RESERVOIR ENGINEERING OF CONVENTIONAL AND UNCONVENTIONAL PETROLEUM RESOURCES

RESERVOIR ENGINEERING OF CONVENTIONAL AND UNCONVENTIONAL PETROLEUM RESOURCES

NNAEMEKA EZEKWE, PhD, PE

TIGA Petroleum, Inc.
Houston, Texas, USA

Library of Congress Cataloging-in-Publication Data

Name: Ezekwe, Nnaemeka
Title: Reservoir Engineering of Conventional and Unconventional Petroleum Resources
Description: Houston; TIGA Petroleum, Inc. 2020.
Identifiers: Library of Congress Control Number (LCCN): 2019919253
ISBN: 978-2-7333890-0-6 (Hardcover)
Subjects: Petroleum Engineering, Reservoir Engineering, Production Engineering

Send all inquiries to:

Dr. Nnaemeka Ezekwe
TIGA Petroleum, Inc.
27 Treasure Cove Drive
Spring, Texas 77381
United States of America
Email: **nezekwe@tigapetroleum.com**
Website: **http://www.tigapetroleum.com**

ISBN-13: 978-1-7333890-0-6
ISBN-13: 978-1-7333890-1-3
ISBN-13: 978-1-7333890-2-0
ISBN-13: 978-1-7333890-3-7
ISBN-13: 978-1-7333890-4-4

Printed in the United States of America.
First Printing, January 2020

Dedication

To the enduring memory of my parents, Vincent Nweke and Rosaline Oriaku, for their commitment, courage, and sacrifices they made toward my education and those of my brother and sisters. This book is truly a testament that the seed they planted and nurtured have borne bountiful fruit for all.

Table of Contents

Title i

Copyright ii

Dedication iii

Preface .... xxii-xxiii

Acknowledgments .... xxiv-xxvi

About the Author xxvii

Chapter 1 **Conventional and Unconventional Reservoirs** 1
- 1.1 Introduction 1
- 1.2 Conventional Reservoirs Versus Unconventional Reservoirs 1
  - 1.2.1 Conventional Reservoirs 2
  - 1.2.2 Unconventional Reservoirs 5
  - 1.2.3 Unconventional Reservoirs in the United States of America (USA) 11
  - 1.2.4 Potential Resources in Unconventional Reservoirs in the World 19
- Abbreviations 20
- References 21

Chapter 2 **Porosity of Reservoir Rocks** 23
- 2.1 Introduction 23
- 2.2 Total Porosity and Effective Porosity 23
- 2.3 Sources of Porosity Data 24
  - 2.3.1 Direct Methods for Measurement of Porosity 24
  - 2.3.2 Indirect Methods for Derivation of Porosity 25
- 2.4 Applications of Porosity Data 28
  - 2.4.1 Volumetric Calculation 29
  - 2.4.2 Calculation of Fluid Saturations 29
  - 2.4.3 Reservoir Characterization 29
- Nomenclature 30
- Abbreviations 30
- References 30
- Additional References 31
- Problems 31

Chapter 3 **Permeability and Relative Permeability** 33
3.1 Introduction 33
3.2 Sources of Permeability Data 33
3.2.1 Permeability from Core Samples 34
3.2.2 Permeability from Pressure Transient Tests 34
3.2.3 Permeability from Well Logs Based on Empirical Correlations 34
3.3 Relative Permeability 37
3.4 Sources of Relative Permeability Data 38
3.4.1 Laboratory Measurements of Relative Permeability Data 38
3.4.2 Estimations from Field Data 38
3.4.3 Empirical Correlations 38
3.5 Three-Phase Relative Permeability 41
3.6 Applications of Permeability and Relative Permeability Data 41
Nomenclature 42
Abbreviations 42
References 43
Additional References 44
Problems 45

Chapter 4 **Reservoir Fluid Saturations** 47
4.1 Introduction 47
4.2 Determination of Water Saturations 47
4.2.1 Clean Sands 48
4.2.2 Shaly Sands 49
4.2.3 Carbonate Rocks 50
4.2.4 Water Saturations from Nuclear Magnetic Resonance Logs 50
4.2.5 Uncertainties in Estimation of Water Saturation 51
4.3 Determination of Reservoir Productive Intervals 51
4.3.1 Net Sands, Net Reservoir, and Net Pay 52
Nomenclature 53
Abbreviations 53
References 53
Additional References 54
Problems 54

Chapter 5 **Pressure-Volume-Temperature (PVT) Properties of Reservoir Fluids** 55
5.1 Introduction 55
5.2 Phase Diagrams 55
5.2.1 Single-Component Systems 55
5.2.2 Binary Systems 56
5.2.3 Multicomponent Systems 57
5.2.4 Retrograde Behavior of Gas-Condensate Systems 58

5.3 Gas and Gas-Condensate Properties 59
5.3.1 Ideal Gas Equation 59
5.3.2 Real Gas Equation 60
5.3.3 Gas Gravity 60
5.3.4 Reduced Temperature and Pressure 61
5.4 Pseudo-Critical Properties of Gas Mixtures 62
5.4.1 Composition of Gas Mixtures Known 62
5.4.2 Correction for non-Hydrocarbon Gas Impurities 63
5.4.3 Composition of Gas Mixture Unknown 63
5.5 Wet Gas and Gas Condensate 64
5.5.1 Recombination Method 64
5.5.2 Correlation Method 65
5.6 Correlations for Gas Compressibility Factor 68
5.7 Gas Formation Volume Factor (FVF) 70
5.8 Gas Density 70
5.9 Gas Viscosity 70
5.10 Gas Coefficient of Isothermal Compressibility 71
5.11 Correlations for Calculation of Oil PVT Properties 77
5.11.1 Bubble Point Pressure 77
5.11.2 Solution Gas-Oil Ratio (GOR) 78
5.11.3 Oil Formation Volume Factor (FVF) 78
5.11.4 Coefficient of Isothermal Compressibility of Oil 79
5.11.5 Oil Viscosity 80
5.12 Correlations for Calculation of Water PVT Properties 83
5.12.1 Water Formation Volume Factor (FVF) 83
5.12.2 Density of Formation Water 83
5.12.3 Coefficient of Isothermal Compressibility of Formation Water 83
5.12.4 Viscosity of Formation Water 83
Nomenclature 84
Subscripts 84
References 85
Additional References 86
Problems 86

Chapter 6 **Reservoir Fluid Sampling and PVT Laboratory Measurements** 89
6.1 Overview of Reservoir Fluid Sampling 89
6.2 Reservoir Type and State 92
6.2.1 Undersaturated Oil Reservoirs 92
6.2.2 Undersaturated Gas Condensate Reservoirs 92
6.2.3 Saturated Oil Reservoirs 92
6.2.4 Saturated Gas Condensate Reservoirs 93

6.3 Well Conditioning 93
6.4 Subsurface Sampling Methods and Tools 93
6.4.1 Conventional Bottomhole Samplers 93
6.4.2 Pistonned Bottomhole Samplers 94
6.4.3 Single Phase Samplers 94
6.4.4 Exothermic Samplers 94
6.5 Wireline Formation Testers 94
6.5.1 Oil-Based Mud Contamination of WFT Samples 94
6.5.2 Formation Pressures from WFT 96
6.5.3 Capillary Effects on WFT Formation Pressures 96
6.5.4 Effects of Supercharging on WFT Formation Pressures 99
6.5.5 Comments on Applications of WFT Pressure Data 99
6.6 PVT Laboratory Measurements 100
6.6.1 Fluid Composition 100
6.6.2 Constant Composition Expansion (CCE) 100
6.6.3 Differential Liberation (DL) 101
6.6.4 Constant Volume Depletion (CVD) 101
6.6.5 Separator Tests 102
6.6.6 Viscosity Measurements 103
6.7 Applications of Laboratory PVT Measurements 103
6.7.1 Calculation of Oil FVF and Solution GOR 103
6.7.2 Calculation of Gas Compressibility Factor, Gas FVF, and Total FVF 103
6.7.3 Calculation of Oil Compressibility Factor 104
Nomenclature 105
Subscripts 105
Abbreviations 105
References 106
Additional References 106
Problems 107
Appendix 6A Typical Reservoir Fluid Study for a Black Oil Sample 109
Appendix 6B Typical Reservoir Fluid Study for a Gas Condensate Sample 112

Chapter 7 **PVT Properties Predictions from Equations of State** 117
7.1 Historical Introduction to Equations of State (EOS) 117
7.2 van der Waals (vdW) EOS 117
7.3 Soave-Redlich-Kwong (SRK) EOS 118
7.4 Peng-Robinson (PR) EOS 119
7.5 Phase Equilibrium of Mixtures 119
7.6 Roots from Cubic EOS 120
7.7 Volume Translation 121
7.8 Two-Phase Flash Calculation 122
7.8.1 Generalized Procedure for Two-Phase Flash Calculations 122

7.9 Bubble Point and Dew Point Pressure Calculations 123
7.10 Characterization of Hydrocarbon Plus Fractions 123
7.11 Phase Equilibrium Predictions with Equations of State 125
Nomenclature 128
Subscripts 128
Superscripts 128
Abbreviations 128
References 129
Problems 130

Chapter 8 **The General Material Balance Equation** 131
8.1 Introduction 131
8.2 Derivation of the General Material Balance Equation (GMBE) 131
8.2.1 Development of Terms in the Expression of Eq. (8.1) 131
8.3 The GMBE for Gas Reservoirs 133
8.4 Discussion on the Application of the GMBE 133
Nomenclature 134
Subscripts 134
Abbreviations 134
References 134
Problems 135

Chapter 9 **Gas Reservoirs** 137
9.1 Introduction 137
9.2 Volumetric Gas Reservoirs 137
9.2.1 Volumetric Calculations for Dry Gas Reservoirs 137
9.2.2 Volumetric Calculations for Wet Gas and Retrograde Gas Condensate Reservoirs 138
9.2.3 Material Balance for Volumetric Dry Gas, Wet Gas, and Retrograde Gas Condensate Reservoirs 139
9.3 Gas Reservoirs with Water Influx 141
9.3.1 Volumetric Approach 141
9.3.2 Material Balance Approach 142
9.3.3 The Cole Plot 142
9.3.4 The Havlena-Odeh Straight Line Method 142
9.4 Water Influx Models 143
9.4.1 Fetkovich Aquifer Model 144
9.4.2 Carter-Tracy Aquifer Model 146
9.5 Geopressured Gas Reservoirs 149
9.5.1 The Ramagost and Farshad Method 150
9.5.2 The Roach Method 152
9.6 Case Histories of Two Gas Reservoirs 154
9.6.1 The Case History of Red Hawk Reservoir 154
9.6.2 The Case History of West Cameron 580 Reservoirs 164

Nomenclature ... 173
Subscripts ... 173
Abbreviations ... 173
References ... 174
Additional References ... 174
Problems ... 175
Appendix 9A Correlations for Estimating Residual Gas Saturations for Gas Reservoirs Under Water Influx ... 177
Appendix 9B Dimensionless Pressure for Finite and Infinite Aquifers ... 178
Appendix 9C Dimensionless Pressure for Infinite Aquifers ... 179

Chapter 10 **Oil Reservoirs** ... 181
10.1 Introduction ... 181
10.2 Oil Reservoir Drive Mechanisms ... 181
10.3 Gravity Drainage Mechanism ... 182
10.4 Volumetric Undersaturated Oil Reservoirs ... 183
10.4.1 Volume Calculations Above Bubble Point Pressure ... 183
10.4.2 Volume Calculations Below Bubble Point Pressure ... 185
10.5 Undersaturated Oil Reservoirs with Water Influx ... 186
10.5.1 Volumetric Method ... 186
10.5.2 Material Balance Method ... 186
10.6 Volumetric Saturated Oil Reservoirs ... 192
10.6.1 Volumetric Method ... 192
10.6.2 Material Balance Method ... 193
10.7 Material Balance Approach for Saturated Oil Reservoirs with Water Influx ... 193
10.8 Case History of Manatee Reservoirs ... 194
10.8.1 Reservoir Geology ... 196
10.8.2 Rock and Fluid Properties ... 200
10.8.3 Reservoir Pressure and Production Data ... 200
Nomenclature ... 203
Subscripts ... 203
Abbreviations ... 203
References ... 204
Problems ... 204

Chapter 11 **Production Forecasting for Conventional and Unconventional Reservoirs** ... 207
11.1 Introduction ... 207
11.2 Flow Regimes ... 207
11.3 Arps' Decline Equations ... 209
11.3.1 Range of Values for Arps' b Parameter ... 210
11.3.2 Effective and Nominal (Exponential) Decline Rates ... 210
11.3.3 Type Curves for Analysis of Production Data ... 211

11.3.4 Normalized Rates and Pressures ... 212
11.3.5 Flowing Material Balance ... 212
11.3.6 Diagnostic Plots ... 213
11.3.7 Transient Decline Models ... 214
11.3.8 Construction of Type Wells for Production Forecasting ... 215
11.3.9 Analyses of Well Production Data Using a Step-by-Step Procedure ... 216
Nomenclature ... 229
Subscripts ... 229
Abbreviations ... 229
References ... 230
Problems ... 230
Appendix 11 Data for Examples 11.1 and 11.2 ... 236

Chapter 12 **Fluid Flow in Petroleum Reservoirs** ... **253**
12.1 Introduction ... 253
12.2 Fluid Types ... 253
12.2.1 Incompressible Fluids ... 253
12.2.2 Slightly Compressible Fluids ... 253
12.2.3 Compressible Fluids ... 254
12.3 Definition of Fluid Flow Regimes ... 254
12.3.1 Transient Flow ... 254
12.3.2 Pseudosteady-State (PSS) Flow ... 255
12.3.3 Steady-State (SS) Flow ... 255
12.4 Darcy Fluid Flow Equation ... 255
12.5 Radial Forms of the Darcy Equation ... 256
12.5.1 Steady-State Flow, Incompressible Fluids ... 256
12.5.2 Average Permeability of Parallel Beds ... 257
12.5.3 Average Permeability of Serial Concentric Segments ... 258
12.5.4 Pseudosteady State, Incompressible Fluids ... 259
12.5.5 Steady-State Flow, Compressible Fluids ... 260
12.6 Derivation of the Continuity Equation in Radial Form ... 261
12.7 Derivation of Radial Diffusivity Equation for Slightly Compressible Fluids ... 261
12.8 Solutions of the Radial Diffusivity Equation for Slightly Compressible Fluids ... 262
12.8.1 Constant Terminal Rate Solution ... 262
12.8.2 Constant Terminal Pressure Solution ... 265
12.9 Derivation of the Radial Diffusivity Equation for Compressible Fluids ... 267
12.10 Transformation of the Gas Diffusivity Equation with Real Gas Pseudo-Pressure Concept ... 267
12.11 The Superposition Principle ... 270
12.11.1 Applications of Constant Terminal Rate Solutions with Superposition Principle ... 270
12.11.2 Applications of Constant Terminal Pressure Solution with Superposition Principle ... 273

12.12 Well Productivity Index 276
12.13 Well Injectivity Index 277
Nomenclature 277
Subscripts 278
Abbreviations 278
References 278
Additional References 278
Problems 278
Appendix 12A Chart for Exponential Integral 281
Appendix 12B Tabulation of p_D vs t_D for Radial Flow, Infinite Reservoirs with Constant Terminal Rate at Inner Boundary 282
Appendix 12C Tabulation of p_D vs t_D for Radial Flow, Finite Reservoirs with Closed Outer Boundary and Constant Terminal Rate at Inner Boundary 283
Appendix 12D Tabulation of p_D vs t_D for Radial Flow, Finite Reservoirs with Constant Pressure Outer Boundary and Constant Terminal Rate at Inner Boundary 285
Appendix 12E Tabulation of Q_D vs t_D for Radial Flow, Infinite Reservoirs with Constant Terminal Pressure at Inner Boundary 291
Appendix 12F Tabulation of Q_D vs t_D for Radial Flow, Finite Reservoirs with Closed Outer Boundary and Constant Terminal Pressure at Inner Boundary 293

Chapter 13 **Well Test Analysis: Straightline Methods** 297
13.1 Introduction 297
13.2 Basic Concepts in Well Test Analysis 297
13.2.1 Radius of Investigation 297
13.2.2 Skin and Skin Factor 298
13.2.3 Flow Efficiency and Damage Ratio 299
13.2.4 Effective Wellbore Radius 300
13.2.5 Drawdown Well Tests 300
13.2.6 Buildup Well Tests 301
13.2.7 Wellbore Storage 302
13.3 Line Source Well, Infinite Reservoir Solution of the Diffusivity Equation with Skin Factor 302
13.4 Well Test Analyses with Straightline Methods 304
13.4.1 Slightly Compressible Fluids 304
13.4.2 Compressible Fluids 319
13.5 Special Topics in Well Test Analyses 336
13.5.1 Multiphase Flow 336
13.5.2 Wellbore Storage Effects 338
13.5.3 Wellbore Phase Redistribution Effects 338
13.5.4 Boundary Effects 338
13.5.5 Multilayered Reservoirs 339

Nomenclature 340
Subscripts 341
Abbreviations 341
References 341
Additional References 343
Problems 343

Chapter 14 **Well Test Analysis: Type Curves** 297
14.1 Introduction 347
14.2 What Are Type Curves? 347
14.3 Gringarten Type Curves 347
14.3.1 Unit-Slope Line 348
14.4 Bourdet Derivative Type Curves 349
14.5 Agarwal Equivalent Time 349
14.6 Type-Curve Matching 350
14.7 Procedures for Manual Application of Type-Curve Matching in Well Test Analysis 350
14.8 Stages of the Type-Curve Matching Procedures 351
14.8.1 Identification of the Interpretation Model 351
14.8.2 Calculation from Interpretation Model Parameters 352
14.8.3 Validation of the Interpretation Model Results 352
Nomenclature 355
Subscripts 355
Abbreviations 355
References 355
Problems 356
Appendix 14A Characteristic Shapes of Pressure and Pressure-Derivative Curves for Selected Well, Reservoir, and Boundary Models 357
Appendix 14B Data for Example 14.1 359
Appendix 14C Calculation of Pressure Derivatives 362

Chapter 15 **Well Test Analysis: Hydraulically Fractured Wells and Naturally Fractured Reservoirs** 363
15.1 Introduction 363
15.2 Hydraulically Fractured Wells 363
15.3 Definition of Dimensionless Variables for Fractured Wells 363
15.4 Flow Regimes in Fractured Wells 363
15.4.1 Fracture Linear Flow 364
15.4.2 Bilinear Flow 364
15.4.3 Formation Linear Flow 364
15.4.4 Pseudo-radial Flow 364
15.5 Fractured Well Flow Models 364
15.5.1 Finite Conductivity Vertical Fracture 364
15.5.2 Infinite Conductivity Vertical Fracture 364
15.5.3 Uniform Flux Vertical Fracture 365

15.6 Fractured Well Test Analysis: Straightline Methods 365
15.6.1 Bilinear Flow 365
15.6.2 Procedure for Application of Straightline Methods on Well Test Data During Bilinear Flow Regime 366
15.6.3 Formation Linear Flow 366
15.6.4 Procedure for Application of Straightline Methods on Well Test Data During Formation Linear Flow Regime 367
15.6.5 Pseudo-Radial Flow 368
15.6.6 Procedure for Application of Straightline Methods on Well Test Data During Pseudo-Radial Flow Regime 368
15.7 Fractured Well Test Analysis: Type-Curve Matching 368
15.7.1 Identification of the Interpretation Model 369
15.7.2 Calculation from Interpretation Model Parameters 369
15.7.3 Validation of the Interpretation Model Results 370
15.7.4 Procedure for Analysis of Well Test from Hydraulically Fractured Wells 370
15.8 Naturally Fractured Reservoirs 374
15.9 Naturally Fractured Reservoir Models 374
15.9.1 Homogeneous Reservoir Model 374
15.9.2 Multiple Region or Composite Reservoir Model 375
15.9.3 Anisotropic Reservoir Model 376
15.9.4 Single Fracture Model 376
15.9.5 Double Porosity Model 377
15.10 Well Test Analysis in Naturally Fractured Reservoirs Based on Double Porosity Model 378
15.11 Well Test Analysis in NFRs: Straightline Methods 378
15.12 Well Test Analysis in NFRs: Type Curves 380
15.13 Procedure for Analysis of Well Test from NFRs Assuming Double Porosity Behavior 381
15.13.1 Identification of Flow Periods 381
15.13.2 Calculation of Fracture and Reservoir Parameters from Type Curves 381
15.13.3 Validation of Results with Straightline Methods 381
Nomenclature 386
Subscripts 386
Abbreviations 386
References 387
Additional References 387
Problems 388

Chapter 16 Well Test Analysis: Deconvolution Concepts 389
16.1 Introduction 389
16.2 What Is Deconvolution? 389
16.3 The Pressure-Rate Deconvolution Model 390
16.3.1 The von Schroeter et al.[3] Deconvolution Algorithm 390

16.4 Application of Deconvolution to Pressure-Rate Data 390
16.5 Examples on the Application of the von Schroeter Deconvolution Algorithm to Real Well Test Data 391
16.6 General Guidelines for Application of von Schroeter Deconvolution Algorithm to Pressure-Rate Data from Well Tests 395
References 396
Additional References 396
Problems 396

Chapter 17 **Well Test Analysis: Diagnostic Fracture Injection Tests in Unconventional Reservoirs** **397**
17.1 Introduction 397
17.2 Typical DFIT Profile 397
17.3 Simple Fracture Orientation Models 398
17.4 Design, Planning, and Execution of DFIT 399
17.4.1 Selection of Test Interval 399
17.4.2 Injection Rates and Volumes 399
17.4.3 Selection of Shut-in Methods 400
17.4.4 Selection of Equipment for DFIT 400
17.4.5 Modeling DFIT with a Hydraulic Fracture Model 401
17.4.6 Other DFIT Execution Guidelines 401
17.5 Analyses of Diagnostic Fracture Injection Tests (DFITs) 401
17.5.1 Diagnostic Analyses 401
17.5.2 Before Closure Analysis (BCA) 407
17.5.3 After Closure Analysis (ACA) 408
17.6 Procedure for DFIT Analyses 411
Nomenclature 420
Subscripts 420
Abbreviations 420
References 421
Additional References 421
Problems 421
Appendix 17A Input Data For Example 17.1 422
Appendix 17B Diagnostic Plots Data For Example 17.1 427
Appendix 17C BCA and ACA Plots Data For Example 17.1 429

Chapter 18 **Immiscible Fluid Displacement** **435**
18.1 Introduction 435
18.2 Basic Concepts in Immiscible Fluid Displacement 435
18.2.1 Rock Wettability 435
18.2.2 Capillary Pressure 436
18.2.3 Relative Permeability 438

18.2.4 Mobility and Mobility Ratio 438
18.2.5 Fluid Displacement Efficiency 438
18.2.6 Volumetric Displacement Efficiency 438
18.2.7 Total Recovery Efficiency 438
18.3 Fractional Flow Equations 438
18.3.1 Fractional Flow Equation for Oil Displaced by Water 439
18.3.2 Fractional Flow Equation for Oil Displaced by Gas 440
18.4 The Buckley-Leverett Equation 441
18.5 The Welge Method 443
18.5.1 Water Saturation at the Flood Front 443
18.5.2 Average Water Saturation Behind the Flood Front 443
18.5.3 Average Water Saturation after Water Breakthrough 444
18.6 Summary 446
Nomenclature 447
References 447
Additional References 448
Problems 448

Chapter 19 **Secondary Recovery Methods** 451
19.1 Introduction 451
19.2 Waterflooding 451
19.2.1 Waterflood Patterns 451
19.2.2 Waterflood Design 452
19.2.3 Recommended Steps in Waterflood Design 453
19.2.4 Waterflood Management 453
19.2.5 Management of Waterflooded Reservoirs 457
19.3 Gasflooding 458
19.3.1 Applications of Gasflooding 458
19.3.2 Gasflood Design 458
19.3.3 Recommended Steps in Gasflood Design 459
19.3.4 Gasflood Management 459
19.3.5 Management of Gasflood Reservoirs 460
Nomenclature 461
Abbreviations 461
References 461
Additional References 462
Problems 462

Chapter 20 **Low Salinity Waterflooding** 465
20.1 Introduction 465
20.2 Laboratory Displacement Experiments Using Sandstone Cores 466
20.3 Field Tests in Sandstone Reservoirs 466
20.4 Laboratory Displacement Experiments Using Carbonate Cores 466

20.5 Field Tests in Carbonate Reservoirs 466
20.6 Mechanisms for LSE 467
20.6.1 Mechanisms for LSE in Sandstone Cores and Reservoirs 467
20.6.2 Mechanisms for LSE in Carbonate Cores/Reservoirs 468
20.7 Guidelines for Planning, Designing, and Installing LSWF Projects 469
20.8 Brief Case History of Claire Ridge LSWF Project 476
20.8.1 An Overview of Clair Ridge Field 476
20.8.2 Low Salinity Coreflood Tests 476
20.8.3 Reservoir Modeling and Analysis 477
20.8.4 Selection of Water Injection Facilities 477
Nomenclature 477
Abbreviations 477
References 478
Additional References 480
Problems 480

Chapter 21 **Enhanced Oil Recovery** **481**
21.1 Introduction 481
21.2 EOR Processes 481
21.3 EOR Screening Criteria 483
21.3.1 EOR Screening Criteria for Miscible Gas Injection Processes 484
21.3.2 EOR Screening Criteria for Chemical Flooding Processes 484
21.3.3 EOR Screening Criteria for Thermal Processes 484
21.4 Miscible Gas Injection Processes 485
21.4.1 Basic Concepts on Miscibility for Gas Displacement Processes 485
21.4.2 First-Contact Miscibility (FCM) 486
21.4.3 Multiple-Contact Miscibility (MCM) 486
21.4.4 Vaporizing Gas Drive MCM Process 486
21.4.5 Condensing Gas Drive MCM Process 487
21.4.6 Combined Condensing/Vaporizing (CV) Gas Drive MCM Process 488
21.5 Methods for Determination of MMP or MME for Gasfloods 488
21.5.1 Analytical Techniques for Estimation of MMP or MME 488
21.5.2 Experimental Methods 496
21.6 Types of Miscible Gas Flooding 497
21.6.1 Nitrogen/Flue-gas Miscible Gas Flooding 497
21.6.2 Hydrocarbon (HC) Miscible Gas Flooding 498
21.6.3 Carbon Dioxide Gas Flooding 498
21.6.4 Types of Miscible Gas Injection Strategies 499
21.7 Chemical Flooding Processes 499
21.7.1 Polymer/Surfactant Flooding 499
21.7.2 Alkali/Surfactant/Polymer (ASP) Flooding 499
21.7.3 Polymer Flooding 500
21.7.4 Microbial Enhanced Oil Recovery (MEOR) 500

21.8 Thermal Processes 500
21.8.1 Steamflooding Methods 501
21.8.2 Steamflood Models 502
21.8.3 Management of Steamflood Projects 503
21.8.4 In-Situ Combustion (ISC)/High Pressure Air Injection (HPAI) 504
21.9 Implementation of EOR Projects 504
21.9.1 Process Screening and Selection 505
21.9.2 Quick Economic Evaluation of Selected Processes 505
21.9.3 Geologic and Reservoir Modeling of Selected Processes 506
21.9.4 Expanded Economic Evaluation of Selected Processes 506
21.9.5 Pilot Testing 506
21.9.6 Upgrade Geologic and Reservoir Models with Pilot Test Data/Results 507
21.9.7 Final Detailed Economic Evaluation 507
21.9.8 Field-Wide Project Implementation 507
21.9.9 EOR Process Project Management 507
Nomenclature 508
Abbreviations 508
References 508
Additional References 511
Problems 512

Chapter 22 **Geologic Modeling and Reservoir Characterization** 515
22.1 Introduction 515
22.2 Sources of Data for Geologic Modeling and Reservoir Characterization 515
22.2.1 Seismic Data 515
22.2.2 Outcrop and Basin Studies 515
22.2.3 Well Log Data 516
22.2.4 Core Data 516
22.2.5 Formation Pressures and Fluid Properties Data 516
22.2.6 Pressure Transient Test Data 516
22.2.7 Reservoir Performance Data 516
22.3 Data Quality Control and Quality Assurance 516
22.4 Scale and Integration of Data 517
22.5 General Procedure for Geologic Modeling and Reservoir Characterization 517
22.5.1 Generation of Geologic Surfaces or Horizons 517
22.5.2 Structural Modeling 517
22.5.3 Stratigraphic Modeling 518
22.5.4 Correlation and Assignment of Well Log Data 518
22.5.5 Property Data Modeling 518
22.5.6 Uncertainty Analysis 524

22.5.7 Upscaling of Geologic Model to Reservoir Flow Model 525
Nomenclature 533
Abbreviations 533
References 533
Additional References 534
Problems 535

Chapter 23 **Reservoir Simulation** 537
23.1 Introduction 537
23.2 Derivation of the Continuity Equation in Rectangular Form 539
23.3 Flow Equations for Three-Phase Flow of Oil, Water, and Gas 539
23.4 Basic Concepts, Terms, and Methods in Reservoir Simulation 541
23.4.1 Grid Systems 541
23.4.2 Timesteps 545
23.4.3 Formulations of Simulator Equations 545
23.4.4 Material Balance Errors and Other Convergence Criteria 546
23.4.5 Numerical Dispersion 546
23.4.6 Well Model 547
23.4.7 Model Initialization 548
23.4.8 History Matching 548
23.4.9 Predictions 550
23.4.10 Uncertainty Analysis 550
23.5 General Structure of Reservoir Flow Models 550
23.5.1 Definition of Model and Simulator 551
23.5.2 Geologic Model Data 551
23.5.3 Fluid Properties Data 551
23.5.4 Rock/Fluid Properties Data 551
23.5.5 Model Equilibration Data 551
23.5.6 Well Data 551
23.5.7 Simulator Data Output 551
Nomenclature 552
Subscripts 552
Abbreviations 552
References 553
Additional References 554
Problems 555

Chapter 24 **Reservoir Management** 559
24.1 Introduction 559
24.2 Reservoir Management Principles 559
24.2.1 Conservation of Reservoir Energy 560
24.2.2 Early Implementation of Simple, Proven Strategies 560
24.2.3 Systematic and Sustained Practice of Data Collection 560
24.2.4 Application of Emerging Technologies for Improved Hydrocarbon Recovery 560
24.2.5 Long-Term Retention of Staff in Multi-Disciplinary Teams 560
24.3 Case Histories Demonstrating Applications of Reservoir Management Principles 561
24.3.1 The Case History of 26R Reservoir (1976–1996) 561
24.3.2 Application of Reservoir Management Principles to 26R Reservoir 564
24.3.3 The Case History of MBB/W31S Reservoirs (1976–1999) 566
24.3.4 Application of Reservoir Management Principles to MBB/W31S Reservoirs 568
24.3.5 The Case History of the Shaybah Field 571
24.3.6 Application of Reservoir Management Principles to the Shaybah Field 572
References 575
Additional References 577
Problems 577

Chapter 25 **Economic Evaluation of Petroleum Projects and Property** 589
25.1 Introduction 589
25.2 Classification of Oil and Gas Resources 589
25.2.1 Definition of Terms in the SPE PRMS 590
25.2.2 Estimation of Recoverable Resources 591
25.2.3 Analytic Methods for Resource Determination 591
25.2.4 Resource Assessment Techniques 593
25.3 Time Value of Money 593
25.3.1 Future Worth (Value) of a Single Payment 594
25.3.2 Present Worth (Value) of a Single Payment 595
25.3.3 Uniform Series Compound Amount 595
25.3.4 Sinking Fund Deposit Amount 596
25.3.5 Capital Recovery Amount 597
25.3.6 Uniform Series Present Worth Amount 597
25.3.7 Timing of Payments (or Investments) Considerations 598
25.3.8 Non-Uniform Series of Payments (or Investments) 600
25.3.9 Period, Nominal, and Effective Interest Rates 602
25.3.10 The Rule of 72 603
25.3.11 The Rule of 78 604
25.3.12 Cash Flow Diagram 606

25.4 Economic Valuation Parameters 607
25.4.1 Valuation Parameters Not Based on the Time Value of Money 608
25.4.2 Valuation Parameters Based on the Time Value of Money 609
25.4.3 Recommended Practice for Economic Valuations 619
25.5 Depreciation, Depletion, and Amortization 619
25.5.1 Depreciation Methods 619
25.5.2 Depletion Methods 622
25.5.3 Amortization Methods 622
25.6 Economic Evaluation of a Petroleum Property 623
25.6.1 Economic Evaluation of an Onshore Petroleum Property in the USA 623
25.6.2 Economic Evaluation of a Petroleum Property in Other Parts of the World 624
25.7 Decision and Risk Analysis Applied to Petroleum Evaluations 626
Nomenclature 627
Abbreviations 627
References 628
Problems 628
Appendix 25A Economic Evaluation of a Drilling Prospect from paper SPE 68588 632
Appendix 25B Cash Flow Model for a Multi-Well Development Project 635
Appendix 25C Worksheet for a PSC with Fixed Scale 642

INDEX 645-670

Preface

This book was envisioned as a partial but important update of the original book, "Petroleum Reservoir Engineering Practice," published in 2010 by Prentice Hall. The vision was to add several chapters on analyses of unconventional reservoirs and a chapter on economic evaluation of petroleum projects to the original book. I think this book provides adequate coverage of both subjects which will enhance its value to engineers. A chapter was added on low salinity waterflooding as an important technology for improved oil recovery. Another important expansion of the book is the addition of problems at the end of each chapter to help readers test their understanding of the subject matter presented in the chapter. The addition of problems will also assist lecturers to assign homework from the book. Except in these four main areas, this book retains substantial portions of material covered in the original book although important changes in wording and presentation were made in this version.

Unconventional reservoirs have completely changed the landscape in American petroleum industry. United States of America (USA) is currently one of the largest oil producers in the world surpassing Saudi Arabia and Russia. This tremendous reversal in oil productivity trend is driven by revolution in the productivity of unconventional reservoirs in the USA. In Chapter 1, the differences between conventional and unconventional reservoirs are presented. Also, in this chapter, unconventional reservoirs were shown to hold the future of hydrocarbon supply in the USA and the rest of the world. The potential of unconventional resources are almost limitless as the technologies developed in the USA are applied to similar resources found around the world. Engineers pursuing a career in the petroleum industry are advised to develop knowledge and skills on the engineering of unconventional reservoirs.

In Chapters 2 to 7, the sources and applications of basic rock and fluid properties data that are fundamental for all petroleum reservoir engineering calculations are presented. Chapter 2 presents the sources and applications of porosity. Chapter 3 covers sources and applications of permeability and relative permeability. Chapter 4 discusses methods and models for determination of fluid saturations, and classification of reservoir rocks for volumetric calculations. These topics are presented at the introductory to intermediate levels. The main objective is to emphasize the importance of these sources of data as basic inputs for most reservoir engineering calculations. Chapter 5 was devoted to rigorous calculations of Pressure-Volume-Temperature (PVT) properties of reservoir fluids from correlations. In concert with Chapter 5, Chapter 6 presents routine reservoir fluid sampling methods, and laboratory measurements of PVT properties from reservoir fluid samples. In Chapter 7, the prediction of PVT properties from equations of state is presented. The application of equations of state in compositional simulation justified the presentation of this subject at the intermediate-to-advance levels for many engineers who are involved in compositional reservoir simulation work.

The fundamentals of petroleum reservoir engineering are treated from Chapters 8 to 10. The general material balance equation is developed from basic concepts in Chapter 8 and applied as a fundamental tool for basic reservoir engineering analysis. In Chapter 9, volumetric and graphical methods for calculation of gas-in-place for different types of gas reservoirs are discussed. This approach is extended to oil reservoirs in Chapter 10. The use of case histories to illustrate analytic methods for evaluation of performance of gas and oil reservoirs is demonstrated in Chapters 9 and 10.

Production forecasting for conventional and unconventional reservoirs is discussed in Chapter 11. This is based primarily on Arps' decline equations which have become the workhorse of the petroleum industry for production forecasting on conventional and unconventional reservoirs. The methodologies (decline curve analysis or DCA) for using Arps' equation for production forecasting on conventional reservoirs are applicable to unconventional reservoirs although rate transient analysis (RTA) techniques should be applied to account for production characteristics of unconventional reservoirs such as relatively long transient flow periods and influence of boundary dominated flows (BDF). The Stretched Exponential Decline Model (SEDM) and the Duong Model are presented in the chapter as two transient decline models which can be coupled with the modified Arps' decline equations to improve production forecasting for unconventional reservoirs. These applications are illustrated with examples.

Fluid flow in petroleum reservoirs is introduced in Chapter 12 with the derivation of the continuity equation and the radial diffusivity equation. In Chapter 12, the fundamental equations that form the bases for pressure transient analysis (PTA) by straightline methods are developed and applied later in Chapter 13. The use of type curves in well test analysis, especially Gringarten and Bourdet type curves, are presented in Chapter 14 with emphasis on procedures for type-curve matching. Well test analysis methods for hydraulically fractured wells and naturally fractured conventional reservoirs are presented in Chapter 15. Deconvolution concepts for well test analysis are covered in Chapter 16. This is followed with the presentation of well test analysis in unconventional reservoirs in Chapter 17 based mainly on diagnostic fracture injection tests (DFIT). Note that many characteristic behaviors of fracture flow are applicable to both conventional and unconventional reservoirs.

Basic concepts in immiscible fluid displacement are discussed in Chapter 18. These include derivations of the fractional flow equation, the Buckley-Leverett equation, and the Welge method. This is followed with the introduction of secondary recovery methods in Chapter 19 focused mainly on waterflooding and gasflooding.

Low salinity waterflooding is presented in Chapter 20 as an improved oil recovery method distinct from secondary recovery methods of Chapter 19 or enhanced oil recovery methods presented in Chapter 21. This definition of low salinity waterflooding is solely at the discretion of this author since it can be categorized as a secondary recovery method or an enhanced oil recovery method depending on any particular perspective. Either categorization is completely acceptable to the author.

Enhanced oil recovery methods are discussed in Chapter 21. In the chapter, special emphasis is placed on screening criteria and field implementation of enhanced oil recovery processes because many engineers should expect to be involved in such activities someday in their careers.

Chapters 18 to 21 are designed to introduce practicing engineers to fundamental methodologies for application of secondary, improved, and enhanced oil recovery processes, and also to develop practical procedures for field implementation of these processes.

Geologic modeling and reservoir characterization methods and procedures are presented in Chapter 22. This is followed with concepts in reservoir simulation in Chapter 23. In both chapters, the focus was on presenting methods and procedures for applying these tools on the practical aspects of building reservoir models rather than on theory.

The principles of reservoir management that I first enunciated in 2003 are presented in Chapter 24. These principles of reservoir management were developed from my experience from managing various types of reservoirs around the world. The principles are simple and practical and can be applied to any reservoir anywhere in the world. The application of the five principles of reservoir management are illustrated with case histories in Chapter 24.

Economic evaluation is key to the success of any petroleum project. Every project in the petroleum industry must be shown to be economic and profitable before it can be sanctioned and implemented. Chapter 25 presents fundamental criteria that can be used to judge the economic profitability of most projects. The chapter also discusses concepts, contracts, and economic models used in many petroleum operations involving host countries and international oil companies.

This book could not have been written without the support of my wife and children who endured long hours of my seclusion over many years to work on material for the book. I am very grateful for their patience and understanding. I give special thanks to my wife, Anulika, and my children (Nkemdirim, Chukwuemeka, Chioma, Ifeoma, Obinna, and Ezenwanyi). The odyssey of writing this book made my love for them much stronger.

Acknowledgments

The author is grateful to the Society of Petroleum Engineers for permissions to reprint the following tables and figures published in this book:

Chapter 2

Figure 2.1

Al-Ruwaili, S.A., and Al-Waheed, H.H.: "Improved Petrophysical Methods and Techniques for Shaly Sands Evaluation," paper SPE 89735 presented at the 2004 SPE International Petroleum Conference in Puebla, Mexico (November 8–9, 2004).

Chapter 4

Figure 4.4

Worthington, P.F., and Cosentino, L.: "The Role of Cutoffs in Integrated Reservoir Studies," *SPEREE* (August 2005) 276–290.

Chapter 5

Figures 5.9, 5.10

Sutton, R.P.: "Compressibility Factors for High-Molecular-Weight Reservoir Gases," paper SPE 14265 presented at the 60th Annual Technical Conference and Exhibition, Las Vegas, Nevada, (September 22–25, 1985).

Figures 5.11, 5.12

Mattar, L., Brar, G.S., and Aziz, K.: "Compressibility of Natural Gases," *J. Cdn. Pet. Tech.* (Oct-Dec. 1975) 77–80.

Chapter 6

Figure 6.1

Moffatt, B.J., and Williams, J.M.: "Identifying and Meeting the Key Needs for Reservoir Fluid Properties – A Multi-Disciplinary Approach," paper SPE 49067 presented at the 1998 SPE Annual Technical Conference and Exhibition, New Orleans, Louisiana, (September 27–30, 1998).

Figures 6.4, 6.5, 6.6, 6.7, 6.9

Elshahawi, H., Samir, M., and Fathy, K.: "Correcting for Wettability and Capillary Pressure Effects on Formation Tester Measurements," paper SPE 63075 presented at the 2000 SPE Annual Technical Conference and Exhibition, Dallas, Texas, (October 1–4, 2000).

Chapter 7

Tables 7.1, 7.2

Jhaveri, B.S., and Youngren, G.K.: "Three-Parameter Modification of the Peng-Robinson Equation of State to Improve Volumetric Predictions," *SPERE* (August 1988) 1033–1040.

Figures 7.1, 7.2

Wang, P., and Pope, G.A.: "Proper Use of Equations of State for Compositional Reservoir Simulation," SPE 69071, Distinguished Author Series, (July 2001) 74–81.

Chapter 9

Figure 9.4

Havlena, D., and Odeh, A.S.: "The Material Balance as an Equation of a Straight Line," *JPT* (August 1963) 896–900.

Figure 9.11

Hammerlindl, D.J.: "Predicting Gas Reserves in Abnormally Pressured Reservoirs," paper SPE 3479 presented at the 1971 SPE Fall Meeting, New Orleans, (October 3-5, 1971).

Chapter 10

Figures 10.1, 10.4, 10.5

Havlena, D., and Odeh, A.S.: "The Material Balance as an Equation of a Straight Line," *JPT* (August 1963) 896-900.

Chapter 11

Figure 11.1

Chen, H., and Teufel, L.W.: "A New Rate-Time Type Curve for Analysis of Tight-Gas Linear and Radial Flow", paper SPE 63094 presented at the 2000 SPE Annual Technical Conference and Exhibition, Dallas, Texas, (October 1-4 2000).

Chapter 12

Appendices 12B, 12C, 12D, 12E, 12F

van Everdingen, A.F., and Hurst, W.: "The Application of the Laplace Transformation to Flow Problems in Reservoirs," *Trans.*, AIME (1949) 186, 305-324.

Chapter 13

Table 13.2

Tiab, D., Ispas, I.N., Mongi, A., and Berkat, A.: "Interpretation of Multirate Tests by the Pressure Derivative- 1. Oil Reservoirs," paper SPE 53935 presented at the 1999 SPE Latin American and Caribbean Petroleum Engineering Conference, Caracas, Venezuela, (April 21-23, 1999).

Chapter 14

Figure 14.1

Agarwal, R.G., Al-Hussainy, R., and Ramey, H.J., Jr.: "An Investigation of Wellbore Storage and Skin Effect in Unsteady Liquid Flow: I. Analytical Treatment," *SPEJ* (September 1970) 279-290.

Figure 14.2

Gringarten, A.C., Bourdet, D.P., Landel, P.A., and Kniazeff, V.J.: "A Comparison Between Different Skin and Wellbore Storage Type-Curves for Early-Time Transient Analysis," paper SPE 8205 presented at the SPE Annual Technical Conference and Exhibition, Las Vegas, Nevada, (September 23-26, 1979).

Chapter 15

Figures 15.1, 15.3, 15.4, 15.5, 15.6, 15.7

Cinco-Ley, H., and Samaniego-V, F.: "Transient Pressure Analysis for Fractured Wells," *JPT,* (September 1981) 1749-1766.

Figures 15.8, 15.9, 15.11

Cinco-Ley, H.: "Evaluation of Hydraulic Fracturing By Transient Pressure Analysis Methods," paper SPE 10043 presented at the International Petroleum Exhibition and Technical Symposium, Beijing, China, (March 18-26, 1982).

Figure 15.10

Cinco-L, H., Samaniego-V, F., and Dominguez-A, N.: "Transient Pressure Behavior for a Well With a Finite-Conductivity Vertical Fracture," *SPEJ* (August 1978) 253-264.

Figures 15.16, 15.17a, 15.17b, 15.18, 15.19a, 15.19b, 15.20, 15.21, 15.22, 15.23, 15.24, 15.25

Cinco-Ley, H.: Well-Test Analysis for Naturally Fractured Reservoirs," *JPT* (January 1996) 51-54.

Figure 15.26

Warren, J.E., and Root, P.J.: "The Behavior of Naturally Fractured Reservoirs," *SPEJ* (September 1963) 245-255.

Figure 15.27

Serra, K., Reynolds, A.C., and Raghavan, R.: "New Pressure Transient Analysis Methods for Naturally Fractured Reservoirs," *JPT* (December 1983) 2271-2283.

Figure 15.32, 15.33, 15.34

Gringarten, A.C.: "Interpretation of Tests in Fissured and Multilayered Reservoirs with Double-Porosity Behavior: Theory and Practice," *JPT* (April 1984) 549-564.

Chapter 16

Figures 16.1, 16.2, 16.3, 16.4, 16.5, 16.6, 16.7, 16.8

Levitan, M.M., Crawford, G.E., and Hardwick, A.: "Practical Considerations for Pressure-Rate Deconvolution of Well-Test Data," *SPEJ* (March 2006) 35-46.

Chapter 17

Figures 17.4, 17.5

Barree, R.D., Miskimins, J.L., and Gilbert, J.V.: "Diagnostic Fracture Injection Tests: Common Mistakes, Misfires, and Misdiagnoses", *SPE Production & Operations* (May 2015) 84-98; SPE 169539-PA.

Chapter 23

Figures 23.2, 23.7, 23.9, 23.10

Aziz, K.: "Reservoir Simulation Grids: Opportunities and Problems," *JPT* (July 1993) 658-663.

Chapter 24

Figures 24.16, 24.17, 24.18, 24.21, 24.22

Salamy, S.P., Al-Mubarak, H.K., Ghamdi, M.S., and Hembling, D.: "Maximum-Reservoir-Contact-Wells Performance Update: Shaybah Field, Saudi Arabia," *SPE Production & Operation* (November 2008) 439-443.

Figures 24.19, 24.20

Saleri, N.G., Salamy, S.P., Mubarak, H.K., Sadler, R.K., Dossary, A.S., and Muraikhi, A.J.: "Shaybah-220: A Maximum-Reservoir-Contact (MRC) Well and its Implications for Developing Tight-Facies Reservoirs," *SPEREE* (August 2004) 316-321.

Chapter 25

Figures 25.1

Society of Petroleum Engineers (SPE)-Petroleum Resources Management Systems (PRMS)- 2018 Update

Figures 25.13, 25.14

Feng, Zhuo, Shui-Bo Zhang, and Ying Gao: "On Oil Investment and Production: A Comparison of Production Sharing Contracts and Buyback Contracts." *Energy Economics* 42 (2014): 395-NE402.

About the Author

Dr. Nnaemeka Ezekwe is a world renowned reservoir engineer who has practiced in all the major oil producing regions in Africa, Asia, Europe, Middle East, North America, and South America. This vast experience on various types of reservoirs was applied in crafting this book as a guide for practicing engineers or a textbook for college students. Dr. Ezekwe worked for sixteen years in key supervisory and managerial roles for Bechtel Petroleum Operations. In 1998, he joined Pennzoil Exploration and Production Company as a senior petroleum engineer advisor responsible for providing technical guidance on domestic projects in the Permian Basin and Gulf of Mexico, and also on international projects in the Caspian Sea, Egypt, Equatorial Guinea, and Venezuela. In 2000, as a senior reservoir engineer advisor at Devon Energy Corporation, he led pioneering analyses on the appraisal and development of several major Lower Tertiary deepwater reservoirs in the Gulf of Mexico. He also led appraisal and development of pre-salt reservoirs in the Campos and Santos Basins in Brazil, and deepwater reservoirs in Angola. Dr. Ezekwe is the author of *Petroleum Reservoir Engineering Practice*, a best-selling book on reservoir engineering published by Prentice Hall in 2010. In 2011, he joined BP America as EOR Deployment Manager responsible for implementation of low salinity waterflooding for improved oil recovery on all BP reservoirs worldwide. In 2015, he was contracted by Petrobras Oil & Gas B.V. as a reservoir engineer expert to provide technical guidance on appraisal and development of all petroleum assets in Africa including producing deepwater reservoirs in West Africa. In 2017, Dr. Ezekwe founded TIGA Petroleum, Inc., as a reservoir engineering consulting firm based in Houston, Texas. He holds BS, MS, and PhD degrees in chemical and petroleum engineering, and an MBA degree, all from the University of Kansas. He has published many technical papers in chemical and petroleum engineering journals. Dr. Ezekwe served twice as an SPE Distinguished Lecturer in 2004 and in 2012. In this role, he lectured on reservoir management strategies and intelligent well/field technologies in more than 100 countries around the world. He was elected SPE Distinguished Member in 2013. Dr. Ezekwe is a registered professional engineer in California and Texas, USA.

CHAPTER 1 Conventional and Unconventional Reservoirs

1.1 Introduction

The technologies applied on reservoirs to produce oil and gas are being revolutionized due to successes achieved with recovering large quantities of hydrocarbons from reservoirs classified as unconventional. Consequently, the knowledge gained on reservoir characteristics, behaviors, and performances since the first oil well was drilled near Titusville, Pennsylvania, United States of America (USA) in 1859, has undergone considerable revisions because those earlier reservoirs are now classified as conventional. This book starts with classifying reservoirs into two broad categories termed conventional and unconventional. This classification, based initially on rock properties, should not be considered strict boundaries. There are some overlaps between conventional and unconventional reservoirs classified on the basis of rock/fluid properties, drilling and completion practices, and reservoir productivity. The objective of classifying reservoirs into conventional and unconventional categories is to provide a rough guideline based primarily on basic rock properties that could be used to distinguish between these reservoirs. Generally, conventional reservoirs relatively have higher matrix permeabilities and porosities while unconventional reservoirs have lower matrix permeabilities and porosities. Conventional reservoirs may or may not have natural fractures. Most unconventional reservoirs have natural fractures that are critical for them to produce fluids at commercial rates. This is a generalized rule-of-thumb kind of classification. More detailed basis for classifying conventional and unconventional reservoirs is presented in this chapter.

1.2 Conventional Reservoirs Versus Unconventional Reservoirs

The most direct way to compare conventional and unconventional reservoirs is through the use of the Resource Triangle[1] shown in Figure 1.1. As shown in the Resource Triangle, conventional reservoirs occupy areas around the apex of the triangle. They are harder to find, relatively smaller in volume, easier to produce, have higher permeabilities, require less technology, and less development costs to achieve commercial production. The unconventional reservoirs are located in areas from the middle to the base of the triangle. They are generally on the opposite side of the earlier characterizations of conventional reservoirs. Specifically, unconventional reservoirs are easier to find (occur in extensive basins), larger in volume, harder to produce, have lower permeabilities, require more technology, and more development costs to achieve commercial production.

Further differences between conventional and unconventional reservoirs in terms of rock properties are shown in Table 1.1. In this table, conventional reservoirs, which could be composed of sandstones, limestones, and dolomites, generally have permeabilities that are greater than 0.1 millidarcy (md), and pore throat diameters that are greater than 1 μm. Unconventional reservoirs, represented by tight sands, silts, and shales, generally have permeabilities that are less than 0.1 md [typically in the micro-darcy (μd) and nano-darcy (nd) range], and pore throat sizes that range from 0.1 μm to 10^{-3} μm. The larger pore throat diameters for conventional reservoirs represent higher matrix porosities whereas the smaller pore throat diameters for unconventional reservoirs indicate lower matrix porosities. Some conventional reservoirs have natural fractures whereas most unconventional reservoirs (tight sands and shales) have natural fractures. It is important to note that the natural matrix fractures in unconventional reservoirs help to transport fluids to the induced fractures created by hydraulic fracturing of wells. This enhances the producibility of unconventional reservoirs in spite of the extremely low matrix permeabilities shown in Table 1.1.

TABLE 1.1 Rock Matrix Property Range of Conventional and Unconventional Formations

Reservoir Type	Formation Types	Permeability Range	Pore Throat Size Range
Conventional	Limestones, dolomites, and sandstones	$k \geq 0.1$ md	$d_{pore} \geq 1\ \mu m$
Unconventional	Tight sands/ silts	$0.1 \geq k \geq 10^{-2}\ \mu d$	$1\ \mu m \geq d_{pore} \geq 10^{-3}\ \mu m$
	Shales	$10\ \mu d \geq k \geq 10^{-5}\ \mu d$	$10^{-1}\ \mu m \geq d_{pore} \geq 10^{-2}\ \mu m$

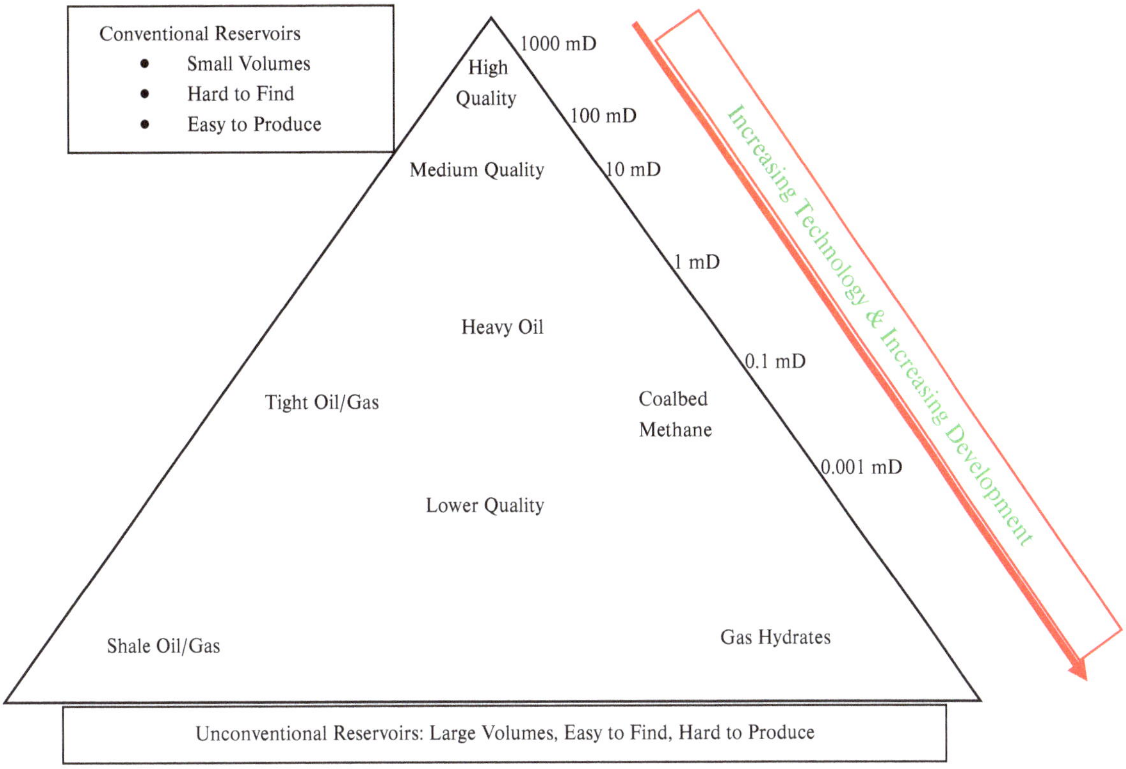

▲ FIGURE 1.1 **Resources Triangle (Adapted from Holditch[1])**

1.2.1 Conventional Reservoirs

The main sources of oil and gas produced around the world have been from conventional reservoirs for over 160 years since the first oil well was drilled near Titusville, Pennsylvania, United States of America (USA) in 1859. Conventional reservoirs were developed primarily with vertical wells and then later with horizontal wells and multilateral wells as drilling technologies were improved and advanced. A high majority of conventional reservoirs did not require special completion technologies such as multistage hydraulic fracturing to achieve commercial productivity because they have high matrix permeabilities. Some conventional reservoirs were hydraulically fractured to improve productivity, but it was not a critical requirement for the wells to produce at commercial rates. Rock and fluid properties of conventional reservoirs were measured in laboratories using specified, verifiable procedures because over many years the petroleum industry had documented reports on the expected ranges of these properties. Analytical tools such as material balance analyses (Chapter 8), volumetric calculations (Chapters 9 and 10), Darcy-based flow equations (Chapter 12), pressure transient analyses (Chapters 13 and 14), analytical models (Chapter 18), numerical models (Chapter 23), etc., were developed for conventional reservoirs. The focus on conventional reservoirs started to change around 2005 as the magnitude of potential resources existing in unconventional reservoirs slowly began to unfold in the petroleum industry, especially in the United States of America.

The global map of giant oil and gas fields is shown in Figure 1.2. Most of the giant fields contain conventional reservoirs. As reported in BP (British Petroleum) Statistical Review of World Energy[3] in 2017, the proved oil reserves in the world at the end of 2016 were 1707 billion barrels (Table 1.2). About 48% of the proved oil reserves were in the Middle East. Most of these proved oil reserves were in conventional reservoirs. Figure 1.3 shows the growth of proved oil reserves in the world from 1996 to 2016. Note that proved oil reserves in the world grew by 318.4 billion barrels in 10 years. The proved gas reserves in the world at the end of 2016 were 6589 trillion cubic feet (tcf) as shown in Table 1.3. About 43% of the proved gas reserves were in the Middle East and existed mainly in conventional reservoirs. The distribution of proved gas reserves in the world from 1996 to 2016 is shown in Figure 1.4.

In the BP Statistical Review of World Energy,[3] Reserves-to-Production (R/P) ratio was calculated by dividing the remaining proved reserves at the end of the year by the production for that year. The result, which was represented as the R/P ratio, was the length of time that the remaining reserves would last if production was continued at that rate into the future. For instance, the remaining total world proved oil reserves at the end of 2016 were 1706.8 billion barrels (Table 1.2). As reported in the BP Statistical Review of World Energy, total world oil production rate in 2016 was 92.15 million barrels per day. The oil R/P ratio for the world at the end of 2016 was 50.7 years. This means that it will take the world 50.7 years to produce the proved oil reserves remaining at the end of 2016 if production were maintained at 92 million barrels per day. Similarly, the total world remaining gas reserves at the end of 2016 were 6588.8 Tcf (Table 1.3). The total world gas production in 2016 was 125.4 Tcf. These numbers gave a gas R/P ratio of 52.5 years. As explained previously, this means that it will take 52.5 years to produce proved gas reserves remaining at the end of 2016 at the 2016 annual gas production rate. In these estimations, it was assumed that all produced oil and gas volumes would be consumed. Since these estimates are based only on proved reserves and does not include unproved or other contingent resources, it is evident that conventional reservoirs would continue to play important roles in the world energy supply for the foreseeable future.

Oil, gas, and coal are the main sources of energy for the world.[4,5] In 2035, 75% of the total energy supply for the world would come from oil, gas, and coal. Out of these three fuels, gas is the fastest growing source of primary energy and will surpass coal by 2035 (Figures 1.5 and 1.6). With the growth of gas as the primary source of energy, the world would rely more on conventional and unconventional reservoirs for the supply of gas. This observation would be discussed further after the potential gas resources recoverable from unconventional reservoirs is presented later in the chapter.

TABLE 1.2 World Proved Oil Reserves at the End of 2016 (from BP Statistical Review[3])

World Geographic Regions	Proved Oil Reserves (Billions of barrels)	Top Three Countries in Region in Order
Middle East	813.5	Saudi Arabia, Iran, Iraq
South and Central America	327.9	Venezuela, Brazil, Ecuador
North America	227.5	Canada, USA, Mexico
Europe And Eurasia	161.5	Russia, Kazakhstan, Azerbaijan
Africa	128	Libya, Nigeria, Algeria
Asia Pacific	48.4	China, India, Vietnam
Total	**1706.8**	

TABLE 1.3 World Proved Gas Reserves at the End of 2016 (from BP Statistical Review[3])

World Geographic Regions	Proved Gas Reserves (Trillion Cubic Feet)	Top Three Countries in Region in Order
Middle East	2803.2	Iran, Qatar, Saudi Arabia
Europe and Eurasia	2002.0	Russia, Turkmenistan, Norway
Asia Pacific	619.3	China, Australia, Indonesia
Africa	503.3	Nigeria, Algeria, Egypt
North America	393.0	USA, Canada, Mexico
South and Central America	268.0	China, India, Vietnam
Total	**6588.8**	

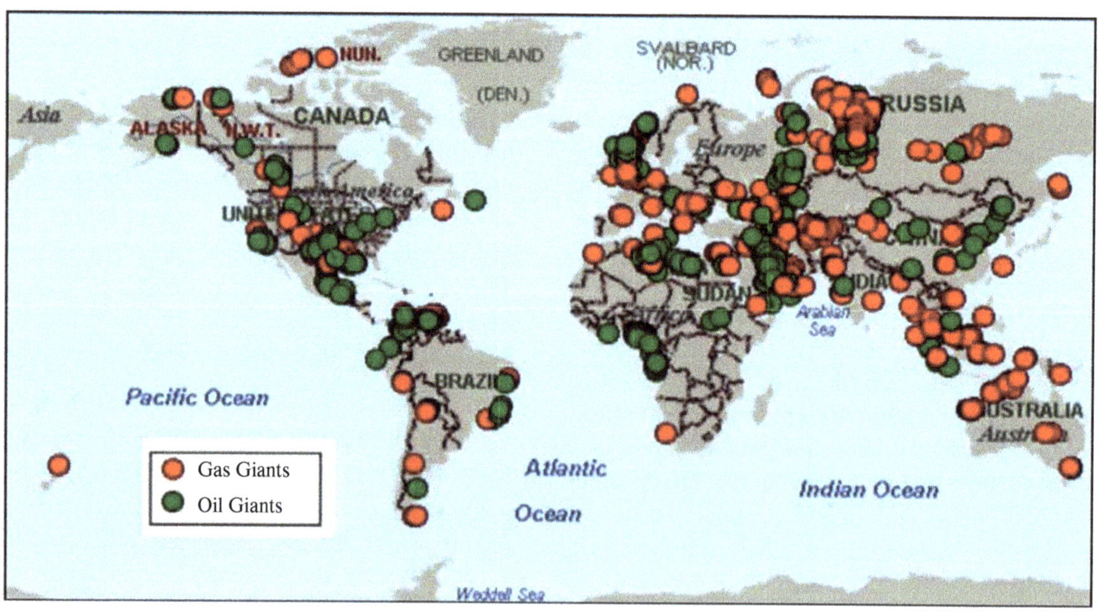

▲ **FIGURE 1.2** **Map of giant oil and gas fields in the world[2]**

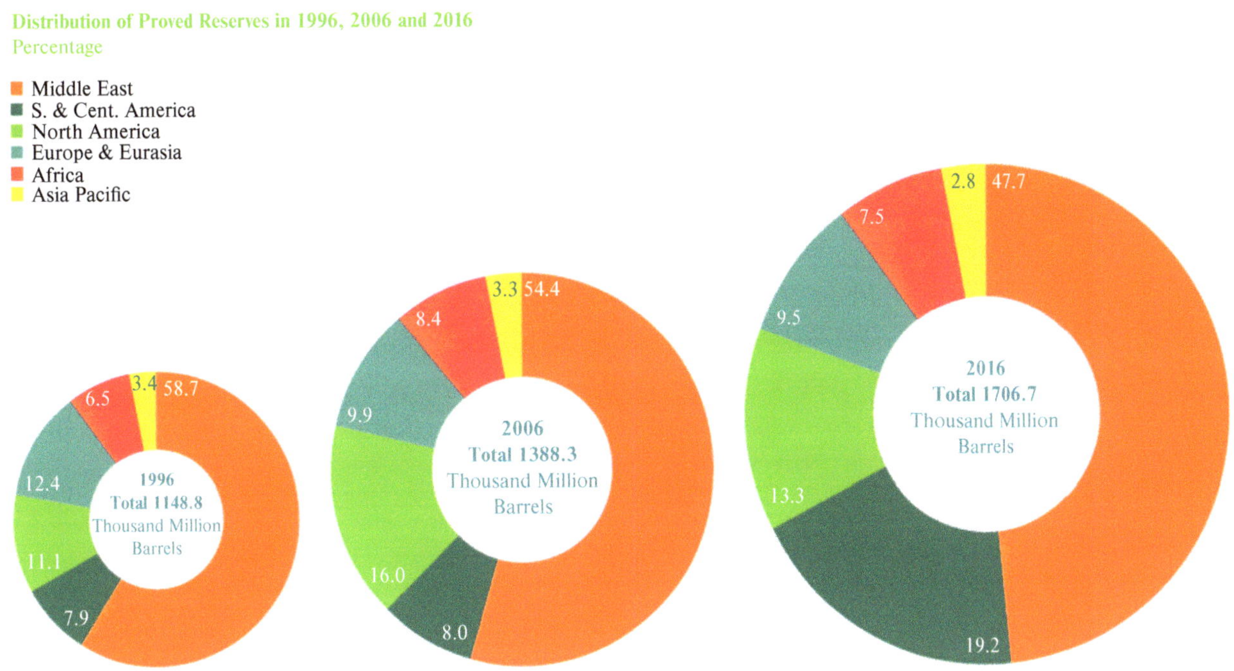

▲ **FIGURE 1.3** World proved oil reserves[3]

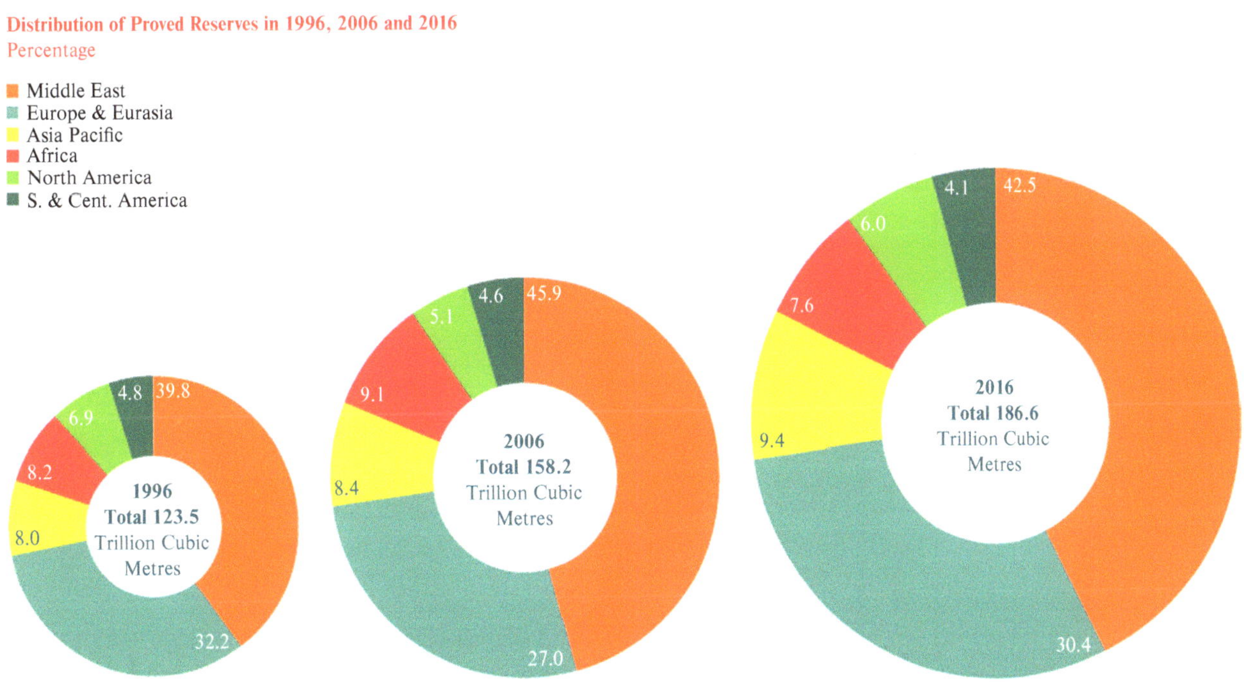

▲ **FIGURE 1.4** World proved gas reserves[3]

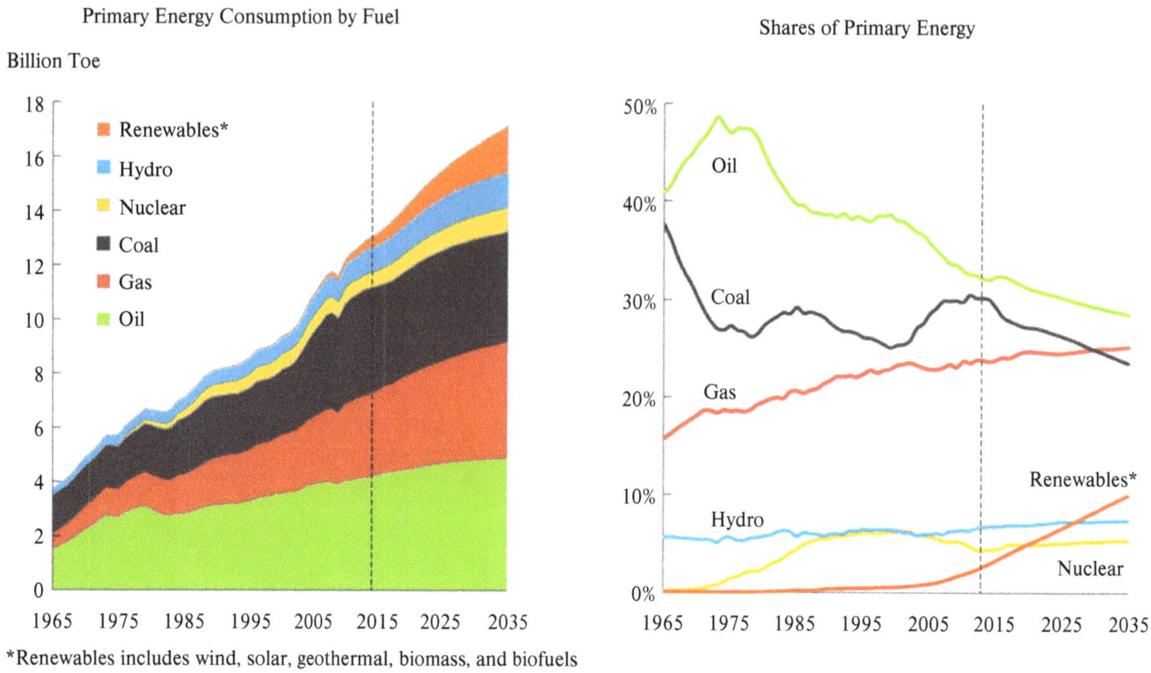

▲ **FIGURE 1.5** **Primary energy consumption in the world**[4]

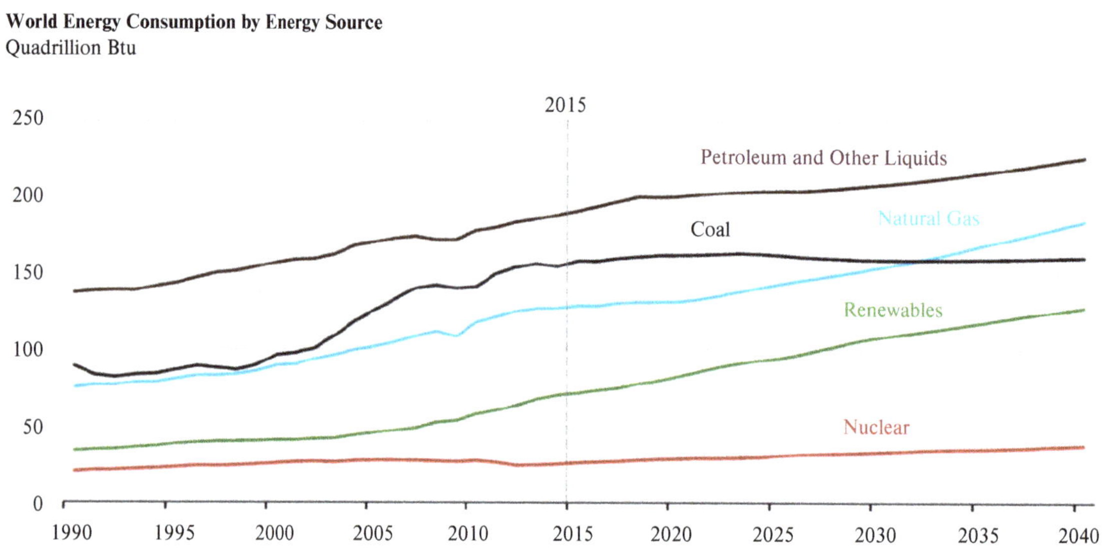

▲ **FIGURE 1.6** **World energy consumption by energy source**[5]

1.2.2 Unconventional Reservoirs

Most reservoirs that exhibit characteristics and behaviors that are different from the hitherto familiar conventional reservoirs are grouped under the term unconventional reservoirs. These reservoirs are considered to be unconventional because they exhibit or possess most of the following characteristics:

1. Low permeability–less than 0.1 md.
2. Low matrix porosity–less than 10%.
3. Require extensive stimulation such as hydraulic fracturing to produce at commercial and economic rates.
4. Require advanced technologies that may include horizontal drilling and microseismic monitoring to improve production and operational efficiencies.

5. The reservoirs may be the source rocks or are close to the source rocks. Unlike conventional reservoirs, a seal or trap may not be present.
6. The reservoirs may not have definable hydrocarbon/water contacts.
7. "Sweet spots" with improved productivity may exist. A "sweet spot" represents concurrence of favorable rock properties (geochemical, geomechanical, and geological) that promote accumulation and producibility of hydrocarbons. In unconventional reservoirs, "sweet spots" are generally isolated and not contiguous.
8. Abnormal reservoir pressures (higher or lower than hydrostatic) are prevalent.
9. Hydrocarbons-in-place are generally large and hard to define and quantify. Recoveries are generally lower in comparison to conventional reservoirs.
10. More wells may be required to develop unconventional reservoirs because the estimated ultimate recovery (EUR) per well is lower compared to conventional reservoirs.
11. Transient (unsteady state) flow in unconventional reservoirs may last for months or even years due to low matrix permeabilities in comparison to conventional reservoirs where transient flows may last for hours or several days but rarely longer than a month.
12. Free, desorbed hydrocarbons may be interacting under forces that are not in accordance with hydrodynamics.
13. Oil and gas wells show steep decline rates within the first 1 or 2 years of production.
14. Initial production (IP) from wells in unconventional reservoirs is generally less than the IP of wells in conventional reservoirs.
15. A dewatering phase may be encountered within IP phase as observed in coalbed methane (CBM) wells.

On the basis of these characteristics, certain reservoir-types are classified as unconventional. Note that other reservoir-types can be considered to be unconventional (such as Heavy oil and/or Natural Bitumen, Gas hydrates, etc.) depending on the criteria used in the definition. The reservoir-types that are discussed in this book under the unconventional term are grouped as follows:

1. Tight gas/oil
2. Shale gas/oil
3. Coalbed methane

1.2.2.1 Tight Gas/Oil Reservoirs

Tight gas and oil reservoirs are described geologically as consisting of fine-grained rock particles composed of siltstones, sandstones, and carbonates. They can be fluvial, laterally discontinuous, or blanket deposits. The formations may be stacked, lenticular, or both. The reservoirs can have high or low pressures and exist in high or low temperature environments. A major distinction between tight and shale reservoirs is that the hydrocarbons found in tight reservoirs were sourced from another formation (likely shale formations) then migrated and became trapped in the tight formations. Most of the characteristics enumerated earlier for unconventional reservoirs apply to tight oil and gas reservoirs. Note that in many books, reports, and technical papers, shale oil and gas reservoirs are classified as tight oil and gas reservoirs because the characteristics and performance of both reservoir types overlap in many basins. Major tight oil reservoirs in the United States of America are found in the Bakken formation, Eagle Ford formation, Monterrey formation, Utica play, and the Niobarra formation, etc. (Figure 1.7).

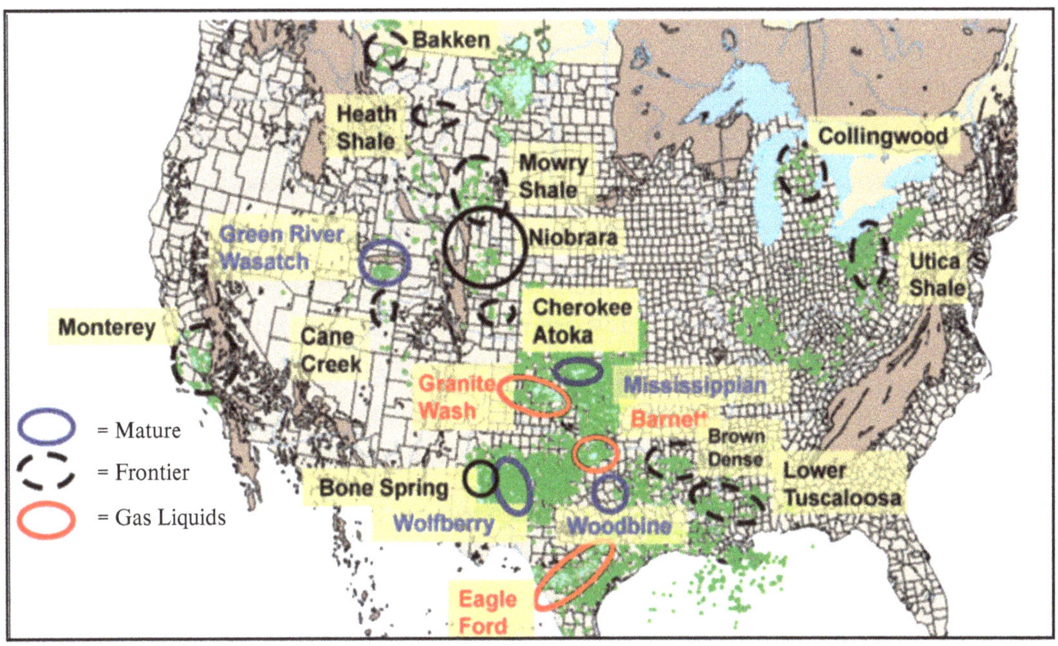

▲ FIGURE 1.7 Map of tight oil plays in the USA[6]

The major tight gas plays in the United States are Anadarko, Deep Bossier, Permian Basin, Appalachian Basin, Austin Chalk, etc. (Figure 1.8). Tight oil and gas resources exist in other basins around the world as shown in Figure 1.9 but are not yet developed commercially as in the United States of America.

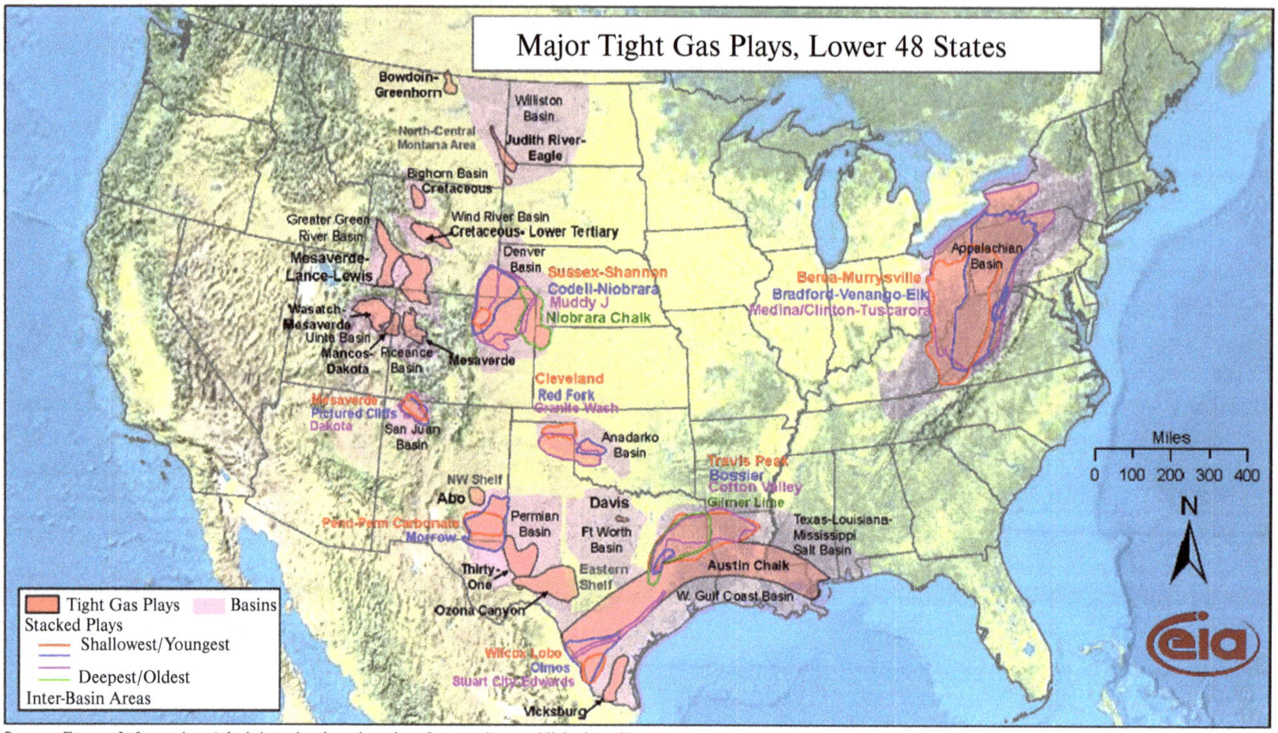

Source: Energy Information Administration based on data from various published studies
Updated: June 6, 2010

▲ **FIGURE 1.8 Map of tight gas plays in the USA**[7]

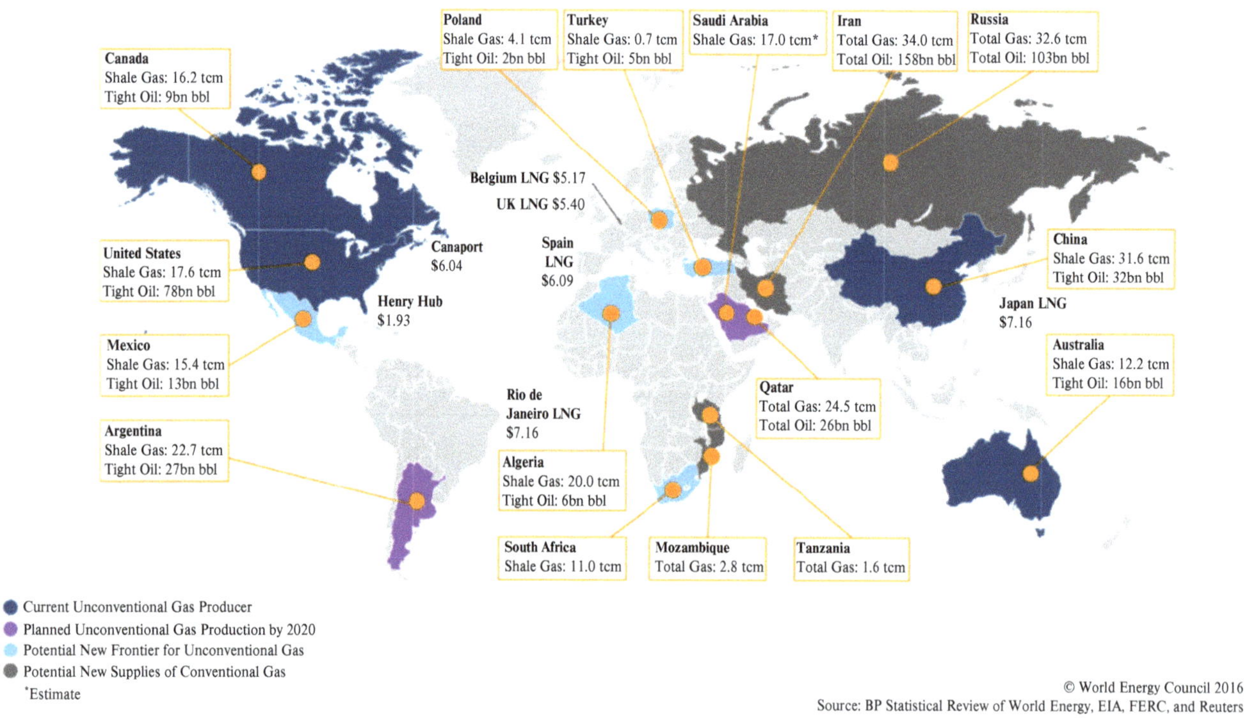

▲ **FIGURE 1.9 World map showing countries with tight oil and gas resources**[8]

1.2.2.2 Shale Gas/Oil Reservoirs

Shale gas and oil reservoirs are finely-grained, sedimentary rocks composed of silts, muds, and clays. Generally, shale reservoirs have low matrix permeability usually in the micro/nano-darcy range (Table 1.1). The matrix porosity is also low and, in most cases, less than 10%. The term shales is widely used in the petroleum industry to describe many rocks that are not typical sandstones or carbonates. Consequently, the variability of rocks described as shale changes from basin to basin and even within the same basin. Shale rocks have high total organic carbon (TOC). TOC is a measure of the amount of organic material available for conversion into hydrocarbons in the transformation processes that generate hydrocarbons. Most of the hydrocarbons that exist in conventional reservoirs were generated in shale formations and then migrated and became trapped in sandstone and carbonate formations. Shale reservoirs act as both the source, container, and the trap for the hydrocarbons that exist in them. Shale reservoirs manifest most of the characteristics of unconventional reservoirs enumerated earlier in this chapter. Hydrocarbon accumulations in shale formations can vary from dry gas to wet gas to condensate and then oil depending on the depth of the reservoir. These hydrocarbon phases exist in the Eagle Ford Shale in South Texas, USA. A map showing shale plays in the USA is presented in Figure 1.10. Figure 1.11 shows worldwide map of shale gas and shale oil basins.

1.2.2.3 Coalbed Methane

Coalbed methane (CBM) or coal seam gas (CSG) is an unconventional source of natural gas consisting mostly of methane that is produced from coal beds and coal seams. Coal as a rock is both the source and reservoir for CBM. CBM is formed during the transformative processes that created the coal. Coal gas is adsorbed on the coal matrix and seams in large quantities, depending on the reservoir pressure of the coal beds. Additional gas is stored in fractures that may exist within the coal beds. Generally, coal matrix permeability is in the millidarcy range. Hence, hydraulic fracturing is generally required to produce the gas at commercial rates. The natural fractures in the coal beds are initially filled with water. This water must be produced during the "dewatering phase" before higher gas production

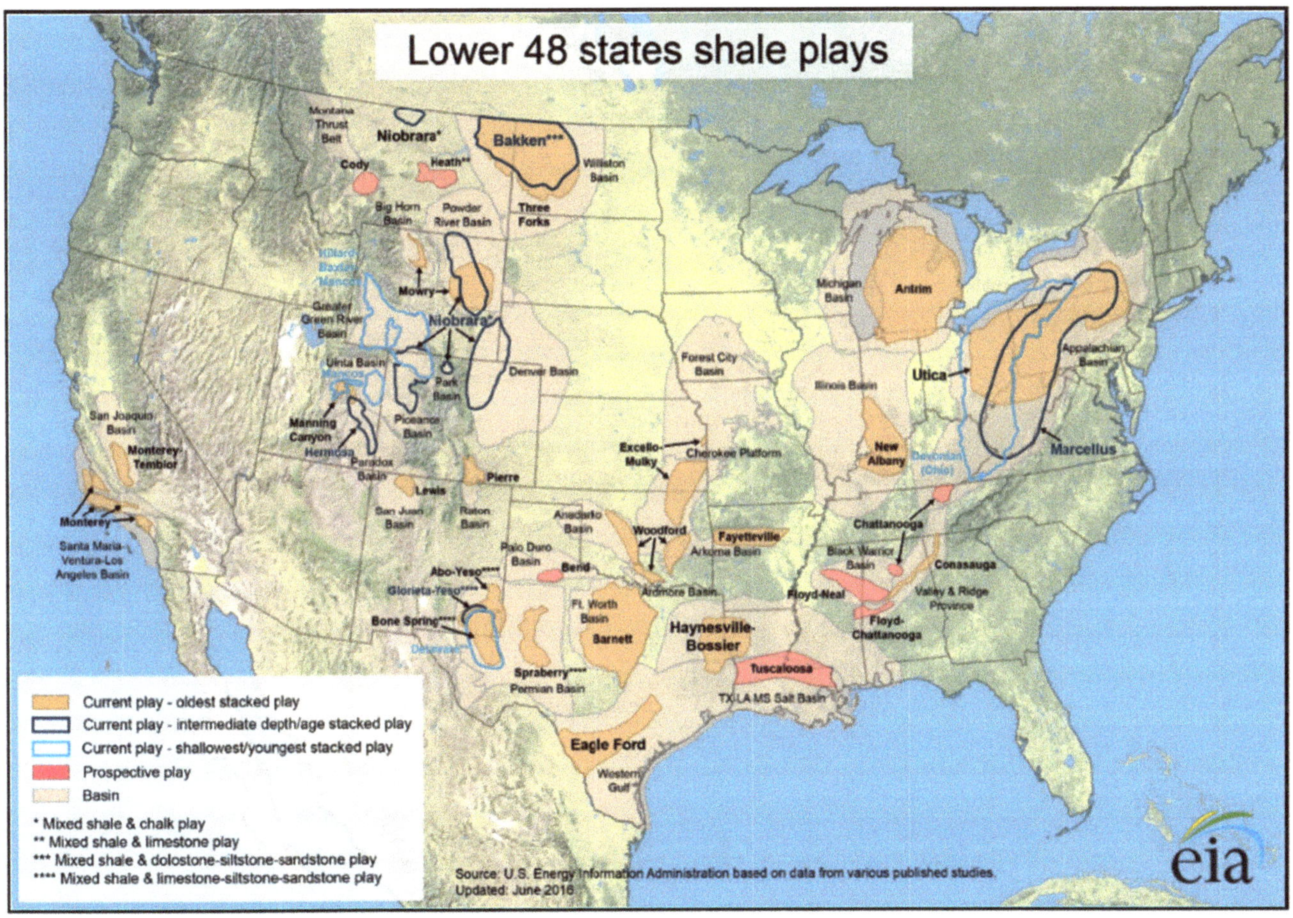

▲ FIGURE 1.10 Map of shale plays in the USA[9]

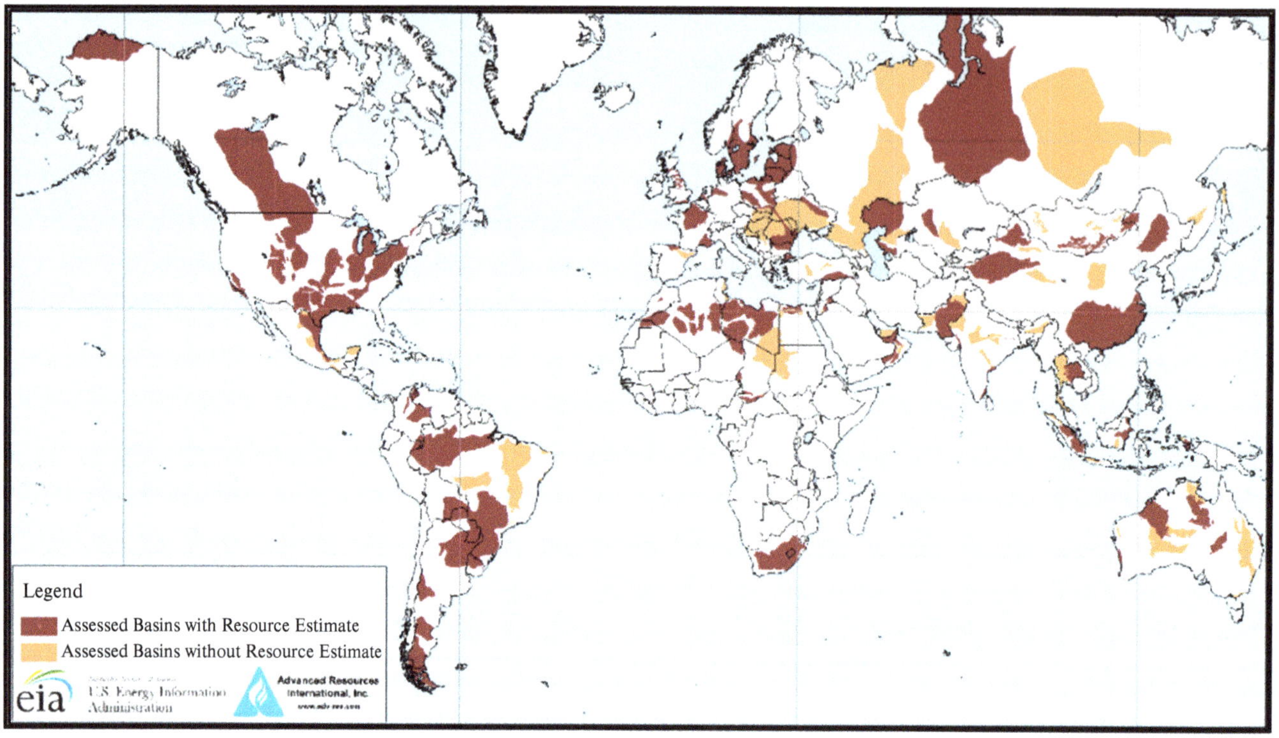

▲ **FIGURE 1.11** **World map of shale oil and gas basins**[10]

rates could be achieved (Figure 1.12). Vertical and/or horizontal wells have been used to produce coal gas (Figure 1.13). It has been estimated that the total gas-in-place in coal deposits worldwide could be as high as 11296 tcf mainly in Russia, China, and United States (Table 1.4). CBM is produced in commercial quantities in United States, China, India, and Australia. Figure 1.14 shows CBM-proved reserves and production in the United States from 1989 to 2016. The map of CBM fields and coal basins in the United States is shown in Figure 1.15.

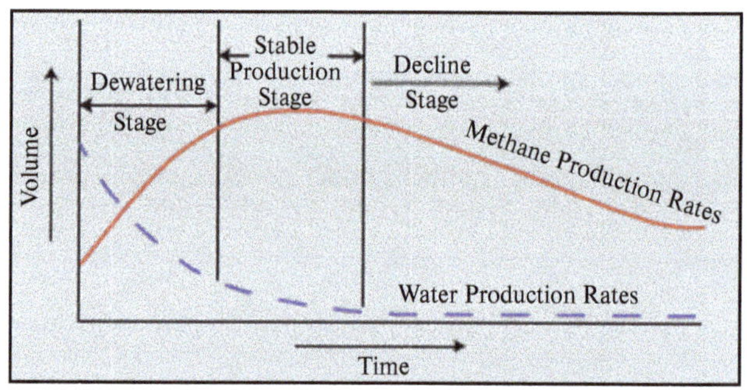

▲ **FIGURE 1.12** **Typical CBM well production profile**[11]

TABLE 1.4 World Estimates of CBM Gas-in-Place Volumes (from Thakur[12])

Country	Estimated CBM Gas-in-Place (Trillion Cubic Feet, tcf)
United States of America	388
Russia	700–5860
China	700–875
Canada	212–2682
Australia	282–2682
Germany	106
India	35
South Africa	35
Poland	106
Other countries	177–353
Total Gas-in-Place	**275–11296**

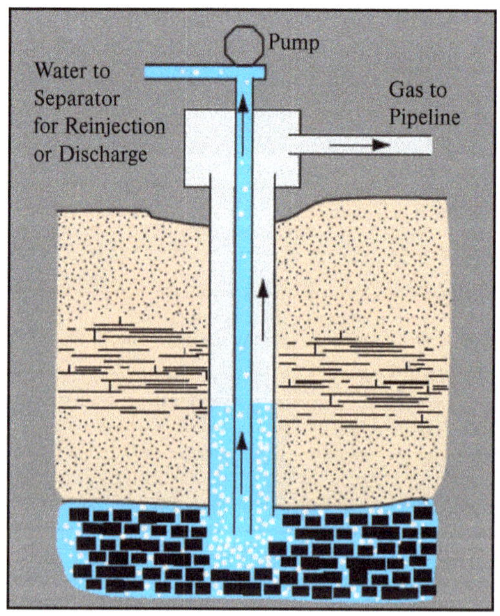

▲ **FIGURE 1.13** **Typical CBM well**[11]

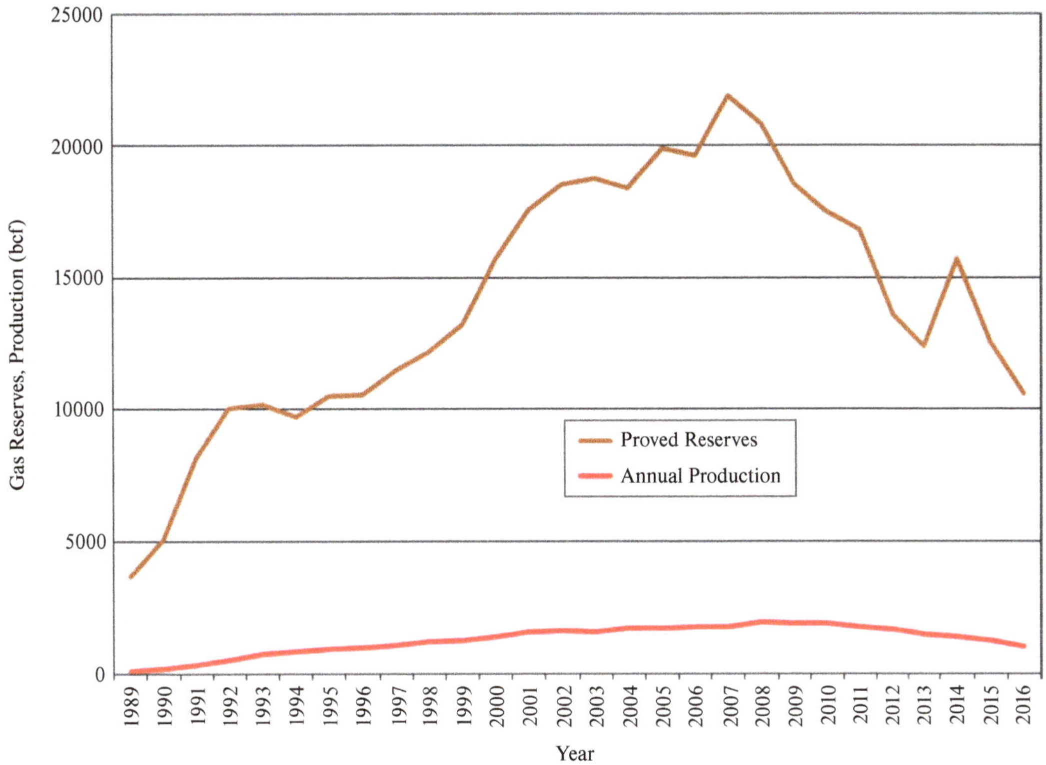

▲ **FIGURE 1.14** **CBM proved reserves and production in the USA from 1986 to 2016**[14]

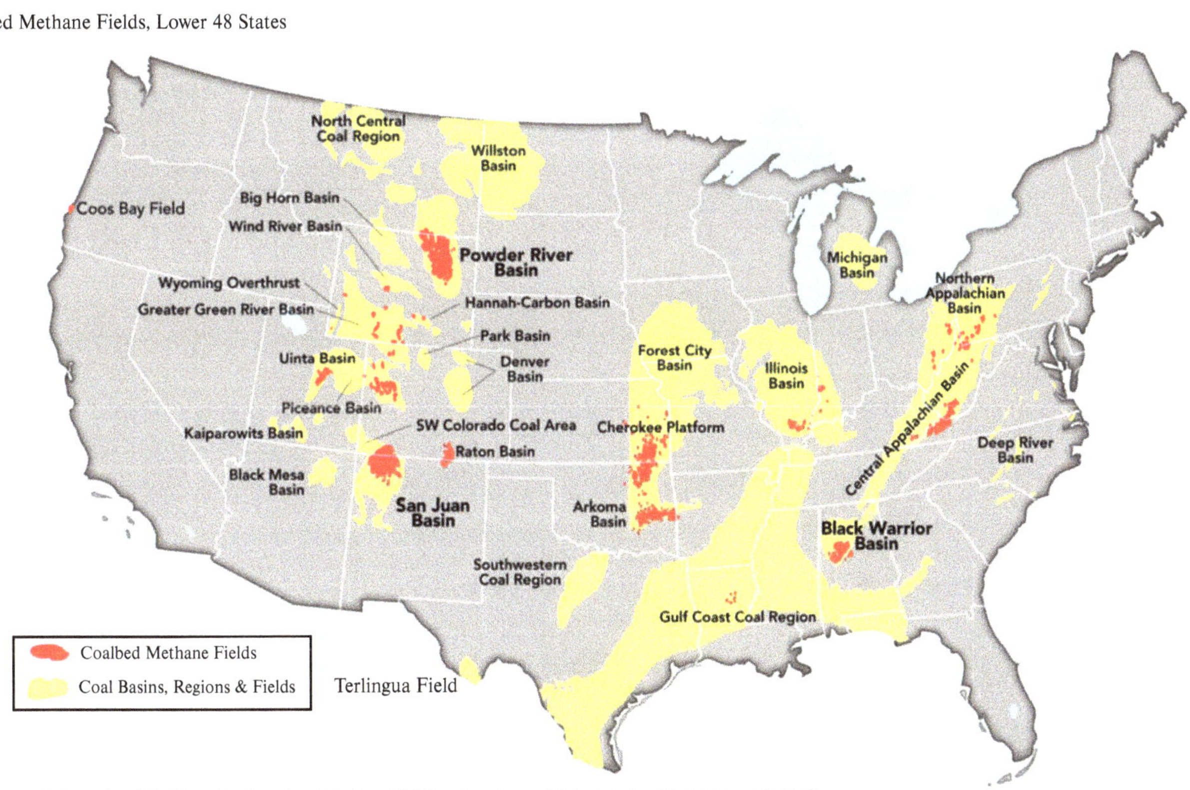

Source: Energy Information Administration based on data from USGS and various published studies. Updated april 8, 2009.

▲ **FIGURE 1.15** **Coalbed methane (CBM) fields and coal basins in the USA**[14]

1.2.3 Unconventional Reservoirs in the United States of America (USA)

The production of oil and gas from unconventional reservoirs has revolutionized the supply and mix of fossil energy in the United States of America (USA). Prior to 2000, production from unconventional reservoirs was virtually nonexistent in the country. But since the early 2000s, technological and operational successes achieved earlier in unconventional shale plays have been repeated across the country on unconventional tight oil/gas plays. As shown in Figure 1.16, projections by the U.S. Energy Information Administration (EIA) show that production from tight oil plays will continue to grow for many years in the future while production from conventional (non-tight) reservoirs will progressively decline over the same period. For example, the EIA[16] projects that tight oil production will account for 65% of the cumulative oil production in the United States of America from 2017 to 2050. The growth in gas production from shale gas plays shows even a higher trend in Figure 1.16. The projected growths in oil and gas production are substantiated by the huge base of recoverable resources. A map of United States of America divided into supply regions by EIA is shown in Figure 1.17. The total technically recoverable crude oil resources by region as reported by EIA[17] in April 2018 were 284.6 billion barrels (Bbbls) (Table 1.5). In Table 1.5, technically recoverable resources (TRR) was defined by EIA as the sum of proved reserves and unproved resources. Similarly, Table 1.6 shows the total technically recoverable dry natural gas resources as 2462.3 tcf. These estimates of TRR as reported by EIA showed that the United States of America has a strong base of resources to meet demand for many years in the future. The strong resource base was built-up in recent years (from 2008) by active appraisal and development of shale and tight basin plays. Table 1.7 is a summary of technical recoverable tight/shale resources by region as reported by EIA in April 2018. In the 2018 report, EIA stated that tight/shale resources accounted for 37% of the total U.S. technically recoverable crude oil resources and 50% of the technically recoverable natural gas resources. This observation underscores the importance of unconventional reservoirs in future supply of oil and gas in the United States of America.

TABLE 1.5 TRR–USA Crude Oil (Jan. 1, 2016)[17]

Regions	Billion Barrels of Oil		
	Proved Reserves	Unproved Reserves	Total Technically Recoverable Reserves
East	0.5	4.2	4.6
Gulf Coast	5.8	32.0	37.8
Midcontinent	2.4	14.1	16.5
Southwest	9.2	71.7	80.9
Rocky Mountain/ Dakotas	8.3	39.2	47.6
West Coast	2.2	4.5	6.6
Total Lower 48 Onshore	**28.4**	**165.7**	**194.1**
Gulf (currently available)	4.3	36.6	40.9
Eastern/Central Gulf (unavailable until 2022)	0.0	3.7	3.7
Pacific	0.4	6.0	6.5
Atlantic	0.0	3.3	3.3
Total Lower 48 Offshore	**4.7**	**49.6**	**54.4**
Alaska (Onshore and Offshore)	2.1	34.0	36.1
Total USA	**35.2**	**249.3**	**284.6**

TABLE 1.6 TRR–USA Dry Natural Gas (Jan. 1, 2016)[17]

Regions	Trillion Cubic Feet of Gas		
	Proved Reserves	Unproved Reserves	Total Technically Recoverable Reserves
East	94.0	613.1	707.1
Gulf Coast	53.6	370.3	423.9
Midcontinent	45.4	143.8	189.2
Southwest	43.8	194.9	238.7
Rocky Mountain/ Dakotas	56.9	255.5	312.4
West Coast	1.8	34.1	35.8
Total Lower 48 Onshore	**295.5**	**1611.6**	**1907.1**
Gulf (currently available)	7.3	209.5	216.8
Eastern/Central Gulf (Unavailable until 2022)	0.0	21.5	21.5
Pacific	0.3	9.3	9.6
Atlantic	0.0	31.7	31.7
Total Lower 48 Offshore	7.6	272.0	279.6
Alaska (Onshore and Offshore)	4.6	271.1	275.6
Total USA	**307.7**	**2154.6**	**2462.3**

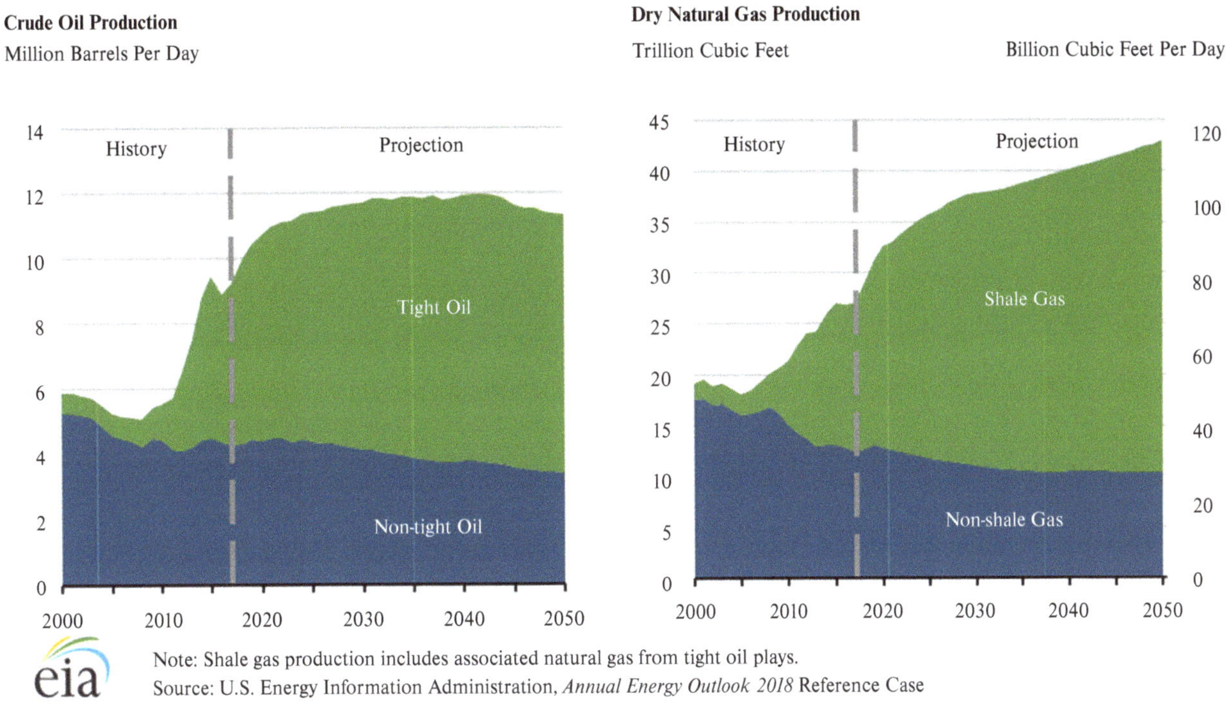

▲ **FIGURE 1.16** **Crude oil and dry gas production in the USA**[15]

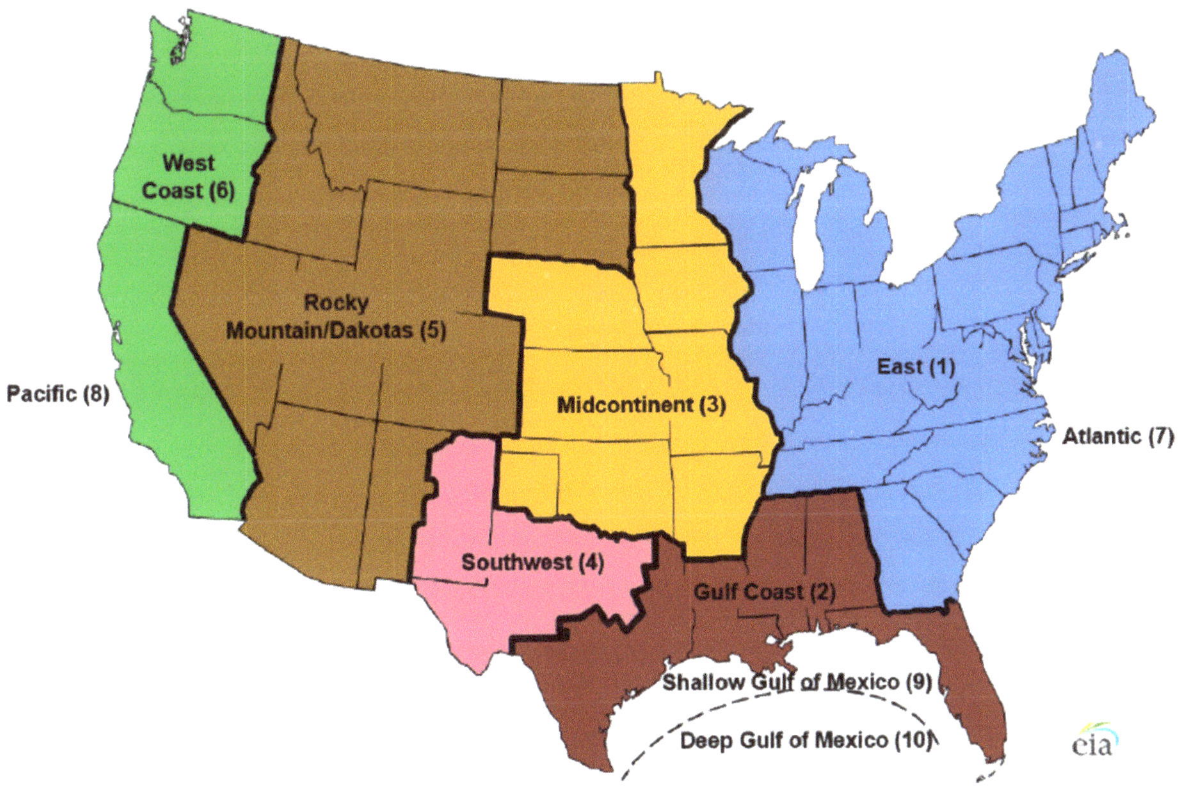

Source: U.S. Energy Information Administration

▲ **FIGURE 1.17** **Map of USA by regions**[17]

TABLE 1.7 USA Unproved TRR–Tight/Shale Oil and Gas (Jan. 1, 2016)[17]

Region	Technical Recoverable Resources			Major Plays
	Crude Oil (Billion bbls)	Natural Gas (Tcf)	Natural Gas Plant Liquids (NGPL) (Billion bbls)	
East	3.3	579.1	22.6	Marcellus, Utica
Gulf Coast	25.3	280.1	10.0	Haynesville, Cotton Valley
Midcontinent	1.6	73.6	1.6	Fayetteville, Woodford
Southwest	73.6	281.9	23.5	Wolfcamp, Niobrara
West Coast	0.2	12.7	0.0	Basin Central
Total	**104.0**	**1227.4**	**57.7**	

The EIA[16] projected in 2018 that natural gas production accounted for the largest proportion of total energy production in the USA (Figure 1.18). Natural gas production will account for almost 39% of total USA energy production by 2050. Most of the growth in natural gas production will come from shale gas and tight oil plays as shown in Figure 1.16. The USA crude oil production in 2018 is projected to exceed 9.6 million BOPD and will level off between 11.5 and 11.9 million BOPD (Figure 1.16). Note that energy production from coal will initially decline and then level off from 2025 to 2050 (Figure 1.18). The fuel mix of energy consumption in the USA has changed as a direct result of abundant and cheap natural gas supply from unconventional reservoirs. Figure 1.19 shows that energy consumption from natural gas will grow the highest in comparison to other sources of energy. Again, note that energy consumption from coal will decline and then level off. As shown also in Figure 1.19, industrial and power sectors account for most of the growth in energy consumption supplied from natural gas.

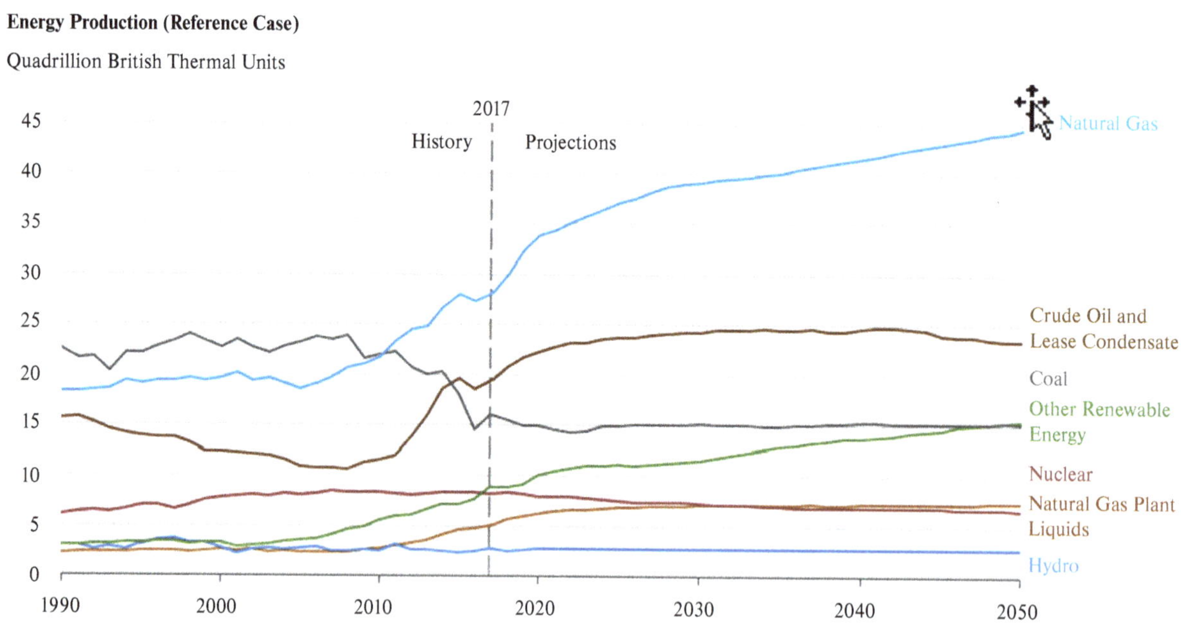

▲ **FIGURE 1.18** Energy production in the USA[16]

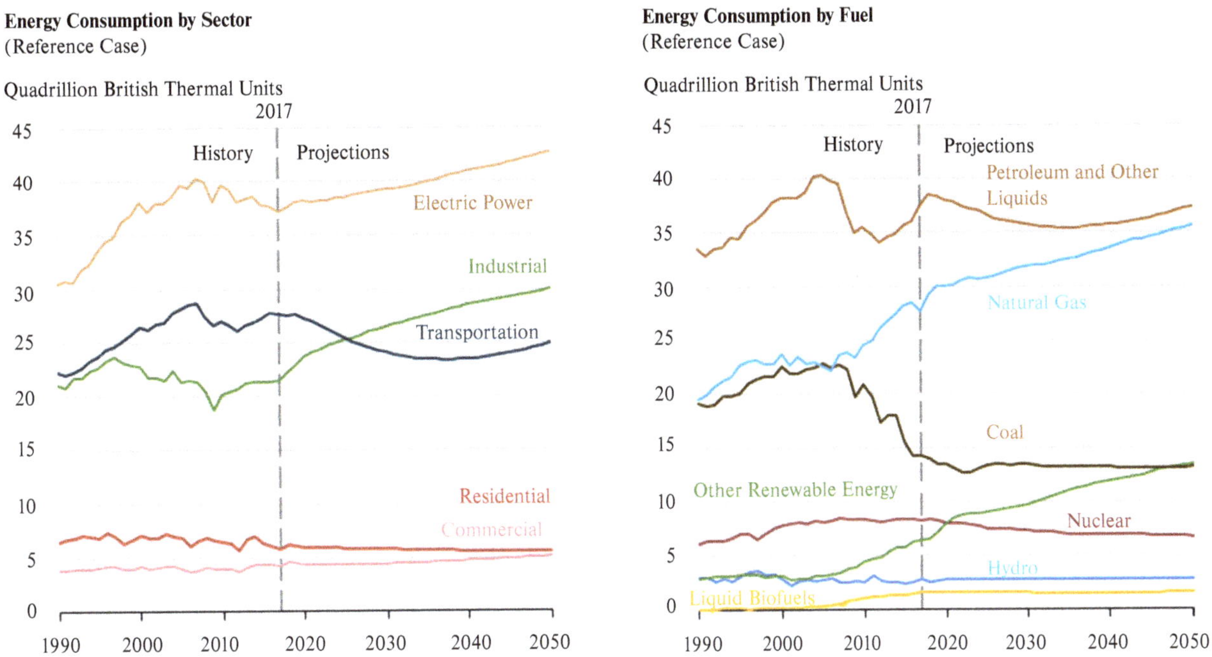

▲ **FIGURE 1.19** **Energy consumption by fuel and sector in the USA**[16]

The United States of America has been a net importer of energy since 1953 according to EIA.[16] It has been projected by EIA that USA will be net exporter of energy in early 2020s (Figure 1.20). This is attributed to increasing export of natural gas and liquefied natural gas (LNG) and decreasing import of crude oil. The USA transiting from an energy importer to an energy exporter is simply another major transformation created by the abundant energy supply from unconventional sources. As shown in Figure 1.16, growth in tight oil and shale gas production are sufficient to meet the energy consumption in the country with some excess left for export. This is the main lesson learned from the transformative change in energy supply created by production from unconventional shale and tight resources in the United States of America. This lesson is bound to spread soon to other parts of the world, leading to abundant energy supply in many countries.

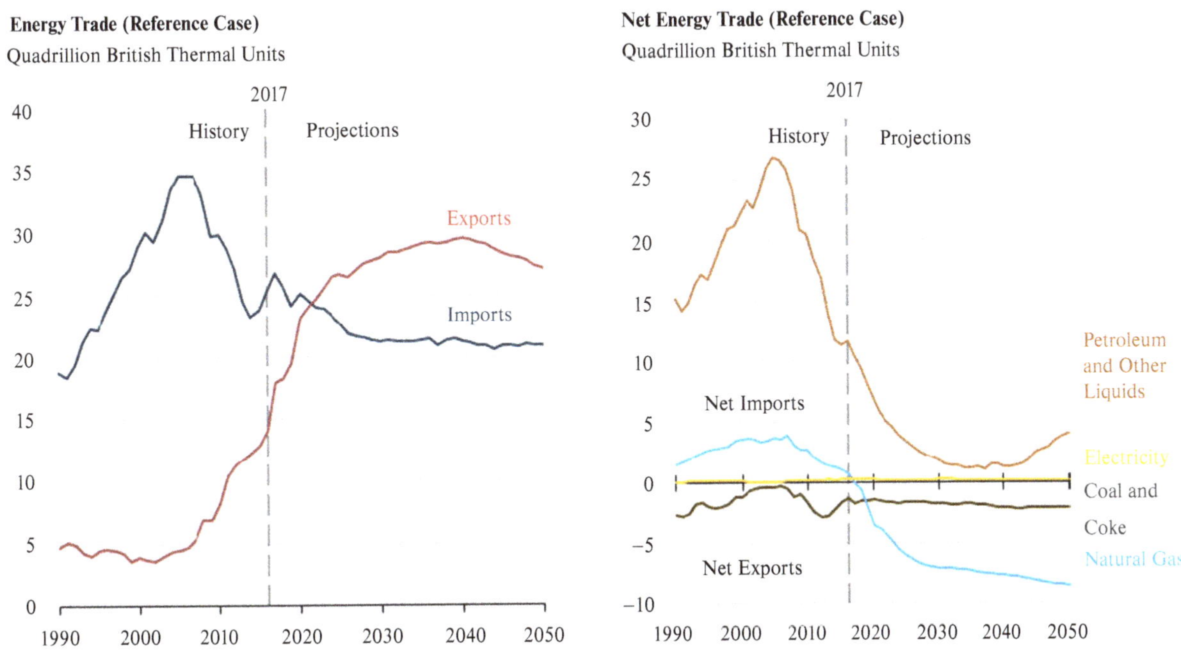

▲ **FIGURE 1.20** **Energy trade for USA**[16]

The successes achieved in exploiting unconventional resources in the United States of America were led by independent oil companies. The successes achieved in the Barnett shale led to the development of the Fayetteville and Bakken formations. These in turn led to successes in the Woodford, Haynesville, Marcellus, Eagle Ford, Horn River, etc., formations.[6] Most of the early successes were achieved with the exploitation of shale gas plays. Figure 1.10 shows a map of the shale gas plays in the United States of America. Production from shale plays (Marcellus and Utica) in the East region is projected to lead the growth in total USA shale gas production followed by the Gulf Coast region (mainly Eagle Ford and Haynesville plays) as shown in Figure 1.21. The shale gas production contributed from Marcellus, Utica, Eagle Ford, Haynesville, and other shale plays are shown in Figure 1.22. In 2015, natural gas production in the United States of America from shale gas plays was 37.4 Bcf/D (Figure 1.22). This represented 50% of total natural gas production in the country as reported by the EIA.[18]

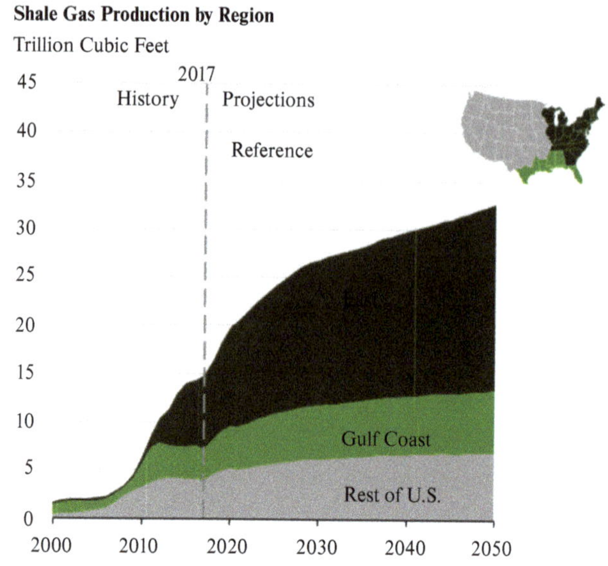

▲ **FIGURE 1.21** **Shale gas production by region in the USA**[16]

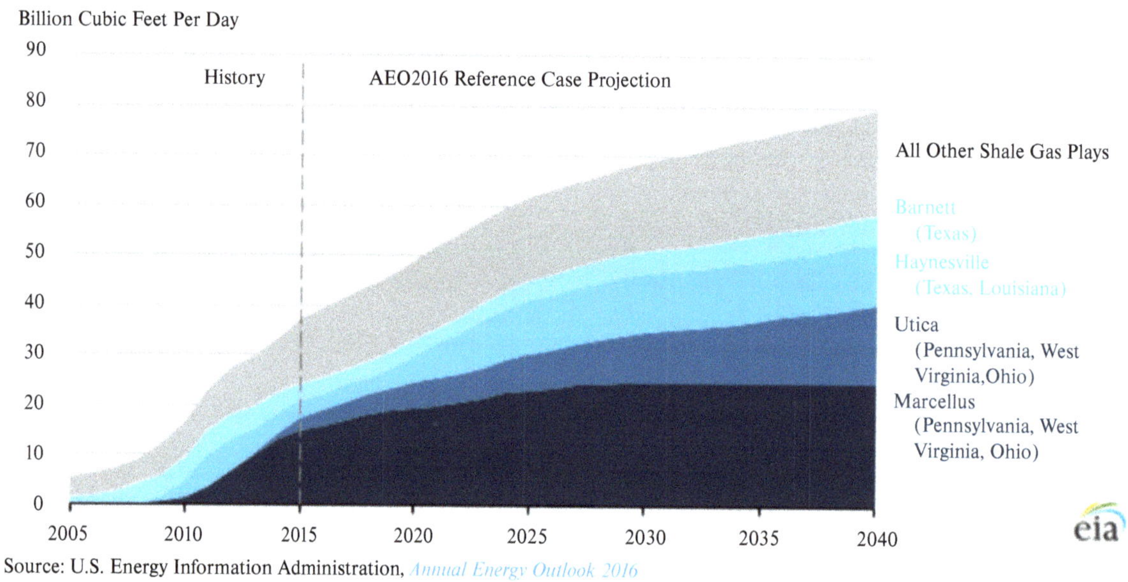

▲ **FIGURE 1.22** **Shale gas production by formation in the USA**[18]

From 2008 to 2009, the sale price of natural gas in the USA slumped to about $3.0 per Mcf due to oversupply and lower demand. In response to the price decline, many independent producers shifted their focus to shale/tight plays containing wet gas or tight oil. The basic technologies (horizontal wells and multistage fracturing) that were used to develop shale/tight gas plays were redirected to developing wet gas and tight oil plays. In January 2009, 80% of the active rigs in the USA were drilling for natural gas and 20% were drilling for oil. By January 2012, the distribution had changed to 60% of the active rigs in the USA drilling for oil and 40% drilling for natural gas.[6] The EIA[19] projects that tight oil production will increase through the early 2040s to surpass 8.2 million BOPD and represent nearly 70% of total oil production in the country (Figure 1.16). In 2017, tight oil production accounted for 54% of total USA oil production. Most of the growth in the USA tight oil production is driven by development of tight oil resources in the Permian basin (Figure 1.23). The most prolific tight oil plays in the Permian Basin are Spraberry, Bone Spring, and Wolfcamp. These three plays are projected to account for 43% of cumulative tight oil production in the USA through 2050. The Bakken and Eagle Ford formations will account for 20% and 17%, respectively, of cumulative tight oil production in the USA through 2050. A map showing most of the tight oil plays in the USA is shown as Figure 1.7.

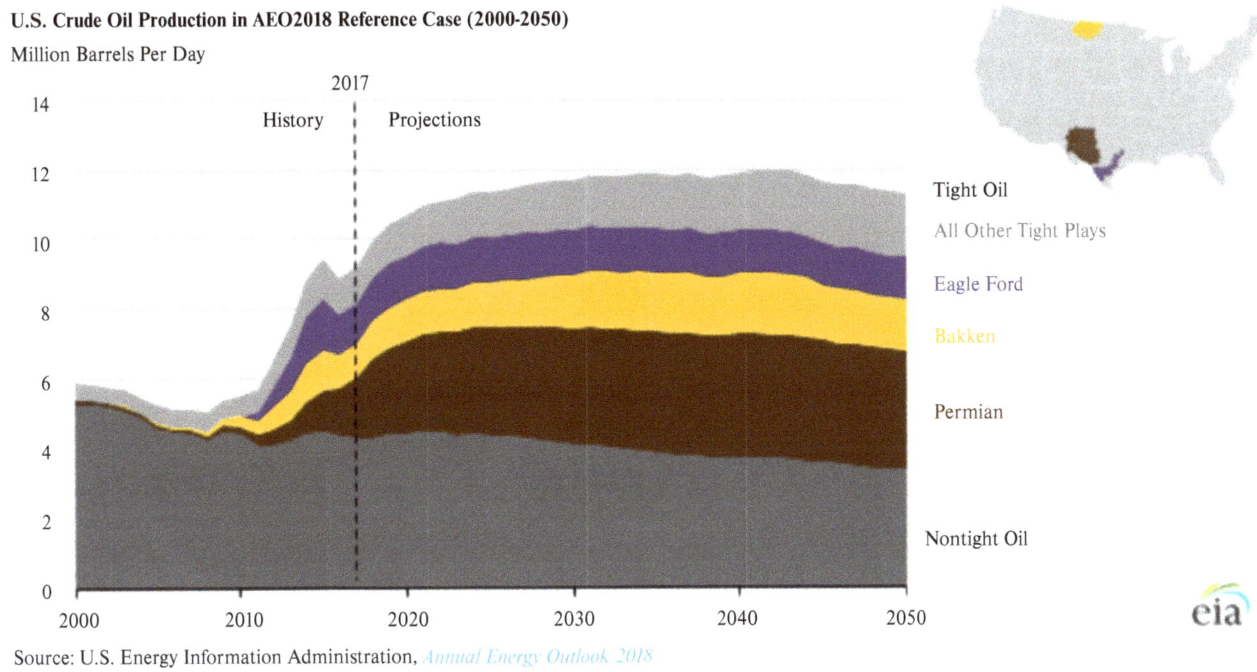

▲ **FIGURE 1.23** **Oil production by formation in the USA**[19]

The rise of unconventional resources in the USA was driven by advances in drilling and completions technology coupled with improvements in operational efficiencies. The major advance in drilling technology is the progression from drilling vertical wells to horizontal wells with very long laterals. The capability to drill horizontal wells with long laterals was accompanied with the technology to complete these wells with multistage hydraulic fracturing. The history of the advancement in the two technologies (horizontal wells and multistage fracturing) is illustrated with the development in the Barnett Shale as shown in Figure 1.24. Development of the Barnett Shale was started with vertical wells then directional wells and at last horizontal wells with multistage fracture completions. Note the improvement in initial production with the advancement in drilling and completion technology as illustrated in Figure 1.24. Figure 1.25 shows the increase in average lateral length of horizontal wells, the increase in the number of fractures stages, and increase in the initial gas production per well from 2006 to 2011. Figures 1.24 and 1.25 demonstrate the advancement of drilling and completion technology, and the impact they had on productivity of wells drilled in USA unconventional gas basins. Similar increases in the number of horizontal wells drilled in the Eagle Ford Shale from 2008 to 2016 were reported by the EIA[21] as shown in Figure 1.26. The EIA[22] reported that in 2016 hydraulically fractured horizontal wells represented 69% of all oil and gas wells drilled in the USA and 83% of the total linear footage drilled (Figure 1.27). In 2011, horizontal fractured wells became the dominant method for development of new oil and gas wells in the USA when the total linear footage drilled and completed became higher than all other drilling and completion methods (Figure 1.27). As of 2016, 670,000 of the 977,000 producing wells in the USA were drilled horizontally and completed with multistage hydraulic fractures.[22] Independent operators of unconventional resources in the USA have continued to drill longer horizontal laterals, increase proppant loading, decrease stage spacing, and reduce the number of days used to drill a well.[23] The increase in proppant loading and decrease of fracture stage spacing have increased stimulated reservoir volume which translates to enhanced productivity. In 2017, an operator[24] in the Utica Shale reported drilling the world's longest onshore lateral with a total measured depth of 27,400 feet and completeable lateral length of 19,300 feet. This well was drilled in less than 17 days.

The other important factor that played a crucial role in the development of unconventional resources in the USA is continuous improvement of operational efficiencies and effectiveness. The USA operators have focused on continuously driving down drilling and completion costs and costs of other phases of the operations. They have also developed technologies such as microseismic sensing to target "sweet spots" in many unconventional resource basins. Figure 1.28 illustrates the efficiency gained by operators in the Bakken and Eagle Ford plays by driving down the average number of days needed to drill a well between 2012 and 2014. Another operational efficiency that operators are using is called pad drilling. **Pad drilling** allows operators to drill multiple wells from a single location. Pad drilling reduces costs by removing the need to build multiple locations especially in rough terrain thereby reducing the time required to drill and complete a well.

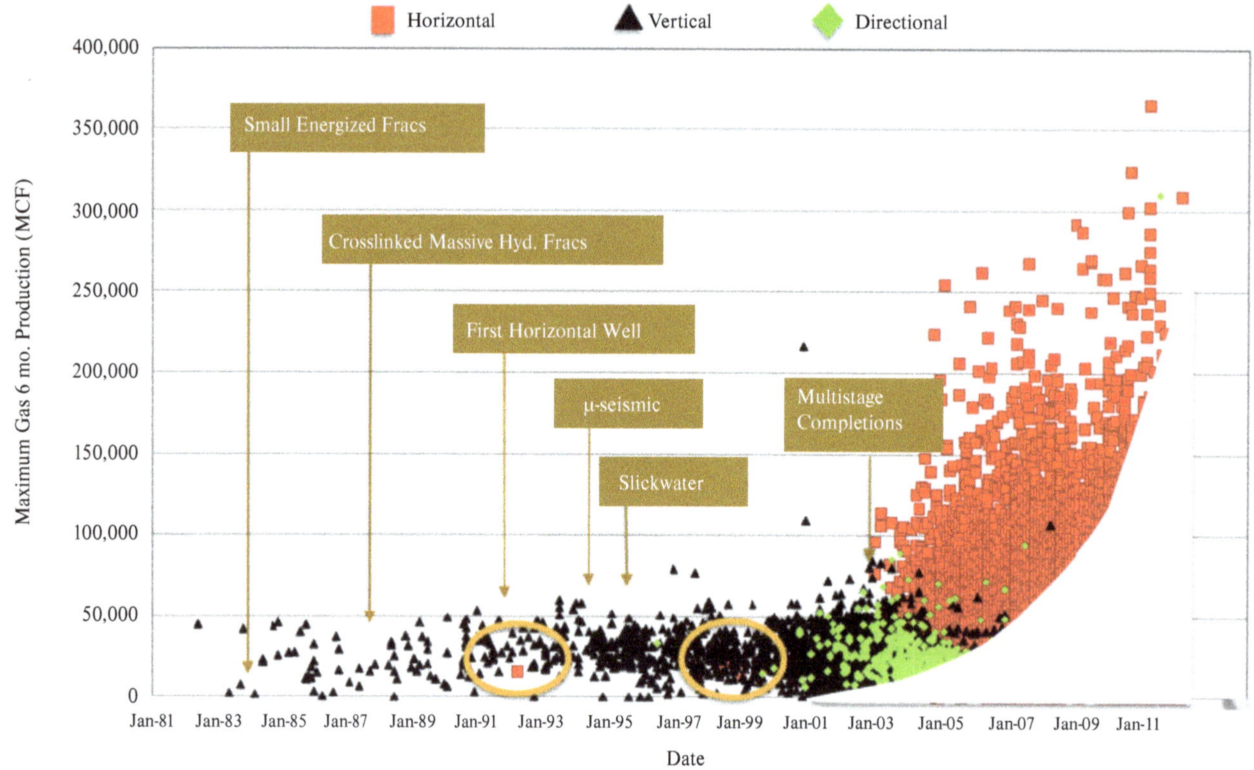

▲ **FIGURE 1.24** Barnett Shale development[20]

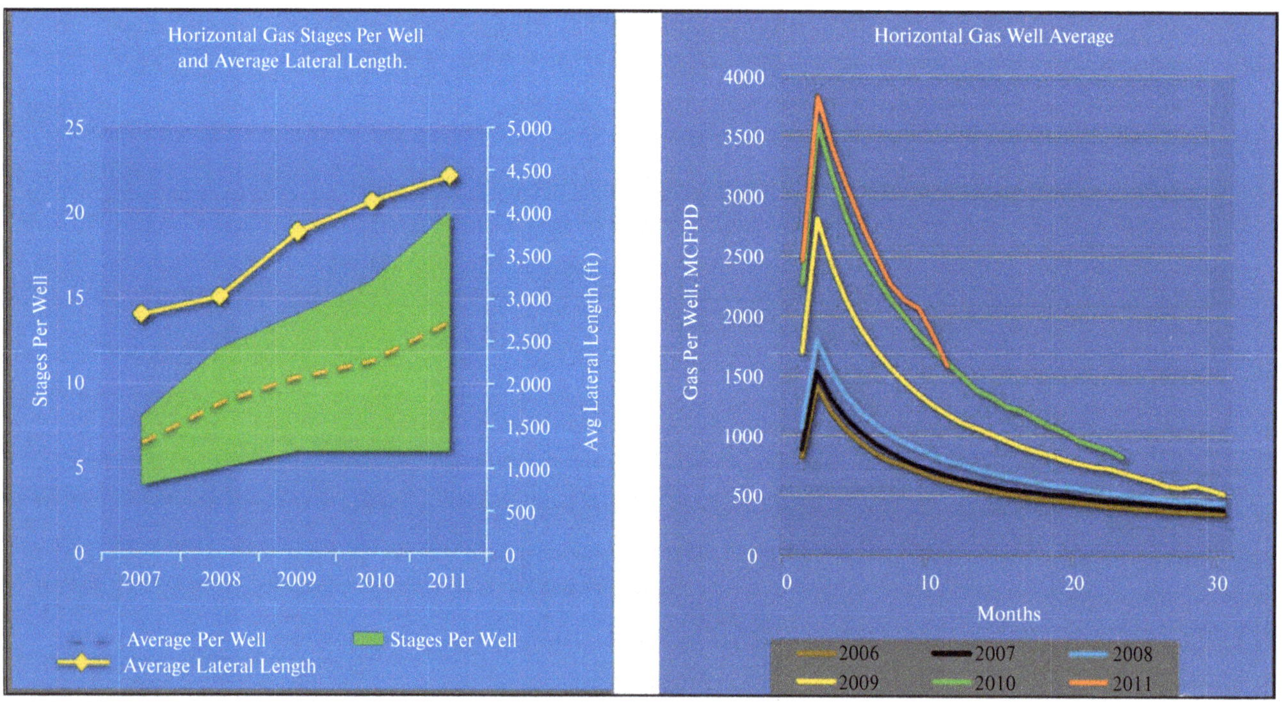

▲ **FIGURE 1.25** Increase in lateral length, fracture stages, and initial production for USA gas basins[20]

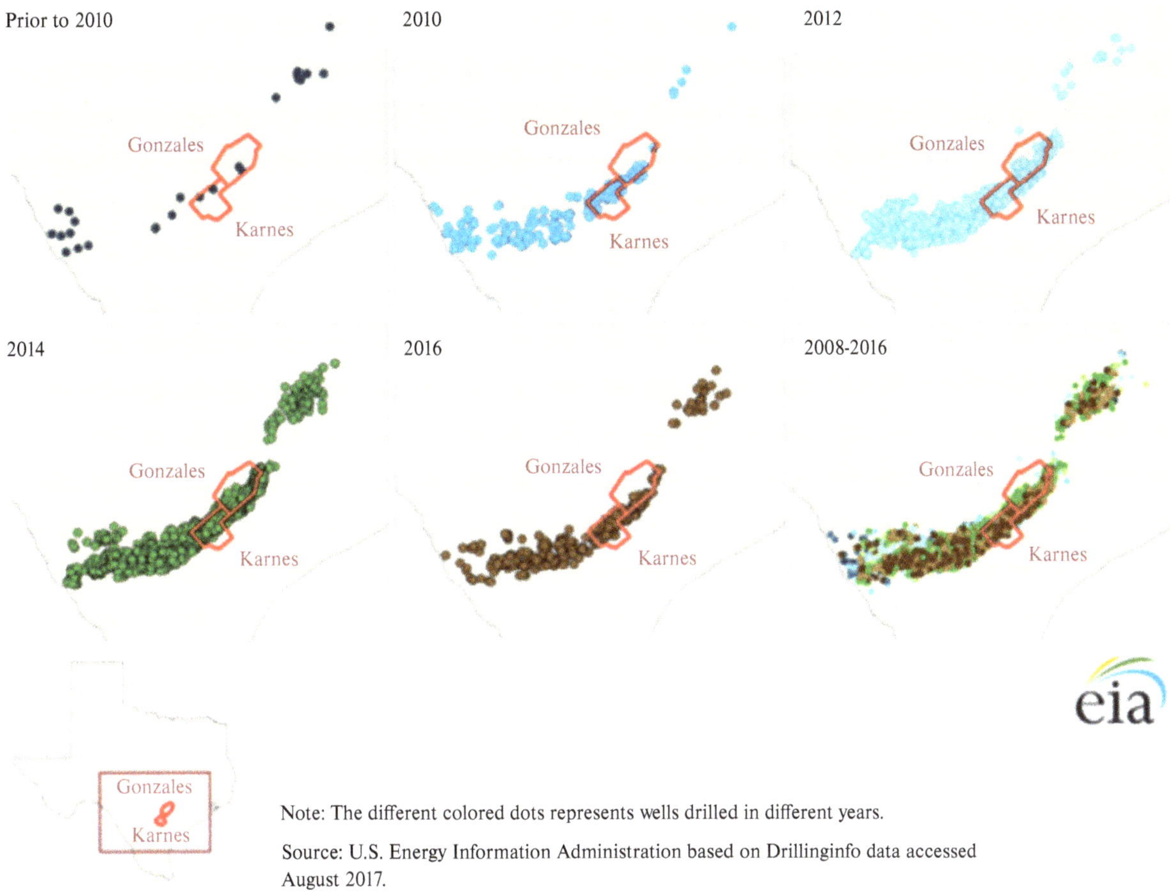

Note: The different colored dots represents wells drilled in different years.

Source: U.S. Energy Information Administration based on Drillinginfo data accessed August 2017.

▲ **FIGURE 1.26** **Horizontal wells drilled in the Eagle Ford Shale, 2008–2016**[21]

Monthly Crude Oil and Natural Gas Well Drilling Footage by Type (2000-2016)

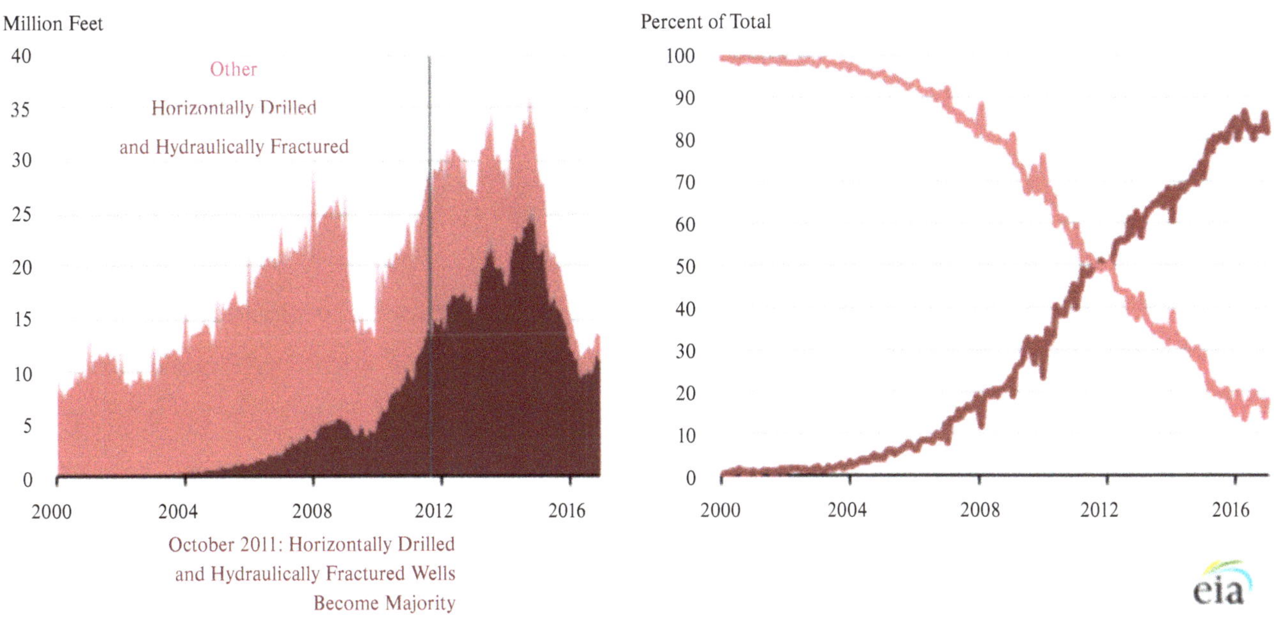

Source: U.S. Energy Information Administration, based on Drillinginfo Inc. and IHS Markit

▲ **FIGURE 1.27** **Drilling footage by type in the USA**[22]

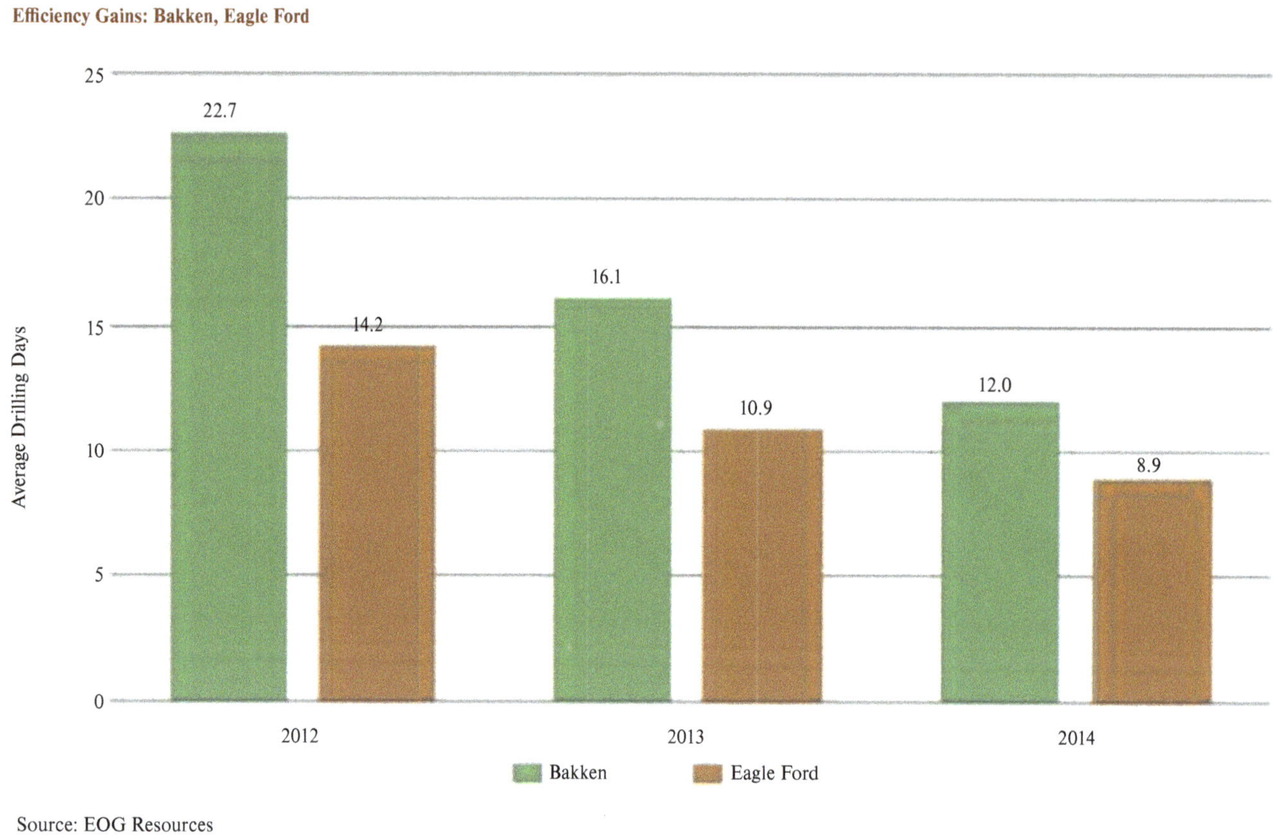

▲ **FIGURE 1.28** **Drilling efficiency gains in Bakken and Eagle Ford plays**[28]

1.2.4 Potential Resources in Unconventional Reservoirs in the World

An assessment of the world's shale gas and shale oil resources was conducted for the U.S. EIA by Advanced Resources International (ARI) in 2013.[25] The assessment covered 26 regions around the world situated in 41 countries (Figure 1.11). An updated summary of the unconventional resources assessed around the world is presented in Table 1.8. The EIA/ARI study (updated in 2015) assessed total risked shale gas-in-place in the world as 35,262 Tcf and the risked technically recoverable shale gas resource as 7576 Tcf. The same updated study assessed the risked total shale oil-in-place in the world as 8430 billion bbls with risked technically recoverable shale oil resources as 419 billion bbls. The EIA/ARI report[25] noted that additional unconventional resources exist in the Middle East, Central Africa, and other countries not yet covered in the study. Bearing this in mind, Table 1.8 demonstrates that considerable unconventional resources exist around the world and could be available to meet future world energy demand. According to the EIA/ARI study, United States, China, Algeria, Canada, and Mexico account for 67% of the risked technically recoverable shale gas resources. For the risked technically recoverable shale oil resources, Russia, United States, China, Argentina, Libya, and Venezuela account for 67% of the total.

In 2014, shale gas and tight oil were commercially produced in only four countries, namely, United States, Canada, China, and Argentina.[26] Most of the unconventional oil and gas production are currently occurring in the United States of America followed distantly by Canada. Future production is expected to spread to Mexico and Algeria. These six countries are projected to produce 70% of global shale production by 2040 as reported by EIA.[27]

The size of assessed technically recoverable unconventional resources sends a clear message to the world that the world will **not** run out of energy resources in the foreseeable future. The technological breakthroughs that have driven the unprecedented revolution underway in the USA unconventional resource plays are bound to spread throughout the world in the near future. The major and independent oil companies that have developed these technologies in the USA have operations in most shale/tight oil regions in the world. Sooner than later, these technologies would be applied to develop these resources outside the

TABLE 1.8 EIA/ARI World Shale Gas and Shale Oil Resources Assessment–Updated Sept. 24, 2015.[5] (Note: North American includes resources from USA obtained from EIA.[6])

Region	Risked Shale Gas in-Place (Tcf)	Technically Recoverable Resources–Wet Shale Gas(Tcf)	Risked Shale Oil in-Place (Billion bbl)	Technically Recoverable Resources–Shale Oil (Billion bbl)
North America	6,962	1,741	2,000	100
Australia	2,008.0	429.3	364.1	15.6
South America	6,402	1,433	1,152	60
Eastern Europe	3,555	628	1,341	79
Western Europe	1,341	255	210	9
North Africa	5,106	972	882	38
Sub-Saharan Africa	1,997	434	393	16
Asia	6,460	1,397	1,372	61
Caspian	253	28	221	11
Middle East	1,178	260	495	29
Total	**35,262**	**7576**	**8430**	**419**

USA even with greater success and efficiency. The lesson to take away from this chapter is that exploration, appraisal, and development of unconventional resources will continue to play a dominant role in the petroleum industry. Future petroleum engineers should develop knowledge and skills on the peculiar characteristics, management, and engineering of unconventional reservoirs.

Abbreviations

ARI	Advanced Resources International
Bcf/D	Billion cubic feet of gas per day
BOPD	barrels of oil per day
BP	British Petroleum
CBM	coalbed methane
CSG	coal seam gas
EIA	Energy Information Administration
EUR	estimated ultimate recovery
LNG	liquefied natural gas
Mcf	Thousand cubic feet of gas
R/P	reserves-to-production ratio
Tcf	Trillion cubic feet of gas
TOC	total organic carbon
TRR	technically recoverable resources
USA	United States of America
U.S.	United States of America

REFERENCES

1. Holditch, S. A.: "Tight Gas Sands," JPT (June 2006) 86-93. SPE 103356
2. Global Map of Giant Oil and Gas Fields. http://holidaymapq.com/map-of-oil-fields.html. Accessed March 30, 2018.
3. Dudley, Bob. "BP Statistical Review of World Energy. June 2017." *London, UK* (2017). https://www.bp.com/content/dam/bp/en/corporate/pdf/energy-economics/statistical-review-2017/bp-statistical-review-of-world-energy-2017-full-report.pdf. Accessed March 28, 2018.
4. Outlook, BP Energy. "edition (2017)." (2017). https://www.bp.com/content/dam/bp/pdf/energy-economics/energy-outlook-2017/bp-energy-outlook-2017.pdf. Accessed March 28, 2018.
5. U.S. Energy Information Administration. "International Energy Outlook 2017." September 14, 2017. https://www.eia.gov/outlooks/ieo/pdf/0484(2017).pdf. Accessed April 2, 2018.
6. Stark, P.H.: "Shale Gas, Tight Oil and EOR Creating Rare Opportunity for Industry and Nation." The American Oil and Gas Reporter, (February, 2012). https://www.aogr.com/magazine/editors-choice/shale-gas-tight-oil-and-eor-creating-rare-opportunity. Accessed April 9, 2018.
7. U.S. Energy Information Administration. Major Tight Gas Plays, Lower 48 States, Updated June 6, 2010. https://www.eia.gov/oil_gas/rpd/tight_gas.jpg. Accessed April 20, 2018.
8. World Energy Council. "Unconventional Gas, a Global Phenomenon." February 2016. https://www.worldenergy.org/wp-content/uploads/2016/02/World-map-Unconventional-gas-a-global-phenomenon-World-Energy-Resources.jpg. Accessed April 3, 2018.
9. U.S. Energy Information Administration. Lower 48 States Shale Plays. Updated June 2016. https://www.eia.gov/maps/images/shale_gas_lower48.pdf. Accessed April 5, 2018.
10. U.S. Energy Information Administration. World Shale Resource Assessments, September 24, 2015. https://www.eia.gov/analysis/studies/worldshalegas/. Accessed April 3, 2018.
11. U.S. Geological Survey. 'Coal-Bed Methane: Potential and Concerns.' Oct. 2000. https://pubs.usgs.gov/fs/fs123-00/fs123-00.pdf. Accessed April 5, 2018.
12. Thakur, Pramod. Chapter 1 - Global Reserves of Coal Bed Methane and Prominent Coal Basins, Elsevier 2017. https://www.sciencedirect.com/science/article/pii/B9780128030950000016. Accessed April 5, 2018.
13. U.S. Energy Information Administration. "Coalbed Methane." https://www.eia.gov/dnav/ng/NG_ENR_COALBED_DCU_NUS_A.htm. Accessed April 5, 2018.
14. U.S. Energy Information Administration. Coalbed Methane Fields, Lower 48 States. Updated April 8, 2009. https://www.eia.gov/oil_gas/rpd/coalbed_gas.pdf. Accessed April 20, 2018.
15. U.S. Energy Information Administration. "Oil and Natural Gas Resources and Technology." March 26, 2018. https://www.eia.gov/outlooks/aeo/section_issues.php#grt. Accessed April 10, 2018.
16. U.S. Energy Information Administration. Annual Energy Outlook 2018. February 6, 2018. https://www.eia.gov/outlooks/aeo/. Accessed April 4, 2018.
17. U.S. Energy Information Administration. Oil and Gas Supply Module. April 2018. file:///C:/Unconventional%20Reservoirs/EIA/oilgas_2018.pdf. Accessed April 10, 2018.
18. U.S. Energy Information Administration. Future U.S. Tight Oil and Shale Gas Production Depends on Resources, Technology, Markets. August 22, 2016. https://www.eia.gov/todayinenergy/detail.php?id=27612. Accessed April 12, 2018.
19. U.S. Energy Information Administration. Tight Oil Remains the Leading Source of Future U.S. Crude Oil Production. February 22, 2018. https://www.eia.gov/todayinenergy/detail.php?id=35052. Accessed April 12, 2018.
20. Usman, Ahmed. Optimized Shale-Resource Development: Well Placement and Hydraulic-Fracture Stages. Society of Petroleum Engineers Distinguished Lecturer Program. 2013-2014. www.spe.org/dl/slides-graphics.php. Accessed April 13, 2018.
21. U.S. Energy Information Administration. Oil and Natural Gas Resources and Technology. March 2018. https://www.eia.gov/outlooks/aeo/section_issues.php#grt. Accessed April 8, 2018.
22. U.S. Energy Information Administration. Hydraulically Fractured Horizontal Wells Accounts for Most New Oil and Natural Gas Wells. January 30, 2018. https://www.eia.gov/todayinenergy/detail.php?id=34732. Accessed April 13, 2018.
23. Collins, G. US Unconventional Oil Responds to Low-Price Environment. June 01, 2015. http://digital.ogj.com/ogjournal/20150601/?pg=44#pg44. Accessed April 13, 2018.
24. Marcellus Drilling News. Great Scott!. Eclipse Drills New Longest Lateral in World-in Utica. https://marcellusdrilling.com/2017/05/great-scott-eclipse-drills-new-longest-lateral-in-world-in-utica/. Accessed April 14, 2018.
25. U.S. Energy Information Administration. "EIA/ARI World Shale Gas and Shale Oil Resource Assessment Technically Recoverable Shale Gas and Shale Oil Resources: An Assessment of 137 Shale Formations in 41 Countries outside the United States." *US Energy Information Administration, www. adv-res. com* (2013). https://www.eia.gov/analysis/studies/worldshalegas/archive/2013/pdf/fullreport_2013.pdf. Accessed April 3, 2018.
26. U.S. Energy Information Administration. Shale Gas and Tight Oil are Commercially Produced in Just Four Countries. February 13, 2015. https://www.eia.gov/todayinenergy/detail.php?id=19991. Accessed April 14, 2018.
27. U.S. Energy Information Administration. Shale Gas Production Drives World Natural Gas Production Growth. August 15, 2016. https://www.eia.gov/todayinenergy/detail.php?id=27512. Accessed April 14, 2018.

CHAPTER

2 Porosity of Reservoir Rocks

2.1 Introduction

Porosity is defined as a measure of the capacity of reservoir rocks to contain or store fluids. The fluids stored in the pore spaces within the reservoir rocks could be gas, oil, and water. High porosity values indicate high capacities of the reservoir rocks to contain these fluids, while low porosity values indicate the opposite. Consequently, porosity data are routinely used qualitatively and quantitatively to assess and estimate the potential volume of hydrocarbons contained in a reservoir. For instance, in a discovery well that shows the presence of hydrocarbons in the reservoir rocks, the set of data that is reviewed at least qualitatively to evaluate reservoir potential is porosity data acquired with either logging-while-drilling (LWD) tools or by running wireline tools. Porosity data are obtained from direct measurements on core samples and/or indirectly from well logs. In most cases, porosity data from core samples are used to validate or calibrate porosity data from well logs. Porosity data are also used in reservoir characterization for the classification of lithological facies, and the assignment of permeabilities using porosity-permeability transforms. Since porosity data are very important in many reservoir engineering calculations, this chapter begins by reviewing basic concepts in the determination of rock porosities. This review is concise and serves to refresh the reader with the many sources of porosity data that exist through applications of different formation evaluation tools.

2.2 Total Porosity and Effective Porosity

The porosity of a rock is a measure of its capacity to contain or store fluids. Porosity is calculated as the pore volume of the rock divided by its bulk volume.

$$\text{Porosity} = \frac{\text{Pore Volume}}{\text{Bulk Volume}} \tag{2.1}$$

Expressed in terms of symbols, Eq. (2.1) is represented as:

$$\phi = \frac{V_P}{V_B} \tag{2.2}$$

In Eq. (2.2), ϕ = porosity; V_p = pore volume; and V_B = bulk volume. Pore volume is the total volume of pore spaces in the rock, and bulk volume is physical volume of the rock, which includes the pore spaces and matrix materials (sand and shale, etc.) that compose the rock.

Two types of porosities can exist in a rock. These are termed primary porosity and secondary porosity. **Primary porosity** is described as the porosity of the rock that formed at the time of its deposition. **Secondary porosity** develops after deposition of the rock. Secondary porosity includes vugular spaces in carbonate rocks created by the chemical process of leaching, or fracture spaces formed in fractured reservoirs. Porosity is further classified as total porosity and effective porosity. **Total porosity** is defined as the ratio of the entire pore space in a rock to its bulk volume. **Effective porosity** is the total porosity less the fraction of the pore space occupied by shale or clay. In very clean sands, total porosity is equal to effective porosity. As shown in Figure 2.1, effective porosity represents pore space that contains hydrocarbon and non-clay water.[1] Free formation water that is bound to neither clay nor shale is called non-clay water. An accurate definition of effective porosity is total porosity minus volume of clay-bound water (Figure 2.1). The relationship between total porosity and effective porosity can be represented for a shaly sand model as:

$$\phi_t = \phi_e + V_{sh} \times \phi_{sh} \tag{2.3}$$

In Eq. (2.3), ϕ_t = total porosity, fraction; ϕ_e = effective porosity, fraction; V_{sh} = volume of shale, fraction; and ϕ_{sh} = shale porosity, fraction. The determination of shale porosity from well logs can be difficult and erroneous because the selection of the 100% shale section can be wrong and subjective.[1] For this reason, an approximate form of Eq. (2.3) is obtained by replacing shale porosity, ϕ_{sh}, with total porosity, ϕ_t, to get:

$$\phi_t = \phi_e + V_{sh} \times \phi_t \tag{2.4}$$

For a clay model, effective porosity is represented as:

$$\phi_t = \phi_e + V_{cbw} \tag{2.5}$$

In Eq. (2.5), V_{cbw} = volume of clay-bound water, fraction. The application of Eq. (2.5) for calculation of accurate effective porosity depends on accurate quantification of the volume of clay-bound water. This can be determined from an elemental capture spectroscopy (ECS) well logs.[1]

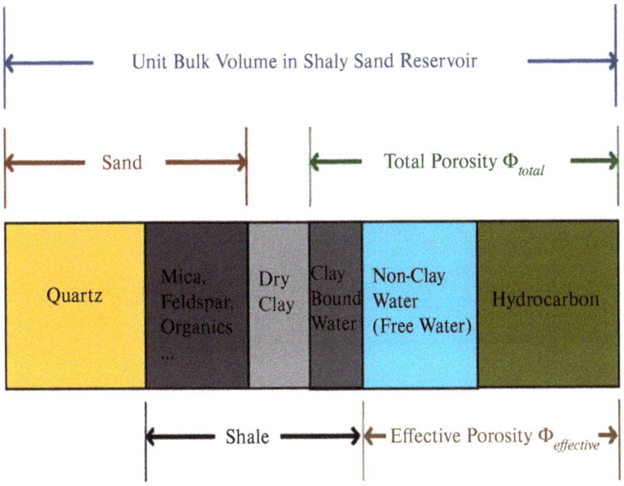

▲ **FIGURE 2.1** Porosity model for a shaly sand reservoir (from Al-Ruwaili et al.[1] © 2004 SPE, Reproduced with permission)

2.3 Sources of Porosity Data

Rock porosity data are obtained by direct or indirect measurements. Laboratory measurements of porosity data on core samples are examples of direct methods. Determinations of porosity data from well log data are considered indirect methods.

2.3.1 Direct Methods for Measurement of Porosity

Direct measurements of porosity data on core samples in a laboratory typically require measurements of bulk and pore volumes of the core samples. For irregular-shaped core samples, the bulk volume is determined by gravimetric or volumetric methods. In gravimetric methods, the apparent loss in weight of the sample when immersed completely in a liquid of known density is measured. Volumetric methods measure the volume of liquid displaced by the rock sample when completely immersed in the liquid. These methods use specially designed equipment so that the liquid is not absorbed by the rock sample. For regular-shaped samples, the bulk volume is calculated from physical measurement of the dimensions of the core sample. For instance, if the core plug is cylindrical in shape, the bulk volume is calculated as:

$$V_B = \pi r^2 l \tag{2.6}$$

In Eq. (2.6), V_B = bulk volume; r = radius of the core plug; and l = length of the core plug.

Other direct methods for measuring the porosity of a rock sample include use of mercury porosimeter or gas expansion porosimeter. The use of mercury porosimeter or gas expansion porosimeter for measurement of porosity is not presented in this book because they are described in many introductory textbooks[2] on petroleum reservoir engineering.

Most laboratory routines based on direct methods measure total porosity. It is important to remember to distinguish between total porosity data obtained from core samples and porosity data derived from well logs, which may include effective porosities. Porosity data obtained from core samples using direct methods are generally considered to be accurate and reliable. They are used to calibrate and validate log-derived porosity data which are based on indirect methods.

EXAMPLE 2.1 Calculation of Porosity from Gravimetric Data

Problem

The dimensions of a cylindrical core sample are 10.16 cm long and 3.81 cm in diameter after it was thoroughly cleaned and dried. The dried core sample weighed 365.0 g. The core sample was then completely (100%) saturated with brine that has specific gravity of 1.04. The weight of the saturated core sample is 390.0 g. Calculate the porosity of the core sample.

Solution

Using Eq. (2.6), the bulk volume of the core sample is:

$$V_B = \pi r^2 l$$

$$= \pi \left(\frac{3.81}{2}\right)^2 \times 10.16 = 115.8333 \text{ cm}^3$$

The pore volume of the core sample is given by:

$$V_p = \frac{\text{wt. of saturated core} - \text{wt. of dried core}}{\text{specific gravity of brine}}$$

$$= \frac{390.0 - 365.0}{1.0400} = 24.0385 \text{ cm}^3$$

Using Eq. (2.2), porosity of the core sample is:

$$\phi = \frac{V_P}{V_B}$$

$$= \frac{24.0385}{115.8333}$$

$$= 0.2075 \text{ or } 20.75\%$$

End of Example 2.1

TABLE 2.1 Summary of Principles, Advantages, and Disadvantages of Porosity Tools

Attributes	Type of Porosity Log			
	Density	Neutron	Sonic	NMR
Basic principle	Gamma ray attenuation	"Slowed" neutrons or Gamma ray capture	Transit times	Excitation of hydrogen in pore spaces
Required data	Matrix and fluid densities	Calibration	Matrix and fluid transit times	Hydrogen index
Advantages	Little effect of presence of gas in formation	Ability to detect presence of gas in formation; can be used in cased hole	Good compensation for environmental effects; combinable with induction logs	Lithology independent
Disadvantages	Shallow depth of investigation; affected by wellbore washouts	Sensitive to irregular borehole; requires calibration	Depth of investigation dependent on type of formation	Environmental corrections; tool run speed affects results

2.3.2 Indirect Methods for Derivation of Porosity

Indirect methods for derivation of porosity data are based on well log data. The well logs generally used for this purpose are density, sonic, neutron, and nuclear magnetic response (NMR) logs. In most formation evaluation programs, density, sonic, and neutron logs are routinely acquired. The NMR log is frequently run in many wells because of its capability of providing other data for formation evaluation, in addition to porosity data. In most deepwater wells, it is common practice to run NMR logs, in addition to density, sonic, and neutron logs. It is important to note that density, sonic, and neutron logs are lithology-dependent, while the NMR logs are lithology-independent for derivation of porosity.[3] NMR data are very sensitive to environmental conditions. It is recommended that NMR tools should be run together with conventional logs, such as density logs or neutron logs for quality control and validation of the NMR data. A summary of the basic principles, data requirements, advantages, and disadvantages of all the porosity tools is provided in Table 2.1.

2.3.2.1 Density Logs

Density logs are based on the attenuation of gamma rays in the formation.[3] Density logging tools measure the attenuation of gamma rays produced by a gamma source of known strength. The attenuation caused by the interaction between gamma ray photons and electrons on the outer shell of electrons (called Compton scattering) is directly proportional to the bulk density, (ρ_b), of the formation. The formation bulk density is related to formation matrix density, (ρ_{ma}), and formation fluid density, (ρ_f), as:

$$\rho_b = (1-\phi)\,\rho_{ma} + \phi\rho_f \tag{2.7}$$

Rearranging Eq. (2.7), density-derived porosity is given by:

$$\phi_d = \frac{\rho_{ma}-\rho_b}{\rho_{ma}-\rho_f} \tag{2.8}$$

In Eq. (2.8), ϕ_d = density-derived porosity; ρ_{ma} = matrix density; ρ_b = bulk density; and ρ_f = fluid density. The porosity data obtained from density logs are considered to be total porosity. This relationship can be represented as:

$$\phi_d = \phi_t \tag{2.9}$$

For density logs, effective porosity is derived from Eq. (2.3) as:

$$\phi_e = \phi_d - V_{sh}\times\phi_{dsh} \tag{2.10}$$

In Eq. (2.10), ϕ_{dsh}, is the shale porosity derived from the density logs. The depth of investigation of density logging tools is shallow and typically within the zone invaded by mud filtrate. For this reason, it is sometimes appropriate to assume that the density of formation fluid is equal to the density of the mud filtrate. However, this assumption may cause errors in the density-derived porosity data, if virgin formation fluid remains within the depth of investigation of the density tool.[4] The matrix density can be determined from elemental capture spectroscopy (ECS) log, if available.

EXAMPLE 2.2 Calculation of Porosity from Density Logs

Problem

The bulk density of a clean, sandy interval saturated with water was measured by the density logging tool to be 2.4 g/cm^3. Assuming that the density of the formation water is 1.04 g/cm^3 and the density of the matrix is 2.67 g/cm^3, calculate the density porosity of this interval.

Solution

Using Eq. (2.8), density porosity is calculated to be:

$$\phi_d = \frac{\rho_{ma}-\rho_b}{\rho_{ma}-\rho_f} = \frac{2.67-2.4}{2.67-1.04} = 0.166 \text{ or } 16.6\%$$

End of Example 2.2

2.3.2.2 Sonic (Acoustic) Logs

In sonic (acoustic) logging, the formation is probed with sound waves. The time sound waves take to travel a given distance is measured. This interval transit time depends on the elastic properties of the rock matrix, the properties of the fluid in the rock, and the porosity of the rock. Wyllie et al.[5] proposed that the interval transit time, (Δt), can be represented as the sum of the transit time in the matrix fraction, (Δt_{ma}), and the transit time in the liquid fraction, (Δt_f), thus:

$$\Delta t = (1-\phi)\Delta t_{ma} + \phi \Delta t_f \tag{2.11}$$

Rearranging Eq. (2.11), sonic-derived porosity is given by:

$$\phi_s = \frac{\Delta t - \Delta t_{ma}}{\Delta t_f - \Delta t_{ma}} \tag{2.12}$$

In Eq. (2.12), ϕ_s = sonic-derived porosity; Δt = transit time; Δt_f = fluid transit time; and Δt_{ma} = transit time for the rock matrix. Total porosity is related to porosity derived from sonic logs as:

$$\phi_s = \phi_t + V_{clay} \times \phi_{scl} \tag{2.13}$$

In Eq. (2.13), V_{clay} = the volume of clay; and ϕ_{scl} = sonic porosity derived in the clay. Effective porosity as calculated from sonic logs as:

$$\phi_e = \phi_s - V_{sh} \times \phi_{ssh} \tag{2.14}$$

In Eq. (2.14), V_{sh} = volume of shale; and ϕ_{ssh} = sonic porosity derived for shale. Analysis of sonic logs based on Eq. (2.12) gives reliable porosity data only for consolidated formations. For unconsolidated sandstones and carbonates, Eq. (2.12) gives porosity values that are too high. Other equations similar to Eq. (2.12) have been proposed for calculation of porosity for unconsolidated formations and carbonates by Raymer et al.[6] These equations should be used for calculations of sonic porosities on unconsolidated formations and carbonates. Note that sonic logs are well-compensated for environmental effects such as mud velocity, borehole diameter, etc., and that its depth of investigation is dependent on the compactness of the formation.

EXAMPLE 2.3 Calculation of Porosity from Sonic Logs

Problem

The transit time for a well-consolidated sandstone interval saturated with brine was measured to be 82×10^{-6} sec/ft. The matrix transit time is 55.5×10^{-6} sec/ft and the brine transit time is 189×10^{-6} sec/ft. Calculate the sonic porosity for the interval.

Solution

Applying Eq. (2.12), sonic porosity is calculated to be:

$$\phi_s = \frac{\Delta t - \Delta t_{ma}}{\Delta t_f - \Delta t_{ma}}$$

$$= \frac{82\times10^{-6} - 55.5\times10^{-6}}{189\times10^{-6} - 55\times10^{-6}}$$

$$= 0.199 \text{ or } 19.9\%$$

End of Example 2.3

2.3.2.3 Neutron Porosity Logs

The first logging tool that was used for the estimation of formation porosity is the neutron logging tool, which was introduced around 1940. The neutron porosity logging tool consists of either a chemical source or an electrical source of fast neutrons, and detectors located some distance from the source. The fast neutrons from the neutron source are slowed down by successive collisions with individual nuclei in the rock, thereby losing most of their energy. The detectors in the neutron tool record either the "slowed" down neutrons directly or capture gamma radiation generated when the neutrons are captured by nuclei. The neutron porosity log is sensitive to the amount of hydrogen in the formation because the neutrons interact most effectively with hydrogen due to the closeness of their masses. Neutron logs estimate the amount of hydrogen in the rock and relate it to the amount of fluid in the formation. From the amount of fluid in the formation, the porosity of the rock is estimated after calibration for different lithologies (sandstone, dolomite, and limestone). Neutron porosity tools are sensitive to borehole conditions, especially variations in the size of the borehole. In combination with density porosity logs, neutron porosity logs can be used to detect the presence of gas in some formations. This known crossover of density porosity log and neutron porosity log in gas-filled formation intervals results from the apparent increase of density-derived porosities and apparent decrease of neutron-derived porosities in gas-filled formation intervals (Figure 2.2).

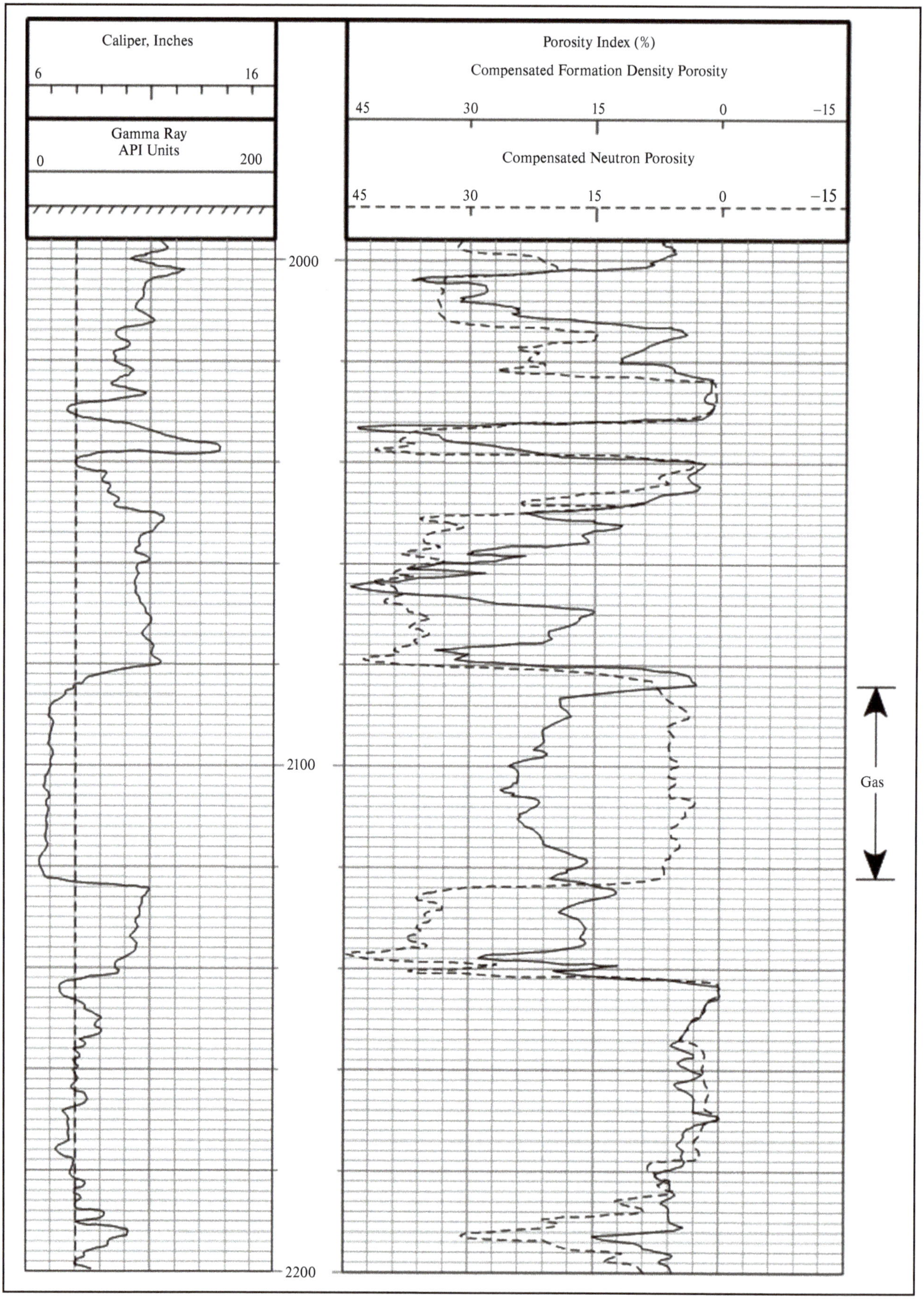

▲ FIGURE 2.2 **Density and neutron well logs showing crossover in a gas interval**

For neutron porosity logs, correction for total porosity is applied as:

$$\phi_n = \phi_t + V_{clay} \times \phi_{ncl} \quad (2.15)$$

In Eq. (2.15), ϕ_n = porosity from neutron logs; and ϕ_{ncl} = neutron porosity for clay. Effective porosity is defined as:

$$\phi_e = \phi_n - V_{sh} \times \phi_{nsh} \quad (2.16)$$

In Eq. (2.16), ϕ_{nsh}, is the neutron porosity for shale.

2.3.2.4 Nuclear Magnetic Resonance Porosity Logs

Nuclear magnetic resonance (NMR) porosity tools have a clear advantage over other porosity tools (density, sonic, and neutron) because their determination of porosity is independent of lithology of the rock. Porosities calculated from density, sonic, and neutron logs depend on "knowing" or estimating the properties of the rock matrix. NMR porosities are calculated from the number of hydrogen atoms in the fluids (hydrocarbon and water) within a specific measurement volume of the tool, and are independent of the lithology of the rock formation.[7] For reservoirs with highly heterogeneous rocks consisting of mixed or unknown lithology, porosity data derived from NMR logs are more consistent and reliable than porosity data from the other porosity tools.[8] NMR logs report porosities in terms of total porosity, bound-fluid porosity, and free-fluid porosity (Figure 2.1). **Free-fluid porosity** (also termed free-fluid index) is a qualitative measure of effective porosity and is linked to the hydrocarbon storage potential of the formation. A comparison of porosities measured with NMR in the laboratory to porosities measured by direct methods on core samples from a reservoir is shown in Figure 2.3. The porosity data plotted in Figure 2.3 show close agreements between NMR porosities and core porosities measured on core samples. It demonstrates a method for calibrating NMR porosities with core porosities, which can then be used to calculate porosities from NMR data in other wells. In addition to measurement of porosity, NMR tools are used for determination of pore size distributions, measurements of permeability (Chapter 3), and fluid saturations (Chapter 4). NMR tools have become standard in most wireline logging operations because they can quickly provide qualitative data on formation porosity, permeability, pore size distributions, and fluid saturations.[9,10] These data are very valuable and useful. They are frequently used to make decisions on selection of fluid sampling points and formation intervals to be tested in discovery, appraisal, and development wells.

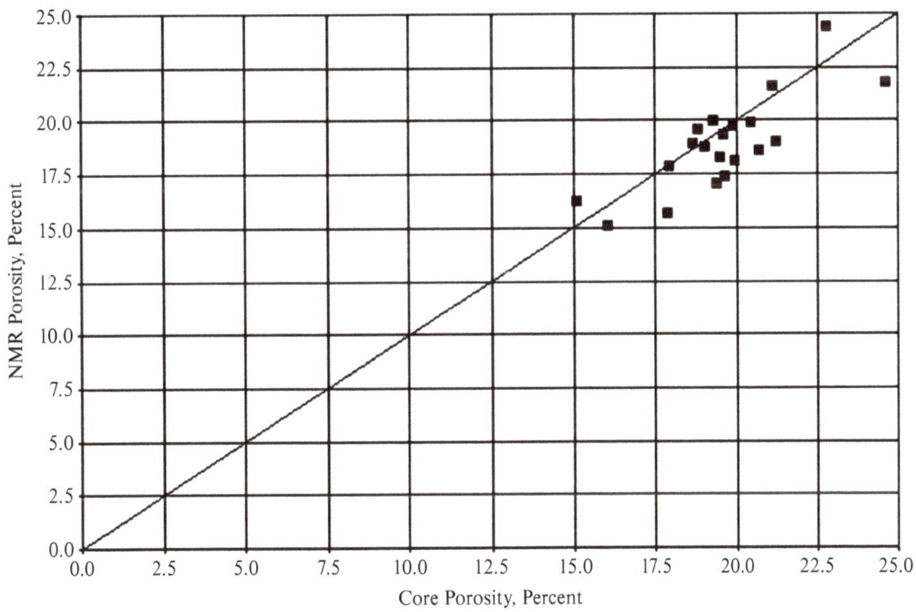

▲ FIGURE 2.3 NMR porosity versus core porosity based on core samples from a reservoir

2.4 Applications of Porosity Data

One of the primary rock property data used in most reservoir evaluation is porosity data. Consequently, it is important that accurate values of porosity data for the reservoir rocks are measured and validated by other independent methods. Porosity data are used in these basic reservoir evaluations:

1. Volumetric calculation of fluids in the reservoir
2. Calculation of fluid saturations
3. Geologic characterization of the reservoir

2.4.1 Volumetric Calculation

A general formula for the calculation of the volume of hydrocarbons in a reservoir is represented as:

$$HCPV = Area \times Thickness \times \phi \times (1 - S_w) \quad (2.17)$$

In Eq. (2.17), HCPV = hydrocarbon pore volume; Area = hydrocarbon-bearing area of the reservoir; Thickness = net productive thickness or pay of the reservoir; ϕ = porosity, fraction; and S_w = water saturation, fraction. The hydrocarbon volumes of specific types of fluids (oil and/or gas) in the reservoir can be calculated with minor modifications of Eq. (2.17) as demonstrated in Chapter 9 (Gas Reservoirs) and Chapter 10 (Oil Reservoirs). Note the prominence of porosity in Eq. (2.17) for the calculation of volumes of hydrocarbons present in a reservoir. It is evident from Eq. (2.17) that inaccurate porosity data can directly cause underestimation or overestimation of the hydrocarbon volumes in the reservoir. For marginal reservoirs, underestimation of in-place hydrocarbon volumes may contribute to a decision not to pursue development of the reservoir. Overestimation of in-place hydrocarbon volumes may lead to economic losses, if projected reserves estimated prior to development are far below actual reservoir performance. Note that there are other geologic and reservoir factors (such as permeability barriers, faults, compartments, recovery mechanisms) which can also cause reservoir performance to be below projected levels. The impact of these other factors on reservoir performance are presented and discussed in more detail in several chapters in this book.

2.4.2 Calculation of Fluid Saturations

For clean, non-shaly rocks, water saturations can be calculated from the Archie equation[1] as:

$$C_t = \phi_t^m \times S_w^n \times C_w \quad (2.18)$$

In Eq. (2.18), C_t = formation conductivity; ϕ_t = total porosity; S_w = water saturation; C_w = formation water conductivity; m = cementation factor; and n = saturation exponent. The parameters, m and n, are also called the electrical properties of the rock.

For shaly sands, water saturations can be calculated from modified forms[1] of the Archie equation, which are shown as:

$$C_t = \phi_t^{m^v} \times S_{wt}^{n^v} \times C_{we} \quad (2.19)$$

$$C_t = \phi_t^{m^v} \times S_{wt}^{n^v} \times (C_w + X) \quad (2.20)$$

In Eqs. (2.19) and (2.20), C_{we} = effective conductivity; and m^v, n^v are general forms of the electrical properties. In Eq. (2.19), C_{we}, is expressed in terms of, C_w, and a function of shale (in the shale model) or a function of clay (in the clay model). In Eq. (2.20), X, is a function that accounts for the conductivity caused by shale or clay that occur in shaly sands. Note that in Eq. (2.20) as, X, approaches zero, Eq. (2.20) becomes equivalent to Eq. (2.18).

The main point to note from Eqs. (2.18), (2.19), and (2.20) is that total porosity is an important data input for calculation of water saturation with water saturation models. If errors exist in the calculations of total porosity, these errors will be transferred to the calculation of water saturations. This could ultimately lead to errors in the estimation of reservoir-in-place hydrocarbon volumes as shown in Eq. (2.17). The calculation of water (fluid) saturation is presented in more detail in Chapter 4.

2.4.3 Reservoir Characterization

Porosities can be measured directly from cores or indirectly determined from well logs as discussed previously in this chapter. On the one hand, rock permeability can be measured most reliably from cores or in aggregate sense from well tests. Indirect methods for acquiring permeability data are discussed in Chapter 3. There are usually more porosity data than permeability data available on a reservoir. A cross-plot of permeability versus porosity data (Figure 2.4) to create a porosity-permeability transform is sometimes used to assign permeability values to areas of the reservoir where permeability data do not exist. The practice of using porosity-permeability transforms in reservoir characterization is presented in Chapter 22.

Facies or rock types can be defined or assigned to parts of a reservoir by using porosity values as part of a system of criteria for rock classification. This process of classifying reservoir rock in terms of facies or rock types is useful in the process of reservoir characterization. This is also presented in Chapter 22.

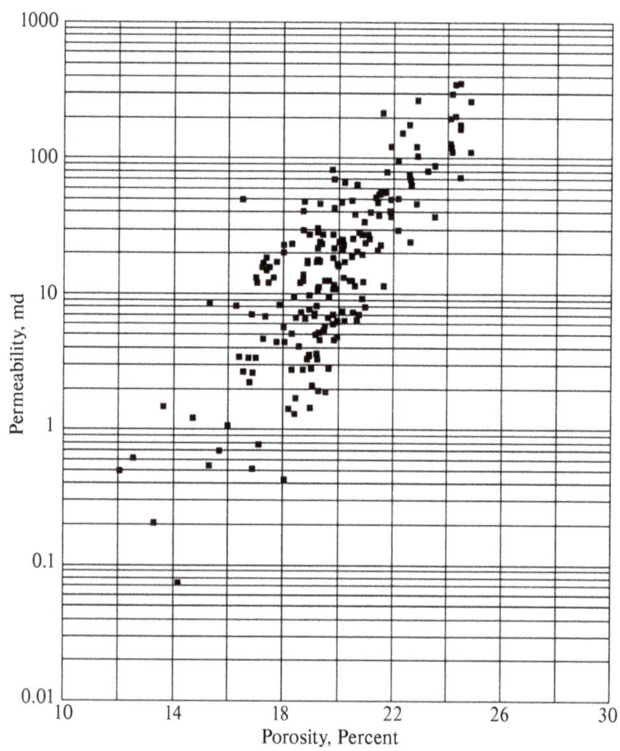

▲ **FIGURE 2.4 Porosity-permeability cross-plot based on core samples from a reservoir**

Nomenclature

C_t	formation conductivity	ϕ_e	effective porosity
C_w	formation water conductivity	ϕ_n	neutron-derived porosity
C_{we}	effective conductivity	ϕ_{ncl}	neutron-derived porosity in clay
l	length of core plug	ϕ_{nsh}	neutron-derived porosity in shale
m	cementation factor	ϕ_s	sonic-derived porosity
n	saturation exponent	ϕ_{scl}	sonic-derived porosity in clay
r	radius of core plug	ϕ_{sh}	shale porosity
S_w	water saturation, fraction	ϕ_{ssh}	sonic-derived porosity in shale
V_B	bulk volume	ϕ_t	total porosity
V_{cbw}	volume of clay-bound water	ρ_b	bulk density
V_{clay}	volume of clay	ρ_f	fluid density
V_p	pore volume	ρ_{ma}	rock matrix density
V_{sh}	shale volume	Δt	formation interval transit time
X	function in Eq. (2.20)	Δt_f	fluid transit time
ϕ	porosity	Δt_{ma}	rock matrix transit time
ϕ_d	density-derived porosity		
ϕ_{dsh}	density-derived shale porosity		

Abbreviations

ECS	elemental capture spectroscopy	LWD	logging-while-drilling
HCPV	hydrocarbon pore volume	NMR	nuclear magnetic resonance

REFERENCES

1. Al-Ruwaili, S.A., and Al-Waheed, H.H.: "Improved Petrophysical Methods and Techniques for Shaly Sands Evaluation," paper SPE 89735 presented at the 2004 SPE International Petroleum Conference in Puebla, Mexico, November 8-9, 2004.
2. Amyx, J.W., Bass, D.M., Jr., and Whiting, R. L.: *Petroleum Reservoir Engineering, Physical Properties,* McGraw-Hill, New York, 1960.
3. Coates, G.R., Menger, S., Prammer, M., and Miller, D.: "Applying NMR Total and Effective Porosity to Formation Evaluation," paper SPE 38736 presented at the 1997 SPE Annual Technical Conference and Exhibition, San Antonio, Texas, October 5-8, 1997.
4. Ellis, D.: "Formation Porosity Estimation from Density Logs," *Petrophysics,* (September-October 2003) 306-316.
5. Wyllie, M.R.J., Gregory, A.R., and Gardner, G.H.F.: "Elastic Wave Velocity in Heterogeneous and Porous Media," *Geophysics,* (1956) 41-70.
6. Raymer, L.L., Hunt, E.R., and Gardner, J.S.: "An Improved Sonic Transit Time-To-Porosity Transform," SPWLA Twenty-First Annual Logging Symposium, July 8-11, 1980.
7. Freedman, R.: "Advances in NMR Logging," SPE 89177, Distinguished Author Series, *JPT* (January 2006) 60-66.
8. Bachman, H.N., Crary, S., Heidler, R., LaVigne, J., and Akkurt, R.: "Porosity Determination from NMR Log Data: The Effects of Acquisition Parameters," paper SPE 110803 presented at the 2007 SPE Annual Technical Conference and Exhibition, Anaheim, California, November 11-14, 2007.

9. Akkurt, R., Kersey, D.G., and Zainalabedin, K.: "Challenges for Everyday NMR: An Operator's Perspective," paper SPE 102247 presented at the 2006 SPE Annual Technical Conference and Exhibition, San Antonio, Texas, September 24-27, 2006.

10. Seifert, D.J., Akkurt, R., Al-Dossary, S., Shokeir, R., and Ersoz, H.: "Nuclear Magnetic Resonance Logging: While Drilling, Wireline, and Fluid Sampling," paper SPE 105605 presented at the 15th SPE Middle East Oil & Gas Show and Conference, Bahrain, Kingdom of Bahrain, March 11-14, 2007.

ADDITIONAL REFERENCES

- Mezzatesta, A., Méndez, F., and Rodriguez, E.: "Effective and Total Porosities: Their Reconciliation in Carbonate and Shaly-Sand Systems," paper SPE 102811 presented at the 2006 SPE Annual Technical Conference and Exhibition, San Antonio, Texas, September 24-27, 2006.
- Towler, B.F.: *Fundamental Principles of Reservoir Engineering,* SPE Textbook Series Vol. 8 (2002).
- Wiltgen, N.: "Shale Volumes Have Large Uncertainties in the Lower Tertiary Deepwater," paper SPE 114775 presented at the 2008 SPE Annual Technical Conference and Exhibition, Denver, Colorado, September 21-24, 2008.
- Wiltgen, N.: "The Contradiction in the Lower Tertiary Deepwater GoM," paper SPE 114758 presented at the 2008 SPE Annual Technical Conference and Exhibition, Denver, Colorado, September 21-24, 2008.

PROBLEMS

2-1. A cylindrical core sample that measured 10 cm long and 3.5 cm in diameter was cleaned and dried in an oven for 7 days. The core sample weighed 362.5 g after the drying process. It was then immersed completely in brine under vacuum for 2 days to achieve 100% saturation. If the weight of the saturated core sample was 385 g, what is its porosity assuming the specific gravity of the brine is 1.03?

2-2. A cleaned, dried core sample weighed 363.75 g in air. The same sample was saturated with brine and weighed 387.5 g in air. The saturated sample was then immersed completely in the same brine and weighed 278.32 g. If the specific gravity of the brine is 1.03, calculate the porosity of the core sample.

2-3. A mercury pycnometer weighed 420 g when filled with mercury only at 68°F. A cleaned, dried core sample that weighed 30 g in air at 68°F was then placed in the mercury-filled pycnometer. The weight of the mercury-filled pycnometer with the core sample inside was measured as 283.08 g at 68°F. The core sample was then removed from the pycnometer and saturated with brine at 68°F. The weight of the saturated core sample in air was 33.75 g. The specific gravity of the saturated brine at 68°F is 1.1999. The density of mercury is 13.545 g/cm^3 at 68°F. Calculate the porosity of the core sample.

2-4. The bulk volume of a cleaned and dried core sample was determined to be 12.375 cm^3 using a mercury pycnometer. The core sample weighed 25 g in air at room temperature. The core sample was then crushed, and its grain density determined to be 2.66 g/cm^3. Calculate the porosity of the core sample.

2-5. The volume of the core chamber of a gas expansion porosimeter was known accurately to be 19.5 cm^3. A core sample was placed in the core chamber to determine its porosity. The volume of air extracted from the core sample by expansion was measured to be 9.1 cm^3. If the bulk volume of the core sample was determined to be 13 cm^3 using a mercury pycnometer, what is the porosity of the core sample?

2-6. The bulk density of a sandstone interval with 100% formation water saturation was measured to be 2.2 g/cm^3 with a density logging tool. The density of the formation water is 1.2 g/cm^3. If the density of the sandstone matrix is 2.65 g/cm^3, calculate the density porosity of the interval.

2-7. A dolomite interval saturated 100% with formation water has a measured bulk density of 2.5 g/cm^3 from a density logging tool. If the matrix density of dolomite is assumed to be 2.88 g/cm^3 and the density of formation water is 1.2 g/cm^3, calculate the density porosity of the interval.

2-8. The bulk density of a sandstone interval was measured as 2.185 g/cm^3. From saturation logs, it was determined that the interval was 60% saturated with oil and 40% saturated with formation water. If the matrix density of sandstone is 2.65 g/cm^3, calculate the density porosity of the interval. Assume density of oil is 0.88 g/cm^3 and density of formation water is 1.1 g/cm^3.

2-9. The bulk density of a sandstone interval measured 1.89 g/cm^3 on a density logging tool. The interval was saturated 90% with gas and 10% with formation water on the basis of saturation logs. Calculate the density porosity of the interval if the matrix density of sandstone is 2.65 g/cm^3. Calculate the density porosity of the interval assuming the interval was 100% saturated with formation water. Assume gas density is 0.0007 g/cm^3 and the formation water density is 1.1 g/cm^3.

2-10. A vuggy dolomite reservoir has an interval which measured a transit time of 58.85×10^{-6} sec/ft. The matrix transit time is 45.6×10^{-6} sec/ft. and the fluid transit time is 144.0×10^{-6} sec/ft. Calculate the sonic porosity for the interval.

2-11. A mercury porosimeter is used to measure porosity by injecting mercury into the pore spaces of the core sample at high pressures. During the injection, some air is trapped within the core spaces. The trapped air is compressed to a small fraction of its original volume, but an error is introduced in the porosity measurement by the entrapment. Assume the initial conditions for a porosity measurement using a mercury porosimeter is 65 °F and 14.7 psia. If a porosity value of 25% was calculated from the volume of mercury injected into the core sample at 65 °F, what is the percentage error in the porosity value when the mercury injection pressure was:

a. 500 psia
b. 1500 psia
c. 2500 psia
d. What is the effect of mercury injection pressure on porosity values measured?

2-12. A gas porosimeter is used to measure the porosity of a core sample. A diagram of a gas porosimeter is represented in Figure 2P.1. A core sample with dimensions of 0.75 in by 0.75 in by 2.5 in (Figure 2P.2) was encapsulated with an epoxy resin to form a cylindrical sample with dimensions of 1.5 in diameter and 2.5 in long (Figure 2P.2). The encapsulated core sample was placed in a core chamber (V_1) (Figure 2P.1). The volume of the core chamber (V_1) is 6 cubic inches. The gas porosimeter was evacuated completely with a vacuum pump. The core chamber (V_1) only was then charged with nitrogen to a stabilized pressure of 1200 psia. All the contents of the core chamber (V_1) was later expanded to fill a second chamber (V_2) connected to core chamber (V_1). The volume of the second chamber (V_2) is 150 cubic inches (Figure 2P.1). The final pressure of the connected porosimeter chambers (V_1 and V_2) was equilibrated to a stabilized pressure of 15.5 psia. Assuming the temperature of the gas porosimeter remained the same throughout the entire experiment, calculate the porosity of the core sample.

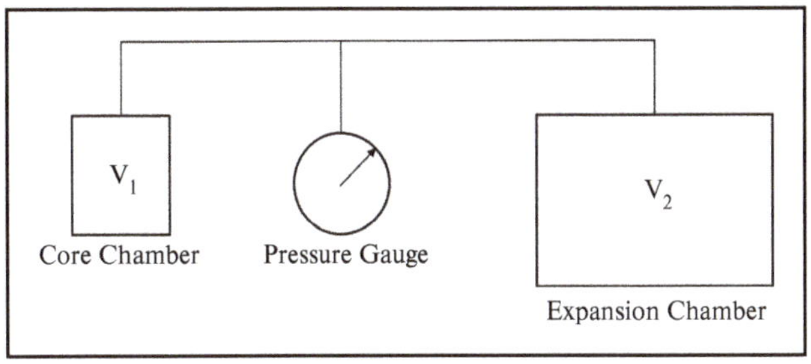

▲ **FIGURE 2P.1** **Schematic diagram of a gas porosimeter**

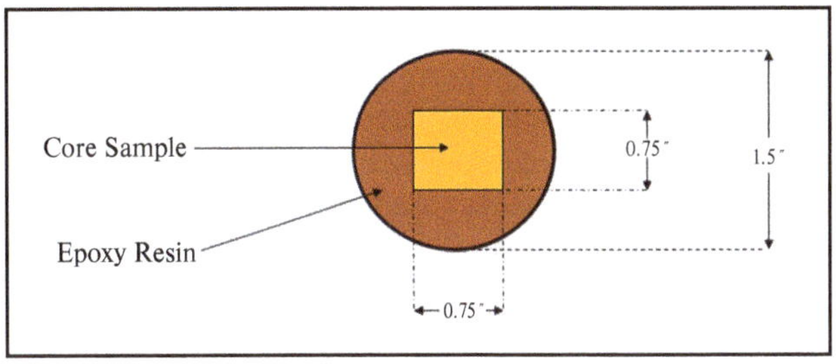

▲ **FIGURE 2P.2** **Cross-sectional diagram of encapsulated core sample**

CHAPTER

3 Permeability and Relative Permeability

3.1 Introduction

Permeability is a measure of the ability of a porous medium, such as reservoir rock, to transmit fluids through its system of interconnected pore spaces. If the porous medium is completely saturated (100% saturated) with a single fluid, the permeability measured is the absolute permeability. **Absolute permeability** is an intrinsic property of the porous medium, and the magnitude of absolute permeability is independent of the type of fluid in the pore spaces. When the pore spaces in the porous medium are occupied by more than one fluid, the permeability measured is the effective permeability of the porous medium to that particular fluid. For instance, the effective permeability of a porous medium to oil is the permeability to oil when other fluids, including oil, occupy the pore spaces. **Relative permeability** is defined as the ratio of effective permeability to absolute permeability of a porous medium. The relationship for relative permeability is represented as:

$$k_{ri} = \frac{k_i}{k_a} \tag{3.1}$$

In Eq. (3.1), k_{ri} = relative permeability of the porous medium to fluid, i; k_i = effective permeability of the porous medium for fluid, i; and k_a = absolute permeability of the porous medium. For instance, the relative permeability of a porous medium to oil is expressed in a form similar to Eq. (3.1) as:

$$k_{ro} = \frac{k_o}{k_a} \tag{3.2}$$

In Eq. (3.2), k_{ro} = relative permeability of the porous medium to oil; k_o = effective permeability of the porous medium to oil; and k_a = absolute permeability of the porous medium. Similarly, the relative permeability of the porous medium to water and gas are expressed in Eqs. (3.3) and (3.4), respectively, as:

$$k_{rw} = \frac{k_w}{k_a} \tag{3.3}$$

$$k_{rg} = \frac{k_g}{k_a} \tag{3.4}$$

In Eq. (3.3), k_{rw} = relative permeability of the porous medium to water; and k_w = effective permeability of the porous medium to water. In Eq. (3.4), k_{rg} = relative permeability of the porous medium to gas; and k_g = effective permeability of the porous medium to gas. Absolute permeability is measured in a laboratory by flowing a fluid of known viscosity through a core sample from a porous medium, while measuring the flow rates and pressure differences across the core sample. The core sample must be totally saturated (100%) with the fluid. By definition, any fluid can be used to measure absolute permeability. In practice, absolute permeability is measured by flowing air through a core sample that has been completely dried. Core permeability measured with air or any other gas must be reviewed and corrected for Klinkenberg[1] effect, if necessary. At low pressures, gas permeability can be higher than absolute permeability due to Klinkenberg effect, as shown in Eq. (3.5):

$$k_{air} = k_a\left(1 + \frac{b}{\bar{P}}\right) \tag{3.5}$$

In Eq. (3.5), k_{air} = air (gas) permeability of core sample as measured; k_a = absolute permeability of the core sample; b = Klinkenberg constant for a given gas in a given core sample; and $\bar{P}$ = mean flowing pressure of the gas at which the gas permeability was measured. Klinkenberg[1] found that at low pressures gas permeability could be higher than absolute permeability of the porous medium due to interactions between gas molecules and the walls of the pore spaces. This interaction is referred to as "slippage," and is related to the molecular size and kinetic energy of the gas. The apparent increase in core air permeabilities at low measurement pressures due to this slippage is called the Klinkenberg effect. The standard unit of measure for permeability in the petroleum industry is the darcy (d). The darcy is the permeability of a porous medium with cross-sectional area of 1 cm^2 through which a fluid with a viscosity of 1 cp is flowing at the volumetric rate of 1 cc/sec under a pressure gradient of 1 atm per 1 cm of length in the direction of flow [see Eq. (3.6)]. The millidarcy (md) is 1/1000th of a darcy.

3.2 Sources of Permeability Data

The main sources of permeability data are:

1. Core samples
2. Pressure transient tests
3. Well logs

These sources of permeability data sample volumes of the reservoir at different scales. Permeability data measured on core samples are at a different scale from permeability data calculated from pressure transient tests or well logs. A typical core plug is about 2 inches in diameter by 2 inches long. It is taken from a full-diameter core sample at every half-foot to 1-foot frequency, depending on the sampling program. This represents a sample volume of 0.0036 ft^3 for each core plug. In contrast, a typical well log, such as an NMR (Nuclear Magnetic Resonance) log, has a vertical resolution of 2 to 5 ft, depending on the configuration of the tool. Consequently, the volume of reservoir rock sampled by an NMR log is several orders of magnitude larger than the volume of reservoir rock sampled by core plugs. Pressure transient tests are capable of achieving radii of investigation that extended several hundred feet away from the wellbore, depending on rock and fluid properties of the reservoir and duration of the test. Clearly, the volume of reservoir rock sampled by pressure transient tests is much larger than the volume of reservoir rock sampled by NMR logs. The differences in scale among the various sources of permeability data should be considered before the data are integrated and used for reservoir analysis. Many techniques have been reported in the literature for integrating permeability data with different scales. A method based on calculation of running averages was reported by Worthington.[2] This method was applied to Saudi Arabian reservoirs by Al-Ali and Worthington.[3] Shafer and Ezekwe[4] demonstrated that core and well log permeability data on different scales could be integrated by applying regression analysis based on the Coates[5] equation.

3.2.1 Permeability from Core Samples

The primary source of reliable permeability data is core samples. Ample permeability data are measured on core plugs taken from whole core samples. In the absence of whole cores, permeability data can be obtained from sidewall core samples. Generally, permeability data from sidewall cores are not as reliable as those measured on core plugs taken from whole core samples and should be used with caution. Core plugs are usually taken parallel to the plane of deposition of the formation. Consequently, the permeability data measured on core plugs are horizontal permeability data. To measure vertical permeability, the core plugs should be taken perpendicular to the plane of deposition. In practice, vertical permeability data are seldom measured. However, if the dominant oil displacement mechanism (such as gravity drainage discussed in Chapter 10) is dependent on vertical permeability, extra effort should be made to obtain vertical permeability data from laboratory core measurements. The typical permeability data measured under a routine core analysis (RCAL) program is air permeability. Air permeability is calculated using the Darcy equation [Eq. (3.6)] from data obtained by placing the core sample in a chamber and measuring the pressure differential across it at stabilized air flow rates under steady-state conditions. Air permeability should be measured at the net confining stress of the reservoir. **Net confining stress** is the pressure difference between the overburden pressure and the pore pressure at the depth of the reservoir. Air permeability data should be corrected for the Klinkenberg effect using Eq. (3.5). As part of a quality control program, the core samples used in core permeability measurements should be examined for the presence of induced fractures, fissures, or any deformations which could influence the magnitude of the permeability data.

3.2.2 Permeability from Pressure Transient Tests

Pressure transient tests are very important sources of permeability data for reservoirs. There are various forms of pressure transient tests that are used to calculate permeability data. These tests can be categorized in two groups, namely: wireline or tubing-conveyed openhole tests and traditional wellhead or down-hole well tests. The wireline or tubing-conveyed openhole tests include wireline formation testers (WFT) (see Chapter 6) and drillstem tests. Traditional wellhead or down-hole well tests comprise of drawdown or buildup tests conducted on active wells, as discussed in Chapters 13, 14, and 15 for conventional reservoirs and Chapter 17 for unconventional reservoirs. Generally, permeability data from WFT or traditional well tests are at different scales from permeability data measured from core samples or obtained from well logs. The volume of reservoir rock sampled by pressure transient tests depends on the extent of the interval tested, rock and fluid properties of the interval, and duration of the test. Pressure transient tests measure effective permeability and should be used to condition permeability data from core samples and well logs as necessary to achieve observed well and reservoir performance.

3.2.3 Permeability from Well Logs Based on Empirical Correlations

The concept of permeability as a proportionality constant that governs the rate of fluid flow is demonstrated by Darcy's equation for single phase, linear, horizontal flow in a porous medium as:

$$k = \frac{q \times \mu \times L}{A \times \Delta P} \tag{3.6}$$

In Eq. (3.6), k = permeability, darcy; q = flow rate, cm^3/s; μ = viscosity, cp; L = length, cm; A = cross-sectional area, cm^2; and ΔP = pressure difference, atm. If a dimensionless analysis is performed on Eq. (3.6) by substituting for the dimensions of permeability, the result is:

$$\begin{aligned} k &= \frac{rate \times viscosity \times length}{area \times pressure} \\ &= \frac{\left(\frac{L^3}{time}\right)\left(\frac{force \times time}{L^2}\right) \times L}{L^2 \times \left(\frac{force}{L^2}\right)} \\ &= L^2 \end{aligned} \tag{3.7}$$

Eq. (3.7) shows that permeability has dimensions of area, (L^2), which are physically related to the surface area of grains exposed to flow. The size and distribution of rock grains in a porous medium are related to surface area, which affects the permeability of the porous medium. This physical aspect of permeability has been used to create empirical equations for prediction of permeabiliy.[6]

An equation that predicts permeability from the porosity and surface area of grains exposed to fluid flow was proposed by Kozeny[7] and Carman[8] as:

$$k = \frac{\phi^3}{F_s \tau^2 A_g^2 (1-\phi)^2} \tag{3.8}$$

Eq. (3.8) is called the Kozeny-Carman equation. In Eq. (3.8), k = permeability, μm²; ϕ = porosity, fraction; F_s = shape factor; τ = tortuosity; and A_g = surface area per unit grain volume, μm^{-1}. The term, $F_s\tau^2$, is called the Kozeny constant. The Kozeny constant is variable and has been shown by Rose and Bruce[9] to vary from 5 to 100 for reservoir rocks. A modification of the Kozeny-Carman equation was proposed by Wyllie and Rose[10] by substituting irreducible water saturation for specific surface area to get:

$$k^{1/2} = \frac{C_1\phi}{S_{wir} - C_2} \tag{3.9}$$

In Eq. (3.9), C_1 and C_2, are constants that vary depending on the rock and fluid properties of the formation; and S_{wir} = irreducible water saturation. A more generalized form of the Wyllie-Rose correlation was proposed by Timur[11] as:

$$k = \frac{P\phi^Q}{S_{wir}^R} \tag{3.10}$$

In Eq. (3.10), P, Q, and R, are regression parameters to be determined by fitting the equation to core data. A form of Eq. (3.10) was developed for sandstones by Timur[11] as:

$$k = \frac{0.136\phi^{4.4}}{S_{wir}^2} \tag{3.11}$$

In Eq. (3.11), k = permeability, md; ϕ = porosity, percent; and S_{wir} = irreducible water saturation, percent. The Timur equation is based on 155 sandstone core samples from three different oil fields in North America. A modified form[12] of the Timur equation which was in general use is represented as:

$$k = \left[\frac{100\phi^{2.25}}{S_{wir}}\right]^2 \tag{3.12}$$

In Eq. (3.12), k = permeability, md; ϕ = porosity, fraction; and S_{wir} = irreducible water saturation, fraction. Eq. (3.12) was used to correlate actual data by adjusting the constant, 100, and the exponent, 2.25, as necessary. Coates and Denoo[13] developed an equation that ensured that permeability is zero at zero effective porosity and 100% irreducible water saturation. The Coates-Denoo[13] equation is represented as:

$$k = \left[\frac{100\phi_e^2(1-S_{wir})}{S_{wir}}\right]^2 \tag{3.13}$$

In Eq. (3.13), k = permeability, md; ϕ_e = effective porosity, fraction; and S_{wir} = irreducible water saturation, fraction. Bulk volume irreducible water can be represented as:

$$V_{bwi} = \phi_t S_{wir} \tag{3.14}$$

In Eq. (3.14), V_{bwi} = bulk volume irreducible water, fraction; and ϕ_t = total porosity, fraction. Eq. (3.13) can be represented in terms of bulk-volume irreducible water by multiplying the numerator and denominator of the equation with total porosity to get:

$$k = \left[\frac{100\phi_e^2(\phi_t - V_{bwi})}{V_{bwi}}\right]^2 \tag{3.15}$$

The improved reliability of nuclear magnetic resonance (NMR) logs led to the reformulation of Eq. (3.15) by Coates et al.[5] in terms of free-fluid index, (*FFI*), and bulk volume irreducible, (*BVI*), saturations. *FFI* is the fractional part of the pore volume occupied by fluids that are free to flow and, *BVI*, is the fractional part of the pore volume occupied by immobile bound water. The Coates equation for calculation of permeability from NMR logs is expressed as:

$$k = \left(\frac{\phi}{C}\right)^m \left(\frac{FFI}{BVI}\right)^n \tag{3.16}$$

In Eq. (3.16), k = permeability, md; ϕ = porosity, percent; C = 10; m = 4; and n = 2. The values given for C, m, and n, parameters are default values. Values of these parameters are dependent on the core samples and can be regressed from actual data. *FFI* is obtained by summing the, T_2, distribution over, T_2, values greater than the, T_2, cutoff. The transverse relaxation time, T_2, is the characteristic time constant that represents the envelope of the spin-echo signal decay in an NMR logging sequence.[14] The main difficulty of using NMR logs to determine, *FFI*, is calibration of, T_2, cutoff. One of the procedures of calibrating, T_2, cutoff is to match irreducible water saturations, (S_{wir}), obtained from an independent method, such as the Dean-Starks method or Centrifugal method, with irreducible water saturations derived from NMR data.[15] The irreducible water saturations of core samples shown in Figure 3.1 were measured by the centrifugal method and matched against irreducible water saturations from NMR core laboratory data based on, T_2, cutoffs. The, T_2, cutoffs were then used to determine, *FFI*, and, *BVI*, for the core samples. NMR permeability for each core sample was then calculated from these data using Eq. (3.16). The comparison of calculated NMR permeability versus core permeability measured by routine core analysis method is shown in Figure 3.2. The approach discussed here is a method regularly used to calibrate NMR calculated permeability with core permeability obtained through routine core analysis methods. The parameters for the Coates equation obtained from regression of the calibrated permeability data are then used to calculate permeability data from NMR data measured in other wells. The Coates equation is widely used for the calculation of permeability data as discussed, if NMR logs are available. This is the method recommended for the calculation of permeability data from NMR and porosity logs.

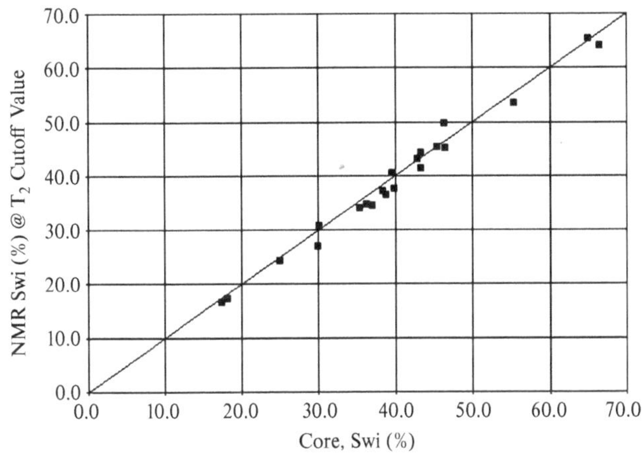

▲ **FIGURE 3.1 Determination of T_2 cutoff from irreducible water saturation comparison**

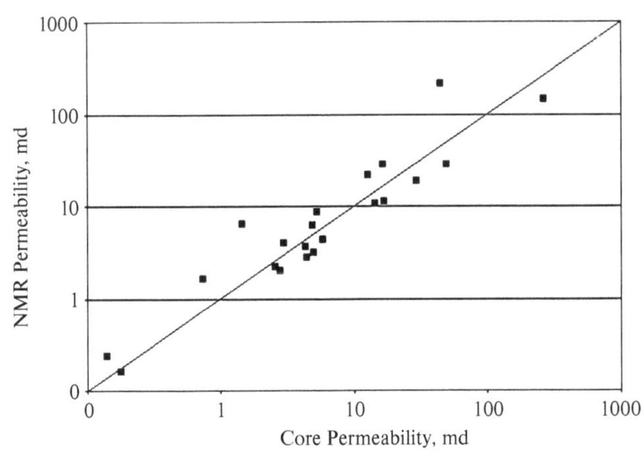

▲ **FIGURE 3.2 NMR permeability calculated from Coates equation versus core permeability**

An important correlation for predicting permeability was developed by Amaefule et al.[16] from the Kozeny-Carman equation [Eq. (3.8)]. By dividing both sides of Eq. (3.8) with effective porosity, ϕ_e, and taking the square root of both sides of the equation, the resulting equation is expressed as:

$$\sqrt{\frac{k}{\phi_e}} = \left[\frac{\phi_e}{1-\phi_e}\right]\left[\frac{1}{\sqrt{F_s \tau A_g}}\right] \tag{3.17}$$

The reservoir quality index (RQI) parameter is defined in terms of permeability in millidarcies (md) as:

$$RQI = 0.0314\sqrt{\frac{k}{\phi_e}} \tag{3.18}$$

In Eq. (3.18), RQI = reservoir quality index, μm; k = permeability, md; and ϕ_e = effective porosity, fraction. The pore volume-to-grain volume ratio, ϕ_z, is defined as:

$$\phi_z = \frac{\phi_e}{1-\phi_e} \tag{3.19}$$

The flow zone indicator (FZI) parameter is designated by the following expression as:

$$FZI = \frac{1}{\sqrt{F_s \tau A_g}} = \frac{RQI}{\phi_z} \tag{3.20}$$

In Eq. (3.20), FZI = flow zone indicator, μm. Taking the logarithm of both sides of Eq. (3.20) gives:

$$\log RQI = \log FZI + \log \phi_z \tag{3.21}$$

As demonstrated by Amaefule et al.,[16] a log-log plot of RQI versus ϕ_z will give a straight line with unit slope for all samples with similar FZI values. Data from core samples with different FZI values will lie on separate, parallel lines. The value of the constant for FZI is determined from the intercept of the unit slope straight line at $\phi_z = 1$. As stated by Amaefule et al.,[16] core data that lie on the same straight line have similar pore throat attributes, and thereby can be categorized as a hydraulic unit. The classification of reservoir rock as hydraulic units is useful as a means of characterizing the reservoir as discussed in Chapter 22. Numerous reservoir studies[17-19] have been conducted by application of the hydraulic unit concepts. However, note that the application of this method for prediction of permeability data requires the availability of significant quantities of core permeability data.

A common empirical relationship that is widely used to correlate permeability and porosity data is a crossplot of the data on a semi-log scale as represented by the equation:

$$\log k = a\phi + b \tag{3.22}$$

In Eq. (3.22), a and b, are regression constants that depend on the data set. Eq. (3.22) has no theoretical basis but has been adopted because the distribution of permeability data appears to be log-normal. An example of a permeability-porosity, $(\phi - k)$, crossplot for core samples from a reservoir is shown in Figure 3.3. The, $(\phi - k)$, crossplot is a convenient method for displaying core or log data but is generally misused as a correlation for predicting permeability data from porosity data. Worthington[2] and Delfiner[20] reported that indiscriminate use of the, $(\phi - k)$, crossplot for predicting permeability data could cause the underestimation of permeability data by at least a factor of 2. The, $(\phi - k)$, crossplot is not recommended as a method for predicting permeability data in the form of a correlation. The crossplot can be used qualitatively to review permeability data predicted using Eq. (3.16) or Eq. (3.21). The application of the, $(\phi - k)$, crossplots in the form of cloud transforms should be used for reservoir characterization, as described in Chapter 22.

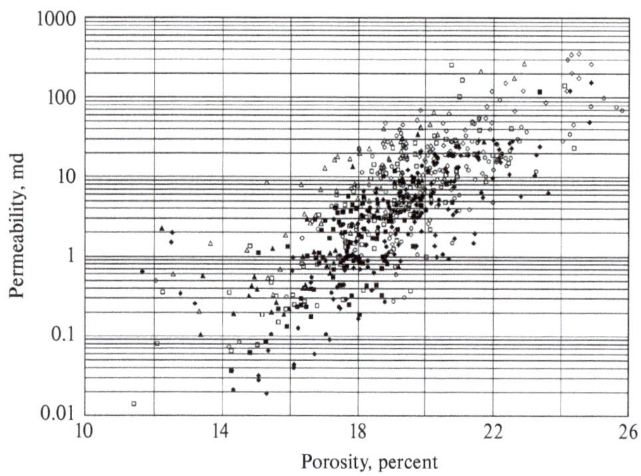

▲ **FIGURE 3.3** **Porosity-permeability crossplot based on core samples from a reservoir**

3.3 Relative Permeability

The relative permeability of a fluid flowing in a porous medium is the ratio of its effective permeability to the absolute permeability of the porous medium as represented in Eq. (3.1). Relative permeability data can be presented graphically in plots called relative permeability curves. A typical relative permeability curve for an oil-water system is shown in Figure 3.4. In Figure 3.4, the range of water saturation is from the initial water saturation, S_{wi}, to water saturation at residual oil saturation, $1 - S_{orw}$. Oil relative permeability, k_{ro}, is highest at, S_{wi}, and declines to zero at, $1 - S_{orw}$. Water relative permeability, k_{rw}, increases from zero at, S_{wi}, to its highest value at, $1 - S_{orw}$. Figure 3.4 shows the location of critical water saturation, S_{wc}. Critical water saturation is the level of water saturation at which water starts to flow in the reservoir. The relative permeability curve for a gas-oil system is shown in Figure 3.5. The range of gas saturation is from initial gas saturation, S_{gi}, to gas saturation at residual oil saturation, $1 - S_{org}$. Oil relative permeability, k_{ro}, is highest at, S_{gi}, and declines to zero at, $1 - S_{org}$. Gas relative permeability, k_{rg}, increases from zero at, S_{gi}, to its highest value at, $1 - S_{org}$. The critical gas saturation, S_{gc}, is the level of gas saturation at which gas starts to flow in the reservoir.

The relative permeability of a fluid in a multi-fluid system is a function of the saturation level of the fluid as shown in Figures 3.4 and 3.5. However, many distributions of the fluid are possible at the same saturation level, depending on the wettability and the direction of saturation changes (or saturation history) of the porous media. For this reason, permeability data measured for drainage displacements (decrease in saturation of the wetting phase) are usually different from permeability data measured for imbibition displacement (increase in saturation of the wetting phase). This difference is described as the **hysteresis** of the permeability curves. It is similar to the hysteresis of the capillary pressure curves discussed in Chapter 18, Section 18.2.2. Basic concepts on rock wettability are presented in Chapter 18, Section 18.2.1.

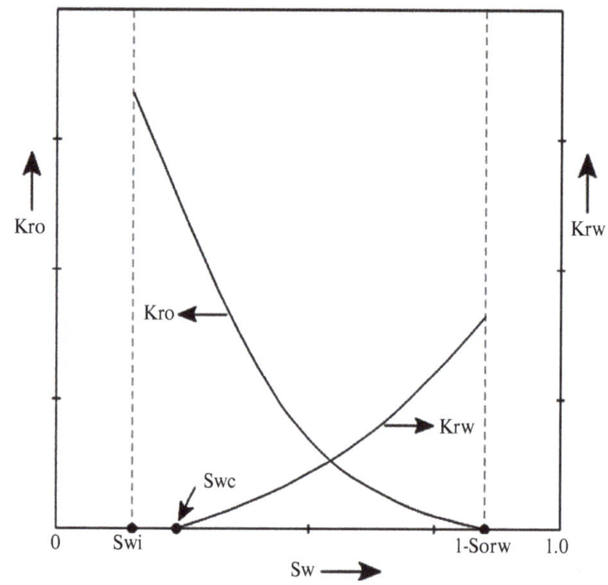

▲ **FIGURE 3.4** **Oil-water relative permeability curves**

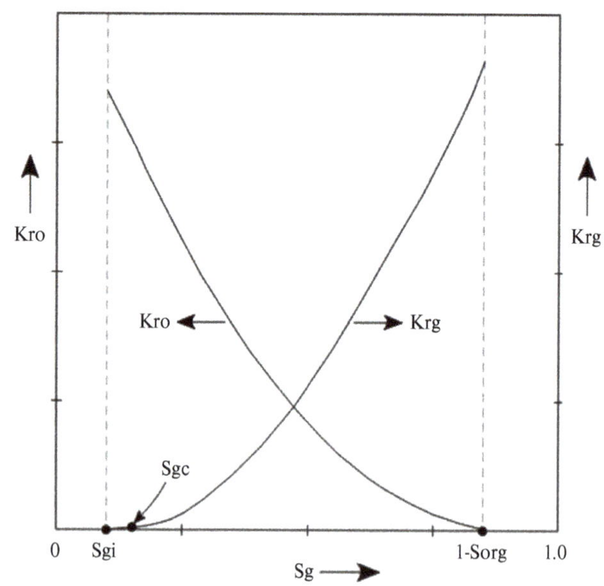

▲ **FIGURE 3.5** **Gas-oil relative permeability curves**

3.4 Sources of Relative Permeability Data

Relative permeability data are obtained from many sources for hydrocarbon reservoirs. These sources include laboratory measurements on core samples, calculations from field production data, and empirical correlations. The most reliable source of relative permeability data are laboratory measurements on core samples.[21] Some success has been reported on estimating relative permeability data from field production data.[22] In the absence of laboratory data, relative permeability data can be calculated from empirical correlations.

3.4.1 Laboratory Measurements of Relative Permeability Data

Laboratory methods used in measuring relative permeability data can be classified under two groups, namely: Steady-state methods and Unsteady-state methods. Steady-state methods are more widely used and are considered more reliable than unsteady-state methods.[21] However, steady-state methods require more experimental time and more elaborate equipment than unsteady-state methods. Some steady-state measurements may require days of experimental time in comparison to hours for unsteady-state methods. Both methods are affected by capillary end effects. **Capillary end effect** is a phenomenon that causes the saturation of the wetting phase to be high close to the inlet and outlet ends of the core sample. The higher saturations at the ends of the cores are caused by the tendency of the wetting phase to remain in the core capillary spaces rather than exit into a non-capillary space. Several measures are employed in relative permeability measurements to reduce or eliminate capillary end effects in core samples. One of the methods used to eliminate capillary end effects is the Hassler's technique that involves the placement of porous plates in contact with both ends of the core.[21] Both steady-state and unsteady-state measurements should be conducted at reservoir conditions.

3.4.1.1 Steady-state methods

Steady-state methods yield the most reliable relative permeability data because capillary equilibrium is achieved in the method, fluid saturations can be measured directly, and the calculation procedure is based on the Darcy equation [Eq. (3.6)]. In the steady-state method, two or three fluids are injected simultaneously at constant rates or pressures to achieve equilibrium inside the porous medium.[21] After equilibrium is achieved, the saturations, flow rates, and pressure gradients of each phase are measured. Using the Darcy equation, the effective permeability of each phase is then calculated. The entire saturation range can be measured in a stepwise fashion by changing the ratio of injection rates and repeating the procedure.[23] Steady-state measurement requires many hours or even days to complete because equilibrium conditions must be attained at each saturation level.

3.4.1.2 Unsteady-state methods

Unlike steady-state methods, relative permeability data can be obtained with unsteady-state methods very quickly. Because unsteady-state methods do not require the attainment of equilibrium in the displacement process, many relative permeability data can be acquired in a few hours. In a typical unsteady-state method, the in-situ fluids are displaced by injection of the displacing fluid at constant rate or pressure, while measuring the produced fluids continuously. The relative permeability data are calculated from the production data on the basis of the Buckley-Leverett equation.[24,25] A widely used unsteady-state method is the use of centrifuge for measurement of relative permeability data. In this method, saturated core samples are rotated at high angular speed, creating a known centrifugal force. The rate of fluid production is measured as a function of time. Relative permeability data are then calculated from the measurements.[26] The centrifuge method is extremely fast and is not affected by the viscous fingering problems associated with other unsteady-state methods. Permeability data from unsteady-state methods are affected by displacement problems, such as viscous fingering and channeling, including instability of the displacement process.[27] The method is recommended to be used to measure relative permeability at the saturation endpoints, while steady-state methods are used to infill the data between the endpoints.[28]

3.4.2 Estimations from Field Data

Relative permeability data can be estimated through history-matching of field production data using reservoir simulators. The generation of relative permeability data from historical production data has been reported by many authors, including Kulkarni and Datta-Gupta,[22] Watson et al.,[29] Yang and Watson,[30] Reynolds et al.,[31] and Eydinov et al.[32] The agreement between permeability data measured on core samples and those estimated from production data is generally not good. This should be expected since cores are samples of very small volumes of the reservoir (microscopic scale), while production data are based on much larger volumes of the reservoir (macroscopic scale). Relative permeability data calculated from field data should be used to adjust laboratory-derived relative permeability data for reservoir simulation applications.

3.4.3 Empirical Correlations

Several correlations are available for estimation of relative permeability data. These correlations should be used in situations where measured relative permeability data are not available. Reasonable caution should be exercised in applying these correlations because they may not be representative of the rock systems that exist in the reservoir. The most widely used correlations for the estimation of relative permeability data are modifications of the Corey equations.[33]

The basic Corey equations are represented as:

$$k_{rwp} = \left(S_{wp}^*\right)^4 \tag{3.23}$$

$$k_{rnwp} = \left(1 - S_{wp}^*\right)^2 \left[1 - \left(S_{wp}^*\right)^2\right] \tag{3.24}$$

In Eqs. (3.23) and (3.24), k_{rwp} = relative permeability of the wetting phase; k_{rnwp} = relative permeability of the non-wetting phase; and S_{wp}^* = normalized saturation of the wetting phase. For a drainage process, the normalized saturation of the wetting phase is expressed as:

$$S_{wp}^* = \frac{S_{wp} - S_{wpr}}{1 - S_{wpr}} \tag{3.25}$$

For an imbibition process, the normalized saturation of the wetting phase is expressed as:

$$S_{wp}^* = \frac{S_{wp} - S_{wpr}}{1 - S_{wpr} - S_{nwpr}} \tag{3.26}$$

In Eqs. (3.25) and (3.26), S_{wp} = saturation of the wetting phase; S_{wpr} = residual saturation of the wetting phase; and S_{nwpr} = residual saturation of the non-wetting phase.

Applying the Corey equations to a water-wet, two-phase oil-water system, and assuming oil displacement process is by imbibition, substituting Eq. (3.26) gives:

$$S_w^* = \frac{S_w - S_{wi}}{1 - S_{wi} - S_{orw}} \tag{3.27}$$

From Eq. (3.27),

$$1 - S_w^* = \frac{1 - S_w - S_{orw}}{1 - S_{wi} - S_{orw}} \tag{3.28}$$

In Eqs. (3.27) and (3.28), S_w = water saturation, fraction; S_{orw} = residual oil saturation to water, fraction; and S_{wi} = initial water saturation, fraction. Substituting Eq. (3.27) into Eq. (3.23) gives:

$$k_{rw} = \left(\frac{S_w - S_{wi}}{1 - S_{wi} - S_{orw}}\right)^4 \tag{3.29}$$

In Eq. (3.29), k_{rw} = relative permeability of water. A generalized form of the equation can be written by expressing the Corey exponent as, N, and including the endpoint of relative permeability of water at residual oil saturation, $k_{rw@S_{orw}}$, such that Eq. (3.29) becomes:

$$k_{rw} = k_{rw@S_{orw}} \left(\frac{S_w - S_{wi}}{1 - S_{wi} - S_{orw}}\right)^N \tag{3.30}$$

Eq. (3.24) can be simplified by dropping the second term in the equation to get:

$$k_{rnwp} = \left(1 - S_{wp}^*\right)^2 \tag{3.31}$$

Substituting Eq. (3.28) into Eq. (3.31) yields:

$$k_{ro} = \left(\frac{1 - S_w - S_{orw}}{1 - S_{wi} - S_{orw}}\right)^2 \tag{3.32}$$

In Eq. (3.32), k_{ro} = relative permeability of oil. A generalized form of the equation can be written by expressing the Corey exponent as N and including the endpoint of relative permeability of oil at initial water saturation, $k_{ro@S_{wi}}$, such that Eq. (3.32) becomes:

$$k_{ro} = k_{ro@S_{wi}} \left(\frac{1 - S_w - S_{orw}}{1 - S_{wi} - S_{orw}}\right)^N \tag{3.33}$$

Corey equations can also be derived for a water-wet, gas-oil system as follows. From Eq. (3.27), it can be deduced that:

$$S_w^* = \frac{1 - S_g - S_{wi} - S_{org}}{1 - S_{wi} - S_{org}} \tag{3.34}$$

Substituting Eq. (3.34) into Eq. (3.23) gives:

$$k_{ro} = \left(\frac{1 - S_g - S_{wi} - S_{org}}{1 - S_{wi} - S_{org}}\right)^4 \tag{3.35}$$

A generalized form of the equation can be written by expressing the Corey exponent as N and including the endpoint of relative permeability of oil at initial gas saturation, $k_{ro@S_{gi}}$, such that Eq. (3.35) becomes:

$$k_{ro} = k_{ro@S_{gi}} \left(\frac{1 - S_g - S_{wi} - S_{org}}{1 - S_{wi} - S_{org}}\right)^N \tag{3.36}$$

The gas relative permeability, k_{rg}, can be derived by substituting Eq. (3.28) into Eq. (3.31) to get:

$$k_{rg} = \left(\frac{1 - S_w - S_{org} - S_{gc}}{1 - S_{wi} - S_{org} - S_{gc}}\right)^2 = \left(\frac{S_g - S_{gc}}{1 - S_{wi} - S_{org} - S_{gc}}\right)^2 \tag{3.37}$$

A generalized form of the equation can be written by expressing the Corey exponent as N and including the endpoint of relative permeability of gas at residual oil saturation, $k_{rg@S_{org}}$, such that Eq. (3.37) becomes:

$$k_{rg} = k_{rg@S_{org}} \left(\frac{S_g - S_{gc}}{1 - S_{wi} - S_{org} - S_{gc}} \right)^N \tag{3.38}$$

Note that similar modified Corey equations can be derived for an oil-wet system assuming drainage displacement process from Eqs. (3.23), (3.24), and (3.25).

In the absence of data, Corey exponent can be estimated for different rock formations as follows: well consolidated sandstones, N = 3.0; poorly consolidated sandstones, N = 3.5; cemented sandstones and limestones, N = 4.0; and fractured formations, N = 1.0.

EXAMPLE 3.1 Calculation of Oil-Water and Gas-Oil Relative Permeability Curves Using the Modified Corey Equations

Problem

Calculate the oil-water and gas-oil relative permeability data for a water-wet, poorly consolidated rock with the following rock and fluid properties:

Initial water saturation, S_{wi}	0.2
Residual oil saturation to water, S_{orw}	0.2
Residual oil saturation to gas, S_{org}	0.2
Critical gas saturation, S_{gc}	0.05
Oil curve endpoint for water, $k_{ro@S_{wi}}$	0.9
Oil curve endpoint for gas, $k_{ro@S_{gi}}$	0.9
Water curve endpoint, $k_{rw@S_{orw}}$	0.15
Gas curve endpoint, $k_{rg@S_{org}}$	1.0
Corey exponent	3.5

Solution

Using Eq. (3.30), the relative permeability of water at water saturation, $S_w = 0.3$, is calculated as:

$$k_{rw} = k_{rw@S_{orw}} \left(\frac{S_w - S_{wi}}{1 - S_{wi} - S_{orw}} \right)^N$$

$$= 0.15\left(\frac{0.3 - 0.2}{1 - 0.2 - 0.2} \right)^{3.5}$$

$$= 0.000284$$

Similarly, the remaining water relative permeabilities at other water saturations were calculated as shown in Table 3.1.

Using Eq. (3.33), the relative permeability of oil at water saturation, $S_w = 0.3$, is calculated as:

$$k_{ro} = k_{ro@S_{wi}} \left(\frac{1 - S_w - S_{orw}}{1 - S_{wi} - S_{orw}} \right)^N$$

$$= 0.9\left(\frac{1 - 0.3 - 0.2}{1 - 0.2 - 0.2} \right)^{3.5}$$

$$= 0.4755$$

Similarly, the remaining oil relative permeabilities at other water saturations were calculated as shown in Table 3.1. The calculated relative permeability data for oil and water are shown in Figure 3.6.

Using Eq. (3.36), the relative permeability of oil at gas saturation, $S_g = 0.4$, is calculated as:

$$k_{ro} = k_{ro@S_{gi}} \left(\frac{1 - S_g - S_{wi} - S_{org}}{1 - S_{wi} - S_{org}} \right)^N$$

$$= 0.9\left(\frac{1 - 0.4 - 0.2 - 0.2}{1 - 0.2 - 0.2} \right)^{3.5}$$

$$= 0.01925$$

Similarly, the remaining oil relative permeabilities at other gas saturations were calculated as shown in Table 3.2.

Using Eq. (3.38), the relative permeability of gas at gas saturation, $S_g = 0.4$, is calculated as:

$$k_{rg} = k_{rg@S_{org}} \left(\frac{S_g - S_{gc}}{1 - S_{wi} - S_{org} - S_{gc}} \right)^N$$

$$= 1.0 \times \left(\frac{0.4 - 0.05}{1 - 0.2 - 0.2 - 0.05} \right)^{3.5}$$

$$= 0.2056$$

The remaining gas relative permeabilities at other gas saturations were calculated as shown in Table 3.2. The calculated relative permeability data for gas and oil are shown in Figure 3.7.

End of Example 3.1

TABLE 3.1 Oil-Water Relative Permeabilities Calculated for Example 3.1

S_w	k_{rw}	k_{ro}
0.20	0.0000	0.9000
0.25	0.0000	0.6637
0.30	0.0003	0.4755
0.35	0.0012	0.3288
0.40	0.0032	0.2177
0.45	0.0070	0.1364
0.50	0.0133	0.0795
0.55	0.0227	0.0420
0.60	0.0363	0.0192
0.65	0.0548	0.0070
0.70	0.0792	0.0017
0.75	0.1106	0.0002
0.80	0.1500	0.0000

TABLE 3.2 Gas-Oil Relative Permeabilities Calculated for Example 3.1

S_g	k_{ro}	k_{rg}
0.05	0.6637	0.0000
0.10	0.4755	0.0002
0.15	0.3288	0.0026
0.20	0.2177	0.0106
0.25	0.1364	0.0290
0.30	0.0795	0.0633
0.35	0.0420	0.1199
0.40	0.0192	0.2056
0.45	0.0070	0.3281
0.50	0.0017	0.4954
0.55	0.0002	0.7164
0.60	0.0000	1.0000

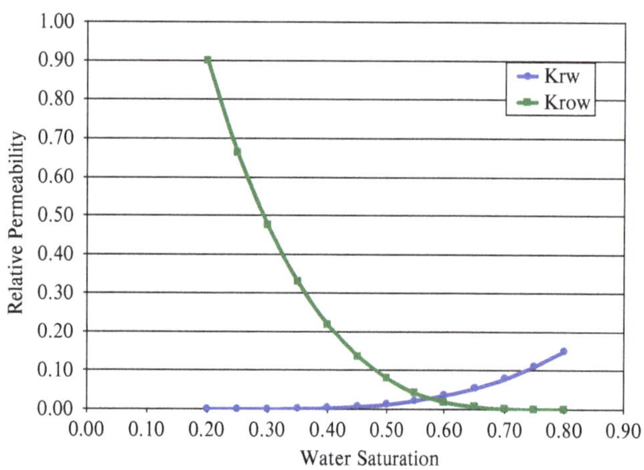

▲ FIGURE 3.6 Oil-water relative permeability plot for Example 3.1

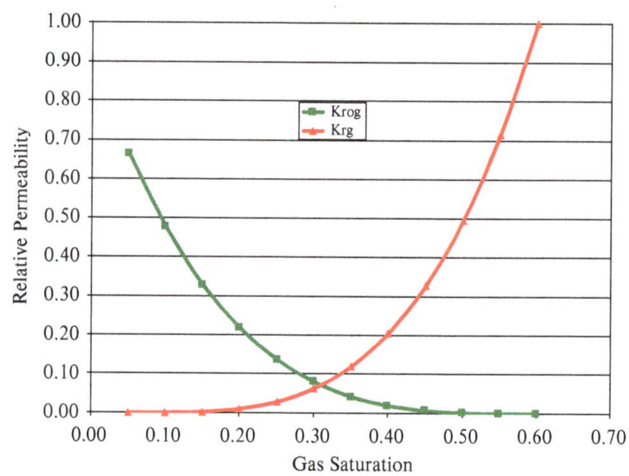

▲ FIGURE 3.7 Gas-oil relative permeability plot for Example 3.1

3.5 Three-Phase Relative Permeability

The relative permeability curves shown in Figures 3.4 and 3.5 are based on two-phase systems. However, in many reservoirs, the flow of three phases (oil, water, and gas) is present during primary, secondary, and enhanced oil recovery processes. The simultaneous flow of three phases requires the use of three-phase relative permeability data to represent fluid flow in the reservoir. Three-phase relative permeability data are usually not measured in the laboratory because they are expensive, tedious, and in some cases impractical. The current accepted engineering practice for prediction of three-phase relative permeability is based on empirical models. Most of the empirical models use data from two-phase relative permeabilities, capillary pressure, and saturation history to predict three-phase relative permeability data. Stone's empirical models (Stone I[34] and Stone II[35]) are the two empirical models widely used in the petroleum industry for the prediction of three-phase relative permeability data. Numerous other empirical models that appear to be superior to Stone's models have been published in the literature.[36-38] However, reviews of these models available in the literature indicate that the empirical models should be tested for accuracy before using them for a specific application.[36-38]

3.6 Applications of Permeability and Relative Permeability Data

Permeability and relative permeability data are two of the most important data that determine the flow of fluids in the reservoir. Consequently, permeability and relative permeability data are very important input in flow calculations, including reservoir simulation. As discussed in Chapter 23, it is common practice in reservoir simulation to adjust permeability and relative permeability data to match performance history. The history-matched model is then used in prediction mode to predict future production targets and reservoir performance. This practice should be conducted with careful consideration of the reasonableness and soundness of the data generated by this process. Permeability and relative permeability data should not be adjusted to the extent that the resulting data violate known physical properties of the rocks and the fluids that are resident in the reservoir.[39,40]

Nomenclature

A	area, cm^2
A_g	surface area
b	Klinkenberg constant
F_s	shape factor
k	permeability
k_a	absolute permeability
k_{air}	permeability measured with air
k_g	effective permeability for gas
k_i	effective permeability for fluid, i
k_o	effective permeability for oil
k_{rg}	relative permeability for gas
k_{ri}	relative permeability for fluid, i
k_{ro}	relative permeability for oil
k_{rw}	relative permeability for water
k_{rwp}	relative permeability of the wetting phase
k_{rnwp}	relative permeability of the non-wetting phase
k_w	effective permeability for water
L	length, cm
N	Corey's exponent
$\bar{p}$	mean flowing pressure
Δp	pressure difference, atm
q	flow rate, cc/sec
S_g	gas saturation, fraction
S_{gc}	critical gas saturation, fraction
S_{nwpr}	residual saturation of the non-wetting phase
S_{org}	residual oil saturation to gas
S_{orw}	residual oil saturation to water
S_w	water saturation, fraction
S_w^*	normalized water saturation
S_{wc}	critical water saturation, fraction
S_{wi}	initial water saturation, fraction
S_{wir}	irreducible water saturation, fraction
S_{wp}	saturation of the wetting phase
S_{wpr}	residual saturation of the wetting phase
S_{wp}^*	normalized saturation of the wetting phase
T_2	NMR characteristic time constant, milliseconds
V_{bwi}	bulk volume irreducible water
ϕ	porosity
ϕ_e	effective porosity
ϕ_t	total porosity
ϕ_z	pore volume-to-grain volume ratio
μ	viscosity, cp
τ	tortuosity

Abbreviations

BVI	bulk volume irreducible
FFI	free-fluid index
FZI	flow zone indicator
NMR	nuclear magnetic resonance
RCAL	routine core analysis
RQI	reservoir quality index
WFT	wireline formation tester

REFERENCES

1. Klinkenberg, L.J.: "The Permeability of Porous Media to Liquids and Gases," *Drilling and Production Practices*, American Petroleum Institute, Washington, DC, 1941, 200-213.
2. Worthington, P.F.: "The Effect of Scale on the Petrophysical Estimation of Intergranular Permeability," *Petrophysics* (January-February 2004), 45, 1, 59-72.
3. Al-Ali, H.A., and Worthington, P.F.: "Applications of Petrophysical Scale Reconciliation to Saudi Arabian Reservoirs," paper IPTC 10726 presented at the International Petroleum Technology Conference, Doha, Qatar, November 21-23, 2005.
4. Shafer, J.L., and Ezekwe, N.J.: "Methods for Upscaling Diverse Rock Permeability Data for Reservoir Characterization and Modeling," paper SPE 109128 presented at the Annual Technical Conference and Exhibition, Anaheim, California, November 11-14, 2007.
5. Coates, G.R., Miller, M., and Henderson, G.: "The MRIL in Conoco 33–1: An Investigation of a New Magnetic Resonance Imaging Log," paper DD presented at the 32nd SPWLA Annual Logging Symposium, Midland, Texas, June 16-19, 1991.
6. Shang, B.Z., Hamman, J.G., Chen, H., and Caldwell, D.H.: "A Model to Correlate Permeability with Efficient Porosity and Irreducible Water Saturation," paper SPE 84303 presented at SPE Annual Technical Conference and Exhibition, Denver, Colorado, October 5-8, 2003.
7. Kozeny, J.: Uber Kapillare Leitung des Wassers im Boden: Sitzungsber. Akad. Wiss. Wien, 1927, 136, 271-306.
8. Carman, P.C.: "Fluid Flow through a Granular Bed," *Trans. Inst. Chem. Eng. London,* 1937, 15, 150-156.
9. Rose, W., and Bruce, W.A.: "Evaluation of Capillary Character in Petroleum Reservoir Rock," *Trans., AIME* (1949) 127-142.
10. Wyllie, M.R.J., and Rose, W.D.: "Some Theoretical Considerations Related to the Quantitative Evaluation of the Physical Characteristics of Reservoir Rock from Electrical Log Data," *Trans., AIME* (1950) 105-118.
11. Timur, A: "An Investigation of Permeability, Porosity, and Residual Water Saturation Relationship for Sandstone Reservoirs," *The Log Analyst* (July-August 1968) 9, 3-5.
12. Coates, G.R., and Dumanoir, J.L.: "A New Approach to Improved Log-Derived Permeability," *Proceedings,* SPWLA 14th Annual Logging Symposium, Lafayette, Louisiana (May 6-9, 1973).
13. Coates, G., and Denoo, S.: "The Producibility Answer Product," *The Technical Review,* Schlumberger, Houston, Texas (June 1981) 29, 2, 55-63.
14. Freedman, R.: "Advances in NMR Logging," Distinguished Author Series, *JPT* (January 2006) 60-66.
15. Shafer, J., Chen, S., Georgi, D., and Chen, J.: "Methods and Procedures for Calibrating NMR Log Derived Permeabilities," The 11th Formation Evaluation Symposium of Japan, Chiba, Japan (October 5-6, 2005) 1-15.
16. Amaefule, J.O., Altunbay, M., Tiab, D., Kersey, D.G., and Keelan, D.K.: "Enhanced Reservoir Description: Using Core and Log Data to Identify Hydraulic (Flow) Units and Predict Permeability in Uncored Intervals/Wells," paper SPE 26436 presented at the SPE 68th Annual Technical Conference and Exhibition, Houston, Texas, October 3-6, 1993.
17. Abbaszadeh, M., Fujii, H., and Fujimoto, F.: "Permeability Prediction by Hydraulic Flow Units–Theory and Applications," *SPEFE* (December 1996) 263-271.
18. Soto B.R., Garcia, J.C., Torres, F., and Perez, G.S.: "Permeability Prediction Using Hydraulic Flow Units and Hybrid Soft Computing Systems," paper SPE 71455 presented at the 2001 SPE Annual Technical Conference and Exhibition, New Orleans, Louisiana, September 30-October 3, 2001.
19. Shenawi, S.H., White, J.P., Elrafie, E.A., and Kilany, A.: "Permeability and Water Saturation Distribution by Lithologic Facies and Hydraulic Units: A Reservoir Simulation Case Study," paper SPE 105273 presented at the 15th SPE Middle East Oil & Gas Show and Conference, Bahrain, March 11-14, 2007.
20. Delfiner, P.: "Three Statistical Pitfalls of Phi-K Transforms," *SPERE&E* (December 2007) 609-617.
21. Honarpour, M., and Mahmood, S.M.: "Relative Permeability Measurements: An Overview," *JPT* (August 1988) 963-966.
22. Kulkarni, K.N., and Datta-Gupta, A.: "Estimating Relative Permeability From Production Data: A Streamline Approach," *SPEJ* (December 2000) 402-411.
23. Braun, E.M., and Blackwell, R.J.: "A Steady-State Technique for Measuring Oil-Water Relative Permeability Curves at Reservoir Conditions," paper SPE 10155 presented at the 56th Annual Technical Conference and Exhibitions, San Antonio, Texas, October 5-7, 1981.
24. Johnson, E.F., Bossler, D.P., and Naumann, V.O.: "Calculation of Relative Permeability from Displacement Experiments," SPE 1023G, *Trans. AIME* (1959) 370-372.
25. Batycky, J.P., McCaffery, F.G., Hodgins, P.K., and Fisher, D.B.: "Interpreting Relative Permeability and Wettability from Unsteady-State Displacement Measurements," *SPEJ* (June 1981) 296-308.
26. Spronsen, E.V.: "Three-Phase Relative Permeability Measurements Using Centrifuge Method," paper SPE/DOE 10688 presented at the SPE/DOE Third Joint Symposium on Enhanced Oil Recovery, Tulsa, Oklahoma, April 4-7, 1982.
27. Peters, E.J., and Khataniar, S.: "The Effect of Instability on Relative Permeability Curves Obtained by the Dynamic-Displacement Method," *SPEFE* (December 1987) 469-474.
28. Shafer, J.L., Braun, E.M., Wood, A.C., and Wooten, J.M.: "Obtaining Relative Permeability Data Using a Combination of Steady-State and Unsteady-State Corefloods," paper 9009 presented at the 1990 SCA Conference, Houston, Texas, August 14-16, 1990.
29. Watson, A.T, Seinfeld, J.H., Gavalas, G.R., and Woo, P.T.: "History Matching in Two-Phase Petroleum Reservoirs," *SPEJ* (December 1980) 521-532.
30. Yang, P-H., and Watson, A.T.: "A Bayesian Methodology for Estimating Relative Permeability Curves," *SPERE* (May 1991) 259-265.

31. Reynolds, A.C., Li, R., and Oliver, D.S.: "Simultaneous Estimation of Absolute and Relative Permeability by Automatic History Matching of Three-Phase Flow Production Data," *JCPT* (2004) 43, 3, 37-46.

32. Eydinov, D., Gao, G., Li, G., and Reynolds, A.C.: "Simultaneous Estimation of Relative Permeability and Porosity/Permeability Fields by History Matching Production Data," *JCPT* (December 2009) 48, 12, 13-25.

33. Corey, A.T.: "The Interrelation between Gas and Oil Relative Permeabilities," *Production Monthly* (1954) 19, 1, 38-41.

34. Stone, H.L.: "Probability Model for Estimating Three-Phase Relative Permeability," *JPT* (February 1970) 214-218.

35. Stone, H. L.: "Estimation of Three-Phase Relative Permeability and Residual Oil Data," *JCPT* (1973) 12, 53-61.

36. Manjnath, A., and Honarpour, M.M.: "An Investigation of Three-Phase Relative Permeability," paper SPE 12915 presented at the 1984 Rocky Mountain Regional Meeting, Casper, Wyoming, May 21-23, 1984.

37. Blunt, M.J.: "An Empirical Model for Three-Phase Relative Permeability," *SPEJ* (December 2000) 435-445.

38. Ahmadloo, F., Asghari, K., and Jamaloei, B.Y.: "Experimental and Theoretical Studies of Three-Phase Relative Permeability," paper SPE 124538 presented at the 2009 SPE Annual Technical Conference and Exhibition, New Orleans, Louisiana, October 4-7, 2009.

39. Coats, K.H.: "Use and Misuse of Reservoir Simulation Models," *JPT* (1969) 183-190.

40. Wang, K., Killough, J.E., and Sepehrnoori, K.: "A New Upscaling Method of Relative Permeability Curves for Reservoir Simulation," paper SPE 124819 presented at the 2009 SPE Annual Technical Conference and Exhibition, New Orleans, Louisiana, October 4-7, 2009.

ADDITIONAL REFERENCES

- Ahmed, U., Crary, S.F., and Coates, G.R.: "Permeability Estimation: The Various Sources and Their Interrelationships," *JPT* (May 1991) 578-587.
- Amabeoku, M.O., Kersey, D.G., BinNasser, R.H., and Belowi, A.R.: "Relative Permeability Coupled Saturation-Height Models Based on Hydraulic (Flow) Units in a Gas Field," *SPEREE* (December 2008) 1013-1028.
- App, J.F., and Mohanty, K.K.: "The Benefit of Local Saturation Measurements in Relative Permeability Estimation from Centrifuge Experiments," *SPEJ* (September 2002) 288-298.
- Babadagli, T., and Al-Salmi, S.: "A Review of Permeability-Prediction Methods for Carbonate Reservoirs Using Well-Log Data," *SPEREE* (April 2004) 75-88.
- Blunt, M.J.: "Effects of Heterogeneity and Wetting on Relative Permeability Using Pore Level Modeling," *SPEJ* (March 1997) 70-87.
- Corey, A.T., Rathjens, C.H., Henderson, J.H., and Wyllie, M.R.J.: "Three-Phase Relative Permeability," *JPT* (November 1956) 63-65.
- Grattoni, C.A., and Bidner, M.S.: "History Matching of Unsteady-State Corefloods for Determining Capillary Pressure and Relative Permeabilities," paper SPE 21135 presented at the SPE Latin American Petroleum Engineering Conference, Rio de Janeiro, Brazil, October 14-19, 1990.
- Haddad, S., Cribbs, M., Sagar, R., Tang, Y., Viro, E., and Castelijins, K.: "Integrating Permeabilities from NMR, Formation Tester, Well Test and Core Data," paper SPE 71722 presented at the 2001 SPE Annual Technical Conference and Exhibition, New Orleans, Louisiana, September 30-October 3, 2001.
- Haddad, S., Cribbs, M., Sagar, R., Viro, E., Castelijins, K., and Tang, Y.: "So What is the Reservoir Permeability?" paper SPE 63138 presented at the 2000 SPE Annual Technical Conference and Exhibition, Dallas, Texas, October 1-4, 2000.
- Heaviside, J., Black, C.J.J., and Berry, J.F.: "Fundamentals of Relative Permeability: Experimental and Theoretical Considerations," paper SPE 12173 presented at the 58th Annual Technical Conference and Exhibition, San Francisco, California, October 5-8, 1983.
- Honarpour, M., Koederitz, L.F., and Harvey, A.H.: "Empirical Equations for Estimating Two-Phase Relative Permeability in Consolidated Rock," *JPT* (December 1982) 2905-2908.
- Johnson, C.E., Jr.: "Graphical Determination of the Constants in the Corey Equation for Gas-Oil Relative Permeability Ratio," SPE 2346, *JPT Forum* (October 1968) 1111-1113.
- Jones, S.C., and Roszelte, W.O.: "Graphical Techniques for Determining Relative Permeability From Displacement Experiments," *JPT* (May 1978) 807-817.
- Kjosavik, A., Ringen, J.K., and Skjaeveland, S.M.: "Relative Permeability Correlation for Mixed-Wet Reservoirs," *SPEJ* (March 2002) 49-58.
- Labban, Z.S., and Horne, R.N.: "Estimating Relative Permeabilities by Automated History Matching Using Linear Modeling Techniques," paper SPE 21206 presented at the 11th SPE Symposium on Reservoir Simulation, Anaheim, California, February 17-20, 1991.
- Nelson, P.H.: "Permeability-Porosity Relationships in Sedimentary Rocks," *The Log Analyst* (May-June 1994) 38-62.

PROBLEMS

3-1. A cleaned and dried core plug was used in a permeability measurement at 68 °F. The core plug is 2 cm long and 1.6 cm in diameter. The volume of air that flowed through the core plug was measured to be 1002 cm^3 in 500 seconds at outlet pressure of 1 atm absolute and flowing temperature of 68 °F. The inlet pressure of air was 1.46 atm absolute at 68 °F. Calculate the air permeability of the core plug.

3-2. The permeability measurement in Problem 3-1 was repeated on the same core plug with water as the fluid at 68 °F. The flow rate of water through the core plug was measured to be 0.021 cm^3/s at the same inlet and outlet pressures as in Problem 3-1. Assuming that water has no reactive effects on the core plug, calculate the permeability of the core plug.

3-3. What are the reasons for the difference, if any, between the permeability values calculated for the core plug in Problem 3-1 and Problem 3-2?

3-4. Describe experimental procedures that could be used to conduct and improve air permeability measurements.

3-5. Calculate the oil-water and gas-oil relative permeability data using the Corey correlations for a water-wet, well-consolidated rock with the rock and fluid properties shown below:

Initial water saturation	0.12
Residual oil saturation to water	0.15
Residual oil saturation to gas	0.15
Critical gas saturation	0.03
Oil curve endpoints	1.0
Water curve endpoint	1.0
Gas curve endpoint	1.0

3-6. Calculate the oil-water and gas-oil relative permeability data using the Corey correlations for a water-wet, cemented limestones with the rock and fluid properties shown below:

Initial water saturation	0.25
Residual oil saturation to water	0.25
Residual oil saturation to gas	0.25
Critical gas saturation	0.025
Oil curve endpoints	1.0
Water curve endpoint	0.75
Gas curve endpoint	1.0

CHAPTER

4 Reservoir Fluid Saturations

4.1 Introduction

The pore spaces in reservoir rocks are occupied by fluids. In petroleum reservoirs, the fluids are usually water and hydrocarbons. The relative volumes of water and hydrocarbons in the pore volume of the reservoir rock are designated as saturations. **Water saturation** in the reservoir rock is the fraction of the pore volume occupied by water. By the same definition, **hydrocarbon saturation** in the reservoir rock is the fraction of the pore volume occupied by hydrocarbons. The sum of the water and hydrocarbon saturations in the reservoir rock is equal to unity. This relationship can be expressed simply as:

$$S_h + S_w = 1 \tag{4.1}$$

In Eq. (4.1), S_h = hydrocarbon saturation, fraction; and S_w = water saturation, fraction. If the hydrocarbon in the reservoir rock exists in oil and gas phases, Eq. (4.1) can be written as:

$$S_o + S_g + S_w = 1 \tag{4.2}$$

In Eq. (4.2), S_o = oil saturation, fraction; and S_g = gas saturation, fraction.

The importance of determining accurate data for water saturations in the reservoir, especially at discovery, is evident from either Eq. (4.1) or Eq. (4.2). The existence of low water saturations in the reservoir indicates the existence of high hydrocarbon saturations. Conversely, high water saturations are interpreted as representing the existence of low hydrocarbon saturations. This rule-of-thumb is routinely applied qualitatively to assess the potential hydrocarbon contents of a reservoir after it has been penetrated with a well, especially at discovery. For calculation of the volume of hydrocarbons in a reservoir, a general equation that is usually applied is expressed as:

$$HCPV = Area \times Thickness \times \phi \times (1 - S_w) \tag{4.3}$$

In Eq. (4.3), *HCPV* = hydrocarbon pore volume; *Area* = hydrocarbon-bearing area of the reservoir; *Thickness* = net productive thickness or pay of the reservoir; ϕ = porosity, fraction; and S_w = water saturation, fraction. Water saturation is one of the key data required in Eq. (4.3). If the water saturation data is incorrect, it could result in the over- or under-estimation of the volume of hydrocarbon present in the reservoir. The economic impact of erroneous calculation of in-place-hydrocarbon volume cannot be overstated. It could lead to execution of uneconomic projects with erroneously high estimated in-place-hydrocarbon volumes or lead to the abandonment of projects with erroneously low estimates of in-place-hydrocarbon volumes.

In this chapter, various models and equations used for the calculation of water saturations in clean sands, shaly sands, and carbonate reservoirs are presented. The widely used Archie equation for clean sands is presented. This is followed with the presentation of models for shaly sands, such as the Waxman-Smits model, Simandoux equation, Poupon-Leveaux equation, and the Dual-Water model. The application of the Archie equation to carbonate reservoirs is discussed. This section ends with a presentation on the use of nuclear magnetic resonance (NMR) logs for the determination of in-situ fluid saturations for most types of reservoirs.

Also, in this chapter the use of cutoffs in the determination of net pay thickness is discussed. As Eq. (4.3) shows, net pay thickness is a very important data required in the calculation of in-place-hydrocarbon volumes. The impact of net pay thickness data on the calculation of hydrocarbon volumes is as important as those of porosity and water saturation. The criteria generally used in the application of cutoffs are based on porosity, permeability, and water saturation limiting values. There are no accepted industry-wide limits on the criteria used for the determination of net pay based on these petrophysical properties. The lack of accepted industry practice is because the application of cutoff criteria is by necessity reservoir-specific, since it is rare to encounter two reservoirs that have identical petrophysical properties. Even in the same reservoir owned jointly by different companies, it is not uncommon to find that the cutoff criteria applied by each company are different. This situation has resulted in partners on the same reservoir using different values of net pay in calculating the in-place-hydrocarbon volumes. In this chapter, common practices and definitions used in the application of cutoff criteria to determine net pay are presented.

4.2 Determination of Water Saturations

Many models and equations have been developed over many years for the determination of water saturations in hydrocarbon-bearing formations. The models and equations vary in terms of complexity from the Archie equation developed for clean sands to more elaborate models developed for shaly sands. In this section, the Archie equations for clean sands are presented followed by a selection of models developed for shaly sands. The application of the Archie equations on carbonate reservoirs is discussed. Finally, the utility of NMR logs in the evaluation of water saturation for most types of reservoirs is discussed.

4.2.1 Clean Sands

Clean sands are classified as sands that satisfy the assumptions used in the development of the Archie equations.[1] These are sands that do not contain clays or clay minerals. In such sands, conduction of electricity occurs only through free ions within the formation water. There is an absence of the "shale effect." Archie equations are assumed to apply at these "perfect" rock conditions. In practice, Archie equations are generally applied under rock conditions that do not meet these perfect conditions.

4.2.1.1 Archie equations

The derivation of Archie equations[1] follows from the definition of **resistivity index** (I_R) as the ratio of the resistivity of a reservoir rock partially saturated with water (R_t) to the resistivity of the rock if fully saturated with water (R_o). Thus, the resistivity index is defined as:

$$I_R = \frac{R_t}{R_o} \tag{4.4}$$

In Eq. (4.4), I_R = resistivity index; R_t = formation resistivity or resistivity of partially water-saturated rock, $\Omega \cdot$ m; and R_o = resistivity of fully (100%) water-saturated rock, $\Omega \cdot$ m. Resistivity index can also be related to water saturation as:

$$I_R = S_w^{-n} \tag{4.5}$$

In Eq. (4.5), S_w = water saturation, fraction; and n = saturation exponent. Combining Eqs. (4.4) and (4.5) gives:

$$S_w^{-n} = \frac{R_t}{R_o} \tag{4.6}$$

For a clean rock that is fully saturated with water, **formation resistivity factor** (sometimes called **formation factor**), F, is defined as:

$$F = \frac{R_o}{R_w} \tag{4.7}$$

In Eq. (4.7), F = formation factor; and R_w = resistivity of formation water, $\Omega \cdot$ m. Substituting Eq. (4.7) into Eq. (4.6) gives:

$$S_w^{-n} = \frac{R_t}{FR_w} \tag{4.8}$$

Archie[1] expressed the formation factor, F, as a function of total porosity, ϕ_t, thus:

$$F = \phi_t^{-m} \tag{4.9}$$

In Eq. (4.9), ϕ_t = total porosity, fraction; and m = cementation exponent. Winsauer et al.[2] modified Eq. (4.9) later by inserting the tortuosity factor, a, to get:

$$F = a\phi_t^{-m} \tag{4.10}$$

Substituting Eq. (4.10) into Eq. (4.8) gives:

$$S_w^{-n} = \frac{R_t}{a\phi_t^{-m} R_w} \tag{4.11}$$

Eq. (4.11) can be rearranged to give:

$$S_w = \left(\frac{a}{\phi_t^m} \frac{R_w}{R_t} \right)^{1/n} \tag{4.12}$$

Eq. (4.12) is the Archie equation for the calculation of water saturation in reservoir rocks. Electrical conductivity of a material is the reciprocal of its resistivity. For example, the conductivity of a reservoir rock is the reciprocal of its resistivity. This can be expressed as:

$$C_t = \frac{1}{R_t} \tag{4.13}$$

In Eq. (4.13), C_t = conductivity of the reservoir rock, Sm^{-1}. Expressing Archie equation [Eq. (4.12)] in terms of conductivities yields:

$$S_w = \left(\frac{a}{\phi_t^m} \frac{C_t}{C_w} \right)^{1/n} \tag{4.14}$$

In Eq. (4.14), C_w = conductivity of the formation water, Sm^{-1}. If it is assumed that the parameter, a = 1, then Eq. (4.14) can be rearranged as:

$$C_t = \phi_t^m \times S_w^n \times C_w \tag{4.15}$$

4.2.1.2 Determination of Archie parameters *n*, *m*, and *a*

The Archie parameters, n, m, and a, can be determined by applying the expression derived by rearranging Eq. (4.12) as:

$$\log R_t = -n \log S_w + \log\left(\frac{aR_w}{\phi_t^m} \right) \tag{4.16}$$

A least-squares straight line fitted through a log-log plot of, R_t versus, S_w laboratory data from core samples has a negative slope equal to the parameter, n. An alternative method that is equivalent to Eq. (4.16) can be derived from Eq. (4.5) as:

$$\log I_R = -n \log S_w \tag{4.17}$$

A least-squares straight line fitted through a log-log plot of, I_R versus S_w, laboratory data from core samples yields the same value of parameter, n.

The parameters, m and a, can be determined by application of Eq. (4.10) in the form:

$$\log F = -m\log\phi_t + \log a \tag{4.18}$$

A least-squares straight line drawn through a log-log plot of, F versus ϕ_t, data based on core samples has a negative slope equal to the parameter m, and an intercept equal to the parameter a. Note that the entire core data used in this plot must be obtained at $S_w = 1$, as required by the definition of formation factor in Eq. (4.7).

A mathematically rigorous method for determining Archie parameters was proposed by Maute et al.[3] This method determines Archie parameters, n, m, and a, by minimizing the mean-square water saturation error, ε, represented as:

$$\varepsilon = \sum_j \sum_i \left[S_{wij} - \left(\frac{a}{\phi_{tj}^m} \frac{R_{wj}}{R_{tij}} \right)^{1/n} \right]^2 \tag{4.19}$$

In Eq. (4.19), ε = water saturation error, fraction; j = core sample index; i = index for each core sample, j, data; S_{wij} = ith laboratory-measured water saturation for core sample, j, fraction; R_{tij} = ith laboratory-measured resistivity for core sample, j, $\Omega \cdot$ m; R_{wj} = resistivity of water used in core sample, j, $\Omega \cdot$ m; and ϕ_{tj} = total porosity of core sample, j, fraction. Maute et al.[3] method uses **all** available core sample data in the regression of the Archie parameters, n, m, and a. Detailed procedure for the implementation of this method was provided by Maute et al.[4]

EXAMPLE 4.1 Calculation of the Water Saturation Using the Archie Equation

Problem

Calculate the water saturation of clean consolidated sandstone given the following rock properties and Archie parameters:

Total porosity, ϕ_t	0.26
Resistivity of sandstone, R_t	22.5 $\Omega \cdot$ m
Resistivity of formation water, R_w	0.265 $\Omega \cdot$ m
Archie parameter, m	1.953
Archie parameter, n	1.861
Archie parameter, a	0.838

Solution

Substituting the given data and Archie parameters into Eq. (4.12), water saturation for the clean sandstone is calculated as:

$$S_w = \left(\frac{a}{\phi_t^m} \frac{R_w}{R_t} \right)^{1/n}$$

$$= \left[\left(\frac{0.838}{(0.26)^{1.953}} \right) \left(\frac{0.265}{22.5} \right) \right]^{1/1.861}$$

$$= 0.344 \text{ or } 34.4\%$$

End of Example 4.1

4.2.2 Shaly Sands

Shaly sands can be described as reservoir rocks that contain shales. These sands are sometimes described as "non-Archie" rocks and exhibit the effects of the presence of shales on the electrical conductivity of rocks. The presence of shales causes reservoir rocks to become conductive, which adds to the conductivity of the formation water. Archie equation assumes that the formation water is the only conductive phase in the formation and the reservoir rock is non-conductive. To account for the effects of shales on the extra conductivity of reservoir rocks, many shaly sand models have been proposed in the literature.[5] Most of the shaly sand saturation models have the following forms:[5,6]

$$C_t = \phi_t^{m^v} \times S_w^{n^v} \times C_{we} \tag{4.20}$$

$$C_t = \phi_t^{m^v} \times S_w^{n^v} \times (C_w + X) \tag{4.21}$$

The forms of Eqs. (4.20) and (4.21) are similar to the Archie equation as shown in Eq. (4.15). In Eq. (4.20), C_{we} = effective conductivity that accounts for the extra conductivity caused by the presence of shales in the rocks. In Eq. (4.21), X = a function that accounts for the extra conductivity caused by the presence of shales in the rocks. In Eqs. (4.20) and (4.21), m^v and n^v, are general forms of the electrical properties of the rock that correspond to the Archie parameters of, m and n. They reduce to Archie parameters as the shale function, $X \rightarrow 0$.

4.2.2.1 Waxman-Smits model

The Waxman-Smits[7,8] model is based on the results of an extensive experimental study on the effects of shales on the conductivity of shaly sands. The model takes the form of Eq. (4.21) and is represented as:

$$C_t = \phi_t^{m^*} \times S_w^{n^*} \times \left(C_w + \frac{BQ_v}{S_w} \right) \tag{4.22}$$

In Eq. (4.22), B = specific conductivity of exchangeable cations, mho/m per mEq/cm^3. B is also called counterion equivalent conductance; Q_v = clay cation-exchange-capacity (CEC) in milliequivalents per unit pore volume, mEq/cm^3; The parameters, m^* and n^*, are the electrical parameters for the Waxman-Smits equation. The experimental procedure for determining the parameters, m^* and n^*, is very complicated.[6] An approximate approach is to use Archie parameters, m and n, to replace Waxman-Smits parameters, m^* and n^*, in Eq. (4.22).[6]

4.2.2.2 Simandoux equation

The Simandoux[9] equation takes the form of Eq. (4.20), and is represented as:

$$C_t = C_w \phi_t^m S_w^n + \varepsilon V_{sh} C_{sh} \tag{4.23}$$

In Eq. (4.23), V_{sh} = wetted shale volume of rock, fraction; C_{sh} = conductivity of wetted shale, Sm^{-1}. The shale-term coefficient, ε, has a range of $0 \le \varepsilon \le 1$. As $S_w \rightarrow$ high, $\varepsilon \rightarrow 1$. The parameters, m and n, are the same as Archie parameters.

4.2.2.3 Poupon-Leveaux equation

The Poupon-Leveaux[10] equation is also of the form of Eq. (4.20). It is expressed as:

$$C_t = C_w \phi_t^m S_w^n + C_{sh} S_w^n V_{sh}^{2-V_{sh}} \tag{4.24}$$

This basic equation has been modified by adding additional term in the development of the Indonesian saturation equation represented as:[6]

$$C_t = C_w \phi_t^m S_w^n + C_{sh} S_w^n V_{sh}^{2-V_{sh}} + S_w^n \sqrt{C_w \phi_t^m V_{sh}^{2-V_{sh}} C_{sh}} \tag{4.25}$$

It is also the precursor of the Nigerian saturation equation expressed as:[6]

$$C_t = \left[\sqrt{C_w \phi_t^m} + \sqrt{C_{sh} V_{sh}^{2.8}} \right]^2 S_w^n \tag{4.26}$$

The parameters, m and n, are the same as Archie parameters.

4.2.2.4 The Dual-Water model

The Dual-Water model proposed by Clavier et al.[11] assumes that the clay-bound water and the free non-clay water act as two parallel conductive layers that contribute to the total conductivity, C_t, measured in the formation. The Dual-Water model is expressed as:

$$C_t = \phi_t^m S_w^n \left[C_w + \frac{v_Q Q_v \left(C_{cbw} - C_w \right)}{S_w} \right] \tag{4.27}$$

In Eq. (4.27), C_{cbw} = conductivity of clay-bound water, Sm^{-1}; v_Q = amount of clay water associated with milliequivalents of clay counterions, mEq^{-1}cm^3; and Q_v = cation-exchange- capacity per unit pore volume, mEq cm^{-3}. The parameters, m and n, can be approximated with the Archie parameters.

4.2.3 Carbonate Rocks

Carbonate rocks, unlike sandstones, have complex pore systems. These pore systems may have bi- or trimodal pore size distributions. Pore sizes may range from less than an inch to feet. The pore geometry of carbonate rocks is very heterogeneous and variable. The texture and structure of carbonate rocks are further rendered more complex by the diagenesis caused by chemical dissolution, precipitation, dolomitization, leaching, and fracturing. Due to these reasons, petrophysical models comparable in terms of simplicity to the Archie equation have not been developed for carbonate rocks. In some petrophysical analyses, the Archie equation is used to calculate water saturation in carbonate rocks. This approach could lead to significant errors. An alternative approach is to base the calculation of water saturation of carbonate rocks on data from NMR logs.

4.2.4 Water Saturations from Nuclear Magnetic Resonance Logs

The use of nuclear magnetic resonance (NMR) logging tools for determination of porosity and permeability has been described in Chapters 2 and 3, respectively. Data from NMR logging tools are also used to determine water saturations. Water saturations in reservoir rocks are determined on the basis of T_2 distribution measured with NMR logging tools.[12] The transverse relaxation time, which is named, T_2, is the characteristic time constant that represents the envelope of the spin-echo signal decay in an NMR response sequence.[12] T_2 distributions computed from NMR echo data are used to determine total porosity, effective porosity, permeability, pore size distribution, and water saturation. An example of a T_2 distribution curve for typical fluid-saturated shaly sandstone is shown in Figure 4.1. The T_2 distribution curve can be divided into two regions representing bound water and free fluids by application of a T_2 cutoff (see Figure 4.1). For shaly sandstones, bound water includes clay- and capillary-bound water, and free fluids include non-clay water (or free water) and hydrocarbons as shown in Figure 4.2. A default value of 33 milliseconds is sometimes applied as the T_2 cutoff for sandstones in the absence of laboratory data. For carbonates, the default T_2 cutoff is 100 milliseconds or higher, depending on the heterogeneity of the rock. More accurate T_2 cutoffs for specific rock samples can be measured in the laboratory with NMR tools after the samples are centrifuged. Another method for determining T_2 cutoffs is by matching irreducible water saturation obtained from an independent source (such as using the Dean-Stark method on core samples) with the irreducible water saturation derived from NMR data.[13] The irreducible water saturations of core samples shown in Figure 4.3 were measured by the centrifugal method and matched against irreducible water saturations from NMR core laboratory data for determination of T_2 cutoffs. In many formation evaluation processes, water saturations calculated from Archie-type equations are compared to water saturations derived from NMR data. In highly heterogeneous formations, such as carbonates, NMR-derived water saturations are considered more reliable in comparison to saturations calculated with Archie-type equations.

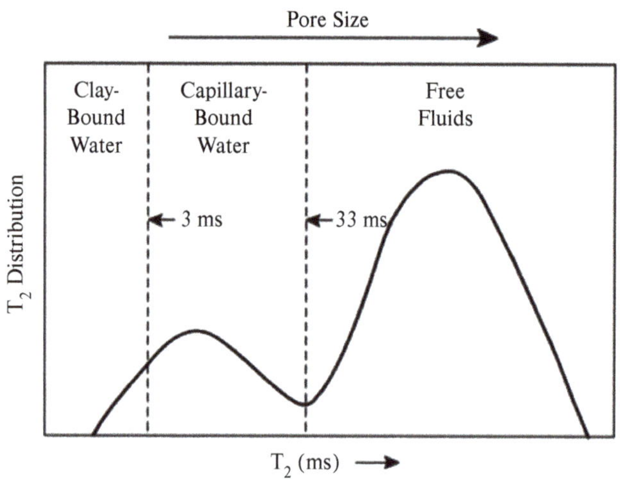

▲ FIGURE 4.1 T_2 distribution curve for typical fluid-saturated shaly sandstone

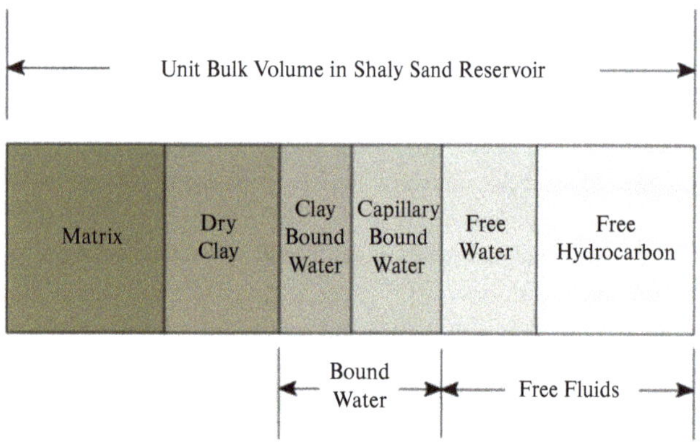

▲ FIGURE 4.2 Fluid saturation distribution in a shaly sand reservoir

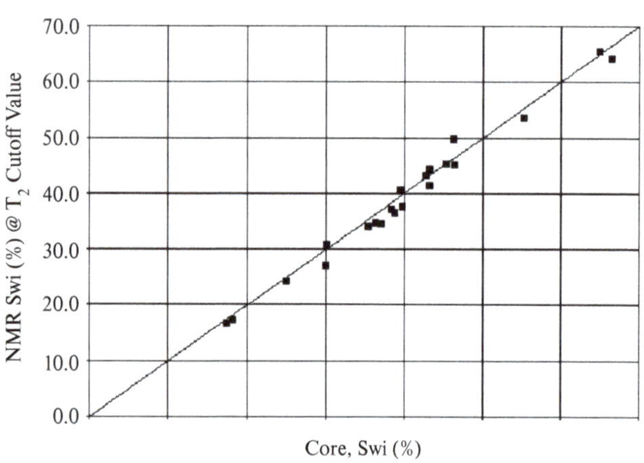

▲ FIGURE 4.3 Determination of T_2 cutoff from irreducible water saturation comparison

4.2.5 Uncertainties in Estimation of Water Saturation

The estimation of water saturation for Archie and non-Archie rocks require input data that have considerable uncertainties associated with them. For instance, calculation of water saturations for Archie rocks (clean sands) using the Archie equation require the following input data: porosity, ϕ; cementation exponent, m; saturation exponent, n; the tortuosity factor, a; formation conductivity, C_t; and formation water conductivity, C_w. For the non-Archie rocks (shaly sands), additional input data required include shale conductivity, C_{sh}; and volume of shale, V_{sh}. In similar manner, water saturations calculated from NMR data are affected by uncertainties related to the computation of the T_2 distribution and accurate values for T_2 cutoffs. The uncertainties associated with these input data have direct impact on the calculated water saturations.[14,15] The degree of uncertainty in calculated water saturations can be minimized by using regression analysis on core data to generate variable parameters, m, n, and a, for Archie or Archie-type equations. The core database can also be classified in terms of facies to improve the correlation of the parameters to rock type.[16]

4.3 Determination of Reservoir Productive Intervals

In Chapter 2, the importance of porosity data as a means of defining the storage capacity of reservoir rocks was discussed. In Chapter 3, permeability was presented as a measure of the ability of reservoir rocks to transmit fluids. Earlier in this chapter, various equations for the calculation of water saturation in different types of formations were presented. As shown in Eq. (4.1), the hydrocarbon saturation is the formation pore volume not occupied by water. As a result, the level of water saturation in a formation is a direct measure of the level of hydrocarbon saturation present in the formation. Porosity, permeability, and water saturation are the three petrophysical properties that constitute the criteria for the application of cutoffs in the determination of productive intervals in a reservoir. Generally, formation cutoffs are applied as limiting values that are used to eliminate formation intervals that appear not to have the potential to contribute significantly to the storage and productive capacity of the reservoir.[17] With this description of cutoffs, it is evident that there are no generally accepted criteria for the application of cutoffs in the petroleum industry.[18] Even on the same reservoirs, petroleum companies have used different cutoff criteria to obtain widely different estimation of productive intervals. And in many organizations, subsurface specialists (engineers, geologists, petrophysicists, etc.) frequently disagree on the cutoffs to be used in the determination of productive intervals. The lack of general criteria for the application of cutoffs can be accommodated by recognition of the fact that reservoirs are rarely identical in terms of their petrophysical properties. Consequently, it is impractical to expect that general cutoff criteria or guidelines can be adequate or applicable to all reservoirs. Due to their arbitrary nature, cutoff criteria should be devised for each reservoir to accommodate the objectives of the reservoir evaluation process in terms of in-place-hydrocarbon volumes, displacement mechanisms, and projected hydrocarbon recoveries.[19]

4.3.1 Net Sands, Net Reservoir, and Net Pay

Net sands, net reservoir, and net pay are terms commonly used to represent different categories of the formation intervals. These terms are defined in this book in accordance with the definitions proposed by Worthington.[17,20] The definitions of gross rock, net sand, net reservoir, and net pay are illustrated in Figure 4.4. **Gross rock thickness** represents the entire thickness of the formation interval. **Net sand** is the summation of the intervals whose sand content is greater than or equal to a limiting value set by the shale cutoff. Shale cutoff represents the limiting fractional volume of the formation that is considered to be shale and non-productive. In some reservoirs, it is preferable to use net sands in the characterization of reservoir models because it includes all the potential sources of drive energy that are present in the reservoir. **Net reservoir** comprises of intervals of net sand that have the capacity to both store and transmit reservoir fluids. This is generally determined by the application of porosity and permeability cutoffs. **Net pay** is the summation of the net reservoir intervals that contain hydrocarbons above the limit set by the water saturation cutoff. Note the definitions of net-to-gross sand, net-to-gross reservoir, and net-to-gross pay ratios as shown in Figure 4.4. Net pay thickness is typically used as shown in Eq. (4.3) for the calculation of in-place-hydrocarbon volumes. But observe that since net pay is derived from the application of arbitrarily set cutoff criteria, it may lead to under- or over-estimation of in-place-hydrocarbon volumes. By extension, this could cause under- or over-estimation of reserves. It is recommended that net pay thicknesses or net-to-gross pay ratios should be subjected to uncertainty analysis, as applied to porosity and water saturation data, in the evaluation of ranges of in-place-hydrocarbon volumes and potential recoveries.[21]

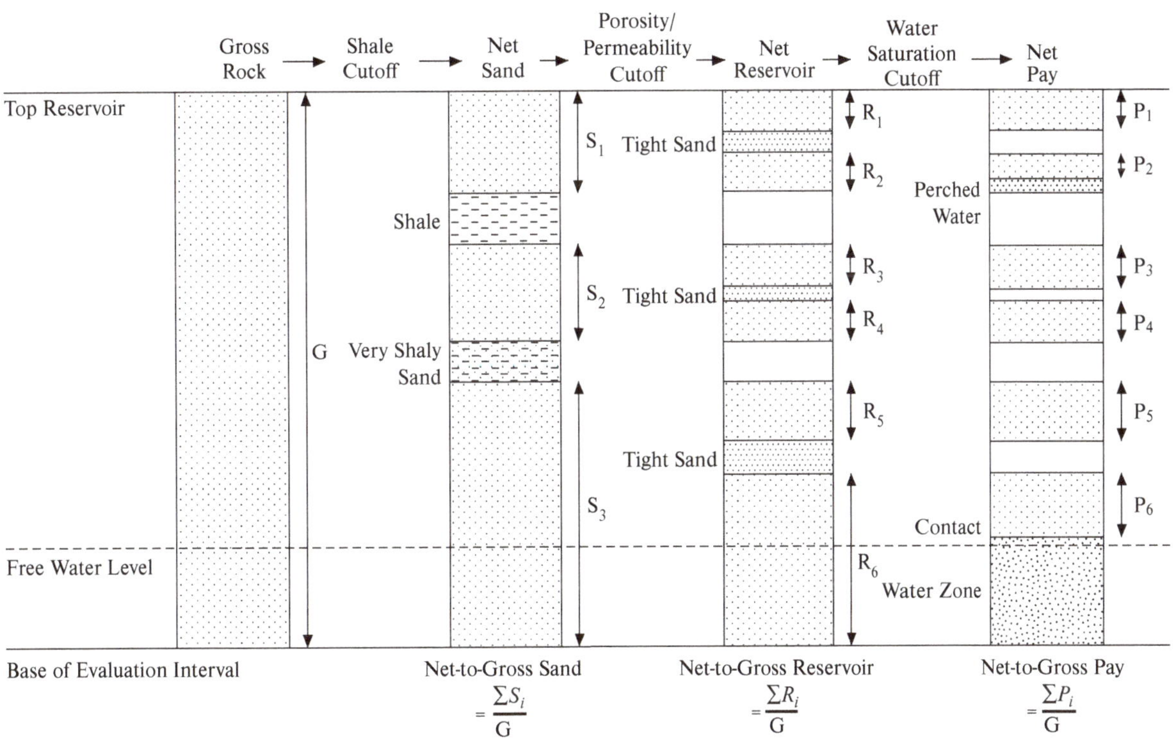

▲ **FIGURE 4.4 Definitions of net sand, net reservoir, and net pay with cutoff criteria based on petrophysical properties (from Worthington and Cosentino[17] © 2005 SPE, Reproduced with permission)**

Nomenclature

a	tortuosity factor
B	specific conductivity of exchangeable cations, mho/m per mEq/cm^3
C_{cbw}	conductivity of clay-bound water, Sm^{-1}
C_o	conductivity of fully water-saturated formation or rock, Sm^{-1}
C_{sh}	conductivity of shale, Sm^{-1}
C_t	conductivity of formation or rock, Sm^{-1}
C_w	conductivity of formation water, Sm^{-1}
C_{we}	effective conductivity of shaly sands, Sm^{-1}
F	formation factor
I_R	resistivity index, ratio
m	cementation factor
m^v	cementation factor for shaly sands
n	saturation exponent
n^v	saturation exponent for shaly sands
Q_v	clay cation-exchange-capacity, mEq/cm^3
R_o	resistivity of fully water-saturated formation or rock, $\Omega \cdot$ m
R_t	resistivity of formation or rock, $\Omega \cdot$ m
R_w	resistivity of formation water, $\Omega \cdot$ m
S_g	gas saturation, fraction
S_h	hydrocarbon saturation, fraction
S_o	oil saturation, fraction
S_w	water saturation, fraction
T_2	NMR characteristic time constant, milliseconds
V_{sh}	rock shale volume, fraction
v_Q	clay water associated with clay counterions, mEq^{-1}cm^3
X	function to account for extra conductivity of shale
ε	mean-square in water saturation regression
ϕ	porosity
ϕ_e	effective porosity
ϕ_t	total porosity

Abbreviations

CEC	cation-exchange-capacity
HCPV	hydrocarbon pore volume
NMR	nuclear magnetic resonance

REFERENCES

1. Archie, G.E.: "The Electrical Resistivity Log as an Aid in Determining Some Reservoir Characteristics," *Trans., AIME* (1942) 146, 54-62.
2. Winsauer, W.O., Shearin, Jr., H.M., Masson, P.H., and Williams, M.: "Resistivity of Brine-Saturated Sands in Relation to Pore Geometry," *AAPG Bulletin* (1952) 36, 2, 253-277.
3. Maute, R.E., Lyle, W.D., and Sprunt, E.S.: "Improved Data-Analysis Method Determines Archie Parameters from Core Data," *JPT* (January 1992) 103-107.
4. Maute, R.E., Lyle, W.D., and Sprunt, E.S.: "Supplement to SPE 19399, Improved Data-Analysis Method Archie Parameters from Core Data," paper SPE 24223.
5. Worthington, P.F.: "The Evolution of Shaly-Sand Concepts in Reservoir Evaluation," *The Log Analyst* (January-February 1985) 26, 1, 23-40.
6. Al-Ruwali, S.B., and Al-Waheed, H.H.: "Improved Petrophysical Methods and Techniques for Shaly Sands Evaluation," paper SPE 89735 presented at the 2004 SPE International Petroleum Conference, Puebla, Mexico, November 8-9, 2004.
7. Waxman, M.H., and Smits, L.J.M.: "Electrical Conductivities in Oil-Bearing Shaly Sands," *SPEJ* (June 1968) 107-122.
8. Waxman, M.H., and Thomas, E.C.: "Electrical Conductivities in Shaly Sands—1. The Relation Between Hydrocarbon Saturation and Resistivity Index; II. The Temperature Coefficient of Electrical Conductivity," *JPT* (February 1974) 213-223.
9. Simandoux, P.: "Dielectric Measurements on Porous Media: Application to the Measurement of Water Saturations: Study of the Behaviour of Argillaceous Formations," *Revue de l'Institu Francais du Petrole* 18, supplementary issue (1963) 193-215 (Translated text in Shaly Sand Reprint Volume, SPWLA, Houston, Texas, pp. IV 97-124).
10. Poupon, A., and Leveaux, J.: "Evaluation of Water Saturation in Shaly Formations," *The Log Analyst* (July-August 1971) 3-8.

11. Clavier, C., Coates, G., and Dumanoir, J.: "Theoretical and Experimental Bases for the Dual-Water Model for Interpretation of Shaly Sands," *SPEJ* (April 1984) 153-168.
12. Freedman, R.: "Advances in NMR Logging," SPE 89177, Distinguished Author Series, *JPT* (January 2006) 60-66.
13. Shafer, J., Chen, S., Georgi, D., and Chen, J.: "Methods and Procedures for Calibrating NMR Log Derived Permeabilities," The 11th Formation Evaluation Symposium of Japan, Chiba, Japan (October 5-6, 2005) 1-15.
14. Worthington, P.F., and Pallatt, N.: "Effect of Variable Saturation Exponent on the Evaluation of Hydrocarbon Saturation," *SPEFE* (December 1992) 331-336.
15. Zeybek, A.D., Onur, M., Tureyen, O.I., Ma, S.M., Al-Shahri, A.M., and Kuchuk, F.J.: "Assessment of Uncertainty in Saturation Estimated From Archie's Equation," paper SPE 120517 presented at the 2009 SPE Middle East Oil & Gas Show and Conference, Bahrain, Kingdom of Bahrain, March 15-18, 2009.
16. Worthington, P.F.: "Improved Quantification of Fit-for-Purpose Saturation Exponents," *SPEREE* (August 2004) 270-284.
17. Worthington, P.F., and Cosentino, L.: "The Role of Cutoffs in Integrated Reservoir Studies," *SPEREE* (August 2005) 276-290.
18. Caldwell, R.H., and Heather, D.I.: "Characterizing Uncertainty in Oil and Gas Evaluations," paper SPE 68592 presented at the SPE Hydrocarbon Economics and Evaluation Symposium, April 2-3, 2001, Dallas, Texas.
19. Worthington, P.F.: "The Application of Cutoffs in Integrated Reservoir Studies," *SPEREE* (December 2008) 968-975.
20. Worthington, P.F.: "Net Pay: What Is It? What Does It Do? How Do We Quantify It? How Do We Use It?" paper SPE 123561 presented at the 2009 SPE Asia Pacific Oil and Gas Conference and Exhibition, Jakarta, Indonesia, August 4-6, 2009.
21. Egbele, E., Ezuka, I., and Onyekonwu, M.: "Net-To-Gross Ratios: Implications in Integrated Reservoir Management Studies," paper SPE 98808 presented at the Nigeria Annual International Conference and Exhibition, August 1-3, 2005, Abuja, Nigeria.

ADDITIONAL REFERENCES

- Juhász, I.: "The Central Role of Q_v and Formation-Water Salinity in the Evaluation of Shaly Formations," *The Log Analyst* (July-August 1979) 3-13.
- Poupon, A., Clavier, C., Dumanoir, J., Gaymard, R., and Misk, A.: "Log Analysis of Sand-Shale Sequences–A Systematic Approach," *JPT* (July 1970) 867-881.
- Worthington, P.F.: "Identifying the Potential Upside of Hydrocarbon Saturation from Electric Logs," *SPEREE* (February 2009) 53-67.
- Worthington, P.F.: "Petrophysical Type Curves for Identifying the Electrical Character of Petroleum Reservoirs," *SPEREE* (December 2007) 711-729.

PROBLEMS

4-1. Calculate the water saturation of a dolomite interval in the Ellenburger formation, Permian Basin of West Texas, USA, given the following rock properties and parameters. What is the hydrocarbon saturation of the interval?

Total porosity, ϕ_t	4%
Formation water resistivity, R_w	0.028 Ω · m
Interval formation resistivity, R_t	40 Ω · m
Formation factor, F	625
Saturation exponent, n	2

4-2. Calculate the water saturation of a sandstone interval in the Wilcox formation of South Texas, USA, given the following rock properties and Archie parameters. What is the hydrocarbon saturation of the interval?

Total porosity, ϕ_t	18%
Formation water resistivity, R_w	0.07 Ω · m
Interval formation resistivity, R_t	9.3 Ω · m
Cementation factor, m	2.15
Tortuosity factor, a	0.62
Saturation exponent, n	2.0

4-3. What is the hydrocarbon saturation of a limestone interval within the Devonian formation in New Mexico, USA, given the following rock properties and resistivity parameters?

Total porosity, ϕ_t	8%
Formation water resistivity, R_w	0.036 Ω · m
Interval formation resistivity, R_t	35 Ω · m
Formation resistivity factor, F	156.25
Saturation exponent, n	2.0

4-4. A clean sandstone interval is 100% saturated with water. Calculate the total porosity of the interval if it has the following resistivity data and Archie parameters.

Formation water resistivity, R_w	0.049 Ω · m
Interval formation resistivity, R_t	0.27 Ω · m
Cementation factor, m	2.15
Tortuosity factor, a	0.62
Saturation exponent, n	2.0

4-5. Calculate the resistivity of an interval within the Burbank formation in Kansas, USA, given the following rock properties and Archie parameters.

Total porosity, ϕ_t	20%
Formation water resistivity, R_w	0.25 Ω · m
Interval formation resistivity index, I_R	4
Cementation factor, m	2.15
Tortuosity factor, a	0.62
Saturation exponent, n	2.0

CHAPTER

5 Pressure-Volume-Temperature (PVT) Properties of Reservoir Fluids

5.1 Introduction

Fluids exist in reservoirs as mixtures of gas, oil, and water. Some reservoirs may contain only gas and water, only oil and water, or mixtures of gas, oil, and water. Irrespective of the proportions of these fluids present in a reservoir, obtaining fluid samples and studying their phase behavior in a laboratory are necessary for establishing reservoir type, devising strategies for reservoir management, and estimating expected hydrocarbon recovery. The importance of collecting representative reservoir fluid samples (preferably early in the life of the reservoir) and having the samples analyzed in a reputable laboratory cannot be over-emphasized. It is one of the essential functions of the engineers working on new or existing reservoirs as an integral part of a comprehensive data collection program. Sampling methods for reservoir fluids and types of laboratory measurements are presented in Chapter 6. This chapter begins by discussing basic concepts of phase diagrams as tools for understanding the phase behavior of reservoir fluids. The chapter provides many correlations for calculation of properties of reservoir fluids especially in circumstances where laboratory data measured on reservoir fluid samples may not be available.

5.2 Phase Diagrams

The PVT properties of reservoir fluids are introduced by reviewing the basic concepts of phase diagrams. **Phase diagrams** are graphical representations that relate the properties of a fluid system. The properties used in the representation could be intensive or extensive properties. **Intensive** properties (pressure, temperature, density) are independent of the extent of the system. **Extensive** properties (mass, volume, heat) depend on the extent of the system. Common forms of phase diagrams show pressure-temperature (P-T), pressure-volume (P-V), and temperature-volume (T-V) relationships. The basic concepts of phase behavior and phase diagrams are discussed in this book by the use of P-T diagrams.

5.2.1 Single-Component Systems

The P-T diagram of a single-component system such as pure water is shown in Figure 5.1. In this diagram, the pure substance can exist as three phases—solid, liquid, and vapor depending on the temperature and pressure of the system. A **phase** is a continuous, homogeneous portion of the system separated by a physical interface if another phase is present.

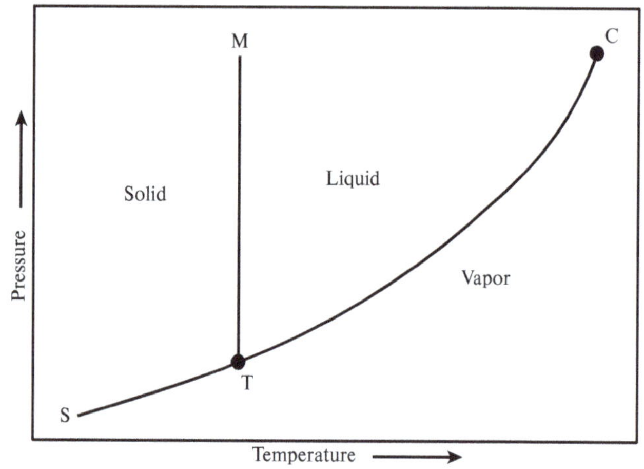

▲ FIGURE 5.1 P-T phase diagram of a single-component system

The three curves in Figure 5.1 represent states of phase equilibrium. The line ST is the **sublimation curve**. Along the sublimation curve, solid and vapor phases are in equilibrium. The line MT is the **melting curve**, along which solid and liquid phases are in equilibrium. Line TC is the **vapor pressure curve,** which represents the locus of equilibrium between liquid and vapor phases. The point T, at which the three curves meet, is the triple point. At the **triple point,** the three phases are in equilibrium. Along the vapor pressure curve, the terminal point C is called the **critical point.** Above the critical point, the liquid and vapor phases are indistinguishable. The physical interface between them disappears. The concept of the critical point and its location on the phase diagram are used later in this chapter to classify the phase behavior of petroleum systems.

The phases within a system composed of one or more components are considered to be in **thermodynamic equilibrium** when the chemical potentials of each component in the phases are equal. Equilibrated phases within a system have the same temperatures and pressures, and there are no net mass transfers between the phases. If a system has two phases denoted as i and j, the condition of equilibrium is denoted as:

$$\psi_i = \psi_j \\ T_i = T_j \text{ and} \\ p_i = p_j \tag{5.1}$$

In Eq. (5.1), ψ is chemical potential, T is temperature, and p is pressure in each phase.

The number of intensive properties required to define a system in equilibrium can be determined by using Gibbs' phase rule.[1] If C is the number of components in a system and P is the number of phases, then the number of intensive properties, F (or degrees of freedom), that must be specified to define its equilibrated state is:

$$F = C - P + 2 \tag{5.2}$$

Applying the Gibbs' phase rule to Figure 5.1, it can be deduced that for a single-component system with two phases in coexistence (such as a liquid phase and a vapor phase), if the temperature is known, then the pressure is fixed by the vapor pressure curve because the number of degrees of freedom, F, is equal to 1.

Another form of a phase diagram representing a single-component system is shown in Figure 5.2. This is the pressure-volume (P-V) phase diagram of a single-component system. The solid curve ACB in Figure 5.2 is called the **phase envelope.** Within the phase envelope, vapor and liquid phases coexist in equilibrium. Outside the phase envelope (left of segment AC in Figure 5.2) is only the liquid phase, and to the right of segment BC is the vapor phase. Line AC is the bubble point line, and line BC is the dew point line. The **bubble point** for any system is the condition of temperature and pressure at which the liquid phase is in equilibrium with an infinitesimal amount of the vapor phase. Similarly, the **dew point** is the temperature and pressure at which the vapor phase is in equilibrium with an infinitesimal amount of the liquid phase. In Figure 5.2, the critical point is represented as C. As was the case in Figure 5.1, beyond the critical point, liquid and vapor phases overlap and cannot be distinguished from each other.

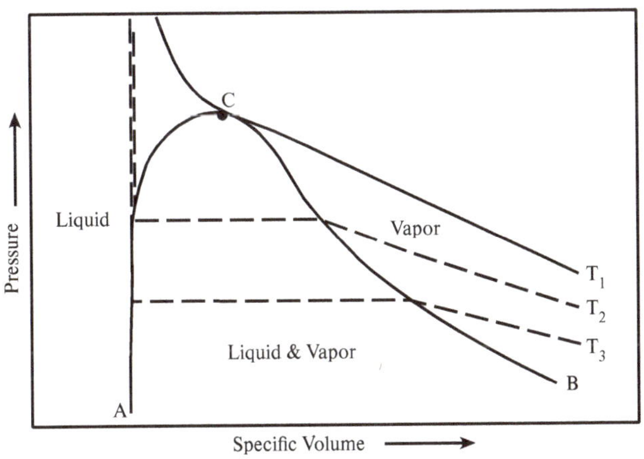

▲ FIGURE 5.2 P-V phase diagram of a single-component system

An important feature of P-V phase diagrams is the use of **constant temperature lines,** or **isotherms.** These are shown as T_1, T_2, and T_3 in Figure 5.2. Along isotherms T_2 and T_3, the specific volume decreases with an increase in pressure within the vapor region. Within the two-phase region, the specific volume decreases at constant pressure. Across the bubble point line, the pressure increases rapidly with a relatively small reduction in the specific volume, because the liquid phase is significantly less compressible than the vapor phase. This concept is used in the measurement of bubble point pressures of most fluid systems. Note that the phase behavior along the T_1 isotherm is significantly different from the phase behavior along the isotherm T_2 or T_3 because it does not traverse the two-phase region within the phase envelope.

A temperature-volume (T-V) phase diagram of a single-component system is shown in Figure 5.3. This diagram is very similar to the P-V diagram shown in Figure 5.2. The main difference is that instead of isotherms shown in Figure 5.2, **constant pressure lines or isobars,** are shown in Figure 5.3.

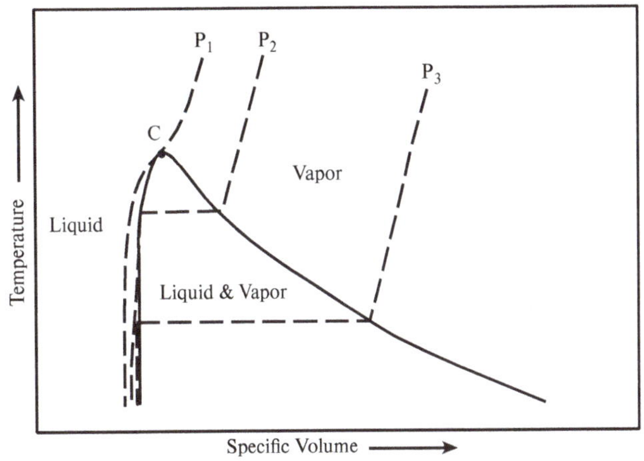

▲ FIGURE 5.3 T-V phase diagram of a single-component system

5.2.2 Binary Systems

Binary systems are mixtures of two components. Figure 5.4 is the P-T phase diagram of a binary mixture of components A and B. The vapor pressure curve of pure component A is shown terminating at the critical point C_A. Similarly, the vapor pressure curve of pure component B terminates at the critical point C_B. The phase diagrams of hypothetical mixtures of components A and B will lie between the vapor pressure curves of the pure components, as shown in Figure 5.4 for the 50-50 mole percent mixture. The area in Figure 5.4 bounded by the solid curve LCV is the phase envelope, which encloses conditions of temperatures and pressures where liquid and vapor phases can coexist in equilibrium. The curved line LC defines the bubble point line, and curved line VC is the dew point line. The bubble point line LC and the dew point line VC meet at point C, which is defined as the critical point. Unlike the critical point that terminates the vapor pressure curve of a pure single-component system (Figure 5.1), the critical point of a binary mixture does not define the highest temperature and pressure at which two phases can coexist. The highest temperature in Figure 5.4 at which two phases can coexist is shown by line EE. This is called the **cricondentherm.** Similarly, the highest pressure at which two phases can coexist is shown by line QQ. This is called the **cricondenbar.** Note that the cricondentherm and the cricondenbar are higher than the critical temperature and pressure of the mixture, respectively. The cricondentherm lies between the critical temperatures of components A and B. The cricondenbar is higher than the critical pressure of either component A or component B.

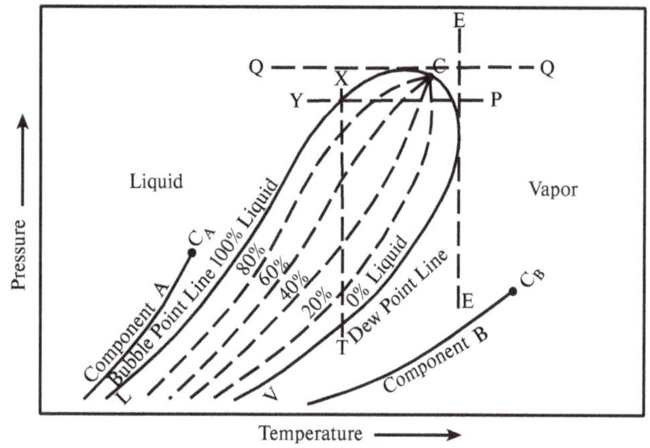

▲ **FIGURE 5.4** **P-T phase diagram of a binary system**

The dashed lines within the phase envelope in Figure 5.4 denote lines of constant volume percentage liquid drawn from the critical point. At the bubble point line, the mixture is 100 percent liquid. At the dew point line, the mixture is 100 percent vapor. Consider a process in which the mixture is taken at constant temperature along the line XT. At point X, the mixture is an undersaturated liquid. As the pressure is reduced, the liquid expands until it reaches the bubble point line, where an infinitesimal amount of bubble will form. This is the bubble point pressure of the mixture at the isotherm XT. As the pressure is reduced along XT, the volume of liquid phase in the mixture will decrease until the dew point line is reached. At the dew point, an infinitesimal amount of liquid will remain. This is the dew point pressure of the mixture at the isotherm XT. Further pressure reduction to point T will create a system composed entirely of the vapor phase.

A process similar to the isothermal expansion of the mixture along XT can be repeated by expanding the mixture at constant pressure (isobaric) along YP. At point Y, the mixture is undersaturated liquid, just as at point X. At the bubble point line, an infinitesimal amount of bubble will form. Further expansion along YP will result in the reduction of the volume of liquid present in the mixture. At the dew point line, a very small amount of liquid will remain. Further temperature increases to point P will create a vapor system.

The P-T phase diagram of ethane/n-heptane mixtures of varying composition of ethane is shown in Figure 5.5. This figure illustrates that mixtures of the same components having varying compositions will have varying phase envelopes. For instance, an ethane/n-heptane mixture with 90.22 percent ethane has a phase envelope shown as C_1. The shape of this phase envelope is different from the phase envelope of the C_2 mixture (50.25% ethane) or that of the C_3 mixture (9.78% ethane). In Figure 5.5 the critical points of the three mixtures are connected with a dashed line, which is the locus of the critical points of all potential mixtures of ethane and n-heptane at temperatures and pressures between the critical points of pure ethane and pure n-heptane.

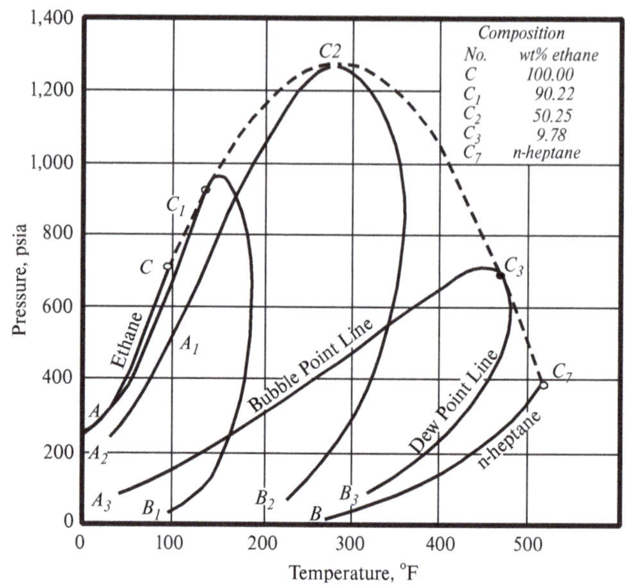

▲ **FIGURE 5.5** **P-T phase diagram for ethane and n-heptane system at various compositions of ethane (from Kay[2])**

5.2.3 Multicomponent Systems

The phase diagram of a binary system (Figure 5.4) introduces several useful concepts, such as the phase envelope, critical point, cricondentherm, and cricondenbar. Figure 5.5 (phase diagrams of ethane/n-heptane at various compositions) also illustrates the effect of the composition of the mixture on the shape of the phase envelope. The shape and nature of phase diagrams for multicomponent systems are similar to those of binary systems. However, it is much more difficult to map out the phase envelope of a multicomponent system. These concepts are used in this section to discuss the phase envelopes of various fluid types that will be encountered in petroleum reservoirs.

The phase envelopes of common reservoir fluid types are shown in Figure 5.6. The reservoir fluid types shown are for a gas, a gas condensate, a volatile oil, and a black oil reservoir. In Figure 5.6, the letter C denotes the location of the critical point on the phase envelopes. Consider a reservoir whose initial pressure and temperature are denoted as p_i,T_i, as shown in Figure 5.6. Note the path of the solid line starting at p_i,T_i, which represents isothermal depletion by reduction in reservoir pressure in relation to the phase envelopes that represent a gas, a gas condensate, a volatile oil, and a black oil reservoir. The phase envelope (A) for the gas reservoir is clearly to the left of the solid line. The critical temperature (point C) and cricondentherm of the phase envelope are lower than reservoir temperature, T_i. As the reservoir pressure is reduced at constant reservoir temperature, there is no chance that the solid line will cross any part of the phase envelope that defines the gas reservoir. Consequently, no liquid will form inside the gas reservoir as its pressure is reduced by production.

Gas reservoirs are further classified as dry or wet. This classification is based on the pressure and temperature of the facilities used in handling and processing the produced fluids. If the pressure and temperature of the flow and processing facilities fall within the two-phase region defined by the phase envelope, then liquids could form within the wellbore, flowlines, and separators. This type of gas is called a **wet gas** (Figure 5.7). Describing the gas as wet does not mean that some of the gas exists as liquid within the reservoir. If the temperature and pressure of the facilities fall outside the two-phase region of the phase envelope, no liquid will condense in the wellbore, flowlines, and separators. This type of gas is described as a **dry gas.**

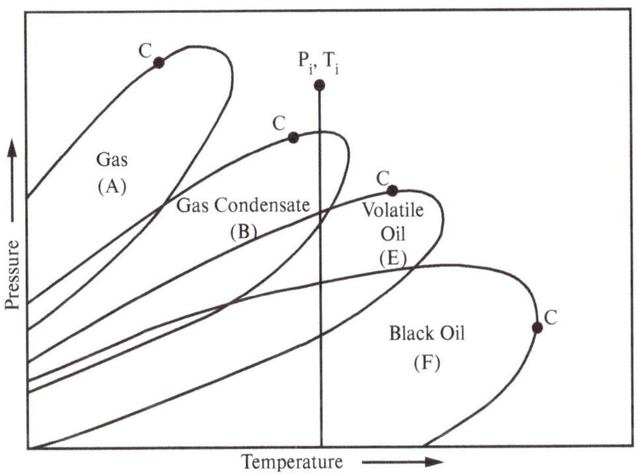

▲ FIGURE 5.6 Phase diagram showing various reservoir fluid types

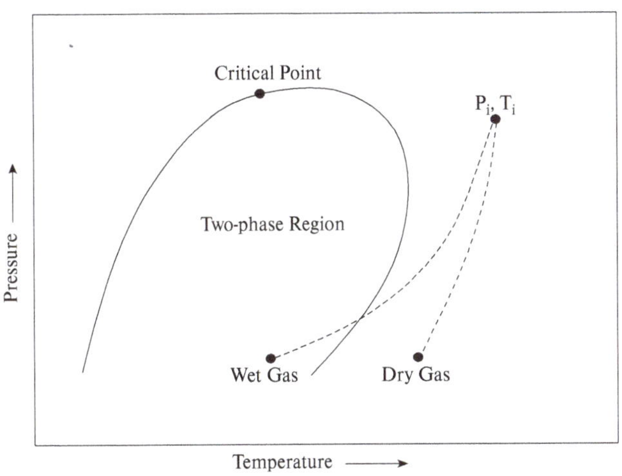

▲ FIGURE 5.7 Phase diagram showing dry gas and wet gas behavior based on the phase envelope

The phase envelope (B) for a gas-condensate reservoir intersects the solid line from p_i, T_i. For the gas-condensate phase envelope, the reservoir temperature, T_i, is higher than the critical temperature (point C) but lower than the cricondentherm. Note that the solid line crosses the phase envelope on the side of the dew point line. As reservoir pressure is reduced along the solid line, a liquid phase will condense inside the reservoir when the dew point line is reached. The nature and behavior of the condensation process are presented in more detail in Section 5.2.4 for retrograde gas-condensate reservoirs.

Volatile oil reservoirs are represented in Figure 5.6 by the phase envelope, E. The critical temperature (point C) and the cricondentherm of the phase envelope are higher than the reservoir temperature, T_i. At initial reservoir pressure and temperature, p_i, T_i, the fluid in the reservoir is a compressed and undersaturated liquid. When reservoir pressure is reduced at constant temperature along the solid line, the reservoir fluid will expand as a single phase until it reaches the bubble point line of the phase envelope. Further reduction of reservoir pressure along the solid line will cause a gas phase to form in the reservoir. Initially, the gas phase exists as isolated bubbles within the liquid phase. As reservoir pressure continues to decline, more gas will evolve until the scattered bubbles will coalesce to form a continuous phase. Generally, the stock-tank oil gravities are higher than 35°API, for volatile oils, resulting in gas-oil ratios that range between 1000 and 3000 scf/STB, and oil formation volume factors that are higher than 1.5 RB/STB. Black oil reservoirs are represented by phase envelope, F, in Figure 5.6. Note the location of the critical point on the phase envelope for black oils compared with its location on the phase envelope for volatile oils. The reservoir temperature, T_i, is much farther from the critical temperature for black oils in comparison with volatile oils. As is the case for volatile oils, a gas phase will form in black oil reservoirs when the reservoir pressure reaches the bubble point pressure on the bubble point line along the solid line of isothermal depletion (p_i, T_i). Generally, the stock-tank oil gravities of black oil are less than 35°API, producing gas-oil ratios are less than 2000 scf/STB, and oil formation volume factors are less than 2 RB/STB.

It is apparent from the stated rules of thumb for volatile and black oils that there is significant overlap in properties between the two types of oils. Actual designation of a particular oil type should be determined from careful PVT laboratory measurements.

5.2.4 Retrograde Behavior of Gas-Condensate Systems

The phase envelope of a gas-condensate reservoir is represented in Figure 5.6 by phase envelope B. A similar phase envelope has been expanded in Figure 5.8 to illustrate the phenomenon of retrograde condensation. The dashed lines within the phase envelope are lines of constant liquid volume varying from 100 percent liquid volume at the bubble point line to 0 percent liquid volume at the dew point line.

The reservoir temperature, T_i, is higher than the critical temperature, C, but lower than the cricondentherm of the phase envelope. This is a necessary condition for the occurrence of retrograde behavior.

Consider an experiment in which reservoir pressure is reduced at constant reservoir temperature, T_i, along the path defined by the solid line ABDE from initial conditions at p_i, T_i. At point A on the dew point line, a liquid phase will begin to form in the reservoir. As reservoir pressure is reduced, more liquid will condense in the reservoir such that at point B, 10 percent of the reservoir volume is occupied by the liquid phase. Further reduction in reservoir pressure will increase the percentage of liquid volume to a maximum of about 12 percent. Beyond this maximum, as pressure is reduced, the volume of liquid phase in the reservoir will begin to shrink until it reaches 10 percent again at point D. From point D to point E, liquid volume will continue to shrink and will disappear completely at the lower dew point at point E. This condensation behavior, which begins with increasing liquid volume followed by a reversal of liquid volume with pressure reduction at constant temperature in the reservoir, is described as **retrograde condensation.**

Generally, the liquid phase that forms in the reservoir rarely reaches the critical liquid saturation necessary to become mobile. The condensed liquid may appear as isolated "islands" of the liquid phase scattered throughout the reservoir. In a strict sense, the process described along the isotherm ABDE never occurs in an actual reservoir. As condensation occurs in a reservoir, the liquid phase will contain the heavier components, and the vapor phase will contain fewer of the heavier components. Because the liquid phase is not mobile, none of these heavier components will be produced. Consequently, the composition of the fluid system remaining in the reservoir is not constant but changes with production. As the composition changes, so does the phase envelope, resulting in retrograde behavior that is different from the process described here, which assumes constant composition of the fluid system in the reservoir.

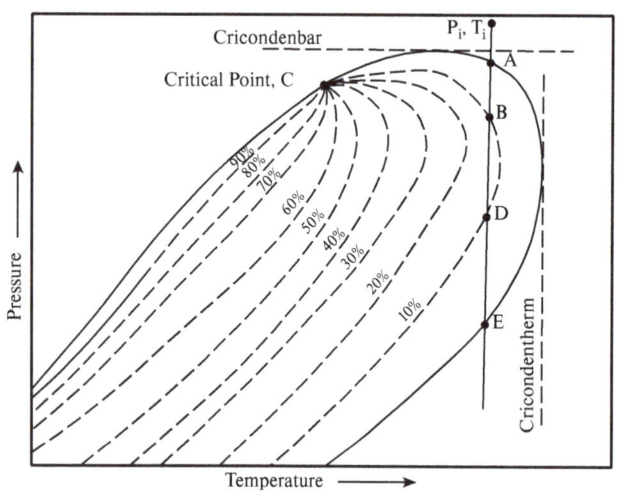

▲ **FIGURE 5.8 Phase diagram showing retrograde behavior**

Proper management of retrograde gas-condensate reservoirs is very important if the goal is to maximize hydrocarbon recovery. The key steps are identifying the gas-condensate system as exhibiting retrograde behavior and quantifying the extent of retrograde condensation that could occur either in the reservoir or within the production facilities. This can be done by taking several representative samples of reservoir fluid as close to initial reservoir conditions as possible. The samples should be analyzed by competent PVT laboratories, especially covering the range of operating conditions expected during depletion. The PVT data should be modeled with an equation of state. Further reservoir work should involve performing compositional simulation by using a geological model of the reservoir. Geologic modeling and reservoir simulation are presented in Chapters 22 and 23, respectively.

Management of gas-condensate reservoirs may require pressure maintenance, usually by gas injection, either to maintain reservoir pressure or to evaporate some of the condensed liquid. Ideally, it will be more beneficial in terms of recovery to maintain reservoir pressure higher than the dew point so that no condensation occurs within the reservoir. This will require initiating the gas injection project as early as possible in the recovery process. The decision to undertake a pressure maintenance or liquid recovery project depends on a number of factors, such as the location of the reservoir, the availability of gas for injection, the cost of injected gas, the availability of facilities for recovery of liquids, and expected recovery from the project. Each of these factors should be evaluated and combined in a comprehensive economic analysis of the project as discussed in Chapter 24, on Reservoir Management strategies. More details in designing gas cycling for gas-condensate reservoirs are presented in Chapter 19, Section 19.3 under Gasflooding.

5.3 Gas and Gas-Condensate Properties

This section presents equations and correlations that can be used to calculate thermodynamic properties of dry, wet, or condensate gases. Recall that a dry gas is one in which no liquid forms either in the reservoir, in flowlines, or in the separators during the entire depletion process (Figure 5.7). A wet gas is one in which a liquid phase forms in the flowlines or separators, but no liquid phase forms in the reservoir. For condensate gas, a liquid phase can precipitate in the reservoir, the flowlines, and the separators (Figure 5.8).

5.3.1 Ideal Gas Equation

The ideal gas equation is derived from the combination of Boyle's law and Charles' law for ideal gases. Boyle's law states that the pressure of a given mass of gas is inversely proportional to its volume if its temperature is kept constant:

$$pV = k, \text{if } T \text{ is constant} \tag{5.3}$$

In Eq. (5.3), p is pressure, V is volume, T is temperature, and k is constant. Another form of Eq. (5.3) that is sometimes useful is:

$$p_1V_1 = p_2V_2, \text{ at constant temperature, } T \tag{5.4}$$

In Eq. (5.4), p_1V_1 and p_2V_2 are pressures and volumes at conditions 1 and 2, respectively.

Charles' law states that the pressure of a given mass of gas is directly proportional to its temperature if its volume is kept constant:

$$\frac{p}{T} = k, \text{ if } V \text{ is constant} \tag{5.5}$$

Another way of expressing Charles' law is:

$$\frac{p_1}{T_1} = \frac{p_2}{T_2} \text{ at constant volume, } V \tag{5.6}$$

Combining Boyle's and Charles' laws gives the ideal gas equation:

$$pV = nRT \tag{5.7}$$

In Eq. (5.7), n is the number of moles, and R is the gas constant. If p is in psia, V is in ft^3, n is in lb-mole, and T is in °R, then the gas constant, $R = 10.732$ (psia · ft^3)/(lb-mole · °R). The ideal gas equation can be used to represent real gas behavior at low temperatures and pressures. The ideal gas equation can be expressed in another useful form by combining Eqs. (5.4) and (5.6) or derived from Eq. (5.7) as:

$$\frac{p_1V_1}{T_1} = \frac{p_2V_2}{T_2} \tag{5.8}$$

In Eq. (5.8), pressures, volumes, and temperatures are denoted at two thermodynamic states, 1 and 2, respectively.

5.3.2 Real Gas Equation

Gases, especially those found at reservoir conditions, do not exhibit ideal behavior. In fact, it can be said that no gases exhibit true ideal behavior, primarily because of the conditions of the kinetic theory (conditions that define ideal gas behavior). At low temperatures and pressures, low molecular weight gases approach ideal behavior. The thermodynamic behavior of real gases can be represented with the real gas equation as:

$$pV = znRT \tag{5.9}$$

In Eq. (5.9), z is sometimes called the **gas factor, gas deviation factor,** or **gas compressibility factor**. The gas compressibility factor, z, is a modifying factor introduced to account for the non-ideal behavior of real gases. When the compressibility factor, z, is equal to unity, the gas behaves as an ideal gas. Eqs. (5.7) and (5.9) become identical. Values of the gas compressibility factor vary over a wide range, depending on the pressure, temperature, and composition of the gas. The gas compressibility factor is one of the key properties necessary to represent the thermodynamic behavior of a gas. In this chapter, methods of calculating the gas compressibility factor using correlations are presented. Calculation of gas compressibility factors using equations of state is treated in Chapter 7.

5.3.3 Gas Gravity

The specific gravity of gas, known simply as **gas gravity,** is defined with reference to air. It is defined as the ratio of the density of the gas to the density of air at the same temperature and pressure:

$$\gamma_g = \frac{\rho_g}{\rho_{air}} \tag{5.10}$$

In Eq. (5.10), γ_g is gas gravity, ρ_g is gas density, and ρ_{air} is density of air.

Most of the petroleum industry has adopted the temperature of 60°F and the pressure of 14.65 psia as standard conditions. Unless otherwise stated, these standard conditions are adopted in this book.

Density is defined as mass per unit volume. Hence, gas density is defined as:

$$\rho_g = \frac{m_g}{V_g} \tag{5.11}$$

In Eq. (5.11), m_g is the mass of the gas, and, V_g is the volume of the gas. $m_g = nM_g$, where n is moles of gas and M_g is its molecular weight. If we assume ideal gas behavior at standard conditions (T_{sc} = 60°F and p_{sc} = 14.65 psia), then we can apply Eq. (5.7) such that Eq. (5.11) becomes:

$$\rho_g = \frac{pM_g}{RT} \tag{5.12}$$

Similarly, the density of air is given by:

$$\rho_{air} = \frac{pM_{air}}{RT} \tag{5.13}$$

In Eq. (5.13), M_{air} is the molecular weight of air, which is equal to 28.9586 (see Table 5.1). Combining Eqs. (5.12) and (5.13) gives:

$$\gamma_g = \frac{M_g}{M_{air}} = \frac{M_g}{28.9586} \tag{5.14}$$

TABLE 5.1 Physical Properties of Compounds

Compound	Formula	Molecular Weight	Critical Pressure (psia)	Critical Temperature (°F)	Critical Volume (ft³/lb)	Liquid Specific Gravity (Water = 1)	Gas Specific Gravity (Air = 1)	Acentric Factor
Methane	CH_4	16.042	667.0	–116.66	0.0985	(0.3)*	0.55400	0.0115
Ethane	C_2H_6	30.069	706.6	89.92	0.0775	0.35643	1.03830	0.0994
Propane	C_3H_8	44.096	615.5	205.92	0.0728	0.50738	1.52270	0.1529
Isobutane	C_4H_{10}	58.122	527.9	274.41	0.0715	0.56295	2.00710	0.1865
n-Butane	C_4H_{10}	58.122	550.9	305.55	0.0703	0.58408	2.00710	0.2003
Isopentane	C_5H_{12}	72.149	490.4	369	0.0685	0.62460	2.49140	0.2284
n-Pentane	C_5H_{12}	72.149	488.8	385.8	0.0676	0.63113	2.49140	0.2515
n-Hexane	C_6H_{14}	86.175	436.9	453.8	0.0688	0.66404	2.97580	0.2993
n-Heptane	C_7H_{16}	100.202	396.8	512.9	0.0682	0.68819	3.46020	0.3483
n-Octane	C_8H_{18}	114.229	360.7	564.2	0.0673	0.70698	3.94450	0.3977
n-Nonane	C_9H_{20}	128.255	330.7	610.8	0.0693	0.72186	4.42890	0.4421
n-Decane	$C_{10}H_{22}$	142.282	304.6	652.2	0.0703	0.73406	4.91330	0.4875
Carbon Monoxide	CO	28.01	506.7	–220.63	0.0527	0.79265	0.96720	0.0510
Carbon Dioxide	CO_2	44.01	1070.0	87.76	0.0343	0.82203	1.51970	0.2239
Hydrogen Sulfide	H_2S	34.082	1306.5	212.81	0.0462	0.80269	1.17690	0.1010
Air	—	28.9586	551.9	–220.97	0.0458	0.87603	1.00000	—
Hydrogen	H_2	2.0159	190.7	–399.9	0.5319	0.07087	0.06961	–0.2140
Oxygen	O_2	31.9988	731.4	–181.43	0.0367	1.14230	1.10500	0.0222
Nitrogen	N_2	28.0135	492.5	–232.53	0.0511	0.80687	0.96740	0.0372
Water	H_2O	18.0153	3200.1	705.1	0.04975	1.00000	0.62210	0.3443

*Apparent value at 60°F

5.3.4 Reduced Temperature and Pressure

The **principle of corresponding states** provides a method for correlating the properties of pure substances and mixtures. It states that two substances of similar chemical nature at the same conditions will exhibit similar physical properties when referenced to the same properties, such as critical temperatures and pressures.[3] The conditions of temperature and pressure referenced to critical temperature and critical pressure are called **reduced temperature,** (T_r) and **reduced pressure** (p_r).

$$T_r = \frac{T}{T_c} \tag{5.15}$$

$$p_r = \frac{p}{p_c} \tag{5.16}$$

In Eqs. (5.15) and (5.16), T_c is critical temperature and, p_c is critical pressure.

For gas mixtures, pseudo-reduced temperature, (T_{pr}), and pseudo-reduced pressure, (p_{pr}), are represented as:

$$T_{pr} = \frac{T}{T_{pc}} \tag{5.17}$$

$$p_{pr} = \frac{p}{p_{pc}} \tag{5.18}$$

In Eqs. 5.17 and 5.18, T_{pc} is the pseudo-critical temperature, and p_{pc} is the pseudo-critical pressure for gas mixtures. Pseudo-critical properties for mixtures do not have real physical meaning as critical properties for pure substances. The main use of pseudo-critical properties for mixtures is as a means of correlating the properties of the mixtures. Section 5.4 presents various methods for estimating pseudo-critical properties for gas mixtures.

5.4 Pseudo-Critical Properties of Gas Mixtures

The methods for calculating the pseudo-critical properties of gas mixtures are presented under two groups. These groups are:

1. Composition of gas mixtures known
2. Composition of gas mixtures unknown

5.4.1 Composition of Gas Mixtures Known

If the composition of the mixture is known, the pseudo-critical properties of gas mixtures can be calculated by using various mixing rules. The simplest mixing rule was proposed by Kay.[2] Kay's mixing rule is the molar weighted average of the critical properties of the pure components of the gas mixture. It is written as:

$$p_{pc} = \sum y_i p_{ci} \tag{5.19}$$

$$T_{pc} = \sum y_i T_{ci} \tag{5.20}$$

In Eqs. (5.19) and (5.20), y_i is the mole fraction of component i; p_{ci} and T_{ci} are the critical pressure and critical temperature, respectively, of component i. Equations (5.19) and (5.20) are suitable for low molecular weight, homologous gas mixtures with specific gravity less than 0.75. These equations can be used for quick estimation of pseudo-critical properties and ultimately for calculation of gas compressibility (z) factors. However, it should be noted that this method underestimates gas compressibility factors, with errors as high as 15%.[3]

The mixing rules proposed by Stewart, Burkhart, and Voo (termed SBV)[4] and modified by Sutton (termed SSBV)[3] improved the accuracy of calculating the pseudo-critical properties of high molecular weight gas mixtures. The mixing rules proposed by Stewart et al.[4] for high molecular weight gas mixtures are:

$$J = \frac{1}{3}\sum_{i=1}^{n_c}\left(\frac{yT_c}{p_c}\right)_i + \frac{2}{3}\left[\sum_{i=1}^{n_c}\left(y\sqrt{\frac{T_c}{p_c}}\right)_i\right]^2 \tag{5.21}$$

$$K = \sum_{i=1}^{n_c}\left(\frac{yT_c}{\sqrt{p_c}}\right)_i \tag{5.22}$$

$$T_{pc} = \frac{K^2}{J} \tag{5.23}$$

$$p_{pc} = \frac{T_{pc}}{J} \tag{5.24}$$

In Eqs. (5.21) and (5.22), y = mole fraction of component i in mixture; T_c = critical temperature of component i in °R; and p_c = critical pressure of component i in psia.

Sutton[3] modified the SBV equations [Eqs. (5.21) to (5.24)] for gas mixtures that contain high concentrations of heptane plus fraction (up to 14.27 mole %). The SSBV modifications are:

$$F_j = \frac{1}{3}\left(\frac{yT_c}{p_c}\right)_{C7+} + \frac{2}{3}\left(\frac{y^2T_c}{p_c}\right)_{C7+} \tag{5.25}$$

$$\varepsilon_j = 0.6081F_j + 1.1325F_j^2 - 14.004F_j y_{C7+} + 64.434F_j y_{C7+}^2 \tag{5.26}$$

$$\varepsilon_k = \left(\frac{T_c}{\sqrt{p_c}}\right)_{C7+} \times \left(0.3129y_{C7+} - 4.8156y_{C7+}^2 + 27.3751y_{C7+}^3\right) \tag{5.27}$$

$$J' = J - \varepsilon_j \tag{5.28}$$

$$K' = K - \varepsilon_k \tag{5.29}$$

$$T_{pc} = \frac{K'^2}{J'} \tag{5.30}$$

$$p_{pc} = \frac{T_{pc}}{J'} \tag{5.31}$$

The critical properties of pure components in a mixture are provided in Table 5.1. The critical properties of the heptanes-plus fraction must be determined from correlations. Sutton[3] reported that the Kessler-Lee[5] equations provide the lowest error in comparison with other methods. The Kessler-Lee[5] equations are:

$$p_{pc,C7+} = \exp\left[\begin{aligned}&8.3634 - \frac{0.0566}{\gamma_{C7+}} - \left(0.24244 + \frac{2.2898}{\gamma_{C7+}} + \frac{0.11857}{\gamma_{C7+}^2}\right)\frac{T_{b,C7+}}{10^3}\\ &+ \left(1.4685 + \frac{3.648}{\gamma_{C7+}} + \frac{0.47227}{\gamma_{C7+}^2}\right)\frac{T_{b,C7+}^2}{10^7}\\ &- \left(0.42019 + \frac{1.6977}{\gamma_{C7+}^2}\right)\frac{T_{b,C7+}^3}{10^{10}}\end{aligned}\right] \tag{5.32}$$

$$T_{pc,C7+} = 341.7 + 811\gamma_{C7+} + (0.4244 + 0.1174\gamma_{C7+})T_{b,C7+} + (0.4669 - 3.2623\gamma_{C7+})\frac{10^5}{T_{b,C7+}} \tag{5.33}$$

The boiling point, $T_{b,c7+}$, is estimated from the Whitson[6] equation:

$$T_{b,C7+} = \left(4.5579 \times M_{C7+}^{0.15178} \times \gamma_{C7+}^{0.15427}\right)^3 \quad (5.34)$$

The application of the SSBV equations for calculation of pseudo-critical properties of a mixture is shown in Example 5.3. The units of the terms in all the equations are as shown in the example.

5.4.2 Correction for non-Hydrocarbon Gas Impurities

Gas mixtures may contain varying amounts of hydrogen sulfide, carbon dioxide, nitrogen, and water vapor as impurities. The critical properties of hydrocarbon mixtures should be corrected for the presence of these substances. Sutton[7] reported that the correlations proposed by Wichert and Aziz[8] for correcting for the presence of hydrogen sulfide and carbon dioxide are superior to other published methods. The Wichert and Aziz correlations are:

$$\xi = 120\left[\left(y_{CO_2} + y_{H_2S}\right)^{0.9} - \left(y_{CO_2} + y_{H_2S}\right)^{1.6}\right] + 15\left(y_{H_2S}^{0.5} - y_{H_2S}^{4}\right) \quad (5.35)$$

$$T_{pc}^{*} = T_{pc} - \xi \quad (5.36)$$

$$p_{pc}^{*} = \frac{p_{pc}\left(T_{pc} - \xi\right)}{T_{pc} + y_{H_2S}\left(1 - y_{H_2S}\right)\xi} \quad (5.37)$$

At present, there are no satisfactory published correlations for correcting for the presence of nitrogen and water vapor. In Section 5.4.3, a method proposed by Standing[9] that could be used to correct for the presence of water and nitrogen in hydrocarbon gas mixtures is presented. The units of the terms in all the equations are as shown in Example 5.3.

5.4.3 Composition of Gas Mixture Unknown

The critical properties of gas mixtures can be calculated from Sutton[7] correlations when the composition of the mixture is not known. Sutton[7] based the correlation on hydrocarbon gas gravity, which is calculated using the method proposed by Standing:[9]

$$y_{HC} = 1 - y_{H_2S} - y_{CO_2} - y_{N_2} \quad (5.38)$$

The original Standing equation [Eq. (5.38)] can be modified slightly to account for the presence of water vapor as with the other impurities:

$$y_{HC} = 1 - y_{H_2S} - y_{CO_2} - y_{N_2} - y_{H_2O} \quad (5.39)$$

The hydrocarbon gas gravity, $\gamma_{g,HC}$, is calculated as:

$$\gamma_{g,HC} = \frac{\gamma_g - \left(y_{H_2S}M_{H_2S} + y_{CO_2}M_{CO_2} + y_{N_2}M_{N_2} + y_{H_2O}M_{H_2O}\right)/M_{AIR}}{y_{HC}} \quad (5.40)$$

According to Standing,[9] the mixture's pseudo-critical properties are calculated as follows:

$$p_{pc} = y_{HC}p_{pc,HC} + y_{H_2S}p_{c,H_2S} + y_{CO_2}p_{c,CO_2} + y_{N_2}p_{c,N_2} + y_{H_2O}p_{c,H_2O} \quad (5.41)$$

$$T_{pc} = y_{HC}T_{pc,HC} + y_{H_2S}T_{c,H_2S} + y_{CO_2}T_{c,CO_2} + y_{N_2}T_{c,N_2} + y_{H_2O}T_{c,H_2O} \quad (5.42)$$

Note that if hydrogen sulfide and carbon dioxide are present in the gas mixture, then p_{pc} and T_{pc} calculated from Eqs. (5.41) and (5.42) should be corrected by using the Wichert and Aziz correlations [Eqs. (5.35), (5.36), and (5.37)].

Sutton[7] proposed two sets of correlations for calculating hydrocarbon pseudo-critical properties based on hydrocarbon gas gravity. The first set of correlations is for associated gas from gas and oil reservoirs. The second set of correlations is to be used for gas condensates.

The correlations to be used for associated hydrocarbon gas gravity as proposed by Sutton[7] are:

$$p_{pc,HC} = 671.1 + 14\gamma_{g,HC} - 34.3\gamma_{g,HC}^2 \quad (5.43)$$

$$T_{pc,HC} = 120.1 + 429\gamma_{g,HC} - 62.9\gamma_{g,HC}^2 \quad (5.44)$$

In Eqs. (5.43) and (5.44), p_{pc} is in psia and T_{pc} is in °R. Equations (5.43) and (5.44) are applicable over a hydrocarbon gas gravity range of $0.554 \le \gamma_{g,HC} \le 1.862$; temperature range of $32 \le T\left({}^{\circ}F\right) \le 460$; and pressure range of $12 \le P\ (\text{psia}) \le 10{,}000$. The equations apply if the gas mixture contains less than 10 mol % hydrogen sulfide, less than 55.8 mol % carbon dioxide, and less than 21.7 mol % nitrogen. Note that the hydrocarbon gas gravity is equal to total gas gravity if the associated gas contained no impurities. The correlations were derived from a database that contained 3256 compositions with 4817 gas compressibility factor measurements. The average absolute error reported by Sutton[7] for compressibility factors calculated with the correlations is 0.80%.

The correlations for condensate hydrocarbon gas gravity proposed by Sutton[7] are:

$$p_{pc,HC} = 744 - 125.4\gamma_{g,HC} + 5.9\gamma_{g,HC}^2 \quad (5.45)$$

$$T_{pc,HC} = 164.3 + 357.7\gamma_{g,HC} - 67.7\gamma_{g,HC}^2 \quad (5.46)$$

In Eqs. (5.45) and (5.46), p_{pc} is in psia and T_{pc} is in °R. Equations (5.45) and (5.46) are applicable over a hydrocarbon gas gravity range of $0.554 \le \gamma_{g,HC} \le 2.819$; temperature range of $0 \le T\,(°F) \le 460$; and pressure range of $12 \le P\,(\text{psia}) \le 17{,}065$. The composition of the gas mixture should contain less than 90 mole % hydrogen sulfide, less than 89.9 mole % carbon dioxide, and less than 33.3 mole % nitrogen. Note that the hydrocarbon gas gravity is equal to total gas gravity if the condensate gas contained no impurities. The database for these correlations contained 2264 compositions with 10,177 compressibility factor measurements. The average absolute error reported by Sutton[7] for compressibility factors calculated with the correlations is 1.11%.

5.5 Wet Gas and Gas Condensate

In this section, methods for calculating the specific gravities of wet and condensate gases are presented. As defined previously in Section 5.3, a wet gas forms no liquids inside the reservoir but condenses liquids in the separators and flowlines. A condensate gas will condense liquids inside the reservoir if reservoir pressure falls below the dew point. The methods presented here are not applicable to a condensate reservoir if reservoir pressure falls below the dew point. Before these methods are applied, it is important to ensure that excessive liquid condensation has not occurred in the flowlines before the sampling point at the separators. The first method is a recombination method that is used when the compositions of the gas and liquid phases sampled at the separators are determined from laboratory measurements. The second method is based on correlations when the compositions of the produced phases are not known.

5.5.1 Recombination Method

The recombination method is illustrated with a two-stage separation system consisting of a separator and a stock tank. The compositions of the separator gas, stock-tank gas, and stock-tank oil are known.

EXAMPLE 5.1 Calculation of Gravity of a Wet Gas by the Recombination Method

Problem

Suppose a gas reservoir produces wet gas through a two-stage separation system consisting of a separator and a stock tank. The gas-oil ratio (GOR) of the separator is given as 62,000 scf/STB. The stock-tank GOR is 350 scf/STB. Stock-tank oil gravity is 50.6°API. Calculate the composition of the reservoir gas given the compositions of the separator and stock-tank fluids as shown in Table 5.2.

Solution

The specific gravity of oil is defined as the density of oil relative to the density of water at the same temperature and pressure:

$$\gamma_o = \frac{\rho_o}{\rho_w} \tag{5.47}$$

In Eq. (5.47), ρ_o and ρ_w are densities of oil and water, respectively, in pounds per cubic foot.

The specific gravity of oil can be related to API gravity (°API) as:

$$\gamma_o = \frac{141.5}{131.5 + {}^\circ API} \tag{5.48}$$

From Eq. (5.48), stock-tank oil specific gravity,

$$\gamma_o = \frac{141.5}{131.5 + 50.6} = 0.7770.$$

Molecular weight of stock-tank oil from Table 5.2 is

$$\sum x_i M_i = 105.7146 \text{ lbm/lb-mole}.$$

Step 1: Convert separator GOR to lb-mole/STB.

Assuming standard conditions of

$T_{sc} = 60°\text{F}$ and $p_{sc} = 14.65$ psia:

Molar volume,

$$V_M = RT_{sc}/p_{sc} =$$

$$= 10.732(60 + 459.67)/14.65$$

$$= 380.69 \text{ scf/lb-mole}$$

Pound moles of separator gas per STB

$$= (62000 \text{ scf/STB}) \times (\text{lb-mole}/380.69 \text{ scf})$$

$$= 162.8622 \text{ lb-mole/STB}$$

Step 2: Convert stock-tank GOR to lb-mole/STB.

Pound moles of stock tank gas per STB

$$= (350 \text{ scf/STB}) \times (\text{lb-mole}/380.69 \text{ scf})$$

$$= 0.9194 \text{ lb-mole/STB}$$

Step 3: Convert stock-tank liquid to lb-mole/STB.

Pound moles of stock tank liquid per STB =

$$(0.777 \times 62.4 \text{ lbm/ft}^3)$$

$$\times (5.615 \text{ ft}^3/\text{STB})$$

$$\times (\text{lb-mole}/105.7146 \text{ lbm})$$

$$= 2.5753 \text{ lb-mole/STB}$$

TABLE 5.2 | Compositions of Separator and Stock-Tank Fluids for Example 5.1

Component	Vapor Composition Separator, Mole Fraction, y_i	Vapor Composition Stock-Tank, Mole Fraction, y_i	Liquid Composition Stock-Tank, Mole Fraction, x_i	Molecular Weight of Components, M_i	$x_i M_i$
N_2	0.0030	0.0010	0.0000	28.0134	0.0000
CO_2	0.0042	0.0014	0.0000	44.0100	0.0000
CH_4	0.7905	0.2700	0.0015	16.0430	0.0241
C_2H_6	0.1230	0.2221	0.0072	30.0700	0.2165
C_3H_8	0.0563	0.2812	0.0328	44.0970	1.4464
i-C_4H_{10}	0.0052	0.0426	0.0126	58.1230	0.7323
n-C_4H_{10}	0.0118	0.1107	0.0138	58.1230	0.8021
i-C_5H_{12}	0.0020	0.0232	0.0409	72.1500	2.9509
n-C_5H_{12}	0.0023	0.0285	0.0267	72.1500	1.9264
C_6H_{14}	0.0014	0.0171	0.0894	86.1770	7.7042
C_{7+}	0.0003	0.0022	0.7751	116.0000	89.9116
Total	**1.0000**	**1.0000**	**1.0000**	—	**105.7146**
Properties of the heptanes plus stock tank liquid:					
Specific gravity		0.802			
Molecular weight		116 lb/lb mole			

End of Example 5.1

The composition of the reservoir gas is computed as shown in Table 5.3 using the values calculated in steps 1-3. Because the reservoir gas composition has been determined, calculation of pseudo-critical properties can proceed as illustrated in Example 5.3. Note that a similar procedure can be used to calculate the specific gravity of the reservoir gas if fluid sampling was conducted only at the primary separator.

5.5.2 Correlation Method

The gas specific gravity of wet or condensate reservoirs can be calculated from production data if the compositions of the fluid phases at the separators are not known. The production data required for the calculation are the specific gravities of the fluids plus gas and liquid production data at the separators. For instance, for a three-stage separation system, the gas gravities and gas-liquid ratios from the primary separator, secondary separator, and stock tank, as well as the stock-tank liquid gravity, are required. Thus, the expression for the gas specific gravity for a three-stage separation system can be derived as reported by Gold et al.:[10]

$$\gamma_w = \frac{R_1\gamma_1 + 4602\gamma_o + R_2\gamma_2 + R_3\gamma_3}{R_1 + \left(133{,}316\gamma_o / M_o\right) + R_2 + R_3} \tag{5.49}$$

In Eq. (5.49), γ_w is wellstream gas gravity (air = 1); R_1, R_2, and R_3 are the gas-liquid ratios in scf/STB for the primary separator, secondary separator, and stock tank, respectively; γ_1, γ_2, and γ_3 are gas gravities for the primary separator, secondary separator, and stock tank, respectively; M_o and γ_o are the molecular weight and specific gravity, respectively, of the stock-tank liquid.

For a two-stage separation system, the expression for the specific gravity is derived as:

$$\gamma_w = \frac{R_1\gamma_1 + 4602\gamma_o + R_3\gamma_3}{R_1 + \left(133{,}316\gamma_o / M_o\right) + R_3} \tag{5.50}$$

Note the absence of the terms R_2 and γ_2 to represent the secondary separator in Eq. (5.50) because the primary separator and stock tank constitute a two-stage separation system. Eqs. (5.49) and (5.50) can be used to calculate gas gravities accurately if all the terms in the equations are available from production data. In most cases, only the primary separator and stock-tank production data are measured. From the production data, the primary separator gas-liquid ratio, R_1, the primary separator gas gravity, γ_1, and the gravity of the stock-tank liquid, γ_o, can be obtained. All the other terms in Eqs. (5.49) and (5.50) are unknown.

TABLE 5.3 Calculation of Gravity of a Wet Gas by the Recombination Method with Two-Stage Separation

Component	Vapor Composition Separator, Mole Fraction, y_{SP}	Vapor Composition Stock Tank, Mole Fraction, y_{ST}	Liquid Composition Stock Tank, Mole Fraction, x_{ST}	Component Moles Separator Gas $y_{SP} \times 162.8622$	Component Moles Stock Tank Gas $y_{ST} \times 0.9194$	Component Moles Stock Tank Liquid $x_{ST} \times 2.5753$	Total	Composition of Reservoir Gas
N_2	0.0030	0.0010	0.0000	0.4886	0.0009	0.0000	0.4895	0.0029
CO_2	0.0042	0.0014	0.0000	0.6840	0.0013	0.0000	0.6853	0.0041
CH_4	0.7905	0.2700	0.0015	128.7426	0.2482	0.0039	128.9947	0.7754
C_2H_6	0.1230	0.2221	0.0072	20.0321	0.2042	0.0185	20.2548	0.1218
C_3H_8	0.0563	0.2812	0.0328	9.1691	0.2585	0.0845	9.5121	0.0572
i-C_4H_{10}	0.0052	0.0426	0.0126	0.8469	0.0392	0.0324	0.9185	0.0055
n-C_4H_{10}	0.0118	0.1107	0.0138	1.9218	0.1018	0.0355	2.0591	0.0124
i-C_5H_{12}	0.0020	0.0232	0.0409	0.3257	0.0213	0.1053	0.4524	0.0027
n-C_5H_{12}	0.0023	0.0285	0.0267	0.3746	0.0262	0.0688	0.4695	0.0028
C_6H_{14}	0.0014	0.0171	0.0894	0.2280	0.0157	0.2302	0.4740	0.0028
C_{7+}	0.0003	0.0022	0.7751	0.0489	0.0020	1.9961	2.0470	0.0123
Total	**1.0000**	**1.0000**	**1.0000**	—	—	—	**166.3569**	**1.0000**
Basis: 1 bbl of stock-tank liquid								

The gas produced at the secondary separator and stock tank (for a three-stage separation system) or the stock tank (for a two-stage separation system) is accounted for as additional gas produced, G_{pa}. Thus, for a three-stage separation system:[10]

$$G_{pa} = R_2\gamma_2 + R_3\gamma_3 \tag{5.51}$$

For a two-stage separation system:

$$G_{pa} = R_3\gamma_3 \tag{5.52}$$

Similarly, the vapor equivalent of the primary separator liquid, V_{eq}, accounts for the stock-tank liquid production and gas production beyond the primary separator. This is represented for a three-stage separation system as:

$$V_{eq} = (133{,}316\gamma_o/M_o) + R_2 + R_3 \tag{5.53}$$

For a two-stage separation system:

$$V_{eq} = (133{,}316\gamma_o/M_o) + R_3 \tag{5.54}$$

Using Eqs. (5.51) and (5.53), Eq. (5.49) can be rewritten for a three-stage separation system as:

$$\gamma_w = \frac{R_1\gamma_1 + 4602\gamma_o + G_{pa}}{R_1 + V_{eq}} \tag{5.55}$$

The expression for a two-stage separation system is identical to Eq. (5.55) except that G_{pa} refers to Eq. (5.52), and V_{eq} refers to Eq. (5.54).

Gold et al.[10] proposed a correlation for estimating molecular weight, M_o, of the stock-tank liquid as:

$$M_o = \frac{5954}{{}^oAPI - 8.811} = \frac{42.43\gamma_o}{1.008 - \gamma_o} \tag{5.56}$$

The additional gas produced, G_{pa}, and vapor equivalent, V_{eq}, for a three-stage or a two-stage separation system can be estimated with correlations proposed by Gold et al.[10] The correlation for additional gas produced in a three-stage separation system is:

$$G_{pa} = \beta_1(p_{s1} - 14.65)^{\beta_2}\gamma_1^{\beta_3}\gamma_{API}^{\beta_4}T_{s1}^{\beta_5}T_{s2}^{\beta_6} \tag{5.57}$$

In Eq. (5.57), $\beta_1 = 2.9922$, $\beta_2 = 0.97050$, $\beta_3 = 6.8049$, $\beta_4 = 1.0792$, $\beta_5 = -1.1960$, and $\beta_6 = 0.55367$.

The correlation for the vapor equivalent in a three-stage separation system is:

$$V_{eq} = \beta_0 + \beta_1 p_{s1}^{\beta_2}\gamma_1^{\beta_3}\gamma_{API}^{\beta_4}T_{s1}^{\beta_5}T_{s2}^{\beta_6} \tag{5.58}$$

For Eq. (5.58), $\beta_0 = 535.92$, $\beta_1 = 2.6231$, $\beta_2 = 0.79318$, $\beta_3 = 4.6612$, $\beta_4 = 1.2094$, $\beta_5 = -0.84911$, and $\beta_6 = 0.26987$. In Eqs. (5.57) and (5.58), p_{s1} is the primary separator pressure in psia; γ_1 is the primary separator gas gravity; T_{s1} is the primary separator temperature in °F; γ_{API} is the API gravity of the stock-tank liquid; and T_{s2} is the secondary separator temperature in °F.

Similar equations for a two-stage separation system were also correlated by Gold et al.[10] The correlation for additional gas produced in a two-stage separation system is:

$$G_{pa} = \beta_1(p_{s1} - 14.65)^{\beta_2}\gamma_1^{\beta_3}\gamma_{API}^{\beta_4}T_{s1}^{\beta_5} \tag{5.59}$$

In Eq. (5.59), $\beta_1 = 1.4599$, $\beta_2 = 1.3394$, $\beta_3 = 7.0943$, $\beta_4 = 1.1436$, and $\beta_5 = -0.93446$.

The correlation for vapor equivalent in a two-stage separation system is:

$$V_{eq} = \beta_0 + \beta_1 p_{s1}^{\beta_2} \gamma_1^{\beta_3} \gamma_{API}^{\beta_4} T_{s1}^{\beta_5} \tag{5.60}$$

In Eq. (5.60), $\beta_0 = 635.53$, $\beta_1 = 0.36182$, $\beta_2 = 1.0544$, $\beta_3 = 5.0831$, $\beta_4 = 1.5812$, and $\beta_5 = -0.79130$. Other terms in Eqs. (5.59) and (5.60) are as defined previously in Eq. (5.58).

According to Gold et al.,[10] the average absolute error calculated for G_{pa} estimated with Eq. (5.57) is 10.5%, and with Eq. (5.59) is 14.9%. The average absolute error calculated for V_{eq} estimated with Eq. (5.58) is 5.8%, and with Eq. (5.60) is 6.3%. Equations (5.57) to (5.60) can be used for reservoir gas with total non-hydrocarbon content up to 20 mole %. For reservoir gas with non-hydrocarbon content between 5 and 20 mole %, the average absolute error of gas gravity calculated with the correlations is 0.95%, and the maximum error is 8.1%. The presence of non-hydrocarbon content up to 20 mole % increased the average absolute error of the V_{eq} correlation by 15 percentage points from 6% to 21%.

The total reservoir fluid production rate for a wet gas or condensate reservoir can be estimated if we have the total gas production rate (separator plus stock-tank gas), stock-tank oil production rate, and properties of the stock-tank oil with Eq. (5.61).

$$q = q_g + 133.316\left(\gamma_o q_o / M_o\right) \tag{5.61}$$

In Eq. (5.61), q = total reservoir fluid rate, Mscf/D; q_g = gas production rate (separator plus stock-tank gas), Mscf/D; γ_o = stock-tank oil gravity; q_o = stock-tank oil rate, STB/D; M_o = molecular weight of stock-tank oil, lbm/lb-mole. Alternatively, if the vapor equivalent of the primary separator liquid, V_{eq} is calculated from prior equations or correlations, then the total reservoir fluid production rate can be estimated by using Eq. (5.62).

$$q = q_{s1}\left(1 + V_{eq}/R_1\right) \tag{5.62}$$

In Eq. (5.62), q = total reservoir fluid rate, Mscf/D; q_{s1} = gas rate of the primary separator, Mscf/D; and R_1 = gas-liquid ratio of the primary separator, scf/STB.

EXAMPLE 5.2 Calculation of Gas Gravity and Total Reservoir Fluid Production Rate

Problem

The separator and stock tank conditions and production data of a condensate gas reservoir in the Gulf of Mexico, USA, are as follows:

Primary separator pressure	1155 psia
Primary separator temperature	92°F
Primary separator gas rate	49,545 Mcf/D
Primary separator GOR	28,721 scf/STB
Primary separator gas gravity	0.6346 (air = 1)
Stock-tank gas rate	778 Mcf/D
Stock-tank GOR	451 scf/STB
Stock-tank gas gravity	0.9394 (air = 1)
Stock-tank liquid gravity	44.8°API
Stock-tank pressure	15.025 psia
Stock-tank temperature	60°F
Condensate production rate	1725 STB/D

Solution

The surface rates and fluid properties of the primary separator and stock-tank are known. The gas gravity can be calculated by using the recombination equation for two-stage separation shown previously as Eq. (5.50):

$$\gamma_w = \frac{R_1\gamma_1 + 4602\gamma_o + R_3\gamma_3}{R_1 + \left(133{,}316\gamma_o/M_o\right) + R_3}$$

From Eq. (5.48),

$$\gamma_o = \frac{141.5}{131.5 + {}^oAPI} = \frac{141.5}{131.5 + 44.8} = 0.8026$$

From Eq. (5.56),

$$M_o = \frac{42.43\gamma_o}{1.008 - \gamma_o} = \frac{42.43(0.8026)}{1.008 - 0.8026} = 165.7951$$

Substituting,

$$\gamma_w = \frac{28721(0.6346) + 4602(0.8026) + 451(0.9394)}{28721 + (133316)\times 0.8026/165.7951 + 451}$$

$$= 0.7493$$

Actual reported reservoir fluid gas gravity is 0.7437. The calculated error is 0.75%.

Assuming that surface rates and fluid properties are not available, gas gravity can still be estimated from the operating parameters of the primary separator by using the correlations from Gold et al.[10]

From Eq. (5.59),

$$G_{pa} = \beta_1\left(p_{s1} - 14.65\right)^{\beta_2} \gamma_1^{\beta_3} \gamma_{API}^{\beta_4} T_{s1}^{\beta_5}$$

$$= 1.4599(1155 - 14.65)^{1.3394}(0.6346)^{7.0943}$$

$$\times(44.8)^{1.1436}(92)^{-0.93446}$$

$$= 814.9231 \text{ scf/STB}$$

From Eq. (5.60),

$$V_{eq} = \beta_0 + \beta_1 p_{s1}^{\beta_2} \gamma_1^{\beta_3} \gamma_{API}^{\beta_4} T_{s1}^{\beta_5}$$

$$= 635.53 + 0.36182(1155)^{1.0544}(0.6346)^{5.0831}$$

$$\times(44.8)^{1.5812}(92)^{-0.79130}$$

$$= 1328.673 \text{ scf/STB}$$

Substituting Eq. (5.55):

$$\gamma_w = \frac{R_1\gamma_1 + 4,602\gamma_o + G_{pa}}{R_1 + V_{eq}}$$
$$= \frac{28721(0.6346) + 4601(0.8026) + 814.9231}{28721 + 1328.673}$$
$$= 0.7566$$

Actual reported reservoir fluid gas gravity is 0.7437. The calculated error is 1.7%.

The total reservoir fluid production rate is estimated by using Eq. (5.62):

$$q = q_{s1}\left(1 + V_{eq}/R_1\right)$$
$$= 49545(1 + 1328.673/28721)$$
$$= 51,837 \text{ Mscf/D}$$

End of Example 5.2

5.6 Correlations for Gas Compressibility Factor

Standing and Katz (SK)[11] developed a generalized chart for predicting the compressibility factors of gases using parameters of reduced pressure, p_r, and reduced temperature, T_r, based on the principle of corresponding states. The SK chart with extended ranges has been reproduced numerically by many authors for computer applications. Takacs[12] reviewed the accuracies of these correlations for numerical representation of the SK chart. On the basis of the errors reported by Takacs, the correlations of Dranchuk and Abou-Kassem (DAK)[13] and Hall and Yarborough (HY)[14] gave the most accurate representation of the SK chart over a wide range of temperatures and pressures. Sutton[7] also compared several methods developed by various authors for numerical representation of the SK chart. According to Sutton,[7] the Dranchuk et al. (DPR)[15] method showed the lowest average error but higher standard deviation, whereas the DAK method had comparably low average error and low standard deviation. On the basis of these studies reported by Takacs[12] and Sutton,[7] the DAK method was selected for the calculations of gas compressibility factors (z factor).

The DAK correlation is based on the 11-parameter Starling equation of state. Using nonlinear regression methods, Dranchuk and Abou-Kassem[13] fitted the Starling equation of state to 1500 data points from the SK chart, with an average absolute error of 0.486%. The DAK correlation for calculating z factors is represented as follows:

$$z = 1 + c_1(T_{pr})\rho_r + c_2(T_{pr})\rho_r^2 - c_3(T_{pr})\rho_r^5 + c_4(\rho_r, T_{pr}) \quad (5.63)$$

In Eq. (5.63):

$$\rho_r = 0.27p_{pr}/(z \times T_{pr}) \quad (5.64)$$

$$c_1(T_{pr}) = A_1 + A_2/T_{pr} + A_3/T_{pr}^3 + A_4/T_{pr}^4 + A_5/T_{pr}^5 \quad (5.65)$$

$$c_2(T_{pr}) = A_6 + A_7/T_{pr} + A_8/T_{pr}^2 \quad (5.66)$$

$$c_3(T_{pr}) = A_9\left(A_7/T_{pr} + A_8/T_{pr}^2\right) \quad (5.67)$$

$$c_4(\rho_r, T_{pr}) = A_{10}\left(1 + A_{11}\rho_r^2\right)\left(\rho_r^2/T_{pr}^3\right)\exp\left(-A_{11}\rho_r^2\right) \quad (5.68)$$

The constants, A_1 through A_{11}, are as follows:

$$A_1 = 0.3265 \quad A_2 = -1.0700 \quad A_3 = -0.5339$$
$$A_4 = 0.01569 \quad A_5 = -0.05165 \quad A_6 = 0.5475$$
$$A_7 = -0.7361 \quad A_8 = 0.1844 \quad A_9 = 0.1056$$
$$A_{10} = 0.6134 \quad A_{11} = 0.7210$$

The DAK correlation is applicable over the following ranges of reduced pressure and temperature: $0.2 \le p_{pr} < 30$, $1.0 < T_{pr} \le 3.0$ and $p_{pr} < 1.0$; $0.7 < T_{pr} \le 1.0$. The correlation gives poor results for $T_{pr} = 1.0$ and $p_{Pr} > 1.0$.

Eq. (5.63) can be solved for z by using iterative techniques such as the Newton-Raphson method. The form of the Newton-Raphson method is:

$$x_{n+1} = x_n - \frac{f_n}{f_n'} \quad (5.69)$$

By rearranging Eq. (5.63), f_n is:

$$F(z) = z - \begin{pmatrix} 1 + c_1(T_{pr})\rho_r + c_2(T_{pr})\rho_r^2 \\ -c_3(T_{pr})\rho_r^5 + c_4(\rho_r, T_{pr}) \end{pmatrix} = 0 \quad (5.70)$$

The derivative, f_n', is the derivative of Eq. (5.70) with respect to z at constant reduced temperature:

$$\left(\partial F(z)/\partial z\right)_{T_{pr}}$$
$$= 1 + c_1(T_{pr})\rho_r/z + 2c_2(T_{pr})\rho_r^2/z - 5c_3(T_{pr})\rho_r^5/z$$
$$+ \frac{2A_{10}\rho_r^2}{T_{pr}^3 z}\left(1 + A_{11}\rho_r^2 - \left(A_{11}\rho_r^2\right)^2\right)\exp\left(-A_{11}\rho_r^2\right) \quad (4.71)$$

The DAK correlations are also provided in Figures 5.9 and 5.10, which express z factors as a function of reduced temperature and pressure. These figures should be used to obtain z factors in manual calculations as shown in Example 5.4.

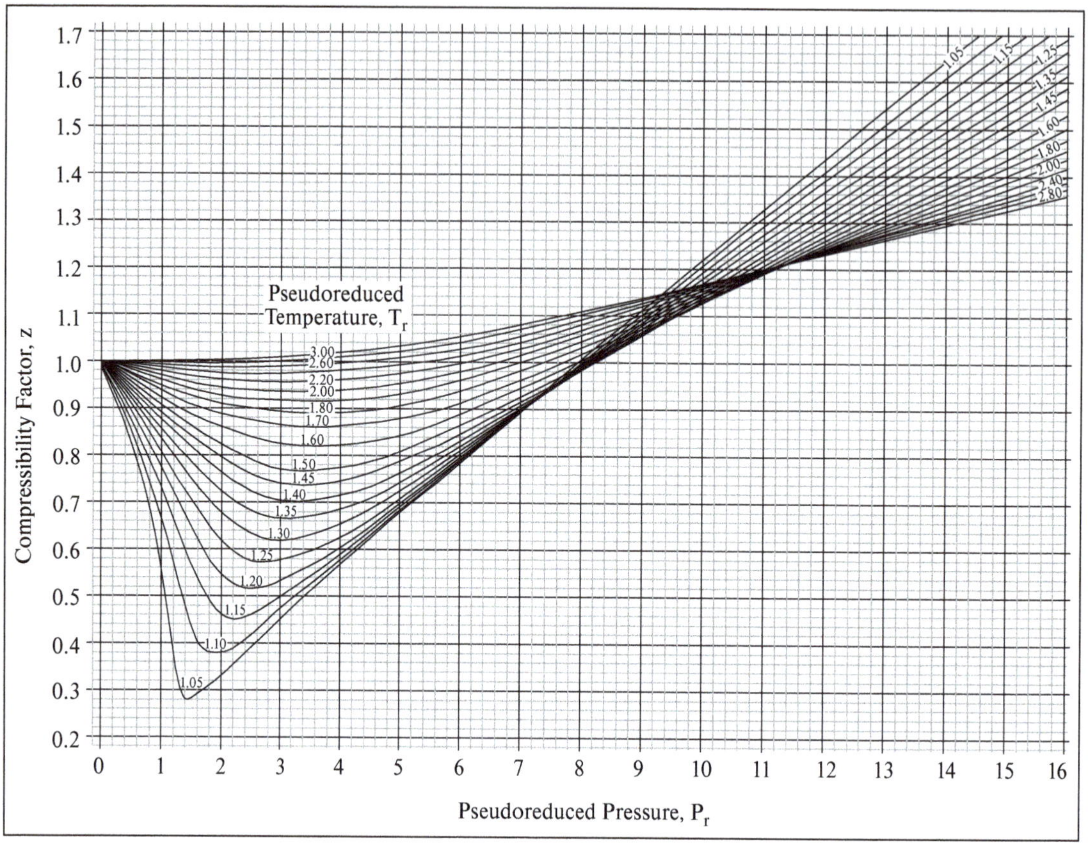

▲ **FIGURE 5.9** **Gas compressibility factor chart for $0 \le p_r \le 16$ (from Sutton[3] © 1985 SPE, Reproduced with permission)**

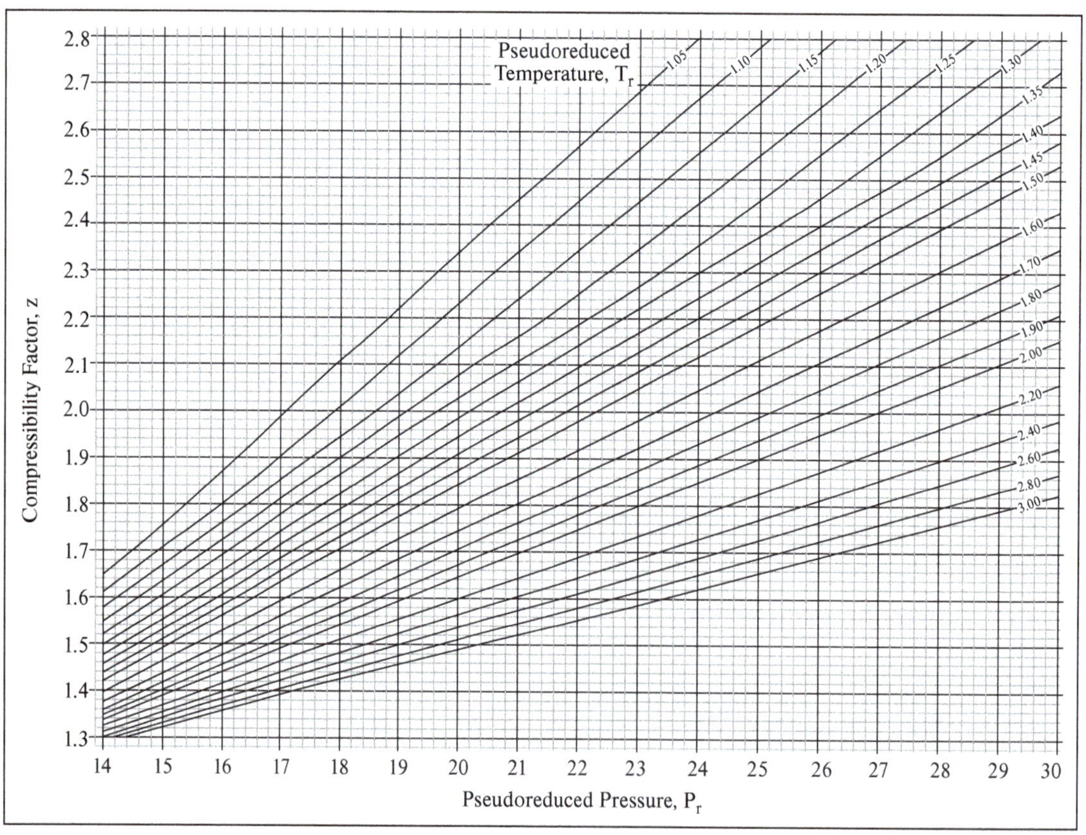

▲ **FIGURE 5.10** **Gas compressibility factor chart for $14 \le p_r \le 30$ (from Sutton[3] © 1985 SPE, Reproduced with permission)**

5.7 Gas Formation Volume Factor (FVF)

The z factor calculated from Eq. (5.63) numerically or by using the graphs in Figures 5.9 and 5.10 can be used to calculate the gas **formation volume factor** (FVF). Gas FVF is defined as its volume at reservoir conditions of temperature and pressure relative to its volume at standard temperature and pressure. Gas FVF, represented as B_g, is thus:

$$B_g = \frac{V_R}{V_{sc}} \tag{5.72}$$

In Eq. (5.72), V_R is the gas volume at reservoir temperature and pressure, and V_{sc} is the gas volume at standard temperature (60°F) and standard pressure (14.65 psia).

Applying Eq. (5.9) for real gases,

$$V_R = \frac{z_R nRT_R}{p_R} \tag{5.73}$$

$$V_{sc} = \frac{z_{sc} nRT_{sc}}{p_{sc}} \tag{5.74}$$

At standard conditions, ideal gas behavior is assumed for most gases, and hence, $z_{sc} = 1.0$. Substituting Eqs. (5.73) and (5.74) into Eq. (5.72) gives:

$$\begin{aligned} B_g &= \frac{z_R T_R (14.65)}{p_R (519.67)} \\ &= \frac{0.02819 z_R T_R}{p_R}, \ \frac{\text{ft}^3}{\text{scf}} \\ &= \frac{0.005021 z_R T_R}{p_R}, \ \frac{\text{RB}}{\text{scf}} \end{aligned} \tag{5.75}$$

Note that in Eq. (5.75), T_{sc} = absolute temperature in °R. This is calculated as: T_{sc} = 60°F + T_{abs} = 60 + 459.67 = 519.67 °R. In this book, absolute zero temperature, T_{abs} = 459.67 °R unless otherwise stated.

5.8 Gas Density

Density is mass per unit volume. Gas density is:

$$\rho_g = \frac{m}{V} \tag{5.76}$$

Assuming real gas behavior, gas volume, V, is given by Eq. (5.9) as:

$$V = \frac{znRT}{p} \tag{5.77}$$

Substituting Eq. (5.77) into Eq. (5.76) gives:

$$\rho_g = \frac{mp}{znRT} \tag{5.78}$$

Because the molecular weight of gas, $M_g = m/n$, where, m is the mass of gas, and, n is the number of moles of gas, Eq. (5.78) can be written as:

$$\rho_g = \frac{M_g p}{zRT} \tag{5.79}$$

From Eq. (5.14), gas gravity, $\gamma_g = M_g/28.9586$. Expressing Eq. (5.79) in terms of gas gravity:

$$\rho_g = \frac{28.9586 \gamma_g p}{zRT} \tag{5.80}$$

5.9 Gas Viscosity

Many correlations are available in the literature for estimation of gas viscosity. These include correlations reported by Carr et al.,[16] Jossi et al.,[17] Dean and Stiel,[18] Lohrenz et al.,[19] Lee et al.,[20] Lucas,[21] Londono et al.,[22] and Sutton.[7] The correlations reported by Lee et al[20] and Sutton[7] for calculations of gas viscosities are recommended.

The expression for the Lee et al.[20] correlation is:

$$\mu_g = 1 \times 10^{-4} K \exp\left(X \rho_g^Y\right) \tag{5.81}$$

where $$K = \frac{(9.379 + 0.01607 M_g) T^{1.5}}{209.2 + 19.26 M_g + T} \tag{5.82}$$

and $$X = 3.448 + \left(\frac{986.4}{T}\right) + 0.01009 M_g \tag{5.83}$$

and $$Y = 2.447 - 0.2224X \tag{5.84}$$

Note that in Eqs. (5.81) to (5.84), ρ_g is gas density in g/cm^3; M_g is gas molecular weight in lbm/lb-mole; T is temperature in °R and μ_g is gas viscosity in centipoise (cp).

The Lee et al.[20] correlation should be used for gases having a specific gravity less than 0.77. The ranges of applicability are as follows: $100 \leq p$ (psia) ≤ 8000; $200 \leq T$ (°F) ≤ 340; $0.55 \leq N_2$ (mol %) ≤ 4.8; and $0.90 \leq CO_2$ (mol %) ≤ 3.20. The reported accuracy of the Lee et al.[20] correlation was a standard deviation of ±2.69% and a maximum deviation of 8.99%.

The Sutton[7] correlation should be used for gases having gravity as high as 1.861. It is a hybrid correlation that uses the Lucas[21] method for low-pressure viscosity, and the Lee et al.[20] method for the viscosity ratio. The Sutton[7] correlation is expressed as follows:

$$\mu_g = \mu_{gsc} \exp\left[X \rho_g^Y\right] \quad (5.85)$$

In Eq. (5.85),

$$X = 3.47 + \frac{1588}{T} + 0.0009 M_g \quad (5.86)$$

$$Y = 1.66378 - 0.04679 X \quad (5.87)$$

$$\mu_{gsc}\xi = 10^{-4}\left[\begin{matrix} 0.807 T_{pr}^{0.618} - 0.357 \exp\left(-0.449 T_{pr}\right) + \\ 0.340 \exp\left(-4.058 T_{pr}\right) + 0.018 \end{matrix}\right] \quad (5.88)$$

$$\xi = 0.9490\left(\frac{T_{pc}}{M_g^3 p_{pc}^4}\right)^{1/6} \quad (5.89)$$

In Eqs. (5.85) to (589), ρ_g is gas density in g/cm^3; M_g is gas molecular weight in lbm/lb-mole; T is temperature in °R; T_{pc} is pseudo-critical temperature in °R; T_{pr} is pseudo-reduced temperature, dimensionless; p_{pc} is pseudo-critical pressure in psia; ξ is viscosity normalizing parameter; μ_{gsc} is low-pressure gas viscosity in centipoise (cp); and μ_g is gas viscosity in centipoise (cp). The Sutton[7] correlation applies under the following conditions: $0 \le H_2S$ (mol%) ≤ 1.7; $0 \le CO_2$ (mol %) ≤ 8.9; $0 \le N_2$ (mol %) ≤ 5.2; $0 \le$ Heptanes-plus (mol %) ≤ 24.3; $0.554 \le \gamma_g \le 1.861$; $14.7 \le p$ (psia) ≤ 20305; $-45.7 \le T$°R ≤ 1112; and $0.008 \le \mu_g$ (cp) ≤ 0.435. The average error of the correlation is minus 0.5%, with a standard deviation of 4.2%.

5.10 Gas Coefficient of Isothermal Compressibility

The **coefficient of isothermal compressibility** of gas (or simply **gas compressibility)** is defined as the change in gas volume per change in pressure at constant temperature. Gas compressibility, c_g, is expressed as:

$$c_g = -\frac{1}{V}\left(\frac{\partial V}{\partial p}\right)_T \quad (5.90)$$

For an ideal gas, $V = nRT/p$, and hence:

$$\left(\frac{\partial V}{\partial p}\right)_T = -\frac{nRT}{p^2} \quad (5.91)$$

Substituting Eq. (5.91) into Eq. (5.90) gives:

$$c_g = \frac{1}{p} \quad (5.92)$$

Thus, the units for c_g are psi^{-1}. Eq. (5.92) is useful for an estimate of the value of gas compressibility. But remember that to arrive at this equation, ideal gas behavior was assumed. The true value of gas compressibility will differ substantially depending on the conditions of pressure and temperature of the gas system being evaluated.

For a real gas, $V = \frac{znRT}{p}$, and hence:

$$\left(\frac{\partial V}{\partial p}\right)_T = \frac{nRT}{p}\left(\frac{\partial z}{\partial p}\right)_T - \frac{znRT}{p^2} = \left(\frac{znRT}{p}\right)\frac{1}{z}\left(\frac{\partial z}{\partial p}\right)_T - \left(\frac{znRT}{p}\right)\left(\frac{1}{p}\right) \quad (5.93)$$

Dividing both sides of Eq. (5.93) by $1/V$, we get:

$$\frac{1}{V}\left(\frac{\partial V}{\partial p}\right)_T = \frac{1}{z}\left(\frac{\partial z}{\partial p}\right)_T - \frac{1}{p} \quad (5.94)$$

From Eqs. (5.90) and (5.94), then,

$$c_g = \frac{1}{p} - \frac{1}{z}\left(\frac{\partial z}{\partial p}\right)_T \quad (5.95)$$

The similarity between Eq. (5.92) for an ideal gas and Eq. (5.95) for a real gas is clearly evident. By definition for an ideal gas, $z = 1$, Eq. (5.95) reduces to Eq. (5.92).

The principle of corresponding states can be used to transform Eq. (5.95) into a reduced form. From Eq. (5.18), $p = p_{pr} \times p_{pc}$, and $dp = p_{pc} dp_{pr}$. Substituting in Eq. (5.95) gives:

$$c_g = \frac{1}{p_{pr} p_{pc}} - \frac{1}{z p_{pc}}\left(\frac{\partial z}{\partial p_{pr}}\right)_{T_{pr}} \quad (5.96)$$

Trube[23] introduced the term **pseudo-reduced compressibility,** c_r, which is defined as:

$$c_r = c_g \times p_{pc} \quad (5.97)$$

Multiplying Eq. (5.96) by p_{pc} gives:

$$c_r = \frac{1}{p_{pr}} - \frac{1}{z}\left(\frac{\partial z}{\partial p_{pr}}\right)_{T_{pr}} \quad (5.98)$$

Mattar et al.[24] applied Eq. (5.63) to obtain Figures 5.11 and 5.12, in which the product, $c_r T_{pr}$, is plotted as a function of reduced pseudo-temperature, T_{pr}, and reduced pseudo-pressure, p_{pr}. The use of Figures 5.11 and 5.12 to estimate gas compressibility is demonstrated in Example 5.4.

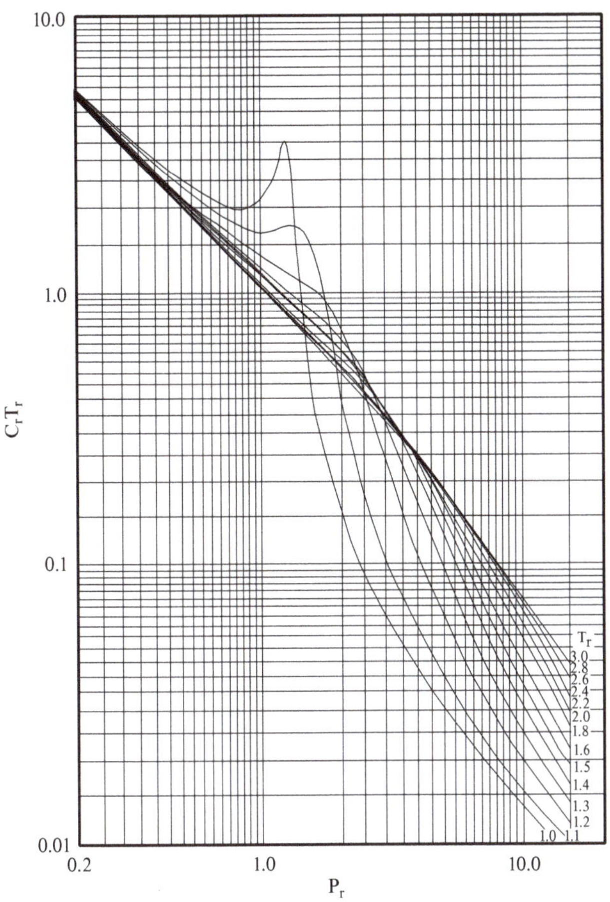

▲ **FIGURE 5.11 Pseudo-reduced compressibility chart for $1.05 \le T_r \le 3.0$ and $0.2 \le p_r \le 15.0$ (from Mattar et al.[24] © 1975 SPE, Reproduced with permission)**

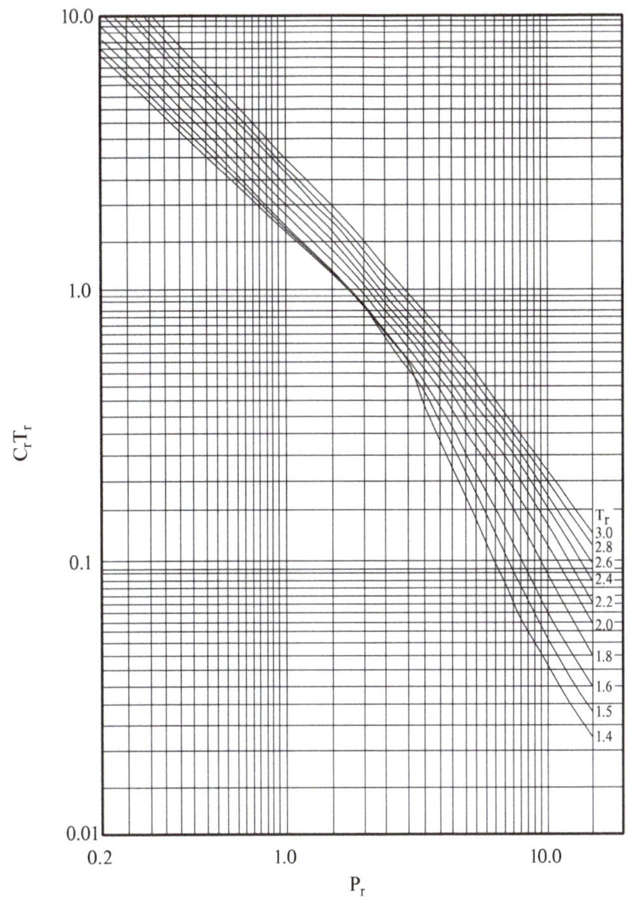

▲ **FIGURE 5.12 Pseudo-reduced compressibility chart for $1.4 \le T_r \le 3.0$ and $0.2 \le p_r \le 15.0$ (from Mattar et al.[24] © 1975 SPE, Reproduced with permission)**

EXAMPLE 5.3 Calculation of Pseudo-critical Properties of a Gas Mixture Using SSBV Mixing Rules

Problem

Calculate the pseudo-critical properties of a gas mixture with composition shown in Table 5.4 using the SSBV mixing rules.

Solution

Step 1: Calculate the boiling point of C_{7+} using Eq. (5.34) from Whitson:[6]

$$T_{b,C7+} = \left(4.5579 \times M_{C7+}^{0.15178} \times \gamma_{C7+}^{0.15427}\right)^3$$

$$= \left(4.5579 \times 193.72^{0.15178} \times 0.816^{0.15427}\right)^3$$

$$= 948.13\ ^\circ\text{R}$$

Step 2: Calculate the critical pressure and temperature of C_{7+} using the Kessler-Lee[5] equations (Eqs. (5.32) and (5.33)):

$$p_{pc,C7+} = \exp\left[\begin{array}{l} 8.3634 - \dfrac{0.0566}{\gamma_{C7+}} - \left(0.24244 + \dfrac{2.2898}{\gamma_{C7+}} + \dfrac{0.11857}{\gamma_{C7+}^2}\right)\dfrac{T_{b,C7+}}{10^3} \\ + \left(1.4685 + \dfrac{3.648}{\gamma_{C7+}} + \dfrac{0.47227}{\gamma_{C7+}^2}\right)\dfrac{T_{b,C7+}^2}{10^7} \\ - \left(0.42019 + \dfrac{1.6977}{\gamma_{C7+}^2}\right)\dfrac{T_{b,C7+}^3}{10^{10}} \end{array}\right]$$

$$= \exp\left[\begin{array}{l} 8.3634 - \dfrac{0.0566}{0.816} - \left(0.24244 + \dfrac{2.2898}{0.816} + \dfrac{0.11857}{0.816^2}\right)\dfrac{948.13}{10^3} \\ + \left(1.4685 + \dfrac{3.648}{0.816} + \dfrac{0.47227}{0.816^2}\right)\dfrac{948.13^2}{10^7} \\ - \left(0.42019 + \dfrac{1.6977}{0.816^2}\right)\dfrac{948.13^3}{10^{10}} \end{array}\right]$$

$$= 264.88 \text{ psia}$$

TABLE 5.4 Properties and Compositions of Gas Mixture for Example 5.3

Component	Mole Fraction	Molecular Weight lbm/ lb-mole	Critical Temperature °R	Critical Pressure psia
N_2	0.0030	28.013	227.16	493.1
CO_2	0.1016	44.010	547.58	1071.0
H_2S	0.0504	34.080	672.12	1300.0
CH_4	0.6632	16.043	343.00	666.4
C_2H_6	0.0387	30.070	549.49	706.5
C_3H_8	0.0283	44.097	665.73	616.0
$i\text{-}C_4H_{10}$	0.0067	58.123	734.13	527.9
$n\text{-}C_4H_{10}$	0.0130	58.123	765.29	550.6
$i\text{-}C_5H_{12}$	0.0060	72.150	828.77	490.4
$n\text{-}C_5H_{12}$	0.0066	72.150	845.47	488.6
C_6H_{14}	0.0096	86.177	913.27	436.9
C_{7+}	0.0729	193.72	—	—
Total	**1.0000**	—	—	—
Molecular weight, C_{7+} =		193.72		
Specific gravity, C_{7+} =		0.816		

$$T_{pc,C7+} = 341.7 + 811\gamma_{C7+} + (0.4244 + 0.1174\gamma_{C7+})T_{b,C7+} + (0.4669 - 3.2623\gamma_{C7+})\frac{10^5}{T_{b,C7+}}$$

$$= 341.7 + 811\times0.816 + (0.4244 + 0.1174\times0.816)948.13 + (0.4669 - 3.2623\times0.816)\frac{10^5}{948.13}$$

$$= 1265.17\ ^\circ R$$

Step 3: Calculate the terms in SBV correlations (Eqs. (5.21) to (5.24)) as shown in Table 5.5:

$$J = \frac{1}{3}\sum_{i=1}^{n_c}\left(\frac{yT_c}{p_c}\right)_i + \frac{2}{3}\left[\sum_{i=1}^{n_c}\left(y\sqrt{\frac{T_c}{p_c}}\right)_i\right]^2$$

$$= \frac{1}{3}[0.8986] + \frac{2}{3}[0.8632]^2 = 0.7963$$

$$K = \sum_{i=1}^{n_c}\left(\frac{yT_c}{\sqrt{p_c}}\right)_i = 20.2428$$

$$T_{pc} = \frac{K^2}{J} = \frac{(20.2428)^2}{0.7963} = 514.59\ ^\circ R$$

$$p_{pc} = \frac{T_{pc}}{J} = \frac{514.59}{0.7963} = 646.23 \text{ psia}$$

Step 4: Apply Sutton's corrections for high concentrations of heptane plus fraction (Eqs. (5.25) to (5.31)):

$$F_j = \frac{1}{3}\left(\frac{yT_c}{p_c}\right)_{C7+} + \frac{2}{3}\left(\frac{y^2T_c}{p_c}\right)_{C7+}$$

$$= \frac{1}{3}(0.3482) + \frac{2}{3}(0.0254) = 0.1330$$

$$\varepsilon_j = 0.6081F_j + 1.1325F_j^2 - 14.004F_jy_{C7+} + 64.434F_jy_{C7+}^2$$

$$= 0.6081(0.1330) + 1.1325(0.1330)^2 - 14.004(0.1330)(0.0729) + 64.434(0.1330)(0.0729)^2$$

$$= 0.0107$$

$$\varepsilon_k = \left(\frac{T_c}{\sqrt{p_c}}\right)_{C7+} \times (0.3129y_{C7+} - 4.8156y_{C7+}^2 + 27.3751y_{C7+}^3)$$

$$= \left(\frac{1265.17}{\sqrt{264.88}}\right) \times \left[0.3129(0.0729) - 4.8156(0.0729)^2 + 27.3751(0.0729)^3\right]$$

$$= 0.6082$$

$$J' = J - \varepsilon_j = 0.7963 - 0.0107 = 0.7856$$

$$K' = K - \varepsilon_k = 20.2428 - 0.6082 = 19.6346$$

$$T_{pc} = \frac{K'^2}{J'} = \frac{(19.6346)^2}{0.7856} = 490.73\ ^\circ R$$

$$p_{pc} = \frac{T_{pc}}{J'} = \frac{490.73}{0.7856} = 624.66 \text{ psia}$$

Step 5: Adjust pseudo-critical properties from step 4 for the presence of hydrogen sulfide and carbon dioxide in the mixture by using the Wichert and Aziz[8] correlation (Eqs. (5.35) to (5.37)):

$$\xi = 120\left[(y_{CO_2} + y_{H_2S})^{0.9} - (y_{CO_2} + y_{H_2S})^{1.6}\right] + 15(y_{H_2S}^{0.5} - y_{H_2S}^4)$$

$$= 120\left[(0.1520)^{0.9} - (0.1520)^{1.6}\right] + 15\left[(0.0504)^{0.5} - (0.0504)^4\right]$$

$$= 19.50$$

$$T_{pc}^* = T_{pc} - \xi = 490.73 - 19.50 = 471.23\ ^\circ R$$

$$p_{pc}^* = \frac{p_{pc}(T_{pc} - \xi)}{T_{pc} + y_{H_2S}(1 - y_{H_2S})\xi}$$

$$= \frac{624.66(471.23)}{490.73 + 0.0504(1 - 0.0504)(19.50)}$$

$$= 598.70 \text{ psia}$$

TABLE 5.5 Calculation of Pseudo-critical Properties of Gas Mixture Using SSBV Mixing Rules in Example 5.3

Component	Mole Fraction y_i	Molecular Weight (lbm/lb mole) M_i	Critical Temperature (°R) T_{ci}	Critical Pressure (psia) p_{ci}	$y_i M_i$	$y_i T_{ci}$	$(T_{ci}/p_{ci})^{0.5}$	$(p_{ci})^{0.5}$	$(y_i T_{ci})/p_{ci}$	$y_i(T_{ci}/p_{ci})^{0.5}$	$y_i T_{ci}/\sqrt{p_{ci}}$
N_2	0.0030	28.013	227.16	493.1	0.0840	0.6815	0.6787	22.2059	0.0014	0.0020	0.0307
CO_2	0.1016	44.010	547.58	1071.0	4.4714	55.6341	0.7150	32.7261	0.0519	0.0726	1.7000
H_2S	0.0504	34.080	672.12	1300.0	1.7176	33.8748	0.7190	36.0555	0.0261	0.0362	0.9395
CH_4	0.6632	16.043	343.00	666.4	10.6397	227.4776	0.7174	25.8147	0.3414	0.4758	8.8119
C_2H_6	0.0387	30.070	549.49	706.5	1.1637	21.2653	0.8819	26.5801	0.0301	0.0341	0.8000
C_3H_8	0.0283	44.097	665.73	616.0	1.2479	18.8402	1.0396	24.8193	0.0306	0.0294	0.7591
$i\text{-}C_4H_{10}$	0.0067	58.123	734.13	527.9	0.3894	4.9187	1.1793	22.9761	0.0093	0.0079	0.2141
$n\text{-}C_4H_{10}$	0.0130	58.123	765.29	550.6	0.7556	9.9488	1.1789	23.4649	0.0181	0.0153	0.4240
$i\text{-}C_5H_{12}$	0.0060	72.150	828.77	490.4	0.4329	4.9726	1.3000	22.1450	0.0101	0.0078	0.2245
$n\text{-}C_5H_{12}$	0.0066	72.150	845.47	488.6	0.4762	5.5801	1.3154	22.1043	0.0114	0.0087	0.2524
C_6H_{14}	0.0096	86.177	913.27	436.9	0.8273	8.7674	1.4458	20.9022	0.0201	0.0139	0.4194
C_{7+}	0.0729	193.72	1265.17	264.88	14.1222	92.2309	2.1855	16.2751	0.3482	0.1593	5.6670
Total	**1.0000**	—	—	—	**36.3281**	—	—	—	**0.8986**	**0.8632**	**20.2428**

End of Example 5.3

EXAMPLE 5.4 Calculation of Gas Compressibility Factor, Gas Formation Volume Factor, Gas Gravity, Gas Density, Gas Viscosity, and Gas Compressibility for the Gas Mixture in Example 5.3

Problem

Calculate gas compressibility factor, gas formation volume factor, gas gravity, gas density, gas viscosity, and gas compressibility for the gas mixture in Example 5.3.

Given:

Reservoir pressure, p = 9620 psia

Reservoir temperature, T = 146 °F

Solution

From Example 5.3, pseudo-critical temperature, $T_{pc} = 471.23\ °R$ and pseudo-critical pressure, $p_{pc} = 598.70$ psia.

From Eq. (5.17), $T_{pr} = \dfrac{T}{T_{pc}} = \dfrac{459.67+146}{471.23} = 1.28$

From Eq. (5.18), $p_{pr} = \dfrac{p}{p_{pc}} = \dfrac{9620}{598.70} = 16.07$

Using Figure 5.9, gas compressibility factor, $z = 1.67$

Gas formation volume factor, B_g, is expressed in Eq. (5.57), as:

$$B_g = \frac{0.005021 z_R T_R}{p_R}$$

$$= \frac{0.005021(1.67)(605.67)}{9620}$$

$$= 0.00053 \text{ RB/scf}$$

From Table 5.5, the molecular weight of the gas mixture, $M_g = 36.328$ lbm/lb-mole. Gas gravity, γ_g, is shown in Eq. (5.14) as

$$\gamma_g = M_g/28.9586$$
$$= 36.328/28.9586$$
$$= 1.254$$

Gas density, ρ_g, is shown in Eq. (5.80) as:

$$\rho_g = 28.9586\gamma_g p/zRT$$
$$= (28.9586\times1.254\times9620)/(1.67\times10.732\times605.67)$$
$$= 32.182 \text{ lbm/ft}^3$$
$$= 0.516 \text{ gm/cm}^3$$

Gas viscosity is calculated using the Sutton[7] correlation because gas gravity is 1.254, which is greater than the limit of 0.77 for the Lee et al.[20] correlation.

From Eqs. (5.85) to (5.89):

$$\xi = 0.9490\left(\frac{T_{pc}}{M_g^{\,3} p_{pc}^4}\right)^{1/6}$$

$$= 0.9490\left[471.44/(36.328)^3(598.74)^4\right]^{1/6}$$

$$= 0.006183$$

$$\mu_{gsc}\xi = 10^{-4}\begin{bmatrix}0.807T_{pr}^{0.618} - 0.357\exp(-0.449T_{pr}) \\ + 0.340\exp(-4.058T_{pr}) + 0.018\end{bmatrix}$$

$$= 10^{-4}\begin{bmatrix}0.807(1.28)^{0.618} - 0.357\exp(-0.449\times1.28) \\ + 0.340\exp(-4.058\times1.28) + 0.018\end{bmatrix}$$

$$= 0.000076$$

$$\mu_{gsc} = 0.000076/0.006183 = 0.01229 \text{ cp}$$

$$X = 3.47 + 1588/T + 0.0009M_g$$

$$= 3.47 + 1588/605.67 + 0.0009\times36.328$$

$$= 6.1246$$

$$Y = 1.66378 - 0.04679X$$

$$= 1.66378 - 0.04679\times6.1246 = 1.3772$$

$$\mu_g = \mu_{gsc}\exp\left[X\rho_g^Y\right]$$

$$= 0.01229\exp\left[6.1246(0.5161)^{1.3772}\right]$$

$$= 0.1443 \text{ cp}$$

The gas compressibility is calculated with the aid of Figure 5.11:

For $p_{pr} = 16.07$ and

$T_{pr} = 1.28$, $c_r T_{pr} = 0.02$ (Fig. 5.11),

$c_r = 0.02/1.28 = 0.01563$

$c_r = c_g \times p_{pc}$ or $c_g = c_r/p_{pc} = 0.01563/598.74$

$= 26\times10^{-6}$ psi^{-1} or 26 microsips

End of Example 5.4

EXAMPLE 5.5 Calculation of Pseudo-critical Properties Using Kay's Mixing Rules

Problem

Calculate the pseudo-critical properties and gas compressibility factor of the gas mixture in Table 5.4 using Kay's mixing rules.

Solution

Step 1: Calculate the boiling point of C_{7+} using Eq. (5.34) from Whitson:[6]

From Example 5.3, $T_{b,C7+} = 948.13$ °R.

Step 2: Calculate the critical pressure and temperature of C_{7+} using the Kessler-Lee[5] equations [Eqs. (5.32) and (5.33)]:

From Example 5.3,

$P_{pc,C7+} = 264.88$ psia and $T_{pc,C7+} = 1265.17$ °R.

Step 3: Calculate pseudo-critical temperature and pressure for mixture.

Using Kay's mixing rules (Eqs. (5.19) and (5.20)):

$p_{pc} = \sum y_i p_{ci} = 702.9$ psia and $T_{pc} = \sum y_i T_{ci}$

$= 484.34$ °R as shown in Table 5.6.

Step 4: Apply corrections for hydrogen sulfide and carbon dioxide.

From Example 5.3, $\xi = 19.50$ °R.

$$T_{pc}^* = T_{pc} - \xi = 484.34 - 19.50 = 464.84 \text{ °R}.$$

$$p_{pc}^* = \frac{p_{pc}(T_{pc} - \xi)}{T_{pc} + y_{H_2S}(1 - y_{H_2S})\xi}$$

$$= \frac{702.9(464.69)}{484.34 + 0.0504(1 - 0.0504)(19.50)}$$

$$= 673.3 \text{ psia}$$

Step 5: Calculate the gas compressibility factor.

Reservoir pressure = 9620 psia Reservoir temperature = 146 °F

$$T_{pr} = \frac{T}{T_{pc}} = \frac{459.67 + 146}{464.84} = 1.30 \text{ and}$$

$$p_{pr} = \frac{p}{p_{pc}} = \frac{9620}{673.3} = 14.29$$

Using Figure 5.9, gas compressibility factor, $z = 1.50$.

The actual measured z factor for this mixture is 1.67. This was the value calculated using SSBV mixing rules in Example 5.4. The error introduced by using Kay's mixing rules is 10%. This example demonstrates that the SSBV method is more accurate than Kay's method. However, Kay's method is relatively simple and can be used for quick estimation.

TABLE 5.6 Calculation of Pseudo-critical Properties of Gas Mixture in Table 5.4 Using Kay's Mixing Rules in Example 5.5

Component	Mole Fraction y_i	Molecular Weight (lbm/lb-mole) M_i	Critical Temperature (°R) T_{ci}	Critical Pressure (psia) p_{ci}	$y_i M_i$	$y_i T_{ci}$	$y_i p_{ci}$
N_2	0.0030	28.013	227.16	493.1	0.084	0.68	1.5
CO_2	0.1016	44.010	547.58	1071.0	4.471	55.63	108.8
H_2S	0.0504	34.080	672.12	1300.0	1.718	33.87	65.5
CH_4	0.6632	16.043	343.00	666.4	10.640	227.48	442.0
C_2H_6	0.0387	30.070	549.49	706.5	1.164	21.27	27.3
C_3H_8	0.0283	44.097	665.73	616.0	1.248	18.84	17.4
i-C_4H_{10}	0.0067	58.123	734.13	527.9	0.389	4.92	3.5
n-C_4H_{10}	0.0130	58.123	765.29	550.6	0.756	9.95	7.2
i-C_5H_{12}	0.0060	72.150	828.77	490.4	0.433	4.97	2.9
n-C_5H_{12}	0.0066	72.150	845.47	488.6	0.476	5.58	3.2
C_6H_{14}	0.0096	86.177	913.27	436.9	0.827	8.77	4.2
C_{7+}	0.0729	193.72	1267.17	264.88	14.122	92.38	19.3
Total	**1.0000**	—	—	—	**36.328**	**484.34**	**702.9**

End of Example 5.5

EXAMPLE 5.6 Calculation of Pseudo-critical Properties and Gas Compressibility Factor of a Gas Mixture if the Composition is Unknown

Problem

The gas gravity of the gas mixture in Example 5.3 was calculated to be 1.254. Assuming that the gas composition is not known, calculate the pseudo-critical properties of the mixture using Sutton's correlation and the gas compressibility factor.

Solution

Step 1: Calculate the hydrocarbon gas composition.

From Eq. (5.39):

$$y_{HC} = 1 - y_{H_2S} - y_{CO_2} - y_{N_2} - y_{H_2O} = 1 - 0.0504 - 0.1016 - 0.003 = 0.8450$$

Step 2: Calculate the hydrocarbon gas gravity.

From Eq. (5.40):

$$\gamma_{g,HC} = \frac{\gamma_g - \left(y_{H_2S}M_{H_2S} + y_{CO_2}M_{CO_2} + y_{N_2}M_{N_2} + y_{H_2O}M_{H_2O}\right)/M_{AIR}}{y_{HC}}$$

$$= \frac{1.254 - \begin{pmatrix}0.0504\times 34.08 + 0.1016 \\ \times 44.01 + 0.003\times 28.013\end{pmatrix}\Big/ 28.963}{0.845}$$

$$= 1.228$$

Step 3: Calculate the hydrocarbon pseudo-critical properties.

Using Eqs. (5.45) and (5.46):

$$p_{pc,HC} = 744 - 125.4\gamma_{g,HC} + 5.9\gamma_{g,HC}^2 = 744 - 125.4\times 1.228 + 5.9(1.228)^2 = 598.91 \text{ psia}$$

$$T_{pc,HC} = 164.3 + 357.7\gamma_{g,HC} - 67.7\gamma_{g,HC}^2 = 164.3 + 357.7\times 1.228 - 67.7(1.228)^2 = 501.47\ ^\circ\text{R}$$

Step 4: Calculate the pseudo-critical properties for the gas mixture.

Using Eqs. (5.41) and (5.42):

$$p_{pc} = y_{HC}p_{pc,HC} + y_{H_2S}p_{c,H_2S} + y_{CO_2}p_{c,CO_2} + y_{N_2}p_{c,N_2} + y_{H_2O}p_{c,H_2O}$$

$$= 0.845\times 598.91 + 0.0504\times 1300 + 0.1016 \times 1071 + 0.003\times 493.1 = 681.89 \text{ psia}$$

$$T_{pc} = y_{HC}T_{pc,HC} + y_{H_2S}T_{c,H_2S} + y_{CO_2}T_{c,CO_2} + y_{N_2}T_{c,N_2} + y_{H_2O}T_{c,H_2O}$$

$$= 0.845\times 501.47 + 0.0504\times 672.12 + 0.1016 \times 547.58 + 0.003\times 227.16 = 513.93\ ^\circ\text{R}$$

Step 5: Apply corrections for hydrogen sulfide and carbon dioxide.

From Example 5.3, $\xi = 19.50\ ^\circ R$.

$$T_{pc}^* = T_{pc} - \xi = 513.93 - 19.50$$
$$= 494.43\ ^\circ R.$$

$$p_{pc}^* = \frac{p_{pc}\left(T_{pc} - \xi\right)}{T_{pc} + y_{H_2S}\left(1 - y_{H_2S}\right)\xi}$$
$$= \frac{681.89(494.43)}{513.93 + 0.0504(1 - 0.0504)(19.50)}$$
$$= 654.83 \text{ psia}$$

Step 6: Calculate the z factor.

Reservoir pressure = 9620 psia Reservoir temperature = 146 °F

$$T_{pr} = \frac{T}{T_{pc}} = \frac{459.67 + 146}{494.43} = 1.22 \text{ and}$$

$$p_{pr} = \frac{p}{p_{pc}} = \frac{9620}{654.83} = 14.69$$

Using Figure 5.10, the gas compressibility factor is calculated to be, $z = 1.56$. The actual measured, z factor, for this mixture is 1.67. The error from using Sutton's correlation for gases with unknown composition is calculated to be 6.6% for this example.

End of Example 5.6

5.11 Correlations for Calculation of Oil PVT Properties

The most reliable source of PVT data for oil mixtures is to conduct laboratory measurements on fluid samples taken from the reservoir. Chapter 6 describes the methods used to obtain reservoir fluid samples and the types of laboratory measurements that are conducted. However, in many circumstances such as new reservoir discoveries, laboratory PVT data may not be available. To calculate early estimates of oil-in-place volumes, oil reserves, oil production rates, recovery factors, etc., the engineer needs empirical correlations to estimate the PVT properties of the reservoir oil. In this section, these empirical correlations are presented. Many empirical correlations for calculation of oil properties exist in the literature.[25,26] Some of the correlations are based on oil samples from specific geographical regions: Africa,[27,28] Gulf of Mexico,[29,30] Middle East,[31,32] North Sea,[33] and South America.[34] According to the analyses conducted by Al-Shammasi[25] and supported by the work of Valko and McCain,[35] correlations based on specific geographical regions are not better than correlations based on a global data set. Consequently, the correlations recommended in this section may have been derived from either a geographical or a global data set but have been deemed most accurate based on published results. However, the user is encouraged to search for other correlations if not satisfied with the estimates calculated from the recommended correlations.

5.11.1 Bubble Point Pressure

Valko and McCain[35] developed correlations for oil bubble point pressures at reservoir conditions. The correlating parameters are solution gas-oil ratio at bubble point, R_{sb}; oil API gravity, γ_{API}; separator gas gravity, γ_{sp}; and reservoir temperature, T_F, in °F.

The Valko and McCain[35] correlations for bubble point pressure are as follows:

$$\ln p_b = 7.475 + 0.713Z + 0.0075Z^2 \tag{5.99}$$

$$Z = \sum_{n=1}^{4} Z_n \tag{5.100}$$

$$Z_1 = -5.48 - 0.0378 \ln R_{sb} + 0.281\left(\ln R_{sb}\right)^2 - 0.0206\left(\ln R_{sb}\right)^3 \tag{5.101}$$

$$Z_2 = 1.27 - 0.0449\gamma_{API} + 4.36\times10^{-4}\left(\gamma_{API}\right)^2 - 4.76\times10^{-6}\left(\gamma_{API}\right)^3 \tag{5.102}$$

$$Z_3 = 4.51 - 10.84\gamma_{gsp} + 8.39\left(\gamma_{gsp}\right)^2 - 2.34\left(\gamma_{gsp}\right)^3 \tag{5.103}$$

$$Z_4 = -0.7835 + 6.23\times10^{-3}T_F - 1.22\times10^{-5}\left(T_F\right)^2 + 1.03\times10^{-8}\left(T_F\right)^3 \tag{5.104}$$

In Eqs. (5.99) to (5.104), p_b = bubble point pressure, psia; R_{sb} = solution gas-oil ratio at bubble point, scf/STB; γ_{API} = stock-tank oil gravity, °API; γ_{gsp} = separator gas gravity (air = 1); and T_F = reservoir temperature, °F.

The average relative error and the absolute relative error reported for the Valko and McCain[35] correlations are 0.0% and 10.9%, respectively. The global data set (1745 records) used to obtain these correlations has the following range:

$$10 \le R_{sb} \le 2216 \text{ scf/STB}$$
$$6.0 \le \gamma_{API} \le 63.7°API$$
$$0.555 \le \gamma_{gsp} \le 1.685 (\text{air} = 1)$$
$$60 \le T_F \le 342°F$$
$$82 \le p_b \le 6700 \text{ psia}$$

A simpler but less accurate correlation developed by Standing[9] can be used to calculate oil bubble point pressures at reservoir conditions. The Standing correlation has wide acceptance in the petroleum industry. It was developed from data based on samples from California reservoirs. The Standing correlations (valid to 325°F) are as follows:

$$p_b = 18.2\left(C_{pb} - 1.4\right) \tag{5.105}$$

and
$$C_{pb} = \left(R_{sb}/\overline{\gamma}_g\right)^{0.83} \times 10^{(0.00091T_F - 0.0125\gamma_{API})} \tag{5.106}$$

In Eqs. (5.105) and (5.106), $\overline{\gamma}_g$ = average gas gravity (separator plus stock-tank vent gas). Average gas gravity can be calculated as shown in Eq. (5.110) or estimated as shown in Eq. (5.111). All other terms have the same units as in the Valko and McCain[35] correlations. The average relative error and the average absolute relative error of the Standing[9] correlation are −2.1% and 12.7%, respectively, when applied to the same global data set used for the Valko and McCain[35] correlations.

The solution gas-oil ratio at bubble point, R_{sb}, in Eqs. (5.101) and (5.106) is the sum of the producing gas-oil ratio from the separators and the stock-tank for a three or a two-stage separation system. Thus, for a two-stage separation system:

$$R_{sb} = R_{sp} + R_{st} \tag{5.107}$$

In most field operations, the separator gas and stock-tank oil production rates are measured. The stock-tank vent gas is rarely measured. As a result, the total solution gas-oil ratio as represented in Eq. (5.107) is underreported. If the stock-tank vent gas is not included in the reported total solution gas-oil ratio, then a simple correlation reported by Valko and McCain[35] can be used to account for the deficiency. The correlation is:

$$R_{sb} = 1.1618R_{sp} \tag{5.108}$$

Equation (5.108) was based on a data set having 881 data points, with an average absolute relative error of 9.9%.

5.11.2 Solution Gas-Oil Ratio (GOR)

At reservoir pressures equal to or less than bubble point pressures, solution gas-oil ratios can be estimated by using Eqs. (5.105) and (5.106). The errors associated with using these equations are the same as reported for bubble point pressure estimation using the same equations.

5.11.3 Oil Formation Volume Factor (FVF)

Oil formation volume factors (FVF) at or less than bubble point pressures can be estimated by using the correlation obtained by Petrosky and Farshad[29] from regression of data:

$$B_o = 1.0113 + 7.2046 \times 10^{-5}\left[R_s^{0.3738}\left(\frac{\left(\overline{\gamma}_g\right)^{0.2914}}{\gamma_o^{0.6265}}\right) + 0.24626T_F^{0.5371}\right]^{3.0936} \tag{5.109}$$

In Eq. (5.109), B_o = FVF, RB/STB; R_s = total solution gas-oil ratio (separator plus stock-tank gas), scf/STB; $\overline{\gamma}_g$ = average gas gravity (separator plus stock tank), air = 1; γ_o = oil specific gravity (water = 1); and T_F = temperature, °F.

The Petrosky and Farshad[29] correlation was regressed from data taken from Gulf of Mexico oil samples. However, the correlation was applied by Al-Shammasi[25] to a global data set and was found to have the lowest average absolute relative error and standard deviation when compared with all other available correlations. Al-Shammasi[25] reported average absolute relative error of 1.728% and standard deviation of 1.92% for the Petrosky and Farshad correlation. The global data set used by Al-Shammasi[25] has 1345 points and the following ranges:

$$6.0 \le R_s \le 3298.6 \text{ scf/STB}$$

$$0.51 \le \gamma_g \le 3.44 \text{ (air = 1)}$$

$$6.0 \le \gamma_{API} \le 63.7°\text{API}$$

$$74 \le T_F \le 341°\text{F}$$

$$1.02 \le B_o \le 2916 \text{ RB/STB}$$

$$31.7 \le p_b \le 7127.0 \text{ psia}$$

The average gas gravity in Eq. (5.109) is calculated as a weighted average as follows:

$$\overline{\gamma}_g = \frac{\gamma_{gsp}R_{sp} + \gamma_{gst}R_{st}}{R_{sp} + R_{st}} \tag{5.110}$$

In Eq. (5.110), $\overline{\gamma}_g$ = average gas gravity; γ_{gsp} = separator gas gravity; γ_{gst} = stock-tank gas gravity; R_{sp} = separator GOR, scf/STB; and R_{st} = stock-tank GOR, scf/STB. In most field operations, stock-tank gas gravity and stock-tank GOR are not measured. In such circumstances, average gas gravity can be estimated by using a correlation suggested by Valko and McCain:[35]

$$\overline{\gamma}_g = 1.066\gamma_{gsp} \tag{5.111}$$

Eq. (5.111) was based on 626 data points, with an average absolute error of 3.8%. Note that Eq. (5.110) or (5.111) is used to estimate average gas gravity where required in this book.

For pressures greater than bubble point pressure, oil FVF is calculated by using:

$$B_o = B_{ob}\exp\left[c_o\left(p_b - p\right)\right] \tag{5.112}$$

Oil FVF at bubble point pressure, B_{ob}, is calculated from Eq. (5.109). The coefficient of oil isothermal compressibility, c_o, is calculated by using the correlation shown in Eq. (5.113).

5.11.4 Coefficient of Isothermal Compressibility of Oil

The coefficient of isothermal compressibility of oil (or simply oil compressibility) at bubble point pressure or greater than bubble point pressure can be calculated by using correlations developed by Spivey et al.:[36]

$$\ln c_{ob} = 2.434 + 0.475Z + 0.048Z^2 \tag{5.113}$$

$$Z = \sum_{n=1}^{6} Z_n \tag{5.114}$$

$$Z_1 = 3.011 - 2.6254\ln\gamma_{API} + 0.497(\ln\gamma_{API})^2 \tag{5.115}$$

$$Z_2 = -0.0835 - 0.259\ln\gamma_{gsp} + 0.382(\ln\gamma_{gsp})^2 \tag{5.116}$$

$$Z_3 = 3.51 - 0.0289\ln p_b - 0.0584(\ln p_b)^2 \tag{5.117}$$

$$Z_4 = 0.327 - 0.608\ln(p/p_b) + 0.0911[\ln(p/p_b)]^2 \tag{5.118}$$

$$Z_5 = -1.918 - 0.642\ln R_{sb} + 0.154(\ln R_{sb})^2 \tag{5.119}$$

$$Z_6 = 2.52 - 2.73\ln T_F + 0.429(\ln T_F)^2 \tag{5.120}$$

In Eqs. (5.113) to (5.120), c_{ob} = oil compressibility in microsips; γ_{API} = stock-tank oil gravity, °API; γ_{gsp} = separator gas gravity (air = 1); p_b = bubble point pressure, psia; p = reservoir pressure, psia; R_{sb} = solution gas-oil ratio at bubble point, scf/STB; and T_F = reservoir temperature, °F.

The average relative error and average absolute relative error of the Spivey et al.[36] correlations are 0.1% and 7.1%, respectively, for a global data set having 2433 data points from 369 reservoir samples. The applicable ranges of the data set are as follows:

$$11.6 \le \gamma_{API} \le 57.7 \text{ °API}$$

$$0.561 \le \gamma_{gsp} \le 1.798 \text{ (air = 1)}$$

$$120.7 \le p_b \le 6658.7 \text{ psia}$$

$$414.7 \le p \le 8014.7 \text{ psia}$$

$$12 \le R_{sb} \le 1808 \text{ scf/STB}$$

$$70.7 \le T_F \le 320 \text{ °F}$$

$$3.6 \le c_o \le 50.3 \text{ microsips}$$

Note that the units of oil compressibility calculated by using Eq. (5.113) are in microsips, or 10^{-6} psi^{-1}.

Al-Marhoun[37] developed a simpler but less accurate correlation for oil compressibility at pressures greater than the bubble point pressure. The Al-Marhoun[37] correlation, which is based on oil samples from the Middle East, is:

$$\ln c_o = -14.1042 + \frac{2.7314}{\gamma_{ob}} - 56.0605\times10^{-6}\frac{(p-p_b)}{\gamma_{ob}^3} - \frac{580.8778}{(T_F+460)} \tag{5.121}$$

$$\gamma_{ob} = \frac{(\gamma_o + 2.18\times10^{-4}R_{sb}\bar{\gamma}_g)}{B_{ob}} \tag{5.122}$$

In Eqs. (5.121) and (5.122), c_o = oil compressibility, psi^{-1}; and $\bar{\gamma}_g$ = average gas gravity, air = 1. All other terms are as previously defined for Eq. (5.113).

Spivey et al.[36] reported average relative error and average absolute error of –3.3% and 12.2%, respectively, when the Al-Marhoun[37] correlation was applied to the same global data set used in their work.

For pressures less than bubble point pressure, McCain et al.[38] developed several correlations for oil compressibility depending on available data. If bubble point pressure and solution gas-oil ratio are available, then oil compressibility can be calculated by using:

$$\ln c_o = -7.573 - 1.450\ln p - 0.383\ln p_b + 1.402\ln T + 0.256\ln\gamma_{API} + 0.449\ln R_{sb} \tag{5.123}$$

If the bubble point pressure is not available, oil compressibility can be calculated by using:

$$\ln c_o = -7.633 - 1.497\ln p + 1.115\ln T + 0.533\ln\gamma_{API} + 0.184\ln R_{sb} \tag{5.124}$$

If bubble point pressure and solution GOR are not available, oil compressibility can be calculated by using:

$$\ln c_o = -7.114 - 1.394\ln p + 0.981\ln T + 0.770\ln\gamma_{API} + 0.446\ln\bar{\gamma}_g \tag{5.125}$$

In Eqs. (5.123) to (5.125), c_o = oil compressibility, psi^{-1}; p = reservoir pressure, psia; p_b = bubble point pressure, psia; T = reservoir temperature, °R; γ_{API} = stock-tank oil gravity, °API; R_{sb} = solution gas-oil ratio at bubble point, scf/STB; and $\bar{\gamma}_g$ = weighted average of separator and stock-tank gas gravity.

The reported maximum absolute deviation for all the equations is 10.2% for pressures greater than 500 psia. The ranges of data (2500 data points) used to generate these correlations are:

$$500 \le p \le 5300 \text{ psia}$$

$$763 \le p_b \le 5300 \text{ psia}$$

$$78 \le T_F \le 330°\text{F}$$

$$0.58 \le \overline{\gamma}_g \le 1.2 \text{ (air = 1)}$$

$$18 \le \gamma_{API} \le 52°\text{API}$$

$$15 \le R_{sb} \le 1947 \text{ scf/STB}$$

$$31 \le c_o \le 6600 \text{ microsips}$$

5.11.5 Oil Viscosity

The calculation of the viscosity of live oil (gas-saturated oil) at bubble point pressure or at pressures less than bubble point pressure is a two-step process. In the first step, the viscosity of the dead oil (no dissolved gas) is calculated by using the correlation by Ng and Egbogah.[39] In the second step, the viscosity of the live oil is calculated by using the correlation by Beggs and Robinson.[40] The Ng-Egbogah[39] dead oil viscosity correlation is:

$$\log_{10}\left[\log_{10}\left(\mu_{od}+1\right)\right] = 1.8653 - 0.025086\gamma_{API} - 0.5644\log_{10}\left(T_F\right) \tag{5.126}$$

For Eq. (5.126), μ_{od} = dead oil viscosity, cp; γ_{API} = stock-tank oil gravity, °API; and T_F = temperature, °F.

The average absolute relative error reported by Ng and Egbogah[39] for the correlation is 6.6% for 394 samples with the following property ranges:

$$59 \le T_R \le 176°\text{F}$$

$$-59 \le T_{pour} \le 59°\text{F}$$

$$5 \le \gamma_{API} \le 58°\text{API}$$

The Beggs and Robinson[40] correlation for gas-saturated oil at or less than bubble point pressures is:

$$\mu_{ob} = a\left(\mu_{od}\right)^b$$

$$\text{where } a = 10.715\left(R_s+100\right)^{-0.515}$$

$$\text{and } b = 5.44\left(R_s+150\right)^{-0.338} \tag{5.127}$$

In Eq. (5.127), μ_{ob} = viscosity of gas-saturated oil, cp; and R_s = solution gas-oil ratio, scf/STB.

The average error reported by Beggs and Robinson[40] for the correlation is −1.83%, with standard deviation of 27.25%. The correlation was developed from a database of 2073 points from 600 samples having the following ranges:

$$20 \le R_s \le 2070 \text{ scf/STB}$$

$$16 \le \gamma_{API} \le 58°\text{API}$$

$$70 \le T_F \le 295°\text{F}$$

$$0 \le p \le 5250 \text{ psia}$$

For pressures greater than the bubble point, Bergman and Sutton[41] developed the following correlation:

$$\mu_o = \mu_{ob} e^{\alpha\left(p-p_b\right)^{\beta}}$$

$$\text{where } \alpha = 6.5698\times10^{-7}\ln\left(\mu_{ob}\right)^2 - 1.48211 \times10^{-5}\ln\left(\mu_{ob}\right) + 2.27877\times10^{-4}$$

$$\text{and } \beta = 2.24623\times10^{-2}\ln\left(\mu_{ob}\right) + 0.873204 \tag{5.128}$$

In Eq. (5.128), μ_o = undersaturated oil viscosity, cp; p = reservoir pressure, psia; and p_b = bubble point pressure, psia.

The Bergman and Sutton[41] correlation is remarkable because of its applicability to severe conditions of high pressures and high temperatures typically encountered in deepwater reservoirs. The database consisted of 1399 oil samples and 10,248 data points having the following ranges:

$$115 \le p \le 25{,}015 \text{ psia}$$

$$14.5 \le p_b \le 11{,}195 \text{ psia}$$

$$32 \le T_F \le 425°\text{F}$$

$$6 \le \gamma_{API} \le 61.2°\text{API}$$

$$0 \le R_s \le 4630.6 \text{ scf/STB}$$

$$0.204 \le \mu_{od} \le 14{,}200 \text{ cp}$$

$$0.063 \le \mu_{ob} \le 14{,}200 \text{ cp}$$

$$0.067 \le \mu_o \le 24{,}180 \text{ cp}$$

The average absolute relative error reported by Berman and Sutton[41] for the correlation is 3.88%, with standard deviation of 4.97%.

EXAMPLE 5.7 Calculation of Bubble Point Pressure (p_b), Oil FVF at Bubble Point (B_{ob}), Oil Compressibility at Bubble Point (c_{ob}), and Oil Viscosity at Bubble Point (μ_{ob})

Problem

Consider an oil sample from a reservoir that has the following properties:

Solution gas-oil ratio at bubble point, R_{sb}	805 scf/STB
Average gas gravity, $\bar{\gamma}_g$	0.929 (air = 1)
Stock-tank oil gravity, γ_{API}	39.1°API
Reservoir temperature, T_R	200°F

Calculate bubble point pressure, oil formation volume factor at bubble point, oil compressibility at bubble point, and oil viscosity at bubble point.

Solution

To calculate bubble point pressure for the oil sample, use the Valko and McCain correlation given in Eqs. (5.99) to (5.104):

$$Z_1 = -5.48 - 0.0378\ln R_{sb} + 0.281(\ln R_{sb})^2 - 0.0206(\ln R_{sb})^3$$
$$= -5.48 - 0.0378\ln 805 + 0.281(\ln 805)^2 - 0.0206(\ln 805)^3$$
$$= 0.6764$$

$$Z_2 = 1.27 - 0.0449\gamma_{API} + 4.36\times10^{-4}(\gamma_{API})^2 - 4.76\times10^{-6}(\gamma_{API})^3$$
$$= 1.27 - 0.0449(39.1) + 4.36\times10^{-4}(39.1)^2 - 4.76\times10^{-6}(39.1)^3 = -0.1036$$

$$Z_3 = 4.51 - 10.84\gamma_{gsp} + 8.39(\gamma_{gsp})^2 - 2.34(\gamma_{gsp})^3$$
$$= 4.51 - 10.84(0.929) + 8.39(0.929)^2 - 2.34(0.929)^3$$
$$= -0.1956$$

$$Z_4 = -0.7835 + 6.23\times10^{-3}T_F - 1.22\times10^{-5}(T_F)^2 + 1.03\times10^{-8}(T_F)^3$$
$$= -0.7835 + 6.23\times10^{-3}(200) - 1.22\times10^{-5}(200)^2 + 1.03\times10^{-8}(200)^3 = 0.0569$$

$$Z = \sum_{n=1}^{n} Z_n = Z_1 + Z_2 + Z_3 + Z_4$$
$$= 0.6764 - 0.1036 - 0.1956 + 0.0569$$
$$= 0.4341$$

$$\ln p_b = 7.475 + 0.7132Z + 0.0075Z^2$$
$$= 7.475 + 0.7132(0.4341) + 0.0075(0.4341)^2$$
$$= 7.7860$$

$$p_b = \exp(7.7860)$$
$$= 2406.7 \text{ psia}$$

For comparison, the bubble point pressure calculations were repeated using the Standing correlation given in Eqs. (5.105) to (5.106):

$$C_{pb} = (R_{sb}/\bar{\gamma}_g)^{0.83} \times 10^{(0.00091T_F - 0.0125\gamma_{API})}$$
$$= (805/0.929)^{0.83} \times 10^{(0.00091(200) - 0.0125(39.1))}$$
$$= 135.3965$$

$$P_b = 18.2(C_{pb} - 1.4) = 18.2(135.3965 - 1.4)$$
$$= 2438.7 \text{ psia}$$

Actual measured bubble point pressure for the oil sample is 2392 psia. The estimation error for the Valko-McCain[35] correlation for this example is 0.61%, and the error for the Standing[9] correlation is 1.95%.

To calculate formation volume factor at bubble point, the correlation by Petrosky and Farshad[29] given in Eq. (5.109) was used:

$$B_o = 1.0113 + 7.2046 \times10^{-5}\left[R_s^{0.3738}\left(\frac{(\bar{\gamma}_g)^{0.2914}}{\gamma_o^{0.6265}}\right) + 0.24626T_F^{0.5371}\right]^{3.0936}$$
$$= 1.0113 + 7.2046 \times10^{-5}\left[(805)^{0.3738}\left(\frac{(0.929)^{0.2914}}{(0.8294)^{0.6265}}\right) + 0.24626(200)^{0.5371}\right]^{3.0936}$$
$$= 1.5304 \text{ RB/STB}$$

Note that oil specific gravity, $\gamma_o = 141.5/(131.5 + \gamma_{API}) = 141.5/(131.5 + 39.1) = 0.8294$.

The actual measured formation volume factor for the oil sample is 1.4790 rb/STB. The error calculated for the Petrosky-Farshad[29] correlation for this example is 3.5%.

The coefficient of isothermal compressibility of oil is calculated by using the Spivey et al.[36] correlations shown in Eqs. (5.113) to (5.120):

$$Z_1 = 3.011 - 2.6254\ln\gamma_{API} + 0.497(\ln\gamma_{API})^2$$
$$= 3.011 - 2.6254\ln 39.1 + 0.497(\ln 39.1)^2$$
$$= 0.0659$$

$$Z_2 = -0.0835 - 0.259\ln\gamma_{gsp} + 0.382(\ln\gamma_{gsp})^2$$
$$= -0.0835 - 0.259\ln 0.929 + 0.382(\ln 0.929)^2$$
$$= -0.0624$$

$$Z_3 = 3.51 - 0.0289 \ln p_b - 0.0584 (\ln p_b)^2$$
$$= 3.51 - 0.0289 \ln 2392 - 0.0584 (\ln 2392)^2$$
$$= -0.2496$$

$$Z_4 = 0.327 - 0.608 \ln(p/p_b) + 0.0911 [\ln(p/p_b)]^2$$
$$= 0.327 - 0.608 \ln(2392/2392) + 0.0911 [\ln(2392/2392)]^2$$
$$= 0.3270$$

$$Z_5 = -1.918 - 0.642 \ln R_{sb} + 0.154 (\ln R_{sb})^2$$
$$= -1.918 - 0.642 \ln 805 + 0.154 (\ln 805)^2$$
$$= 0.6807$$

$$Z_6 = 2.52 - 2.73 \ln T_F + 0.429 (\ln T_F)^2$$
$$= 2.52 - 2.73 \ln 200 + 0.429 (\ln 200)^2$$
$$= 0.0986$$

$$Z = \sum_{n=1}^{6} Z_n = Z_1 + Z_2 + Z_3 + Z_4 + Z_5 + Z_6$$
$$= 0.0659 - 0.0624 - 0.2496 + 0.3270 + 0.6807 + 0.0986$$
$$= 0.8602$$

$$\ln c_{ob} = 2.434 + 0.475Z + 0.048Z^2$$
$$= 2.434 + 0.475(0.8602) + 0.048(0.8602)^2$$
$$= 2.8781$$

$$c_{ob} = \exp(2.8781) = 17.78 \times 10^{-6} \text{ psi}^{-1}$$
$$= 17.78 \text{ microsips}$$

Alternatively, oil compressibility can be calculated by using the Al-Marhoun[37] correlation shown in Eqs. (5.121) and (5.122):

$$\gamma_{ob} = \frac{(\gamma_o + 2.18 \times 10^{-4} R_{sb} \bar{\gamma}_g)}{B_{ob}}$$
$$= \frac{(0.8294 + 2.18 \times 10^{-4} (805 \times 0.929))}{1.479}$$
$$= 0.6710$$

$$\ln c_o = -14.1042 + \frac{2.7314}{\gamma_{ob}}$$
$$-56.0605 \times 10^{-6} \frac{(p - p_b)}{\gamma_{ob}^3} - \frac{580.8778}{(T_F + 460)}$$
$$= -14.1042 + \frac{2.7314}{0.6710} - 56.0605 \times 10^{-6} \frac{(2392 - 2392)}{(0.6710)^3}$$
$$- \frac{580.8778}{(200 + 460)} = -10.9137$$

$$c_{ob} = \exp(-10.9137) = 18.21 \times 10^{-6} \text{ psi}^{-1}$$
$$= 18.21 \text{ microsips}$$

For this example, there is no significant difference between the oil compressibility at bubble point calculated with the two correlations. But note the relative ease of using the Al-Marhoun correlation.

Calculation of oil viscosity for the sample at bubble point pressure is a two-step process. In step 1, dead oil viscosity is calculated by using the Ng-Egbogah[39] correlation. In step 2, oil viscosity at bubble point is calculated by using the Beggs-Robinson[40] correlation.

Step 1: Calculate dead oil viscosity using the Ng-Egbogah correlation.

From Eq. (5.126):

$$\log_{10}[\log_{10}(\mu_{od} + 1)]$$
$$= 1.8653 - 0.025086\gamma_{API} - 0.5644 \log_{10}(T_F)$$
$$= 1.8653 - 0.025086 \times 39.1 - 0.5644 \log_{10}(200)$$
$$= -0.4143$$

$$\mu_{od} = 1.4278 \text{ cp}$$

Step 2: Calculate oil viscosity at bubble point by using the Beggs-Robinson correlation.

From Eq. (5.127):

$$a = 10.715(R_s + 100)^{-0.515}$$
$$= 10.715(805 + 100)^{-0.515} = 0.3216$$
$$b = 5.44(R_s + 150)^{-0.338}$$
$$= 5.44(805 + 150)^{-0.338} = 0.5350$$
$$\mu_{ob} = 0.3216(1.4278)^{0.5350}$$
$$= 0.3891 \text{ cp}$$

An additional exercise is to calculate oil viscosity for the sample at 4000 psia reservoir pressure using the Bergman-Sutton[41] correlation in Eq. (5.128).

$$\alpha = 6.5698 \times 10^{-7} \ln(\mu_{ob})^2 - 1.48211 \times 10^{-5} \ln(\mu_{ob})$$
$$+ 2.27877 \times 10^{-4}$$
$$= 6.5698 \times 10^{-7} \ln(0.3891)^2 - 1.48211 \times 10^{-5} \ln(0.3891)$$
$$+ 2.27877 \times 10^{-4} = 0.00024245$$

$$\beta = 2.24623 \times 10^{-2} \ln(\mu_{ob}) + 0.873204$$
$$= 2.24623 \times 10^{-2} \ln(0.3891) + 0.873204$$
$$= 0.8520$$

Note that $\mu_{ob} = 0.3891$ cp from the Beggs-Robinson correlation, as shown earlier.

$$\mu_o = \mu_{ob} e^{\alpha(p - p_b)^\beta} = 0.3891 \exp^{0.00024245(4000-2392)^{0.8520}}$$
$$= 0.4434 \text{ cp}$$

End of Example 5.7

5.12 Correlations for Calculation of Water PVT Properties

5.12.1 Water Formation Volume Factor (FVF)

The formation volume factor of water, B_w, can be calculated by using a correlation reported by McCain:[26]

$$B_w = (1+\Delta V_{wp})(1+\Delta V_{wT}) \quad (5.129)$$

where
$$\Delta V_{wT} = -1.0001\times10^{-2} + 1.33391\times10^{-4}T + 5.50654\times10^{-7}T^2 \quad (5.130)$$

and
$$\Delta V_{wp} = -1.95301\times10^{-9}pT - 1.72834\times10^{-13}p^2T - 3.58922\times10^{-7}p - 2.25341\times10^{-10}p^2 \quad (5.131)$$

For Eqs. (5.129) to (5.131), p = pressure in psia, and T = temperature in °F. The correlation is valid for temperatures up to 260°F and pressures up to 5000 psia. The reported error for the correlation is within 2%.

5.12.2 Density of Formation Water

The density of formation water at standard conditions is calculated by using the McCain correlation:[26]

$$\rho_w = 62.368 + 0.438603S + 1.60074\times10^{-3}S^2 \quad (5.132)$$

In Eq. (5.132), ρ_w = density in lbm/ft^3, and S = salinity in weight percent. Water density at reservoir temperature and pressure is calculated by dividing the density at standard conditions by the formation volume factor at the same reservoir conditions.

5.12.3 Coefficient of Isothermal Compressibility of Formation Water

At pressures greater than bubble point, water compressibility can be calculated by using a correlation from Osif:[42]

$$\frac{1}{c_w} = 7.033p + 0.5415C - 537T + 403{,}300 \quad (5.133)$$

In Eq. (5.133), c_w = water compressibility, psi^{-1}; p = pressure, psi; C = salinity, mg/L; and T = temperature, °F. Equation (5.133) is valid for temperatures between 200°F and 270°F, pressures of 1000 to 20,000 psi, and salinities up to 200,000 mg/L. The accuracy of this correlation is not reported. Osif[42] reported that water compressibility was independent of dissolved gas.

5.12.4 Viscosity of Formation Water

Water viscosity at reservoir temperature and 1 atm can be calculated by using a correlation from McCain:[26]

$$\mu_{w1} = AT^{-B} \quad (5.134)$$

where
$$A = 109.574 - 8.40564S + 0.313314S^2 + 8.72213\times10^{-3}S^3 \quad (5.135)$$

and
$$B = 1.12166 - 2.63951\times10^{-2}S + 6.79461\times10^{-4}S^2 + 5.47119\times10^{-5}S^3 - 1.55586\times10^{-6}S^4 \quad (5.136)$$

In Eqs. (5.134) to (5.136), μ_{w1} = water viscosity at reservoir temperature and 1 atm, cp; T = temperature, °F; and S = salinity in weight percent. McCain[26] reported that the correlation has an error of 5% when compared with graphical correlations shown by Matthews and Russell[43] for temperatures between 100°F and 200°F and salinities up to 26 weight percent.

Water viscosity at reservoir pressure is calculated by adjusting water viscosity at 1 atm using a correlation from Mathews and Russell:[43]

$$\frac{\mu_w}{\mu_{w1}} = 0.9994 + 4.0295\times10^{-5}p + 3.1062\times10^{-9}p^2 \quad (5.137)$$

McCain[26] reported that Eq. (5.137) had an error of 4% for data in the temperature range of 86°F to 167°F and pressures less than 10,000 psia. The reported error is within 7% for pressures between 10,000 and 15,000 psia.

Nomenclature

B_g	gas formation volume factor, RB/STB
B_o	oil formation volume factor, RB/STB
B_{ob}	oil formation volume factor at bubble point pressure, RB/STB
B_w	water formation volume factor, RB/STB
c_o	isothermal oil compressibility, psi^{-1}
c_{ob}	isothermal oil compressibility at bubble point, psi^{-1}
c_g	isothermal gas compressibility, psi^{-1}
c_r	pseudo-reduced isothermal compressibility
c_w	isothermal water compressibility, psi^{-1}
C	salinity, mg/L
m	mass, lbm or gm
M	molecular weight, lbm/lb-mole
M_i	molecular weight of component *i*, lbm/lb-mole
n	moles of component
p	pressure
p_b	bubble point pressure, psia
p_c	critical pressure
p_i	initial reservoir pressure
p_{pc}	pseudo-critical pressure
p_{pr}	reduced pseudo-pressure, dimensionless
p_r	reduced pressure, dimensionless
P	phases
R	gas constant
R_S	solution gas-oil ratio, scf/STB
R_{sb}	solution gas-oil ratio at bubble point, scf/STB
S	salinity, wt % solids
T	temperature
T_c	critical temperature
T_i	initial reservoir temperature
T_F	reservoir temperature, °F
T_{pc}	pseudo-critical temperature
T_{pr}	reduced pseudo-temperature, dimensionless
T_r	reduced temperature, dimensionless
x_i	mole fraction of component *i* in the liquid phase
y_i	mole fraction of component *i* in the vapor phase
z	gas compressibility factor
Z	correlation parameter in Eq. (5.99) or (5.113)
γ_{API}	stock-tank oil gravity, °API
γ_g	gas gravity, air = 1
$\bar{\gamma}_g$	average gas gravity, air = 1
γ_o	stock-tank oil specific gravity, water = 1
γ_w	wellstream gas gravity, air = 1
μ_g	gas viscosity, cp
μ_o	oil viscosity, cp
μ_{od}	dead-oil viscosity, cp
μ_w	water viscosity, cp
μ_{w1}	water viscosity at 1 atm and reservoir temperature, cp
ρ	density, mass per unit volume
ρ_r	reduced density
ψ	chemical potential

Subscripts

b	bubble point
g	gas
HC	hydrocarbon
o	oil
R	reservoir
sc	standard conditions
sp	separator
st	stock tank
w	water

REFERENCES

1. Gibbs, J.W.: "The Collected Works of J. Willard Gibbs," Yale U. Press, New Haven, CT (1948).
2. Kay, W.B.: "Density of Hydrocarbon Gases and Vapors at High Temperature and Pressure," *Ind. Eng. Chem.* (Sept. 1936) 1014-1019.
3. Sutton, R.P.: "Compressibility Factors for High-Molecular-Weight Reservoir Gases," paper SPE 14265 presented at the 60th Annual Technical Conference and Exhibition, Las Vegas, NV (Sept. 22-25, 1985).
4. Stewart, W.F., Burkhart, S.F., and Voo, D.: "Prediction of Pseudocritical Parameters for Mixtures," paper presented at the AICHE Meeting, Kansas City, MO (May 18, 1959).
5. Kessler, M.G., and Lee, B.I.: "Improve Prediction of Enthalpy of Fractions," *Hyd. Proc.* (March 1976) 153-158.
6. Whitson, C.H.: "Effect of C7+ Properties on Equation-of-State Predictions," *SPEJ* (Dec. 1987) 685-696.
7. Sutton, R.P.: "Fundamental PVT Calculations for Associated and Gas/Condensate Natural Gas Systems," paper SPE 97099 presented at the 2005 SPE Annual Technical Conference and Exhibition, Dallas, TX (Oct. 9-12, 2005).
8. Wichert, E., and Aziz, K.: "Calculate Z's for Sour Gases," *Hyd. Proc.* (May 1972) 119-122.
9. Standing, M.B.: *Volumetric and Phase Behavior of Oil Hydrocarbon Systems*, 9th Printing, Society of Petroleum Engineers of AIME, Dallas, TX (1981).
10. Gold, D.K., McCain, W.D., and Jennings, J.W.: "An Improved Method for the Determination of the Reservoir-Gas Specific Gravity for Retrograde Gases," *JPT* (July 1989) 747-752.
11. Standing, M.B., and Katz, D.L.: "Density of Natural Gases," *Trans. AIME* (1942) 146, 140.
12. Takacs, G.: "Comparisons Made for Computer Z-Factor Calculations," *Oil and Gas J.* (Dec. 20, 1976) 64-66.
13. Dranchuk, P.M., and Abou-Kassem, J.H.: "Calculation of Z Factors for Natural Gases Using Equations of State," *J. Cdn. Pet. Tech.* (July-Sept. 1975) 34-36.
14. Hall, K.R., and Yarborough, L.: "A New Equation of State for Z-Factor Calculations," *Oil and Gas J.* (June 18, 1973) 82-92.
15. Dranchuk, P.M., Purvis, R.A., and Robinson, D.B.: "Computer Calculation of Natural Gas Compressibility Factors Using the Standing and Katz Correlations," Institute of Petroleum Technical Series, No. IP74-008 (1974) 1-13.
16. Carr, N.L., Kobayashi, R., and Burrows, D.B.: "Viscosity of Hydrocarbon Gases Under Pressure," *Trans. AIME* (1954) 201, 264-272.
17. Jossi, J.A., Stiel, L.I., and Thodos, G.: "The Viscosity of Pure Substances in the Dense Gaseous and Liquid Phases," *AICHE Journal* (Mar. 1962) Vol. 8, No. 1, 59-62.
18. Dean, D.E., and Stiel, L.I.: "The Viscosity of Nonpolar Gas Mixtures at Moderate and High Pressures," *AICHE Journal* (May 1965) 526-532.
19. Lohrenz, J., Bray, B.G., and Clark, C.R.: "Calculating Viscosities of Reservoir Fluids from Their Compositions," *JPT* (Oct. 1964) 1171-1176.
20. Lee, A.L, Gonzalez, M.H., and Eakin, B.E.: "The Viscosity of Natural Gas," *Trans. AIME* (1966) Vol. 237, 997-1000.
21. Poling, B.E., Prausnitz, J.M., and O'Connell, J.P.: *The Properties of Gases and Liquids*, 5th Ed., McGraw-Hill (2001), Chapters 5 and 9.
22. Londono, F.E., Archer, R.A., and Blasingame, T.A.: "Correlations for Hydrocarbon-Gas Viscosity and Gas Density-Validation and Correlation of Behavior Using a Large-Scale Database," *SPEREE* (Dec. 2005) 561-572.
23. Trube, A.S. "Compressibility of Natural Gases," *Trans. AIME* (1957), 210, 61.
24. Mattar, L., Brar, G.S., and Aziz, K.: "Compressibility of Natural Gases," *J. Cdn. Pet. Tech.* (Oct-Dec. 1975) 77-80.
25. Al-Shammasi, A.A.: "A Review of Bubblepoint Pressure and Oil Formation Volume Factor Correlations," *SPEREE* (April 2001) 146-160.
26. McCain, W.D. Jr.: "Reservoir-Fluid Property Correlations: State of the Art," *SPERE* (May 1991) 266.
27. Labedi, R.: "Use of Production Data to Estimate Volume Factor Density and Compressibility of Reservoir Fluids," *J. Pet. Sci. Eng.* (1990) 4, 357.
28. Obomanu, D.A., and Okpobori, G.A.: "Correlating the PVT Properties of Nigerian Crudes," *Trans. ASME* (1987) 109, 214.
29. Petrosky, G.E. Jr., and Farshad, F.: "Pressure-Volume-Temperature Correlations for Gulf of Mexico Crude Oils," *SPEREE* (Oct. 1998) 416-420.
30. Sutton, R.P., and Farshad, F.: "Evaluation of Empirically Derived PVT Properties for Gulf of Mexico Crude Oils," *SPERE* (Feb. 1990) 79-86.
31. Al-Marhoun, M.A.: "PVT Correlations for Middle East Crude Oils," *JPT* (May 1988) 650-666.
32. Dokla, M., and Osman, M.: "Correlation of PVT Properties for UAE Crudes," *SPEFE* (March 1992) 41.
33. Glaso, O.: "Generalized Pressure-Volume-Temperature Correlations," *JPT* (May 1980) 785.
34. Farshad, F., LeBlanc, J.L., Garber, J.D., and Osorio, J.G.: "Empirical PVT Correlations for Colombian Crude Oils," paper SPE 36105 presented at the 1996 SPE Latin American and Caribbean Petroleum Engineering Conference, Port of Spain, Trinidad, and Tobago (April 23-26, 1996).
35. Valko, P.P., and McCain, W.D. Jr.: "Reservoir oil bubblepoint pressures revisited; solution gas-oil ratios and surface gas specific gravities," *J. Pet. Sci. Eng.* 37 (2003) 153-169.
36. Spivey, J.P., Valko, P.P., and McCain, W.D. Jr.: "Applications of the Coefficient of Isothermal Compressibility to Various Reservoir Situations with New Correlations for Each Situation," paper SPE 96415 presented at the 2005 SPE Annual Technical Conference and Exhibition, Dallas, TX (Oct. 9-12, 2005).
37. Al-Marhoun, M.A.: "The Coefficient of Isothermal Compressibility of Black Oils," paper SPE 81432 presented at the SPE 13th Middle East Oil Show & Conference, Bahrain (June 9-12, 2003).

38. McCain, W.D. Jr., Rollins, J.B., and Villena Lanzi, A.J.: "The Coefficient of Isothermal Compressibility of Black Oils at Pressures Below the Bubblepoint," *SPEFE* (Sept. 1988) 659-662.

39. Ng, J.T.H., and Egbogah, E.O.: "An Improved Temperature-Viscosity Correlation for Crude Oil Systems," paper CIM 83-34-32 presented at the 1983 Annual Technical Meeting of the Petroleum Soc. Of CIM, Banff, Alta (May 10-13, 1983).

40. Beggs, H.D., and Robinson, J.R.: "Estimating the Viscosity of Crude Oil Systems," *JPT* (Sept. 1975) 1140-1141.

41. Bergman, D.F., and Sutton, R.P.: "Undersaturated Oil Viscosity Correlation for Adverse Conditions," paper SPE 103144 presented at the 2006 SPE Annual Technical Conference and Exhibition, San Antonio, TX (Sept. 24-27, 2006).

42. Osif, T.L.: "The Effects of Salt, Gas, Temperature, and Pressure on the Compressibility of Water," *SPERE* (Feb. 1988) 175-181.

43. Matthews, C.S., and Russell, D.G.: *Pressure Buildup and Flow Tests in Wells*, Monograph Series, SPE, Richardson, TX (1967) 1.

ADDITIONAL REFERENCES

- Bergman, D.F., and Sutton, R.P.: "An Update to Viscosity Correlations for Gas-Saturated Crude Oils," paper SPE 110195 presented at the SPE Annual Technical Conference and Exhibition, Anaheim, CA (Nov. 11-14, 2007).
- Bergman, D.F., and Sutton, R.P.: "A Consistent and Accurate Dead-Oil-Viscosity Method," *SPEREE* (December 2009) 815-840.
- Borges, P.R.: "Correction Improves Z-Factor Values for High Gas Density," *Oil and Gas J.* (Mar. 4, 1991) 55.
- Eilerts, C.K.: *Phase Relations of Gas-Condensate Fluids, Vol. II*, Monograph 10, Bureau of Mines, American Gas Association (1959) 764-770.
- Elsharkawy, A.M., and Elkamel, A.: "Compressibility Factor for Sour Gas Reservoirs," paper SPE 64284 presented at the 2000 SPE Asia Pacific Oil & Gas Conference, Brisbane, Australia (Oct. 16-18, 2000).
- Elsharkawy, A.M., Hashem, Y.Kh., and Alikhan, A.A.: "Compressibility Factor for Gas Condensates," paper SPE 59702 presented at the SPE 2000 Permian Basin Oil and Gas Conference, Midland, TX (Mar. 21-23, 2000).
- Elsharkawy, A.M.: "Efficient Methods for Calculations of Compressibility, Density, and Viscosity of Natural Gases," *Fluid Phase Equilibria 218* (2004) 1-13.
- *Engineering Data Book*, 12th Ed., Gas Processors Suppliers Assn., Tulsa (2004).
- Kay, W.B.: "The Ethane-Heptane System," *Ind. & Eng. Chem.* (1938) 30, 459.
- Lee, J., and Wattenbarger, R.A.: *Gas Reservoir Engineering*, Soc. of Pet. Engrs., Richardson, TX (1996).
- McCain, W.D. Jr.: *The Properties of Petroleum Fluids*, 2nd Ed., PennWell Publishing Co., Tulsa, OK (1990).
- Piper, L.D., McCain, W.D., and Corredor, J.H.: "Compressibility Factors for Naturally Occurring Petroleum Gases," paper SPE 26668 presented at the 68th Annual Technical Conference and Exhibition, Houston, TX (Oct. 3-6, 1993).
- Whitson, C., and Brule, M.: *Phase Behavior, Monograph Vol. 20*, Soc. Of Pet. Engrs., Richardson, TX (2000).
- Yarborough, L., and Hall, K.R.: "How to Solve Equation of State for Z-Factors," Oil and Gas J. (Feb. 18, 1974) 86-88.

PROBLEMS

5-1. The temperature and pressure of a dry gas reservoir were reported as 220°F and 3275 psia, respectively. Calculate the gas compressibility factor of a sample from the reservoir if its gas gravity was measured to be 0.685. Assume the gas sample contained no impurities.

5-2. A gas sample taken from a gas reservoir has a gravity of 0.6955. The gas sample contains impurities of 9.5 mole % carbon dioxide, 4.5 mole % hydrogen sulfide, and 1.0 mole % nitrogen. Calculate the gas compressibility factor at reservoir temperature and pressure of 180°F and 2200 psia, respectively.

5-3. Calculate the gas compressibility factor of the gas sample in Problem 5-2 by using the total gas gravity of 0.6955 directly in the calculation of its critical properties. The gas sample has the same impurities as in Problem 5-2. What is the percentage error of the compressibility factor calculated with this method if the measured compressibility factor of the sample is 0.893? How does the error compare to the compressibility factor calculated in Problem 5-2 ?

5-4. Calculate the gas deviation factor, gas formation volume factor, gas density, gas viscosity, and gas compressibility for a natural gas sample with the following composition and reservoir properties:

Specific gravity	0.65
Carbon dioxide	0.10 mole %
Nitrogen	2.00 mole %
Hydrogen sulfide	1.50 mole %
Reservoir temperature	215°F
Reservoir pressure	3200 psia

5-5. Calculate the gas gravity, gas compressibility factor, gas density, and gas formation volume factor for a natural gas mixture with the following composition at 2400 psia and 200°F.

Component	Mole Percent
Carbon dioxide	0.97
Nitrogen	0.14
Methane	83.49
Ethane	9.29
Propane	4.10
i-Butane	0.46
n-Butane	0.94
i-Pentane	0.20
n-Pentane	0.17
Hexane	0.24

5-6. Calculate gas compressibility factor and gas formation volume factor of a gas mixture at reservoir pressure and temperature of 8820 psia and 175°F, respectively, using:

a. Kay's mixing rules

b. SSBV's mixing rules

The composition of the gas mixture is given as:

Component	Mole Percent
Carbon dioxide	11.34
Nitrogen	0.40
Hydrogen sulfide	4.02
Methane	64.50
Ethane	5.69
Propane	2.52
i-Butane	0.64
n-Butane	1.37
i-Pentane	0.59
n-Pentane	0.67
Hexane	0.90
Heptane plus	7.36

Molecular weight, Heptane plus	195.58
Specific gravity, Heptane plus	0.824

5-7. A condensate reservoir located in the Gulf of Mexico, USA, has a two-stage separation system consisting of a primary separator and a stock tank. The reservoir produces 1844 barrels of condensate per day with operating conditions and other production data given as:

	Primary Separator	Stock Tank
Pressure (psia)	1130	15.025
Temperature (°F)	87	60
Gas rate (Mcf/D)	48,505	680
GOR (scf/STB)	26,300	368.7
Gravity	0.655 (Air = 1)	42°API

Calculate the fluid gas gravity and fluid production rate from the reservoir.

5-8. A PVT laboratory measurement of a downhole sample from an undersaturated oil reservoir reported its properties as:

Solution gas-oil ratio at bubble point	972 scf/STB
Average gas gravity (Air = 1)	0.953
Stock tank oil gravity	34.9°API
Reservoir temperature	220°F

Calculate the bubble point pressure of the sample using:

a. The Standing correlation

b. The Valko-McCain correlation

c. The measured bubble point pressure of the sample at 220°F was reported to be 3037.7 psia. What is the error in the calculated bubble point pressure by using the Standing correlation compared to the Valko-McCain correlation?

5-9. For the downhole sample in Problem 5-8, calculate the following oil properties at measured bubble point pressure of 3037.7 psia.

a. Formation volume factor

b. Oil compressibility

c. Oil viscosity

d. The measured oil properties for the sample at bubble point pressure were reported as:

Formation volume factor = 1.614 RB/STB

Oil compressibility = 17.06×10^{-6} psia^{-1}

Oil viscosity = 0.400 cp

What are the percentage errors for the calculations in a, b, and c?

5-10. For the downhole sample in Problem 5-8, calculate the following oil properties at reservoir pressure of 3322.7 psia.

a. Oil formation volume factor

b. Oil compressibility

c. Oil viscosity

d. The measured oil properties for the sample at reservoir pressure of 3322.7 psia were reported as:

Formation volume factor = 1.606 RB/STB

Oil compressibility = 15.805×10^{-6} psia^{-1}

Oil viscosity = 0.407 cp

What are the percentage errors for the calculations in a, b, and c?

5-11. The undersaturated reservoir in Problem 5-8 was produced by mechanism of solution gas drive to a reservoir pressure of 2414.7 psia. Calculate the following properties for the oil remaining in the reservoir at this pressure.

a. Oil solution gas-oil ratio

b. Oil formation volume factor

c. Oil viscosity

d. Oil compressibility

e. The measured oil properties for the sample at reservoir pressure of 2414.7 psia were reported as:

Solution gas-oil ratio = 769 scf/STB

Oil formation volume factor = 1.515 RB/STB

Oil viscosity = 0.455 cp

What are the percentage errors for the calculations in a, b, and c?

5-12. The salinity of formation water from a deepwater reservoir was measured as 35,500 ppm. Calculate the following properties of the formation water at reservoir pressure and temperature of 4800 psia and 160°F, respectively.

a. Water formation volume factor

b. Formation water density

c. Formation water isothermal compressibility

d. Formation water viscosity

CHAPTER

6 Reservoir Fluid Sampling and PVT Laboratory Measurements

6.1 Overview of Reservoir Fluid Sampling

One of the important activities in reservoir management is to ensure that representative reservoir fluid samples are obtained by sampling and proper laboratory measurements are conducted on the samples. It is also equally important that reservoir fluid samples should be stored in containers and maintained under conditions that retain the composition of the original sample over time. Laboratory measurements on reservoir fluid samples that are not representative of original or current state of reservoir fluids at the time of sampling could yield erroneous data which could adversely impact all engineering calculations on the reservoir. Considerable care and attention should be spent in planning the sampling program, quality checking the samples at the site and laboratory, and monitoring the actual laboratory process of obtaining data from the samples. This chapter is devoted to the entire process of fluid sampling and pressure-volume-temperature (PVT) data measurements on fluid samples to emphasize their importance in the overall process of reservoir management. A summary of guidelines on fluid sampling methods and PVT laboratory measurements for specific reservoir condition is provided in Table 6.1.

The primary objective of sampling fluids from a reservoir is to obtain a sample that is representative of fluids existing in the reservoir at the time the samples were taken. Reservoir fluid samples are used in one or combinations of the following important functions:

1. PVT studies under reservoir and surface operating conditions.
2. Reservoir fluid characterization or modeling.
3. Economic evaluations (studies of in-place volumes, reserves, etc.).
4. Geochemical analyses to determine fluid sources.
5. Areal and vertical stratification or compartmentalization issues (or reservoir continuity).
6. Evaluation of reservoir performance or management strategies.
7. Special core analyses or core flooding.
8. Design of gathering and separation facilities.
9. Flow assurance studies (flow potential, precipitation of wax, asphaltenes, hydrates etc.).
10. Processing plant design.
11. Early identification of corrosive and dangerous fluid components such as hydrogen sulfide, and/or carbon dioxide. These will also influence the choice of materials used for well tubulars (casing, tubing, etc.), surface equipment (pipelines, separators, tanks, etc.), and processing plants.
12. Crude assays for refining plants.

Reservoir fluid sampling methods are generally divided into two categories: subsurface sampling and surface sampling. Subsurface sampling (sometimes called bottomhole sampling) describes the collection of fluid samples down-hole from cased or open-hole wellbores. In surface sampling, fluid samples are collected at various surface locations such as wellhead, pipelines, separators, stock tank, etc. The method selected for collection of reservoir fluid samples is determined mainly by the type of reservoir. In Chapter 5, five reservoir types were identified based on their initial conditions of temperature and pressure relative to their phase diagrams. The five reservoir types were identified as: dry gas, wet gas, gas-condensate, volatile oil, and black oil reservoirs. In general, dry or wet gas reservoirs can be sampled with subsurface or surface methods. Gas-condensate and volatile oil reservoirs are typically sampled with surface methods. Black oil reservoirs are sampled with subsurface and surface methods depending on the state of the reservoir. Note that the sampling methods prescribed at this point for these reservoir types are based on gross generalizations. There are numerous other factors that influence the selection of sampling method for any reservoir, such as potential uses of the samples, the production characteristics of the wells, the type of mechanical equipment in the wellbore, the cost of preparing the well for sampling, mechanical condition of surface separators, availability of sampling equipment, and safety issues. All these factors, in addition to the state of the reservoir, influence the selection of the sampling method for any reservoir.

The entire process of collecting reservoir fluid samples either by surface or subsurface methods is beset with opportunities for errors at every stage. The sources of these potential errors which can affect the quality of the samples

TABLE 6.1 | Guidelines on Fluid Sampling Methods and PVT Laboratory Measurements

Reservoir			Sampling Method				PVT LAB	
Fluid Type	Fluid State	Life Cycle	Surface: Separator	Subsurface: Bottomhole	Subsurface: WFT	QA/QC Issues	Test Types	QA/QC Issues
Dry/Wet Gas	Single Phase	Expl/App			X	Ensure monophasic samples	Composition, CCE, Viscosity, Separator tests	Verify opening pressures Check lab procedures/protocol Check data for consistency
		Prod	X			Ensure monophasic samples Avoid condensation in tubing, flowlines etc.		
Gas Condensate	Undersaturated	Expl/App	X	X	X	Ensure monophasic samples Minimize well drawdown Check liquids drop-out in tubing, flowlines, and separators Check separator condition, calibration, and efficiency	Composition, Sat Press., CCE, CVD, Separator tests	Verify opening pressures Check lab procedures/protocol Check data for consistency
		Prod	X	X				
	Saturated	Expl/App	X		X	Minimize well drawdown Check liquids drop-out in tubing, flowlines, and separators Check separator condition, calibration, and efficiency Well conditioning at stable rates crucial		
		Prod	X					
Volatile Oil	Undersaturated	Expl/App	X	X	X	Minimize well drawdown Check separator condition, calibration, and efficiency Well conditioning at stable low rates crucial	Composition, Sat. Press., CCE, DL, Viscosity, Separator tests	Verify opening pressures Check lab procedures/protocol Check data for consistency
		Prod	X	X				
	Saturated	Expl/App	X		X			
		Prod	X					
Black Oil	Undersaturated	Expl/App	X	X	X	Minimize well drawdown Check separator condition, calibration, and efficiency Well conditioning at stable rates crucial	Composition, Sat. Press., CCE, DL, Viscosity, Separator tests	Verify opening pressures Check lab procedures/protocol Check data for consistency
		Prod	X	X				
	Saturated	Expl/App	X		X			
		Prod	X					

Expl/App = Exploration/Appraisal
Prod = Production
WFT = Wireline Formation Tester

are illustrated in Figure 6.1. Common errors associated with subsurface sampling are:

1. Contamination from drilling fluids, mud filtrates, and completion fluids.
2. Near wellbore phase separation due to excessive drawdown.
3. Commingled reservoir fluids.
4. Two-phase flow in the wellbore.
5. Intermittent flow or "heading" at low flow rates.
6. Loss of reactive components, such as sulfur compounds, to downhole tubular equipment.
7. Separation of phases in static well fluid columns.
8. Transfer and transportation of samples.
9. Errors in recording of subsurface conditions.

Common errors that are encountered in surface sampling are:

1. Fluid stream not equilibrated in the separator due to improper separator size, high throughput rates, insufficient residence time, improper operation, or poor mechanical condition.
2. Entrainment of liquid in the separator exit gas stream.
3. Gas "carry-under" in the separator exit liquid stream.
4. Emulsions in the separator exit liquid stream.
5. Poor sampling practices at the separator.
6. Transfer and transportation of samples.
7. Recording of data at the separator.

The most effective means of eliminating these common errors present in subsurface and surface sampling methods is to devise a comprehensive and detailed sampling program with

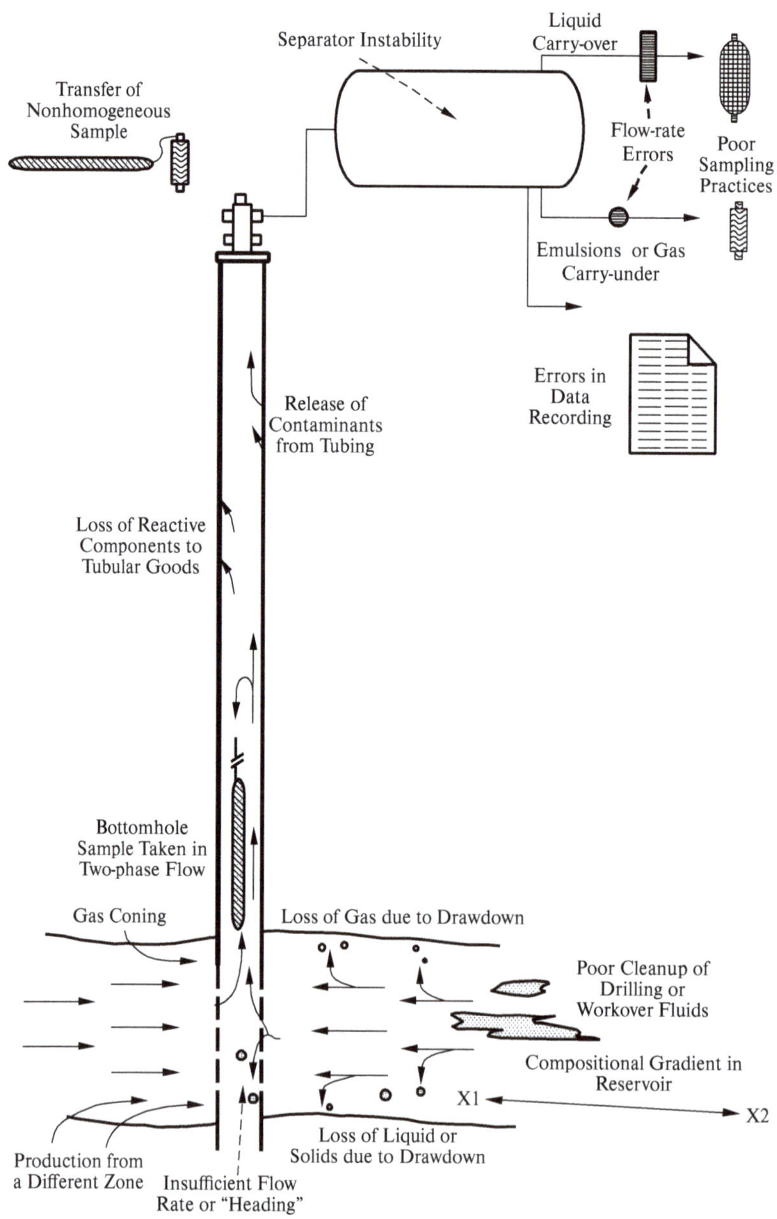

▲ **FIGURE 6.1** **A schematic view of wellsite sampling and measurement errors (from Moffatt et al.[1] © 1998 SPE, Reproduced with permission)**

specific procedures at each major step, identification of well-trained personnel responsible for monitoring activities at the major steps, and verifiable means of quality checking data and samples collected as prescribed in the sampling program.

The probability of capturing a representative reservoir fluid sample is largely influenced by the type and state of the candidate reservoir, the procedures used in well conditioning, sampling methods, and type of sampling tools used. The roles played by these factors are discussed in the following sections.

6.2 Reservoir Type and State

The five reservoir types identified in Chapter 5 are dry gas, wet gas, gas-condensate, volatile oil, and black oil reservoirs. The states of gas-condensate, volatile, and black oil reservoirs are further classified as undersaturated or saturated reservoirs on the basis of initial or current reservoir pressure and temperature. For undersaturated reservoirs, the initial or current reservoir pressure and temperature are above the saturation conditions. In saturated reservoirs, initial or current reservoir pressure and temperature are equal to or less than saturation conditions. It is very important to emphasize that, irrespective of the type and state of a reservoir, the best opportunity to collect the most representative reservoir fluid sample exists early in the life of the reservoir before substantial production has occurred.

6.2.1 Undersaturated Oil Reservoirs

In undersaturated oil reservoirs, initial or current pressures are higher than bubble point pressures of the fluids at reservoir temperature. If the oil reservoir is highly undersaturated, then reservoir pressures are considerably higher than bubble point pressures. For fluids to flow toward a well in a reservoir, a pressure difference must be maintained between the wellbore and the drainage boundary of the well. This pressure difference is called the **drawdown**. High flow rates can be achieved at high drawdowns. Conversely, low flow rates can be maintained at low drawdowns. This concept is utilized in the process of conditioning the well for sampling (termed "well conditioning") and is generally applicable to all reservoir types.

High oil flow rates at high drawdown may cause flowing pressures within the drainage area of the well to fall below bubble point pressure. If this happens, gas bubbles will form within the oil phase in the drainage area. The formation of this gas phase within the drainage area of the well could distort the true gas-oil ratio of the reservoir fluid either by promoting the flow of gas or limiting the flow of oil. In either case, the two-phase system created within the drainage area of the well must be removed before sampling can take place. This is done through well conditioning by systematically reducing the flow rate of the well while monitoring the producing gas-oil ratio. The well is ready for sampling when the producing gas-oil ratio is stable (not changing) after an extended period of flow at approximately constant rate. If there are no other adverse factors such as mechanical condition of the wellbore, prohibitive sampling costs, safety issues, etc., the most representative fluid samples of this type of reservoirs are captured with subsurface samplers.

6.2.2 Undersaturated Gas Condensate Reservoirs

In undersaturated gas condensate reservoirs, initial or current pressures are higher than dew point pressures of the fluids at reservoir temperature. As illustrated in Figure 5.8, a liquid phase will form in the reservoir when reservoir pressure falls below the dew point pressure. This generally will occur if the drawdown applied to the well causes the flowing bottomhole pressure to fall below the dew point pressure. This liquid phase in the near wellbore region must be removed or stabilized by well conditioning prior to sampling. To obtain a sample from an undersaturated gas -condensate reservoir that is representative of the original reservoir fluid, it is important that the following conditions are observed:

1. Reservoir pressure within the drainage area of the well must be above dew point.
2. Samples should be collected as early as possible before substantial production has occurred.
3. Well conditioning should be conducted as long as possible until stable rates (gas and condensate) are achieved at the separator.

The following procedure for sampling gas-condensate reservoirs proposed by McCain and Alexander[2] is recommended:

1. Produce at lowest initial rates to remove liquids from the production string.
2. Maintain that rate reasonably constant until well cleans up.
3. Stabilize separator gas and condensate rates.
4. Then sample.

McCain and Alexander[2] recommended that condensate reservoirs with high permeability which had been produced at high rates should be stabilized over a period of days before sampling to ensure that the spike in heptanes-plus fraction has subsided.

Sampling of gas-condensate reservoirs is typically conducted on the surface at the primary separator. Gas and liquid samples obtained from the separator are recombined at the producing gas-liquid ratio to obtain the composition of a representative reservoir fluid sample as shown in Example 5.1 in Chapter 5.

6.2.3 Saturated Oil Reservoirs

Reservoir pressures in saturated oil reservoirs are below saturation or bubble point pressures at reservoir temperature. In saturated oil reservoirs, a gas cap exists in equilibrium with an oil zone. The pressure at the gas-oil contact is the saturation pressure of the system. It is questionable whether a true representative sample of the original reservoir fluid could be collected from saturated oil reservoirs because of the relative mobilities of gas and oil, the potential effects of gas cusping in the reservoir, and additional evolution of solution gas due to drawdown. Many procedures,[3-6] which are heavily dependent on well conditioning, have been devised for sampling saturated oil reservoirs. These procedures, when properly administered, eliminate gas cusping, replace altered near-wellbore fluids with virgin reservoir fluids, and ensure that oil and gas are flowing at rates that correspond to the gas-oil ratio of the virgin reservoir fluid. The main features of most well conditioning procedures for saturated oil reservoirs are:

1. Reduce well flow rate in a stepwise fashion.
2. Stabilize well at each flow rate and measure producing gas-oil ratio until relatively constant.
3. Continue rate reduction until gas-oil ratio no longer changes with rate reduction.
4. Then sample.

It is considered best practice to sample saturated oil reservoirs at the primary separator. If subsurface sampling is to be conducted, the properly conditioned well should be produced at very low rates during the sampling process.

6.2.4 Saturated Gas Condensate Reservoirs

In saturated gas condensate reservoirs, reservoir pressures are or have declined below dew point pressures to form a mobile or immobile liquid phase. As explained in Chapter 5, retrograde behavior may occur as reservoir pressures continue to decline. The formation of a liquid phase in a gas condensate reservoir continuously alters the composition of fluids remaining in the reservoir because the condensed liquid is composed mostly of heavier hydrocarbons. Since the liquid phase is usually less mobile than the gas phase, the fluids remaining in the reservoir will be richer with heavier hydrocarbons. Consequently, it is doubtful that well conditioning will ever produce a sample that is representative of the original reservoir fluid. Sampling of saturated gas condensate reservoirs is not recommended if the objective of the sampling program is to capture a sample of the original reservoir fluid. However, a sampling program may be devised for other reservoir management purposes, such as monitoring the progress of reservoir depletion.

6.3 Well Conditioning

Well conditioning is the process of preparing a well for sampling by removing near-wellbore fluids which have been altered from virgin reservoir fluids by excessive drawdowns, or by invasion from drilling and remedial fluids. Well conditioning procedures usually (not always) involve stepwise reduction of well flow rates accompanied with measurements of pressures, temperatures, and production rates after stabilization at each step. The well is considered to be properly conditioned when producing gas-oil ratios as measured at appropriate surface locations do not change appreciably with rate change. Generalized well conditioning procedures have been recommended earlier for undersaturated and saturated reservoirs. These generalized procedures must be adapted and possibly revised depending on the specific conditions of the reservoir and the well to be sampled. Detailed well conditioning procedures are also available in *API Recommended Practice 44: Sampling Petroleum Reservoir Fluids.*[3]

Getting a well to be properly conditioned takes time and patience. Conditioning wells in low permeability reservoirs may take several weeks. In some cases, a combination of factors such as well location, lost revenues, and rig cost may prevent adequate conditioning to be achieved on a well. These adverse conditions should be taken into consideration in the preparation of the sampling program with adjustments made to obtain the best possible samples given the prevailing conditions.

The most important aspect of well conditioning is collection of data on the state of the reservoir. The data to be collected should include bottomhole pressures and temperatures, wellhead pressures and temperatures, separator pressures and temperatures, separator gas and liquid flow rates, and stock tank oil, gas and water production rates at each step change. These data are necessary to assess the quality of the conditioning process and hence the quality of the samples that were collected. It is recommended that the data collected during well conditioning should be reviewed thoroughly to ensure that stabilized rates were achieved, and actual sampling was initiated at an appropriate point in the well conditioning process. Specially prepared tables for data collection are normally provided by the service company. Other formats of tables that could be used for data collection during well conditioning and sampling are provided in *API Recommended Practice 44: Sampling Petroleum Reservoir Fluids.*[3]

6.4 Subsurface Sampling Methods and Tools

The subsurface sampling tools that are discussed in this section are:

1. Conventional bottomhole samplers
2. Pistonned bottomhole samplers
3. Single phase samplers
4. Exothermic samplers

6.4.1 Conventional Bottomhole Samplers

Conventional bottomhole samplers are chambers with valves that are used to capture a sample of fluids, typically in cased holes at bottomhole conditions. They are usually run on a wireline. These samplers use either the flow-through or evacuated chamber technique to capture samples. For the flow-through technique, the tool is lowered into the well with the valves open. As the tool descends downhole, the sample chamber is thoroughly flushed with wellbore fluids. At the selected sampling depth, the valves are triggered to close by timing, or mechanical or electrical mechanisms, depending on the type of tool. For the evacuated chamber technique, a vacuumed chamber is positioned on a wireline at the desired sampling depth. The valves are opened, and the chamber is filled with fluids present in the wellbore at that depth. In either technique, the sampler filled with fluids is retrieved to the surface and the fluid sample transferred to another vessel for transportation to the laboratory. In most samplers, the displacing fluid for the transfer has been mercury, water, and glycol. The use of mercury has declined considerably due to safety and environmental concerns. An alternative practice is to transport the sampler directly to the laboratory where better equipment exists for reconstituting the sample before transferring it to another container. This option is suitable for gas condensate samples or oil samples that contain asphaltenes or waxes which can precipitate due to changes in pressures and temperatures.

6.4.2 Pistonned Bottomhole Samplers

Pistonned bottomhole samplers are a newer generation of conventional bottomhole samplers. They are equipped with a displaceable piston in the sample chamber that separates the reservoir fluid samples from a hydraulic fluid on the other side of the piston.[7] This eliminates the use of mercury or any other fluid for transfer of the sample into cylinders for transportation. Piston-type samplers also allow the rate of fluid sampling to be controlled by applying a backpressure during the sampling process. This enhances the chances of collecting the sample in a single phase. However, it is important to point out that this tool can leak hydraulic fluid past the piston and contaminate the fluid sample. This is a potential risk of contamination posed by piston-type bottomhole samplers.

6.4.3 Single Phase Samplers

Single phase samplers are used to capture reservoir fluids which may contain asphaltenes in solution.[5] Asphaltenes may precipitate out of solution if the sampling conditions (temperature and pressure) are reduced. Single phase samplers are pressure-compensated by applying a back pressure with nitrogen against a piston in the sample chamber. By maintaining the back pressure much higher than reservoir pressure, it is expected that monophasic conditions will be retained in the sample chamber. However, since the sampler will undergo temperature reduction as it is retrieved from the well, it is likely that some precipitation of asphaltenes will occur, depending on the concentration of asphaltenes in the sample. The standard practice in most laboratories is to restore the sample by reheating it and maintaining it at reservoir temperature and pressure over several days with agitation to achieve thorough mixing of its contents. The process of asphaltene precipitation may not be completely reversible even after extended period of reheating and mixing. If asphaltenes precipitation could become critical for process design, it might be preferable to collect the sample with exothermic samplers which are temperature-compensated.

6.4.4 Exothermic Samplers

Exothermic samplers are similar to single phase samplers except that they are also designed to maintain the temperature of the sample. They are used mainly for samples that may contain asphaltenes. The goal is to prevent asphaltenes from precipitating by maintaining temperatures and pressures as close as possible to sampling conditions. Exothermic samplers are kept hot with battery-operated heating jackets.

6.5 Wireline Formation Testers

Wireline formation tester (WFT) is the generic name used to describe wireline tools run usually in an open-hole to measure formation pressures and temperatures and collect fluid samples. WFT in their current configurations are very technologically advanced tools. They have become very useful tools for reservoir characterization. In their various industry configurations, the tool consists of a probe and seal assembly that can be extended against the wellbore to create a flow path between the formation and the tool which is isolated from drilling or completion fluids in the wellbore. Fluids can then flow from the formation into several chambers within the tool that can be selectively opened and closed by remote control from the surface. Most modern WFT are equipped with samplers that can apply back pressure on the samples to maintain them in a single phase if desired.

6.5.1 Oil-Based Mud Contamination of WFT Samples

Oil-based mud (OBM) systems are increasingly being used to drill wells because they yield better wellbore quality, increased penetration rate, and reduced drilling time. Drilling of wells with OBM typically causes the invasion of the near wellbore region by OBM filtrates. The depth and extent of the zone invaded by OBM filtrates depend on a number of factors, such as type of rock, formation permeability, buildup of mudcake, drilling operations, etc. Within the invaded zone, formation fluids are contaminated with OBM filtrate. When WFT are used in wells drilled with OBM, contamination of samples by OBM filtrates should be expected. This is the main risk of running WFT in wellbores drilled with OBM.

Several techniques are employed to minimize the level of contamination in fluid samples collected from wellbores drilled with OBM. One of these techniques is the use of pump-out modules in WFT. Pump-out modules provide the option of dumping as much fluids as possible from the invaded zone into the wellbore before samples are collected. The objective of the pump-out process is to replace the contaminated fluid in the invaded zone as much as possible with virgin fluid from the uninvaded parts of the formation. The degree to which the clean-up process is achieved depends on the extent of the invaded zone, the duration of the pump-out period, and mobilities of the fluids in the formation. In many cases, the pump-out module is used in conjunction with an Optical Fluid Analyzer (OFA) to monitor the quality of the fluid before sampling. The OFA monitors fluid flow in the flowline with two sensor systems: an optical spectrometer and an optical gas detector.[8-10] The combined application of pump-out and OFA modules on a sampling operation can be used to reduce contamination in fluid samples to very low levels. Considerable skill is required to decipher the output on OFA time log. The output on OFA time log is subject to a wide range of interpretations. The interpretation of OFA time log is best done by a specialist. Recent advances in OFA technology are leading to in-situ determination of hydrocarbon properties during wireline fluid sampling.[11] This technology will be useful for the in-situ measurement of the properties of fluids in a formation.

Fluid samples collected in wells drilled with OBM should be expected to contain some level of contamination with OBM filtrate. OBM filtrate is miscible with reservoir fluids in most cases. Hence, it is not possible to decontaminate the samples with physical methods. Contamination of fluid samples by OBM filtrates significantly changes the composition and phase behavior of the original reservoir fluids. Even at low levels of contamination, PVT properties of the reservoir fluids such as saturation pressure, formation volume factor, gas-liquid ratio, viscosity, density, and molecular weight can be affected. Because accurate values of these properties are needed in assessment of reserves, reservoir development, and process design, it is extremely important that reliable PVT data of reservoir fluids should be calculated from decontaminated samples.

The two simplest methods for determining the original composition of the reservoir fluid from contaminated samples are called the Subtraction method and the Skimming method. Both methods are based on the observation that there is an exponential relationship between concentrations of components in the C_{n+} of real reservoir fluids and the corresponding carbon number or molecular weight.[12,13] A plot of the composition of the C_{n+} fraction of the contaminated sample against carbon number or molecular weight on a semi-logarithmic scale will not be linear. The departure or "bump" from the straight line is due to contribution from the OBM filtrate. For the Skimming method, the composition of uncontaminated reservoir fluid is determined from a straight line drawn through the plot. This method does not require data on the composition of the OBM or OBM filtrate. An example showing the application of the skimming method to an OBM contaminated sample from a deepwater reservoir in the Gulf of Mexico is shown in Figure 6.2. The Subtraction method requires knowledge of the composition of the OBM or OBM filtrate. The contamination level of the sample is determined on-site or at the laboratory by gas chromatography. The composition of the reservoir fluid is determined by subtracting the mass of OBM or OBM filtrate from the composition of the contaminated fluid sample.

There are other decontamination methods such as the Experimental Method, Scaling Method, and Statistical Method. These methods are more complicated for routine applications but could yield more accurate results. These methods are discussed in detail by Hy-Billiot et al.[13]

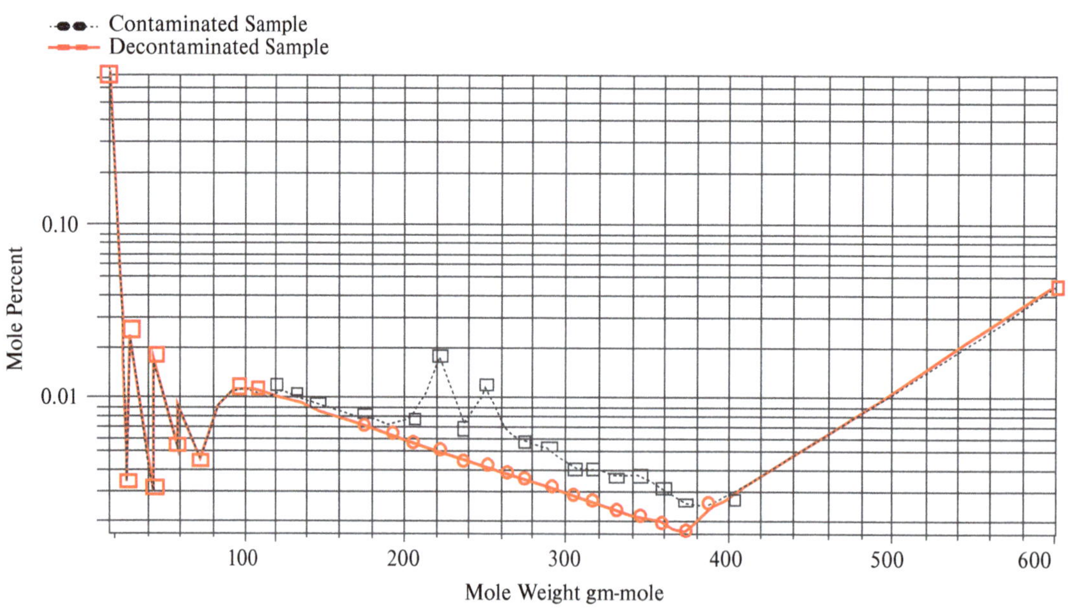

▲ **FIGURE 6.2** **Determination of the composition of the uncontaminated reservoir fluid by the Skimming method**

6.5.2 Formation Pressures from WFT

One of the primary uses of WFT is to measure formation pressures. Formation pressures can be used in a pressure-depth plot to calculate the density of the continuous fluid phase in the formation since:

$$\frac{dp}{dz} = \frac{\rho_f g \cos\theta}{144} \tag{6.1}$$

In Eq. (6.1), p = pressure, psia; z = depth, ft; ρ_f = fluid density, lbm/ft^3; g = acceleration due to gravity, ft/sec^2; and θ = well bore deviation, degrees. An example of pressure-depth plot is shown in Figure 6.3 for a deepwater Gulf of Mexico reservoir. In addition to formation fluid densities, pressure-depth plots can be used to determine fluid contacts, identify existence of reservoir heterogeneities or permeability barriers, and presence of reservoir compartments, and/or multiple reservoirs. However, before formation pressures from WFT are used for these determinations, it is important to recognize that WFT formation pressures are affected by capillary pressure changes and the phenomenon of supercharging within the invaded zone close to the wellbore.

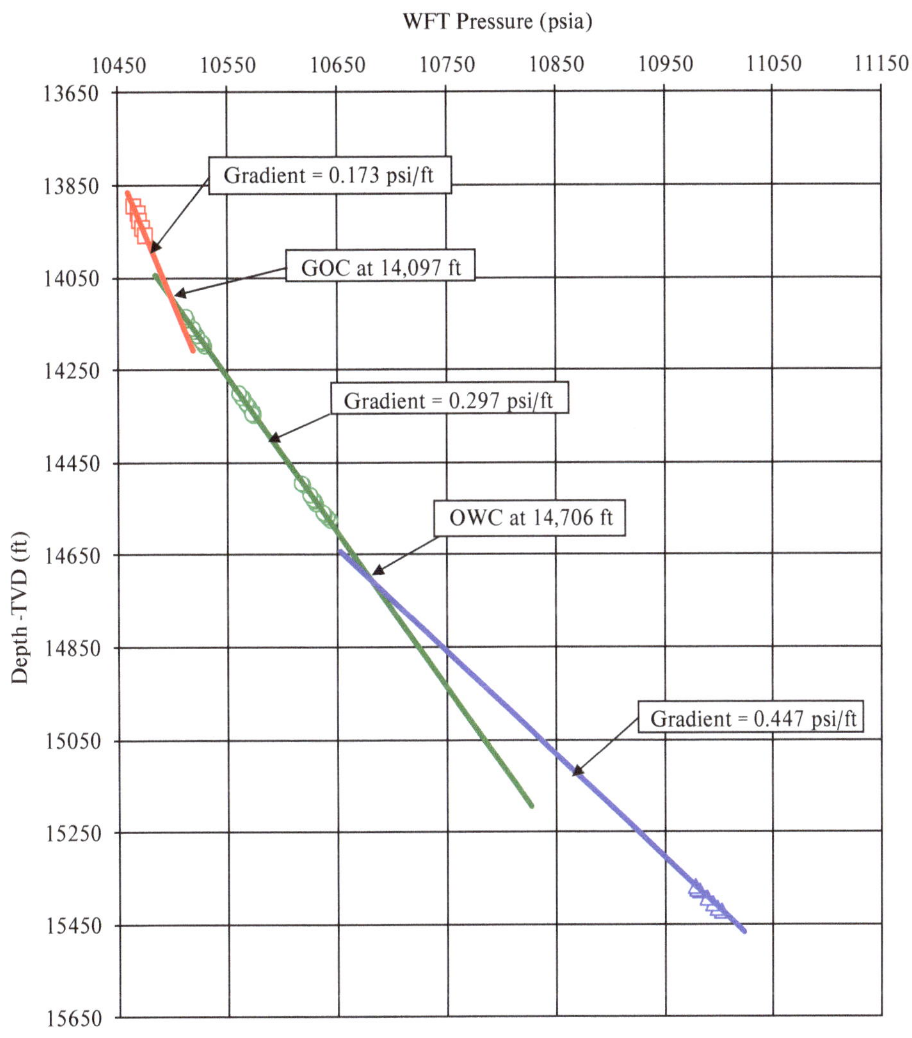

▲ FIGURE 6.3 **Pressure-depth plot for a deepwater Gulf of Mexico reservoir**

6.5.3 Capillary Effects on WFT Formation Pressures

Capillary pressure is defined as the pressure difference between the non-wetting phase and the wetting phase:

$$p_c = p_{nw} - p_w \tag{6.2}$$

In Eq. (6.2), p_c is capillary pressure; p_{nw} is pressure in the non-wetting phase; and, p_w is pressure in the wetting phase. The magnitude of capillary pressure depends on the saturation of each phase, on the nature of the continuous phase, and on the distribution, shape, and size of the pores and pore throats.[14] For a capillary tube, capillary pressure is represented as:

$$p_c = \frac{2\sigma \cos\theta}{r} \tag{6.3}$$

In Eq. (6.3), σ is the interfacial tension between the two fluids; θ is the angle of contact which denotes the wettability of the capillary tube; and r is the radius of the capillary tube. **Rock wettability** is described as the preference of either the water phase or the oil phase to adhere to the surface of the rock. The wettability of a rock is denoted by a measure of its contact angle. Values of contact angles less than 90 degrees indicate a water-wet system, while contact angles greater than 90 degrees indicate an oil-wet system. The capillary pressure profile of a water-wet system is different from that of an oil-wet system as defined by Eq. (6.3). Additional discussion of rock wettability and capillary pressure concepts are presented in Chapter 18, Sections 18.2.1 and 18.2.2, respectively.

The distribution of fluid saturation in a reservoir is governed by capillary pressure, which is dependent on depth. The **Free Water Level** (FWL) in a reservoir is the depth at which oil-water capillary pressure is non-existent, and **Oil-Water Contact** (OWC) is defined as the depth at which oil saturation starts to increase from some level of minimum saturation.[15] The FWL and OWC are not generally at the same depth but are separated by a distance related to the capillary displacement pressure. Capillary displacement is the threshold or entry capillary pressure needed for the non-wetting phase to displace the wetting phase from the largest pores.[14] Capillary displacement pressure is calculated as:

$$p_d = p_{nw} - p_w = \frac{2\sigma\cos\theta}{r_L} \tag{6.4}$$

In Eq. (6.4), p_d is capillary displacement pressure; and r_L is the radius of the largest pore throat. In a water-wet reservoir, the FWL exists at a depth, d_{owc}, below the OWC given by:[15]

$$d_{owc} = \frac{144 \times p_d}{g(\rho_{nw} - \rho_w)} \tag{6.5}$$

In Eq. (6.5), ρ_{nw} and ρ_w = densities of the non-wetting and wetting phases in lbm/ft^3, respectively; p_d = displacement pressure in psi; and d_{owc} is in feet. The location of the OWC and FWL for a water-wet reservoir as illustrated by Elshahawi et al.[15] is shown in Figure 6.4. Note that in Figure 6.4, the FWL is at some distance, d_{owc}, below the OWC. The transition zone as shown in Figure 6.4 is the region above the OWC where water saturation will decrease from a maximum value at the OWC to an irreducible value at some distance above the OWC. The height of the transition zone depends on reservoir wettability, pore size distribution, fluid density difference, and fluid interfacial tension.

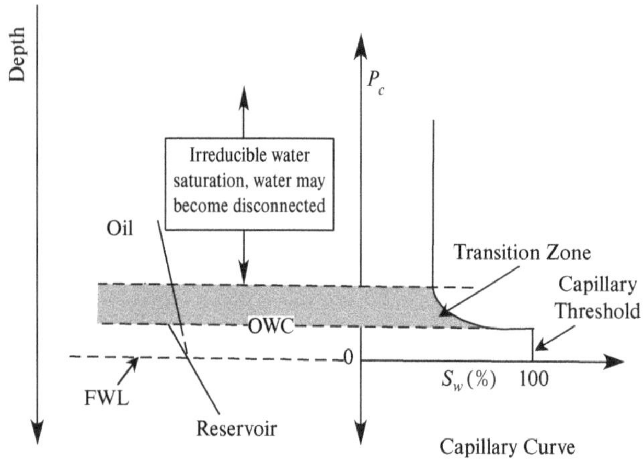

▲ **FIGURE 6.4 Location of OWC and FWL in a water-wet reservoir (from Elshahawi et al.[15] © 2000 SPE, Reproduced with permission).**

The location of the OWC and FWL for an oil-wet reservoir as illustrated by Elshahawi et al.[15] is reproduced in Figure 6.5. Note that for an oil-wet reservoir, the FWL is above the OWC, in contrast to a water-wet reservoir. This is due to the profile of capillary pressure for an oil-wet sand, which is generally negative. The locations of OWC and FWL in Figure 6.5 become evident when the concepts of these fluid levels as defined earlier are applied for an oil-wet reservoir.

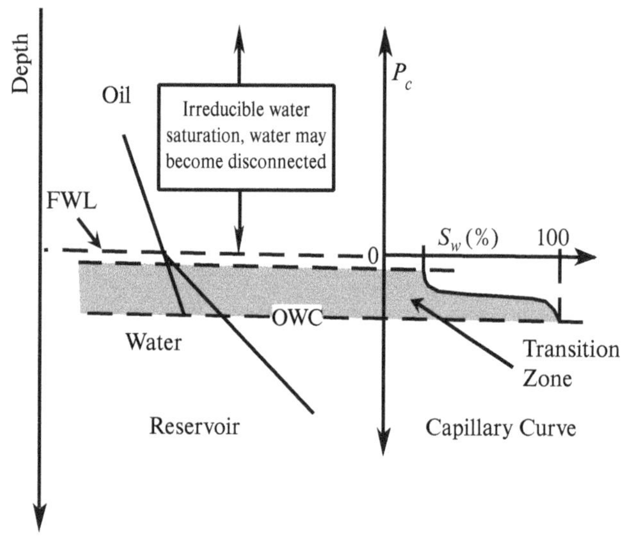

▲ **FIGURE 6.5 Location of OWC and FWL in an oil-wet reservoir (from Elshahawi et al.[15] © 2000 SPE, Reproduced with permission)**

The intersection of the pressure gradient lines drawn through hydrocarbon-bearing, and water-filled (aquifer) sections of a reservoir as measured by WFT indicates an apparent location of the FWL. In general, the apparent FWL location will differ from the true FWL in a direction that reflects the wettability of the rock and by an amount that is dependent on the degree of wettability, magnitude of capillary pressure, and type of mud used in drilling the well.[15] The differences between the apparent FWL from WFT measurements and the true FWL is illustrated in Figures 6.6 and 6.7 for water-wet and oil-wet reservoirs, respectively, drilled with water-based mud (WBM) or OBM. In Figure 6.6, the pressure gradient plot on the left represents water-wet reservoir drilled with WBM. As the plot shows, formation pressures measured by WFT in the oil zone are less (to the left) than true formation pressures because capillary pressure effects in the invaded zone have been eliminated by WBM filtrate. In the water zone, capillary pressure is zero, so WFT measures true formation pressures. Also, in Figure 6.6, the pressure gradient plot on the right denotes a water-wet reservoir drilled with OBM. In this case, WFT measures true formation pressures in the oil zone. In the water zone, WFT pressures are higher than the true formation pressures because OBM filtrates will introduce capillary effects. As Figure 6.6 shows, the summary effect for a water-wet reservoir drilled with either WBM or OBM is that the true FWL is at some level lower than that determined from WFT gradients.

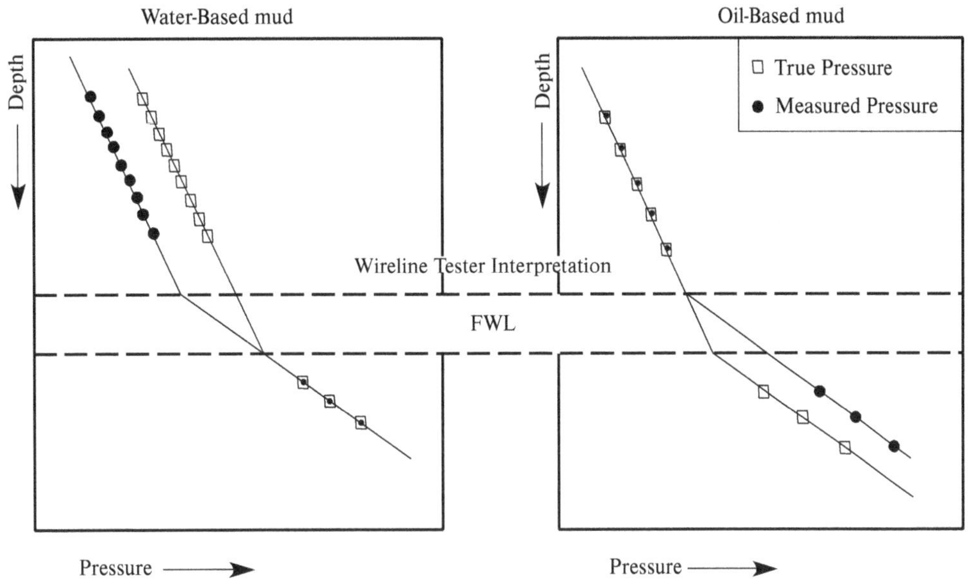

▲ **FIGURE 6.6 Pressure versus depth plot in a water-wet reservoir drilled with a WBM (left) and with OBM (right) (from Elshahawi et al.[15] © 2000 SPE, Reproduced with permission)**

Comparable results can be deduced for an oil-wet reservoir drilled with either WBM or OBM. In Figure 6.7, the plot to the left is for an oil-wet reservoir drilled with WBM. In the oil zone, WFT pressures are higher than true formation pressures due to capillary pressure created by WBM filtrate. In the water zone, there is no change in capillarity. Again, in Figure 6.7, the plot to the right is for an oil-wet reservoir drilled with OBM. There is no difference in formation pressures measured in the oil zone. In the water zone, WFT pressures will be less than the true formation pressures because OBM filtrate will eliminate capillary effects. Generally, for oil-wet sands as Figure 6.7 shows, the true FWL will be higher than the WFT based FWL level for either WBM or OBM.

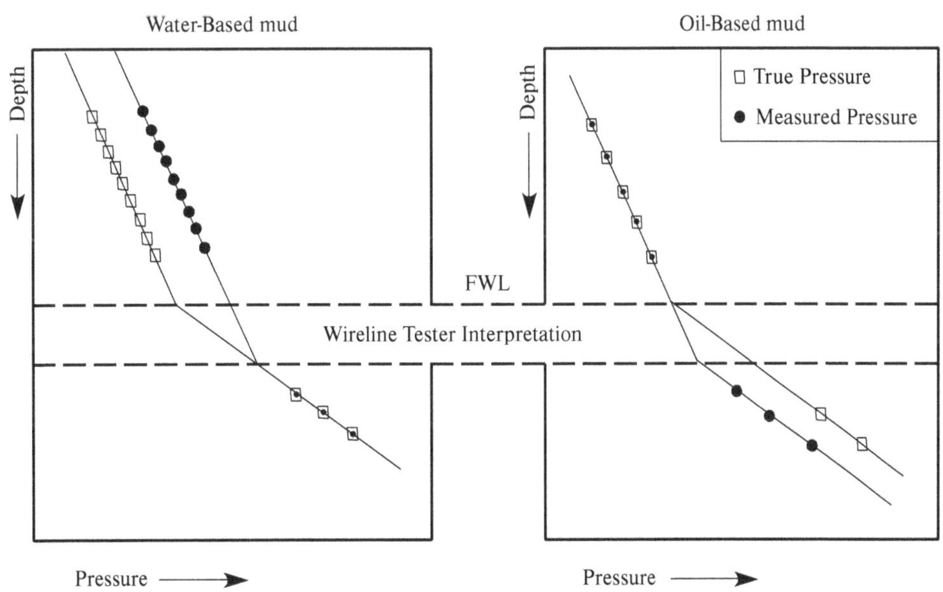

▲ **FIGURE 6.7 Pressure versus depth plot in an oil-wet reservoir drilled with WBM (left) and with OBM (right) (from Elshahawi et al.[15] © 2000 SPE, Reproduced with permission)**

6.5.4 Effects of Supercharging on WFT Formation Pressures

Mud filtrate invasion in the near-wellbore region may create formation pressures that are higher than the true formation pressures. This is called **supercharging**, which is more evident in low permeability reservoirs. High formation pressures due to supercharging may dissipate over time due to deposition of mud cake in the wellbore, which will reduce or eliminate further invasion of mud filtrate. Even if a mud cake is established in the wellbore, higher formation pressures may still exist when WFT is run. Data points affected by supercharging in either the hydrocarbon or water zones will generally be higher and appear to the right side of the true formation pressure line on a pressure-depth plot. This is shown in Figure 6.8 for a deepwater reservoir in the Gulf of Mexico. The primary factors controlling supercharging are pressure differential across the mud cake, properties of the mud cake, and total mobility of the formation.[8] The effects of supercharging can be reduced by delaying running WFT as much as possible, especially after a wiper trip, and by using WFT with pump-out modules to drain the mud filtrate in the invaded zone. A graphical method has been developed that can be used to correct formation pressures affected by supercharging. This is shown in Figure 6.9. This method requires measuring borehole pressures (mud pressures) and formation pressures several times at approximately the same depth. By plotting the formation pressures against borehole pressures, as shown in Figure 6.9, the true formation pressure can be determined at the intersection of a line drawn through the data points and the 45-degree line.

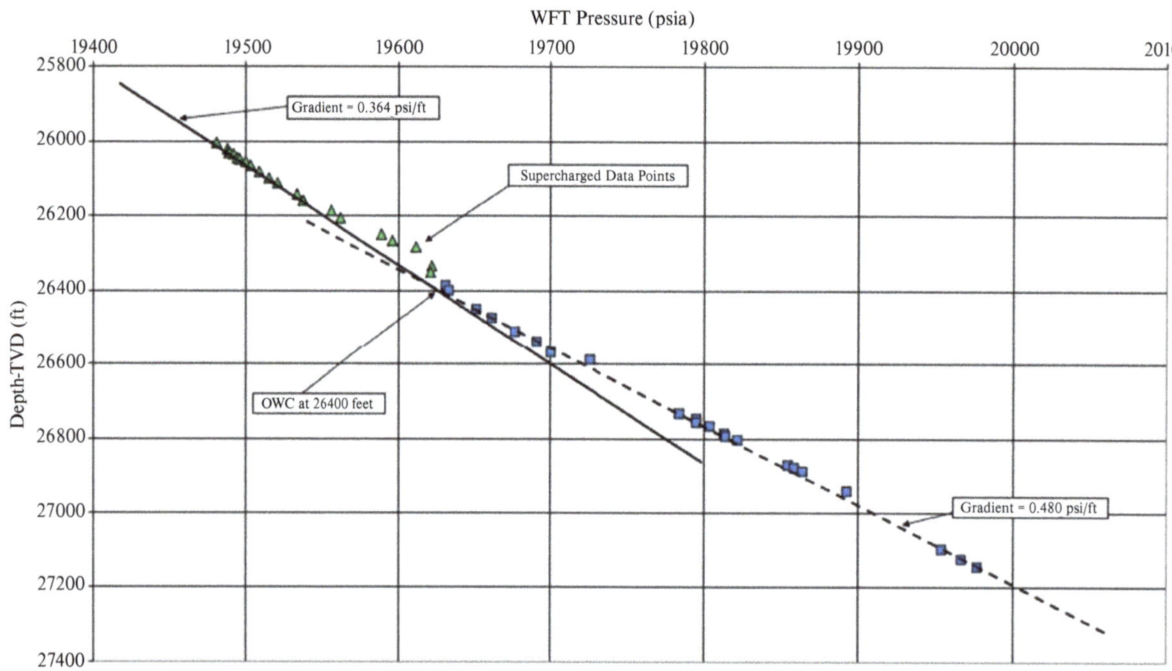

▲ FIGURE 6.8 Pressure-depth plot for a deepwater Gulf of Mexico reservoir with supercharged data points

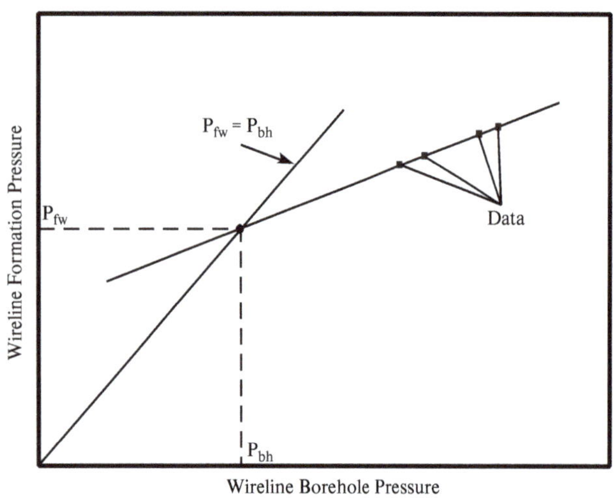

▲ FIGURE 6.9 Correction of supercharged formation pressures (from Elshahawi et al.[15] © 2000 SPE, Reproduced with permission)

6.5.5 Comments on Applications of WFT Pressure Data

WFT pressure data are routinely used in pressure-depth plots to determine the following:

1. Calculation of fluid densities.
2. Estimation of reservoir fluid contact levels.
3. Identification of the existence of reservoir heterogeneities or permeability barriers.
4. Identification of the presence of reservoir compartments and/or multiple reservoirs.

Formation pressures measured with WFT can be affected by capillary pressures, and supercharging as discussed earlier, especially for low permeability reservoirs. Capillary effects can be corrected as suggested by Elshahawi et al.[15] by using core capillary pressure data or capillary pressure data determined from nuclear magnetic resonance (NMR) log. These methods may not be available or at best may be difficult to apply. Supercharging effects can be corrected as shown in Figure 6.9 if required data are available.

Capillary effects and supercharging are factors that should be considered when using WFT data for interpretative purposes. Potential errors in actual measurements (from leaking seals, different hole conditions, accuracy of the pressure gauges, time allowed for pressure stabilization, depth corrections, etc.) must also be taken into consideration. The engineer should avoid the temptation to interpret every deviation from the apparent fluid gradient line as evidence for multiple fluid contacts, reservoir heterogeneities, reservoir compartments, or multiple reservoirs, etc. Interpretation of these geologic features from WFT data must be preceded by elimination of potential effects of capillary pressure, supercharging, measurement errors, depth corrections, etc. It is recommended practice to use independent geological interpretation as supporting evidence and guide in the interpretation of WFT data especially with respect to the presence of compartments, and permeability barriers in the reservoir, or the existence of multiple reservoirs.

6.6 PVT Laboratory Measurements

After a representative sample of the reservoir fluid has been collected, the next step is measurement of the fluid properties in series of pressure-volume-temperature (PVT) experiments. This is a very important operation because the engineer should ascertain that the PVT laboratory chosen for the measurements has the necessary experimental equipment and proper procedures for the prescribed reservoir fluid studies. For measurement of some critical data, it might be advisable to engage two separate PVT laboratories to conduct fluid properties studies on the same samples for verification and quality checking of data reported on the samples.

Reservoir fluid studies typically involve combinations of the following six procedures:

1. Fluid composition.
2. Constant Composition Expansion (CCE).
3. Differential Liberation (DL).
4. Constant Volume Depletion (CVD).
5. Separator tests.
6. Viscosity measurements.

6.6.1 Fluid Composition

The composition of the reservoir fluid sample is usually measured to various levels of detail, depending on the laboratory or as requested. Typically, the mole percent of each component in the C_1 to C_9 range is measured. The average molecular weight and specific gravity of the decane-plus $\left(C_{10}^{+}\right)$ fraction are reported. In some cases, the mole percent of the components in the C_1 to C_6 hydrocarbon range are reported with the molecular weight and specific gravity of the heptane-plus $\left(C_7^{+}\right)$ fraction. The composition of the fluid sample can be used for phase calculations with an equation of state, if needed. The composition of a black oil sample from a typical reservoir fluid study is shown in Appendix 6A. Note that for this study, component composition was determined for hydrocarbons in the C_1 to C_{29} range.

6.6.2 Constant Composition Expansion (CCE)

This laboratory procedure is called by various names, such as constant composition expansion, constant mass expansion, flash expansion, flash vaporization, or pressure-volume relations, etc. Irrespective of what it is called, the actual laboratory procedure is the same. The reservoir fluid sample is placed in a windowed PVT cell maintained at reservoir temperature and pressure. The pressure of the PVT cell is reduced to a predetermined level by increasing the volume of the PVT cell. The contents of the PVT cell are equilibrated at this pressure and volume by agitation. After equilibration, the cell volume is increased until the next predetermined pressure level is reached. This is then followed with equilibration at this pressure level. This

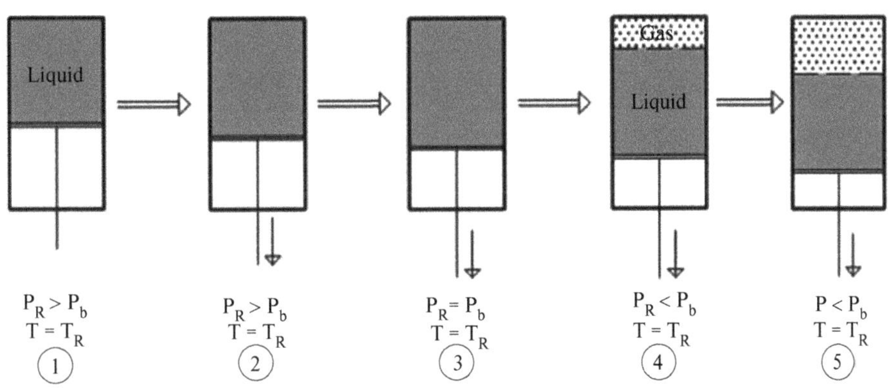

▲ FIGURE 6.10 **Constant composition expansion (CCE) experimental diagram**

procedure is followed until all the pressure stages have been tested. At each equilibrated stage, the pressure and volume of the PVT cell are measured and recorded. Note that no gas or liquid is removed from the PVT cell at any stage. This process is illustrated in Figure 6.10. A typical plot of pressure versus volume that results from this procedure is shown in Figure 6.11. In this case, the bubble point is at the intersection of the compressed liquid line and the two-phase line. Constant composition expansion experiments are conducted for gas condensate and oil samples. Typical reports for CCE experiments are shown in Appendix 6A for an oil sample, and Appendix 6B for a gas condensate sample.

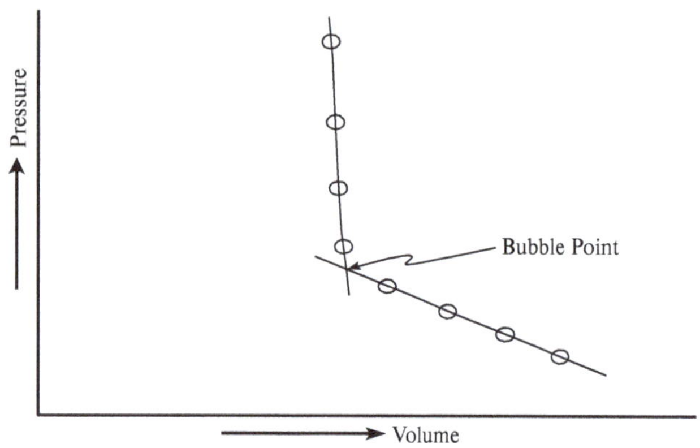

▲ **FIGURE 6.11 Pressure versus volume plot for determination of bubble point pressure from CCE data**

6.6.3 Differential Liberation (DL)

Differential liberation is also called differential vaporization or differential expansion. In this experiment procedure, the reservoir fluid sample is equilibrated in a windowed PVT cell at its bubble point pressure and reservoir temperature. The pressure inside the cell is reduced by increasing its volume. Since the cell pressure is now less than bubble point pressure, a gas phase will form. The gas is equilibrated with the liquid in the cell by agitation. Once equilibrium has been established between the fluid phases, the gas is completely displaced from the cell at constant pressure by slowly reducing the volume of the cell. The volume and specific gravity of the gas expelled from the cell is measured. The volume of liquid left in the cell is also measured. This represents Stage 1, as illustrated in Figure 6.12. The procedure used in Stage 1 is repeated for the next pressure level at Stage 2, as illustrated in Figure 6.12. These stages are repeated until atmospheric pressure is reached as the final stage with the cell still at reservoir temperature. The temperature of the cell is further reduced to 60°F to determine residual oil volume from differential liberation. This is the volume of liquid remaining in the cell at 60°F. A typical report of data from differential liberation is shown in Appendix 6A. From differential liberation data, fluid properties such as oil formation volume factor, gas compressibility factor, gas formation volume factor, solution gas-oil ratio, and oil compressibility are calculated. These data are further adjusted for reservoir engineering applications, as shown in Section 6.7.

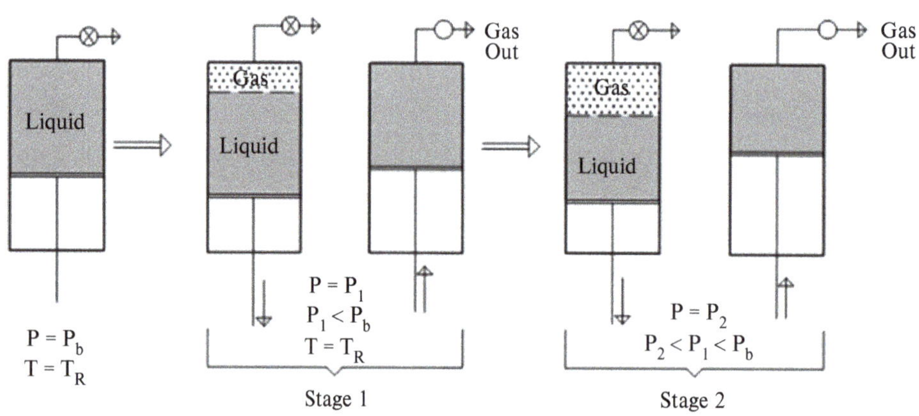

▲ **FIGURE 6.12 Differential liberation experimental diagram**

6.6.4 Constant Volume Depletion (CVD)

Constant volume depletion generally called depletion study is an experimental procedure used to study gas-condensate samples. The fluid sample is equilibrated in a PVT cell at reservoir temperature and pressure. The cell volume is increased, thereby reducing its pressure to some predetermined level. The fluid sample is thoroughly mixed at this lower pressure by agitation. Then a portion of the fluid in the cell is withdrawn slowly while maintaining its pressure constant at the current level until the original cell volume is reached. The composition and specific gravity of the fluid removed from the cell is measured. This procedure is represented as Stage 1 in Figure 6.13. Subsequent stages at lower pressure levels are repeated as described for Stage 1. A typical report from a depletion study is shown in Appendix 6B. The data from a depletion study can be used to fit parameters of any preferred equation of state. The equation of state can then be used in a reservoir simulator to model depletion of a condensate reservoir.

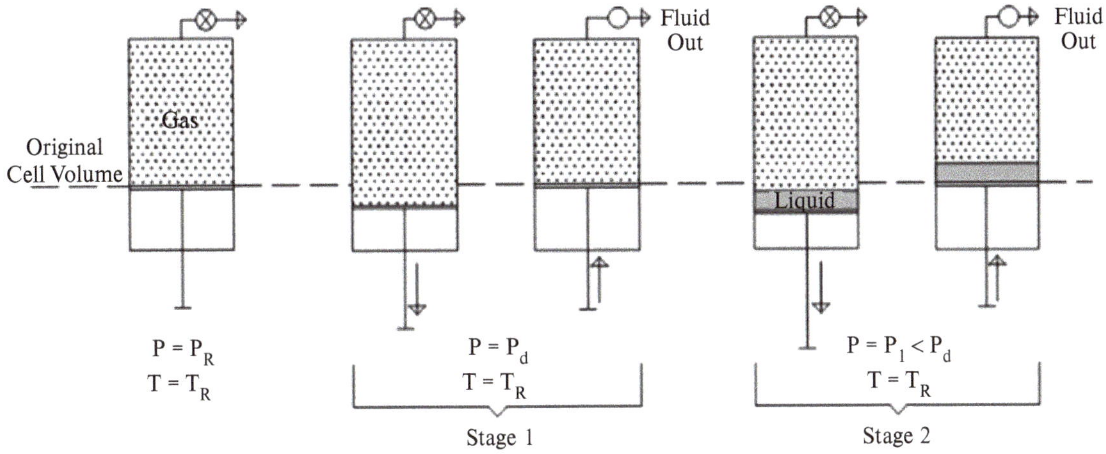

▲ **FIGURE 6.13 Constant volume depletion experimental diagram**

6.6.5 Separator Tests

A separator test is a flash vaporization process. A fluid sample at reservoir temperature and bubble point pressure in a PVT cell is displaced at bubble point pressure through two or more stages of separation. The pressure in the PVT cell is held constant by slowly reducing its volume. The pressures and temperatures of the separators in the laboratory are selected to approximate as closely as possible the expected separator conditions in the field. The stock tank (final stage) is at atmospheric pressure. A three-stage separator test representing this procedure is shown in Figure 6.14. An alternative procedure for separator tests which is used in some laboratories is very similar to the differential liberation experiment. In this procedure, the temperature of fluid sample initially at bubble point pressure and reservoir temperature is reduced to the temperature of the first-stage separator. After equilibration, the pressure of the sample is reduced to the pressure set for the first-stage separator. The liberated gas and the remaining liquid phase are allowed to equilibrate. Then the liberated gas is displaced at constant pressure from the PVT cell into an evacuated chamber for density measurements and compositional analysis. This process is repeated for other separator stages down to stock tank conditions. An example of gas-oil ratio, (R_{sSb}), and formation oil volume factor, (B_{oSb}), measured for a four-stage separation is shown in Appendix 6A. These data are used to adjust fluid properties measured in CCE and DL tests, as demonstrated in Section 6.7.

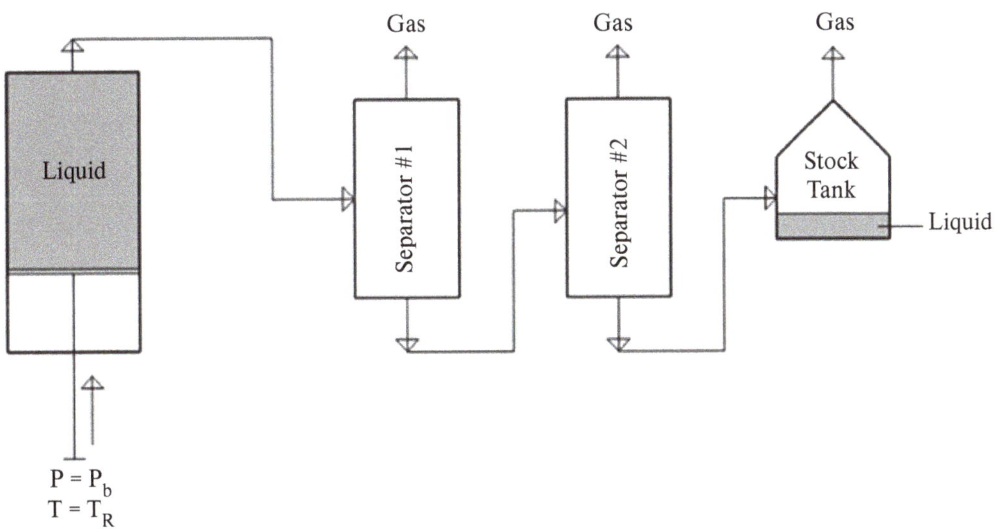

▲ **FIGURE 6.14 A three-stage separator test experimental diagram**

6.6.6 Viscosity Measurements

Oil viscosities are measured at the same pressure levels as the differential liberation tests by obtaining liquid samples devoid of gas at each pressure level in a separate test from the DL test. The equipment used to measure oil viscosities is either a capillary viscometer or electromagnetic (EM) viscometer. Viscosities are calculated from capillary viscometers directly from measured data based on the principles of fluid flow in a tube. Viscosities from the EM viscometer are based on indirect methods. However, either equipment requires viscosity standards for calibration. The reliability of oil viscosities measured with either EM or capillary viscometer depends considerably on the calibration process. Oil viscosities measured for the example fluid study are shown in Appendix 6A. Gas viscosities are calculated from correlations such as Lee et al.[16] correlation.

6.7 Applications of Laboratory PVT Measurements

In laboratory PVT studies of volatile and black oil samples, flash vaporization (CCE) data, differential liberation (DL) data, and separator test data are reported. An example of a reservoir fluid study report is provided in Appendix 6A for a black oil sample. Differential liberation procedures in the laboratory are designed to represent the processes that occur in the reservoir, in which gas released from oil is presumed to migrate away from the oil because of higher mobility of the gas phase. In the reservoir, some of the liberated gas remains in contact with the oil. In the laboratory differential liberation procedure, all the liberated gas is removed from the oil. The actual process that occurs in the reservoir can be reproduced in the laboratory with a procedure called "composite liberation."[17] This procedure is superior to the differential liberation procedure, but it requires a large amount of fluid sample and takes a long time to complete. McCain[18] showed that the differences between oil properties calculated with composite liberation data and differential liberation data are within experimental accuracy. In most applications, differential liberation data in conjunction with separator test data are used for calculation of oil PVT data.

6.7.1 Calculation of Oil FVF and Solution GOR

Oil formation volume factor (FVF) that can be used for material balance and reserves calculations is calculated from flash vaporization, differential liberation, and separator test data. At pressures above bubble point,

$$B_o = \left(\frac{V}{V_{\text{sat}}}\right)_{\text{CCE}} \times B_{oSb} \qquad \text{for } p > p_b \tag{6.6}$$

In Eq. (6.6), B_O = oil FVF; $(V/V_{SAT})_{CCE}$ = relative volume from constant composition expansion; and B_{oSb} = oil FVF at bubble point from separator test.

For pressures below bubble point,

$$B_o = B_{oD}\left(\frac{B_{oSb}}{B_{oDb}}\right) \qquad \text{for } p < p_b \tag{6.7}$$

In Eq. (6.7), B_{oD} = oil FVF from differential liberation; and B_{oDb} is FVF at bubble point from differential liberation.

For pressures below bubble point,

$$R_s = R_{sD}\left(\frac{R_{sSb}}{R_{sDb}}\right) \qquad \text{for } p < p_b \tag{6.8}$$

In Eq. (6.8), R_s = oil solution GOR; R_{sD} = oil solution GOR by differential liberation; R_{sSb} = oil solution GOR at bubble point from separator test; and R_{sDb} = oil solution GOR at bubble point from differential liberation.

6.7.2 Calculation of Gas Compressibility Factor, Gas FVF, and Total FVF

The volume of gas evolved at each stage of the differential liberation process is measured at the temperature and pressure of the PVT cell (reservoir conditions). The volume of the gas is also measured at standard conditions (atmospheric pressure and 60°F). Gas compressibility factor is calculated from these volumes as:

$$z_R = \frac{p_R V_R T_{sc}}{p_{sc} V_{sc} T_R} \tag{6.9}$$

In Eq. (6.9), p_R = pressure of the cell; V_R = volume of the gas at cell conditions; T_R = temperature of the cell; p_{sc} = pressure at standard conditions; V_{sc} = volume of gas at standard conditions; and T_{sc} = temperature at standard conditions (or 60°F).

Gas formation volume factor is calculated from Eq. 5.75 as:

$$\begin{aligned} B_g &= \frac{z_R T_R(14.65)}{p_R(519.67)} \\ &= \frac{0.02819 z_R T_R}{p_R}, \quad \frac{\text{ft}^3}{\text{scf}} \\ &= \frac{0.005021 z_R T_R}{p_R}, \quad \frac{\text{RB}}{\text{scf}} \end{aligned}$$

Note that in Eq. 5.75, pressure at standard conditions is set at 14.65 psia.

Total formation volume factor for differential liberation data is calculated as:

$$B_t = B_{oD} + R_t B_g \tag{6.10}$$

In Eq. (6.10), B_t = total formation volume factor, RB/STB; B_{oD} = oil FVF from differential liberation, RB/STB; R_l = total liberated gas, scf/STB; and B_g = gas FVF, RB/scf.

The calculation of gas properties from differential liberation data is shown in Appendix 6A.

6.7.3 Calculation of Oil Compressibility Factor

If oil compressibility factor is not reported with differential liberation data in Appendix 6A, it can be calculated from constant composition data for pressures above bubble point and differential liberation data for pressures below bubble point.

For pressures above bubble point, oil compressibility is represented as:

$$c_o = -\frac{1}{V}\left(\frac{\partial V}{\partial p}\right)_T \tag{6.11}$$

By integrating Eq. (6.11), assuming c_o is constant as pressure changes, it can be shown that:

$$c_o\left(p_2 - p_1\right) = -\ln\left(\frac{V_2}{V_1}\right) \tag{6.12}$$

Substituting relative volume (V/V_{SAT}) from CCE for volume in Eq. (6.12), we obtain:

$$c_o = \frac{\ln\dfrac{\left(V/V_{sat}\right)_2}{\left(V/V_{sat}\right)_1}}{p_1 - p_2} \tag{6.13}$$

Applying Eq. (6.13) to CCE data in Appendix 6A, $\left(V/V_{sat}\right)_{8000} = 0.9899$ and $\left(V/V_{sat}\right)_{9000} = 0.9797$. Hence,

$$c_o = \frac{\ln\left(\dfrac{0.9899}{0.9797}\right)}{9000 - 8000} = 10.4\times10^{-6}\,\text{psi}^{-1}$$

for 9000 – 8000 psia range

For pressures below bubble point, it was shown by McCain[19] that oil compressibility can be calculated as:

$$c_o = -\frac{1}{B_o}\left[\left(\frac{\partial B_o}{\partial p}\right)_T - B_g\left(\frac{\partial R_s}{\partial p}\right)_T\right] \quad \text{for } p < p_b \tag{6.14}$$

Eq. (6.14) can be rearranged as:

$$c_o = \frac{1}{B_o}\left(\frac{\partial R_s}{\partial p}\right)_T\left[B_g - \left(\frac{\partial B_o}{\partial R_s}\right)_T\right] \tag{6.15}$$

Eq. (6.15) can be rewritten in terms of differential liberation data as:

$$c_o = \frac{1}{B_{oD}}\left(\frac{\partial R_{sD}}{\partial p}\right)_T\left[B_g - \left(\frac{\partial B_{oD}}{\partial R_{sD}}\right)_T\right] \tag{6.16}$$

As an example, Eq. (6.16) can be used to estimate oil compressibility at 5600 psia from differential data in Appendix 6A. From Appendix 6A, $B_{oD} = 1.4232$ RB/bbl and $B_g = 649\times10^{-6}$ RB/scf at 5600 psia. From a plot of B_o versus R_s, $\dfrac{\partial B_o}{\partial R_s} = 0.000405$ RB/scf. From a plot of R_{sD} versus P, $\dfrac{\partial R_{sD}}{\partial p} = 0.1000$ scf/(bbl-psi). Note that the plots are not shown but can be easily generated from the data in Appendix 6A. Substituting in Eq. (6.16),

$$c_o = \frac{1}{1.4232}(0.1000)\left[649\times10^{-6} - 0.000405\right]$$
$$= 17.1\times10^{-6}\text{ psi}^{-1}$$

Note that the calculated oil compressibility is higher the value reported at the same pressure in Appendix 6A. The difference can be traced to the errors associated with calculating the slope of the tangent line at 5600 psia for the $\dfrac{\partial R_{sD}}{\partial P}$ term. But this method could be used to estimate oil compressibility below bubble point pressures from differential liberation data if actual data are not reported.

Nomenclature

B_g	gas formation volume factor, RB/STB
B_o	oil formation volume factor, RB/STB
B_{oD}	oil formation volume factor at $P < P_b$ for DL, RB/residual bbl
B_{oDb}	oil formation volume factor at P_b for DL, RB/residual bbl
B_{oSb}	oil formation volume factor at P_b in separator test, RB/STB
B_w	water formation volume factor, RB/STB
c_o	isothermal oil compressibility, psi^{-1}
c_{ob}	isothermal oil compressibility at bubble point, psi^{-1}
d_{owc}	depth below OWC
p	pressure
p_b	bubble point pressure, psia
p_c	capillary pressure
p_d	capillary displacement pressure or dew point pressure
r	radius of capillary tube
r_L	largest pore throat radius
R	gas constant, (psia × ft)/(1bmol × °R)
R_l	total liberated gas in DL, scf/residual bbl
R_s	solution gas-oil ratio, scf/STB
R_{sSb}	solution gas-oil ratio at P_b in separator test, scf/STB
R_{sD}	solution gas-oil ratio at $P < P_b$ for DL, scf/residual bbl
R_{sDb}	solution gas-oil ratio at P_b for DL, scf/residual bbl
T	temperature
V	volume
V_{sat}	volume at saturated pressure
z	gas compressibility factor
ρ	density, mass per unit volume
σ	Interfacial tension
θ	angle of contact

Subscripts

b	bubble point
f	fluid
n	wetting phase
nw	non-wetting phase
o	oil
R	reservoir
sc	standard conditions
sp	separator
st	stock-tank

Abbreviations

App	appraisal (Table 6.1)
CCE	constant composition expansion
CVD	constant volume depletion
DL	differential liberation
EM	electromagnetic viscometer
Expl	exploration (Table 6.1)
FVF	formation volume factor
FWL	free water level
OBM	oil-based mud
OFA	optical fluid analyzer
OWC	oil-water contact
Prod	production (Table 6.1)
PVT	pressure-volume-temperature
WBM	water-based mud
WFT	wireline formation tester

REFERENCES

1. Moffatt, B.J., and Williams, J.M.: "Identifying and Meeting the Key Needs for Reservoir Fluid Properties–A Multi-Disciplinary Approach," paper SPE 49067 presented at the 1998 SPE Annual Technical Conference and Exhibition, New Orleans, Louisiana (Sept. 27-30, 1998).
2. McCain Jr., W.D., and Alexander, R.A.: "Sampling Gas-Condensate Wells," *SPERE* (August 1992) 358-362.
3. *RP 44, Recommended Practice for Sampling Petroleum Reservoir Fluids*, Second Edition, API, Washington, DC (April 2003).
4. Williams, J.M., and Kokal, S.L.: *Fluid Sampling,* Petroleum Engineering Handbook, General Engineering, Vol. 1, Chapter 4, SPE (2006).
5. Bon, J., Sarma, H.K., Rodrigues, J.T., and Bon, J.G.: "Reservoir Fluid Sampling Revisited-A Practical Perspective," paper SPE 101037 presented at the 2006 SPE Asia Pacific Oil & Gas Conference and Exhibition, Adelaide, Australia (Sept. 11-13, 2006).
6. Towler, B.F.: "Reservoir Engineering Aspects of Bottomhole Sampling of Saturated Oils for PVT Analysis," paper SPE 19438, 1990.
7. Proett, M.A., Gilbert, G.N., Chin, W.C., and Monroe, M.L.: "New Wireline Formation Testing Tool with Advanced Sampling Technology," paper SPE 56711 presented at the 1999 SPE Annual Technical Conference and Exhibition, Houston, Texas (Oct. 3-6, 1999.)
8. *Fundamentals of Formation Testing*, Schlumberger, Sugar Land, Texas (2006).
9. Morris, C.W., Felling, M.M., Butsch, R.J., and Sass, W.: "Using Optical Fluid Analysis to Evaluate Downhole Fluid Sample Contamination," paper SPE 50603 presented at the 1998 SPE European Petroleum Conference, The Hague, The Netherlands (Oct. 20-22, 1998).
10. Smits, A.R., Fincher, D.V., Nishida, K., Mullins, O.C., Schroeder, R.J., and Yamate, T.: "In-situ Optical Fluid Analysis as an Aid to Wireline Formation Sampling," *SPEFE* (June 1995) 91-98.
11. Van Dusen, A., Williams, S., Fadnes, F.H., and Irvine-Fortescue, J.: "Determination of Hydrocarbon Properties by Optical Analysis During Wireline Fluid Sampling," *SPEREE* (August 2003) 286-292.
12. Gozalpour, F., Danesh, A., Tehrani, D.-H., Todd, A.C., and Tohidi, B.: "Predicting Reservoir Fluid Phase and Volumetric Behavior from Samples Contaminated with Oil-Based Mud," *SPEREE* (June 2002) 197-205.
13. Hy-Billiot, J., Bickert, J., Montel, F., and Segalini, G.: "Getting the Best From Formation Tester Sampling," paper SPE 77771 presented at the 2002 SPE Annual Technical Conference and Exhibition, San Antonio, Texas (Sept. 29-Oct. 2, 2002).
14. Elshahawi, H., Fathy, K., and Hiekel, S.: "Capillary Pressure and Rock Wettability Effects on Wireline Formation Tester," paper SPE 56712 presented at the 1999 SPE Annual Technical Conference and Exhibition, Houston, Texas (Oct. 3-6, 1999).
15. Elshahawi, H., Samir, M., and Fathy, K.: "Correcting for Wettability and Capillary Pressure Effects on Formation Tester Measurements," paper SPE 63075 presented at the 2000 SPE Annual Technical Conference and Exhibition, Dallas, Texas (Oct. 1-4, 2000).
16. Lee, A.L, Gonzalez, M.H., and Eakin, B.E.: "The Viscosity of Natural Gas," *Trans. AIME* (1966) Vol. 237, 997-1000.
17. Moses, P.L.: "Engineering Applications of Phase Behavior of Crude Oil and Condensate Systems," *JPT* (July 1986) 38, 715-723.
18. McCain, Jr., W.D.: "Analysis of Black Oil PVT Reports Revisited," paper SPE 77386 presented at the 2002 SPE Annual Technical Conference and Exhibition, San Antonio, Texas (Sept. 29-Oct. 2, 2002).
19. McCain, Jr., W.D.: *The Properties of Petroleum Fluids*, Second Edition, Pennwell Books, Tulsa, OK (1990).

ADDITIONAL REFERENCES

- El-Banbi, A., and McCain, Jr., W.D.: "Sampling Volatile Oil Wells," paper SPE 67232 presented at the SPE Production and Operations Symposium, Oklahoma City, Oklahoma (March 24-27, 2001).
- Felling, M.M. and Morris, C.W.: "Characterization of In-Situ Fluid Responses by Use of Optical Fluid Analysis," *SPEREE* (August 1998) 297-302.
- Michaels, J., Moody, M., and Shwe, T.: "Wireline Fluid Sampling," paper SPE 30610 presented at the SPE Annual Technical Conference & Exhibition, Dallas, Texas (Oct. 22-25, 1995).
- Nagarajan, N.R, Honarpour, M.M., and Sampath, K.: "Reservoir Fluid Sampling and Characterization-Key to Efficient Reservoir Management," paper SPE 101517 presented at the 2006 Abu Dhabi International Petroleum Exhibition and Conference, Abu Dhabi, U.A.E. (November 5-8, 2006).
- Pham, T.R., Al-Afaleg, N.I., Kelder, O., Al-Otaibi, U.F., and Zeybek, M.: "Field Example of Capillary Pressure Effects on Wireline Formation Tester Measurements and OWC Estimation in a Mixed-Wettability Oil Reservoir," paper SPE 93262 presented at the SPE Middle East Oil & Gas Show and Conference, Bahrain (March 12-15, 2005).
- Williams, J.M.: "Getting the Best Out of Fluid Samples," *JPT* (September 1994) 752.

PROBLEMS

6-1. The data from PVT laboratory measurements on a downhole sample from a reservoir in California, USA, are given in the Table 6P.1. From the tabulated data, make the following plots:

a. Oil FVF (B_o) vs. Pressure (p)

b. Solution gas-oil ratio (R_s) vs. Pressure (p)

c. Gas FVF (B_g) vs. Pressure (p)

d. Oil viscosity μ_o vs. Pressure (p)

What is the bubble point pressure of the sample at 220°F?

TABLE 6P.1 | PVT Data for Problem 6-1

Pressure (p) (psig)	Oil FVF (B_o) (RB/STB)	Solution GOR (R_s) (scf/STB)	Gas FVF (B_g) (ft³/scf)	Oil Viscosity (μ_o) (cp)
0	1.073	0	—	1.816
135	1.197	142	0.12593	1.058
300	1.241	207	0.05899	0.937
600	1.289	293	0.02936	0.798
900	1.327	370	0.01922	0.710
1200	1.361	443	0.01418	0.637
1500	1.398	521	0.01113	0.580
1800	1.434	599	0.00919	0.533
2100	1.474	683	0.00780	0.490
2400	1.515	769	0.00680	0.455
2700	1.560	863	0.00604	0.428
3023	1.614	972	—	0.400
3500	1.601	972	—	0.413
4000	1.591	972	—	0.428
4500	1.583	972	—	0.443
5000	1.573	972	—	0.457

6-2. The PVT data from a reservoir fluid study of a black oil sample are provided in Appendix 6A. Make the following plots from the data:

a. Oil FVF (B_o) vs. Pressure (p)

b. Solution gas-oil ratio (R_s) vs. Pressure (p)

c. Gas FVF (B_g) vs. Pressure (p)

d. Total FVF (B_t) vs. Pressure (p)

e. Oil viscosity μ_o vs. Pressure (p)

What is the saturation state of the fluid in the reservoir at reservoir temperature?

6-3. Using the reservoir fluid study report in Appendix 6A, calculate oil FVF (B_o), Solution GOR (R_s), gas FVF (B_g), total FVF (B_t), and oil viscosity (μ_o) at reservoir pressure and temperature of 5950 psia and 212°F, respectively.

6-4. Calculate oil compressibility (c_o) at reservoir pressure of 9500 psia and 5950 psia using the PVT data provided in Appendix 6A. The reservoir temperature is 212°F.

6-5. The formation pressures and depths measured during modular dynamics tester (MDT) runs on a lower tertiary reservoir in deepwater Gulf of Mexico are provided in Table 6P.2. The temperature of the reservoir is 245°F and density of formation water is 1.14 g/cc at 60°F and 14.7 psia. Determine the depth of the oil-water contact based on the data in Table 6P.2.

TABLE 6P.2 | Depth-Pressure Data for Problem 6-5

Depth, TVDSS (feet)	Pressure, (psia)	Depth, TVDSS (feet)	Pressure, (psia)
26004.25	19481.26	26399.19	19633.59
26022.79	19488.12	26451.30	19651.50
26029.81	19488.71	26476.67	19662.34
26030.13	19489.33	26511.07	19677.12
26036.58	19492.16	26540.74	19691.05
26043.19	19494.23	26571.20	19700.39
26049.44	19496.21	26590.17	19726.20
26054.94	19499.80	26732.38	19784.61
26067.53	19503.48	26747.90	19795.55
26080.72	19508.77	26758.45	19795.95
26098.48	19515.23	26767.29	19804.69
26111.47	19520.66	26788.74	19813.84
26143.76	19533.05	26795.58	19814.53
26158.58	19536.92	26806.25	19822.13
26186.27	19556.34	26869.79	19855.13
26205.55	19562.70	26880.48	19858.66
26248.80	19588.96	26891.38	19864.36
26265.66	19596.30	26942.50	19892.97
26284.40	19611.21	27098.25	19954.27
26332.15	19621.10	27125.10	19966.99
26350.64	19620.81	27145.54	19977.85
26388.33	19631.33	—	—

6-6. Modular Dynamic Tester (MDT) runs were conducted on a well drilled in the pre-salt formation in the Campos Basin, Brazil. The formation pressures and depths measured during the survey are presented in Table 6P.3. Assuming a formation water gradient of 0.48 psi/ft, determine the oil-water contact in the formation.

TABLE 6P.3 Depth-Pressure Data for Problem 6-6

Depth, TVDSS (meters)	Pressure, (psia)	Depth, TVDSS (meters)	Pressure, (psia)
7324.93	4762.50	7967.22	4889.70
7331.83	4772.60	7489.41	4891.70
7345.91	4786.70	7486.07	4892.70
7355.55	4795.70	7488.81	4893.70
7369.84	4809.20	7484.79	4893.70
7493.97	4813.20	10203.10	4893.70
9142.40	4815.70	7534.76	4957.80
6629.67	4821.20	8369.64	4965.70
6527.17	4821.60	7627.81	4981.90
5210.58	4828.60	7653.74	4996.60
6199.23	4834.60	7675.14	5004.80
6782.63	4840.70	7732.08	5046.60
7077.70	4841.20	6963.12	5047.60
5221.66	4846.20	7758.35	5064.90
5559.88	4846.70	5857.31	5084.60
7436.68	4858.70	4492.67	5791.00
7617.35	4858.80	8624.16	5825.60

6-7. A discovery well drilled in deepwater Angola, Southwest Africa, was surveyed with a reservoir characterization instrument (RCI) tool assembly. The formation pressures and depths measured during the RCI runs are provided in Table 6P.4. The average pressure of the formation is 6673 psia at formation temperature of 228°F. An oil sample collected during the RCI survey has a bubble point pressure of 3720 psia at formation temperature. The formation water gradient is 0.459 psi/ft at reservoir conditions. What is the apparent location of the OWC in the formation based on the RCI data?

TABLE 6P.4 Depth-Pressure Data for Problem 6-7

Depth, TVDSS (meters)	Pressure, (psia)	Depth, TVDSS (meters)	Pressure, (psia)
6578.19	3476.67	6685.76	3577.97
6585.49	3481.43	6702.76	3590.61
6639.23	3536.54	6716.77	3599.72
6646.10	3542.24	6733.10	3610.16
6649.81	3546.70	6739.00	3614.22
6659.56	3555.78	6747.40	3619.88
6661.98	3558.61	6758.50	3627.21
6665.69	3562.14	6766.33	3632.25
6668.31	3564.88	6775.02	3637.55
6672.95	3568.95	6785.40	3645.06
6675.60	3572.05	6758.10	3626.94

APPENDIX

6A TYPICAL RESERVOIR FLUID STUDY FOR A BLACK OIL SAMPLE

6A.1 Reservoir Fluid Summary

Reservoir temperature, °F	212
Saturation pressure at 212°F, psia	7153
Compressibility of reservoir oil at 212°F, psi^{-1}:	
At static bottomhole pressure of 10,640 psia	8.25E-06
At bubblepoint pressure of 7153 psia	11.96E-06
Saturated oil at 7153 psia, 212°F:	
Density, gm. per cc.	0.6690
Density, lb. per bbl.	234.5
Specific volume, ft^3 per lb.	0.02394
Viscosity, cp.	0.467
Formation volume factor (FVF):	
*FVF at p_b, rb/bbl. residual oil at 60°F	1.5572
**FVF at p_b, rb/stock tank oil at 60°F	1.5186
Solution gas-oil ratio:	
*Standard cu. ft. per bbl. residual oil at 60°F	1153
**Standard cu. ft. per bbl. stock tank oil at 60°F	1049
Reservoir oil at 10460 psia, 212°F:	
Density, gm. per cc.	0.6931
Density, lb. per bbl.	242.9
Specific volume, cu. ft. per lb.	0.02311
Viscosity, cp.	0.676
Formation volume factor:	
*Bbl. per bbl. residual oil at 60°F	1.5031
**Bbl. per bbl. stock tank oil at 60°F	1.4658

*Differential liberation.

**Flash separation at field separator conditions.

6A.2 Calculated Analysis of Reservoir Fluid

Component	Mole %
Hydrogen Sulfide	0.00
Carbon Dioxide	0.23
Nitrogen	0.30
Methane	67.51
Ethane	2.25
Propane	1.87
i-Butane	0.61
n-Butane	1.00
i-Pentane	0.50
n-Pentane	0.55
Hexanes	1.10
Heptanes	1.28
Octanes	1.26
Nonanes	1.38
Decanes	1.28
Undecanes	1.05
Dodecanes	0.99
Tridecanes	0.99
Tetradecanes	0.89
Pentadecanes	0.93
Hexadecanes	1.19
Heptadecanes	1.05
Octadecanes	0.62
Nonadecanes	1.21
Eicosanes	0.69
Heneicosanes	0.58
Docosanes	0.53
Tricosanes	0.49
Tetracosanes	0.44
Pentacosanes	0.43
Hexacosanes	0.32
Heptacosanes	0.35
Octacosanes	0.32
Nonacosanes	0.30
Triacontanes Plus	5.51
Total	**100.00**

Basis of Recombination	= 1167 scf/bbl
Total molecular weight	= 88.37 lb/lbmole
Tricontanes Plus molecular weight	= 610.6
Tricontanes Plus specific gravity	= 0.9847

6A.3 Pressure-Volume Properties at 212°F (Constant Composition Expansion)

Pressure (psia)	Relative Volume (V/V_{sat})	Density (gm/cc)	Compressibility ($\Delta V/V/\Delta psi \times 10^6$)	"Y" Function $(p_{sat} - p)/p(B_t - 1)$
10640*	0.9652	0.6931	8.25	
10000	0.9706	0.6893	8.61	
9000	0.9797	0.6829	9.27	
8000	0.9899	0.6759	10.29	
7153**	1.0000	0.6690	11.96	
Two Phase Region Data Below				
7000	1.0035	0.6667		6.203
6300	1.0230	0.6539		5.876
5600	1.0500	0.6372		5.548
4900	1.0881	0.6148		5.216
4200	1.1437	0.5850		4.893
3500	1.2287	0.5445		4.563
2800	1.3671	0.4894		4.235
2100	1.6155	0.4141		3.909
1400	2.1474	0.3115		3.581
700	3.8323	0.1746		3.255

*Static bottomhole pressure

**Bubble point, p_b

6A.4 Differential Liberation at 212°F

Pressure (psia)	Density (gm/cc)	FVF B_{oD} RB/bbl. residual oil at 60°F	FVF $B_o^\dagger$ (RB/STB)	Compressibility $\frac{\Delta V}{V\Delta p} \times 10^{-6}$ (psi^{-1})	Liberated Gas R_l (scf/bbl)	Solution GOR R_{sD} (scf/bbl)	Solution GOR $R_s^\dagger$ (scf/bbl)
10640*	0.6931	1.5031	1.4658	8.25	0	1153	1049
10000	0.6893	1.5114	1.4739	8.61	0	1153	1049
9000	0.6829	1.5256	1.4877	9.27	0	1153	1049
8000	0.6759	1.5414	1.5032	10.29	0	1153	1049
7153**	0.6690	1.5572	1.5186	11.96	0	1153	1049
Two Phase Region Data Below							
7000	0.6728	1.5436	1.5053	10.64	24	1129	1027
6300	0.6881	1.4803	1.4435	9.49	176	977	889
5600	0.7033	1.4232	1.3879	8.90	310	843	767
4900	0.7175	1.3725	1.3385	8.53	434	719	654
4200	0.7312	1.3263	1.2934	8.20	549	604	550
3500	0.7448	1.2836	1.2517	7.90	656	497	452
2800	0.7584	1.2434	1.2126	7.61	755	398	362
2100	0.7718	1.2056	1.1757	7.35	852	301	274
1400	0.7856	1.1684	1.1394	7.09	947	206	187
700	0.7993	1.1313	—	6.86	1041	112	—
15	0.8215	1.0728	—	—	1153	0	—
15 @ 60°F	0.8813	1.0000	—	—	1153	0	—

*Static bottomhole pressure

**Bubble point pressure, p_b

†Adjusted for field separator conditions per bbl. of stock tank oil at 60°F.

6A.5 Gas Differentially Liberated at 212°F

Pressure (psia)	Specific Gravity (Air = 1.00)		Compressibility Factor (z)	Gas FVF, B_g (RB/MMscf)	Total FVF B_t (RB/STB)	Gas Viscosity (Calculated)* (cp)
	Incremental	Accumulative				
7153	(Saturation Pressure at 212°F)					
7000	0.6030	0.6030	1.133	560	1.5571	0.0309
6300	0.5950	0.5961	1.091	599	1.5853	0.0285
5600	0.5900	0.5935	1.051	649	1.6242	0.0262
4900	0.5861	0.5914	1.013	715	1.6824	0.0241
4200	0.5840	0.5898	0.980	807	1.7688	0.0221
3500	0.5811	0.5884	0.952	941	1.8999	0.0202
2800	0.5820	0.5875	0.935	1154	2.1151	0.0183
2100	0.5832	0.5871	0.928	1528	2.5064	0.0167
1400	0.5920	0.5876	0.936	2313	3.3580	0.0153
700	0.6540	0.5935	0.962	4753	6.0777	0.0140
15	0.9210	0.6254	1.000	230,176	266.4026	0.0122

$B_t = B_o + \left[R_t B_g \right] \times 10^{-6}$

*Based on Lee et al.[16]

6A.6 Viscosity Data at 212°F

Pressure (psia)	Oil Viscosity (cp)	Gas Viscosity (Calculated)† (cp)	Oil/Gas Viscosity Ratio
10640*	0.676		
10000	0.637		
9000	0.579		
8000	0.517		
7153**	0.467		
Two Phase Region Data Below			
7000	0.471	0.0309	15.23
6300	0.507	0.0285	17.81
5600	0.548	0.0262	20.89
4900	0.598	0.0241	24.80
4200	0.671	0.0221	30.36
3500	0.772	0.0202	38.31
2800	0.944	0.0183	51.47
2100	1.286	0.0167	77.05
1400	1.774	0.0153	116.18
700	2.435	0.0140	174.43
15	3.436	0.0122	282.80

*Static bottomhole pressure

**Bubble point, p_b

†Based on Lee et al.[16]

6A.7 Comparison of Reservoir Oil Flash Liberation Tests

Liberation Type At 212°F	Gas-Oil Ratio*	Gas Gravity	Stock Tank Oil Gravity (°API @ 60 °F)	FVF (V_{sat}/V_{st})**
	Total Solution Gas			
Flash separation	1049	0.6079	30.2	1.5186
Differential	1153	0.6254	28.9	1.5572

*Gas-oil ratio is cubic feet of gas at 15.025 psia and 60°F per bbl of stock tank oil at 60°F.

**V_{sat}/V_{st} is the volume of reservoir oil at the saturation pressure and reservoir temperature relative to stock tank oil volume at 60°F.

APPENDIX

6B TYPICAL RESERVOIR FLUID STUDY FOR A GAS CONDENSATE SAMPLE

6B.1 Summary of Reservoir Data and Surface Sampling Conditions

Reservoir Data	
Formation	Unknown
Perforated interval	15,390–15,516 feet
Flowing bottomhole pressure	Not available
Static bottomhole pressure	12,688 psia
Reservoir temperature	267°F
Surface Sampling Summary	
Well status	On production
Choke size	42/64 inch
Date and time sampled	March 2, 1997; 1200 hours
Flowing tubing pressure	9020 psig
Primary separator temperature	92°F
Meter run temperature	92°F
Primary separator pressure	1140 psig
Primary separator gas rate (field)	48,453 Mcf/D
Primary separator gas rate (lab)	49,544.8 Mcf/D
Oil rate measured at primary separator	2120 B/D
Water rate measured at primary separator	40 B/D
Gas-oil ratio	23,370 scf first-stage gas per bbl. separator liquid
Gas-oil ratio	28,721 scf first-stage gas per bbl. stock tank liquid
Standard pressure and temperature	15.025 psia, 60°F

6B.2 Chromatograph Analysis of Separator Gas at 1140 psig and 92°F

Component	Mole (%)	Weight (%)	GPM* @15.025 psia
Nitrogen	0.20	0.30	0.000
Hydrogen Sulfide	0.00	0.00	0.000
Carbon Dioxide	0.94	2.25	0.000
Methane	90.44	79.13	0.000
Ethane	5.28	8.65	1.440
Propane	1.70	4.09	0.478
i-Butane	0.36	1.15	0.122
n-Butane	0.41	1.30	0.132
i-Pentane	0.16	0.65	0.062
n-Pentane	0.12	0.45	0.043
Hexanes	0.16	0.72	0.063
Heptanes Plus	0.23	1.31	0.103
Totals	**100.00**	**100.00**	**2.443**

*GPM is Gallons per Mcf

Calculated Properties of Gas:

Gas specific gravity (Air = 1.00)	= 0.6346
Gas Heat of Combustion (Btu/cu. Ft. @ 15.025 psia & 60°F) Dry	= 1133.7
Gas Heat of Combustion (Btu/cu. Ft. @ 15.025 psia & 60°F) Wet	= 1113.9 water saturated
Gas compressibility factor, z (at 1 atmosphere, 60°F)	= 0.9973

6B.3 Chromatograph Analysis of Separator Liquid at 1140 psig and 92°F

Component	Mole (%)	Liquid Volume (%)	Weight (%)	Mole Weight	Specific Gravity
Hydrogen Sulfide	0.00	0.00	0.00	34.076	0.8014
Carbon Dioxide	0.59	0.23	0.26	44.010	0.8180
Nitrogen	0.04	0.01	0.01	28.013	0.8094
Methane	28.16	9.94	4.49	16.043	0.3324
Ethane	5.89	2.63	1.76	30.070	0.4935
Propane	4.74	3.01	2.08	44.097	0.5070
i-Butane	1.76	1.33	1.02	58.123	0.5629
n-Butane	2.75	2.00	1.59	58.123	0.5840
i-Pentane	2.04	1.72	1.46	72.150	0.6247
n-Pentane	1.84	1.54	1.32	72.150	0.6311
Hexanes	2.08	1.98	1.78	86.177	0.6644
Heptanes Plus	50.11	75.61	84.23	169.100	0.8197
Totals	**100.00**	**100.00**	**100.00**	—	—

Properties	Measured for Heptanes Plus	Calculated for Liquid
Specific gravity	0.8197	0.7358 @ 60/60°F
Molecular weight	169.10	100.60
Cu. Ft/Gal.	15.00	22.63 @ 15.025 psia and 60°F

6B.4 Composition of Reservoir Fluid (Calculated)

Component	Mole (%)
Hydrogen Sulfide	0.00
Carbon Dioxide	0.93
Nitrogen	0.19
Methane	88.01
Ethane	5.30
Propane	1.82
i-Butane	0.42
n-Butane	0.50
i-Pentane	0.23
n-Pentane	0.19
Hexanes	0.24
Heptanes Plus	2.17
Totals	**100.00**

Basis of recombination calculation (1)	= 23,370 scf separator gas per bbl. separator liquid
Basis of recombination calculation (2)	= 42.79 bbl. separator liquid per MMscf separator gas
Total molecular weight	= 21.5 lb/lb-mole
Total specific gravity (Air = 1.0)	= 0.7437
Heptanes Plus molecular weight	= 162.1 lb/lb-mole
Heptanes Plus specific gravity (water = 1.0)	= 0.8125

6B.5 Measured Saturation Pressures from Stepwise Recombinations at 267°F

Gas-Liquid Ratio (scf first-stage gas)/(bbl. first-stage liquid)	Saturation Pressure (Dew Point) (psia)
40,000	9304
30,000	9960
23,000	9912

6B.6 Pressure-Volume Properties of Reservoir Fluid at 267°F (or CCE)

Pressure (psia)	Relative Volume ($V/V_{Res.}$)	Specific Volume (cu. ft./lb)	Reservoir		Deviation Factor (*z*)	Gas Viscosity (cp)	Percent Wellstream Produced††
			bbl/MMcf	Liquid Vol. (%)†			
12668*	1.0000	0.04698	—	—	1.6441	0.0503	0.00
12000	1.0208	0.04796	—	—	1.5898	0.486	2.04
11500	1.0392	0.04883	—	—	1.5511	0.0471	3.78
11000	1.0547	0.04955	—	—	1.5058	0.0460	5.19
10750	1.0643	0.05000	—	—	1.4849	0.0454	6.04
10500	1.0745	0.05048	—	—	1.4643	0.0447	6.93
10000	1.0946	0.05143	—	—	1.4207	0.0434	8.65
9912**	1.0990	0.05164	0.00	0.00	1.4138	0.0432	9.01
9500	1.1187	0.05256	0.16	0.03	1.3793	—	10.61
9000	1.1450	0.05379	0.21	0.04	1.3374	—	12.66
8500	1.1743	0.05517	0.42	0.09	1.2955	—	14.84
8000	1.2085	0.05678	0.77	0.16	1.2547	—	17.25
7500	1.2471	0.05859	1.34	0.28	1.2139	—	19.81
7000	1.2921	0.06071	1.86	0.38	1.1739	—	22.61
6000	1.4078	0.06614	3.05	0.63	1.0963	—	28.97
5000	1.5838	0.07441	4.30	0.89	1.0278	—	36.86
4000	1.8703	0.08787	5.86	1.21	0.9710	—	46.53
3000	2.3994	0.11273	7.16	1.48	0.9342	—	58.32
2000	3.5510	0.16684	7.33	1.51	0.9217	—	71.84
1000	7.2273	0.33957	6.92	1.43	0.9380	—	86.16

*Static bottomhole pressure

**Saturation pressure (dew point)

†Liquid volume is calculated as percent of hydrocarbon pore space at reservoir volume occupied by liquid.

††Wellstream produced is calculated as volume percent of original reservoir fluid.

6B.7 Depletion Study at 267°F: Hydrocarbon Analyses of Produced Wellstream (Mole %)

Component	Reservoir Pressure (psia)						
	Reservoir Fluid						Abandonment Liquid
	12,668	9912	8000	6000	4000	2000	2000
Hydrogen Sulfide	0.00	0.00	0.00	0.00	0.00	0.00	0.00
Nitrogen	0.19	0.19	0.19	0.20	0.19	0.20	0.04
Carbon Dioxide	0.93	0.93	0.94	0.94	0.95	0.93	0.41
Methane	88.01	88.01	88.78	89.29	89.62	88.81	32.10
Ethane	5.30	5.30	5.27	5.25	5.27	5.34	4.41
Propane	1.82	1.82	1.81	1.81	1.81	1.81	2.68
i-Butane	0.42	0.42	0.41	0.40	0.40	0.40	0.88
n-Butane	0.50	0.50	0.49	0.49	0.49	0.48	1.26
i-Pentane	0.23	0.23	0.22	0.21	0.21	0.23	0.88
n-Pentane	0.19	0.19	0.18	0.18	0.18	0.17	0.74
Hexanes	0.24	0.24	0.23	0.22	0.21	0.23	1.64
Heptanes plus	2.17	2.17	1.48	1.01	0.67	1.40	54.96
Totals	**100.00**	**100.00**	**100.00**	**100.00**	**100.00**	**100.00**	**100.00**
Specific Gravity	0.7437	0.7437	0.6976	0.6729	0.6585	0.6942	—
Properties of Heptanes Plus							
Specific Gravity	0.8125	0.8125	0.7890	0.7745	0.7699	0.7908	0.8300
Molecular Weight	162.1	162.1	142.4	132.0	128.9	142.5	181.1
Liquid Content of Well Stream (Gallons/Mscf)	3.914	3.914	3.349	3.026	2.833	3.314	—
Reservoir Fluid Deviation Factor, *z*							
Two-Phase Fluid	—	1.414	1.256	1.101	0.974	0.929	—
Gas Phase Produced	1.644	1.414	1.267	1.110	0.993	0.940	—
Gas FVF, B_g (Res. cu. ft./Mcf)	2.726	2.996	3.326	3.886	5.214	9.871	—
Gas Viscosity (cp)	0.0503	0.0432	0.0348	0.0285	0.0225	0.0173	—
Initial Fluid Produced Cumulative (%)	0.00	9.02	16.53	27.56	45.69	72.05	—

6B.8 Retrograde Condensation During Gas Depletion at 267°F

Pressure (psia)	Reservoir Liquid (Volume %)*
9912	0.00
8000	0.15
6000	0.56
4000	0.82
2000	0.89

*Percent of reservoir hydrocarbon pore space at reservoir pressure.

CHAPTER

7 PVT Properties Predictions from Equations of State

7.1 Historical Introduction to Equations of State (EOS)

The most common and familiar equation of state (EOS) is the ideal gas EOS. This was derived in Chapter 5 from the combination of Boyle's and Charles' laws on gases. The ideal gas EOS was represented in Eq. (5.7) as: $pV = nRT$. The EOS for real gases was represented in Eq. (5.9) as: $pV = znRT$. As discussed in Chapter 5, the gas compressibility factor, z, is used to account for the non-ideality of real gases. The gas compressibility factor is not constant but varies with the temperature and pressure of the system. This imposes mathematical limitations on the direct application of real gas EOS for representation of thermodynamic state of gases, since the gas compressibility factor must be evaluated at each thermodynamic condition.

In 1873, van der Waals[1] proposed an EOS that represented non-ideal behavior by accounting for finite volumes occupied by molecules of substances, and the repulsive and attractive forces between these molecules. The van der Waals (vdW) EOS is shown as:

$$\left(p+\frac{a}{V^2}\right)(V-b)=RT \tag{7.1}$$

In Eq. (7.1), p = pressure; v = volume; T = temperature; a = attraction parameter to correct pressures for attraction between molecules; b = effective molecular volume to correct for volume occupied by the molecules; and R = universal gas constant.

Many EOS of varying degrees of complexity have been proposed since the emergence of the van der Waals (vdW) EOS. A summary of early equations of state as documented by Walas[2] shows the rich history of advancement in the development of equations of state.

The presentation of equations of state is limited to van der Waals–type EOS in this book. These are generally called two-parameter equations of state because they have the form of the original van der Waals (vdW) EOS. In addition to the van der Waals (vdW) EOS, two other equations of state presented in this book are the Soave-Redlich-Kwong (SRK) EOS and the Peng-Robinson (PR) EOS. The reasons for the presentation of only these two other equations of state are that most publications, research work, and software in the petroleum industry are based on the SRK and PR equations of state or variants of these equations of state.

7.2 van der Waals (vdW) EOS

As shown in Eq. (7.1), the van der Waals (vdW) EOS is represented as:

$$\left(p+\frac{a}{V^2}\right)(V-b)=RT$$

In cubic forms, it can be represented as:

$$V^3-\left(b+\frac{RT}{p}\right)V^2+\frac{a}{p}V-\frac{ab}{p}=0 \tag{7.2}$$

$$z^3-z^2\left(\frac{bp}{RT}+1\right)+z\frac{ap}{(RT)^2}-\frac{abp^2}{(RT)^3}=0 \tag{7.3}$$

The above representations of vdW EOS in cubic forms are the bases that these types of equations of state are often referred to as cubic equations of state.

To derive formulas for the constants, a and b, in vdW EOS [Eq. (7.1)], it is useful to observe that at the critical temperature, the first and second derivative of pressure with respect to volume is zero for a pure substance. This means that for a pure substance, the critical temperature isotherm on a pressure-volume (P-V) diagram has a zero gradient as it passes through the critical point. Thus,

$$\left(\frac{\partial p}{\partial V}\right)_{T_c}=-\frac{RT_c}{(V_c-b)^2}+\frac{2a}{V_c^3}=0 \tag{7.4}$$

$$\left(\frac{\partial^2 p}{\partial V^2}\right)_{T_c}=\frac{2RT_c}{(V_c-b)^3}-\frac{6a}{V_c^4}=0 \tag{7.5}$$

Also, vdW EOS can be expressed in terms of critical properties as:

$$\left(p_c+\frac{a}{V_c^2}\right)(V_c-b)=RT_c \tag{7.6}$$

Eqs. (7.4), (7.5), and (7.6) can be solved simultaneously for a and b as:

$$a=3p_cV_c^2=\frac{27R^2T_c^2}{64p_c} \tag{7.7}$$

$$b = \frac{V_c}{3} = \frac{RT_c}{8p_c} \tag{7.8}$$

$$z_c = 0.375 \tag{7.9}$$

For mixtures, the following mixing rules based on the pure component parameters may be applied for vdW EOS:

$$a_m = \sum_i^{N_C} \sum_j^{N_C} y_i y_j \sqrt{a_i a_j} \tag{7.10}$$

$$b_m = \sum_i^{N_C} y_i b_i \tag{7.11}$$

In Eqs. (7.10) and (7.11), a_m and b_m are mixture parameters; y_i and y_j are compositions for mixture components, i and j; a_i, a_j, and b_i are parameters for the pure components, i and j; and N_C, is the total number of components in the mixture.

7.3 Soave-Redlich-Kwong (SRK) EOS

The Soave-Redlich-Kwong (SRK) EOS is a modification of the Redlich-Kwong (RK) EOS. The equations representing RK EOS are presented first followed by the modifications introduced in the SRK EOS.

The RK EOS was proposed in 1949.[3] It is a modification of the vdW EOS with the introduction of a temperature correction for the attraction parameter, a, in the original vdW EOS. The equations representing the RK EOS are as follows:

$$p + \frac{a}{\sqrt{T}V(V+b)} = \frac{RT}{V-b} \tag{7.12}$$

$$a = \Omega_a \frac{R^2 T_c^{2.5}}{p_c} = 0.42748 \frac{R^2 T_c^{2.5}}{p_c} \tag{7.13}$$

$$b = \Omega_b \frac{RT_c}{p_c} = 0.08664 \frac{RT_c}{p_c} \tag{7.14}$$

$$A = \frac{ap}{R^2 T^{2.5}} = 0.42748 \frac{p_r}{T_r^{2.5}} \tag{7.15}$$

$$B = \frac{bp}{RT} = 0.08664 \frac{p_r}{T_r} \tag{7.16}$$

In cubic forms, the RK EOS is shown as:

$$V^3 - V^2 \frac{RT}{p} + \frac{V}{p}\left(\frac{a}{\sqrt{T}} - bRT - pb^2\right) - \frac{ab}{p\sqrt{T}} = 0 \tag{7.17}$$

$$z^3 - z^2 + z(A - B - B^2) - AB = 0 \tag{7.18}$$

For mixtures, the following mixing rules for the parameters are applicable:

$$a_m = \sum_i^{N_C} \sum_j^{N_C} y_i y_j a_{ij}, \text{ where } a_{ij} = \sqrt{a_i a_j} \tag{7.19}$$

$$b_m = \sum_i^{N_C} y_i b_i \tag{7.20}$$

$$A_m = \sum_i^{N_C} \sum_j^{N_C} y_i y_j A_{ij}, \text{ where } A_{ij} = \sqrt{A_i A_j} \tag{7.21}$$

$$B_m = \sum_i^{N} y_i B_i \tag{7.22}$$

In Eqs. (7.19) to (7.22), a_m, b_m, A_m, and B_m are parameters for the mixture; y_i and y_j are compositions of mixture components; a_i b_i, A_i, and B_i are parameters for the pure components; a_{ij} and A_{ij} are the cross-parameter terms between components in the mixture.

In 1972, Soave[4] modified the RK EOS by replacing the temperature dependent term, $a/\sqrt{T}$, in the RK EOS with another term, $a(T, \omega)$ which defined the attraction term, a, as a function of temperature, T, and acentric factor, ω. The SRK EOS is:

$$p + \frac{a(T,\omega)}{V(V+b)} = \frac{RT}{V-b} \tag{7.23}$$

Using the coefficients published by Graboski and Daubert,[5,6] the parameters in the SRK EOS are defined as:

$$a(T,\omega) = a\alpha = 0.42747 \frac{R^2 T_c^2}{p_c} \alpha \tag{7.24}$$

$$\alpha = \left[1.0 + m\left(1.0 - T_r^{0.5}\right)\right]^2 \tag{7.25}$$

$$m = 0.48508 + 1.55171\omega - 0.15613\omega^2 \tag{7.26}$$

$$b = 0.08664 \frac{RT_c}{p_c} \tag{7.27}$$

$$A = \frac{\alpha a p}{R^2 T^2} = \frac{0.42747 \alpha p_r}{T_r^2} \tag{7.28}$$

$$B = \frac{bp}{RT} = \frac{0.08664 p_r}{T_r} \tag{7.29}$$

In Eq. (7.26), ω is the Pitzer acentric factor. Pitzer et al.[7] defined acentric factor as:

$$\omega = -\left(\log\left(\frac{p_v^*}{p_c}\right) + 1\right) \text{ at } T_r = \frac{T}{T_c} = 0.7 \tag{7.30}$$

In Eq. (7.30), p_v^* = vapor pressure at $T = 0.7T_c$; T_c = critical temperature, and p_c = critical pressure. Values of acentric factors for pure substances are given in Table 5.1.

In cubic forms, SRK EOS is represented as:

$$V^3 - V^2\frac{RT}{p} + \frac{V}{p}\left(a\alpha - bRT - pb^2\right) - \frac{a\alpha b}{p} = 0 \tag{7.31}$$

$$z^3 - z^2 + \left(A - B - B^2\right)z - AB = 0 \tag{7.32}$$

For mixtures, as was the case for RK EOS, these mixing rules are applicable:

$$(a\alpha)_m = \sum_i^{N_C}\sum_j^{N_C} y_i y_j (a\alpha)_{ij} \tag{7.33}$$

$$b_m = \sum y_i b_i \tag{7.34}$$

$$A_m = \sum_i^{N_C}\sum_j^{N_C} y_i y_j A_{ij} \tag{7.35}$$

$$B_m = \sum_i^{N_C} y_i B_i \tag{7.36}$$

The cross-parameter terms are expressed as:

$$(a\alpha)_{ij} = \left(1 - k_{ij}\right)\sqrt{(a\alpha)_i (a\alpha)_j} \tag{7.37}$$

$$A_{ij} = \left(1 - k_{ij}\right)\sqrt{A_i A_j} \tag{7.38}$$

The term, k_{ij}, is called the binary interaction parameter (BIP). By definition, $k_{ii} = 0$, and, $k_{ij} = k_{ji}$. Binary interaction parameters are sometimes adjusted during "tuning" of cubic equations of state to match experimental phase equilibria data.

7.4 Peng-Robinson (PR) EOS

In 1976, Peng and Robinson[8] introduced a new two-parameter equation of state developed primarily to improve calculations in the vicinity of the critical region, and also improve calculation of liquid densities. The PR EOS is similar in form to the familiar RK and SRK equations of state. The PR EOS is expressed as:

$$p + \frac{a(T,\omega)}{V(V+b) + b(V-b)} = \frac{RT}{V-b} \tag{7.39}$$

The parameters for the PR EOS are as follows:

$$a(T,\omega) = a\alpha = 0.45724\frac{R^2T_c^2}{p_c}\alpha \tag{7.40}$$

$$\alpha = \left[1.0 + m\left(1.0 - T_r^{0.5}\right)\right]^2 \tag{7.41}$$

$$m = 0.37464 + 1.54226\omega - 0.26992\omega^2 \tag{7.42}$$

$$b = 0.07780\frac{RT_c}{p_c} \tag{7.43}$$

$$A = \frac{\alpha a p}{R^2T^2} = \frac{0.4572\alpha p_r}{T_r^2} \tag{7.44}$$

$$B = \frac{bp}{RT} = \frac{0.07780 p_r}{T_r} \tag{7.45}$$

For heavier components, $\omega > 0.49$:[9,10]

$$m = 0.3796 + 1.485\omega - 0.1644\omega^2 + 0.01667\omega^3 \tag{7.46}$$

In cubic form, PR EOS is represented as:

$$z^3 - (1-B)z^2 + \left(A - 3B^2 - 2B\right)z - \left(AB - B^2 - B^3\right) = 0 \tag{7.47}$$

For mixtures, the same mixing rules presented in Eqs. (7.33) to (7.38) for SRK EOS are applicable to the PR EOS.

7.5 Phase Equilibrium of Mixtures

It was stated in Chapter 5 that the phases in a system composed of one or more components are considered to be in equilibrium if the chemical potential of each component in all the phases are equal at the same temperature and pressure. The chemical potential, μ, of a pure substance is related to temperature and pressure by the expression:

$$d\mu = -sdT + vdp \tag{7.48}$$

In Eq. (7.48), s is the molar entropy, and v is the molar volume. Chemical potential is sometimes called Gibbs energy. For an ideal gas at constant temperature, Eq. (7.48) reduces to:

$$d\mu = vdp = RTd\ln p \tag{7.49}$$

Eq. (7.49) can be applied to a real fluid by replacing the pressure term, p, with a new term, f, called fugacity such that:

$$d\mu = RTd\ln f = vdp \tag{7.50}$$

For an ideal gas, fugacity is equal to pressure and has pressure units. For a real fluid, fugacity approaches pressure at low pressure as defined by the limit:

$$\lim_{p\to 0} f = p \tag{7.51}$$

The ratio of fugacity to pressure for a pure substance is called fugacity coefficient:

$$\phi = \frac{f}{p} \tag{7.52}$$

The fugacity coefficient of a pure substance is expressed by the general thermodynamic relationship:

$$\ln \phi = \ln \frac{f}{p} = \frac{1}{RT} \int_0^p \left(V - \frac{RT}{p} \right) dp \tag{7.53}$$

By substitution of the SRK EOS [Eq. (7.23)] into Eq. (7.53), the fugacity coefficient of a pure substance is expressed as:

$$\ln \phi = z - 1 - \ln(z - B) - \frac{A}{B} \ln \left(\frac{z + B}{z} \right) \tag{7.54}$$

Similarly for the PR EOS, by substituting Eq. (7.39) into Eq. (7.53), the fugacity coefficient for a pure substance is expressed as:

$$\ln \phi = z - 1 - \ln(z - B) - \frac{A}{2\sqrt{2}B} \ln \left(\frac{z + 2.414B}{z - 0.414B} \right) \tag{7.55}$$

For a component in a mixture, the rigorous thermodynamic relationship for the fugacity coefficient is:

$$RT \ln \phi_i = \int_V^\infty \left[\frac{\partial P}{\partial n_i} - \frac{RT}{V} \right] dV - RT \ln z \tag{7.56}$$

In Eq. (7.56), ϕ_i is the partial fugacity coefficient of component i in the mixture where:

$$\phi_i = \frac{f_i}{x_i p} \tag{7.57}$$

In Eq. (7.57), f_i is the partial fugacity of the component in the mixture; x_i is the mole fraction of the component in the mixture; and p is the total pressure of the mixture.

For the SRK EOS, the thermodynamic expression for partial fugacity coefficient of a component in a mixture as derived from Eq. (7.56) is:

$$\ln \phi_i = \frac{B_i}{B}(z - 1) - \ln(z - B) + \frac{A}{B} \left[\frac{B_i}{B} - \frac{2}{a\alpha} \sum_j x_j (a\alpha)_{ij} \right] \ln \left(1 + \frac{B}{z} \right) \tag{7.58}$$

A similar expression for the PR EOS is:

$$\ln \phi_i = \frac{B_i}{B}(z - 1) - \ln(z - B) + \frac{A}{2.828B} \left[\frac{B_i}{B} - \frac{2}{a\alpha} \sum_j x_j (a\alpha)_{ij} \right] \ln \left[\frac{z + 2.414B}{z - 0.414B} \right] \tag{7.59}$$

It was stated earlier that the chemical potentials of all components in all coexisting phases at equilibrium are equal. From Eq. (7.50), chemical potential can be related directly to fugacity. It thus follows that the fugacity of a component in a mixture is equal in all phases of the mixture coexisting at equilibrium. For instance, if the mixture is composed only of a vapor phase and a liquid phase, it means that the fugacity of a component in the vapor phase is equal to its fugacity in the liquid phase. This statement can be expressed simply as:

$$f_i^V = f_i^L \tag{7.60}$$

In Eq. (7.60), f_i is the partial fugacity of component, i; V and L denote the vapor and liquid phases, respectively. Equation (7.60) is the fundamental equation that represents the condition of equilibrium in a mixture. If the mixture consisted of N_C components with N phases in equilibrium, Eq. (7.60) can be extended to illustrate that the partial fugacity of a component in all the N phases are equal:

$$f_i^k = f_i^j, \quad i = 1, \ldots\ldots, N_C; \quad k, j = 1, \ldots\ldots, N \tag{7.61}$$

Applying Eq. (7.57), Eq. (7.60) can be written as:

$$y_i p \phi_i^V = x_i p \phi_i^L \tag{7.62}$$

In Eq. (7.62), p is the pressure of the mixture; y_i and x_i are vapor and liquid phase compositions of component i in equilibrium; and ϕ_i is the corresponding partial fugacity coefficient of the component in each phase. From Eq. (7.62), it is deduced that:

$$\frac{y_i}{x_i} = \frac{\phi_i^L}{\phi_i^V} = K_i \tag{7.63}$$

K_i is the equilibrium constant of component, i.

7.6 Roots from Cubic EOS

The roots of the cubic forms of vdW, RK, SRK, and PR equations of state can be determined by analytical or iterative methods. The solution may yield one or three real roots. When three real roots are obtained, the smallest root is selected for the liquid phase and the largest root is selected for the vapor phase. The middle root is discarded because it has no physical significance. Michelsen[11] devised a method for choosing the correct root for mixtures based on the phase with the lowest normalized Gibbs energy. For the vapor phase, normalized Gibbs energy, g_y^*, is defined as:

$$g_y^* = \sum_{i=1}^{N} y_i \ln f_i(y) \tag{7.64}$$

For the liquid phase, normalized Gibbs energy, g_x^*, is defined as:

$$g_x^* = \sum_{i=1}^{N} x_i \ln f_i(x) \tag{7.65}$$

In Eqs. (7.64) and (7.65), y_i and x_i are mole fractions of vapor and liquid, respectively, and, f_i is the fugacity of the component in the mixture.

7.7 Volume Translation

In a very exhaustive effort to improve volumetric predictions with generalized cubic equations of state, Martin[12] introduced the concept of volume translation. He proved that no other two-parameter cubic equation of state can be developed to be superior to a two-parameter cubic equation with volume translation. Working independently, Peneloux et al.[13] applied volume translation to improve volumetric predictions with SRK EOS with an approach that was directly applicable to phase equilibria and volumetric calculations in the petroleum industry. In 1988, Jhaveri and Youngren[14] extended the work of Peneloux et al.[13] to the PR EOS. The concepts of volume translation are presented as described by Peneloux et al.[13] and extended by Jhaveri and Youngren.[14]

Peneloux et al.[13] introduced a third term, c, into the two-parameter cubic equations of state to correct the molar volume calculated with the cubic equations such that:

$$\tilde{v} = v - c \tag{7.66}$$

In Eq. (7.66), $\tilde{v}$ = corrected (translated) molar volume; v = molar volume calculated with EOS; and c = third (correction) component dependent parameter. For mixtures, the parameter, c, is related to the third parameter, c_i, for each component through a linear mixing rule such that:

$$c = \sum_i c_i z_i \tag{7.67}$$

In Eq. (7.67), z_i is the mole fraction of the component in the mixture. For multicomponent systems, Peneloux et al.[13] showed that volume correction can be applied to the liquid and vapor phases (assuming a biphasic system) as:

$$\tilde{v}_L = v_L - \sum_i^{N_C} c_i x_i \tag{7.68}$$

$$\tilde{v}_V = v_V - \sum_i^{N_C} c_i y_i \tag{7.69}$$

In Eqs. (7.68) and (7.69), $\tilde{v}_L$ and $\tilde{v}_V$ are corrected molar volume for the liquid and vapor phases, respectively; v_L and v_V are EOS calculated molar volumes for the same phases; c_i is the third parameter for component, i; and x_i and y_i are mole fractions of the component in the liquid and vapor phases, respectively.

The important contribution made by Peneloux et al.[13] is that they showed that vapor-liquid equilibrium conditions determined with the unmodified SRK EOS are not altered by the introduction of volume translation parameters. This is seen readily by relating the fugacity of a component of a mixture in the modified EOS to the original EOS in the vapor and liquid phases yielding:

$$\tilde{f}_i^V = f_i^V \exp\left(-\frac{c_i p}{RT}\right) \tag{7.70}$$

$$\tilde{f}_i^L = f_i^L \exp\left(-\frac{c_i p}{RT}\right) \tag{7.71}$$

In Eqs. (7.70) and (7.71), $\tilde{f}_i^V$ and $\tilde{f}_i^L$ are the modified fugacities for the vapor and liquid phase, respectively, and f_i^V and f_i^L are the original fugacities for the vapor and liquid phase. From Eqs. (7.70) and (7.71), the fugacity ratio of the modified EOS is equal to the fugacity ratio of the original EOS not modified with volume shift:

$$\frac{\tilde{f}_i^V}{\tilde{f}_i^L} = \frac{f_i^V}{f_i^L} \tag{7.72}$$

The equivalence of fugacity ratios in Eq. (7.72) demonstrate that equilibrium conditions determined with the original EOS are not affected by volume translation.

For pure components, Peneloux et al.[13] determined the volume shift parameter, c, by calculating the saturated liquid molar volume, v, at the reduced temperature of 0.7 with the unmodified two-parameter EOS and matching it against the experimental volume, v_E, so that:

$$c = v - v_E \tag{7.73}$$

In Eq. (7.73), c = volume shift parameter; v = saturated liquid molar volume calculated with the unmodified two-parameter EOS; v_E = experimentally measured saturated liquid molar volume.

For correlational purposes, Jhaveri and Youngren[14] introduced a dimensionless shift parameter, s_i, for any component defined as:

$$s_i = \frac{c_i}{b_i} \tag{7.74}$$

In Eq. (7.74), s_i = dimensionless shift parameter for component, i; c_i = volume shift parameter for component, i; and b_i = second parameter for component i in the unmodified PR EOS. Values of s_i for light hydrocarbons generated by Jhaveri and Youngren[14] are shown in Table 7.1 for PR EOS. For heavier hydrocarbons (heptanes$^+$), Jhaveri and Youngren[14] developed a correlation that is dependent on molecular weight. The correlation is:

$$s_i = 1 - \frac{d}{M_i^e} \tag{7.75}$$

TABLE 7.1 Volume Shift Parameter for Hydrocarbons (from Jhaveri and Youngren[14] © 1988 SPE, Reproduced with permission)

Component	s_i
Methane	−0.15400
Ethane	−0.10020
Propane	−0.08501
Iso-Butane	−0.07935
n-Butane	−0.06413
Iso-Pentane	−0.04350
n-Pentane	−0.04183
n-Hexane	−0.01478

In Eq. (7.75), M_i = molecular weight of component i; d and e are positive correlation coefficients. Values of d and e for paraffins, naphthenes, and aromatics are shown in Table 7.2 for the PR EOS.

In effect, application of volume translation converts two-parameter cubic equations of state to essentially three-parameter equations. This has been shown by Peneloux et al.[13] and Jhaveri and Youngren[14] to improve volumetric calculations with SRK and PR EOS, respectively.

TABLE 7.2 Volume Shift Parameter Correlation Coefficients for Hydrocarbons Heavier Than Hexane (from Jhaveri and Youngren[14] © 1988 SPE, Reproduced with permission)

Component Type	Correlation Coefficient		Average Error (%)
	d	e	
Paraffins	2.258	0.1823	0.19
Napthenes	3.004	0.2324	0.28
Aromatics	2.516	0.2008	0.24

7.8 Two-Phase Flash Calculation

Consider a biphasic mixture consisting of vapor and liquid phases at equilibrium. Suppose n_t is the total moles in the mixture with n_V moles in the vapor phase and n_L moles in the liquid phase. If, $\mathbf{z}_i$ is the mole fraction of component i in the mixture, y_i is its mole fraction in the vapor phase and x_i is its mole fraction in the liquid phase, the following equations represent the material balance of the component in the mixture:

$$n_t = n_V + n_L \tag{7.76}$$

$$n_t \mathbf{z}_i = n_V y_i + n_L x_i \tag{7.77}$$

For a system in equilibrium, the equilibrium constant, K_i, of a component i is the ratio of the mole fraction of the component in the vapor phase to its mole fraction in the liquid phase, as designated in Eq. (7.77). Thus, as previously shown in Eq. (7.63),

$$K_i = \frac{y_i}{x_i} \tag{7.78}$$

Substituting for y_i in Eq. (7.77) with Eq. (7.78) and rearranging, gives:

$$x_i = \frac{n_t \mathbf{z}_i}{n_V K_i + n_L} \tag{7.79}$$

Substituting for n_L in Eq. (7.79) with Eq. (7.76), we have:

$$x_i = \frac{n_t \mathbf{z}_i}{n_V (K_i - 1) + n_t} \tag{7.80}$$

Dividing Eq. (7.80) by n_t yields:

$$x_i = \frac{\mathbf{z}_i}{F_v (K_i - 1) + 1} \tag{7.81}$$

In Eq. (7.81), F_V is the mole fraction of vapor in the mixture where $F_V = n_V / n_t$.

Similarly, it can be shown that the mole fraction, y_i, of the same component in the vapor phase is represented by:

$$y_i = \frac{K_i \mathbf{z}_i}{F_V (K_i - 1) + 1} \tag{7.82}$$

By definition, the sum of the mole fractions of the equilibrated phases and the entire mixture must be equal to one. Hence for N_C components in the mixture:

$$\sum_{i=1}^{N_C} \mathbf{z}_i = \sum_{i=1}^{N_C} y_i = \sum_{i=1}^{N_C} x_i = 1.0 \tag{7.83}$$

From Eq. (7.83), a constraining equation can be obtained such that:

$$\sum_{i=1}^{N_C} y_i - \sum_{i=1}^{N_C} x_i = 0 \tag{7.84}$$

Substituting Eqs. (7.81) and (7.82) into Eq. (7.84) yields the Rachford-Rice[15] equation:

$$f(F_v) = \sum_{i=1}^{N_C} \frac{\mathbf{z}_i (K_i - 1)}{1 + F_v (K_i - 1)} = 0 \tag{7.85}$$

Eq. (7.85) is a monotonically decreasing convergence function that can be solved by root solving techniques, such as successive substitution method or the Newton-Raphson method.

7.8.1 Generalized Procedure for Two-Phase Flash Calculations

Step 1: Conduct phase stability check on the mixture.

Phase stability of the mixture is based primarily on the minimization of Gibbs free energy as suggested by Michelsen.[11] A procedure for phase stability check based on the Michelsen's method was presented by Whitson and Brule.[16] If a phase is determined not to be stable, then a flash calculation is initiated. Initial estimates of K-values can be obtained from this step. This step may not be necessary if temperature and pressure conditions of the mixture suggest the existence of a biphasic mixture.

Step 2: Estimate initial K-values.

Initial K-values can be obtained from the Wilson[17] equation:

$$K_i = \frac{\exp\left[5.37\left(1 + \omega_i\left(1 - T_{ri}^{-1}\right)\right)\right]}{p_{ri}} \tag{7.86}$$

This step can be skipped if K-values were obtained from Step 1.

Step 3: Calculate F_v from Eq. (7.85).

Solving Eq. (7.85) involves iteration, either by successive substitution or Newton-Raphson method, until convergence is reached at a set tolerance. A faster method for performing flash calculations has been proposed by Li and Johns.[18]

Step 4: Compute phase compositions from Eqs. (7.81) and (7.82).

Step 5: Calculate phase molar volumes from Eqs. (7.32) or (7.47).

Step 6: Compute partial component fugacities from Eqs. (7.58) or (7.59).

Step 7: Check for equivalence of partial component fugacities.

In this step, the equality of partial component fugacity in each phase is checked to a set tolerance level.

Step 8: Convergence check.

If convergence is reached in Step 7, stop. If not, update K-values using Eq. (7.63) for each component. Repeat Steps 3 to 7 until convergence is achieved.

7.9 Bubble Point and Dew Point Pressure Calculations

The bubble point pressure or the dew point pressure of a mixture is defined as the pressure at constant temperature at which the mixture is in equilibrium with an infinitesimal amount of the second phase, assuming a biphasic mixture. At bubble point, the mixture consists entirely of liquid except for an infinitesimal amount of gas (vapor). Conversely, at the dew point, the mixture is mostly gas (vapor) with an infinitesimal amount of liquid. At equilibrium, the sum of the mole fractions of the infinitesimal phase is equal to one. Thus, at bubble point:

$$\sum_{i=1}^{N_C} y_i = 1.0 \tag{7.87}$$

And at dew point:

$$\sum_{i=1}^{N_C} x_i = 1.0 \tag{7.88}$$

For Eqs. (7.87) and (7.88), y_i and x_i are mole fractions of component i in the vapor and liquid phases, respectively, and N_C is the total number of components in the mixture.

Substituting Eq. (7.78) into Eqs. (7.87) and (7.88), gives:

For bubble point:

$$\sum_{i=1}^{N_C} y_i = \sum_{i=1}^{N_C} K_i x_i = \sum_{i=1}^{N_C} K_i \mathbf{z}_i = 1.0 \tag{7.89}$$

In Eq. (7.89), $\mathbf{z}_i$ is the mole fraction of component i in the mixture, and $x_i \approx \mathbf{z}_i$, at the bubble point.

Similarly, at dew point:

$$\sum_{i=1}^{N_C} x_i = \sum_{i=1}^{N_C} \frac{y_i}{K_i} = \sum_{i=1}^{N_C} \frac{\mathbf{z}_i}{K_i} = 1.0 \tag{7.90}$$

In Eq. (7.90), $y_i \approx \mathbf{z}_i$ at the dew point.

Equations (7.89) and (7.90) can be converted to a form similar to Eq. (7.85) for convergence calculations. Thus, Eqs. (7.89) and (7.90) become, respectively:

$$\sum_{i=1}^{N_C} K_i \mathbf{z}_i - 1.0 = 0 \tag{7.91}$$

$$\sum_{i=1}^{N_C} \frac{\mathbf{z}_i}{K_i} - 1.0 = 0. \tag{7.92}$$

7.10 Characterization of Hydrocarbon Plus Fractions

In many laboratory analyses of reservoir fluid samples, the heavy hydrocarbons are typically lumped together and classified as C_n^+ fraction. In some PVT laboratory reports, the C_n^+ fraction is represented as C_7^+ fraction, while higher carbon number such as C_{10}^+ may be chosen to represent the hydrocarbon plus fractions in other reports. Typically, in most of these PVT reports, only the molecular weight and the specific gravity of the C_n^+ fraction are reported. Ideally, a complete True-Boiling-Point (TBP) analyses by distillation should be performed on fluid samples to yield boiling points, specific gravities, and molecular weights for Single-Carbon-Number (SCN) groups which could be used to characterize the fluid samples. TBP data are rarely available in most PVT reports because of the prohibitive expense and time required to measure them.

Although the mole fraction of the hydrocarbon plus fraction in a fluid sample is relatively small, it has a huge impact on the phase behavior of the fluid. In particular, the mole fractions and properties assigned to pseudo-components that result from splitting the plus fraction can significantly influence the phase behavior predicted for the fluid sample. Numerous methods have been used to split the hydrocarbon plus fraction into pseudo-components. In general, these can be grouped into two methods:

Method 1: The hydrocarbon plus fraction is split into many SCN groups using a probability distribution function or some other criteria. The SCN groups are assigned properties from correlations and regrouped into a few Multiple-Carbon-Number (MCN) groups with various averaging techniques. This method was proposed and applied by Whitson.[19,20]

Method 2: The hydrocarbon plus fraction is split into pseudo-components using semi-continuous thermodynamics as proposed by Behrens and Sandler[21] and adapted by Whitson and Brule.[16] The pseudo-components are assigned properties from correlations.

The splitting of the hydrocarbon plus fraction with semi-continuous thermodynamics has wide appeal in the petroleum industry and is available in most commercial PVT software. Since most engineers will most likely encounter this technique in commercial PVT software, the theoretical basis of the method is presented in this book. Semi-continuous thermodynamics is based on representing a fluid mixture with identifiable, discrete components, and using a continuous distribution function to represent the remaining components in the mixture. Behrens and Sandler[21] used a truncated exponential function as the continuous distribution function in their work. The three-parameter gamma function that was first proposed by Whitson[19] is the probability density function used for the description of the method. Note that any other continuous distribution function could be used in place of the three-parameter gamma function.

The three-parameter gamma function is represented as:

$$h(M)=\frac{(M-\eta)^{(\alpha-1)}\exp\left[\dfrac{M-\eta}{\beta}\right]}{\beta^{\alpha}\Gamma(\alpha)} \tag{7.93}$$

For Eq. (7.93), the distribution is defined by the three parameters, α, β, and η. The parameter, α, represents the shape of the distribution. If $\alpha = 1$, Eq. (7.93) reduces to an exponential distribution. The gamma function is represented as Γ. The parameter, β, is the normalized condition, which is given by:

$$\alpha\beta = M_{C_n^+}-\eta \tag{7.94}$$

In Eq. (7.94), $M_{C_n^+}$ is the average molecular weight of the plus fraction; and η is the lowest molecular weight in the plus fraction. Generally, η is approximated from the empirical relation:

$$\eta = 14n-6 \tag{7.95}$$

In Eq. (7.95), n is the first SCN group in the plus fraction.

The cumulative distribution function, $H(x)$, is the integral of $h(x)$ from η to x:

$$H(x)=\int_{\eta}^{x} h(x)\,dx \tag{7.96}$$

The numerical form of Eq. (7.96) is given by:

$$H(x)=e^{-y}\sum_{j=0}^{\infty}\left[\frac{y^{\alpha+j}}{\Gamma(\alpha+j+1)}\right], \tag{7.97}$$

where $y=(x-\eta)/\beta$

The frequency, ξ_i, of a component having molecular weight between boundaries M_{i-1} and M_i is given by the integral:

$$\xi_i=\int_{M_{i-1}}^{M_i} h(M)dM = H(M_i)-H(M_{i-1}) \tag{7.98}$$

The mole fraction, z_i, is calculated by multiplying the mole fraction of the plus fraction, z_n^+, by the respective frequency from Eq. (7.98):

$$z_i = z_n^+\times\xi_i \tag{7.99}$$

The average molecular weight in the same interval is given by:

$$\bar{M}_i=\eta+\alpha\beta\frac{H(M_i,\alpha+1)-H(M_{i-1},\alpha+1)}{H(M_i,\alpha)-H(M_{i-1},\alpha)} \tag{7.100}$$

The above equations represent the basic application of the three-parameter gamma function for splitting of plus fraction into many SCN groups.

Whitson et al.[22] adapted the gamma distribution model with Gaussian quadrature using the principles of semi-continuous thermodynamics to characterize multiple fluid samples from the same reservoir simultaneously. Each fluid sample can have its own molecular weight and specific gravity for the C_n^+ fraction before the split. After the split, each split fraction (pseudo-component) will have the same molecular weight and specific gravity while the mole fraction distribution will be different for each fluid sample. The procedure for applying Gaussian quadrature to the gamma distribution was modified from the procedure proposed by Whitson and Brule.[16]

1. Determine the number of C_n^+ split fractions, N. Obtain the quadrature values, X_i and W_i, from Table 7.3.
2. Specify η and α. Assume $\eta = 92$ and $\alpha - 1$.
3. Specify the heaviest molecular weight of fraction, N, as, M_N. A good estimate is $M_N = 2.5M_{C_n^+}$.
4. Calculate the parameter δ, where $\delta=\exp\left(\dfrac{\alpha\beta^*}{M_{C_n^+}-\eta}-1\right)$ and $\beta^*=(M_N-\eta)/X_N$
5. Calculate the C_n^+ mole fraction, z_i, and molecular weight, M_i, for each fraction

$$z_i = z_{C_n^+}\times\left[W_i f(X_i)\right] \tag{7.101}$$

$$M_i=\eta+\beta^* X_i \tag{7.102}$$

$$f(X)=\frac{(X)^{\alpha-1}}{\Gamma(\alpha)}\frac{(1+\ln\delta)^{\alpha}}{\delta^{X}} \tag{7.103}$$

6. Check whether the condition $\sum_{i=1}^{N} z_i M_i = M_{C_n^+}$ is satisfied. Modify the value of δ and repeat Steps 5 and 6 until a reasonable match is achieved.

TABLE 7.3 Gaussian Quadrature Variables and Weight Factors

Integration Points (Pseudo-components)	Function Variables X_i	Weight Factors W_i
Two Quadrature Points		
1	0.5858	0.8536
2	3.4142	0.1464
Three Quadrature Points		
1	0.4158	0.7111
2	2.2943	0.2785
3	6.2899	0.0104
Five Quadrature Points		
1	0.2636	0.5218
2	1.4134	0.3987
3	3.5964	0.07594
4	7.0858	0.003612
5	12.6408	0.00002337

7. For each split fraction, calculate specific gravity assuming an average Watson Characterization Factor, $\bar{K}_W$. From Whitson:[20]

$$\gamma_i = 6.0108 M_i^{0.17947} K_W^{-1.18241} \tag{7.104}$$

In Eq. (7.104), K_W is obtained from an approximation by Whitson:[19]

$$K_W = 4.5579 M_{C_n^+}^{0.15178} \gamma_{C_n^+}^{-0.84573} \tag{7.105}$$

8. Estimate boiling points for each split fraction from the equation:

$$T_{bi} = \left(K_w \gamma_i\right)^3 \tag{7.106}$$

9. With known boiling points, specific gravities and molecular weights for each split fraction (pseudo-component), the characterization of the hydrocarbon plus fractions is completed by calculating critical properties for each pseudo-component using the Kessler-Lee correlations (Eqs. 5.32 and 5.33). The Riazi-Daubert[23] correlations can also be used. From Riazi-Daubert correlations, the correlation for estimating critical pressure is:

$$p_{ci} = \left(3.12281 \times 10^9\right) \times T_{bi}^{-2.3125} \times \gamma_i^{2.3201} \tag{7.107}$$

For critical temperature, the correlation is:

$$T_{ci} = 24.27871 \times T_{bi}^{0.58848} \times \gamma_i^{0.3596} \tag{7.108}$$

In Eqs. (7.107) and (7.108), T_{bi} is the boiling point of the pseudo-component in degrees Rankine; and, γ_i is the specific gravity of the pseudo-component.

The above procedure is superior to other methods of splitting plus hydrocarbon fractions because multiple fluid samples from the same reservoir could be split together to produce a common set of pseudo-components with the same molecular weights and critical properties but different mole fractions representing each fluid sample. This is essential for the generation of a single "tuned" equation of state model for compositional simulation.

7.11 Phase Equilibrium Predictions with Equations of State

This section discusses the use of equations of state to predict phase equilibrium of reservoir fluids for compositional simulation. The PR EOS and SRK EOS were presented earlier in this chapter because of their simplicity, robustness, and popularity in the petroleum industry, especially in applications that require compositional simulations of petroleum reservoirs. Many commercial software programs that are available for analyses of fluid samples and generation of PVT data input for simulators usually have options for selection of either equation of state. Fundamental topics that laid the foundation for the use of PR and SRK equations of state in phase equilibrium calculations were presented in this chapter to improve understanding of commercial PVT software programs. The key topics that were reviewed are as follows:

1. The concept of fugacity and equivalence of partial fugacity of all components in all phases of a mixture in equilibrium.
2. The significance of calculation of roots of cubic equations of state.
3. Improvement of volumetric predictions with volume shifts from cubic equations of state.
4. The basic equations for flash calculations in compositional simulations.
5. Characterization of the hydrocarbon plus fractions to improve phase equilibrium predictions with equations of state.

Most equations of state used in compositional reservoir simulation are cubic in form. These types of equations of state have been shown to adequately represent the phase behavior of reservoir fluids over **normal** ranges of temperatures and pressures encountered in reservoir processes. But before equations of state can be used in compositional reservoir simulations, they must be adjusted to represent the reservoir fluids under varying processes as temperatures and pressures change. The adjustments, sometimes called "tuning," involve changing the physical properties and other empirical parameters of the components of the fluid system that represent the reservoir fluid mixture. But before getting to "tuning" of equations of state, there are basic steps necessary in preparing an equation of state for use in analyses of PVT data. These steps are summarized as follows:

1. Assemble the PVT data to be used in the analyses.
   a. Evaluate the quality of the PVT data for consistency, errors, and poor laboratory practice.
   b. Check compositions of fluid samples for contamination with mud filtrate, especially if the sample was collected from a wellbore drilled with oil-based mud. Decontaminate the fluid samples if necessary, using the techniques described in Chapter 6.
   c. Tabulate the type of PVT data measured for each fluid sample, such as constant composition expansion (CCE) data, constant volume depletion (CVD) data, differential liberation (DL) data, separator tests data, etc.
   d. Decide the sets of data to be used in "tuning" the EOS, and the sets of data to be used in validating the tuned EOS.
2. Estimate the total number of components to be used to model the reservoir fluid. An initial estimate of twelve components is recommended. Note that the number of components could be increased or reduced during the process of matching experimental PVT data by adjusting the properties of the components as represented in the equation of state fluid model. The effects of number of components on phase equilibria predictions are illustrated in Figures 7.1 and 7.2.
   a. Select the number of pure hydrocarbons to represent the lighter hydrocarbons such as methane, ethane, propane, butane, pentane, and hexane. If non-hydrocarbon gases such as carbon dioxide, nitrogen, and hydrogen sulfide are present, these must be represented as components depending on their concentrations. The intermediate components (C_2 to C_6) can be grouped into two pseudo-components if predictions of liquid production from these groups of hydrocarbons by surface separation are not required.
   b. Select the number of pseudo-components to represent the hydrocarbon plus fraction. This depends on the extent of the compositional analyses of the fluid samples. For instance, if the composition, molecular weight, and specific gravity of C_{10+} were reported, the splitting of the hydrocarbon plus fraction into pseudo-components could be started from this point. Generally, the hydrocarbon plus fraction is split into three or more pseudo-components. For this example with C_{10+} fraction, the hydrocarbons in the C_7 to C_9 range could be grouped into a pseudo-component. It is important to note that selection and grouping of components depend on the surface and reservoir processes that are modeled with compositional simulation. The number of components selected from the non-hydrocarbon group, the pure hydrocarbon group, and the pseudo-components group should be designed to represent the processes that will be modeled.
   c. The technique of semi-continuous thermodynamics with Gaussian quadrature is recommended for splitting the hydrocarbon plus fraction into pseudo-components.
3. Compute the properties (T_c, P_c, ω) for the components.
   a. For the pure hydrocarbons, the properties are taken from published data such as Table 5.1.
   b. For the pseudo-components obtained from splitting of the hydrocarbon-plus fractions, the properties are calculated from correlations as discussed in Section 7.10.
   c. For the grouped components, the properties are calculated from application of one of three averaging methods as presented by Joergensen and Stenby:[24]
      - Weight averaging
      - Molar averaging
      - Mixing rule
   d. Note that the averaging step is embedded in all commercial software and performed almost seamlessly. However, it may be necessary to select the correlations you prefer for calculation of the properties in Step 3b and the type of averaging method to apply in Step 3c.
4. Use the EOS model in current form to compute experimental PVT data and compare results. Determine if tuning EOS parameters is necessary.
5. Tune EOS parameters. This process is generally based on experience since there are many parameters that can be adjusted to tune an EOS.[25,26] The two main approaches are:
   a. Adjustment of critical properties of the pseudo-components.[27,28]
   b. Adjustment of the binary interaction parameters between the light hydrocarbons and the pseudo-components. Also, the binary interaction parameters between the non-hydrocarbons and the pseudo-components may be adjusted.[29,30]

The process of tuning an EOS should proceed in a systematic and organized manner to avoid generating an EOS model that has limited predictive capacity. Most commercial software programs provide computational facilities for performing non-linear regression on the adjusted parameters to minimize the differences between measured and predicted PVT data. The output from the regression analyses should be examined to determine the group of parameters that are most sensitive to adjustments. The most sensitive parameters should be adjusted systematically to improve the match between measured and predicted PVT data.

6. Validate the tuned EOS with other PVT data not used in the tuning process. It is important to test the tuned EOS against other PVT data such as swelling test, multi-contact, and slim-tube data if such data are available.[31]

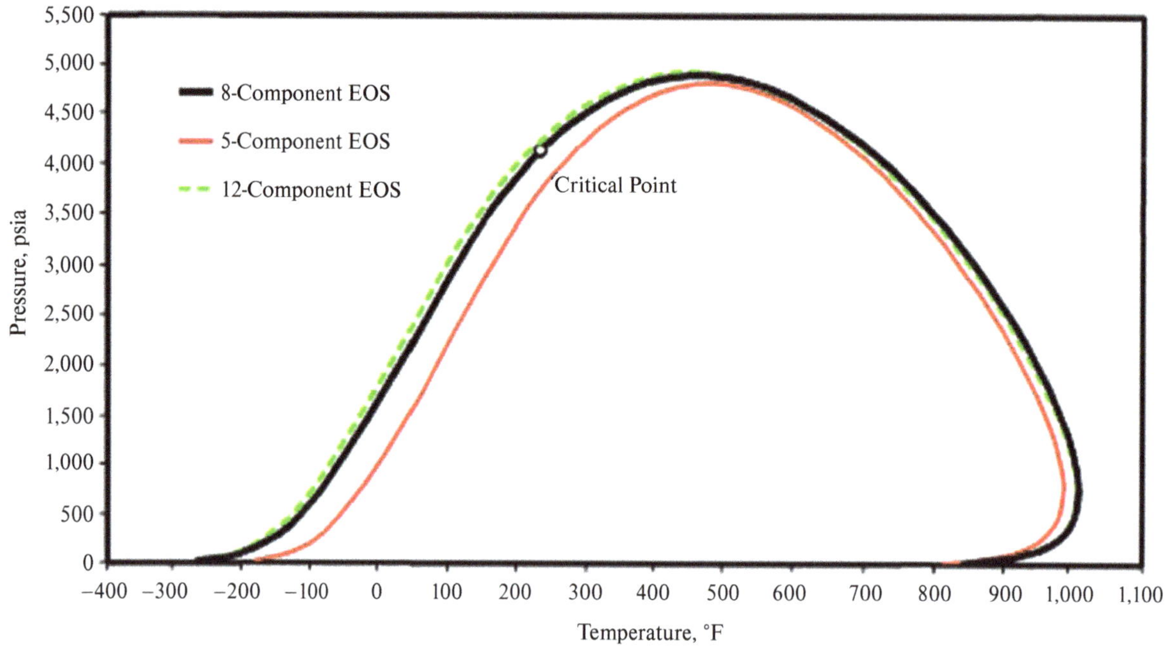

▲ **FIGURE 7.1** **Effects of number of components on phase envelope of an actual gas condensate (from Wang and Pope[28] © 2001 SPE, Reproduced with permission)**

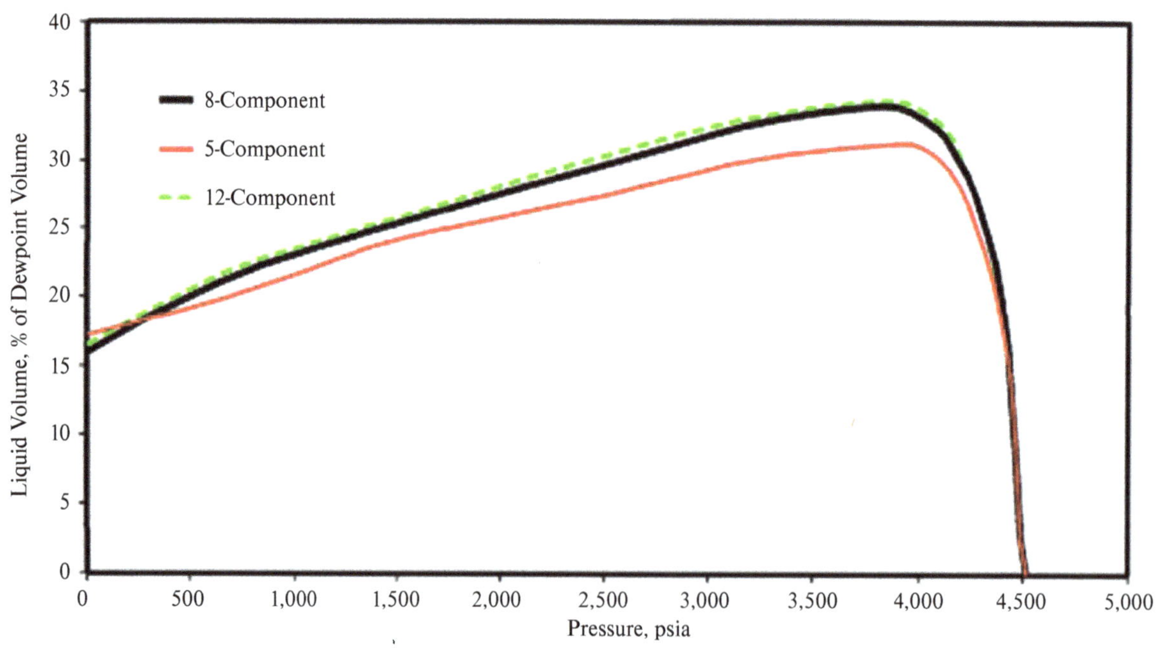

▲ **FIGURE 7.2** **Effects of number of components on liquid dropout (from Wang and Pope[28] © 2001 SPE, Reproduced with permission)**

Nomenclature

c	volume shift parameter
f	fugacity
$\tilde{f}$	modified fugacity
F_v	mole fraction of vapor in mixture
g^*	normalized Gibbs energy
k_{ij}	binary interaction parameter
K	equilibrium constant
K_W	Watson characterization factor
M	molecular weight
n	moles of substance
N	number of phases
N_C	number of components
p	pressure
p_c	critical pressure
p_r	reduced pressure
p_v	vapor pressure
R	universal gas constant
s	molar entropy or dimensionless shift parameter in volume shift
T	temperature
T_b	boiling point temperature
T_c	critical temperature
T_r	reduced temperature
v	molar volume
$\bar{v}$	corrected molar volume
V	volume
V_c	critical volume
x	liquid phase mole fraction
y	vapor phase mole fraction
z	gas compressibility factor
$\mathbf{z}$	total mole fraction of component in mixture
z_c	critical compressibility factor
α	shape parameter in gamma function
β	nomalization parameter in gamma function
β^*	modified normalization parameter in gamma function
η	parameter in gamma function
Γ	gamma function
ϕ	fugacity coefficient
μ	chemical potential
ω	acentric factor
γ	specific gravity
ξ	frequency of a component having molecular weight between certain boundaries

Subscripts

c	critical property
C	component
E	experimental data
i, j	components in mixture
L	liquid phase
V	vapor phase

Superscripts

L	liquid phase
V	vapor phase

Abbreviations

BIP	binary interaction parameter
EOS	equation of state
MCN	multiple carbon number
PR	Peng-Robinson
PVT	pressure-volume-temperature
RK	Redlich-Kwong
SCN	single carbon number
SRK	Soave-Redlich-Kwong
TBP	true-boiling-point
vdW	van der Waals

REFERENCES

1. van der Waals, J.D.: *On the Continuity of the Gas and Liquid State*. Dissertation, Leiden, The Netherlands (1873).
2. Walas, S.M.: *Phase Equilibria in Chemical Engineering*, Butterworth Publishers, Stoneham, MA (1985).
3. Redlich, O., and Kwong, J.N.S.: "On the Thermodynamics of Solutions: V. An Equation of State. Fugacities of Gaseous Solutions," *Chem. Review* (1949) 44, 233-244.
4. Soave, G.: "Equilibrium Constants from a modified Redlich-Kwong Equation of State," *Chem. Eng. Sci.* (1972) 27, 6, 1197-1203.
5. Graboski, M.S., and Daubert, T.E.: "A Modified Soave Equation of State for Phase Equilibrium Calculations. 1. Hydrocarbon Systems," *Ind. Eng. Chem. Process Des. Dev.* (1978) 17, 4, 443-448.
6. Graboski, M.S., and Daubert, T.E.: "A Modified Soave Equation of State for Phase Equilibrium Calculations. 2. Systems Containing CO2, H2S, N2 and CO," *Ind. Eng. Chem. Process Des. Dev.* (1978) 17, 4, 448-454.
7. Pitzer, K.S., Lippmann, D.Z., Curl, R.F., Jr., Huggins, C.M., and Peterson, D.E.: "The Volumetric and Thermodynamic Properties of Fluids. II. Compressibility Factor, Vapor Pressure and Entropy of Vaporization," *J. Am. Chem. Soc.* (1955) 77, 13, 3433-3440.
8. Peng, D.-Y., and Robinson, D.B.: "A New Two-Constant Equation of State," *Ind. Eng. Chem. Fundamentals* (1976) 15, 1, 59-64.
9. Robinson, D.B., Peng, D.-Y., and Ng, H.-Y: "Capabilities of the Peng-Robinson Programs, Part 2: Three-Phase and Hydrate Calculations," *Hydrocarbon Proc.* (1979) 58, 269.
10. Robinson, D.B., and Peng, D.-Y.: "The Characterizations of the Heptanes and Heavier Fractions," Research Report 28, Gas Producers Assn., Tulsa, Oklahoma (1978).
11. Michelsen, M.L.: "The Isothermal Flash Problem. Part II. Phase-Split Calculation," *Fluid Phase Equilibria* (1982) 9, 21.
12. Martin, J.J.: "Cubic Equations of State-Which?" *Ind. & Eng. Chem. Fund.* (1979), 18, 2, 81-97.
13. Peneloux, A., Rauzy, E., and Freze, R: "A Consistent Correction for Redlich-Kwong-Soave Volumes," *Fluid Phase Equilibria* (1982), 8, 7-23.
14. Jhaveri, B.S., and Youngren, G.K.: "Three-Parameter Modification of the Peng-Robinson Equation of State to Improve Volumetric Predictions," *SPERE* (August 1988) 1033-1040.
15. Rachford, H.H., and Rice, J.D.: "Procedure for Use of Electrical Digital Computers in Calculating Flash Vaporization Hydrocarbon Equilibrium," *JPT* (October 1952) 19; *Trans.*, AIME, 195.
16. Whitson, C.H., and Brule, M.R.: *Phase Behavior*, SPE Monograph Volume 20, Society of Petroleum Engineers, Richardson, Texas, 2000.
17. Wilson, G.M.: "A Modified Redlich-Kwong Equation of State, Application to General Physical Data Calculations," paper 15c presented at the 1969 AICHE Natl. Meeting, Cleveland, Ohio.
18. Li, Y., and Johns, R.T.: "Rapid Flash Calculations for Compositional Simulation," *SPEREE* (October 2006) 521-529.
19. Whitson, C.H.: "Characterizing Hydrocarbon Plus Fractions," *SPEJ.* (August 1983) 683-694.
20. Whitson, C.H.: "Effect of C7+ Properties on Equation-of-State Predictions," *SPEJ* (December 1984) 685-696.
21. Behrens, R.A., and Sandler, S.I.: "The Use of Semicontinuous Description to Model the C7+ Fraction in Equation of State Calculations," *SPERE* (August 1988) 1041-1047.
22. Whitson, C.H., Andersen, T.F., and Soreide, I.: "C7+ Characterization of Related Equilibrium Fluids Using the Gamma Distribution," *C7+ Fraction Characterization*, L.G. Chorn, and G.A. Mansoori (eds.), Advances in Thermodynamics, Taylor and Francis, New York (1989) 1, 35-56.
23. Riazi, M.R., and Daubert, T.E.: "Simplify Property Predictions," *Hydro. Proc.* (March 1980) 115.
24. Joergensen, M., and Stenby, E.H.: "Optimization of Pseudo-component Selection for Compositional Studies of Reservoir Fluids," paper SPE 30789 presented at the 1995 SPE Annual Technical Conference and Exhibition, Dallas, Texas, October 22-25.
25. Agarwal, R.K., Li, Y.-K., and Nghiem, L.: "A Regression Technique with Dynamic Parameter Selection for Phase-Behavior Matching," *SPERE*, (February 1990), 115-120.
26. Merrill, R.C., Hartman, K.J., and Creek, J.L.: "A comparison of Equation of State Tuning Methods," paper SPE 28589 presented at the 1994 SPE Annual Technical Meeting and Exhibition, New Orleans, September 26-28.
27. Coats, K.H., and Smart, G.T.: "Application of a Regression-Based EOS PVT Program to Laboratory Data," *SPERE* (May 1986), 277-299.
28. Wang, P., and Pope, G.A.: "Proper Use of Equations of State for Compositional Reservoir Simulation," SPE 69071, Distinguished Author Series, (July 2001) 74-81.
29. Slot-Petersen, C.: "A Systematic and Consistent Approach to Determine Binary Interaction Coefficients for the Peng-Robinson Equation of State," *SPERE* (November 1989) 488-494.
30. Ezekwe, J.N.: "Effect of Paraffinic, Naphthenic and Aromatic Distribution in the Hydrocarbon Mixture and Water on the Phase Equilibria of Carbon Dioxide-Hydrocarbon Systems Over the Temperature Range from 333 K TO 366 K," Ph.D. Dissertation, University of Kansas, 1982.
31. Stalkup, F., and Yuan, H.: "Effect of EOS Characterization on Predicted Miscibility Pressure," paper SPE 95332 presented at the 2005 SPE Annual Technical Conference and Exhibition, Dallas, October 9-12.

PROBLEMS

7-1. A tank contains a mixture of 45 mole percent of propane and 55 mole percent of n-butane at a temperature of 80°F. Assuming a convergence pressure of 800 psia, calculate the dew point and bubble point pressure of the mixture in the tank.

7-2. The tank in Problem 7-1 still contains a mixture of 45 mole percent of propane and 55 mole percent of n-butane. If the pressure and temperature of the tank are 80 psia and 80°F, respectively, calculate the composition of the vapor and liquid phases in the tank assuming the two phases are in equilibrium and a convergence pressure of 800 psia.

7-3. A tank containing a mixture of 70 mole percent methane and 30 mole percent ethane is at a temperature of −60°F. Calculate the dew point and bubble point pressures of the mixture. Assume a convergence pressure of 2000 psia.

7-4. Calculate the composition of the vapor and liquid phases of a multicomponent system in equilibrium at 100 psia and 80°F with the composition shown in Table 7P.1. Assume a convergence pressure of 1000 psia.

TABLE 7P.1 Table for Problem 7-4

Component	Mole Fraction
Ethane	0.20
Propane	0.40
i-Butane	0.10
n-Butane	0.30

7-5. An oil reservoir was discovered with an initial oil composition shown in Table 7P.2. Initial reservoir pressure and temperature were reported as 2500 psia and 120°F, respectively. The initial solution gas-oil ratio was measured to be 3660 scf/STB and oil gravity as 49.9°API. Using Figure 7P.1, determine whether the reservoir is undersaturated or saturated.

TABLE 7P.2 Oil Composition for Problem 7-5

Component	Mole Fraction
Methane	0.3300
Ethane	0.1734
Propane	0.0996
Butane	0.0764
Pentane	0.0384
Hexane	0.0151
Heptanes Plus	0.2671

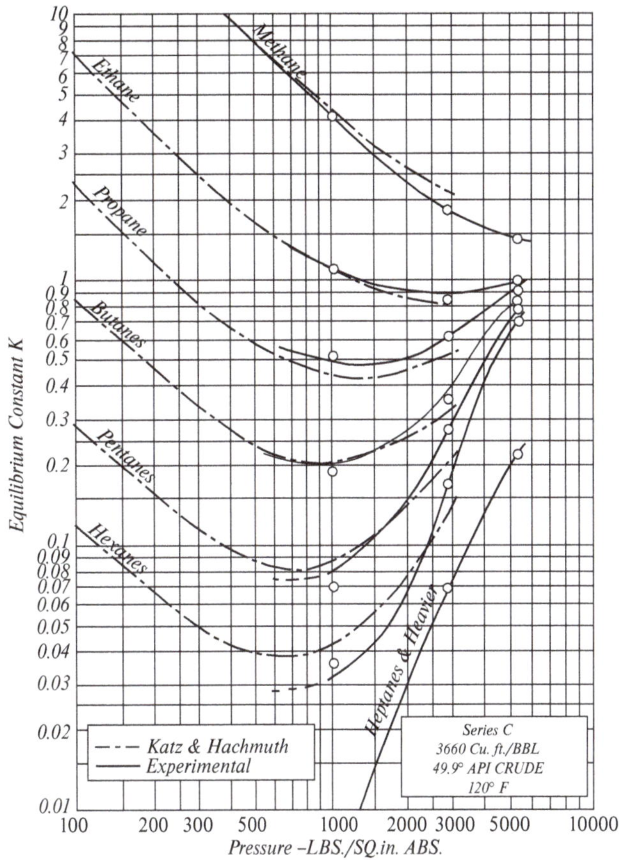

▲ **FIGURE 7P.1** Variation of equilibrium constants with pressure

7-6. The experimental data for a black oil sample were presented in Appendix 6A. Using these data, generate a PVT input data based on an equation of state for a black oil reservoir simulator. (Hint: Use the procedures in Section 7.11 and a PVT software package).

7-7. A reservoir fluid study on a gas condensate sample was presented in Appendix 6B. Generate a PVT input data based on an equation of state for a compositional reservoir simulator using the data in Appendix 6B. (Hint: Use the procedures in Section 7.11 and a PVT software package).

CHAPTER

8 The General Material Balance Equation

8.1 Introduction

The General Material Balance Equation (GMBE) as developed in this book is based on an oil reservoir with a primary gas cap at initial conditions and reservoir pressure designated as p_i. At a later time t, reservoir pressure is assumed to have been reduced from p_i to p due to production of oil, water, and gas. During the production period, it is assumed that there was water influx into the reservoir from an aquifer. It is also assumed that water and/or gas was injected into the reservoir. These conditions are illustrated in Figure 8.1.

The description of the equation that is developed as a material balance equation is in a strict sense not accurate. Actually, it is a volumetric balance based on reservoir barrels instead of a material balance based on mass. But in this book, it is called the GMBE to conform to the generally accepted terminology used in the petroleum industry. The GMBE is derived from the basis that the reservoir pore volume as existing at its initial conditions is constant. Thus, the sum of the volume changes of gas, oil, water, and the formation rock caused by fluid production is equal to zero.

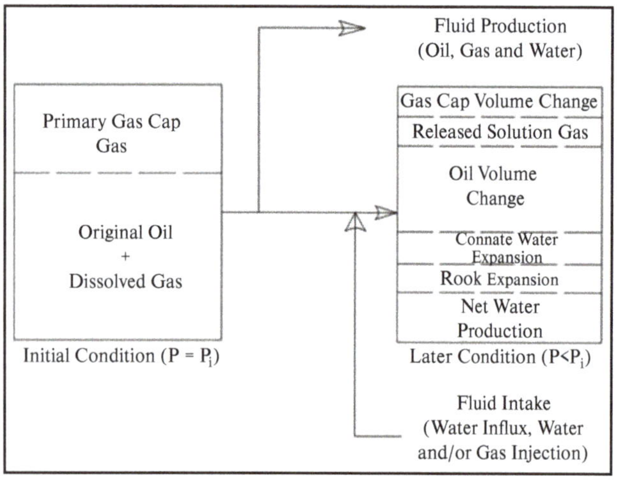

▲ **FIGURE 8.1 Material balance model for an oil reservoir**

8.2 Derivation of the General Material Balance Equation (GMBE)

The GMBE can be represented volumetrically in reservoir barrels as:

Net change in oil and water production volume = Gas cap volume change
+ Released solution gas
+ Oil volume change
+ Connate water expansion
+ Rock expansion (8.1)

Before developing the terms of the expression for the GMBE in Eq. (8.1), it is necessary to derive other expressions that apply to oil reservoirs with primary gas caps. The ratio of original reservoir gas cap volume and the original reservoir oil zone volume is defined as:

$$m = \frac{\text{Original reservoir gas cap volume}}{\text{Original reservoir oil zone volume}} = \frac{GB_{gi}}{NB_{oi}} \quad (8.2)$$

In Eq. (8.2), G = original gas in place (OGIP) in the gas cap, scf; N = original oil in place (OOIP) in the oil zone, STB; B_{gi} = initial gas formation volume factor (FVF), RB/scf; B_{oi} = initial oil FVF, RB/STB. From Eq. (8.2), $NmB_{oi} = GB_{gi}$.

8.2.1 Development of Terms in the Expression of Eq. (8.1)

1. Net change in oil production volume:

   $$\text{Net change in oil production volume (RB)} = N_pB_o \quad (8.3)$$

   In Eq. (8.3), N_p = cumulative oil production, STB; and B_o = oil FVF, RB/STB, at current reservoir pressure, p.

2. Net change in water production volume:

   $$\text{Net change in water production volume (RB)} = W_pB_w - W_e - W_{inj}B_w \quad (8.4)$$

   In Eq. (8.4), W_p = cumulative water production, STB; W_e = cumulative water influx, RB; W_{inj} = cumulative water injection, STB; and B_w = water FVF, RB/STB, at current reservoir pressure, p.

3. Gas cap volume change;

   $$\text{Gas cap volume change (RB)} = \left(G - G_p + G_{inj}\right)B_g - GB_{gi} \quad (8.5)$$

In Eq. (8.5), G_p = cumulative gas production, scf; G_{inj} = cumulative gas injection, scf; and B_g = gas FVF, RB/scf. Expressing Eq. (8.5) in terms of N by using Eq. (8.2) and the relationship $G_p = N_pR_p$:

$$\text{Gas cap volume change (RB)} = \left(\frac{NmB_{oi}}{B_{gi}} - N_pR_p + G_{inj}\right)B_g - NmB_{oi} \tag{8.6}$$

In Eq. (8.6), R_p = cumulative production gas-oil ratio, scf/STB.

4. Solution gas released into the gas cap:

$$\begin{aligned}\text{Released solution gas (RB)} &= \left[NR_{si} - (N - N_p)R_s\right]B_g \\ &= \left[N(R_{si} - R_s) + N_pR_s\right]B_g\end{aligned} \tag{8.7}$$

In Eq. (8.7), R_{si} = initial solution gas-oil ratio (GOR), scf/STB; and R_s = solution GOR, scf/STB, at current reservoir pressure, p.

5. Oil volume change:

$$\text{Oil volume change}(RB) = NB_o - NB_{oi} \tag{8.8}$$

6. Connate water and rock expansion:

$$\begin{aligned}&\text{Reduction in Hydrocarbon Pore Volume (HCPV)} \\ &= \text{Connate water and rock expansion}\end{aligned} \tag{8.9}$$

$$-d(\text{HCPV}) = (c_wV_w + c_fV_p)(p_i - p) \tag{8.10}$$

$$V_w = V_pS_{wi} \tag{8.11}$$

In Eqs. (8.10) and (8.11), c_w = water compressibility, psi^{-1}; c_f = rock compressibility, psi^{-1}; V_w = connate water volume, RB; V_p = total pore volume, RB; p_i = initial reservoir pressure, psi; p = reservoir pressure in psi at time, t; and S_{wi} = initial connate water saturation, fraction.

$$V_p = \frac{\text{HCPV}}{1 - S_{wi}} \tag{8.12}$$

$$\begin{aligned}\text{HCPV} &= \text{initial oil reservoir volume} \\ &\quad + \text{initial reservoir gas volume} \\ &= NB_{oi} + NmB_{oi} = (1+m)NB_{oi}\end{aligned} \tag{8.13}$$

From Eqs. (8.12) and (8.13),

$$V_p = \frac{(1+m)NB_{oi}}{1 - S_{wi}} \tag{8.14}$$

$$V_w = \frac{(1+m)NB_{oi}S_{wi}}{1 - S_{wi}} \tag{8.15}$$

Substituting Eqs. (8.14) and (8.15) into Eq. (8.10) gives;

$$\begin{aligned}-d(\text{HCPV}) &= (1+m)NB_{oi}\left[\frac{c_wS_{wi} + c_f}{1 - S_{wi}}\right](p_i - p) \\ &= (1+m)NB_{oi}\left[\frac{c_wS_{wi} + c_f}{1 - S_{wi}}\right]\Delta p\end{aligned} \tag{8.16}$$

Substituting Eqs. (8.3), (8.4), (8.6), (8.7), (8.8) and (8.16) into Eq. (8.1) gives:

$$\begin{aligned}&N_pB_o + W_pB_w - W_e - W_{inj}B_w \\ &= \left(\frac{NmB_{oi}}{B_{gi}} - N_pR_p + G_{inj}\right)B_g - NmB_{oi} \\ &\quad + \left[N(R_{si} - R_s) + N_pR_s\right]B_g + NB_o - NB_{oi} \\ &\quad + (1+m)NB_{oi}\left[\frac{c_wS_{wi} + c_f}{1 - S_{wi}}\right]\Delta p\end{aligned} \tag{8.17}$$

By expanding and arranging like terms, Eq. (8.17) becomes:

$$\begin{aligned}&NB_{oi} - NB_o + N_pB_o + NmB_{oi} - \frac{NmB_{oi}B_g}{B_{gi}} \\ &\quad + N_pR_pB_g - NR_{si}B_g + NR_sB_g - N_pR_sB_g \\ &= G_{inj}B_g + W_{inj}B_w + W_e - W_pB_w \\ &\quad + (1+m)NB_{oi}\left[\frac{c_wS_{wi} + c_f}{1 - S_{wi}}\right]\Delta p\end{aligned} \tag{8.18}$$

Adding the expression $N_pB_gR_{si}$ to both sides of Eq. (8.18) gives:

$$\begin{aligned}&NB_{oi} - NB_o + N_pB_o + NmB_{oi} - \frac{NmB_{oi}B_g}{B_{gi}} + N_pR_pB_g \\ &\quad - NR_{si}B_g + NR_sB_g - N_pR_sB_g + N_pB_gR_{si} \\ &= G_{inj}B_g + W_{inj}B_w + W_e - W_pB_w \\ &\quad + (1+m)NB_{oi}\left[\frac{c_wS_{wi} + c_f}{1 - S_{wi}}\right]\Delta p + N_pB_gR_{si}\end{aligned} \tag{8.19}$$

Grouping terms in Eq. (8.19) yields:

$$\begin{aligned}&NB_{oi} + NmB_{oi} - N\left[B_o + (R_{si} - R_s)B_g\right] \\ &\quad + N_p\left[B_o + (R_{si} - R_s)B_g\right] \\ &\quad + N_p(R_p - R_{si})B_g - \frac{NmB_{oi}B_g}{B_{gi}} \\ &= G_{inj}B_g + W_{inj}B_w + W_e - W_pB_w \\ &\quad + (1+m)NB_{oi}\left[\frac{c_wS_{wi} + c_f}{1 - S_{wi}}\right]\Delta p\end{aligned} \tag{8.20}$$

The two-phase FVF, B_t, is defined as:

$$B_t = B_o + (R_{si} - R_s) B_g \quad \text{and } B_{ti} = B_{oi}, \text{ at initial conditions} \tag{8.21}$$

Substituting Eq. (8.21) into Eq. (8.20) gives:

$$\begin{aligned} &N(B_{ti} - B_t) + N_p\left[B_t + (R_p - R_{si})B_g\right] + NmB_{ti}\left[1 - \frac{B_g}{B_{gi}}\right] \\ &= G_{inj}B_g + W_{inj}B_w + W_e - W_pB_w \\ &+(1+m)NB_{oi}\left[\frac{c_wS_{wi} + c_f}{1 - S_{wi}}\right]\Delta p \end{aligned} \tag{8.22}$$

Eq. (8.22) is the General Material Balance Equation (GMBE). Further rearrangement of Eq. (8.22) gives a useful form of the GMBE:

$$\begin{aligned} &N(B_t - B_{ti}) + \frac{NmB_{ti}}{B_{gi}}\left[B_g - B_{gi}\right] \\ &+(1+m)NB_{ti}\left[\frac{c_wS_{wi} + c_f}{1 - S_{wi}}\right]\Delta p \\ &+G_{inj}B_g + W_{inj}B_w + W_e \\ &= N_p\left[B_t + (R_p - R_{si})B_g\right] + W_pB_w \end{aligned} \tag{8.23}$$

Another useful form of Eq. (8.22) is rearranged as:

$$\begin{aligned} &N_p\left[B_t + (R_p - R_{si})B_g\right] + W_pB_w - G_{inj}B_g - W_{inj}B_w \\ &= N\left[(B_t - B_{ti}) + (1+m)B_{ti}\left(\frac{c_wS_{wi} + c_f}{1 - S_{wi}}\right)\Delta p \right. \\ &\left. + \frac{mB_{ti}}{B_{gi}}(B_g - B_{gi})\right] + W_e \end{aligned} \tag{8.24}$$

8.3 The GMBE for Gas Reservoirs

The GMBE was developed for a saturated oil reservoir with a primary gas cap. However, it can be modified to apply to gas reservoirs (dry gas, wet gas, and gas-condensate reservoirs). For gas-condensate reservoirs, there is the requirement that the gas initially exists as a single phase (that means no condensation has occurred in the reservoir) before the GMBE can be applied. Starting from Eq. (8.24) and expanding all terms, the result is:

$$\begin{aligned} &N_pB_t + N_pR_pB_g - N_pR_{si}B_g + W_pB_w - G_{inj}B_g - W_{inj}B_w \\ &= NB_t - NB_{ti} + (NB_{ti} + NmB_{ti}) \times \left[\frac{c_wS_{wi} + c_f}{1 - S_{wi}}\right]\Delta p \\ &+ \frac{NmB_{ti}}{B_{gi}}(B_g - B_{gi}) + W_e \end{aligned} \tag{8.25}$$

From Eq. (8.2), $NmB_{ti} = GB_{gi}$. Also, $G_p = N_pR_p$. Substituting in Eq. (8.25) gives:

$$\begin{aligned} &N_pB_t + G_pB_g - N_pR_{si}B_g + W_pB_w - G_{inj}B_g - W_{inj}B_w \\ &= NB_t - NB_{ti} + (NB_{ti} + GB_{gi}) \\ &\times \left[\frac{c_wS_{wi} + c_f}{1 - S_{wi}}\right]\Delta p + G(B_g - B_{gi}) + W_e \end{aligned} \tag{8.26}$$

Since at initial conditions, it is assumed there is no oil in the gas reservoir, $N = 0$, and $N_p = 0$. Also, it is assumed that there is no water or gas injection. Thus, Eq. (8.26) reduces to:

$$\begin{aligned} &G_pB_g + W_pB_w \\ &= GB_{gi}\left(\frac{c_wS_{wi} + c_f}{1 - S_{wi}}\right)\Delta p + G(B_g - B_{gi}) + W_e \end{aligned} \tag{8.27}$$

Eq. (8.27) is the general material balance equation for gas reservoirs. It is applied in Chapter 9 to derive equations for volumetric and geopressured gas reservoirs.

8.4 Discussion on the Application of the GMBE

Schilthius[1] published the first GMBE in 1936. In 1953, van Everdingen et al.[2] applied a linear form of the GMBE to a partial water-drive reservoir. The full potential of the GMBE as a straight line equation was developed by Havlena and Odeh[3,4] in classical papers published in 1963 and 1964. Since then, there has been a large body of work on the applications of the GMBE on many reservoir engineering problems published in the literature.

The GMBE is a valuable, analytical tool for evaluation of reservoir drive mechanisms. Even though it is zero dimensional, it offers tools for gaining diagnostic insights into the processes occurring in the reservoir. Some of these diagnostic applications of the GMBE are presented in Chapter 10 using the straight line method of Havlena and Odeh.[3,4] Computerized applications of the Havlena and Odeh[3] method have been published in the literature by Wang and Teasdale[5] for gas reservoirs, and Wang et al.[6] for oil reservoirs. Computer software for GMBE applications is also available commercially. The applications provide the reservoir engineer with powerful tools to gain in-depth knowledge of the key processes influencing reservoir performance before embarking on more rigorous analyses.

The most rigorous method for reservoir evaluation and analyses is reservoir modeling (also literally termed reservoir simulation). The application of reservoir modeling is widespread in the petroleum industry. In contrast to GMBE, which is zero-dimensional, reservoir modeling is multi-dimensional. The application of reservoir modeling to petroleum reservoirs is presented in Chapters 22 and 23.

With the availability of fast, powerful digital computers with large storage capacity for reservoir modeling, questions have been raised on the utility of material balance methods for reservoir analyses. The debate must be engaged bearing in mind that material balance methods were mostly developed before the advent of modern computers. Furthermore, all the analyses that can be performed with material balance methods can be replicated in many cases more rigorously with reservoir modeling.

However, material balance methods are still very useful and are incredibly simple tools for gaining an understanding of the reservoir processes before undertaking the more elaborate and time-consuming task of reservoir modeling. In practical terms, material balance methods and reservoir modeling should be viewed as complementary tools in the toolkit of the engineer for reservoir analyses. If the reservoir engineer is unsure of the reservoir drive mechanisms, it is advisable to start the process of analyzing the performance of the reservoir with material balance methods, and later advance to reservoir modeling as more knowledge of the reservoir drive mechanisms is gained. In practice, material balance methods can be of assistance in determining the type of reservoir model to construct and in assessing the quality of reservoir performance data to use in conducting reservoir modeling.

Nomenclature

B_g	gas formation volume factor, RB/scf
B_o	oil formation volume factor, RB/STB
B_t	two-phase formation volume factor, RB/STB
B_w	water formation volume factor, RB/STB
c_f	formation compressibility factor, psi^{-1}
c_w	water compressibility factor, psi^{-1}
G	original gas in place, scf
G_{inj}	cumulative gas injection, scf
G_p	cumulative gas produced, scf
m	ratio original reservoir gas volume to original reservoir oil volume
N	original oil in place, STB
N_p	cumulative oil production, STB
p	pressure, psi
R_p	cumulative production gas-oil ratio, scf/STB
R_s	solution gas-oil ratio, scf/STB
S_w	water saturation, fraction
V_p	total pore volume, RB
V_w	connate water volume, RB
W_e	cumulative water influx, RB
W_{inj}	cumulative water injection, STB
W_P	cumulative water production, STB

Subscripts

i	initial
o	oil or condensate
p	produced

Abbreviations

FVF	Formation volume factor
GMBE	General Material Balance Equation
HCPV	Hydrocarbon pore volume
OGIP	Original gas in place
OOIP	Original oil in place

REFERENCES

1. Schilthius, R.J.: "Active Oil and Reservoir Energy," *Trans.* AIME (1936) 148, 33-51.
2. van Everdingen, A.F., Timmerman, E.H., and McMahon, J.J.: "Application of the Material Balance Equation to a Partial Water-Drive Reservoir," *Trans.* AIME (1953) 198, 51-60.
3. Havlena, D., and Odeh, A.S.: "The Material Balance as an Equation of a Straight Line," *JPT* (August 1963) 896-900.
4. Havlena, D., and Odeh, A.S.: "The Material Balance as an Equation of a Straight Line–Part II, Field Cases," *JPT* (July 1964) 815-822.
5. Wang, B., and Teasdale, T.S.: "GASWAT-PC: A Microcomputer Program for Gas Material Balance with Water Influx," paper SPE 16484 presented at the 1987 SPE Petroleum Industry Applications of Microcomputers, Lake Conroe, Montgomery, Texas, (June 23-26, 1987).
6. Wang, B., Litvak, B.L., and Bowman, G.W.: "OILWAT: Microcomputer Program for Oil Material Balance with Gas Cap and Water Influx," paper SPE 24437 presented at the 1992 SPE Petroleum Computer Conference, Houston, Texas, (July 19-22, 1992).

PROBLEMS

8-1. The historical production data including average reservoir pressure for a shaly, turbite reservoir in Kern County, California, USA, are given in Table 8P.1. The PVT data of reservoir fluids are given in Table 8P.2. Other rock and fluid properties for the reservoir are given as follows:

Initial reservoir pressure, p_i	3200 psia
Bubble point pressure, p_{bub}	2995 psia
Initial water saturation, S_{wi}	57%
Water formation volume factor, B_w	1.02 RB/STB
Water compressibility factor, c_w	3.0E-06 psi^{-1}
Rock compressibility factor, c_f	1.1E-05 psi^{-1}

TABLE 8P.1 | Historical Production Data Including Reservoir Pressure for Problem 8-1

DATE	Pressure (psia)	Cum. Oil (MSTB)	Cum. Water (MSTB)	Cum. Gas (MMscf)
Dec-76	3150	935	7	771
Dec-77	3145	2394	25	2107
Dec-78	3080	4222	69	4264
Dec-79	3000	7139	118	8714
Dec-80	2900	11134	387	18295
Dec-81	2800	14965	1245	30574
Dec-82	2700	18056	2421	44210
Dec-83	2595	20540	3609	54690
Dec-84	2500	22892	5012	67177
Dec-85	2400	24860	6614	79766
Dec-86	2300	26412	7916	90878
Dec-87	2220	27733	9520	100973
Dec-88	2330	28905	11159	109425
Dec-89	2260	30034	13215	119121
Dec-90	2230	31011	15071	128718
Dec-91	2200	31927	16747	137483
Dec-92	2185	32749	18786	147115
Dec-93	2200	33618	20722	157869
Dec-94	2180	34644	23000	170442
Dec-95	2080	35864	25596	188577
Jul-96	2060	36515	27259	203075

TABLE 8P.2 | PVT Data for the Reservoir in Problem 8-1

Pressure (psia)	Oil FVF, Bo (RB/STB)	Soln. GOR, Rs (scf/STB)	Gas FVF, Bg (scf/RB)
3200	1.56	950	1100
3100	1.57	950	1067
3000	1.58	950	1050
2900	1.57	933	1025
2800	1.56	900	1000
2700	1.55	866	967
2600	1.53	833	933
2500	1.50	800	880
2400	1.48	766	850
2300	1.46	733	800
2200	1.45	700	780
2000	1.43	638	700
1500	1.38	500	500
1000	1.32	375	320
500	1.24	250	140
250	1.20	175	80

Using the data given, perform the following analyses:

a. Material balance calculations on the reservoir assuming no water influx into the reservoir.

b. Material balance calculations on the reservoir assuming no water influx and negligible water and rock compressibility effects.

c. Compare the original oil in place (OOIP) calculated from a. and b. What is the impact of ignoring water and rock compressibility on estimation of OOIP?

d. What is the likely range of OOIP in the reservoir?

e. Is material balance method suitable for estimating OOIP for the reservoir? Explain your answer?

CHAPTER

9 Gas Reservoirs

9.1 Introduction

This chapter presents methods for calculations of in-place volumes and reserves for gas reservoirs. Gas reservoirs are classified as dry, wet, or condensate depending on path of depletion (in terms of pressure and temperature) relative to the phase envelope of the reservoir fluids. In Chapter 5, the characteristics of dry, wet, and retrograde condensate gas reservoirs were presented. For dry gas reservoirs, no liquids are formed either in the reservoir or production facilities because the path of depletion is completely outside the phase envelope (Figure 5.7). For wet gas reservoirs, no liquids are condensed within the reservoir, but liquids may condense in the production facilities because temperature and pressure conditions in the production facilities may fall within the two-phase region (Figure 5.7). Retrograde gas-condensate reservoirs may initially exist as single-phase fluid systems but condense liquid phases in the reservoir as pressure is reduced by production (Figure 5.8). In this chapter, gas reservoirs are further classified as volumetric or non-volumetric reservoirs. **Volumetric gas reservoirs** are defined as completely isolated, closed systems with approximately constant hydrocarbon pore volumes. Volumetric gas reservoirs are presumed not to gain significant pressure support or fluid influx from outside sources, such as water influx from aquifers or neighboring shale (non-reservoir) layers. On the other hand, **non-volumetric gas reservoirs** exhibit evidence of pressure support or influx of fluids (mostly water) from outside sources, such as aquifers or neighboring shale intervals. A special group of gas reservoirs are classified as overpressured gas reservoirs. These reservoirs are sometimes called geopressured or abnormally pressured gas reservoirs. The term **geopressured gas reservoir** is used to represent such reservoirs in this book. Geopressured gas reservoirs have pressure gradients that are sometimes approximately twice as high as those of normally pressured gas reservoirs.

9.2 Volumetric Gas Reservoirs

Volumetric gas reservoirs are considered as totally isolated, closed systems with approximately constant hydrocarbon pore volume. For volumetric gas reservoirs, it is assumed that the reservoir does not receive significant pressure support or fluid from outside sources, such as water influx from aquifers or neighboring shale (non-reservoir) layers. In practice, there are relatively few gas reservoirs that are truly volumetric as defined. The classification of a gas reservoir as volumetric allows the application of simplifying assumptions for the assessment of its in-place-hydrocarbon volumes and prediction of reservoir performance. In this section, volumetric calculations for dry gas, wet gas, and retrograde gas condensate reservoirs are presented. This is then followed by material balance on volumetric gas reservoirs.

9.2.1 Volumetric Calculations for Dry Gas Reservoirs

The original gas-in-place (OGIP) denoted as G in a dry gas reservoir can be calculated in terms of standard cubic feet (scf) with a simple volumetric equation:

$$G = \frac{7758 \times A \times h \times \phi \times (1 - S_{wi})}{B_{gi}} \tag{9.1}$$

In Eq. (9.1), A = reservoir area in acres; h = net sand thickness (or net pay), feet; ϕ = porosity, fraction; S_{wi} = initial water saturation, fraction (and $S_{gi} = 1 - S_{wi}$, where S_{gi} is initial gas saturation, fraction); and B_{gi} = initial gas formation volume factor (FVF), RB/scf. In the units of RB/scf, initial gas FVF is shown in Eq. (5.75) as:

$$B_{gi} = \frac{0.005021 z_i T_R}{p_i} \tag{9.2}$$

Methods for calculating B_{gi} for a gas mixture from its composition or gas gravity are shown in example calculations in Chapter 5.

At any intermediate pressure, the cumulative volume of gas produced from the reservoir is the difference between the OGIP and the volume of gas remaining in the reservoir. This is represented as:

$$G_p = G - G_r \tag{9.3}$$

In Eq. (9.3), G_p is the cumulative volume of produced gas, and G_r is the volume of gas remaining in the reservoir at the intermediate pressure. From Eq. (9.1),

$$G_r = \frac{7758 \times A \times h \times \phi \times (1 - S_{wi})}{B_g} \tag{9.4}$$

Note that in Eq. (9.4), B_g is the gas FVF at intermediate pressure, p. At an abandonment pressure, p_a, Eq. (9.3) is represented as:

$$G_p = G - G_a \tag{9.5}$$

In Eq. (9.5), G_a is the gas remaining in the reservoir at abandonment pressure, p_a. Substituting Eqs. (9.1) and (9.4) into Eq. (9.5) gives:

$$G_p = 7758 \times A \times h \times \phi \times (1 - S_{wi}) \left[\frac{1}{B_{gi}} - \frac{1}{B_{ga}} \right] \tag{9.6}$$

The ultimate recovery factor, R, for a volumetric gas reservoir is defined as:

$$R = \frac{G - G_a}{G} = \frac{G_P}{G} \tag{9.7}$$

Substituting Eqs. (9.1) and (9.6) into Eq. (9.7), the ultimate gas recovery factor, R, is:

$$\begin{aligned} R &= 1 - \frac{B_{gi}}{B_{ga}} \\ &= 1 - \frac{p_a z_i}{p_i z_a} \end{aligned} \tag{9.8}$$

Note that the basic assumption in developing the equations for a volumetric dry gas reservoir is that initial water saturation does not change over the productive life of the reservoir. Gas recovery factors for volumetric dry gas reservoirs are usually high, approaching 90% in many cases. Even at low reservoir pressures, recovery from volumetric gas reservoirs can be improved by utilizing compressors to reduce well head pressures. However, it is important to be aware that there are few true volumetric gas reservoirs. As reservoir pressures are reduced through production, limited migration of fluids, especially water from neighboring shale layers or aquifers, can occur. These migrations may increase water saturations, and pressures in gas reservoirs. But if fluid influx is insignificant relative to the size of the reservoir, the assumption of volumetric behavior is considered valid and applicable even in such cases.

9.2.2 Volumetric Calculations for Wet Gas and Retrograde Gas Condensate Reservoirs

The original gas-in-place for a wet or a retrograde gas condensate reservoir can be calculated with the same equations as for a dry gas reservoir, but an important difference should be noted. The equations that follow will apply to a retrograde condensate reservoir if reservoir pressure is **above** its dew point pressure and **no condensation** has occurred in the reservoir. If the pressure of a retrograde gas condensate reservoir is below its dew point pressure, volumetric calculation of in-place volumes and reserves will be in error since condensation must have taken place in the reservoir. Liquid condensation changes the volume and composition of gas remaining in the reservoir. This affects the uniformity of the composition of gas in the reservoir, which is the basic assumption in volumetric calculations. The best approach for estimating reserves for condensate reservoirs in this condition is to conduct thorough constant volume depletion (CVD) experiments (described in Chapter 6) with fluid samples obtained from the reservoir, and by performing compositional simulation of the depletion process with the CVD data.

The original wet gas or retrograde condensate gas in place, G_c, can be calculated volumetrically as:

$$G_c = \frac{7758 \times A \times h \times \phi \times (1 - S_{wi})}{B_{gi}} \tag{9.9}$$

All the terms in Eq. (9.9) are as previously defined for Eq. (9.1). The main difference between the application of the volumetric equation to a dry gas reservoir versus a wet gas or a retrograde gas condensate reservoir is calculation of the initial gas FVF, B_{gi}. For wet or retrograde condensate gases, calculation of wellstream compositions should include all the liquids condensed in the production facilities. The best approach is to sample the gas and liquid streams at the separators and stock tank. The total composition of the reservoir gas (wet or condensate) is then determined by the recombination method. This method was demonstrated in Example 5.1 in Chapter 5. If the compositions of the streams at the separators and stock tank are not available, the correlation method proposed by Gold et al.[1] should be applied. This method was used in Example 5.2 in Chapter 5 to calculate the gas gravity of a retrograde gas condensate reservoir. With the gas gravity calculated from either example, the gas compressibility factor can be calculated as shown in Example 5.6 in Chapter 5. The cumulative wet or retrograde gas condensate produced, G_{cp}, is given by:

$$G_{cp} = G_p + \frac{133{,}316(\gamma_o N_p)}{M_o} \tag{9.10}$$

In Eq. (9.10), G_p is cumulative dry gas produced in scf; γ_o is the specific gravity of stock tank oil (condensate); N_p is cumulative stock oil (condensate) produced in STB; and M_o is the molecular weight of stock tank oil (condensate), lbm/lb-mole. Equations for calculating specific gravity of the oil and molecular weight of the oil are presented in Chapter 5 as Eqs. (5.48) and (5.56), respectively.

EXAMPLE 9.1 Calculation of Gas-in-Place for the Condensate Reservoir in Example 5.2

Problem

Calculate the gas-in-place for the condensate reservoir in Example 5.2 in Chapter 5. Additional data for the condensate reservoir in Example 5.2 are as follows:

Initial reservoir pressure, p_i	12,688 psia
Reservoir temperature, T_R	267°F
Area, A	297 acres
Net thickness, h	79 feet
Average porosity, ϕ	21%
Initial connate water saturation, S_{wi}	0.301
Well stream gas gravity, γ_g	0.7437

Solution

Step 1: Calculate pseudo-critical pressure and temperature of the gas.

From Eqs. (5.45) and (5.46) in Chapter 5:

$$p_{pc,HC} = 744 - 125.4\gamma_{g,HC} + 5.9\gamma_{g,HC}^2$$
$$= 744 - 125.4 \times 0.7437 + 5.9 \times (0.7437)^2$$
$$= 654.0 \text{ psia}$$

$$T_{pc,HC} = 164.3 + 357.7\gamma_{g,HC} - 67.7\gamma_{g,HC}^2$$
$$= 164.3 + 357.7 \times 0.7437 - 67.7 \times (0.7437)^2$$
$$= 392.9\ ^oR$$

Compositional analysis of the gas shows negligible amounts of nitrogen and carbon dioxide. Consequently, no corrections are applied to the calculated critical properties due to these non-hydrocarbons.

Step 2: Calculate pseudo-reduced pressure and temperature.

The pseudo-reduced pressure and temperature are calculated as:

$$p_{pr} = \frac{p}{p_{pc}} = \frac{12{,}688}{654} = 19.4$$

$$T_{pr} = \frac{T}{T_{pc}} = \frac{459.67 + 267}{392.9} = 1.85$$

Step 3: Calculate gas compressibility factor.

From Figure 5.10, gas compressibility factor, $z = 1.66$. The initial gas formation volume factor is calculated from Eq. (9.2):

$$B_{gi} = \frac{0.005021 z_i T_R}{p_i}$$
$$= \frac{0.005021 \times 1.66 \times (459.67 + 267)}{12{,}688}$$
$$= 4.77 \times 10^{-4} \text{ RB/scf}$$

Step 4: Calculate condensate gas in place.

The condensate gas in place is calculated from Eq. (9.9) as:

$$G_c = \frac{7758 \times A \times h \times \phi \times (1 - S_{wi})}{B_{gi}}$$
$$= \frac{7758 \times 297 \times 79 \times 0.21 \times (1 - .301)}{4.77 \times 10^{-4}}$$
$$= 56.0 \times 10^9 \text{ scf or 56 Bcf}$$

End of Example 9.1

9.2.3 Material Balance for Volumetric Dry Gas, Wet Gas, and Retrograde Gas Condensate Reservoirs

The material balance equations developed in this section will apply to volumetric dry gas, wet gas, and retrograde gas condensate reservoirs. It is important to note that these equations are not applicable to a retrograde gas condensate reservoir if reservoir pressure is **below** dew point pressure and **condensation** has occurred in the reservoir. In addition, for wet and retrograde gas condensate reservoirs, extreme care must be taken to ensure that the total gas produced from the reservoir includes all the liquids condensed and produced from the production facilities and converted to its vapor equivalent volume. The equations for converting produced liquid volumes to their vapor equivalent volumes were presented as Eqs. (5.53) and (5.54) in Chapter 5.

For any volumetric gas reservoir, a simple material balance equation for the reservoir can be written as:

$$\text{Initial Gas Moles} = \text{Gas Moles Remaining} + \text{Gas Moles Produced} \quad (9.11)$$

Eq. (9.11) can be written as:

$$n_i = n_r + n_p \quad (9.12)$$

In Eq. (9.12), n = moles of gas; and the subscripts, i = initial, r = remaining, and p = produced. Re-arranging Eq. (9.12), gives:

$$n_p = n_i - n_r \quad (9.13)$$

In terms of surface gas units (scf), Eq. (9.13) becomes:

$$G_p = G - G\left(\frac{B_{gi}}{B_g}\right) \tag{9.14}$$

Alternatively, Eq. (9.14) can be derived from the General Material Balance Equation (GMBE) for gas reservoirs developed in Chapter 8. The GMBE for gas reservoirs is shown in Eq. (8.27) as:

$$G_pB_g + W_pB_w = GB_{gi}\left(\frac{c_wS_{wi} + c_f}{1 - S_{wi}}\right)\Delta p + G\left(B_g - B_{gi}\right) + W_e \tag{9.15}$$

Since for volumetric reservoirs, water influx and water production are assumed to be zero, and water and formation compressibility are considered negligible, Eq. (9.15) can be reduced to Eq. (9.14). From Eq. (9.2):

$$\frac{B_{gi}}{B_g} = \frac{z_i p}{z p_i} \tag{9.16}$$

Substituting Eq. (9.16) into Eq. (9.14), yields:

$$G_p = G - G\left(\frac{z_i p}{z p_i}\right) \tag{9.17}$$

From Eq. (9.17), it can be derived easily that:

$$\frac{p}{z} = \frac{p_i}{z_i} - \frac{p_i}{z_i}\left(\frac{G_p}{G}\right) \tag{9.18}$$

Equation (9.18) is the basic equation derived from application of material balance to volumetric gas reservoirs. A plot of p/z vs. G_p yields a straight line. When $G_p = 0$, the intercept on the y-axis is equal to p_i/z_i from which initial pressure can be estimated if not available. When $p/z = 0$, the intercept on the x-axis is equal to the original gas-in-place, G. The p/z vs. G_p plot is widely misapplied to all types of gas reservoirs. It is important to remember that it applies in a strict sense only to volumetric gas reservoirs. Volumetric reservoirs by definition have no fluid influxes either from aquifers or any other external sources. However, the p/z vs. G_p plot can be used in a diagnostic sense to qualitatively demonstrate the existence of potential energy support for the reservoir from external sources. Some of the characteristic shapes of the plot for different reservoir drive mechanisms are shown in Figure 9.1 for volumetric, geopressured, weak, and strong water-drive gas reservoirs. It is sometimes permissible to apply the p/z vs. G_p plot to a retrograde gas condensate reservoir at pressures below the dew point if the amount of liquid condensation in the reservoir is presumed to be less than 10% of the hydrocarbon pore volume. However, limited valid results can be derived from the application of this plotting technique on a retrograde gas condensate reservoir below dew point pressures since in a practical sense, it is difficult to determine with sufficient degree of accuracy, the actual amount of condensation that may have occurred in the reservoir at these pressures.

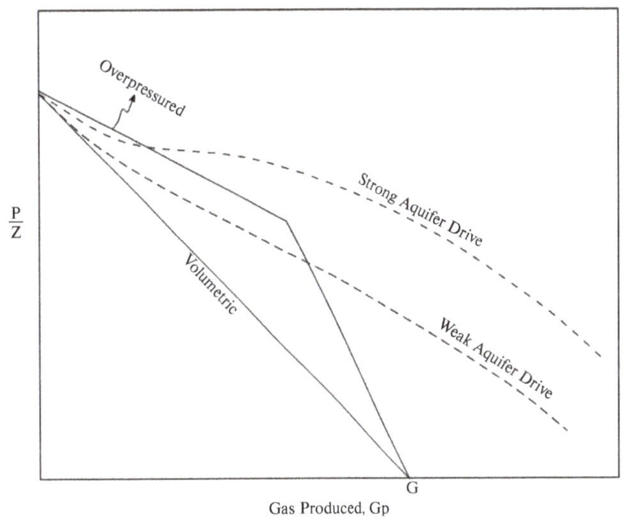

▲ **FIGURE 9.1** Sketch of p/z plots for various gas reservoir drive mechanisms

EXAMPLE 9.2 Calculation of the OGIP for a Volumetric, Dry Gas Reservoir Using the p/z vs. G_p Plot

Problem

Calculate the OGIP for a volumetric, dry gas reservoir from the pressure and cumulative gas production data in Table 9.1. The gas compressibility factors at each reservoir pressure are also shown in Table 9.1.

TABLE 9.1 Data for p/z vs. G_p Plot for a Volumetric, Dry Gas Reservoir

Pressure (psia)	Compressibility Factor, z	Cum. Gas Production G_p (MMscf)	p/z
3600	0.85	0.0	4235.3
3450	0.83	47.8	4156.6
3300	0.82	126.5	4024.4
3150	0.81	204.8	3888.9
2850	0.79	382.5	3607.6
2685	0.77	440.0	3487.0

Solution

From the data in Table 9.1, the p/z vs. G_p plot can be readily performed on a spreadsheet and plotted as shown in Figure 9.2. Extrapolation of the straight line to $p/z = 0$ gives the original gas-in-place (OGIP) as: $G = 2400$ MMscf or 2.4 Bcf.

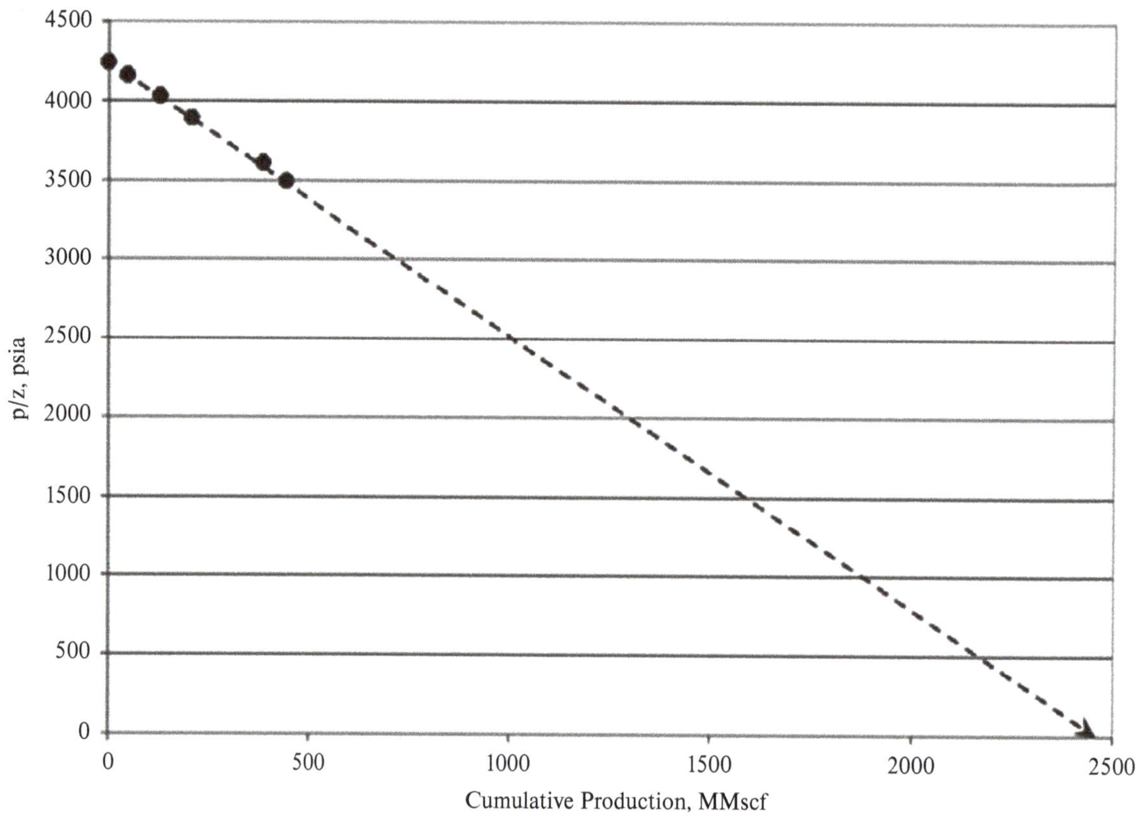

▲ FIGURE 9.2 p/z vs. G_p plot for Example 9.2

End of Example 9.2

9.3 Gas Reservoirs with Water Influx

Gas reservoirs with water influx are non-volumetric reservoirs that have water invading the reservoir mainly from an adjoining aquifer. The strength of the aquifer can be classified qualitatively as weak or strong depending on the rate of water influx into the reservoir. Weak aquifer influx indicates that relatively the rate of water influx is low, while strong aquifer influx indicates a high rate of water influx. Two methods are presented in this section for assessing the performance of gas reservoirs with water influx. These methods are the volumetric approach and the material balance approach. The volumetric method is presented first followed by the material balance method.

9.3.1 Volumetric Approach

Suppose water had invaded a gas reservoir from an adjoining aquifer. Let us assume that the entire hydrocarbon pore volume of the gas reservoir had been occupied by the volume of water that invaded the reservoir and that water was not produced. Based on these assumptions, volumetric methods can be used to estimate the total volume of gas produced from the reservoir and its recovery factor. From Eq. (9.1), the original volume of gas in the reservoir is:

$$G = \frac{7758 \times A \times h \times \phi \times (1 - S_{wi})}{B_{gi}} \tag{9.19}$$

At abandonment, the volume of gas remaining in the reservoir, G_a, is given as:

$$G_a = \frac{7758 \times A \times h \times \phi \times S_{gr}}{B_{ga}} \quad (9.20)$$

In Eq. (9.20), S_{gr} = residual gas saturation at abandonment; and B_{ga} = gas FVF at abandonment pressure. The cumulative gas produced, G_p, is equal to:

$$G_p = G - G_a \quad (9.21)$$

Substituting with Eqs. (9.19) and (9.20):

$$G_p = 7758 \times A \times h \times \phi \times \left[\frac{S_{gi}}{B_{gi}} - \frac{S_{gr}}{B_{ga}} \right] \quad (9.22)$$

In Eq. (9.22), $S_{gi} = 1 - S_{wi}$, from Eq. (9.19). Gas recovery factor, R, is given by:

$$R = \frac{G_p}{G} = \frac{\frac{S_{gi}}{B_{gi}} - \frac{S_{gr}}{B_{ga}}}{\frac{S_{gi}}{B_{gi}}} = \left[1 - \frac{S_{gr} B_{gi}}{S_{gi} B_{ga}} \right] \quad (9.23)$$

Residual gas saturations are difficult to determine either from laboratory measurements or correlations. At best, it can be estimated from laboratory core flood experiments. In absence of any laboratory data, it can be estimated from Agarwal[2] correlations provided in Appendix 9A. These correlations should be used with extreme caution since they may not be representative of the particular reservoir under analysis. Recoveries from gas reservoirs with water influx are generally less than recoveries from volumetric gas reservoirs. The recovery factors for gas reservoirs with water influx range from 50% to 70%, while volumetric gas reservoirs could have recovery factors as high as 90%. The lower recoveries from gas reservoirs with water influx are due to trapped or bypassed gas as water invades the reservoir. The general strategy for managing a gas reservoir with strong water drive is to produce the reservoir at high gas production rates in an attempt to "out run" the invading water. In some cases, this strategy may require producing gas wells at high gas-water ratios until the wells can no longer flow on their own energy. Even under such circumstances, de-watering practices could be used to prolong the productive life of the gas wells.

9.3.2 Material Balance Approach

The material balance equation for a gas reservoir with water influx can be derived readily from the GMBE shown as Eq. (9.15). Assuming water and formation compressibilities are negligible, the material balance equation for a gas reservoir with water influx can be re-arranged from Eq. (9.15) as:

$$GB_{gi} - (G - G_p) B_g = W_e - W_p B_w \quad (9.24)$$

9.3.3 The Cole Plot

Eq. (9.24) can be re-arranged into the following form:

$$\frac{G_p B_g}{B_g - B_{gi}} = G + \frac{W_e - W_p B_w}{B_g - B_{gi}} \quad (9.25)$$

Cole[3] plotted the expression, $G_p B_g / (B_g - B_{gi})$, on the y-axis, and cumulative gas production, G_p, on the x-axis to develop a graphic representation called the Cole plot. A sketch of the Cole plot for various reservoir drive mechanisms is shown in Figure 9.3. As shown in Figure 9.3, the Cole plot for a volumetric depletion drive gas reservoir is a horizontal line. The characteristic shapes of the Cole plot for weak, moderate, and strong water-drive reservoirs are also shown in Figure 9.3. The Cole plot is most useful as a diagnostic plot that will indicate the presence of water influx much earlier in the production life of the reservoir than the p/z plot.[4]

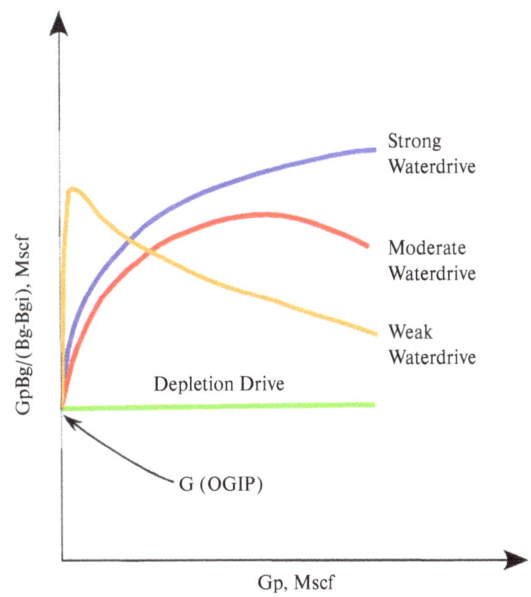

▲ FIGURE 9.3 **Sketch of the Cole plot for gas reservoirs**

9.3.4 The Havlena-Odeh Straight Line Method

Havlena and Odeh[5,6] re-arranged Eq. (9.24) in the form of a straight line equation as follows:

$$\frac{G_p B_g - W_p B_w}{B_g - B_{gi}} = G + \frac{W_e}{B_g - B_{gi}} \quad (9.26)$$

Equation (9.26) can be written in terms of aquifer function such that:

$$\frac{G_p B_g - W_p B_w}{B_g - B_{gi}} = G + C_{aq} \frac{\Phi(p,t)}{B_g - B_{gi}} \quad (9.27)$$

In Eq. (9.27), $W_e = C_{aq}\Phi(p,t)$ where C_{aq} = aquifer constant; and $\Phi(p,t)$ = aquifer function. The definitions of the aquifer constant and aquifer function depend on the type of aquifer. A plot of $(G_pB_g - W_pB_w)/(B_g - B_{gi})$ on the y-axis and $\Phi(p,t)/(B_g - B_{gi})$ on the x-axis yields a straight line with the intercept on the y-axis equal to G and the slope equal to the aquifer constant, C_{aq}. Figure 9.4 is a representation of potential shapes of curves that may result from the Havlena and Odeh (HO) plot. As shown in Figure 9.4, if the size of aquifer is correct, the resulting plot will be a straight line. If the aquifer size is too small, the plot will curve upward. If the size of the aquifer is too large, the plot will curve downward. An S-shaped curve indicates that a linear aquifer might be more appropriate. The Havlena-Odeh Straight Line method is an important tool that can be used to assess the type and strength of the aquifer and gain knowledge on the performance of the reservoir before using more sophisticated techniques, such as reservoir modeling.

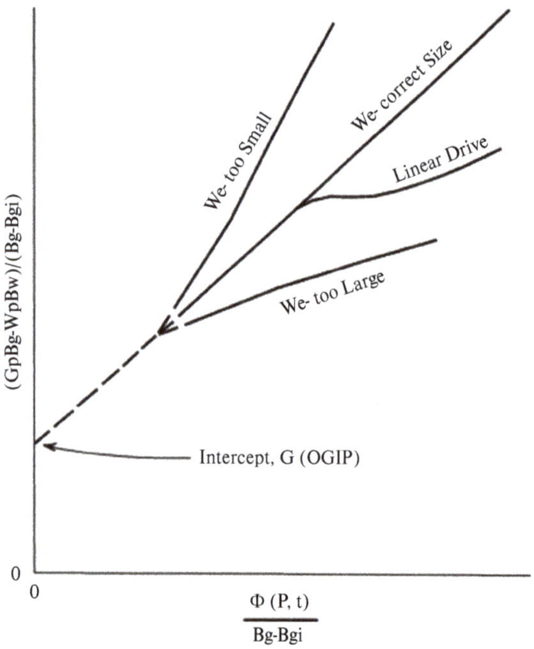

▲ FIGURE 9.4 Sketch of the Havlena-Odeh[5] plot for a gas reservoir (© 1963 SPE, Reproduced with permission)

9.4 Water Influx Models

Water influx into reservoirs (oil and gas) can be estimated with steady-state, unsteady-state, pseudosteady-state, and approximation models. The Schilithuis[7] aquifer model is an example of a steady-state model. Steady-state models assume that the rate of water influx is directly proportional to the pressure drop between the original oil-water contact and the external aquifer boundary. For steady-state models, the pressure at the external aquifer boundary is assumed to be constant. Steady-state models are not presented in this book because they have limited applications for calculating water influx, especially for reservoirs with short production history. The van Everdingen and Hurst[8] aquifer model is an unsteady-state model which is based on the radial diffusivity equation. (The derivation and applications of the radial diffusivity equation are presented in Chapter 12.) The van Everdingen-Hurst aquifer model uses the principle of superposition to calculate cumulative water influx. The process of calculating water influx using the van Everdingen-Hurst method is also presented in Chapter 12. The Fetkovich[9] aquifer model is a pseudosteady-state model. The Fetkovich aquifer model was developed from a combination of an inflow equation and a material balance model based on the aquifer. The Carter-Tracy[10] aquifer model is an approximate model which is based on the van Everdingen-Hurst unsteady-state model. The Fetkovich and Carter-Tracy aquifer models do not require superposition to calculate cumulative water influx. Both methods provide simple and more direct methods for calculating cumulative water influx with accuracy that is comparable to the more tedious van Everdingen-Hurst method. Furthermore, most of the data on the aquifer (such as porosity, net thickness, permeability, compressibility, pressure, etc.) required for calculation of water influx are usually not measured or known with any reasonable level of accuracy. This is mainly due to common industry practice of **not** drilling wells intentionally into the aquifer to acquire these data. The rock properties of the aquifer are often based on the rock properties of the reservoir. Consequently, calculation procedures such as Fetkovich and Carter-Tracy methods are adequate for predicting cumulative water influx for most engineering applications.

The Fetkovich model applies to finite-acting aquifers, while the Carter-Tracy aquifer models can be applied to both finite and infinite-acting aquifers. Both models are applicable to both radial and linear aquifers. The geometrical representations of radial and linear aquifer geometries are shown in Figures 9.5 and 9.6, respectively. The Fetkovich aquifer model applies to edge-water and bottom-water drive reservoirs, while the Carter-Tracy aquifer model applies to edge-water drive reservoirs. In edge-water drive, water influx occurs around the flanks of the reservoir. In bottom-water drive, the reservoir is underlain by the aquifer which influxes vertically into the reservoir. Figures 9.7 and 9.8 are sketches of edge-water drive and bottom-water drive reservoirs, respectively.

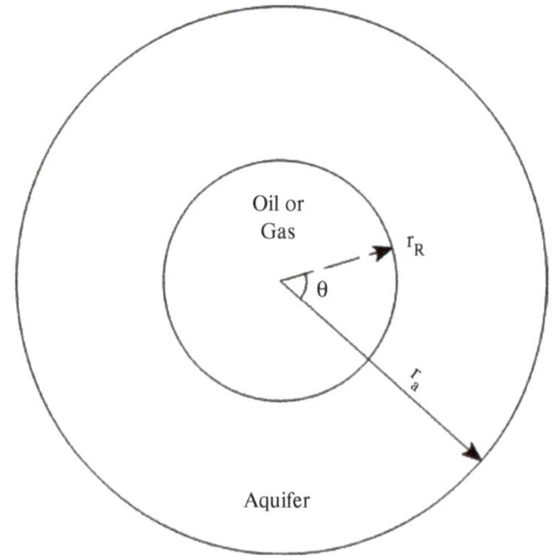

▲ FIGURE 9.5 Sketch of a reservoir with radial aquifer

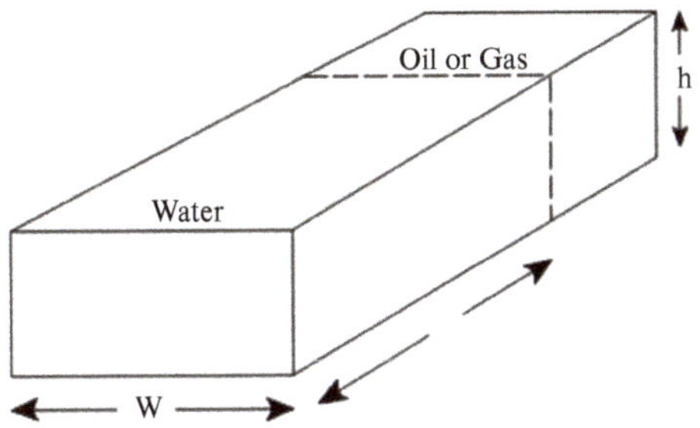

▲ **FIGURE 9.6** Sketch of a reservoir with linear aquifer

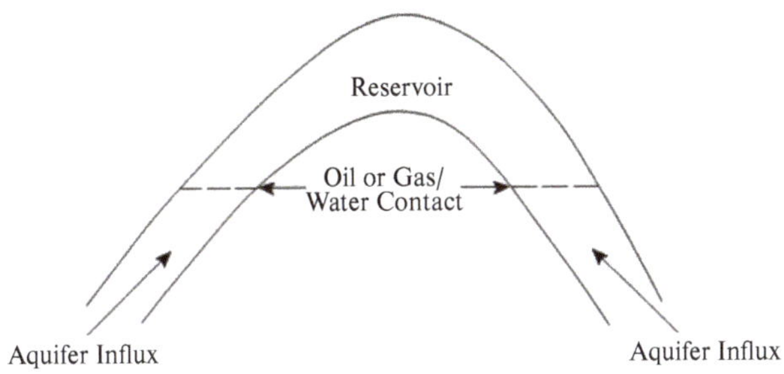

▲ **FIGURE 9.7** Sketch of edge-water drive reservoir

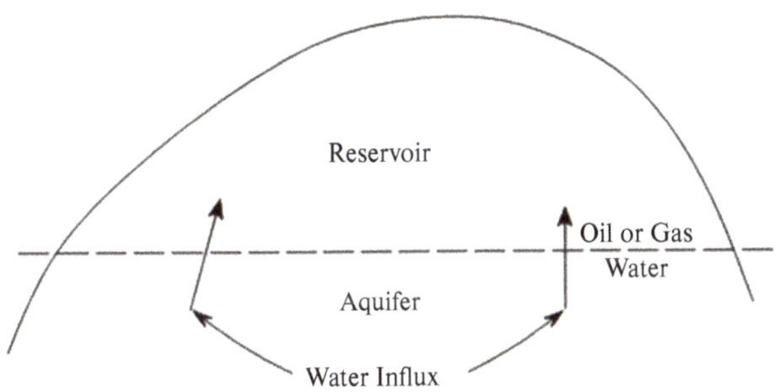

▲ **FIGURE 9.8** Sketch of bottom-water drive reservoir

9.4.1 Fetkovich Aquifer Model

The Fetkovich[9] aquifer model for calculating water influx is based on equating an inflow equation for the aquifer to a material balance model on the aquifer. Because pseudosteady-state is assumed in developing this model, early transient effects on cumulative water influx are ignored. Note that for this reason, cumulative water influx calculated with the Fetkovich method is usually less than cumulative water influx calculated with either the van Everdingen-Hurst or Carter-Tracy methods. However, Fetkovich[9] demonstrated in his paper that this "simplified approach is accurate enough for engineering purposes."

The generalized inflow equation for an aquifer can be written as:

$$q_w = J\left(\bar{p}_a - p_R\right)^n = \frac{dW_e}{dt} \tag{9.28}$$

In Eq. (9.28), q_w = water influx rate; J = aquifer productivity index; $\bar{p}_a$ = average aquifer pressure; p_R = pressure at the reservoir/aquifer boundary (or original hydrocarbon/aquifer contact); n = inflow equation exponent. For Darcy flow at pseudosteady-state or steady-state conditions, $n = 1$, and for turbulent flow conditions, $n = 0.5$; and W_e = cumulative water influx.

The cumulative water influx can be represented in a simple material balance equation based on the expansion of the aquifer as:

$$W_e = c_t W_i\left(p_{a,i} - \bar{p}_a\right) \tag{9.29}$$

In Eq. (9.29), c_t = total aquifer compressibility; W_i = initial aquifer water volume; $p_{a,i}$ = initial aquifer pressure; and $\bar{p}_a$ = average aquifer pressure. Initial aquifer pressure is assumed equal to initial reservoir pressure in most cases. Eq. (9.29) can be expressed in an equivalent form as:

$$\bar{p}_a = p_{a,i}\left(1 - \frac{W_e}{c_t W_i p_{a,i}}\right) = p_{a,i}\left(1 - \frac{W_e}{W_{ei}}\right) \tag{9.30}$$

In Eq. (9.30), $W_{ei} = c_t W_i p_{a,i}$. W_{ei} is the maximum amount of influx water at initial conditions. Differentiating Eq. (9.30) with respect to time, t and re-arranging gives:

$$\frac{dW_e}{dt} = -\frac{W_{ei}}{p_{a,i}}\frac{d\bar{p}_a}{dt} \tag{9.31}$$

Equating Eqs. (9.28) and (9.29) followed by integrating between initial conditions and time, t gives:

$$\int_{p_{a,i}}^{p_a} \frac{d\bar{p}_a}{\bar{p}_a - p_R} = -\int_0^t \frac{Jp_{a,i}}{W_{ei}}dt \tag{9.32}$$

After integration of Eq. (9.32), the result is:

$$\bar{p}_a - p_R = \left(p_{a,i} - p_R\right)\exp\left(\frac{-Jp_{a,i}t}{W_{ei}}\right) \tag{9.33}$$

Substituting Eq. (9.28) into Eq. (9.33) and re-arranging results in:

$$\frac{dW_e}{dt} = J\left(p_{a,i} - p_R\right)\exp\left(\frac{-Jp_{a,i}t}{W_{ei}}\right) \tag{9.34}$$

Integrating Eq. (9.34) for cumulative water influx yields:

$$W_e = \frac{W_{ei}}{p_{a,i}}\left(p_{a,i} - p_R\right)\left[1 - \exp\left(\frac{-Jp_{a,i}t}{W_{ei}}\right)\right] \tag{9.35}$$

Fetkovich[9] approximated constant pressure at the reservoir/aquifer boundary by assuming that the pressure history at the boundary can be divided into small finite number of time intervals such that the incremental water influx at the nth interval can be represented thus:

$$\Delta W_{en} = \frac{W_{ei}}{p_{a,i}}\left(\bar{p}_{a,n-1} - \bar{p}_{R,n}\right)\left[1 - \exp\left(\frac{-Jp_{a,i}\Delta t_n}{W_{ei}}\right)\right] \quad (9.36)$$

For Eq. (9.36),

$$\bar{p}_{a,n-1} = p_{a,i}\left(1 - \frac{W_{e,n-1}}{W_{ei}}\right) \quad (9.37)$$

$$\bar{p}_{R,n} = \frac{p_{R,n-1} + p_{R,n}}{2} \quad (9.38)$$

The Fetkovich method was developed for finite aquifer systems. The productivity indices for various aquifer flow geometries and boundary conditions are provided in Table 9.2.

TABLE 9.2 Aquifer Productivity Indices

Type of Outer Aquifer Boundary	Radial Flow* (STB/D-psi)	Linear Flow* (STB/D-psi)
Finite-no flow	$J = \frac{0.00708kh\left(\frac{\theta}{360}\right)}{\mu\left[\ln\left(\frac{r_a}{r_R}\right) - 0.75\right]}$	$J = \frac{0.003381kwh}{\mu L}$
Finite-constant pressure	$J = \frac{0.00708kh\left(\frac{\theta}{360}\right)}{\mu\left[\ln\left(\frac{r_a}{r_R}\right)\right]}$	$J = \frac{0.001127kwh}{\mu L}$

*All variable are in field units.

EXAMPLE 9.3 Calculation of Water Influx Using the Fetkovich Method

Problem

The properties of a radial aquifer are shown in Table 9.3. The pressure history at the reservoir-aquifer boundary is also provided in Table 9.3. Calculate the water influx using the Fetkovich method assuming a finite aquifer with no flow at the outer aquifer boundary.

TABLE 9.3 Aquifer/Reservoir Data and Pressure History for Example 9.3

Initial reservoir pressure	3665 psia
Porosity, ϕ	0.225
Permeability, k	669 md
Aquifer thickness, h	45 feet
Water viscosity, μ	0.35 cp
Total aquifer compressibility, c_t	6×10^{-6} psia^{-1}
Reservoir radius, r_R	8210 feet
Aquifer radius, r_a	123,150 feet
Encroachment angle, θ	360°
Pressure History at Reservoir/Aquifer Boundary	
Time, t (days)	**Pressure, p (psia)**
0	3665
365	3586
730	3543
1095	3520
1460	3491

Solution

Step 1: Calculate the maximum water influx from the aquifer, W_{ei}.

Initial volume of water in the aquifer, W_i, is calculated as:

$$W_i = \frac{\pi\left(r_a^2 - r_R^2\right)h\phi\left(\frac{\theta}{360}\right)}{5.615}$$

$$= \frac{\pi\left[(123150)^2 - (8210)^2\right]\times 45 \times 0.225 \times 1}{5.615}$$

$$= 85.53\times 10^9 \text{ RB}$$

$$W_{ei} = c_t W_i p_{a,i} = 6\times 10^{-6} \times 85.53\times 10^9 \times 3665$$

$$= 1.88\times 10^9 \text{ RB}$$

Step 2: Calculate aquifer productivity index, J.

From Table 9.2,

$$J = \frac{0.00708kh\frac{\theta}{360}}{\mu\left[\ln\left(\frac{r_a}{r_R}\right) - 0.75\right]}$$

$$= \frac{0.00708\times 669\times 45}{0.35\left[\ln\left(\frac{123150}{8210}\right) - 0.75\right]}$$

$$= 311.0141 \text{ STB/D-psi}$$

Step 3: Calculate incremental water influx for each time period, n.

From Eq. (9.36):

$$\Delta W_{en} = \frac{W_{ei}}{p_{a,i}}\left(\bar{p}_{a,n-1} - \bar{p}_{R,n}\right)\left[1-\exp\left(\frac{-Jp_{a,i}\Delta t_n}{W_{ei}}\right)\right]$$

From Eq. (9.37):

$$\bar{p}_{a,n-1} = p_{a,i}\left(1-\frac{W_{e,n-1}}{W_{ei}}\right)$$

And from Eq. (9.38):

$$\bar{p}_{R,n} = \frac{p_{R,n-1}+p_{R,n}}{2}$$

Substituting above equations at $n = 1$ gives:

$$\bar{p}_{a,0} = 3665\left(1-\frac{0}{1.88\times10^9}\right) = 3665 \text{ psia}$$

$$\bar{p}_{R,1} = \frac{3665+3586}{2} = 3625.5 \text{ psia}$$

$$\bar{p}_{a,0} - \bar{p}_{R,1} = 3665 - 3625.5 = 39.5 \text{ psia}$$

$$\Delta W_{e1} = \frac{1.88\times10^9}{3665}\times39.5 \times\left[1-\exp\left(\frac{-311.0141\times3665\times365}{1.88\times10^9}\right)\right] = 4.023\times10^6 \text{ RB}$$

At $n = 2$:

$$\bar{p}_{a,1} = 3665\left(1-\frac{4.023\times10^6}{1.88\times10^9}\right) = 3657.16 \text{ psia}$$

$$\bar{p}_{R,2} = \frac{3586+3543}{2} = 3564.5 \text{ psia}$$

$$\bar{p}_{a,1} - \bar{p}_{R,2} = 3657.16 - 3564.5 = 92.66 \text{ psia}$$

$$\Delta W_{e2} = \frac{1.88\times10^9}{3665}\times92.66 \times\left[1-\exp\left(\frac{-311.0141\times3665\times365}{1.88\times10^9}\right)\right] = 9.436\times10^6 \text{ RB}$$

Cumulative water influx at $n = 2$ is:

$$W_{e2} = W_{e1} + \Delta W_{e2} = 4.023\times10^6 + 9.436\times10^6 = 1.346\times10^7 \text{ RB}$$

The calculations for the remaining time periods are summarized in Table 9.4.

TABLE 9.4 Water Influx Calculations Using the Fetkovich Method for Example 9.3

n	t (days)	$p_{R,n}$ (psia)	$\bar{p}_{R,n}$ (psia)	$\bar{p}_{a,n}$ (psia)	$\bar{p}_{a,n-1} - \bar{p}_{R,n}$ (psia)	J (STB/D-psi)	ΔW_e (RB)	W_e (RB)
0	0	3665	3665.0	3665.0	0.0	311.0141	0	0
1	365	3586	3625.5	3657.2	39.5	311.0141	4.02E+6	4.02E+6
2	730	3543	3564.5	3638.8	92.7	311.0141	9.44E+6	1.35E+7
3	1095	3520	3531.5	3617.5	107.3	311.0141	1.09E+7	2.44E+7
4	1460	3491	3505.5	3595.2	112.0	311.0141	1.14E+7	3.58E+7

End of Example 9.3

9.4.2 Carter-Tracy Aquifer Model

The Carter-Tracy[10] method is an approximate method that calculates aquifer influx with results that are comparable to the exact solutions of the van Everdingen-Hurst method. The Carter-Tracy method is recommended in preference to the van Everdingen-Hurst method because it does not use superposition which makes the calculations less tedious especially if performed manually on a calculator.

The equations for the Carter-Tracy[10] method were developed by expressing cumulative water influx as a function of varying pressure with the aid of a convolution integral:

$$W_e\left(t_{Dj}\right) = B\int_0^{t_{Dj}} \Delta p(\lambda) Q_D'\left(t_{Dj}-\lambda\right)d\lambda \qquad (9.39)$$

In Eq. (9.39), W_e = cumulative water influx; B = constant; p = pressure; Q_D = dimensionless cumulative water influx; and t_D = dimensionless time. In terms of field units,

$$B = 1.119\phi c_t h r_R^2\left(\frac{\theta}{360}\right) \quad \text{for radial flow} \qquad (9.40)$$

$$B = 0.178\phi c_t hL \quad \text{for linear flow} \qquad (9.41)$$

Also, in terms of field units, dimensionless time, t_D, is defined as:

$$t_D = \frac{0.00633kt}{\phi\mu c_t r_R^2} \quad \text{for radial flow} \tag{9.42}$$

$$t_D = \frac{0.00633kt}{\phi\mu c_t L^2} \quad \text{for linear flow} \tag{9.43}$$

In Eqs. (9.40) to (9.43), ϕ = porosity, fraction; c_t = total aquifer compressibility, psi^{-1}; h = net aquifer thickness, feet; r_R = reservoir radius, feet; L = length of reservoir, feet; θ = angle of encroachment, degrees; k = permeability, md; and t = time, days.

By approximating water influx by a series of constant rate intervals, Eq. (9.39) can be represented with the following alternate forms:

$$W_e\left(t_{Dj}\right) = \sum_{n=0}^{j-1} q_{Dn}\left(t_{Dn+1} - t_{Dn}\right) \tag{9.44}$$

$$W_e\left(t_{Dj}\right) = W_e\left(t_{Di}\right) + \sum_{n=i}^{j-1} q_{Dn}\left(t_{Dn+1} - t_{Dn}\right) \tag{9.45}$$

Combining Eqs. (9.39) and (9.45), Laplace transform techniques can be used to solve for cumulative water influx in terms of pressure drop, Δp_n:

$$W_e\left(t_{Dn}\right) = W_e\left(t_{Dn-1}\right) + \left[\frac{B\Delta p\left(t_{Dn}\right) - W_e\left(t_{Dn-1}\right) p_D'\left(t_{Dn}\right)}{p_D\left(t_{Dn}\right) - t_{Dn-1} p_D'\left(t_{Dn}\right)}\right] \times \left(t_{Dn} - t_{Dn-1}\right) \tag{9.46}$$

The intermediate equations before arriving at Eq. (9.46) are shown in the paper by Carter and Tracy.[10] In Eq. (9.46),

$$\Delta p\left(t_{Dn}\right) = p_0 - p\left(t_{Dn}\right) \tag{9.47}$$

In Eq. (9.46), n and $n-1$ represent current and previous time intervals, respectively. Also, $p\left(t_{Dn}\right)$ represents pressure at dimensionless time interval, t_{Dn}, and $W_e\left(t_{Dn-1}\right)$, represents cumulative water influx at dimensionless time interval, t_{Dn-1}. In Eq. (9.47), p_0 is initial aquifer pressure at $t=0$. In Eq. (9.46), $p_D\left(t_{Dn}\right)$ represents dimensionless pressure as a function of dimensionless time, t_{Dn}, and p_D', represents the derivative of p_D. p_D and p_D' for finite and infinite acting aquifers can be obtained from polynomials developed by Fanchi.[11] The polynomials are provided in Appendix 9B. Similar polynomials for infinite acting aquifers only were published by Edwardson et al.[12] These polynomials are also presented in Appendix 9C. Klins et al.[13] published sets of polynomials for finite and infinite acting reservoirs. These were not provided in this book. For infinite acting aquifers, the polynomials reported by Edwardson et al.[12] are recommended because the reported average error of 0.2% is less than the average error of 1.5% reported by Fanchi.[11] However, for consistency and digital applications, the polynomials presented by Fanchi[11] may be preferable since the polynomials are also applicable to finite aquifers.

EXAMPLE 9.4 Calculation of Water Influx Using the Carter-Tracy Method

Problem

Using the same data presented in Table 9.3 for the example on application of the Fetkovich method, calculate the water influx using the Carter-Tracy method.

Solution

Step 1: Calculate the parameter, B, for radial flow.

From Eq. (9.40):

$$B = 1.119\phi c_t h r_R^2\left(\frac{\theta}{360}\right)$$
$$= 1.119 \times 0.225 \times 6 \times 10^{-6} \times 45 \times (8210)^2 \times \left(\frac{360}{360}\right)$$
$$= 4582.08 \text{ RB/psi}$$

Step 2: Calculate dimensionless time, t_D.

From Eq. (9.42):

$$t_D = \frac{0.00633kt}{\phi\mu c_t r_R^2} = \frac{0.00633 \times 669 \times t}{0.225 \times 0.35 \times 6 \times 10^{-6} \times (8210)^2}$$
$$= 0.1323t$$

Step 3: Calculate pressure change, $\Delta p\left(t_{Dn}\right)$.

From Eq. (9.47):

$$\Delta p\left(t_{Dn}\right) = p_0 - p\left(t_{Dn}\right)$$

At $n=1$, $t_{D1} = 48.30$.

$$\Delta p\left(t_{D1}\right) = 3665 - 3586 = 79 \text{ psia.}$$

At $n=2$, $t_{D2} = 96.62$.

$$\Delta p\left(t_{D2}\right) = 3665 - 3543 = 122 \text{ psia.}$$

Step 4: Calculate dimensionless pressure for infinite acting aquifers at $n=1$.

From Appendix 9C using the correlations by Edwardson et al.[12] at $t_{D1} = 48.3$;

$$p_D = \frac{370.529\sqrt{t_D} + 137.582t_D + 5.69549t_D\sqrt{t_D}}{328.834 + 265.488\sqrt{t_D} + 45.2157t_D + t_D\sqrt{t_D}}$$
$$= \frac{370.529\sqrt{48.3} + 137.582 \times 48.3 + 5.69549 \times 48.3\sqrt{48.3}}{328.834 + 265.488\sqrt{48.3} + 45.2157 \times 48.3 + 48.3\sqrt{48.3}}$$
$$= 2.3718$$

Step 5: Calculate dimensionless pressure derivative for infinite acting aquifers at $n = 1$

From Appendix 9C using the correlations by Edwardson et al.[12] at $t_{D1} = 48.3$;

$$p'_D = \frac{716.441 + 46.7984\sqrt{t_D} + 270.038t_D + 71.0098t_D\sqrt{t_D}}{1269.86\sqrt{t_D} + 1204.73t_D + 618.618t_D\sqrt{t_D} + 538.072t_D^2 + 142.410t_D^2\sqrt{t_D}}$$

$$= \frac{716.441 + 46.7984\sqrt{48.3} + 270.038\times 48.3 + 71.0098\times 48.3\sqrt{48.3}}{1269.86\sqrt{48.3} + 1204.73\times 48.3 + 618.618\times 48.3\sqrt{48.3} + 538.072(48.3)^2 + 142.410(48.3)^2\sqrt{48.3}}$$

$$= 0.009878$$

Step 6: Calculate water influx at, $n = 1$.

From Eq. (9.46):

$$W_e(t_{Dn}) = W_e(t_{Dn-1}) + \left[\frac{B\Delta p(t_{Dn}) - W_e(t_{Dn-1})p'_D(t_{Dn})}{p_D(t_{Dn}) - t_{Dn-1}p'_D(t_{Dn})}\right]\times(t_{Dn} - t_{Dn-1})$$

At $t_{D1} = 48.3$;

$$W_e(t_{D1}) = 0 + \left[\frac{4582.08\times 79 - 0}{2.3718 - 0}\right]\times(48.3 - 0)$$
$$= 7{,}371{,}829 \text{ RB}$$

Step 7: Repeat Steps 4, 5, 6 at $n = 2$ and $t_{D2} = 96.61$.

$$p_D = \frac{370.529\sqrt{96.61} + 137.582\times 96.61 + 5.69549\times 96.61\sqrt{96.61}}{328.834 + 265.488\sqrt{96.61} + 45.2157\times 96.61 + 48.3\sqrt{96.61}}$$
$$= 2.7061$$

$$p'_D = \frac{716.441 + 46.7984\sqrt{96.61} + 270.038\times 96.61 + 71.0098\times 96.61\sqrt{96.61}}{1269.86\sqrt{96.61} + 1204.73\times 96.61 + 618.618\times 96.61\sqrt{96.61} + 538.072(96.61)^2 + 142.410(96.61)^2\sqrt{96.61}}$$
$$= 0.005036$$

$$W_e(t_{D2}) = 7{,}371{,}829 + \left[\frac{4582.08\times 122 - 7{,}371{,}829\times 0.005036}{2.7061 - 48.3\times 0.005036}\right]\times(96.61 - 48.3)$$
$$= 17{,}607{,}516 \text{ RB}$$

The calculations for the remaining time periods are summarized in Table 9.5. Note that cumulative water influx predicted with the Fetkovich method is less than the influx calculated with the Carter-Tracy method by comparing Tables 9.4 and 9.5. This is because the Fetkovich method does not include transient effects for unsteady-state conditions in early time periods, and is limited in its application to infinite-acting aquifers.

TABLE 9.5 Water Influx Calculations Using the Carter-Tracy Method for Example 9.4

n	t (days)	$p_{R,n}$ (psia)	t_D (psia)	$\Delta p\ (t_n)$	B (RB/psi)	p_D	p'_D	W_e (RB)
0	0	3665	0.00	0	4582.08	0.0000	0.0000	0
1	365	3586	48.30	79	4582.08	2.3718	0.0099	7.37E+6
2	730	3543	96.61	122	4582.08	2.7061	0.0050	1.76E+7
3	1095	3520	144.91	145	4582.08	2.9043	0.0034	2.89E+7
4	1460	3491	193.21	174	4582.08	3.0457	0.0025	4.20E+7

End of Example 9.4

Generally, it is recommended practice that a water-drive gas reservoir should be modeled with a numerical aquifer in a reservoir simulator. In constructing the reservoir model, the aquifer size should reflect the assumed reservoir-aquifer ratio depending on whether the aquifer is considered to be finite- or infinite-acting. The Havlena-Odeh plot can assist in the decision pertaining to the size and strength of the aquifer. If sufficient production history is available, a reasonable aquifer size will give an acceptable history match. This aquifer size should be reviewed to ensure that it is reasonable and fits with the known geology of the area. Construction of gas reservoir models with numerical aquifers reduces the difficulties associated with determining the rate of water influx using analytical models such as Fetkovich, and Carter-Tracy models. In addition, it provides a versatile tool that can be used to investigate many possible water drive mechanisms that may be occurring in the reservoir. The applications of reservoir simulation in evaluation of reservoir performance are discussed in Chapter 23.

9.5 Geopressured Gas Reservoirs

Geopressured gas reservoirs are abnormally pressured reservoirs which have been encountered all over the world, including the Gulf Coast region of the United States of America. The initial pressure gradients in normally pressured reservoirs range between 0.43 psi/ft and 0.5 psi/ft. In geopressured reservoirs, initial pressure gradients are between 0.6 psi/ft and 1.0 psi/ft. According to Poston and Berg,[14] the higher pressures in geopressured reservoirs are probably caused by inability of excess fluids to leak-off after major tectonic events of compressional folding or rapid deposition of thick sediments in young sedimentary sections. Figure 9.9 shows the occurrences of geopressured zones along the Gulf Coast of the United States of America.

In normally pressured gas reservoirs, gas compressibility is dominant in comparison to water and formation (rock) compressibility as the source of energy for gas production. For this reason, water and formation compressibility were considered negligible in the development of the material balance equation [Eq. (9.14)] for volumetric, normally pressured gas reservoirs. In geopressured reservoirs, water and formation compressibility can be almost as high as gas compressibility at initial pressures. Consequently, it is necessary to include water and formation compressibility in developing material balance equations for geopressured reservoirs. If the conventional p/z versus G_p plot developed for normally pressured gas reservoirs [Eq. (9.18)] is applied to geopressured gas reservoirs, the original gas-in-place will be overestimated at early stages of reservoir depletion. The typical shape of p/z vs. G_p plot for a geopressured gas reservoir is shown in Figure 9.10. Note the typical dual slope

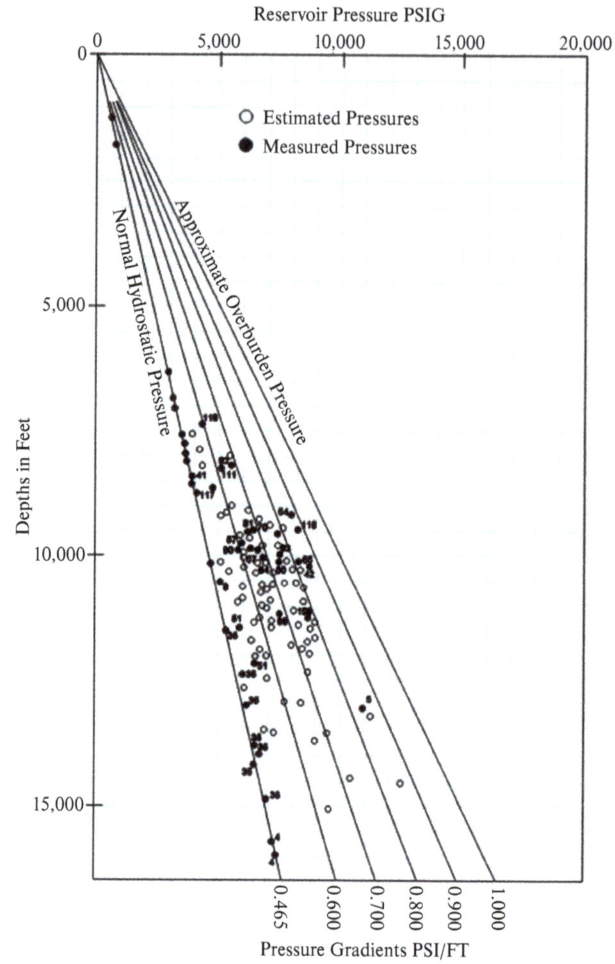

▲ **FIGURE 9.9 Geopressured zones along Gulf Coast, USA (from Dickinson[15])**

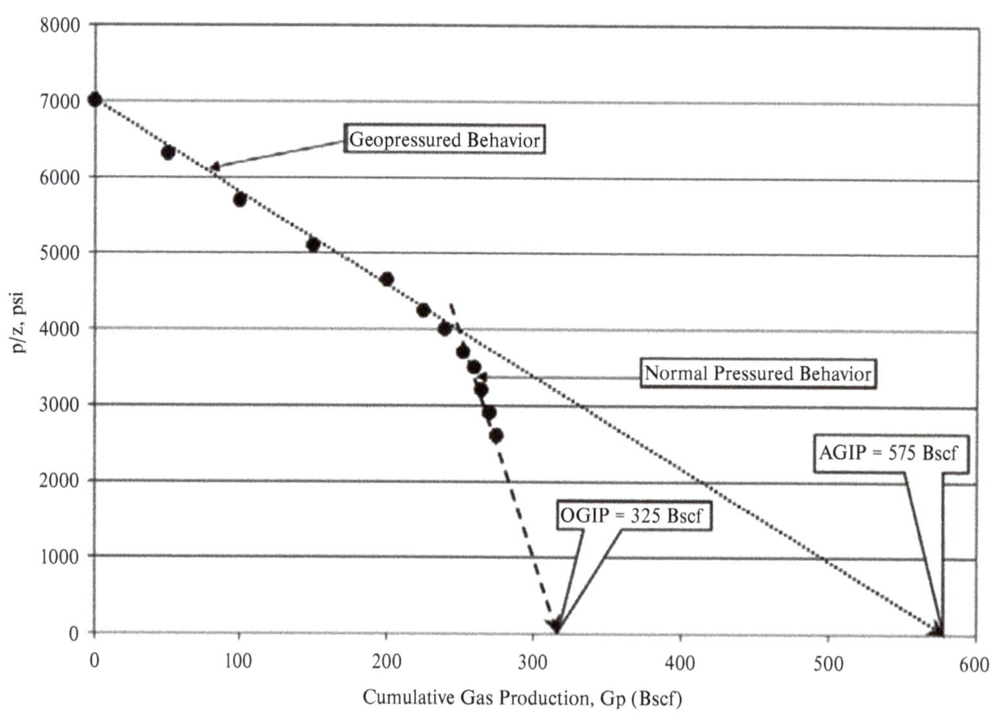

▲ **FIGURE 9.10 Typical p/z vs. G_p plot for a geopressured gas reservoir**

of the plot, representing the flatter slope during the early production history when water and rock compressibility effects are significant and the steeper slope at later stages of depletion when gas compressibility is dominant. Extrapolation of the early straight line will give an apparent gas-in-place (AGIP) which is significantly higher than the actual original gas-in-place (OGIP). This could cause significant errors in the estimation of OGIP and reserves for geopressured reservoirs. To avoid this error, two methods are presented for calculation of OGIP for geopressured gas reservoirs. The methods account for the effects of water and formation compressibility in the development of material balance equations for geopressured reservoirs.

9.5.1 The Ramagost and Farshad Method

The GMBE for a geopressured gas reservoir, assuming no water influx and negligible water production, can be derived from Eq. (9.15) as:

$$G_pB_g = GB_{gi}\left(\frac{c_wS_{wi}+c_f}{1-S_{wi}}\right)\Delta p + G\left(B_g - B_{gi}\right) \quad (9.48)$$

Eq. (9.48) can be rearranged to get:

$$G = \frac{G_pB_g}{B_g - B_{gi} + \left[\dfrac{B_{gi}\left(p_i-p\right)\times\left(c_f+c_wS_{wi}\right)}{1-S_{wi}}\right]} \quad (9.49)$$

Eq. (9.49) was further rearranged by Ramagost and Farshad[16] to develop a useful form of the material balance equation for geopressured reservoirs which is similar in form to Eq. (9.18) for normally pressured gas reservoirs:

$$\frac{p}{z}\left[1-\left(\frac{c_f+c_wS_{wi}}{1-S_{wi}}\right)\left(p_i-p\right)\right] = \frac{p_i}{z_i} - \left(\frac{p_i}{z_iG}\right)G_p \quad (9.50)$$

Equation (9.50) is different from Eq. (9.18) only by the expression: $\left[1-\left\{\left(c_f+c_wS_{wi}\right)/\left(1-S_{wi}\right)\right\}\left(p_i-p\right)\right]$. This expression accounts for the effects of water and rock compressibility on the p/z ratio of geopressured reservoirs. Eq. (9.50) is in the form of the equation for a straight line where:

$$Y \text{ variable} = \frac{p}{z}\left[1-\left(\frac{c_f+c_wS_{wi}}{1-S_{wi}}\right)\left(p_i-p\right)\right] \quad (9.51)$$

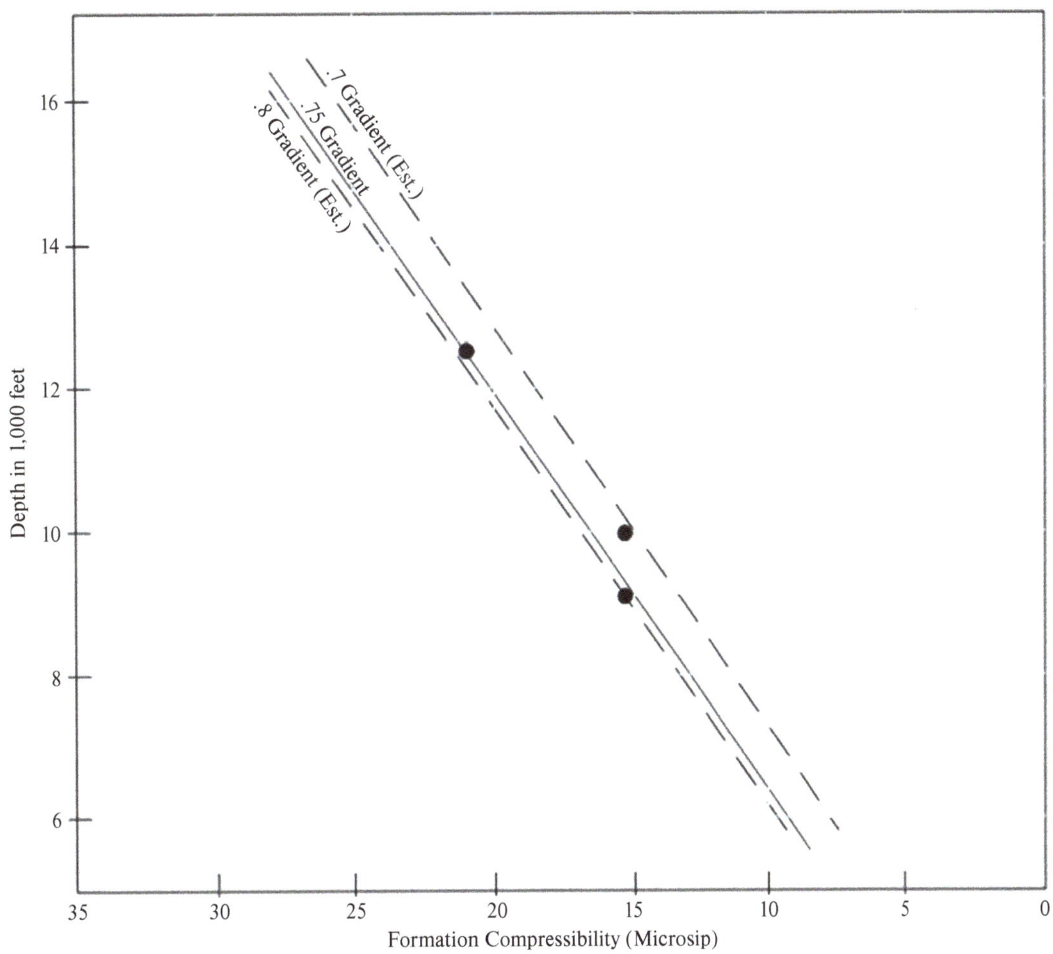

▲ FIGURE 9.11 Correlation for estimation of formation compressibility (from Hammerlindl[19] © 1971 SPE, Reproduced with permission)

$$X \text{ variable} = G_p \tag{9.52}$$

$$\text{slope, } m = -\left(\frac{p_i}{z_i G}\right) \tag{9.53}$$

$$Y \text{ intercept} = \frac{p_i}{z_i} \tag{9.54}$$

The Ramagost and Farshad method can be used to calculate OGIP by making a plot of the production data as shown in Example 9.5 using Eqs. (9.51) to (9.54). This approach assumes that the formation compressibility is known and that it is a single constant value. Formation compressibility is not constant and may be variable, especially at early stages of reservoir depletion. The variability of formation compressibility has been reported by Poston and Chen[17] and Fetkovich et al.[18] Figure 9.11 provides correlations for estimation of formation compressibility. These correlations should only be used if formation compressibility data measured in a laboratory for the reservoir are not available. Note that the formation compressibility obtained from these correlations is most likely not accurate.

EXAMPLE 9.5 Calculation of Original Gas in Place Using the Ramagost-Farshad Method

Problem

Calculate the OGIP with the Ramagost-Farshad method for a geopressured reservoir using the reservoir data and production history of the Anderson "L" reservoir as reported by Duggan.[20] The data summary for the Anderson "L" reservoir is as follows:

Depth	11,167 ft. (subsea)
Porosity, ϕ	24%
Initial water saturation, S_{wi}	35%
Permeability, k	0.7 to 791 md
Estimated dew point pressure	6118 psia
Initial bottomhole pressure	9507 psia
Initial pressure gradient	0.843 psi/ft
Bottom-hole temperature	266°F
Volumetric OGIP	2240 Mcf/ac-ft or 69.552 Bcf

Solution

The formation compressibility was estimated from Figure 9.11 as 19.3 × 10^{-6} psi^{-1} and the compressibility of water was estimated as 3.2 × 10^{-6} psi$^{-1.}$ The calculations were performed on a spreadsheet as shown in Table 9.6 using Eqs. (9.51) to (9.54). The data in Table 9.6 were used to plot Figure 9.12. Note in Figure 9.12, the OGIP was shown to be 68 Bcf. This is comparable to volumetrically calculated OGIP of 69.6 Bcf reported by Duggan.[20] The AGIP was shown to be 107 Bcf based on the p/z plot. This represents an error of about 54% if the p/z method was used early in this reservoir to estimate OGIP.

TABLE 9.6 Data and Calculations for Example 9.5 Using the Ramagost-Farshad method

DATE	Bottomhole Pressure (psia)	Compressibility Factor, z	Cumulative Gas Production (MMscf)	p/z	Y Variable
12/22/65	9507	1.440	0.0	6602.1	6602.1
03/01/66	9292	1.418	392.5	6552.9	6508.6
06/22/66	8970	1.387	1642.2	6467.2	6358.1
09/29/66	8595	1.344	3225.8	6395.1	6211.9
11/17/66	8332	1.316	4260.3	6331.3	6097.6
12/30/66	8009	1.282	5503.5	6247.3	5953.3
03/23/67	7603	1.239	7538.1	6136.4	5769.4
05/15/67	7406	1.218	8749.2	6080.5	5679.1
07/31/67	7002	1.176	10509.3	5954.1	5485.5
09/14/67	6721	1.147	11758.9	5859.6	5346.8
10/19/67	6535	1.127	12789.2	5798.6	5257.2
03/05/68	5764	1.048	17262.5	5500.0	4853.3
09/04/68	4766	0.977	22890.8	4878.2	4151.6
03/19/69	4295	0.928	28144.6	4628.2	3870.4
09/29/69	3750	0.891	32566.7	4208.8	3447.6
03/31/70	3247	0.854	36819.9	3802.1	3054.4

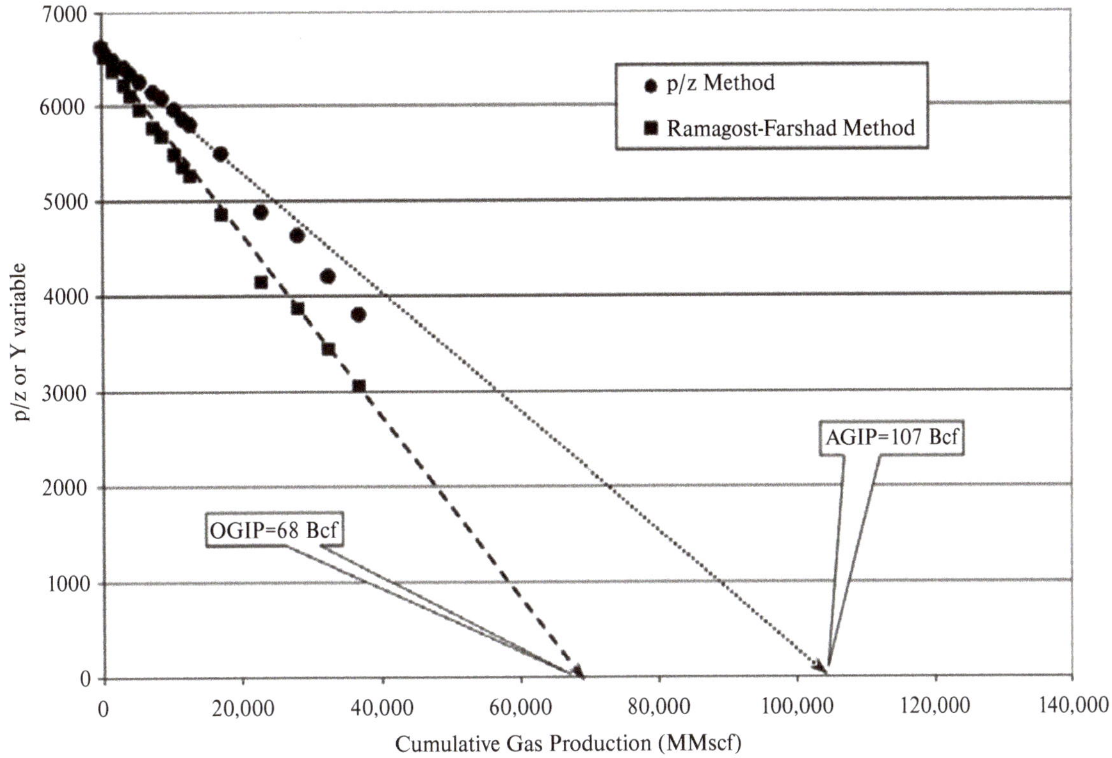

▲ **FIGURE 9.12** **Comparison of *p/z* method and Ramagost-Farshad method for the Anderson "L" Reservoir in Example 9.5**

End of Example 9.5

9.5.2 The Roach Method

Roach[21] re-arranged the familiar material balance equation for geopressured reservoir shown in Eq. (9.49) into an alternative form that treats the formation compressibility and original gas-in-place as the unknown variables. The Roach[21] form of the material balance equation is as follows:

$$\frac{1}{(p_i - p)}\left(\frac{p_i z}{p z_i} - 1\right) = \frac{1}{G}\left[\frac{G_p}{(p_i - p)}\left(\frac{p_i z}{p z_i}\right)\right] - \left[\frac{c_f + c_w S_{wi}}{1 - S_{wi}}\right] \quad (9.55)$$

Again, Eq. (9.55) is in the form of equation for a straight line where:

$$Y \text{ variable} = \frac{1}{(p_i - p)}\left(\frac{p_i z}{p z_i} - 1\right) \quad (9.56)$$

$$X \text{ variable} = \left[\frac{G_p}{(p_i - p)}\left(\frac{p_i z}{p z_i}\right)\right] \quad (9.57)$$

$$\text{slope, } m = \frac{1}{G} \quad (9.58)$$

$$Y \text{ intercept} = -\left[\frac{c_f + c_w S_{wi}}{1 - S_{wi}}\right] \quad (9.59)$$

Example 9.6 is the application of the Roach method to the same data used in Example 9.5. Note that this method assumes that formation compressibility is constant although unknown. The strength of the Roach method is that it can be applied without prior knowledge of the data for formation compressibility.

EXAMPLE 9.6 Calculation of the Original Gas in Place Using the Roach Method

Problem

Calculate the OGIP with the Roach Method using the Anderson "L" reservoir data in Example 9.5.

Solution

The terms for the x-axis and the y-axis were calculated in a spreadsheet as shown in Table 9.7. Figure 9.13 is the plot of the Roach method based on the data in Table 9.7. The slope of the straight line in Figure 9.13 is $m = 1.4082 \times 10^{-5}$ MMscf^{-1}. From Eq. (9.58):

$$\text{slope, } m = \frac{1}{G}$$

$$\text{or } G = \frac{1}{m} = \frac{1}{1.4082 \times 10^{-5}}$$

$$= 71 \text{ Bcf}$$

The OGIP calculated with the Roach method compares favorably to the volumetric OGIP (69 Bcf) reported by Duggan[20] and the OGIP of 68 Bcf calculated in Example 9.5 using the Ramagost-Farshad method.

The y-intercept in Figure 9.13 is -2.2×10^{-5} psi^{-1}. From the y-intercept, the formation compressibility factor can be estimated from Eq. (9.59):

$$Y_{\text{intercept}} = -\left[\frac{c_f + c_w S_{wi}}{1 - S_{wi}}\right]$$

$$-2.2 \times 10^{-5} = -\left[\frac{c_f + 3.2 \times 10^{-6} \times 0.35}{1 - 0.35}\right]$$

$$c_f = (1 - 0.35) \times 2.2 \times 10^{-5} - 3.2 \times 10^{-6} \times 0.35$$

$$= 13.18 \times 10^{-6} \text{ psi}^{-1}$$

Note that the formation compressibility estimated from Figure 9.11 in Example 9.5 is 19.3×10^{-6} psi^{-1}. The difference illustrates the fact that formation compressibility data exhibit a wide range of values and should be used with caution.

TABLE 9.7 Data and Calculations for Example 9.6 Using the Roach Method

DATE	Bottomhole Pressure (psia)	Compressibility Factor, *z*	Cumulative Gas Production (MMscf)	X Variable	Y Variable
12/22/65	9507	1.440	0.0	0.00	0.00E+00
03/01/66	9292	1.418	392.5	1.84	3.49E-05
06/22/66	8970	1.387	1642.2	3.12	3.88E-05
09/29/66	8595	1.344	3225.8	3.65	3.55E-05
11/17/66	8332	1.316	4260.3	3.78	3.64E-05
12/30/66	8009	1.282	5503.5	3.88	3.79E-05
03/23/67	7603	1.239	7538.1	4.26	3.99E-05
05/15/67	7406	1.218	8749.2	4.52	4.08E-05
07/31/67	7002	1.176	10509.3	4.65	4.34E-05
09/14/67	6721	1.147	11758.9	4.76	4.55E-05
10/19/67	6535	1.127	12789.2	4.90	4.66E-05
03/05/68	5764	1.048	17262.5	5.54	5.35E-05
09/04/68	4766	0.977	22890.8	6.53	7.45E-05
03/19/69	4295	0.928	28144.6	7.70	8.18E-05
09/29/69	3750	0.891	32566.7	8.87	9.88E-05
03/31/70	3247	0.854	36819.9	10.21	1.18E-04

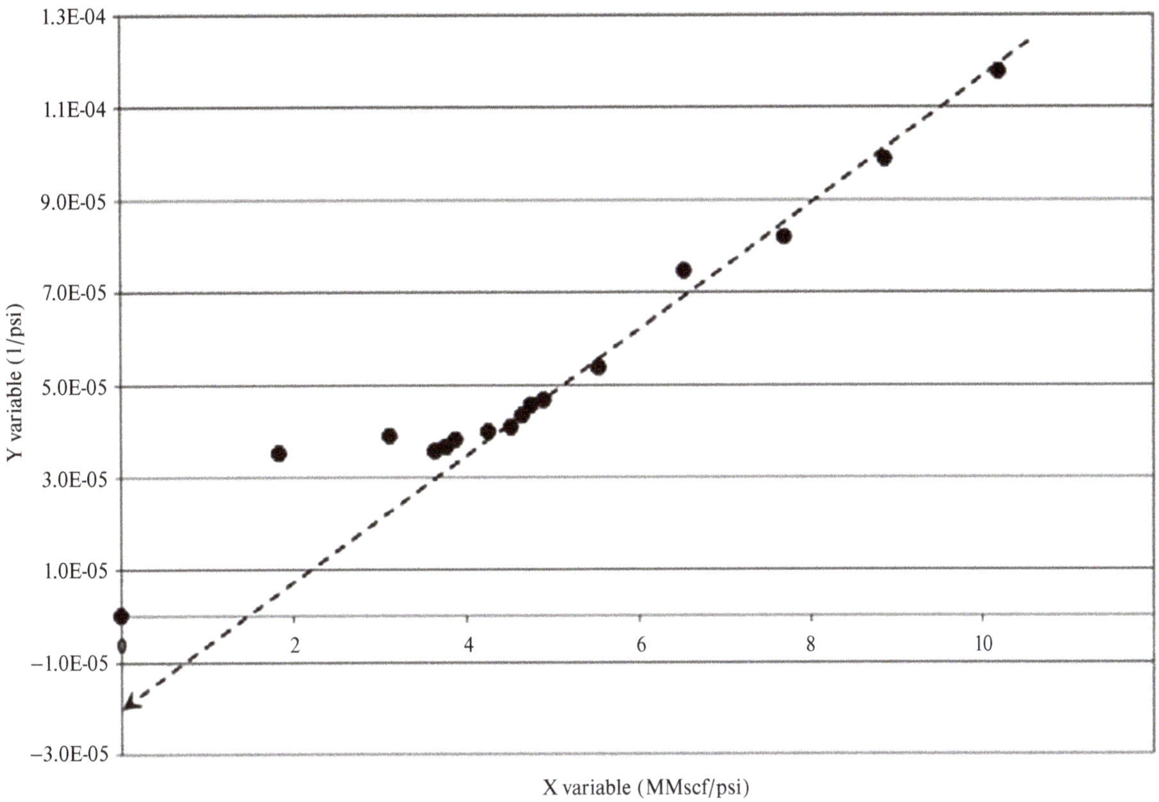

▲ FIGURE 9.13 **Plot of the Roach method for the Anderson "L" reservoir in Example 9.6**

End of Example 9.6

9.6 Case Histories of Two Gas Reservoirs

The case histories of two gas reservoirs are presented to serve as examples for discussions on performance of gas reservoirs. These reservoirs were selected because they have been depleted in terms of economic production, and no further production is expected. One of the reservoirs is classified as a dry gas reservoir, and the other is classified as a retrograde gas condensate reservoir. The case histories have been condensed from all the activities that occurred in these reservoirs and only main events that could have affected reservoir performance are presented. The data presented are considered sufficient for routine reservoir analyses and should provide some useful insights on the performance of the reservoirs. The case histories are presented under the following categories: reservoir geology, reservoir rock and fluid properties, and reservoir pressure and production data. For each case history, review problems have been constructed in the Problems section (Problems 9-5 and 9-6) at the end of this chapter to assist in the discussion and analysis of reservoir performance.

9.6.1 The Case History of Red Hawk Reservoir

The Red Hawk gas reservoir is located in Garden Banks (GB) Block 877 in the Gulf of Mexico (GOM), USA as shown in Figure 9.14. The water depth at GB Block 877 is about 5300 ft, which is classified as deep water in the GOM. The Red Hawk gas reservoir was discovered in August 2001. Gas production began in July 2004 through the first cell spar facility built in the world. The two main productive intervals in the Red Hawk reservoir were depleted in April 2008.

9.6.1.1 Reservoir Geology

The Red Hawk reservoir is composed of sands trapped on the west flank of a large salt dome. The Upper Pliocene sands can be described as slope fans with imbedded channel/levee complex. The three productive intervals present in the reservoir are named the Brouweri, the Surculus, and the Tamalis, after the nanofossils found in each interval. The

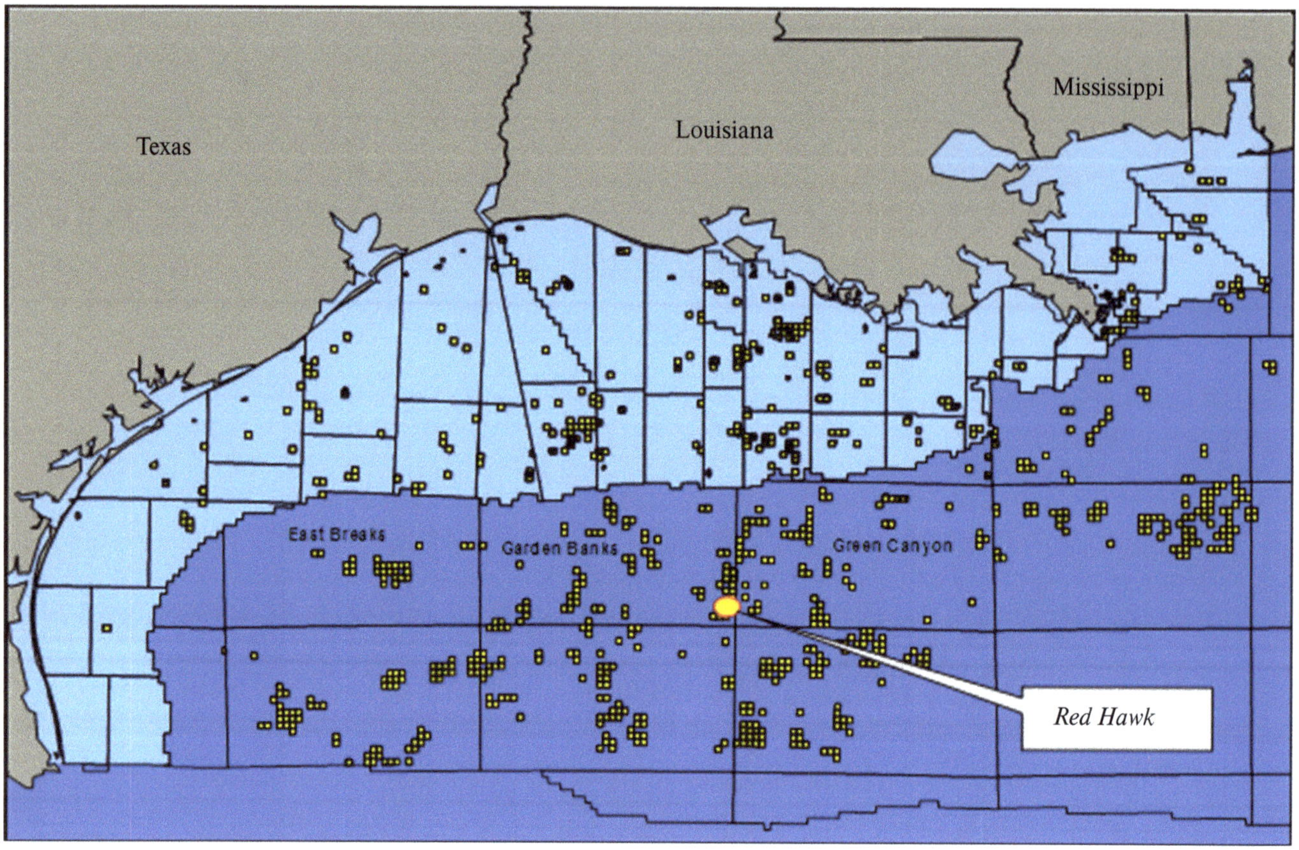

▲ **FIGURE 9.14** **Location of Red Hawk reservoir in Gulf of Mexico, USA**

TABLE 9.8 | Summary of Red Hawk Well Drilling History

Block Number	Borehole #	Date Drilled	Measured Depth (ft)	True Vertical Depth (ft)	Intervals Penetrated	Classification
877	1 OH	August 2001	23,518	23,500	D. Brouweri	Wet
					D. Surculus	Pay
					D. Tamalis	Pay
877	1 ST1	Sept. 2001	20,994	20,500	All 3 Sands	Wet
877	1 ST2	Oct. 2001	21,345	20,419	All 3 Sands	Pay
921	1	Dec. 2001	17,814	16,600	D. Brouweri	Pay
					D. Surculus	Wet
					D. Tamalis	Wet
877	2 OH	Sept. 2002	18,206	16,551	D. Brouweri	Wet
					D. Surculus	Absent
					D. Tamalis	Absent
877	2 ST1	Jan. 2003	18,153	17,186	D. Brouweri	Absent
					D. Surculus	Pay
					D. Tamalis	Faulted out
877	2 ST2	Jan. 2003	18,933	17,832	D. Brouweri	Absent
					D. Surculus	Pay
					D. Tamalis	Pay

type log for the Red Hawk reservoir is shown in Figure 9.15. The Brouweri sands appear to contain limited reserves and were not part of the initial development. Structure maps on the Surculus and the Tamalis intervals are shown as Figure 9.16 and Figure 9.17, respectively. As shown in the structure maps, the two productive intervals are rotated at steep angles near the salt dome. The structure maps show the presence of faults which appear to isolate portions of the reservoir. The Red Hawk reservoirs were penetrated with seven boreholes during discovery, appraisal, and development phases. A cross-section through Well 877 #1 boreholes is in Figure 9.18. The results of drilling at the locations of all boreholes are summarized in Table 9.8. The gas-water contact was located at 18,250 feet TVD subsea for the Surculus and Tamalis intervals on the basis of wireline formation tester data.

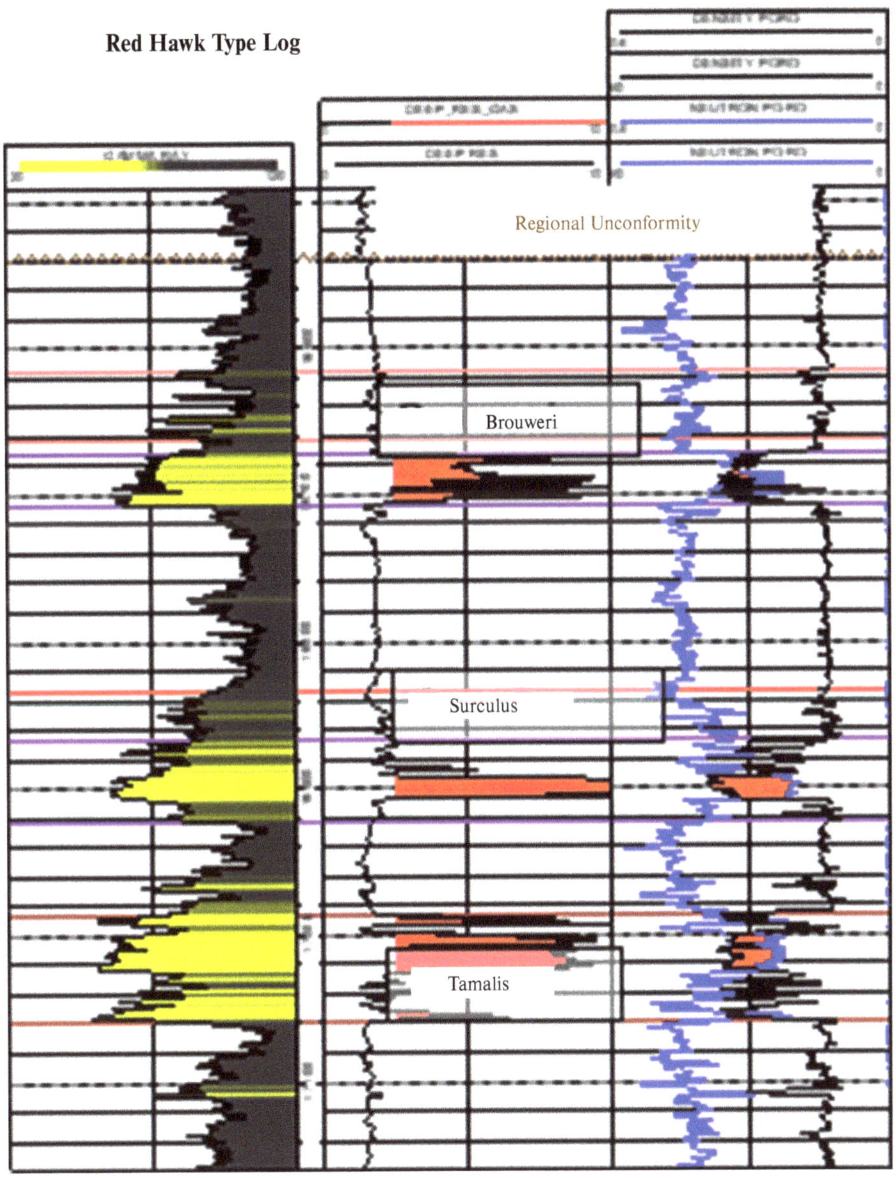

▲ FIGURE 9.15 Type log for Red Hawk reservoir

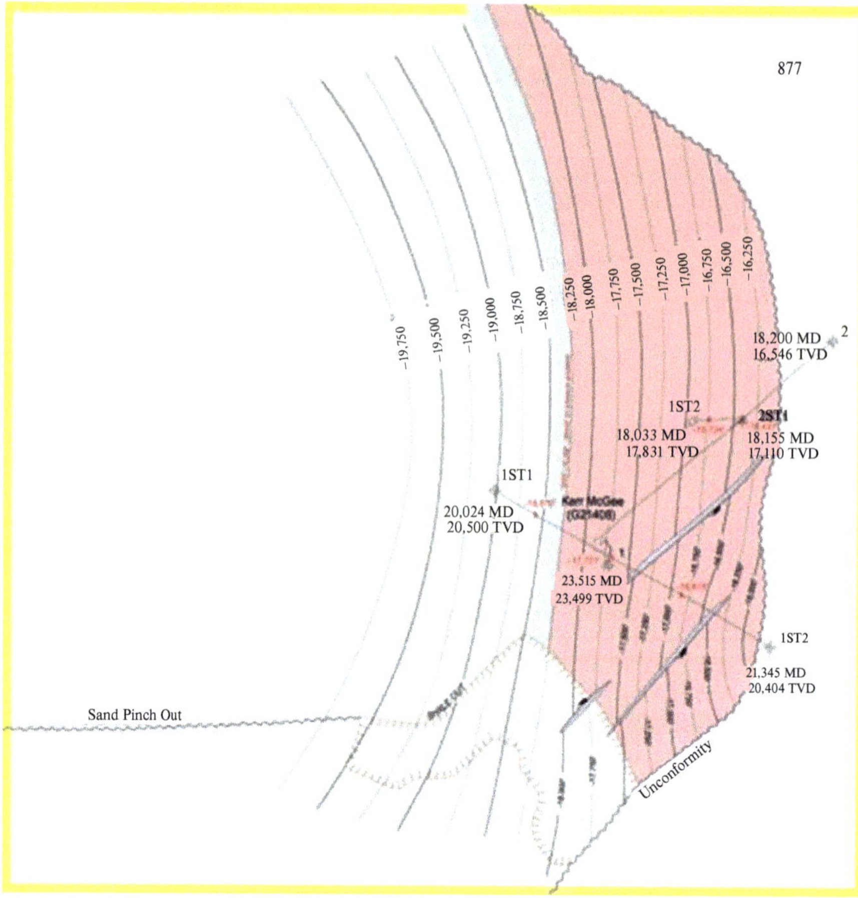

FIGURE 9.16
Structure map for Surculus interval

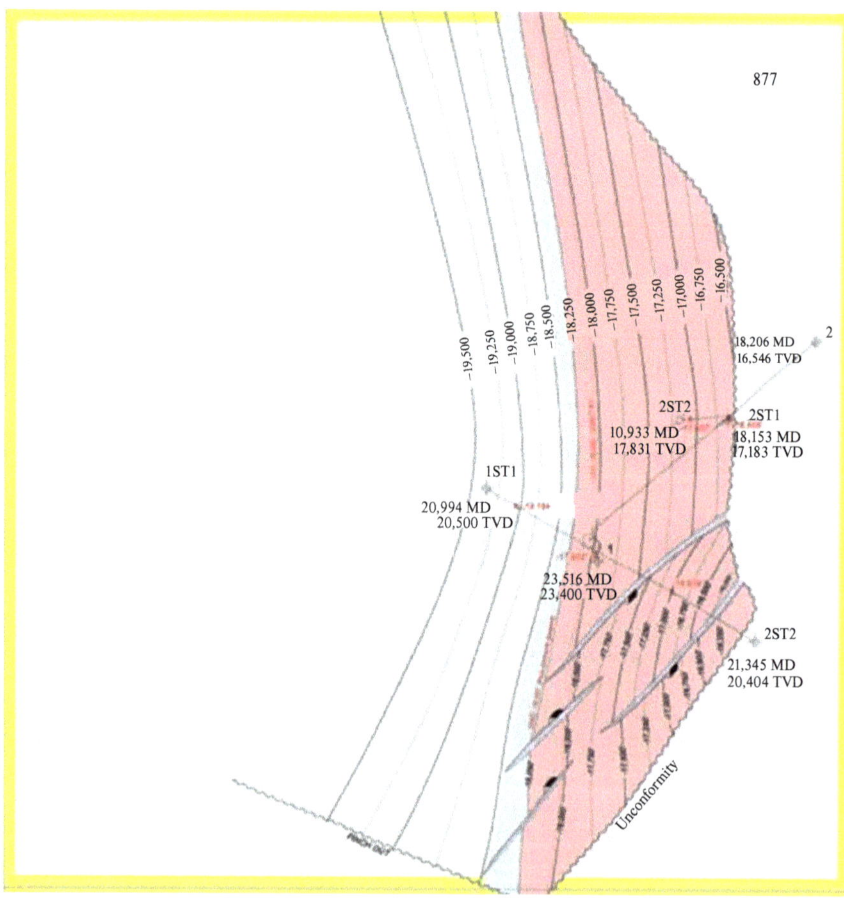

FIGURE 9.17
Structure map for Tamalis interval

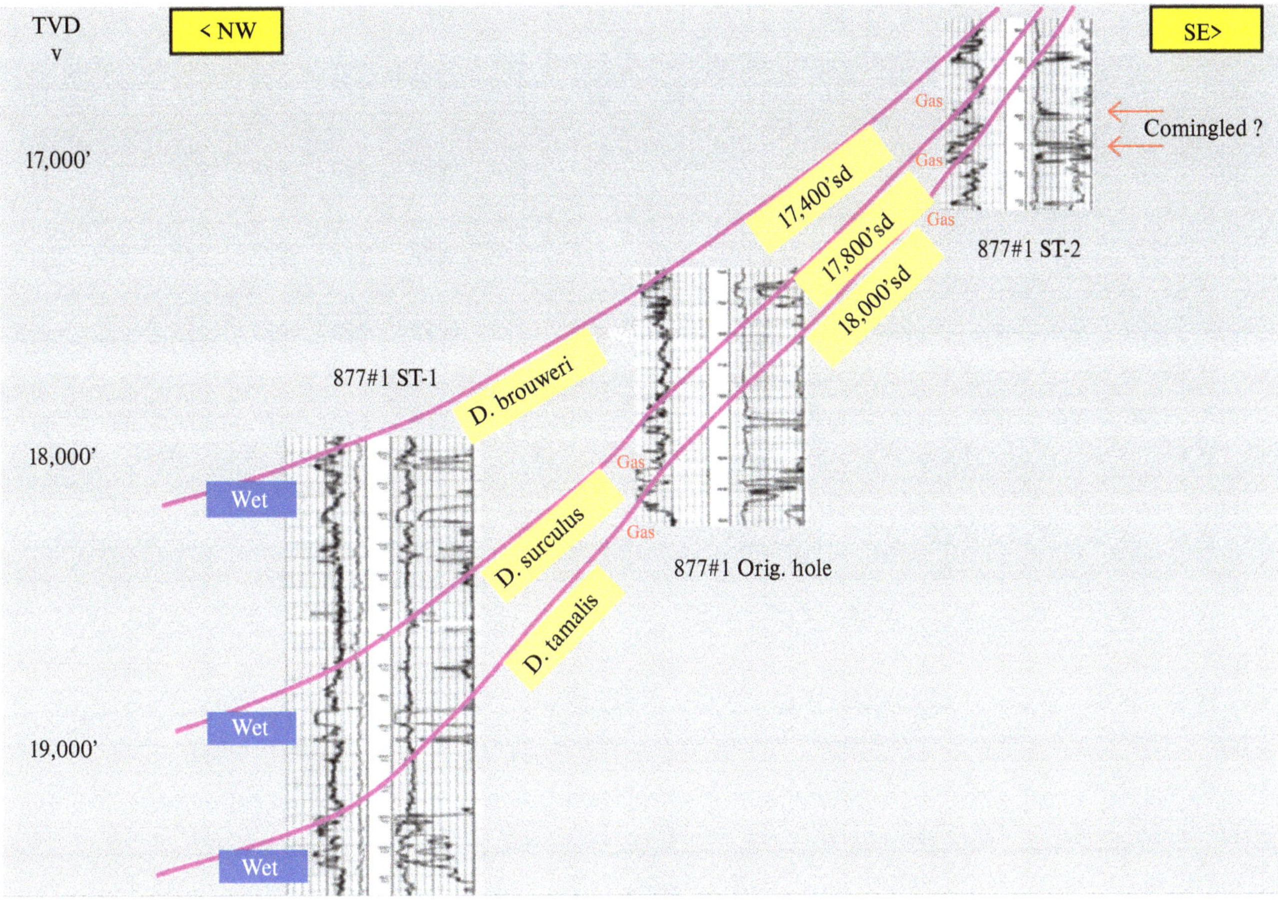

▲ **FIGURE 9.18** **Cross-section of Well 877#1 boreholes**

9.6.1.2 Reservoir Rock and Fluid Properties

The average net pay thickness determined for the Surculus and Tamalis intervals are 45 feet and 99 feet, respectively. The average rock properties for each interval are summarized in Table 9.9. The composition of the reservoir fluid from a typical bottomhole sample obtained from laboratory analysis is shown in Table 9.10. The gas-liquid ratio for the sample is 234,000 scf/STB. This low liquid yield essentially led to the classification of Red Hawk as a dry gas reservoir. Also, note the high composition of methane in the sample. The reservoir temperature for the Surculus and Tamalis intervals is 144°F.

TABLE 9.9 Average Rock Properties of Intervals in Red Hawk Reservoir

Rock Property	Intervals	
	Surculus	Tamalis
Porosity (%)	24	23
Permeability (md)	165	190
Water Saturation (%)	33	26
Total Compressibility (1/psi)	2.35E-04	2.35E-04

9.6.1.3 Reservoir Pressure and Production Data

The initial reservoir pressures of the Surculus and Tamalis intervals are 8923 psia and 9000 psia, respectively. The gas reserves in the Surculus and Tamalis intervals were recovered through two wells. Well 1ST2 was completed in the Surculus and Tamalis intervals, and Well 2ST2 was completed only in the Tamalis intervals (Figures 9.16, 9.17, and 9.18). A permanent downhole pressure gauge was installed in each well. These gauges were used to measure flowing bottomhole pressures during production, and static bottomhole pressures during periodic shut-ins. From these sources, historical static bottomhole pressures (SBHP) were estimated. The historical SBHP and monthly gas production rates are presented in Table 9.11 and Figure 9.19 for the Well 1ST2, and Table 9.12 and Figure 9.20 for Well 2ST2. Gas production was initiated in the two wells in mid-July 2004 and climbed quickly to average above 56 MMCFPD in 1 month (Figures 9.19 and 9.20). These high gas production rates were maintained in the both wells until mid-September 2005, when both wells were shut-in due to damage to production facilities caused by the hurricane named Rita. The wells were returned to production in mid-May 2006 after about 8 months of shut-in. There were signs of aquifer influx into the two producing intervals during

TABLE 9.10 | **Composition of Red Hawk Reservoir Fluid**

Sample Depth 17,340 Feet. Reservoir Fluid Flashed from 11,000 psia and 137°F to 15.025 psia and 60°F		
Sampling Conditions:	11,000 psia at 137°F	
Gas-Liquid Ratio	234,016	ft³ of Stock Tank Vapors / bbl Stock Tank Liquid
API Gravity of Liquid	43.7	@ 60°F
Color of Stock Tank Liquid:	Clear	
Gas-liquid ratio is cubic feet of gas at 15.025 psia and 60°F per barrel of stock tank liquid at 60°F.		

Chromatograph Analysis of Flash Gas

Component	Mole %	GPM @ 15.025 psia	Wt %	Mole Weight
Nitrogen	0.249	0.000	0.428	28.013
Carbon Dioxide	0.026	0.000	0.071	44.010
Hydrogen Sulfide	0.000	0.000	0.000	34.076
Methane	99.258	0.000	97.648	16.043
Ethane	0.150	0.041	0.277	30.070
Propane	0.067	0.019	0.182	44.097
Iso-Butane	0.026	0.009	0.094	58.123
N-Butane	0.032	0.010	0.116	58.123
Iso-Pentane	0.022	0.008	0.098	72.150
N-Pentane	0.021	0.008	0.092	72.150
Hexanes	0.034	0.014	0.179	85.817
Heptanes	0.036	0.015	0.208	96.798
Octanes	0.038	0.019	0.258	111.708
Nonanes	0.021	0.011	0.160	120.668
Decane Plus	0.020	0.014	0.189	142.285
Totals	**100.000**	**0.168**	**100.000**	

Calculated Properties of Gas			
Gas Specific Gravity	(Air = 1.00)	=	0.5641
Net Heat of Combustion	(Btu/Cu.Ft. @ 15.025 Psia @ 60°F)	Dry =	940.8 Real
Gross Heat of Combustion	(Btu/Cu.Ft. @ 15.025 Psia @ 60°F)	Dry =	1044.4 Real
Gross Heat of Combustion—Sat.	(Btu/Cu.Ft. @ 15.025 Psia @ 60°F)	Wet =	1026.1 Water Sat.
Gas Compressibility		Z =	0.9978

the shut-in period. These signs were reflected mostly in higher flowing bottomhole pressures measured after the wells were returned to production. However, no sharp increases in water production rates were observed in the two wells (Figures 9.21 and 9.22) after the shut-in period. Well 1ST2 showed steeper decline in reservoir pressures and productivity following the shut-in period due to hurricane Rita. This was attributed to the possibility that the aquifer water moved updip and isolated Well 1ST2 within a compartment created by faults (Figure 9.16). In effect, the updip movement of the aquifer potentially reduced the effective drainage area of the well. There was no evidence of communication between Well 1ST2 and 2ST2 based on interference tests. A compressor was installed in October 2007 to assist production from both wells.

Gas production from Well 1ST2 in the Surculus and Tamalis intervals was terminated in mid-February 2008. Cumulative gas production from Well 1ST2 is 49.84 Bcf. Also at this time, Well 2ST2 was showing increased water production from the Tamalis interval. The processing facilities were limited to handling no more than 250 BWPD. In March 2008, the Surculus interval was tested in Well 2ST2. The tests showed that the Surculus interval in this well was completely swept by aquifer influx because it produced only water. Attempts to close the valve across the Surculus interval and re-open the valve across the Tamalis interval were not successful. Gas production from Well 2ST2 in the Tamalis interval was terminated in mid-April 2008. Cumulative gas production from Well 2ST2 is 73.02 Bcf.

TABLE 9.11 | Historical Performance Data for Well 1ST2

Date	SBHP (PSIA)	Gas Rate (MCFPD)	Liquids Rate (BOPD)	Water Rate (BWPD)
Jun-04	8923	0	0	0
Jul-04	8903	15,843	30	0
Aug-04	8640	55,955	34	0
Sep-04	8354	52,822	63	0
Oct-04	8100	56,730	68	0
Nov-04	7829	59,449	78	0
Dec-04	7585	60,677	80	0
Jan-05	7403	60,111	71	0
Feb-05	7265	62,611	71	0
Mar-05	6994	64,462	80	0
Apr-05	6805	62,332	102	5
May-05	6693	62,059	94	23
Jun-05	6515	59,756	126	29
Jul-05	6411	63,145	141	21
Aug-05	6221	53,395	181	27
Sep-05	6105	37,224	134	20
Oct-05	6105	0	0	0
Nov-05	6105	0	0	0
Dec-05	6105	0	0	0
Jan-06	6105	0	0	0
Feb-06	6105	0	0	0
Mar-06	6105	0	0	0
Apr-06	6105	0	0	0
May-06	6090	34,297	157	12
Jun-06	6075	63,666	176	31
Jul-06	6050	65,891	175	18
Aug-06	5778	64,536	161	16
Sep-06	5553	57,591	148	15
Oct-06	5350	55,544	136	14
Nov-06	5083	52,000	150	15
Dec-06	4903	50,711	163	16
Jan-07	4563	49,775	220	20
Feb-07	4339	53,606	235	18
Mar-07	4090	52,675	217	17
Apr-07	3767	45,654	176	13
May-07	3347	38,639	120	9
Jun-07	2969	31,274	79	7
Jul-07	2642	25,619	50	5
Aug-07	2396	20,082	34	4
Sep-07	2171	15,767	29	3
Oct-07	2019	11,031	23	3
Nov-07	1915	7614	21	2
Dec-07	1815	6597	18	2
Jan-08	1730	6921	9	1
Feb-08	1654	2414	8	3
Mar-08	1685	0	0	0

TABLE 9.12 | Historical Performance Data for Well 2ST2

Date	SBHP (PSIA)	Gas Rate (MCFPD)	Liquids Rate (BOPD)	Water Rate (BWPD)
Jun-04	9000	0	0	0
Jul-04	8950	14,816	45	0
Aug-04	8900	56,565	19	0
Sep-04	8608	53,749	37	0
Oct-04	8461	60,109	49	0
Nov-04	8238	66,777	58	0
Dec-04	8096	66,458	64	0
Jan-05	7966	63,298	77	0
Feb-05	7872	66,243	82	0
Mar-05	7754	64,965	87	0
Apr-05	7643	64,211	60	3
May-05	7541	64,386	65	10
Jun-05	7449	61,170	28	15
Jul-05	7396	59,994	11	6
Aug-05	7302	49,311	19	15
Sep-05	7250	39,395	2	1
Oct-05	7249	0	0	0
Nov-05	7249	0	0	0
Dec-05	7249	0	0	0
Jan-06	7249	0	0	0
Feb-06	7249	0	0	0
Mar-06	7249	0	0	0
Apr-06	7249	660	0	0
May-06	7200	41,773	29	22
Jun-06	7150	63,575	13	27
Jul-06	7076	64,454	8	3
Aug-06	6928	62,573	9	5
Sep-06	6866	58,353	16	8
Oct-06	6777	60,198	18	9
Nov-06	6665	64,363	12	6
Dec-06	6520	65,995	14	7
Jan-07	6378	62,319	19	10
Feb-07	6249	73,354	18	9
Mar-07	6138	76,566	21	11
Apr-07	6056	75,150	19	10
May-07	5950	75,687	22	12
Jun-07	5859	73,029	18	10
Jul-07	5764	72,957	21	11
Aug-07	5670	73,819	21	12
Sep-07	5589	74,352	20	11
Oct-07	5557	70,249	18	10
Nov-07	5487	69,730	18	10
Dec-07	5415	74,225	17	9
Jan-08	5331	71,106	20	22
Feb-08	5345	62,926	30	155
Mar-08	5334	46,391	31	224
Apr-08	5317	12,992	24	271
May-08	5317	0	0	0

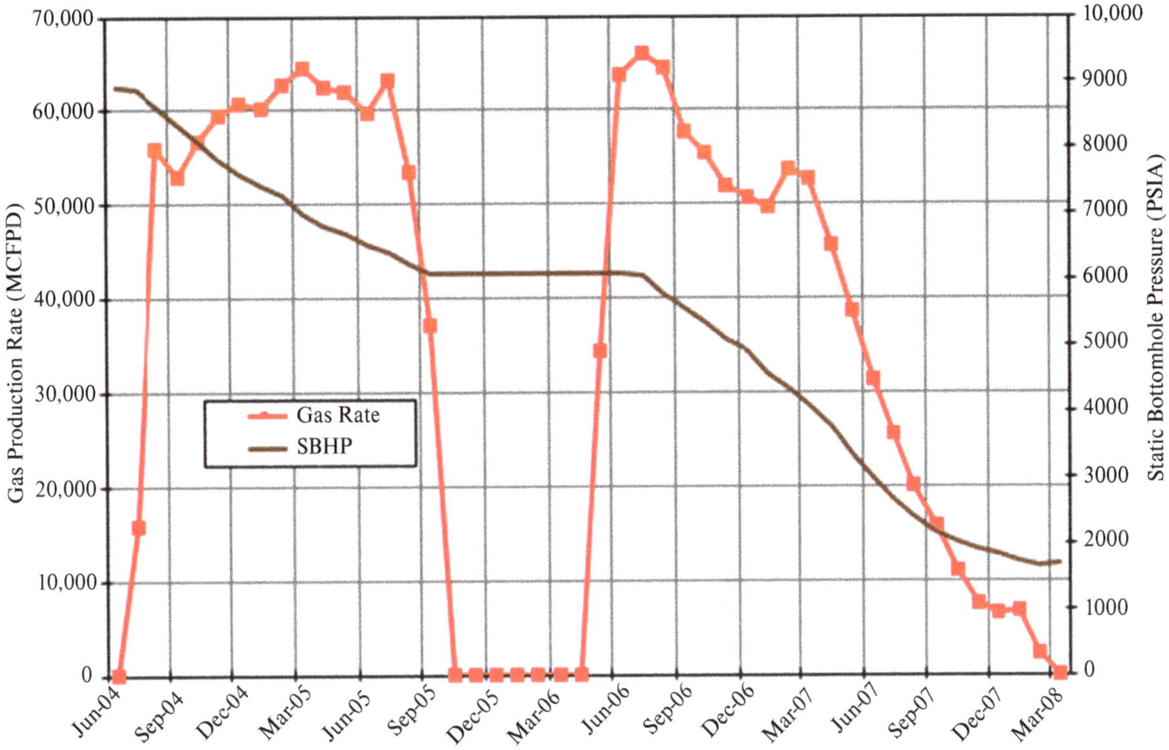

▲ **FIGURE 9.19** Historical pressure and gas production for Well 1ST2

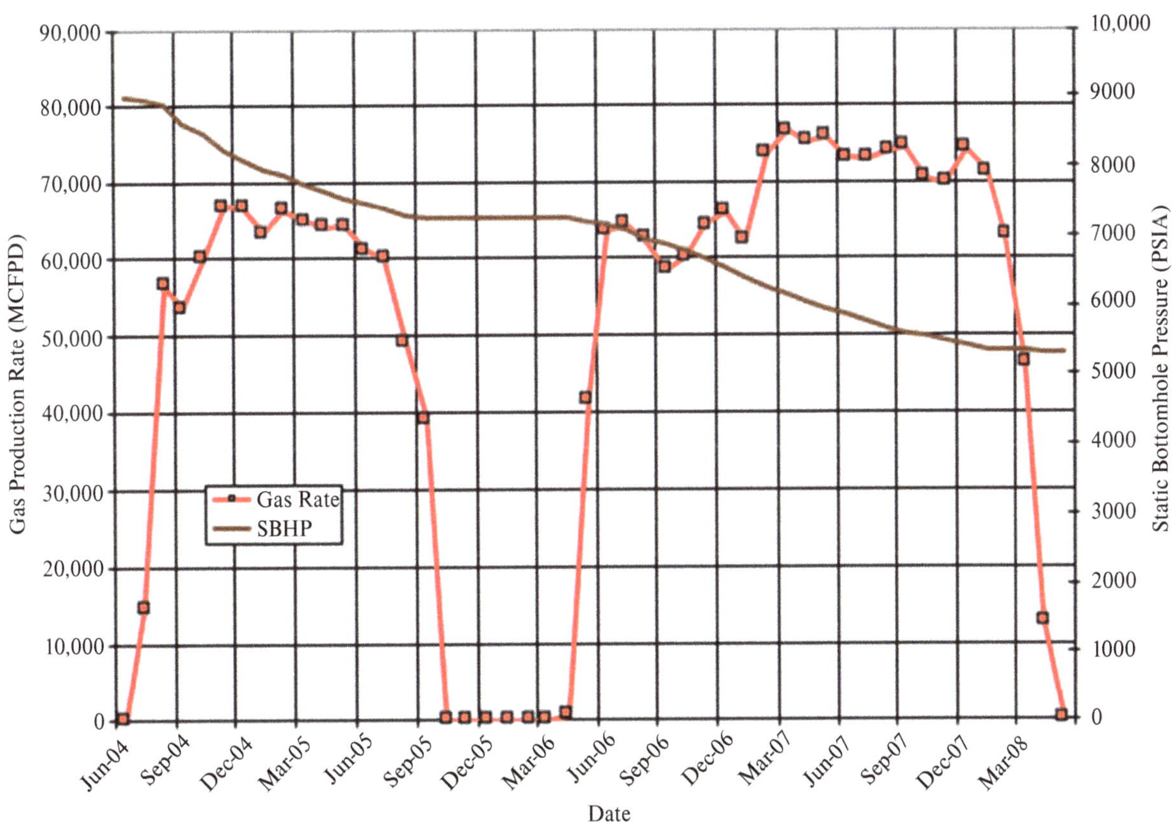

▲ **FIGURE 9.20** Historical pressure and gas production for Well 2ST2

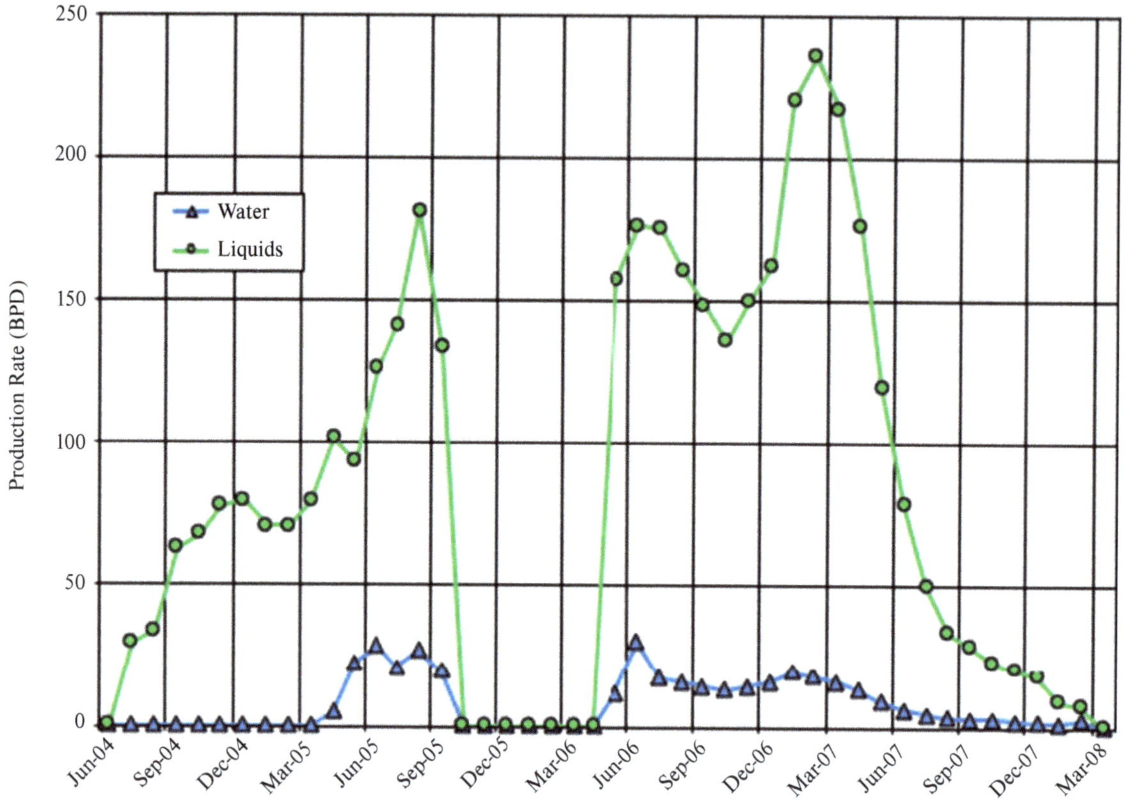

▲ **FIGURE 9.21** **Historical liquids and water production for Well 1ST2**

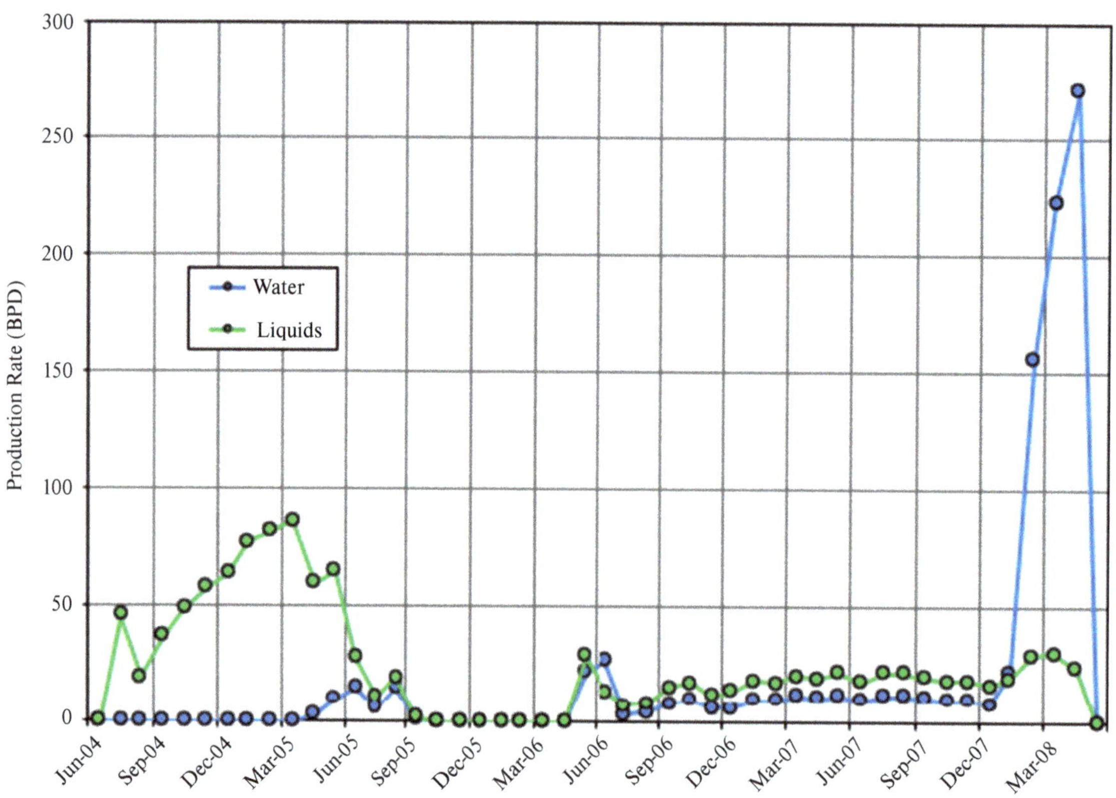

▲ **FIGURE 9.22** **Historical liquids and water production for Well 2ST2**

9.6.2 The Case History of West Cameron 580 Reservoirs

The West Cameron Block 580 (WC 580) field consists of mid-sized, over-pressured, and moderately rich retrograde gas condensate reservoirs located in the Gulf of Mexico at a water depth of 240 feet. The reservoirs are located mostly within WC 580 Block but extend partly into East Cameron Block 334 (EC 334). These blocks are located about 140 miles south of Lake Charles in Louisiana, USA. The WC 580 reservoir consists of the M-4/M-4A and the M-6 reservoirs that extend over WC 580 and EC 334 blocks. These reservoirs were produced through separate wells.

9.6.2.1 Reservoir Geology

The sequences of sands that constitute the M-4/M-4A and the M-6 reservoirs are illustrated in a cross-section drawn in Figure 9.23. The M-4/M-4A reservoir is structurally higher than the M-6 reservoir. The structure map of the M-4/M-4A reservoir is shown as Figure 9.24. The structure map of the M-6 reservoir is shown as Figure 9.25. Both reservoirs were trapped against a fault. A summary of the structural data for the M-4/M-4A and M-6 reservoirs is provided in Table 9.13.

TABLE 9.13 Summary of Reservoir Structure Data for WC 580 Reservoirs

Structure Parameter	Reservoir Name	
	M-4/M-4A	M-6
Reservoir average TVD (ft, ss)	15,283	15,525
Lowest known gas (ft, ss)	15,087	15,558
Gas-water contact (ft, ss)	15,450	15,590
Total productive volume (ac.-ft)	31,956	35,886

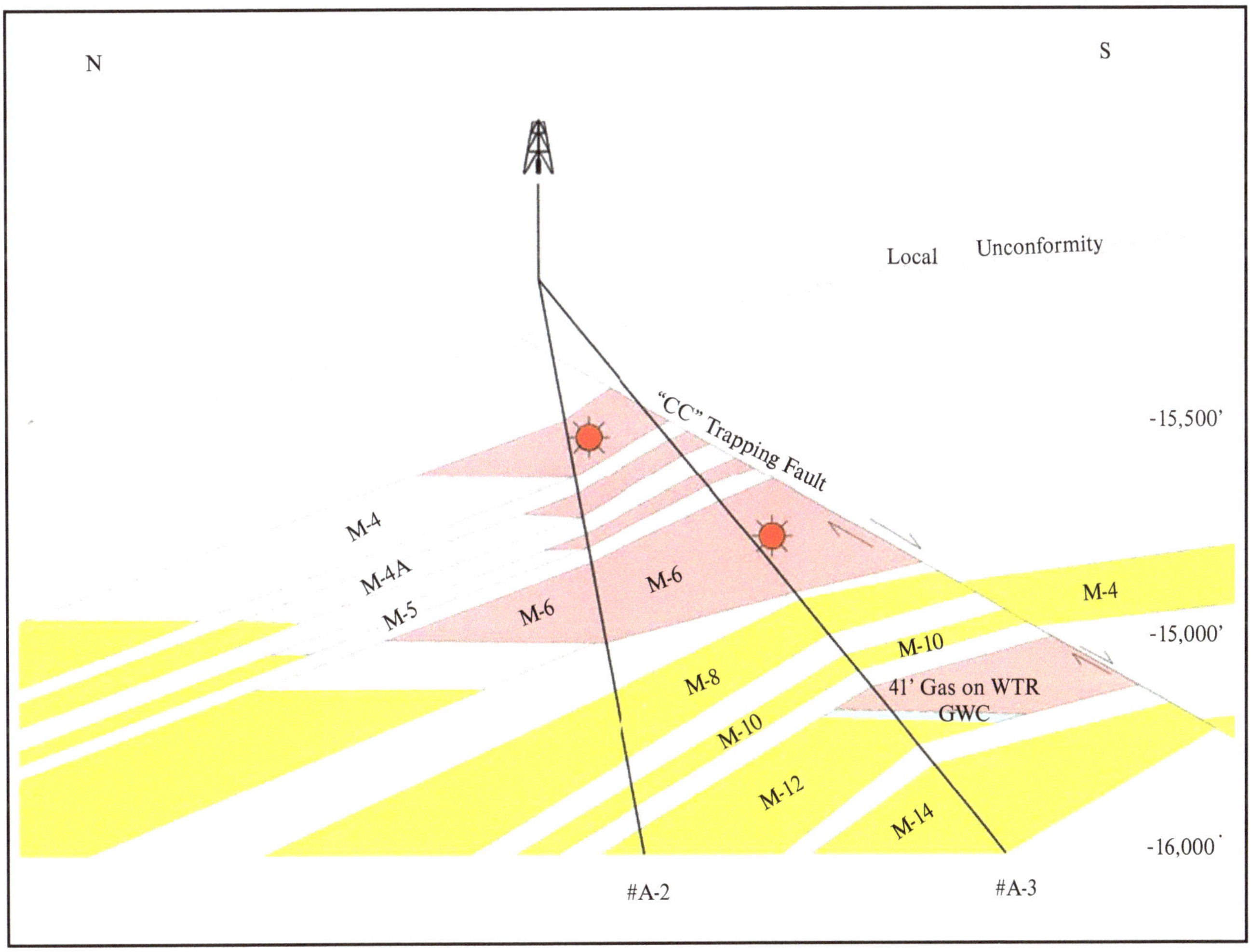

▲ FIGURE 9.23 Cross-section of WC 580 reservoirs

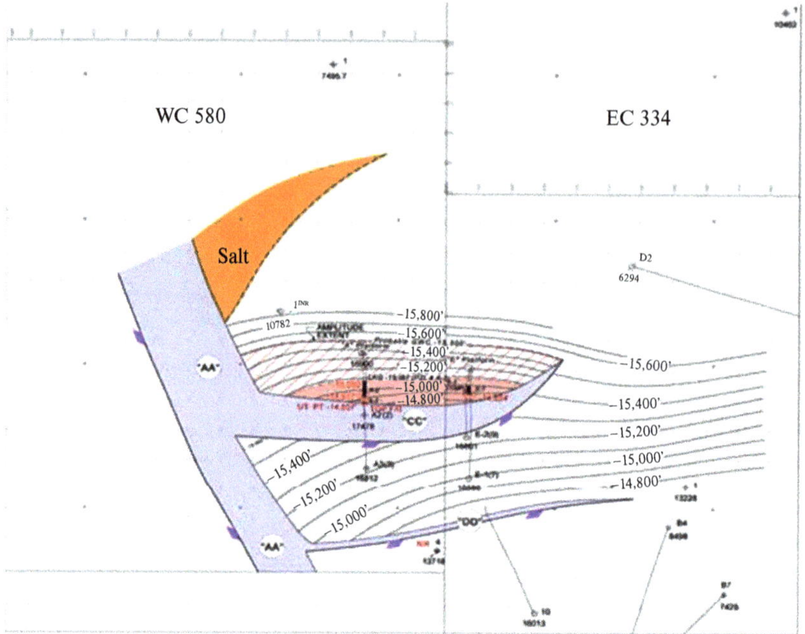

▲ **FIGURE 9.24** **Structure map of the M-4/M-4A reservoirs**

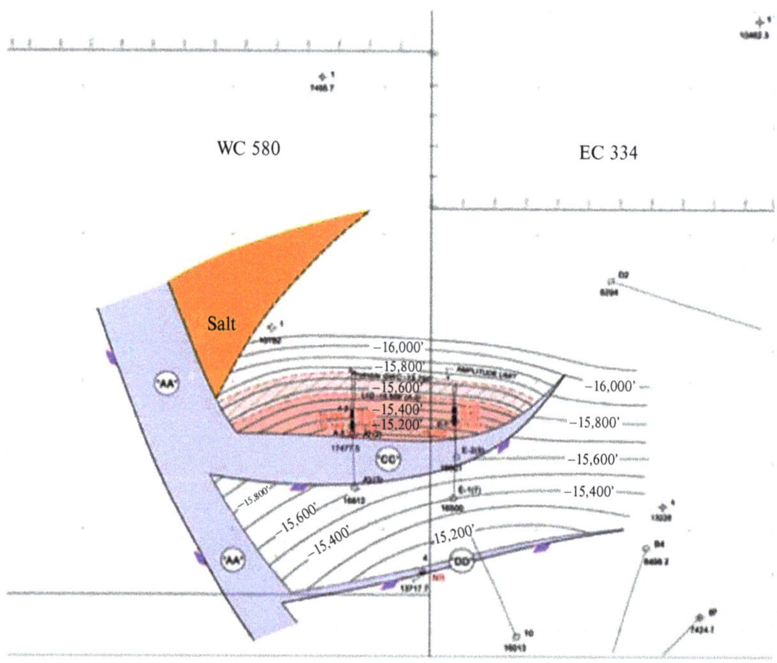

▲ **FIGURE 9.25** **Structure map of the M-6 reservoir**

9.6.2.2 Reservoir Rock and Fluid Properties

The average rock properties for the M-4/M-4A and M-6 reservoirs are presented in Table 9.14. The fluid properties data for a recombined sample taken at the separator from the M-4/M-4A reservoir is given in Table 9.15. Similar data for a recombined sample obtained at the separator for the M-6 reservoir are shown in Table 9.16. Note the high yield of condensate liquids obtained from both samples. The two reservoirs are classified as retrograde gas condensate reservoirs.

TABLE 9.14 Average Rock Properties for WC 580 Reservoirs

Rock Property	Reservoirs	
	M-4/M-4A	M-6
Porosity (%)	21.3	23.0
Permeability (md)	58.6	117.5
Initial water saturation (%)	28.6	20.0
Rock compressibility (1/psi)	7E-06	7E-06

TABLE 9.15 Fluid Properties for M-4/M-4A Reservoirs

Sample type:	Recombined fluid sample at primary separator
Separator temperature:	92°F
Separator pressure:	1155 psia
GOR at separator:	23,370 scf/bbl separator liquid
Stock tank pressure:	15.025 psia
Stock tank temperature:	60°F
GOR at stock tank:	451 scf/STB
Condensate yield:	36.43 STB/MMscf
Condensate gravity:	44.8°API
Initial reservoir pressure:	12,721 psia
Reservoir temperature:	267°F
Dew point pressure at 267°F:	9912 psia

Composition of Reservoir Fluid

Component	Mole (%)
Hydrogen Sulfide	0.00
Carbon Dioxide	0.93
Nitrogen	0.19
Methane	88.01
Ethane	5.30
Propane	1.82
i-Butane	0.42
n-Butane	0.50
i-Pentane	0.23
n-Pentane	0.19
Hexanes	0.24
Heptanes Plus	2.17
Total	**100.00**

Specific gravity of reservoir fluid:	0.7437 (Air = 1.0)
Heptanes Plus molecular weight:	162.1 lb/lb-mole
Heptanes Plus specific gravity:	0.8125 (Water = 1.0)

TABLE 9.16 Fluid Properties for M-6 Reservoirs

Sample type:	Recombined fluid sample at primary separator
Separator temperature:	70°F
Separator pressure:	1075 psia
GOR at separator:	17,899 scf/bbl separator liquid
Stock tank pressure:	15.025 psia
Stock tank temperature:	60°F
GOR at stock tank:	417 scf/STB
Condensate yield:	43.48 STB/MMscf
Condensate gravity:	45.2°API
Initial reservoir pressure:	12,825 psia
Reservoir temperature:	272°F
Dew point pressure at 267°F:	9850 psia

Composition of Reservoir Fluid

Component	Mole (%)
Hydrogen Sulfide	0.00
Carbon Dioxide	0.74
Nitrogen	0.17
Methane	87.85
Ethane	4.84
Propane	1.93
i-Butane	0.47
n-Butane	0.57
i-Pentane	0.25
n-Pentane	0.23
Hexanes	0.38
Heptanes Plus	2.57
Total	**100.00**

Specific gravity of reservoir fluid:	0.7714 (Air = 1.0)
Heptanes Plus molecular weight:	166.9 lb/lb-mole
Heptanes Plus specific gravity:	0.8140 (Water = 1.0)

TABLE 9.17 Historical Production Data for M-4/M-4A Reservoirs

DATE	Gas Rate (MCFPD)	Liquids Rate (BOPD)	Water Rate (BWPD)
Feb-97	12,381	577	1
Mar-97	42,633	1784	1
Apr-97	137,111	5054	9
May-97	163,106	6409	4
Jun-97	166,299	5790	20
Jul-97	163,112	5586	27
Aug-97	148,001	4829	24
Sep-97	128,068	4216	17
Oct-97	81,525	2859	87
Nov-97	73,940	2573	1
Dec-97	64,649	2229	48
Jan-98	76,924	2284	363
Feb-98	45,577	1307	352
Mar-98	75,475	2348	885
Apr-98	71,929	1962	1018
May-98	73,953	1896	1115
Jun-98	57,066	1460	1419
Jul-98	27,560	856	1691
Aug-98	26,414	810	1908
Sep-98	22,468	651	2063
Oct-98	23,469	683	2535
Nov-98	22,937	726	3026
Dec-98	21,803	691	3437
Jan-99	19,751	649	3797
Feb-99	18,721	636	4345
Mar-99	19,374	612	5402
Apr-99	18,666	599	6415
May-99	14,438	584	6883
Jun-99	6493	231	3758
Jul-99	3144	140	3848
Aug-99	2845	80	3751
Sep-99	3149	117	3781
Oct-99	3994	62	4629
Nov-99	2480	65	4489
Dec-99	1970	68	5012
Jan-00	2136	30	5626
Feb-00	2076	42	5607
Mar-00	1952	42	4423
Apr-00	1803	33	3922
May-00	2475	43	4259
Jun-00	2444	37	4,105
Jul-00	1348	13	2835
Aug-00	2131	13	3442
Sep-00	2146	11	3570
Oct-00	1792	10	3933
Nov-00	0	12	4418

9.6.2.3 Reservoir Pressure and Production Data

Initial production started in the M-4/M-4A reservoirs in February 1997 through one well located in WC 580. In April 1997, a second well was added in EC 334. Peak production was achieved quickly in June 1997 at the total rate of 166,299 MCFPD. Gas production declined rapidly from the peak rate with corresponding steep decline in reservoir pressure. Water production started to increase in January 1998 and rose quickly to a high rate of 6883 BWPD in May 1999. The monthly production data for the M-4/M-4A reservoirs are provided in Table 9.17 and shown as Figure 9.26. The average static bottomhole pressures for the M-4/M-4A reservoirs are given in Table 9.18 and shown in Figure 9.27. Note the coincidence between the decline in gas production rates and reservoir pressures, and the rise in water production. Cumulative gas, liquids, and water production from the M-4/M-4A reservoirs are 56.8 Bcf, 1.9 MMBO, and 3.7 MMBW, respectively.

Initial production in the M-6 reservoir started in October 1996 through a single well located in WC 580. A second well was put in production in EC 334 in April 1997. Peak production at the rate of 165,663 MCFPD was achieved in September 1997. The performance of the M-6 reservoirs is very similar to the performance of the M-4/M-4A reservoirs. The M-6 reservoirs also shows rapid decline in gas production which is accompanied with rapid rise in water production. The monthly production data for the M-6 reservoir are provided in Table 9.19 and shown as Figure 9.28. The average static bottomhole pressures for the M-6 reservoirs are given in Table 9.20 and shown in Figure 9.29. Cumulative gas, liquids, and water production from the M-6 reservoirs are 74.2 Bcf, 3.3 MMBO, and 13.0 MMBW, respectively.

TABLE 9.18 Average Reservoir Pressures for M-4/M-4A Reservoirs

Date	Pressure (psia)	Date	Pressure (psia)
Oct-96	12,721	Dec-98	4032
Nov-96	12,721	Jan-99	3925
Dec-96	12,721	Feb-99	3823
Jan-97	12,721	Mar-99	3736
Feb-97	12,555	Apr-99	3634
Mar-97	12,327	May-99	3532
Apr-97	11,710	Jun-99	3456
May-97	10,396	Jul-99	3452
Jun-97	9366	Aug-99	3459
Jul-97	8492	Sep-99	3467
Aug-97	7654	Oct-99	3476
Sep-97	6819	Nov-99	3486
Oct-97	6455	Dec-99	3496
Nov-97	6211	Jan-00	3506
Dec-97	5866	Feb-00	3516
Jan-98	5751	Mar-00	3526
Feb-98	5916	Apr-00	3538
Mar-98	5620	May-00	3550
Apr-98	5259	Jun-00	3563
May-98	4867	Jul-00	3576
Jun-98	4593	Aug-00	3590
Jul-98	4583	Sep-00	3603
Aug-98	4467	Oct-00	3615
Sep-98	4345	Nov-00	3627
Oct-98	4251	Dec-00	3640
Nov-98	4140		

TABLE 9.19 Historical Production Data for M-6 Reservoirs

Date	Gas Rate (MCFPD)	Liquids Rate (BOPD)	Water Rate (BWPD)
Oct-96	44,452	3035	1
Nov-96	60,025	3561	1
Dec-96	0	0	1
Jan-97	0	0	1
Feb-97	0	0	1
Mar-97	28,766	1584	1
Apr-97	131,272	6503	3
May-97	149,188	7101	11
Jun-97	155,106	7129	36
Jul-97	154,536	7292	35
Aug-97	154,004	7056	33
Sep-97	165,663	7585	32
Oct-97	161,093	7353	176
Nov-97	157,661	6998	746
Dec-97	132,358	5947	1669
Jan-98	116,736	5231	4607
Feb-98	55,396	2180	6,583
Mar-98	111,881	4841	7570
Apr-98	103,241	4440	7432
May-98	71,407	3044	7996
Jun-98	37,024	1362	9612
Jul-98	32,449	1191	10,735
Aug-98	31,924	1157	12,116
Sep-98	12,540	482	7702
Oct-98	21,303	1015	11,901
Nov-98	30,851	1118	14,402
Dec-98	19,748	792	13,352
Jan-99	19,449	878	13,417
Feb-99	19,993	812	13,087
Mar-99	19,843	758	12,955
Apr-99	17,947	571	11,458
May-99	9385	282	7330
Jun-99	8290	283	10,078
Jul-99	14,590	467	12,401
Aug-99	15,939	628	13,335
Sep-99	6290	212	10,315
Oct-99	13,161	393	12,253
Nov-99	13,263	395	12,590
Jan-00	12,864	413	13,099
Feb-00	12,606	396	13,203
Mar-00	7462	272	7307
Apr-00	6537	245	7246
May-00	6461	235	7622
Jun-00	6349	232	7359

Jul-00	7431	266	8336
Aug-00	7025	246	8222
Sep-00	6470	224	8300
Oct-00	6745	225	8266
Nov-00	6240	207	7770
Dec-00	2765	93	4400
Jan-01	4144	143	8868
Feb-01	6047	228	9023
Mar-01	5889	218	9108
Apr-01	5402	199	8701
May-01	5345	204	9728
Jun-01	4748	159	9057
Jul-01	4144	125	7101

TABLE 9.20 | Average Reservoir Pressures for M-6 Reservoirs

Date	Pressure (psia)	Date	Pressure (psia)
Sep-96	12,825	Mar-99	5793
Oct-96	12,584	Apr-99	5722
Nov-96	12,533	May-99	5665
Dec-96	12,482	Jun-99	5669
Jan-97	12,482	Jul-99	5668
Feb-97	12,482	Aug-99	5598
Mar-97	12,417	Sep-99	5545
Apr-97	12,099	Oct-99	5501
May-97	11,582	Nov-99	5463
Jun-97	10,622	Dec-99	5435
Jul-97	10,188	Jan-00	5412
Aug-97	9734	Feb-00	5390
Sep-97	9183	Mar-00	5404
Oct-97	8614	Apr-00	5416
Nov-97	8227	May-00	5425
Dec-97	7903	Jun-00	5433
Jan-98	7636	Jul-00	5440
Feb-98	7528	Aug-00	5446
Mar-98	7053	Sep-00	5451
Apr-98	6719	Oct-00	5455
May-98	6499	Nov-00	5459
Jun-98	6404	Dec-00	5463
Jul-98	6296	Jan-01	5466
Aug-98	6188	Feb-01	5469
Sep-98	6185	Mar-01	5471
Nov-98	6,119	Apr-01	5474
Dec-98	5997	May-01	5476
Jan-99	5927	Jun-01	5478
Feb-99	5858	Jul-01	5480

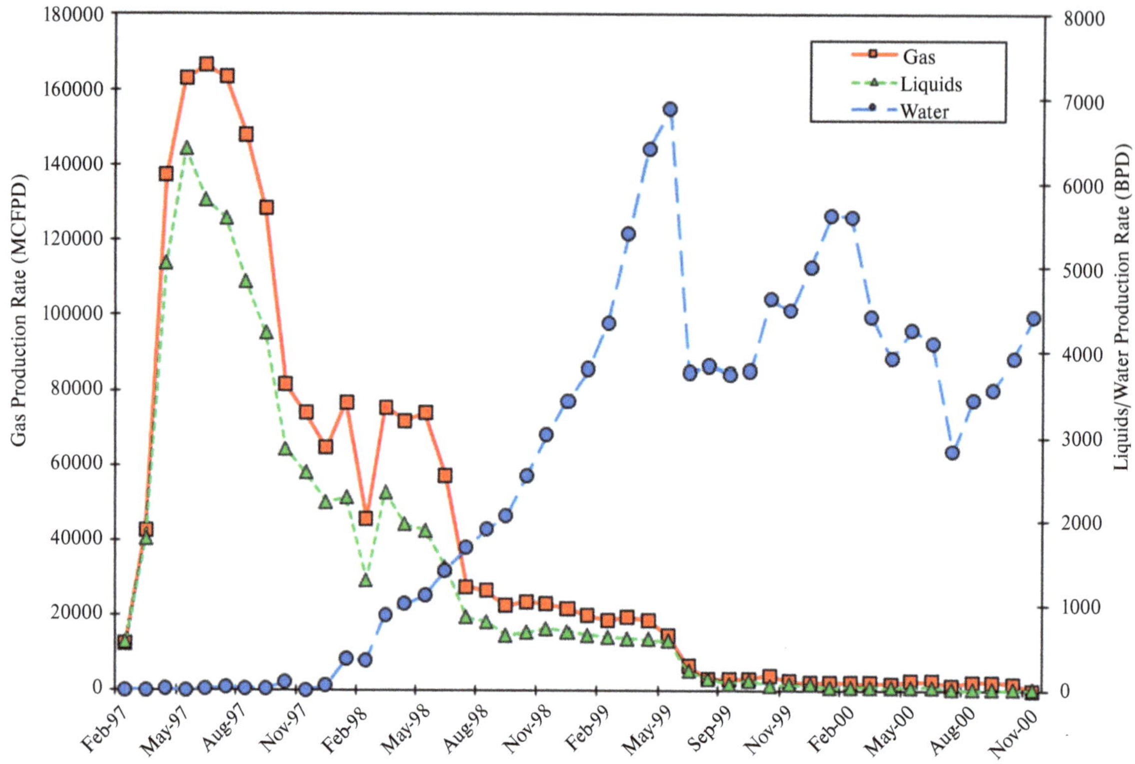

▲ **FIGURE 9.26** Historical production data for M-4/M-4A reservoirs

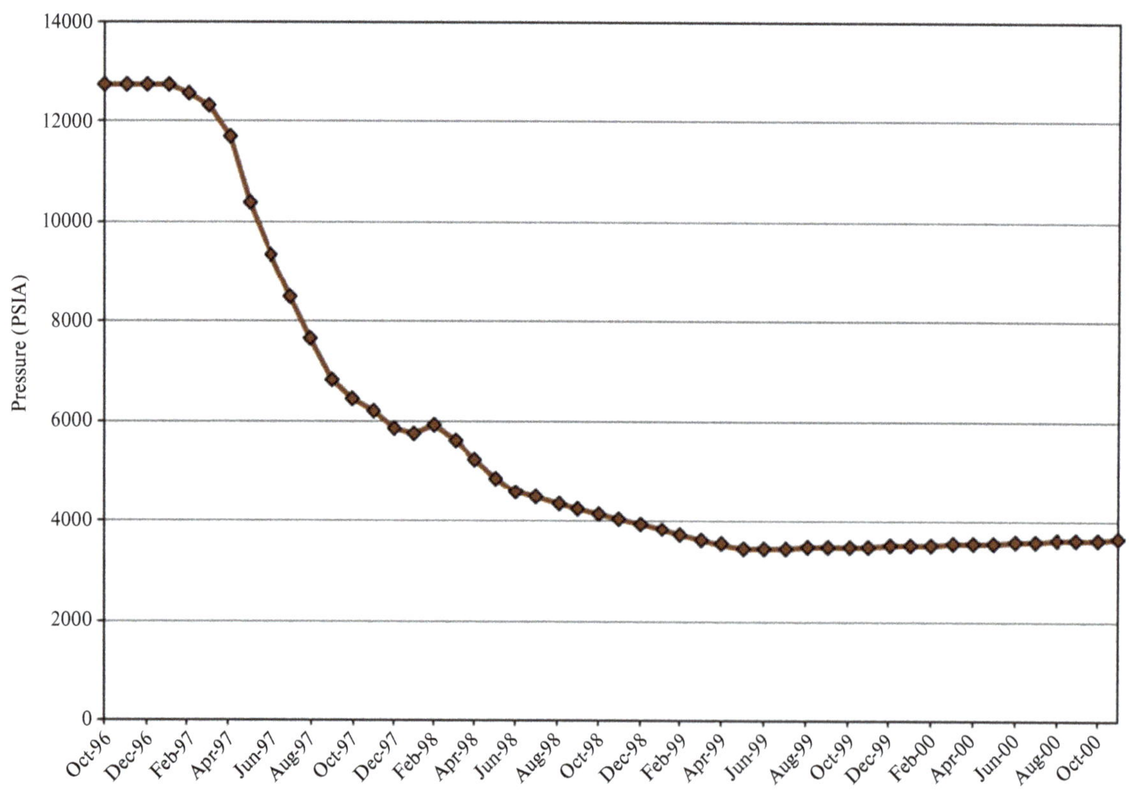

▲ **FIGURE 9.27** Average reservoir pressures for M-4/M-4A reservoirs

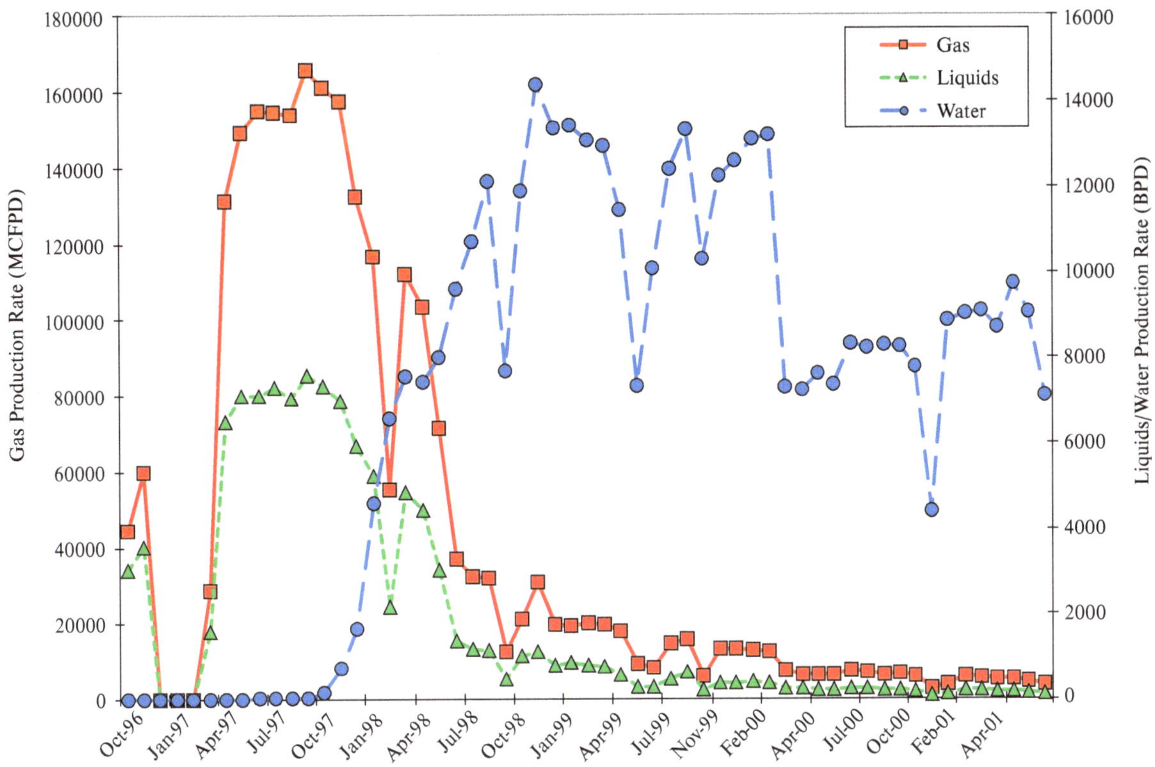

▲ **FIGURE 9.28** Historical production data for M-6 reservoir

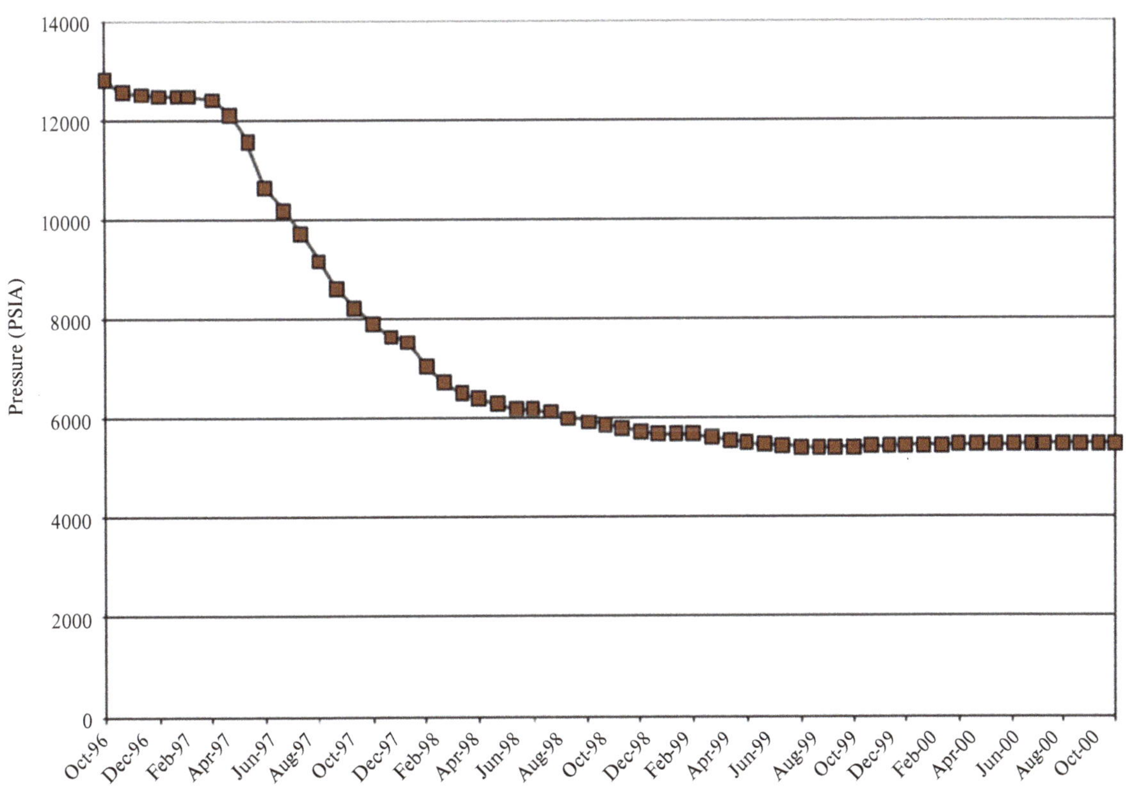

▲ **FIGURE 9.29** Average reservoir pressures for M-6 reservoir

Nomenclature

A	reservoir area, acres
B	constant in Eq. 8.39, RB/psi
B_g	gas formation volume factor, RB/scf
B_w	water formation volume factor, RB/STB
c_f	formation compressibility, psi^{-1}
c_w	water compressibility, psi^{-1}
c_t	total compressibility, psi^{-1}
G	original gas in place, scf
G_{cp}	cumulative wet or condensate gas produced
G_p	cumulative gas produced, scf
G_r	gas remaining in reservoir, scf
h	net sand thickness, feet
k	permeability, md
J	aquifer productivity index, STB/day-psi
L	length of linear reservoir or aquifer, feet
M_o	molecular weight of stock tank oil, lbm/lbmole
n	moles of gas
N_P	cumulative condensate produced, STB
p	pressure, psia
$p_{a,i}$	initial aquifer pressure, psia
$\bar{p}_a$	average aquifer pressure, psia
p_D	dimensionless pressure
p'_D	derivative of p_D
p_0	initial aquifer pressure at $t = 0$, psia
p_R	pressure at reservoir/aquifer boundary, psia
q_w	water influx rate, STB/day
Q_D	dimensionless cumulative water influx
q_D	dimensionless aquifer rate
r	radius, feet
R	recovery factor
S_g	gas saturation, fraction
S_{gi}	initial gas saturation, fraction
S_{gr}	residual gas saturation, fraction
S_w	initial water saturation, fraction
t	time, days
T_R	reservoir temperature, $°R$
t_D	dimensionless time
W_e	cumulative water influx, RB
W_{ei}	Maximum volume of encroachable water at initial conditions, RB
W_i	initial aquifer water volume, RB
W_p	cumulative water produced, STB
w	width of linear reservoir or aquifer, feet
z	gas compressibility factor
θ	aquifer encroachment angle
ϕ	porosity, fraction
μ	viscosity, cp
γ_o	specific gravity of stock tank oil

Subscripts

a	abandonment or aquifer
c	condensate
g	gas
i	initial
o	oil or condensate
p	produced
r	remaining or residual
R	reservoir
w	water

Abbreviations

AGIP	Apparent Gas-In-Place
GMBE	General Material Balance Equation
GOR	Gas-Oil Ratio
HO	Havlena-Odeh
OGIP	Original Gas-In-Place
SS	Sub-sea
TVD	True Vertical Depth

REFERENCES

1. Gold, D.K., McCain, W.D., and Jennings, J.W.: "An Improved Method for the Determination of the Reservoir-Gas Specific Gravity for Retrograde Gases," *JPT* (July 1989) 747-752.
2. Agarwal, R.G.: "Unsteady-State Performance of Water-Drive Gas Reservoir," PhD dissertation, Texas A&M University, College Station, Texas (1967).
3. Cole, F.W.: *Reservoir Engineering Manual,* Gulf Publishing Co., Houston, Texas (1969) 285.
4. Pletcher, J.L.: "Improvements to Reservoir Material-Balance Methods," *SPEREE* (February 2002) 49-59.
5. Havlena, D., and Odeh, A.S.: "The Material Balance as an Equation of a Straight Line," *JPT* (August 1963) 896-900.
6. Havlena, D., and Odeh, A.S.: "The Material Balance as an Equation of a Straight Line- Part II, Field Cases," *JPT* (July 1964) 815-822.
7. Schilthuis, R.J.: "Active Oil and Reservoir Energy," *Trans.* AIME (1936) 148, 33-51.
8. van Everdingen, A.F., and Hurst, W.: "The Application of the Laplace Transformation to Flow Problems in Reservoirs," *Trans.* AIME (1949) 186, 305-324.
9. Fetkovich, M.J.: "A Simplified Approach to Water Influx Calculations-Finite Aquifer Systems," *JPT* (July 1971) 814-828.
10. Carter, R.D., and Tracy, G.W.: "An Improved Method for Calculating Water Influx," *Trans.*, AIME (1960), 219, 415-417.
11. Fanchi, J.R.: "Analytical Representation of the van Everdingen-Hurst Aquifer Influence Functions for Reservoir Simulation," *SPEJ* (June 1985) 405-406.
12. Edwardson, M.J., Girner, H.M., Parkison, H.R., Williams, C.D., and Matthews, C.S.: "Calculation of Formation Temperature Disturbances Caused by Mud Circulation," *JPT* (April 1962) 416-426.
13. Klins, M.A., Bouchard, A.J., and Cable, C.L.: "A Polynomial Approach to the van Everdingen-Hurst Dimensionless Variables for Water Encroachment," *SPERE* (February 1988) 320-326.
14. Poston, S.W., and Berg, R.R.: *Overpressured Gas Reservoirs*, SPE, Richardson, Texas (1997).
15. Dickinson, G.: "Geological Aspects of Abnormal Reservoir Pressures in Gulf Coast Louisiana," *AAPG Bull.* (February 1953), 37, 2, 410.
16. Ramagost, B.P., and Farshad, F.F.: "P/Z Abnormally Pressured Gas Reservoirs," paper SPE 10125 presented at the SPE 56th Annual Fall Technical Conference and Exhibition, San Antonio, Texas, (October 5-7, 1981).
17. Poston, S.W., and Chen, H.Y.: "Case History Studies: Abnormal Pressured Gas Reservoirs," paper SPE 18857 presented at the SPE Production Operations Symposium, Oklahoma City, Oklahoma, (March 13-14, 1989).
18. Fetkovich, M.J., Reese, D.E., and Whitson, C.H.: "Application of a General Material Balance for High-Pressure Gas Reservoirs," *SPEJ* (March 1998) 3-13.
19. Hammerlindl, D.J.: "Predicting Gas Reserves in Abnormally Pressured Reservoirs," paper SPE 3479 presented at the 1971 SPE Fall Meeting, New Orleans, Louisiana, (October 3-5, 1971).
20. Duggan, J.O.: "The Anderson "L"–An Abnormally Pressured Gas Reservoir in South Texas," *JPT* (February 1972) 132-138.
21. Roach, R.H.: "Analyzing Geopressured Reservoirs–A Material Balance Technique," paper SPE 9968 available from SPE, Richardson, Texas, (August 1981).

ADDITIONAL REFERENCES

- Ambastha, A.K.: "Evaluation of Material Balance Analysis Methods for Volumetric, Abnormally-pressured Gas Reservoirs," *JCPT* (October 1993) 32, 8, 19-24.
- Bass, D.M.: "Analysis of Abnormally Pressured Gas Reservoirs with Partial Water Influx," paper SPE 3850 presented at the Abnormal Subsurface Pressure Symposium, Baton Rouge, Louisiana, (May 15-16, 1972).
- Begland, T.F., and Whitehead, W.R.: "Depletion Performance of Volumetric High-Pressured Gas Reservoirs," *SPERE* (August 1989) 279-282.
- Bernard, W.J.: "Gulf Coast Geopressured Gas Reservoirs: Drive Mechanism and Performance Prediction," paper SPE 14362 presented at the 1985 Annual Technical Conference, Las Vegas, Nevada, (September 22-25, 1985).
- Bruns, J.R., Fetkovich, M.J., and Meitzen, V.C.: "The Effect of Water Influx on p/z-Cumulative Gas Production Curves," *JPT* (March 1965) 287-291.
- Lee, J., and Wattenbarger, R.A.: *Gas Reservoir Engineering*, SPE Textbook Series, Vol. 5, SPE, Richardson, Texas, (1996).
- Wang, B., and Teasdale, T.S.: "GASWAT-PC: A Microcomputer Program for Gas Material Balance with Water Influx," paper SPE 16484 presented at the 1987 SPE Petroleum Industry Applications of Microcomputers, Lake Conroe, Montgomery, Texas, (June 23-26, 1987).

PROBLEMS

9-1. The initial pressure of a volumetric, dry gas reservoir is 3200 psia at reservoir temperature of 250°F. The gas gravity of a reservoir gas sample is 0.646 (Air = 1.0). The gas sample contained no impurities. The physical and petrophysical parameters of the gas reservoir are:

Area = 5970 acres

Net pay thickness = 75 feet

Average porosity = 18%

Average initial connate water saturation = 23%

Calculate:

a. The original gas-in-place (OGIP) at standard conditions.

b. The percentage recovery at reservoir pressure of 1600 psia.

9-2. A volumetric, dry gas reservoir at reservoir temperature of 215 °F contains gas that has 0.75 gas gravity (Air = 1.0) and no impurities. The production history of the reservoir was reported as:

Reservoir Pressure (psia)	Cumulative Gas Production (scf)
3973	3.27 × 109
2346	8.61 × 109

Determine:

a. Initial reservoir pressure.

b. Original gas-in-place (OGIP).

c. Reservoir area if the net formation thickness is 73 feet, average porosity is 18.0%, and water saturation is 22%.

d. What volume of gas will be produced from 2346 psia to an abandonment pressure of 100 psia?

e. What volume of gas will be left in the reservoir at abandonment pressure of 100 psia?

9-3. A radial reservoir is surrounded by a larger radial aquifer. The areas of the reservoir and the aquifer are 5500 acres and 792,000 acres, respectively. The pressure history at the reservoir/aquifer boundary is shown in Table 9P.1. The petrophysical properties of the aquifer are given in Table 9P.1. Calculate the volume of water influx into the reservoir by:

a. The Fetkovich method assuming a finite aquifer with constant pressure at the outer aquifer boundary.

b. The Fetkovich method assuming a finite aquifer with no flow across the outer aquifer boundary.

c. The Carter-Tracy method assuming an infinite aquifer.

d. Compare the volumes of water influx calculated from a, b, and c. What are the reasons for any differences?

TABLE 9P.1 Aquifer/Reservoir Data and Pressure History

Initial reservoir pressure	5225 psia
Porosity, ϕ	18%
Permeability, k	270 md
Aquifer thickness, h	50 ft
Water viscosity, μ	0.70 cp
Total aquifer compressibility, c_t	6.8×10^{-6} psi^{-1}
Encroachment angle, θ	360°

Pressure History at Reservoir/Aquifer Boundary

Time, t (days)	Pressure, p (psia)
0	5225
365	5150
730	5100
1095	5060
1460	5025

9-4. A geopressured dry gas reservoir in the Gulf Coast, USA, has its subsurface data summarized below. The production data of the reservoir are given in Table 9P.2.

Depth	10,000 ft (subsea)
Porosity	27%
Initial water saturation	30%
Permeability	100–600 md
Initial bottomhole pressure	8600 psia
Initial pressure gradient	0.86 psi/ft
Bottomhole temperature	236°F
Water compressibility	2.9×10^{-6} psia^{-1}
Formation compressibility	17.2×10^{-6} psia^{-1}

TABLE 9P.2 Production Data for Problem 9-4

Cumulative Gas Production (MMscf)	Reservoir Pressure (psia)	Gas Compressibility Factor, Z
0.00	8600.00	1.3038
400.40	8497.58	1.2934
1650.20	8182.01	1.2618
3216.70	7795.29	1.2240
4252.60	7544.95	1.1999
5495.90	7250.16	1.1721
7529.30	6781.36	1.1289
8742.80	6509.47	1.1045
10,493.80	6127.53	1.0709
11,561.40	5900.68	1.0514
12,775.70	5648.19	1.0300
17,255.20	4767.80	0.9585
22,903.10	3772.17	0.8832
28,165.40	2959.34	0.8261
32,606.90	2359.49	0.7866
36,885.20	1856.29	0.7551
40,200.50	1516.71	0.7346
45,000.60	1102.95	0.7107
50,010.30	769.42	0.6921

Calculate the original gas-in-place (OGIP) by:

a. The p/z method.

b. The Ramagost-Farshad method.

c. The Roach method.

d. Compare the OGIP calculated by the three methods and calculate the percentage errors relative to the p/z method.

9-5. The case history of Red Hawk reservoir was presented in Section 9.6.1 of this chapter. Based on the production history of the reservoir shown in Tables 9.11 and 9.12, analyze the performance of the reservoir as follows:

a. Calculate the original gas-in-place (OGIP) using the p/z method.

b. Is the p/z method applicable to the reservoir?

c. Determine OGIP using the Havlena-Odeh method.

d. Compare the OGIP from the p/z method to the OGIP from the Havlena-Odeh method. What method is more reliable for the reservoir?

e. Discuss reservoir management strategies that could be applied to a gas reservoir under an active water drive.

f. How do these strategies compare to the management of a volumetric dry gas reservoir?

9-6. The production performance data of the M4/M-4A and M-6 reservoirs in WC 580/EC 334 were presented in Section 9.6.2 of this chapter. The reservoirs are structurally separate and were produced through separate wells as stated in the case histories. Perform the following analyses using the data in the case histories:

a. Calculate the original gas-in-place (OGIP) for each reservoir by volumetric method.

b. Calculate the OGIP for each reservoir using the p/z method.

c. Calculate the OGIP for each reservoir using the Ramagost-Farshad method and the Roach method.

d. Determine the OGIP for each reservoir using material balance methods.

e. Compare the OGIPs calculated using the methods stated above. Discuss the applicability of each method for calculation of OGIP for the two reservoirs.

f. What method is most reliable for calculation of OGIP considering the thermodynamic state of the reservoirs?

APPENDIX 9A CORRELATIONS FOR ESTIMATING RESIDUAL GAS SATURATIONS FOR GAS RESERVOIRS UNDER WATER INFLUX (FROM AGARWAL[2])

Agarwal[2] developed correlations for the estimation of residual gas saturations in gas reservoirs under water influx. Note that these correlations are to be used if no laboratory data are available for the reservoir.

For unconsolidated sandstones:

$$S_{gr} = \frac{a_1\left(10^2 S_{gi}\right) + a_2\left(10^4 \phi S_{gi}\right) + a_3\left(10^2 \phi\right) + a_4}{100} \tag{9A-1}$$

In Eq. (9A-1),

$a_1 = -0.51255987,\ a_2 = 0.026097212,$

$a_3 = -0.26769575,\ a_4 = 14.796539.$

For consolidated sandstones:

$$S_{gr} = \frac{a_1\left(10^2 S_{gi}\right) + a_2\left(10^2 S_{gi}\right)^2}{100} \tag{9A-2}$$

In Eq. (9A-2), $a_1 = 0.80841168,\ a_2 = -0.0063869116$

For limestones:

$$S_{gr} = \frac{a_1\left(10^2 \phi\right) + a_2 \log(k) + a_3\left(10^2 S_{gi}\right) + a_4}{100} \tag{9A-3}$$

Eq. (9A-3),

$a_1 = -0.53482234,\ a_2 = 3.3555165,$

$a_3 = 0.15458573,\ a_4 = 14.403977.$

APPENDIX 9B DIMENSIONLESS PRESSURE FOR FINITE AND INFINITE AQUIFERS (FROM FANCHI[11])

The regression equation is given as:

$$p_D = a_0 + a_1 t_D + a_2 \ln t_D + a_3 (\ln t_D)^2 \quad (9B\text{-}1)$$

The derivative, p_D', can be obtained from differentiation of Eq. (9B-1).

Case	Regression Coefficients				Correlation Range of	Standard Error of Estimate	Average Deviation From Actual
r_a / r_R	a_0	a_1	a_2	a_3	t_D	σ	(%)
1.5	0.10371	1.66657	−0.04579	−0.01023	0.06 to 0.6	0.00032	0.06
2.0	0.30210	0.68178	−0.01599	−0.01356	0.22 to 5.0	0.00140	0.16
3.0	0.51243	0.29317	0.01534	−0.06732	0.52 to 5.0	0.00091	0.09
4.0	0.63656	0.16101	0.15812	−0.09104	1.5 to 10.0	0.00058	0.03
5.0	0.65106	0.10414	0.30953	−0.11258	3.0 to 15.0	0.00041	0.02
6.0	0.63367	0.06940	0.41750	−0.11137	4.0 to 30.0	0.00081	0.04
8.0	0.40132	0.04104	0.69592	−0.14350	8.0 to 45.0	0.00047	0.02
10.0	0.14386	0.02649	0.89646	−0.15502	12.0 to 70.0	0.00043	0.01
∞	0.82092	−3.68E-4	0.28908	0.02882	0.01 to 1000.0	0.01710	1.50

APPENDIX 9C DIMENSIONLESS PRESSURE FOR INFINITE AQUIFERS (FROM EDWARDSON ET AL.[12])

For $0.01 < t_D < 500$;

$$p_D = \frac{370.529\sqrt{t_D} + 137.582t_D + 5.69549t_D\sqrt{t_D}}{328.834 + 265.488\sqrt{t_D} + 45.2157t_D + t_D\sqrt{t_D}} \tag{9C-1}$$

$$p'_D = \frac{716.441 + 46.7984\sqrt{t_D} + 270.038t_D + 71.0098t_D\sqrt{t_D}}{1269.86\sqrt{t_D} + 1204.73t_D + 618.618t_D\sqrt{t_D} + 538.072t_D^2 + 142.410t_D^2\sqrt{t_D}} \tag{9C-2}$$

CHAPTER

10 Oil Reservoirs

10.1 Introduction

One of the important functions of a reservoir engineer is the estimation of oil-in-place (OIP) volumes, and the calculation of oil reserves thereafter. In this chapter, methods for estimating OIP volumes are presented for volumetric reservoirs, reservoirs with primary gas caps, and reservoirs with water influx. As was the case for gas reservoirs (Chapter 9), volumetric oil reservoirs are assumed to have constant hydrocarbon pore volume. Oil reservoirs are discovered at undersaturated or saturated conditions. As discussed in Chapter 5, an undersaturated oil reservoir has no initial gas cap at discovery, and reservoir pressure is at or above bubble point pressure. A saturated oil reservoir has a primary gas cap at discovery, and reservoir pressure is below bubble point pressure. The behavior of undersaturated reservoirs is remarkably different from those of saturated reservoirs in terms of performance and oil recovery. For these reasons, oil reservoirs are classified as undersaturated or saturated for purposes of the calculation methods presented in the chapter. However, it is worthy to note that most undersaturated reservoirs will ultimately become saturated reservoirs possibly with secondary gas caps, if reservoir pressure is allowed to fall below bubble point pressure as a result of production of reservoir fluids. Reservoir fluids are produced either through the release of energy stored in the fluids and the formation or supply of external energy from influx of fluids, represented as water influx from aquifers or injection of fluids like water and/or gas. The relative importance of these sources of drive energy varies over the life of the reservoir. The different drive mechanisms that can exist in an oil reservoir are presented in the chapter.

10.2 Oil Reservoir Drive Mechanisms

Reservoir drive mechanism is a term that is used to describe the predominant force or forces that enable the production of reservoir fluids. For volumetric, undersaturated reservoirs, the predominant forces that drive production are the expansion of the fluids and the formation rocks. For volumetric, saturated reservoirs, the predominant forces are the expansion of the gas caps (primary or secondary), and the expansion of the gas in solution in the oil (or solution gas). For non-volumetric reservoirs, the strength of the aquifer determines relative contributions from expansion of the solution gas, the gas cap, and water influx. These drive mechanisms have been reduced to indices as a means of quantifying their contributions to the production of fluids from the reservoir. Using the scheme devised by Pirson,[1] and the General Material Balance Equation (GMBE), four drive indices are defined for an oil reservoir. The GMBE as derived in Chapter 8 for an oil reservoir (Eq. 8.23) is:

$$\begin{aligned}&N\left(B_t-B_{ti}\right)+\frac{NmB_{ti}}{B_{gi}}\left[B_g-B_{gi}\right]\\&+\left(1+m\right)NB_{ti}\left[\frac{c_wS_{wi}+c_f}{1-S_{wi}}\right]\Delta p\\&+G_{inj}B_g+W_{inj}B_w+W_e\\&=N_p\left[B_t+\left(R_p-R_{si}\right)B_g\right]+W_pB_w\end{aligned}\tag{10.1}$$

Assuming no gas or water injection and rearranging, Eq. (10.1) becomes:

$$\begin{aligned}&N\left(B_t-B_{ti}\right)+\frac{NmB_{ti}}{B_{gi}}\left[B_g-B_{gi}\right]\\&+\left(1+m\right)NB_{ti}\left[\frac{c_wS_{wi}+c_f}{1-S_{wi}}\right]\Delta p+W_e-W_pB_w\\&=N_p\left[B_t+\left(R_p-R_{si}\right)B_g\right]\end{aligned}\tag{10.2}$$

Dividing Eq. (10.2) by the term on the right-side of the equation gives:

$$\begin{aligned}&\frac{N\left(B_t-B_{ti}\right)}{N_p\left[B_t+\left(R_p-R_{si}\right)B_g\right]}+\frac{\frac{NmB_{ti}}{B_{gi}}\left[B_g-B_{gi}\right]}{N_p\left[B_t+\left(R_p-R_{si}\right)B_g\right]}\\&+\frac{\left(1+m\right)NB_{ti}\left[\frac{c_wS_{wi}+c_f}{1-S_{wi}}\right]\Delta p}{N_p\left[B_t+\left(R_p-R_{si}\right)B_g\right]}\\&+\frac{\left(W_e-W_pB_w\right)}{N_p\left[B_t+\left(R_p-R_{si}\right)B_g\right]}=1\end{aligned}\tag{10.3}$$

The first term in Eq. (10.3), which represents drive by expansion of oil, is the oil drive index, I_{OD}:

$$I_{OD} = \frac{N(B_t - B_{ti})}{N_p\left[B_t + (R_p - R_{si})B_g\right]} \tag{10.4}$$

The second term in Eq. (10.3), which represents expansion of the gas cap, is the gas cap drive index, I_{GCD}:

$$I_{GCD} = \frac{\frac{NmB_{ti}}{B_{gi}}\left[B_g - B_{gi}\right]}{N_p\left[B_t + (R_p - R_{si})B_g\right]} \tag{10.5}$$

The third term in Eq. (10.3) is the formation and connate water compressibility drive index, I_{FWCD}:

$$I_{FWCD} = \frac{(1+m)NB_{ti}\left[\frac{c_w S_{wi} + c_f}{1 - S_{wi}}\right]\Delta p}{N_p\left[B_t + (R_p - R_{si})B_g\right]} \tag{10.6}$$

The fourth term in Eq. (10.3) is the water drive index, I_{WD}:

$$I_{WD} = \frac{(W_e - W_p B_w)}{N_p\left[B_t + (R_p - R_{si})B_g\right]} \tag{10.7}$$

As shown in Eq. (10.3), the sum of the indices is equal to one. Hence,

$$I_{OD} + I_{GCD} + I_{FWCD} + I_{WD} = 1 \tag{10.8}$$

For oil reservoirs with gas cap and water influx, the formation and water compressibility drive index is negligible. For such reservoirs, Eq. (10.8) reduces to:

$$I_{OD} + I_{GCD} + I_{WD} = 1 \tag{10.9}$$

Note that it is usually necessary to include formation and water compressibility drive index in calculations involving undersaturated reservoirs, especially high pressure, undersaturated oil reservoirs.

10.3 Gravity Drainage Mechanism

The gravity drainage mechanism is a very effective and efficient means of oil production in many reservoirs with steep dip angles of the formation. In this process, oil drains downdip from the gas cap and gas migrate updip from the oil column in a system of counter-flow assuming the reservoir has sufficient vertical permeability for vertical flow. Gravity drainage mechanism is an effective means of oil production if most of the following conditions exist in the reservoir:

1. High formation dip angle
2. High effective oil permeability
3. Low oil viscosity
4. Low residual oil saturation
5. Large gas cap

Reservoirs under gravity drainage mechanism can achieve very high oil recoveries. Dykstra[2] reported an oil recovery of 64% for the Lakeview Pool in Midway Sunset Field after 40 years of production. Oil production from the Lakeview Pool was by free-fall gravity drainage because the gravity head was the only remaining source of energy after extensive blow-out of the discovery well. The key reservoir and fluid data for the Lakeview Pool are as follows: average formation dip angle = 24 degrees; effective oil permeability = 1100 md; average oil viscosity = 19 cp; and estimated residual oil saturation = 0.10. The gas cap was considered to be large. Dykstra[2] also reported oil recovery of 58.5% for the Mile Six Pool. Reservoir pressure was maintained at about 800 psi during the reported 25-year history of the reservoir. The key reservoir and fluid data for the Mile Six Pool are as follows: formation dip angle = 17.5 degrees; oil permeability = 300 md; oil viscosity = 1.05 cp; residual oil saturation = 0.20; and a large gas cap. Wei et al.[3] estimated ultimate oil recovery for the 26R Reservoir in Kern County, California, USA, at 50%. The 26R reservoir is a gravity drainage reservoir with pressure maintenance by crestal gas injection. At the top of the 26R Reservoir, the average formation dip is 23 degrees and increases to 62 degrees at the oil-water contact. Other key reservoir and fluid properties for the 26R Reservoir are as follows: average formation permeability = 88 md; oil viscosity = 0.42 cp; and residual oil saturation = 0.49. The 26R reservoir had a very large gas cap. The 26R Reservoir is an example of forced or assisted gravity drainage because of gas injection at the crest of the structure for pressure maintenance.

The rate of gravity drainage is affected by reservoir pressure since fluid properties depend on pressure. Essley et al.[4] suggest that the benefits of gravity drainage can be improved by maintaining reservoir pressure through gas injection in the crestal areas of the reservoir. This mode of operation can be described as gravity drainage with complete pressure maintenance. Gravity drainage reservoirs can also be operated with declining gas-cap pressure either due to no gas injection or insufficient gas injection into the gas-cap to maintain pressure (Hall[5]). Analyses of the performance of gravity drainage reservoirs should consider the influence of pressure maintenance by gas injection on oil recoveries.

Many analytical models have been published on gravity drainage mechanism (Li and Horne[6]). Most of these analytical models are complicated and do not adequately represent oil production by gravity drainage. For these reasons, none of these analytical models is recommended or presented in this book. Rather, it is considered best practice to analyze the performance of reservoirs whose recovery mechanism is dominated by gravity drainage with reservoir models on numerical simulators. The principles and applications of reservoir simulation are presented in Chapter 23.

10.4 Volumetric Undersaturated Oil Reservoirs

Volumetric undersaturated oil reservoirs are defined as reservoirs that are completely isolated with constant hydrocarbon pore volume and whose pressures are above bubble point pressures. Volumetric reservoirs may be considered closed systems that do not receive significant pressure support or fluid influx from outside sources, such as water influx from aquifers. In this section, the discussions and equations that follow apply **only** to undersaturated, volumetric reservoirs.

10.4.1 Volume Calculations Above Bubble Point Pressure

The original oil-in-place (OOIP), represented as N in an undersaturated oil reservoir, can be calculated in terms of stock tank barrels (STB) with the equation:

$$N = \frac{7758 \times A \times h \times \phi \times (1 - S_{wi})}{B_{oi}} \quad (10.10)$$

In Eq. (10.10), A = area in acres; h = average net thickness in feet; ϕ = porosity, fraction; S_{wi} = water saturation, fraction; and B_{oi} = initial formation volume factor (FVF) in RB/STB. Suppose the reservoir is depleted to a lower pressure which is still above bubble point pressure. The cumulative oil production, N_p, is the difference between the OOIP and the oil remaining in the reservoir, N_r:

$$N_p = N - N_r \quad (10.11)$$

The oil remaining in the reservoir at this lower pressure is given by:

$$N_r = \frac{7758 \times A \times h \times \phi \times (1 - S_{wi})}{B_o} \quad (10.12)$$

In Eq. (10.12), B_o is the FVF at the lower reservoir pressure. Note that the lower reservoir pressure is still above bubble point pressure for this derivation. Substituting Eqs. (10.10) and (10.12) into Eq. (10.11) gives:

$$N_p = 7558 \times A \times h \times \phi \times (1 - S_{wi}) \left[\frac{1}{B_{oi}} - \frac{1}{B_o} \right] \quad (10.13)$$

The oil recovery factor, R, at pressures above bubble point is given by:

$$R = \frac{N_p}{N} = \left[1 - \frac{B_{oi}}{B_o} \right] \quad (10.14)$$

Equation (10.14) can also be derived starting from the GMBE shown as Eq. (10.2). Since this is a volumetric reservoir, there is no water influx, (W_e = 0), and water production is negligible, ($W_p = 0$). Also assume that water and formation compressibilities are negligible, ($c_w = 0$, and $c_f = 0$). (It is demonstrated later in this chapter that this assumption is not recommended, especially for reservoirs with high formation compressibility.) With these assumptions, the GMBE reduces to:

$$N(B_t - B_{ti}) = N_p \left[B_t + (R_p - R_{si}) B_g \right] \quad (10.15)$$

Above bubble point pressure, $R_p = R_{si}$, since all produced gas comes from the gas in solution. Also, above bubble point pressure, $B_{ti} = B_{oi}$, and, $B_t = B_o$, since the amount of solution gas is constant. Thus, Eq. (10.15) becomes:

$$N(B_t - B_{ti}) = N_p B_t \text{ and transforms to: } \quad N(B_o - B_{oi}) = N_p B_o \quad (10.16)$$

From Eq. (10.16), oil recovery factor, R, above bubble point pressure is:

$$R = \left[1 - \frac{B_{oi}}{B_o} \right] \text{ same as Eq. (10.14)}$$

To estimate oil recovery above bubble point pressure based on oil expansion alone, contributions to oil recovery caused by formation and water expansion were ignored. This could lead to large errors in the estimation of oil recovery at pressures above bubble point. Inclusion of formation and water compressibilities is necessary in the calculation of oil recovery above bubble point pressure, especially for reservoirs with high formation compressibility. This type of reservoir is encountered frequently in high pressure, deepwater environments. The expression for oil recovery above bubble point pressure that includes formation and water compressibilities can be derived from the GMBE. Again since there is no water influx (volumetric reservoir), and water production is negligible, Eq. (10.2) becomes:

$$N(B_t - B_{ti}) + NB_{ti} \left[\frac{c_w S_{wi} + c_f}{1 - S_{wi}} \right] \Delta p = N_p \left[B_t + (R_p - R_{si}) B_g \right] \quad (10.17)$$

Since the reservoir is above bubble point, Eq. (10.17) can be written in equivalent form as:

$$N(B_o - B_{oi}) + NB_{oi} \left[\frac{c_w S_{wi} + c_f}{1 - S_{wi}} \right] \Delta p = N_p B_o \quad (10.18)$$

Oil compressibility, c_o, can be calculated from the expression:

$$c_o = \frac{B_o - B_{oi}}{B_{oi} \Delta p} \quad (10.19)$$

Substituting Eq. (10.19) into Eq. (10.18) gives:

$$N c_o B_{oi} \Delta p + NB_{oi} \left[\frac{c_w S_{wi} + c_f}{1 - S_{wi}} \right] \Delta p = N_p B_o \quad (10.20)$$

Since for volumetric, undersaturated reservoirs, $S_o = 1 - S_{wi}$, Eq. (10.20) can be written as:

$$NB_{oi}\left[\frac{c_oS_o + c_wS_{wi} + c_f}{1 - S_{wi}}\right]\Delta p = N_pB_o \tag{10.21}$$

The reservoir effective compressibility, c_e, is defined as:

$$c_e = \left[\frac{c_oS_o + c_wS_{wi} + c_f}{1 - S_{wi}}\right] \tag{10.22}$$

Substituting Eq. (10.22) into Eq. (10.21) gives:

$$NB_{oi}c_e\Delta p = N_pB_o \tag{10.23}$$

By including formation and water compressibilities, oil recovery factor is expressed as:

$$R = \frac{N_p}{N} = c_e\Delta p\left[\frac{B_{oi}}{B_o}\right] \tag{10.24}$$

EXAMPLE 10.1 Calculation of Oil Recovery Factor for an Undersaturated Reservoir

Problem

Estimate oil recovery factors for a super-undersaturated reservoir in ultra-deepwater Gulf of Mexico, USA by neglecting formation and water compressibility, and including formation and water compressibility. The volumetric reservoir was produced from initial pressure and abandoned at average reservoir pressure of 12,000 psia. The average reservoir properties are as follows:

Initial reservoir pressure, p	19,600 psia
Temperature, T	252°F
Porosity, ϕ	0.22
Initial water saturation, S_{wi}	0.38
Permeability, k	42 md
Water compressibility, c_w	4×10^{-6} psi^{-1}
Formation compressibility, c_f	15×10^{-6} psi^{-1}
Bubble point pressure, p_b	1200 psia

The oil PVT properties are presented in Table 10.1.

TABLE 10.1 PVT Data of the Reservoir in Example 10.1

Pressure (psia)	FVF (RB/STB)	Rs (scf/STB)
19564	1.123	254
15000	1.143	254
14000	1.148	254
13000	1.153	254
12000	1.159	254
11000	1.165	254
10000	1.172	254
9000	1.178	254
8000	1.186	254
7000	1.193	254
6000	1.201	254
5000	1.210	254
4000	1.219	254
3000	1.229	254
2000	1.240	254
1240	1.248	254

Solution

Using Eq. (10.14) which neglects formation and water compressibility, and PVT data in Table 10.1:

$$R = \left[1 - \frac{B_{oi}}{B_o}\right] = 1 - \frac{1.123}{1.159}$$

$$= 0.031 \text{ or } 3.1\%$$

Using Eq. (10.24) which includes formation and water compressibility:

$$R = c_e\Delta p\left[\frac{B_{oi}}{B_o}\right]$$

From Eq. (10.22), the reservoir effective compressibility is given by:

$$c_e = \left[\frac{c_oS_o + c_wS_{wi} + c_f}{1 - S_{wi}}\right]$$

From Eq. (10.19), oil compressibility is estimated as:

$$c_o = \frac{B_o - B_{oi}}{B_{oi}\Delta p} = \frac{1.159 - 1.123}{1.123 \times (19600 - 12000)}$$

$$= 4.218 \times 10^{-6} \text{ psi}^{-1}$$

By substituting with reservoir data, the reservoir effective compressibility is:

$$c_e = \left[\frac{c_oS_o + c_wS_{wi} + c_f}{1 - S_{wi}}\right]$$

$$= \frac{4.218 \times 10^{-6} \times 0.62 + 4 \times 10^{-6} \times 0.38 + 15 \times 10^{-6}}{(1 - 0.38)}$$

$$= 3.086 \times 10^{-5} \text{ psi}^{-1}$$

The recovery factor is:

$$R = c_e \Delta p \left[\frac{B_{oi}}{B_o} \right]$$

$$= 3.086 \times 10^{-5} \times (19600 - 12000) \times \left[\frac{1.123}{1.159} \right]$$

$$= 0.23 \text{ or } 23\%$$

This rather simple example illustrates the impact of formation and fluid compressibilities on oil recovery, especially for reservoirs with high formation compressibility. These types of reservoirs are encountered in high pressure conditions, which are prevalent in deepwater environments. The effect of formation compressibility on oil recovery is comparable to the effect it has on the performance of geopressured gas reservoirs. This was discussed in Chapter 9. The intent of this example is to emphasize the importance of including formation and fluid compressibilities in the estimation of oil recovery for **all** undersaturated reservoirs.

End of Example 10.1

10.4.2 Volume Calculations Below Bubble Point Pressure

Suppose the pressure of an oil reservoir at initial conditions was above bubble point pressure. The OOIP can be calculated with Eq. (10.10). If the reservoir was produced until the pressure falls below bubble point, gas will evolve and form a gas phase dispersed within the oil phase. As reservoir pressure continues to fall due to production, gas saturation will continue to increase. When the critical gas saturation is reached, gas will begin to flow either toward the producing wells or migrate toward the top of the structure to form a secondary gas cap. Assuming the reservoir is volumetric (no water influx), the oil remaining in the reservoir at any pressure below the bubble point can be estimated as:

$$N_r = \frac{7758 \times A \times h \times \phi \times (1 - S_{wi} - S_g)}{B_o} \qquad (10.25)$$

In Eq. (10.25), S_g is the gas saturation at any pressure below the bubble point. From Eqs. (10.10), (10.11), and (10.25), the oil recovery factor for the reservoir below bubble point is:

$$R = \frac{N_p}{N}$$

$$= \frac{\left[\left(\frac{1 - S_{wi}}{B_{oi}} \right) - \left(\frac{1 - S_{wi} - S_g}{B_o} \right) \right]}{\left[\frac{1 - S_{wi}}{B_{oi}} \right]} \qquad (10.26)$$

$$= \left[1 - \left(\frac{1 - S_{wi} - S_g}{1 - S_{wi}} \right) \left(\frac{B_{oi}}{B_o} \right) \right]$$

Equation (10.26) is not in a very useful form because of the difficulty of determining gas saturation in the reservoir at any pressure below the bubble point. In some cases, gas saturations can be estimated from core studies or material balance calculations. Note however, that these sources for gas saturation are unreliable and may lead to large errors in the application of Eq. (10.26).

The oil recovery factor for a reservoir below bubble point pressure can be determined from the GMBE. Again, assuming the reservoir is volumetric (no water influx), and water production is negligible, Eq. (10.2) reduces to Eq. (10.17):

$$N(B_t - B_{ti}) + NB_{ti} \left[\frac{c_w S_{wi} + c_f}{1 - S_{wi}} \right] \Delta p$$

$$= N_p \left[B_t + (R_p - R_{si}) B_g \right]$$

Since gas is present in the reservoir, the compressibility of gas is much higher than the combined compressibility of the formation and water. Hence, formation and water compressibility are considered negligible and Eq. (10.17) becomes:

$$N(B_t - B_{ti}) = N_p \left[B_t + (R_p - R_{si}) B_g \right] \qquad (10.27)$$

The oil recovery factor, R, can be derived from Eq. (10.27) as:

$$R = \frac{(B_t - B_{ti})}{\left[B_t + (R_p - R_{si}) B_g \right]} \qquad (10.28)$$

In Eq. (10.28), R_p is the cumulative produced gas-oil ratio. R_p is calculated as:

$$R_p = \frac{G_p}{N_p} \qquad (10.29)$$

Also, in Eq. (10.29), G_p = cumulative gas production, scf; and N_p = cumulative oil production, STB. Note that all the other terms in Eq. (10.28) except, R_p, are based on the properties of the reservoir fluid and are dependent on the pressure of the reservoir. In contrast to Eq. (10.26), it is much easier to calculate the recovery factor for a reservoir below its bubble point by using Eq. (10.28). The cumulative production gas-oil ratio can be calculated from production data, although serious errors might occur if accurate production data are not maintained. The other terms in Eq. (10.28) can be determined from PVT data on the reservoir fluids. Again, if representative samples were not taken from the reservoir and reliable PVT data were not measured, application of Eq. (10.28) may lead to considerable errors.

By examining Eq. (10.28), it can be readily deduced that reducing the cumulative produced gas-oil ratio, R_p, can lead to higher recovery factors. The cumulative produced gas-oil ratio can be reduced in practice by isolating high gas producing intervals with packers, partially shutting-off high gas producing intervals with cement, or simply shutting in high gas-oil ratio wells. If the facilities are available, some or all of the produced gas can be returned to the reservoir through the process of gas re-injection. The practice of reservoir management by gas re-injection for pressure maintenance is discussed in Chapter 24.

10.5 Undersaturated Oil Reservoirs with Water Influx

Undersaturated oil reservoirs with water influx are treated differently from volumetric, undersaturated oil reservoirs. Unlike volumetric reservoirs, pressures in reservoirs with water influx could be supported completely or partially, depending on the strength of the aquifer. Fluid saturations will also change within areas of the reservoir that had been invaded by water influx from the aquifer.

10.5.1 Volumetric Method

At initial conditions, the OOIP in undersaturated reservoirs with water influx is determined with Eq. (10.10). Suppose the aquifer influx is strong, and reservoir pressure is maintained at initial levels, the oil remaining in the reservoir after a period of production is given by:

$$N_r = \frac{7758 \times A \times h \times \phi \times (1 - S_{wi} - S_{or})}{B_{oi}} \tag{10.30}$$

Combining Eqs. (10.10) and (10.30), oil recovery factor for this special case of no pressure decline in the undersaturated reservoir is given by:

$$\begin{aligned} R &= \frac{N_p}{N} = \frac{N - N_r}{N} = 1 - \frac{N_r}{N} \\ &= 1 - \left[\frac{1 - S_{wi} - S_{or}}{1 - S_{wi}}\right] \end{aligned} \tag{10.31}$$

In Eqs. (10.30) and (10.31), S_{or} is the residual oil saturation in the areas of the reservoir invaded by water. Residual oil saturation can be determined from special core analyses. If it is assumed that reservoir pressure declined from initial levels but still was maintained above the bubble point, the oil remaining in the reservoir after a period of production will be:

$$N_r = \frac{7758 \times A \times h \times \phi \times (1 - S_{wi} - S_{or})}{B_o} \tag{10.32}$$

For this case, the oil recovery factor is derived as:

$$R = 1 - \frac{N_r}{N} = 1 - \left[\frac{1 - S_{wi} - S_{or}}{1 - S_{wi}}\right]\left[\frac{B_{oi}}{B_o}\right] \tag{10.33}$$

Eqs. (10.31) and (10.33) are to be used only for estimation of oil recovery factors in undersaturated reservoirs with water influx. The reasons are due to difficulties of obtaining reliable values for residual oil saturations, and the likelihood that some oil bearing intervals may have been bypassed due to poor sweep efficiency. Consequently, these equations may yield high estimates of oil recovery factors.

10.5.2 Material Balance Method

The GMBE can be applied to undersaturated oil reservoirs with water influx. Starting with Eq. (10.2), the GMBE for an undersaturated reservoir with water influx is:

$$\begin{aligned} &N(B_t - B_{ti}) + NB_{ti}\left[\frac{c_w S_{wi} + c_f}{1 - S_{wi}}\right]\Delta p + W_e - W_p B_w \\ &= N_p\left[B_t + (R_p - R_{si})B_g\right] \end{aligned} \tag{10.34}$$

Formation and water compressibility are considered negligible for undersaturated reservoirs with strong water influx. Hence, Eq. (10.34) becomes:

$$\begin{aligned} &N(B_t - B_{ti}) + W_e - W_p B_w \\ &= N_p\left[B_t + (R_p - R_{si})B_g\right] \end{aligned} \tag{10.35}$$

Re-arranging Eq. (10.35) gives:

$$\begin{aligned} &N_p\left[B_t + (R_p - R_{si})B_g\right] + W_p B_w \\ &= N(B_t - B_{ti}) + W_e \end{aligned} \tag{10.36}$$

The left side of Eq. (10.36) represents cumulative voidage from the reservoir and the right side represents energy from oil expansion and water influx. If cumulative reservoir voidage is denoted as F, then:

$$F = N_p\left[B_t + (R_p - R_{si})B_g\right] + W_p B_w \tag{10.37}$$

Also, if oil expansion is represented as E_o, then:

$$E_o = B_t - B_{ti} \tag{10.38}$$

Substituting Eqs. (10.37) and (10.38) into Eq. (10.36) yields:

$$F = NE_o + W_e \tag{10.39}$$

Dividing both sides of Eq. (10.39) by E_o gives:

$$\frac{F}{E_o} = N + \frac{W_e}{E_o} \tag{10.40}$$

Eq. (10.40) is the Havlena-Odeh[7] straight line form of the material balance equation for an undersaturated oil reservoir with water influx. A plot of F/E_o vs. W_e/E_o will give a straight line with the intercept on the y-axis equal to N (the OOIP), and the slope equal to unity. Note that for the special case in which reservoir pressure is still above bubble point, $F = N_pB_t + W_pB_w$ in Eq. (10.37). The shape of F/E_o vs. W_e/E_o plot depends on the values of W_e. If correct values of W_e are used, the plot will be a straight line. If values of W_e are too small, the plot will curve upward. If the values are too large, the plot will curve downward. A linear aquifer will yield an S-shaped curve. These possible shapes of the F/E_o vs. W_e/E_o plots are shown in Figure 10.1. Note that Figure 10.1 is similar to Figure 9.4 for a gas reservoir with water influx.

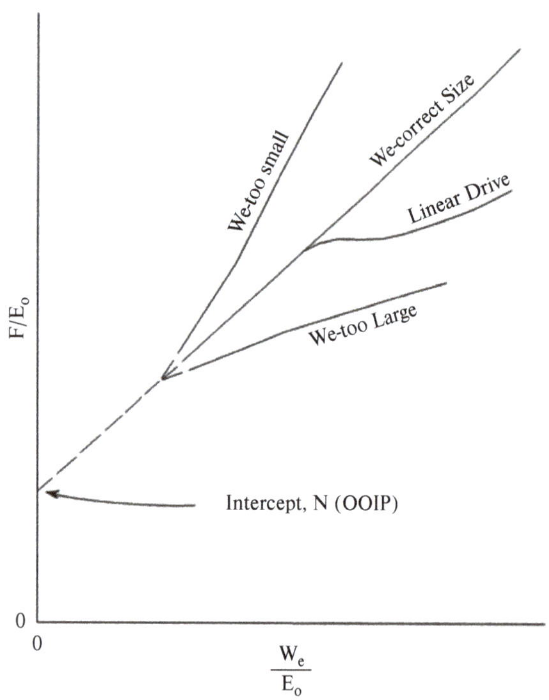

▲ **FIGURE 10.1 Oil reservoir with water influx (from Havlena and Odeh[7] © 1963 SPE, Reproduced with permission)**

It is important to note that Eq. (10.40) was derived by assuming that formation and water compressibility are negligible. This will apply to relatively low-pressure reservoirs with low formation and water compressibility. However, this assumption will not apply to high-pressure reservoirs with high formation and water compressibility, such as reservoirs encountered in a deepwater environment. For reservoirs with high formation and water compressibility, Eq. (10.40) will not be applicable. To derive an equation similar in form to Eq. (10.40), the GMBE for such reservoirs is the same as Eq. (10.34):

$$N\left(B_t - B_{ti}\right) + NB_{ti}\left[\frac{c_wS_{wi} + c_f}{1 - S_{wi}}\right]\Delta p + W_e - W_pB_w$$
$$= N_p\left[B_t + \left(R_p - R_{si}\right)B_g\right]$$

As shown previously in Eqs. (10.37) and (10.38):

$$F = N_p\left[B_t + \left(R_p - R_{si}\right)B_g\right] + W_pB_w$$

$$E_o = B_t - B_{ti}$$

A new term is defined to represent formation and water expansion as:

$$E_{fw} = B_{ti}\left[\frac{c_wS_w + c_f}{1 - S_{wi}}\right]\Delta p \tag{10.41}$$

Substituting these terms into Eq. (10.34) and re-arranging gives:

$$\frac{F}{E_o + E_{fw}} = N + \frac{W_e}{E_o + E_{fw}} \tag{10.42}$$

Note that the only difference between Eqs. (10.40) and (10.42) is the addition of the term, E_{fw}, for formation and water expansion. A plot of $F/\left(E_o + E_{fw}\right)$ vs. $W_e/\left(E_o + E_{fw}\right)$ will give a straight line with the intercept on the y-axis equal to N (the OOIP), and the slope equal to unity. The characteristics and shapes of this line will be the same as described previously for the plot of F/E_o vs. W_e/E_o in Figure 10.1.

EXAMPLE 10.2 Calculation of OOIP for an Undersaturated Reservoir with Water Influx

Problem

Calculate the OOIP and water influx volumes for an undersaturated reservoir with water influx. Calculate water influx volumes with the Carter-Tracy method assuming an infinite-acting aquifer. The properties of the reservoir and aquifer are as follows:

Initial pressure, p_i	10,535 psia
Current reservoir pressure, p	9806 psia
Temperature, T	212°F
Porosity, ϕ	0.275
Permeability, k	36.5 md
Net thickness, h	12.5 ft
Water compressibility, c_w	3×10^{-6} psi^{-1}
Water viscosity, μ_w	0.588 cp
Water formation volume factor, B_w	1.02 RB/STB
Formation compressibility, c_f	20×10^{-6} psi^{-1}
Total compressibility, c_t	23×10^{-6} psi^{-1}
Aquifer encroachment angle, θ	180°

Reservoir Properties:

Area of reservoir, A	1022 acres
Radius of reservoir, r_R	3764.39 ft
Initial water saturation, S_{wi}	0.30
Bubble point pressure, p_b	8104 psia

The PVT properties of the oil above bubble point pressure are presented in Table 10.2 and Figure 10.2. The production and pressure history of the reservoir are presented in Table 10.3.

TABLE 10.2 PVT Data of the Reservoir in Example 10.2

Pressure (psia)	FVF (RB/STB)	Rs (scf/STB)
10535	1.552	1417
10000	1.561	1417
9500	1.568	1417
9000	1.576	1417
8500	1.584	1417
8104	1.592	1417

TABLE 10.3 Production Data for the Reservoir in Example 10.2

Date	Cum (days)	Res. Press, (psia)	Cum Oil (STB)	Cum Gas (Mcf)	Cum Water (STB)	Cum GOR (scf/STB)
1-May-2003	0	10,535	0	0	0	0
1-Jun-2003	30	10,527	75,926	46,419	1,907	611
1-Jul-2003	61	10,519	151,993	123,750	1,913	814
1-Aug-2003	92	10,502	211,109	188,814	1,922	894
1-Sep-2003	122	10,293	363,314	360,419	1,927	992
1 -Oct-2003	153	10,076	544,988	580,480	1,939	1,065
1-Nov-2003	183	9,806	719,095	802,639	1,956	1,116
1-Dec-2003	214	9,784	902,185	1,044,303	2,018	1,158
1-Jan-2004	245	9,637	1,084,186	1,290,928	2,078	1,191
1-Feb-2004	274	9,497	1,264,710	1,539,785	2,124	1,218
1-Mar-2004	305	9,356	1,453,826	1,803,340	2,137	1,240
1-Apr-2004	335	9,142	1,638,530	2,092,009	2,167	1,277
1-May-2004	366	9,079	1,803,146	2,353,667	2,195	1,305
1-Jun-2004	396	9,035	1,951,444	2,589,641	2,206	1,327
1-Jul-2004	427	9,003	2,097,122	2,822,767	2,228	1,346
1-Aug-2004	458	9,013	2,226,579	3,024,247	2,241	1,358
1-Sep-2004	488	9,023	2,306,688	3,152,648	2,241	1,367
1 -Oct-2004	519	8,912	2,458,169	3,387,957	2,248	1,378
1-Nov-2004	549	8,800	2,611,623	3,640,493	2,248	1,394
1-Dec-2004	580	8,689	2,774,991	3,908,663	2,248	1,409
1-Jan-2005	611	8,697	2,940,599	4,174,634	2,248	1,420
1-Feb-2005	639	8,704	3,089,323	4,417,571	2,270	1,430
1-Mar-2005	670	8,712	3,215,581	4,613,268	2,435	1,435
1-Apr-2005	700	8,709	3,369,394	4,860,906	2,588	1,443
1-May-2005	731	8,705	3,471,539	5,103,481	5,254	1,470
1-Jun-2005	761	8,597	3,581,969	5,275,471	11,974	1,473
1-Jul-2005	792	8,488	3,660,120	5,394,232	23,754	1,474
1-Aug-2005	823	8,380	3,731,048	5,498,702	36,340	1,474

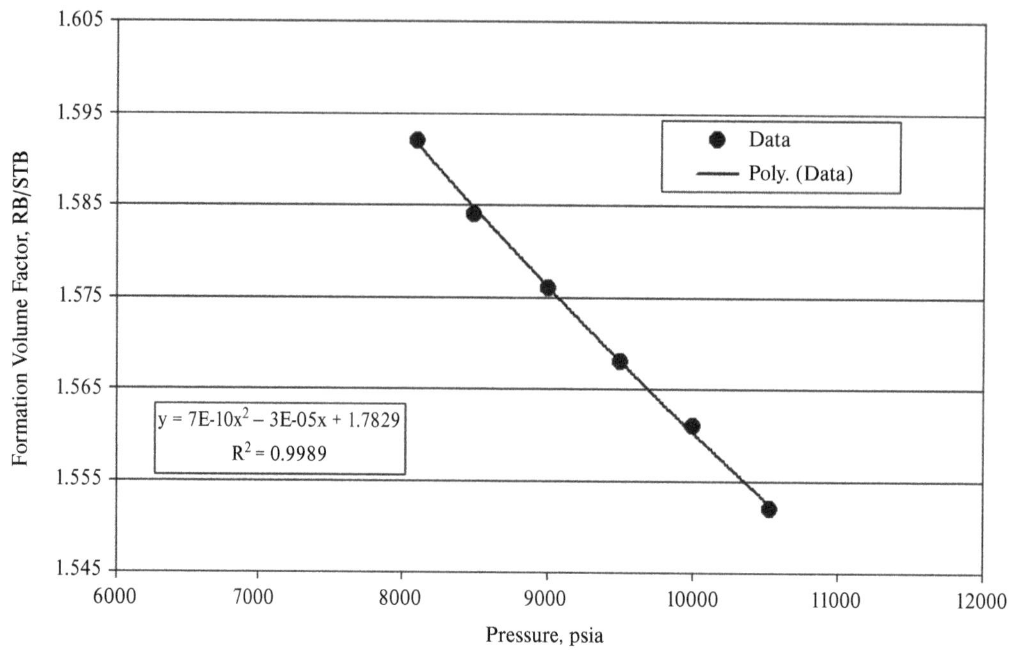

▲ FIGURE 10.2 **PVT data for the reservoir in Example 10.2**

Solution

Step 1: Calculate the parameter, B, for the Carter-Tracy method (see Example 9.4 in Chapter 9).

$$B = 1.119\phi c_t h r_R^2 \left(\frac{\theta}{360}\right)$$
$$= 1.119 \times 0.275 \times 23 \times 10^{-6}$$
$$\times 12.5 \times (3764.39)^2 \times \left(\frac{180}{360}\right)$$
$$= 626.8446 \text{ RB/psi}$$

Step 2: Calculate the dimensionless time, t_D:

$$t_D = \frac{0.00633kt}{\phi \mu c_t r_R^2}$$
$$= \frac{0.0063 \times 36.5 \times t}{0.275 \times 0.588 \times 23 \times 10^{-6} \times (3764.39)^2}$$
$$= 0.00436t$$

Step 3: Calculate water influx volumes using the Carter-Tracy method as shown in Example 9.4 (Chapter 9):

The remaining procedure for calculating the water influx volumes are shown in Example 9.4 in Chapter 9. The water influx volumes are calculated as shown in Table 10.4. The results in Table 10.4 were generated in a spreadsheet and are similar to Table 9.5 in Chapter 9.

Step 4: Calculation of the terms in Eq. (10.42):

For the production data of November 1, 2003 as example:

From Eq. (10.38):

$$E_o = B_t - B_{ti}$$
$$= 1.5560 - 1.5445 = 0.0115 \text{ RB/STB}$$

The values for B_t and B_{ti} were calculated from a polynomial fit of the data shown in Table 10.2 and Figure 10.2. The equation for the polynomial is: $FVF = 7E - 10p^2 - 3E - 05p + 1.7829$.

From Eq. (10.41):

$$E_{fw} = B_{ti}\left[\frac{c_w S_w + c_f}{1 - S_{wi}}\right]\Delta p$$
$$= 1.5445\left[\frac{3 \times 10^{-6} \times 0.3 + 20 \times 10^{-6}}{1 - 0.3}\right](10535 - 9806)$$
$$= 0.0336 \text{ RB/STB}$$

$W_e = 235,206$ RB, as shown in Table 10.4.

From Eq. (10.37):

$$F = N_p B_t + W_p B_w$$
$$= 719,095 \times 1.5560 + 1956 \times 1.02$$
$$= 1,120,928 \text{ RB}$$

Values for the y-axis and the x-axis were calculated as follows:

$$y_{axis} = \frac{F}{(E_o + E_{fw})}$$

$$= \frac{1{,}120{,}928}{(0.0115 + 0.0336)}$$

$$= 24{,}850{,}381 \text{ STB}$$

$$x_{axis} = \frac{W_e}{(E_o + E_{fw})}$$

$$= \frac{235{,}206}{(0.0115 + 0.0336)}$$

$$= 5{,}214{,}383 \text{ STB}$$

A plot of y_{axis} vs. x_{axis} for the results in Table 10.5 is shown in Figure 10.3. The intercept on the y_{axis} shows the OOIP is estimated to be 21.5 MMBO. The volumetrically calculated OOIP for this example is 21 MMBO. Thus, the OOIP calculated from material balance, and volumetrically from geologic maps are remarkably close for this example. It is important to note that the material balance calculations required many iterations on the estimated cumulative volume of water influx from the aquifer. The cumulative water influx volume was the key variable that was systematically changed iteratively before achieving the solution presented for this example. These iterations were not shown in Table 10.4 and Table 10.5 due to limitations of space. Note that the early data points (June to October 2003) were not included in Figure 10.3 because of the large variability of the data. These might have been caused by instability of production data, especially during the early stages of initiating production from the reservoir. Note also that toward the end of the production history (May to August 2005), the plotted data points appear to have decreased. This may have been caused by using constant formation compressibility for the entire calculations. Formation compressibility is not typically constant but varies with depletion of the reservoir. The variability of formation compressibility was reported by Poston and Chen[8] and Fetkovich et al.[9] An important requirement for using the material balance method for estimating OOIP and W_e is that the plotted points must fall on the unit slope line as evident in Eq. (10.42). This condition is a necessary check at each iteration step.

TABLE 10.4 Calculation of Water Influx Volumes for the Reservoir in Example 10.2 Using the Carter-Tracy Method

n	t (days)	$p_{R,n}$ (psia)	t_D	$\Delta p\ (t_D)$ (psia)	$\sqrt{t_D}$	p_D	p'_D	W_e(RB)
0	0	10535	0.00	0	0.0000	0.0000	0.0000	0
1	30	10527	0.13	8	0.3617	0.3535	1.1766	1,856
2	61	10519	0.27	16	0.5157	0.4780	0.7451	4,927
3	92	10502	0.40	33	0.6333	0.5652	0.5644	10,756
4	122	10293	0.53	242	0.7293	0.6320	0.4635	53,778
5	153	10076	0.67	459	0.8167	0.6897	0.3943	128,834
6	183	9806	0.80	729	0.8932	0.7379	0.3461	235,206
7	214	9784	0.93	751	0.9659	0.7820	0.3082	335,610
8	245	9637	1.07	898	1.0335	0.8216	0.2784	448,551
9	274	9497	1.19	1038	1.0930	0.8553	0.2558	564,958
10	305	9356	1.33	1179	1.1532	0.8885	0.2356	699,870
11	335	9142	1.46	1393	1.2086	0.9182	0.2191	850,070
12	366	9079	1.60	1456	1.2632	0.9468	0.2044	1,004,142
13	396	9035	1.73	1500	1.3140	0.9727	0.1921	1,150,891
14	427	9003	1.86	1532	1.3644	0.9979	0.1810	1,299,186
15	458	9013	2.00	1522	1.4131	1.0217	0.1712	1,439,851
16	488	9023	2.13	1512	1.4587	1.0435	0.1627	1,569,719
17	519	8912	2.26	1623	1.5043	1.0650	0.1548	1,712,000
18	549	8800	2.39	1735	1.5471	1.0847	0.1479	1,857,497
19	580	8689	2.53	1846	1.5902	1.1043	0.1414	2,015,367
20	611	8697	2.66	1838	1.6322	1.1230	0.1355	2,167,624
21	639	8704	2.79	1831	1.6691	1.1392	0.1306	2,301,018
22	670	8712	2.92	1823	1.7092	1.1565	0.1256	2,444,072
23	700	8709	3.05	1826	1.7470	1.1727	0.1211	2,579,621
24	731	8705	3.19	1830	1.7853	1.1887	0.1168	2,716,984
25	761	8597	3.32	1938	1.8215	1.2038	0.1130	2,857,732
26	792	8488	3.45	2047	1.8583	1.2188	0.1092	3,010,987
27	823	8380	3.59	2155	1.8943	1.2333	0.1058	3,171,731

TABLE 10.5 Tabulation of the Results from Eq. (10.42) Used in Plotting Figure 10.3 for Example 10.2

Date	Cum Oil (STB)	Cum Water (STB)	FVF (RB/STB)	Δp (PSIA)	E_o (RB/STB)	E_{fw} (RB/STB)	W_e (RB)	F (RB)	$\frac{W_e}{(E_o+E_{fw})}$ (STB)	$\frac{F}{(E_o+E_{fw})}$ (STB)
01-May-2003	0	0	1.5445	0	0.0000	0.0000	0	0	0	0
01-Jun-2003	75,926	1,907	1.5447	8	0.0001	0.0004	1,856	119,225	3,779,867	242,837,944
01-Jul-2003	151,993	1,913	1.5448	16	0.0002	0.0007	4,927	236,747	5,016,846	241,082,255
01-Aug-2003	211,109	1,922	1.5450	33	0.0005	0.0015	10,756	328,133	5,309,683	161,976,375
01-Sep-2003	363,314	1,927	1.5483	242	0.0037	0.0112	53,778	564,474	3,611,365	37,906,243
01-Oct-2003	544,988	1,939	1.5517	459	0.0071	0.0212	128,834	847,629	4,550,169	29,936,727
01-Nov-2003	719,095	1,956	1.5560	729	0.0115	0.0336	235,206	1,120,928	5,214,383	24,850,381
01-Dec-2003	902,185	2,018	1.5564	751	0.0118	0.0346	335,610	1,406,209	7,220,536	30,254,112
01-Jan-2004	1,084,186	2,078	1.5588	898	0.0143	0.0414	448,551	1,692,149	8,057,286	30,395,922
01-Feb-2004	1,264,710	2,124	1.5611	1,038	0.0166	0.0479	564,958	1,976,537	8,765,694	30,667,239
01-Mar-2004	1,453,826	2,137	1.5635	1,179	0.0190	0.0544	699,870	2,275,228	9,545,107	31,030,480
01-Apr-2004	1,638,530	2,167	1.5671	1,393	0.0226	0.0642	850,070	2,570,022	9,788,953	29,594,997
01-May-2004	1,803,146	2,195	1.5682	1,456	0.0237	0.0671	1,004,142	2,829,987	11,055,009	31,156,483
01-Jun-2004	1,951,444	2,206	1.5690	1,500	0.0245	0.0692	1,150,891	3,064,051	12,292,894	32,727,716
01-Jul-2004	2,097,122	2,228	1.5695	1,532	0.0250	0.0706	1,299,186	3,293,807	13,582,123	34,434,558
01-Aug-2004	2,226,579	2,241	1.5694	1,522	0.0248	0.0702	1,439,851	3,496,622	15,153,289	36,799,156
01-Sep-2004	2,306,688	2,241	1.5692	1,512	0.0247	0.0697	1,569,719	3,621,942	16,631,162	38,374,457
01-Oct-2004	2,458,169	2,248	1.5711	1,623	0.0266	0.0748	1,712,000	3,864,413	16,877,095	38,095,822
01-Nov-2004	2,611,623	2,248	1.5731	1,735	0.0286	0.0800	1,857,497	4,110,659	17,107,893	37,859,934
01-Dec-2004	2,774,991	2,248	1.5751	1,846	0.0305	0.0851	2,015,367	4,373,124	17,424,141	37,808,471
01-Jan-2005	2,940,599	2,248	1.5749	1,838	0.0304	0.0848	2,167,624	4,633,550	18,823,757	40,237,983
01-Feb-2005	3,089,323	2,270	1.5748	1,831	0.0303	0.0844	2,301,018	4,867,418	20,060,119	42,433,819
01-Mar-2005	3,215,581	2,435	1.5747	1,823	0.0301	0.0841	2,444,072	5,065,961	21,402,668	44,362,478
01-Apr-2005	3,369,394	2,588	1.5747	1,826	0.0302	0.0842	2,579,621	5,308,500	22,551,799	46,408,458
01-May-2005	3,471,539	5,254	1.5748	1,830	0.0303	0.0844	2,716,984	5,472,317	23,699,689	47,733,888
01-Jun-2005	3,581,969	11,974	1.5767	1,938	0.0322	0.0894	2,857,732	5,659,996	23,509,887	46,563,459
01-Jul-2005	3,660,120	23,754	1.5787	2,047	0.0342	0.0944	3,010,987	5,802,432	23,423,186	45,138,495
01-Aug-2005	3,731,048	36,340	1.5807	2,155	0.0361	0.0994	3,171,731	5,934,574	23,408,921	43,800,051

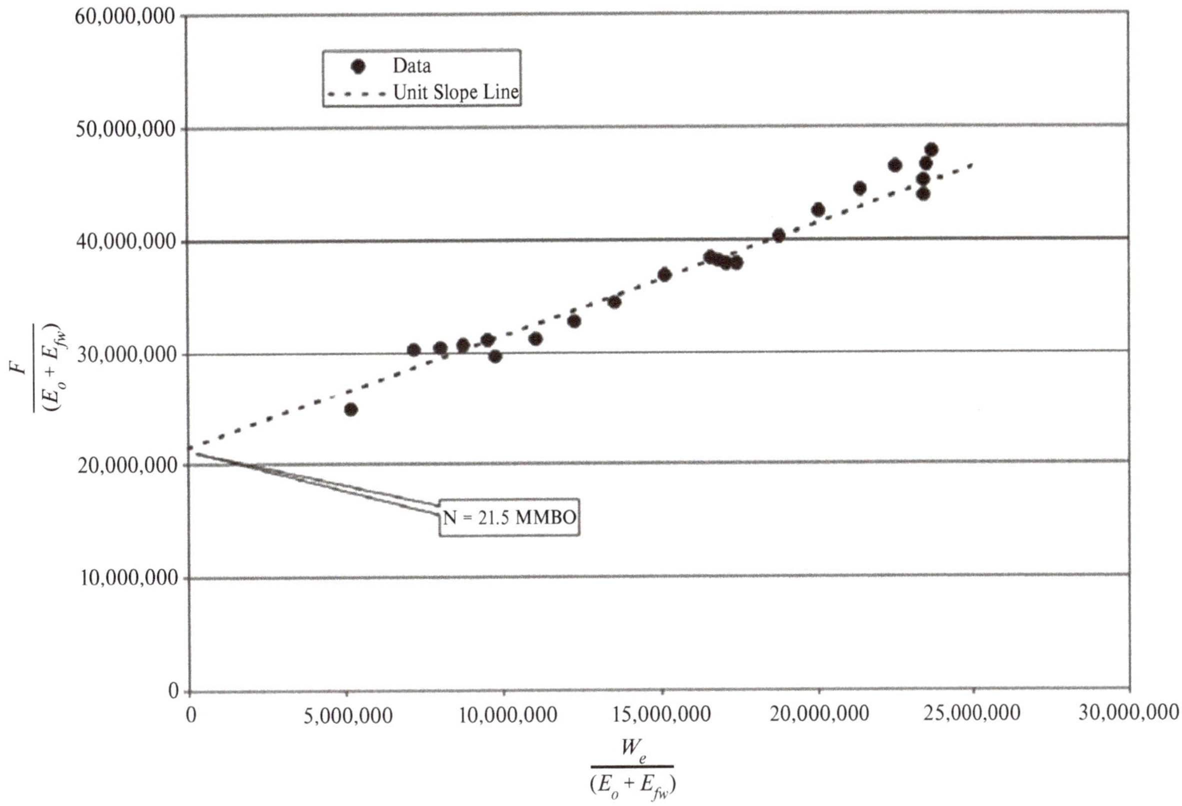

▲ **FIGURE 10.3** **Plot of material balance calculations from Table 10.5 for Example 10.2**

End of Example 10.2

10.6 Volumetric Saturated Oil Reservoirs

A volumetric saturated oil reservoir is described as a reservoir with a primary gas cap in contact with the oil zone at discovery and negligible influence from external energy sources such as an aquifer. At initial conditions, the oil zone is assumed to be in thermodynamic equilibrium with the gas cap so that gas saturation within the oil zone is essentially zero. Note that the concept of volumetric reservoirs described previously as isolated, closed systems for other types of reservoirs is applicable to saturated oil reservoirs.

10.6.1 Volumetric Method

The original gas-in-place (OGIP) in the primary gas cap of a saturated oil reservoir can be calculated as:

$$G = \frac{7758 \times A_{gc} \times h_{gc} \times \phi_{gc} \times \left(1 - S_{wigc}\right)}{B_{gi}} \qquad (10.43)$$

For Eq. (10.43), A_{gc} = area of gas cap, acres; h_{gc} = average net thickness of gas cap, feet; ϕ_{gc} = average porosity of gas cap, fraction; and S_{wigc} = initial average connate water saturation in the gas cap, fraction. Similarly, the OOIP in the oil zone of a saturated reservoir can be calculated as:

$$N = \frac{7758 \times A_{oz} \times h_{oz} \times \phi_{oz} \times \left(1 - S_{wioz}\right)}{B_{oi}} \qquad (10.44)$$

For Eq. (10.44), A_{oz} = area of the oil zone, acres; h_{oz} = average net thickness of the oil zone, feet; ϕ_{oz} = average porosity in oil zone, fraction; and S_{wioz} = average connate water saturation in the oil zone. For simplicity, it can be assumed that the initial connate water saturations in the gas cap and oil zone are equal. Also, it can be assumed that the porosity of the formation is the same in the gas cap and in the oil zone. With both assumptions, the ratio of the initial gas cap volume to the initial oil volume can be derived from Eqs. (10.43) and (10.44) as:

$$\begin{aligned} m &= \frac{A_{gc} h_{gc}}{A_{oz} h_{oz}} \\ &= \frac{G B_{gi}}{N B_{oi}} \end{aligned} \qquad (10.45)$$

10.6.2 Material Balance Method

The GMBE for a volumetric saturated oil reservoir is obtained from Eq. (10.2) as:

$$N(B_t - B_{ti}) + \frac{NmB_{ti}}{B_{gi}}[B_g - B_{gi}] + (1+m)NB_{ti}\left[\frac{c_w S_{wi} + c_f}{1 - S_{wi}}\right]\Delta p_e - W_p B_w = N_p[B_t + (R_p - R_{si})B_g] \quad (10.46)$$

For reservoirs with gas caps, formation and water compressibility are generally negligible because of the higher compressibility of the gas in the gas cap. By re-arrangement, Eq. (10.46) becomes:

$$N_p[B_t + (R_p - R_{si})B_g] + W_p B_w = N(B_t - B_{ti}) + \frac{NmB_{ti}}{B_{gi}}[B_g - B_{gi}] \quad (10.47)$$

As in Eq. (10.37), the cumulative voidage, $F = N_p[B_t + (R_p - R_{si})B_g] + W_p B_w$. The oil expansion is represented in Eq. (10.38) as, $E_o = B_t - B_{ti}$. The gas expansion term is expressed as:

$$E_g = B_g - B_{gi} \quad (10.48)$$

Substituting Eq. (10.47) becomes:

$$F = NE_o + Nm\left(\frac{B_{ti}}{B_{gi}}\right)E_g = N\left[E_o + m\left(\frac{B_{ti}}{B_{gi}}\right)E_g\right] \quad (10.49)$$

As demonstrated by Havlena and Odeh,[7] if m is known, a plot of F vs. $\left[E_o + m(B_{ti}/B_{gi})E_g\right]$ will give a straight line passing through the origin with the slope equal to N. This is shown as Figure 10.4. If N and m are unknown, a plot of F vs. $\left[E_o + m(B_{ti}/B_{gi})E_g\right]$ for various values of m will produce a plot similar to Figure 10.5. If m is too small, the line will curve upward. If m is too large, the line will curve downward. If the correct value of m is assumed, a straight line that passes through the origin will be obtained as shown in Figure 10.5.

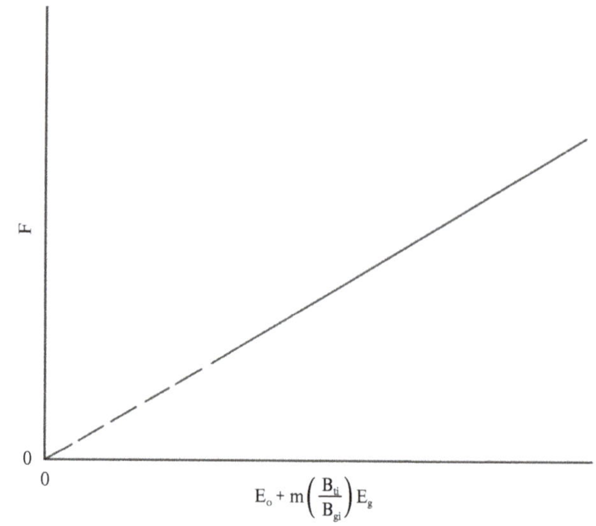

▲ **FIGURE 10.4** **Volumetric saturated oil reservoir with *m* known and *N* unknown (from Havlena and Odeh[7] © 1963 SPE, Reproduced with permission)**

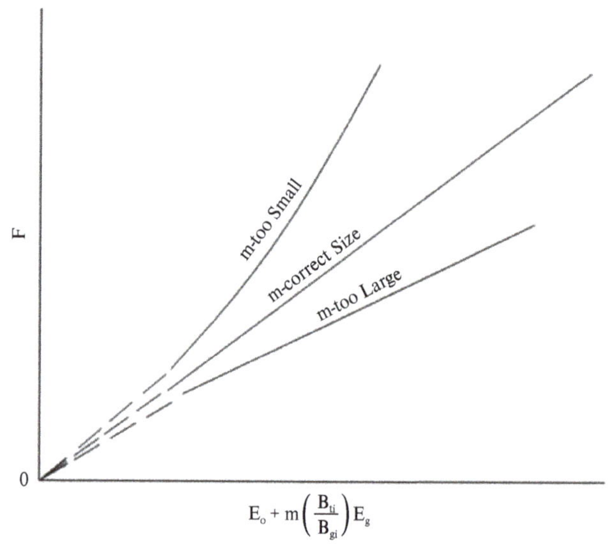

▲ **FIGURE 10.5** **Volumetric saturated oil reservoir with *m* and *N* as unknowns (from Havlena and Odeh[7] © 1963 SPE, Reproduced with permission)**

10.7 Material Balance Approach for Saturated Oil Reservoirs with Water Influx

The GMBE for a saturated oil reservoir with water influx is as shown in Eq. (10.2):

$$N(B_t - B_{ti}) + \frac{NmB_{ti}}{B_{gi}}[B_g - B_{gi}] + (1+m)NB_{ti}\left[\frac{c_w S_{wi} + c_f}{1 - S_{wi}}\right]\Delta p + W_e - W_p B_w = N_p[B_t + (R_p - R_{si})B_g]$$

Since the reservoir has a gas cap and water influx, the effects of formation and water compressibility are negligible. Thus, Eq. (10.2) can be reduced and re-arranged into the following form:

$$N_p\left[B_t+\left(R_p-R_{si}\right)B_g\right]+W_pB_w$$
$$=N\left(B_t-B_{ti}\right)+\frac{NmB_{ti}}{B_{gi}}\left[B_g-B_{gi}\right]+W_e \tag{10.50}$$

Using the symbols, F, E_o, E_g, as defined in Eqs. (10.37), (10.38), and (10.48) in Eq. (10.50) gives:

$$F=N\left[E_o+m\left(\frac{B_{ti}}{B_{gi}}\right)E_g\right]+W_e \tag{10.51}$$

Eq. (10.51) can be re-arranged in the form of a straight line equation for application of the Havlena-Odeh[7] method:

$$\frac{F}{\left[E_o+m\left(\frac{B_{ti}}{B_{gi}}\right)E_g\right]}=N+\frac{W_e}{\left[E_o+m\left(\frac{B_{ti}}{B_{gi}}\right)E_g\right]} \tag{10.52}$$

A plot of $F/\left[E_o+m\left(B_{ti}/B_{gi}\right)E_g\right]$ on the y-axis against $W_e/\left[E_o+m\left(B_{ti}/B_{gi}\right)E_g\right]$ on the x-axis will give a straight line with the intercept of the y-axis equal to N, and slope equal to unity if the correct values of W_e are used for known gas cap size, m. If W_e is too small, the line will curve upward. If W_e is too large, the line will curve downward. A linear aquifer will yield an S-shaped curve. The shapes of the curves that could result from this plot are as shown in Figure 10.1. Note the similarity of the plots resulting from applying the Havlena-Odeh[7] method to an undersaturated reservoir with water influx, and a saturated reservoir with water influx. The main difference between Eqs. (10.40) and (10.52) is the inclusion of the gas expansion term in Eq. (10.52) for the gas cap in saturated reservoirs.

10.8 Case History of Manatee Reservoirs

The Manatee reservoirs are located in the Gulf of Mexico, USA, about 120 miles southwest of New Orleans, Louisiana, in Green Canyon Block 155 (GC 155) and in the western part of the same basin that contain the Angus and Troika reservoirs (Figure 10.6). The water depth at this location is 1900 feet. There are three reservoirs that constitute the Manatee reservoirs. These are the S10A1, S10A2, and S10B reservoirs. The S10A1 reservoir is relatively small in terms of oil-in-place volumes and was not part of initial development. The type log of the sands in the Manatee reservoirs is shown as Figure 10.7. This case history is based on the S10A2 and S10B reservoirs.

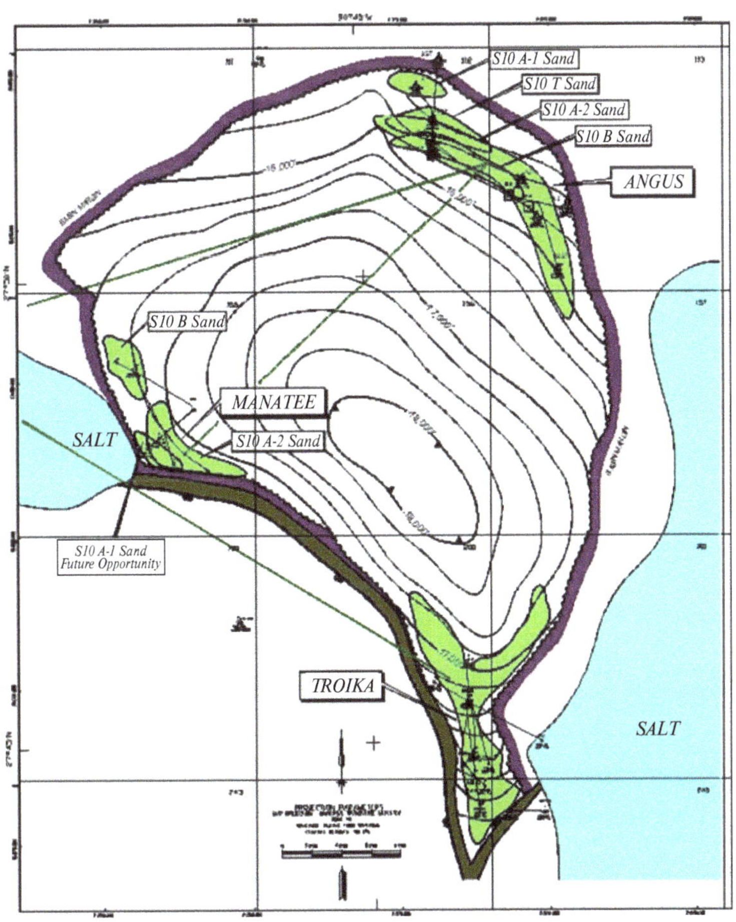

FIGURE 10.6 Basin map showing Manatee, Angus, and Troika reservoirs

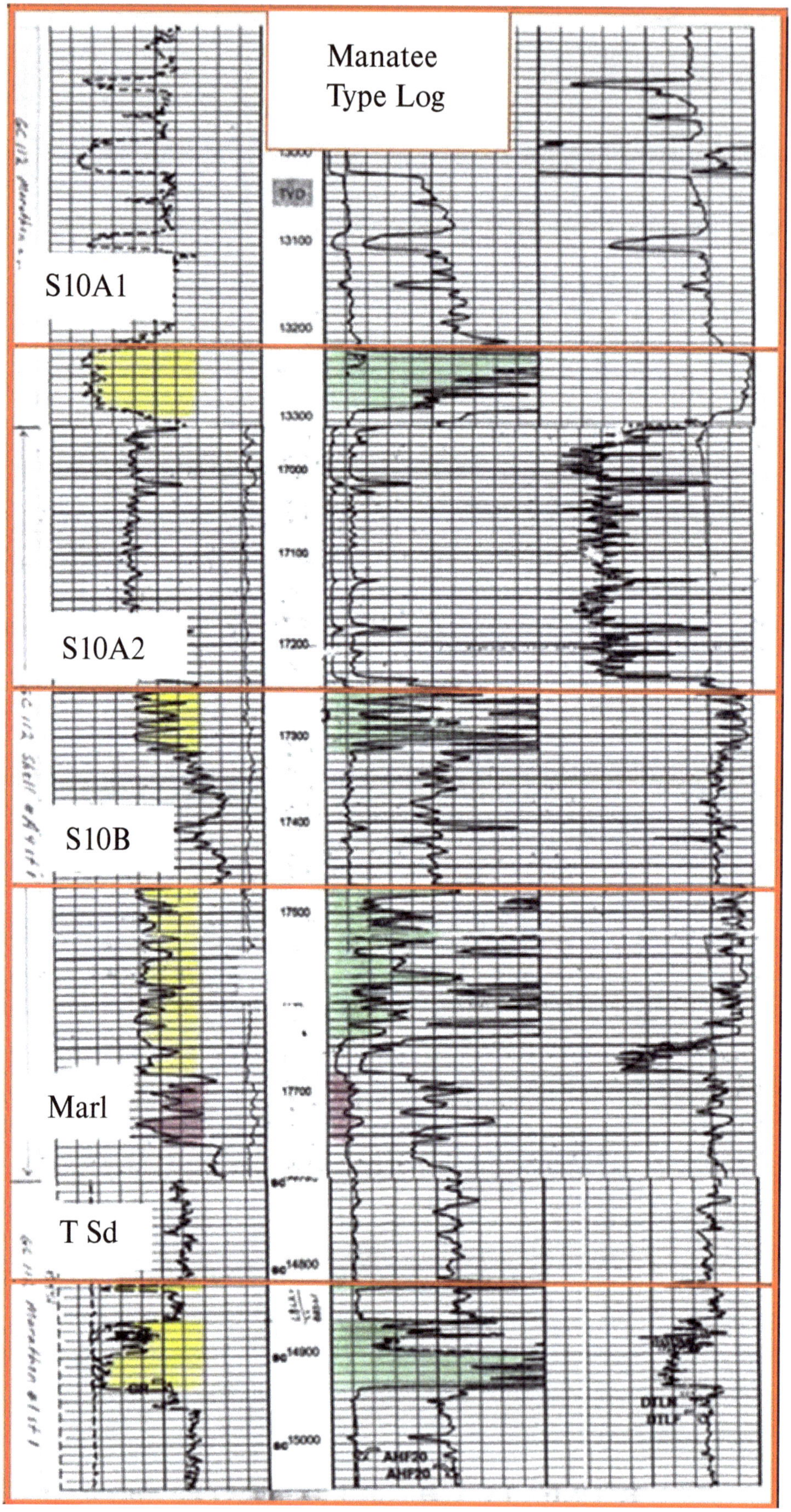

▲ **FIGURE 10.7** Type log of Manatee reservoirs

10.8.1 Reservoir Geology

The S10A2 and S10B reservoirs consist of turbidite sands trapped against marl and salt in a mini-basin setting (Figure 10.6). The age of the pay intervals is lower Pliocene. The structure and net pay isopach maps of the S10A2 reservoir are shown as Figures 10.8 and 10.9, respectively. The planimetered reservoir volume of the S10A2 reservoir is 12,140 acre-ft. The structure and net pay isopach maps of the S10B reservoir are shown as Figures 10.10 and 10.11, respectively. The planimetered reservoir volume of the S10B reservoir is 13,707 acre-ft. The Manatee sands have five well penetrations (Wells #1, #1ST1, #1ST2, #1ST3, and #2) as shown in the maps. The oil-water contacts in the S10A2 and S10B reservoirs were determined to be at 17,000 feet and 17,100 feet, respectively. In June 1998, original pressures measured in the S10A2 and S10B reservoirs at the oil-water contact were 11,225 psia and 11,237 psia, respectively.

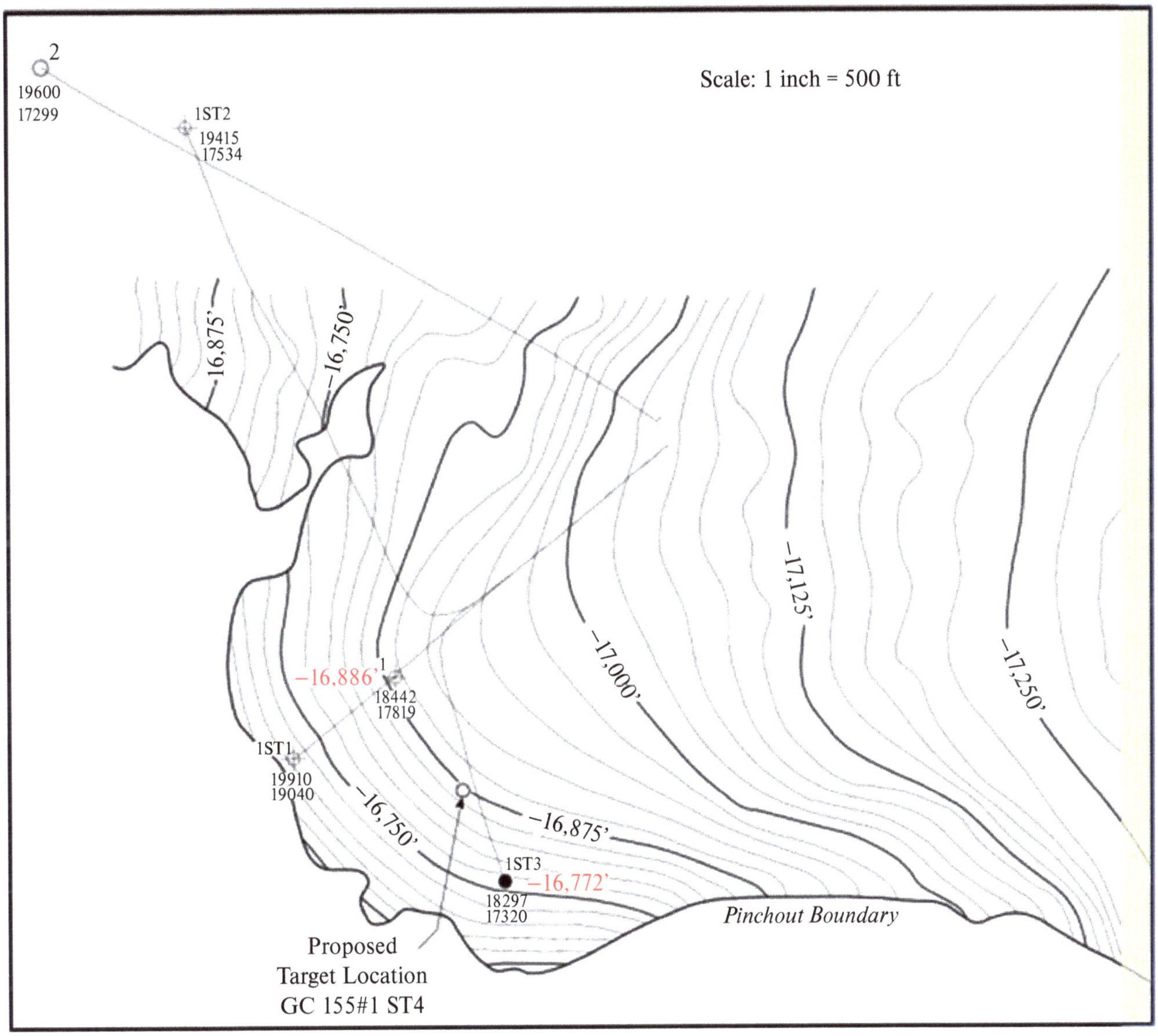

▲ **FIGURE 10.8** Structure map of S10A2 reservoir

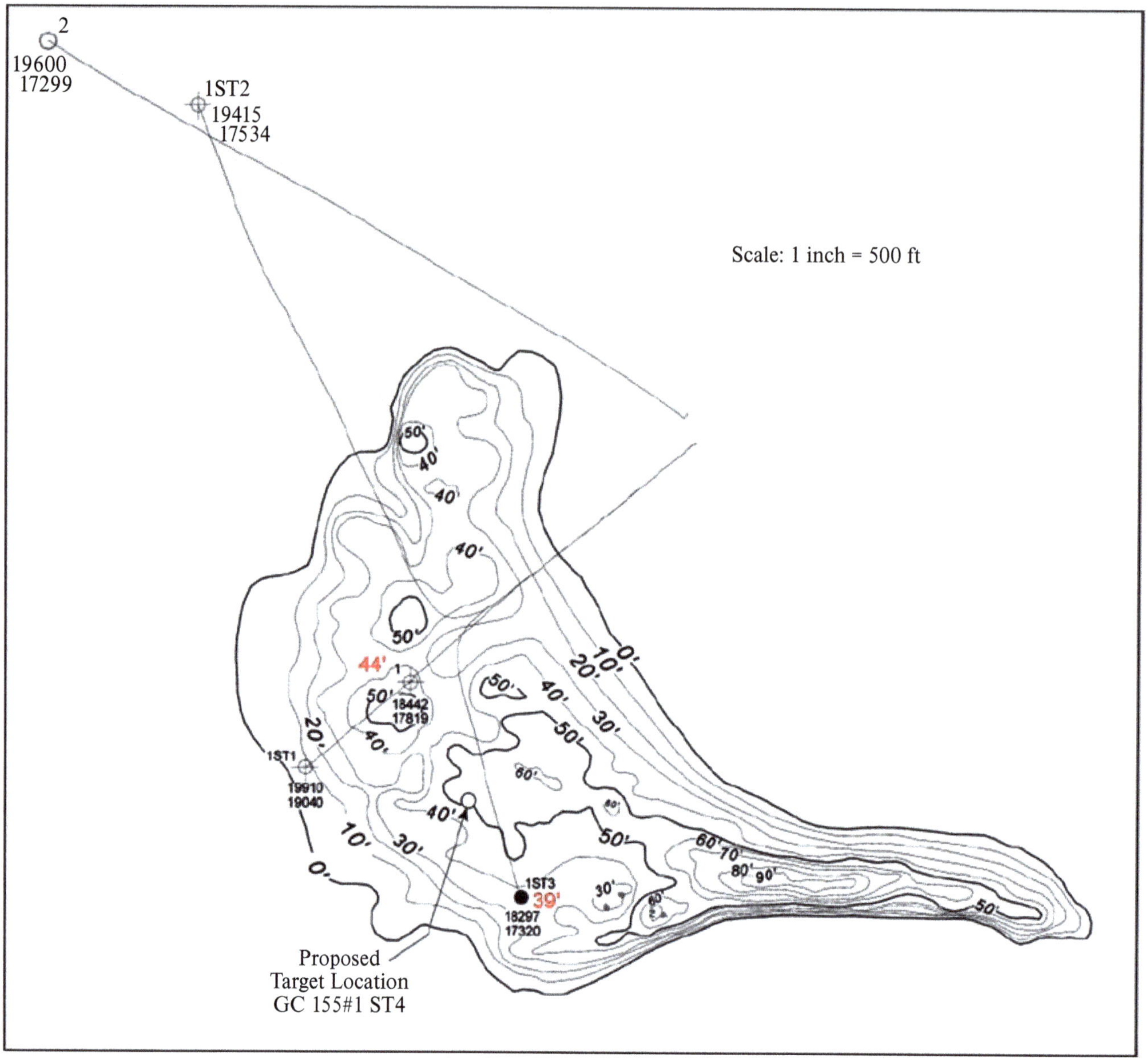

▲ FIGURE 10.9 **Net pay isopach map for S10A2 reservoir**

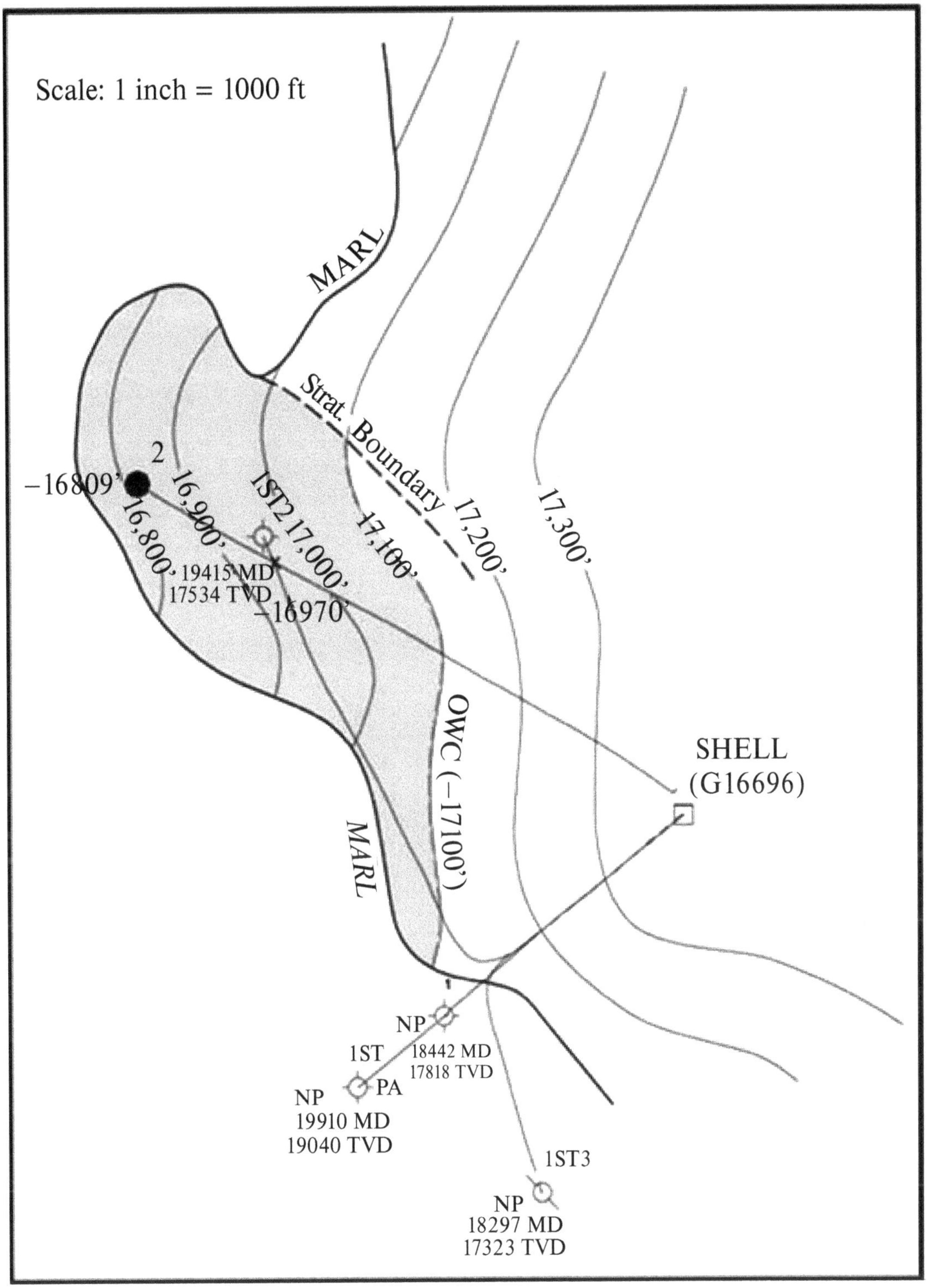

▲ FIGURE 10.10 **Structure map of S10B reservoir**

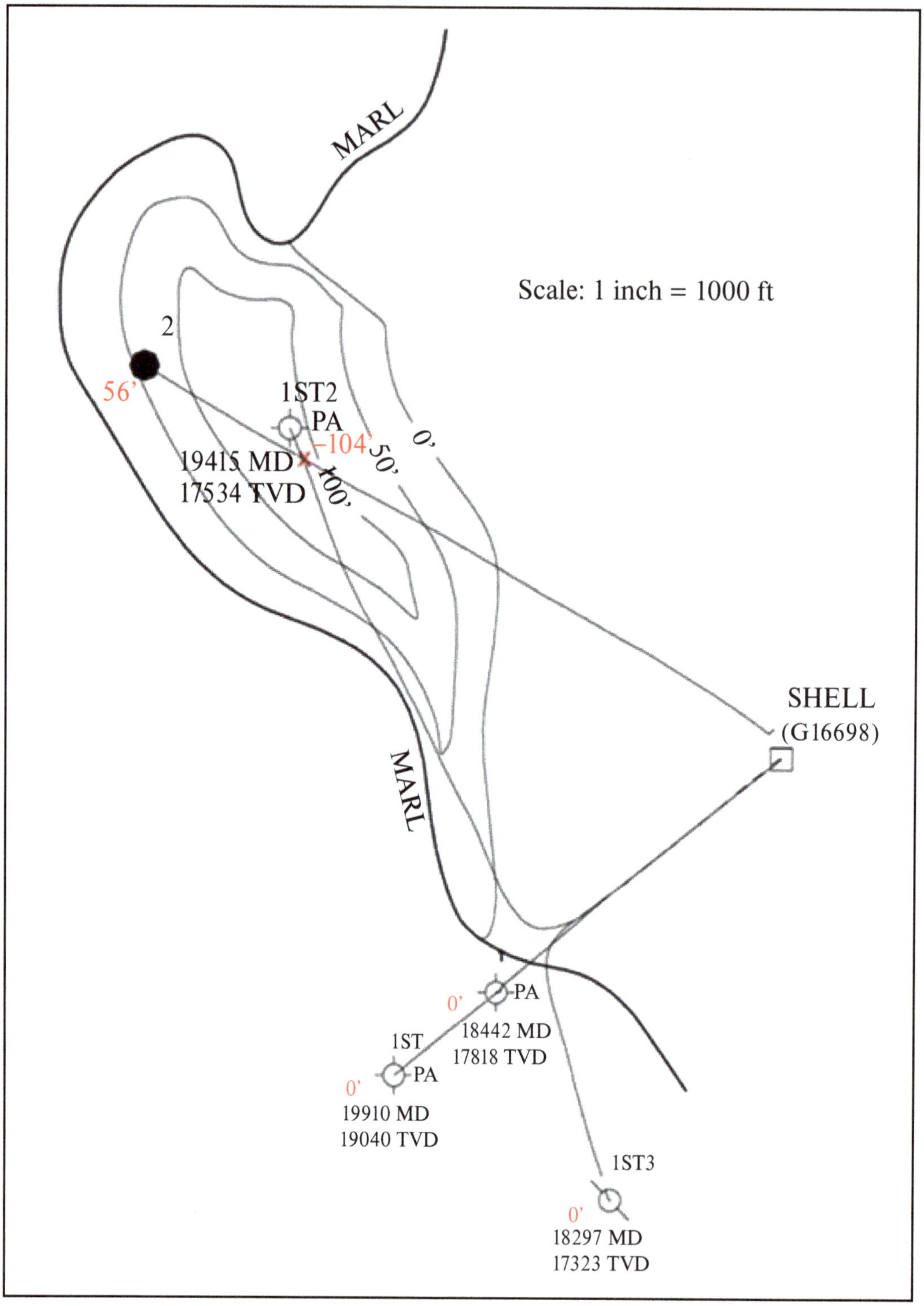

▲ **FIGURE 10.11** **Net pay isopach map for S10B reservoir**

10.8.2 Rock and Fluid Properties

The rock properties of the S10A2 and S10B sands are summarized in Table 10.6. These averages are based on three well penetrations of the S10A2 sands, and two well penetrations of the S10B sands. The PVT properties of reservoir fluids for the S10A2 and S10B reservoirs are given in Table 10.7 for oil and Table 10.8 for associated gas.

TABLE 10.6 Rock/Fluid Properties for S10A2 and S10B Reservoirs

Rock/Fluid Property	Reservoirs	
	S10A2	S10B
Porosity (%)	27.3	28.0
Permeability (md)	165	400
Initial water saturation (%)	21.7	24.6
Oil compressibility (1/psi)	6.6E-06	6.6E-06
Water compressibility (1/psi)	3.0E-06	3.0E-06
Formation compressibility (1/psi)	15E-06	15E-05
Total compressibility (1/psi)	21E-06	21E-06
Temperature (°F)	185	188
Gas Gravity (Air = 1)	0.71	0.70
Oil Gravity (°API)	36.5	35.4

TABLE 10.7 Oil PVT Data for S10A2 and S10B Reservoirs

Pressure (PSIA)	FVF (RB/STB)	Rs (scf/STB)	Viscosity (cp)
12,000	1.843	1805	0.372
11,581	1.847	1805	0.363
11,161	1.850	1805	0.354
10,742	1.855	1805	0.345
10,323	1.859	1805	0.336
9904	1.864	1805	0.327
9484	1.869	1805	0.317
9065	1.875	1805	0.308
8646	1.882	1805	0.298
8227	1.889	1805	0.289
7807	1.897	1805	0.280
7388	1.905	1805	0.271
6969	1.915	1805	0.262
6549	1.927	1805	0.254
6130	1.939	1805	0.246
5273	1.777	1464	0.274
4326	1.614	1123	0.315
3247	1.452	781	0.378
1938	1.289	440	0.498
15	1.126	99	0.884

TABLE 10.8 Gas PVT Data for S10A2 and S10B Reservoirs

Pressure (PSIA)	Gas FVF (RB/scf)	Viscosity (cp)
12,000	0.000446	0.0446
11,369	0.000454	0.0434
10,738	0.000462	0.0422
10,108	0.000471	0.0409
9477	0.000482	0.0395
8846	0.000493	0.0381
8215	0.000507	0.0366
7584	0.000523	0.0351
6954	0.000542	0.0335
6323	0.000566	0.0317
5692	0.000595	0.0299
5061	0.000633	0.0279
4430	0.000685	0.0258
3800	0.000760	0.0236
3169	0.000876	0.0212
2538	0.001072	0.0188
1907	0.001438	0.0166
1276	0.002230	0.0148
646	0.004674	0.0136
15	0.219731	0.0129

10.8.3 Reservoir Pressure and Production Data

The Manatee Field was discovered in June 1998 and oil production started in July 2002. The S10A2 reservoir was produced from a single well designated as Well #1ST3 (Figure 10.8). The S10B reservoir was also produced from a single well shown as Well #2 in Figure 10.10. The production history of the S10A2 reservoir is shown in Table 10.9 and Figure 10.12. The production history of the S10B reservoir is shown in Table 10.10 and Figure 10.13. The historical average reservoir pressures are also provided in Table 10.9 and Table 10.10 for S10A2 and S10B reservoirs, respectively.

TABLE 10.9 Historical Production Data for S10A2 Reservoir

Date	Oil Rate (BOPD)	Gas Rate (MCFPD)	Water Rate (BWPD)	Reservoir Pressure (PSIA)
Jan-02	0	0	0	11,225
Feb-02	0	0	0	
Mar-02	0	0	0	
Apr-02	0	0	0	
May-02	0	0	0	
Jun-02	0	0	0	
Jul-02	214	239	0	
Aug-02	362	243	0	
Sep-02	54	119	0	
Oct-02	135	326	0	9600
Nov-02	71	155	0	
Dec-02	0	0	0	
Jan-03	0	0	0	
Feb-03	0	0	0	
Mar-03	249	466	0	
Apr-03	570	951	0	
May-03	1104	1975	0	
Jun-03	927	1592	0	
Jul-03	250	389	0	
Aug-03	878	1331	0	
Sep-03	163	266	0	
Oct-03	1135	1214	0	8400
Nov-03	815	1214	0	
Dec-03	494	724	0	
Jan-04	448	952	0	
Feb-04	488	787	0	
Mar-04	193	197	0	
Apr-04	181	423	0	
May-04	235	371	0	8400
Jun-04	46	72	0	
Jul-04	194	318	0	
Aug-04	644	1072	0	
Sep-04	349	502	0	
Oct-04	103	172	0	
Nov-04	180	277	0	
Dec-04	337	640	0	
Jan-05	322	628	0	
Feb-05	0	0	0	
Mar-05	0	0	0	

TABLE 10.10 Historical Production Data for S10B Reservoir

Date	Oil Rate (BOPD)	Gas Rate (MCFPD)	Water Rate (BWPD)	Reservoir Pressure (PSIA)
Jan-02	0	0	0	11,237
Feb-02	0	0	0	
Mar-02	0	0	0	
Apr-02	0	0	0	
May-02	0	0	0	
Jun-02	0	0	0	
Jul-02	1431	1193	0	
Aug-02	3267	1212	0	
Sep-02	3637	6049	0	
Oct-02	1245	2052	0	9417
Nov-02	4587	7406	0	
Dec-02	5702	9213	0	
Jan-03	6238	10,265	0	
Feb-03	6900	10,611	0	
Mar-03	8012	13,838	0	
Apr-03	6700	11,190	0	
May-03	6311	11,291	0	
Jun-03	6680	11,457	0	
Jul-03	7491	12,199	0	
Aug-03	6643	10,491	0	
Sep-03	6645	10,825	0	
Oct-03	5562	8005	0	8350
Nov-03	5376	8005	0	
Dec-03	5904	8653	0	
Jan-04	4826	10260	0	
Feb-04	3935	6871	0	
Mar-04	4512	7337	0	8050
Apr-04	3613	5851	638	
May-04	3544	5591	625	
Jun-04	2986	4647	1538	
Jul-04	1810	2969	1157	
Aug-04	816	1359	1224	
Sep-04	563	809	844	
Oct-04	802	1353	1556	
Nov-04	851	1340	1651	
Dec-04	831	1699	1613	
Jan-05	988	1901	1918	
Feb-05	959	1810	1862	
Mar-05	1046	1875	2030	

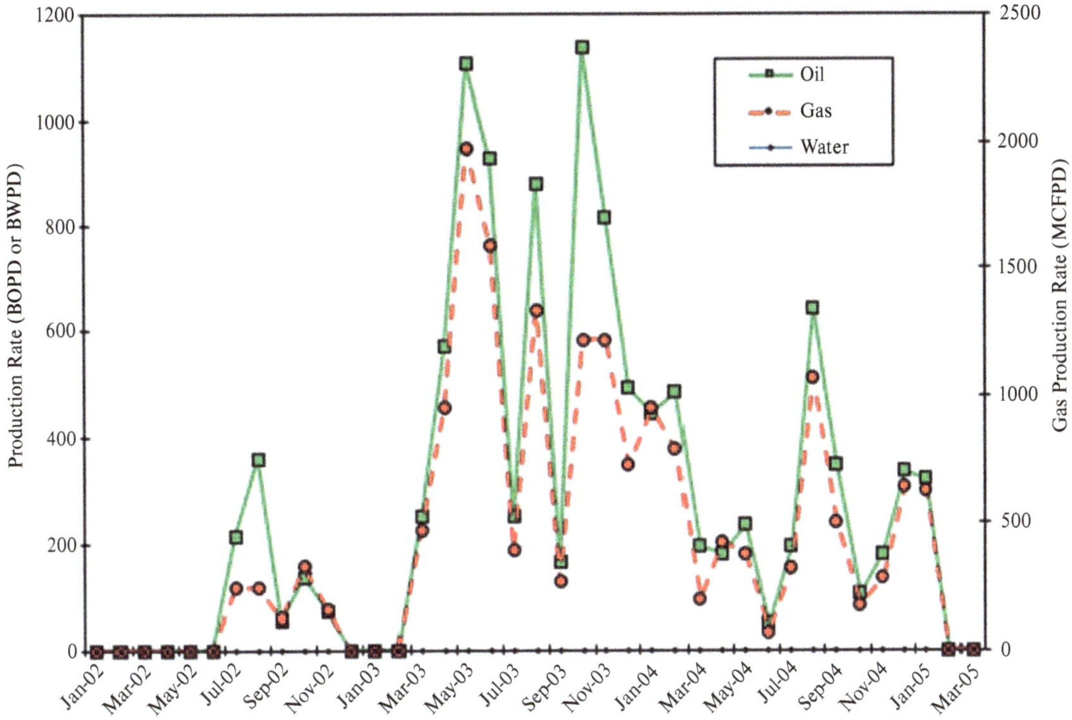

▲ FIGURE 10.12 Historical production data for S10A2 reservoir

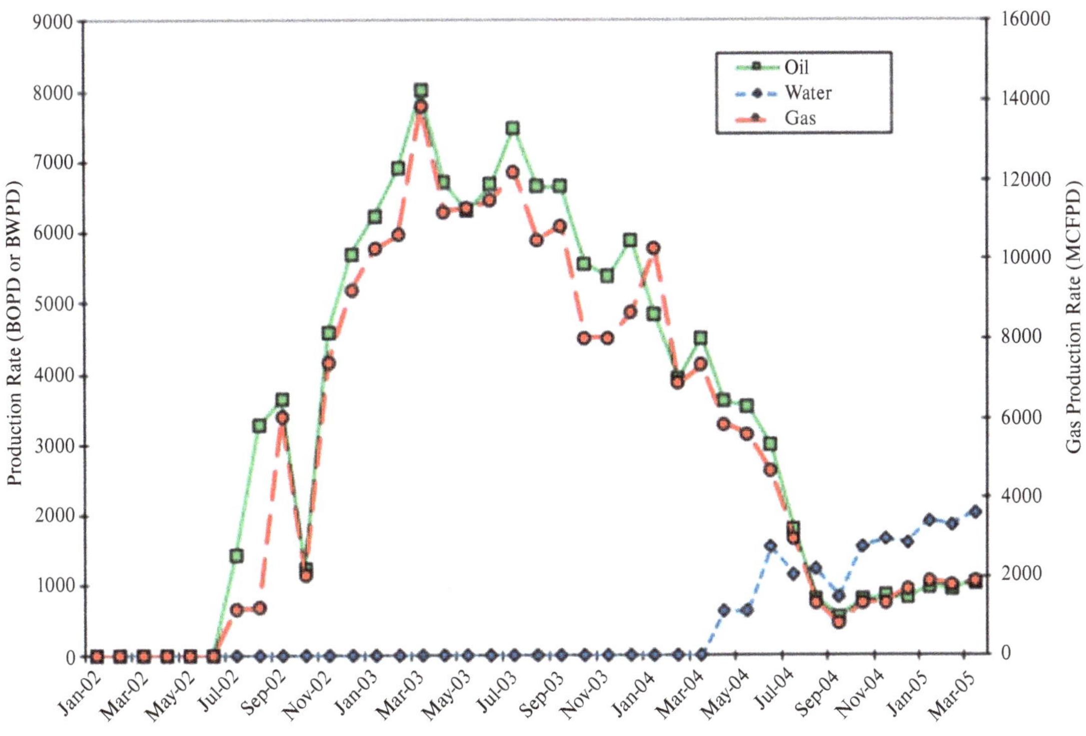

▲ FIGURE 10.13 Historical production data for S10B reservoir

Nomenclature

A	reservoir area, acres	I	drive index
B_g	gas formation volume factor, RB/scf	m	ratio of original reservoir gas cap volume to original reservoir oil volume
B_o	formation volume factor, RB/STB	N	original oil in place, STB
B_t	two-phase formation volume factor, RB/STB	N_p	cumulative oil production, STB
B_w	water formation volume factor, RB/STB	N_r	oil remaining in reservoir, STB
c_e	reservoir effective compressibility, psi^{-1}	p	pressure, psia
c_f	formation compressibility, psi^{-1}	p_b	bubble point pressure, psia
c_o	oil compressibility, psi^{-1}	R	oil recovery factor
c_w	water compressibility, psi^{-1}	R_p	cumulative production gas-oil ratio, scf/STB
E_{fw}	formation and water expansion term in material balance equation	R_s	solution gas-oil ratio, scf/STB
E_g	gas expansion term in material balance equation	S_g	gas saturation, fraction
E_o	oil expansion term in material balance equation	S_o	oil saturation, fraction
F	cumulative reservoir voidage, RB	S_w	water saturation, fraction
G	original gas in place, scf	W_e	cumulative water influx, RB
G_{inj}	cumulative gas injection, scf	W_{inj}	cumulative water injection, STB
G_p	cumulative gas produced, scf	W_p	cumulative water production, STB
h	average net sand thickness, feet	ϕ	porosity, fraction

Subscripts

FWCD	formation and water compressibility drive	OD	oil drive
g	gas	p	produced
gc	gas cap	r	remaining or residual
GCD	gas cap compressibility drive	w	water
i	initial	WD	water drive
o	oil		
oz	oil zone		

Abbreviations

FVF	Formation Volume Factor	OIP	Oil-In-Place
GMBE	General Material Balance Equation	OOIP	Original Oil-In-Place
OGIP	Original Gas-In-Place	STB	Stock Tank Barrels

REFERENCES

1. Pirson, S.J.: *Elements of Oil Reservoir Engineering*, 2nd Ed., McGraw-Hill, New York (1958).
2. Dykstra, H.: "The Prediction of Oil Recovery by Gravity Drainage," *JPT* (May 1978) 818–830.
3. Wei, M.H., Yu, J.P., Moore, D.M., Ezekwe, N., Querin, M.E., and Williams, L.L.: "Case History of Pressure Maintenance by Crestal Gas Injection in the 26R Gravity Drainage Reservoir," paper SPE 24035 presented at the Western Regional Meeting, Bakersfield, California, (March 30–April 1, 1992).
4. Essley, P.L., Hancock, G.L., and Jones, K.E.: "Gravity Drainage Concepts in a Steeply Dipping Reservoir," paper SPE 1029-G presented at the SPE-AIME Petroleum Production and Reservoir Engineering Conference, Tulsa, Oklahoma, (March 20–21, 1958).
5. Hall, H.N.: "Analysis of Gravity Drainage." *JPT* (September 1961) 927–936.
6. Li, K., and Horne, R.N.: "Prediction of Oil Production by Gravity Drainage," paper SPE 84184 presented at the SPE Annual Technical Conference and Exhibition in Denver, Colorado, (October 5–8, 2003).
7. Havlena, D., and Odeh, A.S.: "The Material Balance as an Equation of a Straight Line," *JPT* (August 1963) 896–900.
8. Poston, S.W., and Chen, H.Y.: "Case History Studies: Abnormal Pressured Gas Reservoirs," paper SPE 18857 presented at the SPE Production Operations Symposium, Oklahoma City, Oklahoma, (March 13–14, 1989).
9. Fetkovich, M.J., Reese, D.E., and Whitson, C.H.: "Application of a General Material Balance for High-Pressure Gas Reservoirs," *SPEJ* (March 1998) 3–13.

PROBLEMS

10-1. Calculate the OOIP of an undersaturated sandstone reservoir with the following parameters:

Area of reservoir, A	5500 acres
Net pay thickness, h	65 feet
Average porosity, ϕ	30%
Initial water saturation, S_{wi}	27%
Initial oil formation volume factor, B_{oi}	1.44 RB/STB

10-2. The OOIP of a volumetric, undersaturated reservoir was calculated to be 275 × 10^6 STB by volumetric methods at initial reservoir pressure of 3600 psia. The PVT data of the oil are given in Table 10P.1. If the current reservoir is 3400 psia after 732,000 STB of oil have been produced, calculate the cumulative oil that would be produced when the reservoir pressure has declined to 2700 psia. Assume water and rock compressibility are negligible.

TABLE 10P.1 PVT for Problem 10-2

Pressure (PSIA)	Solution GOR (scf/STB)	Oil FVF (RB/STB)
3600	567	1.310
3200	567	1.317
2800	567	1.325
2500	567	1.333
2400	554	1.310
1800	434	1.263
1200	337	1.210
600	223	1.140
200	143	1.070

10-3. The oil PVT data of a super-undersaturated volumetric reservoir in deepwater, Gulf of Mexico, USA, are given in Table 10P.2. A summary of other reservoir properties is provided in Table 10P.3. Calculate the reservoir recovery factors at abandonment pressure of 8000 psia assuming:

a. Water and formation rock compressibility are negligible
b. Water and formation rock compressibility are not negligible
c. Compare the recovery factors from a and b. Discuss the impact of water and formation rock compressibility factors on recovery factors.

TABLE 10P.2 PVT Data for Problem 10-3

Pressure (PSIA)	Oil FVF (RB/STB)	Soln. GOR (scf/STB)
15,000	1.152	535
14,000	1.158	535
13,000	1.164	535
12,000	1.170	535
11,000	1.176	535
10,000	1.182	535
9000	1.188	535
8000	1.194	535
7000	1.200	535
6000	1.206	535
5000	1.212	535
4000	1.218	535

TABLE 10P.3 Reservoir Properties for Problem 10-3

Initial reservoir pressure, p (psia)	15,000
Bubble point pressure, (psia)	1500
Reservoir temperature, T (°F)	245
Porosity, ϕ	0.27
Permeability, k (md)	105
Initial water saturation, S_{wi}	0.35
Water compressibility, c_w (psi^{-1})	4.5×10^{-6}
Formation compressibility, c_f (psi^{-1})	16.7×10^{-6}

10-4. A Stevens Zone reservoir in Kern County, California, USA, was discovered with initial reservoir pressure of 2200 psia, and bubble point pressure of 1850 psia. The production history of the reservoir showed negligible water production and no evidence of water influx from an aquifer. The water and formation rock compressibility are small and negligible. Total production from the reservoir was 180×10^6 STB oil and 147.65×10^9 scf gas as reservoir pressure declined from initial pressure of 1850 psia to 1300 psia. The PVT data for the oil are given in Table 10P.4. Calculate:

a. Oil-in-place in the reservoir at 1850 psia

b. Average net formation thickness of the reservoir given an average porosity of 11%, a connate water saturation of 29%, and a productive area of 49,000 acres.

TABLE 10P.4 PVT Data for Problem 10-4

Pressure (PSIA)	Oil FVF (RB/STB)	Soln. GOR (scf/STB)	Gas FVF (RB/scf)
2200	1.383	727	—
1850	1.388	727	0.00130
1600	1.358	654	0.00150
1300	1.321	563	0.00182
1000	1.280	469	0.00250
700	1.241	374	0.00375
400	1.199	277	0.00691
100	1.139	148	0.02495
40	1.100	78	0.05430

10-5. The case history of the S10A2 reservoir was presented in Section 10.8. Using the data given in the case history, calculate:

a. OOIP by volumetric methods

b. OOIP using the GMBE. Assume infinite-acting aquifer

c. OOIP using the Havlena-Odeh method. Assume infinite-acting aquifer

d. Compare OOIPs from a, b, and c. What could be the reasons for differences, if any?

e. The cumulative production from the S10A2 reservoir is low compared to the OOIP. What are the potential factors related to the reservoir that could have limited the productivity of the reservoir?

f. Is the S10A2 reservoir a good candidate for a new well or a side-track well? (Consider oil recovery only as the basis for your answers. No economic evaluation is required).

g. What are the potential effects of the Manatee reservoirs located in the same basin as the Angus and Troika reservoirs (Figure 10.6)? Could the reservoirs communicate hydraulically?

10-6. The case history of the S10B reservoir was presented in Section 10.8. Applying the data given in the case history, calculate:

a. OOIP by volumetric methods

b. OOIP using the GMBE. Assume infinite-acting aquifer

c. OOIP using the Havlena-Odeh method. Assume infinite-acting aquifer

d. Compare OOIPs from a, b, and c. What could be the reasons for any differences, if any?

e. Is the S10B reservoir depleted? Are there resources left for additional recovery?

f. What are the potential effects of the Manatee reservoirs located in the same basin as the Angus and Troika reservoirs (Figure 10.6)? Could the reservoirs communicate hydraulically?

CHAPTER

11 Production Forecasting for Conventional and Unconventional Reservoirs

11.1 Introduction

Production forecasting is a practice undertaken by many engineers to estimate expected production from a well, groups of wells, reservoirs, and fields. The forecasts are used with economic analyses to justify development of assets, drilling of a single well or multiple wells, remediation of a well or groups of wells, facilities improvement, abandonment decisions, buying or selling of assets, exploration of new areas, reserves determination and classification, etc. The main point is that production forecasting plays a central and crucial role in all aspects of finding, developing, producing, buying, selling, and abandoning of all types of petroleum reservoirs/resources. Production forecasting is a very important function performed regularly by petroleum professionals in various capacities depending on the type and size of the organization. The degree and depth of participation by the petroleum professional depends on whether the organization is a major oil company, an independent oil company, a governmental agency, a country, or a regulatory agency. Irrespective of the type and size of the organization, the practice of production forecasting uses similar processes although the level and sophistication of the methods employed vary between organizations.

The most common method of production forecasting is described as Decline Curve Analysis (DCA). The foundations of most DCA can be traced back to decline equations published by Arps[1] in 1944. Arps' decline equations are empirical equations verified with actual production data published earlier by Cutler.[2] Other DCA methods based on type curves were published later notably by Fetkovich et al.[3] But DCA based on Arps' equations remain the workhorse of the petroleum profession for production forecasting of conventional and unconventional reservoirs. As will be shown later in the chapter, different forms of Arps' equations work for conventional reservoirs and separate forms of the equations work for unconventional reservoirs.

In this chapter, Arps' decline equations will be presented. Other decline models such as the Duong model and Stretch Exponential Decline Model (SEDM) that work better for unconventional wells during transient linear flow will also be presented. Most of the material on DCA will be presented with focus on unconventional wells. But these materials can be adapted and applied to conventional wells as well since they exhibit boundary dominated flow (BDF) characteristics in most cases.

11.2 Flow Regimes

In Chapter 12, Section 12.3, the definitions and characteristics of three flow regimes are presented. These are transient (or unsteady state) flow, pseudosteady-state flow, and steady-state flow. The reader may review the full description of these flow regimes presented there if needed. The characteristics of transient flow are that the pressure transients created by initial constant rate production at the well have not reached the outer boundary of the reservoir, that the pressure distribution in the reservoir is not constant but depends on elapsed time and distance from the well since production was initiated. For pseudosteady-state (PSS) flow, the characteristics are that the pressure transients have reached **all** the outer boundaries of the reservoir, and that the pressure distributions are changing at **constant** rate. Pseudosteady-state flow is related to BDF. PSS assumes constant flow rate at the well with varying bottomhole flowing pressure (BHFP) while BDF assumes variable well flow rate with constant BHFP. The relationship between PSS and BDF will be discussed later in this chapter. In steady-state flow regime, the characteristics are that the pressure distribution at any point within the established drainage area of the well is constant and does not change with time, that the pressure at the reservoir boundary is constant, and fluid flow into the reservoir (or drainage area of the well) is equal to constant production rate at the well.

The radial flow equation for a well producing a slightly compressible fluid at constant rate under transient flow conditions in an infinite-acting reservoir is derived in Section 13.3 as Eq. (13.25) and repeated here as:

$$p_{(w,t)} = p_i - \frac{162.6qB\mu}{kh}\left[\log\frac{kt}{\phi\mu c_t r_w^2} - 3.23 + 0.8691s\right] \tag{11.1}$$

A similar equation for a well producing a slightly compressible fluid at constant rate under PSS flow is derived in Section 12.8.1.2 as Eq. (12.75) and represented here as:

$$\bar{p} = p_w + \frac{141.2qB\mu}{kh}\left[\ln\left(\frac{r_e}{r_w}\right) - 0.75 + s\right] \tag{11.2}$$

In Eqs. (11.1) and (11.2), p = pressure, psia; q = flow rate, STB/D; B = formation volume factor, RB/STB; μ = viscosity, cp; k = permeability, md; h = formation thickness, ft;

ϕ = porosity, fraction; c_t = total compressibility, psi^{-1}; r_w = wellbore radius, ft; r_e = reservoir boundary, ft; t = time, hr; s = skin factor; and $\bar{p}$ = average pressure, psia.

For compressible fluids such as gases, transient radial flow equation at constant rate was derived in terms of normalized pseudo-pressure in Section 13.4.2.2 as Eq. (13.79). The same equation is reproduced here as:

$$p_{pnwf} = p_{pni} - \frac{162.6 q_g \bar{B} \bar{\mu}}{kh}\left[\log\frac{kt}{\phi\bar{\mu}\bar{c}_t r_w^2} - 3.23 + 0.8691 s'\right] \tag{11.3}$$

In Eq. (11.3), p_{pnwf} = normalized flowing bottomhole pseudo-pressure, psia; p_{pni} = normalized initial reservoir pseudo-pressure, psia; q_g = gas flow rate, Mcf/D; $\bar{B}$ = gas formation volume factor at average reservoir pressure, RB/Mcf; $\bar{\mu}$ = gas viscosity at average reservoir pressure, cp; k = permeability, md; h = formation thickness, ft; t = time, hr; ϕ = porosity, fraction; $\bar{c}_t$ = total compressibility at average reservoir pressure, psi^{-1}; r_w = wellbore radius, ft; $s' = s + Dq_g$, as defined in Eq. (13.73).

For PSS flow of compressible fluids such as gases at constant rate, Eq. (12.94) is reproduced here as:

$$\bar{p}_p - p_{pwf} = \frac{q_g T p_{sc}}{1.98\times10^{-3} khT_{sc}}\left[\ln\frac{0.472 r_e}{r_w}\right] \tag{11.4}$$

In Eq. (11.4), $\bar{p}_p$ = average reservoir pseudo-pressure, psia; p_{pwf} = bottomhole flowing pseudo-pressure, psia; q_g = gas flow rate, Mcf/D; T = reservoir temperature, °R; p_{sc} = standard pressure, psia; k = permeability, md; h = formation thickness, ft; T_{sc} = standard temperature, °R; r_e = reservoir boundary, ft; and r_w = wellbore radius, ft.

For Eqs. (11.2) and (11.4), average reservoir pressure can be calculated using either Eq. (12.95) or Eq. (12.96).

The time to reach pseudo-steady state assuming radial flow can be estimated from Eq. (12.8) as:

$$t_{pss} = \frac{948\phi\mu c_t r_e^2}{k} \tag{11.5}$$

The units for Eq. (11.5) are as stated for Eq. (12.8).

Equations (11.1) to (11.4) were derived assuming constant rate production at the well. Except under special circumstances, most wells are produced at constant bottomhole pressure. This condition is the same as the constant terminal pressure solution presented in Section 12.8.2. There is a technical distinction drawn when describing pressure distribution in the reservoir depending on whether the well is produced at constant rate or constant pressure. If the well is produced at constant rate, the pressure response is described as pseudo-steady state whereas pressure response from constant pressure production is described as boundary-dominated flow (BDF). Anderson and Mattar[4] showed that the concept of **Material Balance Time** (MBT) as developed by Blasingame et al.[5] can be used to perform exact transformation of constant pressure data to constant rate data during BDF regime. Material balance time is calculated as cumulative production divided by production rate in corresponding units:

For an oil well:

$$\text{MBT} = \frac{N_p}{q_o} \tag{11.6}$$

For a gas well:

$$\text{MBT} = \frac{G_p}{q_g} \tag{11.7}$$

In Eqs. (11.6) and (11.7), N_p = cumulative oil production; q_o = oil production rate; G_p = cumulative gas production; and q_g = gas production rate, are all in corresponding units. The application of material balance time in making diagnostic plots is presented in Section 11.3.6 .

Fetkovich[6] developed an equation for variable rate, constant bottomhole pressure shown here as:

$$\frac{q}{q_i} = \exp\left[\frac{-q_i t}{N_{pi}\left(1 - \frac{p_{wf}}{p_i}\right)}\right] \tag{11.8}$$

where N_{pi} is:

$$N_{pi} = \frac{\pi\left(r_e^2 - r_w^2\right)\phi c_t h p_i}{5.615 B_o} \tag{11.9}$$

Equation (11.8) can be reduced to a form similar to Arps' exponential decline equation [Eq. (11.17)] by setting flowing bottomhole pressure equal to zero, ($p_{wf} = 0$), which then gives:

$$\frac{q}{q_i} = \exp\left[\frac{-(q_i)_{\max} t}{N_{pi}}\right] \tag{11.10}$$

In Chapter 15, Section 15.4, four flow regimes that can occur in a fractured well have been identified as:

1. Fracture linear flow
2. Bilinear flow
3. Formation linear flow
4. Pseudo-radial flow

More details on the duration and characteristics of these flow regimes are presented in Section 15.4. As stated there, fracture linear flow is short-lived and does not have much practical application in analyzing production from fractured wells. Bilinear and formation linear flows can occur in fractured wells producing from unconventional reservoirs. Pseudo-radial flow is likely to occur in conventional reservoirs and less likely to occur in unconventional reservoirs. The flow regime that may occur in fractured wells producing from unconventional reservoirs is boundary dominated flow which was discussed earlier.

The equation for bilinear flow in an oil well producing at constant rate has been derived in Section 15.6.1 as Eq. (15.14) and represented here as:

$$\Delta p = \frac{44.1 qB\mu}{h\left(k_f w_f\right)^{1/2}\left(\phi\mu c_t k\right)^{1/4}} t^{1/4} \tag{11.11}$$

The units for Eq. (11.11) are as described for Eq. (15.14) in Section 15.6.1. A similar equation for gas wells is presented as Eq. (15.15). Note that Eq. (11.11) shows that a log-log plot of Δp vs. t will give a straight line with a quarter-slope as shown in Figure 15.4. This is a diagnostic plot that can be used to show the presence of bilinear flow in a fractured well.

The equation for linear flow in an oil well producing at constant rate was derived in Section 15.6.3 as Eq. (15.22) and reproduced here as:

$$\Delta p = \frac{4.064 qB}{hL_f}\sqrt{\frac{\mu t}{k\phi c_t}} \tag{11.12}$$

The units for Eq. (11.12) are the same as for Eq. (15.22) in Section 15.6.3. A similar equation for gas wells is presented there as Eq. (15.23). A log-log plot of Δp vs. t will yield a half-slope straight line as shown in Figure 15.7. This is a diagnostic plot that can be used to identify linear flow in a fractured well.

11.3 Arps' Decline Equations

In 1945, Arps[1] proposed a set of equations that have been widely used to forecast production for wells producing from conventional and unconventional reservoirs. For all the equations, Arps defined the decline rate, D, as:

$$D = -\frac{dq/dt}{q} \tag{11.13}$$

Arps further defined an approximately constant positive parameter, b, as:

$$b = -\frac{d\left(\dfrac{q}{dq/dt}\right)}{dt} \tag{11.14}$$

Substituting Eq. (11.13) into Eq. (11.14) gives:

$$b = \frac{d\left(1/D\right)}{dt} \tag{11.15}$$

Assuming that the parameter b is constant, Arps integrated Eqs. (11.13) and (11.15) to get:

$$q(t) = \frac{q_i}{\left(1 + bD_i t\right)^{1/b}} \tag{11.16}$$

Equation (11.16) is the Arps hyperbolic decline equation. Note that this is an empirical equation based on well production data analyzed by Arps. It has been observed later that the parameter b is not constant and may vary over the productive life of a well.

For the special case where the parameter $b = 0$, Eq. (11.16) reduces to:

$$q(t) = q_i e^{-D_i t} \tag{11.17}$$

Eq. (11.17) is the Arps exponential decline equation.

For another special case where the parameter $b = 1$, Eq. (11.16) becomes:

$$q(t) = \frac{q_i}{\left(1 + D_i t\right)} \tag{11.18}$$

Eq. (11.18) is the Arps harmonic decline equation.

The Arps[7] equation for cumulative production for the hyperbolic decline case is:

$$N_p = \frac{q_i^b}{(1-b)D_i}\left(q_i^{1-b} - q_t^{1-b}\right) \tag{11.19}$$

The cumulative production equation for the exponential decline case is:

$$N_p = \frac{q_i - q_t}{D_i} \tag{11.20}$$

For the harmonic decline case, the cumulative production equation is:

$$N_p = \frac{q_i}{D_i}\ln\left(\frac{q_i}{q_t}\right) \tag{11.21}$$

In Eqs. (11.19) to (11.21), cumulative oil production, N_p, may be replaced with, G_p, for cumulative gas production.

It is important to note that Arps' decline equations were not derived from any fluid flow or material balance equations that are routinely applied in many reservoir engineering analyses. Again, Arps' decline equations were derived on the basis of empirical analyses of production data published earlier by Cutler[2] as reported by Arps.[1] Since the publication of Arps' decline equations, many researchers have attempted to link these equations to familiar reservoir engineering analytical equations. One of the noteworthy effort in this area was undertaken by Fetkovich. Earlier, it was noted that Eq. (11.10) developed by Fetkovich[6] is similar in form to Arps' exponential decline equation. In this case, the Arps' decline rate, D_i, is defined as:

$$D_i = \frac{(q_i)\max}{N_{pi}} \quad (11.22)$$

Under strict constraints, Arps' decline models can be theoretically applicable if the following conditions are satisfied:

1. Constant flowing bottomhole pressure
2. Boundary dominated flow (BDF) is present
3. Constant fluid and rock properties
4. Homogeneous, isotropic formations
5. Parameter b is constant
6. Fully bounded reservoir with no influx
7. Constant productivity index

Most wells producing from conventional or unconventional reservoirs do not meet these conditions. For instance, it is highly unlikely that producing wells can maintain constant flowing bottomhole pressure for any reasonable length of time. Further fluid properties change as reservoir pressure is depleted especially for gas reservoirs. For wells producing from unconventional reservoirs, BDF may take many months or even years to be reached or in some cases BDF is never attained. For such wells in very low permeability reservoirs, the wells could produce under the transient flow regime for a very long time and never reach BDF.

Even though **all** these conditions are never met in actual producing wells, Arps' decline model remains the most common method used by many engineers to forecast oil and gas production for wells producing from conventional and unconventional reservoirs. The reasons for the popularity of Arps' equations are that they are very easy to apply and can be fitted to many production profiles. They have also been used in many commercial software for forecasting reserves and performing economic analyses. Other forecasting tools, such as numerical simulation, are readily available for the same purposes but they do not match the simplicity of the Arps' models. These empirical models have been in use in the industry since 1945 and are expected to be used in the petroleum industry for many more years in the future.

11.3.1 Range of Values for Arps' *b* Parameter

Arps[1] reported in his original work that the range of values for the parameter b was 0.0 to 0.7 for all the 149 wells that were analyzed from the data assembled by Cutler.[2] The range of values for the parameter b for 121 wells was between 0.0 and 0.4. None of the wells had a value of parameter b higher than 0.7. Arps reported that a parameter b value equal to 1.0 is very rare but can occur. It is important to note that the well data used in Arps' analyses were most likely obtained from reservoirs that are now classified as conventional due to high formation permeabilities. During that era, only conventional reservoirs were considered commercial to develop and produce. Unconventional reservoirs were not usually produced due to low permeabilities and non-commercial production rates.

The range of values for the parameter b reported by Arps and subsequent investigators has fueled the discussion on whether the value of parameter b can be greater than one (1.0). Later investigators such as Long and Davis[8] have reported that parameter b values greater than one have been calculated from many reservoirs. The trend in parameter b values higher than one seems to follow the development of low permeability formations found in unconventional reservoirs. Most wells producing from unconventional reservoirs exhibit transient behavior for a long time (for months and even years) and it has been observed that parameter b values higher than one is needed to fit their production data. Theoretically, parameter b value of 2 would fit transient linear flow, and parameter b value of 4 would fit transient bilinear flow. The point here is that there is no actual limit on the value of parameter b necessary to fit transient production data. However, it is advisable to employ a range of parameter b values that is reasonable and customary for wells producing from a reservoir or a basin.

The implication of using parameter b values greater than one in forecasting cumulative production with hyperbolic equation is that the result is unbounded as rate declines [see Eq. (11.19)]. Even for harmonic decline model where parameter b is equal to one, cumulative production is also unbounded as rate declines [see Eq. (11.21)]. To avoid this problem in making long-term forecast, Long and Davis[8] recommended switching from hyperbolic decline to exponential decline at a predetermined minimum terminal decline rate. This is the reason for the wide application of minimum terminal decline rate, (D_{min}), in making long-term forecasts for wells producing from unconventional reservoirs. Selection of minimum terminal decline rate in production forecasting is arbitrary and varics among organizations. However, Hale et al.[9] have published a review of historical terminal decline rates for unconventional reservoirs in most producing basins in the United States of America. This study should be used as a source of minimum terminal decline rates if more reliable data specific to the reservoir or basin are not available.

11.3.2 Effective and Nominal (Exponential) Decline Rates

It is important to draw a distinction between effective and nominal decline rates. Effective decline rate, D_e, is defined as:

$$D_e = \frac{q_i - q_t}{q_i} \quad (11.23)$$

In Eq. (11.23), q_i is the initial production rate, and q_t is the production rate at a future time, t. Thus, effective decline rate is the decline in production over a time period. It can also be considered the proportion by which production rate has decreased over a stated time period. Nominal decline rate, D, is defined with the aid of the Arps' exponential equation as:

$$q_t = q_i \exp(-Dt) \tag{11.24}$$

Nominal decline rate is a continuous function as defined by Eq. (11.24). For an exponential decline, it is constant over time. A relationship can be developed between effective and nominal decline rates as follows. Equation (11.23) can be rearranged simply as:

$$D_e = 1 - \frac{q_t}{q_i} \tag{11.25}$$

Substituting Eq. (11.24) into Eq. (11.25) at $t = 1$ gives:

$$D_e = 1 - \exp(-D) \tag{11.26}$$

Equation (11.26) is rearranged to give:

$$D = -\ln(1 - D_e) \tag{11.27}$$

This equation can be used to convert effective decline rate into nominal decline rate and vice-versa. Note that effective decline rates are frequently used in most petroleum industry financial reports.

11.3.3 Type Curves for Analysis of Production Data

The concept and applications of type curves will be introduced and discussed in Chapter 14 for well test analysis. In Chapter 14, the focus is on the use of type curves to analyze pressure transient data. Here the application of type curves is for analyses of production data. There are several versions of type curves used to analyze production data. The prominent and widely used type curves are:

1. Fetkovich[6] type curves
2. Blasingame[10] type curves
3. Agarwal-Gardner[11] type curves

The main characteristics of type curves will be described generally with the Fetkovich type curves. Specific features of Blasingame et al.[10] and Agarwal-Gardner[11] type curves are covered in more details in the cited references. A version of the Fetkovich type curve updated by Chen and Teufel[12] is provided in Figure 11.1 for the discussions. The axes of Figure 11.1 are dimensionless decline flow rate, q_{dD}, and dimensionless decline time, t_{dD}. The dimensionless decline flow rate is expressed as:

$$q_{dD} = \frac{q(t)}{q_i} = q_D \left[\ln\left(\frac{r_e}{r_w}\right) - \frac{1}{2} \right] \tag{11.28}$$

where,

$$q_D = \frac{141.3 q(t) \mu B}{kh(p_i - p_{wf})} \tag{11.29}$$

Substituting Eq. (11.29) into Eq. (11.28) gives:

$$q_{dD} = \frac{q(t)}{\dfrac{kh(p_i - p_{wf})}{141.3\mu B \left[\ln\left(\dfrac{r_e}{r_w}\right) - \dfrac{1}{2} \right]}} \tag{11.30}$$

and dimensionless decline time is expressed as:

$$t_{dD} = \frac{t_D}{\dfrac{1}{2}\left[\left(\dfrac{r_e}{r_w}\right)^2 - 1\right]\left[\ln\left(\dfrac{r_e}{r_w} - \dfrac{1}{2}\right)\right]} \tag{11.31}$$

where,

$$t_D = \frac{0.00634 kt}{\phi \mu c_t r_w^2} \tag{11.32}$$

In Eqs. (11.28) to (11.32), p_i = initial reservoir pressure, psia; p_{wf} = flowing bottomhole pressure, psia; $q(t)$ = flow rate, STB/D; B = formation volume factor, RB/STB; μ = viscosity, cp; k = permeability, md; h = formation thickness, ft; ϕ = porosity, fraction; c_t = total compressibility, psi^{-1}; r_w = wellbore radius, ft; r_e = reservoir radius, ft; and t = time, days.

The left side of Figure 11.1 represents transient flow plotted on various values of r_{eD} stems. The dimensionless parameter, r_{eD}, is defined as:

$$r_{eD} = \frac{r_e}{r_w} \tag{11.33}$$

To account for wellbore damage, this is sometimes defined as:

$$r_{eD} = \frac{r_e}{r_{wa}} \tag{11.34}$$

where,

$$r_{wa} = r_w e^{-s} \tag{11.35}$$

In Eq. (11.35), r_{wa} = apparent wellbore radius, ft; and s = skin factor.

The left-side stems are analytical solutions for transient radial flow charactèrized by the r_{eD} stems. Note that the transient stems are concave upward and coalesces to a single stem at approximately $t_{dD} = 0.3$. Chen and Teufel[12] modified the original Fetkovich type curves by including a stem that represents transient linear flow.

The right side of Figure 11.1 represents BDF. The stems are characterized by Arps' parameter b values of 0 to 1. Note that BDF stems are concave downward. The importance of Fetkovich type curves is that transient and BDF stems are represented on the same plot and can be used to match production data that include transient and BDF regimes.

This was the fundamental work by Fetkovich[6] that led to the use of type curves for production data analyses. Later investigators have followed this approach.

Most type curve matching are done on computer screens by using readily available commercial software. It is highly unlikely that any engineer will perform manual type curve matching in the current computerized era (same as in pressure transient analysis as pointed out in Chapter 14.7). Type curve matching on computer screens is fast, visual, and efficient. It allows for matching the same data with other type curves. This approach gives results that can be compared leading to better interpretation of flow regimes, reservoir properties, and production forecasts. Computer-based type curve matching follows specific steps which are similar to manual type curve matching steps. For this reason, new practitioners of type curve matching are advised to review the approach (steps) recommended by Mattar and Anderson.[13]

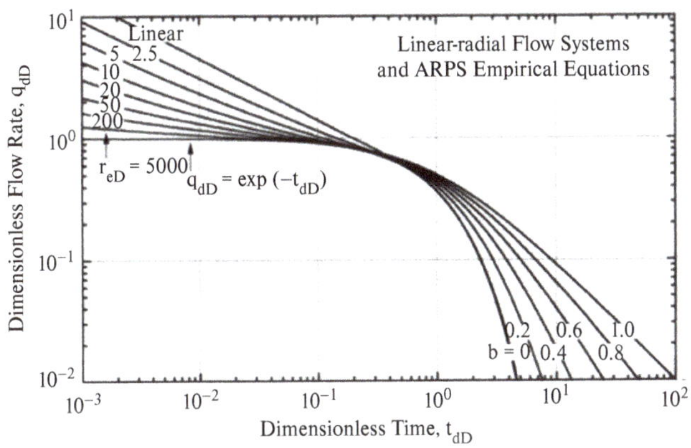

▲ **FIGURE 11.1 Fetkovich type curve (from Chen and Teufel[12] © 2000 SPE, Reproduced with permission)**

11.3.4 Normalized Rates and Pressures

The most common parameters used in the analyses of production data are normalized rates and normalized pressures. Normalized rate is simply calculated as production rate divided by pressure difference, thus:

$$\text{Normalized Rate} = \frac{q}{p_i - p_{wf}} \tag{11.36}$$

Normalized pressure, which is the inverse of normalized rate, is then calculated as:

$$\text{Normalized Pressure} = \frac{p_i - p_{wf}}{q} \tag{11.37}$$

In Eqs. (11.36) and (11.37), p_i = initial reservoir pressure; p_{wf} = flowing bottomhole pressure; and q = production rate. Note that for gas wells, pressures are transformed to the corresponding real gas pseudo-pressures. Real gas pseudo-pressure has been defined in Eq. (12.86) as:

$$p_p = 2\int_{p_b}^{p} \frac{p}{\mu(p)z(p)}\,dp \tag{11.38}$$

In terms of real gas pseudo-pressures, Eq. (11.36) becomes:

$$\frac{q_g}{p_{pi} - p_{pwf}} \tag{11.39}$$

It is recommended practice to use normalized rates and normalized pressures in making diagnostic plots for production data analyses. Normalization of rates and pressures accounts for the effects of declining flowing bottomhole pressures on the variability of production rate. However, for many wells, reliable pressure data are not measured or not reported. Consequently, for many diagnostic plots only the production rate data are available. Note that the use of only rate data could introduce variability and errors in the diagnostic plots.

11.3.5 Flowing Material Balance

The Flowing Material Balance method (FMB) is a production data analysis technique derived from the Agarwal-Gardner Rate Cumulative type curves.[11] FMB can be used to estimate original fluids-in-place (oil or gas) by extrapolating the linear data trend on Cartesian plot of normalized rate versus normalized cumulative production to intersect the *x*-axis. The intersection point on the *x*-axis (which is normalized cumulative production) yields the original fluids-in-place. The method does not require shut-in pressure data except initial reservoir pressure. A derivation of the FMB is provided by Mattar and Anderson.[13]

The FMB for a well producing undersaturated oil is:

$$\frac{q_o}{\Delta p} = -\frac{Q}{\Delta p c_t} \times \frac{1}{N b_{pss}} + \frac{1}{b_{pss}} \tag{11.40}$$

where b_{pss} is:

$$b_{pss} = \frac{141.2 B\mu}{kh}\left(\ln\frac{r_e}{r_w} - \frac{3}{4}\right) \tag{11.41}$$

A plot of $q/\Delta p$ versus $Q/(c_t \Delta p)$ gives a straight line with an *x*-intercept of N as the oil-in-place. In Eqs. (11.40) and (11.41), q_o = oil production rate in STB/D; Δp = flowing pressure drop, psia; Q = cumulative production, STB; c_t = total compressibility, psi^{-1}; N = Original oil-in-place, bbl; B = formation volume factor, RB/STB; μ = oil viscosity, cp; k = permeability, md; h = formation thickness, ft; r_e = reservoir radius, ft; and r_w = wellbore radius, ft.

Similarly, for a gas well:

$$\frac{q_g}{\Delta p_p} = -\frac{2 q_g t_{ca} p_i}{(c_t \mu z)\Delta p_p} \times \frac{1}{G b'_{pss}} + \frac{1}{b'_{pss}} \tag{11.42}$$

where b'_{pss} is:

$$b'_{pss} = \frac{1.417 \times 10^6 \times T}{kh}\left(\ln\frac{r_e}{r_w} - \frac{3}{4}\right) \tag{11.43}$$

and material balance pseudo-time, t_{ca}, is defined as:

$$t_{ca} = \frac{\mu c_t}{q_g} \int_0^t \frac{q_g dt}{\bar{\mu} \bar{c}_t} \quad (11.44)$$

The parameters $\bar{\mu}$ and $\bar{c}_t$ are evaluated at average reservoir pressure.

A plot of $q_g/\Delta p$ versus $2q_g t_{ca} p_i / (c_{ti} \mu_i z_i \Delta p_p)$ gives a straight line with an x-intercept equal to G_i as the initial gas-in-place. In Eqs. (11.42) to (11.44), q_g = gas production rate, MMscf/D; μ = gas viscosity, cp; z = gas compressibility factor; Δp_p = flowing pseudo-pressure drop, psi^2/cp; t_{ca} = material balance pseudo-time, days; G = original gas-in-place, MMscf; and T = reservoir temperature, °R.

11.3.6 Diagnostic Plots

The most common diagnostic plot used to identify flow regimes from well production data is a log-log plot of normalized rates versus material balance times. The equations for calculating normalized rates and material balance times were shown as Eqs. (11.36) and (11.6), respectively. From this plot, bilinear flow regime with quarter slope, linear flow regime with half slope, and BDF flow regime with unit slope can be identified. It is important that the material balance times used in making the plot should be increasing monotonically. Erratic data created by operational changes such as low production rates resulting from partial shut-ins should be filtered out from the production data. For many wells, flowing bottomhole pressures (FBHP) are not measured or not reported. In that case, an alternative plot of production rate versus material balance times should be used instead. An example plot of rate versus material balance time for a well in the Barnett Shale is shown in Figure 11.2. Note that in this plot, bilinear flow denoted with a quarter slope line was not observed. Bilinear flows have short duration and rarely occur in actual data. In many cases, appearance of bilinear flow is caused by unloading of completion fluids (clean-up period) and rapidly decreasing flowing bottomhole pressures. This apparent appearance of bilinear flow should be ignored unless it can be shown from production data that the well has cleaned up sufficiently and FBHP has stabilized. It is important to emphasize that the appropriate parameter to use in making diagnostic plots is normalized rates if pressure data are available. Normalized rates account for changes in production rates caused by changing FBHP.

Another diagnostic plot that can be made is a Cartesian plot of normalized pressures versus the square root of time. Equation (11.12) shows the resulting plot will give a straight line that passes through the origin (see Figure 15.8). If we denote the slope of the straight line as m_{Lf} then:

$$m_{Lf} = \frac{4.064B}{hL_f} \sqrt{\frac{\mu}{k\phi c_t}} \quad (11.45)$$

If the area of the fracture can be estimated, permeability of the system can be calculated from Eq. (11.45).

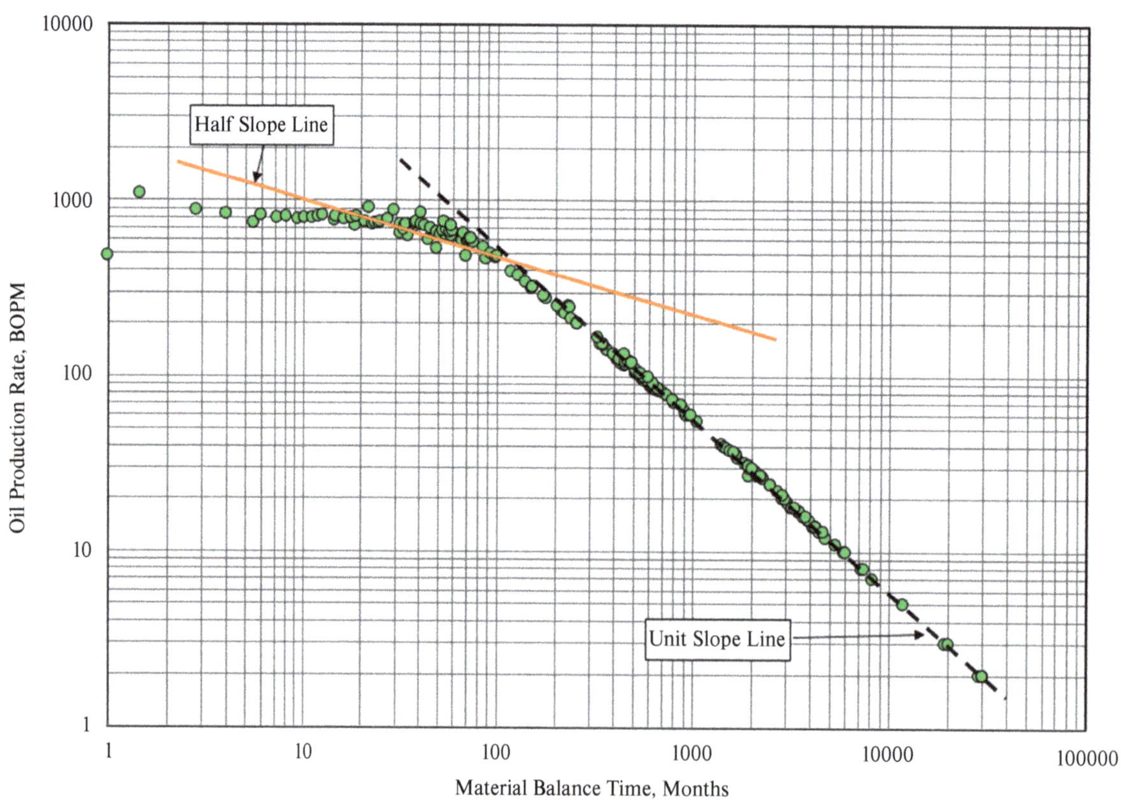

▲ **FIGURE 11.2 Diagnostic log-log plot of oil rate vs. material balance time**

11.3.7 Transient Decline Models

Since earlier Arps' decline models, transient decline models have been proposed recently to improve matching of data primarily from the transient flow regimes. The prominent models in this group are:

1. The Power Law Loss Ratio Model by Ilk et al.[14]
2. The Stretched Exponential Decline Model by Valko[15]
3. The Duong Model by Duong[16]

The Power Law Loss Ratio Model is complex, difficult to use, and does not have an analytical solution for cumulative production (Russel et al.[18]). For these reasons, it is not presented in this book.

11.3.7.1 The Stretched Exponential Decline Model (SEDM)

The SEDM proposed by Valko[15] in 2009 is a three-parameter equation represented as:

$$q = q_i \exp\left[-\left(\frac{t}{\tau}\right)^n\right] \tag{11.46}$$

Rearranging Eq. (11.46) gives:

$$\ln\left(\frac{q}{q_i}\right) = -\left(\frac{t}{\tau}\right)^n \tag{11.47}$$

Taking logarithm of both sides of Eq. (11.47) gives:

$$\log\left[\ln\left(\frac{q_i}{q}\right)\right] = \log\left[\left(\frac{t}{\tau}\right)^n\right] \tag{11.48}$$

Expanding Eq. (11.48) yields:

$$\log\left[\ln\left(\frac{q_i}{q}\right)\right] = n\log[t] - n\log[\tau] \tag{11.49}$$

In Eqs. (11.46) to (11.49), $q =$ production rate; $q_i =$ production rate at $t = 0$; $\tau =$ characteristic time with the same units as t; and $n =$ model parameter. From Eq. (11.49), a log-log plot of $\ln(q_i/q)$ vs. t may yield a straight line. The slope of the straight line gives the parameter n and the intercept of the line can be used to calculate τ.

For cumulative production, Eq. (11.46) is integrated to give:

$$Q = \frac{q_i \tau}{n}\left\{\Gamma\left[\frac{1}{n}\right] - \Gamma\left[\frac{1}{n}, \left(\frac{t}{\tau}\right)^n\right]\right\} \tag{11.50}$$

In Eq. (11.50), $Q = N_p$ or G_p, for cumulative oil or gas production, respectively. In Eq. (11.50), $\Gamma\left[\frac{1}{n}\right]$ is the "complete" gamma function, and $\Gamma\left[\frac{1}{n}, \left(\frac{t}{\tau}\right)^n\right]$, is the "incomplete" gamma function. The gamma functions are available from Microsoft Excel™ spreadsheet and other mathematical software packages. The SEDM is complex and difficult to use especially for calculating cumulative production using the gamma functions.

A simplified procedure for using SEDM proposed by Yu et al.[17] is summarized as follows:

1. Assume a value for q_i close to stabilize peak production.
2. Plot $\log\left[\ln(q_i/q)\right]$ *vs.* $\log t$.
3. If a linear correlation is evident from the plot, the slope of the line is the parameter n and the intercept of the line is equal to the term, $-n\left[\log(\tau)\right]$. Calculate the parameter τ from the equation.

Application of the SEDM will be illustrated with Examples 11.1 and 11.2.

11.3.7.2 The Duong Model

Duong[16] observed that fracture flow regimes in a well can be modeled with an equation of the form:

$$q_o = q_1 t^{-n} \tag{11.51}$$

In Eq. (11.51), n = 0.5 for linear flow or n = 0.25 for bilinear flow, and q_1 is the flow rate at Day 1.

Oil cumulative production, N_p, is given by:

$$N_p = \int_0^t q_o dt = q_1 \frac{t^{1-n}}{(1-n)} \tag{11.52}$$

Substituting Eq. (11.51) into Eq. (11.52) gives:

$$\frac{q_o}{N_p} = \frac{(1-n)}{t} \tag{11.53}$$

Duong observed that field production data are better represented by an equation of the form:

$$\frac{q_o}{N_p} = a t^{-m} \tag{11.54}$$

In Eq. (11.54), a is a fitting parameter and m is always positive and close to unity. Eq. (11.54) indicates that a log-log plot of q_o/N_p vs. t will give a straight line with slope m and intercept a.

From Eq. (11.54), Duong[16] derived equations for rate q and cumulative production N_p as follows:

$$\frac{q_o}{q_1} = t^{-m} \exp\left[\frac{a}{1-m}\left(t^{1-m} - 1\right)\right] \tag{11.55}$$

$$N_p = \frac{q_1}{a} \exp\left[\frac{a}{1-m}\left(t^{1-m} - 1\right)\right] \tag{11.56}$$

From Eq. (11.55), a plot of q_o vs. $t^{-m} \exp\left[\frac{a}{1-m}\left(t^{1-m}-1\right)\right]$ yields q_1 as the slope of a line passing through the origin. With the parameters, a, m, and q_1 determined, Eq. (11.55) can be used to forecast future production and cumulative production can be calculated from Eq. (11.56). Note that, q_g and G_p for gas can be substituted for q_o and N_p respectively, in all of the above equations. Application of the Duong model will be illustrated with Examples 11.1 and 11.2.

11.3.8 Construction of Type Wells for Production Forecasting

A type well can be described as a representative well modeled to be a typical sample well producing from a reservoir, a geologic structure or formation, a geographical area or region, or a resource play. The production profile of the type well is assumed to represent the production profiles of wells producing or planned to be drilled in the target area. Type wells are used in making investment decisions especially on new resource plays, in planning of new developments, and in assignment/classification of reserves. A type well is assumed to represent the production profile of a group of wells producing from reservoirs or geologic structures with the same or similar characteristics. The type well should be constructed from a significantly large group of wells to maintain statistical validity. It is also equally important that the wells in the group have similar characteristics such as well type (vertical or horizontal), well location (sweet spots or not), vintage (close first oil dates), lateral well length (for horizontal wells), fracture size (number of fracture stages), fracture fluid type, completion technique, operational practices, etc. All these factors can affect the production profiles of wells in a group, so it is important to ensure that the wells have similar characteristics as much as possible.

The general industry practice for constructing a type well starts with creating a group of wells producing from a formation or region with similar geologic properties and similar reservoir characteristics. The well group is filtered down further by applying other factors such as well type, fracture size, first oil/gas date, completion technique, and other factors that could affect well performance as discussed earlier. The simplest approach used in constructing a type well from the group is to shift the production profile of each well to a common time frame which is termed time zero. After shifting all wells to the common reference time, the average production of the wells in the group at successive time steps is assigned as the production of the type well. Russell et al.[18] recommended that all wells in the group should be used in the averaging process including wells that have been shut-in or no longer in production to avoid constructing incorrect and optimistic type wells. The recommended method for construction of type wells could be represented in equation form as:

$$\text{Type well rate at time step} = \frac{\text{Sum of production rates of wells in group at time step}}{\text{Total number of wells in group at time step}} \qquad (11.57)$$

In Eq. (11.57), the time step typically represents months of production (1, 2, 3, ... months of production) assuming that the common reference time is zero month for all wells in the group. Note that other time step such as days or date could be used as long as all wells in the group have the same reference time or date.

Freeborn et al.[19] reported that type wells constructed as described with Eq. (11.57) could have implicit errors resulting from sequence bias. Sequence bias is created inadvertently when new wells are drilled in locations likely to have better productivity in early stages of development of an area. As the project progresses, new wells are likely to be drilled in less productive areas. This type of sequence bias is likely to create type wells that are optimistic since they are based on the production profile of the more productive wells. Another source of sequence bias is improvement in technology (drilling, fracturing, and completion technologies) as latter wells are drilled leading to better productivity. In this case, sequence bias will create type wells that are pessimistic since production profiles are based on earlier, less productive wells. If the drilling of new wells is relatively random, the sequence bias will have little or no influence on the type well curve generated with this process. However, the effect of sequence bias should be considered, and the type well constructed as described should be modified as necessary to avoid understating or overstating expected ultimate recovery (EUR).

Another source of errors in the construction of type wells as described by Freeborn et al.[19] is the presence of "Gap" wells. Gap wells are wells whose production history does not cover the entire time span over which the type well is being constructed. For instance, if the type well is being constructed over a 10-year time span, gap wells are wells whose production history may not cover the entire 10-year period mainly because they started producing latter than other wells in the group. The effect of gap wells can be eliminated by including the predicted production profile of such wells in the construction of type wells as recommended by Freeborn et al.[19]

Another type well that is commonly used in production forecasting is called percentile type well. Percentile type wells are type wells whose rate-time profile and EUR are referenced to a certain probability of occurrence. For instance, a P90 Type Well is a representative well that implies that there is a 90% probability that the well will recover its assigned EUR or higher. Similarly, a P10 Type Well represents a well with a 10% probability of recovering its designated EUR or higher. Percentile type wells can be generated by several methods. One method is to group wells with rate/time profiles and EURs that reflect the targeted EUR at the desired percentile. Using a large sample of wells through the process of aggregation in constructing percentile type wells creates a more representative type well as reported by Freeborn and Russell.[20]

11.3.9 Analyses of Well Production Data Using a Step-by-Step Procedure

This section gives a concise step-by-step procedure that can be used in the analyses of production data from unconventional wells. The section will be followed with two examples in which these steps were used to analyze production data from a gas well and an oil well. Note that these steps should be used as guides and should be modified as necessary to suit specific circumstances. The analyses of production data of unconventional wells leading ultimately to forecasting of future performance could be an arduous process depending on many variables. In the examples that follow, most of the key processes have been covered and should be sufficient to prepare the reader to undertake these tasks.

Step 1: Review production, pressure and all available data from the well for any anomalies and inconsistencies. This step is a methodical quality check of all data related to the well. The review should be comprehensive and thorough. For instance, instances of shut-in or disruption of production should be investigated. The impact of such disruptions on the production profile should be noted and understood. The review should include well history starting from drilling and completion of the well, initial production period, and any operator's practices that could influence the performance of the well. In some circumstances, detailed data on well history may not be available especially for data retrieved from public records. The point to keep in mind is that good knowledge of the well history and associated data helps in understanding the results generated from analysis of the production data.

Step 2: Prepare diagnostic plots from the production data, and pressure data if available. The recommended practice in making diagnostic plots is to plot normalized rate versus material balance time as discussed in Section 11.3.6. *Diagnostic Plots.* In many cases, pressure data may not be available. In such cases, a plot of rate versus material balance time could be used. From the diagnostic plots, flow regimes representing bilinear flow, linear flow, and BDF could be identified. Knowing the start and end of these flow regimes is very useful in the further analyses of the production data.

Step 3: Match the production data with Arps' models and modified Arps models. Theoretically, Arps hyperbolic model is capable of modelling transient bilinear flow with parameter, $b = 4$, and transient linear flow with parameter, $b = 2$. BDF could be modeled with parameter, b, range of $0 \leq b \leq 1$. This was discussed in Section 11.3.1. *Range of Values for Arps' b Parameter.* In this step, the goal is to match the production data as close as possible with Arps' models including modified Arps' models. The modified Arps' model could include an exponential tail incorporating the minimum terminal decline rate as discussed in Section 11.3.1. In some cases, Arps' modified model may give a better match of the production data than either the SEDM or Duong model. In this step, other models such as the flowing material balance model could be explored also if pressure data are available.

Step 4: Fit the production data with Duong's model as discussed in Section 11.3.7.2. *Duong's Model.* The key point to note in this step is that Duong's model is a transient model designed to fit production data during transient linear flow period. It is **not** designed to fit production data during bilinear flow or BDF periods. From the diagnostic plot, define the linear flow period if present. Then fit the Duong's model to production data **within** the linear flow period to determine Duong's parameters a, m, and q_1 as discussed in Section 11.3.7.2. Make sure that the straight line used to determine, q_1, is drawn through the origin as recommended.

Step 5: Fit the production data to SEDM as discussed in Section 11.3.7.1. *SEDM.* As pointed out in Step 4, SEDM is a transient flow model just as the Duong's model. Consequently, the model should be used to fit production data during linear flow as determined from the diagnostic plot.

Note that in Steps 4 and 5, Arps' exponential decline model could be combined with Duong's model or SEDM by transitioning to the exponential model using the concept of minimum terminal decline rate as the transition point to represent flow during BDF. This approach is illustrated in Example 11.2 with SEDM.

EXAMPLE 11.1 Analyses of Historical Production Data from a Dry Gas Well in the Permian Basin

The historical production data for a gas well producing from the Permian Basin was obtained from public records. The production data consisted only of monthly dry gas production rates. Water production was negligible. There were no pressure data (wellhead or reservoir) available from the public records. The well is a horizontal well. The public records did not have the lateral length of the well or any data on how the well was completed. It is, however, assumed that the well was hydraulically fractured. There is no information on the number of fracture stages used. The lease name is West Ranch, field name is Pinon, and the production zone is named Dimple. The total well depth is 7354 feet, top perforation depth is at 5875 feet, and bottom perforation depth is at 6840 feet. Gas production started on September 1, 2010 and ended on October 31, 2016. The lease operator is Occidental West Texas Overthrust Inc. The production data used in the analyses are provided in Table 11.1 of Appendix 11.

Problem

Conduct production data analyses on the gas well using Arps' decline models, Duong model, and Stretch Exponential Decline model.

Solution

Step 1: Review production data for inconsistencies, shut-in periods, partial shut-ins, and any other anomalies. Remove any spurious data. Inspect the data for reasonableness and consistency. Make a plot of production data versus time as shown in Figure 11.3.

Step 2: Make a diagnostic plot from the production data as described in Section 11.3.6. *Diagnostic Plots.* The recommended practice is to plot normalized rates versus material balance time. Note that in this example, pressure data were not available. In such circumstances, a plot of production rate versus material balance time is considered adequate and acceptable. The diagnostic plot for the gas well is shown in Figure 11.4. The diagnostic plot indicates the presence of linear flow represented by the half-slope line and BDF represented by the unit slope line.

Step 3: Fit production data to Arps' hyperbolic model using Eq. (11.16). A reasonable fit of the production data was obtained with Arps' parameters, q_i, b, and D_i of 48,650 MCFPM, 0.8, and 0.15 per month, respectively, as shown in Figure 11.5. A modified Arps' model was also used to fit the production data. The modification of the Arps' hyperbolic model was introduced by switching to Arps' exponential model at the beginning of BDF at approximately, Time = 37 months or November 2013. The decline rate at the switch, D_{sw}, was calculated from the equation that relates decline rate to the other Arps' parameters as:

$$D = \frac{D_i}{\left(1 + bD_i t\right)} \tag{11.58}$$

Substituting Eq. (11.58) for, D_{sw}, gives:

$$D_{sw} = \frac{D_i}{\left(1 + bD_i t_{sw}\right)} \tag{11.59}$$

Equation (11.59) is a very useful equation that can be rearranged to calculate the time to switch from hyperbolic to exponential decline models if D_{sw} is given instead. The rearranged equation is:

$$t_{sw} = \frac{\frac{D_i}{D_{sw}} - 1}{bD_i} \tag{11.60}$$

Using Eq. (11.59), for D_i = 0.15 month^{-1}, $b = 0.8$, and $t_{sw} = 37$ months as:

$$D_{sw} = \frac{0.15}{1 + 0.8 \times 0.15 \times 37} = 0.02757 \text{ month}^{-1}$$

The fit of the production data with the Arps' hyperbolic model and with modified Arps' model (hyperbolic model switched to exponential model) is also shown in Figure 11.5. The actual cumulative production is 723,161 MCF, hyperbolic model predicted 730,519 MCF, and modified Arps' model predicted 719,497 MCF. There is a slight improvement in matching actual cumulative production by using the modified Arps' model. Switching from hyperbolic model (or any other transient model) to Arps' exponential model during BDF is the recommended industry practice for matching production data and more importantly, in the prediction phase. This practice leads to more conservative estimates of reserves and prevents gross over-estimation of reserves especially when using the hyperbolic model. The results of applying the Arps' hyperbolic model and the modified Arps' model are shown in Table 11.1.

Step 4: Fit production data to Duong's model. From Eq. (11.54), make a log-log plot of q_g/G_p vs. t, as shown in Figure 11.6. From Figure 11.6, the intercept is $a = 3.7$ month^{-1} and the slope of the line is $m = 1.6726$. Next from Eq. (11.55), a plot of q_g vs. $t^{-m} \exp\left[\frac{a}{1-m}\left(t^{1-m} - 1\right)\right]$, gives a straight line as shown in Figure 11.7. It is important that this line should be drawn through the origin and if possible, fitted through the data points that represent the linear flow period. Note that this is necessary because the Duong's model is a transient flow model and not designed to represent data in the BDF regime. The slope of the line drawn through the origin is equal to: q_1 = 14,000 MCFPM. With the Duong's parameters determined as: $a = 3.7$, $m = 1.6726$, and q_1 = 14,000, the Duong rate equation can be written as:

$$q_g = 14000t^{-1.6726}\exp\left[\left(\frac{3.7}{1-1.6726}\right)\times\left(t^{1-1.6726}-1.0\right)\right]$$

This rate equation was used to predict the gas rates shown in Table 11.2 (Appendix 11) and Figure 11.8. The actual cumulative gas production is 723,161 MCF and the cumulative gas production predicted with the Duong's model is 685,418 MCF for a difference of −37,743 MCF or −5.2%. Figure 11.8 shows a comparison of actual versus predicted production rates and cumulative production using Duong's model.

Step 5: Fit production data to SEDM. Using Eq. (11.49), make a log-log plot of $\ln q_i/q_g$ vs. t. A plot for the gas well is shown in Figure 11.9. From the plot, the slope of the straight line is $n = 1.30014$ and the intercept equal to 0.034. From the intercept it was calculated that $\tau = 13.4744$. The initial production, q_i, is assumed equal to 45,930 MCFPM. This is the maximum production rate reported for the gas well. Substituting these parameters into Eq. (11.46) gives the rate equation as:

$$q_g = 45930\exp\left[-\left(\frac{t}{13.4744}\right)^{1.30014}\right]$$

A comparison of the actual versus predicted production rates and cumulative gas production is shown in Table 11.2 (Appendix 11) and Figure 11.10. Note that the SEDM under-predicted gas production beyond the transient period. This should be expected because the SEDM is designed to model production during the transient period. The recommended industry practice is to switch from SEDM to perhaps Arps' exponential model after the transient period and definitely during BDF period. The time to switch could be determined from the end of the transient period or the start of BDF using the diagnostic plot. The effective decline rate at the switch point could be assigned from accepted basin rates or company designated rates. An excellent start, if no other data are available, is to use the basin decline rates published by Hale et al.[9]

The equation for calculation of cumulative production based on SEDM is shown as Eq. (11.50). As pointed earlier, this equation is complex and not easily adaptable for calculation of cumulative production especially if analyses of multiple wells are required.

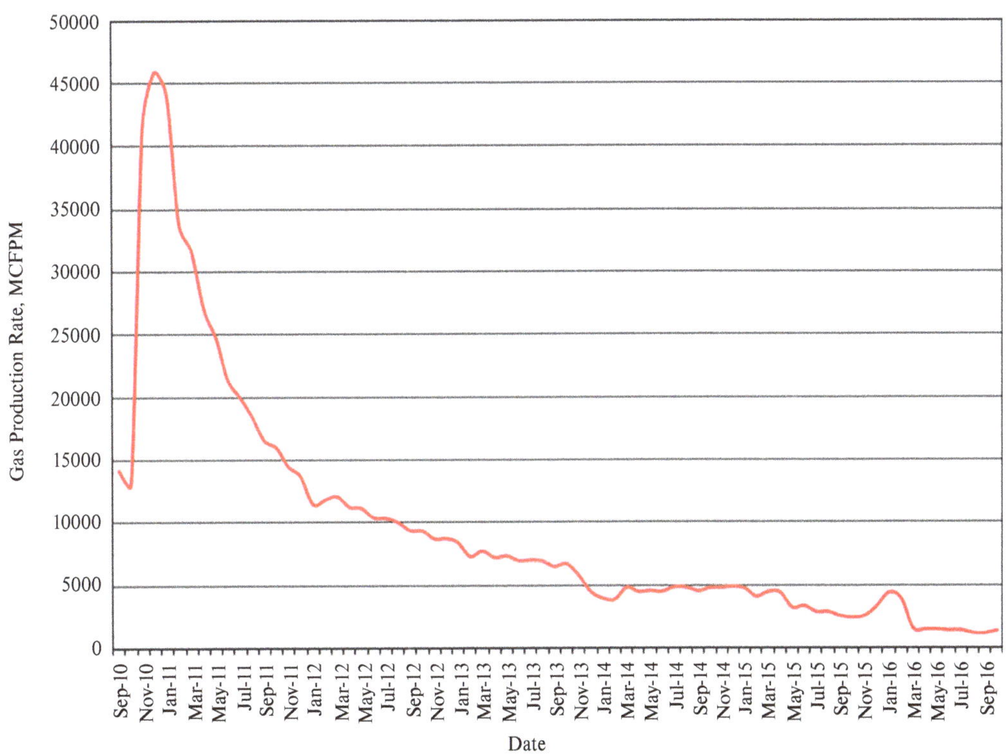

FIGURE 11.3 Production data for gas well in Example 11.1

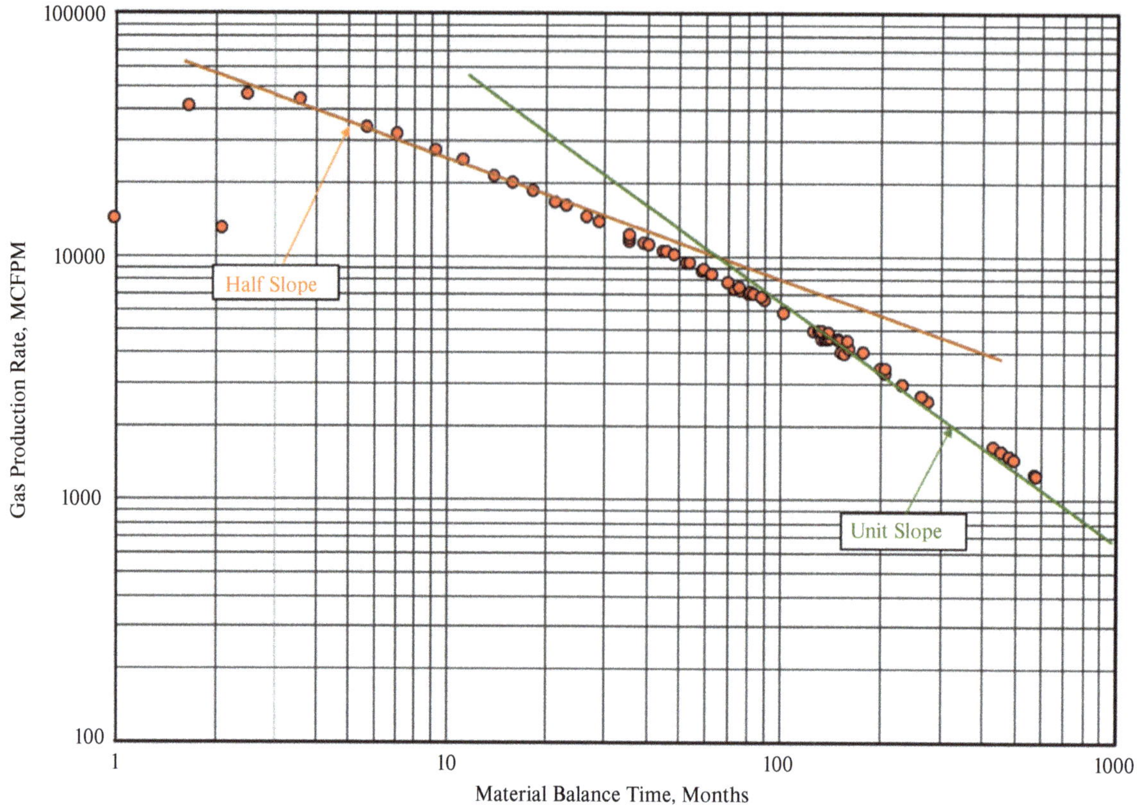

▲ **FIGURE 11.4** **Diagnostic plot for gas well in Example 11.1**

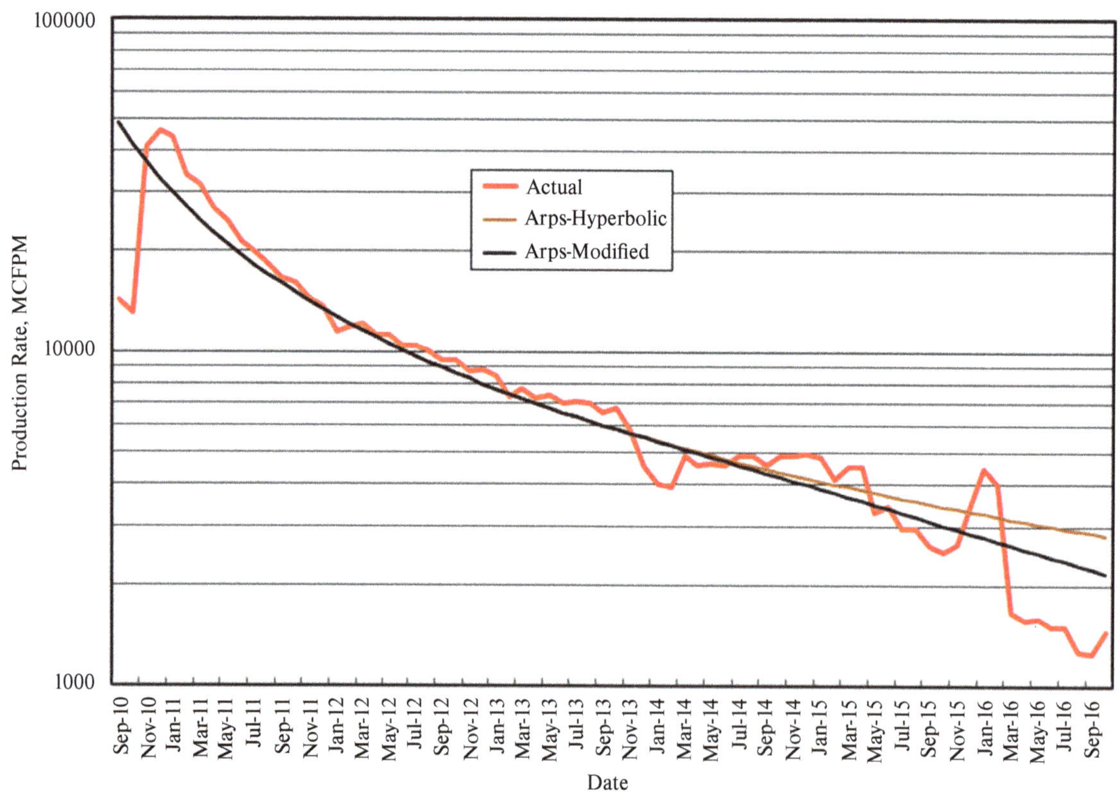

▲ **FIGURE 11.5** **Production history fitted with hyperbolic decline model and modified Arps' model**

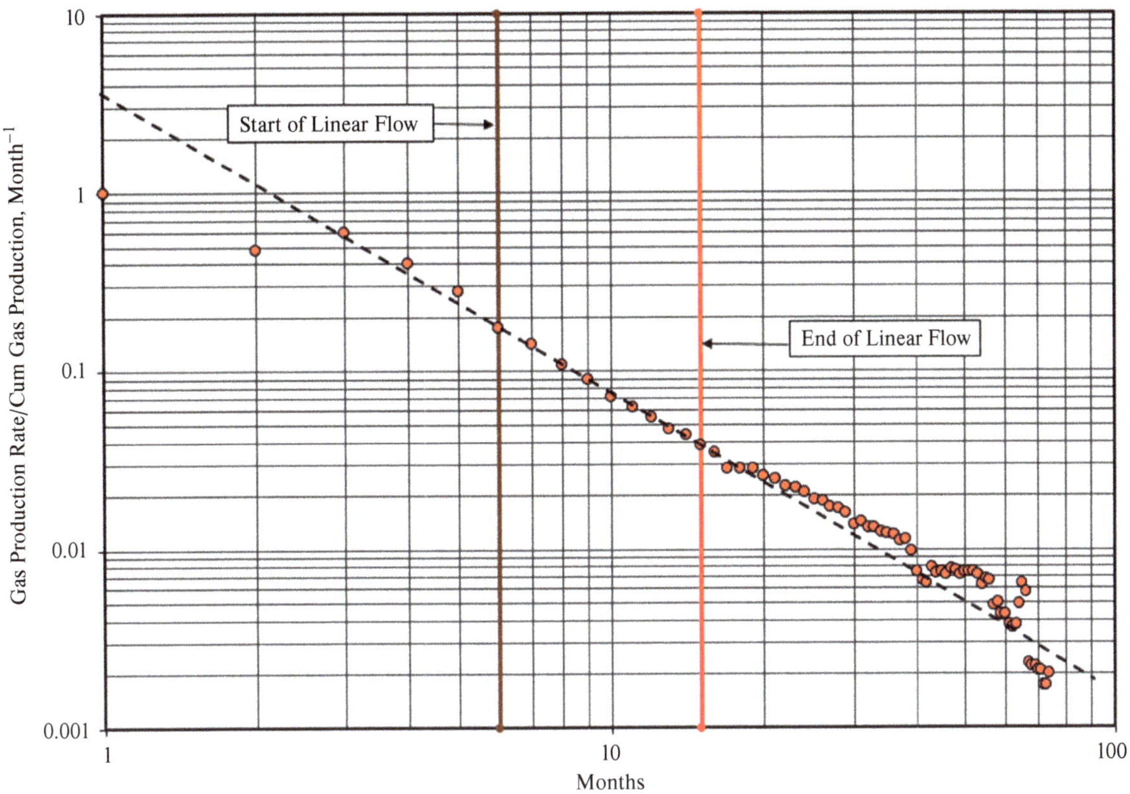

▲ **FIGURE 11.6** Duong plot of q_g/G_p vs. t for Example 11.1

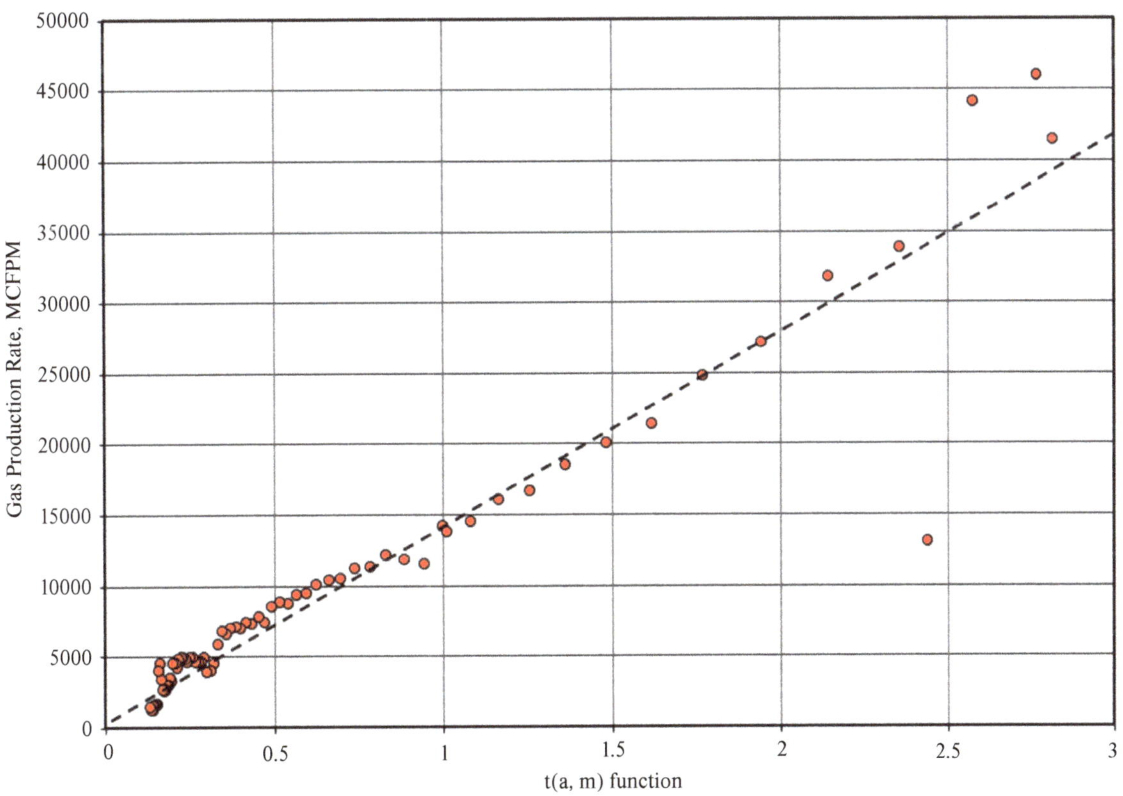

▲ **FIGURE 11.7** Duong plot of q_g vs. $t(a, m)$ function for Example 11.1

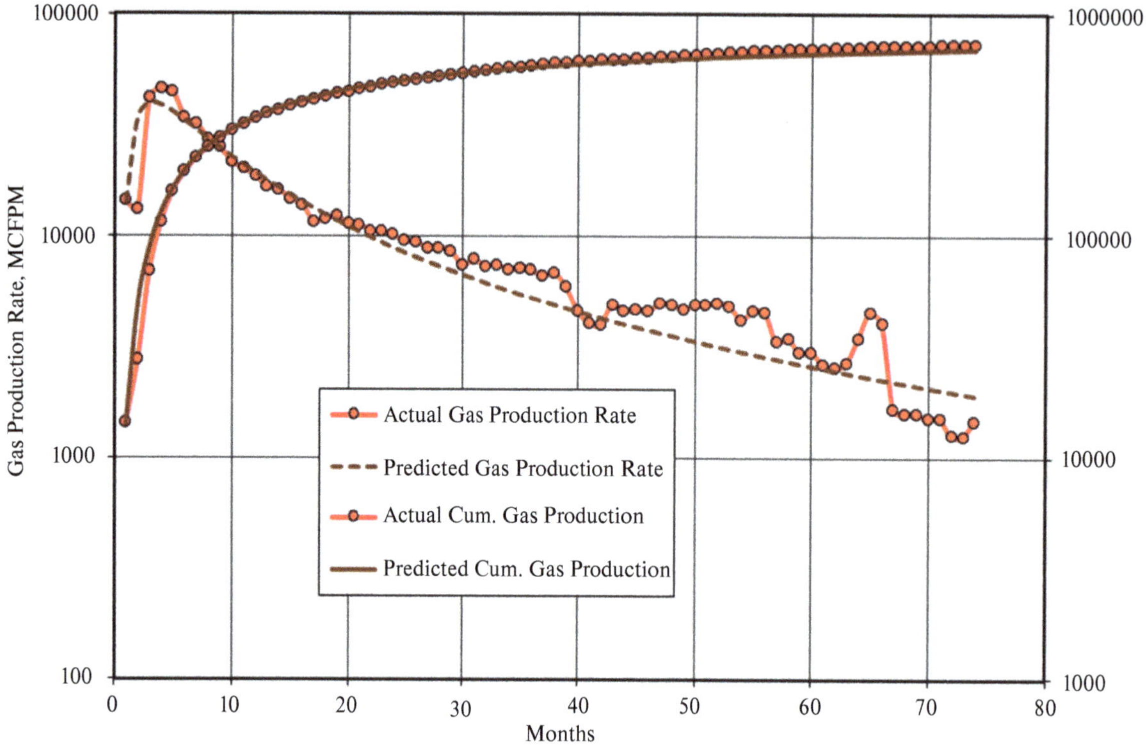

▲ **FIGURE 11.8** **Comparison of actual vs. predicted using Duong model for gas well in Example 11.1**

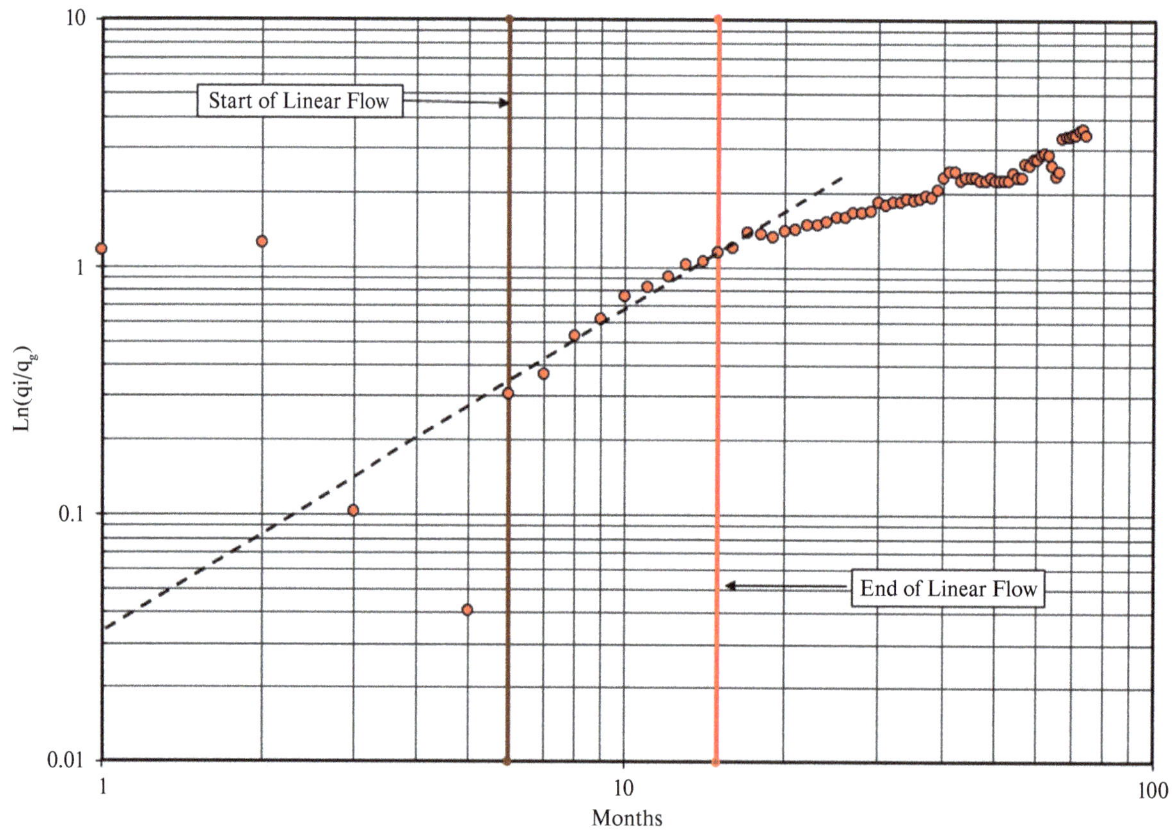

▲ **FIGURE 11.9** **SEDM log-log plot of ln(q_i/q_g) vs. t for Example 11.1**

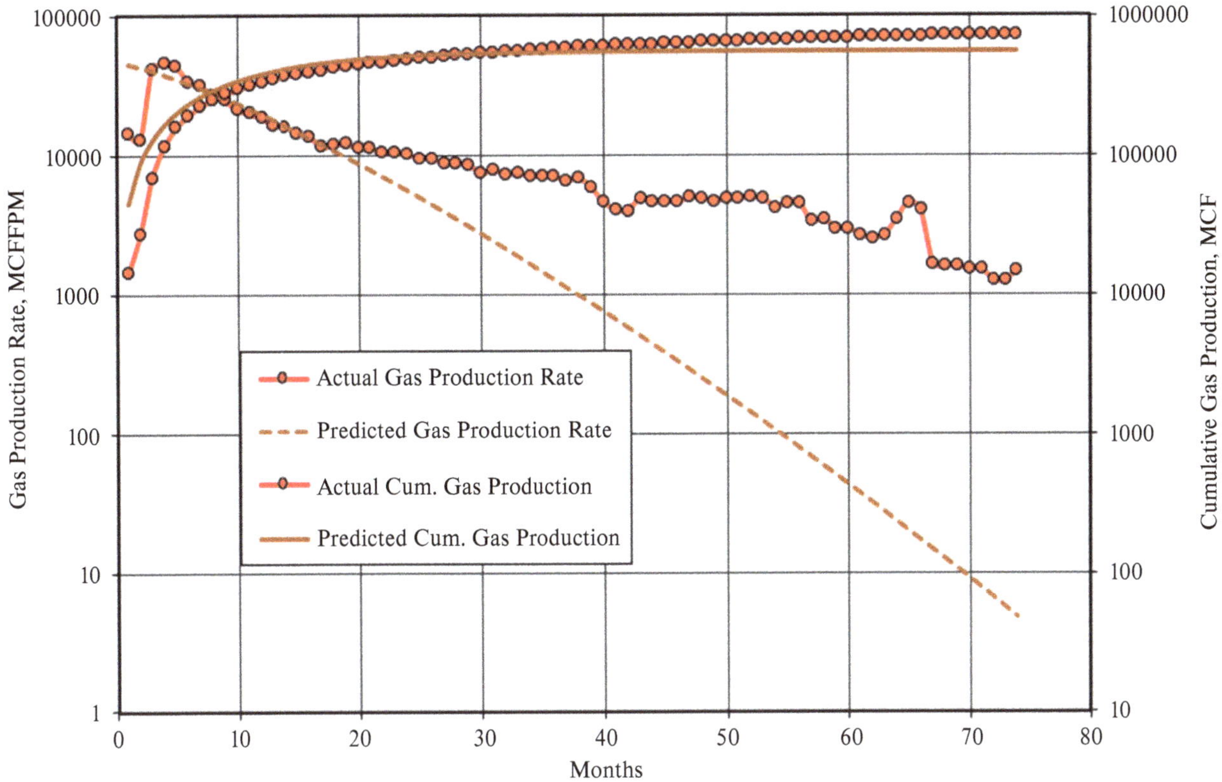

▲ **FIGURE 11.10** **Comparison of actual vs. predicted using SEDM model for gas well in Example 11.1**

End of Example 11.1

EXAMPLE 11.2 Analyses of Historical Production Data from an Oil Well in the Barnett Shale

The analyses were based on the historical production data of a vertical oil well completed in the Barnett Shale in the Fort Worth Basin. The well was spudded on August 24, 1994, and completed on September 20, 1994. It is assumed that the well was hydraulically fractured before completion. Production data retrieved from public records indicate that the well had produced mainly oil with negligible gas production and no reported water production. The gravity of produced oil was reported as 40°API. Note that no pressure data were available for the well through public records. Total well depth was 4605 feet at elevation of 1316 feet KB. The depth of top perforations was 4595 feet with depth of bottom perforations at 4605 feet. First oil production date was October 1, 1994, and last production date was March 1, 2013, for a total production life of 218 months. Cumulative oil production was 59,911 bbls. The lease name is Putnam "M" and the field is Throck Morton County Regular. The operator was reported as Bengal Petroleum, USA.

Problem

Conduct production data analyses on the oil well using Arps' decline models, Duong model, and Stretch Exponential Decline (SEDM) model.

Solution

Step 1: This well was selected because it has long production history of 218 months. The production data were reviewed for inconsistencies, shut-in periods, partial shut-ins, and any other anomalies. The data were also inspected for reasonableness and consistency. A plot of production data versus time is shown as Figure 11.11. It appears from a review of the publicly available data that this well produced over a long time with no protracted shut-in periods.

Step 2: Make a diagnostic plot from the production data as described in Section 11.3.6. *Diagnostic Plots.* A diagnostic plot of the production data is shown as Figure 11.12. Note that actual rates were used instead of normalized rates because pressure data were not available from public records. The recommended practice is to use normalized rates if pressure data are available. The diagnostic plot in Figure 11.12 indicate the presence of linear flow (with half-slope line) and BDF (with unit-slope line).

Step 3: Fit production data to Arps' exponential, hyperbolic, and modified Arps' segment models. In this step, the production data was fit to Arps' exponential model, then Arps' hyperbolic model, and then to segmented Arps' models consisting initially of hyperbolic model, followed by an exponential model for the middle segment, and finally a terminal exponential model for the last segment. The practice of using different models for segments of the production history is sometimes used in cases where a single model would not represent or fit the entire production history especially if the production history is long. For the initial hyperbolic segment, the model fitting parameters were $q_i = 1082$ BOPM, $b = 4$, and $D_i = 0.4$ per yr. For the first exponential model, the fitting parameters were, $q_i = 622.4$ BOPM and nominal decline rate, $D_i = 0.6931$ per yr. For the terminal exponential model, the fitting parameters were, $q_i = 146.87$ BOPM and $D_i = 0.25$ per yr. A comparison of the actual production data to Arps' exponential, hyperbolic, and modified segment models is shown in Table 11.3 (Appendix 11) and Figure 11.13. The actual cumulative production is 59,911 BO compared to 47,543 BO for the exponential model, 128,444 BO for the hyperbolic model, and 61,875 for the modified segmented models. Clearly, the exponential model grossly under-predicted cumulative production while the hyperbolic model largely over-predicted cumulative production. The modified segment model matched cumulative production closely and over-predicted it by just 1964 BO. This illustration justifies the use of the segmented modeling approach in cases where a single decline model would not represent wells with long historical production data.

Step 4: Use the Duong's model to fit the production data. From Eq. (11.54), make a log-log plot of q_o/N_p vs. t as shown in Figure 11.14. From Figure 11.14, the intercept, $a = 2.0$ month^{-1} and the slope of the line, $m = 1.3027$. Next from Eq. (11.55), a plot of q_o vs. $t^{-m} \exp\left[\frac{a}{1-m}\left(t^{1-m}-1\right)\right]$ gives a straight line as shown in Figure 11.15. It is important that this line should be drawn through the origin. Figure 11.15 shows considerable scatter of data points within the linear flow period. A straight line representing the start and end of linear flow was drawn through the origin. The slope of this line is $q_1 = 666.67$ BOPM. This value of q_1 gave the closest fit of the production data. With the Duong's parameters determined as $a = 2.0$, $m = 1.3027$, and $q_1 = 666.67$, the Duong rate equation can be written as:

$$q_o = 666.67t^{-1.3027} \exp\left[\left(\frac{2.0}{1-1.3027}\right) \times \left(t^{1-1.3027} - 1.0\right)\right]$$

This rate equation was used to predict the oil rates shown in Table 11.4 (Appendix 11) and Figure 11.16. The actual cumulative oil production is 59,911 BO and the cumulative prediction from Duong's model is 67,636 BO for a difference of +7,725 BO or +12.9%. Figure 11.16 shows a comparison of actual versus predicted production rates and cumulative production using the Duong model.

Step 5: Fit the production data to SEDM and a hybrid (combined) SEDM and exponential model. On the basis of Eq. (11.49), make a log-log plot of $\ln q_i/q_o$ vs. t. A plot for the oil well is shown in Figure 11.17. From the plot, the slope of the straight line is $n = 1.0416$ and the intercept = 0.003. From the intercept, $\tau = 264.2476$. The initial production, q_i, is assumed equal to 782 BOPM. This is the rate that gave the closest match of early production data reported for the well. Substituting these parameters into Eq. (11.46) gives the rate equation as:

$$q_o = 782 \exp\left[-\left(\frac{t}{264.2476}\right)^{1.0416}\right]$$

This SEDM rate equation was used to predict the rates shown in Figure 11.18. Note that the rates predicted with this equation did not match the actual rates of the well during BDF. This should be expected because SEDM is a transient flow model and may not match production during BDF regime. For this reason, Arps' exponential model was fitted to production data during the BDF regime. The nominal decline rate for the exponential model is $D_i = 0.05776$ per month, and initial production is $q_i = 629.347$ BOPM. This is the production rate predicted with the SEDM at the end of linear flow. This methodology is similar to the use of a minimum terminal decline rate widely used in the industry to forecast the tail end of a production profile as discussed in

Section 11.3.1. A comparison of actual production rates versus predicted rates using SEDM and the hybrid model consisting of SEDM and Arps' exponential model is shown in Table 11.4 (Appendix 11) and Figure 11.18. The actual cumulative oil production of the well is 59,911 BO compared to 117,204 predicted with the SEDM, and 53,480 BO predicted with the hybrid model (SEDM plus Exponential model). It can be seen that the use of the hybrid model improved matching of the actual production data of the well.

The procedures for analyzing production data from unconventional wells have been illustrated with Examples 11.1 and 11.2. These examples have been used here to demonstrate concepts and recommended practices based on data from single wells. But the development of many unconventional resources may involve hundreds and sometimes thousands of wells producing from the same formation or reservoir. The engineers working on such projects with hundreds of wells are routinely required to provide forecasts for new wells, budgets, and booking of reserves. It is obvious that performing analyses manually on these number of wells could be a daunting and extremely time-consuming task. For this reason, software packages are available commercially to assist in preparing forecasts based on large number of wells. This process has been named **auto forecasting**. Auto forecasting is efficient, objective, eliminates bias, and yields results that can be consistent and repeatable (Russell and Freeborn[21]). It is recommended to use software that have algorithm to perform auto forecasting especially if hundreds of wells will need to be analyzed and used to prepare forecasts for a project or a resource play.

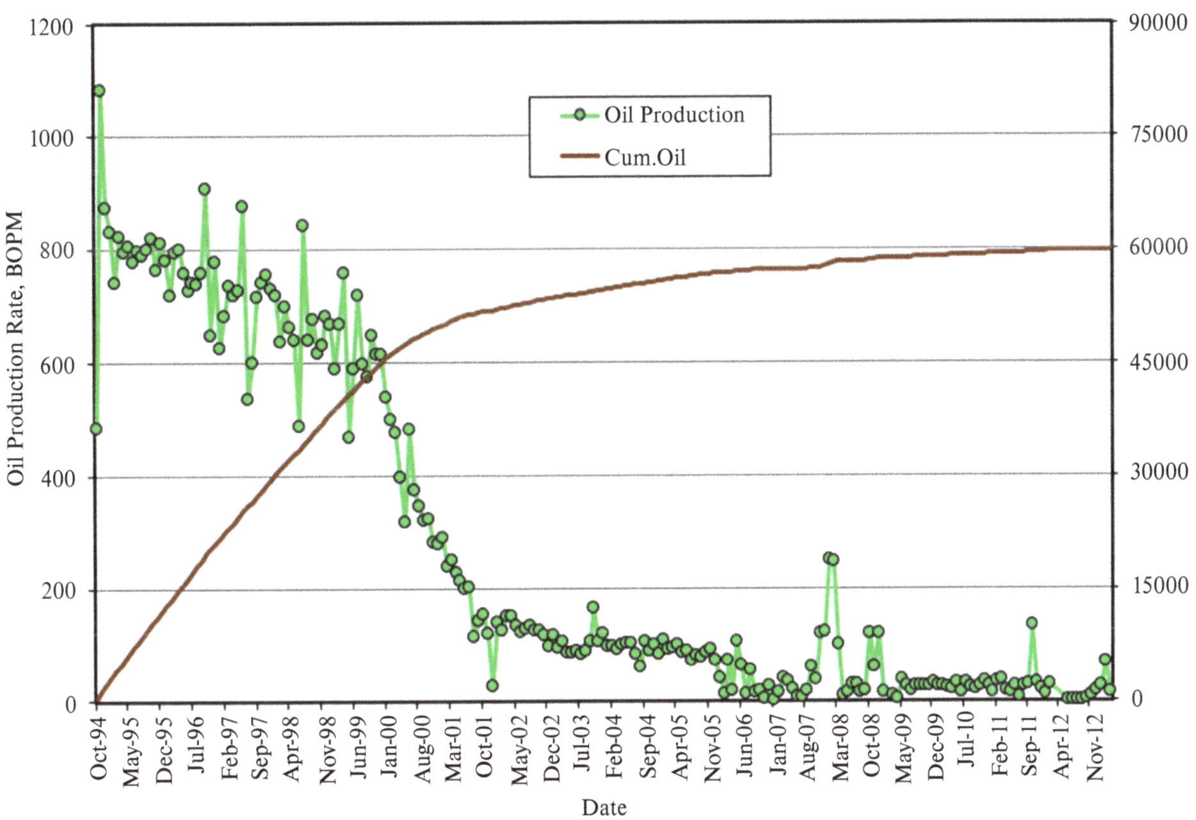

▲ **FIGURE 11.11 Historical production data for oil well in Example 11.2**

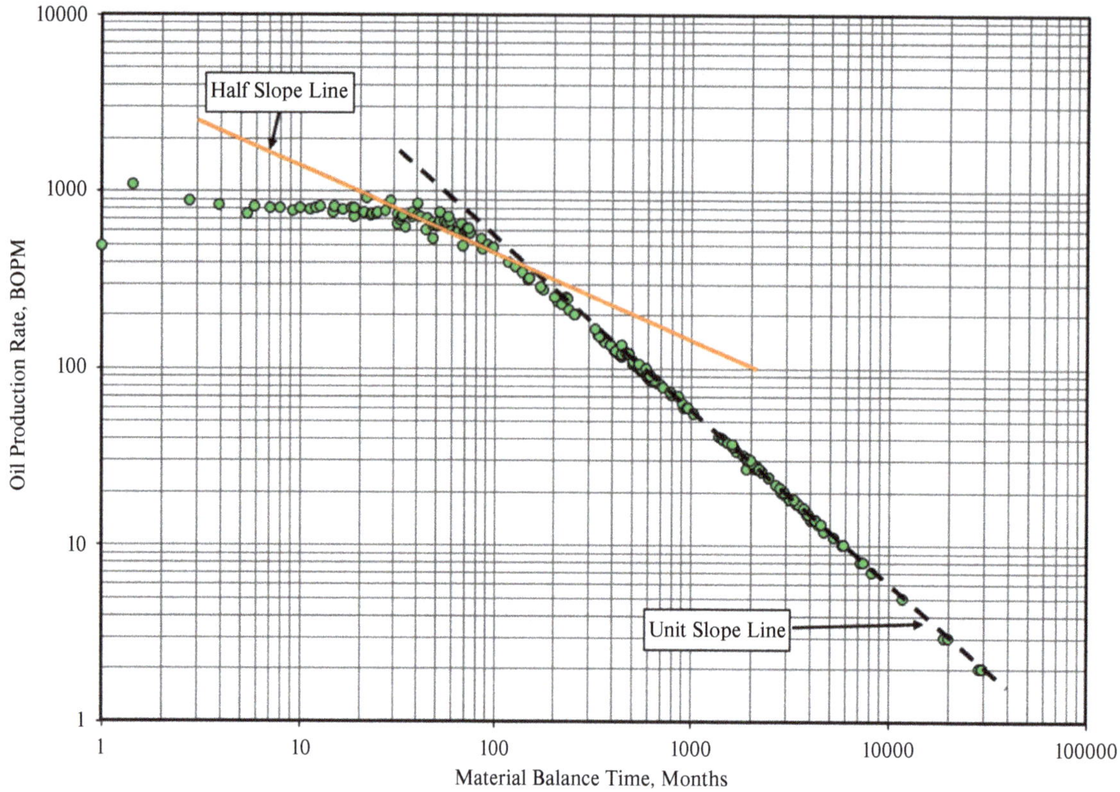

▲ **FIGURE 11.12** **Diagnostic plot for oil well in Example 11.2**

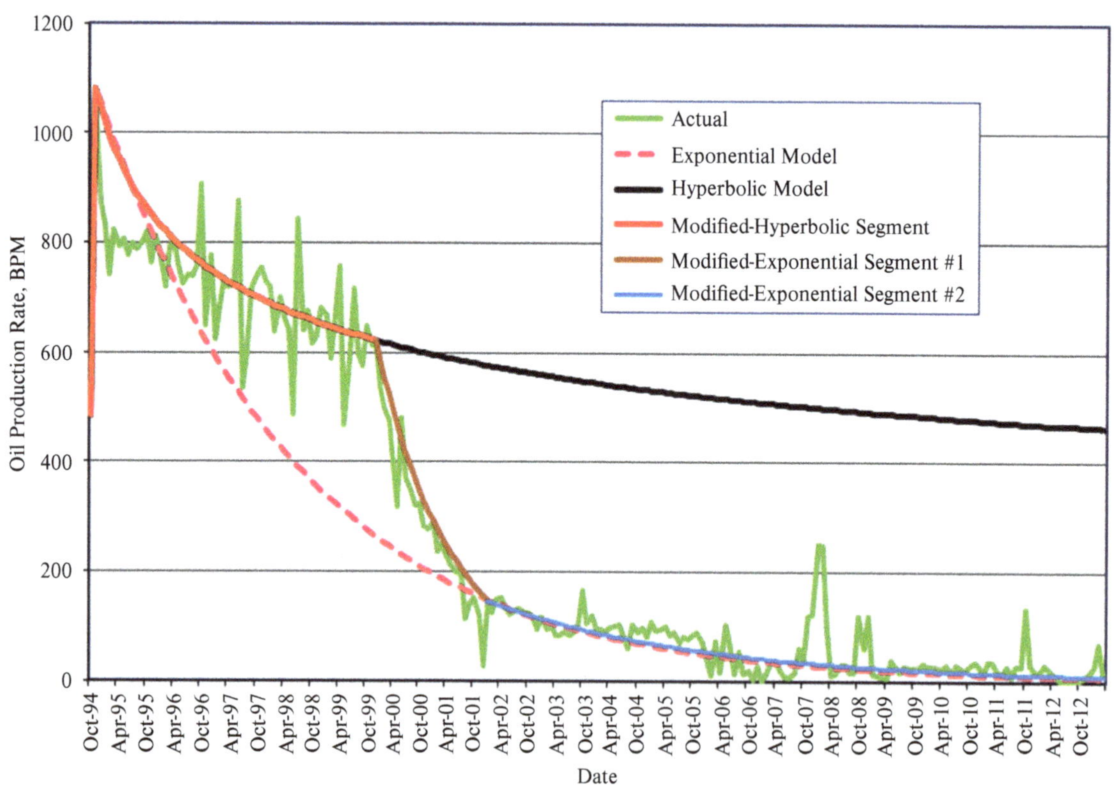

▲ **FIGURE 11.13** **Production history fitted with Arps' model for Example 11.2**

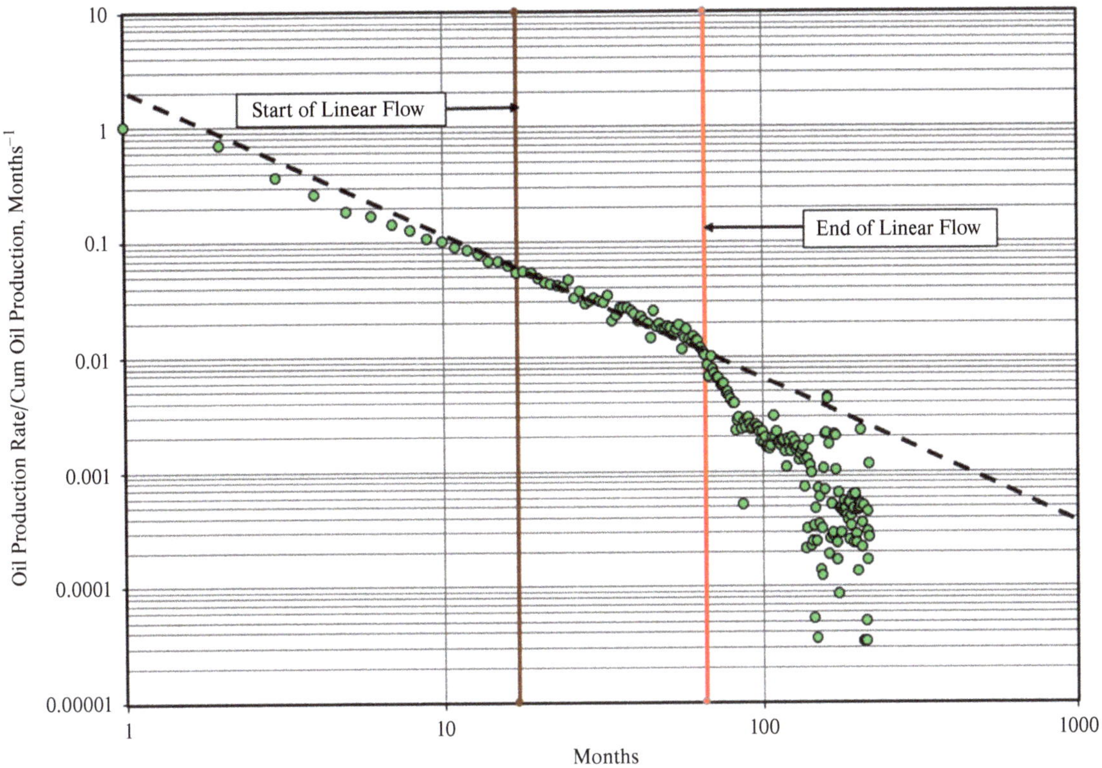

▲ **FIGURE 11.14** Duong plot of q_o/N_p vs. t for Example 11.2

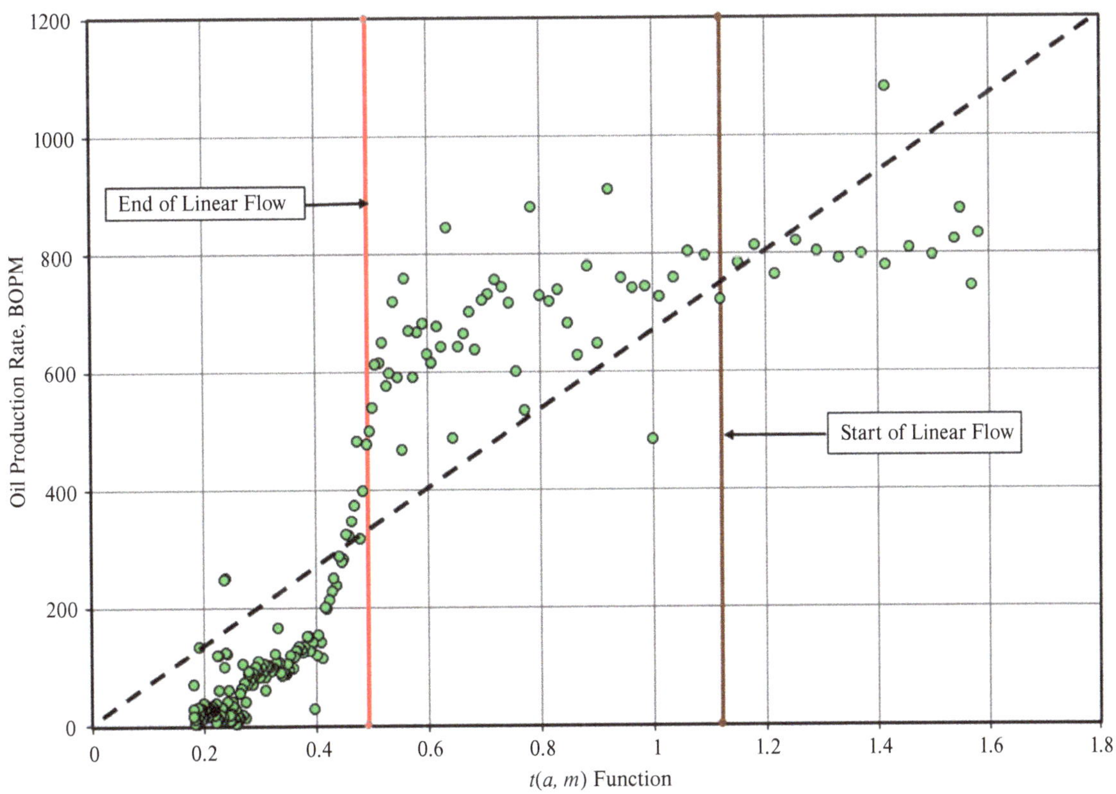

▲ **FIGURE 11.15** Duong plot of q_o vs. $t(a, m)$ function for Example 11.2

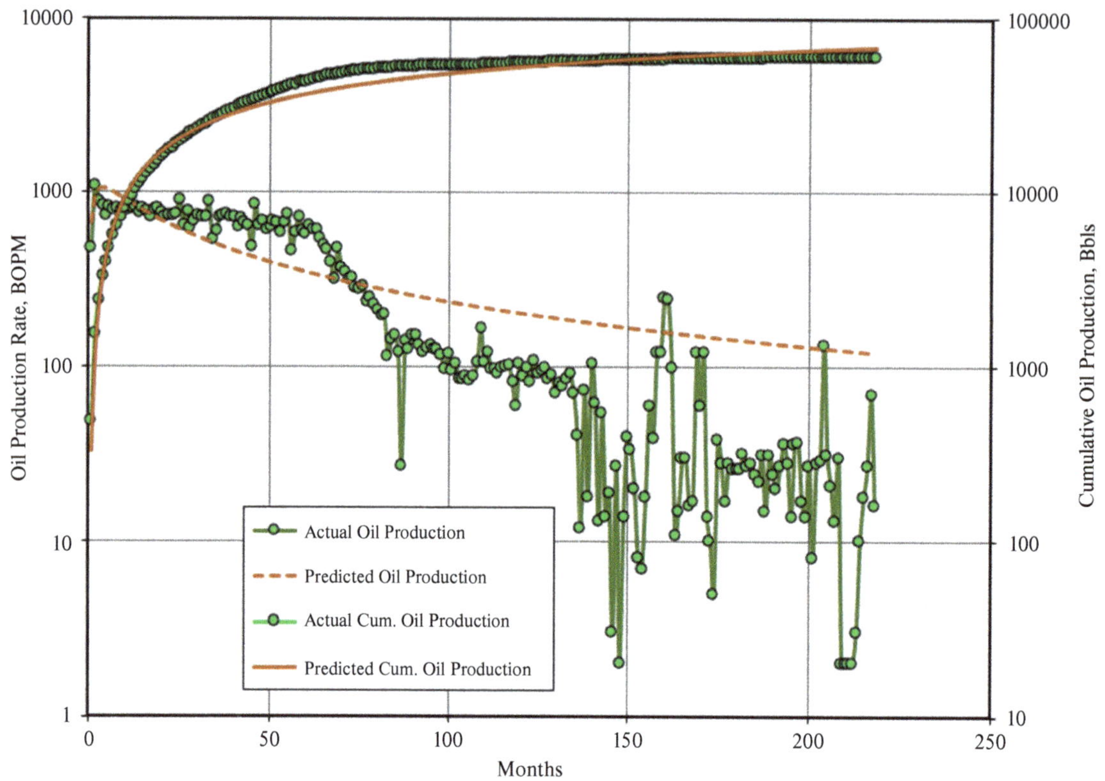

▲ **FIGURE 11.16** **Comparison of actual vs. predicted using Duong model for Example 11.2**

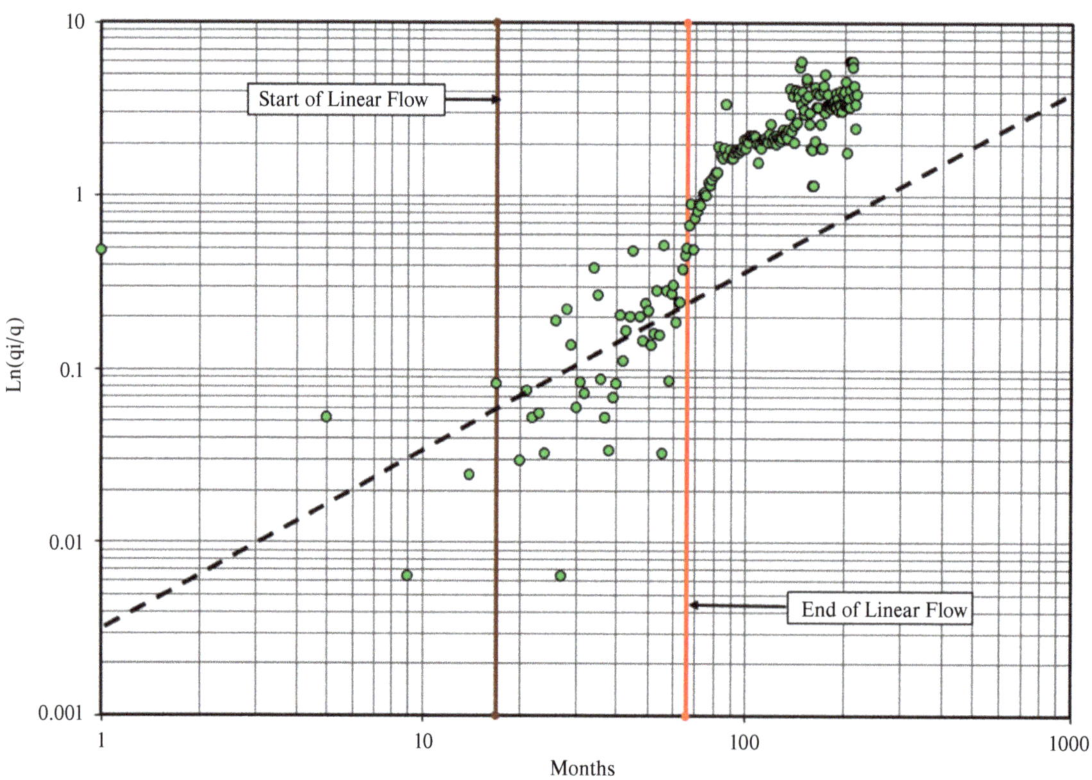

▲ **FIGURE 11.17** **SEDM plot of ln q_i/q_o vs. t for Example 11.2**

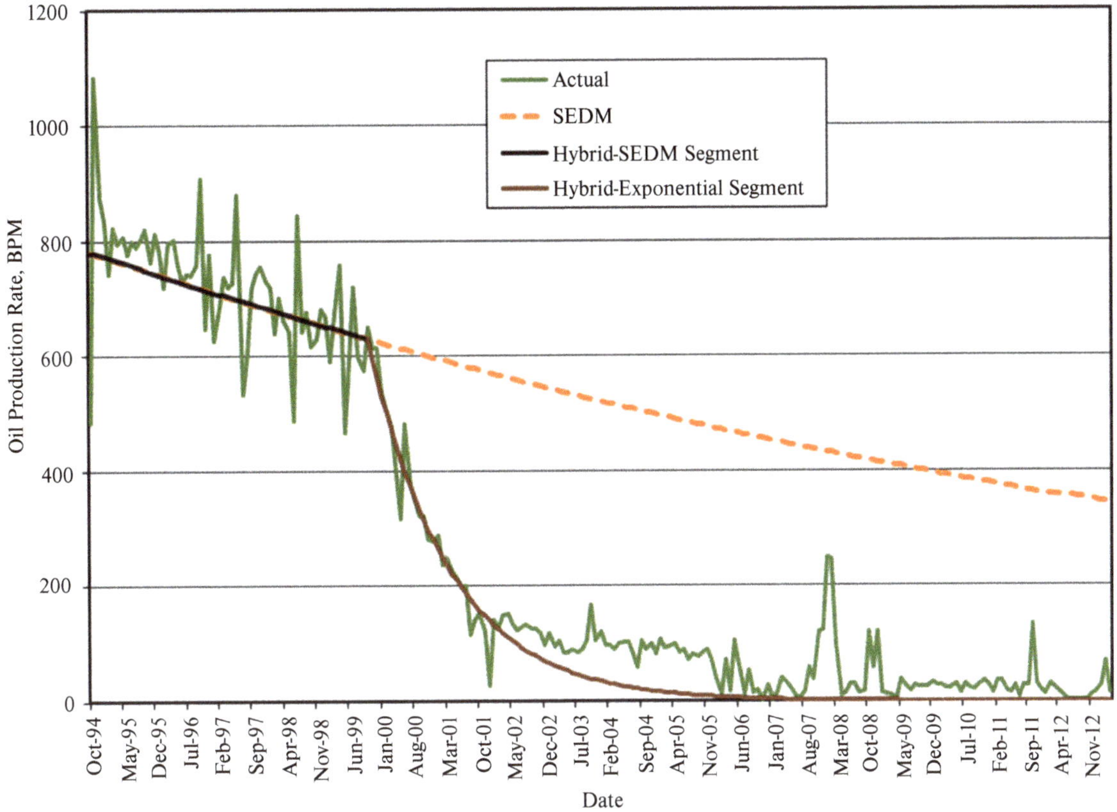

▲ **FIGURE 11.18 Production history fitted with SEDM model and hybrid model for Example 11.2**

End of Example 11.2

Nomenclature

b	Arps' parameter
B	formation volume factor, RB/STB
c_t	total compressibility, psi^{-1}
D	Arps' nominal decline rate
D_e	effective decline rate
D_{min}	minimum terminal decline rate
h	reservoir or formation thickness, feet
k	permeability, md
k_f	fracture permeability, md
L_f	fracture half-length, feet
Δp	pressure change, psia
p	pressure, psia
p_D	dimensionless pressure
p_p	real gas pseudo-pressure, psia^2/cp
p_{wf}	well bottom-hole flowing pressure, psia
p_{ws}	shut-in bottomhole pressure, psia
q	volume flow rate, STB/D
q_D	dimensionless flow rate
q_{dD}	dimensionless decline flow rate
q_g	gas volume flow rate, Mcf/D
r_{eD}	dimensionless reservoir radius
r_w	wellbore radius, feet
r_{wa}	effective wellbore radius, feet
s	skin factor
s'	apparent skin factor.
t	time, hours
t_D	dimensionless time
t_{dD}	dimensionless decline time
T	temperature, °R
w_f	fracture width, feet
ϕ	porosity, fraction
μ	fluid viscosity, cp
γ	exponential Euler's constant approximately 1.781

Subscripts

f	fracture
g	gas
i	initial conditions
o	oil
w	water

Abbreviations

API	American Petroleum Institute
BDF	Boundary Dominated Flow
BHFP	Bottom Hole Flowing Pressure
DCA	Decline Curve Analysis
FMB	Flowing Material Balance
KB	Kelly Bushing
PSS	Pseudosteady-state
SEDM	Stretched Exponential Decline Model

REFERENCES

1. Arps, J.J.: "Analysis of Decline Curves," *Trans. AIME* (1945), 160, 228-247, SPE 945228-G.
2. Cutler, W.W.: "Estimation of Underground Oil Reserves by Well Production Curves." *U.S. Bur. Mines Bull.* (1924) 228.
3. Fetkovich, M.J., Vienot, M.E., Bradley, M.D., and Kiesow, U.G.: "Decline-Curve Analysis Using Type Curves-Case Histories," *SPE Formation Evaluation* (December 1987) 637-656, SPE 13169-PA.
4. Anderson, D.M., and Mattar, L.: "Material-Balance-Time During Linear and Radial Flow," Petroleum Society of Canada (2003), PETSOC-2003-201.
5. Blasingame, T.A., McCray, T.L., and Lee, W.J.: "Decline Curve Analysis for Variable Pressure Drop/Variable Flowrate Systems," paper SPE 21513 presented at the SPE Gas Technology Symposium, January 23-24, 1991.
6. Fetkovich, M.J.: "Decline Curve Analysis Using Type Curves," *JPT* (June 1980), 1065-1077, SPE 4629.
7. Arps, J.J.: "Estimation of Primary Oil Reserves," *Trans. AIME* (1956), 207182-191, SPE 627-G.
8. Long, D.R., and Davis, M.J.: "A New Approach to the Hyperbolic Curve," *JPT* (1988), 40, 7, 909-912, SPE 16237-PA.
9. Hale, B.W., Cobb, W.M., and Hall, R.K.: "Historical Terminal Decline Rates Review of Unconventional Reservoirs in the United States," paper SPE 185056 presented at the SPE Unconventional Resources Conference, Calgary, Alberta, Canada, February15-16, 2017.
10. Blasingame, T.A., Johnston, J.L., and Lee, W.J.: "Type-Curve Analysis Using the Pressure Integral Method," paper SPE 18799 presented at the SPE California Regional Meeting, Bakersfield, April 5-7, 1989.
11. Agarwal, R.G., Gardner, D.C., Kleinsteiber, S.W., and Fussel, D.D.: "Analyzing Well Production Data Using Combined Type Curve and Decline Curve Concepts," paper SPE 57916 presented at the 1998 SPE Annual Technical Conference and Exhibition, New Orleans, September 27-30.
12. Chen, H., and Teufel, L.W.: "A New Rate-Time Type Curve for Analysis of Tight-Gas Linear and Radial Flow," paper SPE 63094 presented at the 2000 SPE Annual Technical Conference and Exhibition, Dallas, Texas, October 1-4, 2000.
13. Mattar, L., and Anderson, D.M.: "A Systematic and Comprehensive Methodology for Advanced Analysis of Production Data," paper SPE 84472 presented at the SPE Technical Conference and Exhibition, Denver, Colorado, October 5-8, 2003.
14. Ilk, D., Rushing, J.A., Perego, A.D., and Blasingame, T.: "Exponential vs. Hyperbolic Decline in Tight Sands-Understanding the Origin and Implications for Reserves Estimates Using Arps' Decline Curves," paper SPE 116731 presented at the SPE Annual Technical Conference and Exhibition, Denver, September 21-24, 2008.
15. Valko, P.P.: "Assigning Value to Stimulation in the Barnett Shale: A Simultaneous Analysis of 7000 Plus Production Histories and Well Completion Records," paper SPE 119369 presented at the 2009 SPE Hydraulic Fracturing Technology Conference, The Woodlands, Texas, January 19-21.
16. Duong, A.N.: "Rate-Decline Analysis for Fracture-Dominated Shale Reservoirs," *SPE RE&E* (June 2011) 14, 3, 377-387, SPE 137748.
17. Yu, S., Lee, W.J., Miocevic, D.J., Dong, L., and Seth, H.: Estimating Proved Reserves in Tight/Shale Wells Using the Modified SEPD Method," paper SPE 166198 presented at the SPE Annual Technical Conference and Exhibition, New Orleans, Louisiana, 30 Sept.-2 Oct. 2013.
18. Russell, B., Freeborn, R., and Keinick, W.: "A Practical Guide to Unconventional Petroleum Evaluation," paper SPE 158867 presented at the SPE Asia Pacific Oil and Gas Conference and Exhibition, Perth, Australia, October 22-24, 2012.
19. Freeborn, R., Russell, B., and Keinick, W.: "Creating Analogs, Fact and Fiction," paper SPE 162630 presented at the SPE Hydrocarbon, Economics, and Evaluation Symposium, Calgary, Alberta, September 24-25, 2012.
20. Freeborn, R., and Russell, B.: "Creating More-Representative Type Wells," *SPE Economics and Management* (April 2016), 8, 2, 50-58, SPE 175967-PA.
21. Russell, B., and Freeborn, R.: "A practical Guide to Unconventional Petroleum Evaluation Part 2," SPE 167215 presented at the SPE Unconventional Resources Conference, Calgary, Alberta, Canada, November 5-7, 20013.

PROBLEMS

11-1. The historical production data of a vertical well producing from the Barnett Shale in the Fort Worth basin are given in Table 11P.1. The production data were retrieved from public records. No pressure data were available through the public records. The production data in Table 11P.1 were from March 1986 to September 1993. The well was spudded on January 14, 1986 and completed on March 20, 1986. Total depth of the well was 4450 ft at elevation of 1091 ft KB. Top perforation depth was at 3220 ft and bottom perforation depth was at 3234 ft. The gravity of produced oil was 38.5°API. Assume the well was hydraulically fractured.

Using the data in Table 11P.1, perform production data analyses on the well using:

a. Diagnostic plots
b. Arps' decline models
c. Duong model
d. SEDM
e. Hybrid models
f. Identify the model that gave the best fit of the production data

TABLE 11P.1 Oil Production Data for Problem 11-1

Date	Oil (BOPM)	Water (BWPM)	Gas (MCFPM)	Date	Oil (BOPM)	Water (BWPM)	Gas (MCFPM)
Mar-86	287	0	0	Jan-90	183	0	89
Apr-86	594	0	0	Feb-90	159	0	83
May-86	315	0	54	Mar-90	166	0	101
Jun-86	306	0	168	Apr-90	164	0	90
Jul-86	315	0	178	May-90	167	0	78
Aug-86	313	0	202	Jun-90	153	0	85
Sep-86	294	0	179	Jul-90	153	0	75
Oct-86	307	0	174	Aug-90	154	0	92
Nov-86	293	0	169	Sep-90	145	0	82
Dec-86	308	0	170	Oct-90	149	0	91
Jan-87	318	0	201	Nov-90	135	0	73
Feb-87	286	0	185	Dec-90	134	0	62
Mar-87	311	0	189	Jan-91	144	0	46
Apr-87	303	0	172	Feb-91	129	0	53
May-87	307	0	174	Mar-91	136	0	47
Jun-87	300	0	122	Apr-91	129	0	55
Jul-87	319	0	164	May-91	141	0	43
Aug-87	315	0	139	Jun-91	127	0	1
Sep-87	304	0	107	Jul-91	129	0	1
Oct-87	331	0	126	Aug-91	127	0	1
Nov-87	313	0	135	Sep-91	123	0	1
Dec-87	330	0	111	Oct-91	127	0	1
Jan-88	339	0	125	Nov-91	125	0	1
Feb-88	312	0	111	Dec-91	129	0	1
Mar-88	324	0	109	Jan-92	128	0	1
Apr-88	332	0	123	Feb-92	119	0	1
May-88	330	0	123	Mar-92	123	0	1
Jun-88	327	0	130	Apr-92	117	0	1
Jul-88	332	0	103	May-92	121	0	1
Aug-88	324	0	130	Jun-92	117	0	1
Sep-88	312	0	105	Jul-92	117	0	1
Oct-88	321	0	111	Aug-92	116	0	1
Nov-88	305	0	125	Sep-92	113	0	1
Dec-88	319	0	93	Oct-92	120	0	1
Jan-89	306	0	115	Nov-92	114	0	1
Feb-89	284	0	93	Dec-92	120	0	1
Mar-89	310	0	112	Jan-93	118	0	1
Apr-89	286	0	111	Feb-93	110	0	1
May-89	283	0	113	Mar-93	122	0	1
Jun-89	267	0	117	Apr-93	115	0	1
Jul-89	240	0	125	May-93	120	0	1
Aug-89	217	0	108	Jun-93	109	0	1
Sep-89	196	0	111	Jul-93	115	0	1
Oct-89	202	0	69	Aug-93	117	0	1
Nov-89	186	0	97	Sep-93	118	0	1
Dec-89	188	0	71	—	—	—	—

11-2. The historical production data for a gas well in the Permian Basin is given in Table 11P.2. The production data were obtained from public records with no pressure data. Water production was not reported. This was a horizontal well and it was assumed to have been hydraulically fractured. Total well depth was 19,532 ft. Top perforation was at 11,938 ft and bottom perforation was at 12,015 ft. Using the data in Table 11P.2, perform production data analyses on the well using:

a. Diagnostic plots
b. Arps' decline models
c. Duong model
d. SEDM
e. Hybrid models
f. Identify the model that gave the best fit of the production data

TABLE 11P.2 | Gas Production Data for Problem 11-2

Date	Gas (MCFPM)	Date	Gas (MCFPM)
Dec-10	1674	Nov-13	1490
Jan-11	2199	Dec-13	1485
Feb-11	2032	Jan-14	1052
Mar-11	2569	Feb-14	956
Apr-11	2255	Mar-14	1271
May-11	2325	Apr-14	1206
Jun-11	2211	May-14	792
Jul-11	2167	Jun-14	1334
Aug-11	2084	Jul-14	1446
Sep-11	1912	Aug-14	1190
Oct-11	1830	Sep-14	393
Nov-11	1817	Oct-14	162
Dec-11	1483	Nov-14	325
Jan-12	1035	Dec-14	75
Feb-12	1004	Jan-15	100
Mar-12	2394	Feb-15	556
Apr-12	1860	Mar-15	765
May-12	1558	Apr-15	725
Jun-12	1411	May-15	353
Jul-12	401	Jun-15	788
Aug-12	802	Jul-15	234
Sep-12	1417	Aug-15	172
Oct-12	783	Sep-15	0
Nov-12	1341	Oct-15	125
Dec-12	1670	Nov-15	894
Jan-13	1907	Dec-15	716
Feb-13	1396	Jan-16	653
Mar-13	1641	Feb-16	408
Apr-13	1628	Mar-16	221
May-13	1260	Apr-16	0
Jun-13	1453	May-16	652
Jul-13	1281	Jun-16	889
Aug-13	865	Jul-16	853
Sep-13	835	Aug-16	941
Oct-13	1055	Sep-16	45

11-3. The gas production data for a well in the Permian Basin are provided in Table 11P.3. The production data were taken from public records with no associated pressure data reported. Water and liquid hydrocarbon production were negligible. The well was a horizontal well and presumed to be hydraulically fractured. Total well depth was 7254 ft. Top and bottom perforations were 5187 ft and 7012 ft, respectively. Using the data in Table 11P.3, perform production data analyses on the well using:

a. Diagnostic plots
b. Arps' decline models
c. Duong model
d. SEDM
e. Hybrid models
f. Identify the model that gave the best fit of the production data

TABLE 11P.3 Gas Production Data for Problem 11-3

Date	Gas (MCFPM)	Date	Gas (MCFPM)
Mar-10	28923	Jul-13	9625
Apr-10	71115	Aug-13	9195
May-10	69992	Sep-13	8525
Jun-10	61770	Oct-13	9992
Jul-10	57078	Nov-13	8511
Aug-10	51109	Dec-13	9556
Sep-10	44368	Jan-14	9093
Oct-10	38681	Feb-14	7557
Nov-10	37196	Mar-14	8805
Dec-10	35539	Apr-14	8237
Jan-11	31798	May-14	8284
Feb-11	26266	Jun-14	7897
Mar-11	28120	Jul-14	7939
Apr-11	25544	Aug-14	7710
May-11	25306	Sep-14	7289
Jun-11	22450	Oct-14	7155
Jul-11	22079	Nov-14	7135
Aug-11	22021	Dec-14	7257
Sep-11	19852	Jan-15	6941
Oct-11	19269	Feb-15	6261
Nov-11	18560	Mar-15	6745
Dec-11	17350	Apr-15	6411
Jan-12	14698	May-15	6537
Feb-12	15623	Jun-15	6364
Mar-12	16460	Jul-15	6291
Apr-12	15430	Aug-15	6534
May-12	15407	Sep-15	6030
Jun-12	13792	Oct-15	5792
Jul-12	14050	Nov-15	5957
Aug-12	14017	Dec-15	5962
Sep-12	12712	Jan-16	5763
Oct-12	12209	Feb-16	5369
Nov-12	12607	Mar-16	2324
Dec-12	12312	Apr-16	2181
Jan-13	11673	May-16	2196
Feb-13	10105	Jun-16	2046
Mar-13	11244	Jul-16	2174
Apr-13	10489	Aug-16	2092
May-13	10879	Sep-16	1995
Jun-13	10051	Oct-16	2125

11-4. The historical production data for an oil well in the Permian Basin is given in Table 11P.4. The production data were obtained from public records with no pressure data. Water production was not reported. This was a horizontal well and it was assumed to have been hydraulically fractured. Total well depth was 11,530 ft. Top perforation was at 7599 ft and bottom perforation was at 11,490 ft. Using the data in Table 11P.4, perform production data analyses on the well using:

a. Diagnostic plots
b. Arps' decline models
c. Duong model
d. SEDM
e. Hybrid models
f. Identify the model that gave the best fit of the production data

TABLE 11P.4 | Oil Production Data for Problem 11-4

Date	Oil BOPM	Date	Oil BOPM
Aug-12	3075	Oct-14	516
Sep-12	4759	Nov-14	439
Oct-12	2340	Dec-14	528
Nov-12	2960	Jan-15	406
Dec-12	3248	Feb-15	365
Jan-13	2367	Mar-15	427
Feb-13	2774	Apr-15	271
Mar-13	2110	May-15	348
Apr-13	1703	Jun-15	291
May-13	1563	Jul-15	192
Jun-13	1494	Aug-15	238
Jul-13	1638	Sep-15	248
Aug-13	1291	Oct-15	341
Sep-13	665	Nov-15	308
Oct-13	681	Dec-15	184
Nov-13	534	Jan-16	134
Dec-13	460	Feb-16	135
Jan-14	305	Mar-16	264
Feb-14	258	Apr-16	207
Mar-14	406	May-16	264
Apr-14	624	Jun-16	320
May-14	297	Jul-16	322
Jun-14	377	Aug-16	294
Jul-14	337	Sep-16	274
Aug-14	574	Oct-16	310
Sep-14	486	—	—

11-5. The gas production data for a well in the Permian Basin are provided in Table 11P.5. The production data were taken from public records with no associated pressure data. Water and liquid hydrocarbon production were reported. Water and liquid hydrocarbon production were negligible. The well was a horizontal well and presumed to be hydraulically fractured. Total well depth was 16,069 ft. Top and bottom perforations were 12,100 ft and 14,970 ft, respectively. Using the data in Table 11P.5, perform production data analyses on the well using:

a. Diagnostic plots
b. Arps' decline models
c. Duong model
d. SEDM
e. Hybrid models
f. Identify the model that gave the best fit of the production data

TABLE 11P.5 | Gas Production Data for Problem 11-5

Date	Gas (MCFPM)	Date	Gas (MCFPM)
Jan-10	50884	Jun-13	35630
Feb-10	139252	Jul-13	35392
Mar-10	165197	Aug-13	33205
Apr-10	139722	Sep-13	17377
May-10	122918	Oct-13	36567
Jun-10	94651	Nov-13	34472
Jul-10	107861	Dec-13	29794
Aug-10	96107	Jan-14	32821
Sep-10	103766	Feb-14	27768
Oct-10	99927	Mar-14	27559
Nov-10	91250	Apr-14	27451
Dec-10	83610	May-14	34021
Jan-11	77548	Jun-14	27679
Feb-11	64067	Jul-14	27913
Mar-11	66847	Aug-14	29259
Apr-11	63164	Sep-14	27049
May-11	64468	Oct-14	27019
Jun-11	64605	Nov-14	25632
Jul-11	62113	Dec-14	29668
Aug-11	58700	Jan-15	27479
Sep-11	54492	Feb-15	25170
Oct-11	53770	Mar-15	26601
Nov-11	52387	Apr-15	24890
Dec-11	51442	May-15	27278
Jan-12	49334	Jun-15	24125
Feb-12	44552	Jul-15	24938
Mar-12	41171	Aug-15	25244
Apr-12	40539	Sep-15	16747
May-12	43615	Oct-15	2096
Jun-12	40381	Nov-15	24393
Jul-12	45098	Dec-15	33750
Aug-12	40673	Jan-16	27379
Sep-12	37313	Feb-16	24685
Oct-12	36591	Mar-16	22741
Nov-12	38425	Apr-16	20595
Dec-12	38784	May-16	19498
Jan-13	39628	Jun-16	17739
Feb-13	30649	Jul-16	22383
Mar-13	35008	Aug-16	23495
Apr-13	31758	Sep-16	20992
May-13	33296	Oct-16	20810

APPENDIX

11 DATA FOR EXAMPLES 11.1 AND 11.2

TABLE 11.1 Actual Production Rates Compared to Arps' Model Predicted Rates for Example 11.1

Date	Actual Gas Prod. Rate (MCFPM)	Cumulative Gas Prod. (MCF)	Material Balance Time (Months)	Predicted-Arps Model	
				Hyperbolic (MCFPM)	Modified (MCFPM)
1-Sep-10	14,222	14,222	1.00	48,650	48,650
1-Oct-10	13,079	27,301	2.09	42,224	42,224
1-Nov-10	41,460	68,761	1.66	37,180	37,180
1-Dec-10	45,930	114,691	2.50	33,125	33,125
1-Jan-11	44,086	158,777	3.60	29,803	29,803
1-Feb-11	33,771	192,548	5.70	27,035	27,035
1-Mar-11	31,725	224,273	7.07	24,699	24,699
1-Apr-11	27,105	251,378	9.27	22,702	22,702
1-May-11	24,807	276,185	11.13	20,978	20,978
1-Jun-11	21,365	297,550	13.93	19,476	19,476
1-Jul-11	20,019	317,569	15.86	18,157	18,157
1-Aug-11	18,482	336,051	18.18	16,991	16,991
1-Sep-11	16,595	352,646	21.25	15,953	15,953
1-Oct-11	16,043	368,689	22.98	15,024	15,024
1-Nov-11	14,514	383,203	26.40	14,188	14,188
1-Dec-11	13,735	396,938	28.90	13,432	13,432
1-Jan-12	11,507	408,445	35.50	12,745	12,745
1-Feb-12	11,860	420,305	35.44	12,120	12,120
1-Mar-12	12,135	432,440	35.64	11,547	11,547
1-Apr-12	11,276	443,716	39.35	11,021	11,021
1-May-12	11,190	454,906	40.65	10,537	10,537
1-Jun-12	10,434	465,340	44.60	10,090	10,090
1-Jul-12	10,399	475,739	45.75	9,676	9,676
1-Aug-12	10,077	485,816	48.21	9,292	9,292
1-Sep-12	9413	495,229	52.61	8,934	8,934
1-Oct-12	9386	504,615	53.76	8,600	8,600
1-Nov-12	8756	513,371	58.63	8,288	8,288
1-Dec-12	8785	522,156	59.44	7,996	7,996
1-Jan-13	8470	530,626	62.65	7,722	7,722
1-Feb-13	7348	537,974	73.21	7,464	7,464
1-Mar-13	7786	545,760	70.10	7,222	7,222
1-Apr-13	7266	553,026	76.11	6,993	6,993

Date	Actual Gas Prod. Rate (MCFPM)	Cumulative Gas Prod. (MCF)	Material Balance Time (Months)	Predicted-Arps Model	
				Hyperbolic (MCFPM)	Modified (MCFPM)
1-May-13	7403	560,429	75.70	6,777	6,777
1-Jun-13	7013	567,442	80.91	6,573	6,573
1-Jul-13	7067	574,509	81.29	6,379	6,379
1-Aug-13	6985	581,494	83.25	6,196	6,196
1-Sep-13	6540	588,034	89.91	6,021	6,021
1-Oct-13	6774	594,808	87.81	5,856	5,856
1-Nov-13	5845	600,653	102.76	5,698	5,697
1-Dec-13	4553	605,206	132.92	5,548	5,542
1-Jan-14	4035	609,241	150.99	5,405	5,391
1-Feb-14	3941	613,182	155.59	5,268	5,244
1-Mar-14	4894	618,076	126.29	5,138	5,102
1-Apr-14	4562	622,638	136.48	5,013	4,963
1-May-14	4,639	627,277	135.22	4,894	4,828
1-Jun-14	4,565	631,842	138.41	4,779	4,697
1-Jul-14	4,916	636,758	129.53	4,670	4,569
1-Aug-14	4,896	641,654	131.06	4,564	4,445
1-Sep-14	4,605	646,259	140.34	4,463	4,324
1-Oct-14	4,865	651,124	133.84	4,366	4,206
1-Nov-14	4,875	655,999	134.56	4,273	4,092
1-Dec-14	4,949	660,948	133.55	4,183	3,980
1-Jan-15	4,815	665,763	138.27	4,096	3,872
1-Feb-15	4,171	669,934	160.62	4,013	3,767
1-Mar-15	4,533	674,467	148.79	3,933	3,664
1-Apr-15	4,513	678,980	150.45	3,855	3,565
1-May-15	3,308	682,288	206.25	3,781	3,468
1-Jun-15	3,443	685,731	199.17	3,708	3,373
1-Jul-15	2,962	688,693	232.51	3,639	3,282
1-Aug-15	2,967	691,660	233.12	3,571	3,192
1-Sep-15	2,635	694,295	263.49	3,506	3,106
1-Oct-15	2,518	696,813	276.73	3,443	3,021
1-Nov-15	2,658	699,471	263.16	3,382	2,939
1-Dec-15	3,419	702,890	205.58	3,323	2,859
1-Jan-16	4,491	707,381	157.51	3,265	2,781
1-Feb-16	4,028	711,409	176.62	3,210	2,706
1-Mar-16	1,649	713,058	432.42	3,156	2,632
1-Apr-16	1,575	714,633	453.74	3,104	2,561
1-May-16	1,578	716,211	453.87	3,053	2,491
1-Jun-16	1,495	717,706	480.07	3,004	2,423
1-Jul-16	1,500	719,206	479.47	2,956	2,357
1-Aug-16	1,254	720,460	574.53	2,909	2,293
1-Sep-16	1,242	721,702	581.08	2,864	2,231
1-Oct-16	1,459	723,161	495.66	2,820	2,170

TABLE 11.2 | Actual Production Rates Compared to Predicted Rates with Duong Model and SEDM for Example 11.1

Date	Actual Gas Prod. Rate (MCFPM)	Cumulative Gas Prod. (MCF)	Duong Model		SEDM Model	
			Predicted Rates (MCFPM)	Cumulative (MCF)	Predicted Rates (MCFPM)	Cumulative (MCF)
1-Sep-10	14,222	14,222	14,000	14,000	44,395	44,395
1-Oct-10	13,079	27,301	34,108	48,108	42,241	86,636
1-Nov-10	41,460	68,761	39,453	87,561	39,856	126,492
1-Dec-10	45,930	114,691	38,713	126,274	37,373	163,865
1-Jan-11	44,086	158,777	36,043	162,317	34,867	198,732
1-Feb-11	33,771	192,548	32,944	195,261	32,389	231,121
1-Mar-11	31,725	224,273	29,944	225,205	29,974	261,095
1-Apr-11	27,105	251,378	27,211	252,416	27,644	288,738
1-May-11	24,807	276,185	24,781	277,197	25,416	314,155
1-Jun-11	21,365	297,550	22,640	299,837	23,301	337,456
1-Jul-11	20,019	317,569	20,757	320,594	21,306	358,762
1-Aug-11	18,482	336,051	19,100	339,694	19,433	378,195
1-Sep-11	16,595	352,646	17,637	357,331	17,684	395,879
1-Oct-11	16,043	368,689	16,341	373,672	16,056	411,935
1-Nov-11	14,514	383,203	15,189	388,862	14,548	426,484
1-Dec-11	13,735	396,938	14,160	403,022	13,156	439,639
1-Jan-12	11,507	408,445	13,238	416,260	11,874	451,513
1-Feb-12	11,860	420,305	12,408	428,668	10,697	462,210
1-Mar-12	12,135	432,440	11,658	440,326	9,620	471,830
1-Apr-12	11,276	443,716	10,979	451,305	8,637	480,467
1-May-12	11,190	454,906	10,361	461,666	7,741	488,208
1-Jun-12	10,434	465,340	9,797	471,464	6,928	495,136
1-Jul-12	10,399	475,739	9,282	480,745	6,191	501,327
1-Aug-12	10,077	485,816	8,808	489,553	5,524	506,850
1-Sep-12	9,413	495,229	8,373	497,926	4,921	511,771
1-Oct-12	9,386	504,615	7,971	505,897	4,379	516,150
1-Nov-12	8,756	513,371	7,600	513,497	3,890	520,040
1-Dec-12	8,785	522,156	7,255	520,752	3,452	523,492
1-Jan-13	8,470	530,626	6,936	527,688	3,059	526,552
1-Feb-13	7,348	537,974	6,638	534,326	2,708	529,259
1-Mar-13	7,786	545,760	6,361	540,687	2,394	531,653
1-Apr-13	7,266	553,026	6,102	546,790	2,113	533,766
1-May-13	7,403	560,429	5,860	552,649	1,864	535,630
1-Jun-13	7,013	567,442	5,633	558,282	1,642	537,272
1-Jul-13	7,067	574,509	5,420	563,702	1,445	538,717
1-Aug-13	6,985	581,494	5,219	568,921	1,270	539,986
1-Sep-13	6,540	588,034	5,031	573,952	1,115	541,101
1-Oct-13	6,774	594,808	4,853	578,805	978	542,079

Date	Actual Gas Prod. Rate (MCFPM)	Cumulative Gas Prod. (MCF)	Duong Model		SEDM Model	
			Predicted Rates (MCFPM)	Cumulative (MCF)	Predicted Rates (MCFPM)	Cumulative (MCF)
1-Nov-13	5,845	600,653	4,685	583,490	857	542,936
1-Dec-13	4,553	605,206	4,526	588,016	750	543,685
1-Jan-14	4,035	609,241	4,376	592,392	656	544,341
1-Feb-14	3,941	613,182	4,234	596,627	573	544,914
1-Mar-14	4,894	618,076	4,099	600,726	500	545,413
1-Apr-14	4,562	622,638	3,971	604,697	436	545,849
1-May-14	4,639	627,277	3,850	608,547	379	546,228
1-Jun-14	4,565	631,842	3,734	612,281	330	546,559
1-Jul-14	4,916	636,758	3,624	615,904	287	546,846
1-Aug-14	4,896	641,654	3,519	619,423	249	547,095
1-Sep-14	4,605	646,259	3,418	622,841	216	547,312
1-Oct-14	4,865	651,124	3,323	626,164	188	547,499
1-Nov-14	4,875	655,999	3,231	629,396	163	547,662
1-Dec-14	4,949	660,948	3,144	632,540	141	547,803
1-Jan-15	4,815	665,763	3,061	635,601	122	547,924
1-Feb-15	4,171	669,934	2,980	638,581	105	548,029
1-Mar-15	4,533	674,467	2,904	641,485	91	548,120
1-Apr-15	4,513	678,980	2,830	644,315	78	548,199
1-May-15	3,308	682,288	2,760	647,074	68	548,266
1-Jun-15	3,443	685,731	2,692	649,766	58	548,324
1-Jul-15	2,962	688,693	2,627	652,393	50	548,375
1-Aug-15	2,967	691,660	2,564	654,957	43	548,418
1-Sep-15	2,635	694,295	2,504	657,461	37	548,455
1-Oct-15	2,518	696,813	2,446	659,907	32	548,487
1-Nov-15	2,658	699,471	2,390	662,297	27	548,514
1-Dec-15	3,419	702,890	2,336	664,633	23	548,537
1-Jan-16	4,491	707,381	2,284	666,917	20	548,557
1-Feb-16	4,028	711,409	2,234	669,151	17	548,574
1-Mar-16	1,649	713,058	2,186	671,337	15	548,589
1-Apr-16	1,575	714,633	2,139	673,477	13	548,602
1-May-16	1,578	716,211	2,094	675,571	11	548,612
1-Jun-16	1,495	717,706	2,051	677,622	9	548,622
1-Jul-16	1,500	719,206	2,009	679,631	8	548,629
1-Aug-16	1,254	720,460	1,968	681,599	7	548,636
1-Sep-16	1,242	721,702	1,929	683,527	6	548,642
1-Oct-16	1,459	723,161	1,891	685,418	5	548,647

TABLE 11.3 Actual Production Rates Compared to Arps' Models Predicted Rates for Example 11.2

Date	Actual Oil Prod. Rate (BOPM)	Cumulative Oil Prod. (BO)	Material Balance Time (Months)	Arps Models-Predicted Rates		
				Exponential (BOPM)	Hyperbolic (BOPM)	Modified (BOPM)
1-Oct-94	485	485	1.00	485	485	485
1-Nov-94	1,082	1,567	1.45	1,082	1,082	1,082
1-Dec-94	874	2,441	2.79	1,057	1,049	1,049
1-Jan-95	833	3,274	3.93	1,033	1,020	1,020
1-Feb-95	742	4,016	5.41	1,010	995	995
1-Mar-95	822	4,838	5.89	986	972	972
1-Apr-95	794	5,632	7.09	964	952	952
1-May-95	806	6,438	7.99	942	934	934
1-Jun-95	777	7,215	9.29	920	918	918
1-Jul-95	798	8,013	10.04	899	902	902
1-Aug-95	789	8,802	11.16	879	888	888
1-Sep-95	802	9,604	11.98	859	875	875
1-Oct-95	820	10,424	12.71	839	863	863
1-Nov-95	763	11,187	14.66	820	852	852
1-Dec-95	811	11,998	14.79	801	841	841
1-Jan-96	782	12,780	16.34	783	832	832
1-Feb-96	720	13,500	18.75	765	822	822
1-Mar-96	794	14,294	18.00	748	813	813
1-Apr-96	802	15,096	18.82	731	805	805
1-May-96	759	15,855	20.89	714	797	797
1-Jun-96	726	16,581	22.84	698	789	789
1-Jul-96	742	17,323	23.35	682	782	782
1-Aug-96	740	18,063	24.41	666	775	775
1-Sep-96	757	18,820	24.86	651	768	768
1-Oct-96	908	19,728	21.73	636	762	762
1-Nov-96	648	20,376	31.44	621	756	756
1-Dec-96	777	21,153	27.22	607	750	750
1-Jan-97	626	21,779	34.79	593	744	744
1-Feb-97	682	22,461	32.93	580	739	739
1-Mar-97	737	23,198	31.48	567	734	734
1-Apr-97	719	23,917	33.26	554	728	728
1-May-97	728	24,645	33.85	541	724	724
1-Jun-97	878	25,523	29.07	529	719	719
1-Jul-97	534	26,057	48.80	517	714	714
1-Aug-97	600	26,657	44.43	505	710	710
1-Sep-97	717	27,374	38.18	493	705	705

Date	Actual Oil Prod. Rate (BOPM)	Cumulative Oil Prod. (BO)	Material Balance Time (Months)	Arps Models-Predicted Rates		
				Exponential (BOPM)	Hyperbolic (BOPM)	Modified (BOPM)
1-Oct-97	742	28,116	37.89	482	701	701
1-Nov-97	756	28,872	38.19	471	697	697
1-Dec-97	730	29,602	40.55	460	693	693
1-Jan-98	720	30,322	42.11	450	689	689
1-Feb-98	638	30,960	48.53	439	686	686
1-Mar-98	700	31,660	45.23	429	682	682
1-Apr-98	663	32,323	48.75	420	679	679
1-May-98	641	32,964	51.43	410	675	675
1-Jun-98	486	33,450	68.83	401	672	672
1-Jul-98	844	34,294	40.63	391	668	668
1-Aug-98	641	34,935	54.50	383	665	665
1-Sep-98	676	35,611	52.68	374	662	662
1-Oct-98	616	36,227	58.81	365	659	659
1-Nov-98	630	36,857	58.50	357	656	656
1-Dec-98	681	37,538	55.12	349	653	653
1-Jan-99	667	38,205	57.28	341	650	650
1-Feb-99	589	38,794	65.86	333	647	647
1-Mar-99	669	39,463	58.99	325	645	645
1-Apr-99	757	40,220	53.13	318	642	642
1-May-99	467	40,687	87.12	311	639	639
1-Jun-99	589	41,276	70.08	304	637	637
1-Jul-99	718	41,994	58.49	297	634	634
1-Aug-99	597	42,591	71.34	290	632	632
1-Sep-99	576	43,167	74.94	283	629	629
1-Oct-99	649	43,816	67.51	277	627	627
1-Nov-99	614	44,430	72.36	270	625	625
1-Dec-99	613	45,043	73.48	264	622	622
1-Jan-00	539	45,582	84.57	258	620	587
1-Feb-00	499	46,081	92.35	252	618	554
1-Mar-00	476	46,557	97.81	247	616	523
1-Apr-00	398	46,955	117.98	241	614	494
1-May-00	317	47,272	149.12	235	612	466
1-Jun-00	481	47,753	99.28	230	609	440
1-Jul-00	373	48,126	129.02	225	607	415
1-Aug-00	347	48,473	139.69	220	605	392
1-Sep-00	321	48,794	152.01	215	603	370

Date	Actual Oil Prod. Rate (BOPM)	Cumulative Oil Prod. (BO)	Material Balance Time (Months)	Arps Models-Predicted Rates		
				Exponential (BOPM)	Hyperbolic (BOPM)	Modified (BOPM)
1-Oct-00	323	49,117	152.07	210	602	349
1-Nov-00	282	49,399	175.17	205	600	330
1-Dec-00	278	49,677	178.69	200	598	311
1-Jan-01	288	49,965	173.49	196	596	294
1-Feb-01	238	50,203	210.94	191	594	277
1-Mar-01	250	50,453	201.81	187	592	262
1-Apr-01	228	50,681	222.29	183	591	247
1-May-01	213	50,894	238.94	178	589	233
1-Jun-01	199	51,093	256.75	174	587	220
1-Jul-01	201	51,294	255.19	170	585	208
1-Aug-01	115	51,409	447.03	167	584	196
1-Sep-01	142	51,551	363.04	163	582	185
1-Oct-01	153	51,704	337.93	159	581	175
1-Nov-01	120	51,824	431.87	155	579	165
1-Dec-01	27	51,851	1920.41	152	577	156
1-Jan-02	141	51,992	368.74	148	576	147
1-Feb-02	126	52,118	413.63	145	574	144
1-Mar-02	150	52,268	348.45	142	573	141
1-Apr-02	152	52,420	344.87	138	571	138
1-May-02	134	52,554	392.19	135	570	135
1-Jun-02	123	52,677	428.27	132	568	132
1-Jul-02	128	52,805	412.54	129	567	130
1-Aug-02	134	52,939	395.07	126	566	127
1-Sep-02	126	53,065	421.15	123	564	124
1-Oct-02	126	53,191	422.15	120	563	122
1-Nov-02	117	53,308	455.62	118	561	119
1-Dec-02	97	53,405	550.57	115	560	117
1-Jan-03	118	53,523	453.58	112	559	114
1-Feb-03	95	53,618	564.40	110	557	112
1-Mar-03	105	53,723	511.65	107	556	110
1-Apr-03	86	53,809	625.69	105	555	107
1-May-03	86	53,895	626.69	102	554	105
1-Jun-03	89	53,984	606.56	100	552	103
1-Jul-03	84	54,068	643.67	98	551	101
1-Aug-03	89	54,157	608.51	96	550	99
1-Sep-03	106	54,263	511.92	93	549	97

Date	Actual Oil Prod. Rate (BOPM)	Cumulative Oil Prod. (BO)	Material Balance Time (Months)	Arps Models-Predicted Rates		
				Exponential (BOPM)	Hyperbolic (BOPM)	Modified (BOPM)
1-Oct-03	166	54,429	327.89	91	547	95
1-Nov-03	106	54,535	514.48	89	546	93
1-Dec-03	121	54,656	451.70	87	545	91
1-Jan-04	97	54,753	564.46	85	544	89
1-Feb-04	97	54,850	565.46	83	543	87
1-Mar-04	91	54,941	603.75	81	542	85
1-Apr-04	99	55,040	555.96	79	540	84
1-May-04	102	55,142	540.61	78	539	82
1-Jun-04	103	55,245	536.36	76	538	80
1-Jul-04	82	55,327	674.72	74	537	79
1-Aug-04	60	55,387	923.12	72	536	77
1-Sep-04	105	55,492	528.50	71	535	75
1-Oct-04	89	55,581	624.51	69	534	74
1-Nov-04	99	55,680	562.42	68	533	72
1-Dec-04	83	55,763	671.84	66	532	71
1-Jan-05	109	55,872	512.59	65	531	69
1-Feb-05	92	55,964	608.30	63	530	68
1-Mar-05	95	56,059	590.09	62	529	67
1-Apr-05	100	56,159	561.59	60	528	65
1-May-05	85	56,244	661.69	59	527	64
1-Jun-05	90	56,334	625.93	58	526	63
1-Jul-05	71	56,405	794.44	56	525	61
1-Aug-05	81	56,486	697.36	55	524	60
1-Sep-05	78	56,564	725.18	54	523	59
1-Oct-05	86	56,650	658.72	52	522	58
1-Nov-05	91	56,741	623.53	51	521	56
1-Dec-05	71	56,812	800.17	50	520	55
1-Jan-06	41	56,853	1386.66	49	519	54
1-Feb-06	12	56,865	4738.75	48	518	53
1-Mar-06	73	56,938	779.97	47	517	52
1-Apr-06	18	56,956	3164.22	46	516	51
1-May-06	105	57,061	543.44	45	516	50
1-Jun-06	63	57,124	906.73	44	515	49
1-Jul-06	13	57,137	4395.15	43	514	48
1-Aug-06	55	57,192	1039.85	42	513	47
1-Sep-06	14	57,206	4086.14	41	512	46

Date	Actual Oil Prod. Rate (BOPM)	Cumulative Oil Prod. (BO)	Material Balance Time (Months)	Arps Models-Predicted Rates		
				Exponential (BOPM)	Hyperbolic (BOPM)	Modified (BOPM)
1-Oct-06	19	57,225	3011.84	40	511	45
1-Nov-06	3	57,228	19076.00	39	510	44
1-Dec-06	27	57,255	2120.56	38	510	43
1-Jan-07	2	57,257	28628.50	37	509	42
1-Feb-07	14	57,271	4090.79	36	508	41
1-Mar-07	40	57,311	1432.78	35	507	40
1-Apr-07	34	57,345	1686.62	35	506	40
1-May-07	20	57,365	2868.25	34	505	39
1-Jun-07	8	57,373	7171.63	33	505	38
1-Jul-07	7	57,380	8197.14	32	504	37
1-Aug-07	18	57,398	3188.78	32	503	36
1-Sep-07	60	57,458	957.63	31	502	36
1-Oct-07	39	57,497	1474.28	30	502	35
1-Nov-07	121	57,618	476.18	29	501	34
1-Dec-07	123	57,741	469.44	29	500	33
1-Jan-08	251	57,992	231.04	28	499	33
1-Feb-08	248	58,240	234.84	27	498	32
1-Mar-08	99	58,339	589.28	27	498	31
1-Apr-08	11	58,350	5304.55	26	497	31
1-May-08	15	58,365	3891.00	26	496	30
1-Jun-08	30	58,395	1946.50	25	496	30
1-Jul-08	30	58,425	1947.50	24	495	29
1-Aug-08	16	58,441	3652.56	24	494	28
1-Sep-08	17	58,458	3438.71	23	493	28
1-Oct-08	120	58,578	488.15	23	493	27
1-Nov-08	60	58,638	977.30	22	492	27
1-Dec-08	120	58,758	489.65	22	491	26
1-Jan-09	14	58,772	4198.00	21	491	26
1-Mar-09	10	58,782	5878.20	21	490	25
1-Apr-09	5	58,787	11757.40	20	489	24
1-May-09	38	58,825	1548.03	20	489	24
1-Jun-09	28	58,853	2101.89	19	488	23
1-Jul-09	17	58,870	3462.94	19	487	23
1-Aug-09	28	58,898	2103.50	19	487	23
1-Sep-09	26	58,924	2266.31	18	486	22
1-Oct-09	26	58,950	2267.31	18	485	22

Date	Actual Oil Prod. Rate (BOPM)	Cumulative Oil Prod. (BO)	Material Balance Time (Months)	Arps Models-Predicted Rates		
				Exponential (BOPM)	Hyperbolic (BOPM)	Modified (BOPM)
1-Nov-09	26	58,976	2268.31	17	485	21
1-Dec-09	32	59,008	1844.00	17	484	21
1-Jan-10	27	59,035	2186.48	17	483	20
1-Feb-10	28	59,063	2109.39	16	483	20
1-Mar-10	24	59,087	2461.96	16	482	19
1-Apr-10	22	59,109	2686.77	15	481	19
1-May-10	31	59,140	1907.74	15	481	19
1-Jun-10	15	59,155	3943.67	15	480	18
1-Jul-10	31	59,186	1909.23	14	479	18
1-Aug-10	24	59,210	2467.08	14	479	18
1-Sep-10	20	59,230	2961.50	14	478	17
1-Oct-10	27	59,257	2194.70	13	478	17
1-Nov-10	36	59,293	1647.03	13	477	16
1-Dec-10	28	59,321	2118.61	13	476	16
1-Jan-11	14	59,335	4238.21	13	476	16
1-Feb-11	36	59,371	1649.19	12	475	15
1-Mar-11	37	59,408	1605.62	12	475	15
1-Apr-11	17	59,425	3495.59	12	474	15
1-May-11	14	59,439	4245.64	11	473	15
1-Jun-11	27	59,466	2202.44	11	473	14
1-Jul-11	8	59,474	7434.25	11	472	14
1-Aug-11	28	59,502	2125.07	11	472	14
1-Sep-11	29	59,531	2052.79	10	471	13
1-Oct-11	134	59,665	445.26	10	471	13
1-Nov-11	31	59,696	1925.68	10	470	13
1-Dec-11	21	59,717	2843.67	10	470	13
1-Jan-12	13	59,730	4594.62	9	469	12
1-Feb-12	30	59,760	1992.00	9	468	12
1-Jun-12	2	59,762	29881.00	9	468	12
1-Jul-12	2	59,764	29882.00	9	467	12
1-Aug-12	2	59,766	29883.00	9	467	11
1-Sep-12	2	59,768	29884.00	8	466	11
1-Oct-12	3	59,771	19923.67	8	466	11
1-Nov-12	10	59,781	5978.10	8	465	11
1-Dec-12	18	59,799	3322.17	8	465	10
1-Jan-13	27	59,826	2215.78	8	464	10
1-Feb-13	69	59,895	868.04	8	464	10
1-Mar-13	16	59,911	3744.44	7	463	10

TABLE 11.4 Actual Production Rates Compared to Predicted Rates with Duong Model, SEDM, and Hybrid Models for Example 11.2

Date	Actual Oil Prod. Rate (BOPM)	Cumulative Oil Prod. (BO)	Duong Model		SEDM		Hybrid Model	
			Predicted Rates (BOPM)	Cumulative (BO)	Predicted Rates (BOPM)	Cumulative (BO)	Predicted Rates (BOPM)	Cumulative (BO)
1-Oct-94	485	485	667	333	780	780	780	780
1-Nov-94	1,082	1,567	944	1,164	777	1,557	777	1,557
1-Dec-94	874	2,441	1,033	2,161	775	2,332	775	2,332
1-Jan-95	833	3,274	1,054	3,208	772	3,104	772	3,104
1-Feb-95	742	4,016	1,047	4,261	770	3,873	770	3,873
1-Mar-95	822	4,838	1,027	5,298	767	4,640	767	4,640
1-Apr-95	794	5,632	1,001	6,313	764	5,405	764	5,405
1-May-95	806	6,438	972	7,299	762	6,166	762	6,166
1-Jun-95	777	7,215	944	8,257	759	6,926	759	6,926
1-Jul-95	798	8,013	915	9,187	757	7,682	757	7,682
1-Aug-95	789	8,802	888	10,088	754	8,436	754	8,436
1-Sep-95	802	9,604	861	10,962	751	9,188	751	9,188
1-Oct-95	820	10,424	836	11,811	749	9,936	749	9,936
1-Nov-95	763	11,187	812	12,635	746	10,683	746	10,683
1-Dec-95	811	11,998	789	13,435	744	11,426	744	11,426
1-Jan-96	782	12,780	768	14,214	741	12,167	741	12,167
1-Feb-96	720	13,500	747	14,971	738	12,905	738	12,905
1-Mar-96	794	14,294	728	15,708	736	13,641	736	13,641
1-Apr-96	802	15,096	709	16,427	733	14,374	733	14,374
1-May-96	759	15,855	692	17,127	731	15,105	731	15,105
1-Jun-96	726	16,581	675	17,810	728	15,833	728	15,833
1-Jul-96	742	17,323	659	18,477	725	16,559	725	16,559
1-Aug-96	740	18,063	644	19,128	723	17,281	723	17,281
1-Sep-96	757	18,820	629	19,765	720	18,002	720	18,002
1-Oct-96	908	19,728	616	20,388	718	18,719	718	18,719
1-Nov-96	648	20,376	602	20,997	715	19,435	715	19,435
1-Dec-96	777	21,153	590	21,593	713	20,147	713	20,147
1-Jan-97	626	21,779	578	22,176	710	20,857	710	20,857
1-Feb-97	682	22,461	566	22,748	708	21,565	708	21,565
1-Mar-97	737	23,198	555	23,309	705	22,270	705	22,270
1-Apr-97	719	23,917	544	23,859	702	22,972	702	22,972
1-May-97	728	24,645	534	24,398	700	23,672	700	23,672

Date	Actual Oil Prod. Rate (BOPM)	Cumulative Oil Prod. (BO)	Duong Model		SEDM		Hybrid Model	
			Predicted Rates (BOPM)	Cumulative (BO)	Predicted Rates (BOPM)	Cumulative (BO)	Predicted Rates (BOPM)	Cumulative (BO)
1-Jun-97	878	25,523	524	24,927	697	24,370	697	24,370
1-Jul-97	534	26,057	515	25,447	695	25,064	695	25,064
1-Aug-97	600	26,657	506	25,957	692	25,757	692	25,757
1-Sep-97	717	27,374	497	26,458	690	26,447	690	26,447
1-Oct-97	742	28,116	488	26,951	687	27,134	687	27,134
1-Nov-97	756	28,872	480	27,435	685	27,819	685	27,819
1-Dec-97	730	29,602	472	27,911	682	28,501	682	28,501
1-Jan-98	720	30,322	465	28,379	680	29,181	680	29,181
1-Feb-98	638	30,960	457	28,840	677	29,858	677	29,858
1-Mar-98	700	31,660	450	29,294	675	30,533	675	30,533
1-Apr-98	663	32,323	443	29,740	672	31,206	672	31,206
1-May-98	641	32,964	436	30,180	670	31,876	670	31,876
1-Jun-98	486	33,450	430	30,613	668	32,544	668	32,544
1-Jul-98	844	34,294	424	31,040	665	33,209	665	33,209
1-Aug-98	641	34,935	417	31,460	663	33,871	663	33,871
1-Sep-98	676	35,611	411	31,875	660	34,532	660	34,532
1-Oct-98	616	36,227	406	32,283	658	35,189	658	35,189
1-Nov-98	630	36,857	400	32,686	655	35,845	655	35,845
1-Dec-98	681	37,538	395	33,084	653	36,498	653	36,498
1-Jan-99	667	38,205	389	33,476	651	37,149	651	37,149
1-Feb-99	589	38,794	384	33,862	648	37,797	648	37,797
1-Mar-99	669	39,463	379	34,244	646	38,443	646	38,443
1-Apr-99	757	40,220	374	34,621	643	39,086	643	39,086
1-May-99	467	40,687	370	34,993	641	39,727	641	39,727
1-Jun-99	589	41,276	365	35,360	639	40,366	639	40,366
1-Jul-99	718	41,994	360	35,723	636	41,002	636	41,002
1-Aug-99	597	42,591	356	36,081	634	41,636	634	41,636
1-Sep-99	576	43,167	352	36,435	632	42,268	632	42,268
1-Oct-99	649	43,816	348	36,784	629	42,897	629	42,897
1-Nov-99	614	44,430	343	37,130	627	43,524	594	43,491
1-Dec-99	613	45,043	339	37,471	625	44,149	561	44,052
1-Jan-00	539	45,582	336	37,809	622	44,772	529	44,581
1-Feb-00	499	46,081	332	38,142	620	45,392	500	45,081
1-Mar-00	476	46,557	328	38,472	618	46,009	471	45,552

Date	Actual Oil Prod. Rate (BOPM)	Cumulative Oil Prod. (BO)	Duong Model		SEDM		Hybrid Model	
			Predicted Rates (BOPM)	Cumulative (BO)	Predicted Rates (BOPM)	Cumulative (BO)	Predicted Rates (BOPM)	Cumulative (BO)
1-Apr-00	398	46,955	324	38,798	615	46,625	445	45,997
1-May-00	317	47,272	321	39,121	613	47,238	420	46,417
1-Jun-00	481	47,753	317	39,440	611	47,849	396	46,814
1-Jul-00	373	48,126	314	39,756	609	48,458	374	47,188
1-Aug-00	347	48,473	311	40,068	606	49,064	353	47,541
1-Sep-00	321	48,794	307	40,377	604	49,668	333	47,875
1-Oct-00	323	49,117	304	40,683	602	50,270	315	48,189
1-Nov-00	282	49,399	301	40,985	600	50,870	297	48,486
1-Dec-00	278	49,677	298	41,285	597	51,467	280	48,767
1-Jan-01	288	49,965	295	41,581	595	52,062	265	49,031
1-Feb-01	238	50,203	292	41,875	593	52,655	250	49,281
1-Mar-01	250	50,453	289	42,166	591	53,246	236	49,517
1-Apr-01	228	50,681	286	42,453	588	53,834	223	49,739
1-May-01	213	50,894	284	42,738	586	54,420	210	49,949
1-Jun-01	199	51,093	281	43,021	584	55,004	198	50,148
1-Jul-01	201	51,294	278	43,300	582	55,586	187	50,335
1-Aug-01	115	51,409	276	43,577	580	56,166	177	50,511
1-Sep-01	142	51,551	273	43,852	578	56,744	167	50,678
1-Oct-01	153	51,704	271	44,123	575	57,319	157	50,835
1-Nov-01	120	51,824	268	44,393	573	57,892	149	50,984
1-Dec-01	27	51,851	266	44,660	571	58,463	140	51,124
1-Jan-02	141	51,992	263	44,924	569	59,032	132	51,256
1-Feb-02	126	52,118	261	45,186	567	59,599	125	51,381
1-Mar-02	150	52,268	259	45,446	565	60,164	118	51,499
1-Apr-02	152	52,420	256	45,704	563	60,726	111	51,610
1-May-02	134	52,554	254	45,959	560	61,287	105	51,715
1-Jun-02	123	52,677	252	46,212	558	61,845	99	51,814
1-Jul-02	128	52,805	250	46,463	556	62,401	94	51,908
1-Aug-02	134	52,939	248	46,712	554	62,955	88	51,996
1-Sep-02	126	53,065	246	46,959	552	63,507	83	52,080
1-Oct-02	126	53,191	244	47,204	550	64,057	79	52,158
1-Nov-02	117	53,308	242	47,446	548	64,605	74	52,233
1-Dec-02	97	53,405	240	47,687	546	65,151	70	52,303
1-Jan-03	118	53,523	238	47,926	544	65,694	66	52,369

Date	Actual Oil Prod. Rate (BOPM)	Cumulative Oil Prod. (BO)	Duong Model		SEDM		Hybrid Model	
			Predicted Rates (BOPM)	Cumulative (BO)	Predicted Rates (BOPM)	Cumulative (BO)	Predicted Rates (BOPM)	Cumulative (BO)
1-Feb-03	95	53,618	236	48,163	542	66,236	62	52,431
1-Mar-03	105	53,723	234	48,398	540	66,776	59	52,490
1-Apr-03	86	53,809	232	48,631	538	67,313	56	52,546
1-May-03	86	53,895	230	48,862	536	67,849	53	52,598
1-Jun-03	89	53,984	229	49,091	534	68,382	50	52,648
1-Jul-03	84	54,068	227	49,319	531	68,914	47	52,695
1-Aug-03	89	54,157	225	49,545	529	69,443	44	52,739
1-Sep-03	106	54,263	223	49,769	527	69,971	42	52,781
1-Oct-03	166	54,429	222	49,992	525	70,496	39	52,820
1-Nov-03	106	54,535	220	50,213	523	71,020	37	52,857
1-Dec-03	121	54,656	218	50,432	521	71,541	35	52,892
1-Jan-04	97	54,753	217	50,650	520	72,061	33	52,925
1-Feb-04	97	54,850	215	50,866	518	72,578	31	52,956
1-Mar-04	91	54,941	214	51,080	516	73,094	29	52,986
1-Apr-04	99	55,040	212	51,293	514	73,607	28	53,014
1-May-04	102	55,142	211	51,505	512	74,119	26	53,040
1-Jun-04	103	55,245	209	51,715	510	74,629	25	53,065
1-Jul-04	82	55,327	208	51,923	508	75,137	23	53,088
1-Aug-04	60	55,387	206	52,130	506	75,643	22	53,110
1-Sep-04	105	55,492	205	52,335	504	76,146	21	53,131
1-Oct-04	89	55,581	203	52,540	502	76,648	20	53,151
1-Nov-04	99	55,680	202	52,742	500	77,149	19	53,169
1-Dec-04	83	55,763	201	52,944	498	77,647	18	53,187
1-Jan-05	109	55,872	199	53,143	496	78,143	17	53,203
1-Feb-05	92	55,964	198	53,342	494	78,637	16	53,219
1-Mar-05	95	56,059	197	53,539	493	79,130	15	53,234
1-Apr-05	100	56,159	195	53,735	491	79,621	14	53,247
1-May-05	85	56,244	194	53,930	489	80,109	13	53,261
1-Jun-05	90	56,334	193	54,123	487	80,596	12	53,273
1-Jul-05	71	56,405	191	54,315	485	81,081	12	53,285
1-Aug-05	81	56,486	190	54,506	483	81,564	11	53,296
1-Sep-05	78	56,564	189	54,696	481	82,046	10	53,306
1-Oct-05	86	56,650	188	54,884	479	82,525	10	53,316
1-Nov-05	91	56,741	187	55,072	478	83,003	9	53,325

Date	Actual Oil Prod. Rate (BOPM)	Cumulative Oil Prod. (BO)	Duong Model		SEDM		Hybrid Model	
			Predicted Rates (BOPM)	Cumulative (BO)	Predicted Rates (BOPM)	Cumulative (BO)	Predicted Rates (BOPM)	Cumulative (BO)
1-Dec-05	71	56,812	185	55,258	476	83,479	9	53,334
1-Jan-06	41	56,853	184	55,443	474	83,953	8	53,342
1-Feb-06	12	56,865	183	55,626	472	84,425	8	53,350
1-Mar-06	73	56,938	182	55,809	470	84,895	7	53,357
1-Apr-06	18	56,956	181	55,990	469	85,364	7	53,364
1-May-06	105	57,061	180	56,171	467	85,831	7	53,371
1-Jun-06	63	57,124	179	56,350	465	86,296	6	53,377
1-Jul-06	13	57,137	178	56,528	463	86,759	6	53,383
1-Aug-06	55	57,192	177	56,705	461	87,220	6	53,389
1-Sep-06	14	57,206	176	56,881	460	87,680	5	53,394
1-Oct-06	19	57,225	174	57,056	458	88,138	5	53,399
1-Nov-06	3	57,228	173	57,230	456	88,594	5	53,403
1-Dec-06	27	57,255	172	57,403	454	89,048	4	53,408
1-Jan-07	2	57,257	171	57,575	453	89,501	4	53,412
1-Feb-07	14	57,271	170	57,746	451	89,952	4	53,416
1-Mar-07	40	57,311	169	57,916	449	90,401	4	53,419
1-Apr-07	34	57,345	168	58,085	447	90,849	3	53,423
1-May-07	20	57,365	168	58,253	446	91,294	3	53,426
1-Jun-07	8	57,373	167	58,420	444	91,738	3	53,429
1-Jul-07	7	57,380	166	58,586	442	92,181	3	53,432
1-Aug-07	18	57,398	165	58,752	441	92,621	3	53,435
1-Sep-07	60	57,458	164	58,916	439	93,060	3	53,438
1-Oct-07	39	57,497	163	59,079	437	93,498	2	53,440
1-Nov-07	121	57,618	162	59,242	436	93,933	2	53,442
1-Dec-07	123	57,741	161	59,403	434	94,367	2	53,445
1-Jan-08	251	57,992	160	59,564	432	94,799	2	53,447
1-Feb-08	248	58,240	159	59,724	431	95,230	2	53,449
1-Mar-08	99	58,339	159	59,883	429	95,659	2	53,450
1-Apr-08	11	58,350	158	60,041	427	96,086	2	53,452
1-May-08	15	58,365	157	60,198	426	96,511	2	53,454
1-Jun-08	30	58,395	156	60,354	424	96,935	2	53,455
1-Jul-08	30	58,425	155	60,510	422	97,358	1	53,457
1-Aug-08	16	58,441	154	60,665	421	97,778	1	53,458
1-Sep-08	17	58,458	154	60,819	419	98,197	1	53,459

Date	Actual Oil Prod. Rate (BOPM)	Cumulative Oil Prod. (BO)	Duong Model		SEDM		Hybrid Model	
			Predicted Rates (BOPM)	Cumulative (BO)	Predicted Rates (BOPM)	Cumulative (BO)	Predicted Rates (BOPM)	Cumulative (BO)
1-Oct-08	120	58,578	153	60,972	417	98,615	1	53,461
1-Nov-08	60	58,638	152	61,124	416	99,030	1	53,462
1-Dec-08	120	58,758	151	61,276	414	99,445	1	53,463
1-Jan-09	14	58,772	150	61,426	413	99,857	1	53,464
1-Mar-09	10	58,782	150	61,576	411	100,268	1	53,465
1-Apr-09	5	58,787	149	61,726	409	100,678	1	53,466
1-May-09	38	58,825	148	61,874	408	101,086	1	53,467
1-Jun-09	28	58,853	147	62,022	406	101,492	1	53,468
1-Jul-09	17	58,870	147	62,169	405	101,897	1	53,468
1-Aug-09	28	58,898	146	62,315	403	102,300	1	53,469
1-Sep-09	26	58,924	145	62,461	402	102,701	1	53,470
1-Oct-09	26	58,950	144	62,605	400	103,101	1	53,470
1-Nov-09	26	58,976	144	62,750	398	103,500	1	53,471
1-Dec-09	32	59,008	143	62,893	397	103,897	1	53,472
1-Jan-10	27	59,035	142	63,036	395	104,292	1	53,472
1-Feb-10	28	59,063	142	63,178	394	104,686	1	53,473
1-Mar-10	24	59,087	141	63,319	392	105,078	0	53,473
1-Apr-10	22	59,109	140	63,460	391	105,469	0	53,474
1-May-10	31	59,140	140	63,600	389	105,858	0	53,474
1-Jun-10	15	59,155	139	63,739	388	106,246	0	53,474
1-Jul-10	31	59,186	138	63,878	386	106,632	0	53,475
1-Aug-10	24	59,210	138	64,016	385	107,017	0	53,475
1-Sep-10	20	59,230	137	64,153	383	107,400	0	53,476
1-Oct-10	27	59,257	136	64,290	382	107,782	0	53,476
1-Nov-10	36	59,293	136	64,426	380	108,162	0	53,476
1-Dec-10	28	59,321	135	64,561	379	108,541	0	53,476
1-Jan-11	14	59,335	135	64,696	377	108,919	0	53,477
1-Feb-11	36	59,371	134	64,830	376	109,294	0	53,477
1-Mar-11	37	59,408	133	64,964	374	109,669	0	53,477
1-Apr-11	17	59,425	133	65,097	373	110,042	0	53,477
1-May-11	14	59,439	132	65,229	372	110,413	0	53,478
1-Jun-11	27	59,466	131	65,361	370	110,783	0	53,478
1-Jul-11	8	59,474	131	65,492	369	111,152	0	53,478
1-Aug-11	28	59,502	130	65,623	367	111,519	0	53,478

Date	Actual Oil Prod. Rate (BOPM)	Cumulative Oil Prod. (BO)	Duong Model		SEDM		Hybrid Model	
			Predicted Rates (BOPM)	Cumulative (BO)	Predicted Rates (BOPM)	Cumulative (BO)	Predicted Rates (BOPM)	Cumulative (BO)
1-Sep-11	29	59,531	130	65,753	366	111,885	0	53,478
1-Oct-11	134	59,665	129	65,882	364	112,249	0	53,479
1-Nov-11	31	59,696	129	66,011	363	112,612	0	53,479
1-Dec-11	21	59,717	128	66,139	362	112,974	0	53,479
1-Jan-12	13	59,730	127	66,267	360	113,334	0	53,479
1-Feb-12	30	59,760	127	66,394	359	113,693	0	53,479
1-Jun-12	2	59,762	126	66,521	357	114,050	0	53,479
1-Jul-12	2	59,764	126	66,647	356	114,406	0	53,479
1-Aug-12	2	59,766	125	66,772	355	114,760	0	53,480
1-Sep-12	2	59,768	125	66,897	353	115,114	0	53,480
1-Oct-12	3	59,771	124	67,022	352	115,465	0	53,480
1-Nov-12	10	59,781	124	67,146	350	115,816	0	53,480
1-Dec-12	18	59,799	123	67,269	349	116,165	0	53,480
1-Jan-13	27	59,826	123	67,392	348	116,513	0	53,480
1-Feb-13	69	59,895	122	67,514	346	116,859	0	53,480
1-Mar-13	16	59,911	122	67,636	345	117,204	0	53,480

CHAPTER

12 Fluid Flow in Petroleum Reservoirs

12.1 Introduction

In previous chapters, various methods for the calculation of volumes of hydrocarbons in-place in petroleum reservoirs and potential recoveries were presented. These methods do not provide the engineer with techniques for estimating the rates at which the hydrocarbon reserves could be produced. Determination of achievable production rates is a critical element in the economic evaluation of hydrocarbon resources. If the predicted production rates are low due to geologic barriers and/or rock and fluid properties, economic development of the hydrocarbon resource may be delayed or in some cases abandoned if these factors that limit capacity to produce are not resolved successfully in favor of the project. This requirement is especially important in high cost environments, such as deepwater reservoirs, where high production rates are generally required for economic viability of the project. The rate of fluid production at the wells is controlled by the rate of fluid flow in the reservoir. The main factors that control fluid flow in petroleum reservoirs are the permeability of the formation, the viscosity of the fluid, and the pressure difference between the well and the reservoir. (This pressure difference is called the **drawdown.**) The relationship between these factors and fluid flow in reservoirs was established in 1856 by Henry Darcy in an equation that still bears his name. Most equations that represent fluid flow in petroleum reservoirs are based on the Darcy equation. In this chapter, the radial form of the Darcy flow equation is presented for single-phase flow of three main types of fluids. The three main types of fluids that are considered are incompressible fluids, slightly compressible fluids, and compressible fluids. During fluid flow, the state of pressure distribution in the reservoir depends on duration of the flow (time) and distance from the well. Depending on flow time and size of the reservoir, three flow regimes can be distinguished. These are transient flow (or unsteady-state flow), pseudosteady-state flow (or semi-steady state flow), and steady-state flow. The characteristics of each flow regime are presented in this chapter. The continuity equation that governs fluid flow in a porous medium such as petroleum reservoirs is derived. From the continuity equations, diffusivity equations for slightly compressible fluids and compressible fluids are derived. Mathematical solutions to the diffusivity equations for the two types of fluids are presented for transient and pseudosteady-state flow regimes. The theory and application of the principle of superposition are illustrated with examples for the single-well, multi-production rates case; the multi-well reservoir system case; and the single well near a sealing boundary case.

12.2 Fluid Types

The three types of fluids discussed in this section are incompressible fluids, slightly compressible fluids, and compressible fluids. These classifications are based on the response of the fluid types to pressure changes at constant temperature.

12.2.1 Incompressible Fluids

Incompressible fluids are described as fluids whose volume does not change with pressure at constant temperature. However, all fluids are compressible to some degree. Hence, incompressible fluids do not exist in practice. This class of fluids has been defined to simplify the derivation of certain forms of the Darcy equation for radial flow. These equations are widely used to represent fluid flow and are sufficiently accurate for most engineering applications.

12.2.2 Slightly Compressible Fluids

The coefficient of isothermal compressibility, c, of any fluid is expressed as:

$$c = -\frac{1}{V}\left(\frac{\partial V}{\partial p}\right)_T \tag{12.1}$$

For slightly compressible fluids, it is assumed that the compressibility coefficient is small and constant over the range of pressures under consideration. Equation (12.1) can be integrated between a low base pressure, p_b, where the volume is V_b and the pressure of interest, p, at volume V to give the expression:

$$\int_{p_b}^{p} -c dp = \int_{V_b}^{V} \frac{dV}{V} \tag{12.2}$$

Integrating Eq. (12.2) gives:

$$c\left(p_b - p\right) = \ln\left(\frac{V}{V_b}\right) \tag{12.3}$$

Rearranging Eq. (12.3) gives:

$$V = V_b\left[e^{c(p_b - p)}\right] \tag{12.4}$$

The exponential function, e^x, can be defined by a power series as:

$$e^x = \sum_{n=0}^{\infty}\frac{x^n}{n!} = 1 + x + \frac{x^2}{2!} + \frac{x^3}{3!} + \frac{x^4}{4!} + \ldots.. \tag{12.5}$$

Applying the expression in Eq. (12.5), for small values of x, Eq. (12.4) can be written as:

$$V = V_b\left[1 + c\left(p_b - p\right)\right] \tag{12.6}$$

Eq. (12.6) is applicable to slightly compressible fluids with small and constant compressibility coefficient. Most liquids can be classified as slightly compressible, especially undersaturated oil.

12.2.3 Compressible Fluids

Generally, all gases can be classified as compressible fluids. As shown in Eq. (5.90), gas compressibility, c_g, can be expressed as:

$$c_g = -\frac{1}{V}\left(\frac{\partial V}{\partial p}\right)_T$$

Note that this equation is equivalent to Eq. (12.1) defined above for slightly compressible fluids. Applying real gas equation of state, it was shown in Eq. (5.95) that gas compressibility for real gases can be expressed as:

$$c_g = \frac{1}{p} - \frac{1}{z}\left(\frac{\partial z}{\partial p}\right)_T$$

By comparing the above equation for gas compressibility to Eq. (12.3) for liquid compressibility, it is evident that gas compressibility is several times higher than liquid compressibility for the same pressure change.

12.3 Definition of Fluid Flow Regimes

The three types of fluid flow regimes discussed in this section are transient flow, pseudosteady-state flow, and steady-state flow. The flow regimes depend on elapsed time and distance from the wellbore and have characteristic pressure distributions in the reservoir.

12.3.1 Transient Flow

Suppose you are standing at the banks of a calm, motionless body of water such as a small lake. If you picked up a piece of rock and threw it into this body of water, a radial system of waves will radiate toward you from the point of entry of the rock. Before the waves reach the banks of this body of water, pressure distributions (or pressure transients) caused by your action are changing rapidly and are not constant. They depend only on distance from the point of entry of the rock and the elapsed time since the rock pierced the surface of the water. This is the transient phase of wave motion caused by the rock you threw into the body of water. This analogy can be extended to a circular reservoir with uniform thickness and homogeneous, isotropic properties (means constant rock and fluid properties in all directions). Suppose a single well exists at the center of this circular reservoir. The radius of the single wellbore is r_w, and the radius of the outer boundary of the circular reservoir is r_e. A schematic diagram of this simple reservoir is represented as Figure 12.1. If fluid production at a constant rate, q_w, is initiated at this well at time $t = 0$, pressure transients will radiate from the well and travel toward the outer boundaries of the reservoir. These pressure transients are similar to the waves caused by the rock you threw travelling toward the banks of the lake. The rate of travel of the pressure transient is independent of the production rate at the well but is proportional to a property of the reservoir termed formation diffusivity. The formation diffusivity, η, is defined as:

$$\eta = \frac{k}{\phi\mu c_t} \tag{12.7}$$

In Eq. (12.7), k = formation permeability; ϕ = formation porosity; μ = fluid viscosity; and c_t = total compressibility of the system. As long as the leading front of the pressure transient has not reached the outer boundaries of the reservoir, fluid flow toward the well is in the transient state, and the reservoir acts as if it is infinite in size in response to the well. During the transient flow phase, pressure distribution in the reservoir is not constant and depends on elapsed time and distance from the well. This transient flow regime is illustrated in Figure 12.2 for the circular reservoir in Figure 12.1. Note in Figure 12.2 that reservoir pressure at the outer boundary is still at initial pressure, p_i. Also note that transient flow is also called unsteady-state flow.

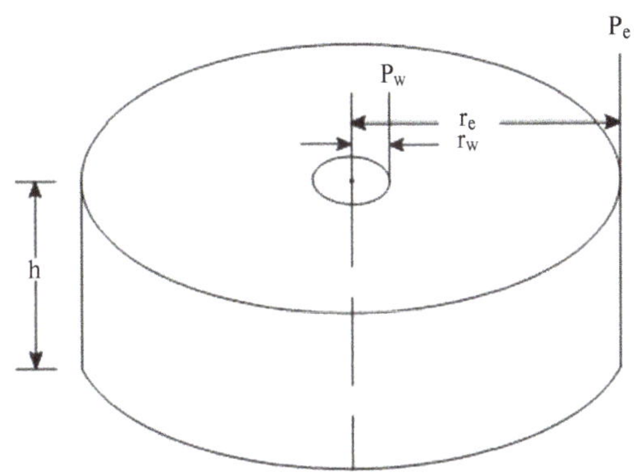

▲ **FIGURE 12.1** Diagram of a circular reservoir with a well at center

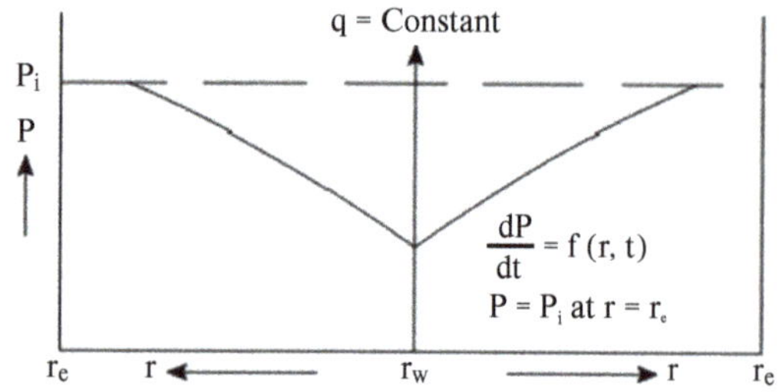

▲ **FIGURE 12.2 Transient flow diagram showing pressure distribution**

12.3.2 Pseudosteady-State (PSS) Flow

The pseudosteady-state (PSS) flow regime is sometimes called semi-steady-state flow. Using the illustration in Figure 12.1, PSS flow is achieved in the reservoir after the pressure transient has reached the outer boundaries of the reservoir and the pressure distribution is changing at a constant rate. This is illustrated in Figure 12.3. Note that as shown in Figure 12.3, pressure gradient at the outer boundary is zero. Hence, fluid flow across the outer boundary is also zero. The time, t_{pss}, for the start of PSS flow is given by:

$$t_{pss} = \frac{948\phi\mu c_t r_e^2}{k} \tag{12.8}$$

In Eq. (12.8), t_{pss} = time, hours; ϕ = formation porosity, fraction; μ = fluid viscosity, cp; c_t = total compressibility of system, psi^{-1}; r_e = outer boundary radius, feet; and k = formation permeability, md.

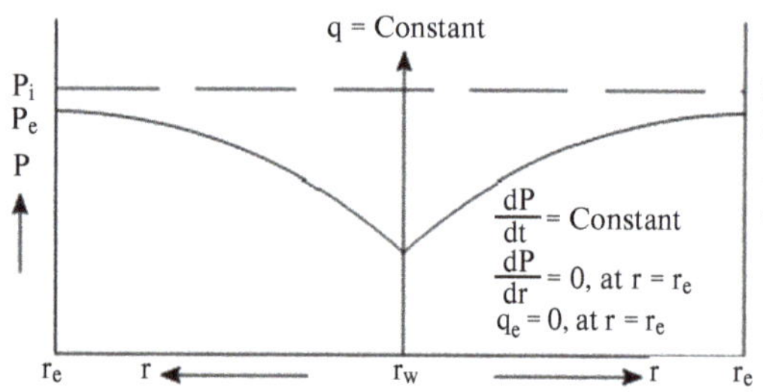

▲ **FIGURE 12.3 Pseudosteady-state flow diagram showing pressure distribution**

Note that there is a short time period under most flow conditions between the end of the transient flow regime and the establishment of the PSS flow regime. This time period has been called the Late Transient Flow (LTF) regime. The duration of the LTF regime is very short and is not considered in most flow systems. For these reasons, it is not covered in this book. However, the time to reach the start of the late transient flow period can be estimated with the following equation:

$$t_{LTF} = \frac{240\phi\mu c_t r_e^2}{k} \tag{12.9}$$

The units of the terms in Eq. (12.9) are the same as in Eq. (12.8). Equations (12.8) and (12.9) apply to reservoirs containing slightly compressible fluids such as oil. For gas reservoirs, the start of PSS flow is given by:

$$t_{pss} = \frac{948\phi\mu r_e^2}{k\overline{p}} \tag{12.10}$$

Also, for gas reservoirs the start of the late transient flow period can be estimated with:

$$t_{LTF} = \frac{240\phi\mu r_e^2}{k\overline{p}} \tag{12.11}$$

The average pressure, $\overline{p}$, in Eqs. (12.10) and (12.11) can be estimated as the arithmetic average of initial pressure and flowing well pressure. Other methods for estimating average pressure are presented later in Eqs. (12.95) and (12.96). If the flowing well pressure is not available, average pressure can be assumed to be equal to initial reservoir pressure if the reservoir has not been depleted significantly. In Eqs. (12.10) and (12.11), μ = gas viscosity, cp; ϕ = formation porosity, fraction; r_e = outer boundary radius, feet; k = formation permeability, md; and $\overline{p}$ = average reservoir pressure, psi.

12.3.3 Steady-State (SS) Flow

When the circular reservoir in Figure 12.1 reaches steady-state (SS) flow conditions, the pressure distribution at any point in the reservoir will not change with time. This condition is illustrated in Figure 12.4. Note that as shown in Figure 12.4 at SS flow conditions, the reservoir pressure at the outer boundary is constant, and the rate of fluid flow into the reservoir at the outer boundary is equal to the flow rate at the well. The SS flow regime is best represented by aquifer influx cases where the rate of water influx into the reservoir is considered equal to the total rate of fluid withdrawal from the reservoir.

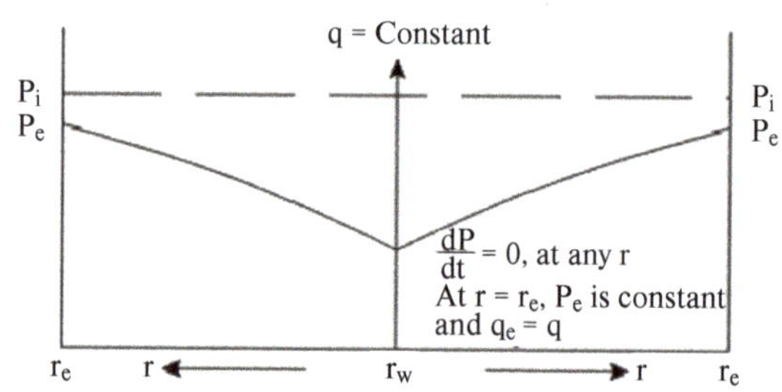

▲ **FIGURE 12.4 Steady-state flow diagram showing pressure distribution**

12.4 Darcy Fluid Flow Equation

In 1856, Henry Darcy demonstrated through a series of experiments that the flow velocity of a homogenous fluid through a porous medium under laminar (non-turbulent) conditions is proportional to the potential gradient. Darcy's law is expressed for linear flow as:

$$v = -\frac{k\rho}{\mu}\frac{d\Phi}{dl} \tag{12.12}$$

In Eq. (12.12), k = permeability; ρ = fluid density; μ = fluid viscosity; Φ = potential; and l = length of flow path. If the fluid is assumed to be incompressible, the potential can be represented as:

$$\Phi = \frac{p}{\rho} + gZ \tag{12.13}$$

In Eq. (12.13), p = pressure; Z = elevation above a datum plane; and g = acceleration due to gravity. Substituting Eq. (12.13) into Eq. (12.12) gives:

$$v = -\frac{k}{\mu}\left(\frac{dp}{dl} + \rho g \frac{dZ}{dl}\right) \tag{12.14}$$

In terms of Darcy units, Eq. (12.14) becomes:

$$v = -\frac{k}{\mu}\left(\frac{dp}{dl} + \frac{\rho g}{1.0133\times10^{6}}\frac{dZ}{dl}\right) \tag{12.15}$$

In Eq. (12.15), v = apparent fluid velocity, cm/sec; k = permeability, darcy; μ = fluid viscosity, cp; p = pressure, atm; l = length, cm; ρ = fluid density, gm/cm^3; g = acceleration due to gravity, cm/sec^2; and Z = elevation, cm. In field units, Eq. (12.15) converts to:

$$q = -1.127\times10^{-3}\frac{kA}{\mu B}\left[\frac{dp}{dl} + 0.4335\gamma\sin\theta\right] \tag{12.16}$$

There are several important terms to note in Eq. (12.16). The volumetric flow rate, q, is in STB/D. The volumetric flow rate is related to the apparent fluid velocity by the expression:

$$q = \frac{vA}{B} \tag{12.17}$$

In Eqs. (12.16) and (12.17), v = apparent fluid velocity, res bbl/day/ft^2; A = total cross-sectional area, ft^2; and B = formation volume factor, RB/STB. Also in Eq. (12.16), k = permeability, md; μ = fluid viscosity, cp; p = pressure, psia; l = flow path length, ft; γ = fluid specific gravity (water = 1); and θ = dip angle of the reservoir or formation measured counter-clockwise from the horizontal to the positive flow path. The negative sign in Eq. (12.16) accounts for the sign convention that flow is considered positive in the positive direction of the flow path length, l, and pressure decreases in the direction of flow. The constant, 1.127×10^{-3}, is the conversion term from Darcy units to field units. The gravity term, $(0.4335\gamma\sin\theta)$, in Eq. (12.16) is small in comparison to the pressure gradient, (dp/dl), in most petroleum engineering applications of the linear form of the Darcy flow equation. For this reason, it is generally ignored. By ignoring the gravity term, Eq. (12.16) becomes:

$$q = -1.127\times10^{-3}\frac{kA}{\mu B}\left[\frac{dp}{dl}\right] \tag{12.18}$$

Assuming all the terms in Eq. (12.18) are constant except pressure, p, and length, l, of the linear flow system and integrating between its limits, Eq. (12.18) becomes:

$$q\int_0^L dl = -1.127\times10^{-3}\frac{kA}{\mu B}\int_{p_1}^{p_2} dp \text{ where } p_1 > p_2 \tag{12.19}$$

Integration of Eq. (12.19) and rearrangement of the terms gives:

$$q = 1.127\times10^{-3}\frac{kA(p_1 - p_2)}{\mu BL} \tag{12.20}$$

Eq. (12.20) is the steady-state, linear flow equation for incompressible fluids.

12.5 Radial Forms of the Darcy Equation

In this section, radial forms of the Darcy equation for incompressible and compressible fluids are presented for PSS and SS flows. Also, methods for averaging permeabilities for parallel and serial beds are derived.

12.5.1 Steady-State Flow, Incompressible Fluids

Starting from Eq. (12.16) and assuming the gravity term is negligible, the Darcy equation can be represented in radial form in field units as:

$$\frac{qB}{2\pi rh} = 1.127\times10^{-3}\frac{k}{\mu}\frac{dp}{dr} \tag{12.21}$$

Note that in Eq. (12.21), the negative sign was dropped because pressure is decreasing in the same direction as decreasing radius (see Figure 12.1). Separating the variables and integrating between the wellbore radius, r_w, and the external reservoir boundary, r_e, within the corresponding pressure limits, p_w and p_e, respectively, gives:

$$q\int_{r_w}^{r_e}\frac{dr}{r} = \frac{1.127\times10^{-3}\times2\pi kh}{\mu B}\int_{p_w}^{p_e} dp \tag{12.22}$$

After integration and simplification, Eq. (12.22) reduces to:

$$q = \frac{0.00708kh(p_e - p_w)}{\mu B\ln\left(\frac{r_e}{r_w}\right)} \tag{12.23}$$

Eq. (12.23) is the radial flow equation for an incompressible fluid under steady-state conditions. In Eq. (12.23), q = well production rate, STB/D; k = average permeability of the drainage area of the well, md; h = average reservoir thickness, feet; p_e = pressure at external reservoir boundary or drainage area, psia; p_w = pressure at the wellbore, psia; μ = fluid viscosity, cp; B = fluid formation volume factor, RB/STB; r_e = external radius, feet; and r_w = wellbore radius, feet. Equation (12.23) is generally used to calculate flow rates in reservoirs where the conditions are assumed to be close to steady state. These are mainly reservoirs where pressure is maintained by water or gas injection or reservoirs with active water influx from aquifers. Note that p_e can be estimated as the average static pressure of the reservoir or drainage area of the well.

12.5.2 Average Permeability of Parallel Beds

Consider the sketch in Figure 12.5, which represents a formation consisting of sand beds of varying thickness and permeability producing into a common wellbore. If it is assumed that the fluids flowing in the sand beds have the same properties and the same drawdown was applied across all the sand beds, then the total flow rate of the well is the sum of the flow rate from each sand bed. This statement can be expressed as:

$$q_t = q_1 + q_2 + q_3 + \ldots\ldots + q_n, \text{ if there are } n \text{ sand beds in the formation} \quad (12.24)$$

Applying Eq. (12.23) to Eq. (12.24) gives:

$$\frac{0.00708 k_{avg} h_t (p_e - p_w)}{\mu B \ln(r_e/r_w)} = \frac{0.00708 k_1 h_1 (p_e - p_w)}{\mu B \ln(r_e/r_w)} + \frac{0.00708 k_2 h_2 (p_e - p_w)}{\mu B \ln(r_e/r_w)} + \ldots + \frac{0.00708 k_n h_n (p_e - p_w)}{\mu B \ln(r_e/r_w)} \quad (12.25)$$

Cancelling like terms in Eq. (12.25) yields:

$$k_{avg} h_t = k_1 h_1 + k_2 h_2 + \ldots.. + k_n h_n$$

$$k_{avg} = \frac{\sum_i^n k_i h_i}{\sum_i^n h_i} \quad (12.26)$$

The product (kh) is a measure of the productive capacity of a sand bed or an interval. Assuming the intervals contain fluids with similar properties, the (kh) product is frequently used in a qualitative sense to assess the relative productivity of intervals in a well or a reservoir.

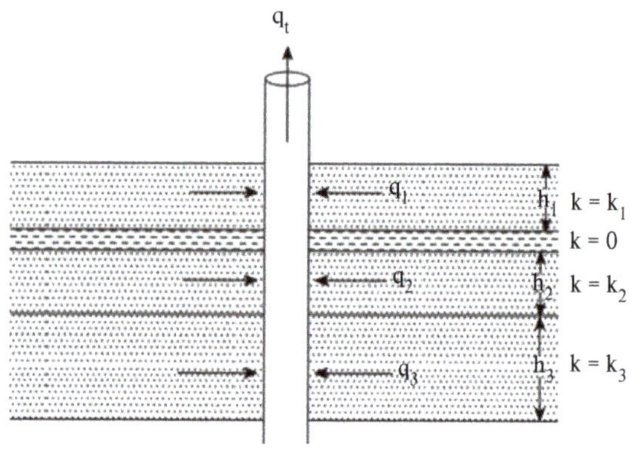

▲ **FIGURE 12.5 Radial flow in parallel beds of different permeabilities**

EXAMPLE 12.1 Calculation of Average Permeability for Parallel Beds

Problem

Calculate the average permeability of the reservoir in Figure 12.5 if the permeability and thickness of the three beds (intervals) are as follows:

Interval	Permeability (md)	Thickness (Feet)
1	500	15
2	350	17
3	200	22

Solution

Using Eq. (12.26), the average permeability of the reservoir is calculated as:

$$k_{avg} = \frac{\sum_i^n k_i h_i}{\sum_i^n h_i}$$

$$= \frac{500 \times 15 + 350 \times 17 + 200 \times 22}{15 + 17 + 22}$$

$$= 330.56 \text{ md}$$

End of Example 12.1

12.5.3 Average Permeability of Serial Concentric Segments

Consider the diagram in Figure 12.6, which shows radial flow in serial, concentric segments with different permeabilities but the same thickness. The flow through each segment is the same but the pressure drop across each segment is different. The total pressure drop across all the segments is equal to the sum of the pressure drop across each segment. This statement is represented in the form of an equation as:

$$(p_e - p_w) = (p_e - p_2) + (p_2 - p_1) + (p_1 - p_w) \qquad (12.27)$$

Substituting Eq. (12.27) with Eq. (12.23) gives:

$$\frac{\mu B \ln(r_e/r_w) q}{0.00708 k_{avg} h} = \frac{\mu B \ln(r_e/r_2) q}{0.00708 k_e h} + \frac{\mu B \ln(r_2/r_1) q}{0.00708 k_2 h} + \frac{\mu B \ln(r_1/r_w) q}{0.00708 k_1 h} \qquad (12.28)$$

Note as shown in Figure 12.6, $k_e = k_3$, and $r_e = r_3$.

Eliminating similar terms in Eq. (12.28) results in:

$$\frac{\ln(r_e/r_w)}{k_{avg}} = \frac{\ln(r_e/r_2)}{k_e} + \frac{\ln(r_2/r_1)}{k_2} + \frac{\ln(r_1/r_w)}{k_1} \qquad (12.29)$$

Eq. (12.29) can be rearranged so that:

$$k_{avg} = \frac{\ln(r_e/r_w)}{\frac{1}{k_e}\ln\left(\frac{r_e}{r_2}\right) + \frac{1}{k_2}\ln\left(\frac{r_2}{r_1}\right) + \frac{1}{k_1}\ln\left(\frac{r_1}{r_w}\right)} \qquad (12.30)$$

A general form of Eq. (12.30) for n segments in series is:

$$k_{avg} = \frac{\ln\left(\frac{r_e}{r_w}\right)}{\sum_{i=1}^{n} \frac{1}{k_i}\ln\left(\frac{r_i}{r_{i-1}}\right)} \qquad (12.31)$$

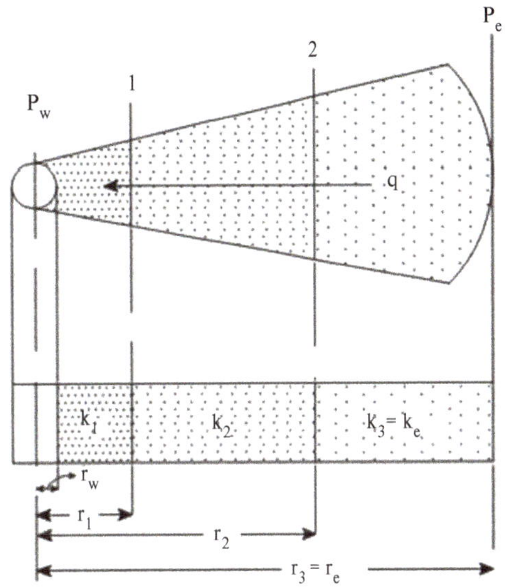

▲ **FIGURE 12.6** Radial flow in serial concentric segments

EXAMPLE 12.2 Calculation of Average Permeability for Concentric Segments in Series

Problem

Suppose the drainage area of a well consists of three segments with different permeabilities. A damaged zone extending 10 feet from the wellbore has a reduced permeability of 25 md. This is followed by another segment extending 300 feet from the wellbore with a permeability of 350 md. Beyond this segment, the permeability of the formation is 250 md. The drainage radius of the well is 1000 feet, and the wellbore radius is 0.5 feet. Calculate the average permeability of the drainage area.

Solution

From Eq. (12.31), the average permeability of the drainage area is:

$$k_{avg} = \frac{\ln\left(\frac{r_e}{r_w}\right)}{\sum_{i=1}^{n} \frac{1}{k_i}\ln\left(\frac{r_i}{r_{i-1}}\right)}$$

$$= \frac{\ln(r_e/r_w)}{\frac{1}{k_e}\ln\left(\frac{r_e}{r_2}\right) + \frac{1}{k_2}\ln\left(\frac{r_2}{r_1}\right) + \frac{1}{k_1}\ln\left(\frac{r_1}{r_w}\right)}$$

$$= \frac{\ln\left(\frac{1000}{0.5}\right)}{\frac{1}{250}\ln\left(\frac{1000}{300}\right) + \frac{1}{350}\ln\left(\frac{300}{10}\right) + \frac{1}{25}\ln\left(\frac{10}{0.5}\right)}$$

$$= 56.57 \text{ md}$$

End of Example 12.2

EXAMPLE 12.3 Calculation of Well Production Rate in Steady-State Flow

Problem

Calculate the oil production rate of a well in reservoir whose pressure is maintained at 3500 psia by water injection. The flowing well pressure is 500 psia. Other fluid and reservoir data are as given:

Wellbore radius, r_w	0.5 ft
External radius of reservoir, r_e	2500 ft
Average permeability, k	300 md
Average reservoir thickness, h	25 ft
Oil viscosity, μ_o	4.5 cp
Oil FVF, B_o	1.25 RB/STB

Solution

Note that since reservoir pressure was maintained at 3500 psia, fluid properties (oil viscosity and oil FVF) should be evaluated at this pressure and not at the average pressure of 2000 psia between the well and the reservoir. Most of the pressure drop in a flowing well occurs close to the wellbore. Hence, the volumetric average pressure of the reservoir is closer to 3500 psia than 2000 psia. Substituting Eq. (12.23) with the given data, the oil production rate of the well is:

$$q = \frac{0.00708 \times 300 \times 25 \times (3500 - 500)}{4.5 \times 1.25 \times \ln\left(\dfrac{2500}{0.5}\right)}$$

$$= 3325 \text{ STB/D}$$

End of Example 12.3

12.5.4 Pseudosteady State, Incompressible Fluids

The radial form of the Darcy flow equation can be expressed as:

$$q = \frac{2\pi rkh}{\mu}\frac{dp}{dr} \tag{12.32}$$

As discussed previously for PSS flow, the flow rate is zero at the external boundary of the reservoir and increases to the constant flow rate at the wellbore (Figure 12.3). Since the compressibility of the fluid is constant, the flow rate at any radius r in the reservoir is proportional to the reservoir volume between the external radius r_e and radius r. The flow rate q at radius r can be related to the flow rate q_w at the wellbore with radius r_w by the expression:

$$\frac{q}{q_w} = \frac{\left(r_e^2 - r^2\right)}{\left(r_e^2 - r_w^2\right)} \tag{12.33}$$

Substituting Eq. (12.33) into Eq. (12.32) gives:

$$q_w\left(\frac{r_e^2 - r^2}{r_e^2 - r_w^2}\right) = \frac{2\pi rkh}{\mu}\frac{dp}{dr} \tag{12.34}$$

The term r_w^2 is very small in comparison to the term r_e^2. Neglecting the term r_w^2, Eq. (12.34) becomes:

$$q_w\left(1 - \frac{r^2}{r_e^2}\right) = \frac{2\pi rkh}{\mu}\frac{dp}{dr} \tag{12.35}$$

Separating the variables in Eq. (12.35) and showing the limits of integration gives:

$$q_w\int_{r_w}^{r_e}\frac{dr}{r} - \frac{q_w}{r_e^2}\int_{r_w}^{r_e} rdr = \frac{2\pi kh}{\mu}\int_{p_w}^{p_e} dp \tag{12.36}$$

Integrating between the limits as shown in Eq. (12.36) gives:

$$q_w \ln\frac{r_e}{r_w} - \frac{q_w}{r_e^2}\left(\frac{r_e^2}{2} - \frac{r_w^2}{2}\right) = \frac{2\pi kh(p_e - p_w)}{\mu} \tag{12.37}$$

Again, the term r_w^2 is small compared with r_e^2. Neglecting r_w^2, Eq. (12.37) becomes:

$$q_w\left(\ln\frac{r_e}{r_w} - \frac{1}{2}\right) = \frac{2\pi kh(p_e - p_w)}{\mu} \tag{12.38}$$

From Eq. (12.38), the flow rate at the wellbore is given by:

$$q_w = \frac{2\pi kh(p_e - p_w)}{\mu\left[\ln\left(\dfrac{r_e}{r_w}\right) - \dfrac{1}{2}\right]} \tag{12.39}$$

In field units, Eq. (12.39) is converted to:

$$q_w = \frac{0.00708kh(p_e - p_w)}{\mu B\left[\ln\left(\dfrac{r_e}{r_w}\right) - \dfrac{1}{2}\right]} \tag{12.40}$$

The units of the variables in Eq. (12.40) are the same as the units of the variables in Eq. (12.23). Note that Eq. (12.40) is very similar to Eq. (12.23) except for the constant term, $(-1/2)$, that occurs in the denominator of Eq. (12.40). Because of the presence of this term in the denominator, it is obvious that well production rates calculated with Eq. (12.40) will be slightly higher that the rates calculated with Eq. (12.23) using the same data. Equation (12.40) is applicable to depletion-drive reservoirs producing by fluid and rock expansion above the bubblepoint at constant compressibility of the fluid.

EXAMPLE 12.4 Calculation of Well Production Rate in Pseudosteady-State Flow

Problem

Calculate the oil production rate of a well in reservoir whose pressure at the reservoir boundary is estimated to be 3500 psia. The flowing well pressure is 500 psia. Other fluid and reservoir data are as given:

Wellbore radius, r_w	0.5 ft
External radius of reservoir, r_e	2500 ft
Average permeability, k	300 md
Average reservoir thickness, h	25 ft
Oil viscosity, μ_o	4.0 cp
Oil FVF, B_o	1.26 RB/STB
Bubblepoint pressure, p_b	1800 psia

Solution

Note that fluid properties in this example were evaluated at an arithmetic average reservoir pressure of 2000 psia. This approach is a quick approximation since reservoir pressure varies from the wellbore to the boundary of the drainage area. Substituting Eq. (12.40) with the data provided, the well production rate is calculated as:

$$q = \frac{0.00708 \times 300 \times 25 \times (3500 - 500)}{4.0 \times 1.26 \times \left[\ln\left(\frac{2500}{0.5}\right) - \frac{1}{2}\right]}$$

$$= 3{,}942 \text{ STB/D}$$

End of Example 12.4

12.5.5 Steady-State Flow, Compressible Fluids

Gas is the reservoir fluid that is highly compressible at reservoir conditions. Gas flow rate is typically measured in units of standard cubic feet per day (scf/D). This can be converted into flow rate at reservoir conditions using the gas formation volume factor. Gas formation volume factor was derived in Eq. (5.75) as:

$$B_g = \frac{zTp_{sc}}{T_{sc}p} \text{ in } \frac{\text{ft}^3}{\text{scf}}$$

$$= \frac{zTp_{sc}}{5.615T_{sc}p} \text{ in } \frac{\text{RB}}{\text{scf}}$$

Thus, gas flow rate, q, in scf/D is converted to RB/D by:

$$qB_g = q\left(\frac{zTp_{sc}}{5.615T_{sc}p}\right) \tag{12.41}$$

Substituting Eq. (12.41) into Eq. (12.21) gives:

$$\frac{q}{2\pi rh}\left(\frac{zTp_{sc}}{5.615T_{sc}p}\right) = 1.127 \times 10^{-3} \frac{k}{\mu}\frac{dp}{dr} \tag{12.42}$$

Rearranging Eq. (12.42) gives:

$$\frac{q(z\mu)Tp_{sc}}{0.03976T_{sc}pkhr} = \frac{dp}{dr} \tag{12.43}$$

Assuming the product $(z\mu)$ is constant and integrating between the limits, r_e and r_w, (Figure 12.1) gives:

$$\frac{q(z\mu)Tp_{sc}}{0.03976T_{sc}kh}\int_{r_w}^{r_e}\frac{dr}{r} = \int_{p_w}^{p_e} pdp$$

$$\frac{q(z\mu)Tp_{sc}}{0.03976T_{sc}kh}\ln\left(\frac{r_e}{r_w}\right) = \frac{1}{2}\left(p_e^2 - p_w^2\right) \tag{12.44}$$

Rearranging Eq. (12.44) gives:

$$q = \frac{0.01988T_{sc}kh\left(p_e^2 - p_w^2\right)}{(z\mu)Tp_{sc}\ln\left(\frac{r_e}{r_w}\right)} \tag{12.45}$$

In Eq. (12.45), q = gas flow rate, scf/D; T_{sc} = standard temperature, °R; k = permeability, md; h = formation thickness, feet; p_e = external boundary pressure, psia; p_w = wellbore flowing pressure, psia; z = compressibility factor; μ = gas viscosity, cp; T = reservoir temperature, °R; p_{sc} = standard pressure, psia; r_e = external radius, feet; and r_w = wellbore radius, feet. Note that the assumption that the product $(z\mu)$ is constant is generally valid for pressures up to 2000 psia. Consequently, Eq. (12.45) is only applicable for this pressure range.

12.6 Derivation of the Continuity Equation in Radial Form

Consider the volume element of thickness, Δr, in Figure 12.7 located at a distance, r, from a well at the center of the circular system. Suppose there is radial fluid flow through the element toward the well at the center. The reservoir system is assumed to have constant thickness, h, and constant rock and fluid properties. Over a time period, Δt, the principle of mass conservation can be applied over the volume element such that:

Mass flowing into volume element in Δt – Mass flowing out of volume element in Δt = Rate of mass accumulation in volume element in Δt (12.46)

$$\text{Mass rate IN} = 2\pi\left(r+\Delta r\right)h\rho v\Big|_{r+\Delta r} \tag{12.47}$$

$$\text{Mass rate OUT} = 2\pi rh\rho v\Big|_{r} \tag{12.48}$$

Rate of Accumulation

$$= \frac{2\pi r\Delta rh\left[\{\phi\rho\}_{t+\Delta t} - \{\phi\rho\}_{t}\right]}{\Delta t} \tag{12.49}$$

Combining Eqs. (12.47), (12.48), and (12.49) into Eq. (12.46) gives:

$$2\pi\left(r+\Delta r\right)h\rho v\Big|_{r+\Delta r} - 2\pi rh\rho v\Big|_{r} = \frac{2\pi r\Delta rh\left[\{\phi\rho\}_{t+\Delta t} - \{\phi\rho\}_{t}\right]}{\Delta t} \tag{12.50}$$

Dividing both sides of Eq. (12.50) by $2\pi r\Delta rh$ and taking limits as Δr and Δt approach zero gives:

$$\frac{\partial}{\partial r}(\rho v) + \frac{1}{r}(\rho v) = \frac{\partial}{\partial t}(\phi\rho) \tag{12.51}$$

Rearranging Eq. (12.51) gives:

$$\frac{1}{r}\left[\frac{\partial(r\rho v)}{\partial r}\right] = \frac{\partial(\phi\rho)}{\partial t} \tag{12.52}$$

Eq. (12.52) is the continuity equation which represents the conservation of mass for flow in radial systems. In Eq. (12.52), ρ = fluid density; v = fluid flow velocity; and ϕ = porosity of the system.

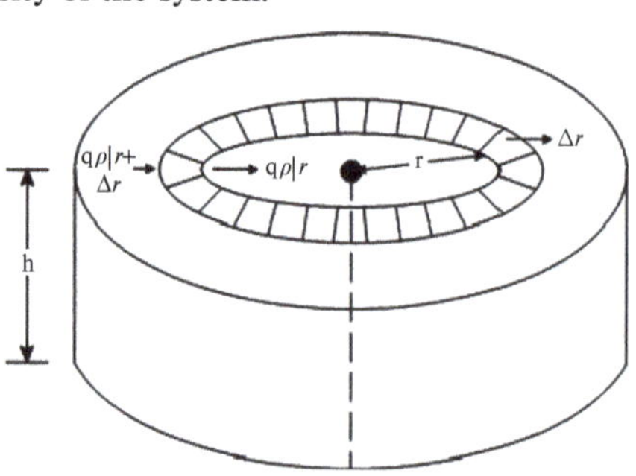

▲ **FIGURE 12.7 Volume element diagram for derivation of the radial continuity equation**

12.7 Derivation of Radial Diffusivity Equation for Slightly Compressible Fluids

In deriving the diffusivity equation for slightly compressible fluids, the following assumptions are made:

1. Compressibility of the fluid is small and constant over the pressure range of interest.
2. Viscosity of the fluid is constant and independent of pressure.
3. Permeability of the medium is constant and isotropic.

The above assumptions are necessary to simplify the derivation of the diffusivity equation for slightly compressible fluids.

The Darcy flow equation expressed in radial form in terms of fluid velocity is represented as:

$$v = 2.637\times10^{-4}\frac{k}{\mu}\frac{\partial p}{\partial r} \tag{12.53}$$

Note that in Eq. (12.53) the negative sign was not included because the pressure gradient necessary for flow to the well decreases as radius r decreases. In Eq. (12.53), v = volumetric fluid velocity, ft^3/hr-ft^2; k = permeability, md; μ = fluid viscosity, cp; p = pressure, psia; and r = radius, ft.

Substituting Eq. (12.53) into Eq. (12.52) and rearranging gives:

$$\frac{1}{r}\frac{\partial}{\partial r}\left(r\rho\frac{\partial p}{\partial r}\right) = \frac{1}{2.637\times10^{-4}}\frac{\mu}{k}\frac{\partial(\phi\rho)}{\partial t} \tag{12.54}$$

Applying the chain rule to the right side and left side of Eq. (12.54) yields:

$$\frac{\rho}{r}\frac{\partial}{\partial r}\left(r\frac{\partial p}{\partial r}\right) + \frac{\partial p}{\partial r}\frac{\partial\rho}{\partial p}\frac{\partial p}{\partial r} = \frac{1}{2.637\times10^{-4}}\frac{\mu}{k}\phi\rho\left(\frac{1}{\phi}\frac{\partial\phi}{\partial p}\frac{\partial p}{\partial t} + \frac{1}{\rho}\frac{\partial\rho}{\partial p}\frac{\partial p}{\partial t}\right) \tag{12.55}$$

From Eq. (12.1), fluid compressibility is expressed as:

$$c = -\frac{1}{V}\left(\frac{\partial V}{\partial p}\right)_T = \frac{1}{\rho}\left(\frac{\partial\rho}{\partial p}\right)_T$$

For an isothermal system, the preceding equation can be rearranged so that:

$$\frac{\partial\rho}{\partial p} = c\rho \tag{12.56}$$

Note that the fluid compressibility, c, in Eq. (12.56) is assumed to be small and constant.

Pore volume compressibility, c_f, is defined by the following equation:

$$c_f = \frac{1}{\phi}\frac{\partial \phi}{\partial p} \tag{12.57}$$

The total compressibility, c_t, of the system is the sum of the fluid compressibility and formation compressibility:

$$c_t = c + c_f \tag{12.58}$$

Substituting Eqs. (12.56), (12.57), and (12.58) into Eq. (12.55) and dividing by ρ, gives:

$$\frac{1}{r}\frac{\partial}{\partial r}\left(r\frac{\partial p}{\partial r}\right) + c\left(\frac{\partial p}{\partial r}\right)^2 = \frac{\phi \mu c_t}{2.637\times 10^{-4} k}\frac{\partial p}{\partial t} \tag{12.59}$$

The pressure gradient is assumed to be small, so the term $(\partial p/\partial r)^2$ can be ignored. Hence, Eq. (12.59) reduces to:

$$\frac{1}{r}\frac{\partial}{\partial r}\left(r\frac{\partial p}{\partial r}\right) = \frac{\phi \mu c_t}{2.637\times 10^{-4} k}\frac{\partial p}{\partial t} \tag{12.60}$$

Eq. (12.60) is the radial diffusivity equation for liquid flow in a porous medium. The coefficient $\left(\frac{\phi \mu c_t}{2.637\times 10^{-4} k}\right)$ is the reciprocal of the formation diffusivity, η, as defined in Eq. (12.7). The units of η in Eq. (12.60) are feet^2 per hour.

12.8 Solutions of the Radial Diffusivity Equation for Slightly Compressible Fluids

There are two main solutions to the diffusivity equation for slightly compressible fluids that have wide applications in reservoir engineering. These are called the constant terminal rate solution and the constant terminal pressure solution. These designations are based on the initial and boundary conditions imposed on the diffusivity equation to achieve solutions. Note that these solutions apply to a bounded, cylindrical (circular) reservoir with a well located at its center (Figure 12.1). This model of the reservoir is modified as necessary for special cases.

12.8.1 Constant Terminal Rate Solution

For all the cases under the constant terminal rate solution, the following initial and boundary conditions apply:

1. At time, $t = 0$, the reservoir is at uniform pressure, $p = p_i$, for all r.
2. At time, $t > 0$, a single well of radius, r_w, at the center of the cylinder (reservoir) produces at a constant rate, q.

12.8.1.1 Case 1: Line Source Well, Infinite Reservoir

This is a special case of the constant terminal rate solution. This case assumes that the well located at the center of the cylinder has zero radius and the reservoir acts as if it is infinite in size. Hence, in addition to the conditions stated for the constant terminal rate solution, the following two conditions apply to this special case:

1. The well has zero radius, $r_w = 0$.
2. The reservoir is infinite in size (i.e., $p \rightarrow p_i$, as $r \rightarrow \infty$, for all t).

A solution to the diffusivity equation for a line source well in an infinite reservoir with the stated initial and boundary conditions was given by Matthews and Russell[1] as:

$$p(r,t) = p_i - \frac{q\mu}{2\pi kh}\left[-\frac{1}{2}Ei\left(-\frac{\phi \mu c_t r^2}{4kt}\right)\right] \tag{12.61}$$

The exponential integral, Ei, function is defined as:

$$-Ei(-x) = \int_x^{\infty} \frac{e^{-u}}{u} du \tag{12.62}$$

For $x < 0.02$,

$$Ei(-x) = \ln(1.781x) \tag{12.63}$$

Other values of the exponential integral can be read from a chart in Appendix 12A.

In field units, Eq. (12.61) can be written as:

$$p(r,t) = p_i - \frac{70.6qB\mu}{kh}\left[-Ei\left(-\frac{948\phi \mu c_t r^2}{kt}\right)\right] \tag{12.64}$$

In Eq. (12.64), p = pressure, psia; q = flow rate, STB/D; B = formation volume factor, RB/STB; μ = fluid viscosity, cp; k = permeability, md; h = formation thickness, ft; ϕ = porosity, fraction; c_t = total compressibility, psi^{-1}; r = radius, ft; and t = time, hr. Note that Eq. (12.64) can be used to calculate pressures at any distance, r, from the wellbore if:

$$t > \frac{3.79\times 10^5 \phi \mu c_t r_w^2}{k} \tag{12.65}$$

Also, Eq. (12.64) can be used to calculate pressure at any distance, r, from the wellbore if the reservoir is assumed to be infinite-acting. The reservoir is assumed to be infinite-acting if:

$$t < \frac{948\phi \mu c_t r_e^2}{k} \tag{12.66}$$

EXAMPLE 12.5 Calculation of Reservoir Pressure Assuming Line Source Well

Problem

A single well located in a deepwater reservoir is producing oil at a constant rate of 5500 STB/D. Calculate the pressure at a location 500 feet away from the well after 5 days and 10 days of production. The drainage boundary of the well is located 1000 feet away. Other rock and fluid properties of the reservoir are as follows:

Initial pressure, p_i	19,250 psia
Formation thickness, h	345 ft
Formation permeability, k	7 md
Formation porosity, ϕ	0.19
Total compressibility, c_t	4×10^{-6} psi^{-1}
Oil viscosity, μ	4.5 cp
Oil FVF, B_o	1.1 RB/STB
Reservoir boundary, r_e	1000 ft
Wellbore radius, r_w	0.5 ft

Solution

Step 1: Check that the conditions set by Eqs. (12.65) and (12.66) are satisfied.

Before using Eq. (12.64), it is important to check that the conditions set by Eqs. (12.65) and (12.66) are met. Equation (12.65) is used to check that the well has been on production for sufficient length of time so that the accuracy of Eq. (12.64) is not affected. Equation (12.66) is used to check that the reservoir is still infinite-acting.

The length of time the well is on production must be greater than the time calculated from Eq. (12.65):

$$t > \frac{3.79\times10^{5}\phi\mu c_t r_w^2}{k}$$

$$> \frac{3.79\times10^{5}\times0.19\times4.5\times4\times10^{-6}\times(0.5)^2}{7}$$

$$> 0.05 \text{ hrs}$$

Typically, the time required to exceed this limit as is the case in this example is very small. Since the calculations are performed after 5 and 10 days of production, the application of Eq. (12.64) is valid.

Next, it is important to check that the reservoir is infinite-acting after 10 days of production. This time of production must be less than the time calculated from Eq. (12.66):

$$t < \frac{948\phi\mu c_t r_e^2}{k}$$

$$< \frac{948\times0.19\times4.5\times4\times10^{-6}\times(1000)^2}{7}$$

$$< 463 \text{ hrs or } 19.3 \text{ days}$$

The reservoir will cease to be infinite-acting after about 19.3 days of production. Hence, after 10 days of production, the reservoir is still infinite-acting.

Step 2: Calculate reservoir pressure after 5 days, and 10 days of production at 500 feet from well.

Substituting Eq. (12.64) with the data provided gives:

$$p(r,t) = p_i - \frac{70.6qB\mu}{kh}\left[-Ei\left(-\frac{948\phi\mu c_t r^2}{kt}\right)\right]$$

$$= 19250 - \frac{70.6\times5500\times1.1\times4.5}{7\times345}$$

$$\left[-Ei\left(\frac{-948\times0.19\times4.5\times4\times10^{-6}\times(500)^2}{7\times24\times t}\right)\right]$$

$$= 19250 - 795.89\left[-Ei\left(\frac{-4.8246}{t}\right)\right]$$

At 5 days of production, the $-Ei(-x)$ integral is:

$$-Ei\left(\frac{-4.8246}{t}\right) = -Ei\left(\frac{-4.8246}{5}\right)$$

$$= -Ei(-0.96)$$

From the chart in Appendix 12A, $-E_i(-0.96) = 0.24$

$$p(r,t) = 19250 - 795.8944\times0.24$$

$$= 19{,}059 \text{ psia}$$

At 10 days of production, the $-Ei(-x)$ integral is:

$$-Ei\left(\frac{-4.8246}{t}\right) = -Ei\left(\frac{-4.8246}{10}\right)$$

$$= -Ei(-0.48)$$

From Appendix 12A, $-Ei(-0.48) = 0.585$

$$p(r,t) = 19250 - 795.8944\times0.585$$

$$= 18{,}784 \text{ psia}$$

End of Example 12.5

12.8.1.2 Case 2: Bounded Cylindrical Reservoir, No Flow at Outer Boundary

This case is a variation of the constant terminal rate solution. In addition to the initial and boundary conditions stated earlier for the constant terminal rate solution, the following conditions apply to this case:

1. The well has a radius, r_w.
2. There is no flow across the outer boundary. Hence, at $r = r_e$, $q = 0$.
3. Since there is no flow across the outer boundary, the pressure gradient at the outer boundary is zero. Thus, at $r = r_e$, $\dfrac{\partial p}{\partial r} = 0$.

The solution to the diffusivity equation at constant terminal rate under the stated initial and boundary conditions was reported in the classical work of van Everdingen and Hurst.[2] The solution is presented in terms of dimensionless variables of pressure, time, and radius. The dimensionless pressure variable, p_D, is defined as:

$$p_D = \frac{0.007082kh(p_i - p)}{qB\mu} \tag{12.67}$$

In Eq. (12.67), k = permeability, md; h = formation thickness, ft; p_i = initial pressure, psia: p = pressure at radius, r, and time, t; q = flow rate, STB/D; B = FVF, RB/STB; and μ = fluid viscosity, cp.

The dimensionless time variable, t_D, is defined as:

$$t_D = \frac{0.0002637kt}{\phi\mu c_t r_w^2} \tag{12.68}$$

In Eq. (12.68), the same units apply as for Eq. (12.67). The units for time t are in hours.

The outer boundary radius as a dimensionless variable, r_{eD}, is defined as:

$$r_{eD} = \frac{r_e}{r_w} \tag{12.69}$$

The dimensionless radius variable, r_D, is defined as:

$$r_D = \frac{r}{r_w} \tag{12.70}$$

Note that at the wellbore, $r_D = 1$.

An important solution by van Everdingen and Hurst[2] is for pressure at the wellbore. This is expressed as:

$$(p_D)_{r_D=1} = f(t_D, r_{eD}) \tag{12.71}$$

The tabulations of p_D as functions of t_D at fixed r_{eD} are provided in Appendix 12C for finite systems. Before the pressure transient caused by production at the wellbore reaches the outer boundary, the reservoir is acting as if it is infinite in size. For this condition, values of p_D as functions of t_D are provided in Appendix 12B. In using Appendix 12B to obtain values of p_D, two possible extensions of the tabulations are worthy of note. For $t_D < 0.01$, p_D can be estimated with the expression:

$$p_D \cong 2\sqrt{\frac{t_D}{\pi}} \tag{12.72}$$

For $100 < t_D < 0.25r_{eD}^2$, p_D can be estimated from the expression:

$$p_D \cong 0.5(\ln t_D + 0.80907) \tag{12.73}$$

Note that the reservoir is infinite-acting at a given r_{eD}, if t_D values are smaller than the values listed in Appendix 12C. If this situation occurs, the tabulations in Appendix 12B should be used instead.

An approximate solution for pseudosteady-state flow of Eq. (12.60) under initial and boundary conditions stated in this section is given as:

$$p_{wf} = \bar{p} - \frac{70.6q\mu B}{kh}\left[\ln\left(\frac{r_e^2}{r_w^2} - 1.5\right)\right] \tag{12.74}$$

Rearranging Eq. (12.74) gives:

$$q = \frac{0.00708kh(\bar{p} - p_{wf})}{\mu B\left[\ln\left(\frac{r_e}{r_w}\right) - 0.75\right]} \tag{12.75}$$

Equation (12.75) is the radial flow equation for a slightly compressible fluid under pseudosteady-state conditions. The units of the variables in Eq. (12.75) are the same as the units in Eq. (12.23) except that, $\bar{p}$ = average reservoir pressure, psia. The average reservoir pressure can be calculated as shown in Eqs. (12.95) or (12.96). Note that fluid properties should be calculated at average reservoir pressure. Equation (12.75) can be used to calculate the flow rate of a well in a reservoir under solution-gas drive.

EXAMPLE 12.6 Calculation of Wellbore Pressure for the Well in Example 12.5

Problem

For the well in Example 12.5, calculate flowing wellbore pressure after 5 days and 10 days of production. The reservoir properties and conditions are as listed in Example 12.5.

Solution

Step 1: Calculate t_D.

From Eq. (12.68),

$$t_D = \frac{0.0002637kt}{\phi\mu c_t r_w^2}$$

$$= \frac{0.0002637\times7\times24\times t}{0.19\times4.5\times4\times10^{-6}\times(0.5)^2}$$

$$= 51{,}814.74t$$

At $t = 5$ days, $t_D = 259{,}074$

At $t = 10$ days, $t_D = 518{,}147$

Step 2: Calculate r_{eD}.

From Eq. (12.69),

$$r_{eD} = \frac{r_e}{r_w} = \frac{1000}{0.5} = 2000.$$

Step 3: Calculate p_D.

Since $100 < t_D < 0.25r_{eD}^2$, use Eq. (12.73) to estimate p_D:

$$p_D \cong 0.5(\ln t_D + 0.80907)$$

At $t_D = 259{,}074$, $p_D = 6.6370$

At $t_D = 518{,}147$, $p_D = 6.9835$

Step 4: Calculate p_{wf}.

From Eq. (12.67),

$$p_{wf} = p_i - \frac{141.2qB\mu}{kh}\times p_D$$

$$= 19{,}250 - \frac{141.2\times5500\times1.1\times4.5}{7\times345}\times p_D$$

$$= 19{,}250 - 1591.7888\times p_D$$

At $t = 5$ days, $p_D = 6.6370$, $p_{wf} = 8685$ psia

At $t = 10$ days, $p_D = 6.9835$, $p_{wf} = 8134$ psia

End of Example 12.6

12.8.1.3 Case 3: Bounded Cylindrical Reservoir, Constant Pressure at Outer Boundary

This is another variation of the constant terminal rate solution. In addition to the initial and boundary conditions stated earlier for the constant terminal rate solution, the following conditions apply to this case:

1. The well has a radius, r_w.
2. At outer boundary where, $r = r_e$, $p = p_i$.

The dimensionless variables, p_D, t_D, r_{eD}, and r_D, as defined in Case 2 apply to Case 3. The solution of the diffusivity equation for these conditions by van Everdingen and Hurst[2] in terms of p_D at $r_D = 1$ as a function of t_D for fixed r_{eD} are tabulated in Appendix 12D. The reservoir is infinite-acting at values of t_D less than those listed in Appendix 12D. In such cases, p_D should be obtained from Appendix 12B. Case 3 conditions are applicable to waterflood reservoirs where the pressure at reservoir boundary or drainage volume of the well can be assumed to be constant.

12.8.2 Constant Terminal Pressure Solution

For the constant terminal pressure solution, the following initial and boundary conditions apply:

1. At time, $t = 0$, the reservoir is at uniform pressure, $p = p_i$, for all r.
2. At time, $t > 0$, a single well of radius, r_w, at the center of the cylinder (reservoir) produces at a constant pressure, p_{wf}.
3. There is no flow across the outer boundary. Hence, at $r = r_e$, $q = 0$.
4. Since there is no flow across the outer boundary, the pressure gradient at the outer boundary is zero. Thus, at $r = r_e$, $\partial p/\partial r = 0$.

van Everdingen and Hurst[2] published a solution of the diffusivity equation for slightly compressible fluids for the constant terminal pressure conditions as stated above. The solution provides the means for calculating the cumulative volume of fluid that will flow across any specified inner radius over a specified time interval in response to a specified drop in pressure between an outer boundary and the specified inner radius. The solution is expressed in terms of dimensionless variables, Q_D, t_D, and r_{eD}. Q_D is the dimensionless cumulative flow across the inner radius and is expressed as:

$$Q_D = \frac{QB}{1.119 h\phi c_t r^2 \left(p_i - p_{\text{at radius r}}\right)} \tag{12.76}$$

In Eq. (12.76), Q = cumulative flow across radius r, STB; B = FVF, RB/STB; h = formation thickness, ft.; ϕ = porosity, fraction; c_t = total compressibility, psi^{-1}; r = inner radius, ft.; p_i = initial pressure, psia; and $p_{\text{at radius r}}$ = pressure at radius r, psia. Note that for a well producing at the center of a circular reservoir, pressure at radius r is the same as the flowing bottomhole pressure p_{wf}.

The remaining dimensionless variables, t_D and r_{eD}, are as defined previously in Eqs. (12.68) and (12.69), respectively. Tabulations of Q_D as functions of t_D for infinite systems are presented in Appendix 12E. Further tabulations of Q_D as functions of t_D for specific r_{eD} are provided in Appendix 12F. For $r_{eD} = \infty$ (essentially $r_{eD} > 20$) and $t_D \geq 200$, Q_D can be estimated with a correlation developed by Edwardson et al.[3] as:

$$Q_D = \frac{\left(-4.29881 + 2.02566 t_D\right)}{\ln t_D} \tag{12.77}$$

The constant terminal pressure method can be used to calculate cumulative production at a well for a specified pressure drop. This is illustrated in Example 12.7. However, this method is used primarily to calculate cumulative water influx from an aquifer into a reservoir. In this case, the inner radius is the reservoir boundary and the outer boundary is the external boundary of the aquifer. The cumulative flow, Q, is the same as the cumulative water influx, W_e, as designated in Chapters 9 and 10. The use of Eq. (12.76) for calculating cumulative water influx into a reservoir is presented in more detail under application of the superposition principle later in Section 12.11.2.

EXAMPLE 12.7 Calculation of Cumulative Oil Production for the Well in Example 12.5

Problem

Suppose the well in Example 12.5 was produced at a constant flowing bottom-hole pressure of 13,500 psia starting from initial reservoir pressure of 19,250 psia. Calculate cumulative oil production after 5 days and 10 days of production. The reservoir properties and conditions are listed in Example 12.5.

Solution

Step 1: Calculate t_D.

From Eq. (12.68),

$$t_D = \frac{0.0002637 kt}{\phi\mu c_t r_w^2}$$

$$= \frac{0.0002637 \times 7 \times 24 \times t}{0.19 \times 4.5 \times 4 \times 10^{-6} \times (0.5)^2}$$

$$= 51{,}814.74t$$

At t = 5 days, t_D = 259,074

At t = 10 days, t_D = 518,147

Step 2: Calculate r_{eD}.

From Eq. (12.69),

$$r_{eD} = \frac{r_e}{r_w}$$

$$= \frac{1000}{0.5} = 2000.$$

Step 3: Calculate Q_D.

Since $r_{eD} > 20$ and $t_D > 200$, use Eq. (12.77):

$$Q_D = \frac{\left(-4.29881 + 2.02566 t_D\right)}{\ln t_D}$$

For $t_D = 259{,}074$, $Q_D = 42{,}101.65$

For $t_D = 518{,}147$, $Q_D = 79{,}767.76$

Step 4: Calculate cumulative production, Q.

From Eq. (12.76),

$$Q = \frac{1.119 h\phi c_t \left(r_w^2\right)\left(p_i - p_{wf}\right) \times Q_D}{B}$$

$$= \frac{1.119 \times 345 \times 0.19 \times 4 \times 10^{-6} \times (0.5)^2 \times (19250 - 13500) \times Q_D}{1.1}$$

$$= 0.3834 \times Q_D$$

At t = 5 days, cumulative production:

$$Q = 0.3834 \times Q_D$$

$$= 0.3834 \times 42{,}101.65 = 16{,}142 \text{ STB.}$$

At t = 10 days, cumulative production:

$$Q = 0.3834 \times Q_D$$

$$= 0.3834 \times 79{,}767.76 = 30{,}583 \text{ STB.}$$

End of Example 12.7

12.9 Derivation of the Radial Diffusivity Equation for Compressible Fluids

In this derivation, compressible fluids are assumed to be gases. From Eq. (5.12), the real gas equation of state is:

$$\rho_g = \frac{pM_g}{zRT} \tag{12.78}$$

In terms of gas density, ρ_g, Eq. (12.54) can be written as:

$$\frac{1}{r}\frac{\partial}{\partial r}\left(r\rho_g\frac{\partial p}{\partial r}\right) = \frac{1}{2.637\times10^{-4}}\frac{\mu}{k}\frac{\partial(\phi\rho_g)}{\partial t} \tag{12.79}$$

Substituting Eq. (12.78) into Eq. (12.79) gives:

$$\begin{aligned}&\frac{1}{r}\frac{\partial}{\partial r}\left[r\left(\frac{pM_g}{zRT}\right)\frac{\partial p}{\partial r}\right]\\&=\frac{1}{2.637\times10^{-4}}\frac{\mu}{k}\frac{\partial\left[\phi\left(\frac{pM_g}{zRT}\right)\right]}{\partial t}\end{aligned} \tag{12.80}$$

Eliminating M_g/RT in Eq. (12.80) gives:

$$\begin{aligned}&\frac{1}{r}\frac{\partial}{\partial r}\left[r\left(\frac{p}{z}\right)\frac{\partial p}{\partial r}\right]\\&=\frac{1}{2.637\times10^{-4}}\frac{\mu}{k}\frac{\partial\left[\phi\left(\frac{p}{z}\right)\right]}{\partial t}\end{aligned} \tag{12.81}$$

Applying the product rule to right-side of Eq. (12.81) gives:

$$\begin{aligned}\frac{\partial}{\partial t}\left[\phi\left(\frac{p}{z}\right)\right]&=\phi\frac{\partial}{\partial t}\left(\frac{p}{z}\right)+\left(\frac{p}{z}\right)\frac{\partial\phi}{\partial t}\\&=\phi\frac{\partial}{\partial p}\left(\frac{p}{z}\right)\frac{\partial p}{\partial t}+\left(\frac{p}{z}\right)\frac{\partial\phi}{\partial p}\frac{\partial p}{\partial t}\\&=\phi\left(\frac{p}{z}\right)\frac{\partial p}{\partial t}\left[\left(\frac{z}{p}\right)\frac{\partial}{\partial p}\left(\frac{p}{z}\right)+\frac{1}{\phi}\frac{\partial\phi}{\partial p}\right]\end{aligned} \tag{12.82}$$

Gas compressibility was derived in Eq. (5.95) as:

$$c_g = \frac{1}{p}-\frac{1}{z}\left(\frac{\partial z}{\partial p}\right)_T$$

Expansion of one of the terms in Eq. (12.82) gives:

$$\begin{aligned}\left(\frac{z}{p}\right)\frac{\partial}{\partial p}\left(\frac{p}{z}\right)&=\left(\frac{z}{p}\right)\left(\frac{1}{z}-\frac{p}{z^2}\frac{\partial z}{\partial p}\right)\\&=\frac{1}{p}-\frac{1}{z}\frac{\partial z}{\partial p}=c_g\end{aligned} \tag{12.83}$$

Pore volume compressibility was defined in Eq. (12.57) as:

$$c_f = \frac{1}{\phi}\frac{\partial\phi}{\partial p}$$

Replacing the terms in Eq. (12.82) with the expressions for c_g and c_f from Eqs. (12.83) and (12.57), respectively, gives:

$$\begin{aligned}\frac{\partial}{\partial t}\left[\phi\left(\frac{p}{z}\right)\right]&=\phi\left(\frac{p}{z}\right)\frac{\partial p}{\partial t}(c_g+c_f)\\&=\phi c_t\left(\frac{p}{z}\right)\frac{\partial p}{\partial t}\end{aligned} \tag{12.84}$$

Substituting Eq. (12.84) into Eq. (12.81) gives:

$$\begin{aligned}&\frac{1}{r}\frac{\partial}{\partial r}\left[r\left(\frac{p}{z}\right)\frac{\partial p}{\partial r}\right]\\&=\frac{\phi\mu c_t}{2.637\times10^{-4}k}\left(\frac{p}{z}\right)\frac{\partial p}{\partial t}\end{aligned} \tag{12.85}$$

Eq. (12.85) is the radial diffusivity equation in field units for compressible fluids such as gases.

12.10 Transformation of the Gas Diffusivity Equation with Real Gas Pseudo-Pressure Concept

Al-Hussainy et al.[4] introduced the concept of real gas pseudo-pressure to transform the gas diffusivity equation shown as Eq. (12.85) into a form that is suitable for solution and applicable to gas reservoir engineering problems. The real gas pseudo-pressure approach has two major advantages over previous methods used to linearize the gas diffusivity equation. These advantages are:

1. Real gas flow equations based on real gas pseudo-pressure approach account for pressure dependent gas properties such as gas viscosity and gas compressibility factor.
2. No assumption of small pressure gradients in the entire flow system is necessary. This means that the second degree pressure gradient terms, $(\partial p/\partial r)^2$, are not ignored as was the case in the derivation of the diffusivity equation for liquids [see Eq. (12.59)].

Al-Hussainy et al.[4] defined the real gas pseudo-pressure, p_p, as:

$$p_p = 2\int_{p_b}^{p} \frac{p}{\mu(p)z(p)}dp \tag{12.86}$$

In Eq. (12.86), gas viscosity, μ, and gas compressibility factor, z, are shown as functions of pressure, and p_b is low reference base pressure, usually close to atmospheric pressure. Writing Eq. (12.86) in differential form gives:

$$\frac{\partial p_p}{\partial p} = \frac{2p}{\mu z} \tag{12.87}$$

Applying the chain rule to Eq. (12.87) yields the following equations:

$$\frac{\partial p_p}{\partial r} = \frac{\partial p_p}{\partial p}\frac{\partial p}{\partial r} \tag{12.88}$$

$$\frac{\partial p_p}{\partial t} = \frac{\partial p_p}{\partial p}\frac{\partial p}{\partial t} \tag{12.89}$$

Substituting for $\partial p_p / \partial p$ in Eqs. (12.88) and (12.89) with Eq. (12.87) give:

$$\frac{\partial p_p}{\partial r} = \frac{2p}{\mu z}\frac{\partial p}{\partial r} \tag{12.90}$$

$$\frac{\partial p_p}{\partial t} = \frac{2p}{\mu z}\frac{\partial p}{\partial t} \tag{12.91}$$

Substituting Eqs. (12.90) and (12.91) into Eq. (12.85) gives:

$$\frac{\partial^2 p_p}{\partial r^2} + \frac{1}{r}\frac{\partial p_p}{\partial r} = \frac{\phi \mu c_t}{2.637\times 10^{-4} k}\frac{\partial p_p}{\partial t} \tag{12.92}$$

Eq. (12.92) is the gas diffusivity equation in terms of real gas pseudo-pressure. This equation is still non-linear since gas viscosity and gas compressibility are functions of pressure. Al-Hussainy and Ramey[5] achieved an approximate solution to Eq. (12.92) which can be used for gas flow problems. For a radial infinite-acting reservoir with a well producing at a constant rate at the center, the flowing pressure at the well in terms gas pseudo-pressure is given by:

$$p_{pwf} = p_{pi} - \frac{1637qT}{kh}\left[\log t_D + 0.3513\right] \tag{12.93}$$

In Eq. (12.93), q = gas flow rate, Mscf/D, at standard conditions of 60°F and 14.7 psia; T = reservoir temperature, °R; k = permeability, md; h = formation thickness, ft; and t_D = dimensionless time. The dimensionless time, t_D, was defined in Eq. (12.68) as:

$$t_D = \frac{0.0002637kt}{\phi \mu c_t r_w^2}$$

Note that to calculate the dimensionless time, t_D, from the above equation, gas viscosity and total compressibility should be evaluated at the initial pressure, p_i.

For pseudosteady-state flow in a radial gas reservoir with closed outer boundary and a well producing at the center at constant rate, the gas flow rate can be calculated with an approximate equation from Al-Hussainy and Ramey[5] as:

$$q = \frac{1.987\times 10^{-3} khT_{sc}}{Tp_{sc}}\left[\frac{\overline{p}_p - p_{pwf}}{\ln \frac{0.472 r_e}{r_w}}\right] \tag{12.94}$$

The units in Eq. (12.94) are the same as in Eq. (12.93). The average pressure, $\overline{p}$, can be evaluated rigorously from reservoir isobaric maps as the volume-weighted average pressure, thus:

$$\overline{p} = \frac{\sum_{j=1}^{n} \overline{p}_j V_j}{\sum_{j=1}^{n} V_j} \tag{12.95}$$

In Eq. (12.95), $\overline{p}_j$ is the arithmetic average of pressure data in the *jth* drainage volume, V_j. The average pressure, $\overline{p}$, can also be calculated from a less rigorous method suggested by Carter:[6]

$$\overline{p} = \sqrt{\frac{\left(p_i^2 + p_{wf}^2\right)}{2}} \tag{12.96}$$

EXAMPLE 12.8 Calculation of the Flowing Bottomhole Pressure of a Gas Well

Problem

Calculate the flowing bottomhole pressure of a gas well on production for 35 days at a constant rate of 250 Mscf/D. The reservoir and fluid properties data are given below:

Initial pressure, p_i	3000 psia
Reservoir temperature, T	121°F
Formation thickness, h	125 ft
Formation permeability, k	4 md
Formation porosity, ϕ	0.24
Total compressibility, c_t	1.8×10^{-4} psi^{-1}
Gas viscosity, μ	0.021 cp

Reservoir boundary, r_e 3100 ft

Wellbore radius, r_w 0.5 ft

The gas viscosity and compressibility data for the reservoir are provided in Table 12.1.

TABLE 12.1 Gas Properties for Example 12.8

Pressure, p (psia)	Gas Compressibility factor, z	Gas Viscosity, μ_g (cp)
500	0.9380	0.01200
1000	0.8800	0.01300
1500	0.8230	0.01413
2000	0.7630	0.01600
2500	0.7018	0.01859
3000	0.6843	0.02086
3500	0.6818	0.02285

Solution

Step 1: Check that the reservoir is infinite-acting after 35 days of production.

Using Eq. (12.11), time for start of late transient flow is:

$$t_{LTF} = \frac{240\mu\phi r_e^2}{k\bar{p}}$$

$$= \frac{240\times0.021\times0.24\times(3100)^2}{4\times3000}$$

$$= 968.69 \text{ hours or } 40.36 \text{ days}$$

Since the reservoir was on production at this rate for 35 days and start of late transient was calculated as 40.4 days, the reservoir was still infinite-acting. Hence, Eq. (12.93) can be used to calculate flowing well pressure. Note that in estimating the time for start of late transient flow, it was assumed that average reservoir pressure was equal to initial reservoir pressure.

Step 2: Calculate real gas pseudo-pressure, p_p.

From Eq. (12.86),

$$p_p = 2\int_{p_b}^{p} \frac{p}{\mu(p)z(p)}dp$$

The real gas pseudo-pressure is calculated as shown in Table 12.2 using the trapezoidal rule for integration.

Step 3: Calculate dimensionless time, t_D:

From Eq. (12.68):

$$t_D = \frac{0.0002637kt}{\phi\mu c_t r_w^2}$$

$$= \frac{0.0002637\times4\times24\times35}{0.24\times0.021\times1.8\times10^{-4}\times(0.5)^2}$$

$$= 3.907\times10^6$$

Step 4: Calculate flowing well pressure, p_{wf}:

From Eq. (12.93),

$$p_{pwf} = p_{pi} - \frac{1637qT}{kh}[\log t_D + 0.3513]$$

$$= 721{,}353{,}662 - \frac{1637\times250\times581}{4\times125}\left[\log\left(3.907\times10^6\right) + 0.3513\right]$$

$$= 718{,}051{,}878$$

Note that p_{pi} was obtained from Table 12.2. Also, by interpolation in Table 12.2, the flowing bottomhole pressure at $p_{pwf} = 718{,}051{,}878$ was transformed to be:

$$p_{wf} = 2991.8 \text{ psia}$$

TABLE 12.2 Calculation of Real-Gas Pseudo-Pressure in Example 12.8

Pressure, p (psia)	Gas Compressibility Factor, z	Gas Viscosity, μ_g (cp)	$2p/(\mu z)$ (psia/cp)	Δp (psia)	P_p (psia²/cp)
500	0.9380	0.01200	88,842	500	22,210,377
1000	0.8800	0.01300	174,825	500	88,127,047
1500	0.8230	0.01413	257,976	500	196,327,334
2000	0.7630	0.01600	327,654	500	342,734,826
2500	0.7018	0.01859	383,246	500	520,459,742
3000	0.6843	0.02086	420,330	500	721,353,662
3500	0.6818	0.02285	449,319	500	938,765,931

End of Example 12.8

12.11 The Superposition Principle

The superposition principle states that a new solution (or equation) to a fluid flow problem can be derived by the addition of solutions (or equations) to the original linear differential equation that represents the fluid flow problem. Thus, the superposition principle can be used in conjunction with the constant terminal rate solutions or the constant terminal pressure solutions to develop solutions to fluid flow problems on petroleum reservoirs.

12.11.1 Applications of Constant Terminal Rate Solutions with Superposition Principle

For the purposes of the discussions that follow, the exponential integral solution [Eq. (12.64)] to the diffusivity equation for slightly compressible fluids is used. Remember that this particular solution assumes that the reservoir is infinite-acting. This assumption is also made here. However, other constant terminal rate solutions could be used instead of the exponential integral solution in the application of the superposition principle. In the application of the superposition principle, three special cases are presented for illustrations. These are:

1. Single well, multi-production rates case.
2. Multi-well reservoir system case.
3. Single well near a boundary (Image well) case.

12.11.1.1 Single Well, Multi-Production Rates Case

Suppose there is a single well producing at a constant rate, q_1, at time, $t = 0$. At time, $t = t_1$, the production rate was increased to q_2. The well was produced at this rate until time, $t = t_2$, when its rate was reduced to q_3. The well was then kept at q_3 for the total time, $t = t$. These flow conditions are illustrated in Figure 12.8. Assume there is no near-wellbore damage in this well. Applying the superposition principle, the total pressure drop in this well at time t is the sum of the pressure drop caused by the production rate at q_1 over time t, and the pressure drop caused by the production rate at $(q_2 - q_1)$ over time $(t - t_1)$, and the pressure drop caused by the production rate at $(q_3 - q_2)$ over time $(t - t_2)$. This is represented in the form of an equation as:

$$\Delta p_t = \Delta p_{q_1 @ t} + \Delta p_{(q_2 - q_1)@(t-t_1)} + \Delta p_{(q_3 - q_2)@(t-t_2)} \tag{12.97}$$

From Eq. (12.64),

$$\Delta p_{q_1 @ t} = \left(p_i - p_{wf}\right)_{q_1 @ t} = \frac{70.6 q_1 B\mu}{kh}\left[-Ei\left(-\frac{948\phi\mu c_t r^2}{kt}\right)\right] \tag{12.98}$$

Similarly,

$$\Delta p_{(q_2 - q_1)@(t-t_1)} = \left(p_i - p_{wf}\right)_{(q_2 - q_1)@(t-t_1)} = \frac{70.6(q_2 - q_1)B\mu}{kh}\left[-Ei\left(-\frac{948\phi\mu c_t r^2}{k(t-t_1)}\right)\right] \tag{12.99}$$

$$\Delta p_{(q_3 - q_2)@(t-t_2)} = \left(p_i - p_{wf}\right)_{(q_3 - q_2)@(t-t_2)} = \frac{70.6(q_3 - q_2)B\mu}{kh}\left[-Ei\left(-\frac{948\phi\mu c_t r^2}{k(t-t_2)}\right)\right] \tag{12.100}$$

Substituting Eqs. (12.98), (12.99), and (12.100) into Eq. (12.97) gives:

$$\Delta p_t = \frac{70.6 q_1 B\mu}{kh}\left[-Ei\left(-\frac{948\phi\mu c_t r^2}{kt}\right)\right] + \frac{70.6(q_2 - q_1)B\mu}{kh}\left[-Ei\left(-\frac{948\phi\mu c_t r^2}{k(t-t_1)}\right)\right] + \frac{70.6(q_3 - q_2)B\mu}{kh}\left[-Ei\left(-\frac{948\phi\mu c_t r^2}{k(t-t_2)}\right)\right] \tag{12.101}$$

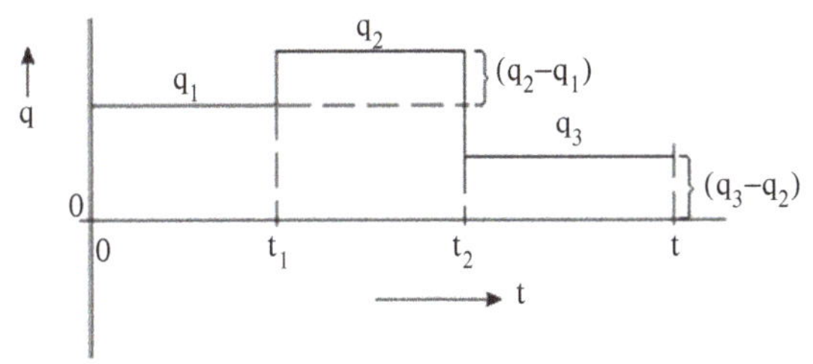

▲ FIGURE 12.8 Diagram for a single well with multi-production rates

EXAMPLE 12.9 Calculation of Total Pressure Drop due to Multi-Production Rates

Problem

Calculate the total pressure drop at an observation well located 1500 feet from a producing well in a virgin reservoir. The single production well was maintained at a constant production rate of 300 STB/D for 10 days; then the rate was increased to a constant rate of 500 STB/D for an additional 25 days. The initial reservoir pressure was 2500 psia. Assume the reservoir was infinite-acting during this test. Other reservoir and fluid properties are as follows:

Initial pressure, p_i	2500 psia
Formation thickness, h	15 ft
Formation permeability, k	300 md
Formation porosity, ϕ	0.24
Total compressibility, c_t	6×10^{-6} psi^{-1}
Oil viscosity, μ	4.2 cp
Oil FVF, B_o	1.2 RB/STB
Wellbore radius, r_w	0.5 ft

Solution

Step 1: Calculate pressure drop caused by initial production at 300 STB/D.

For this event, $q_1 = 300$ STB/D and total time $t = 35 \times 24 = 840$ hrs. Substituting Eq. (12.98), the pressure drop is:

$$\Delta p_{q_1 @ t}$$

$$= (p_i - p_{wf})_{q_1 @ t}$$

$$= \frac{70.6 q_1 B\mu}{kh}\left[-Ei\left(-\frac{948\phi\mu c_t r^2}{kt}\right)\right]$$

$$= \frac{70.6 \times 300 \times 1.2 \times 4.2}{300 \times 15}$$

$$\times\left[-Ei\left(-\frac{948 \times 0.24 \times 4.2 \times 6 \times 10^{-6} \times (1500)^2}{300 \times 840}\right)\right]$$

$$= 23.7216\left[-Ei(-0.0512)\right]$$

From the chart in Appendix 12A, $-Ei(-0.0512) = 2.5$.

Thus, $\Delta p_{q_1 @ t} = 23.7216 \times 2.5 = 59.3$ psia.

Step 2: Calculate the pressure drop caused by production increase to 500 STB/D after 10 days.

The production increment, $q_2 - q_1 = 500 - 300 = 200$ STB/D, and the duration is $t - t_1 = (35 - 10) \times 24 = 600$ hrs. Substituting Eq. (12.99), the pressure drop is:

$$\Delta p_{(q_2 - q_1)@(t - t_1)}$$

$$= (p_i - p_{wf})_{(q_2 - q_1)@(t - t_1)}$$

$$= \frac{70.6 (q_2 - q_1) B\mu}{kh}\left[-Ei\left(-\frac{948\phi\mu c_t r^2}{k(t - t_1)}\right)\right]$$

$$= \frac{70.6 \times 200 \times 1.2 \times 4.2}{300 \times 15}$$

$$\times\left[-Ei\left(-\frac{948 \times 0.24 \times 4.2 \times 6 \times 10^{-6} \times (1500)^2}{300 \times 600}\right)\right]$$

$$= 15.8144 \times \left[-Ei(-0.0717)\right]$$

From the chart in Appendix 12A, $-E_i(-0.0717) = 2.2$.

Thus, $\Delta p_{(q_2 - q_1)@(t - t_1)} = 15.8144 \times 2.2 = 34.8$ psia.

Step 3: Calculate the total pressure drop at the observation well.

The total pressure drop is:

$$\Delta p_t = \Delta p_{q_1 @ t} + \Delta p_{(q_2 - q_1)@(t - t_1)}$$

$$= 59.3 + 34.8 \text{ psia.}$$

$$= 94.1 \text{ psia}$$

End of Example 12.9

12.11.1.2 Multi-Well Reservoir System Case

Consider three wells (A, B, C) in a reservoir (Figure 12.9) producing at constant rates, q_A, q_B, and q_C, respectively, from time $t = 0$, and located at distances, r_{OA}, r_{OB}, and r_{OC}, from an observation well at location O. By the superposition principle, the total pressure drop at the observation well located at O after time t is the sum of the pressure drops caused by Wells A, B, and C, due to production at rates, q_A, q_B, and q_C. In the form of an equation, this can be expressed as:

$$\Delta p_t = \Delta p_A + \Delta p_B + \Delta p_C \tag{12.102}$$

Applying Eq. (12.64) and using the notation in Figure 12.9, Eq. (12.102) becomes:

$$\Delta p_t = \frac{70.6 q_A B\mu}{kh}\left[-Ei\left(-\frac{948\phi\mu c_t r_{OA}^2}{kt}\right)\right]$$

$$+ \frac{70.6 q_B B\mu}{kh}\left[-Ei\left(-\frac{948\phi\mu c_t r_{OB}^2}{kt}\right)\right]$$

$$+ \frac{70.6 q_C B\mu}{kh}\left[-Ei\left(-\frac{948\phi\mu c_t r_{OC}^2}{kt}\right)\right] \tag{12.103}$$

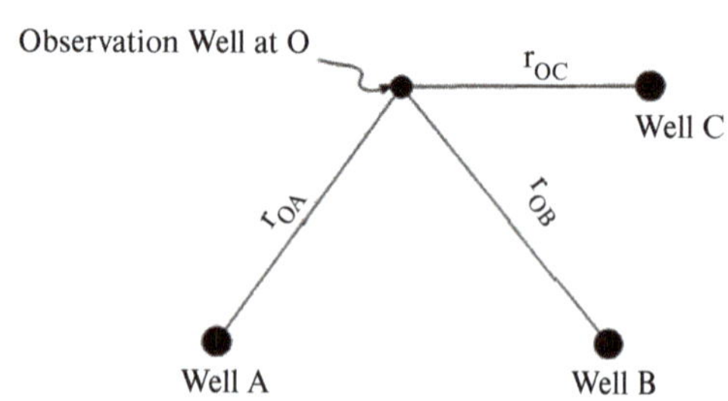

▲ **FIGURE 12.9** Diagram for a multi-well reservoir system

EXAMPLE 12.10 Calculation of Total Pressure Drop in a Multi-Well Reservoir System

Problem

A reservoir at initial pressure of 2500 psia has two wells (Well A and Well B) on production for 35 days from initial production. Well A produced at a constant rate of 500 STB/D and is located 750 feet from an observation well. Well B produced at a constant rate of 300 STB/D and is located 1000 feet from the same observation well. If the reservoir is infinite in size, calculate the total pressure drop measured at the observation well at the end of 35 days. This reservoir has the same rock and fluid properties as the reservoir in Example 12.9.

Solution

Step 1: Calculate the pressure drop caused by Well A.

From Eq. (12.103):

$$\Delta p_A = \frac{70.6 q_A B\mu}{kh}\left[-Ei\left(-\frac{948\phi\mu c_t r_{OA}^2}{kt}\right)\right]$$

$$= \frac{70.6\times 500\times 1.2\times 4.2}{300\times 15}\times\left[-Ei\left(-\frac{948\times 0.24\times 4.2\times 6\times 10^{-6}\times(750)^2}{300\times 35\times 24}\right)\right]$$

$$= 39.5360\times\left[-Ei(-0.0128)\right]$$

Since $x < 0.02$, Eq. (12.63) is used for calculation of the exponential integral. From Eq. (12.63):

$$Ei(-x) = \ln(1.781x)$$

$$= \ln(1.781\times 0.0128) = -3.7811$$

Hence,

$$\Delta p_A = 39.5360\times\left[-(-3.7811)\right]$$

$$= 149.49 \text{ psia}$$

Step 2: Calculate the pressure drop caused by Well B.

From Eq. (12.103):

$$\Delta p_B = \frac{70.6 q_B B\mu}{kh}\left[-Ei\left(-\frac{948\phi\mu c_t r_{OB}^2}{kt}\right)\right]$$

$$= \frac{70.6\times 300\times 1.2\times 4.2}{300\times 15}\times\left[-Ei\left(-\frac{948\times 0.24\times 4.2\times 6\times 10^{-6}\times(1000)^2}{300\times 35\times 24}\right)\right]$$

$$= 23.7216\times\left[-Ei(-0.0228)\right]$$

From the chart in Appendix 12A, $-Ei(-0.0228) = 3.1$.

Hence,

$$\Delta p_B = 23.7216\times 3.1$$

$$= 73.53 \text{ psia}$$

Step 3: Calculate the total pressure drop at the observation well.

The total pressure drop at the observation well is:

$$\Delta p_t = \Delta p_A + \Delta p_B = 149.49 + 73.53$$

$$= 223 \text{ psia}$$

End of Example 12.10

12.11.1.3 Single Well Near A Sealing Boundary Case

Suppose Well A producing at a constant rate, q_A, is located at a distance, r, from a sealing fault as shown in Figure 12.10. There is no flow across the sealing fault (i.e., it is a closed boundary). The pressure drop at observation Well O at a distance, r_{OA}, from Well A, is the sum of the pressure drop caused by Well A and an image Well I located behind the sealing fault at an equal distance, r, from the fault. If the distance from the image well to the observation well is r_{OI}, then the total pressure drop at the observation well, assuming the reservoir is infinite-acting is:

$$\Delta p_{t@well\ O} = \Delta p_A + \Delta p_I = \frac{70.6 q_A B\mu}{kh}\left[-Ei\left(-\frac{948\phi\mu c_t r_{OA}^2}{kt}\right)\right] + \frac{70.6 q_A B\mu}{kh}\left[-Ei\left(-\frac{948\phi\mu c_t r_{OI}^2}{kt}\right)\right] \quad (12.104)$$

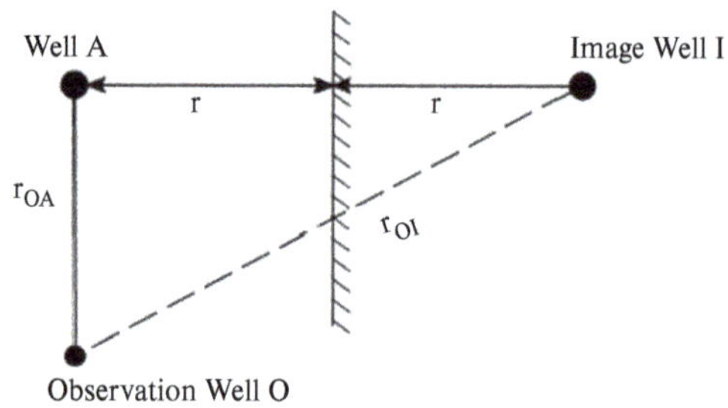

▲ **FIGURE 12.10** Diagram of a well near a sealing boundary

12.11.2 Applications of Constant Terminal Pressure Solution with Superposition Principle

The constant terminal pressure solution coupled with the superposition principle is used primarily to calculate cumulative water influx into a reservoir from an aquifer. Consider a radial hydrocarbon reservoir surrounded by a radial aquifer as shown in Figure 12.11. Applying the terminology that was used to define the initial and boundary conditions of the constant terminal pressure solution, the hydrocarbon reservoir becomes the well with constant pressure at its boundary, r_R, which separates it from the aquifer. The extent of the aquifer designated as, r_a, is the outer boundary of the system. For this reservoir-aquifer system, the dimensionless radius, r_{aD}, is defined as:

$$r_{aD} = \frac{r_a}{r_R} \tag{12.105}$$

The dimensionless time, t_D, is defined as:

$$t_D = \frac{0.0002637kt}{\phi\mu c_t r_R^2} \tag{12.106}$$

Rewriting Eq. (12.76) in terms of the reservoir-aquifer system, the cumulative water influx, W_e, in reservoir barrels (RB) is given by:

$$W_e = 1.119h\phi c_t r_R^2 \Delta p Q_D \tag{12.107}$$

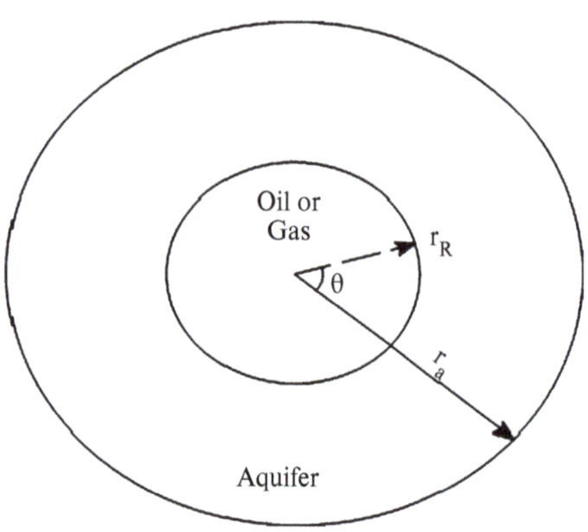

▲ **FIGURE 12.11** Diagram of a reservoir with a radial aquifer

In Eqs. (12.105) to (12.107), r = radius, ft; k = permeability, md; t = time, hours; ϕ = porosity, fraction; μ = fluid viscosity, cp; c_t = total compressibility, psi^{-1}; h = formation thickness, ft; Δp = pressure drop, psi; and Q_D = dimensionless cumulative water influx evaluated as a function of r_{aD} and t_D. Note that all the fluid and rock properties in Eqs. (12.106) and (12.107) are based on the aquifer.

In hydrocarbon reservoirs, pressure change occurs continuously over time. Sudden or step pressure change as required by Eq. (12.107) does not occur under normal conditions. For a reservoir with a continuous pressure history as shown in Figure 12.12, the superposition principle is used to transform the pressure history into a series of stair-step pressure changes which is kept constant over a specific time interval. For instance, to calculate the cumulative water influx at the end of 4 years for the reservoir with the pressure history shown in Figure 12.12, the first pressure drop, Δp_1, is active for 4 years, the next pressure drop, Δp_2, is active for 3 years, the next pressure drop, Δp_3, is active for 2 years, and the final pressure drop, Δp_4, is active for 1 year. This concept can be translated into an equation for cumulative water influx after n time periods on the basis of Eq. (12.107) as:

$$W_{en} = 1.119h\phi c_t r_R^2 \sum_{i=1}^{n} \Delta p_i Q_{D@t_{D(n-i+1)}} \tag{12.108}$$

Eq. (12.108) can be written in a more concise form as:

$$W_{en} = \beta \sum_{i=1}^{n} \Delta p_i Q_{D@t_{D(n-i+1)}}$$

$$\text{where } \beta = 1.119h\phi c_t r_R^2 \tag{12.109}$$

In Eqs. (12.108) and (12.109), W_{en} = cumulative water influx at end of n time periods, RB; Δp_i = pressure change at i time period, psi; Q_D = dimensionless cumulative water influx evaluated as a function of r_{aD} and t_D from Appendices 12E and 12F; and $t_{D(n-i+1)}$ = dimensionless time at $(n-i+1)$ time period. If the angle of contact between the reservoir and the aquifer is less than 360°, then β is calculated as:

$$\beta = 1.119h\phi c_t r_R^2 \left(\frac{\theta}{360}\right) \tag{12.110}$$

In Eq. (12.110), θ is the angle of contact between the reservoir and aquifer in degrees (Figure 12.11).

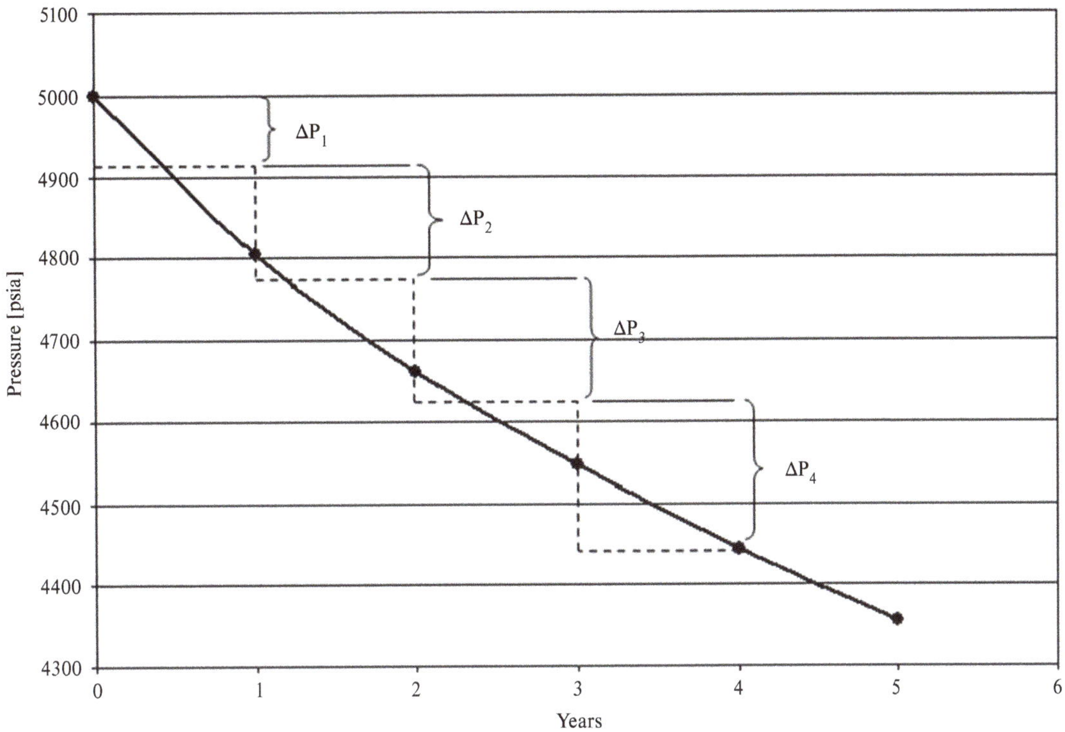

▲ **FIGURE 12.12** Typical pressure history profile at the reservoir/aquifer boundary

The step-wise pressure change as illustrated in Figure 12.12 has been simplified for demonstration purposes. The pressure change for each time period can be calculated with the expression:

$$\Delta p_i = \frac{1}{2}\left(p_{i-2} - p_i\right), \text{ for } i = 1, 2, ..., n \qquad (12.111)$$

For Eq. (12.111), $p_{\leq 0} = p_0 = p_i$, where p_i is the initial pressure. The application of Eq. (12.111) is illustrated in Example 12.11.

The application of the constant terminal pressure solution with superposition principle as developed here is actually the van Everdingen-Hurst Method for calculating cumulative water influx into a reservoir. It was considered appropriate to present this method in this chapter after the introduction of the principle of superposition. It is an alternative method to the Carter-Tracy and Fetkovich methods presented in Chapter 10 for calculating cumulative water influx.

EXAMPLE 12.11 Calculation of Cumulative Water Influx Using the van Everdingen-Hurst Method

Problem

Calculate the cumulative water influx after 5 years for the reservoir with pressure history at the reservoir/aquifer boundary shown in Table 12.3 and Figure 12.13. Assume the aquifer is infinite-acting. Other properties of the aquifer are as follows:

Formation thickness, h	65 ft
Formation permeability, k	150 md
Formation porosity, ϕ	0.20
Total compressibility, c_t	7×10^{-6} psi^{-1}
Water viscosity, μ	1.1 cp
Reservoir radius, r_R	3000 ft
Angle of contact, θ	360°

TABLE 12.3 Pressure History for Example 12.11

Time, t (days)	Pressure, p (psia)
0	5000
365	4804
730	4662
1095	4547
1460	4443
1825	4355

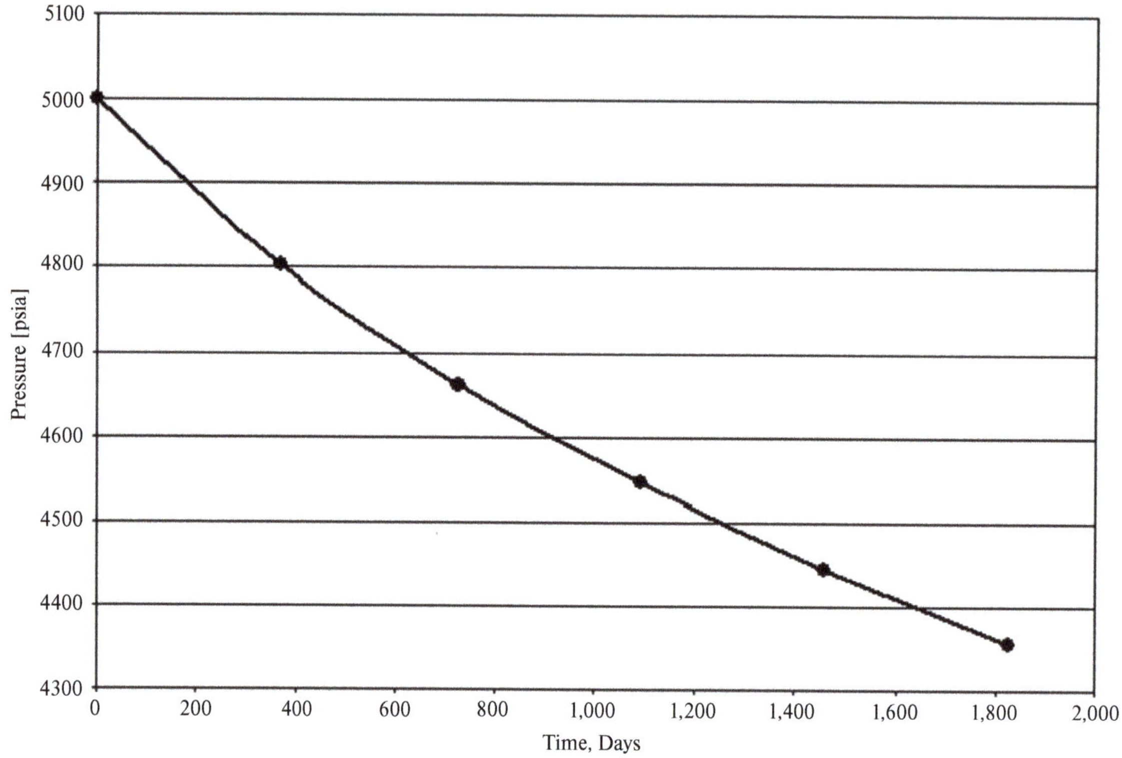

▲ **FIGURE 12.13** **Pressure history at the reservoir/aquifer boundary for Example 12.11**

Solution

Step 1: Calculate the pressure change for each time period.

From Eq. (12.111), the pressure change at each time period is given by:

$$\Delta p_i = \frac{1}{2}\left(p_{i-2} - p_i\right), \text{ for } i = 1, 2, ..., n.$$

Note that for $p_{\le 0} = p_0 = p_i$

For $i = 1$,

$$\Delta p_1 = \frac{1}{2}\left(p_0 - p_1\right)$$

$$= \frac{1}{2}(5000 - 4804) = 98 \text{ psia.}$$

For $i = 2$,

$$\Delta p_2 = \frac{1}{2}\left(p_0 - p_2\right)$$

$$= \frac{1}{2}(5000 - 4662) = 169 \text{ psia.}$$

For $i = 3$,

$$\Delta p_3 = \frac{1}{2}\left(p_1 - p_3\right)$$

$$= \frac{1}{2}(4804 - 4547) = 128.5 \text{ psia.}$$

For $i = 4$,

$$\Delta p_4 = \frac{1}{2}\left(p_2 - p_4\right)$$

$$= \frac{1}{2}(4662 - 4443) = 109.5 \text{ psia.}$$

For $i = 5$,

$$\Delta p_5 = \frac{1}{2}\left(p_3 - p_5\right)$$

$$= \frac{1}{2}(4547 - 4355) = 96 \text{ psia.}$$

Step 2: Calculate dimensionless time for each time period.

From Eq. (12.106), the dimensionless time is calculated as:

$$t_D = \frac{0.0002637kt}{\phi \mu c_t r_R^2}$$

$$= \frac{0.0002637 \times 150 \times 24 \times t}{0.20 \times 1.1 \times 7 \times 10^{-6} \times (3000)^2}$$

$$= 0.0685t$$

For $i = 1$, $t = 365$ days, $t_D = 0.0685 \times 365 =$ 25.00

Values of t_D for subsequent time periods are shown in Table 12.4.

Step 3: Calculate dimensionless cumulative water influx term, Q_D.

From Appendix 12E, obtain values of Q_D for infinite aquifer at each t_D. Values of Q_D are listed in Table 12.4 at each t_D.

Step 4: Calculate cumulative water influx at each time period.

From Eq. (12.110),

$$\beta = 1.119 h\phi c_t r_R^2 \left(\frac{\theta}{360}\right)$$

$$= 1.119 \times 65 \times 0.2 \times 7 \times 10^{-6} \times (3000)^2 \times \frac{360}{360}$$

$$= 916.46 \text{ rb/psi}$$

From Eq. (12.109),

$$W_{en} = \beta \sum_{i=1}^{n} \Delta p_i Q_{D@t_{D(n-i+1)}}$$

For example, at $n = 1$:

$$W_{e1} = \beta \sum_{i=1}^{1} \Delta p_1 Q_{D@t_{D(1-1+1)}} = \beta \times \Delta p_1 \times Q_{D@t_{D1}}$$

$$= 916.46 \times 98 \times 14.55$$

$$= 1,306,780 \text{ rbbls.}$$

Another example, at $n = 3$:

$$W_{e3} = \beta \sum_{i=1}^{3} \Delta p_i Q_{D@t_{D(3-i+1)}}$$

$$= \beta\left[\Delta p_1 Q_{D@t_{D3}} + \Delta p_2 Q_{D@t_{D2}} + \Delta p_3 Q_{D@t_{D1}}\right]$$

$$= 916.46[98 \times 34.14 + 169 \times 24.82 + 129 \times 14.55]$$

$$= 8,630,533 \text{ rbbls.}$$

Similarly, at $n = 5$:

$$W_{e6} = \beta \sum_{i=1}^{5} \Delta p_i Q_{D@t_{D(5-i+1)}}$$

$$= \beta\left[\begin{array}{l}\Delta p_1 Q_{D@t_{D5}} + \Delta p_2 Q_{D@t_{D4}} + \Delta p_3 Q_{D@t_{D3}} \\ + \Delta p_4 Q_{D@t_{D2}} + \Delta p_5 Q_{D@t_{D1}}\end{array}\right]$$

$$= 916.46\left[\begin{array}{l}98 \times 51.41 + 169 \times 43 + 129 \times 34.14 \\ + 109 \times 24.82 + 96 \times 14.55\end{array}\right]$$

$$= 19,072,834 \text{ rbbls.}$$

The cumulative water influxes at time periods $n = 2$ and $n = 4$ are shown in Table 12.4.

TABLE 12.4 Calculation of Cumulative Water Influx in Example 12.11

Time Periods, n	Time, t (days)	Pressure, p (psia)	Δp	t_D	Q_D	Cum. Water Influx, W_e (RB)
0	0	5000	0	0	0.00	0
1	365	4804	98	25	14.55	1,306,780
2	730	4662	169	50	24.82	4,482,690
3	1095	4547	129	75	34.14	8,630,533
4	1460	4443	109	100	43.00	13,537,388
5	1825	4355	96	125	51.41	19,072,834

End of Example 12.11

12.12 Well Productivity Index

The productivity index (PI) of a well is the ratio of its production rate and the pressure drawdown taken at the midpoint of the producing interval. PI is expressed in equation form as:

$$J = \frac{q}{\bar{p} - p_{wf}} \tag{12.112}$$

In Eq. (12.112), J = PI, STB/psia-day; q = production rate, STB/D; $\bar{p}$ = average reservoir pressure, psia; and p_{wf} = flowing bottomhole pressure measured at midpoint of the producing interval, psia. The average reservoir pressure can be estimated quickly as equal to the arithmetic average of all current, static bottomhole pressures corrected to a datum or determined more rigorously with Eq. (12.95). PI should be calculated after production at the well has stabilized. PI is typically used as a measure to monitor the condition of the well. If the production rate of a well is falling, the PI will also be in decline. Factors that affect the PI of a well are:

1. Near-wellbore reduction of permeability due to fines migration, evolution of free gas due to pressure drop, and trapping of condensates for gas reservoirs.

2. Non-Darcy flow near the wellbore due to high-pressure drawdown.
3. Increase in oil viscosity at pressures below bubblepoint for undersaturated reservoirs.
4. Reduction in reservoir pressure for depletion drive reservoirs.
5. Increase in multiphase flow.

Generally, in water drive reservoirs, PI will decline at water breakthrough if calculation of productivity was previously based only on oil production rate. To ensure that the productivity of such well has not declined, water production rate should be included in Eq. (12.112). In reservoir simulation, productivity indices of wells are sometimes modified as necessary to history-match the reported well production data.

12.13 Well Injectivity Index

For injection wells, injectivity index is used in place of PI. Injectivity index (I) is expressed as:

$$I = \frac{q_{inj}}{p_{winj} - \overline{p}} \tag{12.113}$$

In Eq. (12.113), I = injectivity index, STB/psia-day; q_{inj} = injection rate, STB/D; p_{winj} = injection pressure at midpoint of interval, psia; and $\overline{p}$ = average reservoir pressure, psia. Injectivity index is a measure of the ability to inject fluid into a reservoir. A reduction of injectivity index with time is an indicator of near-wellbore reduction in permeability due to formation damage.

Nomenclature

A	cross-sectional area, ft^2
B	formation volume factor, RB/STB
B_g	gas formation volume factor, RB/scf
c	isothermal compressibility, psi^{-1}
c_f	formation compressibility, psi^{-1}
c_g	isothermal gas compressibility, psi^{-1}
c_t	total compressibility, psi^{-1}
g	acceleration due to gravity, ft/sec^2
h	reservoir or formation thickness, ft
I	well injectivity index, STB/psi
J	well productivity index, STB/psi
k	permeability, md
l	length of flow path or linear reservoir
M_g	molecular weight of gas, lbm/lbmole
p_b	base low pressure, psia
p_D	dimensionless pressure
p_p	real gas pseudo-pressure, psia^2/cp
p	pressure, psia
$\overline{p}$	average pressure, psia
p_{wf}	well bottom-hole flowing pressure, psia
Δp	pressure change, psi
q	volume flow rate, STB/D
Q	cumulative flow or production, STB or MMscf
Q_D	dimensionless cumulative flow or production
r	radius or distance, ft
r_{aD}	dimensionless aquifer radius or distance
r_D	dimensionless radius or distance
r_e	reservoir radius or distance, ft
r_{eD}	dimensionless reservoir radius or outer boundary
r_w	wellbore radius, ft
R	gas constant, 10.732 psia-ft^2/lbmol-°R
t	time, hours or days
t_D	dimensionless time
T	temperature, °R
v	apparent fluid velocity, ft/sec
V	volume, ft^2 or RB
V_b	volume at base low pressure, p_b
W_e	cumulative water influx, RB
z	gas compressibility factor
Z	elevation above a datum plane, ft
β	constant in equation in Eq. (10.107)
γ	fluid specific gravity, water = 1
η	formation diffusivity, ft^2/hr
ϕ	porosity, fraction
μ	fluid viscosity, cp
ρ	fluid density, lbm/ft^3
θ	dip angle of formation or contact angle between reservoir and aquifer, degrees
Φ	potential

Subscripts

a	aquifer	LTF	late transient flow
avg	average	pss	pseudo-steady state
g	gas	R	reservoir
i	initial conditions	sc	standard conditions
inj	injection	w	well

Abbreviations

LTF	Late transient flow
PSS	Pseudosteady-state
SS	Steady state

REFERENCES

1. Matthews, C.S., and Russell, D.G.: *Pressure Buildup and Flow Tests in Wells*, Monograph Volume 1, Society of Petroleum Engineering of AIME, Dallas, Texas (1967).
2. van Everdingen, A.F., and Hurst, W.: "The Application of the Laplace Transformation to Flow Problems in Reservoirs," *Trans.* AIME (1949) 186, 305-324.
3. Edwardson, M.J., Girner, H.M., Parkison, H.R., Williams, C.D., and Matthews, C.S.: "Calculation of Formation Temperature Disturbances Caused by Mud Circulation," *JPT* (April 1962) 416-426.
4. Al-Hussainy, R., Ramey, H.J., Jr., and Crawford, P.R.: "The Flow of Real Gases Through Porous Media," *JPT* (May 1966) 624-636.
5. Al-Hussainy, R., and Ramey, H.J., Jr.: "Application of Real Gas Flow Theory to Well Testing and Deliverability Forecasting," *JPT* (May 1966) 637-642.
6. Carter, R.D.: "Solutions of Unsteady-State Radial Gas Flow," *JPT* (May 1962) 549-554.

ADDITIONAL REFERENCES

- Bruce, G.H., Peaceman, D.W., Rachford, Jr., H.H., and Rice, J.D.: "Calculations of Unsteady-State Gas Flow Through Porous Media," *Trans.* AIME (1953) 198, 79-92.
- Russell, D.G., Goodrich, J.H., and Bruskotter, J.F.: "Methods for Predicting Gas Well Performance," *JPT* (January 1966) 99-108.
- Swift, G.W., and Kiel, O.G.: "The Prediction of Gas Well Performance Including the Effect of Non-Darcy Flow," *JPT* (July 1962) 791-798.
- van Everdingen, A.F.: "The Skin Effect and its Influence on the Productive Capacity of a Well," *Trans.* AIME (1953) 198, 171-176.

PROBLEMS

12-1. Calculate the permeability (in darcy) of a water-saturated core if the flow rate of water through the core was measured at 10 cm^3/s. The pressure at the inlet end of the core is 200 psia and the pressure at the outlet end is 100 psia. The core is 3 inches long and 2 inches in diameter. The viscosity of water is 1.0 cp. Assume linear, horizontal flow through the core.

12-2. In a laboratory core test, a sample of filtered crude oil was flowed vertically upward through a core plug that measured 4 cm long at a volume flow rate of 0.0853 cm^3/s. The pressure at the inlet end of the core plug is 3.009 atm and the pressure at the outlet end is 1.002 atm. The cross-sectional area of the core is 4.0 cm^2. The viscosity of the oil is 5 cp, and its density is 0.8 g/cm^3. Calculate the permeability of the core plug.

12-3. In a special core analysis experiment, air was flowed through a cleaned Berea sandstone core plug that measured 5.08 cm in diameter and 30.48 cm in length. The gas permeability of the core plug was determined to be 150 md. The temperature of the air and the core plug was maintained at 20°C. Viscosity of air at 20°C is 0.0083 cp. Assume air behaves as an ideal gas at these conditions.

a. Calculate the mass flow rate of air through the core plug if the pressures at the inlet and outlet ends of the core plug are 3.0 atm and 1.0 atm, respectively.

b. Calculate and compare the volumetric flow rates at the inlet and outlet ends of the core plug.

12-4. Calculate the average permeability of a stratified reservoir if the permeability and thickness of each horizontal strata are given as:

Strata (Interval)	Permeability (md)	Thickness (feet)
1	150	30
2	757	8
3	125	17
4	445	13

12-5. A well is draining four concentric areas with different permeabilities. The segment closest to the well was damaged by drilling mud filtrate. It extends 4 feet from the wellbore and has a reduced permeability of 12.5 md. The next segment extends 175 feet from the wellbore and its permeability is 225 md. This is then followed by another segment that is 230 feet away from the wellbore and has 450 md permeability. The last segment has a permeability of 120 md and extends 700 feet to the drainage boundary of the well from the wellbore. The radius of the wellbore is 0.45 feet. Calculate the average permeability of the drainage area of the well.

12-6. A well is producing only oil from a reservoir at a constant flowing bottomhole pressure of 10,000 psia. Reservoir and fluid data are given as:

Wellbore radius, r_w	0.454 feet
Reservoir radius, r_e	2100 feet
Average formation permeability, k	115 md
Average formation net thickness, h	60 feet
Oil formation volume factor at 12,000 psia, B_o	1.84 RB/STB
Oil formation volume factor at 10,000 psia, B_o	1.86 RB/STB
Oil viscosity at 12,000 psia, μ_o	0.372 cp
Oil viscosity at 10,000 psia, μ_o	0.329 cp

Calculate:

a. Oil production rate. Assume reservoir pressure is maintained at 12,000 psia by water injection, incompressible fluids, and steady-state flow.

b. Oil production rate. Assume reservoir pressure at outer reservoir boundary is 12,000 psia, incompressible fluids, and pseudosteady-state flow.

c. Oil production rate. Assume average reservoir pressure is 11,000 psia, slightly compressible fluids, and pseudosteady-state flow.

d. Compare oil production rates calculated in a, b, and c. Discuss reasons for the differences in oil production rates between the assumed flow conditions.

12-7. A gas well was drilled in the center of a circular dry-gas reservoir, The area of the reservoir was 160 acres and the radius of the wellbore was 0.5 feet. The well produced gas from a 25-feet net thickness interval and the specific gravity of the gas was 0.65. The permeability of the interval was 175 md. The initial pressure of the reservoir was 1950 psia at reservoir temperature of 202°F. There were no impurities in the reservoir gas. Calculate the gas production rate of the well at flowing bottomhole pressure of 1500 psia assuming steady-state flow.

12-8. A well completed in a Gulf of Mexico deepwater reservoir is producing oil at a constant rate of 1400 STB/D. Data on the reservoir are given as follows:

Initial reservoir pressure, p_i	11,237 psia
Formation net thickness, h	56 feet
Formation permeability, k	400 md
Formation porosity, ϕ	0.28
Oil viscosity, μ_o	0.354 cp
Oil FVF, B_o	1.85 RB/STB
Total compressibility, c_t	20E-06 psi^{-1}
Reservoir boundary, r_e	1842 feet
Wellbore radius, r_w	0.5 feet

Calculate:

a. The reservoir pressure at a location 921 feet from the wellbore after 12 hours of initial production assuming the well behaves as a line source.

b. The wellbore pressure after 12 hours of initial production assuming no flow at reservoir boundary.

12-9. Calculate the reservoir pressure at an observation well located 1050 feet from a single well producing oil from a reservoir at its initial pressure. The well produced at a constant rate of 1400 BOPD for 12 days, then the constant production rate was reduced to 1200 BOPD for 19 days. Toward the end of the test, the production rate was increased to 1500 BOPD from 1200 BOPD for 14 days. Reservoir and fluid properties data are given as:

Initial reservoir pressure, p_i	10,205 psia
Formation net thickness, h	60 feet
Formation permeability, k	42 md
Formation porosity, ϕ	0.32
Oil viscosity, μ_o	1.25 cp
Oil FVF, B_o	1.56 RB/STB
Total compressibility, c_t	16E-06 psi^{-1}
Reservoir boundary, r_e	3000 feet
Wellbore radius, r_w	0.5 feet

12-10. Suppose the reservoir in Problem 12-9 had three wells on production at first oil. Well A produced at a constant rate of 1400 BOPD and is located 1050 feet from the observation well. Well B produced at a constant rate of 1200 BOPD and is located 850 feet from the same observation well. And Well C produced at a constant rate of 1500 BOPD and is located 2000 feet from the same observation well. The reservoir has the same rock and fluid properties as in Problem 12-9. Calculate the reservoir pressure at the observation well after 45 days of production.

12-11. Suppose the well in Problem 12-9 was located 2000 feet west of a parallel, sealing fault and 1050 feet north of the observation well (see Figure 12.10). If the well produced at a constant rate of 1400 BOPD for 45 days, calculate the reservoir pressure at the observation well. The reservoir has the same rock and fluid properties as in Problem 12-9.

12-12. A radial reservoir is surrounded by a larger radial aquifer. The areas of the reservoir and the aquifer are 5500 acres and 792,000 acres, respectively. The pressure history at the reservoir/aquifer boundary is shown in Table 12P.1. The petrophysical properties of the aquifer are given also in Table 12P.1. Assuming infinite-acting aquifer, calculate:

a. The volume of water influx into the reservoir from the aquifer using the van Everdingen-Hurst method.

b. Compare the volumes of water influx calculated by the van Everdingen-Hurst method to the volumes of water influx calculated by the Fetkovich and Carter-Tracy methods in Problem 9-3. (Note: The Problem 9-3 and 12-12 are identical). What are the reasons for any differences?

TABLE 12P.1 Aquifer/Reservoir Data and Pressure History for Problem 12-12

Porosity, ϕ	5225 psia
Permeability, k	18%
Aquifer thickness, h	270 md
Water viscosity, μ	50 feet
Total aquifer compressibility, c_t	6.8×10^{-6} psi^{-1}
Encroachment angle, θ	360°

Pressure History at Reservoir/Aquifer Boundary	
Time, t (days)	**Pressure, p (psia)**
0	5225
365	5150
730	5100
1095	5060
1460	5025

APPENDIX 12A CHART FOR EXPONENTIAL INTEGRAL

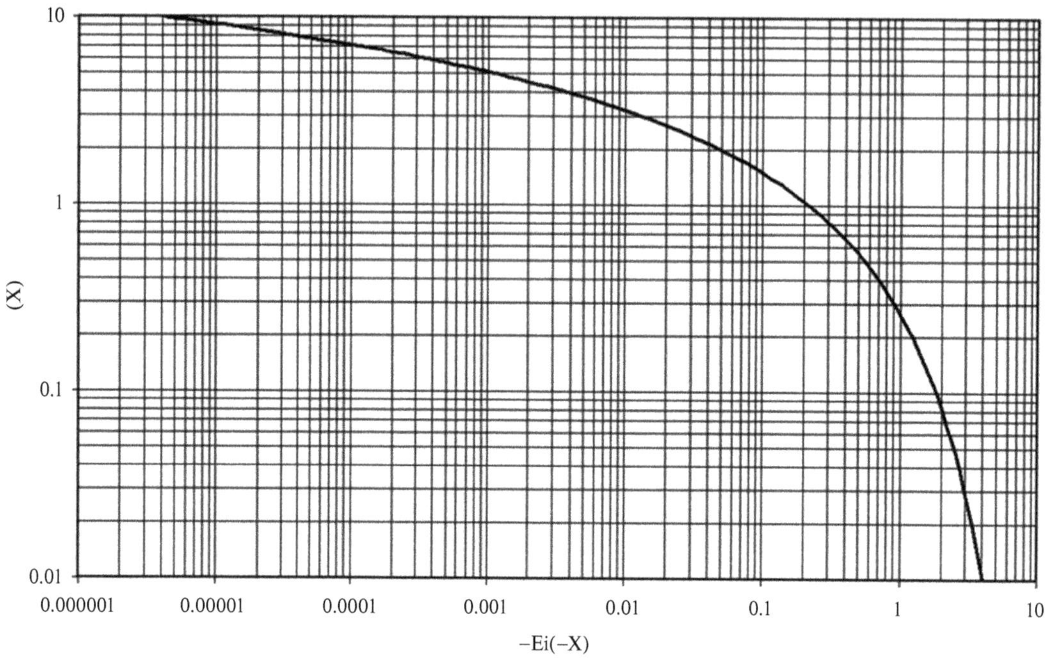

APPENDIX 12B TABULATION OF p_D vs t_D FOR RADIAL FLOW, INFINITE RESERVOIRS WITH CONSTANT TERMINAL RATE AT INNER BOUNDARY (FROM VAN EVERDINGEN AND HURST[2])

t_D	p_D	t_D	p_D
0.01	0.112	15.0	1.829
0.05	0.229	20.0	1.960
0.10	0.315	25.0	2.067
0.15	0.376	30.0	2.147
0.20	0.424	40.0	2.282
0.25	0.459	50.0	2.388
0.30	0.503	60.0	2.476
0.40	0.564	70.0	2.550
0.50	0.616	80.0	2.615
0.60	0.659	90.0	2.672
0.70	0.702	100.0	2.723
0.80	0.735	150.0	2.921
0.90	0.772	200.0	3.064
1.00	0.802	250.0	3.173
1.50	0.927	300.0	3.263
2.00	1.020	400.0	3.406
2.50	1.101	500.0	3.516
3.00	1.169	600.0	3.608
4.00	1.275	700.0	3.684
5.00	1.362	800.0	3.750
6.00	1.436	900.0	3.809
7.00	1.500	1000.0	3.860
8.00	1.556		
9.00	1.604		
10.00	1.651		

APPENDIX

12C TABULATION OF p_D vs t_D FOR RADIAL FLOW, FINITE RESERVOIRS WITH CLOSED OUTER BOUNDARY AND CONSTANT TERMINAL RATE AT INNER BOUNDARY (FROM VAN EVERDINGEN AND HURST[2])

r_{eD} = 1.5		r_{eD} = 2.0		r_{eD} = 2.5		r_{eD} = 3.0		r_{eD} = 3.5		r_{eD} = 4.0		r_{eD} = 4.5	
t_D	p_D	t_D	p_D	t_D	p_D	t_D	p_D	t_D	p_D	t_D	p_D	t_D	p_D
0.06	0.251	0.22	0.443	0.40	0.565	0.52	0.627	1.0	0.802	1.5	0.927	2.0	1.023
0.08	0.288	0.24	0.459	0.42	0.576	0.54	0.636	1.1	0.830	1.6	0.948	2.1	1.040
0.10	0.322	0.26	0.476	0.44	0.587	0.56	0.645	1.2	0.857	1.7	0.968	2.2	1.056
0.12	0.355	0.28	0.492	0.46	0.598	0.60	0.662	1.3	0.882	1.8	0.988	2.3	1.072
0.14	0.387	0.30	0.507	0.48	0.608	0.65	0.683	1.4	0.906	1.9	1.007	2.4	1.087
0.16	0.420	0.32	0.522	0.50	0.618	0.70	0.703	1.5	0.929	2.0	1.025	2.5	1.102
0.18	0.452	0.34	0.536	0.52	0.628	0.75	0.721	1.6	0.951	2.2	1.059	2.6	1.116
0.20	0.484	0.36	0.551	0.54	0.638	0.80	0.740	1.7	0.973	2.4	1.092	2.7	1.130
0.22	0.516	0.38	0.565	0.56	0.647	0.85	0.758	1.8	0.994	2.6	1.123	2.8	1.144
0.24	0.548	0.40	0.579	0.58	0.657	0.90	0.776	1.9	1.014	2.8	1.154	2.9	1.158
0.26	0.580	0.42	0.593	0.60	0.666	0.95	0.791	2.0	1.034	3.0	1.184	3.0	1.171
0.28	0.612	0.44	0.607	0.65	0.688	1.0	0.806	2.25	1.083	3.5	1.255	3.2	1.197
0.30	0.644	0.46	0.621	0.70	0.710	1.2	0.865	2.50	1.130	4.0	1.324	3.4	1.222
0.35	0.724	0.48	0.634	0.75	0.731	1.4	0.920	2.75	1.176	4.5	1.392	3.6	1.246
0.40	0.804	0.50	0.648	0.80	0.752	1.6	0.973	3.0	1.221	5.0	1.460	3.8	1.269
0.45	0.884	0.60	0.715	0.85	0.772	2.0	1.076	4.0	1.401	5.5	1.527	4.0	1.292
0.50	0.964	0.70	0.782	0.90	0.792	3.0	1.328	5.0	1.579	6.0	1.594	4.5	1.349
0.55	1.044	0.80	0.849	0.95	0.812	4.0	1.578	6.0	1.757	6.5	1.660	5.0	1.403
0.60	1.124	0.90	0.915	1.0	0.832	5.0	1.828			7.0	1.727	5.5	1.457
		1.0	0.982	2.0	1.215					8.0	1.861	6.0	1.510
		2.0	1.649	3.0	1.596					9.0	1.994	7.0	1.615
		3.0	2.316	4.0	1.977					10.0	2.127	8.0	1.719
		5.0	3.649	5.0	2.358							9.0	1.823
												10.0	1.927
												11.0	2.031
												12.0	2.135
												13.0	2.239
												14.0	2.343
												15.0	2.447

$r_{eD} = 5.0$		$r_{eD} = 6.0$		$r_{eD} = 7.0$		$r_{eD} = 8.0$		$r_{eD} = 9.0$		$r_{eD} = 10$	
t_D	p_D	t_D	p_D	t_D	p_D	t_D	p_D	t_D	p_D	t_D	p_D
3.0	1.167	4.0	1.275	6.0	1.436	8.0	1.556	10.0	1.651	12.0	1.732
3.1	1.180	4.5	1.322	6.5	1.470	8.5	1.582	10.5	1.673	12.5	1.750
3.2	1.192	5.0	1.364	7.0	1.501	9.0	1.607	11.0	1.693	13.0	1.768
3.3	1.204	5.5	1.404	7.5	1.531	9.5	1.631	11.5	1.713	13.5	1.784
3.4	1.215	6.0	1.441	8.0	1.559	10.0	1.653	12.0	1.732	14.0	1.801
3.5	1.227	6.5	1.477	8.5	1.586	10.5	1.675	12.5	1.750	14.5	1.817
3.6	1.238	7.0	1.511	9.0	1.613	11.0	1.697	13.0	1.768	15.0	1.832
3.7	1.249	7.5	1.544	9.5	1.638	11.5	1.717	13.5	1.786	15.5	1.847
3.8	1.259	8.0	1.576	10.0	1.663	12.0	1.737	14.0	1.803	16.0	1.862
3.9	1.270	8.5	1.607	11.0	1.711	12.5	1.757	14.5	1.819	17.0	1.890
4.0	1.281	9.0	1.638	12.0	1.757	13.0	1.776	15.0	1.835	18.0	1.917
4.2	1.301	9.5	1.668	13.0	1.801	13.5	1.795	15.5	1.851	19.0	1.943
4.4	1.321	10.0	1.698	14.0	1.845	14.0	1.813	16.0	1.867	20.0	1.968
4.6	1.340	11.0	1.757	15.0	1.888	14.5	1.831	17.0	1.897	22.0	2.017
4.8	1.360	12.0	1.815	16.0	1.931	15.0	1.849	18.0	1.926	24.0	2.063
5.0	1.378	13.0	1.873	17.0	1.974	17.0	1.919	19.0	1.955	26.0	2.108
5.5	1.424	14.0	1.931	18.0	2.016	19.0	1.986	20.0	1.983	28.0	2.151
6.0	1.469	15.0	1.988	19.0	2.058	21.0	2.051	22.0	2.037	30.0	2.194
6.5	1.513	16.0	2.045	20.0	2.100	23.0	2.116	24.0	2.090	32.0	2.236
7.0	1.556	17.0	2.103	22.0	2.184	25.0	2.180	26.0	2.142	34.0	2.278
7.5	1.598	18.0	2.160	24.0	2.267	30.0	2.340	28.0	2.193	36.0	2.319
8.0	1.641	19.0	2.217	26.0	2.351	35.0	2.499	30.0	2.244	38.0	2.360
9.0	1.725	20.0	2.274	28.0	2.434	40.0	2.658	34.0	2.345	40.0	2.401
10.0	1.808	25.0	2.560	30.0	2.517	45.0	2.817	38.0	2.446	50.0	2.604
11.0	1.892	30.0	2.846					40.0	2.496	60.0	2.806
12.0	1.975							45.0	2.621	70.0	3.008
13.0	2.059							50.0	2.746		
14.0	2.142										
15.0	2.225										

APPENDIX 12D TABULATION OF p_D vs t_D FOR RADIAL FLOW, FINITE RESERVOIRS WITH CONSTANT PRESSURE OUTER BOUNDARY AND CONSTANT TERMINAL RATE AT INNER BOUNDARY (FROM VAN EVERDINGEN AND HURST[2])

r_{eD} = 1.5		r_{eD} = 2.0		r_{eD} = 2.5		r_{eD} = 3.0		r_{eD} = 3.5		r_{eD} = 4.0		r_{eD} = 6.0	
t_D	p_D	t_D	p_D	t_D	p_D	t_D	p_D	t_D	p_D	t_D	p_D	t_D	p_D
0.050	0.230	0.20	0.424	0.30	0.502	0.50	0.617	0.50	0.620	1.0	0.802	4.0	1.275
0.055	0.240	0.22	0.441	0.35	0.535	0.55	0.640	0.60	0.665	1.2	0.857	4.5	1.320
0.060	0.249	0.24	0.457	0.40	0.564	0.60	0.662	0.70	0.705	1.4	0.905	5.0	1.361
0.070	0.266	0.26	0.472	0.45	0.591	0.70	0.702	0.80	0.741	1.6	0.947	5.5	1.398
0.080	0.282	0.28	0.485	0.50	0.616	0.80	0.738	0.90	0.774	1.8	0.986	6.0	1.432
0.090	0.292	0.30	0.498	0.55	0.638	0.90	0.770	1.0	0.804	2.0	1.020	6.5	1.462
0.10	0.307	0.35	0.527	0.60	0.659	1.0	0.799	1.2	0.858	2.2	1.052	7.0	1.490
0.12	0.328	0.40	0.552	0.70	0.696	1.2	0.850	1.4	0.904	2.4	1.080	7.5	1.516
0.14	0.344	0.45	0.573	0.80	0.728	1.4	0.892	1.6	0.945	2.6	1.106	8.0	1.539
0.16	0.356	0.50	0.591	0.90	0.755	1.6	0.927	1.8	0.981	2.8	1.130	8.5	1.561
0.18	0.367	0.55	0.606	1.0	0.778	1.8	0.955	2.0	1.013	3.0	1.152	9.0	1.580
0.20	0.375	0.60	0.619	1.2	0.815	2.0	0.980	2.2	1.041	3.4	1.190	10.0	1.615
0.22	0.381	0.65	0.630	1.4	0.842	2.2	1.000	2.4	1.065	3.8	1.222	12.0	1.667
0.24	0.386	0.70	0.639	1.6	0.861	2.4	1.016	2.6	1.087	4.5	1.266	14.0	1.704
0.26	0.390	0.75	0.647	1.8	0.876	2.6	1.030	2.8	1.106	5.0	1.290	16.0	1.730
0.28	0.393	0.80	0.654	2.0	0.887	2.8	1.042	3.0	1.123	5.5	1.309	18.0	1.749
0.30	0.396	0.85	0.660	2.2	0.895	3.0	1.051	3.5	1.158	6.0	1.325	20.0	1.762
0.35	0.400	0.90	0.665	2.4	0.900	3.5	1.069	4.0	1.183	7.0	1.347	22.0	1.771
0.40	0.402	0.95	0.669	2.6	0.905	4.0	1.080	5.0	1.215	8.0	1.361	24.0	1.777
0.45	0.404	1.0	0.673	2.8	0.908	4.5	1.087	6.0	1.232	9.0	1.370	26.0	1.781
0.50	0.405	1.2	0.682	3.0	0.910	5.0	1.091	7.0	1.242	10.0	1.376	28.0	1.784
0.60	0.405	1.4	0.688	3.5	0.913	5.5	1.094	8.0	1.247	12.0	1.382	30.0	1.787
0.70	0.405	1.6	0.690	4.0	0.915	6.0	1.096	9.0	1.250	14.0	1.385	35.0	1.789
0.80	0.405	1.8	0.692	4.5	0.916	6.5	1.097	10.0	1.251	16.0	1.386	40.0	1.791
		2.0	0.692	5.0	0.916	7.0	1.097	12.0	1.252	18.0	1.386	50.0	1.792
		2.5	0.693	5.5	0.916	8.0	1.098	14.0	1.253				
		3.0	0.693	6.0	0.916	10.0	1.099	16.0	1.253				

$r_{eD} = 8$		$r_{eD} = 10$		$r_{eD} = 15$		$r_{eD} = 20$		$r_{eD} = 25$		$r_{eD} = 30$		$r_{eD} = 40$	
t_D	p_D	t_D	p_D	t_D	p_D	t_D	p_D	t_D	p_D	t_D	p_D	t_D	p_D
7.0	1.499	10.0	1.651	20.0	1.960	30.0	2.148	50.0	2.389	70.0	2.551	120.0	2.813
7.5	1.527	12.0	1.730	22.0	2.003	35.0	2.219	55.0	2,434	80.0	2.615	140.0	2.888
8.0	1.554	14.0	1.798	24.0	2.043	40.0	2.282	60.0	2.476	90.0	2.672	160.0	2.953
8.5	1.580	16.0	1.856	26.0	2.080	45.0	2.338	65.0	2.514	100.0	2.723	180.0	3.011
9.0	1.604	18.0	1.907	28.0	2.114	50.0	2.388	70.0	2.550	120.0	2.812	200.0	3.063
9.5	1.627	20.0	1.952	30.0	2.146	60.0	2.475	75.0	2.583	140.0	2.886	220.0	3.109
10.0	1.648	25.0	2.043	35.0	2.218	70.0	2.547	80.0	2.614	160.0	2.950	240.0	3.152
12.0	1.724	30.0	2.111	40.0	2.279	80.0	2.609	85.0	2.643	165.0	2.965	260.0	3.191
14.0	1.786	35.0	2.160	45.0	2.332	90.0	2.658	90.0	2.671	170.0	2.979	280.0	3.226
16.0	1.837	40.0	2.197	50.0	2.379	100.0	2.707	95.0	2.697	175.0	2.992	300.0	3.259
18.0	1.879	45.0	2.224	60.0	2.455	105.0	2.728	100.0	2.721	180.0	3.006	350.0	3.331
20.0	1.914	50.0	2.245	70.0	2.513	110.0	2.747	120.0	2.807	200.0	3.054	400.0	3.391
22.0	1.943	55.0	2.260	80.0	2.558	115.0	2.764	140.0	2.878	250.0	3.150	450.0	3.440
24.0	1.967	60.0	2.271	90.0	2.592	120.0	2.781	160.0	2.936	300.0	3.219	500.0	3.482
26.0	1.986	65.0	2.279	100.0	2.619	125.0	2.796	180.0	2.984	350.0	3.269	550.0	3.516
28.0	2.002	70.0	2.285	120.0	2.655	130.0	2.810	200.0	3.024	400.0	3.306	600.0	3.545
30.0	2.016	75.0	2.290	140.0	2.677	135.0	2.823	220.0	3.057	450.0	3.332	650.0	3.568
35.0	2.040	80.0	2.293	160.0	2.689	140.0	2.835	240.0	3.085	500.0	3.351	700.0	3.588
40.0	2.055	90.0	2.297	180.0	2.697	145.0	2.846	260.0	3.107	600.0	3.375	800.0	3.619
45.0	2.064	100.0	2.300	200.0	2.701	150.0	2.857	280.0	3.126	700.0	3.387	900.0	3.640
50.0	2.070	110.0	2.301	220.0	2.704	160.0	2.876	300.0	3.142	800.0	3.394	1000.0	3.655
60.0	2.076	120.0	2.302	240.0	2.706	180.0	2.906	350.0	3.171	900.0	3.397	1200.0	3.672
70.0	2.078	130.0	2.302	260.0	2.707	200.0	2.929	400.0	3.189	1000.0	3.399	1400.0	3.681
80.0	2.079	140.0	2.302	280.0	2.707	240.0	2.958	450.0	3.200	1200.0	3.401	1600.0	3.685
		160.0	2.303	300.0	2.708	280.0	2.975	500.0	3.207	1400.0	3.401	1800.0	3.687
						300.0	2.980	600.0	3.214			2000.0	3.688
						400.0	2.992	700.0	3.217			2500.0	3.689
						500.0	2.995	800.0	3.218				
								900.0	3.219				

$r_{eD} = 50$		$r_{eD} = 60$		$r_{eD} = 70$		$r_{eD} = 80$		$r_{eD} = 90$		$r_{eD} = 100$		$r_{eD} = 200$	
t_D	p_D	t_D	p_D	t_D	p_D	t_D	p_D	t_D	p_D	t_D	p_D	t_D	p_D
200.0	3.064	300.0	3.257	500.0	3.512	600.0	3.603	800.0	3.747	1000.0	3.859	1500.0	4.061
220.0	3.111	400.0	3.401	600.0	3.603	700.0	3.680	900.0	3.803	1200.0	3.949	2000.0	4.205
240.0	3.154	500.0	3.512	700.0	3.680	800.0	3.747	1000.0	3.858	1400.0	4.026	2500.0	4.317
260.0	3.193	600.0	3.602	800.0	3.746	900.0	3.805	1200.0	3.949	1600.0	4.092	3000.0	4.408
280.0	3.229	700.0	3.676	900.0	3.803	1000.0	3.857	1300.0	3.988	1800.0	4.150	3500.0	4.485
300.0	3.263	800.0	3.739	1000.0	3.854	1200.0	3.946	1400.0	4.025	2000.0	4.200	4000.0	4.552
350.0	3.339	900.0	3.792	1200.0	3.937	1400.0	4.019	1500.0	4.058	2500.0	4.303	5000.0	4.663
400.0	3.405	1000.0	3.832	1400.0	4.003	1600.0	4.080	1800.0	4.144	3000.0	4.379	6000.0	4.754
450.0	3.461	1200.0	3.908	1600.0	4.054	1800.0	4.130	2000.0	4.192	3500.0	4.434	7000.0	4.829
500.0	3.512	1400.0	3.959	1800.0	4.095	2000.0	4.171	2500.0	4.285	4000.0	4.478	8000.0	4.894
550.0	3.556	1600.0	3.996	2000.0	4.127	2500.0	4.248	3000.0	4.349	4500.0	4.510	9000.0	4.949
600.0	3.595	1800.0	4.023	2500.0	4.181	3000.0	4.297	3500.0	4.394	5000.0	4.534	10000.0	4.996
650.0	3.630	2000.0	4.043	3000.0	4.211	3500.0	4.328	4000.0	4.426	5500.0	4.552	12000.0	5.072
700.0	3.661	2500.0	4.071	3500.0	4.228	4000.0	4.347	4500.0	4.448	6000.0	4.565	14000.0	5.129
750.0	3.688	3000.0	4.084	4000.0	4.237	4500.0	4.360	5000.0	4.464	6500.0	4.579	16000.0	5.171
800.0	3.713	3500.0	4.090	4500.0	4.242	5000.0	4.368	6000.0	4.482	7000.0	4.583	18000.0	5.203
850.0	3.735	4000.0	4.092	5000.0	4.245	6000.0	4.376	7000.0	4.491	7500.0	4.588	20000.0	5.227
900.0	3.754	4500.0	4.093	5500.0	4.247	7000.0	4.380	8000.0	4.496	8000.0	4.593	25000.0	5.264
950.0	3.771	5000.0	4.094	6000.0	4.247	8000.0	4.381	9000.0	4.498	9000.0	4.598	30000.0	5.282
1000.0	3.787	5500.0	4.094	6500.0	4.248	9000.0	4.382	10000.0	4.499	10000.0	4.601	35000.0	5.290
1200.0	3.833			7000.0	4.248	10000.0	4.382	11000.0	4.499	12500.0	4.604	40000.0	5.294
1400.0	3.862			7500.0	4.248	11000.0	4.382	12000.0	4.500	15000.0	4.605		

$r_{eD} = 300$		$r_{eD} = 400$		$r_{eD} = 500$		$r_{eD} = 600$		$r_{eD} = 700$		$r_{eD} = 800$		$r_{eD} = 900$	
t_D	p_D	t_D	p_D	t_D	p_D	t_D	p_D	t_D	p_D	t_D	p_D	t_D	p_D
6.0×10^3	4.754	1.5×10^4	5.212	2.0×10^4	5.356	4.0×10^4	5.703	5.0×10^4	5.814	7.0×10^4	5.983	8.0×10^4	6.049
8.0×10^3	4.808	2.0×10^4	5.356	2.5×10^4	5.468	4.5×10^4	5.762	6.0×10^4	5.905	8.0×10^4	6.049	9.0×10^4	6.108
10.0×10^3	5.010	3.0×10^4	5.556	3.0×10^4	5.559	5.0×10^4	5.814	7.0×10^4	5.982	9.0×10^4	6.108	10.0×10^4	6.161
12.0×10^3	5.101	4.0×10^4	5.689	3.5×10^4	5.636	6.0×10^4	5.904	8.0×10^4	6.048	10.0×10^4	6.160	12.0×10^4	6.251
14.0×10^3	5.177	5.0×10^4	5.781	4.0×10^4	5.702	7.0×10^4	5.979	9.0×10^4	6.105	12.0×10^4	6.249	14.0×10^4	6.327
16.0×10^3	5.242	6.0×10^4	5.845	4.5×10^4	5.759	8.0×10^4	6.041	10.0×10^4	6.156	14.0×10^4	6.322	16.0×10^4	6.392
18.0×10^3	5.299	7.0×10^4	5.889	5.0×10^4	5.810	9.0×10^4	6.094	12.0×10^4	6.239	16.0×10^4	6.382	18.0×10^4	6.447
20.0×10^3	5.348	8.0×10^4	5.920	6.0×10^4	5.894	10.0×10^4	6.139	14.0×10^4	6.305	18.0×10^4	6.432	20.0×10^4	6.494
24.0×10^3	5.429	9.0×10^4	5.942	7.0×10^4	5.960	12.0×10^4	6.210	16.0×10^4	6.357	20.0×10^4	6.474	25.0×10^4	6.587
28.0×10^3	5.491	10.0×10^4	5.957	8.0×10^4	6.013	14.0×10^4	6.262	18.0×10^4	6.398	25.0×10^4	6.551	30.0×10^4	6.652
30.0×10^3	5.517	11.0×10^4	5.967	9.0×10^4	6.055	16.0×10^4	6.299	20.0×10^4	6.430	30.0×10^4	6.599	40.0×10^4	6.729
40.0×10^3	5.606	12.0×10^4	5.975	10.0×10^4	6.088	18.0×10^4	6.326	25.0×10^4	6.484	35.0×10^4	6.630	45.0×10^4	6.751
50.0×10^3	5.652	12.5×10^4	5.977	12.0×10^4	6.135	20.0×10^4	6.345	30.0×10^4	6.514	40.0×10^4	6.650	50.0×10^4	6.766
60.0×10^3	5.676	13.0×10^4	5.980	14.0×10^4	6.164	25.0×10^4	6.374	35.0×10^4	6.530	45.0×10^4	6.663	55.0×10^4	6.777
70.0×10^3	5.690	14.0×10^4	5.983	16.0×10^4	6.183	30.0×10^4	6.387	40.0×10^4	6.540	50.0×10^4	6.671	60.0×10^4	6.785
80.0×10^3	5.696	16.0×10^4	5.988	18.0×10^4	6.195	35.0×10^4	6.392	45.0×10^4	6.545	55.0×10^4	6.676	70.0×10^4	6.794
90.0×10^3	5.700	18.0×10^4	5.990	20.0×10^4	6.202	40.0×10^4	6.395	50.0×10^4	6.548	60.0×10^4	6.679	80.0×10^4	6.798
100.0×10^3	5.702	20.0×10^4	5.991	25.0×10^4	6.211	50.0×10^4	6.397	60.0×10^4	6.550	70.0×10^4	6.682	90.0×10^4	6.800
120.0×10^3	5.703	24.0×10^4	5.991	30.0×10^4	6.213	60.0×10^4	6.397	70.0×10^4	6.551	80.0×10^4	6.684	100.0×10^4	6.801
140.0×10^3	5.704	26.0×10^4	5.991	35.0×10^4	6.214			80.0×10^4	6.551	100.0×10^4	6.684		
150.0×10^3	5.704			40.0×10^4	6.214								

$r_{eD} = 1000$		$r_{eD} = 1200$		$r_{eD} = 1400$		$r_{eD} = 1600$		$r_{eD} = 1800$		$r_{eD} = 2000$		$r_{eD} = 2200$	
t_D	p_D	t_D	p_D	t_D	p_D	t_D	p_D	t_D	p_D	t_D	p_D	t_D	p_D
1.0×10^5	6.161	2.0×10^5	6.507	2.0×10^5	6.507	2.5×10^5	6.619	3.0×10^5	6.710	4.0×10^5	6.854	5.0×10^5	6.966
1.2×10^5	6.252	3.0×10^5	6.704	2.5×10^5	6.619	3.0×10^5	6.710	4.0×10^5	6.854	5.0×10^5	6.966	5.5×10^5	7.013
1.4×10^5	6.329	4.0×10^5	6.833	3.0×10^5	6.709	3.5×10^5	6.787	5.0×10^5	6.965	6.0×10^5	7.056	6.0×10^5	7.057
1.6×10^5	6.395	5.0×10^5	6.918	3.5×10^5	6.785	4.0×10^5	6.853	6.0×10^5	7.054	7.0×10^5	7.132	6.5×10^5	7.097
1.8×10^5	6.452	6.0×10^5	6.975	4.0×10^5	6.849	5.0×10^5	6.962	7.0×10^5	7.120	8.0×10^5	7.196	7.0×10^5	7.133
2.0×10^5	6.503	7.0×10^5	7.013	5.0×10^5	6.950	6.0×10^5	7.046	8.0×10^5	7.188	9.0×10^5	7.251	7.5×10^5	7.167
2.5×10^5	6.605	8.0×10^5	7.038	6.0×10^5	7.026	7.0×10^5	7.114	9.0×10^5	7.238	10.0×10^5	7.298	8.0×10^5	7.199
3.0×10^5	6.681	9.0×10^5	7.056	7.0×10^5	7.082	8.0×10^5	7.167	10.0×10^5	7.280	12.0×10^5	7.374	8.5×10^5	7.229
3.5×10^5	6.738	10.0×10^5	7.067	8.0×10^5	7.123	9.0×10^5	7.210	15.0×10^5	7.407	14.0×10^5	7.431	9.0×10^5	7.256
4.0×10^5	6.781	12.0×10^5	7.080	9.0×10^5	7.154	10.0×10^5	7.244	20.0×10^5	7.459	16.0×10^5	7.474	10.0×10^5	7.307
4.5×10^5	6.813	14.0×10^5	7.085	10.0×10^5	7.177	15.0×10^5	7.334	30.0×10^5	7.489	18.0×10^5	7.506	12.0×10^5	7.390
5.0×10^5	6.837	16.0×10^5	7.088	15.0×10^5	7.229	20.0×10^5	7.364	40.0×10^5	7.495	20.0×10^5	7.530	16.0×10^5	7.507
5.5×10^5	6.854	18.0×10^5	7.089	20.0×10^5	7.241	25.0×10^5	7.373	50.0×10^5	7.495	25.0×10^5	7.566	20.0×10^5	7.579
6.0×10^5	6.868	19.0×10^5	7.089	25.0×10^5	7.243	30.0×10^5	7.376	51.0×10^5	7.495	30.0×10^5	7.584	25.0×10^5	7.631
7.0×10^5	6.885	20.0×10^5	7.090	30.0×10^5	7.244	35.0×10^5	7.377	52.0×10^5	7.495	35.0×10^5	7.593	30.0×10^5	7.661
8.0×10^5	6.895	21.0×10^5	7.090	31.0×10^5	7.244	40.0×10^5	7.378	53.0×10^5	7.495	40.0×10^5	7.597	35.0×10^5	7.677
9.0×10^5	6.901	22.0×10^5	7.090	32.0×10^5	7.244	42.0×10^5	7.378	54.0×10^5	7.495	50.0×10^5	7.600	40.0×10^5	7.686
10.0×10^5	6.904	23.0×10^5	7.090	33.0×10^5	7.244	44.0×10^5	7.378	56.0×10^5	7.495	60.0×10^5	7.601	50.0×10^5	7.693
12.0×10^5	6.907	24.0×10^5	7.090							64.0×10^5	7.601	60.0×10^5	7.695
14.0×10^5	6.907											70.0×10^5	7.696
16.0×10^5	6.908											80.0×10^5	7.696

$r_{eD} = 2400$		$r_{eD} = 2600$		$r_{eD} = 2800$		$r_{eD} = 3000$	
t_D	p_D	t_D	p_D	t_D	p_D	t_D	p_D
6.0×10^5	7.057	7.0×10^5	7.134	8.0×10^5	7.201	1.0×10^6	7.312
7.0×10^5	7.134	8.0×10^5	7.201	9.0×10^5	7.260	1.2×10^6	7.403
8.0×10^5	7.200	9.0×10^5	7.259	10.0×10^5	7.312	1.4×10^6	7.480
9.0×10^5	7.259	10.0×10^5	7.312	12.0×10^5	7.403	1.6×10^6	7.545
10.0×10^5	7.310	12.0×10^5	7.401	16.0×10^5	7.542	1.8×10^6	7.602
12.0×10^5	7.398	14.0×10^5	7.475	20.0×10^5	7.644	2.0×10^6	7.651
16.0×10^5	7.526	16.0×10^5	7.536	24.0×10^5	7.719	2.4×10^6	7.732
20.0×10^5	7.611	18.0×10^5	7.588	28.0×10^5	7.775	2.8×10^6	7.794
24.0×10^5	7.668	20.0×10^5	7.631	30.0×10^5	7.797	3.0×10^6	7.820
28.0×10^5	7.706	24.0×10^5	7.699	35.0×10^5	7.840	3.5×10^6	7.871
30.0×10^5	7.720	28.0×10^5	7.746	40.0×10^5	7.870	4.0×10^6	7.908
35.0×10^5	7.745	30.0×10^5	7.765	50.0×10^5	7.905	4.5×10^6	7.935
40.0×10^5	7.760	35.0×10^5	7.799	60.0×10^5	7.922	5.0×10^6	7.955
50.0×10^5	7.775	40.0×10^5	7.821	70.0×10^5	7.930	6.0×10^6	7.979
60.0×10^5	7.780	50.0×10^5	7.845	80.0×10^5	7.934	7.0×10^6	7.992
70.0×10^5	7.782	60.0×10^5	7.856	90.0×10^5	7.936	8.0×10^6	7.999
80.0×10^5	7.783	70.0×10^5	7.860	100.0×10^5	7.937	9.0×10^6	8.002
90.0×10^5	7.783	80.0×10^5	7.862	120.0×10^5	7.937	10.0×10^6	8.004
95.0×10^5	7.783	90.0×10^5	7.863	130.0×10^5	7.937	12.0×10^6	8.006
		100.0×10^5	7.863			15.0×10^6	8.006

APPENDIX

12E TABULATION OF Q_D vs t_D FOR RADIAL FLOW, INFINITE RESERVOIRS WITH CONSTANT TERMINAL PRESSURE AT INNER BOUNDARY (FROM VAN EVERDINGEN AND HURST[2])

t_D	Q_D	t_D	Q_D	t_D	Q_D
0.01	0.112	15.0	9.965	3000.0	759.00
0.05	0.278	20.0	12.29	4000.0	975.70
0.10	0.404	25.0	14.55	5000.0	1188.00
0.15	0.520	30.0	16.81	6000.0	1395.00
0.20	0.606	40.0	20.88	7000.0	1599.00
0.25	0.689	50.0	24.82	8000.0	1800.00
0.30	0.758	60.0	28.60	9000.0	1999.00
0.40	0.898	70.0	32.28	10000.0	2196.00
0.50	1.020	80.0	35.99	15000.0	3146.00
0.60	1.140	90.0	39.42	20000.0	4079.00
0.70	1.251	100.0	43.01	25000.0	4994.00
0.80	1.359	150.0	59.80	30000.0	5891.00
0.90	1.469	200.0	75.86	40000.0	7634.00
1.00	1.570	250.0	91.20	50000.0	9342.00
1.50	2.032	300.0	105.80	60000.0	11030.0
2.00	2.442	400.0	134.80	70000.0	12690.0
2.50	2.838	500.0	162.40	80000.0	14330.0
3.00	3.209	600.0	189.70	90000.0	15950.0
4.00	3.879	700.0	216.00	100000.0	17560.0
5.00	4.541	800.0	242.30	150000.0	25380.0
6.00	5.148	900.0	267.70	200000.0	33080.0
7.00	5.749	1000.0	293.10	250000.0	40660.0
8.00	6.314	1500.0	413.60	300000.0	48170.0
9.00	6.861	2000.0	531.50	400000.0	62670.0
10.00	7.417	2500.0	646.60	500000.0	76990.0

t_D	Q_D	t_D	Q_D	t_D	Q_D
6.0×10^{5}	9.113×10^{4}	9.0×10^{7}	9.911×10^{6}	2.0×10^{10}	1.697×10^{9}
7.0×10^{5}	1.051×10^{5}	1.0×10^{8}	10.95×10^{6}	2.5×10^{10}	2.103×10^{9}
8.0×10^{5}	1.189×10^{5}	1.5×10^{8}	1.604×10^{7}	3.0×10^{10}	2.505×10^{9}
9.0×10^{5}	1.326×10^{5}	2.0×10^{8}	2.108×10^{7}	4.0×10^{10}	3.299×10^{9}
1.0×10^{6}	1.462×10^{5}	2.5×10^{8}	2.607×10^{7}	5.0×10^{10}	4.087×10^{9}
1.5×10^{6}	2.126×10^{5}	3.0×10^{8}	3.100×10^{7}	6.0×10^{10}	4.868×10^{9}
2.0×10^{6}	2.781×10^{5}	4.0×10^{8}	4.071×10^{7}	7.0×10^{10}	5.643×10^{9}
2.5×10^{6}	3.427×10^{5}	5.0×10^{8}	5.032×10^{7}	8.0×10^{10}	6.414×10^{9}
3.0×10^{6}	4.064×10^{5}	6.0×10^{8}	5.984×10^{7}	9.0×10^{10}	7.183×10^{9}
4.0×10^{6}	5.313×10^{5}	7.0×10^{8}	6.928×10^{7}	1.0×10^{11}	7.948×10^{9}
5.0×10^{6}	6.544×105	8.0×10^{8}	7.865×10^{7}	1.5×10^{11}	1.170×10^{10}
6.0×10^{6}	7.761×10^{5}	9.0×10^{8}	8.797×10^{7}	2.0×10^{11}	1.550×10^{10}
7.0×10^{6}	8.965×10^{5}	1.0×10^{9}	9.725×10^{7}	2.5×10^{11}	1.920×10^{10}
8.0×10^{6}	1.016×10^{6}	1.5×10^{9}	1.429×10^{8}	3.0×10^{11}	2.290×10^{10}
9.0×10^{6}	1.134×10^{6}	2.0×10^{9}	1.880×10^{8}	4.0×10^{11}	3.020×10^{10}
1.0×10^{7}	1.252×10^{6}	2.5×10^{9}	2.328×10^{8}	5.0×10^{11}	3.750×10^{10}
1.5×10^{7}	1.828×10^{6}	3.0×10^{9}	2.771×10^{8}	6.0×10^{11}	4.470×10^{10}
2.0×10^{7}	2.398×10^{6}	4.0×10^{9}	3.645×10^{8}	7.0×10^{11}	5.190×10^{10}
2.5×10^{7}	2.961×10^{6}	5.0×10^{9}	4.510×10^{8}	8.0×10^{11}	5.890×10^{10}
3.0×10^{7}	3.517×10^{6}	6.0×10^{9}	5.368×10^{8}	9.0×10^{11}	6.580×10^{10}
4.0×10^{7}	4.610×10^{6}	7.0×10^{9}	6.220×10^{8}	1.0×10^{12}	7.280×10^{10}
5.0×10^{7}	5.689×10^{6}	8.0×10^{9}	7.066×10^{8}	1.5×10^{12}	1.080×10^{11}
6.0×10^{7}	6.758×10^{6}	9.0×10^{9}	7.909×10^{8}	2.0×10^{12}	1.420×10^{11}
7.0×10^{7}	7.816×10^{6}	1.0×10^{10}	8.747×10^{8}		
8.0×10^{7}	8.866×10^{6}	1.5×10^{10}	1.288×10^{9}		

APPENDIX 12F TABULATION OF Q_D vs t_D FOR RADIAL FLOW, FINITE RESERVOIRS WITH CLOSED OUTER BOUNDARY AND CONSTANT TERMINAL PRESSURE AT INNER BOUNDARY (FROM VAN EVERDINGEN AND HURST[2])

r_{eD} = 1.5		r_{eD} = 2.0		r_{eD} = 2.5		r_{eD} = 3.0		r_{eD} = 3.5		r_{eD} = 4.0		r_{eD} = 4.5	
t_D	Q_D	t_D	Q_D	t_D	Q_D	t_D	Q_D	t_D	Q_D	t_D	Q_D	t_D	Q_D
0.050	0.276	0.050	0.278	0.10	0.408	0.30	0.755	1.00	1.571	2.00	2.442	2.5	2.835
0.060	0.304	0.075	0.345	0.15	0.509	0.40	0.895	1.20	1.761	2.20	2.598	3.0	3.196
0.070	0.330	0.10	0.404	0.20	0.599	0.50	1.023	1.40	1.940	2.40	2.748	3.5	3.537
0.080	0.354	0.125	0.458	0.25	0.681	0.60	1.143	1.60	2.111	2.60	2.893	4.0	3.859
0.090	0.375	0.150	0.507	0.30	0.758	0.70	1.256	1.80	2.273	2.80	3.034	4.5	4.165
0.10	0.395	0.175	0.553	0.35	0.829	0.80	1.363	2.00	2.427	3.00	3.170	5.0	4.454
0.11	0.414	0.200	0.597	0.40	0.897	0.90	1.465	2.20	2.574	3.25	3.334	5.5	4.727
0.12	0.431	0.225	0.638	0.45	0.962	1.00	1.563	2.40	2.715	3.50	3.493	6.0	4.986
0.13	0.446	0.250	0.678	0.50	1.024	1.25	1.791	2.60	2.849	3.75	3.645	6.5	5.231
0.14	0.461	0.275	0.715	0.55	1.088	1.50	1.997	2.80	2.976	4.00	3.792	7.0	5.464
0.15	0.474	0.300	0.751	0.60	1.140	1.75	2.184	3.00	3.098	4.25	3.932	7.5	5.684
0.16	0.486	0.325	0.785	0.65	1.195	2.00	2.353	3.25	3.242	4.50	4.068	8.0	5.892
0.17	0.497	0.350	0.817	0.70	1.248	2.25	2.507	3.50	3.379	4.75	4.198	8.5	6.089
0.18	0.507	0.375	0.848	0.75	1.229	2.50	2.646	3.75	3.507	5.00	4.323	9.0	6.276
0.19	0.517	0.400	0.877	0.80	1.348	2.75	2.772	4.00	3.628	5.50	4.560	9.5	6.453
0.20	0.525	0.425	0.905	0.85	1.395	3.00	2.886	4.25	3.742	6.00	4.779	10.0	6.621
0.21	0.533	0.450	0.932	0.90	1.440	3.25	2.990	4.50	3.850	6.50	4.982	11.0	6.930
0.22	0.541	0.475	0.958	0.95	1.484	3.50	3.084	4.75	3.951	7.00	5.169	12.0	7.208
0.23	0.548	0.500	0.983	1.0	1.526	3.75	3.170	5.00	4.047	7.50	5.343	13.0	7.457
0.24	0.554	0.550	1.028	1.1	1.605	4.00	3.247	5.50	4.222	8.00	5.504	14.0	7.680
0.25	0.559	0.600	1.070	1.2	1.679	4.25	3.317	6.00	4.378	8.50	5.653	15.0	7.880
0.26	0.565	0.650	1.108	1.3	1.747	4.50	3.381	6.50	4.516	9.00	5.790	16.0	8.060

$r_{eD} = 1.5$		$r_{eD} = 2.0$		$r_{eD} = 2.5$		$r_{eD} = 3.0$		$r_{eD} = 3.5$		$r_{eD} = 4.0$		$r_{eD} = 4.5$	
t_D	Q_D	t_D	Q_D	t_D	Q_D	t_D	Q_D	t_D	Q_D	t_D	Q_D	t_D	Q_D
0.28	0.574	0.700	1.143	1.4	1.811	4.75	3.439	7.00	4.639	9.50	5.917	18.0	8.365
0.30	0.582	0.750	1.174	1.5	1.870	5.00	3.491	7.50	4.749	10.00	6.035	20.0	8.611
0.32	0.588	0.800	1.203	1.6	1.924	5.50	3.581	8.00	4.846	11.00	6.246	22.0	8.809
0.34	0.594	0.900	1.253	1.7	1.975	6.00	3.656	8.50	4.932	12.00	6.425	24.0	8.968
0.34	0.599	1.00	1.295	1.8	2.022	6.50	3.717	9.00	5.009	13.00	6.580	26.0	9.097
0.36	0.603	1.1	1.330	2.0	2.106	7.00	3.767	9.50	5.078	14.00	6.712	28.0	9.200
0.38	0.606	1.2	1.358	2.2	2.178	7.50	3.809	10.00	5.138	15.00	6.825	30.0	9.283
0.40	0.613	1.3	1.382	2.4	2.241	8.00	3.843	11.00	5.241	16.00	6.922	34.0	9.404
0.50	0.617	1.4	1.402	2.6	2.294	9.00	3.894	12.00	5.321	17.00	7.004	38.0	9.481
0.60	0.621	1.6	1.432	2.8	2.340	10.00	3.928	13.00	5.385	18.00	7.076	42.0	9.532
0.70	0.623	1.7	1.444	3.0	2.380	11.00	3.951	14.00	5.435	20.00	7.189	46.0	9.565
0.80	0.624	1.8	1.453	3.4	2.444	12.00	3.967	15.00	5.476	22.00	7.272	50.0	9.586
		2.0	1.468	3.8	2.491	14.00	3.985	16.00	5.506	24.00	7.332	60.0	9.612
		2.5	1.487	4.2	2.525	16.00	3.993	17.00	5.531	26.00	7.377	70.0	9.621
		3.0	1.495	4.6	2.551	18.00	3.997	18.00	5.551	30.00	7.434	80.0	9.623
		4.0	1.499	5.0	2.570	20.00	3.999	20.00	5.579	34.00	7.464	90.0	9.624
		5.0	1.500	6.0	2.599	22.00	3.999	25.00	5.611	38.00	7.481	100.0	9.625
				7.0	2.613	24.00	4.000	30.00	5.621	42.00	7.490		
				8.0	2.619			35.00	5.624	46.00	7.494		
				9.0	2.622			40.00	5.625	50.00	7.497		
				10.0	2.624								

r_{eD} = 5.0		r_{eD} = 6.0		r_{eD} = 7.0		r_{eD} = 8.0		r_{eD} = 9.0		r_{eD} = 10.0	
t_D	Q_D	t_D	Q_D	t_D	Q_D	t_D	Q_D	t_D	Q_D	t_D	Q_D
3.0	3.195	6.0	5.148	9.00	6.861	9	6.861	10	7.417	15	9.965
3.5	3.542	6.5	5.440	9.50	7.127	10	7.398	15	9.945	20	12.32
4.0	3.875	7.0	5.724	10.0	7.389	11	7.920	20	12.26	22	13.22
4.5	4.193	7.5	6.002	11.0	7.902	12	8.431	22	13.13	24	14.09
5.0	4.499	8.0	6.273	12.0	8.397	13	8.930	24	13.98	26	14.95
5.5	4.792	8.5	6.537	13.0	8.876	14	9.418	26	14.79	28	15.78
6.0	5.074	9.0	6.795	14.0	9.341	15	9.895	28	15.59	30	16.59
6.5	5.345	9.5	7.047	15.0	9.791	16	10.361	30	16.35	32	17.38
7.0	5.605	10.0	7.293	16.0	10.23	17	10.82	32	17.10	34	18.16
7.5	5.854	10.5	7.533	17.0	10.65	18	11.26	34	17.82	36	18.91
8.0	6.094	11.0	7.767	18.0	11.06	19	11.70	36	18.52	38	19.65
8.5	6.325	12.0	8.220	19.0	11.46	20	12.13	38	19.19	40	20.37
9.0	6.547	13.0	8.651	20.0	11.85	22	12.95	40	19.85	42	21.07
9.5	6.760	14.0	9.063	22.0	12.58	24	13.74	42	20.48	44	21.76
10.0	6.965	15.0	9.456	24.0	13.27	26	14.50	44	21.09	46	22.42
11.0	7.350	16.0	9.829	26.0	13.92	28	15.23	46	21.69	48	23.07
12.0	7.706	17.0	10.19	28.0	14.53	30	15.92	42	22.26	50	23.71
13.0	8.035	18.0	10.53	30.0	15.11	34	17.22	50	22.82	52	24.33
14.0	8.339	19.0	10.85	35.0	16.39	38	18.41	52	23.36	54	24.94
15.0	8.620	20.0	11.16	40.0	17.49	40	18.97	54	23.89	56	25.53
16.0	8.879	22.0	11.74	45.0	18.43	45	20.26	56	24.39	58	26.11
18.0	9.338	24.0	12.26	50.0	19.24	50	21.42	52	24.88	60	26.67
20.0	9.731	25.0	12.50	60.0	20.51	55	22.46	60	25.36	65	28.02
22.0	10.07	31.0	13.74	70.0	21.45	60	23.40	65	26.48	70	29.29
24.0	10.35	35.0	14.40	80.0	22.13	70	24.98	70	27.52	75	30.49
26.0	10.59	39.0	14.93	90.0	22.63	80	26.26	75	28.48	80	31.61
28.0	10.80	51.0	16.05	100.0	23.00	90	27.28	80	29.36	85	32.67
30.0	10.98	60.0	16.56	120.0	23.47	100	28.11	85	30.18	90	33.66
34.0	11.26	70.0	16.91	140.0	23.71	120	29.31	90	30.93	95	34.60
38.0	11.46	80.0	17.14	160.0	23.85	140	30.08	95	31.63	100	35.48

$r_{eD} = 5.0$		$r_{eD} = 6.0$		$r_{eD} = 7.0$		$r_{eD} = 8.0$		$r_{eD} = 9.0$		$r_{eD} = 10.0$	
t_D	Q_D	t_D	Q_D	t_D	Q_D	t_D	Q_D	t_D	Q_D	t_D	Q_D
42.0	11.61	90.0	17.27	180.0	23.92	160	30.58	100	32.27	120	38.51
46.0	11.71	100.0	17.36	200.0	23.96	180	30.91	120	34.39	140	40.89
50.0	11.79	110.0	17.41	500.0	24.00	200	31.12	140	35.92	160	42.57
60.0	11.91	120.0	17.45			240	31.34	160	37.04	180	44.21
70.0	11.96	130.0	17.46			280	31.43	180	37.85	200	45.36
80.0	11.98	140.0	17.48			320	31.47	200	38.44	240	46.95
90.0	11.99	150.0	17.49			360	31.49	240	39.17	280	47.94
100.0	12.00	160.0	17.49			400	31.50	280	39.56	320	48.54
120.0	12.00	180.0	17.50			500	31.50	320	39.77	360	48.91
		200.0	17.50					360	39.88	400	49.14
		220.0	17.50					400	39.94	440	49.28
								440	39.97	480	49.36
								480	39.98		

CHAPTER

13 Well Test Analysis: Straightline Methods

13.1 Introduction

Well testing by initiation of pressure transients in the wellbore is routinely conducted to collect data which can be analyzed for well and reservoir properties. During drilling operations, well testing is conducted with drillstem tests (DST) or with openhole formation tester tools. For producing wells, testing is typically performed with either flow (drawdown) or shut-in (buildup) tests depending on well conditions, and reservoir properties to be calculated from the test data. For injection wells, test data can be obtained by changing injection rates (injection tests) or by shutting-down injection (falloff tests). Well test data are analyzed with methods that have been developed over time. In this chapter, these methods are reviewed in context of their historical development and significance. The first methods for well test analysis that gained wide acceptance in the petroleum industry are the straightline methods which were presented by Miller et al.[1] and Horner.[2] Straightline methods (sometimes called conventional methods) are based on straight lines drawn over middle time region data on semi-log plots of well test data, and are used primarily to calculate reservoir permeability, skin factor, average reservoir pressure, and estimates of reservoir size or the drainage area of the well. Straightline methods are still in use in well test analysis although as supplementary to other more advanced methods. In 1970, Ramey[3] introduced the use of type curves for well test analysis. Type curves are theoretical solutions to reservoir flow equations under specified initial and boundary conditions. The application of type curves for well test analysis was advanced by Gringarten et al.[4] with the introduction of a unified set of type curves and integrated methodologies for well test analysis that incorporated straightline methods. Further revolutionary advancement in the application of type curves for well test analysis occurred in 1983 with the introduction of derivative type curves by Bourdet et al.[5-7] The combined applications of log-log type curves by Gringarten et al.[4] and derivative type curves from Bourdet et al.[5-7] for well test analysis have enabled identifications of reservoir behaviors and geologic features which were previously undetectable from well test data analysis using straightline methods. Another major advancement in well test analysis was initiated with the introduction of stable deconvolution methods by von Schroeter et al.[8] Deconvolution is the process of converting variable rate pressure data into a constant rate pressure data with duration equal to the duration of the test data. Deconvolution methods provide more data for analyses which aid in the identification of geologic features that would have been remained "hidden" with limited test data. A detailed historical account of the development of well test analysis methods has been documented by Gringarten.[9] In this book, well test analysis methods are presented in the following order starting with straightline methods in this chapter, log-log type curves and derivative type curves in Chapter 14, fractured wells and fractured reservoirs in Chapter 15, deconvolution methods in Chapter 16, and unconventional reservoirs in Chapter 17. The presentation in this chapter covers slightly compressible fluids (liquids) and compressible fluids (gases). Examples are presented to illustrate the application of straightline solution methods. This chapter begins with reviews of basic concepts commonly encountered in well test analysis.

13.2 Basic Concepts in Well Test Analysis

Basic concepts commonly encountered in the acquisition and analyses of pressure transient tests data are reviewed in this section. The concepts that are presented include: radius of investigation, skin and skin factor, flow efficiency and damage ratio, effective wellbore radius, drawdown well tests, buildup well tests, and wellbore storage. These concepts are important in understanding the fundamental principles of pressure transient test analysis discussed in this chapter, and also in Chapters 14, 15, 16, and 17.

13.2.1 Radius of Investigation

When a transient is initiated in a well by changing its production rate or wellbore pressure, the transient will travel away from the wellbore toward the outer boundaries of the reservoir at a rate which depends only on the diffusivity constant of the reservoir. In Chapter 12, the reservoir diffusivity constant, η, was defined in Eq. (12.7) as:

$$\eta = \frac{k}{\phi \mu c_t}$$

In Eq. (12.7) as shown here, k = permeability; ϕ = porosity; μ = fluid viscosity; and c_t = total compressibility. In field units, the reservoir diffusivity constant is given by:

$$\eta = \frac{0.00633k}{\phi \mu c_t} \tag{13.1}$$

In Eq. (13.1), η = diffusivity constant, ft²/day; k = permeability, md; ϕ = porosity, fraction; μ = fluid viscosity, cp; and c_t = total compressibility, psi⁻¹. In well testing, the distance a transient has travelled after it was initiated is termed its radius of investigation. It is a measure of the extent of the reservoir through which the transient has travelled since it was initiated. The radius of investigation, r_i, for an oil reservoir is calculated with the expression:

$$r_i = \sqrt{\frac{kt}{948\phi\mu c_t}} \tag{13.2}$$

In Eq. (13.2), r_i = radius of investigation, feet; k = permeability, md; t = time, hrs.; ϕ = porosity, fraction; μ = fluid viscosity, cp; and c_t = total compressibility, psi⁻¹. For a gas reservoir, the radius of investigation is estimated with the expression:

$$r_i = \sqrt{\frac{kt\bar{p}}{948\phi\mu}} \tag{13.3}$$

Eq. (13.3) has the same units as Eq. (13.2), and $\bar{p}$ is average reservoir pressure in psia. Note that Eq. (13.3) is the same as Eq. (12.10) in Chapter 12. Note also that in Eqs. (13.2) and (13.3), t is total elapsed time since initiation of the transient. The radius of investigation is calculated in most commercial well test analysis reports.

EXAMPLE 13.1 Calculation of Radius of Investigation for an Oil Reservoir

Problem

Calculate the radius of investigation of an oil well producing at a constant rate for 20 hours from initial conditions in a reservoir with properties as follows:

Formation permeability, k	200 md
Oil viscosity, μ	2.5 cp
Formation porosity, ϕ	0.28
Total formation compressibility, c_t	5×10^{-6} psi⁻¹

Solution

By substituting Eq. (13.2) with the given reservoir properties, the radius of investigation is calculated as:

$$r_i = \sqrt{\frac{kt}{948\phi\mu c_t}}$$

$$= \sqrt{\frac{200 \times 20}{948 \times 0.28 \times 2.5 \times 5 \times 10^{-6}}}$$

$$= 1097.97 \text{ feet.}$$

End of Example 13.1

13.2.2 Skin and Skin Factor

The use of the term "skin" to describe a zone close to the wellbore with altered permeability that is different from the permeability of the formation was introduced by van Everdingen.[10] This concept is illustrated in Figure 13.1, showing a wellbore with a zone of altered permeability close to the borehole. The permeability of the zone close to the wellbore can be reduced due to damage from mud filtrate invasion during drilling or completion operations. Near wellbore damage can also result in producing wells from migration and deposition of fine particles from the reservoir or deposition of waxy residue from the produced oil. On the other hand, the permeability in the altered zone could be improved to be higher than formation permeability through acid stimulation and/or hydraulic fracturing. Generally, the primary objective of most stimulative remedial programs is to remove near-wellbore damage, thereby increasing permeability in this zone.

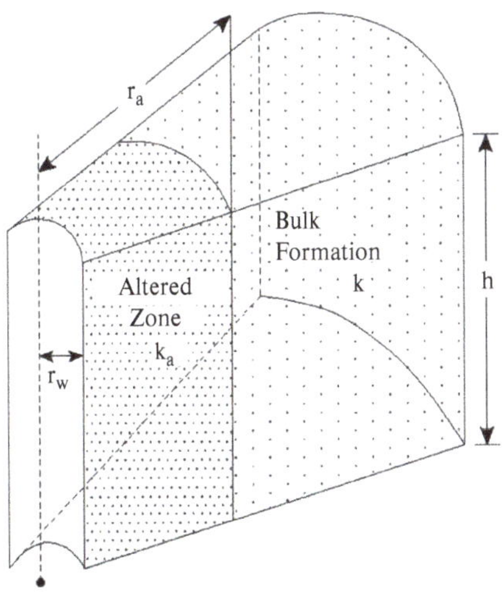

▲ **FIGURE 13.1 Wellbore diagram showing zone with altered permeability**

The presence of formation damage near the wellbore increases the pressure drawdown required for a given production rate. This is illustrated in Figure 13.2. As shown in Figure 13.2, the pressure drop, $(p_a - p_{wf})$, across the damaged zone (from r_a to r_w), is greater than the pressure drop, $(p_a - p'_{wf})$, that would have existed for a given production rate if the well was not damaged. Assuming steady-state radial flow, the pressure drop across the damaged zone can be calculated with Eq. (12.23) as:

$$p_a - p_{wf} = \frac{141.2qB\mu}{k_a h} \ln \frac{r_a}{r_w} \tag{13.4}$$

In Eq. (13.4), p_a is the pressure at the boundary of the damaged zone; p_{wf} is the flowing wellbore pressure; k_a is the permeability of the damaged zone; r_a is the radius of the damaged zone; and r_w is the radius of the wellbore. If the zone was undamaged, the pressure drop across the undamaged zone would have been calculated as:

$$p_a - p'_{wf} = \frac{141.2qB\mu}{kh} \ln \frac{r_a}{r_w} \tag{13.5}$$

In Eq. (13.5), p'_{wf} is the flowing wellbore pressure if the zone was not damaged, and, k is the permeability of the undamaged formation. Subtracting Eq. (13.5) from Eq. (13.4) gives:

$$\begin{aligned} p'_{wf} - p_{wf} &= \frac{141.2qB\mu}{h}\left(\frac{1}{k_a} - \frac{1}{k}\right) \ln \frac{r_a}{r_w} \\ &= \frac{141.2qB\mu}{kh}\left(\frac{k}{k_a} - 1\right) \ln \frac{r_a}{r_w} \end{aligned} \tag{13.6}$$

The pressure difference, $(p'_{wf} - p_{wf})$, is equal to the additional pressure drop, (Δp_s), caused by the presence of damage or skin in the near-wellbore zone. Substituting in Eq. (13.6) gives:

$$\begin{aligned} p'_{wf} - p_{wf} &= \Delta p_s \\ &= \frac{141.2qB\mu}{kh}\left(\frac{k}{k_a} - 1\right) \ln \frac{r_a}{r_w} \end{aligned} \tag{13.7}$$

The zone of altered permeability near the wellbore is often represented in terms of a skin factor. The skin factor, s, is defined as:

$$s = \left(\frac{k}{k_a} - 1\right) \ln \frac{r_a}{r_w} \tag{13.8}$$

With this definition of skin factor in Eq. (13.8), Eq. (13.7) can be written as:

$$\begin{aligned} \Delta p_s &= \frac{141.2qB\mu}{kh} \times s \\ &= \frac{70.6qB\mu}{kh} \times 2s \end{aligned} \tag{13.9}$$

As can be deduced from Eq. (13.8), a positive value for skin factor denotes that the near-wellbore zone is damaged, while a negative value for skin factor indicates that the near-wellbore zone is stimulated. A zero value for skin factor indicates that the near-wellbore zone is neither damaged nor stimulated. It is evident from Eq. (13.8) that skin factors representing damage in wellbores can be very high because the damaged zone permeability is in the denominator. For this reason, the magnitude of skin factors should be used qualitatively rather than quantitatively to assess damage in the near-wellbore zone. (For the remainder of this chapter, the near-wellbore zone is termed the wellbore for discussions of damage or stimulation.)

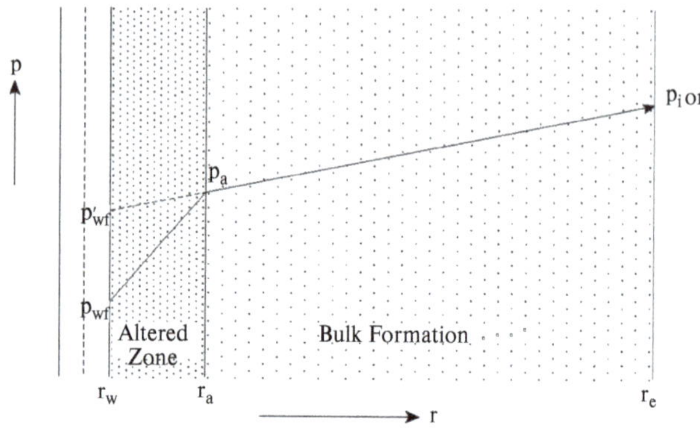

▲ **FIGURE 13.2 Pressure drawdown due to near-wellbore formation damage**

13.2.3 Flow Efficiency and Damage Ratio

Flow efficiency and damage ratio are two parameters that are used to measure the condition of a wellbore. Flow efficiency is the productivity of a damaged wellbore relative to a wellbore that is not damaged. Thus, flow efficiency, E, is calculated as:

$$E = \frac{\bar{p} - p_{wf} - \Delta p_s}{\bar{p} - p_{wf}} \tag{13.10}$$

In Eq. (13.10), $\bar{p}$ = average reservoir pressure; p_{wf} = flowing wellbore pressure; Δp_s = pressure drop due to skin. As is evident from Eq. (13.10), if flow efficiency is less than one, the well appears to be damaged, while flow efficiency greater than one represents a stimulated wellbore. Note that average reservoir pressure can be estimated as the arithmetic average of initial reservoir pressure and flowing bottomhole pressure. However, it can be calculated more rigorously with either Eqs. (12.95) or (12.96).

Damage ratio is the reciprocal of flow efficiency. From Eq. 13.10, damage ratio, DR, is represented as:

$$\begin{aligned} DR &= \frac{1}{E} \\ &= \frac{\bar{p} - p_{wf}}{\bar{p} - p_{wf} - \Delta p_s} \end{aligned} \tag{13.11}$$

In contrast to flow efficiency, a damage ratio that is greater than one indicates a damaged wellbore, whereas a damage ratio that is less than one indicates a stimulated wellbore.

13.2.4 Effective Wellbore Radius

Effective wellbore radius is a measure that can be used to denote the level of damage or stimulation existing in a wellbore. Effective wellbore radius, r_{wa}, is defined as the radius of an undamaged well that exhibits the same pressure drawdown as a damaged or stimulated well. From Eqs. (13.17) and (13.8), additional pressure drop across the damaged zone is given as:

$$\Delta p_s = \frac{141.2qB\mu}{kh} \times s \qquad (13.12)$$

For a reservoir exhibiting homogenous behavior, the same additional pressure drop would be achieved if the wellbore radius was r_{wa} instead of r_w such that:

$$\Delta p_s = \frac{141.2qB\mu}{kh} \ln\frac{r_w}{r_{wa}} \qquad (13.13)$$

From Eqs. (13.12) and (13.13), it follows that:

$$s = \ln\frac{r_w}{r_{wa}} \qquad (13.14)$$

Rearranging Eq. (13.14), effective wellbore radius is given by:

$$r_{wa} = r_w e^{-s} \qquad (13.15)$$

From Eq. (13.15), it can be deduced that the effective wellbore radius of a damaged well is small, while that of a stimulated well is relatively large. For instance, the effective wellbore radius of a vertically fractured well is one-half of the fracture length:

$$r_{wa} = \frac{1}{2} L_f \qquad (13.16)$$

In Eq. (13.16), L_f is the length of one wing of the fracture assuming the fracture has two wings of equal length.

EXAMPLE 13.2 Calculation of Skin Factor and Effective Wellbore Radius

Problem

Calculate the skin factor and effective wellbore radius of a well with the following properties:

Formation permeability, k	250 md
Damaged zone permeability, k_a	50 md
Wellbore radius, r_w	0.5 ft
Damaged zone radius, r_a	3.0 ft

Solution

From Eq. (13.8), skin factor is calculated using the given properties of the wellbore as:

$$s = \left(\frac{k}{k_a} - 1\right)\ln\frac{r_a}{r_w} = \left(\frac{250}{50} - 1\right)\ln\left(\frac{3}{0.5}\right)$$
$$= 7.17$$

Using Eq.(13.15), effective wellbore radius is calculated as:

$$r_{wa} = r_w e^{-s} = 0.5e^{-7.17}$$
$$= 3.85 \times 10^{-4} \text{ ft.}$$

End of Example 13.2

13.2.5 Drawdown Well Tests

One type of flow test that can be conducted on a well is called a drawdown test. In a drawdown test, a well at zero production in a reservoir or drainage volume at initial or stabilized static pressure is placed on production for a specified length of time. There are two main types of drawdown tests, namely, constant single rate tests and multiple rate (or multi-rate) tests. In constant single rate tests, the well is produced at a constant, single rate for the duration of the test. This is shown in Figure 13.3. In multi-rate tests, the production rates of the well are maintained constant at multiple rates over various time periods as illustrated in Figure 13.4. Drawdown tests are affected by following operational problems:

1. It is difficult, sometimes impossible, to maintain the well at a constant production rate over the duration of the test either for single rate or multi-rate tests.
2. Except for virgin reservoirs, the static pressure within the drainage volume of the well may not have been stabilized to a specific average pressure.
3. Early test data may be distorted by unloading of fluids in the wellbore at initiation of production.

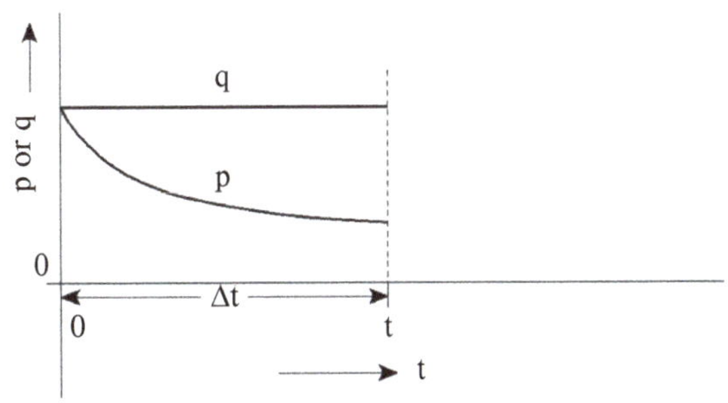

▲ FIGURE 13.3 Constant single rate drawdown test

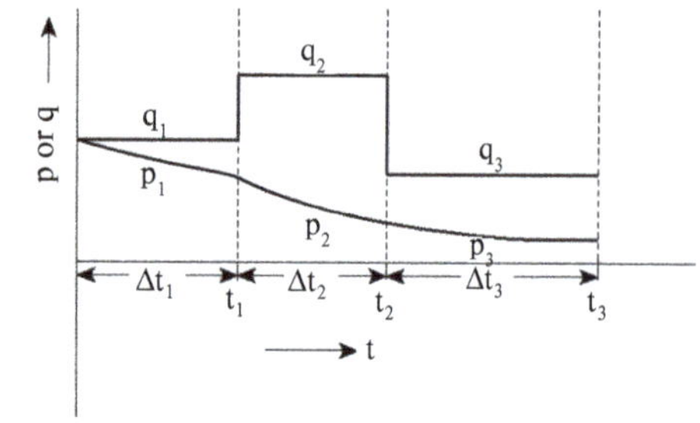

▲ **FIGURE 13.4** **Multiple rates drawdown test**

As shown in Figure 13.5, a typical plot of drawdown pressure data versus time on a semi-log scale can be divided into three time regions: early time region, middle time region, and late time region.

Early Time Region (ETR). In ETR, well test data is influenced by unloading of fluids in the wellbore (wellbore storage effects). The pressure transient is also passing through the near-wellbore zone of altered permeability (damaged or stimulated) shown in Figure 13.1. Well test data collected during ETR can be used to calculate wellbore storage coefficient and wellbore skin factor.

Middle Time Region (MTR). The pressure transient is presumed to have entered MTR after unloading of wellbore fluids have ceased (termination of wellbore storage effects), and the pressure transients have entered zones in the formation with virgin permeability. Well test data collected during MTR can be used to calculate formation permeability and skin factor, as shown later in the chapter.

Late Time Region (LTR). Late time region is reached when the pressure transient is affected by the reservoir or drainage volume boundaries or other severe heterogeneities (such as sealing faults) that may be present in the formation. This is characterized by a change in the slope of the pressure drawdown curve as shown in Figure 13.5.

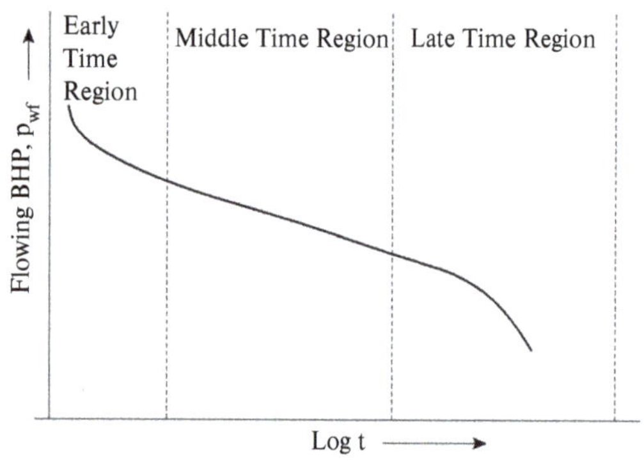

▲ **FIGURE 13.5** **Drawdown test time regions**

13.2.6 Buildup Well Tests

A typical buildup well test is shown in Figure 13.6. A buildup test is characterized by production at a constant rate over a time period denoted as t_p followed by shut-in either at the wellhead or downhole with tools over another time period denoted as Δt. Analyses of buildup test data are affected by the following well conditions:

1. Difficulties of maintaining production at constant rate prior to shut-in.
2. For old producing wells or wells with extended production history, the challenge of calculating production time, t_p, prior to shut-in. The production time before shut-in, t_p, is needed as input for the preparation of the Horner plot.[2]
3. The distortion of early well test data by flows that occur (called afterflow) after the well has been shut in.

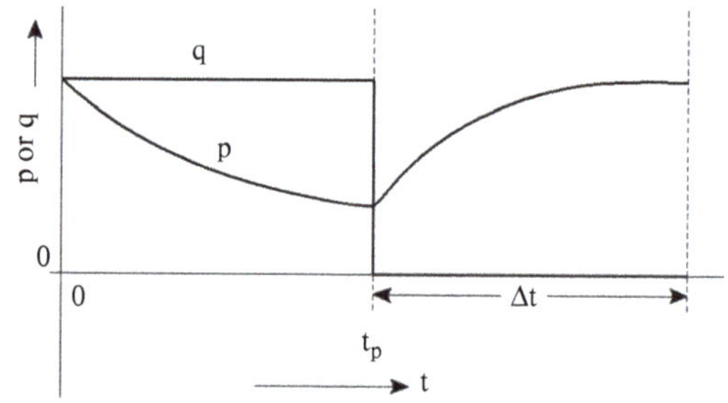

▲ **FIGURE 13.6** **Typical buildup test**

A plot of shut-in pressure versus time on a semi-log scale of buildup test data can be divided into time regions similar to a drawdown test as shown in Figure 13.7. The time regions are also named as in drawdown tests as: early time region, middle time region, and late time region.

Early Time Region (ETR). Buildup data in this time region is affected by afterflow and the altered permeability zone close to the wellbore (Figure 13.1). These data can be used to calculate wellbore storage coefficient and skin factor.

Middle Time Region (MTR). MTR is reached when afterflow ends (end of wellbore storage) and the pressure transient has entered the areas of the formation with virgin permeability. The slope of the straight line drawn through MTR data on a semi-log plot similar to Figure 13.7 is used to calculate formation permeability, skin factor, and an estimate of average reservoir pressure. This application is demonstrated later in the chapter with the Horner Plot.[2]

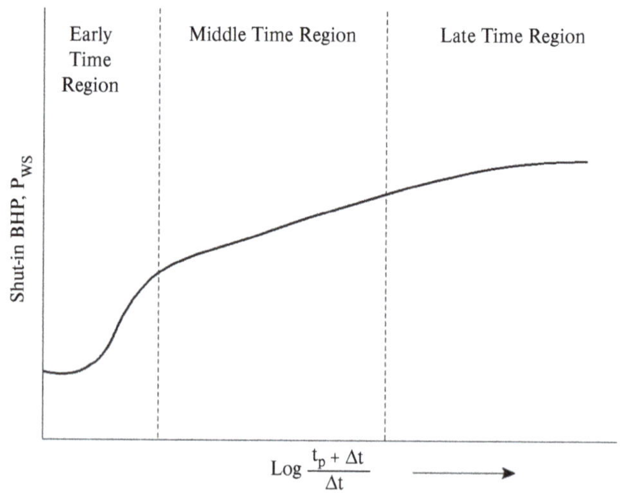

▲ **FIGURE 13.7** **Buildup test time regions**

Late Time Region (LTR). LTR data in buildup tests are affected by reservoir boundaries and other reservoir heterogeneities, such as a sealing fault. The data from LTR can be used to calculate the size of the drainage volume of the well.

13.2.7 Wellbore Storage

In drawdown flow tests, unloading of wellbore fluids at start of production affects early time test data. Similarly, in buildup tests, continued flow of reservoir fluids into the wellbore after the well has been shut in (called afterflow) affects early time test data. Well unloading and afterflow distortions of early time test data are collectively described as wellbore storage effects. Unloading of wellbore fluids in drawdown tests ceases when the rate of fluid production at the wellhead equals the rate of fluid entry from the reservoir. Afterflow in buildup tests ends when hydrostatic pressure of the fluid flowing into the wellbore equals reservoir pressure at the sandface. Wellbore storage coefficient, C, is a parameter used to quantify wellbore storage. It is defined as change in volume of wellbore fluids per unit change in wellbore pressure. This is represented as:

$$C = \frac{\Delta V}{\Delta p} \tag{13.17}$$

In Eq. (13.17), C is wellbore storage coefficient, bbl/psi; ΔV is in barrels; and Δp is in psi. For a well with a rising or falling liquid level, wellbore storage coefficient is calculated as:

$$\begin{aligned} C &= \frac{144 A_{wb}}{\rho_{wb}}, \ \frac{\text{ft}^3}{\text{psi}} \\ &= \frac{25.646 A_{wb}}{\rho_{wb}}, \ \frac{\text{bbl}}{\text{psi}} \end{aligned} \tag{13.18}$$

In Eq. (13.18), A_{wb} is the area of the wellbore, ft^2; and ρ_{wb} is the density of the fluid in the wellbore, Ibm/ft^3. For a wellbore filled with a single phase fluid, wellbore storage coefficient is given by:

$$C = V_{wb} c_{wb} \tag{13.19}$$

For Eq. (13.19), V_{wb} = volume of fluid in wellbore, bbl; and c_{wb} = compressibility of fluid in the wellbore, psi^{-1}. In dimensionless form, wellbore storage coefficient is defined as:

$$C_D = \frac{0.894 C}{\phi c_t h r_w^2} \tag{13.20}$$

In Eq. (13.20), C_D = dimensionless wellbore storage coefficient; C = wellbore storage coefficient, bbl/psi; ϕ = porosity, fraction; c_t = total compressibility, psi^{-1}; h = interval thickness, ft.; and r_w = wellbore radius, ft.

Wellbore storage effects can be reduced in drawdown tests by measuring flow rates downhole close to the producing interval rather at the wellhead. This could increase the cost of conducting drawdown tests. In buildup tests, wellbore storage effects can be minimized by using downhole shut-in tools. This is common practice in conducting buildup tests if there are no obstructions in the wellbore or no substantial risk of losing the tool in the wellbore. The use of downhole shut-in tools reduces the time and cost of buildup tests, especially on wells with high production rates.

13.3 Line Source Well, Infinite Reservoir Solution of the Diffusivity Equation with Skin Factor

In Chapter 12, the constant terminal rate solution of the diffusivity equation for slightly compressible liquids assuming a line source well in an infinite reservoir was presented. The solution presented as Eq. (12.64) in field units is:

$$p(r,t) = p_i - \frac{70.6 qB\mu}{kh}\left[-Ei\left(-\frac{948\phi\mu c_t r^2}{kt}\right)\right]$$

In the above equation, p = pressure, psia; q = flow rate, STB/D; B = formation volume factor, RB/STB; μ = fluid viscosity, cp; k = permeability, md; h = formation thickness, ft; ϕ = porosity, fraction; c_t = total compressibility, psi^{-1}; r = radius, ft; and t = time, hr.

It is important to observe that Eq. (12.64) does not include the pressure drop, Δp_s caused by the zone of altered permeability close to the wellbore as shown in Figure 13.1. If the pressure drop, Δp_s, caused by skin is included for calculation of wellbore pressure, p_w, in a well, Eq. (12.64) can be modified to be:

$$p_{(w,t)} = p_i - \frac{70.6 qB\mu}{kh}\left[-Ei\left(-\frac{948\phi\mu c_t r_w^2}{kt}\right)\right] - \Delta p_s \tag{13.21}$$

Combining Eqs. (13.9) and (13.21) gives:

$$p_{(w,t)} = p_i - \frac{70.6 qB\mu}{kh}\left[-Ei\left(-\frac{948\phi\mu c_t r_w^2}{kt}\right) + 2s\right] \tag{13.22}$$

Introducing the logarithmic approximation to the exponential integral for $x < 0.02$ gives the expression:

$$Ei(-x) \approx \ln(1.781x) \tag{13.23}$$

Assuming the logarithmic approximation to the exponential integral applies, using Eq. (13.23) in Eq. (13.22) yields:

$$p_{(w,t)} = p_i + \frac{70.6 qB\mu}{kh}\left[\ln\frac{1688\phi\mu c_t r_w^2}{kt} - 2s\right] \tag{13.24}$$

Converting from natural logarithm to base-10 logarithm, Eq. (13.24) becomes:

$$p_{(w,t)} = p_i - \frac{162.6qB\mu}{kh}\left[\log\frac{kt}{\phi\mu c_t r_w^2} - 3.23 + 0.8691s\right] \quad (13.25)$$

Eq. (13.25) is the basic equation used in the development of equations for drawdown and buildup analyses. Consequently, note that the assumptions made to achieve the line source solution of the diffusivity equation for slightly compressible fluids for an infinite reservoir apply to drawdown and buildup analyses.

EXAMPLE 13.3 Calculation of Total Pressure Drop of a Well in a Multi-Well Reservoir System

Problem

A reservoir at initial pressure of 2500 psia has three wells (Wells A, B, C) as shown in Figure 13.8 on production for 45 days from initial conditions. Well A was produced at a constant rate of 600 STB/D. Well B was produced at a constant rate of 350 STB/D and is located 1000 feet from Well A. Well C was produced at a constant rate of 150 STB/D and is located 750 feet from Well A. Assuming the reservoir is infinite in size, calculate the total pressure drop in Well A at the end of 45 days. Rock and fluid properties for the reservoir, including parameters for Well A, are as follows:

Initial pressure, p_i	2500 psia
Formation thickness, h	15 ft
Formation permeability, k	300 md
Formation porosity, ϕ	0.24
Total compressibility, c_t	6×10^{-6} psi^{-1}
Oil viscosity, μ_o	4.2 cp
Oil FVF, B_o	1.2 RB/STB
Wellbore radius, r_w	0.5 ft
Skin factor, Well A	4.5

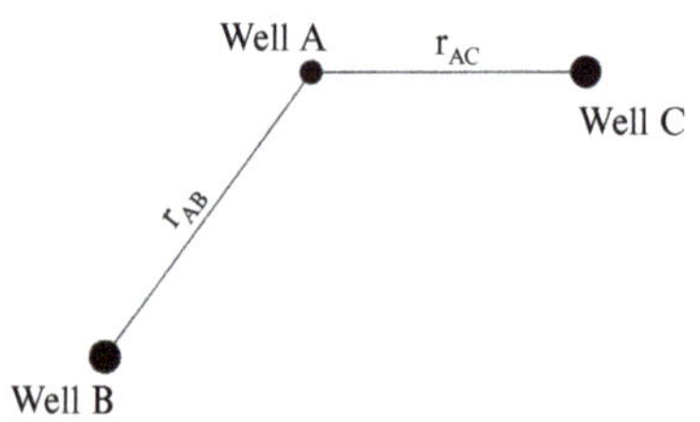

▲ FIGURE 13.8 Diagram of well locations for Example 13.3

Solution

This example is similar to Example 12.10 in Chapter 12. It is presented in this chapter to illustrate the application of the superposition principle for calculation of pressure in a well with a near-wellbore zone of altered permeability. This application of the superposition principle was deferred to this chapter because the concepts of skin and skin factor were not discussed in Chapter 12. To calculate pressure in a well, it is important to include the pressure drop due to presence of skin near the wellbore. This means that Eq. (13.22) should be used in the formulation of superposition equations. For calculations of pressures away from the wellbore, use Eq. (12.64).

Step 1: Calculate the pressure drop caused by Well A. From Eq. (13.22),

$$\Delta p_A = \frac{70.6q_A B\mu}{kh}\left[-Ei\left(-\frac{948\phi\mu c_t r_w^2}{kt}\right) + 2s\right]$$

$$= \frac{70.6\times600\times1.2\times4.2}{300\times15} \times\left[-Ei\left(\frac{948\times0.24\times4.2\times6\times10^{-6}\times(0.5^2)}{300\times45\times24}\right) + (2\times4.5)\right]$$

$$= 47.4432\times\left[-Ei\left(-4.424\times10^{-9}\right)+9\right]$$

Since, $x < 0.2$, use the logarithmic approximation for calculation of the exponential integral.

$$Ei(-x) = \ln(1.781x) = \ln\left(1.781\times4.424\times10^{-9}\right)$$

$$= -18.659$$

Thus,

$$\Delta p_A = 47.4432\times\left[-(-18.659)+9\right]$$

$$= 1312.2337 \text{ psia}$$

Note that Eq. (13.24) could have been used instead to calculate, Δp_A.

Step 2: Calculate the pressure drop caused by Well B. From Eq. (12.64),

$$\Delta p_B = \frac{70.6q_B B\mu}{kh}\left[-Ei\left(-\frac{948\phi\mu c_t r_{AB}^2}{kt}\right)\right]$$

$$= \frac{70.6\times350\times1.2\times4.2}{300\times15} \times\left[-Ei\left(-\frac{948\times0.24\times4.2\times6\times10^{-6}\times(1000)^2}{300\times45\times24}\right)\right]$$

$$= 27.6752\times\left[-Ei(-0.017696)\right]$$

Since $x < 0.02$, $Ei(-x) = \ln(1.781x) = \ln(1.781 \times 0.017696) = -3.4572$

Hence,

$$\Delta p_B = 27.6752\times\left[-(-3.4572)\right]$$

$$= 95.6799 \text{ psia.}$$

Step 3: Calculate the pressure drop caused by Well C.

From Eq. (12.64),

$$\Delta p_C = \frac{70.6 q_C B\mu}{kh}\left[-Ei\left(-\frac{948\phi\mu c_t r_{AC}^2}{kt}\right)\right]$$

$$= \frac{70.6\times150\times1.2\times4.2}{300\times15}$$

$$\times\left[-Ei\left(-\frac{948\times0.24\times4.2\times6\times10^{-6}\times(750)^2}{300\times45\times24}\right)\right]$$

$$=11.8608\times\left[-Ei(-0.009954)\right]$$

Since $x < 0.02$, $Ei(-x) = \ln(1.781x)) = \ln(1.781 \times 0.009954) = -4.0326$

Hence,

$$\Delta p_C = 11.8608\times\left[-(-4.0326)\right]$$

$$= 47.8299 \text{ psia.}$$

The total pressure drop at Well A is:

$$\Delta p_t = \Delta p_A + \Delta p_B + \Delta p_C$$

$$= 1312.23 + 95.68 + 47.83 \text{ psia.}$$

$$= 1455.74 \text{ psia.}$$

End of Example 13.3

13.4 Well Test Analyses with Straightline Methods

In keeping with historical perspective, it is instructive to begin presentation of well test analysis methods with straightline methods. The earliest techniques used to analyze well test data were based on straightline methods. These methods gained wide acceptance in the petroleum industry with publications of the procedures by Miller et al.[1] and Horner.[2] Straightline methods are based on the ability to identify the presence of a straight line on a plot of pressure data versus prescribed function of elapsed time within the dominant flow regime. For instance, the slope of the straight line representing radial flow regime (middle time region) in a drawdown test can be used to calculate reservoir permeability, reservoir pressure, and skin factor, as shown later in this chapter. Straightline methods are applicable to slightly compressible fluids (liquids) and compressible fluids (gases). The application of the method to slightly compressible fluids is presented first, followed by application to compressible fluids.

13.4.1 Slightly Compressible Fluids

In Chapter 12, the derivation of the diffusivity equation for slightly compressible fluids was presented. The constant terminal rate solution to the diffusivity equation assuming line source well in an infinite reservoir was also presented. The key initial and boundary conditions for the solution are repeated in this section as follows:

1. The well has zero radius, $r_w = 0$.
2. The reservoir is infinite in size, (i.e., $p \to p_i$ as $r \to \infty$ for all t).
3. At time, $t = 0$, the reservoir is at uniform pressure, $p = p_i$, for all r.
4. At time, $t > 0$, a single well of radius, r_w, at the center of the cylinder (reservoir) produces at a constant rate, q.

It is important to remember the above initial and boundary conditions (especially items 2, 3, and 4) in the application of the solution of the diffusivity equation to drawdown and buildup analyses.

13.4.1.1 Constant Rate Drawdown Tests

Consider a single well in a reservoir producing a single phase fluid at constant rate from initial conditions. The flowing bottomhole pressure of the well, p_{wf}, can be calculated with Eq. (13.25) as:

$$p_{wf} = p_i - \frac{162.6qB\mu}{kh}\left[\log\frac{kt}{\phi\mu c_t r_w^2} - 3.23 + 0.8691s\right] \quad (13.26)$$

Eq. (13.26) can be rearranged as:

$$p_{wf} = p_i - \frac{162.6qB\mu}{kh}\left[\log t + \log\frac{k}{\phi\mu c_t r_w^2} - 3.23 + 0.8691s\right] \quad (13.27)$$

Eq. (13.27) is in the form of the equation for a straight line: $y = b + mx$.

Assigning terms in Eq. (13.27) to representative terms in the straight line equation gives:

$$y \equiv p_{wf} \quad (13.28)$$

$$x \equiv \log t \quad (13.29)$$

$$m \equiv \left|\frac{162.6qB\mu}{kh}\right| \quad (13.30)$$

$$b \equiv p_i - \frac{162.6qB\mu}{kh}\left[\log\frac{k}{\phi\mu c_t r_w^2} - 3.23 + 0.8691s\right] \quad (13.31)$$

A plot of p_{wf} vs. log t produces a curve similar to Figure 13.5. The part of this curve in the ETR is affected by wellbore storage and the altered permeability zone close to the wellbore. The slope of the straight line representing the MTR is equal to m. Note that by convention the absolute value of m is used as the slope of the line. The slope of the MTR line is simply determined as the absolute value of the difference between two pressure points, (p_{wf2} and p_{wf1}), exactly one log cycle apart. Reservoir permeability, k, or its productive capacity, kh, can be calculated from Eq. (13.30) given the slope of the MTR line as m. Generally, it is frequently difficult to determine precisely the end of wellbore storage effects or end of ETR. A rule of thumb generally used in analyses of drawdown tests is that MTR begins approximately one-and-one-half log cycles from the apparent end of ETR.

By substituting Eq. (13.30) into Eq. (13.26) and rearranging terms, skin factor, s, can be represented as:

$$s = 1.151\left[\frac{p_i - p_{wf}}{m} - \log\frac{kt}{\phi\mu c_t r_w^2} + 3.23\right] \quad (13.32)$$

Using Eq. (13.32), the skin factor, s, can be calculated by selecting any point on the straight line representing the MTR on the plot of p_{wf} vs. log t. By convention however, p_{wf} is determined at $t = 1$ hour sometimes by extrapolation of the MTR straight line. With this approach, Eq. (13.32) is re-written as:

$$s = 1.151\left[\frac{p_i - p_{1hr}}{m} - \log\frac{k}{\phi\mu c_t r_w^2} + 3.23\right] \quad (13.33)$$

It is important to reiterate some of the assumptions (constant flow rate, infinite-acting reservoir, and uniform initial pressure) made in achieving the exponential integral solution to the diffusivity equation, which is the basis for the equations used in the analyses of constant rate drawdown flow tests. In practice, some of these assumptions are difficult to achieve or maintain for the duration of the test. Consequently, analyses of drawdown tests are affected because these conditions are rarely achieved in an actual test.

EXAMPLE 13.4 Calculation of Permeability and Skin Factor from Analyses of a Single, Constant Rate Drawdown Test

Problem

The drawdown data of a discovery well in an oil reservoir are shown in Table 13.1. The well was produced at a single, constant rate of 550 STB/D for a total time of 725 hours. Other reservoir and fluid properties data are as follows:

Formation thickness, h	20 ft
Formation porosity, ϕ	0.23
Initial water saturation, S_{wi}	0.26
Total compressibility, c_t	6×10^{-5} psi^{-1}
Oil viscosity, μ_o	0.42 cp
Oil FVF, B_o	1.45 RB/STB
Wellbore radius, r_w	0.5 ft

TABLE 13.1 Drawdown Flow Test Data for Example 13.4

Time, t (hrs.)	p_{wf} (psia)	Time, t (hrs.)	p_{wf} (psia)
0.00	5,665	48.40	5,405
0.12	5,645	72.50	5,385
0.22	5,615	97.00	5,368
0.42	5,590	122.00	5,359
0.62	5,577	146.00	5,350
1.10	5,566	196.00	5,335
2.05	5,548	244.00	5,320
4.20	5,525	300.00	5,305
8.40	5,493	370.00	5,293
10.20	5,481	485.00	5,265
15.50	5,462	608.00	5,237
24.50	5,438	725.00	5,209
36.20	5,417		

Solution

Step 1: Plot p_{wf} vs. $\log t$.

From the data in Table 13.1, make a plot of p_{wf} vs. t on a semi-log graph. A plot of p_{wf} vs. t is shown as Figure 13.9. This can be done easily on an electronic spreadsheet or with any graphing software.

Step 2: Determine the slope, m, of the MTR straight line.

By examining Figure 13.9, it is evident that wellbore storage effects ended approximately after 1 hour of flow. Applying the one-and-half cycle rule of thumb, the flow test was definitely in the MTR after 50 hours of flow. Boundary effects appear to start after 200 hours of flow. Thus, the MTR straight line was drawn between flow times of 50 hours and 200 hours. The slope of the MTR straight line was determined one cycle apart between $t = 10$ hrs and $t = 100$ hrs as:

$$m = \left|p_{wf@10hrs.} - p_{wf@100hrs.}\right|$$
$$= 5486 - 5371$$
$$= 115 \text{ psi/cycle}$$

Step 3: Calculate formation permeability, k, from m.

Using Eq. (13.30):

$$k = \frac{162.6 qB\mu}{mh}$$

$$= \frac{162.6 \times 550 \times 1.45 \times 0.42}{115 \times 20}$$

$$= 23.68 \text{ md.}$$

Step 4: Calculate skin factor, s.

From Figure 13.9, $p_{wf@1hr}$ = 5595 psia on the extrapolated MTR straight line. Using Eq. (13.33):

$$s = 1.151\left[\frac{p_i - p_{1hr}}{m} - \log\frac{k}{\phi\mu c_t r_w^2} + 3.23\right]$$

$$= 1.151\left[\frac{5665 - 5595}{115} - \log\frac{23.68}{0.23 \times 0.42 \times 6 \times 10^{-5} \times (0.5)^2} + 3.23\right]$$

$$= -3.88$$

The negative skin indicates that this well may have a stimulated zone near the wellbore.

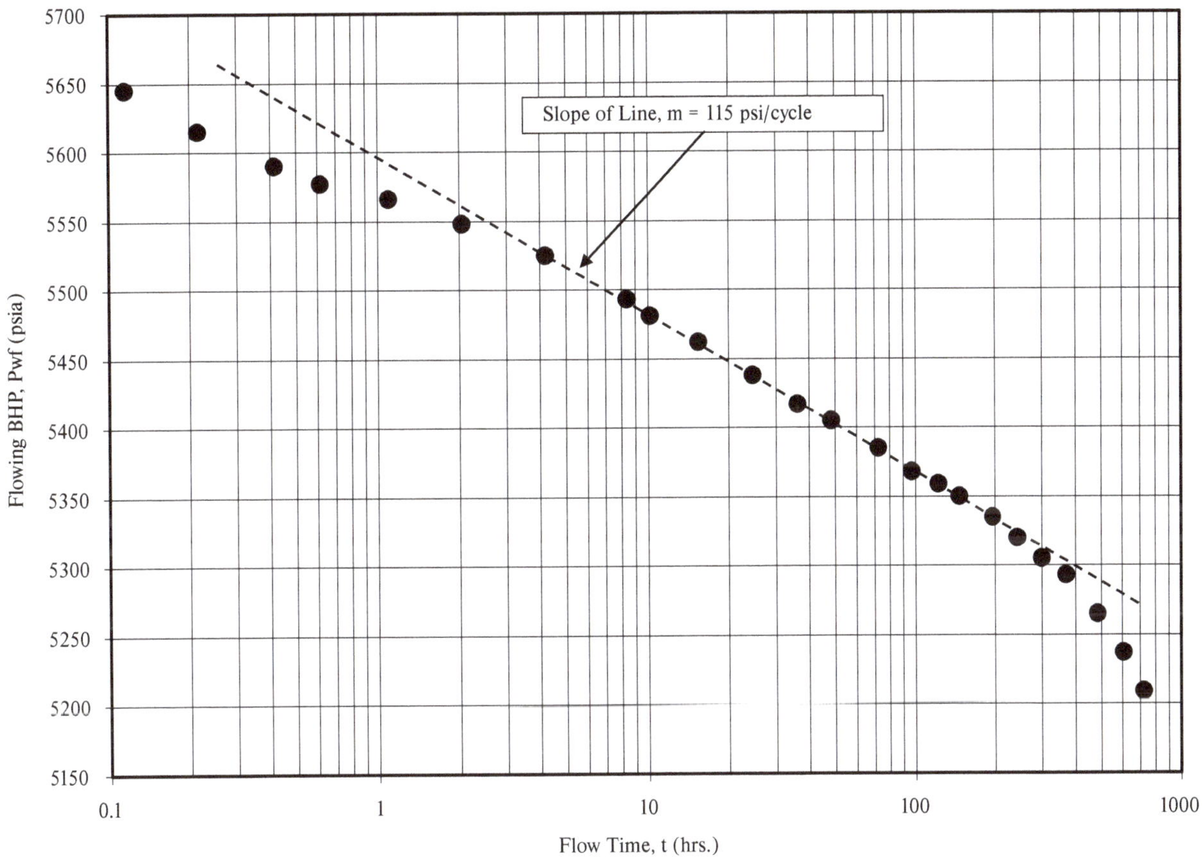

▲ **FIGURE 13.9** **Plot of p_{wf} vs. t for Example 13.4**

End of Example 13.4

13.4.1.2 Multi-Rate Drawdown Tests

It is difficult to maintain a single, constant production rate on a well during the duration of a drawdown test. For this reason, it is more desirable to test a well at multiple flow rates of varying durations. This type of test at multiple flow rates is represented in Figure 13.10. As shown in Figure 13.10, flow at constant rate, q_1, starts at time, $t = 0$; increases to q_2 at t_1; changes to q_3 at t_2; and finally changes to q_n at t_{n-1}. Note that the flow rate is constant at each flow period. An equation can be derived to represent the pressure at the well at any time period using the principle of superposition. To simplify the application of superposition principle, Eq. (13.26) can be written in the following form:

$$p_i - p_{wf} = m_m q\left[\log(t) + b_m\right] \tag{13.34}$$

where $m_m = \dfrac{162.6B\mu}{kh}$ (13.35)

and $b_m = \log\left(\dfrac{k}{\phi\mu c_t r_w^2}\right) - 3.23 + 0.8691s$ (13.36)

Applying the superposition principle on the flow rates shown in Figure 13.10 for n rates gives:

$$\begin{aligned} p_i - p_{wf} &= m_m q_1\left[\log(t) + b_m\right] \\ &+ m_m(q_2 - q_1)\left[\log(t - t_1) + b_m\right] \\ &+ m_m(q_3 - q_2)\left[\log(t - t_2) + b_m\right] + \ldots\ldots \\ &\ldots. + m_m(q_n - q_{n-1})\left[\log(t - t_{n-1}) + b_m\right] \end{aligned} \quad (13.37)$$

Eq. (13.37) can be written as:

$$\begin{aligned} \frac{p_i - p_{wf}}{q_n} &= m_m \sum_{j=1}^{n}\left[\frac{(q_j - q_{j-1})}{q_n}\log(t_n - t_{j-1})\right] \\ &+ m_m b_m \quad \text{for } q_n \neq 0 \end{aligned} \quad (13.38)$$

Substituting for b_m in Eq. (13.38) yields:

$$\begin{aligned} \frac{p_i - p_{wf}}{q_n} &= m_m \sum_{j=1}^{n}\left[\frac{(q_j - q_{j-1})}{q_n}\log(t_n - t_{j-1})\right] \\ &+ m_m\left[\log\left(\frac{k}{\phi\mu c_t r_w^2}\right) - 3.23 + 0.8691s\right] \end{aligned} \quad (13.39)$$

Eq. (13.39) can be adapted for the analyses of multi-rate tests as illustrated with Example 13.5. Note that because this equation is based on the line source solution for an infinite-acting reservoir, the condition of a reservoir or drainage volume that is infinite in size applies for the entire duration of the test for the flow test to be valid. Also, note that Eq. (13.39) is in the form of the equation for a straight line. If Y_F is used to represent values on the y-axis, and X_F for values on the x-axis, then Eq. (13.39) can be written as:

$$Y_F \equiv \frac{p_i - p_{wf}}{q_n} \quad (13.40)$$

$$X_F \equiv \sum_{j=1}^{n}\left[\frac{(q_j - q_{j-1})}{q_n}\log(t_n - t_{j-1})\right] \quad (13.41)$$

$$slope,\ m_m = \frac{162.6B\mu}{kh} \quad (13.42)$$

$$Y_{intercept} \equiv b_I = m_m\left[\log\left(\frac{k}{\phi\mu c_t r_w^2}\right) - 3.23 + 0.8691s\right] \quad (13.43)$$

A Cartesian plot of Y_F vs. X_F results in a straight line with slope, m_m, and intercept, b_I, if the reservoir is infinite-acting. This type of plot is sometimes called rate-normalized plot or multi-rate superposition plot.[11] An expression for calculating skin factor can be derived by rearranging Eq. (13.43) to get:

$$s = 1.151\left[\frac{b_I}{m_m} - \log\left(\frac{k}{\phi\mu c_t r_w^2}\right) + 3.23\right] \quad (13.44)$$

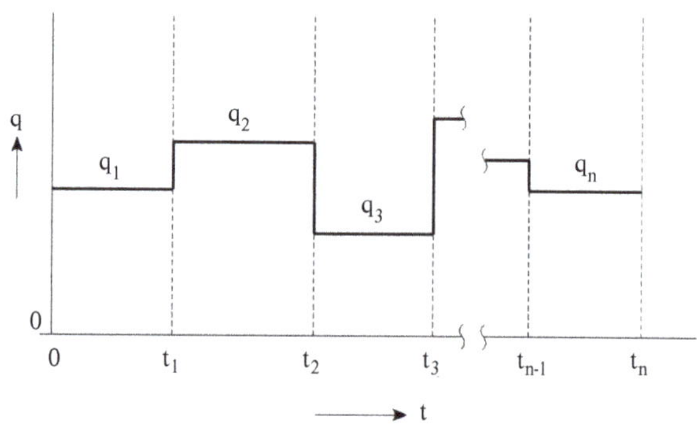

▲ **FIGURE 13.10 Multiple rate flow test**

EXAMPLE 13.5 Calculation of Permeability and Skin Factor from a Multi-Rate Drawdown Test

Problem

A multi-rate test was conducted on an oil well at production rates and flowing bottomhole pressures shown in Table 13.2. Other reservoir properties and well data are listed as follows:

Initial reservoir pressure, p_i	2906 psia
Formation thickness, h	40 ft
Formation porosity, ϕ	0.24
Total compressibility, c_t	5×10^{-5} psi^{-1}
Oil viscosity, μ_o	0.60 cp
Oil FVF, B_o	1.27 RB/STB
Wellbore radius, r_w	0.5 ft

TABLE 13.2 Multi-Rate Test Data and Plotting Functions for Example 13.5 (from Tiab et al.[12] © 1999 SPE, Reproduced with permission)

n	t (hrs.)	q_n (STB/D)	p_{wf} (psi)	Y_F (psi/ STB-D)	X_F
1	0	0	2906	—	—
1	1.00	1580	2023	0.5589	0.0000
1	1.50	1580	1968	0.5937	0.1761
1	1.89	1580	1941	0.6108	0.2765
1	2.40	1580	—	—	—
2	3.00	1490	1892	0.6805	0.5193
2	3.45	1490	1882	0.6872	0.5690
2	3.98	1490	1873	0.6933	0.6241
2	4.50	1490	1867	0.6973	0.6732
2	4.80	1490	—	—	—
3	5.50	1440	1853	0.7313	0.7870
3	6.05	1440	1843	0.7382	0.8193
3	6.55	1440	1834	0.7444	0.8485
3	7.00	1440	1830	0.7472	0.8739
3	7.20	1440	—	—	—
4	7.50	1370	1827	0.7876	0.9737
4	8.95	1370	1821	0.7920	1.0091
4	9.60	1370	—	—	—
5	10.00	1300	1815	0.8392	1.1242
5	12.00	1300	1797	0.8531	1.1535
6	14.40	1260	—	—	—
7	15.00	1190	1775	0.9504	1.3374
7	18.00	1190	1771	0.9538	1.3553
7	19.20	1190	—	—	—
8	20.00	1160	1772	0.9776	1.4225
8	21.60	1160	—	—	—
9	24.00	1137	1756	1.0114	1.4851
10	28.80	1106	—	—	—
11	30.00	1080	1751	1.0694	1.6067
11	33.60	1080	—	—	—
12	36.00	1000	—	—	—
13	36.20	983	1756	1.1699	1.7883
13	48.00	983	1743	1.1831	1.7995

Solution

Step 1: Calculate the Y_F function.

From Eq. (13.40), $Y_F = \dfrac{p_i - p_{wf}}{q_n}$

As example, at $t = 1.89$ hrs, $p_i = 2906$ psi; $p_{wf} = 1941$ psi; $q_n = 1580$ STB/D

$$Y_F = \frac{2906 - 1941}{1580} = 0.6108 \text{ psi/STB-D}$$

Similarly, at $t = 20$ hrs, $p_{wf} = 1772$ psi; $q_n = 1160$ STB/D

$$Y_F = \frac{2906 - 1772}{1160} = 0.9776 \text{ psi/STB-D}$$

The remaining values for the Y_F function are shown in Table 13.2 for other test times.

Step 2: Calculate the X_F function.

From Eq. (13.41),

$$X_F = \sum_{j=1}^{n}\left[\frac{(q_j - q_{j-1})}{q_n}\log(t_n - t_{j-1})\right]$$

As example, at $t = 1.89$ hrs,

$$X_F = \left(\frac{1580 - 0}{1580}\right)\log(1.89 - 0)$$
$$= 0.2765$$

At $t = 20$ hrs,

$$X_F = \frac{1}{1160} \times \begin{bmatrix} (1580-0)\log(20-0) \\ +(1490-1580)\log(20-2.4) \\ +(1440-1490)\log(20-4.8) \\ +(1370-1440)\log(20-7.2) \\ +(1300-1370)\log(20-9.6) \\ +(1260-1300)\log(20-12) \\ +(1190-1260)\log(20-14.4) \\ +(1160-1190)\log(20-19.2) \end{bmatrix}$$
$$= 1.4225$$

Values for X_F are shown in Table 13.2 for the remaining data points. Calculation of X_F (superposition time) can be tedious if done manually. However, the calculation can be performed quite readily on an electronic spreadsheet.

Step 3: Plot X_F vs. Y_F on a Cartesian graph.

A plot of X_F vs. Y_F based on the data in Table 13.2 is shown in Figure 13.11.

Step 4: Determine the slope, m_m.

Note that Figure 13.11 has two straight lines. The slope of the first straight line is determined thus:

$$m_m = \frac{Y_{F2} - Y_{F1}}{X_{F2} - X_{F2}}$$

Using the values for Y_F and X_F at $t = 6.55$ hrs and $t = 3.00$ hrs give:

$$m_m = \frac{0.7444 - 0.6805}{0.8485 - 0.5193}$$
$$= 0.194 \text{ psi/(STB/D-cycle)}$$

Similarly, the slope of the second line is determined to be 0.526 psi/(STB/D-cycle). It is apparent that this second straight line represents late time region data that may have been affected by boundary effects. The slope of this line is not used in the calculations of permeability and skin factor since it is influenced by boundary effects.

Step 5: Calculate formation permeability.

From Eq. (13.42),

$$k = \frac{162.6 B\mu}{m_m h}$$
$$= \frac{162.6 \times 1.27 \times 0.6}{0.194 \times 40} = 15.97 \text{ md.}$$

Step 6: Calculate skin factor.

Using Eq. (13.44),

$$s = 1.151\left[\frac{b_I}{m_m} - \log\left(\frac{k}{\phi \mu c_t r_w^2}\right) + 3.23\right]$$

From Figure 13.11, the intercept on the y-axis is $b_I = 0.5589$. Substituting gives:

$$s = 1.151\left[\frac{0.5589}{0.194} - \log\left(\frac{15.97}{0.24 \times 0.6 \times 5 \times 10^{-5} \times (0.5)^2}\right) + 3.23\right]$$
$$= -0.96$$

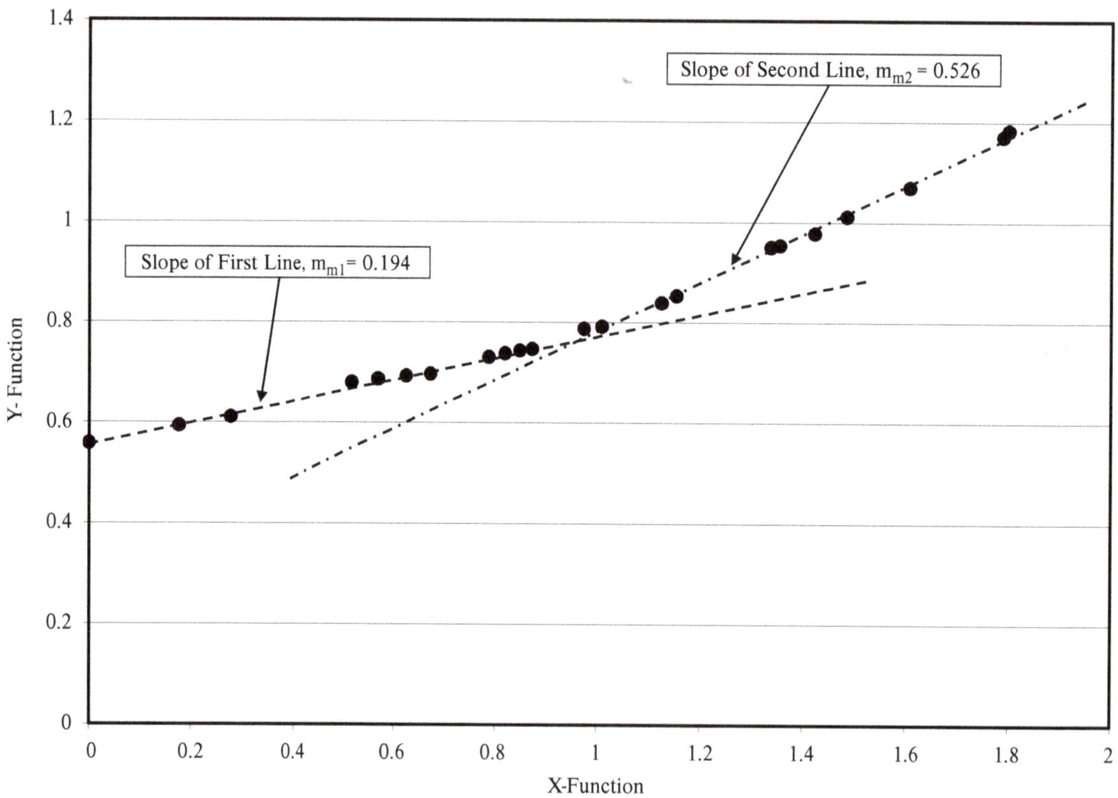

▲ **FIGURE 13.11 Plot of Y_F vs. X_F for Example 13.5**

End of Example 13.5

13.4.1.3 Two-Rate Drawdown Tests

Russell[13] proposed the two-rate test as a substitute for buildup tests, especially in wells where long shut-in periods may result in severe economic losses. Other benefits of the two-rate test are elimination of the effects of afterflow and phase redistribution in buildup tests. The equations representing the two-rate test can be developed directly from Eq. (13.39). If q_1 represents the production rate of the first test of duration, t, as shown in Figure 13.12, and q_2 is the production rate of the second test which starts at t_1, substitution of respective terms in Eq. (13.39) gives:

$$p_i - p_{wf} = \frac{162.6 q_2 B\mu}{kh}\left[\left(\frac{q_1}{q_2}\right)\log(t) + \left(\frac{q_2 - q_1}{q_2}\right)\log(t - t_1)\right] + \frac{162.6 q_2 B\mu}{kh}\left[\log\left(\frac{k}{\phi\mu c_t r_w^2}\right) - 3.23 + 0.8691 s\right] \tag{13.45}$$

By setting $\Delta t = t - t_1$, and rearranging, Eq. (13.45) becomes:

$$p_{wf} = p_i - \frac{162.6 q_1 B\mu}{kh}\left[\log\left(\frac{t_1 + \Delta t}{\Delta t}\right) + \left(\frac{q_2}{q_1}\right)\log(\Delta t)\right] - \frac{162.6 q_2 B\mu}{kh}\left[\log\left(\frac{k}{\phi\mu c_t r_w^2}\right) - 3.23 + 0.8691 s\right] \tag{13.46}$$

Eq. (13.46) is in the form of a straight line equation such that the following assignments can be made for y and x axes:

$$Y_F \equiv p_{wf} \tag{13.47}$$

$$X_F \equiv \left[\log\left(\frac{t_1 + \Delta t}{\Delta t}\right) + \left(\frac{q_2}{q_1}\right)\log(\Delta t)\right] \tag{13.48}$$

$$slope,\ m = -\frac{162.6 q_1 B\mu}{kh} \tag{13.49}$$

$$Y_{intercept} = p_i + m\left(\frac{q_2}{q_1}\right)\left[\log\left(\frac{k}{\phi\mu c_t r_w^2}\right) - 3.23 + 0.8691 s\right] \tag{13.50}$$

As shown by Russell,[13] the skin factor for a two-rate test is calculated from the equation:

$$s = 1.151\left[\frac{q_1}{(q_1 - q_2)}\left(\frac{p_{wf1} - p_{1hr}}{m}\right) - \log\left(\frac{k}{\phi\mu c_t r_w^2}\right) + 3.23\right] \tag{13.51}$$

Note that in Eq. (13.51), p_{1hr} is the pressure at $\Delta t = 1$ hr on the straight line drawn on the plot or extrapolation of the straight line. p_{wf1} is the flowing pressure at $\Delta t = 0$ or the time of the rate change to q_2. The two-rate test is valid only if the rate of the second test, q_2, is maintained constant for the duration of the test. This requirement is difficult to achieve in most tests. Odeh and Jones[14] proposed a method that can handle variable flow rate in the second flow period. Note also that the two-rate flow test is accurate when the reservoir is assumed to be infinite-acting for the entire duration of the test.

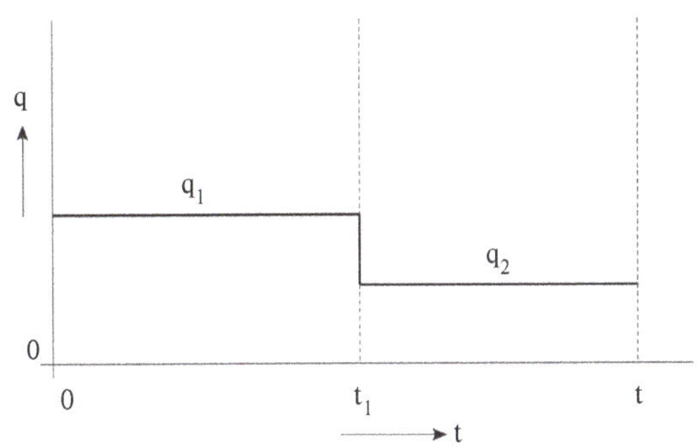

▲ **FIGURE 13.12** Two-rate flow test

EXAMPLE 13.6 Calculation of Permeability and Skin Factor from Two-Rate Drawdown Test

Problem

A two-rate drawdown test was conducted on a low permeability oil reservoir. The reservoir was produced at the first rate of 220 STB/D. The cumulative production at the first rate is 50,985 STB. The production rate of the well was then reduced to 143 STB/D. The flowing bottom-hole pressure of the well at this second rate versus time is shown in Table 13.3. Using data given for this reservoir, calculate formation permeability and skin factor of the well.

Formation thickness, h	26 ft
Formation porosity, ϕ	0.09
Total compressibility, c_t	24×10^{-6} psi^{-1}
Oil viscosity, μ_o	0.76 cp
Oil FVF, B_o	1.24 RB/STB
Wellbore radius, r_w	0.25 ft
Flowing BHP at $\Delta t = 0$	2925 psia

TABLE 13.3 Two-Rate Test Data and Plotting Function for Example 13.6

Time, Δt (hrs)	p_{wf} (psia)	X_F
1.00	3075	3.7453
1.50	3082	3.6837
2.00	3088	3.6400
2.50	3100	3.6061
3.00	3110	3.5785
4.00	3125	3.5348
5.00	3138	3.5010
6.00	3148	3.4733
7.00	3155	3.4500
8.00	3162	3.4298
9.00	3169	3.4119
10.00	3175	3.3960
15.00	3197	3.3348
20.00	3212	3.2914
25.00	3225	3.2579

Solution

Step 1: Calculate production time, t_1, at first rate.

$$t_1 = \frac{24N_p}{q_1} = \frac{24\times 50,985}{220} = 5562 \text{ hrs.}$$

Step 2: Calculate plotting function, X_F.

From Eq. (13.48),

$$X_F = \left[\log\left(\frac{t_1+\Delta t}{\Delta t}\right)+\left(\frac{q_2}{q_1}\right)\log(\Delta t)\right]$$

As an example, at $\Delta t = 1.50$, $t_1 = 5{,}562$, $q_1 = 220$ STB/D, $q_2 = 143$ STB/D:

$$X_F = \left[\log\left(\frac{5562+1.50}{1.50}\right)+\left(\frac{143}{220}\right)\log(1.5)\right]$$

$$= 3.6837$$

Step 3: Plot p_{wf} vs. X_F.

A plot of p_{wf} vs. X_F based on the data in Table 13.3 is shown in Figure 13.13. From the plot, the slope of the line, m, is:

$$m = \frac{p_{wf2} - p_{wf1}}{X_{F2} - X_{F1}}$$

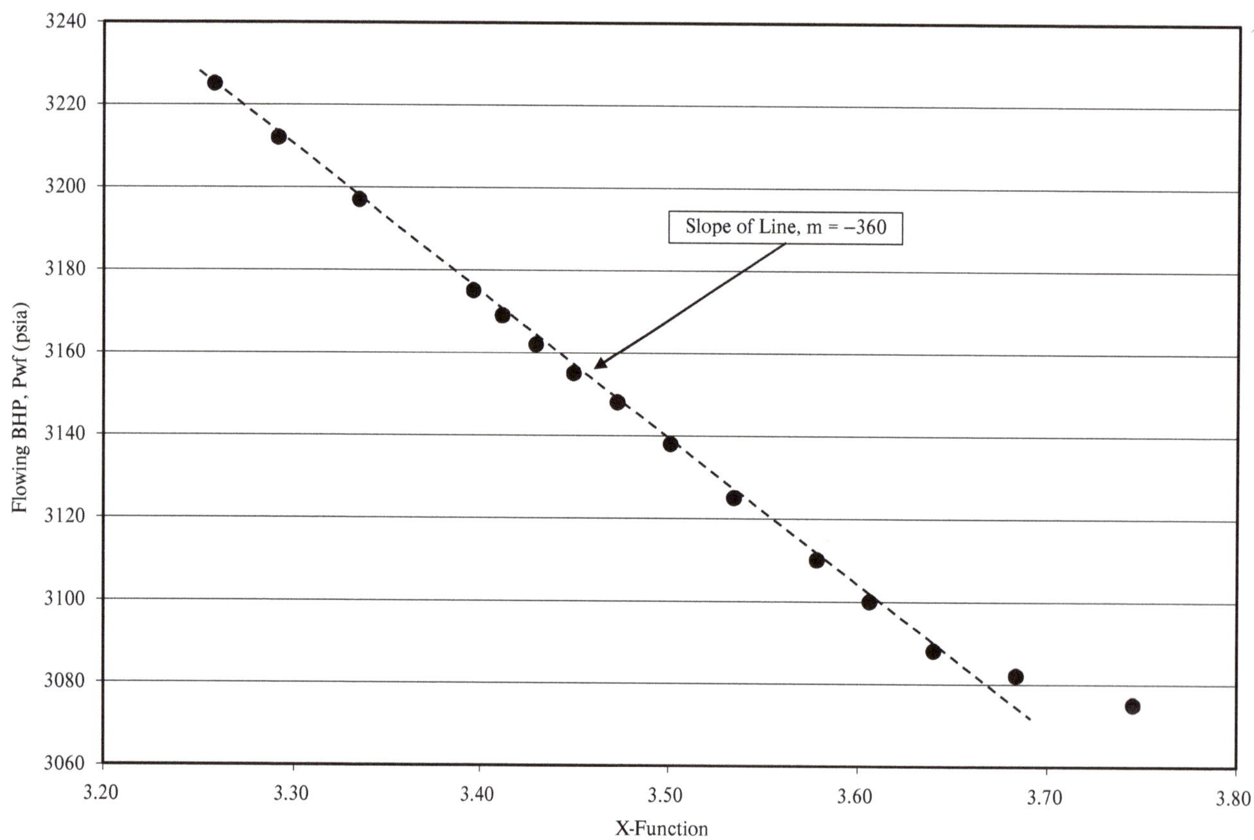

▲ **FIGURE 13.13** Two-rate flow test

Using data points at $\Delta t = 4$ hrs and $\Delta t = 15$ hrs,

$$m = \frac{3{,}197 - 3{,}125}{3.3348 - 3.5348} = -360 \text{ psi/(STB/D-cycle)}$$

Step 4: Calculate formation permeability, k.

From Eq. (13.49),

$$k = -\frac{162.6 q_1 B \mu}{mh} = -\frac{162.6 \times 220 \times 1.24 \times 0.76}{(-360) \times 26} = 3.60 \text{ md.}$$

Step 5: Calculate skin factor, s.

From Eq. (13.51),

$$s = 1.151 \left[\frac{q_1}{(q_1 - q_2)} \left(\frac{p_{wf1} - p_{1hr}}{m} \right) - \log \left(\frac{k}{\phi \mu c_t r_w^2} \right) + 3.23 \right]$$

Substituting with p_{wf1} = 2925 psia and p_{1hr} = 3075 psia gives:

$$s = 1.151 \left[\frac{220}{220 - 143} \left(\frac{2925 - 3075}{-360} \right) - \log \frac{3.6}{0.09 \times 0.76 \times 24 \times 10^{-6} \times (0.25)^2} + 3.23 \right] = -3.60$$

The negative skin factor calculated for this example indicates that the near-wellbore region may be stimulated.

End of Example 13.6

13.4.1.4 Buildup Tests

Most engineers have encountered buildup tests rather than drawdown tests. In a typical buildup test, a producing well is shut in and its bottomhole pressure (BHP) is measured as a function of time. The production profile and pressure response of a typical buildup test was shown in Figure 13.6. The ideal buildup test represented in Figure 13.6 assumes that the production rate before the well was shut in is constant. This is rarely the case since it is extremely difficult to maintain the production rate of a well constant for an extended period of time as was noted for drawdown tests. The equations for analysis of buildup tests are derived based on the assumption of constant production rate before the well was shut in. This is followed with equations for buildup tests with variable production rates before shut-in. Modification of the buildup analysis for wells with variable production rates or long production history before shut-in introduced by Horner[2] is then presented.

13.4.1.5 Buildup Tests with Constant Production Rate before Shut-in

For this case, the basic assumption is that the well is producing at a constant rate before it was shut in. The equation representing pressure drop in a well, assuming the logarithmic approximation to the line source solution of the radial diffusivity equation for slightly compressible fluids applies, is obtained by rearranging Eq. (13.25) as:

$$p_i - p_{(w,t)} = \frac{162.6 q B \mu}{kh} \left[\log \frac{kt}{\phi \mu c_t r_w^2} - 3.23 + 0.8691 s \right] \quad (13.52)$$

Referring to Figure 13.6, suppose a well was produced at a constant rate, q, for a time period, t_p, before it was shut in for a duration represented as Δt. Applying the principle of superposition in time, the production at rate, q, is presumed to have occurred for the total duration of the test $t_p + \Delta t$, and the shut-in at a rate equal to $(-q)$ has a duration equal to the shut-in time, Δt. The total pressure drop at the wellbore, $(p_i - p_{ws})$, is the sum of the pressure drop caused by production at rate q over a time period of $t_p + \Delta t$, and the pressure drop caused by production at rate $(-q)$ over a time period of Δt. Expressing this statement in equation form gives:

$$p_i - p_{ws} = \Delta p_{@q(t_p + \Delta t)} + \Delta p_{@-q(\Delta t)} \quad (13.53)$$

In Eq. (13.53), p_{ws} is the shut-in BHP measured over time Δt.

Substituting Eq. (13.52) into Eq. (13.53) gives:

$$p_i - p_{ws} = \frac{162.6 q B \mu}{kh} \left[\log \frac{k(t_p + \Delta t)}{\phi \mu c_t r_w^2} - 3.23 + 0.8691 s \right] + \frac{162.6(-q) B \mu}{kh} \left[\log \frac{k \Delta t}{\phi \mu c_t r_w^2} - 3.23 + 0.8691 s \right] \quad (13.54)$$

Grouping and cancelling like terms in Eq. (13.54) yields:

$$p_i - p_{ws} = \frac{162.6 q B \mu}{kh} \left[\log \frac{(t_p + \Delta t)}{\Delta t} \right] \quad (13.55)$$

Rearranging Eq. (13.55) gives:

$$p_{ws} = p_i - \frac{162.6 q B \mu}{kh} \left[\log \frac{(t_p + \Delta t)}{\Delta t} \right] \quad (13.56)$$

Eq. (13.56) is the basic equation for buildup analysis assuming a constant production rate before shut-in.

Eq. (13.56) is in the form of an equation for a straight line, $y = mx + b$, where:

$$y \equiv p_{ws} \tag{13.57}$$

$$x \equiv \log \frac{\left(t_p + \Delta t\right)}{\Delta t} \tag{13.58}$$

$$m = \left| \frac{162.6 qB\mu}{kh} \right| \tag{13.59}$$

$$b \equiv p_i \tag{13.60}$$

As the above equations suggest, a plot of p_{ws} vs. $\log \frac{t_p + \Delta t}{\Delta t}$ gives a straight line with slope, m, and intercept equal to p_i at infinite shut-in time. This plot is widely known as the Horner[2] plot and the ratio, $(t_p + \Delta t)/\Delta t$, is called the Horner time ratio. The slope, m, is determined from the MTR straight line when the pressure transient is in the infinite-acting radial flow (IARF) regime. On a semi-log plot, this slope is the pressure difference between two points one cycle apart on the MTR straight line. The permeability of the formation is calculated from the slope of the MTR line as shown in Eq. (13.56).

Skin factor for the well can also be calculated from buildup tests. Starting with Eq. (13.25), the equation for BHP at 1 hour of shut-in time is:

$$p_{ws@1hr} = p_{wf} + \frac{162.6 qB\mu}{kh}\left[\log \frac{k}{\phi \mu c_t r_w^2} - 3.23 + 0.8691 s\right] \tag{13.61}$$

Rearranging Eq. (13.61) and solving for skin factor, s, gives:

$$s = 1.151\left[\frac{p_{1hr} - p_{wf}}{m} - \log \frac{k}{\phi \mu c_t r_w^2} + 3.23\right] \tag{13.62}$$

Note that in Eq. (13.62), p_{1hr} is the shut-in pressure at $\Delta t = 1$ hr on the MTR straight line or extrapolation of the MTR straight line representing IARF regime.

Theoretically, if the reservoir is shut in for a very long time such that shut-in time approaches infinity, $(\Delta t \rightarrow \infty)$, then the Horner time ratio approaches unity as: $\left[\left\{\left(t_p + \Delta t\right)/\Delta t\right\} \rightarrow 1\right]$. Under this condition, the shut-in pressure approaches initial reservoir pressure, p_i, for a new reservoir or extrapolated pressure, p^*, for a partially depleted reservoir. The average reservoir pressure, $\overline{p}$, can be determined from the extrapolated pressure, p^*, with a method developed by Matthews et al.[15] Using correlations by Matthews et al.,[15] the average reservoir pressure is calculated as:

$$\overline{p} = p^* - \frac{m \times f_{pf}}{2.303} \tag{13.63}$$

In Eq. (13.63), m = slope of MTR line; and f_{pf} = pressure function for the drainage area of the well. Matthews et al.[15] published pressure functions for well locations in various shapes of the drainage area of the well. These pressure functions have not been reproduced in this book because they are rarely used by engineers, since it is difficult to determine the appropriate shape of the drainage area of the well. However, the pressure functions can be obtained from Matthews et al.[15] if needed. Another method for determining average reservoir pressure was proposed by Muskat[16] and later modified by Larson.[17] A robust method that does require prior knowledge of the reservoir and its fluid properties was proposed by Crump and Hite.[18] These methods and other similar methods[19-22] in the petroleum literature provide alternative means of rigorously calculating average reservoir pressure from buildup data. However, for many applications, many engineers assume that the extrapolated pressure, p^*, is a good estimate of the average reservoir pressure for the drainage area of the well. This approximation may be permissible depending on the application of the results from analysis of the buildup test, but it can be improved with published methods[15-22] as suggested earlier.

EXAMPLE 13.7 Calculation of Permeability, Reservoir Pressure, and Skin Factor from Buildup Tests

Problem

The discovery well in an oil reservoir was shut in for a buildup test after producing 20,000 STB at a constant rate of 480 STB/D. The shut-in pressures measured versus time are shown in Table 13.4. Calculate the effective permeability of the formation, average reservoir pressure, and the skin factor of the well using the following reservoir and well parameters:

Formation thickness, h	35 ft
Formation porosity, ϕ	0.18
Total compressibility, c_t	2.6×10^{-5} psi^{-1}
Oil viscosity, μ_o	1.5 cp
Oil FVF, B_o	2.2 RB/STB
Wellbore radius, r_w	0.50 ft
Flowing BHP at $\Delta t = 0$	2832 psia

TABLE 13.4 | **Buildup Test Data for Example 13.7**

Δt (hrs)	p_{ws} (psia)	$(t_p + \Delta t)/\Delta t$
0.00	2,832	—
0.10	2,842	10,001
0.35	2,867	2,858
0.80	2,912	1,251
1.40	2,972	715
2.00	3,032	501
3.00	3,092	334
3.50	3,112	287
4.00	3,132	251
5.00	3,152	201
6.00	3,172	168
7.00	3,182	144
8.00	3,197	126
9.00	3,207	112
10.00	3,216	101
15.00	3,249	68
20.00	3,275	51

Solution

Step 1: Calculate the total production time, t_p.

The total production time before shut-in is:

$$t_p = \frac{24N_p}{q} = \frac{24 \times 20,000}{480} = 1000 \text{ hrs.}$$

Step 2: Calculate Horner time ratio.

The Horner time ratio is $\frac{(t_p + \Delta t)}{\Delta t}$. For example, at shut-in time, $\Delta t = 0.35$ hrs., Horner time ratio is:

$$\frac{(t_p + \Delta t)}{\Delta t} = \frac{(1,000 + 0.35)}{0.35} = 2858.1$$

Horner time ratios for other shut-in times are shown in Table 13.4.

Step 3: Plot p_{ws} vs. $\log\left(\frac{t_p + \Delta t}{\Delta t}\right)$ and determine the slope of MTR straight line, m.

A plot of p_{ws} vs. $\log\left(\frac{t_p + \Delta t}{\Delta t}\right)$ is shown in Figure 13.14. By examining Figure 13.14, wellbore storage ended after $\Delta t = 3.5$ hrs. The slope of the MTR straight line representing infinite-acting radial flow (IARF) regime is determined to be 196 psi per cycle. That is, $m = 196$ psi per cycle.

Step 4: Calculate effective formation permeability, k.

From Eq. (13.59),

$$k = \frac{162.6 qB\mu}{mh} = \frac{162.6 \times 480 \times 2.2 \times 1.5}{196 \times 35} = 37.54 \text{ md.}$$

Step 5: Calculate average reservoir pressure.

By extrapolation of the MTR straight line in Figure 13.14,

$$p_i = \bar{p} = p^* = 3,608 \text{ psia.}$$

Note that it has been assumed that initial reservoir pressure is equal to average reservoir pressure which is approximated as equal to extrapolated pressure. This approximation is not accurate but is permissible for many applications. Use the methods suggested by Matthews et al.,[15] Muskat,[16] Larson,[17] Crump and Hite,[18] and other authors[19-22] to determine average reservoir pressure if more accurate value is required.

Step 6: Calculate skin factor, s, for the well.

From Eq. (13.62),

$$s = 1.151\left[\frac{p_{1hr} - p_{wf}}{m} - \log\frac{k}{\phi\mu c_t r_w^2} + 3.23\right]$$

By extrapolation of the MTR straight line in Figure 13.14, $p_{1hr} = 3019$ psia. Substituting above equation gives:

$$s = 1.151\left[\frac{3,019 - 2,832}{196} - \log\left(\frac{37.54}{0.18 \times 1.5 \times 2.6 \times 10^{-5} \times (0.5)^2}\right) + 3.23\right] = -3.62$$

The negative skin factor indicates that this well may be stimulated.

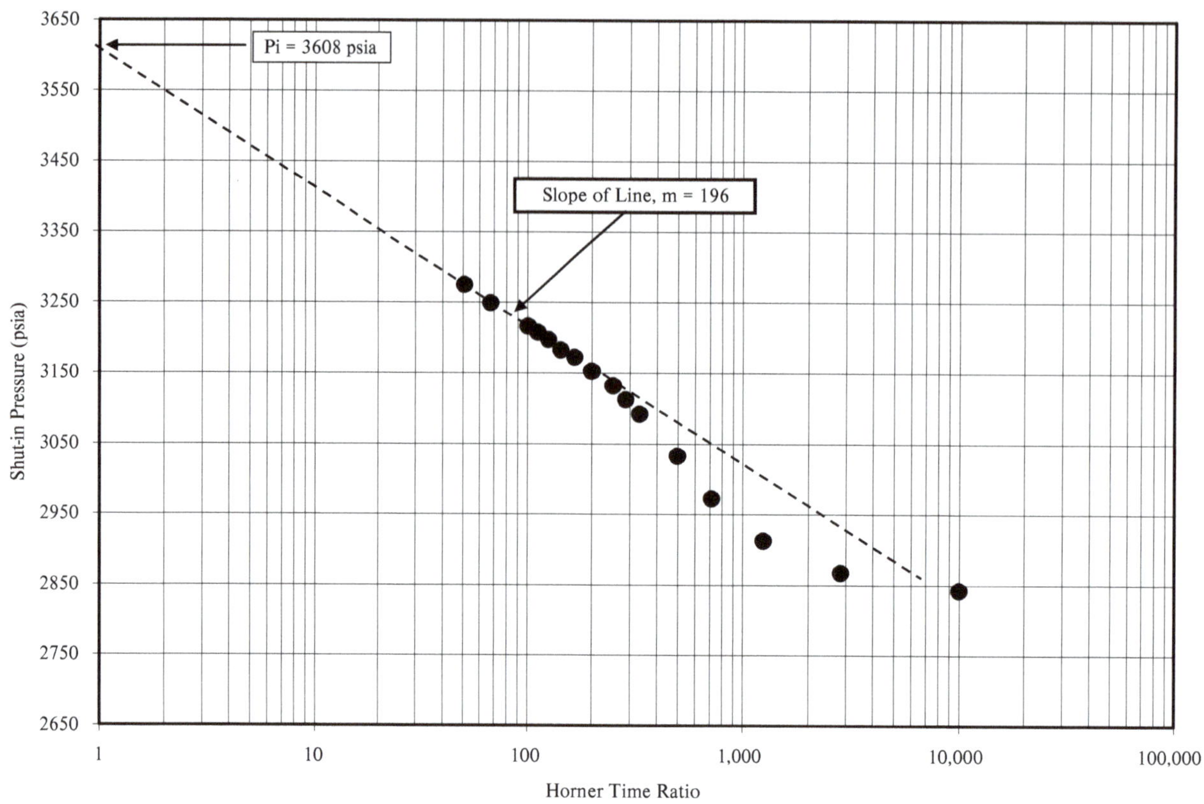

▲ **FIGURE 13.14** **Horner plot for Example 13.7**

End of Example 13.7

13.4.1.6 MDH Plot

Miller, Dyes, and Hutchinson[1] (MDH) proposed a method for analyzing buildup tests conducted on old wells with long production period. The theoretical basis of the MDH method can be derived from Eq. (13.56) which can be written as:

$$p_{ws} = p_i - m\left[\log\frac{(t_p + \Delta t)}{\Delta t}\right] \tag{13.64}$$

Expanding Eq. (13.64) gives:

$$p_{ws} = p_i - m\log\,(t_p + \Delta t) + m\log\,(\Delta t) \tag{13.65}$$

If $t_p >> \Delta t$, then $\log\left(t_p + \Delta t\right) \simeq \log\left(t_p\right) =$ constant. Eq. (13.65) can then be written as:

$$p_{ws} = \kappa + m\log\Delta t,$$
$$\text{where } \kappa = p_i - m\log\left(t_p\right) = \text{constant}. \tag{13.66}$$

A plot of p_{ws} vs. log Δt yields a straight line in the MTR region representing IARF regime. The slope of this line on the MDH plot is equal to the slope of the line on the Horner plot over the same time-function on the MTR straight line.

The MDH plot can be used to determine effective formation permeability and skin factors for wells with long production history but cannot be used to obtain extrapolated pressure, p^*. The MDH plot should be used as the last resort if it is difficult to estimate the length of time the well has been on production before the buildup test.

EXAMPLE 13.8 Calculation of Permeability and Skin Factor from Buildup Tests in Example 13.7 Using the MDH Plot

Problem

Suppose the cumulative production of the well in Example 13.7 is not known. Calculate permeability of the formation and skin factor of the well using the buildup test data in Table 13.4. Reservoir and fluid properties data are as given in Example 13.7.

Solution

Step 1: Plot p_{ws} vs. log Δt.

A plot of p_{ws} vs. log Δt is shown in Figure 13.15.

Step 2: Determine the slope of the MTR straight line.

From Figure 13.15, the slope of the MTR line, $m = 194$ psi/cycle. Note that this is very close to the slope of the MTR line of the Horner plot in Example 13.7.

Step 3: Calculate formation permeability, k.

Using Eq. (13.59),

$$k = \frac{162.6 q B \mu}{m h} = \frac{162.6 \times 480 \times 2.2 \times 1.5}{194 \times 35} = 37.93 \text{ md.}$$

Step 4: Calculate skin factor, s, for the well.

From Eq. (13.62),

$$s = 1.151 \left[\frac{p_{1hr} - p_{wf}}{m} - \log \frac{k}{\phi \mu c_t r_w^2} + 3.23 \right]$$

By extrapolation of the MTR straight line in Figure 13.15, $p_{1hr} = 3018$ psia. Substituting above equation gives:

$$s = 1.151 \left[\frac{3,018 - 2,832}{194} - \log \left(\frac{37.93}{0.18 \times 1.5 \times 2.6 \times 10^{-5} \times (0.5)^2} \right) + 3.23 \right] = -3.62$$

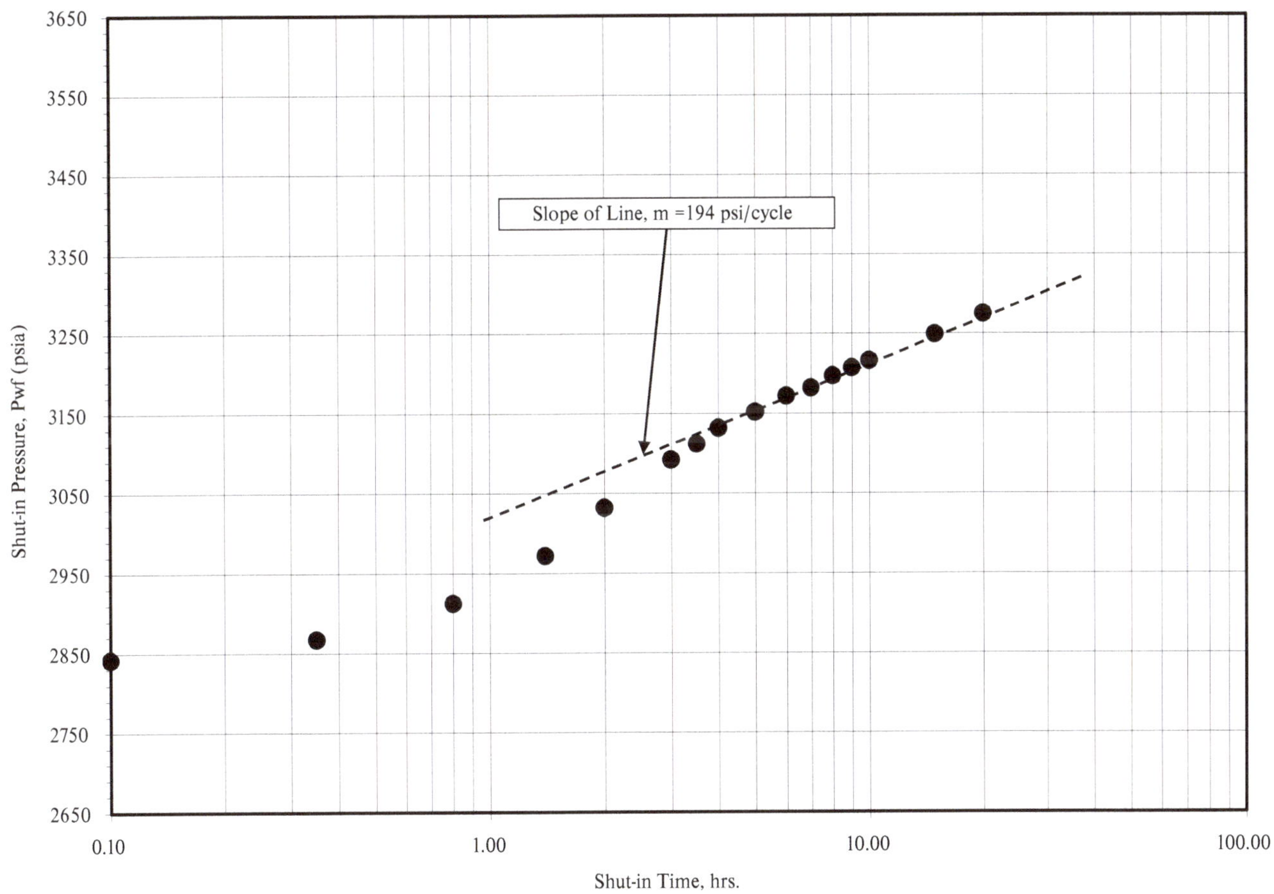

▲ **FIGURE 13.15** **MDH plot for Example 13.8**

End of Example 13.8

13.4.1.7 Horner's Time Approximation

In developing the equations for analyzing buildup tests, it was assumed that the well was produced at a constant rate throughout its production history before shut-in. In practice, it is very difficult to maintain the production rate of a well relatively constant over a reasonable time period. The production rates of most wells vary due to changing downhole and surface conditions. These conditions are usually difficult to control over time in order to maintain a constant production rate at the well. As a result, most wells have variable production rates over time. The production histories of wells with variable production rates can be modeled accurately by application of superposition principle as shown in Eq. (13.39). This approach can be very tedious, especially if the well has a long production history with many variable rates. To simplify estimation of total producing time in such situations, Horner[2] introduced an approximation which is widely used in place of superposition time. Horner's pseudo-producing time, t_{pH} is defined as:

$$t_{pH} = \frac{24N_p}{q_{last}} \qquad (13.67)$$

In Eq. (13.67), N_p = cumulative oil production before shut-in, STB; q_{last} = production rate of the well just before shut-in, STB/D. In practice, q_{last} should be maintained constant and stabilized as long as possible before starting the buildup phase of the test. Note that by definition, Horner's pseudo-producing time, t_{pH}, is different from, t_p, used in Eq. (13.56). This t_p assumes a constant producing rate over the entire history of the well. However, t_{pH} can be used to replace t_p, and q_{last} to replace q in Eq. (13.56) to yield:

$$p_{ws} = p_i - \frac{162.6q_{last}B\mu}{kh}\left[\log\frac{\left(t_{pH}+\Delta t\right)}{\Delta t}\right] \qquad (13.68)$$

A plot of p_{ws} vs. $\log\frac{t_{pH}+\Delta t}{\Delta t}$ gives a straight line with slope, m, and intercept equal to p_i, at infinite shut-in time. This is the same as the Horner plot. The slope of the MTR straight line in the Horner plot, m, is given as:

$$m = \frac{162.6q_{last}B\mu}{kh} \qquad (13.69)$$

All other equations derived previously for the ideal case Horner plot apply to this case with the Horner pseudo-producing time.

EXAMPLE 13.9 Calculation of Permeability, Reservoir Pressure, and Skin Factor from Buildup Tests Using Horner Pseudo-Producing Time Approximation

Problem

The discovery well in an oil reservoir was tested at production rates and duration shown in Table 13.5. The well was then shut in for a buildup test. The pressures versus time data recorded during the shut-in period are shown in Table 13.6. If the flowing bottomhole pressure of the well before it was shut in was 3535 psia, calculate the formation permeability, initial reservoir pressure, and skin factor of the well from the buildup test data. Other reservoir and fluid properties data are given as follows:

Formation thickness, h	50 ft
Formation porosity, ϕ	0.20
Total compressibility, c_t	2.4×10^{-5} psi^{-1}
Oil viscosity, μ_o	0.75 cp
Oil FVF, B_o	1.25 RB/STB
Wellbore radius, r_w	0.50 ft
Flowing BHP at $\Delta t = 0$	3535 psia

TABLE 13.5 Production History of the Well in Example 13.9

Production Rate (STB/D)	Duration (days)
450	7
600	4
500	10

TABLE 13.6 Buildup Test Data for Example 13.9

Δt (hrs)	p_{ws} (psia)	$(t_{pH} + \Delta t)/\Delta t$
0.25	3,605	2026.60
0.50	3,632	1013.80
1.00	3,795	507.40
1.50	3,960	338.60
2.00	4,230	254.20
3.00	4,320	169.80
4.00	4,370	127.60
7.00	4,390	73.34
10.00	4,395	51.64
15.00	4,410	34.76
24.00	4,422	22.10

Solution

Step 1: Calculate Horner pseudo-producing time, t_{pH}.

Using Eq. (13.67),

$$t_{pH} = \frac{24N_p}{q_{last}}$$

$$= \frac{24\left(450\times 7+600\times 4+500\times 10\right)}{500}$$

$$= 506.4 \text{ hrs.}$$

Step 2: Calculate Horner time ratio.

The Horner time ratio is $\left(t_{pH}+\Delta t\right)/\Delta t$. For example, at shut-in time, $\Delta t = 0.50$ hrs, Horner time ratio is:

$$\frac{(t_{pH}+\Delta t)}{\Delta t}=\frac{(506.4+0.50)}{0.50}=1013.8$$

Horner time ratios for other shut-in times are shown in Table 13.6.

Step 3: Plot p_{ws} vs. $\log\left(\frac{t_{pH}+\Delta t}{\Delta t}\right)$ and determine the slope of MTR straight line, m.

A plot of p_{ws} vs. $\log\left(\frac{t_{pH}+\Delta t}{\Delta t}\right)$ is shown in Figure 13.16. By examining Figure 13.16, wellbore storage ended after $\Delta t = 3.0$ hrs. The slope of the MTR straight line representing infinite-acting radial flow (IARF) regime is determined to be 71.4 psi per cycle. That is, $m = 71.4$ psi per cycle.

Step 4: Calculate effective formation permeability, k.

From Eq. (13.69),

$$k=\frac{162.6q_{last}B\mu}{mh}$$
$$=\frac{162.6\times500\times1.25\times0.75}{71.4\times50}$$
$$=21.35\text{ md.}$$

Step 5: Calculate average reservoir pressure, $\bar{p}$.

By extrapolation of the MTR straight line in Figure 13.16, $p_i=\bar{p}=p^*=4514$ psia.

Step 6: Calculate skin factor, s, for the well.

From Eq. (13.62),

$$s=1.151\left[\frac{p_{1hr}-p_{wf}}{m}-\log\frac{k}{\phi\mu c_t r_w^2}+3.23\right]$$

By extrapolation of the MTR straight line in Figure 13.16, $p_{1hr}=4329$ psia. Substituting above equation gives:

$$s=1.151\left[\frac{4329-3535}{71.4}-\log\left(\frac{21.35}{0.20\times0.75\times2.4\times10^{-5}\times(0.5)^2}\right)+3.23\right]$$
$$=8.03$$

The positive skin factor calculated for the well indicates that it may be damaged in the near-wellbore region.

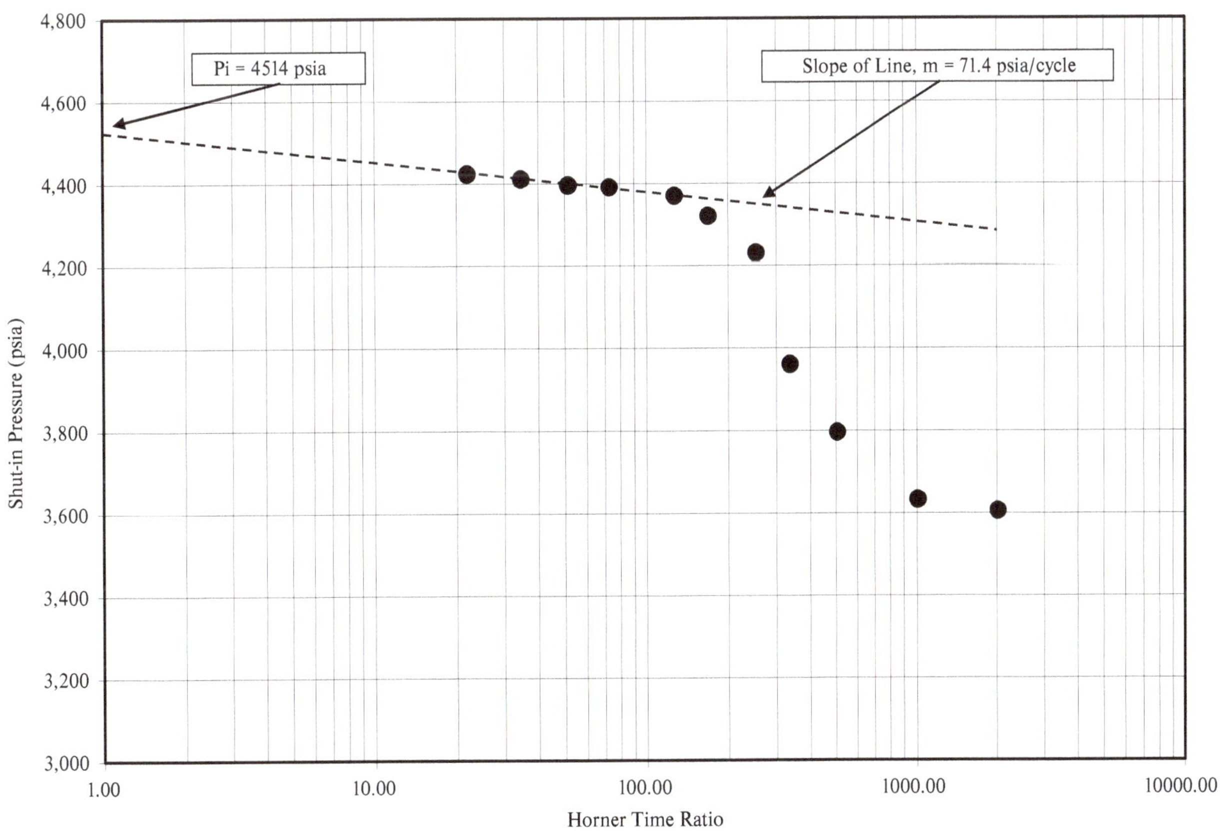

▲ **FIGURE 13.16 Plot of p_{ws} vs. $\{(t_{pH}+\Delta t)/\Delta t\}$ for Example 13.9**

End of Example 13.9

13.4.2 Compressible Fluids

In this section, the application of straightline methods for analyses of drawdown and buildup tests to gas reservoirs is presented. It is worthwhile to note that notional forms of equations developed for liquid reservoirs are applicable to gas reservoirs with slight or no modifications by using transforming pressure functions which account for the variation of gas properties with pressure. In Chapter 12, the concept of pseudo-pressure was introduced and used to linearize to some extent the gas diffusivity equation. In this section, two additional concepts are introduced that will assist in the manipulation of the equations that are used to represent gas flow equations for drawdown and buildup tests.

13.4.2.1 Real Gas Pseudo-Pressure and Pseudo-Time

Real gas pseudo-pressure, p_p, as defined by Al-Hussainy et al.[23] is:

$$p_p = 2\int_{p_b}^{p} \frac{p}{\mu(p)z(p)}dp \tag{13.70}$$

Note that Eq. (13.70) is the same as Eq. (12.86). An analogous function called real gas pseudo-time, t_A, was defined by Agarwal[24] as:

$$t_A = \int_0^t \frac{1}{\mu(p)c_t(p)}dt \tag{13.71}$$

The combined use of these pseudo-variables provides an "effective" linearization of the gas diffusivity equations, especially in applications to type-curve analyses. The use of type curves in well test analysis is discussed in Chapter 14.

Al-Hussainy and Ramey[25] achieved an approximate solution to the gas diffusivity equation expressed in terms of pseudo-pressure as:

$$p_{pwf} = p_{pi} - \frac{5.792\times10^4 q_g p_{sc} T}{khT_{sc}} \times\left[\log\frac{kt}{\phi\mu c_t r_w^2} - 3.23 + 0.8691s'\right] \tag{13.72}$$

In Eq. (13.72), s' is the apparent skin factor which is defined as:

$$s' = s + Dq_g \tag{13.73}$$

In Eq. (13.73), s = skin factor due to formation damage or stimulation; D = non-Darcy flow coefficient; and q_g = gas flow rate.

If standard conditions are set at $T_{sc} = 60°\text{F}$ and $p_{sc} = 14.7$ psia, Eq. (13.72) becomes:

$$p_{pwf} = p_{pi} - \frac{1637 q_g T}{kh}\left[\log\frac{kt}{\phi(\mu c_t)_i r_w^2} - 3.23 + 0.8691s'\right] \tag{13.74}$$

13.4.2.2 Normalized Gas Pseudo-Pressure and Gas Pseudo-Time

Meunier et al.[26] proposed normalization of gas pseudo-pressure and pseudo-time at any arbitrary reference pressure to produce pseudo-variables whose magnitudes are close to those of untransformed pressure and time and have the same units. For gas pseudo-pressure, the normalized pseudo-pressure is expressed as:

$$\begin{aligned} p_{pn} &= \frac{1}{2}\left(\frac{\mu z}{p}\right)_r p_p \\ &= \left(\frac{\mu z}{p}\right)_r \int_{p_b}^{p} \frac{p}{\mu(p)z(p)}dp \end{aligned} \tag{13.75}$$

For the gas pseudo-time, the normalized pseudo-time is expressed as:

$$\begin{aligned} t_{An} &= (\mu c_t)_r \int_0^t \frac{1}{\mu(p)c_t(p)}dt \\ &= (\mu c_t)_r t_A \end{aligned} \tag{13.76}$$

The choice of the reference pressure for the normalization of the pseudo-variables is completely arbitrary. Typically, the initial reservoir pressure or the average reservoir pressure is selected if known. If the average reservoir pressure is the reference pressure, then Eqs. (13.75) and (13.76) can be written as:

$$p_{pn} = \left(\frac{\overline{\mu z}}{\bar{p}}\right)_{\bar{p}} \int_{p_b}^{p} \frac{p}{\mu(p)z(p)}dp \tag{13.77}$$

and

$$t_{An} = (\overline{\mu c_t})_{\bar{p}} \int_0^t \frac{dt}{\mu c_t} \tag{13.78}$$

Using Eq. (13.77), Eq. (13.74) can be written in terms of normalized pseudo-variables as:

$$p_{pnwf} = p_{pni} - \frac{162.6 q_g \bar{B}\bar{\mu}}{kh} \times\left[\log\frac{kt}{\phi\overline{\mu c_t} r_w^2} - 3.23 + 0.8691s'\right] \tag{13.79}$$

13.4.2.3 Pressure and Pressure-Squared as Variables in Gas Flow Equations

In Eqs. (13.74) and (13.79), gas flow equations were expressed as functions of pseudo-pressures and normalized pseudo-pressures, respectively. As explained earlier, this transformation was necessary because gas properties such as viscosity and compressibility factor vary with pressure. However, under certain conditions, some assumptions can be made with respect to the variation of gas viscosity and gas compressibility factor with pressure. The first assumption that can be made is that at pressures above 3500 psia, the gas property, $p/(\mu z)$, is constant. With this assumption, it can be written that at any pressure above the 3500 psia threshold that:

$$\frac{p}{\mu z} = \frac{\bar{p}}{\bar{\mu}\bar{z}}, \text{ at the average reservoir pressure, } \bar{p} \tag{13.80}$$

If the condition represented in Eq. (13.80) applies, then it can be derived from Eq. (13.70) that:

$$p_p = \frac{2\bar{p}p}{\bar{\mu}\bar{z}} \tag{13.81}$$

Substituting Eq. (13.81) into Eq. (13.74) and applying some conversions gives:

$$p_{wf} = p_i - \frac{162.6 q_g \bar{B}\bar{\mu}}{kh} \times \left[\log\frac{kt}{\phi\bar{\mu}\bar{c}_t r_w^2} - 3.23 + 0.8691 s'\right] \tag{13.82}$$

Note that with the stated assumption, Eq. (13.82) is similar in form to Eq. (13.79) developed for normalized pseudo-pressures.

The second assumption that is applicable at low pressures, ($p < 2000$ psia), is that the gas property, (μz), is constant. With this assumption, it can be written that:

$$(\mu z) = (\bar{\mu}\,\bar{z}), \text{ at the average reservoir pressure, } \bar{p} \tag{13.83}$$

If the condition expressed in Eq. (13.83) applies, then it can be derived from Eq. (13.70) that:

$$p_p = \frac{p^2}{\bar{\mu}\bar{z}} \tag{13.84}$$

Substituting Eq. (13.84) into Eq. (13.74) gives:

$$p_{wf}^2 = p_i^2 - \frac{1637 q_g \bar{\mu}\bar{z}T}{kh} \times \left[\log\frac{kt}{\phi\bar{\mu}\bar{c}_t r_w^2} - 3.23 + 0.8691 s'\right] \tag{13.85}$$

13.4.2.4 Constant Rate Drawdown Tests for Gas Wells

In the previous section, several equations were derived for gas flow in porous media. It is important to note the various assumptions that were used to derive each equation. The most important assumption that applies to all the gas flow equations is that the reservoir is radial and infinite-acting. This is the underlying assumption that applies to the solution of the radial diffusivity equation for compressible fluids.

In this section, the gas flow equations derived in the previous section are applied in the analysis of gas drawdown tests at constant rate. When drawdown tests at constant rate for oil wells were discussed, it was pointed out that it is very difficult to maintain the production rate of an oil well constant for the entire duration of a test. It is even more difficult to maintain the production rate of a gas well constant for the duration of a test. Typically, gas wells are produced at conditions that attempt to achieve constant pressures either at the wellhead or bottomhole. These conditions will cause the production rate to vary with time. In the next section, methods for analyzing gas wells with varying production rates are presented. However, it is instructive to review methods for analyzing gas wells at constant production rates. These methods are applicable if the production rate of the gas well is assumed to be approximately constant.

In terms of pseudo-pressure, the expressions for analyses of drawdown tests on gas wells at constant rate can be deduced from Eq. (13.74). On the basis of Eq. (13.74), a plot of p_{pwf} vs. $\log t$ gives a straight line. The slope of the straight line, m, is:

$$m = \left|\frac{1637 q_g T}{kh}\right| \tag{13.86}$$

It can be derived from Eq. (13.74) that the apparent skin is given by:

$$s' = 1.151\left[\frac{p_{pi} - p_{p,1hr}}{m} - \log\left(\frac{k}{\phi(\mu c_t)_i r_w^2}\right) + 3.23\right] \tag{13.87}$$

In terms of normalized pseudo-pressure, the expressions for analysis of drawdown tests on gas wells at constant rate can be deduced from Eq. (13.79). From Eq. (13.79), a plot of p_{pnwf} vs. $\log t$ gives a straight line. The slope of the straight line, m, is:

$$m = \left|\frac{162.6 q_g \bar{B}\bar{\mu}}{kh}\right| \tag{13.88}$$

It can be derived from Eq. (13.79) that the apparent skin is given by:

$$s' = 1.151\left[\frac{p_{pni} - p_{pn,1hr}}{m} - \log\left(\frac{k}{\phi(\mu c_t)_i r_w^2}\right) + 3.23\right] \tag{13.89}$$

In terms of pressure, the expressions for analyses of drawdown tests on gas wells at constant rate can be deduced from Eq. (13.82). From Eq. (13.82), a plot of p_{wf} vs. $\log t$ gives a straight line. The slope of the straight line, m, is:

$$m = \left|\frac{162.6 q_g \bar{B} \bar{\mu}}{kh}\right| \tag{13.90}$$

It can be derived from Eq. (13.82) that the apparent skin is given by:

$$s' = 1.151\left[\frac{p_i - p_{1hr}}{m} - \log\left(\frac{k}{\phi(\mu c_t)_i r_w^2}\right) + 3.23\right] \tag{13.91}$$

Similarly in terms of pressure-squared, the expressions for analyses of drawdown tests on gas wells at constant rate can be deduced from Eq. (13.85). From Eq. (13.85), a plot of p_{wf}^2 vs. $\log t$ gives a straight line. The slope of the straight line, m, is:

$$m = \left|\frac{1{,}637 q_g \bar{\mu} \bar{z} T}{kh}\right| \tag{13.92}$$

It can be derived from Eq. (13.85) that the apparent skin is given by:

$$s' = 1.151\left[\frac{p_i^2 - p_{1hr}^2}{m} - \log\left(\frac{k}{\phi(\mu c_t)_i r_w^2}\right) + 3.23\right] \tag{13.93}$$

Note that by convention, the absolute values of m are used as the slopes of the straight line in Eqs. (13.86), (13.88), (13.90), and (13.92). It is important to observe that in all preceding equations for gas flow, reservoir pressure was assumed to be at initial pressure, p_i. Under this condition, reservoir properties such as gas viscosity, gas formation volume factor, etc., should be evaluated at initial reservoir pressure. In many applications, average reservoir pressure, $\bar{p}$, at the start of the flow test is used. In such cases, reservoir properties should be evaluated at the average reservoir pressure.

EXAMPLE 13.10 Calculation of Permeability and Skin Factor from Analyses of Single, Constant Rate Drawdown Test Using Pseudo-Pressure, Normalized Pseudo-Pressure, Pressure, and Pressure-Squared Methods

Problem

The discovery well in a dry gas reservoir was produced at a constant rate of 6000 Mcf/D. The flowing BHP measured at various time intervals are shown in Table 13.7. Initial reservoir pressure before the test was measured at 4275 psia. Reservoir temperature is 215°F. Assuming the reservoir gas has no non-hydrocarbon impurities, gas properties are summarized as follows:

Gas gravity, γ_g	0.70
Critical Pressure, p_c	660 psia
Critical Temperature, T_c	−78.5°F
Gas FVF, B_{gi}	0.7553 RB/Mcf
Gas compressibility, c_g	1.71×10^{-4} psi^{-1}
Total compressibility, c_t	1.71×10^{-4} psi^{-1}

Reservoir and well data are as follows:

Porosity, ϕ	0.15
Well radius, r_w	0.29 ft
Formation thickness, h	24 ft

TABLE 13.7 Drawdown Flow Test Data for Example 13.10

Flowing Time t (hrs)	Flowing BHP p_{wf} (psia)	Flowing Time t (hrs)	Flowing BHP p_{wf} (psia)
0.0	4,275	20.0	3,867
0.5	4,260	25.0	3,858
0.9	4,245	30.0	3,850
1.3	4,230	35.0	3,844
1.7	4,180	40.0	3,840
2.1	4,130	50.0	3,831
2.4	4,087	60.0	3,824
2.7	4,047	70.0	3,819
3.2	4,010	80.0	3,815
4.0	3,991	90.0	3,810
4.8	3,975	100.0	3,805
6.1	3,950	110.0	3,802
8.0	3,934	120.0	3,798
10.1	3,914	130.0	3,793
12.7	3,895	140.0	3,791
15.2	3,885		

Solution

Step 1: Construct a table of gas properties consisting of gas compressibility factor and gas viscosity versus pressure using the procedure shown in Example 5.4. The calculated gas properties are shown in Table 13.8.

Step 2: Calculate pseudo-pressure from data obtained in Step 1 using Eq. (13.70) as illustrated in Example 12.8. The calculated pseudo-pressures are shown in Table 13.9.

Step 3: Calculate normalized pseudo-pressure from Eq. (13.75). The calculated normalized pseudo-pressures are also shown in Table 13.9.

Step 4: Pseudo-Pressure Plot: Plot p_{pwf} vs. log *t*.

A plot of p_{pwf} vs. log *t* from the data in Table 13.9 is shown in Figure 13.17. From the plot, the slope of the MTR straight line is determined as $m = 32.5 \times 10^6$ psia²/cp-cycle. From Eq. (13.86), formation permeability, *k*, is calculated as:

$$k = \left|\frac{1637 q_g T}{mh}\right|$$

$$= \frac{1637 \times 6000 \times 675}{32.5 \times 10^6 \times 24}$$

$$= 8.50 \text{ md}$$

The apparent skin factor, *s'*, from Eq. (13.87) is:

$$s' = 1.151\left[\frac{p_{pi} - p_{p,1hr}}{m} - \log\left(\frac{k}{\phi(\mu c_t)_i r_w^2}\right) + 3.23\right]$$

$$= 1.151\left[\frac{(191.9 - 85.5)\times 10^6}{32.5 \times 10^6}\right.$$

$$\left. - \log\left(\frac{8.50}{0.15 \times 0.0248 \times 1.71 \times 10^{-4} \times (0.29)^2}\right) + 3.23\right]$$

$$= -1.95$$

Step 5: Normalized Pseudo-Pressure Plot: Plot p_{pnwf} vs. log *t*.

A plot of p_{pnwf} vs. log *t* from the data in Table 13.9 is shown in Figure 13.18. From the plot, the slope of the MTR straight line is determined as m = 85 psia/cycle. From Eq. (13.88), formation permeability, *k*, is calculated as:

$$k = \left|\frac{162.6 q_g \bar{B} \bar{\mu}}{mh}\right|$$

$$= \frac{162.6 \times 6000 \times 0.7553 \times .0248}{85 \times 24}$$

$$= 8.96 \text{ md}$$

The apparent skin factor, *s'*, from Eq. (13.89) is:

$$s' = 1.151\left[\frac{p_{pni} - p_{pn,1hr}}{m} - \log\left(\frac{k}{\phi(\mu c_t)_i r_w^2}\right) + 3.23\right]$$

$$= 1.151\left[\frac{528.69 - 235}{85}\right.$$

$$\left. - \log\left(\frac{8.96}{0.15 \times 0.0248 \times 1.71 \times 10^{-4} \times (0.29)^2}\right) + 3.23\right]$$

$$= -1.77$$

Step 6: Pressure Plot Method: Plot Plot p_{wf} vs. log *t*.

A plot of p_{wf} vs. log *t* from the data in Table 13.9 is shown in Figure 13.19. From the plot, the slope of the MTR straight line is determined as m = 91 psia/cycle. From Eq. (13.90), formation permeability, *k*, is calculated as:

$$k = \left|\frac{162.6 q_g \bar{B} \bar{\mu}}{mh}\right|$$

$$= \frac{162.6 \times 6000 \times 0.7553 \times .0248}{91 \times 24}$$

$$= 8.37 \text{ md.}$$

The apparent skin factor, *s'*, from Eq. (13.91) is:

$$s' = 1.151\left[\frac{p_i - p_{1hr}}{m} - \log\left(\frac{k}{\phi(\mu c_t)_i r_w^2}\right) + 3.23\right]$$

$$= 1.151\left[\frac{4275 - 3986}{91}\right.$$

$$\left. - \log\left(\frac{8.37}{0.15 \times 0.0248 \times 1.71 \times 10^{-4} \times (0.29)^2}\right) + 3.23\right]$$

$$= -2.06$$

Step 7: Pressure-Squared Plot: Plot p^2_{wf} vs. log t.

A plot of p^2_{wf} vs. log t from the data in Table 13.9 is shown in Figure 13.20. From the plot, the slope of the MTR straight line is determined as m = 727,272.73 psia²/cycle. From Eq. (13.92), formation permeability, k, is calculated as:

$$k = \left|\frac{1637 q_g \overline{\mu z} T}{mh}\right|$$

$$= \frac{1637 \times 6000 \times .0248 \times 0.95 \times 675}{727{,}272.73 \times 24}$$

$$= 8.95 \text{ md}$$

The apparent skin factor, s', from Eq. (13.93) is:

$$s' = 1.151\left[\frac{p_i^2 - p_{1hr}^2}{m} - \log\left(\frac{k}{\phi(\mu c_t)_i r_w^2}\right) + 3.23\right]$$

$$= 1.151\left[\frac{(4275)^2 - (3988.62)^2}{727{,}272.73} - \log\left(\frac{8.95}{0.15 \times 0.0248 \times 1.71 \times 10^{-4} \times (0.29)^2}\right) + 3.23\right]$$

$$= -2.00$$

A summary of the results from the four methods is shown in Table 13.10. The results in Table 13.10 indicate that there are insignificant differences between the four methods for this example problem. However, prior work reported by Al-Hussainy and Ramey,[25] Meunier et al.,[26] Aziz et al.,[27] Reynolds et al.,[28] and Wattenbarger and Ramey[29] showed that use of pseudo-pressures or normalized pseudo-pressures provides more accurate analyses of gas drawdown test data. Note also for the four methods, semi-log analyses were based on actual flow time instead of pseudo-time. The application of pseudo-pressure or normalized pseudo-pressure versus actual flow time is the recommended practice for analysis of gas well drawdown data.

TABLE 13.8 | Calculated Gas Properties for Example 13.10

Pressure (psia)	z-Factor z	Gas Viscosity μ_g (cp)	Pressure (psia)	z-Factor z	Gas Viscosity μ_g (cp)
3750	0.91	0.0234	4050	0.93	0.0244
3775	0.91	0.0235	4075	0.93	0.0245
3800	0.91	0.0236	4100	0.93	0.0246
3825	0.91	0.0237	4125	0.93	0.0247
3850	0.92	0.0237	4150	0.93	0.0247
3875	0.92	0.0238	4175	0.94	0.0247
3900	0.92	0.0239	4200	0.94	0.0247
3925	0.92	0.0241	4225	0.95	0.0247
3950	0.92	0.0241	4250	0.95	0.0248
3975	0.92	0.0242	4275	0.95	0.0248
4000	0.92	0.0243	4300	0.96	0.0248
4025	0.93	0.0243			

TABLE 13.9 | Plot Data for Example 13.10

t (hrs)	p_{wf} (psia)	p_{wf}^2 (psia)^2	p_{pwf} (psia^2/cp)	p_{pnwf} (psia)
0.0	4,275	18,275,625	191,862,124	528.69
0.5	4,260	18,147,600	186,434,492	513.73
0.9	4,245	18,020,025	181,013,842	498.79
1.3	4,230	17,892,900	175,607,153	483.89
1.7	4,180	17,472,400	157,566,955	434.18
2.1	4,130	17,056,900	139,546,662	384.53
2.4	4,087	16,703,569	124,121,092	342.02
2.7	4,047	16,378,209	109,821,590	302.62
3.2	4,010	16,080,100	96,621,360	266.24
4.0	3,991	15,928,081	89,833,919	247.54
4.8	3,975	15,800,625	84,114,523	231.78
6.1	3,950	15,602,500	75,197,224	207.21
8.0	3,934	15,476,356	69,514,380	191.55
10.1	3,914	15,319,396	62,419,441	172.00
12.7	3,895	15,171,025	55,686,205	153.45
15.2	3,885	15,093,225	52,142,781	143.68
20.0	3,867	14,953,689	45,770,988	126.12
25.0	3,858	14,884,164	42,589,073	117.36
30.0	3,850	14,822,500	39,760,705	109.56
35.0	3,844	14,776,336	37,637,140	103.71
40.0	3,840	14,745,600	36,221,431	99.81
50.0	3,831	14,676,561	33,036,084	91.03
60.0	3,824	14,622,976	30,558,224	84.20
70.0	3,819	14,584,761	28,786,744	79.32
80.0	3,815	14,554,225	27,369,560	75.42
90.0	3,810	14,516,100	25,598,080	70.54
100.0	3,805	14,478,025	23,826,600	65.66
110.0	3,802	14,455,204	22,763,712	62.73
120.0	3,798	14,424,804	21,348,185	58.83
130.0	3,793	14,386,849	19,580,849	53.96
140.0	3,791	14,371,681	18,873,914	52.01

TABLE 13.10 | Summary of Results for Example 13.10

Method	Effective Gas Permeability, k (md)	Skin Factor, s
p_{wf} vs. log t	8.37	−2.06
p_{wf}^2 vs. log t	8.95	−2.00
p_{pwf} vs. log t	8.50	−2.02
p_{pnwf} vs. log t	8.96	−1.77

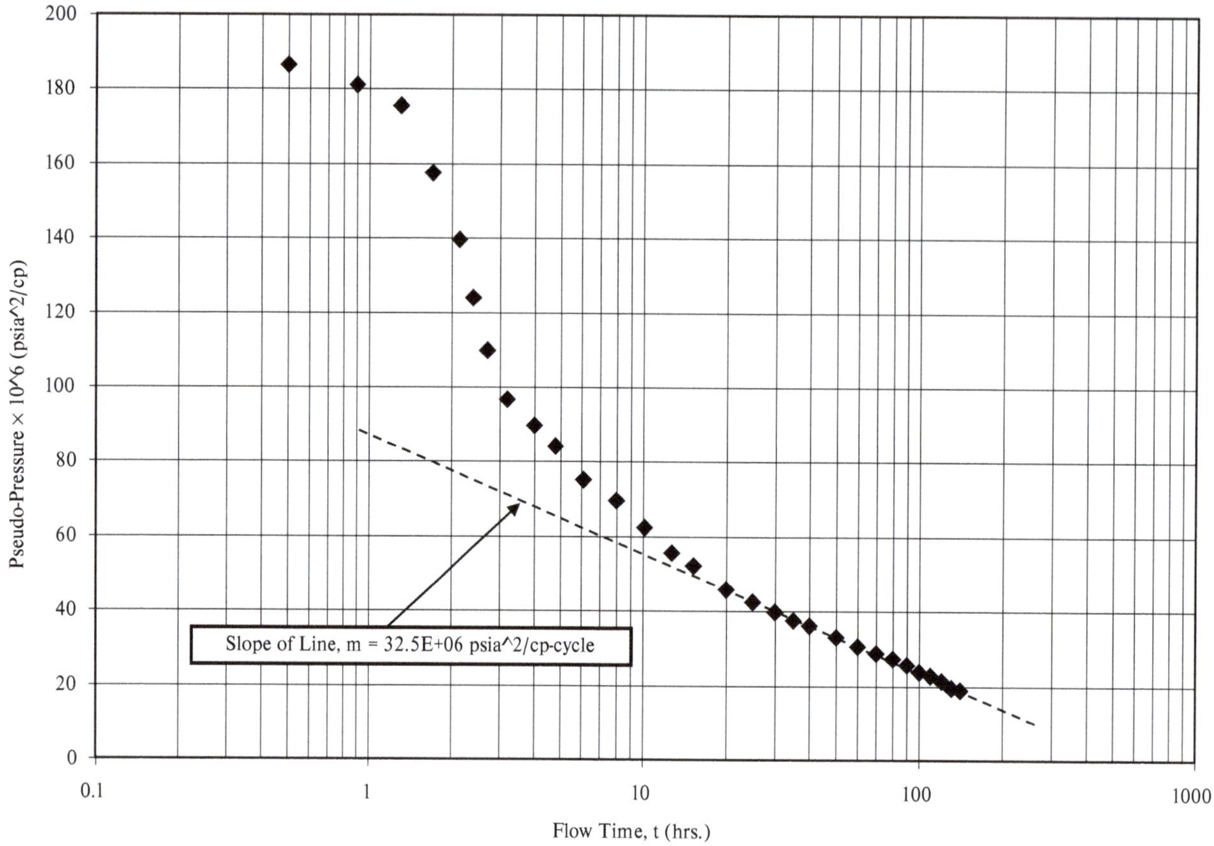

▲ **FIGURE 13.17** **Plot of p_{pwf} vs. t for Example 13.10**

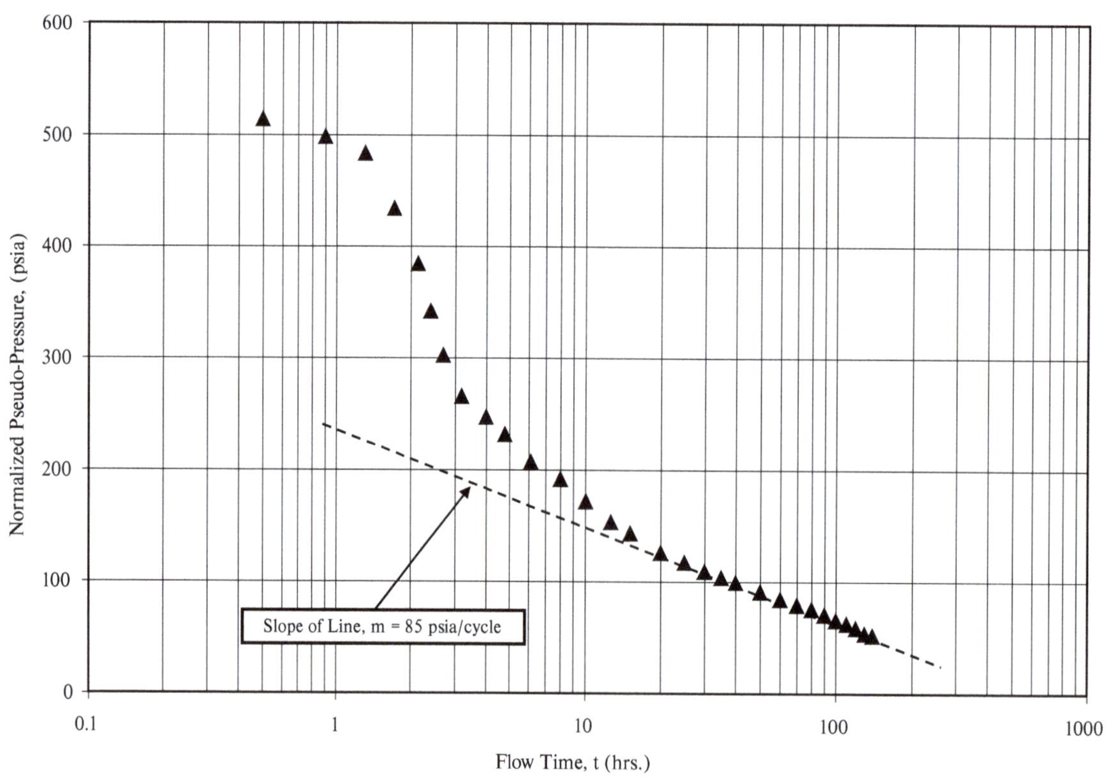

▲ **FIGURE 13.18** **Plot of p_{pnwf} vs. t for Example 13.10**

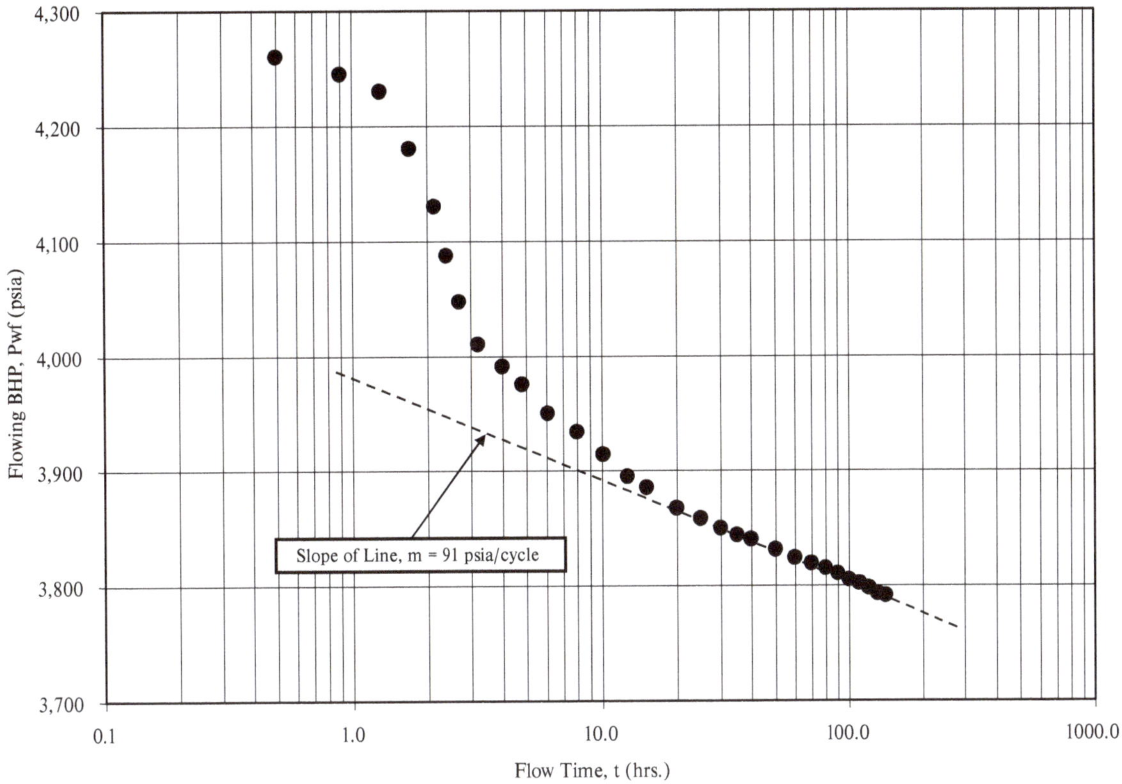

▲ FIGURE 13.19 Plot of p_{wf} vs. t for Example 13.10

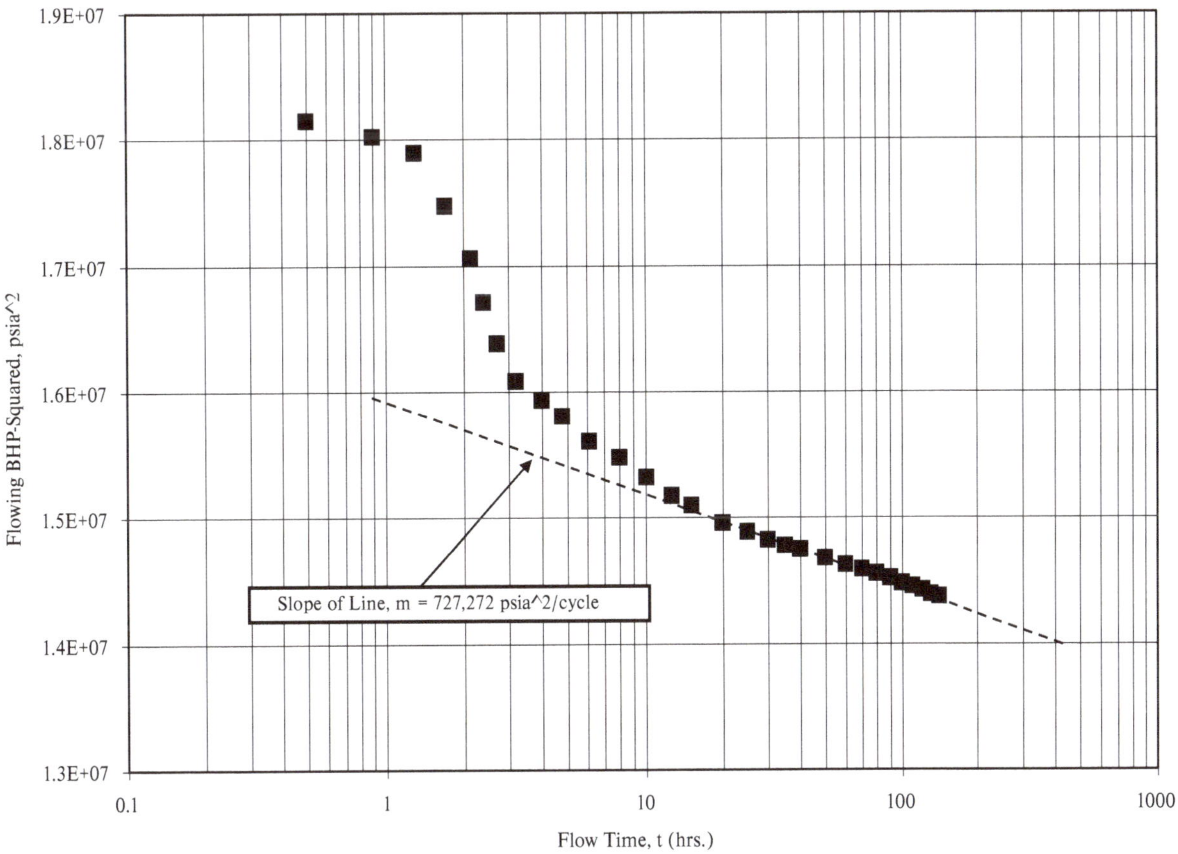

▲ FIGURE 13.20 Plot of p_{wf}^2 vs. t for Example 13.10

End of Example 13.10

13.4.2.5 Multi-Rate Drawdown Tests for Gas Wells

Multi-rate drawdown tests in gas wells are similar to the same type of tests in oil wells. As was the case for oil wells, these multi-rate tests are conducted in gas wells to overcome the difficulty of maintaining the rate constant in a single rate test. A typical multi-rate test is shown in Figure 13.10. As shown in Figure 13.10, flow at constant rate, q_1, starts at time, $t = 0$; increases to rate, q_2, at time, t_1; changes to rate, q_3, at time, t_2; and finally changes to rate, q_n, at time, t_{n-1}. Note that the flow rate is maintained constant at each flow period. The gas flow equation written in the form of normalized pseudo-pressures [Eq. (13.79)] is used in the derivation of a general equation for calculation of the pressure at the well at any time period by applying the principle of superposition. To simplify the application of the superposition principle, Eq. (13.79) can be written in the following form:

$$p_{pni} - p_{pnwf} = m'_m q_g \left[\log(t) + b'_m\right] \tag{13.94}$$

$$\text{where } m'_m = \frac{162.6\bar{B}\bar{\mu}}{kh} \tag{13.95}$$

$$\text{and } b'_m = \log\left(\frac{k}{\phi\bar{\mu}\bar{c}_t r_w^2}\right) - 3.23 + 0.8691 s' \tag{13.96}$$

Applying the superposition principle on the flow rates shown in Figure 13.10 for n rates gives:

$$\begin{aligned} p_{pni} - p_{pnwf} = {} & m'_m q_1\left[\log(t) + b'_m\right] \\ & + m'_m(q_2 - q_1)\left[\log(t - t_1) + b'_m\right] \\ & + m'_m(q_3 - q_2)\left[\log(t - t_2) + b'_m\right] + \ldots\ldots \\ & \ldots. + m'_m(q_n - q_{n-1})\left[\log(t - t_{n-1}) + b'_m\right] \end{aligned} \tag{13.97}$$

Eq. (13.97) can be written as:

$$\frac{p_{pni} - p_{pnwf}}{q_n} = m'_m \sum_{j=1}^{n}\left[\frac{(q_j - q_{j-1})}{q_n}\log(t_n - t_{j-1})\right] + m'_m b'_m \quad \text{for } q_n \neq 0 \tag{13.98}$$

Substituting for b'_m in Eq. (13.98) gives:

$$\frac{p_{pni} - p_{pnwf}}{q_n} = m'_m \sum_{j=1}^{n}\left[\frac{(q_j - q_{j-1})}{q_n}\log(t_n - t_{j-1})\right] + m'_m\left[\log\left(\frac{k}{\phi\mu c_t r_w^2}\right) - 3.23 + 0.8691 s'\right] \tag{13.99}$$

Eq. (13.99) is similar in form to Eq. (13.39) derived for multi-rate tests in oil wells. Note that Eq. (13.99) is valid only for an infinite-acting reservoir. Eq. (13.99) is in the form of the equation for a straight line. If Y_F is used to represent values on the y-axis and X_F for values on the x-axis, then Eq. (13.99) can be written as:

$$Y_F \equiv \frac{p_{pni} - p_{pnwf}}{q_n} \tag{13.100}$$

$$X_F \equiv \sum_{j=1}^{n}\left[\frac{(q_j - q_{j-1})}{q_n}\log(t_n - t_{j-1})\right] \tag{13.101}$$

$$\text{slope, } m'_m = \frac{162.6\bar{B}\bar{\mu}}{kh} \tag{13.102}$$

$$Y_{intercept} = b'_I = m'_m\left[\log\left(\frac{k}{\phi\bar{\mu}\bar{c}_t r_w^2}\right) - 3.23 + 0.8691 s'\right] \tag{13.103}$$

A Cartesian plot of Y_F vs. X_F results in a straight line with slope, m'_m, and intercept, b'_I, if the reservoir is infinite-acting. This type of plot is sometimes called the rate-normalized plot or multi-rate superposition plot.[11] An expression for calculating skin factor, s', can be derived by rearranging Eq. (13.103) to get:

$$s' = 1.151\left[\frac{b'_I}{m'_m} - \log\left(\frac{k}{\phi\bar{\mu}\bar{c}_t r_w^2}\right) + 3.23\right] \tag{13.104}$$

The procedures for applying the above equations to multi-rate gas flow tests are the same as were illustrated in Example 13.5 for multi-rate tests in oil wells.

13.4.2.6 Non-Darcy Flow Coefficient

In Eq. (13.73), the apparent skin factor, s', in gas flow equations was represented as $s' = s + Dq_g$, where s is the skin factor due to formation damage or stimulation, D is the non-Darcy flow coefficient, and, q_g is the gas flow rate. Non-Darcy flow coefficient is used to account for pressure loss due to non-Darcy (or turbulent) flow near the wellbore caused by high gas flow velocities. The effects of non-Darcy flow depend on gas flow rate as shown in the above expression for apparent skin factor. The apparent skin factor can be separated into its components parts by conducting two flow tests of equal duration (called isochronal tests) at different flow rates. This yields two equations which can be solved simultaneously for the skin factor and non-Darcy flow coefficient. This method is illustrated in Example 13.11. If only one flow-rate test is available, the non-Darcy coefficient can be estimated with the expression:[30]

$$D = \frac{2.715 \times 10^{-15}\beta k M p_{sc}}{h T_{sc} r_w \mu_{@p_{wf}}} \tag{13.105}$$

In Eq. (13.105), B = turbulence parameter; k = gas permeability, md; M = gas molecular weight, lbm/lb-mole; p_{sc} = pressure at standard conditions, psia; h = formation thickness, feet; T_{sc} = temperature at standard conditions, °R; r_w = wellbore radius, feet; and $\mu_{@pwf}$ gas viscosity, cp, evaluated at flowing BHP. The turbulence parameter, β, can be determined experimentally or estimated from correlations[31] as:

$$\ln\beta = -1.201\ln k + 23.83 \qquad (13.106)$$

The non-Darcy coefficient was assumed to be constant for the preceding presentation. However, studies performed by Wattenbarger et al.,[29] Firoozabadi et al.,[31] and Swift and Kiel[32] show that non-Darcy coefficient varies with gas flow rates. Rigorous procedures for calculating the variation of non-Darcy coefficients with rate have been published by Fligelman et al.[30] These procedures can be used in analyzing gas flow tests if warranted by flow conditions.

EXAMPLE 13.11 Calculation of Non-Darcy Flow Coefficient from Isochronal Tests

Problem

An isochronal flow test was conducted on a well in a dry gas reservoir at two flow rates of 2000 Mcf/D and 4000 Mcf/D. The measured flowing BHP at the two flow rates are shown in Table 13.11. The properties of the reservoir gas are shown in Table 13.12. Calculate the formation gas permeability, skin factor, and non-Darcy flow coefficient of the well. Assume non-Darcy flow coefficient for the tests is constant. Other reservoir data are provided as follows:

Initial reservoir pressure, p_i	3000 psia
Reservoir temperature, T	121°F
Formation thickness, h	15 ft
Formation porosity, ϕ	0.20
Total compressibility, c_t	4.5×10^{-4} psi^{-1}
Wellbore radius, r_w	0.50 ft

TABLE 13.11 Isochronal Flow Test Data for Example 13.11

Time, t (hrs.)	Flow #1 q_1 = 2000 Mcf/D	Flow #2 q_2 = 4000 Mcf/D
	p_{wf} (psia)	p_{wf} (psia)
0.25	2,576	1,170
0.40	2,555	1,085
0.60	2,533	1,019
0.80	2,524	971
1.00	2,515	925
1.50	2,495	815
2.00	2,479	755
3.00	2,456	650
4.00	2,445	570
5.00	2,434	495
6.00	2,425	435
7.00	2,416	385

TABLE 13.12 Gas Properties for Example 13.11

Pressure, p (psia)	Gas Compressibility factor, z	Gas Viscosity, μ_g (cp)
500	0.9380	0.01200
1,000	0.8800	0.01300
1,500	0.8230	0.01413
2,000	0.7630	0.01600
2,500	0.7018	0.01859
3,000	0.6843	0.02086
3,500	0.6818	0.02285

Solution

Step 1: Using gas properties in Table 13.12, calculate pseudo-pressures for the reservoir pressure range using Eq. (13.70). The results are shown in Table 13.13.

Step 2: Calculate pseudo-pressures for the two flow tests using Table 13.13. The results are shown in Table 13.14.

Step 3: For each flow rate, plot p_{pwf} vs. log t. The plot for the flow rate at 2000 Mcf/D is shown in Figure 13.21, and the plot for the 4000 Mcf/D is shown in Figure 13.22.

Step 4: Determine slope, m, of the MTR straight line for each plot. Calculate gas permeability at each flow rate.

From Figure 13.21, $m = 40 \times 10^6$ psia^2/cp-cycle for Flow #1 at 2000 Mcf/D.

From Figure 13.22, $m = 87 \times 10^6$ psia^2/cp-cycle for Flow #2 at 4000 Mcf/D.

For Flow #1, using Eq. (13.86):

$$k = \frac{1637 q_g T}{mh} = \frac{1637 \times 2000 \times 581}{40 \times 10^6 \times 15} = 3.17 \text{ md.}$$

For Flow #2, using Eq. (13.86):

$$k = \frac{1637 q_g T}{mh} = \frac{1637 \times 4000 \times 581}{87 \times 10^6 \times 15} = 2.91 \text{ md.}$$

Step 5: Calculate apparent skin factor, s'.

For Flow #1, applying Eq. (13.87),

$$s' = 1.151\left[\frac{p_{pi} - p_{p,1hr}}{m} - \log\left(\frac{k}{\phi(\mu c_t)_i r_w^2}\right) + 3.23\right]$$

$$= 1.151\left[\frac{(721.4 - 520.0)\times 10^6}{40 \times 10^6} - \log\left(\frac{3.17}{0.20 \times 0.02086 \times 4.5 \times 10^{-4} \times (0.50)^2}\right) + 3.23\right]$$

$$= 1.6522$$

For Flow #2, applying Eq. (13.87),

$$s' = 1.151\left[\frac{p_{pi} - p_{p,1hr}}{m} - \log\left(\frac{k}{\phi(\mu c_t)_i r_w^2}\right) + 3.23\right]$$

$$= 1.151\left[\frac{(721.4 - 80.0)\times 10^6}{87 \times 10^6} - \log\left(\frac{2.91}{0.20 \times 0.02086 \times 4.5 \times 10^{-4} \times (0.50)^2}\right) + 3.23\right]$$

$$= 4.3853$$

Step 6: Determine skin factor, s, and non-Darcy coefficient, D.

Using Eq. (13.73), the equations for the two flow rates are:

Flow#1: $s + q_1 D = 1.6522$ or $s + 2000D = 1.6522$

Flow#2: $s + q_2 D = 4.3853$ or $s + 4000D = 4.3853$

Solving the two equations simultaneously gives: $s = -1.0809$, and $D = 1.37 \times 10^{-3}$ $(\text{Mcf/D})^{-1}$

From the results of this test, it appears that the well is not damaged and all pressure loss near the wellbore is caused by turbulent or non-Darcy flow.

TABLE 13.13 Calculation of Real Gas Pseudo-Pressure in Example 13.11

Pressure, p (psia)	Gas Compressibility factor, z	Gas Viscosity, μ_g (cp)	$2p/(\mu z)$	Δp	p_p
500	0.9380	0.01200	88,842	500	22,210,377
1,000	0.8800	0.01300	174,825	500	88,127,047
1,500	0.8230	0.01413	257,976	500	196,327,334
2,000	0.7630	0.01600	327,654	500	342,734,826
2,500	0.7018	0.01859	383,246	500	520,459,742
3,000	0.6843	0.02086	420,330	500	721,353,662
3,500	0.6818	0.02285	449,319	500	938,765,931

TABLE 13.14 | **Pseudo-Pressures for Flow Test Data for Example 13.11**

Time, *t* (hrs)	Flow #1 q_1 = 2000 Mcf/D		Flow #2 q_2 = 4000 Mcf/D	
	p_{wf} (psia)	p_p (psia)^2/cp	p_{wf} (psia)	p_p (psia)^2/cp
0.25	2,576	544,393,512	1,170	130,336,695
0.40	2,555	536,431,402	1,085	113,051,234
0.60	2,533	528,147,966	1,019	100,238,733
0.80	2,524	524,776,341	971	91,255,016
1.00	2,515	521,414,618	925	82,909,917
1.50	2,495	513,979,573	815	64,003,115
2.00	2,479	508,066,744	755	54,313,768
3.00	2,456	499,621,880	650	38,416,368
4.00	2,445	495,605,892	570	27,208,683
5.00	2,434	491,604,696	495	17,412,033
6.00	2,425	488,341,992	435	10,069,810
7.00	2,416	485,089,190	385	4,287,467

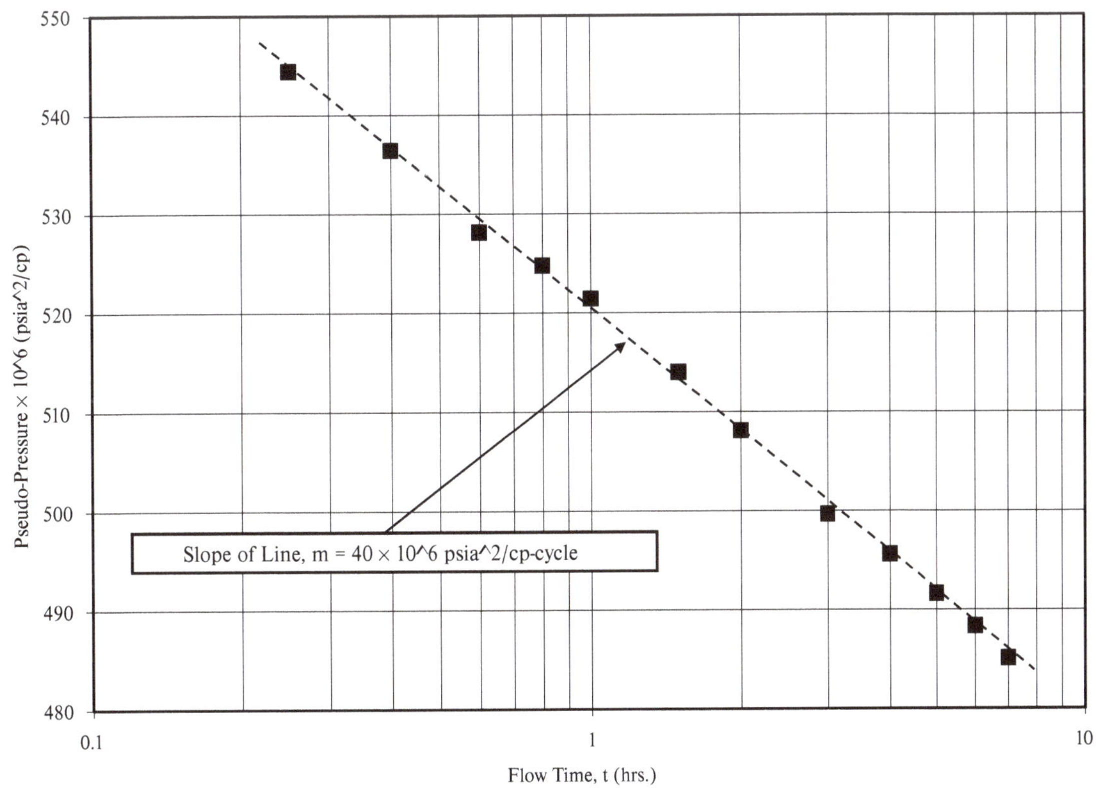

▲ **FIGURE 13.21** **Plot of p_{pwf} vs. *t* for Flow #1 (2000 Mcf/D) in Example 13.11**

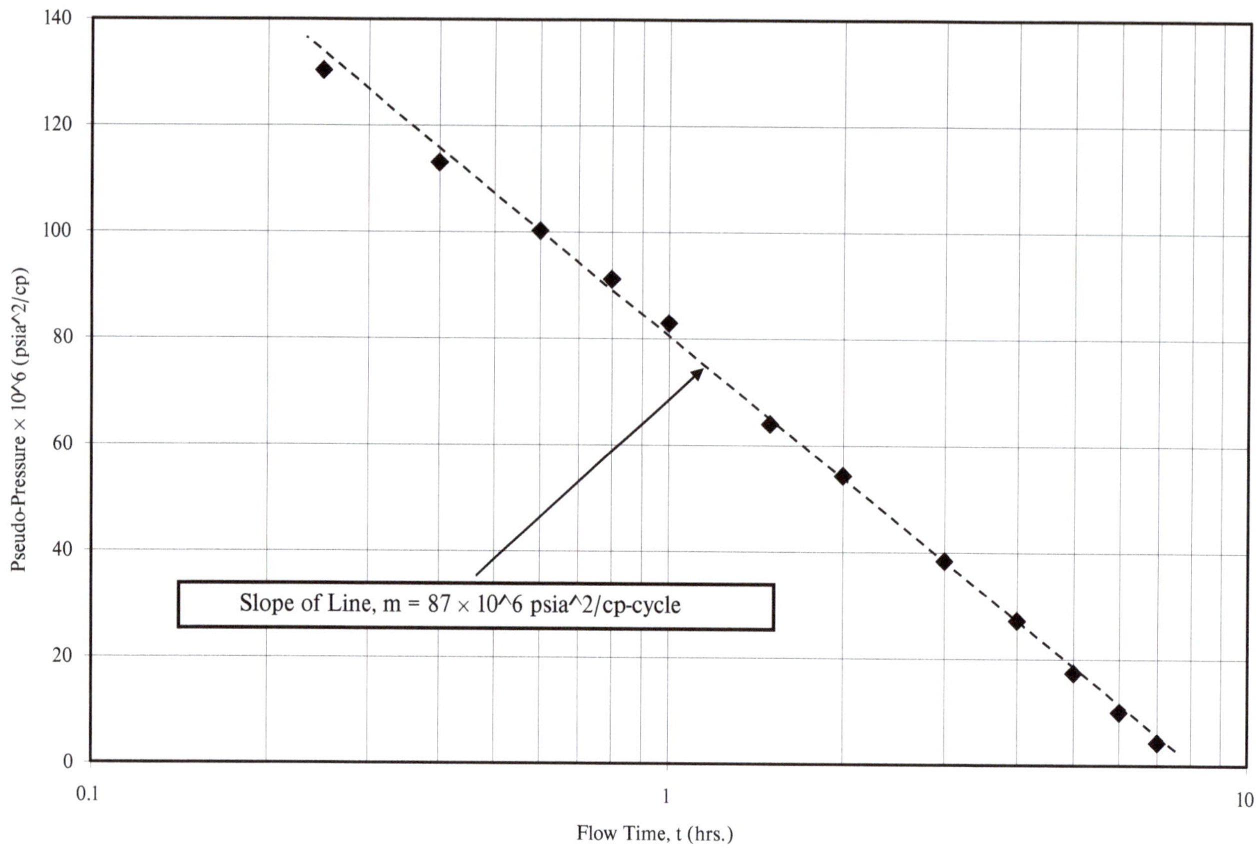

▲ **FIGURE 13.22 Plot of p_{pwf} vs. t for Flow #2 (4000 Mcf/D) in Example 13.11**

End of Example 13.11

13.4.2.7 Buildup Tests for Gas Wells with Constant Production Rate before Shut-in

The equations for analysis of buildup tests in gas wells with constant production rate before shut-in are similar in form to the equations developed previously for oil wells. The equations for buildup analyses in gas wells are developed initially in terms of pseudo-pressure. Other forms of the equations in terms of normalized pseudo-pressure, pressure, and pressure-squared are presented later.

From Eq. (13.74), the gas flow equation in terms of pseudo-pressure can be rearranged as:

$$p_{pi} - p_{pwf} = \frac{1637 q_g T}{kh}\left[\log\frac{kt}{\phi(\mu c_t)_i r_w^2} - 3.23 + 0.8691 s'\right] \tag{13.107}$$

As shown in Figure 13.6, if a gas well was produced at constant rate, q_g, for time period, t_p, before it was shut in for the duration designated as Δt, the total pressure drop that occurred in the wellbore can be represented by application of the superposition principle in time as:

$$p_{pi} - p_{pws} = \Delta p_{p@q_g(t_p+\Delta t)} + \Delta p_{p@-q_g(\Delta t)} \tag{13.108}$$

In Eq. (13.108), p_{pws} is the shut-in pseudo-pressure measured over time Δt.

Substituting Eq. (13.107) into Eq. (13.108) yields:

$$p_{pi} - p_{pws} = \frac{1637 q_g T}{kh}\left[\log\frac{k(t_p+\Delta t)}{\phi(\mu c_t)_i r_w^2} - 3.23 + 0.8691 s'\right] + \frac{1637(-q_g)T}{kh}\left[\log\frac{k(\Delta t)}{\phi(\mu c_t)_i r_w^2} - 3.23 + 0.8691 s'\right] \tag{13.109}$$

Grouping and eliminating similar terms in Eq. (13.109) and rearranging gives:

$$p_{pws} = p_{pi} - \frac{1637 q_g T}{kh}\left[\log\frac{(t_p+\Delta t)}{\Delta t}\right] \tag{13.110}$$

Eq. (13.110) is in the form of an equation for a straight line, $y = mx + b$, where:

$$y \equiv p_{pws} \tag{13.111}$$

$$x \equiv \log \frac{(t_p + \Delta t)}{\Delta t} \quad (13.112)$$

$$m = \left| \frac{1637 q_g T}{kh} \right| \quad (13.113)$$

$$b = p_{pi} \quad (13.114)$$

As shown by the above equations, a plot of p_{pws} vs. $\log \frac{t_p + \Delta t}{\Delta t}$ yields a straight line with slope, m, and intercept, p_{pi}, at infinite shut-in time.

The equation for calculating the apparent skin factor, s', for the well can be derived from Eq. (13.107) as:

$$s' = 1.151 \left[\frac{p_{p,1\text{hr}} - p_{pwf}}{m} - \log \left(\frac{k}{\phi (\mu c_t)_i r_w^2} \right) + 3.23 \right] \quad (13.115)$$

The analysis of the buildup test for a gas well is identical to that of an oil well by using the preceding equations and applying the procedures illustrated in Example 13.7.

The equations for analyses of buildup test in gas wells can be developed in terms of normalized pseudo-pressures. Starting with Eq. (13.79), the gas flow equation in terms of normalized pseudo-pressure can be rearranged as:

$$p_{pni} - p_{pnwf} = \frac{162.6 q_g \bar{B} \bar{\mu}}{kh} \times \left[\log \frac{kt}{\phi \bar{\mu} \bar{c}_t r_w^2} - 3.23 + 0.8691 s' \right] \quad (13.116)$$

Applying the same procedures used previously for derivation of buildup equations in terms of pseudo-pressure, it can be shown that the corresponding equation in terms of normalized pseudo-pressure is:

$$p_{pnws} = p_{pni} - \frac{162.6 q_g \bar{B} \bar{\mu}}{kh} \left[\log \frac{(t_p + \Delta t)}{\Delta t} \right] \quad (13.117)$$

A plot of p_{pnws} vs. $\log \left(\frac{t_p + \Delta t}{\Delta t} \right)$ yields a straight line with slope, m, and intercept, p_{pni}, at infinite shut-in time. The slope of the straight line, m, is represented as:

$$m = \left| \frac{162.6 q_g \bar{B} \bar{\mu}}{kh} \right| \quad (13.118)$$

The apparent skin factor, s', of the well is calculated from the expression:

$$s' = 1.151 \left[\frac{p_{pn,1\text{hr}} - p_{pnwf}}{m} - \log \left(\frac{k}{\phi (\mu c_t)_i r_w^2} \right) + 3.23 \right] \quad (13.119)$$

Analysis of buildup data from gas wells in terms of pseudo-pressure and normalized pseudo-pressure have been presented by plotting, p_{pws} or p_{pnws} vs. $\log \left(\frac{t_p + \Delta t}{\Delta t} \right)$. The ratio, $(t_p + \Delta t)/\Delta t$, is the standard Horner time ratio. Wattenbarger and Ramey[29] showed that buildup data on gas wells can be analyzed accurately by plotting p_{pws} or p_{pnws} vs. $\log \left(\frac{t_p + \Delta t}{\Delta t} \right)$, if production rate before shut-in is low. Production rate before shut-in is considered low if the dimensionless flow rate, $q_D \simeq 0.01$. If the production rate before shut-in is high, the semi-log slope, m, based on the standard Horner time ratio could be approximately ten percent higher than the "correct" slope. Production rate before shut-in is considered high if the dimensionless flow rate, $q_D \simeq 0.1$. According to Wattenbarger and Ramey,[29] dimensionless flow rate is defined as:

$$q_D = \frac{q_g p_{sc} T}{1.987 \times 10^{-5} kh T_{sc} p_{pi}} \quad (13.120)$$

If p_{sc} = 14.7 psia and T_{sc} = 520°R (T_{sc} = 60°F), Eq. (13.120) becomes:

$$q_D = \frac{1422.7 q_g T}{khp_{pi}} \quad (13.121)$$

Reynolds et al.[28] recommended application of two types of Horner time ratios that could be used to achieve accurate analysis of gas well buildup data. The first ratio is based on normalized shut-in time and represented as:

$$\text{Horner Time Ratio} = \frac{(t_p + \Delta t)/(\mu c_t)_i}{\Delta t/(\mu c_t)} \quad (13.122)$$

In Eq. (13.122), $(\mu c_t)_i$ is evaluated at initial pressure, p_i, or average pressure, $\bar{p}$. The term in the denominator, (μc_t), is evaluated at the shut-in pressure, p_{ws}, that is measured at shut-in time, Δt. The second ratio is based on shut-in pseudo-time and expressed as:

$$\text{Horner Time Ratio} = \frac{(t_p + \Delta t)/(\mu c_t)_i}{\Delta t_A} \quad (13.123)$$

Eqs. (13.122) and (13.123) yield almost identical results, although it is easier in terms of computation to apply Eq. (13.122).

Other equations for the analyses of gas well buildup data can be expressed in terms of pressure or pressure-squared from Eqs. (13.82) and (13.85), respectively. In terms of pressure, a plot of p_{ws} vs. $\log \left(\frac{t_p + \Delta t}{\Delta t} \right)$ yields a straight line with slope m. The slope of the straight line is represented as:

$$m = \left| \frac{162.6 q_g \overline{B} \overline{\mu}}{kh} \right| \quad (13.124)$$

The apparent skin factor, s', is determined from the equation:

$$s' = 1.151 \left[\frac{p_{1\,hr} - p_{wf}}{m} - \log \left(\frac{k}{\phi (\mu c_t)_i r_w^2} \right) + 3.23 \right] \quad (13.125)$$

In terms of pressure-squared, a plot of p_{ws}^2 vs. log $\left(\frac{t_p + \Delta t}{\Delta t} \right)$ gives a straight line with slope, m. The slope of the straight line is expressed as:

$$m = \left| \frac{1637 q_g T \overline{\mu z}}{kh} \right| \quad (13.126)$$

The apparent skin factor, s', for the well is calculated from the equation:

$$s' = 1.151 \left[\frac{p_{1\,hr}^2 - p_{wf}^2}{m} - \log \left(\frac{k}{\phi (\mu c_t)_i r_w^2} \right) + 3.23 \right] \quad (13.127)$$

Analyses of gas well buildup data using the pressure, or the pressure-squared technique is not recommended. Studies performed by Al-Hussainy and Ramey,[25] Meunier et al.,[26] Aziz et al.,[27] Reynolds et al.,[28] and Wattenbarger and Ramey[29] showed that use of the pseudo-pressure or normalized pseudo-pressure techniques yield more accurate analyses of gas buildup data in comparison to either pressure or pressure-squared methods. The application of pseudo-pressure or normalized pseudo-pressure is the practice recommended for analyses of gas well buildup data, irrespective of the gas flow rate before shut-in.

13.4.2.8 Horner's Time Approximation

The equations developed previously for the analysis of buildup tests in gas wells assumed that the wells were produced at constant rates before shut-in. As was discussed previously for oil wells, it is even more difficult to maintain the production rate of a gas well constant over extended time intervals. Following the approach developed for oil wells by Horner,[2] an approximation of total producing time for gas wells, also termed Horner's pseudo-producing time, t_{pH}, is defined as:

$$t_{pH} = \frac{24 G_p}{q_{g,last}} \quad (13.128)$$

In Eq. (13.128), G_p = cumulative gas production before shut-in, Mscf; and $q_{g,last}$ = gas production rate just before shut-in, Mcf/D. In practice, q_{last}, should be maintained constant and stabilized as long as possible before starting the buildup phase of the test. Note that by definition, Horner's pseudo-producing time, t_{pH}, is different from t_p used in previous equations such as Eq. (13.110). This t_p assumes a constant producing rate over the entire history of the well. However, t_{pH} can be used to replace t_p and q_{last} in place of q in Eq. (13.110) to yield:

$$p_{pws} = p_{pi} - \frac{1637 q_{g,last} T}{kh} \left[\log \frac{(t_{pH} + \Delta t)}{\Delta t} \right] \quad (13.129)$$

A plot of p_{pws} vs. $\log \frac{t_{pH} + \Delta t}{\Delta t}$ gives a straight line with slope, m, and intercept equal to, p_{pi}, at infinite shut-in time. This is the same as the Horner plot. The slope of the Horner plot, m, is given as:

$$m = \left| \frac{1637 q_{g,last} T}{kh} \right| \quad (13.130)$$

All other equations derived previously for the ideal Horner plot can be modified accordingly as shown for the pseudo-pressure case with the Horner pseudo-producing time.

EXAMPLE 13.12 Calculation of Permeability, Reservoir Pressure, and Skin Factor from Gas Well Buildup Tests Using Horner Pseudo-Producing Time Approximation

Problem

A buildup test was conducted on a discovery well in a dry gas reservoir. The production history of the well before the test is shown in Table 13.15. The data taken during the buildup test are shown in Table 13.16 as shut-in pressures versus time. The gas PVT data of the reservoir are the same as in Example 13.11. Other reservoir data are presented as follows:

Reservoir temperature, T	121°F
Formation thickness, h	30 ft
Formation porosity, ϕ	0.20
Total compressibility, c_t	4.5×10^{-4} psi^{-1}
Wellbore radius, r_w	0.50 ft

TABLE 13.15 Production History of Gas Well in Example 13.12

Flow Time (days)	Production Rate (Mcf/D)
0.5	3500
0.5	0
0.5	4500
0.5	0
0.5	9600
0.5	0
3.0	9000

TABLE 13.16 Buildup Data for Example 13.12

Δt (hrs)	p_{ws} (psia)
0	2130
0.25	2540
0.5	2580
1	2630
2	2680
4	2740
8	2798
12	2830
24	2880
36	2909
48	2925
72	2948
96	2960

Solution

Step 1: Calculate Horner's pseudo-time, t_{pH}.

Using Eq. (13.128) and production history in Table 13.15:

$$t_{pH} = \frac{24G_p}{q_{g,last}}$$

$$= \frac{\begin{pmatrix} 3500\times0.5+0\times0.5+4500 \\ \times0.5+0\times0.5+9600\times0.5 \\ +0\times0.5+9000\times3 \end{pmatrix}\times24}{9000}$$

$$= 95.47 \text{ hrs.}$$

Step 2: Plot p_{pws} vs. $\log\left(\frac{t_{pH}+\Delta t}{\Delta t}\right)$ from data in Table 13.17. Determine slope of MTR line.

From Figure 13.23, slope $m = 85.72 \times 10^6$ psia^2/cp-cycle.

Step 3: Calculate formation permeability, k, and apparent skin factor, s'.

Using Eq. (13.130), formation permeability, k, is:

$$k = \left|\frac{1637 q_{g,last} T}{mh}\right|$$

$$= \frac{1637\times9000\times581}{85.72\times10^6\times30}$$

$$= 3.33 \text{ md.}$$

From Eq. (13.115), apparent skin factor, s', is calculated as:

$$s' = 1.151\left[\frac{p_{p,1\text{hr}} - p_{pwf}}{m} - \log\left(\frac{k}{\phi(\mu c_t)_i r_w^2}\right) + 3.23\right]$$

$$= 1.151\left[\frac{(559.52-386.88)\times10^6}{85.72\times10^6}\right.$$

$$\left.-\log\left(\frac{3.33}{0.20\times0.02086\times4.5\times10^{-4}\times(0.5)^2}\right)+3.23\right]$$

$$= -1.85$$

Step 4: Calculate initial reservoir pressure, p_i.

From Figure 13.23, by extrapolation of the MTR straight line to Horner time ratio = 1, the intercept is, p_{pi} = 718 × 10^6 (psia)^2/cp. This translates to initial reservoir pressure, p_i = 2993 psia.

TABLE 13.17 Plot Data for Figure 13.23 of Example 13.12

$\frac{(t_{pH}+\Delta t)}{\Delta t}$	$p_{pws} \times 10^6$ (psia)^2/cp
382.88	530.78
191.94	545.92
96.47	565.12
48.74	584.62
24.87	608.43
12.93	631.86
8.96	644.97
4.98	665.69
3.65	677.86
2.99	684.61
2.33	694.37
1.99	699.49

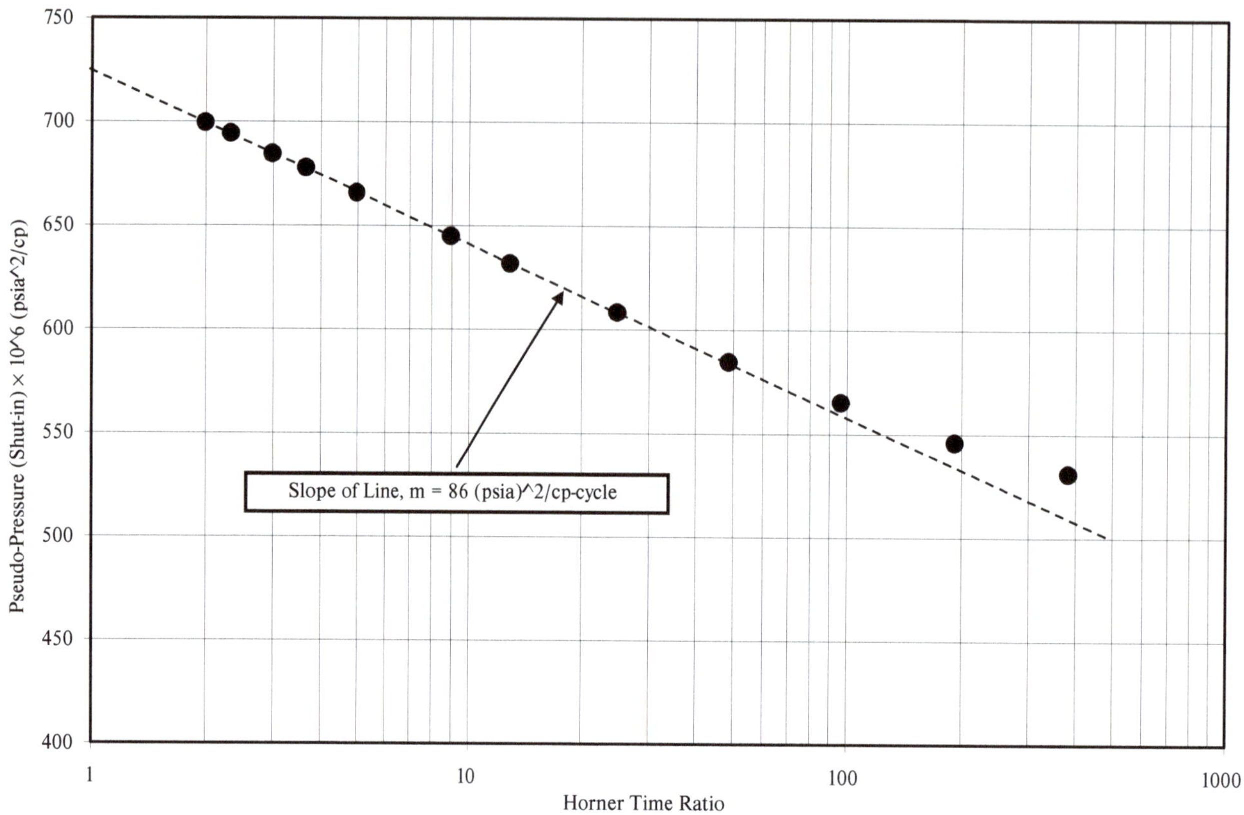

▲ **FIGURE 13.23** **Plot of p_{pws} vs. $\{(t_{pH} + \Delta t)/\Delta t\}$ for Example 13.12**

Alternative Method: Using Normalized Horner Time Ratio.

Step 1: Same as in previous method.

Step 2: Plot p_{pws} vs. $\log \dfrac{(t_p + \Delta t)/(\mu c_t)_i}{\Delta t/(\mu c_t)}$ from data in Table 13.18. Determine slope of MTR line.

From Figure 13.24, slope, m = 82.43 × 10^6 psia^2/cp-cycle.

Step 3: Calculate formation permeability, k, and apparent skin factor, s'.

Using Eq. (13.130), formation permeability, k, is:

$$k = \left|\frac{1637 q_{g,last} T}{mh}\right| = \frac{1637 \times 9000 \times 581}{82.43 \times 10^6 \times 30}$$

$$= 3.46 \text{ md.}$$

From Eq. (13.115), apparent skin factor, s', is calculated as:

$$s' = 1.151\left[\frac{p_{p,1hr} - p_{pwf}}{m} - \log\left(\frac{k}{\phi(\mu c_t)_i r_w^2}\right) + 3.23\right]$$

$$= 1.151\left[\frac{(559.46 - 386.88) \times 10^6}{82.43 \times 10^6} - \log\left(\frac{3.46}{0.20 \times 0.02086 \times 4.5 \times 10^{-4} \times (0.5)^2}\right) + 3.23\right]$$

$$= -1.78$$

Step 4: Calculate initial reservoir pressure, p_i

From Figure 13.24, by extrapolation of the MTR straight line to Horner time ratio = 1, the intercept is, p_{pi} = 718 × 10^6 (psia)^2/cp. This translates to initial reservoir pressure, p_i = 2993 psia.

For this example, using the standard Horner time ratio or the normalized Horner time ratio gave almost identical results. However, the normalized Horner time ratio was shown by Reynolds et al.[28] to give more accurate results. The application of normalized Horner time ratio based on Eq. (13.122) or Eq. (13.123) is recommended.

TABLE 13.18 Plot Data for Figure 13.24 of Example 13.12

$\frac{(t_{pH}+\Delta t)/(\mu c_t)_i}{\Delta t/(\mu c_t)_i}$	$p_{pws} \times 10^6$ (psia)^2/cp	$\frac{(t_{pH}+\Delta t)/(\mu c_t)_i}{\Delta t/(\mu c_t)_i}$	$p_{pws} \times 10^6$ (psia)^2/cp
405.86	530.78	9.12	644.97
202.18	545.92	5.04	665.69
100.85	565.12	3.69	677.86
50.59	584.62	3.01	684.61
25.61	608.43	2.34	694.37
13.22	631.86	2.00	699.49

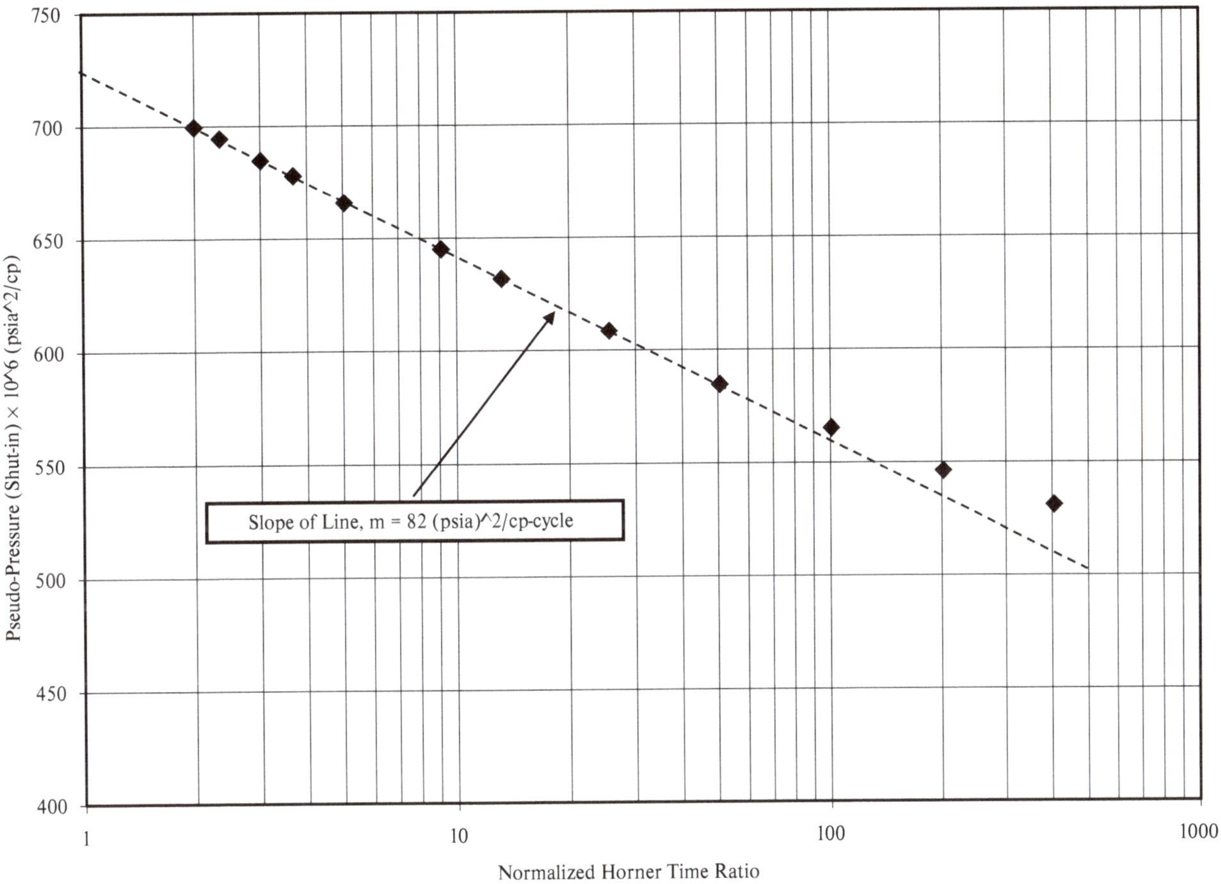

▲ **FIGURE 13.24** Plot of p_{pws} vs. $\{(t_{pH}+\Delta t)/(\mu c_t)_i\}/(\Delta t/\mu c_t)$ for Example 13.12

End of Example 13.12

13.5 Special Topics in Well Test Analyses

13.5.1 Multiphase Flow

All the equations developed previously in this chapter for analyses of drawdown and buildup tests have been based on the assumption that only a single phase fluid (oil or gas) was flowing in the reservoir. This could be the case if, for an oil reservoir, the pressure conditions during the test were above the bubble point pressure. This single phase condition also applies to a gas reservoir if dry gas is present in the reservoir or for a condensate gas reservoir where the flowing pressure conditions are outside the retrograde condensation region. In either situation described above for oil or gas reservoirs, connate water is assumed to be immobile in the reservoirs. When more than one fluid phase is flowing in the reservoir at the same time, then multiphase interactions occur between the fluids and the equations developed assuming single phase flow are rendered invalid and are no longer applicable. Multiphase flow occurs in reservoirs under following conditions:

1. Solution gas drive reservoirs below bubble point pressure with gas and/or water saturations above critical saturations. That is, gas and/or water are mobile in the reservoir.
2. Oil reservoirs above bubble point with mobile water saturations.
3. Gas condensate reservoirs with condensate and/or water saturations above critical saturations. That is, condensate and/or water are mobile in the reservoir.

If a reservoir is under multiphase flow, the flow capacity of each fluid phase is reduced due to relative permeability effects since the flow of any fluid phase impedes the flow of the other fluid phases. As discussed in Chapter 3, the relative permeability of a fluid is defined as the ratio of its effective permeability to the absolute permeability of the porous medium. This can be expressed for any fluid as:

$$k_{ri} = \frac{k_i}{k} \tag{13.131}$$

In Eq. (13.131), k_{ri} = relative permeability of the fluid, i; k_i = effective permeability of the fluid, i; and k = absolute permeability of the porous medium. For instance, if the fluid is oil, the relative permeability of oil is given by, $k_{ro} = k_o/k$. As discussed in Chapter 3, relative permeability is a function of fluid saturation. Hence, effective permeability also depends on saturation. In a well test, the distribution of fluid saturations changes with time. Thus, effective permeability, which affects flow, also changes with time. Furthermore, mobility of a fluid in a porous medium is defined as the ratio of its effective permeability to its viscosity. In equation form, this is expressed as:

$$\lambda_i = \frac{k_i}{\mu_i} \tag{13.132}$$

In Eq. (13.132), λ_i = mobility of the fluid, i; and μ_i = viscosity of the fluid, i. If the fluid is oil, the mobility of oil is expressed as: $\lambda_o = k_o/\mu_o$. A fluid with high mobility flows faster than a fluid with less mobility. Since fluid mobilities depend on effective permeability which varies with saturation and time, it is important to consider multiphase flow in well test analyses.

Multiphase flow well test analysis is presented on the basis of an approximate method developed by Perrine[33] and later verified theoretically by Martin.[34] This approach uses the concepts of total mobility, λ_t, and total compressibility, c_t. Total mobility is defined as:

$$\lambda_t = \frac{k_o}{\mu_o} + \frac{k_w}{\mu_w} + \frac{k_g}{\mu_g} \tag{13.133}$$

Total compressibility is defined as:

$$c_t = c_o S_o + c_w S_w + c_g S_g + c_f \tag{13.134}$$

where

$$c_o = \left(-\frac{1}{B_o}\frac{dB_o}{dp} \right) + \frac{B_g}{1000 B_o}\frac{dR_s}{dp} \tag{13.135}$$

and

$$c_w = \left(-\frac{1}{B_w}\frac{dB_w}{dp} \right) + \frac{B_g}{1000 B_w}\frac{dR_{sw}}{dp} \tag{13.136}$$

In Eqs. (13.135) and (13.136), the units of B_g are RB/Mcf.

The total flow rate, q_t, under reservoir conditions is defined thus:

$$(qB)_t = q_o B_o + q_w B_w + \left(q_g - \frac{q_o R_s}{1000} - \frac{q_w R_{sw}}{1000} \right) B_g \tag{13.137}$$

Application of the Perrine's method to multiphase well test analyses makes the following assumptions:

1. Pressure gradients in the drainage area of the well are small.
2. Saturation gradients in the same area are small.
3. Saturation changes during the test are negligible.

Assuming the above conditions are applicable, well test analyses equations and methods developed earlier for single phase fluid systems are applicable to multiphase fluid systems. Thus, to analyze test data from a well producing two or three phases at the same time, prepare a plot of the test data as was described earlier for a single phase system. For instance, for a buildup test, prepare a plot of p_{pws} vs. $\log \frac{(t_p + \Delta t)/(\mu c_t)_i}{\Delta t/(\mu c_t)}$ as shown in Example 13.12. From the plot, determine the slope, m, of the MTR straight line. From the slope, calculate the effective permeabilities of each fluid phase as follows:

For oil:

$$k_o = \frac{162.6 q_o B_o \mu_o}{mh} \tag{13.138}$$

For water:

$$k_w = \frac{162.6 q_w B_w \mu_w}{mh} \tag{13.139}$$

For gas:

$$k_g = \frac{162.6\left[q_g - (q_o R_s/1000) - (q_w R_{sw}/1000) \right] B_g \mu_g}{mh} \tag{13.140}$$

The total mobility, λ_t, is calculated from:

$$\lambda_t = \frac{162.6 (qB)_t}{mh} \tag{13.141}$$

The absolute permeability, k, can be calculated from total mobility as:

$$k = \frac{\lambda_t}{\left[\frac{k_{ro}}{\mu_o} + \frac{k_{rw}}{\mu_w} + \frac{k_{rg}}{\mu_g}\right]} \tag{13.142}$$

The skin factor, *s*, is calculated in terms of total mobility from:

$$s = 1.151\left[\frac{p_{1hr} - p_{wf}}{m} - \log\frac{\lambda_t}{\phi c_t r_w^2} + 3.23\right] \tag{13.143}$$

Perrine's method for analysis of multiphase flow tests is simple and straight-forward but approximate. There are other methods that have been proposed by Raghavan[35,36] and Al-Khalifah et al.[37] that claim to be improvements over Perrine's method. These other methods can be useful depending on the state of the reservoir. However, Perrine's method is widely used in most applications of well test analysis.

13.5.2 Wellbore Storage Effects

The concept of wellbore storage was discussed earlier in this chapter as the unloading of wellbore fluids at the start of drawdown tests or the continued flow of reservoir fluids into the wellbore (called afterflow) after a well has been shut in at the start of buildup tests. The main effects of wellbore storage on well test data are the distortion of early time data and delay of the start of middle time region data. For wells with high wellbore storage coefficient, most of the MTR data could be lost due to wellbore storage effects. A critical step in the application of straightline methods in well test analysis is determination of the time at which wellbore storage effects have ended. This time marks the beginning of MTR data which in turn establishes the slope of the MTR straight line. Since this step is subject to the interpretation of the data by the analyst, it is reasonable to expect that different analysts will pick different points as the start of the MTR depending on the quality of the well test data being analyzed. This was a persistent problem in the application of straightline methods until the introduction of type curves in the analysis of well test data. In Chapter 14, the application of type curves as tools to assist in the identification of end of wellbore storage effects are presented. The joint application of straightline methods and type curves improved the interpretation of well test data.[9] It should be noted that in all the examples shown in this chapter, the start of MTR straight line appears to have been "magically" determined without further discussion on the procedures that could be used to select the end of wellbore storage. The procedures for determination of the end of wellbore storage are discussed in Chapter 14 in conjunction with the application of type curves for well test analysis.

13.5.3 Wellbore Phase Redistribution Effects

Another wellbore phenomenon which is closely tied to wellbore storage is wellbore phase redistribution effects. Wellbore phase redistribution effects occur in wells with multiphase flow of gas and liquid. When such a well is shut in for buildup tests, gravity effects cause the gas to rise toward the top of the wellbore and the liquid to fall toward the bottom of the wellbore.[38-40] Since the liquid phase is almost incompressible, and the gas has no room to expand in such a closed system, a high-pressure gas column develops at the top of the wellbore with a column of liquid below it. In some cases, the sum of the pressure due to the gas column and the pressure due to the liquid column will exceed the formation pressure at the sandface and cause liquid to flow back into the formation until equilibrium is eventually established. At early times, this phenomenon shows up in buildup data as an anomalous pressure hump as shown in Figure 13.25. In Figure 13.25, pressure curve A represents normal wellbore storage which affects only ETR portion of the buildup test. Pressure curve B represents a buildup test in which ETR and MTR are completely distorted by phase redistribution in the wellbore. If the pressure hump caused by phase redistribution is large, it will delay and sometime distort the MTR data completely. This makes it difficult to determine properties such as formation permeability and skin factor from the buildup data.

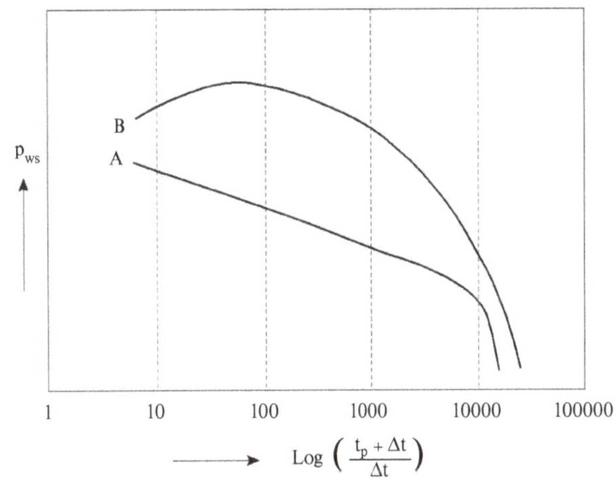

▲ **FIGURE 13.25 Pressure hump caused by phase redistribution in the wellbore**

13.5.4 Boundary Effects

The flow equations that form the bases of drawdown and buildup analyses assume an infinite-acting reservoir. This assumption does not apply in some tests. Depending on the size of the drainage area of the well, the presence of a boundary can influence the pressure response in the wellbore. The boundary can be a no-flow boundary (or closed boundary) created by a sealing fault or impermeable barriers. The boundary can also be a constant pressure boundary imposed by an aquifer, a large gas cap, or fluid injectors. These boundary effects generally terminate the MTR straight line and generate characteristic shapes of the pressure response curves. One of the major advances in well test analyses is the use of derivative type curves to identify specific types of boundary effects. These characteristic type curves are presented in Chapter 14.

13.5.5 Multilayered Reservoirs

It is necessary to restate that all equations developed previously in this chapter for the analyses of drawdown and buildup tests are based on the flow of a single phase fluid in a single layer with homogeneous properties. Most reservoirs consist of more than one layer, and in many cases the rock and fluid properties of the layers are very different. In this section, some observations on the analyses of well test data from multilayered reservoirs are presented. The discussions on multilayered reservoirs proceed by defining two main reservoir systems commonly used in describing these reservoirs. These are commingled reservoirs and crossflow reservoirs. Commingled reservoirs are multilayered reservoirs in which fluid flow between layers takes place only through the wellbore (Figure 13.26). Essentially, the vertical permeabilities across layer boundaries are zero for commingled reservoirs. Crossflow reservoirs are multilayered reservoirs in which fluid flow can occur between layers within the reservoir (Figure 13.27). In crossflow reservoirs, interlayer communication or fluid transfer between the layers exist inside the reservoir and also through the wellbore. Note that the term crossflow as used here is different from common industry usage of crossflow as fluid transfer between layers only through the wellbore. This is sometimes described as backflow.

The pressure response of commingled and crossflow reservoirs can be examined by comparing the transmissivities (kh/μ) and storativities $(\phi c_t h)$ of the composite layers. If the transmissivities and storativities of the composite layers are nearly the same, the reservoirs will exhibit homogeneous behavior. On the other hand, if these properties are remarkably different, the reservoirs will exhibit heterogeneous behavior. The derivative type curve from a test on a reservoir that shows homogeneous behavior is identical to that of single layer reservoir. For a reservoir that shows heterogeneous behavior, the derivative type curve has a characteristic shape. The application of derivative type curves as diagnostic plots for well test analysis of heterogeneous reservoirs are presented in Chapter 14.

Numerous studies[41-45] on commingled reservoirs concluded that conventional straightline methods can be used to determine thickness-averaged formation flow capacity, $\overline{kh}$, and equivalent total skin factor, s_t. The thickness-averaged formation permeability, $\bar{k}$, is defined as:

$$\bar{k} = \frac{\sum_{j=1}^{n} k_j h_j}{\sum_{j=1}^{n} h_j} \tag{13.144}$$

Tariq[46] showed that the equivalent total skin factor of well producing a commingled reservoir with different skin factor for each layer can be approximated by the following expression:

$$s_t = \sum_{j=1}^{n} \frac{k_j h_j s_j}{\overline{kh}} \tag{13.145}$$

Published studies[47-51] on crossflow reservoirs show that early-time well responses of such reservoirs are identical to that of comparably located commingled reservoirs. Russell and Prats[47] reported that a multilayered crossflow reservoir will ultimately behave as a single layer reservoir with equivalent pore volume and drainage area. Furthermore, the total flow capacity is equal to the sum of the flow capacities of the composite layers [same as Eq. (13.144) for commingled reservoirs]. The equivalent total skin factor for crossflow reservoirs can be estimated as reported by Prijambodo et al.[50] as:

$$s_t = \sum \frac{s_j q_j}{q_t} \tag{13.146}$$

In Eq. (13.146), s_j and q_j are the skin factor and flow rate for layer j respectively, and q_t is the total flow rate.

The important decision to note on the testing of multilayered reservoir is whether to measure the total flow rate of all the layers or the flow rate of each layer. If total flow rate is measured, then only average reservoir properties for the multilayer system can be determined from the well test. Testing individual layers will yield determination of individual layer properties such as layer permeability and skin factor. Many studies[52-54] have been published on methods to test and analyze data from individual layers in a multilayer system. These methods are generally expensive but could be justified if the results will lead to improved and accelerated depletion of multilayered reservoirs.

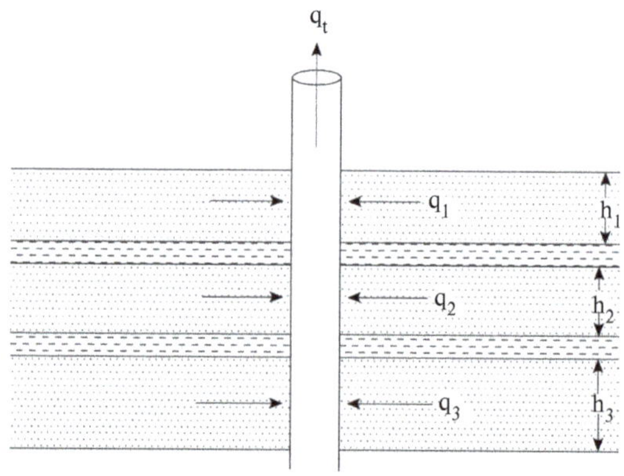

▲ **FIGURE 13.26** Multilayered commingled reservoirs

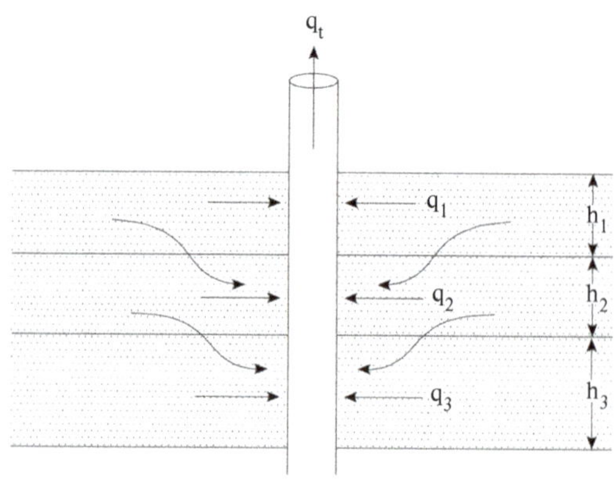

▲ **FIGURE 13.27** Multilayered crossflow reservoirs

Nomenclature

A_{wb}	area of wellbore, feet^2
B	formation volume factor, RB/STB
c	compressibility, psi^{-1}
c_f	formation compressibility, psi^{-1}
c_t	total compressibility, psi^{-1}
c_{wb}	compressibility of wellbore fluid, psi^{-1}
C	wellbore storage coefficient, bbl/psi
C_D	dimensionless wellbore storage coefficient
D	non-Darcy flow coefficient, D/Mcf
E	flow efficiency
f_{pf}	pressure function of the drainage area as defined by Matthews et al.[15]
G_p	cumulative gas production, Mcf
h	reservoir or formation thickness, feet
k	permeability, md
$\overline{k}$	thickness-averaged formation permeability, md
k_a	permeability of the altered zone, md
k_i	effective permeability of fluid, i
k_{ri}	relative permeability of fluid, i
L_f	length of fracture wing, feet
m	slope of MTR straight line
M_g	molecular weight of gas, lbm/lb-mole
N_p	cumulative oil production, STB
Δp	pressure change
Δp_s	pressure change due to skin, psia
p^*	extrapolated pressure in buildup test, psia
p	pressure, psia
$\overline{p}$	average pressure, psia
p_a	pressure at boundary of damaged zone, psia
p_b	low base pressure or reference pressure
p_p	real gas pseudo-pressure, psia^2/cp
p_{pn}	normalized gas pseudo-pressure, psia
p_{pnwf}	normalized flowing bottomhole-pseudo-pressure, psia
p_{pnws}	normalized shut-in bottomhole pseudo-pressure, psia
p'_{wf}	flowing bottomhole pressure of undamaged zone
p_{wf}	well bottom-hole flowing pressure
p_{pws}	shut-in bottomhole pseudo-pressure, psia^2/cp
p_{ws}	shut-in bottomhole pressure, psia
q	volume flow rate, STB/D
q_D	dimensionless flow rate
q_{last}	production rate before shut-in, STB/D
$q_{g,last}$	gas production rate before shut-in, Mcf/D
r	radius or distance, feet
r_a	radius of damaged zone, feet
r_i	radius of investigation, feet
r_w	wellbore radius, feet
r_{wa}	effective wellbore radius, feet
R_s	solution gas-oil ratio, scf/STB
R_{sw}	solution gas-water ratio, scf/STB
s	skin factor
s'	apparent skin factor
S	fluid saturation
Δt	shut-in time in buildup test, hours
t	time, hours
t_A	real gas pseudo-time, hr-psi/cp
t_{An}	normalized real gas pseudo-time, hr
t_p	total producing time before shut-in, hours
t_{pH}	Horner's pseudo-producing time, hours
T	temperature, °F or °R
V	volume, feet^3 or barrels
V_{wb}	volume of wellbore, barrels
z	gas compressibility factor
γ_g	gas gravity, air = 1
η	formation diffusivity, ft^2/D
ϕ	porosity
μ	fluid viscosity, cp
ρ_{wb}	density of fluid in wellbore, lbm/ft^3
β	turbulence parameter
λ	mobility, md/cp

Subscripts

f	formation	sc	standard conditions
g	gas	t	total
i	initial conditions	w	water
o	oil		

Abbreviations

BHP	Bottomhole Pressure	IARF	Infinite Acting Radial Flow
DR	Damage Ratio	LTR	Late Time Region
ETR	Early Time Region	MTR	Middle Time Region
FBHP	Flowing Bottomhole Pressure	OOIP	Original-oil-in-place
FVF	Formation Volume Factor	SBHP	Shut-in Bottomhole Pressure

REFERENCES

1. Miller, C.C., Dyes, A.B., and Hutchinson, C.A.: "The Estimation of Permeability and Reservoir Pressure From Bottom Hole Pressure Build-up Characteristics," *Trans.* AIME (1950) 189, 91-104.
2. Horner, D.R.: "Pressure Buildup in Wells," Pressure Analysis Methods, Reprint Series, SPE, Richardson, Texas, (1967) 9, 25-43.
3. Ramey, H.J.: "Short-Time Well Test Data Interpretation in the Presence of Skin Effect and Wellbore Storage," *JPT* (January 1970) 97-104.
4. Gringarten, A.C., Bourdet, D.P., Landel, P.A., and Kniazeff, V.J.: "A Comparison Between Different Skin and Wellbore Storage Type-Curves for Early-Time Transient Analysis," Paper SPE 8205 presented at the SPE Annual Technical Conference and Exhibition, Las Vegas, Nevada, (September 23-26, 1979).
5. Bourdet, D., Whittle, T.M., Douglas, A.A., and Pirard, Y.M.: "A New Set of Type Curves Simplifies Well Test Analysis," *World Oil* (May 1983) 95-106.
6. Bourdet, D., Ayoub, J.A., Whittle, T.M., Pirard, Y.M., and Kniazeff, V.: "Interpreting Well Tests in Fractured Reservoirs," *World Oil* (October 1983) 77-87.
7. Bourdet, D., Ayoub, J.A., and Pirard, Y.M.: "Use of Pressure Derivative in Well-Test Interpretation," *SPEFE* (June 1989) 293-302.
8. von Schroeter, T., Hollaender, F., and Gringarten, A.C.: "Deconvolution of Well-Test Data as a Nonlinear Total Least-Squares Problem," *SPEJ* (December 2004) 375-390.
9. Gringarten, A.C.: "From Straight Lines to Deconvolution: The Evolution of the State of the Art in Well Test Analysis," *SPEREE* (February 2008) 41-62.
10. van Everdingen, A.F.: "The Skin Effect and its Influence on the Productive Capacity of a Well," *Trans.* AIME (1953) 198, 171-176.
11. Horne, R.N.: *Modern Well Test Analysis–A Computer-Aided Approach*, 2nd Ed., Petroway, Inc., California (2008).
12. Tiab, D., Ispas, I.N., Mongi, A., and Berkat, A.: "Interpretation of Multirate Tests by the Pressure Derivative–1. Oil Reservoirs," Paper SPE 53935 presented at the 1999 SPE Latin American and Caribbean Petroleum Engineering Conference, Caracas, Venezuela, (April 21-23, 1999).
13. Russell, D.G.: "Determination of Formation Characteristics from Two-Rate Flow Tests," *JPT* (December 1963) 1347-1355.
14. Odeh, A.S., and Jones, L.G.: "Two-Rate Flow Test, Variable-Rate Case–Application to Gas-Lift and Pumping Wells," *JPT* (January 1974) 93-99.
15. Matthews, C.S., Brons, F., and Hazebroek, P.: A Method for Determination of Average Pressure in a Bounded Reservoir," *Trans.* AIME (1954) 201, 182-191.
16. Muskat, M.: "Use of Data on the Build-up of Bottom-Hole Pressures," *Trans.* AIME (1937) 123, 44-48.
17. Larson, V.C.: "Understanding the Muskat Method of Analyzing Pressure Buildup Curves," *J. Cdn. Pet. Tech.* (Fall 1963) 2, 136.
18. Crump, J.G., and Hite, R.H.: "A New Method for Estimating Average Reservoir Pressure: The Muskat Plot Revisited," *SPEREE* (April 2008) 298-306.
19. Dietz, D.N.: "Determination of Average Reservoir Pressure from Buildup Surveys," *JPT* (1965) 955-959.
20. Mead, H.N.: "A Practical Approach to Transient Pressure Behavior," Paper SPE 9901 presented at the SPE California Regional Meeting, Bakersfield, California, March 25-27, 1981.

21. Kabir, C.S., and Hasan, A.R.: "Estimating Average Reservoir Pressure Using the Hyperbola Approach: New Algorithm and Field Examples," Paper SPE 36255 presented at the SPE Abu Dhabi International Petroleum Exhibition and Conference, Abu Dhabi, United Arab Emirates, (October 13-16, 1996).
22. Chacon, A., Djebrouni, A., and Tiab, D.: "Determining the Average Reservoir Pressure from Vertical and Horizontal Well Test Analysis Using the Tiab's Direct Synthesis Technique," Paper SPE 88619 presented at the SPE Asia Pacific Oil and Gas Conference and Exhibition, Perth, Australia, (October 18-20, 2004).
23. Al-Hussainy, R., Ramey, H.J., Jr., and Crawford, P.R.: "The Flow of Real Gases Through Porous Media," *JPT* (May 1966), 624-636.
24. Agarwal, Ram G.: "Real Gas Pseudo-Time–A New Function for Pressure Buildup Analysis of MHF Gas Wells," Paper SPE 8279 presented at the 54th Annual Fall Technical Conference and Exhibition of SPE, Las Vegas, Nevada, (September 23-26, 1979).
25. Al-Hussainy, R., and Ramey, H.J., Jr.: "Application of Real Gas Flow Theory to Well Testing and Deliverability Forecasting," *JPT* (May 1966) 637-642.
26. Meunier, D.F., Kabir, C.S., and Wittmann, M.J.: "Gas Well Test Analysis: Use of Normalized Pseudovariables" *SPEFE* (December 1987) 629-636.
27. Aziz, K., Mattar, L., Ko, S., and Brar, G.S.: "Use of Pressure, Pressure-Squared or Pseudo-Pressure in the Analysis of Transient Pressure Drawdown Data from Gas Wells," *J. Cdn. Pet. Tech.* (April-June 1976) 58-65.
28. Reynolds, A.C., Bratvold, R.B., and Ding, W.: "Semilog Analysis of Gas Well Drawdown and Buildup Data," *SPEFE* (December 1987) 657-670.
29. Wattenbarger, R.A., and Ramey, H.J.: "Gas Well Testing with Turbulence, Damage and Wellbore Storage," *JPT* (August 1968) 877-887.
30. Fligelman, H., Cinco-Ley, H., Ramey Jr., H.J., Braester, C., and Couri, F.: "Pressure-Drawdown Test Analysis of a Gas Well-Application of New Correlations," *SPEFE* (September 1989) 406-412.
31. Firoozabadi, A., and Katz, D.L.: "An Analysis of High Velocity Gas Flow Through Porous Media," *JPT* (February 1979) 211-216.
32. Swift, G.W., and Kiel, O.G.: "The Prediction of Gas Well Performance Including the Effect of Non-Darcy Flow," *JPT* (July 1962) 791-798.
33. Perrine, R.L.: "Analysis of Pressure Buildup Curves," *Drill. and Prod. Prac.,* API (1956), 482-509.
34. Martin, J.C.: "Simplified Equations of Flow in Gas Drive Reservoirs and the Theoretical Foundation of Multiphase Pressure Buildup Analyses," *Trans.* AIME (1959) 216, 321-323.
35. Raghavan, R.: "Well Test Analysis: Wells Producing by Solution Gas Drive," *SPEJ* (August 1976) 196-208.
36. Raghavan, R.: "Well Test Analysis for Multiphase Flow," *SPEFE* (December 1989) 585-594.
37. Al-Khalifah, A-J.A., Aziz, K., and Horne, R.N.: "A New Approach to Multiphase Well Test Analysis," paper SPE 16473 presented at the 62nd Annual SPE Technical Conference and Exhibition, Dallas, Texas, (September 27-30, 1987).
38. Stegemeier, G.L., and Mathews, C.S.: "A Study of Anomalous Pressure Build-up Behavior," *Trans.* AIME (1958) 213, 44-50.
39. Fair, W.B.: "Pressure Buildup Analysis with Wellbore Phase Redistribution," *SPEJ* (April 1981) 259-270.
40. Hegeman, P.S., Hallford, D.L., and Joseph, J.A.: "Well Test Analysis with Changing Wellbore Storage," *SPEFE* (September 1993) 201-207.
41. Lefkovits, H.C., Hazebroek, P., Allen, E.E., and Matthews, C.S.: "A Study of the Behavior of Bounded Reservoirs Composed of Stratified Layers," *SPEJ* (March 1961) 43-58.
42. Cobb, W.M., Ramey, H.J. Jr., and Miller, F.G.: "Well-Test Analysis for Wells Producing Commingled Zones," *JPT* (January 1972) 27-37.
43. Raghavan, R., Topaloglu, H.N., Cobb, W.M., and Ramey, H.J. Jr.: "Well Test Analysis for Wells Producing From Two Commingled Zones of Unequal Thickness," *JPT* (September 1974) 1035-1043.
44. Earlougher, R.C. Jr., Kersch, K.M., and Kunzman, W.J.: "Some Characteristics of Pressure Buildup Behavior in Bounded Multiple-Layer Reservoirs Without Crossflow," *JPT* (October 1974) 1178-1186.
45. Raghavan, R.: "Behavior of Wells Completed in Multiple Producing Zones," *SPEFE* (June 1989) 219-230.
46. Tariq, S.M.: "A Study of the Behavior of Layered Reservoirs with Wellbore Storage and Skin Effect," PhD dissertation, Stanford University, Stanford, California, (September 1977).
47. Russell, D.G., and Prats, M.: "Performance of Layered Reservoirs with Crossflow-Single-Compressible-Fluid Case," *SPEJ* (March 1962) 53-67.
48. Katz, M.L., and Tek, M.R.: "A Theoretical Study of Pressure Distribution and Fluid Flux in Bounded Stratified Porous Systems with Crossflow," *SPEJ* (March 1962) 68-82.
49. Pendergrass, J.D., and Berry, V.J. Jr.: "Pressure Transient Performance of a Multilayered Reservoir with Crossflow," *SPEJ* (December 1962) 347-354.
50. Prijambodo, R., Raghavan, R., and Reynolds, A.C.: "Well Test Analysis for Wells Producing Layered Reservoirs with Crossflow," *SPEJ* (June 1985) 380-396.
51. Bourdet, D.: "Pressure Behavior of Layered Reservoirs with Crossflow," paper SPE 13628 presented at the SPE California Regional Meeting, Bakersfield, California, (March 27-29, 1985).
52. Dogru, A.H., and Seinfeld, J.H.: "Design of Well Tests to Determine the Properties of Stratified Reservoirs," paper SPE 7694 presented at the SPE Symposium on Reservoir Simulation, Denver, Colorado, (February 1-2, 1979).
53. Kuchuk, F., Karakas, M., and Ayestaran, L.: "Well Testing and Analysis Techniques for Layered Reservoirs," *SPEFE* (August 1986) 342-354.
54. Ehlig-Economides, C.A., and Joseph, J.: "A New Test for Determination of Individual Layer Properties in a Multilayered Reservoir," *SPEFE* (September 1987) 261-283.

ADDITIONAL REFERENCES

- Earlougher, R.C.: *Advances in Well Test Analysis*, SPE Monograph Volume 5, SPE, Richardson, Texas (1977).
- Lee, J., Rollins, J.B., and Spivey, J.P.: *Pressure Transient Testing*, SPE Textbook Series Vol. 9, SPE, Richardson, Texas (2003).

PROBLEMS

13-1. A drill stem test (DST) is planned for a discovery well drilled into the pre-salt formation of the Campos Basin, offshore Brazil. The duration of the main flow of the DST is scheduled for 52 hours. Rock and fluid properties of the formation are given as:

Net pay thickness, h	66.42 ft
Formation porosity, ϕ	0.08
Initial water saturation, s_{wi}	0.257
Oil viscosity, μ_o	0.56 cp
Oil FVF, B_O	1.56 RB/STB
Total compressibility, c_t	1.38E-05 psi^{-1}

Assuming radial drainage area for the well, calculate:

a. Connected original oil-in-place (OOIP) tested if the permeability of the formation is 100 md.

b. Connected OOIP tested if the permeability of the formation is 300 md.

c. Is the well likely to be commercial based on the volumes of OOIP calculated in a and b?

13-2. A reservoir under development in deepwater, Nigeria, has the following formation and fluid parameters:

Permeability, k	1340 md
Net pay thickness, h	210 ft
Formation porosity, ϕ	0.20
Initial water saturation, s_{wi}	0.23
Oil viscosity, μ_o	1.40 cp
Oil FVF, B_o	1.27 RB/STB
Total compressibility, c_t	1.58E-05 psi^{-1}
Initial reservoir pressure, p_i	4375 psia
Bubble point pressure, p_b	4229 psia

A DST is being designed for a development well in the reservoir. If the main objective of the DST is to prove-up a minimum 100 MMSTB of connected oil volume, what should be the minimum duration of the main flow period of the test? Assume a radial drainage area for the well.

13-3. A new well producing at a constant rate of 1000 BOPD was determined to be damaged by drilling mud filtrate. The damaged zone was estimated to be 1.5 feet thick with a reduced permeability of 30 md. Other parameters of the well are given as:

Permeability, k	225 md
Net pay thickness, h	45 ft
Oil viscosity, μ_o	1.25 cp
Oil FVF, B_o	1.54 RB/STB
Total compressibility, c_t	1.58E-05 psi^{-1}
Wellbore radius, r_w	0.50 ft
Average reservoir pressure, $\overline{p}$	2500 psia
Flowing bottomhole pressure, p_{wf}	1200 psia

Calculate the following parameters for the well:

a. Additional pressure drop due to the damage

b. Flow efficiency

c. Damage ratio

d. Skin factor

e. Effective wellbore radius

13-4. A well in a new reservoir was produced at a constant rate of 375 BOPD for 35 days. The recorded flowing bottomhole pressures of the well versus times are given in Table 13P.1. Other reservoir and fluid parameters are given as:

TABLE 13P.1 Drawdown Data for Problem 13-4

Time, t (hrs.)	FBHP, p_{wf} (psia)	Time, t (hrs)	FBHP, p_{wf} (psia)
0.00	4500	36.00	4253
0.05	4496	48.00	4239
0.10	4474	72.00	4219
0.15	4461	96.00	4205
0.20	4452	120.00	4194
0.35	4434	144.00	4185
0.50	4423	168.00	4177
0.65	4414	192.00	4171
1.00	4400	216.00	4165
1.50	4387	240.00	4160
2.00	4378	300.00	4145
2.50	4371	360.00	4132
3.00	4365	420.00	4118
3.50	4360	480.00	4105
4.00	4356	540.00	4091
6.00	4342	600.00	4078
8.00	4328	660.00	4065
10.00	4317	720.00	4051
12.00	4308	780.00	4038
24.00	4273	840.00	4024

Net pay thickness, h	16 ft
Porosity, ϕ	0.21
Water saturation, s_w	0.23
Oil viscosity, μ_o	0.38 cp
Oil FVF, B_o	1.41 RB/STB
Total compressibility, c_t	6.5E-05 psi^{-1}
Wellbore radius, r_w	0.45 ft

Calculate:

a. Permeability of the reservoir

b. Skin factor for the well

13-5. The pressure in a reservoir was determined to be above its bubble point pressure. A well completed in the reservoir produced oil at a stabilized rate of 275 BOPD. The well was then shut-in for a buildup test after 3 months of production. The data for the buildup test are given in Table 13P.2. Other reservoir parameters are given as:

Net pay thickness, h	70 ft
Porosity, ϕ	0.15
Water saturation, s_w	0.25
Oil viscosity, μ_o	0.85 cp
Oil FVF, B_o	1.25 RB/STB
Total compressibility, c_t	1.6E-05 psi^{-1}
Wellbore radius, r_w	0.25 ft

Calculate:

a. Formation permeability

b. Skin factor for the well

c. Average reservoir pressure

TABLE 13P.2 Buildup Data for Problem 13-5

Time, Δt (hrs)	SBHP, p_{ws} (psia)	Time, Δt (hrs)	SBHP, p_{ws} (psia)
0.00	3534	8.00	4350
0.15	3680	12.00	4364
0.20	3723	16.00	4373
0.30	3800	20.00	4379
0.40	3866	24.00	4384
0.50	3920	30.00	4393
1.00	4103	40.00	4398
2.00	4250	50.00	4402
4.00	4320	60.00	4405
6.00	4340	72.00	4407
7.00	4344	—	—

13-6. Calculate the formation permeability and skin factor for the well using the Miller-Dyes-Hutchinson (MDH) method based on the buildup data and reservoir parameters given in Problem 13-5. Compare the results from the MDH method to the results from the Horner method used in Problem 13-5.

13-7. A buildup test was conducted on a discovery well in a reservoir under evaluation for future development. The well was tested at a constant rate of 1750 BOPD for 10 days before it was shut-in for the buildup test. The data recorded during the buildup test are given in Table 13P.3. Additional reservoir data are given as:

Net pay thickness, h	25 feet
Porosity, ϕ	0.25
Water saturation, s_w	0.30
Oil viscosity, μ_o	2.50 cp
Oil FVF, B_o	1.45 RB/STB
Total compressibility, c_t	23E-06 psi^{-1}
Wellbore radius, r_w	0.454 feet

Calculate:

a. Formation permeability

b. Skin factor for the well

c. Initial reservoir pressure

d. Flow efficiency for the well

e. Damage ration for the well

TABLE 13P.3 Buildup Data for Problem 13-7

Time, Δt (hrs)	SBHP, p_{ws} (psia)	Time, Δt (hrs)	SBHP, p_{ws} (psia)
0.0000	3372.0	3.0	3892.9
0.0083	3554.2	4.0	3942.2
0.0167	3556.0	5.0	3980.4
0.0250	3557.7	6.0	4011.7
0.0333	3559.4	8.0	4061.0
0.0417	3561.1	10.0	4099.2
0.0500	3562.8	12.0	4130.5
0.0667	3566.2	16.0	4179.8
0.0833	3569.6	20.0	4218.0
0.1000	3572.9	24.0	4249.3
0.1333	3579.6	30.0	4287.5
0.1667	3586.1	36.0	4318.8
0.2000	3592.5	48.0	4388.1
0.2500	3602.0	60.0	4426.3
0.5000	3646.4	72.0	4457.6
1.0000	3719.8	84.0	4484.0
2.0000	3804.8	—	—

13-8. Determine the formation permeability and skin factor for the well by applying the MDH method to the buildup data and reservoir parameters presented in Problem 13-7. Compare the results from the MDH method to the results from the Horner method used in Problem 13-7.

13-9. A well on production in a dry gas reservoir was shut-in for a buildup test. The well produced dry gas at a stabilized rate of 9300 Mscf/D for 120 hours before it was shutin for the buildup test. The buildup data recorded during the test are provided in Table 13P.4. The properties of the gas are given in Table 13P.5. Other parameters for the reservoir are given as follows:

Net pay thickness, h	35 feet
Porosity, ϕ	0.21
Wellbore radius, r_w	0.50 feet
Reservoir temperature, T_R	118°F

Calculate reservoir pressure $\left(p^*\right)$, formation permeability, and the apparent skin factor for the well using the following methods:

a. Shut-in pressure versus Horner Time ratio
b. Shut-in pressure squared versus Horner Time ratio
c. Shut-in pseudo-pressure versus Horner Time ratio
d. Shut-in pseudo-pressure versus normalized Horner Time ratio
e. Compare the results from the four methods. Which method is recommended for gas buildup analyses?

TABLE 13P.4 Buildup Data for Problem 13-9

Shutin Time, Δt (hrs)	SBHP, p_{ws} (psia)	Shutin Time, Δt (hrs)	SBHP, p_{ws} (psia)
0.0000	1805.0	4.0	2417.9
0.0167	2013.7	5.0	2435.0
0.0250	2042.9	6.0	2449.0
0.0333	2063.7	7.0	2460.8
0.0667	2113.7	8.0	2471.0
0.1000	2142.9	10.0	2488.1
0.1333	2163.7	12.0	2502.0
0.1667	2179.8	18.0	2533.1
0.2000	2192.9	24.0	2555.1
0.2500	2209.0	30.0	2572.2
0.3333	2229.8	36.0	2586.1
0.4167	2245.8	42.0	2597.9
0.5000	2259.0	48.0	2600.7
0.7500	2288.2	60.0	2612.0
1.0000	2309.0	66.0	2616.9
1.5000	2338.2	72.0	2621.3
2.0000	2359.0	84.0	2629.1
3.0000	2388.2	96.0	2635.9
3.5000	2399.4	120.0	2647.3

TABLE 13P.5 Gas Properties for Problem 13-9

Pressure, p (psia)	Gas Compressibility factor, z	Gas Viscosity, μ_g (cp)
1000	0.8381	0.01200
1250	0.8050	0.01225
1500	0.7756	0.01250
1750	0.7500	0.01275
2000	0.7281	0.01300
2250	0.7100	0.01325
2500	0.6956	0.01350
2750	0.6850	0.01375
3000	0.6781	0.01400
3250	0.6750	0.01425
3500	0.6756	0.01450
3750	0.6800	0.01475
4000	0.6881	0.01500

13-10. A well producing from a dry gas reservoir was the subject of an isochronal test. The flow rates of the first and second tests were 73.41 Mscf/D and 409 Mscf/D, respectively. The duration of both tests was 4 hours. The flowing bottomhole pressures (FBHP) measured at the two flow rates are provided in Table 13P.6. Other reservoir parameters are given as:

Net pay thickness, h	385 feet
Porosity, ϕ	0.07
Wellbore radius, r_w	0.21 feet
Initial reservoir pressure, p_i	450 psia
Reservoir temperature, T_R	94°F
Total compressibility, c_t	22E-04 psia^{-1}
Gas gravity, γ_g	0.59 (Assume no gas impurities)

Calculate:

a. Average formation permeability
b. Skin factor for the well
c. Non-darcy flow coefficient (Assume constant for the flow tests)

TABLE 13P.6 Isochronal flow test data for Problem 13-10

Time, (hrs)	Flow Test # 1 q_1 = 73.41 Mcf/D	Flow Test # 2 q_2 = 409 Mcf/D
	p_{wf} (psia)	p_{wf} (psia)
1.00	432	324
1.50	431	317
2.00	430	314
2.50	430	310
3.00	429	308
3.50	429	306
4.00	428	304

13-11. A two-rate test was conducted on a well in a dry gas reservoir. The well was produced at a constant rate of 1100 Mscf/D for 24 hours during the first flow test. Then the production rate was increased to a constant rate of 2100 Mscf/ for another 24 hours for the second flow test. Note that there was no shut-in of the well between the two flow tests. The FBHP was 950 psia after 24 hours at the end of the first flow test and was 550 psia after 48 hours at the end of the second flow test. Other reservoir parameters are given as:

Net pay thickness, h	76 feet
Porosity, ϕ	0.07
Gas viscosity, μ_g	0.012 cp
Wellbore radius, r_w	0.21 feet
Initial reservoir pressure, p_i	1175 psia
Reservoir temperature, T_R	140°F
Standard pressure	14.7 psia
Standard temperature	60°F

Assuming ideal gas behavior and negligible non-Darcy flow, calculate:

a. The permeability of the formation

b. The skin factor of the well

13-12. The gas well in Problem 13-11 was then shut-in for a buildup test at the end of the 48-hour flow test. Calculate the shut-in bottomhole pressure (SBHP) in the well after 48 hours.

CHAPTER 14 Well Test Analysis: Type Curves

14.1 Introduction

In Chapter 13, well test analysis with straightline methods was presented. In that chapter, the difficulties associated with correctly identifying the time region over which straightline methods are applicable were discussed. In some cases, engineers analyzing the same well test data could identify different times for the start of the infinite-acting radial flow (IARF) regime, which is crucial to the application of straightline methods for calculation of well and reservoir properties. Well test analysis was improved tremendously by combined application of straightline methods and type curves on the same well test data. In this chapter, representative type curves, such as Ramey type curve, Gringarten type curve, and Bourdet derivative type curve are presented. The chapter concludes with a manual illustration of the application of type curves in well test analysis.

14.2 What Are Type Curves?

Type curves are graphic plots of theoretical solutions to flow equations under specific initial and boundary conditions of the interpretation model representing a reservoir-well system.[1] The most common forms of type curves are presented in terms of dimensionless pressure versus dimensionless time. The use of type curves for well test analysis was introduced by Ramey[2] in 1970. Ramey type curves[3] for a single well with skin and wellbore storage in an infinite system is shown in Figure 14.1. Numerous type curves for various well and reservoir boundary conditions were subsequently published by other authors. The best known of these type curves are the McKinley[4] type curves for wellbore storage effects, type curves for a well with infinite conductivity vertical fracture by Gringarten et al.[5] and type curves for a well with a finite conductivity vertical fracture by Cinco-Ley et al.[6] The application of type curves for well test analysis was advanced by Gringarten et al.[7] with the introduction of a unified set of type curves shown in Figure 14.2. The most important advancement that transformed the application of type curves for well test analysis began in 1983 with the introduction of derivative type curves by Bourdet et al.[8-10] The Bourdet derivative type curves are shown in Figure 14.3. The combined application of Gringarten type curves and Bourdet derivative type curves (Figure 14.4) in a single plot for well test analysis has enabled identification of reservoir behaviors and geologic features which were previously undetectable from well test data using only straightline methods.[11]

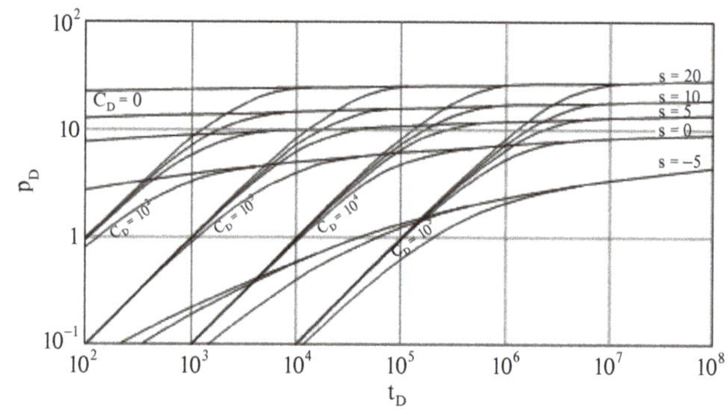

▲ **FIGURE 14.1** **Type curves for a single well with skin and wellbore storage in an infinite reservoir (from Agarwal et al.[3] © 1970 SPE, Reproduced with permission)**

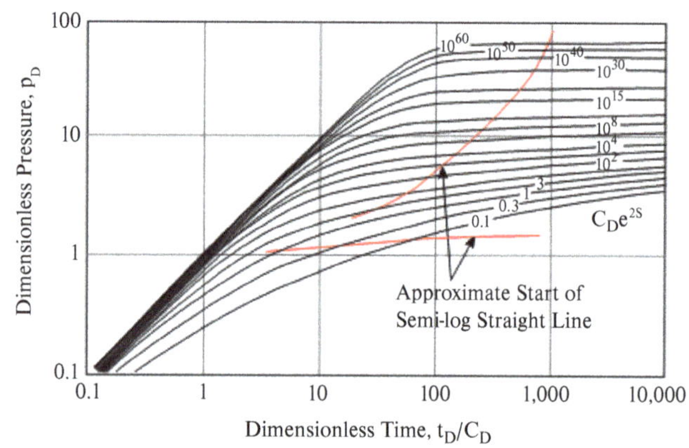

▲ **FIGURE 14.2** **Type curves for a single well with skin and wellbore storage in an infinite reservoir (from Gringarten et al.[7] © 1979 SPE, Reproduced with permission)**

14.3 Gringarten Type Curves

The Gringarten[7] type curves were derived from solutions to the diffusivity equation of a slightly compressible fluid produced from a well with skin and wellbore storage in an infinite-acting reservoir exhibiting homogeneous behavior. In Gringarten type curves (Figure 14.2), dimensionless pressure, p_D is plotted against dimensionless variables, t_D/C_D, with C_De^{2s} as the correlating parameter. The range of C_De^{2s} as the correlating parameter includes damaged and stimulated wells. Also shown in Figure 14.2 are two lines denoting the loci of approximate start of radial flow for an infinite-acting reservoir. This is the approximate location of the middle time region (MTR) or more appropriately the location of the infinite-acting radial flow (IARF) regime.

The MTR or IARF regime on the Gringarten type curves corresponds to the MTR straight line region identified on semi-log plots using straightline methods as shown in Chapter 13. Consequently, Gringarten type curves can be used with straightline methods to determine more accurately the start of IARF flow regime. This diagnostic application of Gringarten type curves was further enhanced by the introduction of derivative type curves as discussed later.

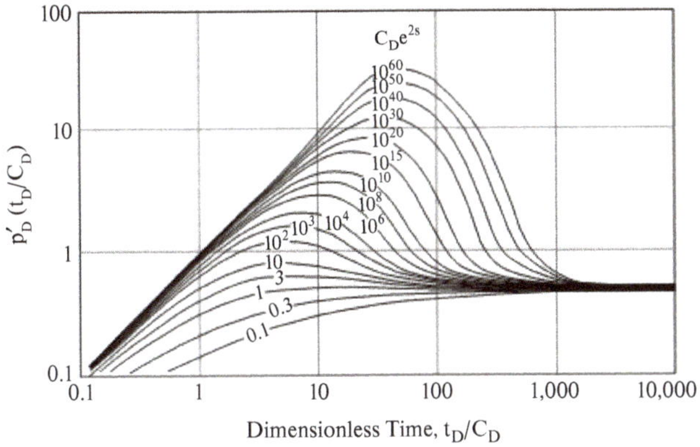

▲ **FIGURE 14.3** **Bourdet derivative type curves (from Bourdet et al.[8])**

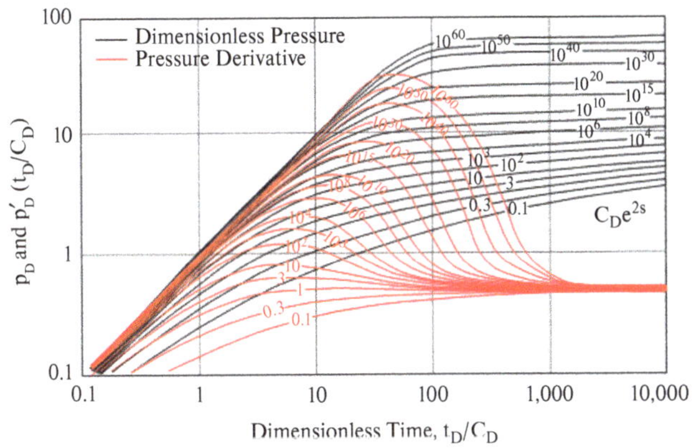

▲ **FIGURE 14.4** **Combined Gringarten type curves and Bourdet derivative type curves**

14.3.1 Unit-Slope Line

At early times, fluid flow is dominated by wellbore storage. During this flow regime, the type curves merge into a straight line whose slope is equal to unity. This line is termed the **unit-slope line**. Note that some of the Gringarten type curves with low values of C_De^{2s} have slopes which are less than one. This indicates stimulation, either by acidization or hydraulic fracturing. All points on the unit slope line imply that:

$$p_D = \frac{t_D}{C_D} \tag{14.1}$$

In Eq. (14.1), p_D is the dimensionless pressure defined as:

$$p_D = \frac{kh(p_i - p)}{141.2qB\mu} \tag{14.2}$$

For Eq. (14.2), p = pressure, psia; k = permeability, md; h = formation thickness, ft.; q = flow rate, STB/D; B = formation volume factor, RB/STB; and μ = viscosity, cp.

Also in Eq. (14.1), t_D is the dimensionless time defined as:

$$t_D = \frac{0.0002637kt}{\phi\mu c_t r_w^2} \tag{14.3}$$

In Eq. (14.3), k = permeability, md; t = time, hrs.; ϕ = porosity, fraction; μ = viscosity, cp; c_t = total compressibility, psi^{-1}; and r_w = wellbore radius, ft.

Finally in Eq. (14.1), C_D is the dimensionless wellbore storage coefficient defined as:

$$C_D = \frac{0.894C}{\phi c_t h r_w^2} \tag{14.4}$$

In Eq. (14.4), the units of all terms are as previously defined for Eqs. (14.2) and (14.3). The term, C, is the wellbore storage coefficient in bbl/psi which can be calculated for a rising or falling fluid level in the wellbore as:

$$C = \frac{25.65A_{wb}}{\rho_{wb}} \tag{14.5}$$

If the wellbore is filled with a single phase fluid, then C can be calculated as:

$$C = V_{wb}c_{wb} \tag{14.6}$$

In Eqs. (14.5) and (14.6), A_{wb} = wellbore area, ft^2; ρ_{wb} = density of wellbore fluid, lbm/ft^3; V_{wb} = volume of fluid in wellbore, ft^3; and c_{wb} = compressibility of wellbore fluid, psi^{-1}.

From Eq. (14.1):

$$C_D = \frac{t_D}{p_D} \tag{14.7}$$

Substituting Eq. (14.7) with Eqs. (14.2) and (14.3), it can be shown that for any point on the unit-slope line:

$$C_D = \frac{0.03723qB}{\phi c_t h r_w^2}\left(\frac{\Delta t}{\Delta p}\right)_{USL} \tag{14.8}$$

Furthermore, substituting Eq. (14.4) into Eq. (14.8) gives:

$$C = 0.04165qB\left(\frac{\Delta t}{\Delta p}\right)_{USL} \tag{14.9}$$

Equations (14.8) and (14.9) can be applied by making a log-log plot of Δp vs. Δt for any test data. If a unit-slope line exists for the test data, any point on the line can be used to determine $(\Delta t/\Delta p)_{USL}$.

14.4 Bourdet Derivative Type Curves

The Bourdet[8-10] derivative type curves were developed from pressure derivatives of the analytical solutions of the same flow equations used in the generation of the Gringarten type curves. Taking derivatives of Eq. (14.1) gives:

$$\frac{dp_D}{d(t_D/C_D)} = p_D' = 1 \tag{14.10}$$

Also, differentiating dimensionless pressure with respect to logarithm of dimensionless time yields:

$$\begin{aligned}\frac{dp_D}{\left[d\ln(t_D/C_D)\right]} &= (t_D/C_D)\left\{dp_D/\left[d(t_D/C_D)\right]\right\} \\ &= (t_D/C_D)\,p_D'\end{aligned} \tag{14.11}$$

Combining Eqs. (14.10) and (14.11) gives:

$$\left(\frac{t_D}{C_D}\right)p_D' = \frac{t_D}{C_D} \tag{14.12}$$

As can be determined from Eq. (14.12), a plot of $(t_D/C_D)\,p_D'$ vs. (t_D/C_D) gives a line with slope equal to unity within the early time region (ETR). Thus, at early times when wellbore storage is dominant, the Bourdet derivative type curves coincides with the unit-slope lines of the Gringarten type curves for damaged wellbores. Stimulated wellbores have low values of the correlating parameter, C_De^{2s}, which indicate stimulation, either by acidization or hydraulic fracturing.

The line-source solution of the diffusivity equation for slightly compressible fluids [Eq. (13.24)] can be transformed in terms of dimensionless variables into:

$$p_D = 0.5\left[\ln t_D + 0.80907 + 2s\right] \tag{14.13}$$

By introducing the term, $\ln C_D$, into Eq. (14.13), it can be derived that:

$$p_D = 0.5\left[\ln(t_D/C_D) + 0.80907 + \ln\left(C_De^{2s}\right)\right] \tag{14.14}$$

Taking derivative of Eq. (14.14) with respect to (t_D/C_D) gives:

$$\frac{dp_D}{d(t_D/C_D)} = p_D' = \frac{0.5}{(t_D/C_D)} \tag{14.15}$$

Thus, when infinite-acting radial flow (IARF) has been reached, the Bourdet derivative type curves become horizontal at constant value of $p_D'(t_D/C_D) = 0.5$. Thus, for Bourdet derivative type curves, there are two asymptotes defined by the unit-slope line and the horizontal line at a constant value of 0.5. Between these two asymptotes, the Bourdet derivative type curves exhibit characteristic shapes which depend on values of the correlating parameter C_De^{2s}.

The Bourdet derivative type curves are shown in Figure 14.3 as a log-log plot of $p_D'(t_D/C_D)$ vs. (t_D/C_D) with C_De^{2s} as the correlating parameter. In practice, the Bourdet derivative type curves are combined with Gringarten type curves in a single plot as shown in Figure 14.4. When both type curves are used together in type-curve matching of test data, it helps to eliminate most of the non-uniqueness of a type-curve match that may occur if only the Gringarten type curves are used. Further, the Bourdet derivative type curves have been used to identify characteristic response shapes representing various conditions of well models, reservoir models, and reservoir boundaries. These have led to significant advances in the interpretation of well test data. With the aid of derivative type curves, it is possible to identify fracture response in wells, double porosity behavior in naturally fractured reservoirs, and boundary effects in reservoirs. A selection of some of the characteristic shapes of derivative type curves for various well, reservoir, and boundary models are shown in Appendix 14A.

14.5 Agarwal Equivalent Time

The Gringarten type curves and Bourdet derivative type curves are based on analytical solutions to the diffusivity equations of slightly compressible fluids at constant production rates. Essentially, these are pressure drawdown type curves. As shown in Figure 14.5, drawdown and buildup type curves are not identical because buildup type curves are affected by the duration of the producing time, t_p, before shut-in. If the duration of the producing time, t_p, before shut-in is relatively short, the deviation between drawdown and buildup type curves tends to be more pronounced. This could lead to erroneous interpretation of buildup tests with drawdown type curves. To reduce the effects of producing time before shut-in and allow the use of drawdown type curves for analysis of buildup tests, Agarwal[12] proposed an equivalent drawdown time which is called Agarwal equivalent time. Agarwal[12] equivalent time, Δt_e, is defined as:

$$\Delta t_e = \frac{t_p\Delta t}{t_p + \Delta t} \tag{14.16}$$

Agarwal equivalent time is strictly applicable to infinite-acting radial flow reservoirs with no wellbore storage. However, it has been shown to be applicable to wells with wellbore storage and fracture effects.[12]

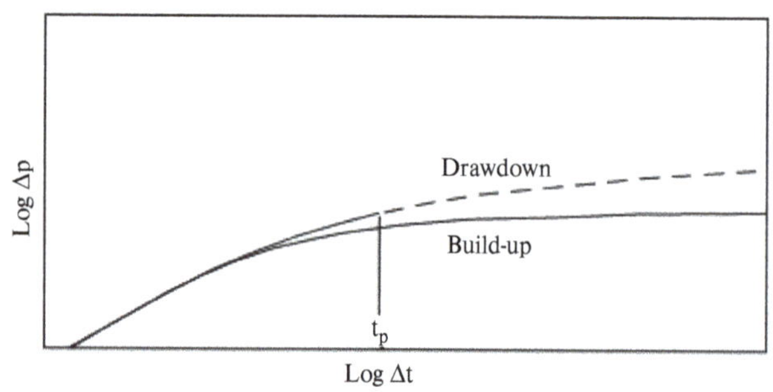

▲ **FIGURE 14.5** Drawdown and buildup type curves comparison

14.6 Type-Curve Matching

Type-curve matching is the methodical process of superimposing a plot of the test data on a set of type curves on the same scale and using the characteristic shapes of the type curves to identify an interpretation model for the test data. The process of matching a plot of the test data to a type curve can be done manually as demonstrated in Example 14.1 or by using graph fitting techniques based on regression analysis as implemented in most commercial software for well test analyses. The identification of an interpretation model (or type curve) for the test data during type-curve matching was enhanced considerably with the introduction of derivative type curves. With application of derivative curves, an interpretation model can be selected that encompasses the entire flow regimes of the test data. This means that by selecting an interpretation model that fits the test data, the interpretation model also includes a well model that represents flow regimes at the early times (ETR), a reservoir model for the flow regime at the intermediate times (MTR), and a boundary model for the late times (LTR). Thus, the test data can be analyzed at various flow regimes if the flow regimes are present. Note that selection of an interpretation model during type-curve matching is a non-unique solution because it is possible to find another interpretation model that can match the test data. In well test analysis using type-curve matching, it is important to include a validation step that checks the results calculated from type curves against other methods such as straightline methods.[7,11] The validation step may incorporate data from other external sources such as geological, petrophysical, and seismic data.

Once a match has been achieved with a type curve, well and reservoir properties corresponding to the test data can be calculated from the dimensionless parameters of the type curve. From the unit-slope line, it was shown previously in Eq. (14.8) that dimensionless wellbore storage coefficient can be calculated as:

$$C_D = \frac{0.03723qB}{\phi c_t h r_w^2}\left(\frac{\Delta t}{\Delta p}\right)_{USL}$$

Similarly, wellbore storage coefficient can be determined from Eq. (14.9) as:

$$C = 0.04165qB\left(\frac{\Delta t}{\Delta p}\right)_{USL}$$

Starting with Eq. (14.3), it can be derived that at the Time Match Point (TMP) along the x-axis, the dimensionless wellbore storage coefficient can be calculated from:

$$C_D = \frac{0.0002637k}{\phi \mu c_t r_w^2}\left[\frac{\Delta t}{t_D/C_D}\right]_{TMP} \tag{14.17}$$

Note that it is common practice to check the values for C_D calculated from Eqs. (14.8) and (14.17) for consistency.

By rearranging Eq. (14.2), formation permeability, k, at the Pressure Match Point (PMP) along the y-axis can be calculated from the expression:

$$k = \frac{141.2qB\mu}{h}\left[\frac{p_D}{\Delta p}\right]_{PMP} \tag{14.18}$$

As part of the validation process, the formation permeability calculated from Eq. (14.18) should be compared to the formation permeability calculated from straightline methods for the same test data.

From the correlating parameter, $C_D e^{2s}$, for the type curve, the skin factor for the test data can be calculated from the relationship:

$$s = 0.5\ln\left(\frac{C_D e^{2s}}{C_D}\right) \tag{14.19}$$

Also, as part of the validation process, the skin factor calculated from Eq. (14.19) should be compared to the skin factor calculated from straightline methods.

14.7 Procedures for Manual Application of Type-Curve Matching in Well Test Analysis

The procedures presented here apply only to manual use of type-curve matching in analysis of well test data. Since the advent of computer-aided well test analysis,[13-15] type-curve matching is generally accomplished in all commercial software through automatic fitting of the test data using linear or non-linear regression techniques. Most engineers may never have to manually plot the test data and match them against a set of type curves. However, these procedures are presented here to enable engineers to gain thorough understanding of the processes that have been automated in computer-aided well test analysis. The procedures presented here are based on the use of Gringarten-Bourdet combined type curves as shown in Figure 14.4. The procedures for manual type-curve matching are as follows:

1. Prepare the test data for analysis by tabulating pressure change versus time. For drawdown test data, tabulate pressure change as $\Delta p = (p_i \text{ or } \bar{p}) - p_{wf}$ versus flowing time, Δt. For buildup test data, tabulate pressure change as $\Delta p = (p_{ws} - p_{wf@\Delta t=0})$ versus equivalent shut-in time, Δt_e, calculated from Eq. (14.16). Plot Δp vs. Δt or Δt_e.
2. If a unit slope line is present on the plot of the test data, select any point $(\Delta p,\ \Delta t \text{ or } \Delta t_e)_{USL}$ on the unit slope line. Calculate dimensionless wellbore storage coefficient from Eq. (14.8) as:

$$C_D = \frac{0.03723qB}{\phi c_t h r_w^2}\left(\frac{\Delta t \text{ or } \Delta t_e}{\Delta p}\right)_{USL}$$

From Eq. (14.9), calculate wellbore storage coefficient as:

$$C = 0.04165qB\left(\frac{\Delta t \text{ or } \Delta t_e}{\Delta p}\right)_{USL}$$

3. For the derivative curve, calculate derivative of pressure with respect to natural logarithm of time. For drawdown test data, calculate the derivative as:

$$-\frac{dp_{wf}}{d(\ln t)} = -t\left[\frac{dp_{wf}}{dt}\right] = t\Delta p' \quad (14.20)$$

For buildup test data, calculate the derivative as:

$$\frac{dp_{ws}}{d(\ln \Delta t_e)} = \Delta t_e\left[\frac{dp_{ws}}{d(\Delta t_e)}\right] = \Delta t_e \Delta p' \quad (14.21)$$

Plot $t\Delta p'$ (or $\Delta t_e \Delta p'$) versus Δt (or Δt_e) on the same graph plotted in Step 1. An algorithm for calculation of the derivatives is provided in Appendix 14C.

4. Select a set of Gringarten-Bourdet type curves on the same scale as the graph plotted in Steps 1 and 2. Note that it is important that the two graphs are on the same scale.
5. Superimpose the plot of the test data on the Gringarten-Bourdet type curves. Achieve a match in the vertical direction by aligning the IARF stabilization line of the test data with similar line on the type curve indicated by $p'_D(t_D/C_D) = 0.5$.
6. If a unit slope line is present on the plot of the test data, achieve a match in the horizontal direction by aligning the unit-slope line of the test data with the unit slope line of the type curve.
7. From the pressure match point $(p_D, \Delta p)_{PMP}$ in Step 5, calculate formation permeability from Eq. (14.18) as:

$$k = \frac{141.2qB\mu}{h}\left[\frac{p_D}{\Delta p}\right]_{PMP}$$

8. From the time match point in Step 6, use the match point $(t_D/C_D, \Delta t \text{ or } \Delta t_e)_{TMP}$ to calculate dimensionless wellbore storage coefficient from Eq. (14.17) as:

$$C_D = \frac{0.0002637k}{\phi\mu c_t r_w^2}\left[\frac{\Delta t \text{ or } \Delta t_e}{t_D/C_D}\right]_{TMP}$$

Compare the values of C_D calculated from Steps 2 and 8. If inconsistent, repeat match and re-calculate.

9. Using the correlating parameter, $C_D e^{2s}$, of the match type curve and value for C_D from Step 8, calculate skin factor for the test data from Eq. (14.19) as:

$$s = 0.5\ln\left(\frac{C_D e^{2s}}{C_D}\right)$$

10. Use straightline methods presented in Chapter 13 to prepare a plot of p_{wf} vs. $\log t$ for drawdown data, and p_{ws} vs. $\log(t_p + \Delta t/\Delta t)$ for buildup data. Determine formation permeability from the time region defined by the stabilization line identified with the derivative type curve in Step 4. Also, calculate skin factor from this MTR line. Compare results from type curve matching and straightline methods. If results are inconsistent, repeat the entire process until reasonable consistency is achieved.

14.8 Stages of the Type-Curve Matching Procedures

The procedures presented earlier for type-curve matching can be grouped into three stages as suggested by Gringarten.[11] These stages are:

1. Identification of interpretation model (Steps 1 to 6).
2. Calculation from interpretation model parameters (Steps 7 to 9).
3. Validation of the interpretation model (Step 10).

14.8.1 Identification of the Interpretation Model

The key step in well test analysis is the identification of the interpretation model based on the characteristic shapes of Bourdet derivative type curves. A sample of these characteristic shapes is shown in Appendix 14A. In ETR, the characteristic shape may indicate a damaged well with wellbore storage and skin or a fractured well with no wellbore storage. During the MTR flow regime, the characteristic shape of the derivative curve may indicate double porosity reservoir or double permeability reservoir. In the LTR, reservoir boundary effects may indicate a sealing fault or a constant pressure boundary. Each of these characteristic shapes may indicate an interpretation model that could assist in the complete analysis of the test data. However, it is important to note that none of these characteristic shapes is unique to a particular well, reservoir, or boundary model. A characteristic shape of the derivative type curve may potentially represent two or more well, reservoir, or boundary responses that are similar in appearance. For instance, the characteristic shape of a double porosity (fractured) reservoir closely resembles the response of a double permeability (multilayered) reservoir. Consequently, it is often necessary to incorporate external data from geological, geophysical, and petrophysical sources to assist in the identification of an interpretation model.

14.8.2 Calculation from Interpretation Model Parameters

After an interpretation model has been selected and fitted to the test data either by type-curve matching or regression analysis, the reservoir parameters can be calculated from the interpretation model parameters using equations developed in this chapter and Chapter 15. The results from the analysis are independent of calculation methods as long as the calculations are based on the same interpretation model.[11] Consequently, calculation methods such as straightline methods and type-curve matching should yield the same result as long as they are based on the same interpretation model.

14.8.3 Validation of the Interpretation Model Results

It is important to validate the results calculated with the interpretation model. This could be done by comparing the results calculated from type-curve matching for instance, to results calculated from straightline methods based on the same interpretation model. If the results are consistent, then the interpretation model should be accepted as being representative of the test data at least for that test period. However, if the results are grossly in disagreement, the entire process of analyzing the test data should be repeated.

EXAMPLE 14.1 Calculation of Reservoir Parameters from Buildup Test Using the Gringarten-Bourdet Type Curves

Problem

A well in a reservoir above its bubble point pressure was producing oil at a constant rate of 185 BOPD before it was shut-in for a buildup test. The buildup test data are given in Appendix 14B as Table 14B.1 showing elapsed shut-in time, Δt, and shut-in pressure, p_{ws}. Other reservoir and well data are given below:

Formation thickness, h	114 ft
Formation porosity, ϕ	0.28
Total compressibility, c_t	4.1×10^{-6} psi^{-1}
Oil viscosity, μ_o	2.2 cp
Oil FVF, B_o	1.1 RB/STB
Wellbore radius, r_w	0.50 ft
Flowing BHP, p_{wf} at $\Delta t = 0$	2820 psia
Producing time before shut-in, t_p	540 hrs

Calculate dimensionless wellbore storage, C_D; wellbore storage coefficient, C; formation permeability, k; and skin factor, s. Use the technique of type curve matching with the combined Gringarten-Bourdet type curve. Compare the results from type curves to the results obtained from using straightline methods based on the Horner plot.

Solution

Step 1: Prepare the test data for analysis.

Calculate Agarwal equivalent shut-in time, Δt_e, and pressure change, $\Delta p = \left(p_{ws} - p_{wf\,@\,\Delta t=0}\right)$. The calculated data are shown in Appendix 14B as Table 14B.2. Plot Δp vs. Δt_e on a log-log scale as shown in Figure 14.6.

Step 2: Calculate pressure derivative with respect to natural logarithm of equivalent time.

Calculate pressure derivatives with the algorithm shown in Appendix 14C. The calculated derivatives are shown in Table 14B.2. Plot $\Delta p'$ vs. Δt_e on a log-log scale as shown in Figure 14.6.

Step 3: Calculate C_D and C from the unit-slope line.

Select any point on the unit-slope line from Figure 14.6. One such point on the unit-slope line is: $(\Delta t_e / \Delta p)_{USL} = (0.06374/41.13)$.

Substituting Eq. (14.8) gives:

$$C_D = \frac{0.03723qB}{\phi c_t h r_w^2}\left(\frac{\Delta t_e}{\Delta p}\right)_{USL}$$

$$= \frac{0.03723 \times 185 \times 1.1}{0.28 \times 4.1 \times 10^{-6} \times 114 \times (0.5)^2}\left[\frac{0.06374}{41.13}\right]$$

$$= 358.86$$

From Eq. (14.9):

$$C = 0.04165qB\left(\frac{\Delta t_e}{\Delta p}\right)_{USL}$$

$$= 0.04165 \times 185 \times 1.1 \times \left(\frac{0.06374}{41.13}\right)$$

$$= 0.013 \text{ RB/psi}$$

Step 4: Perform type-curve matching using Figure 14.6 and Gringarten-Bourdet type curve.

From Figure 14.7, the pressure match point (PMP) is $\left(\frac{p_D}{\Delta p}\right)_{PMP} = \left(\frac{4.5}{100}\right)$. Substituting Eq. (14.18) gives:

$$k = \frac{141.2qB\mu}{h}\left[\frac{p_D}{\Delta p}\right]_{PMP}$$

$$= \frac{141.2 \times 185 \times 1.1 \times 2.2}{114}\left(\frac{4.5}{100}\right)$$

$$= 24.95 \text{ md.}$$

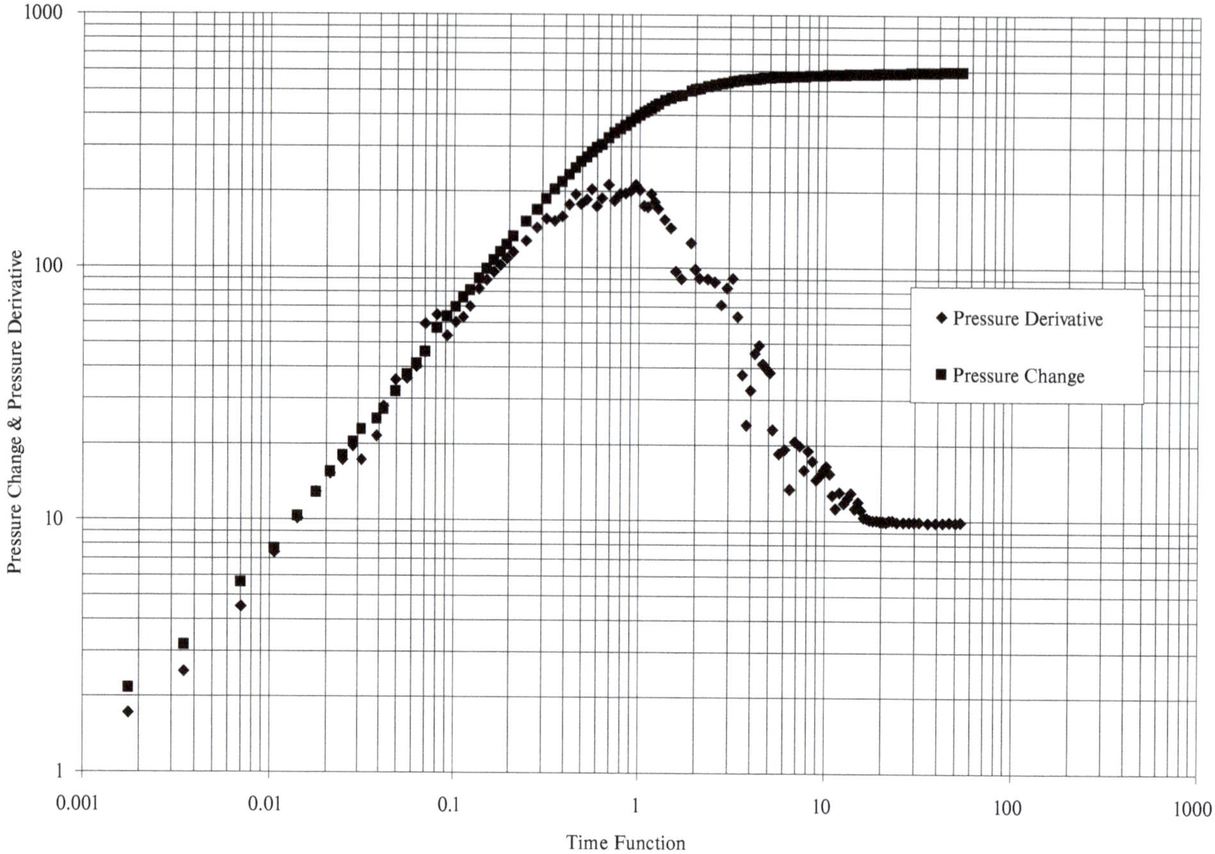

▲ **FIGURE 14.6** Plot of Δp & $\Delta p'$ vs. Δt_e for Example 14.1

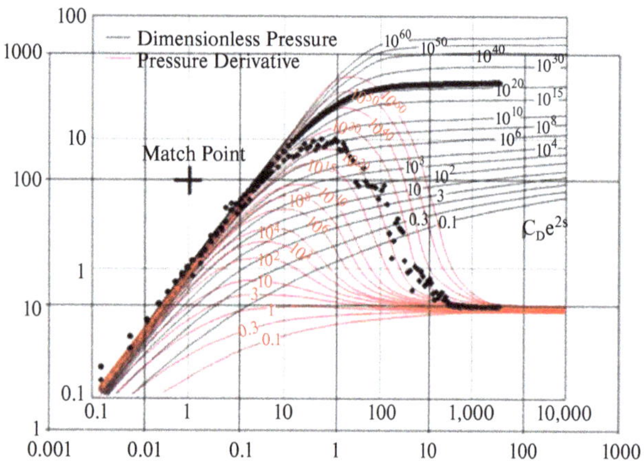

▲ **FIGURE 14.7** Type-Curve match for Example 14.1

Step 5: Calculate, C_D, from time match point (TMP).

From Figure 14.7, the time match point (TMP) is: $\left(\dfrac{\Delta t_e}{t_D/C_D}\right)_{TMP} = \left(\dfrac{0.03}{1}\right)$. Substituting into Eq. (14.17) gives:

$$C_D = \frac{0.0002637k}{\phi \mu c_t r_w^2}\left[\frac{\Delta t_e}{t_D/C_D}\right]_{TMP}$$

$$= \frac{0.0002637 \times 24.95}{0.28 \times 2.2 \times 4.1 \times 10^{-6} \times (0.5)^2}\left(\frac{0.03}{1}\right)$$

$$= 312.61$$

Note the value for dimensionless storage coefficient calculated from TMP by type-curve matching is reasonably close to the value calculated in Step 3 from unit-slope line.

Step 6: Calculate skin factor, s.

From Eq. (14.19):

$$s = 0.5 \ln\left(\frac{C_De^{2s}}{C_D}\right)$$

$$= 0.5 \ln\left(\frac{10^{21}}{312.61}\right), \text{ where } C_De^{2s}$$

$$= 10^{21} \text{ from Figure 12.7.}$$

$$= 22.46$$

Step 7: For validation, calculate permeability and skin factor using straightline methods.

The straightline method applied is the Horner plot. A tabulation of p_{ws} and $(t_p + \Delta t)/\Delta t$ is shown in Appendix 14B as Table 14B.3.

Figure 14.8 is a plot of p_{ws} vs. $\left(t_p + \Delta t/\Delta t\right)$ on a semilog scale. From Figure 14.8, slope of the IARF line is: $m =$ 3413.25 – 3384.46 = 28.79 psi/cycle. From Eq. 13.59,

$$k = \frac{162.6qB\mu}{mh}$$

$$= \frac{162.6\times185\times1.1\times2.2}{28.79\times114}$$

$$= 22.18 \text{ md.}$$

Note the formation permeability calculated here is close to the value of 24.95 md calculated from type-curve matching in Step 4.

From Figure 14.8, $p_{1hr} = 3360$ psia. From Eq. 13.62,

$$s = 1.151\left[\frac{p_{1hr} - p_{wf}}{m} - \log\frac{k}{\phi\mu c_t r_w^2} + 3.23\right]$$

$$= 1.151\left[\frac{3360 - 2820}{28.79} - \log\left(\frac{22.18}{0.28\times2.2\times4.1\times10^{-6}\times(0.5)^2}\right) + 3.23\right]$$

$$= 16.63$$

Again, note the skin factor calculated here is close to the value of 22.46 from type-curve matching. This validation process indicates that the results obtained from analysis of the test data are consistent.

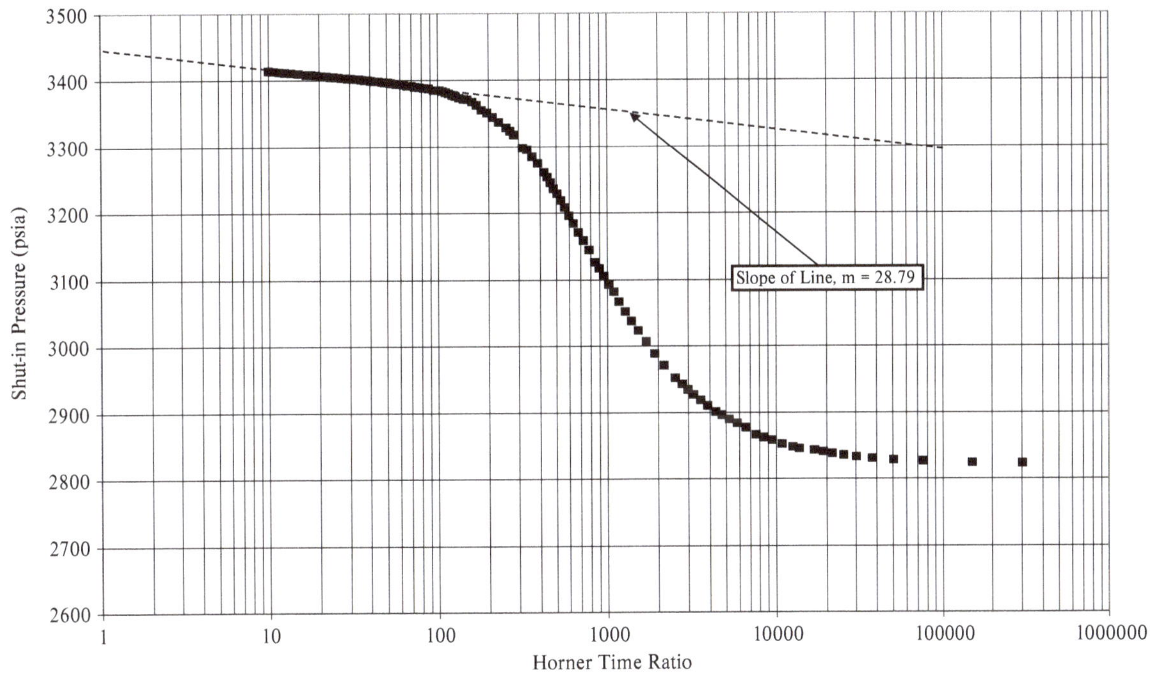

▲ **FIGURE 14.8 Plot of p_{ws} vs. $(t_p + \Delta t/\Delta t)$ for Example 14.1**

End of Example 14.1

Nomenclature

A_{wb}	area of wellbore, feet^2	p_D	dimensionless pressure
B	formation volume factor, RB/STB	p_{wf}	well bottomhole flowing pressure, psia
c_f	formation compressibility, psi^{-1}	p_{ws}	shut-in bottomhole pressure, psia
c_t	total compressibility, psi^{-1}	q	volume flow rate, STB/D
C_{wb}	compressibility of wellbore fluid, psi^{-1}	r_w	wellbore radius, feet
C	wellbore storage coefficient, bbl/psi	s	skin factor
C_D	dimensionless wellbore storage coefficient	Δt	shut-in time in buildup test, hours
h	reservoir or formation thickness, feet	Δt_e	Agarwal equivalent time
k	permeability, md	t	time, hours
L	lateral length of horizontal well, feet	t_D	dimensionless time
m	slope of MTR straight line	t_p	total producing time before shut-in, hours
Δp	pressure change	V_{wb}	volume of wellbore, barrels
$\Delta p'$	pressure change derivative	ϕ	porosity, fraction
p'_D	dimensionless pressure derivative	μ	fluid viscosity, cp
p	pressure, psia	ρ_{wb}	density of fluid in wellbore, lbm/ ft^3

Subscripts

f	formation	o	oil
g	gas	t	total
i	initial conditions	w	water

Abbreviations

BHP	Bottomhole Pressure	MTR	Middle Time Region
DST	Drill Stem Test	PMP	Pressure Match Point
ETR	Early Time Region	PTA	Pressure Transient Analysis
FVF	Formation Volume Factor	TMP	Time Match Point
IARF	Infinite-Acting Radial Flow	USL	Unit Slope Line
LTR	Late Time Region		

REFERENCES

1. Gringarten, A.C.: "Type-Curve Analysis: What It Can and Cannot Do," *JPT* (January 1987), 11-13.
2. Ramey, H.J.: "Short-Time Well Test Data Interpretation in the Presence of Skin Effect and Wellbore Storage," *JPT* (January 1970) 97-104.
3. Agarwal, R.G., Al-Hussainy, R., and Ramey, H.J., Jr.: "An Investigation of Wellbore Storage and Skin Effect in Unsteady Liquid Flow: I. Analytical Treatment," *SPEJ* (September 1970) 279-290.
4. McKinley, R.M.: "Wellbore Transmissibility from Afterflow-Dominated Pressure Buildup Data," *JPT* (July 1971) 863-872.
5. Gringarten, A.C., Ramey, H.J., Jr., and Raghavan, R.: "Unsteady-State Pressure Distribution Created by a Well with a Single Infinite-Conductivity Vertical Fracture," *SPEJ* (August 1974) 347-360.
6. Cinco-Ley, H., Samaniego-V., F., and Dominguez-A., N.: "Transient Pressure Behavior for a Well with a Finite-Conductivity Vertical Fracture," *SPEJ* (August 1978) 253-264.

7. Gringarten, A.C., Bourdet, D.P., Landel, P.A., and Kniazeff, V.J.: "A Comparison Between Different Skin and Wellbore Storage Type-Curves for Early-Time Transient Analysis," Paper SPE 8205 presented at the SPE Annual Technical Conference and Exhibition, Las Vegas, Nevada, September 23-26, 1979.
8. Bourdet, D., Whittle, T.M., Douglas, A.A., and Pirard, Y.M.: "A New Set of Type Curves Simplifies Well Test Analysis," *World Oil* (May 1983) 95-106.
9. Bourdet, D., Ayoub, J.A., Whittle, T.M., Pirard, Y.M., and Kniazeff, V.: "Interpreting Well Tests in Fractured Reservoirs," *World Oil* (October 1983) 77-87.
10. Bourdet, D., Ayoub, J.A., and Pirard, Y.M.: "Use of Pressure Derivative in Well-Test Interpretation," *SPEFE* (June 1989) 293-302.
11. Gringarten, A.C.: "From Straight Lines to Deconvolution: The Evolution of the State of the Art in Well Test Analysis," *SPEREE* (February 2008) 41-62.
12. Agarwal, Ram G.: "A New Method to Account for Producing Time Effects When Drawdown Type Curves Are Used to Analyze Pressure Buildup and Other Test Data," paper SPE 9289 presented at the 55th Annual Fall Technical Conference and Exhibition, Dallas, Texas, (September 21-24, 1980).
13. Gringarten, A.C.: "Computer-Aided Well Test Analysis," paper SPE 14099 presented at the SPE International Meeting on Petroleum Engineering, Beijing, China, (March 17-20, 1986).
14. Ehlig-Economides, C.A., Joseph, J.A., Ambrose, R.W., Jr., and Norwood, C.: "A Modern Approach to Reservoir Testing," *JPT* (December 1990) 1554-1563.
15. Horne, R.N.: *Modern Well Test Analysis-A Computer-Aided Approach*, 2nd Ed., Petroway, Inc., California (2008).

PROBLEMS

Problems

14-1. A Drill Stem Test (DST) was conducted on a discovery well in a deepwater reservoir, offshore West Africa. The reservoir fluid and rock properties are given as:

Net pay thickness, h	64.36 ft
Formation porosity, ϕ	0.18
Initial water saturation, s_{wi}	0.22
Oil viscosity, μ_o	0.36 cp
Oil FVF, B_o	1.52 RB/STB
Oil gravity	39°API
Total compressibility, c_t	1.62E-05 psi^{-1}
Initial reservoir pressure, p_i	8043 psia
Bubblepoint pressure, p_b	3850 psia
Reservoir temperature, T_R	282°F
Wellbore radius, r_w	0.454 ft

The actual DST data of the well can be downloaded from: http://www.tigapetroleum.com/. Using the DST data, calculate:

a. Permeability of the reservoir

b. Skin factor of the well

Hint: This problem should be solved with the aid of a commercial pressure transient analysis (PTA) software.

14-2. A new reservoir in deepwater, offshore Nigeria, is being developed for first oil. A DST was conducted on one of the development wells to assess its productive potential. This is a horizontal well with 984 feet of lateral length. The DST data can be downloaded from: http://www.tigapetroleum.com/.

The reservoir rock and fluid properties are given as:

Net pay thickness, h	98.43 ft
Lateral horizontal length, L	984 ft
Formation porosity, ϕ	0.21
Initial water saturation, s_{wi}	0.20
Oil viscosity, μ_o	1.82 cp
Oil FVF, B_o	1.21 RB/STB
Total compressibility, c_t	1.37E-05 psi^{-1}
Initial reservoir pressure, p_i	4125 psia
Bubble point pressure, p_b	3578 psia
Reservoir temperature, T_R	121°F
Wellbore radius, r_w	0.23 ft
Top reservoir depth	6000 ft

Using the DST data, determine:

a. Permeability of the reservoir

b. Skin factor for the well

c. Productive potential of the well

Hint: This problem should be solved with the aid of a commercial pressure transient analysis (PTA) software.

APPENDIX 14A CHARACTERISTIC SHAPES OF PRESSURE AND PRESSURE-DERIVATIVE CURVES FOR SELECTED WELL, RESERVOIR, AND BOUNDARY MODELS

1. Well Models

A. Wellbore Storage and Skin.

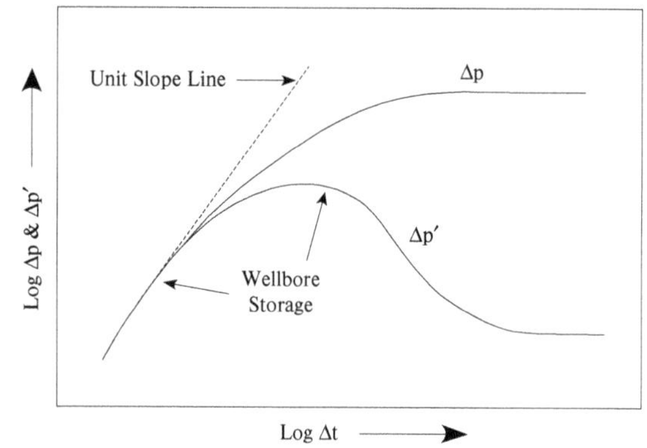

B. Infinite Conductivity Fracture.

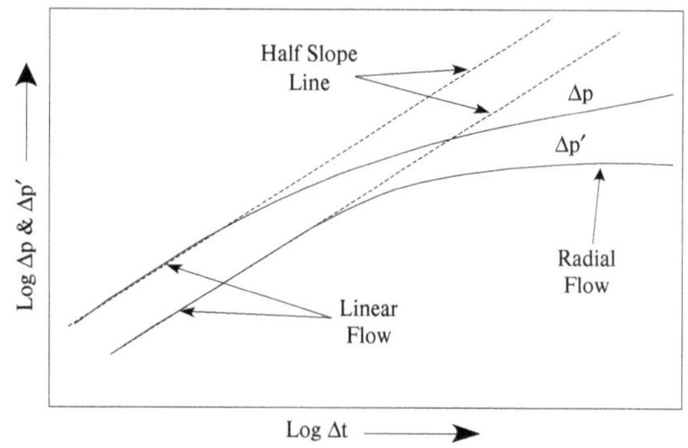

C. Finite Conductivity Fracture.

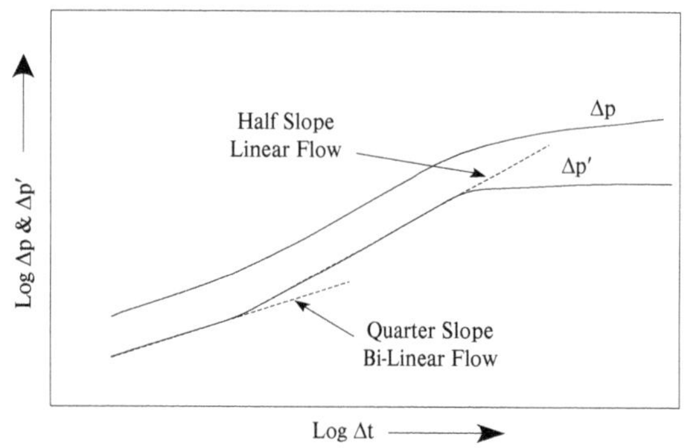

2. Reservoir Models

A. Homogeneous Behavior- Infinite-Acting Radial Flow.

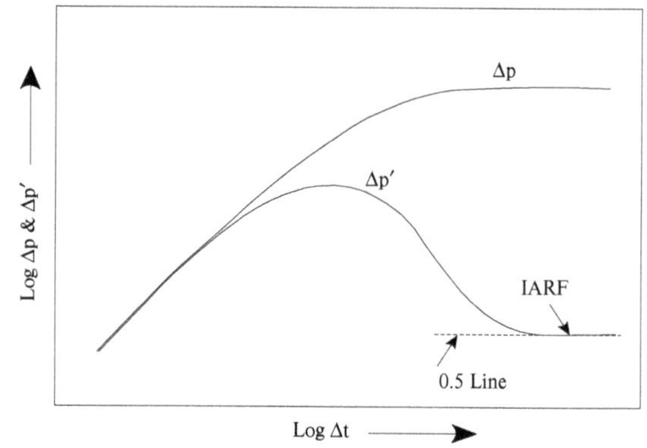

B. Heterogeneous Behavior- Pseudosteady-State Interporosity Flow, Double Porosity Behavior.

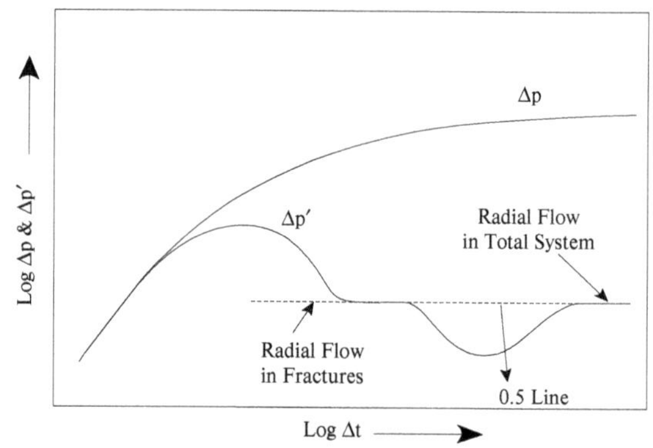

C. Heterogeneous Behavior- Transient Interporosity Flow, Double Porosity Behavior.

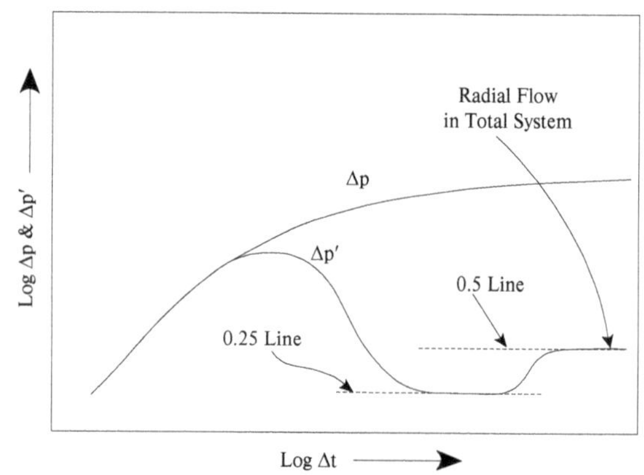

D. Heterogeneous Behavior- Double Permeability Behavior.

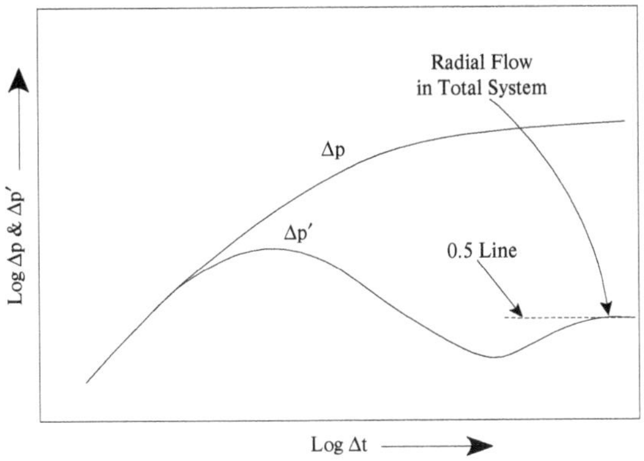

3. Boundary Models

A. Infinite-Acting Reservoir–No Boundary Effects.

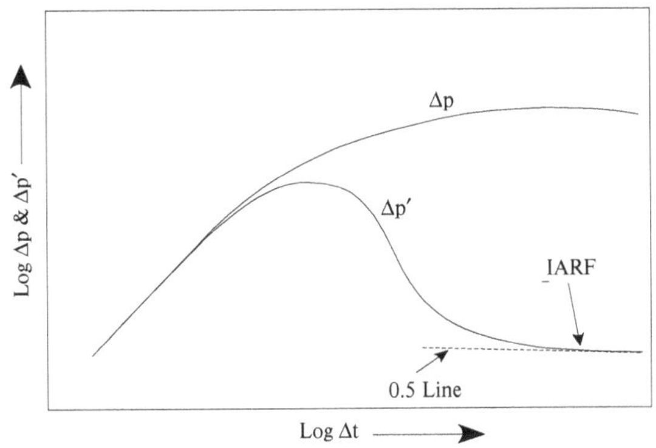

B. Sealing Fault or Single No-Flow Boundary.

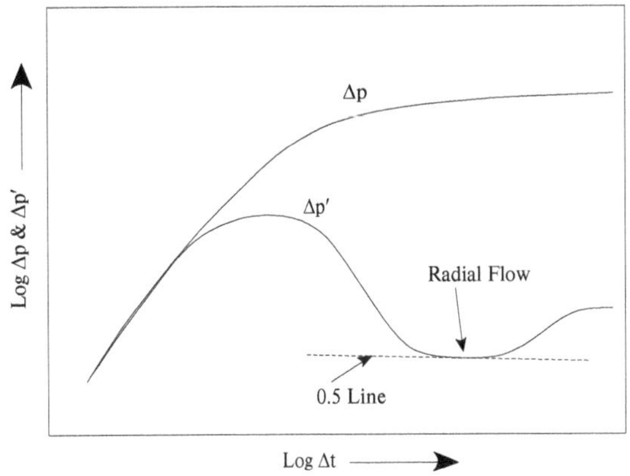

C. Bounded Reservoir or Constant Pressure Boundary Effects.

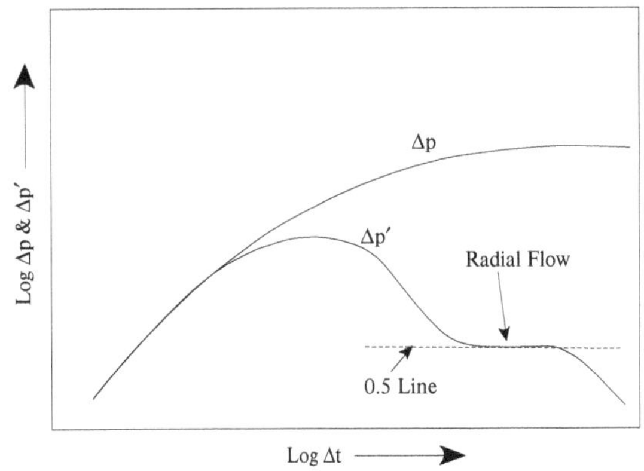

APPENDIX

14B DATA FOR EXAMPLE 14.1

TABLE 14B.1 Buildup Test Data for Example 14.1

Δt (hrs)	p_{ws} (psia)	Δt (hrs)	p_{ws} (psia)	Δt (hrs)	p_{ws} (psia)	Δt (hrs)	p_{ws} (psia)
0.0000	2820.00	0.3188	3006.01	2.5500	3344.09	12.9625	3398.19
0.0018	2822.15	0.3542	3022.46	2.7625	3350.80	13.6000	3398.75
0.0035	2823.18	0.3896	3036.49	2.9750	3355.10	14.2375	3399.31
0.0071	2825.61	0.4250	3051.08	3.1875	3362.38	14.8750	3399.86
0.0106	2827.67	0.4604	3065.68	3.4000	3367.23	15.5125	3400.25
0.0142	2830.28	0.4958	3080.63	3.6125	3370.40	16.1500	3400.80
0.0177	2832.71	0.5313	3091.29	3.8250	3371.72	16.7875	3401.10
0.0213	2835.33	0.5667	3104.95	4.0375	3373.03	17.4250	3401.55
0.0248	2837.76	0.6021	3116.73	4.2500	3375.09	18.0625	3401.83
0.0283	2840.19	0.6375	3125.52	4.4625	3377.52	18.9125	3402.39
0.0319	2842.62	0.6906	3144.23	4.6750	3379.73	19.7625	3402.71
0.0390	2844.86	0.7438	3158.26	4.8875	3381.28	20.6125	3403.21
0.0425	2847.11	0.7969	3170.42	5.1000	3383.15	21.4625	3403.51
0.0496	2851.97	0.8500	3184.07	5.3125	3384.46	22.3125	3403.96
0.0567	2857.19	0.9031	3194.93	5.7375	3384.80	23.1625	3404.24
0.0638	2861.13	0.9563	3207.47	6.1625	3387.00	24.2250	3404.78
0.0708	2865.98	1.0094	3218.31	6.5875	3387.51	25.5000	3405.13
0.0815	2876.72	1.0625	3228.62	7.0125	3388.64	27.6250	3406.22
0.0921	2883.26	1.1156	3235.90	7.4375	3389.96	29.7500	3406.62
0.1027	2889.05	1.1688	3244.89	7.8625	3390.89	31.8750	3407.52
0.1133	2895.59	1.2219	3253.49	8.2875	3391.64	34.0000	3407.90
0.1240	2900.93	1.2750	3260.61	8.7125	3392.76	38.2500	3409.71
0.1381	2909.72	1.3813	3274.46	9.1375	3393.32	42.5000	3410.10
0.1523	2917.76	1.4875	3284.39	9.5625	3394.07	46.7500	3411.42
0.1665	2926.17	1.5938	3294.70	9.9875	3394.63	51.0000	3411.83
0.1806	2934.03	1.7000	3297.68	10.4125	3395.39	55.2500	3412.86
0.1948	2942.06	1.9125	3317.54	10.8375	3395.94	59.5000	3413.25
0.2125	2951.60	2.0188	3323.15	11.2625	3396.56	—	—
0.2479	2970.66	2.1250	3327.83	11.6875	3396.88	—	—
0.2833	2987.87	2.3375	3336.24	12.3250	3397.63	—	—

TABLE 14B.2 Plot Data for Figure 14.6 in Example 14.1

Δt_e	Δp (psi)	$\Delta t_e \Delta p'$	Δt_e	Δp (psi)	$\Delta t_e \Delta p'$	Δt_e	Δp (psi)	$\Delta t_e \Delta p'$
0.0018	2.15	1.7098	0.7427	338.26	182.7992	8.5742	572.76	17.1367
0.0035	3.18	2.4958	0.7957	350.42	194.7352	8.9855	573.32	14.4583
0.0071	5.61	4.4916	0.8487	364.07	195.1176	9.3961	574.07	14.9275
0.0106	7.67	7.4107	0.9016	374.93	200.1915	9.8061	574.63	15.8172
0.0142	10.28	10.0963	0.9546	387.47	210.0963	10.2155	575.39	16.2879
0.0177	12.71	12.8207	1.0075	398.31	201.1232	10.6243	575.94	15.2523
0.0212	15.33	15.1382	1.0604	408.62	174.7776	11.0324	576.56	12.5318
0.0248	17.76	17.0691	1.1133	415.90	172.1645	11.4399	576.88	11.1707
0.0283	20.19	19.4921	1.1662	424.89	193.8354	12.0500	577.63	12.8853
0.0319	22.62	17.1346	1.2191	433.49	180.5007	12.6586	578.19	11.6352
0.0390	24.86	21.4190	1.2720	440.61	169.6267	13.2659	578.75	12.2291
0.0425	27.11	27.9062	1.3777	454.46	153.1716	13.8718	579.31	12.8085
0.0496	31.97	35.5838	1.4834	464.39	142.3950	14.4762	579.86	11.1516
0.0567	37.19	36.0873	1.5891	474.70	96.3010	15.0793	580.25	11.8288
0.0637	41.13	40.0996	1.6947	477.68	89.7894	15.6810	580.80	11.0220
0.0708	45.98	59.2751	1.9058	497.54	124.5963	16.2813	581.10	10.2644
0.0814	56.71	64.3431	2.0112	503.15	97.7732	16.8803	581.55	10.2107
0.0921	63.26	53.1749	2.1167	507.83	90.6307	17.4779	581.83	10.0441
0.1027	69.05	60.0963	2.3274	516.24	89.6178	18.2725	582.39	10.0083
0.1133	75.59	62.9027	2.5380	524.09	87.2892	19.0648	582.71	9.9814
0.1239	80.93	69.4071	2.7484	530.80	70.8360	19.8546	583.21	9.9659
0.1381	89.72	81.8344	2.9587	535.10	83.0795	20.6421	583.51	9.9290
0.1522	97.76	88.7436	3.1688	542.38	90.3366	21.4271	583.96	9.8879
0.1664	106.17	95.4434	3.3787	547.23	63.7965	22.2098	584.24	9.9722
0.1806	114.03	101.4855	3.5885	550.40	37.4576	23.1849	584.78	9.9844
0.1947	122.06	107.8807	3.7981	551.72	23.8049	24.3501	585.13	9.8985
0.2124	131.60	114.7506	4.0075	553.03	32.6983	26.2806	586.22	9.8435
0.2478	150.66	126.5068	4.2168	555.09	45.4847	28.1966	586.62	9.8889
0.2832	167.87	142.3060	4.4259	557.52	49.0577	30.0984	587.52	9.8851
0.3186	186.01	155.2165	4.6349	559.73	41.3656	31.9861	587.90	9.8516
0.3539	202.46	151.5460	4.8437	561.28	39.7761	35.7198	589.71	9.8177
0.3893	216.49	158.0667	5.0523	563.15	38.2492	39.3991	590.10	9.7815
0.4247	231.08	175.4780	5.2607	564.46	22.7654	43.0251	591.42	9.8236
0.4600	245.68	192.5590	5.6772	564.80	18.3097	46.5990	591.83	9.8396
0.4954	260.63	177.4633	6.0930	567.00	19.0236	50.1218	592.86	9.8026
0.5307	271.29	184.2295	6.5081	567.51	13.2130	53.5947	593.25	9.8448
0.5661	284.95	202.9277	6.9226	568.64	20.5387	—	—	—
0.6014	296.73	173.6664	7.3365	569.96	19.7070	—	—	—
0.6367	305.52	187.2953	7.7497	570.89	15.7560	—	—	—
0.6897	324.23	210.9340	8.1622	571.64	18.7560	—	—	—

TABLE 14B.3 Plot Data for Figure 14.8 in Example 14.1

$\frac{(t_p+\Delta t)}{\Delta t}$	p_{ws} (psia)	$\frac{(t_p+\Delta t)}{\Delta t}$	p_{ws} (psia)	$\frac{(t_p+\Delta t)}{\Delta t}$	p_{ws} (psia)	$\frac{(t_p+\Delta t)}{\Delta t}$	p_{ws} (psia)
304990.97	2822.15	1525.70	3022.46	196.48	3350.80	40.71	3398.75
152471.10	2823.18	1387.10	3036.49	182.51	3355.10	38.93	3399.31
76236.32	2825.61	1271.59	3051.08	170.41	3362.38	37.30	3399.86
50824.53	2827.67	1173.85	3065.68	159.82	3367.23	35.81	3400.25
38118.57	2830.28	1090.08	3080.63	150.48	3370.40	34.44	3400.80
30495.17	2832.71	1017.47	3091.29	142.18	3371.72	33.17	3401.10
25412.76	2835.33	953.94	3104.95	134.75	3373.03	31.99	3401.55
21782.41	2837.76	897.89	3116.73	128.06	3375.09	30.90	3401.83
19059.84	2840.19	848.06	3125.52	122.01	3377.52	29.55	3402.39
16942.18	2842.62	782.90	3144.23	116.51	3379.73	28.32	3402.71
13861.97	2844.86	727.05	3158.26	111.49	3381.28	27.20	3403.21
12706.88	2847.11	678.65	3170.42	106.88	3383.15	26.16	3403.51
10891.76	2851.97	636.29	3184.07	102.65	3384.46	25.20	3403.96
9530.39	2857.19	598.92	3194.93	95.12	3384.80	24.31	3404.24
8471.59	2861.13	565.71	3207.47	88.63	3387.00	23.29	3404.78
7624.53	2865.98	535.98	3218.31	82.97	3387.51	22.18	3405.13
6630.16	2876.72	509.24	3228.62	78.01	3388.64	20.55	3406.22
5865.27	2883.26	485.03	3235.90	73.61	3389.96	19.15	3406.62
5258.61	2889.05	463.03	3244.89	69.68	3390.89	17.94	3407.52
4765.72	2895.59	442.94	3253.49	66.16	3391.64	16.88	3407.90
4357.31	2900.93	424.53	3260.61	62.98	3392.76	15.12	3409.71
3910.50	2909.72	391.95	3274.46	60.10	3393.32	13.71	3410.10
3546.82	2917.76	364.03	3284.39	57.47	3394.07	12.55	3411.42
3245.06	2926.17	339.82	3294.70	55.07	3394.63	11.59	3411.83
2990.62	2934.03	318.65	3297.68	52.86	3395.39	10.77	3412.86
2773.19	2942.06	283.35	3317.54	50.83	3395.94	10.08	3413.25
2542.18	2951.60	268.49	3323.15	48.95	3396.56	—	—
2179.14	2970.66	255.12	3327.83	47.20	3396.88	—	—
1906.88	2987.87	232.02	3336.24	44.81	3397.63	—	—
1695.12	3006.01	212.76	3344.09	42.66	3398.19	—	—

APPENDIX

14C CALCULATION OF PRESSURE DERIVATIVES

The calculation of pressure derivatives is based on an algorithm recommended by Bourdet et al.[1] The pressure derivative at a point of interest denoted as I in Figure 14C.1 is determined by using a preceding point denoted as 1 and a succeeding point denoted as 2 to calculate a weighted mean derivative which is placed at point I. Using the notation in Figure 14C.1, the algorithm can be written as:

$$\left(\frac{dp}{dX}\right)_I = \frac{\left[\left(\frac{\Delta p_1}{\Delta X_1}\right)\Delta X_2 + \left(\frac{\Delta p_2}{\Delta X_2}\right)\Delta X_1\right]}{(\Delta X_1 + \Delta X_2)} \tag{14-C.1}$$

This algorithm is illustrated by using a subset of the data from Example 14.1 shown in Table 14B.2. This is shown in Table 14C.1. At $\Delta t = 0.1523$, substituting Eq. (14-C.1) gives:

$$\left(\frac{dp}{dX}\right)_{@\Delta t=0.1523} = \frac{\left[\left(\frac{8.04}{0.0976}\right)\times 0.0889 + \left(\frac{8.41}{0.0889}\right)\times 0.0976\right]}{(0.0976+0.0889)}$$
$$= 88.7436$$

Bourdet et al.[1] recommends using a smoothing parameter, L, to reduce the effects of noise associated with pressure transient data on calculation of pressure derivatives. The smoothing parameter L can be defined as $\Delta(\ln t)$ for drawdown test data and $\Delta(\ln t_e)$ for buildup test data. The range of the smoothing parameter is $0 \le L \le 0.5$. For the pressure derivative calculated as shown earlier, no smoothing was applied, hence $L = 0$. The procedure for applying the smoothing parameter is as follows: For values of $L > 0$, subtract L from I and add L to I which creates a "window" around the point of interest I. Then use the data points just above and below this "window" to calculate the derivative at I. The implementation of the procedure for calculation of pressure derivatives with smoothing parameter, $L > 0$, requires additional programming steps especially for large data sets.

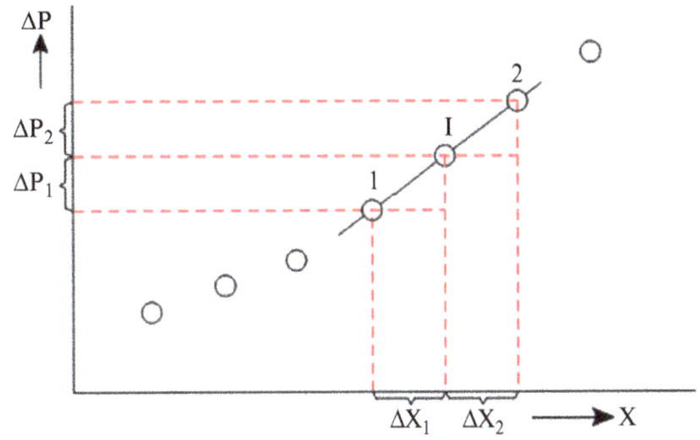

▲ FIGURE 14C.1 A sketch for calculation of pressure derivative

Reference

1. Bourdet, D., Ayoub, J.A., and Pirard, Y.M.: "Use of Pressure Derivative in Well-Test Interpretation," *SPEFE* (June 1989) 293–302.

TABLE 14C.1 Pressure Derivative Calculation (Subset of Data in Table 14B.2)

Δt (hrs)	Δp (psia)	Δt_e	$\ln \Delta t_e$	ΔX_1	ΔX_2	Δp_1	Δp_2	$\Delta t_e \Delta p'$
0.0921	63.26	0.0921	−2.3852	0.1226	0.1092	6.55	5.78	53.1749
0.1027	69.05	0.1027	−2.2761	0.1092	0.0984	5.78	6.55	60.0963
0.1133	75.59	0.1133	−2.1776	0.0984	0.0896	6.55	5.34	62.9027
0.1240	80.93	0.1239	−2.0880	0.0896	0.1082	5.34	8.79	69.4071
0.1381	89.72	0.1381	−1.9799	0.1082	0.0976	8.79	8.04	81.8344
0.1523	97.76	0.1522	−1.8822	0.0976	0.0889	8.04	8.41	88.7436
0.1665	106.17	0.1664	−1.7933	0.0889	0.0817	8.41	7.86	95.4434
0.1806	114.03	0.1806	−1.7117	0.0817	0.0755	7.86	8.03	101.4855
0.1948	122.06	0.1947	−1.6362	0.0755	0.0870	8.03	9.54	107.8807
0.2125	131.60	0.2124	−1.5492	0.0870	0.1541	9.54	19.07	114.7506
0.2479	150.66	0.2478	−1.3951	0.1541	0.1335	19.07	17.21	126.5068
0.2833	167.87	0.2832	−1.2617	0.1335	0.1177	17.21	18.14	142.3060

CHAPTER

15 Well Test Analysis: Hydraulically Fractured Wells and Naturally Fractured Reservoirs

15.1 Introduction

This chapter presents basic concepts on well test analysis of hydraulically fractured wells and naturally fractured reservoirs. Hydraulically fractured wells and naturally fractured reservoirs form a large part of hydrocarbon production capacity around the world. The well test analysis methods presented for these wells and reservoir types emphasize applications of combined straightline and type-curve matching techniques as previously recommended for unfractured wells and reservoirs. In this chapter, well test analysis for hydraulically fractured wells is presented first followed by well test analysis for naturally fractured reservoirs. Additional methods for analysis of diagnostic fracture injection tests (DFIT) in unconventional reservoirs are presented in Chapter 17.

15.2 Hydraulically Fractured Wells

Most wells completed in low permeability sand and shale formations (classified as unconventional reservoirs) in most regions of the United States of America are hydraulically fractured to achieve commercial production. In fact, many gas wells in these areas of United States of America (USA) could be categorized as unproductive without the application of hydraulic fracturing. Also, oil production has advanced into the deepwater regions of the Gulf of Mexico, USA. Deepwater reservoirs are also on production in offshore Brazil, West Africa, and other deepwater sectors around the world. Most of the wells completed in these deepwater reservoirs are hydraulically fractured to achieve and maintain production at high commercial rates in consideration of the high cost of operations in the deepwater environment. Many other oil producing regions in the world (especially in unconventional reservoirs in the USA) are also using hydraulic fracturing to improve well productivity. Consequently, it is important to understand how to analyze pressure transient tests from hydraulically fractured wells to evaluate the effectiveness of the post-fracture operations, predict expected productivity of the well, or evaluate the condition of a fractured well to determine whether additional stimulation is needed.

15.3 Definition of Dimensionless Variables for Fractured Wells

A number of dimensionless variables are defined specifically for fractured wells. The dimensionless fracture conductivity, C_{fD}, is defined as:

$$C_{fD} = \frac{k_f w_f}{kL_f} \tag{15.1}$$

In Eq. (15.1), k_f = fracture permeability, md; w_f = fracture width, ft; k = formation permeability, md; and L_f = fracture half-length, ft.

The dimensionless time, t_{DL_f}, in terms of fracture half-length is defined as:

$$t_{DL_f} = \frac{0.0002637kt}{\phi \mu c_t L_f^2} \tag{15.2}$$

Note that t_{DL_f} is related to the familiar dimensionless time, t_D, defined in Chapter 14, Eq. (14.3) by the expression:

$$t_{DL_f} = t_D \frac{r_w^2}{L_f^2} \tag{15.3}$$

In Eqs. (15.2) and (15.3), t = time, hr; ϕ = porosity, fraction; μ = viscosity, cp; c_t = total compressibility, psi^{-1}; and r_w = wellbore radius, ft.

15.4 Flow Regimes in Fractured Wells

Fluid flow in vertically fractured wells can be represented by four flow regimes as shown in Figure 15.1. The four flow regimes are:

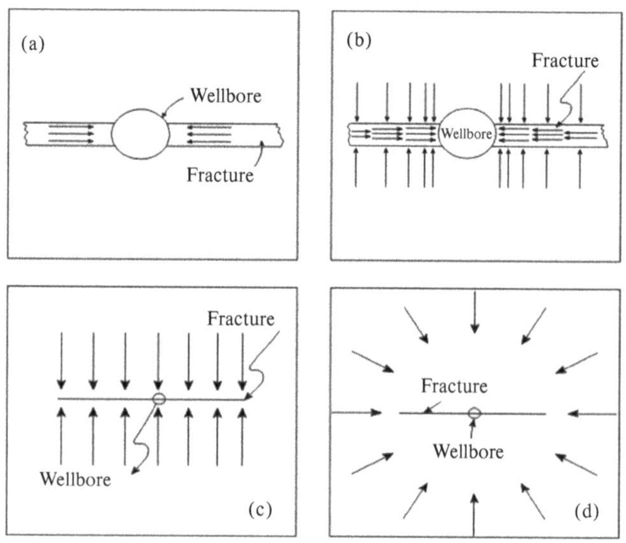

▲ FIGURE 15.1 Fracture flow regimes (from Cinco-Ley and Samaniego[1] © 1981 SPE, Reproduced with permission)

1. Fracture linear flow
2. Bilinear flow
3. Formation linear flow
4. Pseudo-radial flow

15.4.1 Fracture Linear Flow

Initial fluid flow in fractured wells is dominated by fracture linear flow regime (Figure 15.1a). During this flow regime, most of the fluid flow toward the wellbore is caused by the expansion of the fluid in the fracture. The fluid flow is linear and very short-lived. Fracture linear flow has no practical application in well test analysis because its duration is very short. The end of fracture linear flow can be estimated with the expression:[1]

$$t_{DL_f} = \frac{0.01C_{fD}^2}{\eta_{fD}^2} \tag{15.4}$$

In Eq. (15.4), η_{fD} is the dimensionless hydraulic diffusivity which is defined as:

$$\eta_{fD} = \frac{k_f \phi c_t}{k \phi_f c_{ft}} \tag{15.5}$$

The new terms in Eq. (15.5) are defined as follows: ϕ_f = fracture porosity, fraction; and c_{ft} = total compressibility of fluid in the fracture, psi^{-1}.

15.4.2 Bilinear Flow

Bilinear flow, if present in a fracture, begins at the end of fracture linear flow. The flow is described as bilinear because two linear flows occur at the same time. One linear flow occurs in the fracture and the other linear flow occurs in the formation (Figure 15.1b). Bilinear flow occurs in fractures with $C_{fD} < 300$. During bilinear flow regime, most of the fluid entering the wellbore comes from the formation until fracture-tip effects begin to affect well behavior. The time for bilinear flow to end depends on dimensionless fracture conductivity. The dimensionless times for bilinear flow to end for various ranges of dimensionless fracture conductivity are given as follows:[1]

$$t_{DL_f} = \frac{0.1}{C_{fD}^2}, \quad \text{for } C_{fD} \geq 3 \tag{15.6}$$

$$t_{DL_f} = 0.0205\left[C_{fD} - 1.5\right]^{-1.53} \quad \text{for } 1.6 \leq C_{fD} \leq 3 \tag{15.7}$$

$$t_{DL_f} = \left[\frac{4.55}{\sqrt{C_{fD}}} - 2.5\right]^{-4} \quad \text{for } C_{fD} \leq 1.6 \tag{15.8}$$

15.4.3 Formation Linear Flow

Highly conductive fractures with $C_{fD} \geq 300$ exhibit formation linear flow (Figure 15.1c). Bilinear flow does not occur in these high conductive fractures. Formation linear flow is estimated to begin at:

$$t_{DL_f} \simeq \frac{100}{C_{fD}^2} \tag{15.9}$$

The end of formation linear flow period is given by:

$$t_{DL_f} \simeq 0.016 \tag{15.10}$$

15.4.4 Pseudo-radial Flow

All fractured wells eventually exhibit pseudo-radial flow (Figure 15.1d) provided the flow period is sufficiently long and the fracture penetrates less than one-third of the drainage radius of the well.[2] Pseudo-radial flow is essentially equivalent to the radial flow used previously to represent fluid flow in unfractured wells. Hence, all equations developed for radial flow also apply to pseudo-radial flow. Pseudo-radial flow is estimated to begin for highly conductive fractures at:[3]

$$t_{DL_f} \simeq 5 \;\text{ for } C_{fD} \geq 300 \tag{15.11}$$

The range of dimensionless time for start of pseudo-radial flow for less conductive fractures is given by: $0.1 \leq C_{fD} \leq 300$ and $0.1 \leq t_{DL_f} \leq 5$.

15.5 Fractured Well Flow Models

There are three main flow models used to represent fractured wells. These are:

1. Finite conductivity vertical fracture
2. Infinite conductivity vertical fracture
3. Uniform flux vertical fracture

15.5.1 Finite Conductivity Vertical Fracture

Fractures with $C_{fD} < 300$ are considered to be finite conductivity fractures. Finite conductivity fractures can exhibit the four fluid flow regimes which are fracture linear flow, bilinear flow, formation linear flow, and pseudo-radial flow with intervening transition periods. As shown earlier, the duration of fracture linear flow is so short that it has no practical use in well test analysis. Generally, finite conductivity fractures transits from bilinear flow to pseudo-radial flow with the intermediate formation linear flow not fully developed.

15.5.2 Infinite Conductivity Vertical Fracture

Fractures with $C_{fD} \geq 300$ are considered to be infinite conductivity fractures. Infinite conductivity fractures do not exhibit bilinear flow regime. The flow regimes that occur in infinite conductivity fractures are formation linear flow followed by pseudo-radial flow after a transition period. The main characteristic of infinite conductivity fractures is that pressure drop along the length of the fracture is negligible.

15.5.3 Uniform Flux Vertical Fracture

Uniform flux fracture model assumes that the flow into the fracture is uniform along its length.[4] Uniform flux vertical fracture is similar to infinite conductivity vertical fracture in terms of flow behavior. As shown in Figure 15.2, there is little difference between the two fracture models. Uniform flux vertical fractures also exhibit only formation linear and pseudo-radial flow regimes.

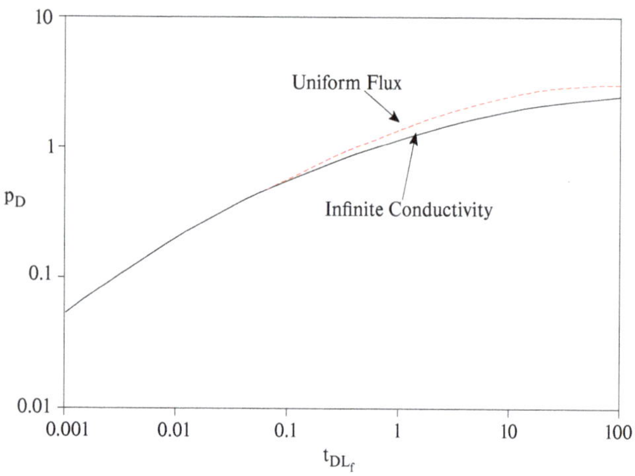

▲ **FIGURE 15.2** **Infinite conductivity vertical fracture versus uniform flux vertical fracture**

15.6 Fractured Well Test Analysis: Straightline Methods

Well test analysis in fractured wells is used to evaluate the effectiveness of hydraulic fracturing by determining fracture parameters, such as fracture half-length, fracture conductivity, and permeability of the formation. The flow regimes that exist during a test are determined by the conductivity of the fracture and the duration of the test. The procedures for fractured well test analysis are presented according to the following flow regimes: bilinear flow, formation linear flow, and pseudo-radial flow.

15.6.1 Bilinear Flow

Bilinear flow occurs in vertical fractures with $C_{fD} < 300$ at early flow times. The duration of bilinear flow for various ranges of C_{fD} are given in Eqs. (15.6) to (15.8). The dimensionless wellbore pressure for bilinear flow is expressed as:

$$p_D = \frac{2.45}{\sqrt{C_{fD}}} t_{DL_f}^{1/4} \tag{15.12}$$

Taking derivatives with respect to dimensionless time t_{DL_f} Eq. (15.12) becomes:

$$t_{DL_f} \frac{dp_D}{dt_{DL_f}} = \frac{0.6125}{\sqrt{C_{fD}}} t_{DL_f}^{1/4} \tag{15.13}$$

For an oil well, Eq. (15.12) can be transformed into:

$$\Delta p = \frac{44.1 qB\mu}{h\left(k_f w_f\right)^{1/2}\left(\phi\mu c_t k\right)^{1/4}} t^{1/4} \tag{15.14}$$

In Eq. (15.14), Δp = pressure change, psia; q = oil rate, STB/D; B = formation volume factor of oil, RB/STB; and h = net formation thickness, ft. Other terms are as defined in previous equations in this chapter.

Similarly, for a gas well, Eq. (15.12) can be transformed into:

$$\Delta p_p = \frac{444.75 q_g T}{h\left(k_f w_f\right)^{1/2}\left(\phi\mu c_t k\right)^{1/4}} t^{1/4} \tag{15.15}$$

In Eq. (15.15), Δp_p = real gas pseudo-pressure change, psia²/cp; q_g = gas rate, Mcf/D; and T = temperature, °R. It can be deduced from Eqs. (15.14) and (15.15) that a plot of Δp or Δp_p vs. $\sqrt[4]{t}$ on Cartesian coordinates will give a straight line with a slope m_{bf} that passes through the origin (Figure 15.3). From this plot, the fracture conductivity can be calculated as follows:

For an oil well:

$$k_f w_f = \left(\frac{44.1 qB\mu}{m_{bf} h}\right)^2 \left(\frac{1}{\phi\mu c_t k}\right)^{1/2} \tag{15.16}$$

For a gas well:

$$k_f w_f = \left(\frac{444.75 q_g T}{m_{bf} h}\right)^2 \left(\frac{1}{\phi\mu c_t k}\right)^{1/2} \tag{15.17}$$

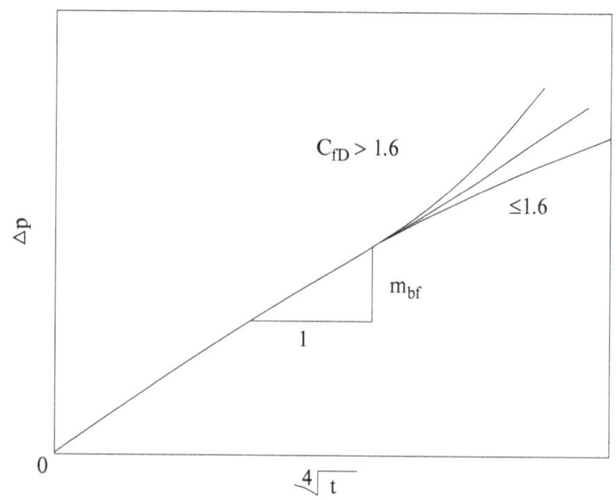

▲ **FIGURE 15.3** **Plot of Δp vs. $\sqrt[4]{t}$ for bilinear flow (from Cinco-Ley and Samaniego[1] © 1981 SPE, Reproduced with permission)**

Note that in Eqs. (15.16) and (15.17), other formation properties such as porosity (ϕ), fluid viscosity (μ), and total compressibility (c_t) are required. The formation permeability, k, can be determined from prefracture well test data.

As Eqs. (15.14) and (15.15) imply, bilinear flow is evident in a log-log plot of Δp vs. t which yields a straight line with a quarter-slope as shown in Figure 15.4. This is a diagnostic plot that can be used to demonstrate the presence of bilinear flow within the test period.

Bilinear flow regime may be distorted by effects of wellbore storage as shown in Figure 15.5 or flow restriction within the fracture which causes additional pressure drop as shown in Figure 15.6.

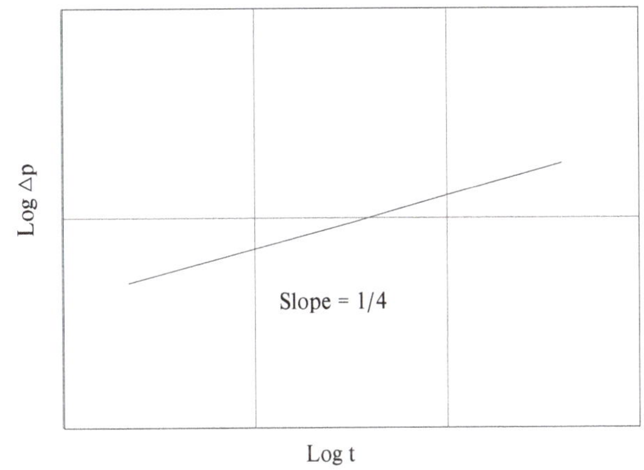

▲ **FIGURE 15.4** **Plot of log Δp vs. log t for bilinear flow (from Cinco-Ley and Samaniego[1] © 1981 SPE, Reproduced with permission)**

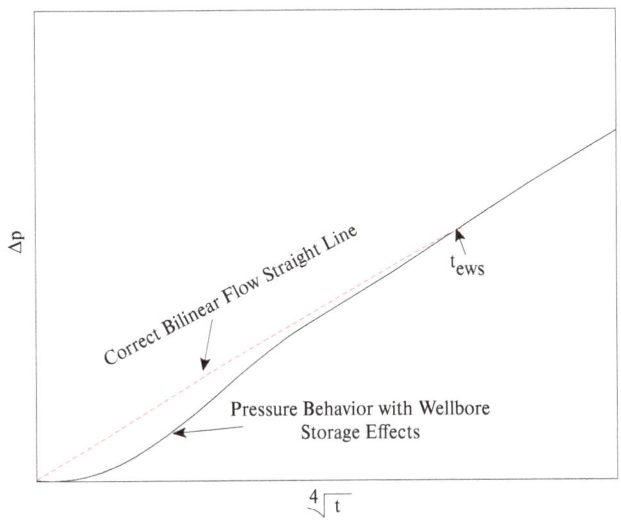

▲ **FIGURE 15.5** **Effects of wellbore storage on Δp vs. $\sqrt[4]{t}$ plot (from Cinco-Ley and Samaniego[1] © 1981 SPE, Reproduced with permission)**

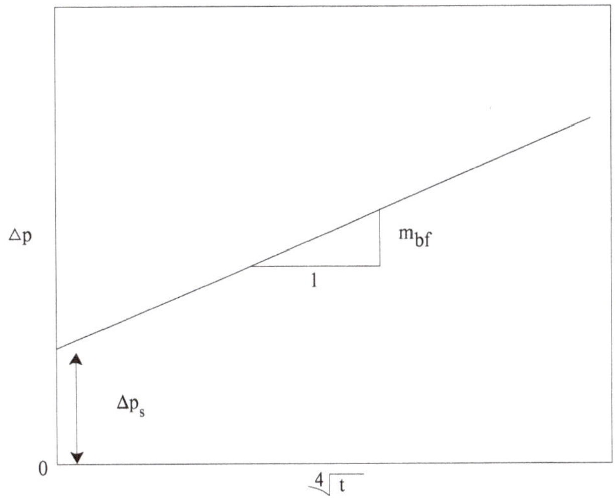

▲ **FIGURE 15.6** **Additional pressure drop Δp_s due to fracture restriction on Δp vs. $\sqrt[4]{t}$ plot (from Cinco-Ley and Samaniego[1] © 1981 SPE, Reproduced with permission)**

15.6.2 Procedure for Application of Straightline Methods on Well Test Data During Bilinear Flow Regime

1. Plot p_{wf} vs. $\sqrt[4]{t}$ for a constant rate flow test or p_{ws} vs. $\sqrt[4]{\Delta t}$ for a buildup test. For gas wells, use appropriate pressure transformation functions, such as real gas pseudo-pressure, as shown in Chapter 13.
2. Determine the slope, m_{bf}, from the straight-line region of the plot.
3. Calculate fracture conductivity from Eqs. (15.16) or (15.17).

Note that it is important to validate the results obtained with straightline methods with the results obtained from application of type-curve matching. This is discussed further under type-curve matching of fractured well tests.

15.6.3 Formation Linear Flow

Formation linear flow occurs in highly conductive fractures with $C_{fD} \geq 300$. The duration of formation linear flow is estimated with Eqs. (15.9) and (15.10). The dimensionless wellbore pressure change for formation linear flow is expressed as:

$$p_D = \sqrt{\pi t_{DL_f}} \tag{15.18}$$

By logarithmic expansion, Eq. (15.18) gives:

$$\log p_D = \frac{1}{2}\log\left(t_{DL_f}\right) + \frac{1}{2}\log(\pi) \tag{15.19}$$

Taking derivative of p_D in Eq. (15.18) with respect to t_{DL_f} yields:

$$t_{DL_f}\frac{dp_D}{dt_{DL_f}} = \frac{1}{2}\sqrt{\pi t_{DL_f}} \tag{15.20}$$

Again, applying logarithmic expansion to Eq. (15.20) gives:

$$\log\left(t_{DL_f}\frac{dp_D}{dt_{DL_f}}\right)=\frac{1}{2}\log\left(t_{DL_f}\right)+\frac{1}{2}\log(\pi) \quad (15.21)$$

From Eq. (15.19), it is evident that a log-log plot of Δp vs. t will yield a half-slope straight line (Figure 15.7). Similarly from Eq. (15.21), a log-log plot of the pressure derivative function versus time will also yield a half-slope straight line.

For an oil well, Eq. (15.18) can be expanded to give:

$$\Delta p=\frac{4.064qB}{hL_f}\sqrt{\frac{\mu t}{k\phi c_t}} \quad (15.22)$$

Similarly for a gas well, Eq. (15.18) can be expanded to give:

$$\Delta p_p=\frac{40.925q_gT}{hL_f}\sqrt{\frac{t}{k\phi\mu c_t}} \quad (15.23)$$

From Eqs. (15.22) and (15.23), it is evident that a plot of Δp or Δp_p vs. $\sqrt{t}$ on Cartesian coordinates will give a straight line with a slope, m_{Lf}, that passes through the origin (Figure 15.8). The fracture half-length, L_f, can be calculated from the slope m_{Lf} as follows:

For an oil well:

$$L_f=\frac{4.064qB}{m_{Lf}h}\sqrt{\frac{\mu}{k\phi c_t}} \quad (15.24)$$

For a gas well:

$$L_f=\frac{40.925q_gT}{m_{Lf}h}\sqrt{\frac{1}{k\phi\mu c_t}} \quad (15.25)$$

Formation linear flow regime can be distorted by wellbore storage effects as shown in Figure 15.9, or flow restrictions within the fracture that will cause additional pressure drop close to the wellbore.

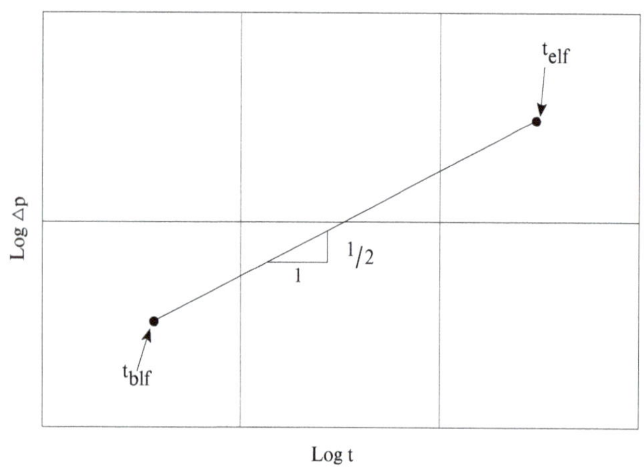

▲ **FIGURE 15.7** **Plot of log Δp vs. log t for linear flow (from Cinco-Ley and Samaniego[1] © 1981 SPE, Reproduced with permission)**

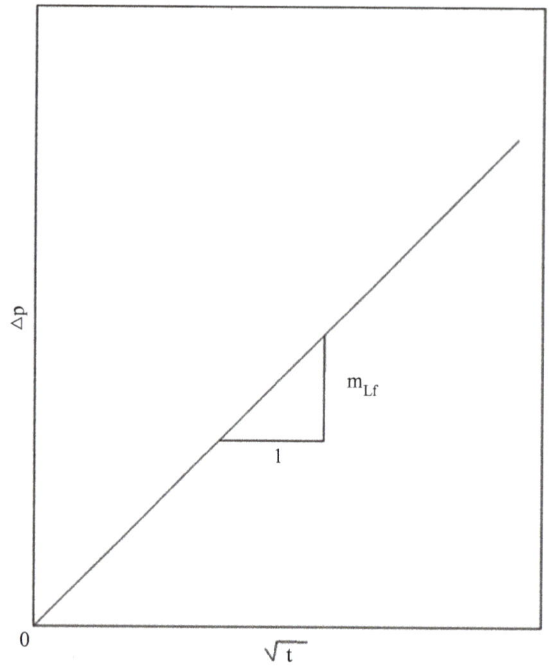

▲ **FIGURE 15.8** **Plot of Δp vs. $\sqrt{t}$ for linear flow (from Cinco-Ley[2] © 1982 SPE, Reproduced with permission)**

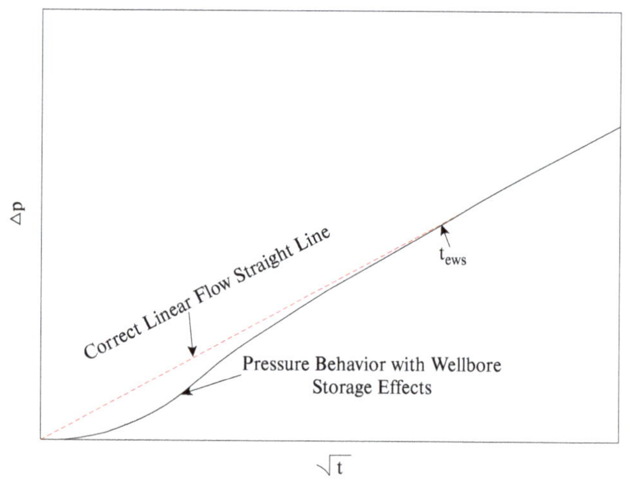

▲ **FIGURE 15.9** **Effects of wellbore storage on Δp vs. $\sqrt{t}$ plot (from Cinco-Ley[2] © 1982 SPE, Reproduced with permission)**

15.6.4 Procedure for Application of Straightline Methods on Well Test Data During Formation Linear Flow Regime

1. Plot p_{wf} vs. $\sqrt{t}$ for a constant rate flow test or p_{ws} vs. $\sqrt{\Delta t}$ for a buildup test. For gas wells, use appropriate pressure transformation functions such as real gas pseudo-pressure, as shown in Chapter 13.
2. Determine the slope, m_{Lf}, from the straight-line region of the plot.
3. Calculate fracture half-length, L_f, from Eqs. (15.24) or (15.25).

Note that it is important to validate the results obtained with straightline methods with the results obtained from application of type-curve matching. This is discussed further under type-curve matching of fractured well tests.

15.6.5 Pseudo-Radial Flow

Finite and infinite conductivity fractures eventually exhibit pseudo-radial flow if the duration of the test is sufficiently long. Infinite conductivity fractures with $C_{fD} \geq 300$ will reach pseudo-radial flow when $t_{DL_f} \simeq 5$. Pseudo-radial flow for finite conductivity fractures is estimated to begin at:[2]

$$t_{DL_f} \simeq 5\exp\left[-0.5\left(C_{fD}\right)^{-0.6}\right] \text{ for } 0.1 \leq C_{fD} < 300. \tag{15.26}$$

Pseudo-radial flow is theoretically equivalent to the radial flow used previously to represent fluid flow in unfractured wells. Hence, all equations developed for radial flow and well test analysis in Chapters 12 and 13, respectively, apply to pseudo-radial flow in fractured wells. Consequently, a semi-log plot of p_{wf} vs. t for a drawdown test or p_{ws} vs. $\left(t_p + \Delta t\right)/\Delta t$ for a buildup test will yield a straight line. For a drawdown test in an oil well, the slope, m, of the straightline can be used to calculate the following formation properties as shown in Chapter 13:

$$k = \left|\frac{162.6qB\mu}{mh}\right| \tag{15.27}$$

$$s = 1.151\left[\frac{\bar{p} - p_{1\text{hr}}}{m} - \log\left(\frac{k}{\phi\mu c_t r_w^2}\right) + 3.23\right] \tag{15.28}$$

For a buildup test in an oil well:

$$k = \left|\frac{162.6qB\mu}{mh}\right| \tag{15.29}$$

$$s = 1.151\left[\frac{p_{1\text{hr}} - p_{wf}}{m} - \log\left(\frac{k}{\phi\mu c_t r_w^2}\right) + 3.23\right] \tag{15.30}$$

For a drawdown test in a gas well using real gas pseudo-pressure, p_p:

$$k = \left|\frac{1,637q_g T}{mh}\right| \tag{15.31}$$

$$s' = 1.151\left[\frac{\bar{p}_p - p_{p,1\text{hr}}}{m} - \log\left(\frac{k}{\phi\bar{\mu}\bar{c}_t r_w^2}\right) + 3.23\right] \tag{15.32}$$

For a buildup test in a gas well using real gas pseudo-pressure, p_p:

$$k = \left|\frac{1,637q_g T}{mh}\right| \tag{15.33}$$

$$s' = 1.151\left[\frac{p_{p,1\text{hr}} - p_{p,wf}}{m} - \log\left(\frac{k}{\phi\bar{\mu}\bar{c}_t r_w^2}\right) + 3.23\right] \tag{15.34}$$

In Eqs. (15.32) and (15.34), $s' = s + Dq_g$, where s' = apparent skin factor; s = skin factor due to damage or stimulation; D = non-Darcy flow coefficient; and q_g = gas flow rate, Mcf/D. Gas properties are evaluated either at average reservoir pressure or initial pressure (Chapters 12 and 13).

The effective wellbore radius, r_{wa}, is related to skin factor, s, by:

$$r_{wa} = r_w e^{-s} \tag{15.35}$$

The half-length of a vertical fracture, L_f, can be estimated from effective wellbore radius as:

$$L_f = 2r_{wa} \tag{15.36}$$

Note that Eqs. (15.27) to (15.36) were previously derived and presented in Chapter 13.

15.6.6 Procedure for Application of Straightline Methods on Well Test Data During Pseudo-Radial Flow Regime

1. Plot p_{wf} vs. t for a constant rate drawdown test or p_{ws} vs. $\left(t_p + \Delta t\right)/\Delta t$ for a buildup test on a semi-log scale. For gas wells, use appropriate pressure transformation functions, such as real gas pseudo-pressure, as shown in Chapter 13.
2. Determine the slope, m, from the straight-line region of the plot.
3. Calculate formation properties from Eqs. (15.27) to (15.34) depending on the type of test and the type of fluid in the reservoir.

Note that it is important to validate the results obtained with straightline methods with the results obtained from application of type-curve matching. This is discussed further under type-curve matching of fractured well tests.

15.7 Fractured Well Test Analysis: Type-Curve Matching

In Chapter 14, analysis of well test data by type-curve matching was presented as consisting of three main stages:

1. Identification of interpretation model.
2. Calculations from interpretation model parameters.
3. Validation of the interpretation model.

These stages are also applicable to analysis of test data from fractured wells by type-curve matching. Fractured well type curves have similar shapes (Figures 15.10 and 15.11) and tend to result in a match that is non-unique. Thus, a systematic application of the above three stages in analyzing fractured well test data in an iterative mode leads ultimately to achievement of results that are consistent with current knowledge of the well and the reservoir.

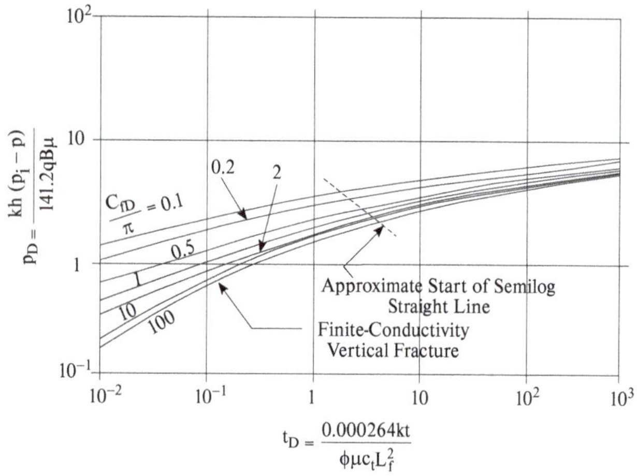

▲ **FIGURE 15.10 Type curve for finite conductivity vertical fracture (from Cinco-Ley et al.[3] © 1978 SPE, Reproduced with permission)**

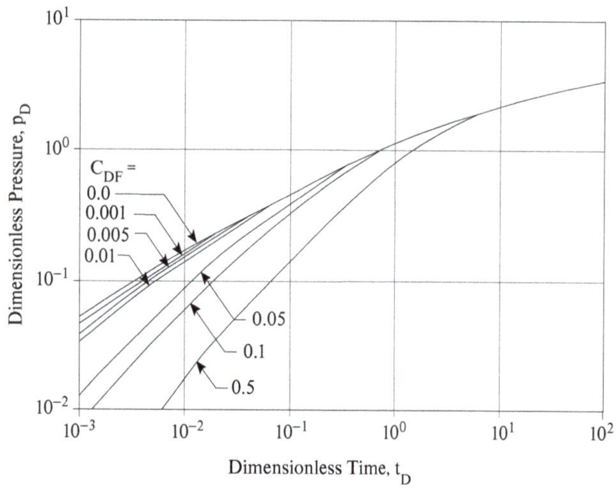

▲ **FIGURE 15.11 Type curve for an infinite conductivity vertical fracture with wellbore storage (from Cinco-Ley[2] © 1982 SPE, Reproduced with permission)**

15.7.1 Identification of the Interpretation Model

Earlier in this chapter, three flow regimes that could be present in testing a fractured well were discussed. These flow regimes were identified as bilinear flow, formation linear flow, and pseudo-radial flow. The fourth flow regime which is fracture linear flow has such a very short duration at early times that it has little or no application in well test analysis. The presence of bilinear flow, formation linear flow, and pseudo-radial flow can be identified with type curves. Note that all these flow regimes may not be present in a test either because of the duration of the test or the conductivity of the fracture. Typically, finite conductivity fractures $(C_{fD} < 300)$ will exhibit bilinear flow followed by transition period before reaching pseudo-radial flow if the duration of the test is sufficiently long. The formation linear flow regime is not fully developed before the emergence of pseudo-radial flow in most finite conductivity fractures. On the other hand, infinite conductivity fractures $(C_{fD} \geq 300)$ exhibit formation linear flow before transiting to pseudo-radial flow depending on the duration of the test. Bilinear flow regime is typically absent in tests of infinite conductive fractures.

In analyzing fractured well test data, type curves could be used initially to identify the presence of these flow regimes. For instance, the presence of bilinear flow is indicated by the presence of the quarter-slope line in a log-log plot of Δp vs. t in a drawdown test or Δp vs. Δt_e for a buildup test. The quarter-slope line should match the quarter-slope line on a type curve. Similarly, the presence of formation linear flow is indicated by the presence of half-slope line in a log-log plot of Δp vs. t in a drawdown test or Δp vs. Δt_e for a buildup test. The half-slope line should match the half-slope line on a type curve. Note that derivative type curves could also be plotted. In that case, on the same log-log plot, plot $t\Delta p'$ vs. t for a drawdown test or plot $\Delta t_e \Delta p'$ vs. Δt_e for a buildup test. Typically for fractured wells, the pressure curve and the derivative curve have the same slope as demonstrated in Eqs. (15.13) and (15.21). The derivative plot is useful in the identification of attainment of pseudo-radial flow.[5] Note that the presence of wellbore storage and fracture damage could distort the early flow regimes.[6,7]

15.7.2 Calculation from Interpretation Model Parameters

After achieving a match with a type curve, fracture and reservoir parameters can be calculated from the match points. From the pressure match point (PMP), the formation permeability, k, can be calculated from the following equations as:

$$k = \frac{141.2qB\mu}{h}\left(\frac{p_D}{\Delta p}\right)_{PMP} \quad \text{for an oil well} \qquad (15.37)$$

$$k = \frac{1422q_g T}{h}\left(\frac{p_D}{\Delta p_p}\right)_{PMP} \quad \text{for a gas well} \qquad (15.38)$$

The units for the terms in Eqs. (15.37) and (15.38) are as follows: k = permeability, md; q = oil rate, STB/D; B = FVF, RB/STB; μ = oil viscosity, cp; h = formation thickness, ft; q_g = gas flow rate, Mcf/D; and T = temperature, °R.

Equation (15.38) is based on real gas pseudo-pressure, p_p. From the time match point (TMP), the fracture half-length, L_f, can be calculated from the following equations as:

$$L_f = \sqrt{\left(\frac{0.0002637k}{\phi\mu c_t}\right)\left(\frac{\Delta t \text{ or } \Delta t_e}{t_{DL_f}}\right)_{TMP}} \quad \text{for an oil well} \tag{15.39}$$

$$L_f = \sqrt{\left(\frac{0.0002637k}{\phi\bar{\mu}\bar{c}_t}\right)\left(\frac{\Delta t \text{ or } \Delta t_e}{t_{DL_f}}\right)_{TMP}} \quad \text{for an gas well} \tag{15.40}$$

In Eqs. (15.39) and (15.40), L_f = fracture half-length, ft; ϕ = porosity, fraction; and c_t = total compressibility, psi^{-1}. Gas properties are evaluated either at average reservoir pressure or initial pressure (Chapters 12 and 13).

From the match of the dimensionless fracture conductivity, C_{fD}, the fracture conductivity can be calculated as:

$$w_f k_f = C_{fD} \times kL_f \tag{15.41}$$

15.7.3 Validation of the Interpretation Model Results

The validation step involves using straightline methods to calculate fracture and reservoir properties as previously discussed and comparing them to the results obtained from type-curve matching. If the results are reasonably close, then it can be presumed that the interpretation model is representative of the test data, and that the results are inherently consistent. If the results are widely divergent, then the analysis should begin again from type-curve matching by choosing another interpretation model and repeating the entire process. This process is highly iterative and should be continued until reasonable agreement is achieved with the results from type-curve matching and straightline methods.

For instance, if the interpretation model had indicated bilinear flow, use the procedures presented earlier to calculate fracture conductivity. Compare it to the fracture conductivity calculated from type-curve matching. If the results are in reasonable agreement, then the analysis is terminated. If not, repeat the entire process starting from selection of another interpretation model.

Similarly, if the interpretation model had indicated formation linear flow, use the procedures enumerated earlier to calculate the fracture half-length. Compare it to the fracture half-length from type-curve matching. If the results are in reasonable agreement, then the analysis is terminated. If not, repeat the entire process starting from selection of another interpretation model.

If the duration of the test is sufficiently long to reach pseudo-radial flow, use the procedures enumerated above to calculate formation permeability. Compare it to the formation permeability from type-curve matching. If the results are in reasonable agreement, then the analysis is terminated. If not, repeat the entire process starting from selection of another interpretation model. From pseudo-radial flow analysis, other well parameters such as skin factor and fracture half-length can be estimated as shown in the procedures for analysis of pseudo-radial flow using straightline methods.

15.7.4 Procedure for Analysis of Well Test from Hydraulically Fractured Wells

The procedure for analysis of well test from hydraulically fractured wells is summarized in the following steps:

1. Plot log Δp vs. log t for a drawdown test or log Δp vs. log Δt_e for a buildup test on the same scale size as the type curves graph.
2. Select either finite conductivity or infinite conductivity type curves. Most wells exhibit finite conductivity behavior. For this reason, make the first pass with finite conductivity type curves.
3. Match your plot to one of the type curves. Identify potential flow regimes on the type curve. Select a match point. Determine pressure and time match points.
4. Calculate reservoir and well parameters from the match points.
5. Make appropriate plots of p_{wf} or p_{ws} vs. $\sqrt[4]{t}$ for bilinear flow; p_{wf} or p_{ws} vs. $\sqrt{t}$ for formation linear flow; p_{wf} vs. log t for drawdown test or $\left[p_{ws} \text{ vs. } \log\left\{(t_p + \Delta t)/\Delta t\right\}\right]$ for a buildup test, if pseudo-radial is indicated.
6. Calculate reservoir and well parameters from straightline methods of Step 5.
7. Compare results from Step 4 to results from Step 6.
8. If results are in reasonable agreement, terminate the analysis. If not, repeat Steps 2 to 7.

Note that the above procedure is for manual analysis of fractured well tests. Most modern well test analysis is computer-aided including non-linear regression matching of the test data to an interpretation model. For this reason, the well test analyst may never have to conduct manual type-curve matching of the test data. As pointed out previously, this procedure was included to illustrate the steps actually involved in computer-aided well test analysis. The goal for this approach is to provide the reader with thorough understanding of the processes embedded in computer-aided analysis of well tests.

EXAMPLE 15.1 Calculation of Well and Reservoir Parameters from a Drawdown Test Data Using Finite Conductivity Type Curves

Problem

A drawdown test was conducted on a hydraulically fractured well. The test data are shown in Table 15.1. Calculate fracture half-length and reservoir properties from the test data. Well and other reservoir properties are given as follows:

TABLE 15.1 Test and Plot Data for Example 15.1

t (hrs)	p_{wf} (psia)	$\sqrt{t}$ (hrs.)$^{0.5}$	Δp (psia)	t (hrs)	p_{wf} (psia)	$\sqrt{t}$ (hrs)$^{0.5}$	Δp (psia)
0.3	2488.64	0.548	61.36	60.0	2182.91	7.746	367.09
0.5	2477.09	0.707	72.91	65.0	2172.85	8.062	377.15
1.0	2457.86	1.000	92.14	70.0	2163.30	8.367	386.70
1.5	2444.35	1.225	105.65	75.0	2154.18	8.660	395.82
2.0	2433.57	1.414	116.43	80.0	2145.46	8.944	404.54
2.5	2424.46	1.581	125.54	85.0	2137.10	9.220	412.90
3.0	2416.49	1.732	133.51	90.0	2129.05	9.487	420.95
3.5	2409.35	1.871	140.65	95.0	2121.30	9.747	428.70
4.0	2402.87	2.000	147.13	100.0	2113.81	10.000	436.19
4.5	2396.90	2.121	153.10	105.0	2106.56	10.247	443.44
5.0	2391.36	2.236	158.64	110.0	2099.54	10.488	450.46
5.5	2386.17	2.345	163.83	115.0	2092.73	10.724	457.27
6.0	2381.28	2.449	168.72	120.0	2086.11	10.954	463.89
6.5	2376.66	2.550	173.34	130.0	2078.51	11.402	471.49
7.0	2372.27	2.646	177.73	140.0	2070.91	11.832	479.09
7.5	2368.08	2.739	181.92	150.0	2063.31	12.247	486.69
8.0	2364.07	2.828	185.93	160.0	2055.71	12.649	494.29
8.5	2360.23	2.915	189.77	170.0	2048.11	13.038	501.89
9.0	2356.53	3.000	193.47	180.0	2040.51	13.416	509.49
9.5	2352.97	3.082	197.03	190.0	2032.91	13.784	517.09
10.0	2349.53	3.162	200.47	200.0	2025.31	14.142	524.69
11.0	2342.97	3.317	207.03	210.0	2020.11	14.491	529.89
12.0	2336.80	3.464	213.20	220.0	2014.91	14.832	535.09
13.0	2330.96	3.606	219.04	230.0	2009.71	15.166	540.29
14.0	2325.41	3.742	224.59	240.0	2004.51	15.492	545.49
15.0	2320.12	3.873	229.88	250.0	1999.31	15.811	550.69
16.0	2315.05	4.000	234.95	260.0	1994.11	16.125	555.89
17.0	2310.19	4.123	239.81	270.0	1988.91	16.432	561.09
18.0	2305.52	4.243	244.48	280.0	1983.71	16.733	566.29
19.0	2301.02	4.359	248.98	290.0	1978.51	17.029	571.49
20.0	2296.67	4.472	253.33	300.0	1973.31	17.321	576.69
22.0	2288.38	4.690	261.62	310.0	1968.71	17.607	581.29
24.0	2280.58	4.899	269.42	320.0	1964.11	17.889	585.89
26.0	2273.20	5.099	276.80	330.0	1959.51	18.166	590.49
28.0	2266.19	5.292	283.81	340.0	1954.91	18.439	595.09
30.0	2259.50	5.477	290.50	350.0	1950.31	18.708	599.69
35.0	2243.98	5.916	306.02	360.0	1945.71	18.974	604.29
40.0	2229.87	6.325	320.13	370.0	1941.11	19.235	608.89
45.0	2216.88	6.708	333.12	380.0	1936.51	19.494	613.49
50.0	2204.82	7.071	345.18	390.0	1931.91	19.748	618.09
55.0	2193.53	7.416	356.47	400.0	1927.31	20.000	622.69

Formation thickness, h	60 ft
Formation porosity, ϕ	0.20
Total compressibility, c_t	20×10^{-6} psi^{-1}
Oil viscosity, μ_o	2.0 cp
Oil FVF, B_o	1.5 RB/STB
Wellbore radius, r_w	0.25 ft
Production rate, q_o	200 STB/D
Initial average pressure, $\bar{p}$	2550 psia

Solution

Step 1: Plot log Δp vs. log t.

The data in Table 15.1 were used to make a log-log plot of Δp vs. t as shown in Figure 15.12.

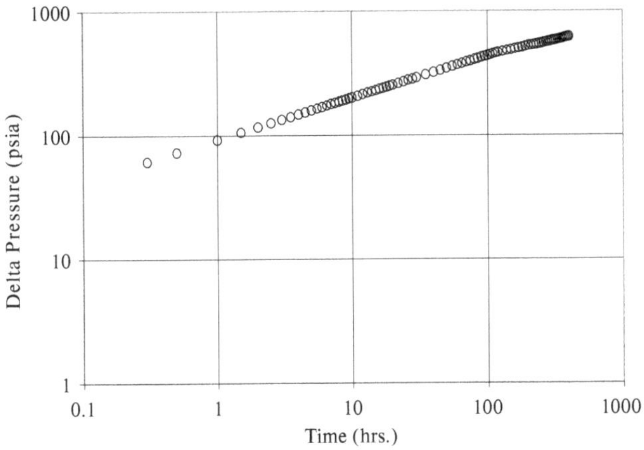

▲ FIGURE 15.12 Log-Log plot of Δp vs. t for Example 15.1

Step 2: Type-curve matching.

The type curve shown as Figure 15.10 was used in the type-curve matching procedure. Figure 15.12 was superimposed on Figure 15.10. The matched (superimposed) plot of Figures 15.10 and 15.12 is shown as Figure 15.13. The matched points are as follows:

Pressure match point (PMP), $(p_D/\Delta p)_{PMP} = (0.6/100)$

Time match point (TMP), $\left(\Delta t/t_{DL_f}\right)_{TMP} = (100/3.0)$

Using Eq. (15.37),

$$k = \frac{141.2qB\mu}{h}\left(\frac{p_D}{\Delta p}\right)_{PMP}$$

$$= \frac{141.2\times200\times1.5\times2}{60}\times\left(\frac{0.6}{100}\right)$$

$$= 8.47 \text{ md}$$

From Eq. (15.39),

$$L_f = \sqrt{\left(\frac{0.0002637k}{\phi\mu c_t}\right)\left(\frac{\Delta t}{t_{DL_f}}\right)_{TMP}}$$

$$= \sqrt{\frac{0.0002637\times8.47}{0.2\times2\times20\times10^{-6}}\times\left(\frac{100}{3}\right)}$$

$$= 96.47 \text{ ft.}$$

From Figure 15.13, the matched type curve is designated as: $C_{fD} = 10\pi$.

From Eq. (15.41), fracture conductivity is calculated as:

$$w_f k_f = C_{fD}\times kL_f$$

$$= 10\pi\times8.47\times96.47$$

$$= 25{,}670 \text{ md-ft}$$

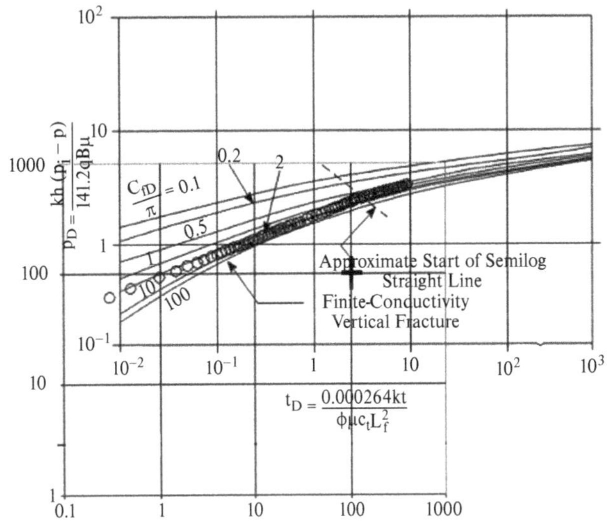

▲ FIGURE 15.13 Type curve match for Example 15.1

Step 3: Plot Δp vs. $\sqrt{t}$.

The data in Table 15.1 were used to make a plot of Δp vs. $\sqrt{t}$ as shown in Figure 15.14. The slope of the straight line is:

$$m_{Lf} = 54 \text{ psia/hrs}^{0.5}$$

Applying Eq. (15.24),

$$L_f = \frac{4.064qB}{m_{Lf}h}\sqrt{\frac{\mu}{k\phi c_t}}$$

$$= \frac{4.064\times200\times1.5}{54\times60}\sqrt{\frac{2}{8.47\times0.2\times20\times10^{-6}}}$$

$$= 91.43 \text{ ft.}$$

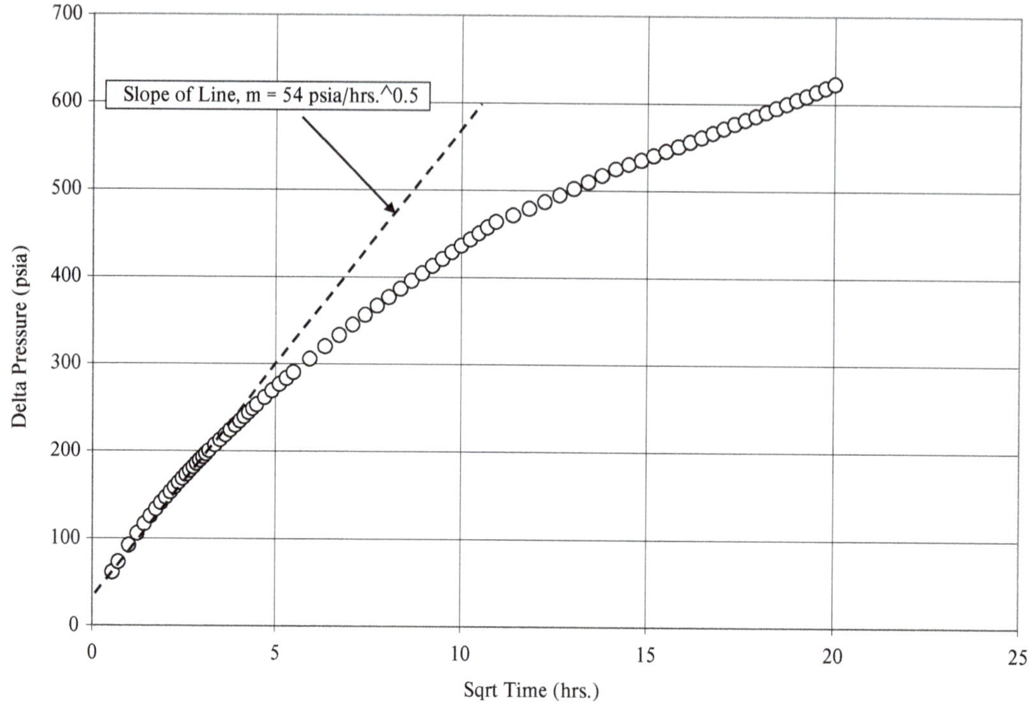

▲ **FIGURE 15.14** **Plot of Δp vs. $\sqrt{t}$ for Example 15.1**

Step 4: Plot p_{wf} vs. $\log t$.

The data in Table 15.1 were used to prepare a semi-log plot of p_{wf} vs. t as shown in Figure 15.15. The slope of the semi-log straight line is: $m = 280$ psi/cycle.

Applying Eq. (15.27),

$$k = \frac{162.6 qB\mu}{mh}$$

$$= \frac{162.6 \times 200 \times 1.5 \times 2}{280 \times 60} = 5.81 \text{ md}$$

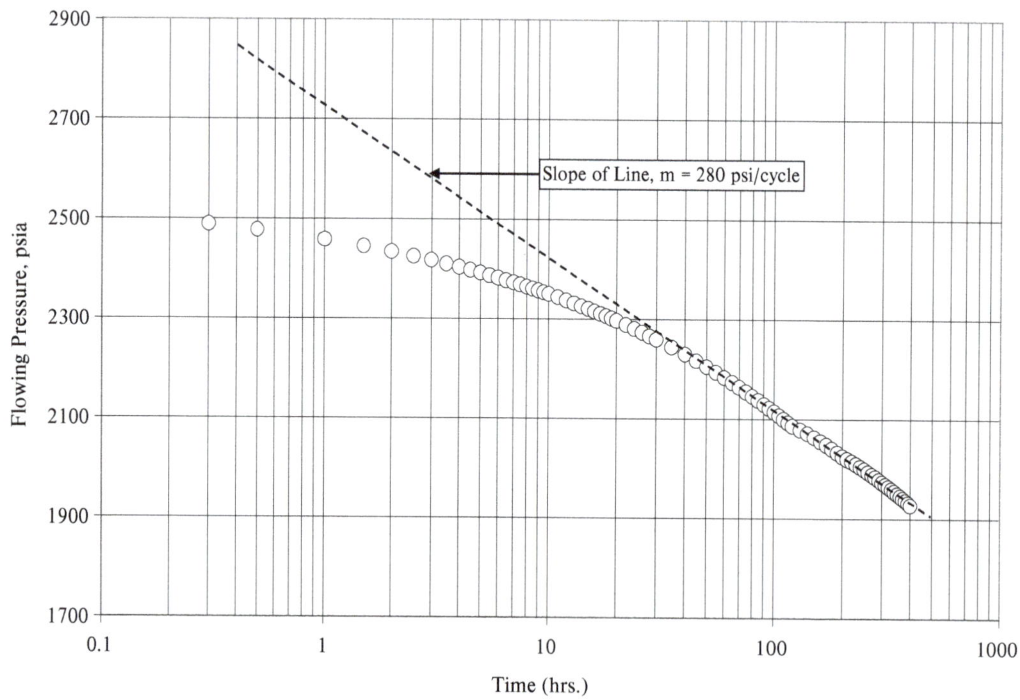

▲ **FIGURE 15.15** **Semi-log plot of p_{wf} vs. t for Example 15.1**

Also, from Figure 15.15, $p_{1hr} = 2738$ psia.

Applying Eq. (15.28),

$$s = 1.151\left[\frac{\bar{p} - p_{1\text{hr}}}{m} - \log\left(\frac{k}{\phi\mu c_t r_w^2}\right) + 3.23\right]$$

$$= 1.151\left[\frac{2550 - 2738}{280} - \log\left(\frac{5.81}{0.2\times 2\times 20\times 10^{-6}\times (0.25)^2}\right) + 3.23\right]$$

$$= -5.19$$

Applying Eq. (15.35),

$$r_{wa} = r_w e^{-s} = (0.25)e^{-(-5.19)} = 44.7 \text{ ft.}$$

Further by applying Eq. (15.36),

$$L_f = 2r_{wa} = 2\times 44.7 = 89.4 \text{ ft.}$$

There are reasonable agreements between well and reservoir parameters calculated from type-curve matching and straightline methods. For instance, the formation permeabilities calculated from the two methods are reasonably close. This indicates that the results obtained from analyzing the well test data with the two methods are reasonably consistent.

End of Example 15.1

15.8 Naturally Fractured Reservoirs

Many reservoirs in different petroleum producing formations across the world are naturally fractured. In a majority of these cases, the productivity of these reservoirs depends on the presence of these fractures. The fractures act as conduits that convey fluids from the matrix to the wellbore. Production rates of wells in naturally fractured reservoirs are relatively high, although these wells may have short lives due to channeling of gas and/or water through the fracture system to the wellbore. The pressure response of naturally fractured reservoirs during a well test is different from that of unfractured reservoirs. In this section, well test analysis is presented as a tool that can be applied in the diagnosis of naturally fractured reservoirs.

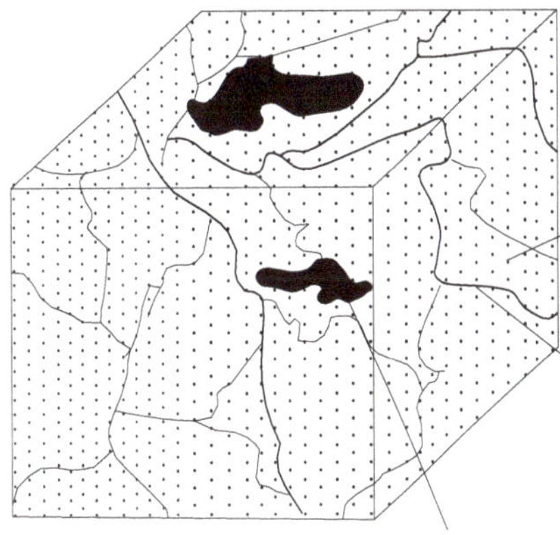

▲ FIGURE 15.16 Naturally fractured rock (from Cinco-Ley[8] © 1996 SPE, Reproduced with permission)

15.9 Naturally Fractured Reservoir Models

Naturally fractured reservoirs (NFRs) are composed of random distributions of fractures, vugs, and matrices, as shown in Figure 15.16. The distribution of fractures in the reservoir can be massive, localized, oriented, or clustered along a fault. The nature of the fractures and their distribution has considerable influence on the pressure response of a well in a test. In well test analysis, naturally fractured reservoirs can be represented with different models based on the type and distribution of the fractures within the system. These representative models are:[8]

1. Homogeneous reservoir
2. Multiple region or composite reservoir
3. Anisotropic reservoir
4. Single fracture reservoir
5. Double-porosity reservoir

15.9.1 Homogeneous Reservoir Model

Homogeneous reservoir model can be used to represent massively fractured formations with small matrix blocks (Figure 15.17a) or fractured formations in which most of the fluids are stored within the fractures (Figure 15.17b). In these types of NFRs, the fractures and the rock matrix behave as a single indistinguishable system. Such NFRs exhibit homogeneous type of behavior in flow tests. Their pressure responses in well tests are similar to those of other homogeneous-acting reservoirs previously described in Chapters 13 and 14. Consequently, all well test analysis techniques composed of straightline methods and type-curve matching developed earlier in Chapters 13 and 14 are equally applicable to these types of NFRs. For instance, a plot of p_{wf} vs. $\log t$ for a drawdown test or a Horner plot of p_{ws} vs. $\log\{(t_p + \Delta t)/\Delta t\}$ for a buildup test yields a straightline with slope, m, from which the flow capacity, kh, of the entire system (fracture plus matrix) can be calculated.

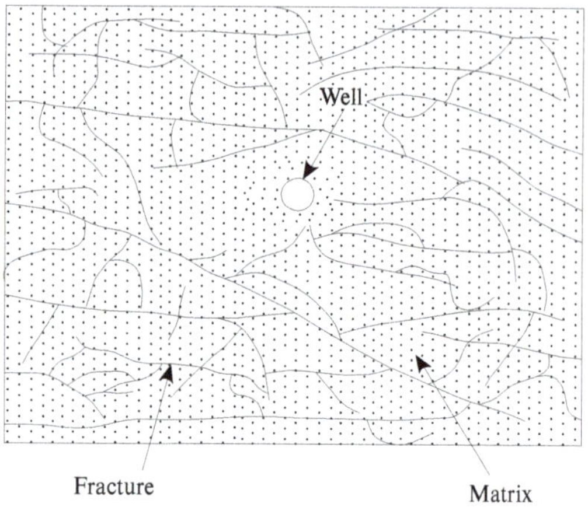

▲ **FIGURE 15.17A** **Massively fractured reservoir (from Cinco-Ley[8] © 1996 SPE, Reproduced with permission)**

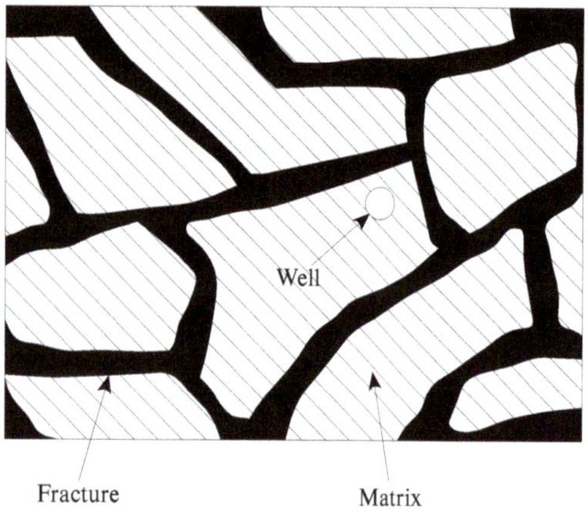

▲ **FIGURE 15.17B** **Naturally fractured reservoir with low porosity/permeability matrix (from Cinco-Ley[8] © 1996 SPE, Reproduced with permission)**

15.9.2 Multiple Region or Composite Reservoir Model

In Figure 15.18, a fractured reservoir composed of two regions is shown. The rock formation in one region is fractured, while the rest of the reservoir is not fractured. The fractured region will most likely exhibit higher transmissibility than the unfractured region. This NFR type can be represented with a composite radial model. Wells completed in the fractured region of the reservoir will tend to exhibit higher production rates than wells completed in the unfractured region. The flow capacity of the composite system depends on the flow capacity of the fractured and unfractured regions. If a test is conducted on a well located in the fractured region, the pressure response will be influenced initially by nearby fractures, and later by the unfractured region if the duration of the test is sufficiently long. Figure 15.19a is schematic log-log plot of the pressure derivative. The first horizontal line in Figure 15.19a represents radial flow from the fracture system after wellbore storage effects have ended. After a transition period, the second horizontal line represents radial flow from the fractured and unfractured regions (composite system) of the reservoir. Figure 15.19b is a schematic semi-log plot of the pressure response. The first straight line in Figure 15.19b represents the fracture system and the second straight line represents the composite reservoir. The flow capacities of the fractured region and the composite reservoir can be calculated from the slopes of these straight lines (m_1 and m_2), respectively.

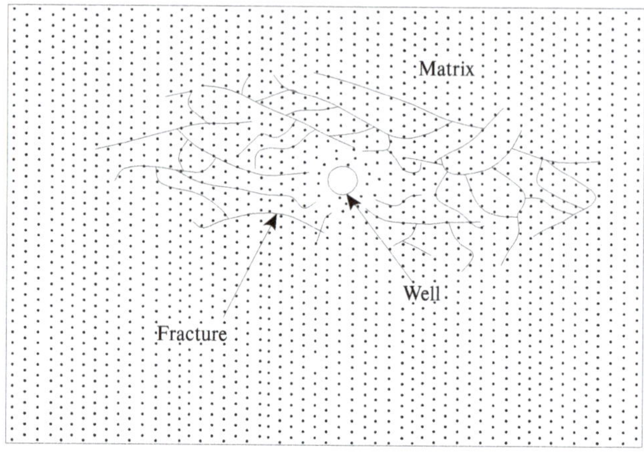

▲ **FIGURE 15.18** **Regionally fractured reservoir (from Cinco-Ley[8] © 1996 SPE, Reproduced with permission)**

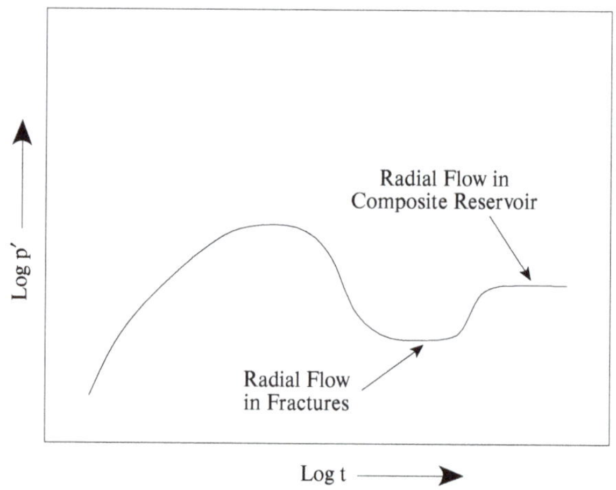

▲ **FIGURE 15.19A** **Derivative plot for a composite reservoir model (from Cinco-Ley[8] © 1996 SPE, Reproduced with permission)**

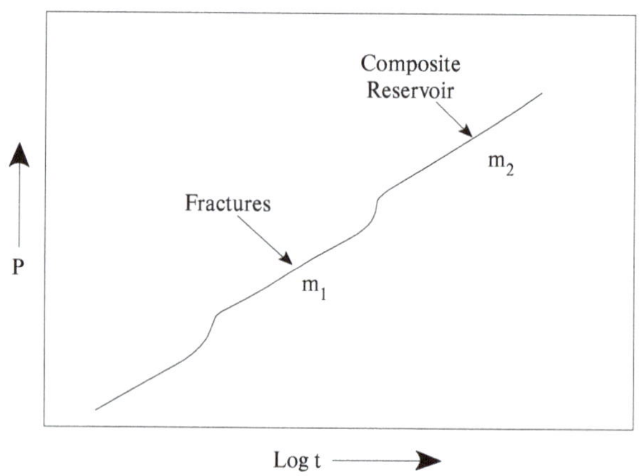

▲ **FIGURE 15.19B** **Semi-log plot for a composite reservoir model (from Cinco-Ley[8] © 1996 SPE, Reproduced with permission)**

15.9.3 Anisotropic Reservoir Model

The fractures in some NFRs are aligned in a particular direction as shown in Figure 15.20. In these reservoirs, the composite permeability in the direction of the fractures is higher than the composite permeability in the direction normal to the fractures. Consequently, these reservoirs exhibit anisotropic behavior. The reservoir exhibits a maximum permeability, k_{max}, and a minimum permeability, k_{min}, as shown in Figure 15.21. In NFR exhibiting anisotropic behavior, interference tests are the best method for determining k_{max} and k_{min} plus the orientation of the principal axis of permeability.[8] The application of interference testing in wells is not covered in this book.

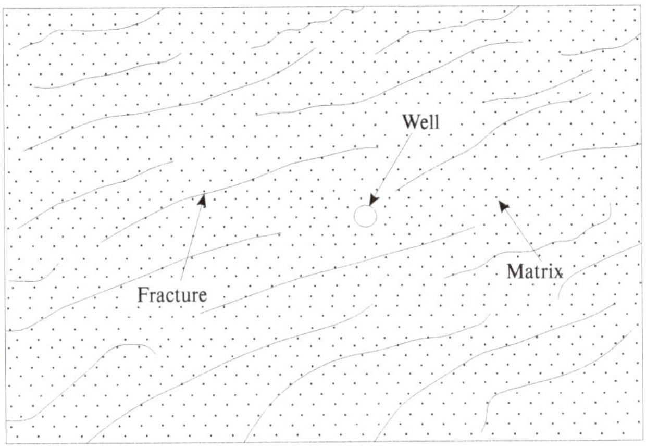

▲ **FIGURE 15.20** Naturally fractured reservoir with directional fractures (from Cinco-Ley[8] © 1996 SPE, Reproduced with permission)

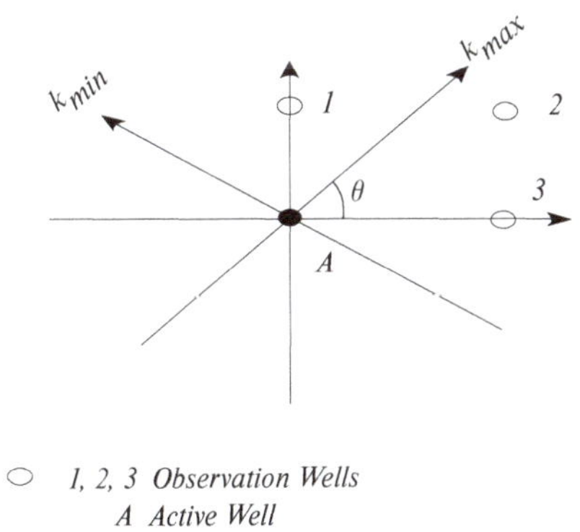

▲ **FIGURE 15.21** Orientation of fracture permeability for anisotropic NFR (from Cinco-Ley[8] © 1996 SPE, Reproduced with permission)

15.9.4 Single Fracture Model

In Figure 15.22, a well located near a high conductivity fracture system is shown. This fracture system may consist of a permeable fault through which fluids can flow to the well from other parts of the reservoir. Wells located in this region of the reservoir are capable of producing at very high, sustained rates. They are also subject to premature water breakthrough if the permeable fault traverses the aquifer. Several reservoir and well parameters can be calculated from analysis of a well test in this reservoir. Figure 15.23 is a log-log pressure derivative plot of a well test for such reservoir. As shown in Figure 15.23, there is a radial flow regime representing the entire system. After a transition period, the response to the permeable fault appears to indicate a constant pressure boundary. This is then followed by bilinear flow representing the flow through the fracture system. Several specialized plots (Figure 15.24) are useful for analysis of flow tests on the well. These include a semi-log plot to calculate the flow capacity, kh, of the total system and a plot of p vs. $\sqrt[4]{t}$ for the bilinear flow regime to calculate fracture conductivity, $w_f k_f$.

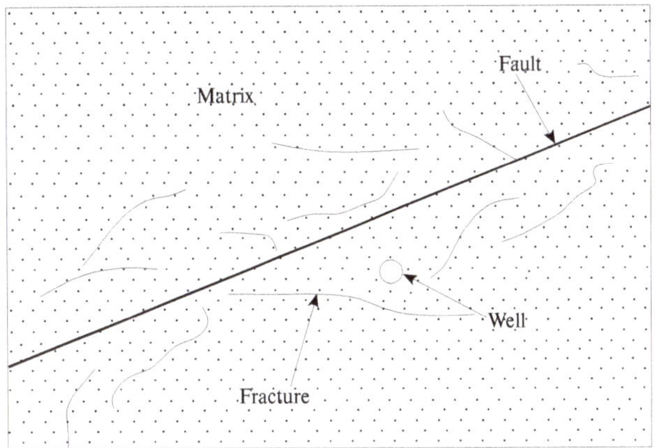

▲ **FIGURE 15.22** Naturally fractured reservoir with highly conductive fault (from Cinco-Ley[8] © 1996 SPE, Reproduced with permission)

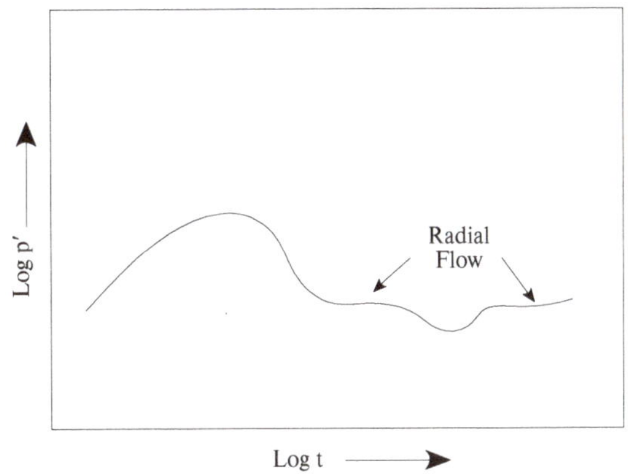

▲ **FIGURE 15.23** Derivative plot for NFR with highly conductive fault (from Cinco-Ley[8] © 1996 SPE, Reproduced with permission)

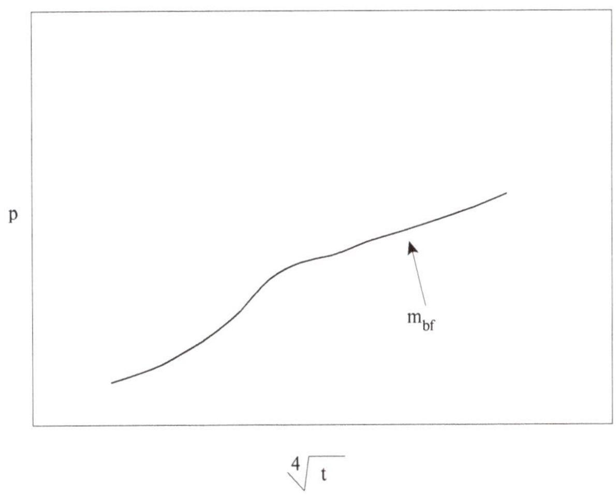

▲ **FIGURE 15.24** Specialized plot for a well near a highly conductive fault (from Cinco-Ley[8] © 1996 SPE, Reproduced with permission)

15.9.5 Double Porosity Model

NFRs have traditionally been described as consisting of a matrix system and a fracture system as shown in Figure 15.25. Reservoirs fluids flow from the matrix system to the fracture system and the fracture system delivers the fluids to the wellbore. The capacity of the matrix system to store and transport fluids is measured by its porosity, ϕ_{ma}, and permeability, k_{ma}, respectively. Similarly, the storage and transport capacity of the fracture system is measured by its porosity, ϕ_f, and permeability, k_f, respectively.

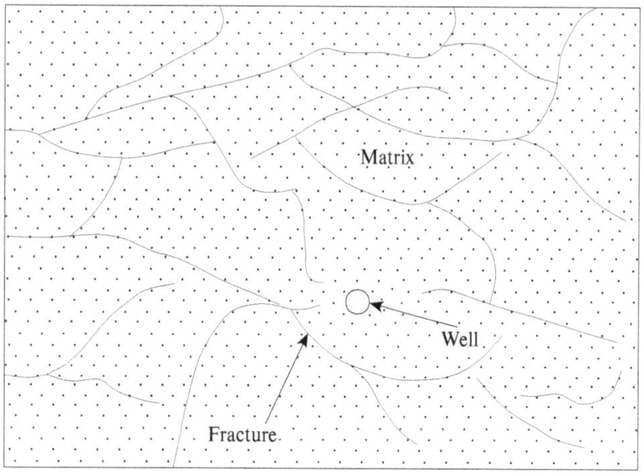

▲ FIGURE 15.25 NFR with fracture and matrix permeability (from Cinco-Ley[8] © 1996 SPE, Reproduced with permission)

Since reservoir fluids are stored in both the matrix and fractures, the double porosity model is used to represent the total system. Two flow conditions are used to describe the flow of fluids from the matrix to the fractures. These are pseudosteady-state flow[9,10] and transient flow[11-13] conditions. It is not readily apparent which flow condition is prevalent in a flow test. Some tests appear to indicate pseudosteady-state interporosity flow behavior while others indicate transient interporosity flow behavior.[14] In the analysis of a flow test, pseudosteady-state flow should be assumed on a trial basis. If analysis of the test data does not conform to this model, then transient flow behavior can be assumed as an alternative model.

The term interporosity flow describes the exchange of fluids between the fracture and the matrix systems. A measure of fluid transfer between the matrix and the fracture systems is called the interporosity flow coefficient. **Interporosity flow coefficient**, λ, is defined as:[14]

$$\lambda = \alpha r_w^2 \frac{k_{ma}}{k_f} \tag{15.42}$$

In Eq. (15.42), k_{ma} = matrix permeability; k_f = fracture permeability; and α = a parameter characteristic of the geometry of the system. The parameter, α, is defined as:

$$\alpha = \frac{4n(n+2)}{l^2} \tag{15.43}$$

In Eq. (15.43), n = the number of normal sets of planes limiting the least permeable medium; and l = characteristic dimension of the matrix block.[14] Figure 15.26 shows an idealized matrix block model of a fractured reservoir. For this representative model, $n = 3$. Figure 15.27 shows a slab model of a fractured reservoir. For this model, $n = 1$. The range of values for λ is from 10^{-9} to 10^{-4}. A high value for λ indicates fast interaction between the fracture system and the matrix system.[8]

The term storativity represents the storage and expansion capacity of the fluid in the particular system. The storativity of the fractures relative to the storativity of the total system is called the storativity ratio. It is a measure of the relative storage and expansion capacity of the fractures in the entire system. **Storativity ratio,** ω, is defined as:

$$\omega = \frac{(\phi V c_t)_f}{(\phi V c_t)_f + (\phi V c_t)_{ma}} \tag{15.44}$$

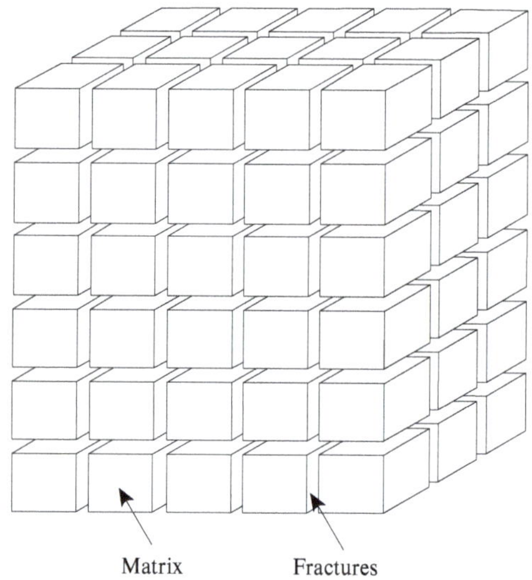

▲ FIGURE 15.26 Idealized matrix block model for NFRs (from Warren and Root[10] © 1963 SPE, Reproduced with permission)

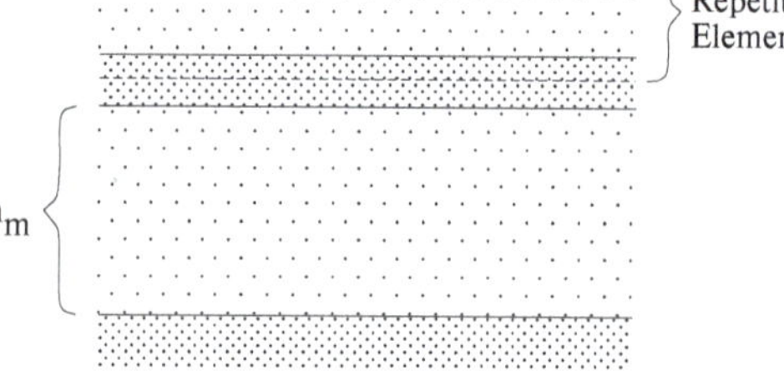

▲ FIGURE 15.27 Slab model for NFRs (from Serra et al.[13] © 1983 SPE, Reproduced with permission)

In Eq. (15.44), V = ratio of the total volume of medium to the bulk volume of the total system. The subscripts, f and ma, denote fracture and matrix, respectively. The range of typical values for ω is from 0.001 to 0.5.[8]

15.10 Well Test Analysis in Naturally Fractured Reservoirs Based on Double Porosity Model

Many NFRs exhibit pseudosteady-state flow from the matrix to the fractures.[14] Pseudosteady-state flow implies that the rate of pressure change in the matrix is constant. Initially, fluid flow to the wellbore is from the fracture system. This followed by a transition period during which fluids begin to flow from the matrix. Finally, an equilibrium condition is reached at which fluids flow from the matrix to the fractures, and then from the fractures to the wellbore. A semi-log plot of Δp vs. t representing this behavior is shown in Figure 15.28. In Figure 15.28, Line 1 represents flow in the fracture system exhibiting radial, homogeneous behavior, and Line 2 represents radial, homogeneous flow in the total system (fracture plus matrix). Line 1 is parallel to Line 2. For this reason, both lines have the same slope. From the slope, the flow capacity, $(kh)_{f+ma}$, of the total system can be calculated. The flow capacity, $(kh)_{f+ma}$, is essentially equal to the flow capacity of the fracture, $(kh)_f$, since the permeability of the fracture is much greater than the permeability of the matrix $\left(\text{i.e., } k_f \gg k_m\right)$.

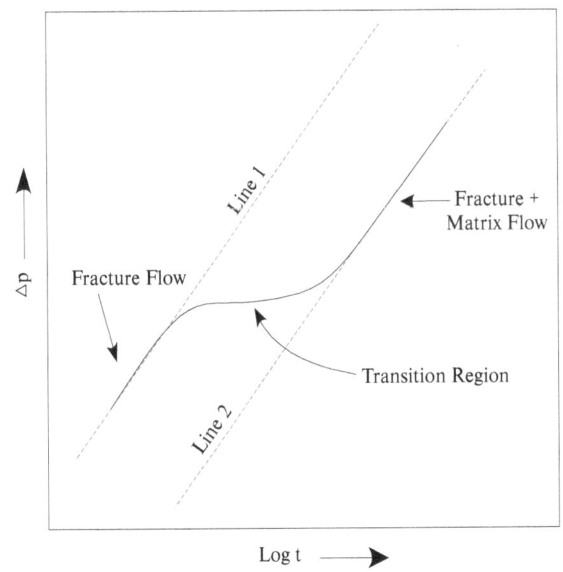

▲ **FIGURE 15.28 Semi-log pressure response of a NFR assuming pseudosteady-state flow**

Some naturally fractured reservoirs exhibit transient flow from the matrix to the fractures.[13] Transient flow conditions imply the rate of pressure change varies with time and location in the matrix. As previously described for pseudosteady-state flow, flow originates in the fractures under transient conditions. This is followed by transition period before stabilized flow goes from the matrix to the fractures, and finally from the fractures to the wellbore. A semi-log plot of Δp vs. t representing this transient behavior is shown in Figure 15.29. In Figure 15.29, Line 1 represents flow in the fracture system exhibiting radial, homogeneous behavior, and Line 3 represents radial, homogeneous flow in the total system (fracture plus matrix). A third straight line, Line 2, represents transitional flow regime from the matrix to the fractures. Lines 1 and 3 have the same slope. The slope of Line 2 is equal to half the slope of either Lines 1 or 3.

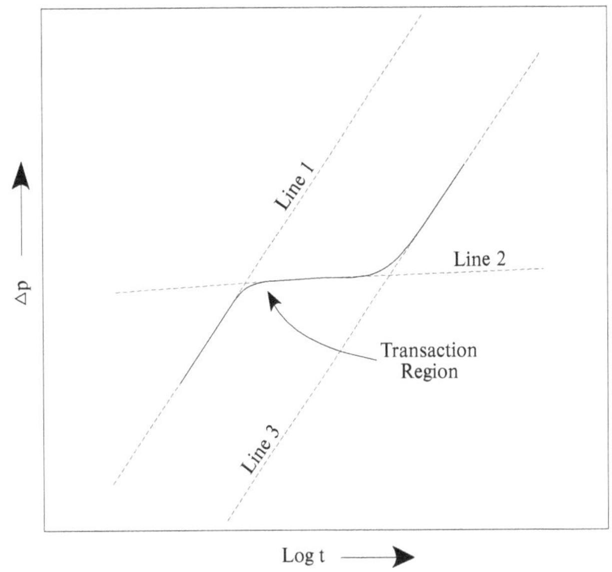

▲ **FIGURE 15.29 Semi-log pressure response of a NFR assuming transient flow**

15.11 Well Test Analysis in NFRs: Straightline Methods

Fluid flow in the matrix of NFRs is described under the double porosity model either by assuming pseudosteady-state conditions or transient state conditions. Most flow tests in NFRs appear to exhibit pseudosteady-state behavior.[14] Consequently, the application of straightline methods in the analysis of well tests in NFRs is based primarily on the assumption of pseudosteady-state flow conditions. However, the equations presented here are applicable if transient flow conditions are assumed.

A semi-log plot of pressure change versus time (Δp vs. log t) is shown in Figure 15.30, assuming pseudosteady-state flow. Lines 1 and 2 are two parallel lines representing homogeneous flow in the fractures and the total system (fractures plus matrix), respectively. The slope, m, of either straight line can be used to calculate the permeability-thickness product, kh, of the fracture or the total system, thus:

$$
\begin{aligned}
(kh)_f &\simeq (kh)_{total} \\
&= \frac{162.6 qB\mu}{m}
\end{aligned}
\tag{15.45}
$$

In Eq. (15.45), $(kh)_{total} = \left[(kh)_f + (kh)_m\right]$. Since fracture permeability, k_f, is much greater than matrix permeability, k_m, $\left(i.e., k_f \gg k_m\right)$, then $(kh)_f \simeq (kh)_{total}$ as indicated in Eq. (15.45).

The storativity ratio, ω, can be calculated from the vertical pressure displacement, δp, between Lines 1 and 2 as shown in Figure 15.30. Thus,

$$\omega = 10^{-\left(\frac{\delta p}{m}\right)} \tag{15.46}$$

As shown in Figure 15.30, a straight line drawn through the middle of the transition curve will intersect Line 1 at t_1, and Line 2 at t_2. The interporosity flow coefficient, λ, can be calculated using either t_1 or t_2 from the following equations:[14]

$$\lambda = \frac{(\phi V c_t)_f \mu r_w^2}{\gamma k_f t_1} = \frac{(\phi V c_t)_{f+m} \mu r_w^2}{\gamma k_f t_2}, \text{ for drawdown tests} \tag{15.47}$$

$$\lambda = \frac{(\phi V c_t)_f \mu r_w^2 \left(t_p + \Delta t_1\right)}{\gamma k_f t_p \Delta t_1} = \frac{(\phi V c_t)_{f+m} \mu r_w^2 \left(t_p + \Delta t_2\right)}{\gamma k_f t_p \Delta t_2}, \text{ for buildup tests} \tag{15.48}$$

In Eqs. (15.47) and (15.48), γ is the exponential of Euler's constant, which is approximately 1.781.

It is clearly evident from Figure 15.30 that these methods used earlier to calculate the storativity ratio, ω, and the interporosity flow coefficient, λ, depend on the presence of the two semi-log straight lines, Line 1 and Line 2. Line 1, which is based on early flow in the fractures, is usually distorted by wellbore storage and may not be observed. Consequently, parameters needed for Eqs. (15.46) to (15.48) may not be obtainable from a semi-log plot similar to Figure 15.30.

An alternative method for calculating the storativity ratio and interporosity flow coefficient is based on the characteristic minimum point that is typically present on the pressure derivative plots of flow tests from NFRs, as shown in Figure 15.31. The dimensionless time value, $t_{D\min}$, at which this minimum point occurs is given as:[15]

$$t_{D\min} = \frac{\omega}{\lambda}\left[\ln\left(\frac{1}{\omega}\right)\right] \tag{15.49}$$

This minimum point also corresponds to the dimensionless pressure derivative value derived as:[16]

$$\left[t_D \frac{\partial p_D}{\partial t_D}\right]_{\min} = 0.5\left[1 + \omega^{\{1/(1-\omega)\}} - \omega^{\{\omega/(1-\omega)\}}\right] \tag{15.50}$$

Eq. (15.50) can be solved for ω by using either successive substitution methods or Newton-Raphson numerical analysis techniques. An initial estimate for ω can be obtained from the expression:[16]

$$\log \omega = \frac{\left[0.01765 + \log\left(t_D \frac{\partial p}{\partial t_D}\right)_{\min}\right]}{0.94903} \tag{15.51}$$

The interporosity flow coefficient is calculated from Eq. (15.49) by substituting the value of the storativity ratio calculated from Eq. (15.50).

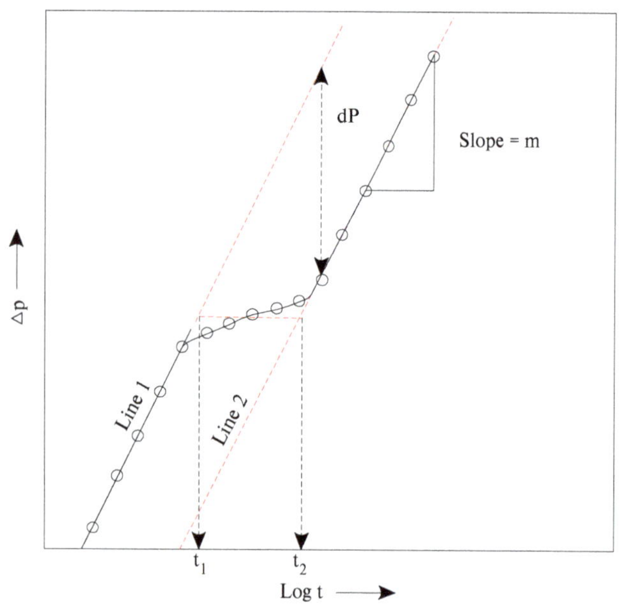

▲ **FIGURE 15.30** **Semi-log plot of Δp vs. *t* for an NFR flow test**

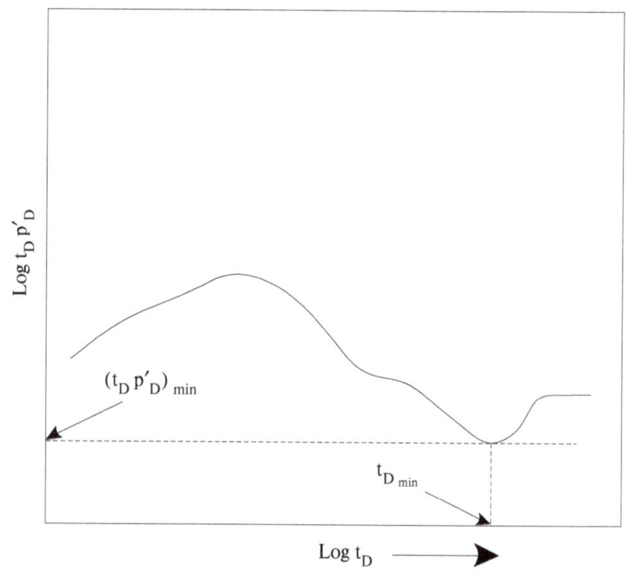

▲ **FIGURE 15.31** **NFR derivative plot**

The skin factor, s, is calculated by extrapolating the second semi-log straight line (Line 2, Figure 15.30) to p_{1hr}. The skin factor is calculated from the conventional expression:

$$s = 1.151\left[\frac{\Delta p_{1hr}}{m} - \log\left(\frac{k}{\phi\mu c_t r_w^2}\right) + 3.23\right] \quad (15.52)$$

Note that in Eq. (15.52), $\Delta p_{1hr} = (p_i - p_{1hr})$ for a drawdown test or $\Delta p_{1hr} = (p_{1hr} - p_{wf@\Delta t=0})$ for a buildup test.

15.12 Well Test Analysis in NFRs: Type Curves

Type curves, in conjunction with straightline methods, are useful in well test analysis of NFRs, especially in cases where early time flows in the fractures are distorted by wellbore storage. A typical analysis using type curves is illustrated in Figure 15.32, which shows the pressure behavior of a well test in a double porosity reservoir. During early time flow from the fractures, the pressure response follows one of the homogeneous curves with $C_De^{2s} = (C_De^{2s})_f$. This is shown as the heavy dark line between points A and B in Figure 15.32. When interporosity flow begins from the matrix to the fractures (termed the transition period), the pressure response curve departs from the C_De^{2s} curve and follow the transition curve characterized by the parameter λ^{e-2s}. This is represented by the segment between points B and C in Figure 15.32. At later times, when both the fractures and matrix are contributing to production, the pressure curve leaves the transition curve and follows a new C_De^{2s} that represents the entire system. This is shown as the segment between points C and D in Figure 15.32.

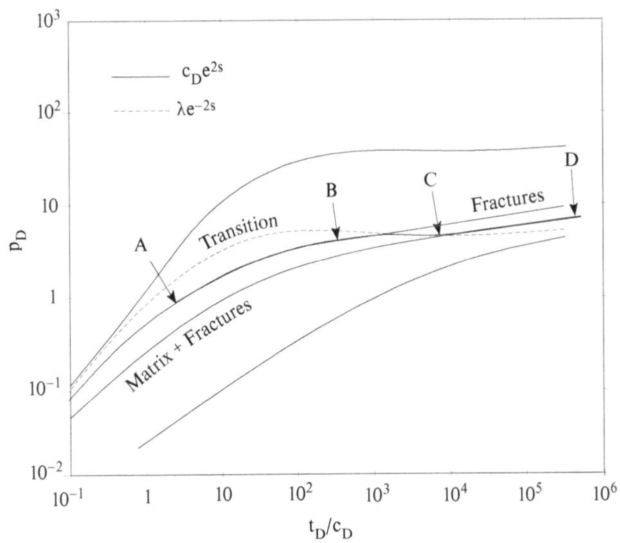

▲ **FIGURE 15.32 Typical type curve analysis of flow test in a NFR (from Gringarten[14] © 1984 SPE, Reproduced with permission)**

The process of determining reservoir and fracture parameters by type-curve matching is the same as discussed previously in Chapter 14. The process begins by making a log-log plot of pressure change versus time on the same scale as the type curve. By examining the plotted data, it is possible to identify different flow periods such as fracture flow, transition flow, and total system flow. This plot is then superimposed on a set of type curves such as shown in Figure 15.33 for pseudosteady-state flow, and Figure 15.34 for transient flow. By shifting the two plots in the manner described in Chapter 14 for type-curve matching, a match point is obtained. From the pressure match point (PMP), the permeability-thickness product for the total system (fracture + matrix), $(kh)_{f+ma}$ is calculated as:

$$(kh)_{f+ma} = 141.2qB\mu\left[\frac{p_D}{\Delta p}\right]_{PMP} \quad (15.53)$$

The permeability of the fracture system is much higher than the permeability of the matrix (i.e., $k_f >> k_{ma}$). For this reason, $(kh)_{f+ma}$ is essentially equal to $(kh)_f$.

From the time match point (TMP), the wellbore storage coefficient, C_D, can be calculated as:

$$C_D = \frac{0.0002637k}{\phi\mu c_t r_w^2}\left(\frac{\Delta t}{t_D/C_D}\right)_{TMP} \quad (15.54)$$

From the fit of the early data that characterize the fracture system, determine the value of $(C_De^{2s})_f$ from the match. Also, determine the value of λe^{-2s} that characterizes the transition region. Finally, determine the value of $(C_De^{2s})_{f+m}$ that characterizes flow of the total system in the later flow period. The storativity ratio, ω, is calculated from these matched values as:

$$\omega = \frac{(C_De^{2s})_{f+m}}{(C_De^{2s})_f} \quad (15.55)$$

The skin factor, s, is calculated as:

$$s = 0.5\ln\left[\frac{(C_De^{2s})_{f+m}}{C_D}\right] \quad (15.56)$$

The interporosity flow coefficient, λ, is calculated from:

$$\lambda = (\lambda e^{-2s})e^{2s} \quad (15.57)$$

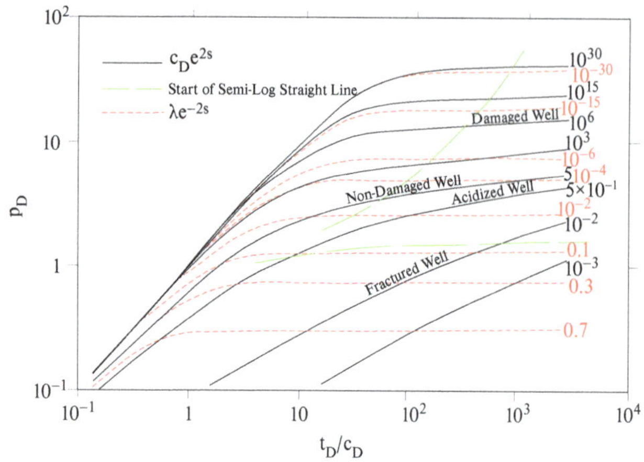

▲ **FIGURE 15.33 Type curve for NFRs under pseudosteady-state flow (from Gringarten[14] © 1984 SPE, Reproduced with permission)**

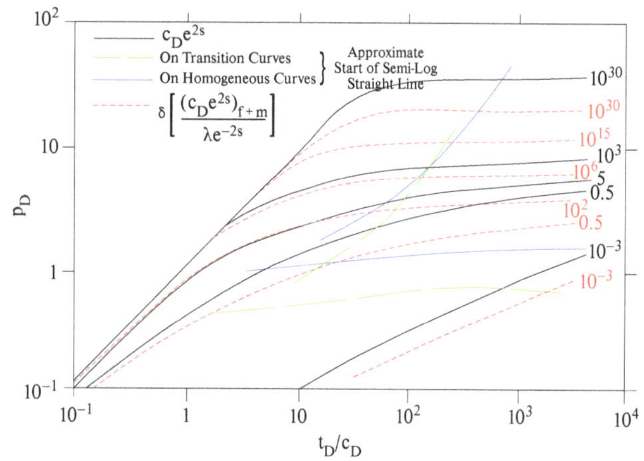

▲ **FIGURE 15.34 Type curve for NFRs under transient flow (from Gringarten[14] © 1984 SPE, Reproduced with permission)**

15.13 Procedure for Analysis of Well Test from NFRs Assuming Double Porosity Behavior

The procedure for analysis of well test data from NFRs can be classified as consisting of three main processes:

1. Identification of flow periods
2. Calculation of fracture and reservoir parameters from type curves
3. Validation of results with straightline methods

15.13.1 Identification of Flow Periods

As discussed earlier, there are potentially three flow periods in the test data from NFRs. These are fracture flow at early times, transition flow period marking the onset of fluid flow from the matrix to the fractures, and stabilized flow from both fracture and matrix at later times. In many cases, fracture flow and transition flow periods may be distorted or masked by wellbore storage. The presence of these flow regimes can be identified on a plot of pressure change and pressure-derivative versus time. The pressure-derivative plot yields the characteristic dip below the homogeneous behavior level which is indicative of double porosity behavior.

1. Plot log Δp vs. log t for flow test or log Δp vs. log Δt_e for a buildup test. Plot $t\Delta p'(or\ \Delta t_e\Delta p')$ vs. $t(or\ \Delta t_e)$ on the same graph. Use the same scale size as the type curves graph to be used later for type-curve matching.
2. Identify any flow periods that may be present. Note if double porosity behavior is present.

15.13.2 Calculation of Fracture and Reservoir Parameters from Type Curves

Fracture and reservoir parameters can be calculated by matching the plot from Step 1 to the type curves in Figure 15.33 or Figure 15.34. Since most test data from NFRs appear to indicate pseudosteady-state flow conditions, use Figure 15.33 for the first trial.

1. Match the flow test data plot of log Δp vs. log t for flow test or log Δp vs. log Δt_e for a buildup test against type curves of Figure 15.33.
2. Determine pressure and time match points (PMP and TMP, respectively). Also determine $\left(C_De^{2s}\right)_f$ for the fracture flow period, λe^{-2s} for the transition flow period, and $\left(C_De^{2s}\right)_{f+m}$ for the stabilized flow from the combined system (fracture plus matrix).
3. Calculate fracture and reservoir parameters using Eqs. (15.53) to (15.57).

15.13.3 Validation of Results with Straightline Methods

The validation step involves using straightline methods to calculate fracture and reservoir parameters and comparing them to the results obtained with type-curve matching. If the results are reasonably close, then it can be presumed that the type curves selected from type-curve matching are representative of the test data and that the results are inherently consistent. If the results are widely divergent, then the analysis should begin again from type-curve matching by choosing another set of type curves and repeating the entire process. This process is highly iterative and should be continued until reasonable agreement is achieved with the results from type-curve matching and straightline methods.

1. Plot p_{wf} vs. t for a drawdown test or p_{ws} vs. $\left(t_p+\Delta t\right)/\Delta t$ for a buildup test on a semi-log scale.
2. Identify flow periods present in plot using the knowledge gained from type-curve matching.
3. If the flow periods illustrated in Figure 15.30 are present, use Eqs. (15.45) to (15.48) to calculate fracture and reservoir parameters.
4. Use the derivative plot to obtain $t_{D\min}$. Calculate ω and λ from Eqs. (15.49) and (15.50).
5. Calculate skin factor from Eq. (15.52).

6. Compare the results obtained from straightline methods to the results obtained from type-curve analysis.
7. If the results are in reasonable agreement, then the analysis is internally consistent. If not, repeat process by using another interpretation model.

The discussions on analysis of test data from NFRs have been conducted on the basis of slightly compressible fluids such as oil reservoirs. For gas reservoirs, it is necessary to use appropriate pressure transformation functions such as real gas pseudo-pressure, as demonstrated in Chapter 13.

EXAMPLE 15.2 Calculation of Fracture and Reservoir Parameters from the Test Data of a Well in a Naturally Fractured Reservoir (NFR)

Problem

A drawdown test was conducted on a well in a naturally fractured reservoir. The test data are shown in Table 15.2. Calculate fracture and reservoir parameters from the test data. Well and other reservoir properties are given as follows:

Formation thickness, h	30 ft
Formation porosity, ϕ_m	0.12
Total compressibility, c_t	14×10^{-6} psi^{-1}
Oil viscosity, μ_o	1.2 cp
Oil FVF, B_o	1.3 RB/STB
Wellbore radius, r_w	0.5 ft
Production rate, q_o	2950 STB/D
Initial average pressure, $\bar{p}$	4480 psia

Solution

Step 1: Plot log Δp vs. log t.

Using the data in Table 15.2 and plotted in Figure 15.35, make a log-log plot of Δp vs. t as shown in Figure 15.36. Also, plot the derivative curve of $t\Delta p'$ vs. t on the same graph using the data in Table 15.2. The derivative curve clearly shows evidence of double porosity behavior which is indicated by the characteristic dip below the homogeneous behavior level. Also, the derivative curve in Figure 15.36 shows pseudosteady-state flow from the combined system of fracture and matrix which is represented by the stabilized leveling of the derivative curve after the dip.

TABLE 15.2 Test and Plot Data for Example 15.2

Time, t (hrs)	p_{wf} (psia)	Δp (psia)	$t\Delta p'$	Time, t (hrs)	p_{wf} (psia)	Δp (psia)	$t\Delta p'$
0.00	4480.00	0.00	—	2.30	4002.00	478.00	72.41
0.05	4462.00	18.00	128.87	2.93	3984.07	495.93	78.05
0.06	4440.00	40.00	136.28	3.93	3959.73	520.27	85.48
0.08	4403.00	77.00	147.34	4.93	3939.90	540.10	89.70
0.09	4379.20	100.80	141.03	5.93	3923.00	557.00	95.16
0.12	4338.50	141.50	135.82	6.93	3907.69	572.31	115.62
0.17	4296.50	183.50	137.56	7.93	3890.08	589.92	134.18
0.20	4270.00	210.00	138.68	8.50	3880.64	599.36	137.50
0.25	4240.00	240.00	145.16	10.50	3850.62	629.38	138.42
0.33	4196.00	284.00	144.70	11.60	3837.00	643.00	144.72
0.40	4170.00	310.00	141.94	13.40	3814.45	665.55	140.49
0.51	4135.00	345.00	149.17	15.00	3800.00	680.00	144.40
0.57	4117.00	363.00	139.44	16.30	3787.00	693.00	146.00
0.65	4100.00	380.00	146.64	17.40	3778.00	702.00	144.80
0.74	4080.00	400.00	120.92	18.35	3770.00	710.00	146.74
0.92	4070.00	410.00	99.51	19.20	3763.50	716.50	144.14
1.00	4060.00	420.00	101.18	20.50	3754.00	726.00	143.71
1.10	4052.40	427.60	76.42	21.50	3747.20	732.80	144.00
1.40	4036.00	444.00	67.38	22.50	3740.60	739.40	143.85
1.65	4025.00	455.00	67.00	23.50	3734.40	745.60	143.76
1.93	4014.49	465.51	69.02	25.00	3725.40	754.60	148.33

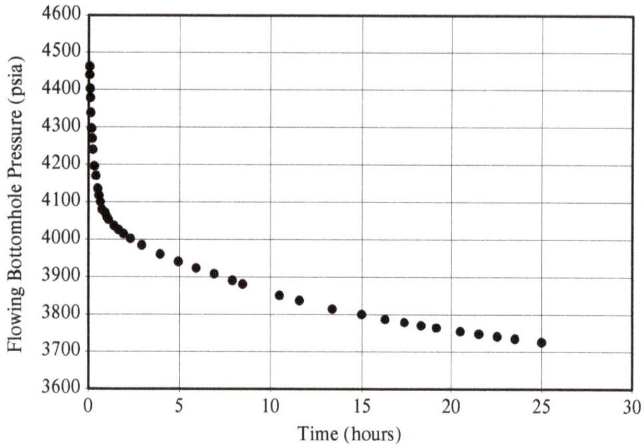

▲ **FIGURE 15.35 Plot of flow test data for Example 15.2**

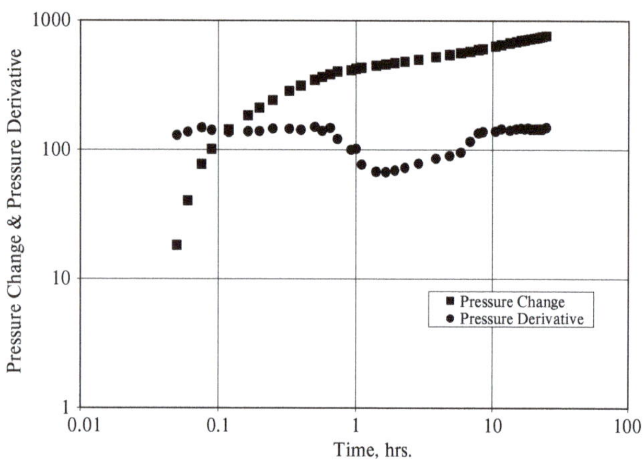

▲ **FIGURE 15.36 Log-Log plot of Δp & $t\Delta p'$ vs. t for Example 15.2**

Step 2: Type-curve matching.

The type curve for NFRs under pseudosteady-state flow shown as Figure 15.33 was used in the type-curve matching procedure. Figure 15.36 was matched against Figure 15.33. The matched plot of Figures 15.33 and 15.36 together is shown as Figure 15.37. The matched data points are as follows:

Pressure match point (PMP), $(p_D/\Delta p)_{PMP} = (0.35/100)$

Time match point (TMP), $\left(\frac{\Delta t}{t_D/C_D}\right)_{TMP} = \left(\frac{1}{10}\right)$

For the pressure match point, applying Eq. (15.53),

$$(kh)_{f+m} = 141.2qB\mu\left[\frac{p_D}{\Delta p}\right]_{PMP}$$

$$= 141.2\times2950\times1.3 \times1.2\times\left(\frac{0.35}{100}\right)$$

$$= 2274.31 \text{ md-ft.}$$

Since the matrix permeability is relatively small in comparison to fracture permeability, $(kh)_{f+ma}$, is essentially equal to, $(kh)_f$.

For the time match point, applying Eq. (15.54),

$$C_D = \frac{0.0002637k}{\phi\mu c_t r_w^2}\left(\frac{\Delta t}{t_D/C_D}\right)_{TMP}$$

$$= \frac{0.0002637\times75.81}{0.12\times1.2\times14\times10^{-6}\times(0.5)^2}\left(\frac{1}{10}\right)$$

$$= 3966.49$$

From Figure 15.37, type-curve match parameters were estimated as follows:

$$\left(C_De^{2s}\right)_{f+ma} = 0.4;\ \left(C_De^{2s}\right)_f = 2.0$$

and $\left(\lambda e^{-2s}\right) = 0.06$

Applying Eq. (15.55), the storativity ratio, ω, is calculated to be:

$$\omega = \frac{\left(C_De^{2s}\right)_{f+m}}{\left(C_De^{2s}\right)_f}$$

$$= \frac{0.4}{2} = 0.2$$

The skin factor, s, is calculated from Eq. (15.56) as:

$$s = 0.5\ln\left[\frac{\left(C_De^{2s}\right)_{f+m}}{C_D}\right]$$

$$= 0.5\ln\left(\frac{0.4}{3966.49}\right)$$

$$= -4.601$$

The interporosity flow coefficient, λ, is calculated from Eq. (15.57) to be:

$$\lambda = \left(\lambda e^{-2s}\right)e^{2s}$$

$$= (0.06)e^{2(-4.601)}$$

$$= 6.05\times10^{-6}$$

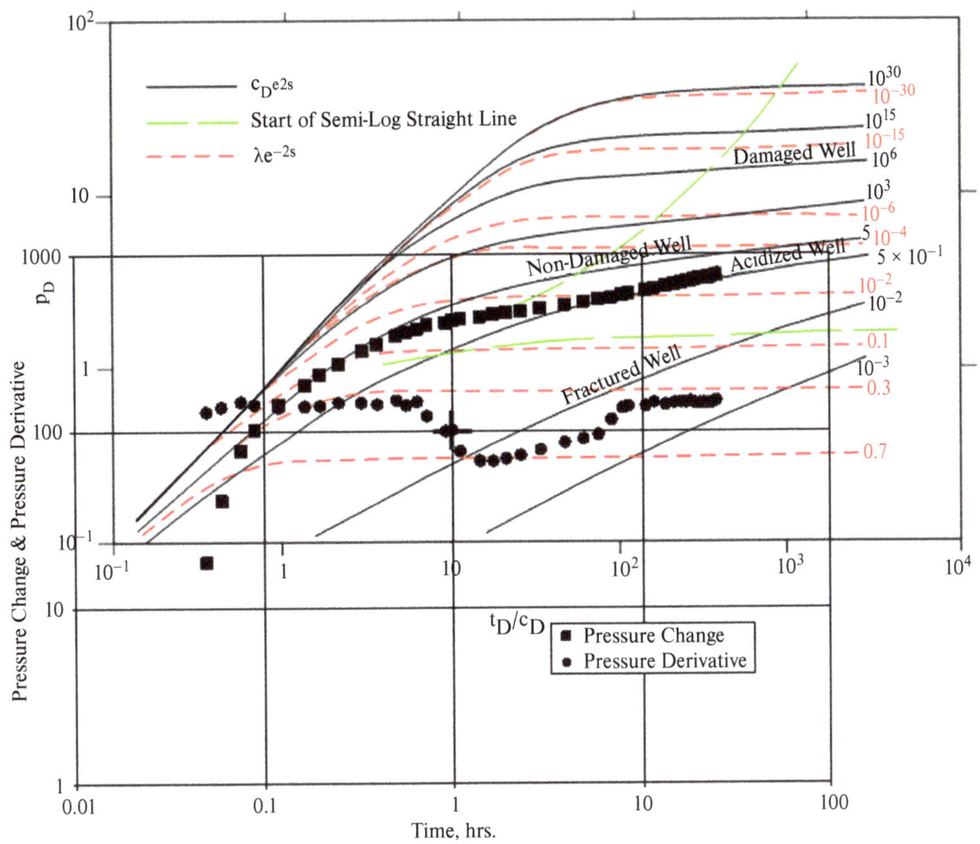

▲ **FIGURE 15.37** Type curve match for Example 15.2

Step 3: Plot p_{wf} vs. log t

Prepare a semi-log plot of p_{wf} vs. t from the data in Table 15.2. The semi-log plot is shown as Figure 15.38. From Figure 15.38, it is obvious that fracture flow lasted for about 1 hour. This is followed by a transition period to matrix flow of duration of 1 to 9 hours. Pseudosteady-state flow from both matrix and fractures started approximately after 10 hours of flow. These flow periods conform to the flow periods identifiable on the derivative curve of Figure 15.36.

The slope of the semi-log straightline in Figure 15.38 is $m = 311$ psi/cycle.

Applying Eq. (15.45), the total flow capacity of the formation is:

$$(kh)_{total} = \frac{162.6qB\mu}{m}$$
$$= \frac{162.6\times2950\times1.3\times1.2}{311}$$
$$= 2406.06 \text{ md-ft.}$$

From Figure 15.38, the vertical pressure displacement, δp, is 155.56 psia. Applying Eq. (15.46), the storativity ratio, ω, is calculated to be:

$$\omega = 10^{-\left(\frac{\delta p}{m}\right)}$$
$$= 10^{-\left(\frac{155.6}{311}\right)} = 0.32$$

A straight line through the middle of the transition curve intersects the two parallel straight lines as shown in Figure 15.38 at $t_1 = 1.7$ hrs and $t_2 = 3.7$ hrs, respectively. From Eq. (15.47), the interporosity flow coefficient, λ, at t_2 is calculated as:

$$\lambda = \frac{(\phi Vc_t)_{f+m}\mu r_w^2}{\gamma k_f t_2}$$
$$= \frac{0.12\times14\times10^{-6}\times1.2\times(0.5)^2}{1.78\times80.2\times3.7}$$
$$= 9.54\times10^{-10}$$

Note that V_{f+m} is essentially equal to 1.0. Also, $k_f = (kh)_f/h = 2406.1/30 = 80.2$ md.

Rearranging Eq. (15.44) yields:

$$(\phi V c_t)_f = \frac{\omega}{1-\omega}(\phi V c_t)_{ma}$$
$$= \frac{0.32}{1-0.32}\left(0.12\times14\times10^{-6}\right) = 7.9059\times10^{-7}\ \text{psi}^{-1}$$

Applying Eq. (15.47), the interporosity flow coefficient, λ, at t_1 is calculated to be:

$$\lambda = \frac{(\phi V c_t)_f\ \mu r_w^2}{\gamma k_f t_1}$$
$$= \frac{7.9059\times10^{-7}\times1.2\times(0.5)^2}{1.78\times80.2\times1.7} = 9.77\times10^{-10}$$

From Figure 15.38, $p_{1\text{hr}}$ = 4190 psia. The skin factor, s, is given by Eq. (15.52) as:

$$s = 1.151\left[\frac{\Delta p_{1\text{hr}}}{m} - \log\left(\frac{k}{\phi\mu c_t r_w^2}\right) + 3.23\right]$$
$$= 1.151\left[\frac{4480-4190}{311} - \log\frac{80.2}{0.12\times1.2\times14\times10^{-6}\times(0.5)^2} + 3.23\right] = -4.65$$

For this example, the flow capacities calculated from type-curve matching and straightline methods are reasonably in close agreement. The same close agreement was achieved for the skin factors calculated from both methods. However, there are poor agreement between the storativity ratios and interporosity flow coefficients calculated from type-curve matching and straightline methods. The storativity ratio and interporosity flow coefficient calculated from type-curve matching should at best be considered gross estimates because of the difficulties associated with estimating the correct values for the type-curve parameters: $\left(C_D e^{2s}\right)_{f+ma}$, $\left(C_D e^{2s}\right)_f$, and λe^{-2s}. Under these circumstances, the storativity ratio and interporosity flow coefficient calculated from straightline methods should be considered to be more representative of the actual values.

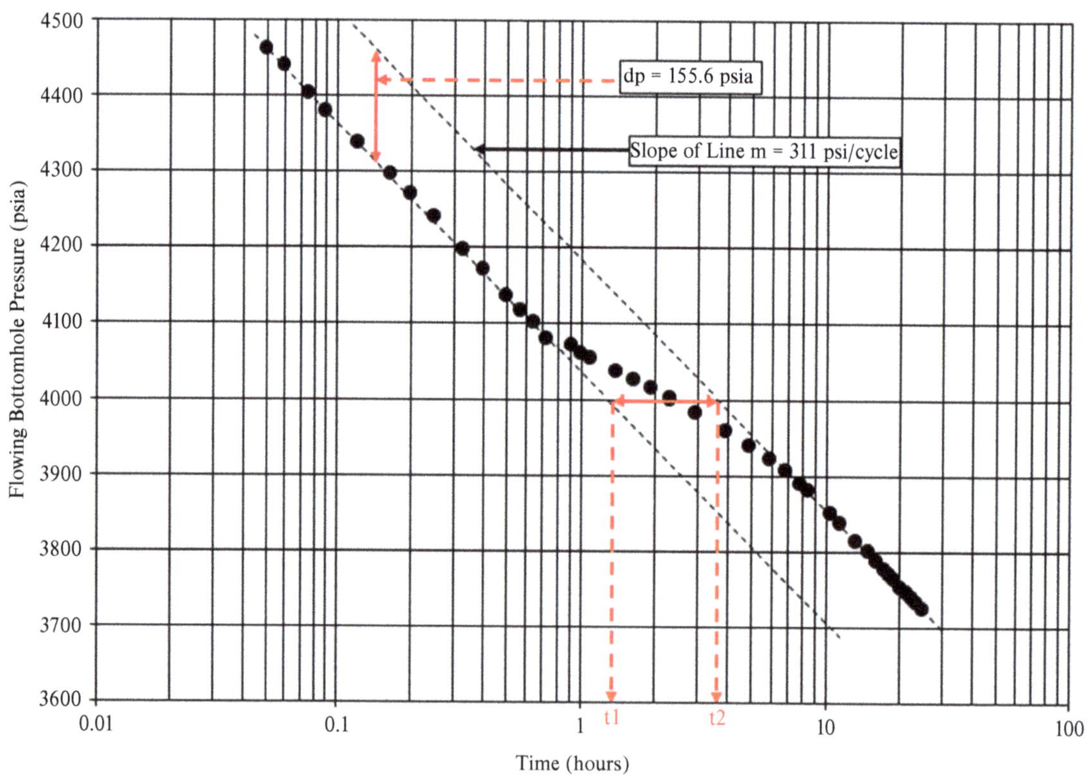

▲ **FIGURE 15.38 Semi-log plot of p_{wf} vs. t for Example 15.2**

Nomenclature

B	formation volume factor, RB/STB
c_{ft}	total compressibility of fluid in fracture, psi^{-1}
c_t	total compressibility, psi^{-1}
C_D	dimensionless wellbore storage coefficient
C_{Df}	dimensionless fracture storage coefficient, $C_{Df} = 0.894C/\left(\phi c_t h L_f^2\right)$
C_{fD}	dimensionless fracture conductivity
h	reservoir or formation thickness, feet
k	permeability, md
k_f	fracture permeability, md
k_{ma}	matrix permeability, md
L_f	fracture half-length, feet
m	slope of straight line during pseudo-radial flow
m_{bf}	slope of straight line during bilinear flow
m_{Lf}	slope of straight line during formation linear flow
Δp	pressure change, psia
Δp_S	pressure drop due to skin, psia
$\Delta p'$	pressure change derivative
p'_D	dimensionless pressure derivative
p	pressure, psia
p_D	dimensionless pressure
p_p	real gas pseudo-pressure, psia^2/cp
p_{wf}	well bottom-hole flowing pressure, psia
p_{ws}	shut-in bottomhole pressure, psia
q	volume flow rate, STB/D
q_g	gas volume flow rate, Mcf/D
r_w	wellbore radius, feet
r_{wa}	effective wellbore radius, feet
s	skin factor
s'	apparent skin factor
Δt	shut-in time in buildup test, hours
Δt_e	Agarwal equivalent time
t	time, hours
t_{blf}	time, begining of linear flow
t_D	dimensionless time
t_{DLf}	dimensionless time defined in terms of L_f
t_{elf}	time, end of linear flow
t_{ews}	time, end of wellbore storage
t_p	total producing time before shut-in, hours
T	temperature, °R
V	ratio of the total volume of medium to the bulk volume of the total system
w_f	fracture width, feet
ϕ	porosity, fraction
ϕ_f	fracture porosity, fraction
μ	fluid viscosity, cp
λ	interporosity flow coefficient
α	characteristic parameter of geometry of system
ω	storativity ratio
η_{fD}	dimensionless hydraulic diffusivity
γ	exponential Euler's constant approximately 1.781

Subscripts

f	fracture
g	gas
i	initial conditions
o	oil
ma	matrix
t	total
w	water

Abbreviations

DST	Drill Stem Test
PMP	Pressure Match Point
PTA	Pressure Transient Analysis
TMP	Time Match Point

REFERENCES

1. Cinco-Ley, H., and Samaniego-V., F.: "Transient Pressure Analysis for Fractured Wells," *JPT* (September 1981) 1749-1766.
2. Cinco-Ley, H.: "Evaluation of Hydraulic Fracturing By Transient Pressure Analysis Methods," paper SPE 10043 presented at the International Petroleum Exhibition and Technical Symposium, Beijing, China, (March 18-26, 1982).
3. Cinco-Ley, H., Samaniego-V., F., and Dominguez-A, N.: "Transient Pressure Behavior for a Well with a Finite-Conductivity Vertical Fracture," *SPEJ* (August 1978) 253-264.
4. Gringarten, A.C., Ramey, H.J., Jr., and Raghavan, R.: "Applied Pressure Analysis for Fractured Wells," *JPT* (July 1975) 887-892.
5. Wong, D.W., Harrington, A.G., and Cinco-Ley, H.: "Application of the Pressure-Derivative Function in the Pressure-Transient Testing of Fractured Wells," *SPEFE* (October 1986) 470-480.
6. Cinco-Ley, H., and Samaniego-V., F.: "Effect of Wellbore Storage and Damage on the Transient Pressure Behavior of Vertically Fractured Wells," paper SPE 6752 presented at the 52nd Annual Technical Conference and Exhibition, Denver, Colorado, (October 9-12, 1977).
7. Cinco-Ley, H., and Samaniego-V., F.: "Transient Pressure Analysis: Finite Conductivity Fracture Case Versus Damaged Fracture Case," paper SPE 10179 presented at the 56th Annual Fall Technical Conference and Exhibition, San Antonio, Texas, (October 5-7, 1981).
8. Cinco-Ley, H.: "Well-Test Analysis for Naturally Fractured Reservoirs," *JPT* (January 1996) 51-54.
9. Barenblatt, G.E., Zheltov, I.P., and Kochina, I.N.: "Basic Concepts in the Theory of Homogeneous Liquids in Fissured Rocks," *J. Appl. Math. Mech.* (1960) 24, 852-864.
10. Warren, J.E., and Root, P.J.: "The Behavior of Naturally Fractured Reservoirs," *SPEJ* (September 1963) 245-255.
11. de Swaan-O., A.: "Analytical Solutions for Determining Naturally Fractured Reservoir Properties by Well Testing," *SPEJ* (June 1976) 117-122.
12. Najurieta, H.L.: "A Theory for Pressure Transient Analysis in Naturally Fractured Reservoirs," *JPT* (July 1980) 1241-1250.
13. Serra, K., Reynolds, A.C., and Raghavan, R.: "New Pressure Transient Analysis Methods for Naturally Fractured Reservoirs," *JPT* (December 1983) 2271-2283.
14. Gringarten, A.C.: "Interpretation of Tests in Fissured and Multilayered Reservoirs with Double-Porosity Behavior: Theory and Practice," *JPT* (April 1984) 549-564.
15. Tiab, D., Restrepo, D.P., and Igbokoyi, A.: "Fracture Porosity of Naturally Fractured Reservoirs," paper SPE 104056 presented at the First International Oil Conference and Exhibition, Cancun, Mexico, (August 31-September 2, 2006).
16. Horne, R.N.: *Modern Well Test Analysis–A Computer-Aided Approach*, 2nd Ed., Petroway, Inc., California (2008).

ADDITIONAL REFERENCES

- Bourdet, D., and Gringarten, A.C.: "Determination of Fissure Volume and Block Size in Fractured Reservoirs by Type-Curve Analysis," paper SPE 9293 presented at the 55th Annual Fall Technical Conference and Exhibition, Dallas, Texas, (September 21-24, 1980).
- Bourdet, D., Ayoub, J.A., and Pirard, Y.M.: "Use of Pressure Derivative in Well-Test Interpretation," *SPEFE* (June 1989) 293-302.
- Cinco-Ley, H.: "Pressure Transient Analysis of Wells with Finite Conductivity Vertical Fractures in Double Porosity Reservoirs," paper SPE 18172 presented at the 63rd Annual Technical Conference and Exhibition, Houston, Texas, (October 2-5, 1988).
- de Swaan, A.: "Analysis of Well Tests in Multiple Fractured Reservoirs–Field Case Applications," paper SPE 59015 presented at the 2000 SPE International Petroleum Conference and Exhibition, Villahermosa, Mexico, (February 1-3, 2000).
- Gringarten, A.C.: "From Straight Lines to Deconvolution: The Evolution of the State of the Art in Well Test Analysis," *SPEREE* (February 2008) 41-62.
- Gringarten, A.C.: "How to Recognize 'Double-Porosity' Systems from Well Tests," *JPT* (June 1987) 631-633.
- Lee, J., Rollins, J.B., and Spivey, J.P.: *Pressure Transient Testing,* SPE Textbook Series Vol. 9, SPE, Richardson, Texas (2003).
- Okoye, C.U., Oshinuga, A.D., and Ghalambor, A.: "Pressure Transient Behavior in Multilayer Vertically Fractured Gas Reservoir with Finite Conductivity," paper SPE 18960 presented at the SPE Joint Rocky Mountain Region/Low Permeability Reservoirs Symposium and Exhibition, Denver, Colorado, (March 6-8, 1989).
- Tiab, D., and Puthigai, S.K.: "Pressure-Derivative Type Curves for Vertically Fractured Wells," *SPEFE* (March 1988) 156-158.
- Tiab, D.: "Multi-Rate Testing for Vertical Wells in Naturally Fractured Reservoirs," paper SPE 88558 presented at the SPE Asia Pacific Oil and Gas Conference and Exhibition, Perth, Australia, (October 18-20, 2004).

PROBLEMS

15-1. A Drill Stem Test (DST) was conducted on a newly discovered reservoir in the pre-salt formation, Campos Basin, offshore Brazil. The DST data can be downloaded for analyses from: http://www.tigapetroleum.com/

There is evidence from core samples that the reservoir could be naturally fractured. Other rock and fluid properties of the reservoir are given as:

Net pay thickness, h	30 ft
Formation porosity, ϕ	0.12
Oil viscosity, μ_o	0.843 cp
Oil FVF, B_o	1.378 RB/STB
Total compressibility, c_t	3.0E-06 psi^{-1}
Reservoir temperature, T_R	282°F
Well bore radius, r_w	0.3 ft

Using the DST data, determine:

a. Formation permeability

b. Skin factor for the well

c. Does the analyses support the presence of natural fractures?

Hint: Use commercial pressure transient analyses (PTA) software to analyze this problem.

CHAPTER

16 Well Test Analysis: Deconvolution Concepts

16.1 Introduction

In the preceding chapters on well test analysis, applications of conventional methods on the analyses of well test data were presented. These conventional methods include straightline methods discussed in Chapter 13 and the application of type curves discussed in Chapter 14. In Chapter 15, analysis of well test data from hydraulically fractured wells and naturally fractured reservoirs using both straightline methods and type curves were presented. In this chapter, deconvolution is introduced as another tool which can be used in the analysis of well test data. It is important to note that deconvolution is not a new interpretation method or a replacement for conventional methods in the analysis of well test data. It is rather a process that is used to convert variable-rate pressure data into constant-rate pressure data, thereby providing more pressure data for the determination of an interpretation model. Deconvolution is rapidly gaining wide application, especially on extensive well test data collected from permanent downhole gauges installed in deepwater wells, and other wells where constant pressure monitoring is desired. In this book, basic concepts of deconvolution are presented with examples and guidelines for application of deconvolution method.

16.2 What Is Deconvolution?

Deconvolution is the process of transforming variable-rate pressure data into a constant-rate drawdown pressure response data with duration equal to the total duration of the test. A corresponding pressure derivative normalized to a unit rate is also obtained during the deconvolution process. The process of deconvolving pressure-rate data is not simply an advanced numerical analysis technique but requires considerable skill and knowledge of the deconvolution algorithm.

In the application of conventional methods (semi-log and derivative type plots), well test analysis is generally limited to transient pressure behavior for a specific test period. This could be either a drawdown or a buildup test period. The rate history of the well is typically accounted for with a superposition-time transform which is meant to remove the effects of variability present in the production rates before the test period. The superposition-time transforms do not remove all the effects due to variability of production rates prior to either drawdown or buildup tests. These residual superposition effects sometimes distort and complicate analysis of well test data with conventional methods. Furthermore, analysis of well test data based on individual test periods using conventional methods are sometimes inconclusive if the test was terminated before characteristic curve shapes evident in conventional plots (semilog or log-log derivative plots) have been reached.

The deconvolution process, on the other hand, incorporates all pressure-rate data obtained during a well test sequence. This includes all the variable rate flow periods and shut-in periods of varying durations. Consequently, the deconvolution process removes the effects of rate variations from the pressure-rate data and yields a constant-rate drawdown pressure response which could be more representative of the well/reservoir system. The deconvolved pressure response typically has a much longer time span and may reveal characteristic features of the reservoir which will not be evident, if only conventional methods had been used in the test analysis. It is important to note that the deconvolution process helps in the identification of an interpretation model for the test data rather than the direct calculation of reservoir parameters. However, once an acceptable interpretation model has been identified through application of deconvolution, reservoir parameters can be estimated based on the interpretation model.

Deconvolution techniques have been applied to variable-rate test data since 1959 with limited success. These techniques can be classified into two methods, namely:[1] Time-domain methods and Spectral methods. **Time-domain methods** use various interpolation schemes to evaluate the convolution integral directly in the time domain.[1] **Spectral methods** are based on Laplace transform techniques.[1] The detailed mathematical application of either of these methods is not covered in this book. However, it is important to observe that application of robust and stable deconvolution techniques were not available until von Schroeter et al.[1-3] introduced a new time-domain based method in 2001. This method is based on a regularized, nonlinear total least squares (TLS) formulation of the deconvolution problem. The von Schroeter et al.[3] deconvolution method was evaluated by Levitan,[4] Levitan et al.,[5] and Onur et al.,[6] and found to be relatively stable under certain guidelines.

16.3 The Pressure-Rate Deconvolution Model

The pressure-rate deconvolution model used by von Schroeter et al.,[1-3] Levitan,[4] and Levitan et al.,[5] is based on the following convolution integral for a single well during a variable well test:

$$p(t) = p_0 - \int_0^t q(\tau)\frac{dp_u(t-\tau)}{dt}d\tau \qquad (16.1)$$

In Eq. (16.1), $q(t)$ is the measured well rate, $p(t)$ is the measured bottomhole pressure, and p_0 is the initial reservoir pressure. Also in Eq. (16.1), $p_u(t)$ is the drawdown pressure response of the well/reservoir system at a constant unit rate production, assuming that the reservoir is in equilibrium and pressure is also uniform throughout the reservoir at start of production. Equation (16.1) is valid only for linear systems (that is, single phase flow of slightly compressible fluid in a porous medium). Extension of Eq. (16.1) to multi-well test data has been reported by Levitan.[7]

Deconvolution is the process of reconstructing the constant-rate drawdown pressure response, $p_u(t)$, together with the initial pressure, p_0, from the pressure, $p(t)$, and rate, $q(t)$, data measured in a variable-rate well test.[4] The deconvolution problem is the same as solving Eq. (16.1) for $p_u(t)$ and p_0, given the measured data of $p(t)$ and $q(t)$. Thus, the deconvolution problem as defined is an inverse problem. The solution of an inverse problem is non-unique.[8] This means that different interpretation models can produce the same constant-rate drawdown pressure response, $p_u(t)$, from the same input of pressure, $p(t)$, and rate, $q(t)$, data measured in a variable-rate well test. It is important to remember this point when evaluating the pressure response data from the deconvolution process.

The deconvolution problem is also ill-conditioned. This means that slight variations in the input data (measured pressure and rate data) can cause large changes in the output data (deconvolved pressure response and initial pressure). The ill-conditioned form of the deconvolution problem combined with errors typically present in measurements of pressure and rate data make the solution of the deconvolution problem very unstable. As reported by von Schroeter et al.,[3] several deconvolution algorithms proposed in the literature were not sufficiently stable in the solution of the deconvolution problem. The algorithm proposed by von Schroeter et al.[1-3] and evaluated by Levitan,[4] Levitan et al.,[5] and Onur et al.[6] have been found to be stable under certain guidelines.

16.3.1 The von Schroeter et al.[3] Deconvolution Algorithm

The von Schroeter algorithm is summarized under three novel concepts as proposed in the work by Levitan:[4]

1. Implicit Constraint Encoding. For this approach, Eq. (16.1) is solved in terms of the function:

$$z(\sigma) = \ln\left[\frac{dp_u(t)}{d\ln t}\right] = \ln\left[\frac{dp_u(\sigma)}{d\sigma}\right] \qquad (16.2)$$

   In Eq. (16.2), $\sigma = \ln(t)$. In terms of the function, $z(\sigma)$, Eq. (16.2) becomes:

$$p(t) = p_0 - \int_{-\infty}^{\ln t} q\left(t-e^{\sigma}\right) e^{z(\sigma)} d\sigma \qquad (16.3)$$

   The selection of $z(\sigma)$ as the new solution variable ensures that the term, $dp_u(t)/d\ln t$, is positive. This condition should be satisfied by the constant-unit-rate response of the system. The use of the variable, $z(\sigma)$, makes Eq. (16.3) nonlinear .
2. Regularization Based on Curvature. The deconvolution problem is ill-conditioned and very sensitive to errors in measured pressure and rate data. Regularization based on the curvature of $z(\sigma)$ is imposed to attain some degree of smoothness of the solution and improve the conditioning of the deconvolution problem.
3. Total Least Squares Error Model. The deconvolution problem is formulated as unconstrained nonlinear minimization with the objective function defined to include errors in the pressure and rate data, residuals of Eq. (16.3), and regularization curvature constraints.

A detailed presentation of the mathematical formulation and implementation of the von Schroeter deconvolution algorithm is not presented in this book. Additional details on the implementation of the algorithm were documented by von Schroeter et al.,[3] and Levitan.[4]

16.4 Application of Deconvolution to Pressure-Rate Data

The convolution integral as written in Eq. (16.1) is valid for a single well producing a single-phase fluid in a variable-rate well test. It is also assumed that wellbore storage and/or skin factor are constant for the entire duration of the well test. This is the model on which the von Schroeter deconvolution algorithm is based. Consequently, for well tests in which these conditions are not met, the deconvolved constant-rate drawdown pressure response may yield erroneous results. In extended well tests, it is possible for wellbore storage and skin factor to change over the duration of the test. If the reservoir pressure is close to saturation pressure of the fluids within the reservoir, multiphase flow will occur when reservoir pressure falls sufficiently below the saturation pressure. These constraining conditions should be observed when subjecting pressure-rate data to the deconvolution process.

Levitan[4] proposed a novel procedure for checking the consistency of the variable-rate test data and ensuring the successful application of the von Schroeter deconvolution algorithm. Inconsistent test data sets are defined as data sets that do not conform to the conditions stipulated for the deconvolution model of Eq. (16.1) as stated earlier. This includes data sets obtained under conditions in which wellbore storage and/or skin factor changed or multiphase flow occurred during the duration of the test. For an inconsistent data set, Levitan[4] applied a variant of the von Schroeter deconvolution algorithm on pressure-rate data from individual flow periods within the test sequence while specifying initial pressure as input. Typically, pressure buildup (PBU) data from the individual flow periods are preferable because PBU data generally have fewer errors associated with them. Levitan[4] found that for a consistent data set, the constant-rate drawdown pressure response deconvolved from PBU data of individual flow periods appear to be identical or show similar trends. This procedure could be used to check variable-rate test data for consistency. Levitan et al.[5] listed three factors that could affect pressure response obtained from applying the von Schroeter deconvolution algorithm on well test data. These are:

1. Accuracy of initial reservoir pressure.
2. Consistency of test data.
3. Effects of local events on deconvolved pressure response.

The deconvolved pressure response is very sensitive to initial reservoir pressure. For this reason, it is crucial to ensure that the source of initial reservoir pressure is reliable and accurate. An inconsistent data set will produce deconvolved system response that is completely misleading. The approach devised by Levitan et al.[5] should be used to check test data for consistency. It was also found by Levitan et al.[5] that localized events such as mismatch of pressure and rate data distort the constant-rate drawdown pressure response at late times.

16.5 Examples on the Application of the von Schroeter Deconvolution Algorithm to Real Well Test Data

The two examples presented in this book are based on the work of Levitan et al.[5] These examples are intended simply to illustrate the application of the von Schroeter algorithm and present the results that could be obtained from deconvolution of well test data.

EXAMPLE 16.1

The production history of a new gas well is shown in Figure 16.1. The total duration of the variable-rate test is 210 hours. Included in the test history are two PBUs labeled PBU 1 and PBU 2 in Figure 16.1. PBU 2 is the longer buildup with duration of 121 hours. The pressure and pressure derivative of the two PBUs (PBU 1 and PBU 2) shown in Figure 16.2 indicate similar trends. However, the pressure derivative curves do not indicate definitive boundary effects. This appears to be in contradiction with the pressure history shown in

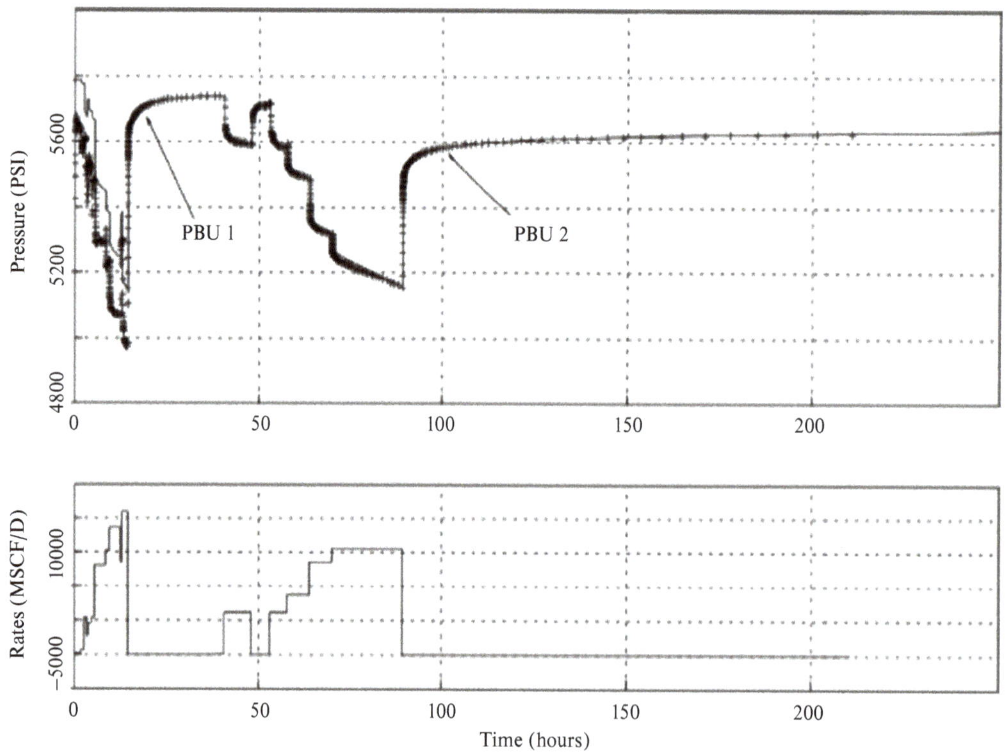

▲ **FIGURE 16.1** Pressure-rate history for Example 16.1 (from Levitan et al.[5] © 2006 SPE, Reproduced with permission)

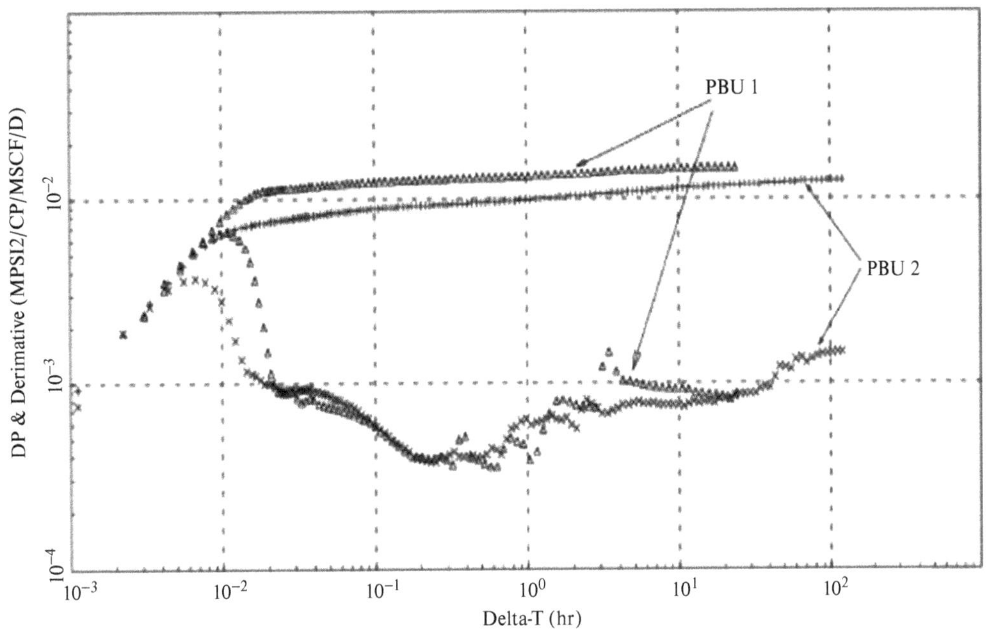

▲ **FIGURE 16.2 Comparison of derivative plots of two PBUs for Example 16.1 (from Levitan et al.[5] © 2006 SPE, Reproduced with permission)**

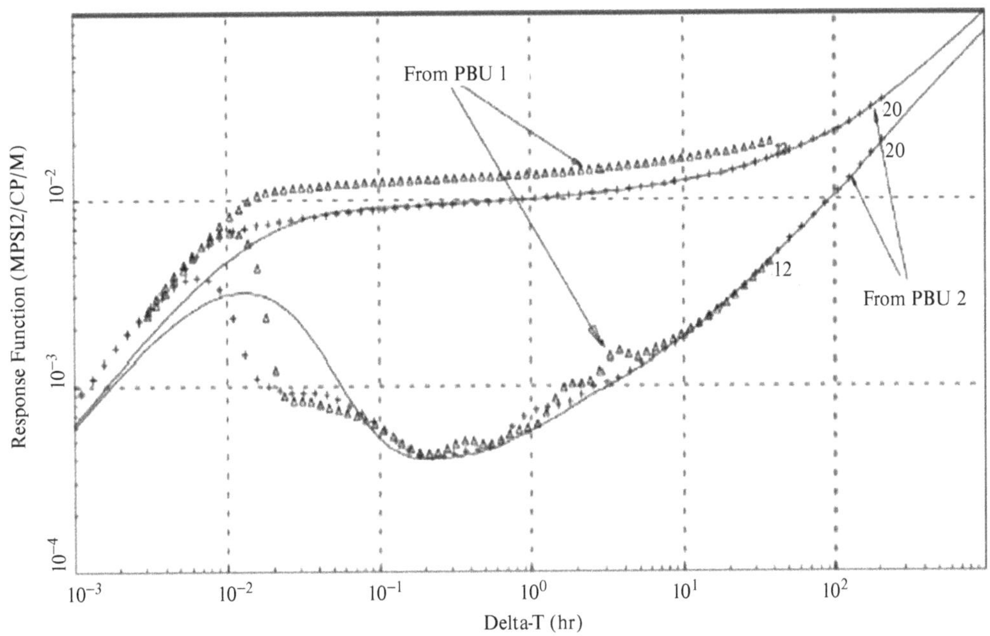

▲ **FIGURE 16.3 Comparison of deconvolved responses for PBU 1 and PBU 2 for Example 16.1 (from Levitan et al.[5] © 2006 SPE, Reproduced with permission)**

Figure 16.1, which shows pressure depletion during PBU 2, indicative of a closed reservoir compartment.[5] Figure 16.3 shows the pressure and pressure derivative of the deconvolved constant-rate drawdown responses derived from PBU 1 and PBU 2 data. The slopes of both pressure derivatives at late times approach unity. This is indicative of a closed reservoir system. Figure 16.4 compares the pressure derivative for PBU 2 obtained by differentiating PBU data with respect to superposition time and the pressure derivative obtained by differentiating the deconvolved pressure response with respect to the logarithm of time. Note the obvious differences between the two curves. The pressure derivative obtained from the deconvolved pressure response is more representative of the interpretation model assumed to be a partially closed system. By assuming a reservoir model that is closed on three sides with the fourth side open for limited pressure support, the pressure buildup data for PBU 2 is matched as shown in Figure 16.5. This model also matches most of the sequence shown in Figure 16.1.

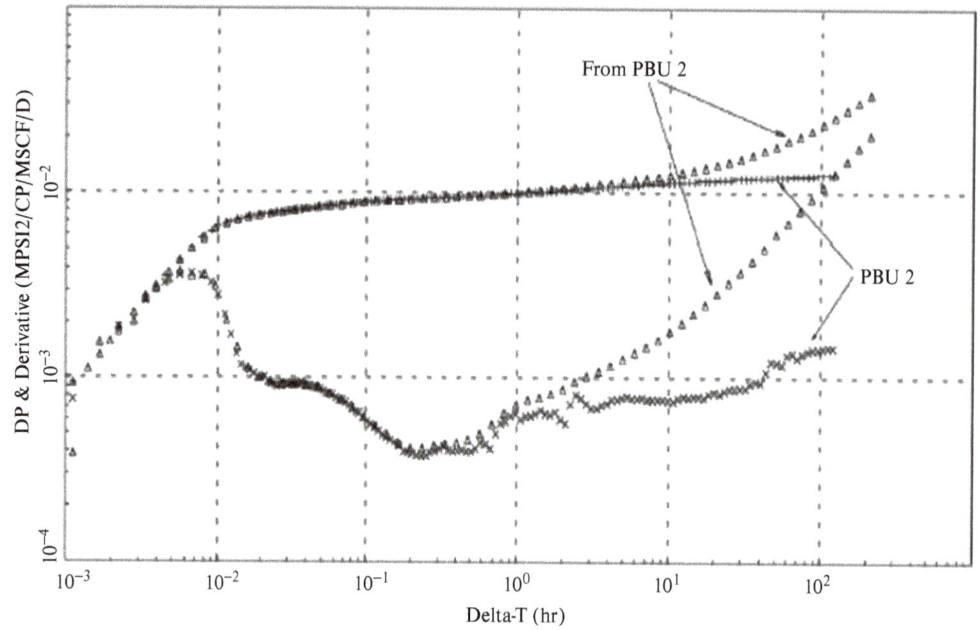

▲ **FIGURE 16.4 Comparison of deconvolved responses for PBU 2 data (triangles) with derivative plot of PBU 2 data (crosses) for Example 16.1 (from Levitan et al.[5] © 2006 SPE, Reproduced with permission)**

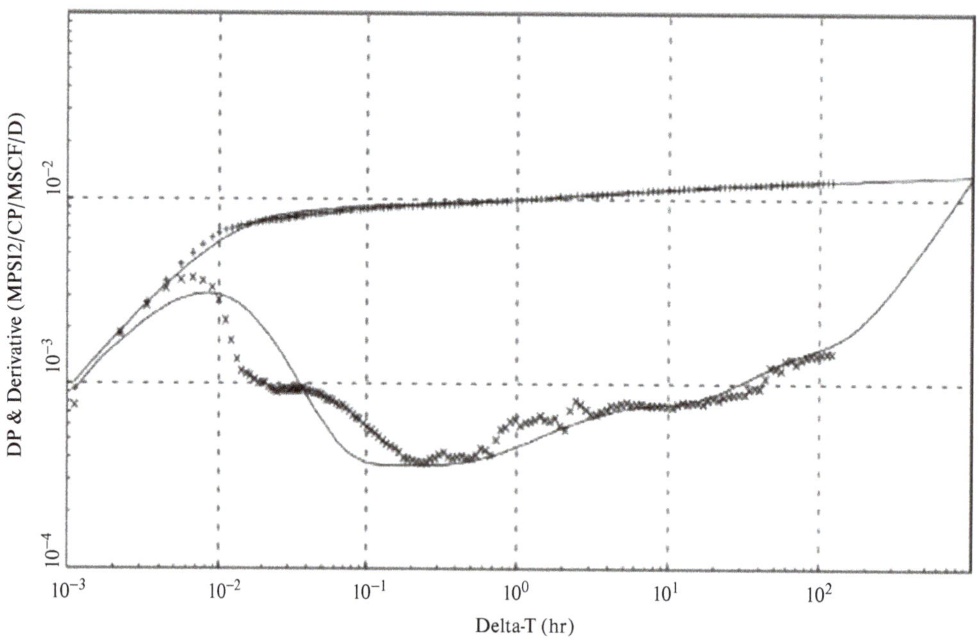

▲ **FIGURE 16.5 Model match of PBU 2 data assuming partially closed reservoir model for Example 16.1 (from Levitan et al.[5] © 2006 SPE, Reproduced with permission)**

EXAMPLE 16.2

The rate and permanent gauge pressure data from a single horizontal well in a gas reservoir is shown in Figure 16.6. The time span of the production history is 12,800 hours. Two PBUs labeled PBU 1 and PBU 2 with durations of 167 hours and 170 hours, respectively, were selected for analysis. The pressure and pressure derivative

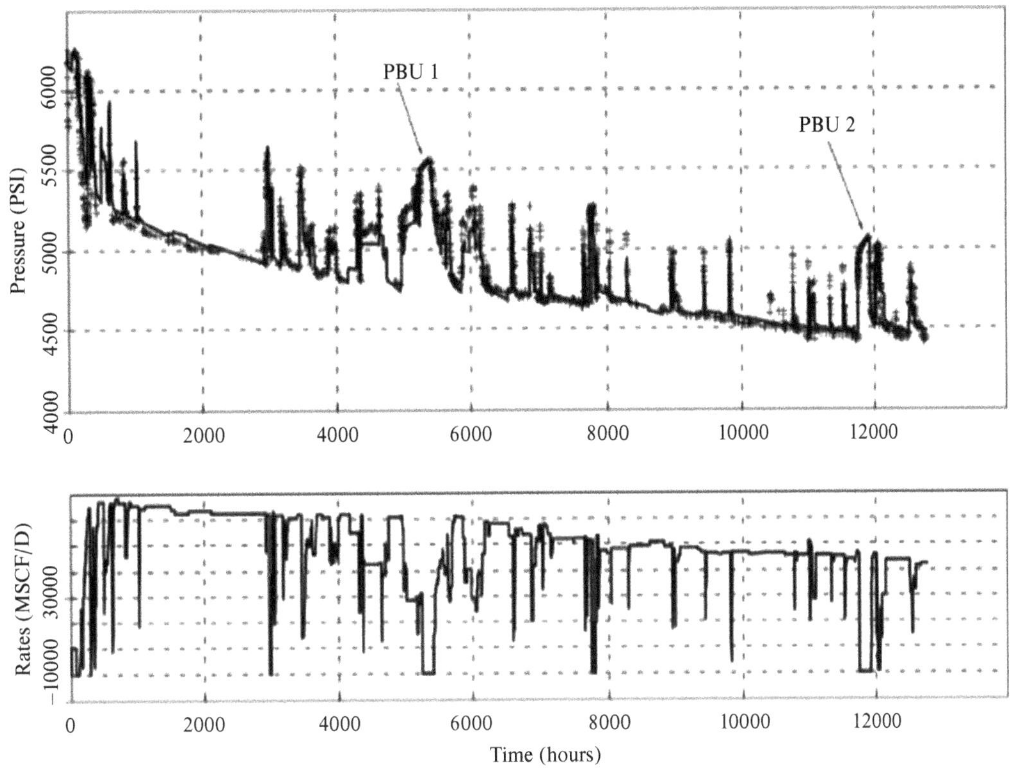

▲ **FIGURE 16.6** **Pressure-rate history for Example 16.2 (from Levitan et al.[5] © 2006 SPE, Reproduced with permission)**

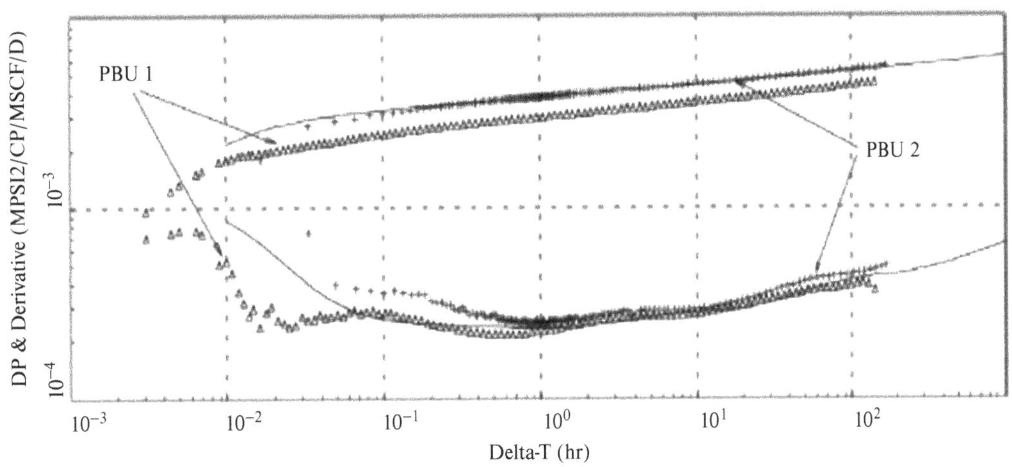

▲ **FIGURE 16.7** **Comparison of derivative plots of two PBUs for Example 16.2 (from Levitan et al.[5] © 2006 SPE, Reproduced with permission)**

plots for PBU 1 and PBU 2 are shown in Figure 16.7. The pressure derivative curves for PBU 1 and PBU 2 show increasing trends at late times. Figure 16.8 shows the pressure and pressure derivative plots for PBU 1 and PBU 2 obtained from constant-rate drawdown pressure response of the two PBUs over time intervals of 5400 and 11,900 hours, respectively. The pressure derivative plots of the two PBUs indicate unit slope at late times. This is indicative of boundary effects. Consequently, a reservoir model of a horizontal well in an elongated rectangular reservoir matched PBU 2 data as shown in Figure 16.8 by the solid curve.

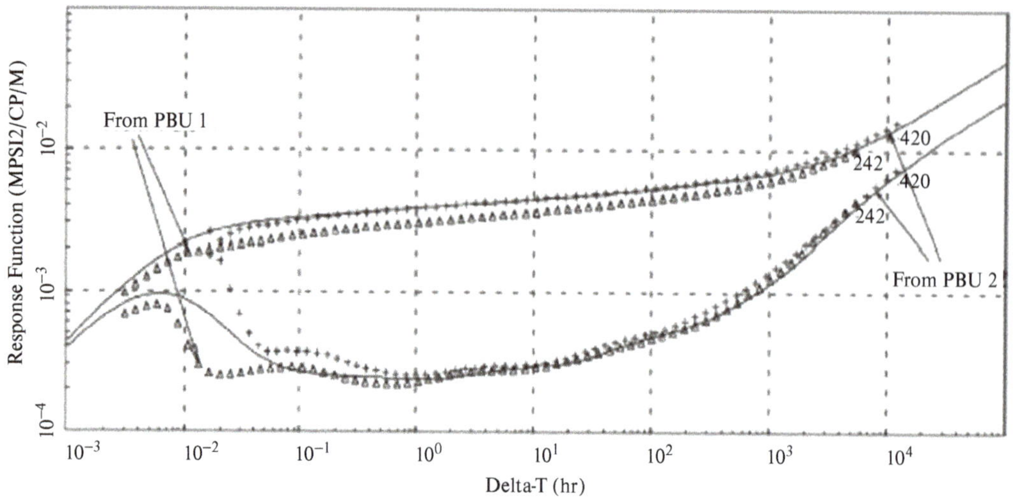

▲ **FIGURE 16.8** **Deconvolved pressure responses for the PBUs in Example 16.2 (from Levitan et al.[5] © 2006 SPE, Reproduced with permission)**

16.6 General Guidelines for Application of von Schroeter Deconvolution Algorithm to Pressure-Rate Data from Well Tests

It is necessary to reiterate that deconvolution is not a new method for interpretation of well test data. It is rather a process for converting variable-rate pressure data into constant-rate pressure data, thereby providing more pressure data for the determination of an interpretation model. Furthermore, deconvolution is not a replacement for conventional methods or type curves in the analysis of well test data. It should be considered as another tool in the toolkit of the analyst for the evaluation of well test data. The application of deconvolution to well test data should yield results that are consistent with results obtained from other methods (such as straight line methods, type curve analysis, etc.), and in conformity with known geological interpretations of the reservoir. The application of deconvolution algorithms to well test data requires considerable skill and knowledge of the technology. As more robust and stable deconvolution algorithms are made available in commercial software for well test analysis, deconvolution methods will become widely used because of its capacity to include all the data recorded in a test sequence. This is especially important in the analysis of data from permanent downhole gauges that are routinely installed in wells located in challenging environments, such as found in deepwater reservoirs.

The following are suggested guidelines for application of the von Schroeter-type deconvolution algorithm to variable-rate well test data:[5]

1. The von Schroeter deconvolution algorithm is based on linear systems. This means that this algorithm is applicable to systems with single phase flow in the reservoir. Reservoir pressure should be above bubble point pressure for oil reservoirs or dew point pressure for gas reservoirs when applying this algorithm to test data.
2. The reservoir should be at equilibrium and at a uniform pressure at the start of the well test.
3. A reliable estimate of the initial reservoir pressure at the start of the test sequence will aid in the deconvolution of the test data.
4. The test data should be evaluated for consistency. This means that the test data should comply with the assumptions of constant wellbore storage and/or skin factor implied in the convolution integral of Eq. (16.1). This can be done by selecting individual pressure buildup data from the test sequence for deconvolution and comparing the results.
5. Check the pressure-rate data for local events such as mismatched pressure-rate data. This type of error may distort the deconvolution pressure transform at late times and create non-existent artifacts in the transformed data.
6. Compare the interpretation model generated from the deconvolved pressure response to other interpretation models based on derivative plots, type curves, or geologic data. All the interpretation models generated from all these methods should be inherently consistent. For instance, if the deconvolved pressure response is indicating a closed reservoir system, this should be confirmed with other interpretations based on derivative plots, type curves, or known geologic data. If the interpretation is inconsistent, the evaluation process should be reexamined in its entirety.
7. Perform a thorough check on the quality of the test data. Poor quality data will generally yield misleading results when subjected to deconvolution.

REFERENCES

1. von Schroeter, T., Hollaender, F., and Gringarten, A.C.: "Deconvolution of Well-Test Data as a Nonlinear Total Least-Squares Problem," *SPEJ* (December 2004) 375-390.
2. von Schroeter, T., Hollaender, F., and Gringarten, A.C.: "Analysis of Well Test Data From Permanent Downhole Gauges by Deconvolution," paper SPE 77688 presented at the 2002 SPE Annual Technical Conference and Exhibition, San Antonio, Texas, (September 29-October 2, 2002).
3. von Schroeter, T., Hollaender, F., and Gringarten, A.C.: "Deconvolution of Well-Test Data as a Nonlinear Total Least-Squares Problem," paper SPE 71574 presented at the 2001 SPE Annual Technical Conference and Exhibition, New Orleans, Louisiana, (September 30-October 3, 2001).
4. Levitan, M.M.: "Practical Application of Pressure/Rate Deconvolution to Analysis of Real Well Tests," *SPEREE* (April 2005) 113-121.
5. Levitan, M.M., Crawford, G.E., and Hardwick, A.: "Practical Considerations for Pressure-Rate Deconvolution of Well-Test Data," *SPEJ* (March 2006) 35-46.
6. Onur, M., Cinar, M., Ilk, D., Valko, P.P., and Blasingame, T.A.: "An Investigation of Recent Deconvolution Methods for Well-Test Data Analysis," *SPEJ* (June 2008) 226-247.
7. Levitan, M.M.: "Deconvolution of Multiwell Test Data," *SPEJ* (December 2007) 420-428.
8. Gringarten, A.C.: "From Straight Lines to Deconvolution: The Evolution of the State of the Art in Well Test Analysis," *SPEREE* (February 2008) 41-62.

ADDITIONAL REFERENCES

- Baygun, B., Kuchuk, F.J., and Arikan, O.: "Deconvolution Under Normalized Autocorrelation Constraints," *SPEJ* (September 1997) 246-253.
- Du, K.: "The Determination of Tested Drainage Area and Reservoir Characterization from Entire Well-Test History by Deconvolution and Conventional Pressure-Transient Analysis Techniques," paper SPE 115720 presented at the 2008 SPE Annual Technical Conference and Exhibition, Denver, Colorado, (September 21-24, 2008).
- Gringarten, A.C., von Schroeter, T., Rolfsvaag, T., and Bruner, J.: "Use of Downhole Permanent Gauge Data to Diagnose Production Problems in a North Sea Horizontal Well," paper SPE 84470 presented at the SPE Annual Technical Conference and Exhibition, Denver, Colorado, (October 5-8, 2003).
- Kuchuk, F.J., Carter, R.G., and Ayestaran, L.: "Deconvolution of Wellbore Pressure and Flow Rate," *SPEFE* (March 1990) 53-59.
- Whittle, T., and Gringarten, A.: "The Determination of Minimum Tested Volume from the Deconvolution of Well Test Pressure Transients," paper SPE 116575 presented at the 2008 SPE Annual Technical Conference and Exhibition, Denver, Colorado, (September 21-24, 2008).

PROBLEMS

16-1. The drillstem test (DST) data in Problem 14-1 included drawdown production data. Repeat the analysis by performing deconvolution of the entire DST data set to evaluate the following:

a. Changes in the interpretation model, if any
b. Changes of reservoir boundaries, if any
c. Improvements in the interpretation by applying deconvolution

Hint: This problem should be solved on a commercial pressure transient analysis (PTA) software that has deconvolution module implemented.

16-2. The DST data in Problem 14-2 included several drawdown production data. Repeat the analysis by performing deconvolution of the entire DST data set to evaluate the following:

a. Changes in the interpretation model, if any
b. Changes of reservoir boundaries, if any
c. Improvements in the interpretation by applying deconvolution

Hint: This problem should be solved on a commercial pressure transient analysis (PTA) software that has deconvolution module implemented.

16-3. The DST data in Problem 15-1 included several drawdown production data. Repeat the analysis by performing deconvolution of the entire DST data set to evaluate the following:

a. Changes in the interpretation model, if any
b. Changes of reservoir boundaries, if any
c. Improvements in the interpretation by applying deconvolution

Hint: This problem should be solved on a commercial pressure transient analysis (PTA) software that has deconvolution module implemented.

CHAPTER

17 Well Test Analysis: Diagnostic Fracture Injection Tests in Unconventional Reservoirs

17.1 Introduction

Unconventional reservoirs were categorized in Chapter 1 as reservoirs that have formations with extremely low rock matrix permeabilities. These formations may also have been naturally fractured. To achieve commercial production from most unconventional reservoirs, hydraulic fracturing of the formation is generally required. Certain data from rock property and geomechanics are necessary in the design and execution of hydraulic fracturing. These rock data cannot be obtained from regular well tests (such as buildups, drawdowns, falloffs, mini-fracs, etc.) used in conventional reservoirs because they will take months (or even years) to reach stabilized flow regimes in unconventional reservoirs. For this reason, diagnostic fracture injection test (DFIT) was developed to acquire these needed rock property and geomechanics data in a much shorter time frame usually in a few days and rarely in more than 2 months. DFIT can be described in general as a low rate, small volume injection of water or similar fluids into a narrow-perforated interval over a short duration of 5 to 10 minutes to create a fracture in the formation. The well is then shut-in and monitored with pressure gauges for specific falloff pressure responses over time. The main objectives of a DFIT are to obtain data such as:

1. The minimum in-situ stress or fracture closure pressure.
2. The nature and complexity of existing natural fractures.
3. Reservoir properties such as permeability and pore pressure.

The basic features of a DFIT are presented in this chapter including key diagnostic plots used in the analysis of a test. Also, the design, planning, and execution of a DFIT are presented. A step-by-step procedure for analysis of a DFIT is presented and illustrated with the solution of a DFIT on a gas well in the Mesaverde formation in the San Juan Basin in the United States of America (USA) as an example.

17.2 Typical DFIT Profile

A sketch of a typical DFIT profile showing pressure and injection trends is shown in Figure 17.1. At the start of a DFIT, the wellbore is filled with water or any other available low compressibility fluid. Care should be taken to remove any trapped air or gas from the fluid column by pressurizing and venting it slowly. With the start of injection, the wellbore pressure would rise rapidly (Point 1) as shown in Figure 17.1 until formation breakdown or breakover pressure is reached denoted as Point 2 in the figure. Formation breakdown pressure is a significant event in a DFIT and denotes the point at which fracture is initiated in the formation. It is marked by a sharp drop in injection pressure. The sequences of a DFIT profile as denoted by annotated points in Figure 17.1 are discussed in detail as follows:

Point 1: Fluid injection initiated at the surface at low rates causing compression of wellbore fluid. Rapid rise of wellbore pressure is observed. Little or none of the injected fluids is expected to penetrate or enter into the formation at these injection pressures.

Point 2: Wellbore pressures reach formation breakdown or breakover pressures. Hydraulic fracturing is initiated and is being propagated through the formation.

Point 3: Fluid injection is stabilized at constant rate to establish Fracture Propagation Pressure (FPP).

Point 4: Fluid injection is terminated.

Point 5: Instantaneous Shut-in Pressure (ISIP) response is registered. Determination of ISIP from the pressure response will be discussed later. ISIP is considered equivalent to the minimum horizontal stress in the formation. It is the minimum pressure required to keep the fracture open.

Point 6: The pressure response from the well is monitored for Fracture Closure Pressure (FCP). FCP is determined from several diagnostic plots which will be presented later. Note that fracture closure marks the end of fracture dominated flow regime and the beginning of reservoir dominated flow regime. The analytic techniques used before fracture closure is termed Before Closure Analysis (BCA). The analytic techniques used after fracture closure is termed After Closure Analysis (ACA). These analytic techniques will be presented later in the chapter. Net Fracture Pressure (NFP) is the difference between ISIP and FCP. This is represented in the form of an equation as:

$$NFP = ISIP - FCP \tag{17.1}$$

Point 7: After fracture closure, the flow regime within the formation is dominated by formation pseudo-linear flow. Formation pseudo-linear flow is identified with certain diagnostic plots which will be demonstrated later. Formation pseudo-linear-flow characteristic plots can be used to determine reservoir pressure.

Point 8: If the well is shut-in for a sufficiently long time, pseudo-radial flow may develop. The length of time required for pseudo-radial flow to develop depends on many factors including the permeability of the formation, injection rate, volume of injected fluid, type of induced fractures, and complexity of existing natural fractures. From the pseudo-radial-flow regime, characteristic plots can be used to determine formation permeability and reservoir pressure.

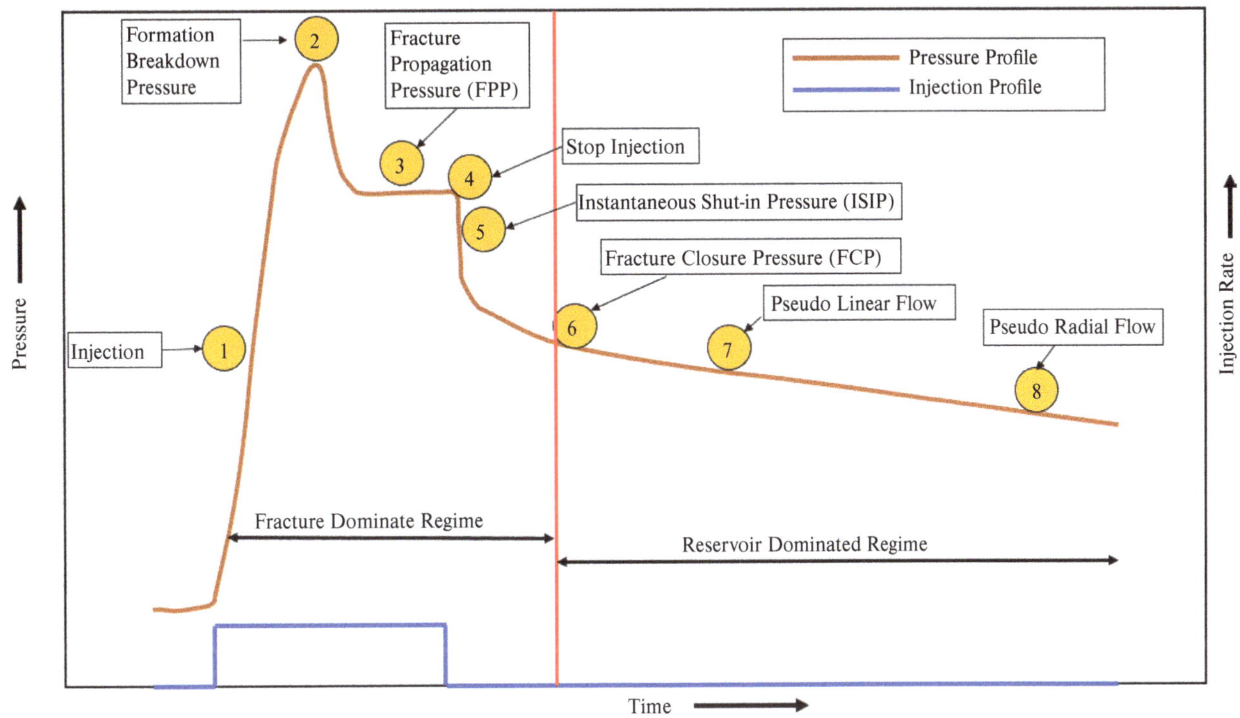

▲ FIGURE 17.1 DFIT profile

17.3 Simple Fracture Orientation Models

A simple rock cube can be used to illustrate in-situ stresses acting on the rock which determine the direction of hydraulically induced fractures. As shown in Figure 17.2, the three principal in-situ stresses acting on the rock cube are the overburden stress, (σ_1), the maximum horizontal stress, (σ_2), and the minimum horizontal stress, (σ_3). The minimum stress is assumed roughly equivalent to fracture closure pressure. For many formations, the overburden stress is higher than the maximum horizontal stress which in turn is higher than the minimum horizontal stress. Consequently, if such a formation is subjected to hydraulic fracturing, the fractures will be vertical and oriented in the direction of maximum horizontal stress as shown in Figure 17.2. Note that a rule-of-the-thumb estimate of overburden pressure (stress) is 1 psi/ft. On the other hand, consider Figure 17.3 that represents a shallow formation (typically at a depth of less than 1000 feet), the overburden pressure is the lowest stress. If such a formation is subjected to hydraulic fracturing, the orientation of the fractures will be horizontal as shown in Figure 17.3.

The importance of knowing the likely orientation of the fractures is that placing the trajectory of the wellbore perpendicular to direction of the fractures promotes well productivity.

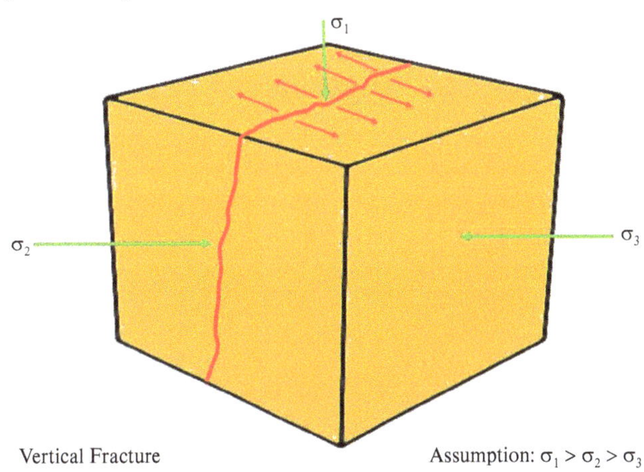

▲ FIGURE 17.2 Orientation of vertical fractures

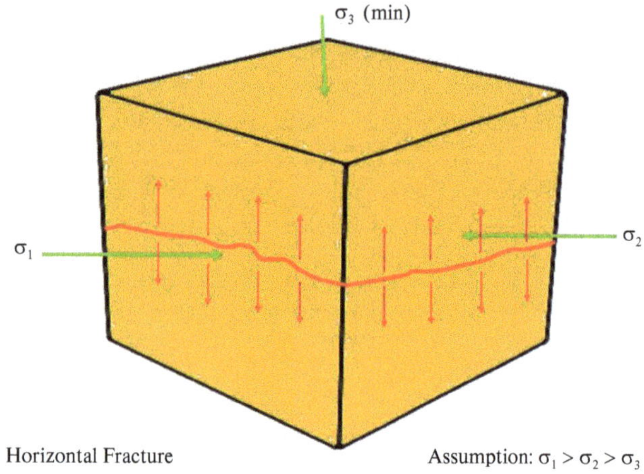

▲ **FIGURE 17.3 Orientation of horizontal fractures**

17.4 Design, Planning, and Execution of DFIT

DFITs involve injection of small volumes of water or similar fluids at low rates into small intervals of a formation that has extremely low permeabilities that range from micro- to nano-darcy. The pressure responses from the formations after the injection are then analyzed to obtain important geomechanical and reservoir rock properties that can be used to design and execute full hydraulic fracturing programs and other field development plans. In this section, the design, planning, and execution of DFITs will be discussed under several sections to address particular items specific to DFITs.

17.4.1 Selection of Test Interval

The interval to be tested should be selected as early as possible. In reviewing a potential test interval, key item to focus on is the quality of the cement bond log around the interval. The cement bond log must be of higher quality to prevent channeling of injected fluids behind the casing into other intervals. Another important item to investigate is the competency of the overlaying and underlaying intervals to withstand the fracture pressures and prevent the fractures from extending out of zone into these intervals.

17.4.2 Injection Rates and Volumes

The characteristics of DFITs in low permeability reservoirs are low injection rates and small injection volumes. Low injection rates and small injection volumes are typically chosen to achieve these three favorable outcomes:

1. Reduce length of the induced fractures.
2. Reduce the time necessary to achieve fracture closure.
3. Reduce the time required to reach pseudo-radial flow.

Consequently, in many DFITs injection rates are typically in the range of 0.1 to 5 bbl/min, and injection volumes are in the range of 1 to 20 barrels of fluid. Since injection rates and pumping times (duration of pumping) are related to injection volume, it is rather obvious that pumping times in most DFITs are kept short to keep injection volumes small. Figure 17.4 shows estimated fracture half-length as a function of pumping times and rates. Note that Figure 17.4 is provided for illustration purposes only. It is based on the assumption of a constant fracture height of 30 feet, fracture width of 0.1 inches, and 100% fracture fluid efficiency. **This graph should not be used to estimate fracture half-length in an actual DFIT.** Fluid efficiency is defined as:

$$\begin{aligned}&\text{Fracture fluid efficiency}\\&=\frac{\text{Total fluids injected} - \text{Fluids leak-off}}{\text{Total fluids injected}}\\&=\frac{\text{Injected fluids remaining in fracture}}{\text{Total fluids injected}}\end{aligned} \tag{17.2}$$

Note that high fracture fluid efficiency implies low leak-off. It also implies that a high percentage of the fracturing fluid was retained in the fracture to promote its growth and propagation into the target formation. Low permeability formations generally exhibit low leak-off characteristics.

In a DFIT, reducing the time to reach fracture closure is desirable because it hastens the initiation of pseudo-radial flow in the formation. The development and identification of pseudo-radial flow is required for the calculation of flow capacity, (kh/μ), and pore pressure, (p_i), under after closure analysis (ACA). An empirical relationship that relates time to attain closure, (t_c), to pumping time, (t_p), and permeability of the leak-off system, (k), is given by Barree et al.[1] as:

$$t_c = \frac{t_p}{3k} \tag{17.3}$$

In Eq. (17.3), t_c is time to achieve fracture closure in minutes; t_p is pumping time in minutes, and k is the permeability of the leak-off system in millidarcy. For example, a formation interval that has a permeability of 0.001 md and 5-minute pumping time during a DFIT will require approximately 1666.7 minutes or 27.8 hours to reach closure using Eq. (17.3). Figure 17.5 is a representation of Eq. (17.3) in a graphical form.

The time to reach pseudo-radial flow is an important consideration in a DFIT design. It is important to shorten this time as much as possible by minimizing fracture length as discussed earlier. The time to reach pseudo-radial flow can be estimated with Eq. (15.2) repeated here in rearranged form as:

$$t = \frac{\phi \mu c_t L_f^2 t_{DL_f}}{0.0002637k} \tag{17.4}$$

The dimensionless time in terms of fracture half-length, t_{DL_f}, is defined in Eq. (15.3). For $t_{DL_f} = 1$, Eq. (17.4) reduces to:

$$t = \frac{\phi \mu c_t L_f^2}{0.0002637k} \tag{17.5}$$

The value of t_{DL_f} depends on the conductivity of the fracture. The range of values of t_{DL_f} is given in Section 15.4.4 of Chapter 15. In Eqs. (17.4) and (17.5), t = time, hr.; ϕ = porosity, fraction; μ = viscosity, cp; c_t = total compressibility, psi^{-1}; L_f = fracture half-length, ft; and k = formation permeability, md.

Applying Eq. (17.5) to a gas reservoir with porosity of 10%, viscosity of 0.0177 cp, total compressibility of 2.26×10^{-4}, fracture half-length of 122 feet, and formation permeability of 0.012 md, gives the time to reach pseudo-radial flow as 1887 hours or 78.6 days. For this example, the calculated time to reach pseudo-radial flow is high and would be even higher for lower permeability formations. Barree et al.[1] argued that most DFITs achieve pseudo-radial flow at times much shorter than the times calculated with Eqs. (17.4) and (17.5) because actual fracture systems are much more complex than the simple planar geometry of the Perkins-Kern-Nordgren (PKN) fracture model that was the basis of the work done by Nolte.[2] The point is that Eqs. (17.4) and (17.5) should be used only as guides with the understanding that actual times to reach pseudo-radial flows could be considerably less depending on the complexity of natural fractures that may pre-exist in the target formation.

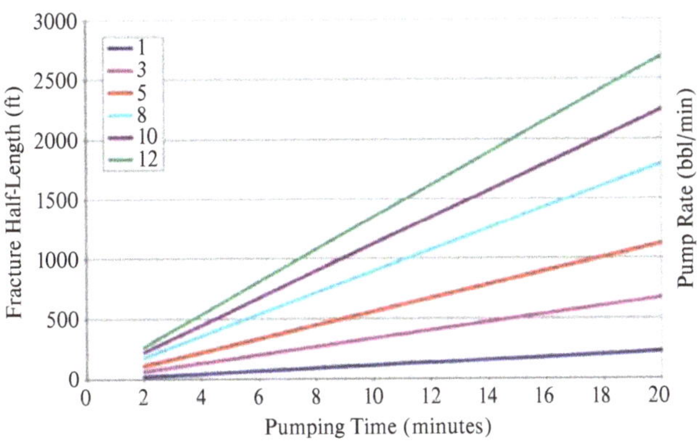

▲ **FIGURE 17.4 Fracture half-lengths as a function of pumping times and rates[1] (from Barree et al.[1] © 2015 SPE, Reproduced with permission)**

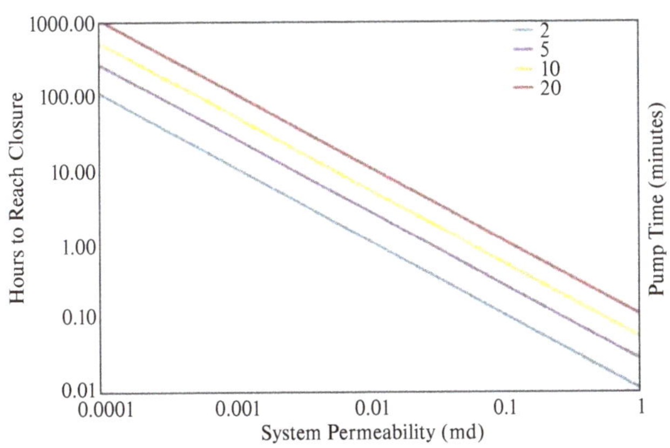

▲ **FIGURE 17.5 Estimated time to reach fracture closure (from Barree, et al[1] © 2015 SPE, Reproduced with permission)**

17.4.3 Selection of Shut-in Methods

At the end of injection, DFITs can be shut in either at the surface or downhole with a retrievable bridge plug. The selection of shut-in method could depend on many factors including the condition of the wellbore, ease of achieving shut in, availability and operability of equipment for downhole shut-in, risks of downhole shut-in, costs of downhole shut-in, time-savings by using downhole shut-in, etc. All these factors should be considered before selection of a shut-in method. However, downhole shut-in has clear advantages over surface shut-in that include reducing wellbore storage effects and reducing the time to reach fracture closure that would result ultimately to quicker initiation of pseudo-radial flow. Reducing wellbore storage by using downhole shut-in reduces afterflow volumes that could exist if the well was shut in at the surface. Afterflow effects are due to expansion flow of the fluid in the wellbore into the fracture and reservoir due to depressurization after shut in. According to Cramer and Nguyen,[3] the afterflow volume can be estimated with the equation:

$$V_{AF} = V_w \times C_w \times \Delta p \tag{17.6}$$

In Eq. (17.6), V_{AF} = afterflow volume, bbls; V_w = wellbore fluid volume, bbls; C_w = wellbore fluid compressibility, psi^{-1}; and Δp = maximum downhole injection pressure minus fracture closure pressure, psi.

Another benefit of selecting downhole shut-in is to prevent pressure hold-up effect especially for sub-pressured reservoirs.[3] Sub-pressured reservoirs are described for this discussion only as reservoirs whose pore pressures are less than the hydrostatic head of the wellbore fluid. And over-pressured reservoirs are reservoirs whose pore pressures are higher than the hydrostatic head of the wellbore fluid column. For sub-pressured reservoirs with surface shut-in and surface pressure gauge, the wellhead pressure must be greater than zero for a valid reading of the pressure gauge. If the wellhead pressure drops below zero (representing vacuum conditions that can create pressure hold-up effect), the pressure readings are no longer accurate and could not be used for DFIT analysis.[4] For sub-pressured reservoirs, the pressure hold-up effect could be prevented by using downhole shut-in with a retrievable bridge plug. In such applications, care should be taken to set the retrievable bridge plug at a depth such that the hydrostatic head of the fluid below the bridge plug is less than reservoir pore pressure. Even for over-pressured formations, use of downhole shut-in with a retrievable bridge plug is preferred because it ensures that wellbore storage effects are minimized, afterflow volume is reduced, and fracture closure time is shortened due to reduced fracture leak-off fluid volumes.[3]

17.4.4 Selection of Equipment for DFIT

By design, DFITs involve injection of small volumes of fluids at relatively low injection rates over short time durations. Consequently, the volume of fluid injected as well as the injection rate should be monitored and measured as precisely as possible. It is recommended that digital measurements of the injection rate, injection volumes, and injection pressures should be used if available, and a precise injection schedule with all these data should be recorded.[1] The pressure gauges used in the test should have resolutions of 0.01 to 0.1 psi with sample rate of one measurement per second.[1] This sampling rate should be maintained during the pumping phase, shut-in and past fracture closure. The sampling rate could be reduced later after fracture closure depending on the length of time the well would be shut in. The objective of any sampling frequency selected should be to acquire as much pressure data as possible depending on the memory capacity of the pressure gauges. Pressure data sampled at high frequencies are more stable for calculation of derivatives during analysis of the DFIT data.

17.4.5 Modeling DFIT with a Hydraulic Fracture Model

The recommended practice is to conduct a simulation of the planned DFIT on a suitable in-house or commercial hydraulic fracturing model. The hydraulic fracture model would require data input of rock properties and fluid saturations of the target formation and the bounding intervals. These properties could be obtained from open-hole and cased-hole logs of the candidate well or from analog wells in the same or similar basins. Using the hydraulic fracture model to simulate the DFIT would provide insights into the following variables of the test:[3]

1. Injection volumes, injection rates, and duration of injection.
2. Effects of perforated interval on propagated fracture length and height.
3. Competency of the bounding intervals to keep the fracture within the target interval.
4. Impact of net fracture pressure on growth of fracture length and height.

The hydraulic fracture model can be used after the DFIT with actual data from the test to validate the results. This practice can be used to refine the fracture model for designing and execution of future DFITs on similar wells in the reservoir.

17.4.6 Other DFIT Execution Guidelines

In some geographical regions, the DFIT may be scheduled for execution under freezing or low temperature conditions. In such circumstances, it is important to insulate the injection lines, injection gauges, and pressure gauges to ensure the readings are not affected by temperature fluctuations either during injection or shut in of the well. In some cases, it might be beneficial to use a low-freeze-point fluid as the injection fluid. An example of a low-freeze-point injection fluid is various mixtures of propylene glycol with water.[3]

It is essential that the injection lines and the wellbore are purged free of air and gas trapped or entrained in the fluid system.[3] This can be accomplished by slowly pressurizing the entire system and allowing trapped/entrained gas (air) to escape by venting. In some cases, it might be beneficial to circulate the well with the injection fluid at least 24 hours before start of injection to allow geothermal gradient to be restored in the wellbore. Making sure the injection fluid column is free of entrained gas or air reduces compressibility of the fluid system and the adverse effects of phase separation doing long periods of shut in.

In most cases, DFITs require the well to be shut in for extended time periods to acquire needed pressure falloff data. These pressure data are very sensitive to any leakages in the system either during injection or shut in. Extreme caution should be taken to ensure there are no leaks either in the surface or downhole equipment. If surface pressure gauges are used for the DFIT, the surface equipment (such as lubricators, pumps, etc.) must be isolated with a valve from the wellhead gauges during the long period of shut-in. In addition, the wellbore should be monitored to ensure there are no leaks or any fluid phase segregation that can compromise the pressure data. If the well is shut-in downhole with a retrievable bridge plug (RBP), pressure tests of the RBP should be conducted at the beginning and end of the test to ensure that it had not leaked.[4]

17.5 Analyses of Diagnostic Fracture Injection Tests (DFITs)

Analyses of DFITs can be conducted by using these three specific steps:

1. Diagnostic analysis.
2. Before closure analysis (BCA).
3. After closure analysis (ACA).

Under diagnostic analyses, the DFIT data are evaluated using characteristic plots to identify fracture leak-off types, and to determine fracture closure using G-function time plots. Other diagnostic plots that give similar results are root time plots. In conducting BCA, the main diagnostic plot is the log-log plot of pressure difference (ISIP minus shut-in pressures) versus shut-in time and the semi-log derivatives versus shut-in time. This plot can be used to identify fracture flow regime before fracture closure, confirm fracture closure, and flow regime after closure. In performing ACA, diagnostic plots of pressure change and its semi-log derivative versus square linear flow time function can be used to identify flow regimes after fracture closure. Other plots that can be used to calculate reservoir properties such as transmissibility and reservoir pressure are pressure change versus radial flow time function and the Horner plot.

17.5.1 Diagnostic Analyses

The main characteristic plot used under diagnostic analyses to identify fracture leak-off type and to determine fracture closure pressure (FCP) are plots of pressure derivative and semi-log of pressure derivative versus G-function time. The G-function is a dimensionless time function that relates elapsed shut-in time to pumping time as defined by Nolte.[2] The basic G-function equations are defined as follows:

$$G(\Delta t_D) = \frac{4}{\pi}\left[g(\Delta t_D) - g_0\right] \tag{17.7}$$

In Eq. (17.7), g_0 = dimensionless loss-volume function at $\Delta t_D = 0$.

For low leak-off, the intermediate function, $g(\Delta t_D)$, is defined as:

$$g(\Delta t_D) = \frac{4}{3}\left[(1+\Delta t_D)^{1.5} - \Delta t_D^{1.5}\right] \tag{17.8}$$

For high leak-off, the intermediate function, $g(\Delta t_D)$, is defined as:

$$g(\Delta t_D) = (1+\Delta t_D)\sin^{-1}(1+\Delta t_D)^{-0.5} + \Delta t_D^{0.5} \tag{17.9}$$

The dimensionless time, Δt_D, is defined as:

$$\Delta t_D = \frac{t - t_p}{t_p} \tag{17.10}$$

Note that in Eq. (17.10), t is total elapsed time from start of fracture initiation, and t_p is elapsed time from fracture initiation to shut-in as defined by Barree et al[5].

The diagnostic plots based on G-function are constructed by plotting three items on the same graph. These three items are: falloff pressure versus G-function, (p vs. G), pressure derivative versus G-function, (dp/dG vs. G), and semi-log pressure derivative versus G-function, (Gdp/dG vs. G). This diagnostic plot based on the G-function can be used to determine fracture closure pressure (FCP), and to identify four common fracture leak-off types. The fracture leak-off types that can be identified using G-function diagnostic plot are:

1. Normal leakoff.
2. Pressure-dependent leakoff.
3. Fracture tip extension.
4. Fracture height recession/transverse storage.

It is important to note that the value of G-function time determined at fracture closure, (G_c), can be used to estimate fracture fluid efficiency with the expression:

$$\eta = \frac{G_c}{G_c + 2} \tag{17.11}$$

In Eq. (17.11), η = fracture fluid efficiency, fraction; and G_c = value of G-function at fracture closure.

Also, the root time plots can be used to determine fracture closure pressure and identify common fracture leakoff types listed earlier. For the root time plots, pressure versus the square root of time, (p vs. $\sqrt{t}$), pressure derivative versus square root of time, ($dp/d\sqrt{t}$ vs. $\sqrt{t}$), and semi-log pressure derivative versus the square root of time, ($\sqrt{t}\,dp/d\sqrt{t}$ vs. $\sqrt{t}$), are plotted on the same graph similar to the G-function plots. Generally, the root time plots are used to verify and confirm fracture closure pressure and fracture leakoff type identified with the G-function plot.

17.5.1.1 Normal Leak-off

Normal leakoff fracture is described as a fracture with constant surface area and constant composite reservoir permeability. The fracture is assumed to stop propagating after shut-in. A sketch of the G-function plots representing normal leakoff is shown as Figure 17.6. In Figure 17.6, p vs. G, dp/dG vs. G, and $G\,dp/dG$ vs. G, are plotted on the same graph. For normal leakoff fracture, the $G\,dp/dG$ vs. G plot is linear close to the origin and lies on a straight line that passes through the origin. Fracture closure is identified as the point of departure of the semi-log curve from the straight line drawn through the origin. This is identified with a dashed line in Figure 17.6. Note that as defined for normal leakoff, the first derivative, dp/dG, should be constant, and p vs. G should form a straight line until fracture closure. In diagnostic plots with actual data, these specific features generally may not be present. A similar diagnostic plot based on square root of shut-in time can be used to determine fracture closure. Figure 17.7 is a plot of shut-in pressure, (p vs. $\sqrt{t}$), first derivative, ($dp/d\sqrt{t}$ vs. $\sqrt{t}$), and semi-log derivative, ($\sqrt{t}\,dp/d\sqrt{t}$ vs. $\sqrt{t}$), all versus the square root of shut-in time. This plot is very similar in form to the G-function plot. It should be used to verify and confirm the determination of fracture closure with the G-function plot.[5]

17.5.1.2 Pressure-Dependent Leakoff

Pressure-dependent leakoff (PDL) occurs when the formation has existing natural (secondary) fracture system that comes into communication with the induced hydraulic fracture. Figure 17.8 is a plot of pressure, its first derivative, and its semi-log derivative versus G-function time for PDL. PDL is identified by the characteristic "hump" above the straight line through the origin developed by the semi-log derivative curve. The end of the "hump" marks the end of PDL and denotes the opening pressure of the secondary fracture system. It also represents the beginning of leakoff dominated by the matrix. Fracture closure is still identified by the departure of the straight line through the origin from the semi-log derivative curve as shown in Figure 17.8. The PDL square-root plot is similar to the PDL G-function plot in many respects as shown in Figure 17.9. The PDL square-root plot should be used to confirm fracture closure determined with the G-function plot.[5]

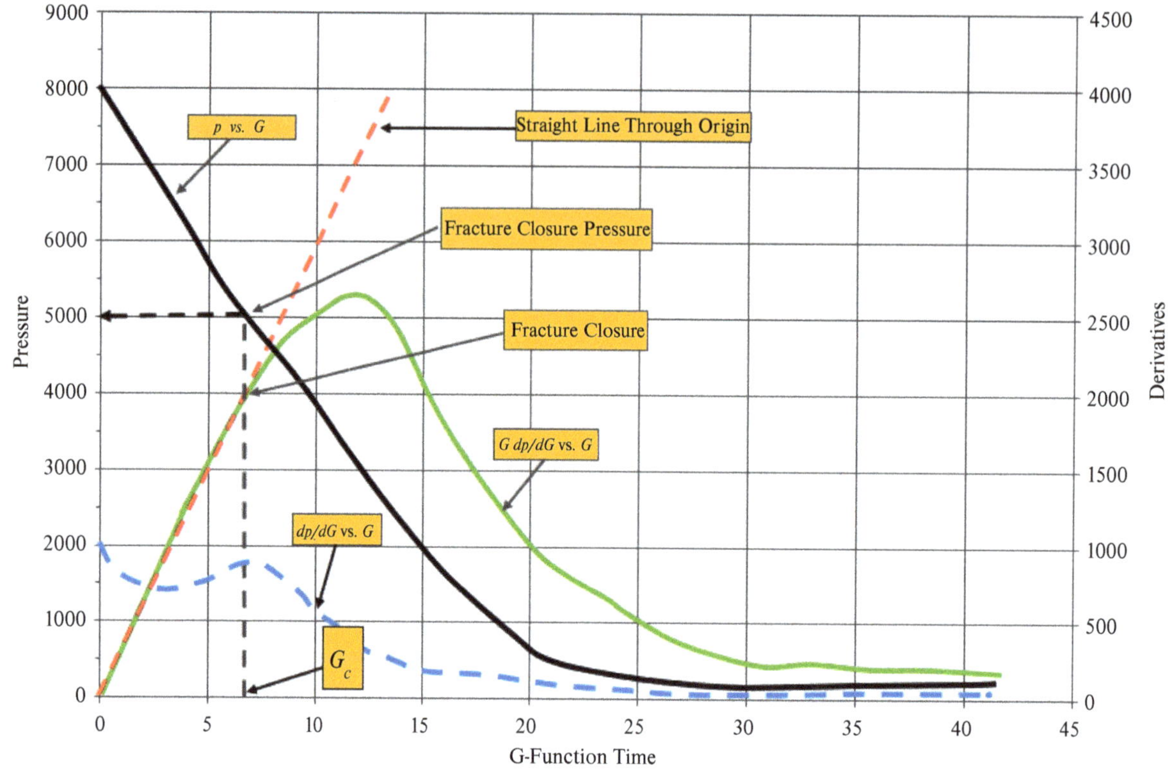

▲ **FIGURE 17.6** Normal leakoff plot based on G-function plot

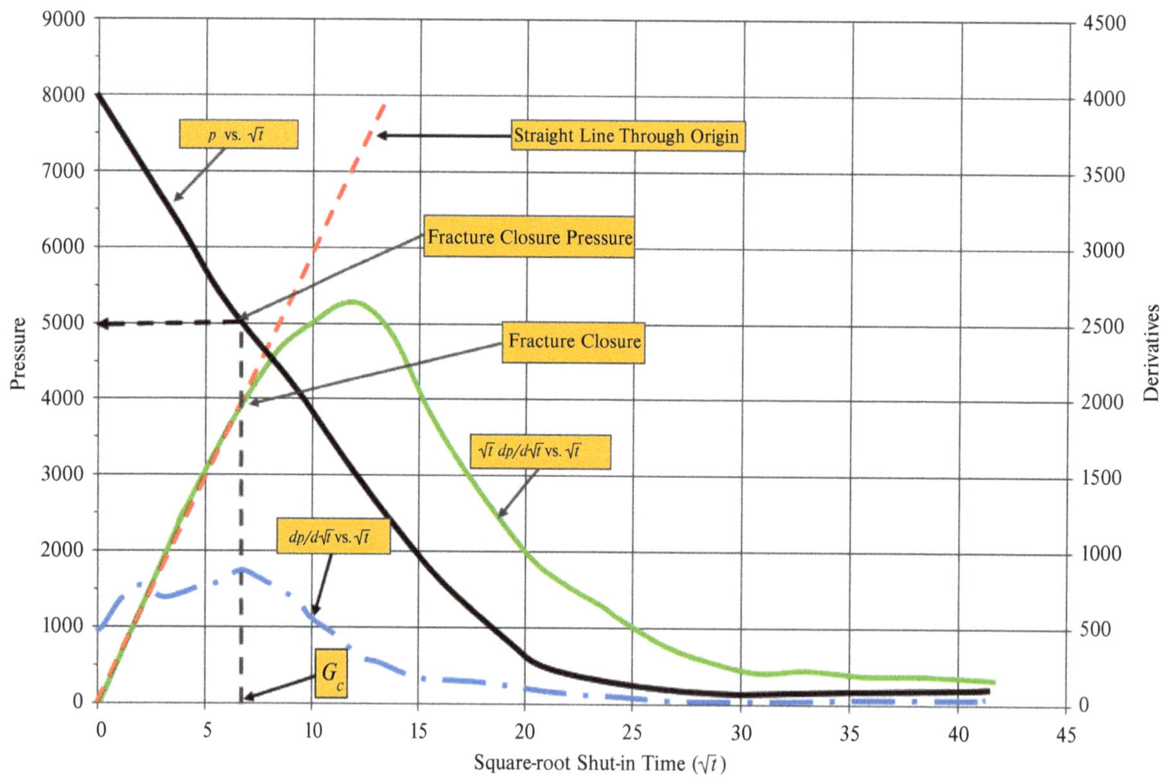

▲ **FIGURE 17.7** Normal leakoff based on square-root time plot

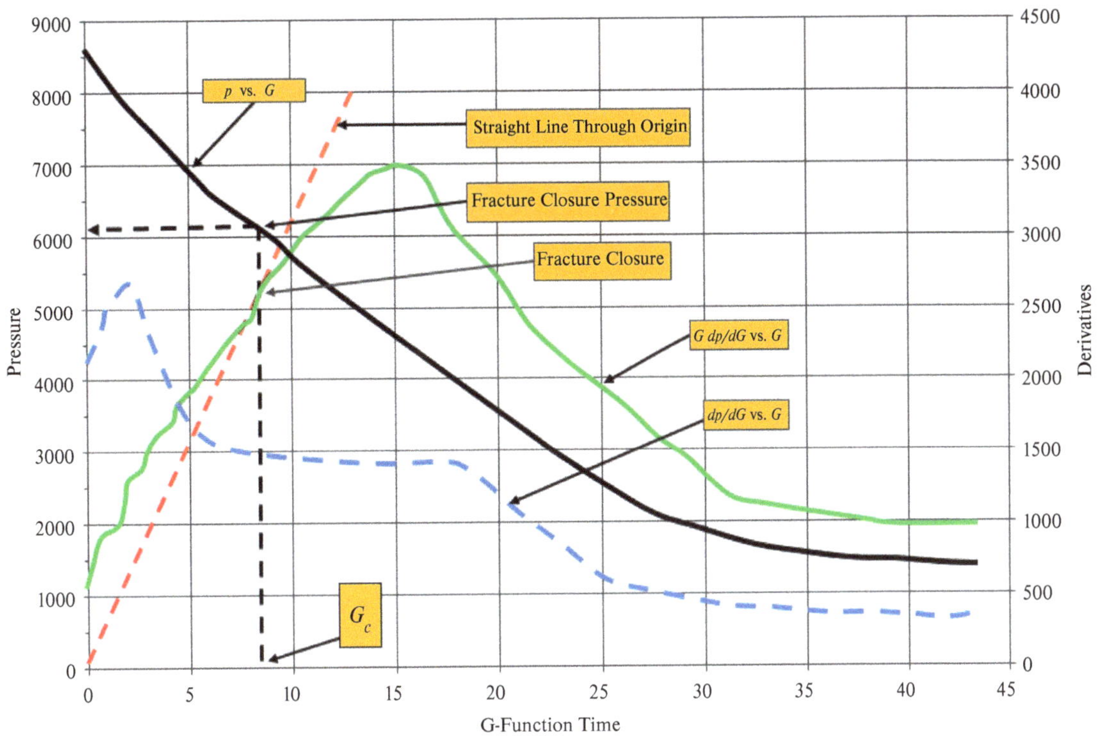

▲ **FIGURE 17.8** **PDL plot based on G-function plot**

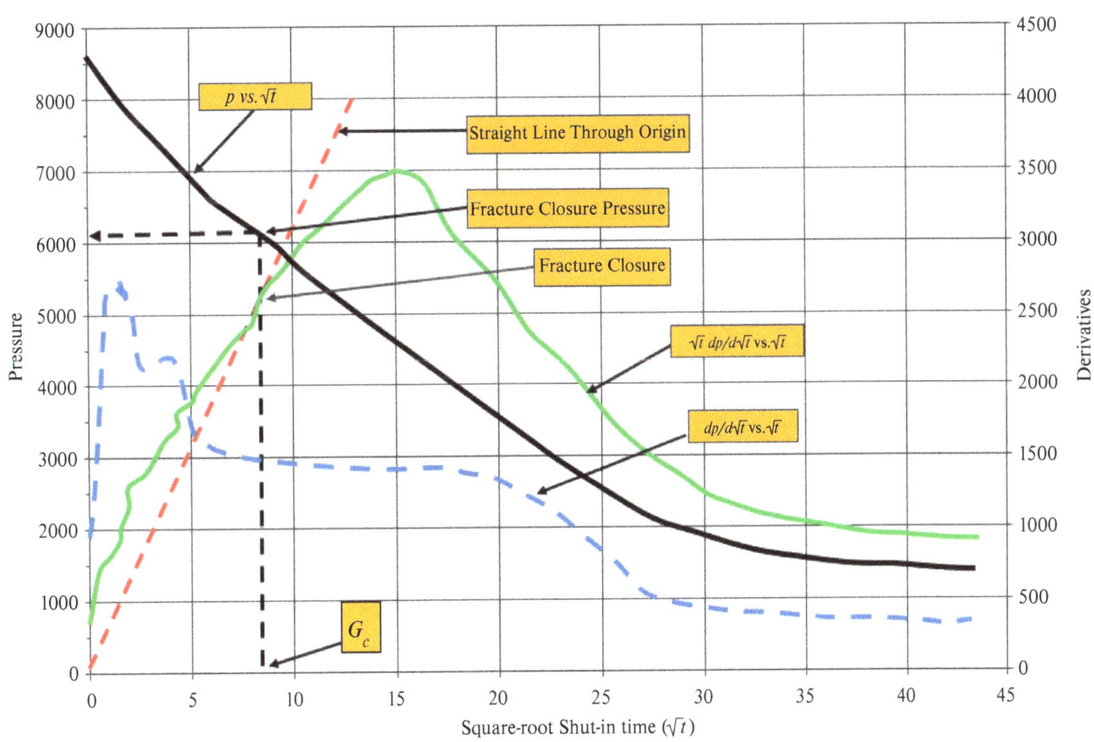

▲ **FIGURE 17.9** **PDL plot based on square-root time plot**

17.5.1.3 Fracture Tip Extension

Fracture tip extension can occur in very low permeability formations when the fracture continues to grow at its tips after the well has been shut-in. Instead of leaking-off, the energy stored in the fracturing fluids continues to extend the fracture along its length after shut-in. Figure 17.10 is a plot of pressure, its first derivative, and its semi-log derivative versus G-function for fracture tip extension. The slope of the semi-log derivative curve is initially large and positive but decreases with time resulting in a curve that is concave downwards. The pressure and first derivative curves are typically concave upwards. Since the semi-long curve continues to increase, fracture closure may not occur and cannot be identified as shown in Figure 17.10. Figure 17.11 is a plot of pressure, its first derivative, and its semi-log derivative versus square root of shut-in time. Note the similarities between this plot and Figure 17.10. Both plots should be used together to diagnose fracture behavior.[5]

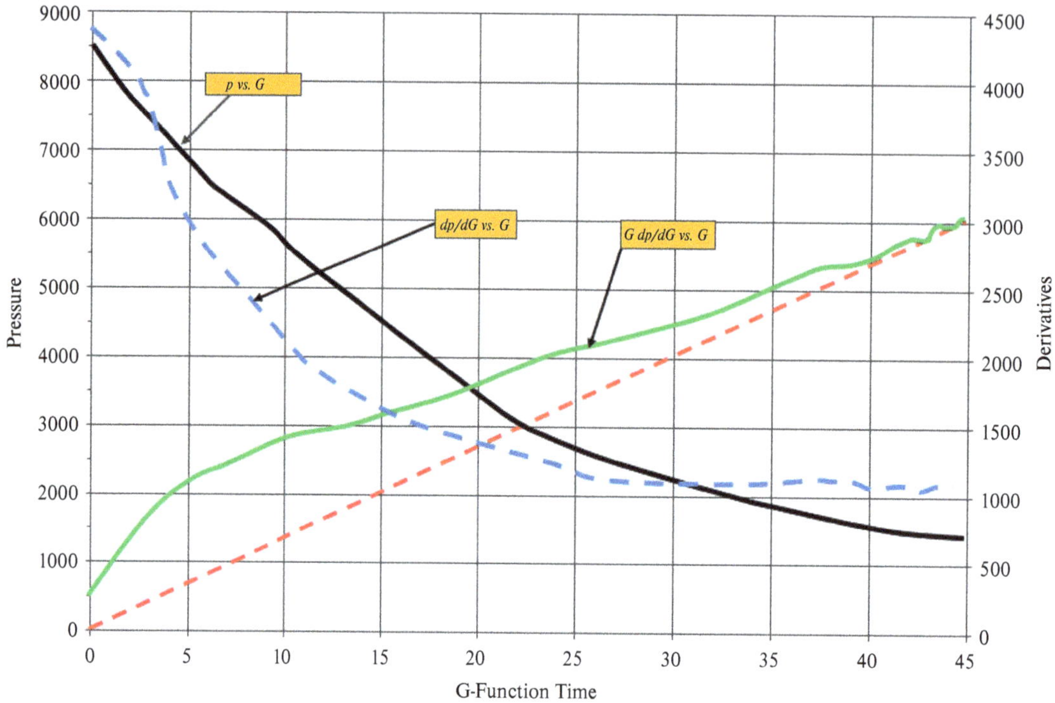

▲ **FIGURE 17.10** **Fracture tip extension based on G-function plot**

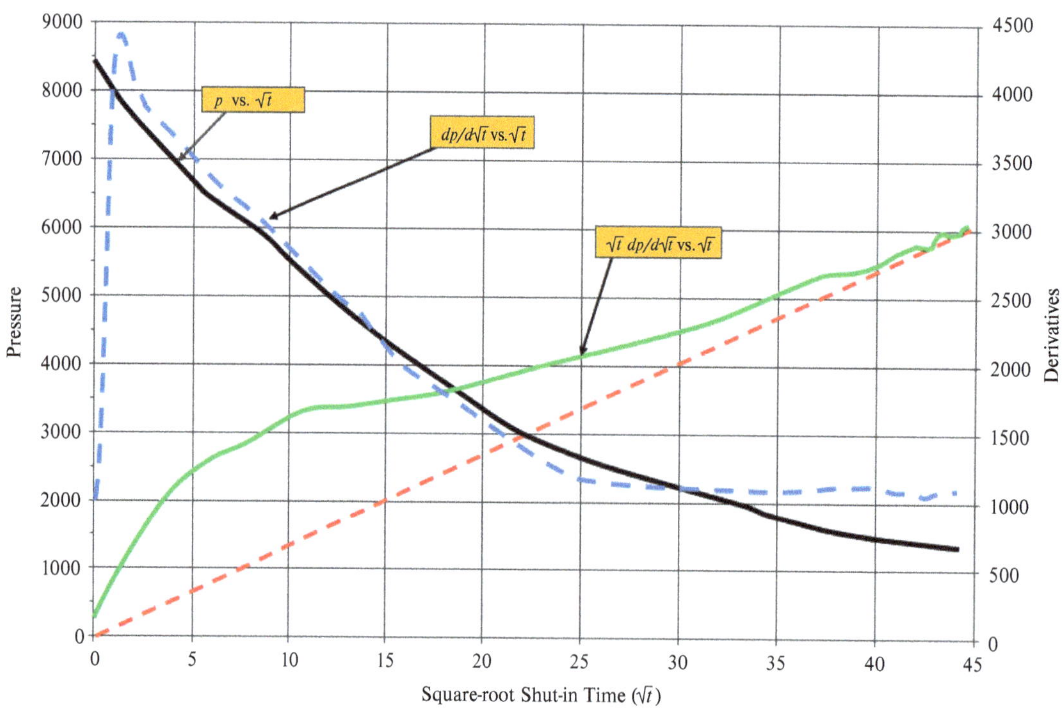

▲ **FIGURE 17.11** **Fracture tip extension based on square-root time plot**

17.5.1.4 Fracture Height Recession/ Transverse Storage

Fracture height recession is caused by the main fracture propagating through overlaying and/or underlaying impermeable layers during injection. Transverse storage occurs when the main fracture intersects a secondary fracture (natural or induced) system that provides pressure support to it. Figure 17.12 shows a plot of pressure, its first derivative, and its semi-log derivative versus G-function for fracture height-recession and transverse storage. Both events are identified by the same characteristic plots. The semi-log derivative curve exhibits a "belly" below the straight line drawn through the origin that is characteristic of fracture height recession and transverse storage. The point at which the straight line is tangent to the semi-log derivative plot is determined as fracture closure as shown in Figure 17.12. A plot of pressure, its first derivative, and its semi-log derivative versus square root of shut-in time is shown as Figure 17.13. Note the

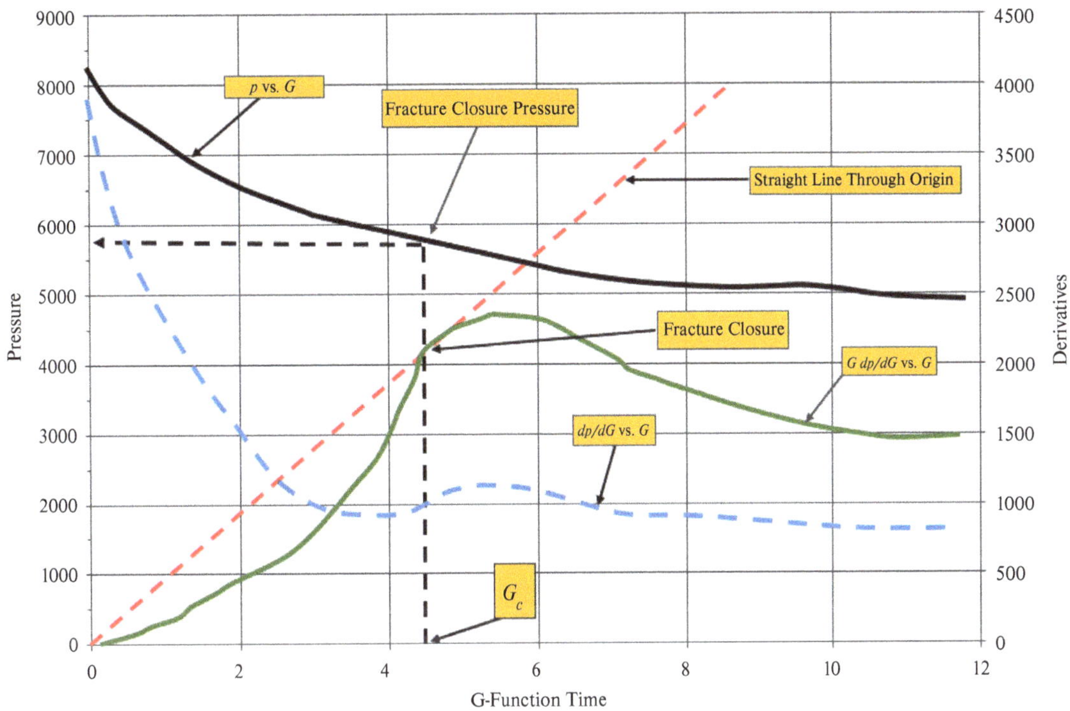

FIGURE 17.12 Fracture height recession based on G-function plot

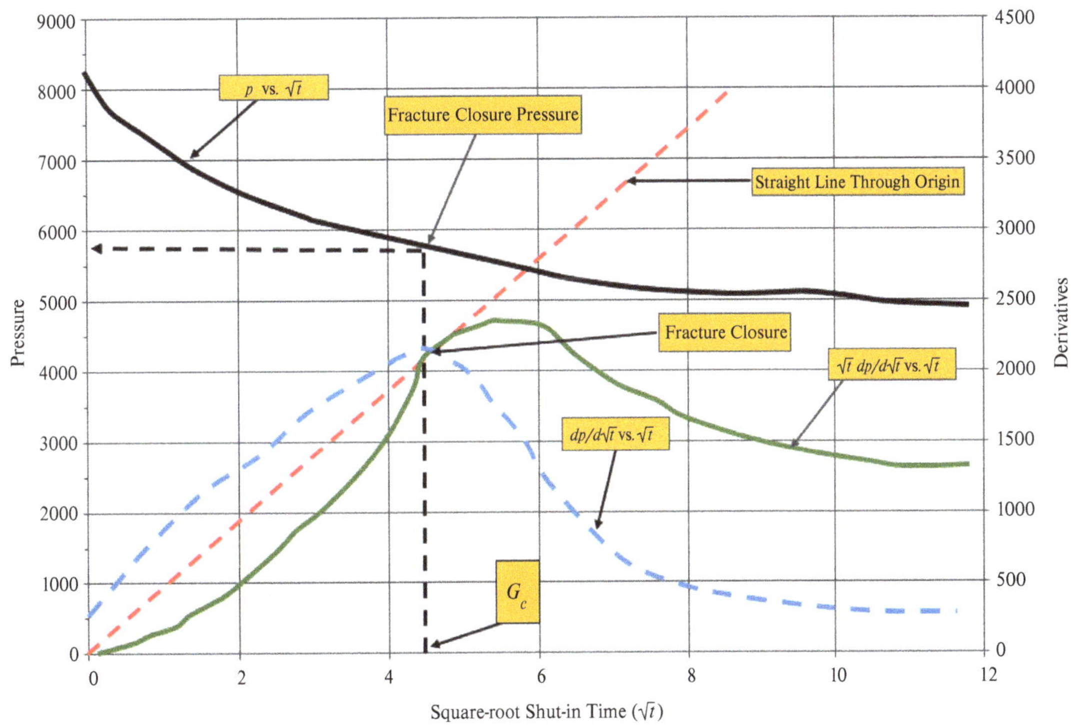

FIGURE 17.13 Fracture height recession based on square-root time plot

similarities between Figures 17.12 and 17.13. Both plots should be used in the determination of fracture closure for fracture height recession.[5]

17.5.2 Before Closure Analysis (BCA)

Before closure analysis (BCA) starts with the determination of instantaneous shut-in pressure (ISIP). ISIP is defined as:

$$\text{ISIP} = \text{Final Injection Pressure} - \text{Pressure drop from frictional losses in wellbore} \quad (17.12)$$

ISIP can be difficult to determine accurately because the pressure in the wellbore drops rapidly due to severe frictional losses after stoppage of injection and shut-in of the well. There are several methods that can be used to determine ISIP. The recommended method is to plot early wellbore pressures after shut in versus the log of shut-in time as shown in Figure 17.14. ISIP is determined as the point of intersection of a vertical line from the final injection pressure and a line tangent to the wellbore pressure curve. This method is likely to yield the lower bound of ISIP. However, the method is recommended because it appears objective and free from bias. There is opportunity to adjust the ISIP derived from this method by using the log-log plot of the pressure difference based on the ISIP and the semi-log derivative of the pressure difference versus the change in shut-in time from ISIP. This plot would yield two curves that are close to parallel before fracture closure. The magnitude of the semi-log derivative curve should be the pressure difference curve multiplied by the slope of the semi-log derivative curve.[5] The assumed ISIP should be adjusted as necessary to satisfy this requirement as close as practical. Note that adjusting the value of ISIP would shift the pressure difference curve but have no effect on semi-log derivative curve. A rough rule of thumb is that the magnitude of the pressure difference curve is about two times the magnitude of the semi-log derivative curve.

A sketch of log-log plot of pressure difference and the semi-log derivative of pressure difference versus time change is shown in Figure 17.15. Note this plot is for illustration purposes only. This plot is similar to the log-log plot used in pressure build-up analyses. Figure 17.15 shows a unit slope line representing wellbore storage. This is followed by half-slope lines indicating the presence of fracture linear flow before fracture closure. After fracture closure, the plot indicates the existence of radial flow after fracture linear flow. Note that formation linear flow is not indicated in this plot. Formation linear flow would have been represented by negative half slope if present. The log-log plot is useful only for diagnostic purposes before proceeding to after closure analysis (ACA).

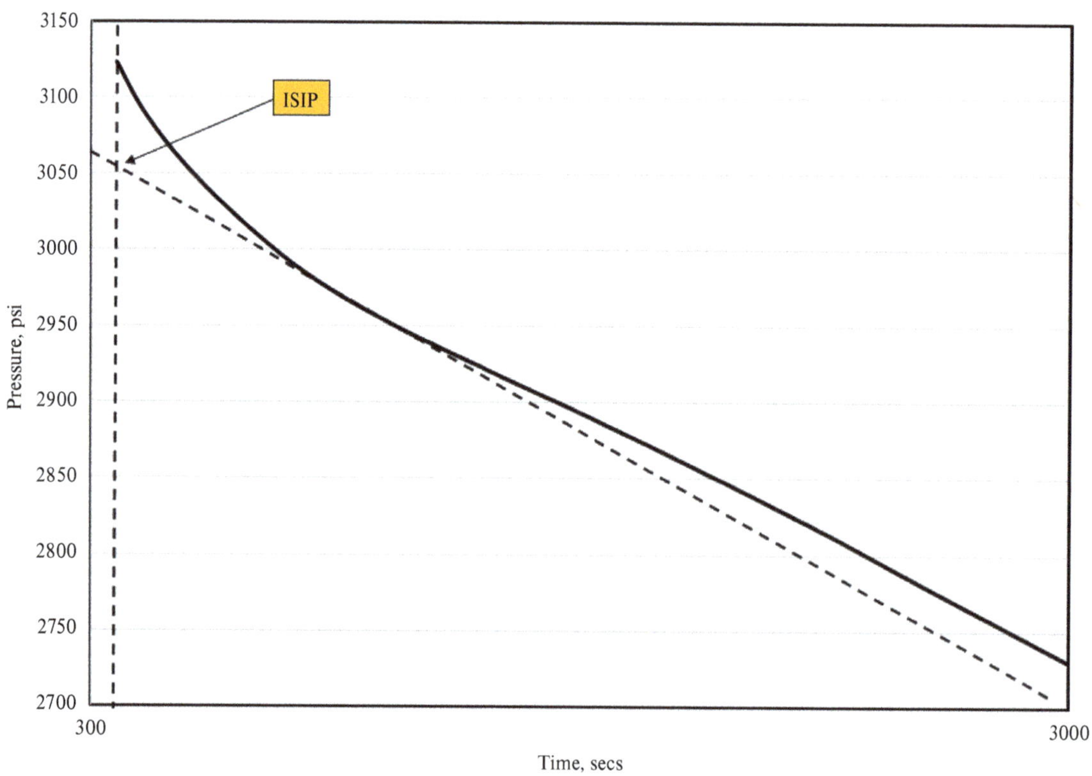

▲ **FIGURE 17.14 Determination of ISIP**

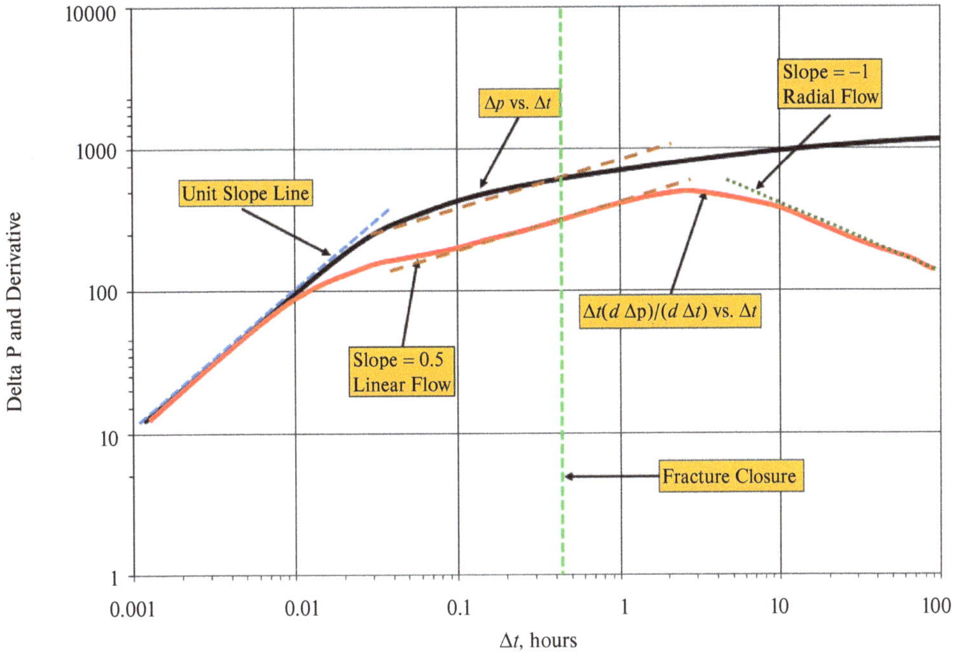

▲ **FIGURE 17.15** Log-log plot of pressure difference and derivative vs. time change

17.5.3 After Closure Analysis (ACA)

The after closure analysis techniques presented in this section is based on a method developed by Talley et al.[6] For this method, flow regimes are identified from a log-log plot of measured falloff pressure minus assumed reservoir pressure versus the square of linear-flow time function and its semi-log derivative versus the same time function. An example of this plot is shown as Figure 17.16. If linear flow is present, the slope of semi-log derivative of the pressure difference would be half (0.5) as shown in Figure 17.16. If radial flow is present, the slope of semi-log derivative curve would be one (1.0) as shown in the same figure. Note that the slope of the pressure difference curve is determined by the assumed value of reservoir pressure used in constructing the plot. This can be adjusted as needed to achieve the desired slope of half. The slope of the semi-log derivative curve is not affected by the assumed value of reservoir pressure.

According to Talley et al.[6], the linear-flow time function is defined as:

$$F_L(t,t_c) = \frac{2}{\pi}\sin^{-1}\sqrt{\frac{t_c}{t}} \quad \text{for } t \geq t_c \tag{17.13}$$

In Eq. (17.13), F_L = linear-flow time function; t_c = fracture closure time; and t = time since closure in corresponding units of time.

When a stable linear flow period is identified with the log-log plot (Figure 17.16), a cartesian plot of falloff pressure versus linear-flow function time would yield a straight line which can be extended to determine reservoir pressure on the pressure axis as shown in Figure 17.17.

If pseudo-radial flow period is identified with the log-log plot (Figure 17.16), a cartesian plot of falloff pressure versus radial-flow time function would yield a straight line that can be extended to determine reservoir pressure on the pressure axis as shown in Figure 17.18.

The radial-flow time function was defined by Talley et al.[6] as follows:

$$p_w(t) - p_r = m_R F_R(t,t_c) \tag{17.14}$$

$$F_R(t,t_c) = \frac{1}{4}\ln\left(1 + \frac{\chi \times t_c}{t - t_c}\right), \quad \chi = \frac{16}{\pi^2} \cong 1.6 \tag{17.15}$$

In Eqs. (17.14) and (17.15), $p_w(t)$ = falloff pressure at time, t; p_r = reservoir pressure; m_R = slope of straight line; F_R = radial flow time function; t = time since fracture closure; and t_c = fracture closure time in corresponding units. The slope of the straight line can be used to determine transmissibility of the formation using the relationship:

$$\frac{kh}{\mu} = \frac{251{,}000 \times Q_t}{m_R t_c} \tag{17.16}$$

In Eq. (17.16), k = permeability in md; h = net pay thickness, ft; μ = native fluid viscosity, cp; Q_t = total volume of fluid injected in bbls; t_c = fracture closure time in minutes; and m_R = slope in psi.

If pseudo-radial flow exists, the familiar Horner plot (Section 13.4.1.5), can be used to determine reservoir pressure and formation transmissibility. A Horner plot is shown as Figure 17.19. In this plot, falloff pressure is plotted versus Horner time, $(t_p + \Delta t)/\Delta t$, where t_p is elapsed time from fracture initiation to shut in and Δt is elapsed time since fracture initiation as defined previously in Eq. (17.10). Extrapolation of a straight line through the pseudo-radial flow period to Horner time ratio of 1.0 gives reservoir pressure, p_r. The slope of the straight line, m_H, can be used to calculate formation transmissibility from the expression:

$$\frac{kh}{\mu} = \frac{162.6 \times 1440 \times q}{m_H} \tag{17.17}$$

In Eq. (17.17), k = permeability, md; h = net pay thickness, ft.; μ = native fluid viscosity, cp; and q = average injection rate in barrels per minute. The formation transmissibility calculated from the Horner plot can be used to corroborate the number obtained from Eq. (17.16) based on linear flow.

An empirical equation has been developed for the estimation of formation permeability based on certain rock properties and G-function time at fracture closure.[5] The equation is given as:

$$k = \frac{0.0086 \mu_f \sqrt{0.01 p_z}}{\phi c_t \left(\frac{G_c E r_p}{0.038} \right)^{1.96}} \tag{17.18}$$

In Eq. (17.18), k = formation permeability, md; μ_f = injected fluid viscosity, cp; $p_z = p_{ISIP} - p_{FCL}$ in psi or net fracture pressure in psi; ϕ = porosity, fraction; c_t = total compressibility, psi^{-1}; G_c = G-function time at closure; E = Young's modulus, Mpsi; and r_p = storage ratio. The storage ratio is equal to 1.0 for normal leak-off. It is less than 1.0 for the other fracture leak-off types. Note that permeabilities calculated with Eq. (17.18) should be used as estimates only.

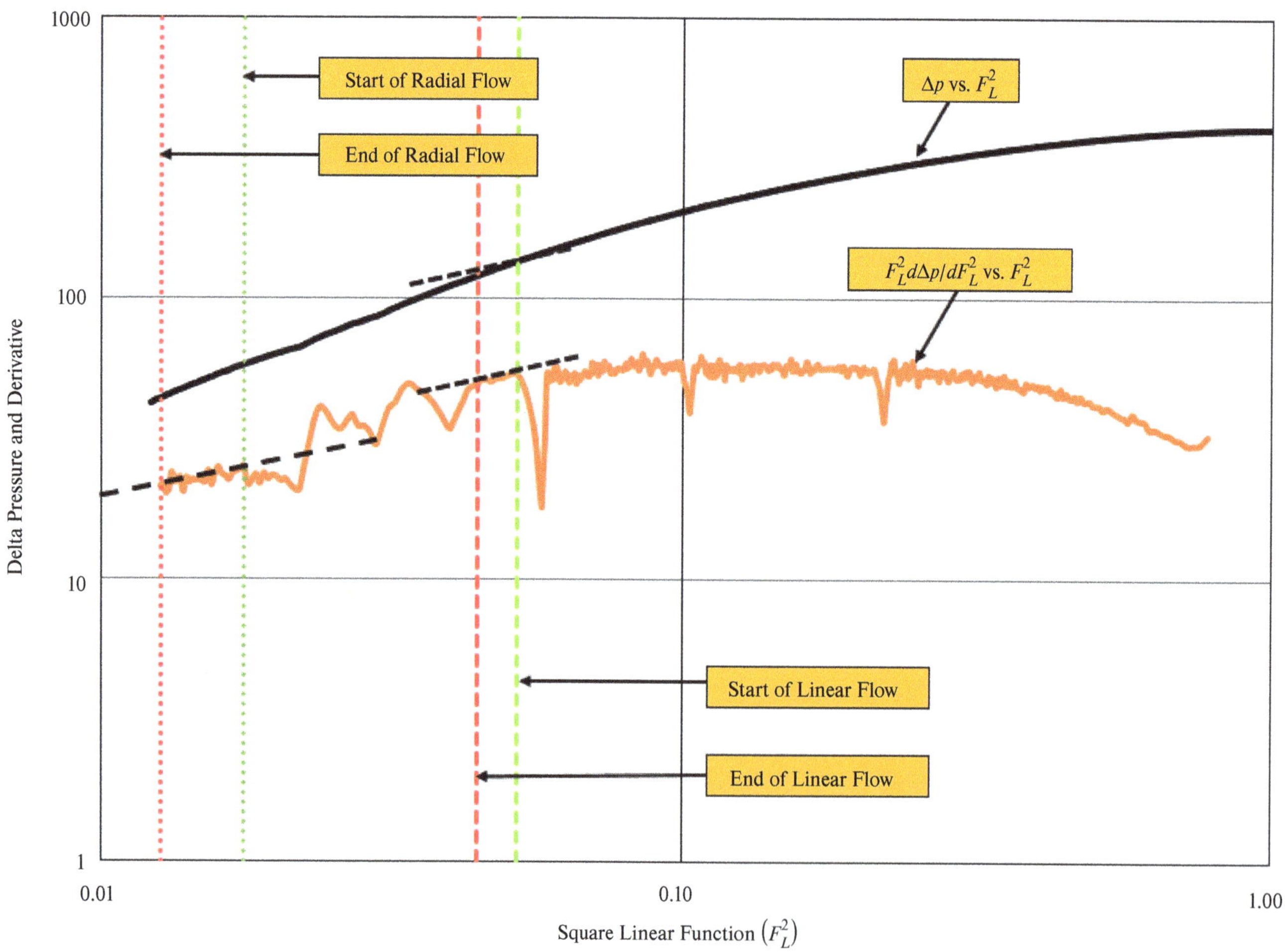

▲ **FIGURE 17.16 Log-log plot of pressure difference and derivative vs. square linear function**

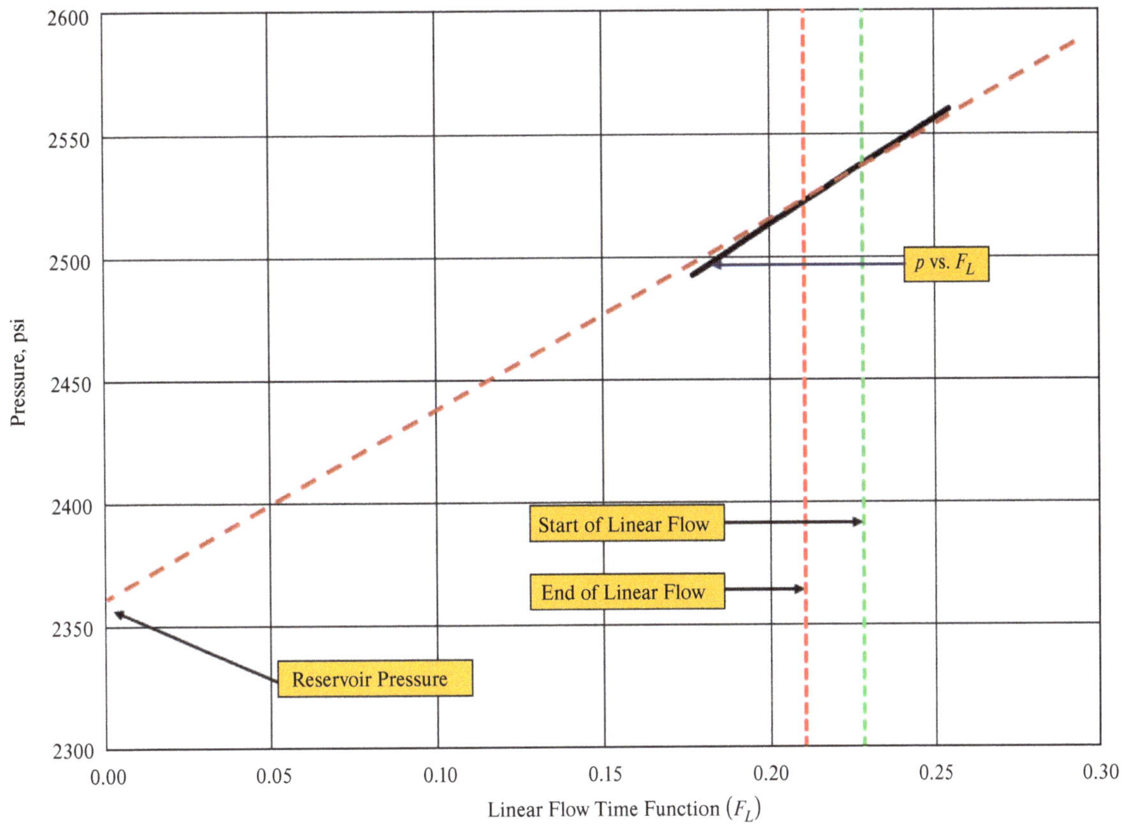

▲ **FIGURE 17.17** **ACA linear flow cartesian plot**

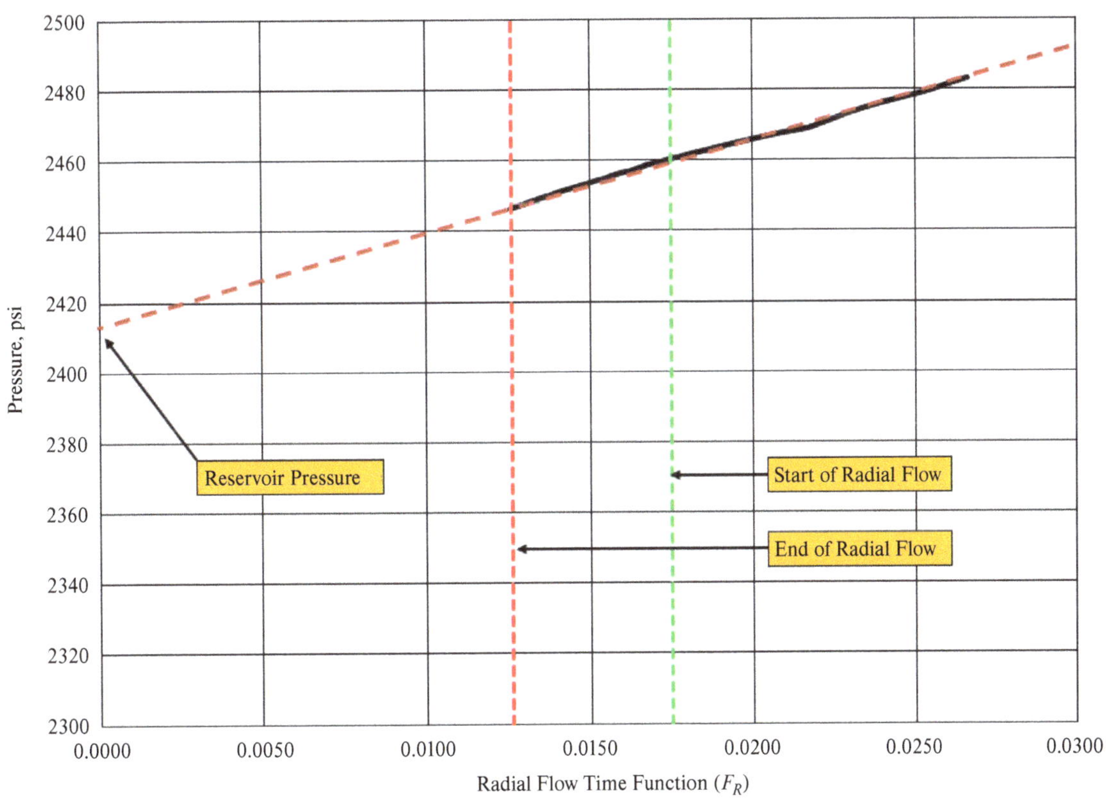

▲ **FIGURE 17.18** **ACA radial flow cartesian plot**

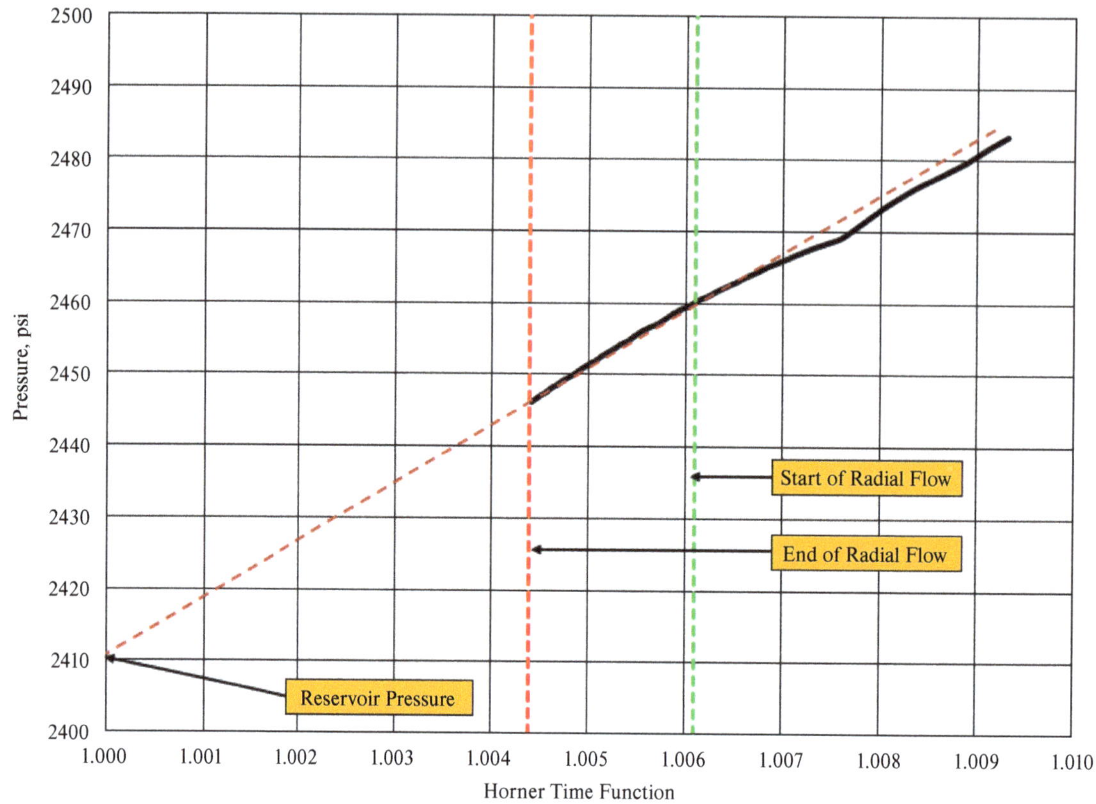

▲ **FIGURE 17.19** Horner time plot

17.6 Procedure for DFIT Analyses

The following steps are recommended for analyses of DFITs. These steps could be modified and adapted for DFIT analyses depending on the quality and quantity of measured injection data and pressure falloff data.

Step 1: Plot injection rate and pressure data versus injection time.

From the plot, determine fracture initiation (breakdown) pressure and time (Figure 17.1). Note time and pressure at which injection was stopped. Examine plot for other important features such as fracture propagation pressure (FPP), instantaneous shut-in pressure (ISIP), and any notable abnormalities based on pressure trends.

Step 2: Plot bottomhole pressure at the end of injection versus shut-in time to end of test.

Examine plot for any pressure irregularities due to leakages either in surface or downhole equipment and pressure hold-up effects. Look for any abnormal pressure trends. This step should be devoted to quality-checking the pressure data before proceeding to analysis.

Step 3: Make the G-function plot: $(p_w \text{ vs. } G), (dp_w/dG \text{ vs. } G)$, and $(G\,dp_w/dG \text{ vs. } G)$.

The equations for making the G-function plot were presented in Section 17.5.1, Diagnostic Analysis. From the G-function plot, determine fracture closure pressure (FCP) and fracture closure G-function time, (G_c). Identify the type of fracture leak-off based on the characteristics of the plot.

Step 4: Make the square root of time plot: $(p_w \text{ vs. } \sqrt{t})$, $(dp_w/d\sqrt{t} \text{ vs. } \sqrt{t})$, $(\sqrt{t}\,dp_w/d\sqrt{t} \text{ vs. } \sqrt{t})$.

Use this plot to verify and confirm the FCP determined in Step 3 from the G-function plot.

Step 5: Plot bottomhole pressure (p_w) versus log of shut-in time (t).

This is a plot of BHP versus log of total shut-in time (Figure 17.14). Determine ISIP from the plot. This plot will generally give a low value for ISIP, but it is objective and free of bias. The ISIP determined from this method can be adjusted higher if necessary, with the aid of the log-log plot in Step 6.

Step 6: Make a log-log plot of Δp vs. Δt and $\Delta t d(\Delta p)/d(\Delta t)$ vs. Δt

The pressure difference is: $\Delta p = ISIP\text{-}p_W$. The time difference is: $\Delta t = t - t_{ISIP}$. An example of plot is shown as Figure 17.15. From plot, identify well-bore storage, fracture linear flow before closure, formation linear flow after closure, and pseudo-radial flow if these events are present. A summary of characteristic slopes identifiable from a log-log plot is shown in Table 17.1.

Note that for gas wells, the log-log diagnostic plot for Step 6 and successive plots in Steps 7 to 10 should be prepared in terms of pseudo-pressure or normalized pseudo-pressure to account for gas compressibility. As defined in Eq. (13.70), real gas pseudo-pressure is represented as:

$$p_p = 2\int_{p_b}^{p} \frac{p}{\mu(p)z(p)} dp \tag{17.19}$$

The equation for normalized pseudo-pressures was defined in Eq. (13.75) as:

$$p_{pn} = \frac{1}{2}\left(\frac{\mu z}{p}\right) p_p \tag{17.20}$$

As discussed in Example 13.10, use of pseudo-pressures or normalized pseudo-pressures yield more accurate analyses of gas falloff pressure data. The use of actual shut-in time instead of pseudo-time was recommended.[8]

The use of normalized pseudo-pressure and actual shut-in time for analysis of a DFIT on a gas well is shown as Example 17.1.

Step 7: Make ACA log-log plot of Δp vs. F_L^2 and $F_L^2\, d\Delta p/dF_L^2$ vs. F_L^2.

From the plot, determine ACA linear flow and radial flow intervals as shown in Figure 17.16.

Step 8: Make ACA linear-flow time function plot: $(p \text{ vs. } F_L)$.

Plot a cartesian plot of bottomhole pressure versus linear-flow time function as shown in Figure 17.17. Extend the line representing the linear flow period to intersect the pressure axis. The intercept is the value of reservoir pressure.

Step 9: Make ACA radial flow cartesian plot: $(p \text{ vs. } F_R)$.

Make a cartesian plot of bottomhole pressure versus radial-flow time function as shown in Figure 17.18. Extend the line representing the radial flow period to intersect the pressure axis. The intercept is the value of reservoir pressure. The slope of the line can be used to calculate formation permeability or transmissibility as given in Eq. (17.16).

Step 10: Make a Horner plot: $\left(p \text{ vs. } \Delta t/(\Delta t + t_p)\right)$.

If pseudo-radial flow exists, a Horner plot of bottomhole pressure versus Horner time function can be made as shown in Figure 17.19. Extend the line representing the radial flow period to intersect the pressure axis. The intercept is the value of reservoir pressure. The slope of the line can be used to calculate formation permeability or transmissibility as given in Eq. (17.17).

TABLE 17.1 Summary of Characteristic Slopes and Flow Regimes in a Log-Log Plot

Type of Analysis	Slope	Flow Regime	Description of Flow Regime
Before closure analysis (BCA)	+0.25	Fracture bilinear	Fluid flow is normal to the fracture and along the fracture (Figure 15.1b)
	+0.50	Fracture linear	Fluid flow is along the fracture (Figure 15.1a)
After closure analysis (ACA)	−0.50	Formation linear	Fluid flow in the formation is normal to the fracture plane (Figure 15.1c)
	−1.00	Formation pseudo-radial	Fluid flow is radial with respect to the formation and wellbore (Figure 15.1d)

EXAMPLE 17.1 Analysis of a DFIT on a Gas Well in Mesaverde Formation, San Juan Basin, USA

Problem

The example is based on DFIT data published by Craig and Blasingame[7] from a gas well in the Mesaverde formation, San Juan Basin, USA. The test was conducted on Well GM 543-33. As reported by Craig and Blasingame,[7] the injection fluid was 17.69 barrels of 1% KCL water and the injection rate averaged 3.30 bbl/min over 5.30 minutes of injection period. The injection data and pressure falloff data recorded during the test are reproduced as Tables 17A.1 and 17A.2, respectively, in Appendix 17A. The properties of the Mesaverde formation were reported as follows:

Reservoir pressure (assumed)	2332 psia
Reservoir temperature	160°F
Gross thickness	14 ft
Net thickness	12 ft
Porosity	10%
Gas saturation	50%
Gas gravity	0.63
Young's modulus	5×10^6 psi
Poisson's ratio	0.20

Solution

Step 1: Plot injection rate and pressure data versus injection time.

Using data from Table 17A.1 in Appendix 17A, a plot of injection rate and pressure versus injection time was made as shown in Figure 17.20. From Figure 17.20, fracture initiation appears to have started after 70 seconds of injection. Fracture propagation pressure was 3908 psi and lasted for about 70 seconds. Elapsed time from initiation of fracture to end of injection denoted as t_p is 248 seconds. Total injection time is 318 seconds.

Step 2: Plot bottomhole pressure at the end of injection versus shut-in time to end of test.

A plot of bottomhole pressure versus shut-in time to the end of test is shown as Figure 17.21. This figure was plotted from data in Table 17A.2 of Appendix 17A. A close examination of the plot indicated that there were no leaks or any other abnormalities in the test data.

Step 3: Make the G-function plot: $(p_w \text{ vs. } G)$, $(dp_w/dG \text{ vs. } G)$, and $(G\, dp_w/dG \text{ vs. } G)$.

To make the G-function plot, values of the G-function time were calculated using the intermediate function for high leak-off (Eq. (17.9) and the basic G-function equation (Eq. (17.7). The dimensionless time, (t_D), was calculated as represented in Eq. (17.10). A table containing data for p_w, t_D, G, dp/dG, and $G\, dp/dG$ is shown in Appendix 17B as Table 17B.1 for a section of the falloff data. A plot of the data in Table 17B.1 is shown as Figure 17.22. From this plot, the fracture closure pressure was determined to be 2805.5 psi and the G-function value at closure was, $G_c = 4.5374$. Note the "hump" of the semi-log derivative curve above the straight line from the origin indicating that the fracture exhibits characteristics of pressure-dependent leak-off (Figure 17.22).

Step 4: Make the square root of time plot: $(p_w \text{ vs. } \sqrt{t})$, $(dp_w/d\sqrt{t} \text{ vs. } \sqrt{t})$, $(\sqrt{t}\, dp_w/d\sqrt{t} \text{ vs. } \sqrt{t})$.

The square root of time plot was made from data shown in Table 17B.2 in Appendix 17B. Figure 17.23 shows the square root of time plot for pressure, pressure derivative, and semi-log pressure derivative. Note the characteristic hump above the straight line as seen in Figure 17.22 for the G-function plot indicating pressure-dependent leak-off. The fracture closure pressure determined from the square-root plot is 2805 psi. This confirms the same value determined with the G-function plot.

Step 5: Plot bottomhole pressure (p_w) versus log of shut-in time (t).

This plot is shown as Figure 17.24. The ISIP was determined from this plot to be 3050 psi.

Step 6: Make a log-log plot of Δp_{pnw} vs. Δt and $\Delta t \times d(\Delta p_{pnw})/d(\Delta t)$ vs. Δt.

Normalized pseudo-pressures were calculated as shown in Table 17C.1 in Appendix 17C.

A plot of Δp_{pnw} vs. Δt and $\Delta t \times d\Delta p_{pnw}/d\Delta t$ vs. Δt was made from the data in Table 17C.2 in Appendix 17C. This plot is shown as Figure 17.25. From this diagnostic plot, wellbore storage, fracture linear flow, formation linear flow, and pseudo-radial flow were identified as shown in Figure 17.25. An expanded diagnostic log-log plot of the same terms is shown as Figure 17.26. In this plot, wellbore storage and fracture linear flow before fracture closure are clearly identified.

Step 7: Make ACA log-log plot of Δp_{pn} vs. F_L^2 and $F_L^2 d\Delta p_{pn}/dF_L^2$ vs. F_L^2.

For the ACA analysis, the linear function values were calculated using Eq. (17.13). These values are shown in Table 17C.3 in Appendix 17C. The log-log plot of Δp_{pn} vs. F_L^2 and $F_L^2 d\Delta p_{pn}/dF_L^2$ vs. F_L^2 shown in Figure 17.27 was prepared from the values in the table. Note that in the figure, the duration of formation linear flow was identified with a slope of 0.5 and that of pseudo-radial flow was shown with a slope of 1.0.

Step 8: Make ACA linear flow plot.

A cartesian plot of normalized pseudo-pressure versus linear flow time function, $(p_{pnw} \text{ vs. } F_L)$, was made as shown in Figure 17.28. Table 17C.4 of Appendix 17C shows the data used in the plot. The intercept on the y-axis is normalized pseudo-pressure value of 262.5 psi. This is transformed to reservoir pressure value of 2356 psi.

Step 9: Make ACA radial flow cartesian plot.

The cartesian plot of normalized pseudo-pressure versus radial flow time function, $(p_{pnw} \text{ vs } F_R)$, shown as Figure 17.29 was prepared using the data contained in Table 17C.5 in Appendix 17C. The intercept of a straight line through the radial flow period on the y-axis gives normalized pseudo-pressure value of 320 psi. This transforms to reservoir pressure value of 2410 psi. The slope of this straight line is 3152.35 psi. Using Eq. (17.16),

$$\frac{kh}{\mu} = \frac{251{,}000 \times Q_t}{m_R t_c}$$

Rearranging the equation gives:

$$k = \frac{251{,}000 \times Q_t \times \mu}{m_R t_c h}$$

Using data derived from the test data, Q_i = 3.30 bbls/min × 4.1333 min = 13.64 bbls of total injection vol. from fracture initiation; t_c = 28.883 mins.; μ = 0.0177 cp; and h = 12 ft.

Substituting these data into the equation gives:

$$k = \frac{251000 \times 13.64 \times 0.0177}{3152.35 \times 28.883 \times 12} = 0.0555 \text{ md.}$$

This is the formation permeability calculated from the radial flow cartesian plot.

Step 10: Make a Horner plot

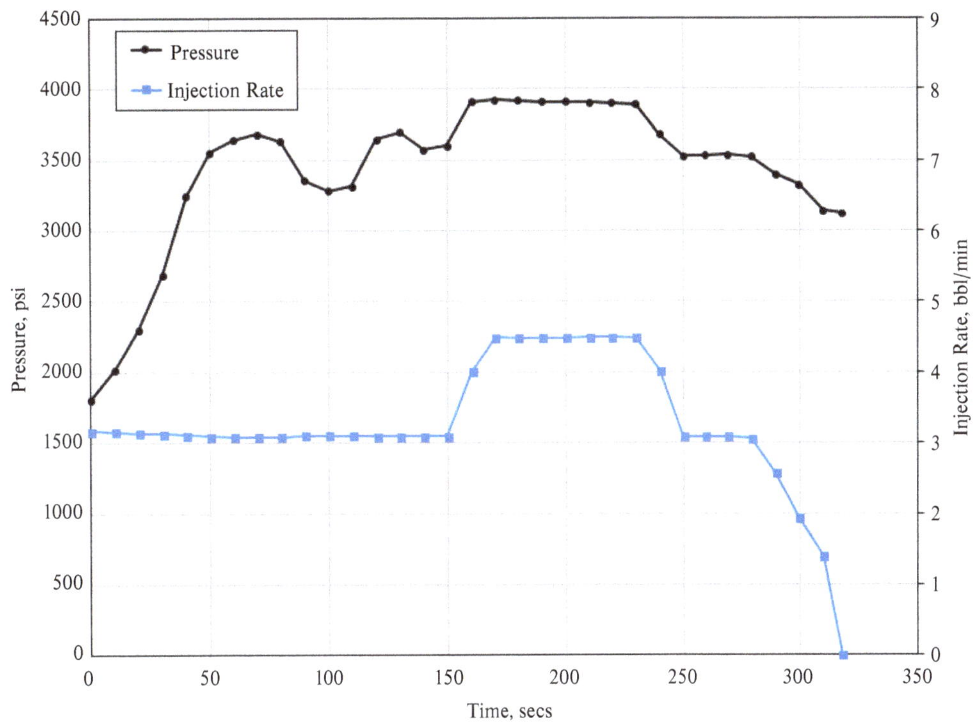

FIGURE 17.20 Injection rate and pressure vs. injection time for Example 17.1

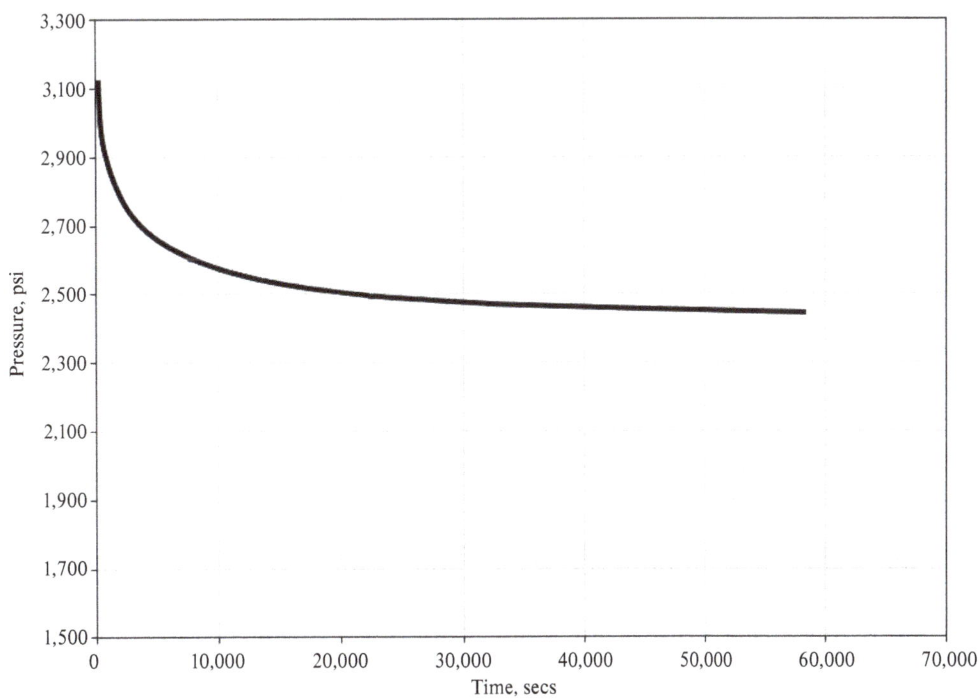

FIGURE 17.21 Falloff pressure data vs. shut-in time for Example 17.1

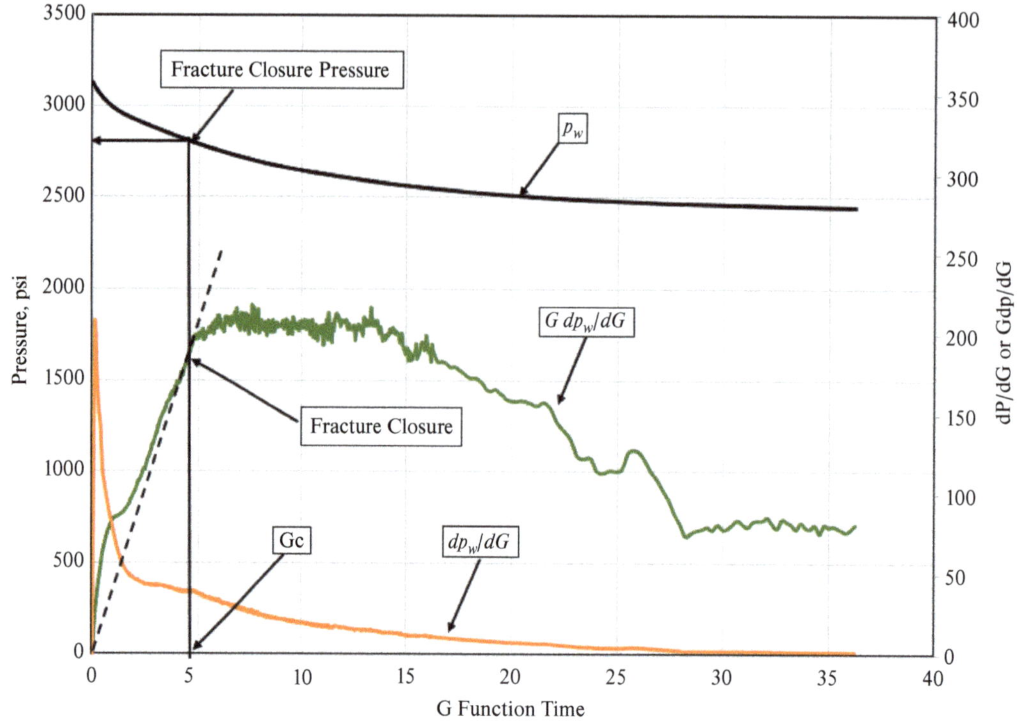

▲ **FIGURE 17.22** **G-function plot for Example 17.1**

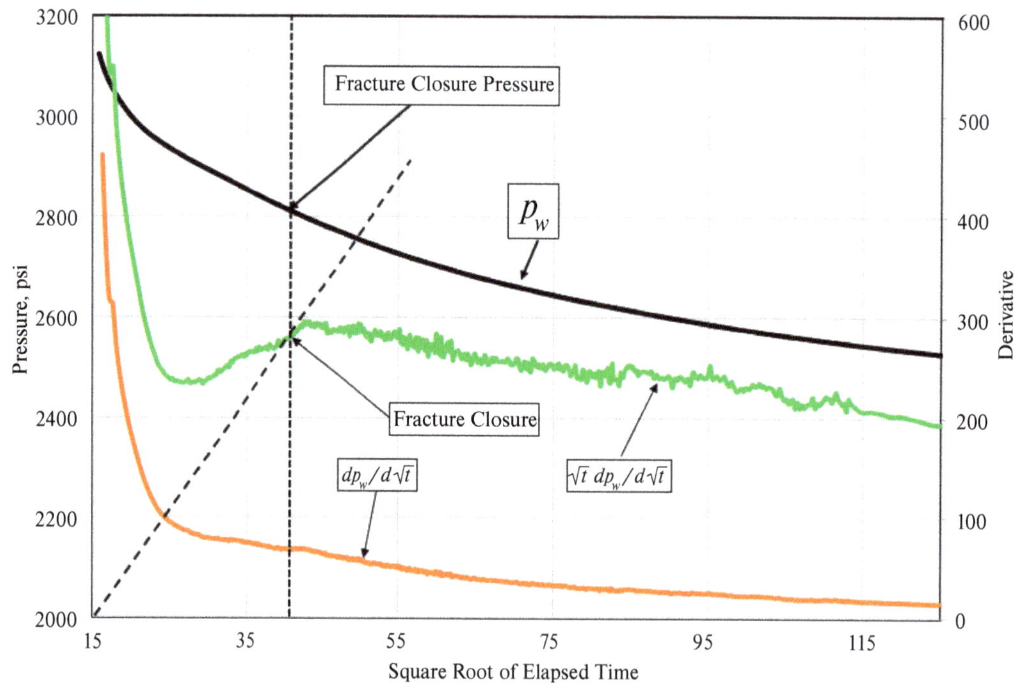

▲ **FIGURE 17.23** **Square-root of time plot for Example 17.1**

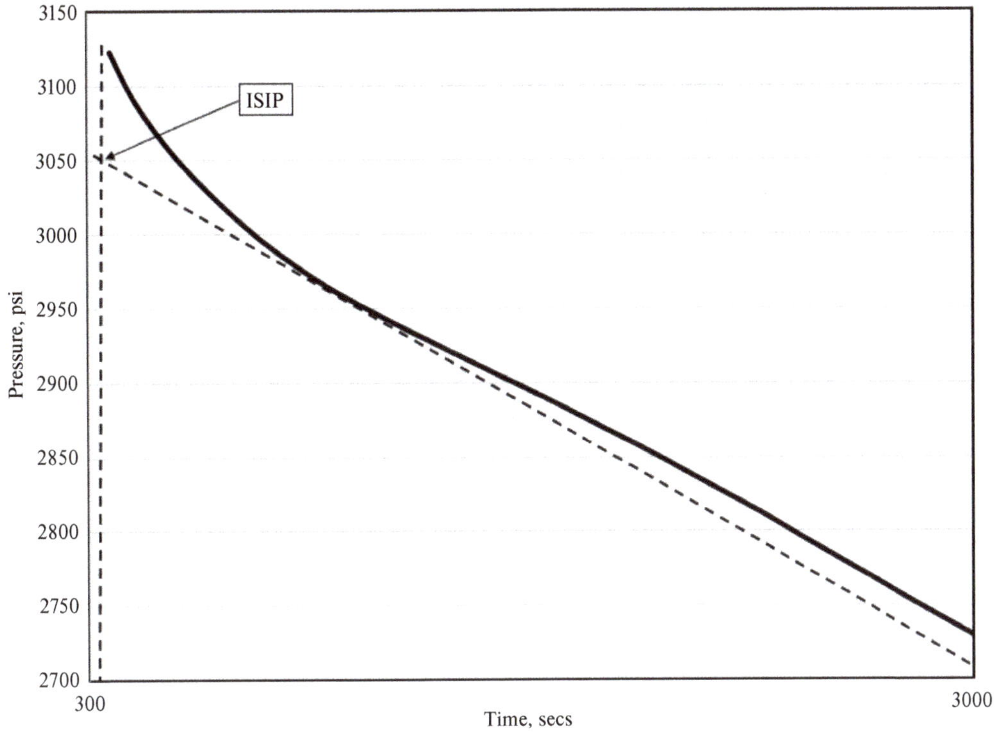

▲ **FIGURE 17.24** p_w **vs. log (*t*) plot for Example 17.1**

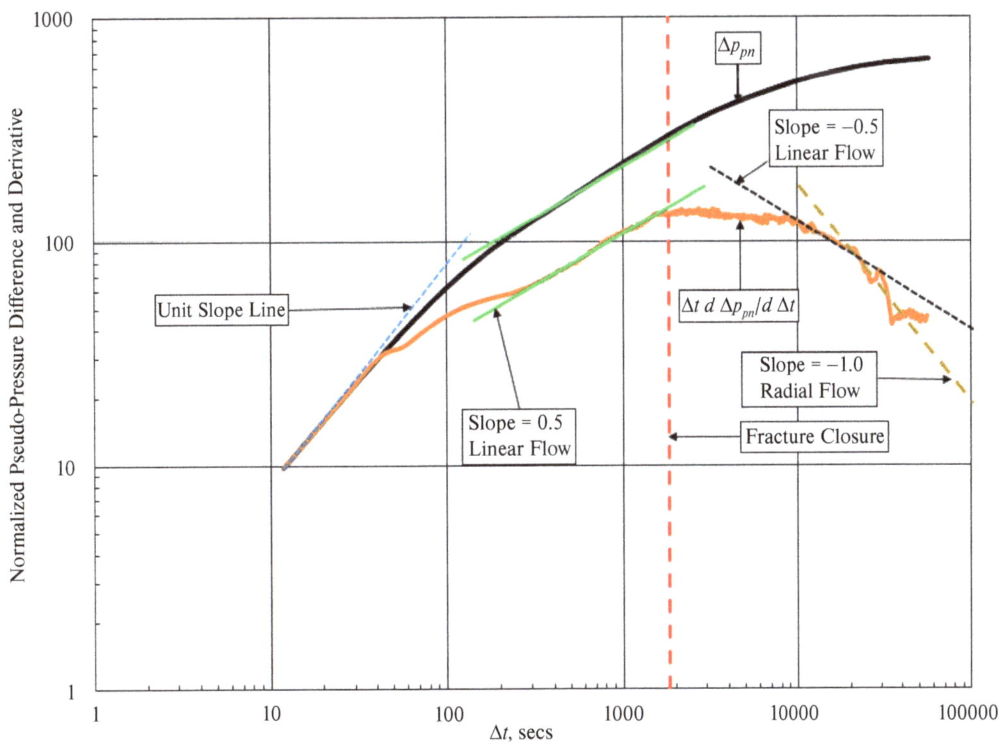

▲ **FIGURE 17.25** **Log-log plot of Δp_{pnw} vs. Δt and $\Delta t \times d(\Delta p_{pnw})/d\,(\Delta t)$ vs. Δt for Example 17.1**

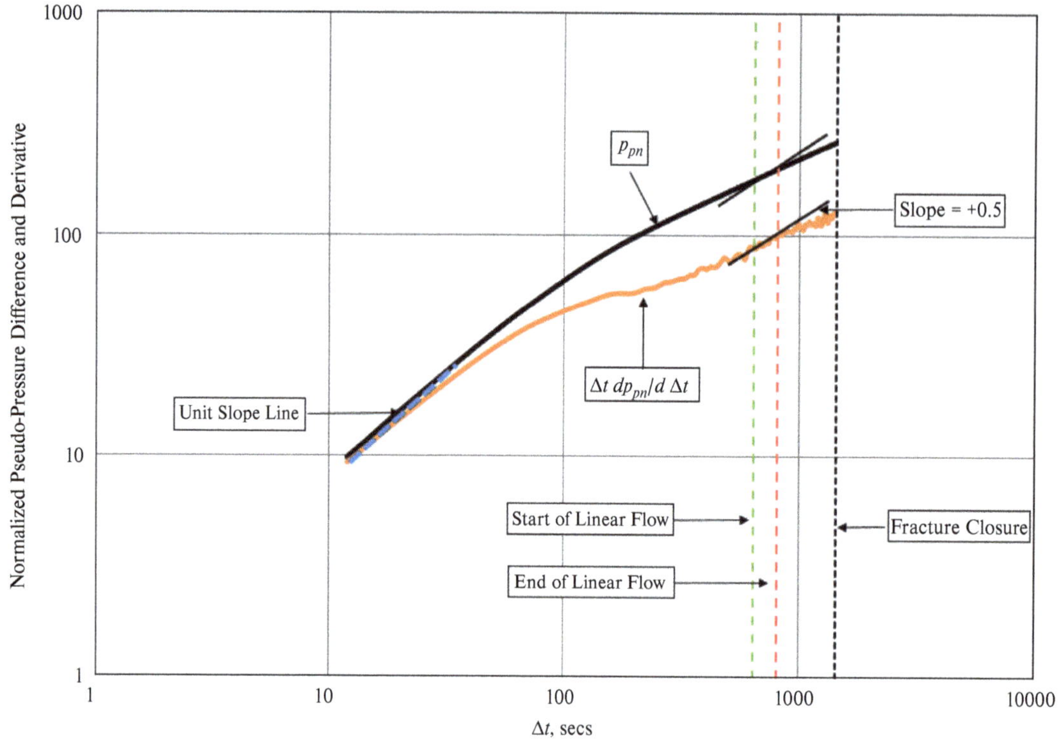

▲ **FIGURE 17.26** **Expanded log-log plot for Example 17.1**

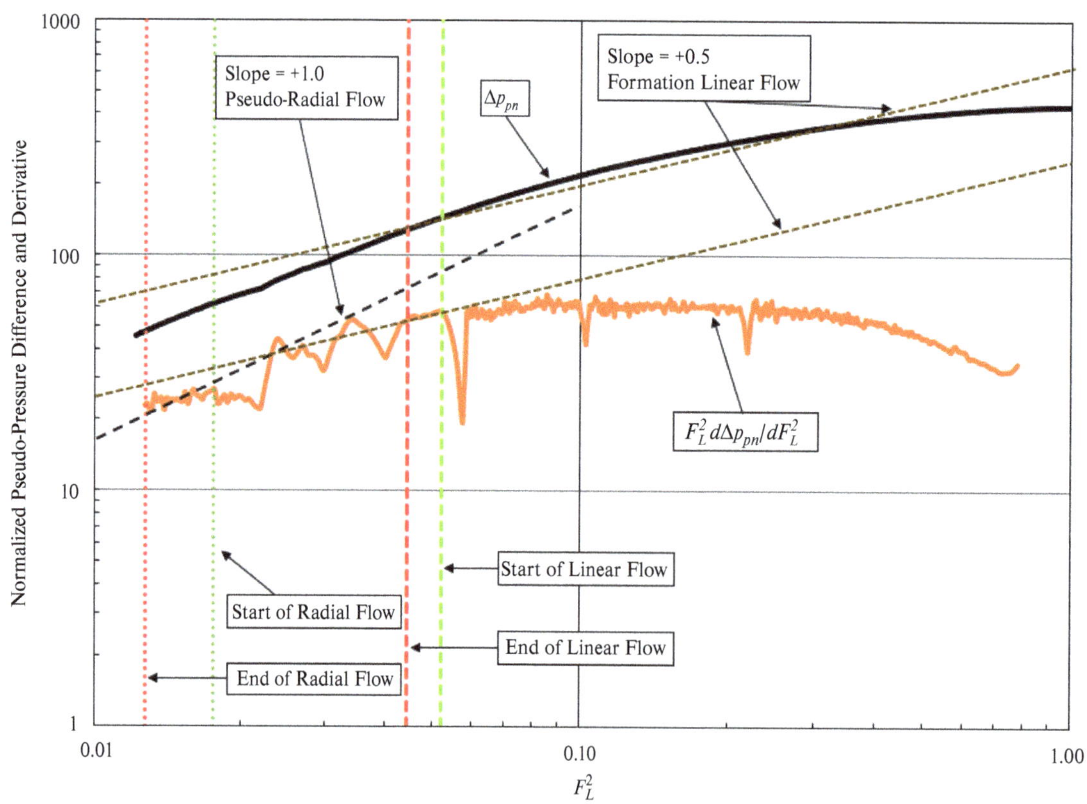

▲ **FIGURE 17.27** **Log-log plot of Δp_{pn} vs. F_L^2 and $F_L^2 d\Delta p_{pn}/dF_L^2$ vs. F_L^2 for Example 17.1**

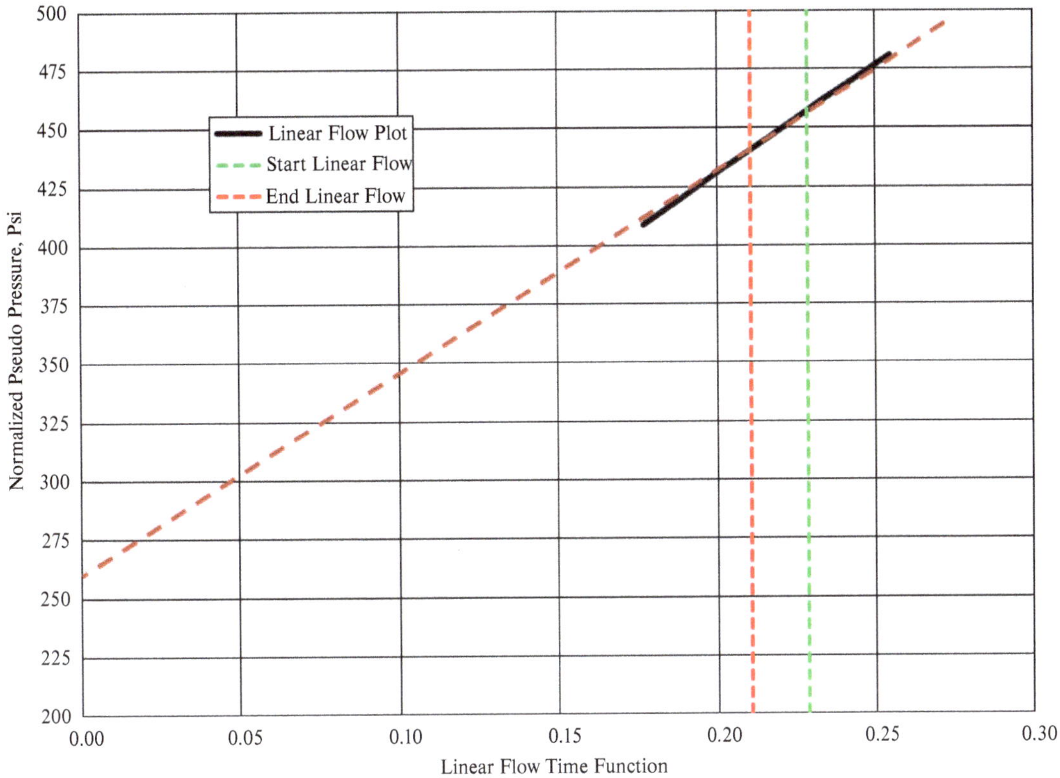

▲ **FIGURE 17.28** **ACA linear flow time function plot (p_{pnw} vs. F_L) for Example 17.1**

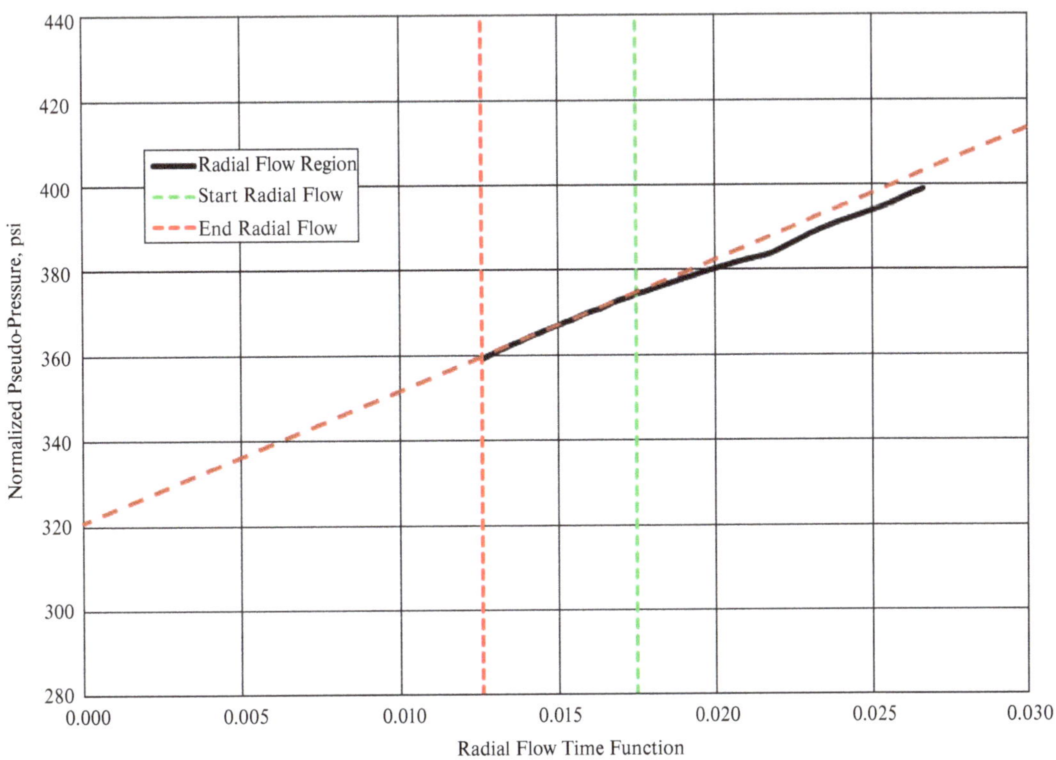

▲ **FIGURE 17.29** **ACA radial flow time function plot (p_{pnw} vs. F_R) for Example 17.1**

It appears that a pseudo-radial flow exists based on plots in Figures 17.27 and 17.29. On this basis, a Horner plot was made using the data in Table 17C.5 of Appendix 17C. The Horner plot is shown as Figure 17.30. A straight line drawn through the radial flow period has an intercept of 320 psi (normalized pseudo-pressure) on the y-axis as shown in Figure 17.30. This transforms to corresponding reservoir pressure of 2410 psi. Note that the same value of reservoir pressure was calculated from the radial time function plot in Step 9. The slope of the Horner straight line, m_H, was calculated to be 8294.2222 psi.

Substituting Eq. (17.17),

$$k = \frac{162.6 \times 1440 \times q \times \mu}{m_H \times h}$$

$$k = \frac{162.6 \times 1440 \times 0.0177 \times 0.7}{8294.2222 \times 12} = 0.0291 \text{ md.}$$

The key to using Eq. (17.17) is estimating injection rate after fracture closure. For this example, injection rate after fracture closure was assumed as the average injection rate just before the termination of injection. The average injection rate used in the calculation of formation permeability was 0.7 bbls/min. Clearly, actual injection rate after fracture closure could be substantially less than this number. A lower formation permeability would be calculated if a lower injection rate was used in the calculation. This is the major deficiency of using the Horner plot to calculate formation permeability from Eq. (17.17).

The results from analyzing the DFIT in Example 17.1 were compared to the results reported by Craig and Blasingame[7] in Table 17.2. The reservoir pressures calculated from analyses of the DFIT data in Example 17.1 are very close to the results reported by Craig and Blasingame.[7] There are some variations in formation permeabilities reported by the two analyses as shown in Table 17.2. It is highly suspected that the main sources of the differences are the actual total injection volumes and fracture injection rates used in the calculations by Craig and Blasingame.[7] Even with these unknown numbers, the formation permeabilities reported in Table 17.2 are reasonably within range.

TABLE 17.2 Comparison of Results for Example 17.1

Parameter	DFIT Analysis (Craig and Blasingame[7])				Example 17.1		
	ACA, Q_t only	ACA, Q_t + Storage	Type Curve	BU Type Curve	ACA Linear	ACA Radial	Horner Plot
p_i, psi	2332	2332	2332	2402	2356	2410	2410
k, md	0.002	0.019	0.011	0.012	—	0.056	0.029

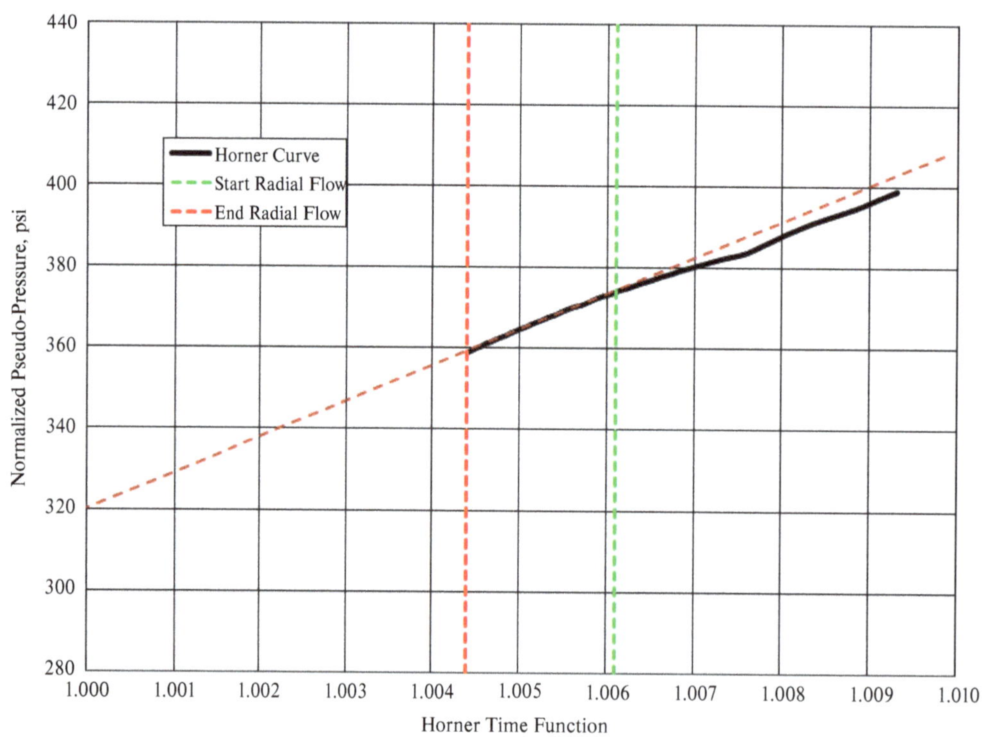

▲ **FIGURE 17.30** **ACA Horner plot for Example 17.1**

End of Example 17.1

Nomenclature

c_t	total compressibility, psi^{-1}	p_z	net fracture pressure, psia
E	Young's Modulus, Mpsi	q	injection rate, bbl/min
F_L	Linear-flow time function	Q_t	total injected fluid volume, bbls.
F_R	Radial-flow time function	r_p	storage ratio
G_c	G-function at fracture closure	Δt	time difference, secs
G	G-function	t	elapsed time from fracture initiation, secs.
h	reservoir or formation thickness, feet	t_c	time to reach closure, secs.
k	permeability, md	t_{DL_f}	dimensionless time in terms of fracture half-length
L_f	fracture half-length, feet	t_p	elapsed time from fracture initiation to end of injection, secs.
m_H	slope of Horner plot straight line	V_{AF}	afterflow volume
m_R	slope of radial-flow straight line	V_w	wellbore fluid volume
Δp	pressure difference, psia	z	gas compressibility factor
p	pressure, psia	γ_g	gas gravity, air = 1
p_p	pseudo-pressure, psia^2/cp	η	fracture fluid efficiency
p_{pn}	normalized pseudo-pressure, psia	ϕ	porosity
p_{pni}	normalized initial pseudo-pressure, psia	σ	formation stress
p_{pnw}	normalized pseudo-pressure-falloff, psi	μ	fluid viscosity, cp
p_r	reservoir pressure, psia		
p_w	fall-off pressure, psi		

Subscripts

c	fracture closure	pn	normalized pseudo-pressure
i	initial conditions	t	total
p	pseudo-pressure	w	water

Abbreviations

ACA	After Closure Analysis	FPP	Fracture Propagation Pressure
BCA	Before Closure Analysis	ISIP	Instantaneous Shut-In Pressure
BHP	Bottomhole Pressure	NFP	Net Fracture Pressure
DFIT	Diagnostic Fracture Injection Test	PKN	Perkins-Kern-Nordgren
FCP	Fracture Closure Pressure	RBP	Retrievable Bridge Plug

REFERENCES

1. Barree, R.D., Miskimins, J.L., and Gilbert, J.V.: "Diagnostic Fracture Injection Tests: Common Mistakes, Misfires, and Misdiagnoses," *SPE Production & Operations,* SPE 169539-PA (May 2015) 84-98.
2. Nolte, K.G.: "Determination of Fracture Parameters from Fracturing Pressure Decline," paper SPE 8341 presented at the SPE Annual Technical Conference and Exhibition, Las Vegas, Nevada, (September 23-26, 1979).
3. Cramer, D.D., and Nguyen, D.H.: "Diagnostic Fracture Injection Testing Tactics in Unconventional Reservoirs," paper SPE 163863 presented at the SPE Hydraulic Fracturing Technology Conference, The Woodlands, Texas, (February 4-6, 2013).
4. Martin, A.R., Cramer, D.D., Nunez, O., and Roberts, N.R.: "A Method to Perform Multiple Diagnostic Fracture Injection Tests Simultaneously in a Single Wellbore," *SPE Production and Operations,* SPE 152019-PA (May 2013) 191-200.
5. Barree, R.D., Barre, V.L., and Craig, D.P.: "Holistic Fracture Diagnostics: Consistent Interpretation of Prefrac Injection Tests Using Multiple Analysis Methods," *SPE Production and Operations,* SPE 107877-PA (August 2009) 396-406.
6. Talley, G.R., Swindell, T.M., Waters, G.A., and Nolte, K.G.: "Field Application of After-Closure Analysis of Fracture Calibration Tests," paper SPE 52220 presented at the 1999 SPE Mid-Continent Operations Symposium, Oklahoma City, Oklahoma, (March 28-31, 1999).
7. Craig, D.P., and Blasingame, T.A.: "Application of a New Fracture-Injection/Fall-off Model Accounting for Propagating, Dilated, and Closing Hydraulic Fractures," paper SPE 100578 presented at the 2006 SPE Gas Technology Symposium, Calgary, Alberta, Canada, (May 15-17).
8. Al-Hussainy, R., and Ramey, H.J., Jr.: "Application of Real Gas Flow Theory to Well Testing and Deliverability Forecasting," *JPT* (May 1966) 637-642.

ADDITIONAL REFERENCES

- Barree, R.D., and Mukherjee, H.: "Determination of Pressure Dependent Leakoff and Its Effects on Fracture Geometry," paper SPE 36424 presented at the 71st Annual Technical Conference and Exhibition, Denver, Colorado, (October 6-9, 1996).
- Nojabaei, B., and Kabir, C.S.: "Establishing Key Reservoir Parameters With Diagnostic Fracture Injection Testing," *SPE Reservoir Evaluation & Engineering,* SPE 153979-PA (October 2012) 563-570.

PROBLEMS

17-1. The diagnostic fracture injection test (DFIT) on a gas well in the Mesaverde formation, San Juan Basin, USA, was analyzed in Example 17.1 to illustrate the step-by-step analytic process for DFIT. The injection data and pressure fall-off data are given in Tables 17A.1 and 17A.2, respectively, of Appendix 17A. Other formation properties of the Mesaverde formation are given in Example 17.1. Using the DFIT data, repeat the analyses with a DFIT analyses software to redetermine the following:

a. Fracture closure pressure (FCP)
b. Instantaneous shut-in pressure (ISIP)
c. Identify fracture linear flow period
d. Identify formation linear flow period
e. Identify pseudo-radial-flow period
f. Reservoir pressure from formation linear flow period
g. Reservoir pressure and permeability from pseudo-radial-flow period
h. Reservoir pressure and permeability from Horner plot
i. Compare the results to the results in Table 17.2

APPENDIX

17A INPUT DATA FOR EXAMPLE 17.1

TABLE 17A.1 Injection Time, Pressure, and Rate for Example 17.1 (Data from Craig and Blasingame[7])

Time, *t* (secs)	p_w (psi)	Inj. Rate, q_{inj} (bbl/min)
0	1808.60	3.17
10	2021.12	3.16
20	2301.28	3.15
30	2687.07	3.14
40	3248.57	3.12
50	3556.64	3.10
60	3640.80	3.09
70	3685.89	3.09
80	3633.14	3.09
90	3357.52	3.11
100	3284.20	3.11
110	3314.79	3.11
120	3647.31	3.10
130	3696.94	3.10
140	3576.05	3.10
150	3602.03	3.10
160	3908.88	4.01
170	3921.19	4.50
180	3919.12	4.49
190	3910.34	4.49
200	3907.54	4.49
210	3905.70	4.50
220	3901.36	4.50
230	3889.46	4.49
240	3680.62	4.02
250	3527.56	3.09
260	3531.31	3.09
270	3535.23	3.09
280	3522.83	3.06
290	3397.62	2.58
300	3324.64	1.93
310	3143.15	1.40
318	3123.09	0.00

TABLE 17A.2 Time and Pressure Fall-Off Data for Example 17.1 (Data from Craig and Blasingame[7])

Time, t (secs)	p_w (psia)	t (secs)	p_w (psia)	t (secs)	p_w (psia)	t (secs)	p_w (psia)
318	3123.09	978	2890.50	1638	2819.58	2298	2769.14
333	3098.72	993	2888.39	1653	2818.22	2313	2768.07
348	3080.28	1008	2886.43	1668	2816.82	2328	2767.22
363	3065.26	1023	2884.50	1683	2815.62	2343	2766.22
378	3052.42	1038	2882.64	1698	2814.37	2358	2765.21
393	3040.86	1053	2880.74	1713	2813.12	2373	2764.35
408	3030.73	1068	2878.77	1728	2811.77	2388	2763.40
423	3021.44	1083	2877.09	1743	2810.58	2403	2762.53
438	3012.94	1098	2875.16	1758	2809.31	2418	2761.54
453	3005.09	1113	2873.43	1773	2808.06	2433	2760.60
468	2997.89	1128	2871.64	1788	2806.77	2448	2759.80
483	2991.38	1143	2869.80	1803	2805.53	2463	2758.74
498	2985.36	1158	2868.05	1818	2804.31	2478	2758.03
513	2979.83	1173	2866.27	1833	2803.19	2493	2757.08
528	2974.66	1188	2864.61	1848	2801.85	2508	2756.22
543	2969.80	1203	2862.76	1863	2800.62	2523	2755.21
558	2965.31	1218	2861.14	1878	2799.50	2538	2754.39
573	2961.19	1233	2859.40	1893	2798.24	2553	2753.57
588	2957.39	1248	2857.76	1908	2796.94	2568	2752.65
603	2953.69	1263	2856.05	1923	2795.83	2583	2751.79
618	2950.18	1278	2854.45	1938	2794.63	2598	2750.88
633	2946.91	1293	2852.79	1953	2793.45	2613	2750.16
648	2943.63	1308	2851.25	1968	2792.27	2628	2749.18
663	2940.51	1323	2849.67	1983	2791.13	2643	2748.39
678	2937.56	1338	2848.07	1998	2790.01	2658	2747.55
693	2934.61	1353	2846.55	2013	2788.82	2673	2746.76
708	2931.86	1368	2844.90	2028	2787.74	2688	2746.04
723	2929.15	1383	2843.36	2043	2786.63	2703	2745.23
738	2926.45	1398	2841.89	2058	2785.55	2718	2744.39
753	2923.97	1413	2840.39	2073	2784.49	2733	2743.61
768	2921.17	1428	2839.02	2088	2783.26	2748	2742.73
783	2916.83	1443	2837.40	2103	2782.21	2763	2742.10
798	2916.29	1458	2836.07	2118	2781.18	2778	2741.29
813	2913.98	1473	2834.49	2133	2780.04	2793	2740.39

Time, t (secs)	p_w (psia)	t (secs)	p_w (psia)	t (secs)	p_w (psia)	t (secs)	p_w (psia)
828	2911.65	1488	2833.11	2148	2779.14	2808	2739.68
843	2909.41	1503	2831.66	2163	2778.07	2823	2738.82
858	2907.03	1518	2830.30	2178	2777.08	2838	2738.14
873	2904.76	1533	2828.87	2193	2775.93	2853	2737.40
888	2902.63	1548	2827.60	2208	2775.07	2868	2736.54
903	2900.59	1563	2826.21	2223	2773.97	2883	2735.93
918	2898.42	1578	2824.76	2238	2773.02	2898	2735.09
933	2896.35	1593	2823.41	2253	2772.01	2913	2734.34
948	2894.24	1608	2822.17	2268	2771.08	2928	2733.61
963	2892.34	1623	2820.88	2283	2770.04	2943	2732.87
2958	2732.25	3618	2703.71	4638	2670.54	5958	2638.44
2973	2731.43	3633	2703.08	4668	2669.70	5988	2637.73
2988	2730.71	3648	2702.51	4698	2668.78	6018	2637.20
3003	2729.99	3663	2702.16	4728	2668.05	6048	2636.48
3018	2729.22	3678	2701.42	4758	2667.20	6078	2635.89
3033	2728.54	3693	2700.93	4788	2666.40	6108	2635.22
3048	2727.89	3708	2700.27	4818	2665.50	6138	2634.66
3063	2727.12	3723	2699.77	4848	2664.81	6168	2634.14
3078	2726.51	3738	2699.14	4878	2663.97	6198	2633.51
3093	2725.64	3753	2698.66	4908	2663.20	6228	2632.84
3108	2724.97	3768	2698.17	4938	2662.32	6258	2632.20
3123	2724.35	3783	2697.55	4968	2661.54	6288	2631.75
3138	2723.65	3798	2697.15	4998	2660.80	6318	2631.20
3153	2722.94	3813	2696.50	5028	2660.13	6348	2630.54
3168	2722.19	3828	2696.05	5058	2659.26	6378	2630.03
3183	2721.71	3843	2695.40	5088	2658.58	6408	2629.35
3198	2721.02	3858	2694.80	5118	2657.66	6438	2628.79
3213	2720.34	3873	2694.33	5148	2656.94	6468	2628.28
3228	2719.56	3888	2693.78	5178	2656.28	6498	2627.71
3243	2718.96	3903	2693.32	5208	2655.63	6528	2627.08
3258	2718.32	3918	2692.79	5238	2654.80	6558	2626.55
3273	2717.76	3948	2691.72	5268	2654.07	6588	2626.01
3288	2717.09	3978	2690.72	5298	2653.41	6618	2625.47
3303	2716.44	4008	2689.64	5328	2652.62	6648	2624.86
3318	2715.71	4038	2688.81	5358	2651.92	6678	2624.34

Time, *t* (secs)	p_w (psia)	*t* (secs)	p_w (psia)	*t* (secs)	p_w (psia)	*t* (secs)	p_w (psia)
3333	2715.22	4068	2687.75	5388	2651.26	6708	2623.75
3348	2714.48	4098	2686.64	5418	2650.34	6738	2623.19
3363	2713.85	4128	2685.73	5448	2649.75	6768	2622.85
3378	2713.33	4158	2684.89	5478	2649.06	6798	2622.09
3393	2712.69	4188	2683.95	5508	2648.41	6828	2621.61
3408	2711.96	4218	2682.86	5538	2647.71	6858	2621.03
3423	2711.40	4248	2682.02	5568	2647.05	6888	2620.53
3438	2710.89	4278	2681.06	5598	2646.32	6918	2619.88
3453	2710.18	4308	2680.19	5628	2645.67	6948	2619.56
3468	2709.55	4338	2679.33	5658	2644.91	6978	2619.03
3483	2709.02	4368	2678.36	5688	2644.32	7008	2618.36
3498	2708.40	4398	2677.36	5718	2643.61	7038	2617.91
3513	2707.76	4428	2676.55	5748	2642.92	7068	2617.45
3528	2707.13	4458	2675.70	5778	2642.28	7098	2616.88
3543	2706.71	4488	2674.81	5808	2641.65	7128	2616.41
3558	2706.04	4518	2674.02	5838	2641.04	7158	2615.88
3573	2705.39	4548	2673.12	5868	2640.33	7188	2615.35
3588	2705.02	4578	2672.14	5898	2639.66	7218	2614.90
3603	2704.40	4608	2671.41	5928	2639.00	7248	2614.27
7278	2613.73	9678	2579.03	12318	2551.97	27918	2479.80
7308	2613.24	9738	2578.36	12378	2551.38	28518	2478.45
7338	2612.68	9798	2577.62	12438	2550.79	29118	2477.21
7368	2612.18	9858	2576.82	12498	2550.27	29718	2476.04
7398	2611.68	9918	2576.17	12558	2549.72	30318	2474.69
7428	2611.20	9978	2575.52	12618	2549.35	30918	2473.33
7458	2610.74	10038	2574.89	12678	2548.77	31518	2471.82
7488	2610.19	10098	2574.11	12738	2548.17	32118	2470.38
7518	2609.69	10158	2573.42	12798	2547.75	32718	2469.03
7578	2608.70	10218	2572.85	12858	2547.24	33318	2468.32
7638	2607.72	10278	2572.07	12918	2546.72	33918	2467.72
7698	2602.69	10338	2571.33	13218	2544.31	34518	2467.03
7758	2605.78	10398	2570.70	13518	2542.02	35118	2466.31
7818	2604.84	10458	2569.99	13818	2539.76	35718	2465.62
7878	2603.98	10518	2569.43	14118	2537.47	36318	2464.97
7938	2603.00	10578	2568.69	14418	2535.33	36918	2464.20
7998	2601.98	10638	2567.96	14718	2533.29	37518	2463.56

Time, t (secs)	p_w (psia)	t (secs)	p_w (psia)	t (secs)	p_w (psia)	t (secs)	p_w (psia)
8058	2601.27	10698	2567.50	15018	2531.25	38118	2462.82
8118	2600.23	10758	2566.72	15318	2529.39	38718	2462.19
8178	2599.38	10818	2566.08	15618	2527.33	39318	2461.53
8238	2598.49	10878	2565.44	15918	2525.58	39918	2460.81
8298	2597.66	10938	2564.79	16218	2523.82	40518	2460.32
8358	2596.77	10998	2564.28	16518	2522.09	41118	2459.54
8418	2595.86	11058	2563.69	16818	2520.34	41718	2459.05
8478	2595.21	11118	2563.00	17118	2518.61	42318	2458.43
8538	2594.26	11178	2562.41	17418	2516.98	42918	2457.62
8598	2593.33	11238	2561.83	17718	2515.59	43518	2456.92
8658	2592.43	11298	2561.31	18018	2514.07	44118	2456.51
8718	2591.67	11358	2560.67	18318	2512.62	44718	2455.96
8778	2590.95	11418	2560.15	18918	2509.64	45318	2455.31
8838	2590.02	11478	2559.60	19518	2506.93	45918	2454.57
8898	2589.35	11538	2559.01	20118	2504.25	46518	2454.15
8958	2588.38	11598	2558.53	20718	2501.64	47118	2453.50
9018	2587.56	11658	2557.83	21318	2499.29	47718	2452.99
9078	2586.77	11718	2557.41	21918	2496.92	48318	2452.49
9138	2585.83	11778	2556.83	22518	2492.66	48918	2451.84
9198	2585.17	11838	2556.29	23118	2492.54	49518	2451.42
9258	2584.52	11898	2555.72	23718	2489.98	50118	2450.92
9318	2583.56	11958	2555.18	24318	2488.24	50718	2450.41
9378	2582.66	12018	2554.61	24918	2487.05	51318	2449.81
9438	2582.09	12078	2554.02	25518	2485.46	51918	2449.39
9498	2581.27	12138	2553.56	26118	2484.00	52518	2448.96
9558	2580.49	12198	2553.06	26718	2483.16	53118	2448.40
9618	2579.69	12258	2552.54	27318	2481.56	53718	2448.04
54318	2447.42	54918	2447.04	55518	2446.54	56118	2446.05
56718	2445.72	57318	2445.39	57918	2444.79	58518	2444.34

APPENDIX

17B DIAGNOSTIC PLOTS DATA FOR EXAMPLE 17.1

TABLE 17B.1 Partial Table of t_D, p_w, G, dp/dG, and $G\ dp/dG$ for Example 17.1

Total Time, *t* (secs)	Elapsed Time, *t* (secs)	t_D (secs)	p_w (psi)	*G*	*dp/dG*	*G dp/dG*
318	248	0.0000	3123.09	0.0000	0.0000	0.0000
333	263	0.0605	3098.72	0.1085	204.9666	22.2368
348	278	0.1210	3080.28	0.2070	174.3816	36.1047
363	293	0.1815	3065.26	0.2995	154.1395	46.1658
378	308	0.2419	3052.42	0.3873	141.8741	54.9477
393	323	0.3024	3040.86	0.4713	116.2932	63.1877
408	338	0.3629	3030.73	0.5520	108.1329	68.3772
423	353	0.4234	3021.44	0.6298	101.3257	72.7097
438	368	0.4839	3012.94	0.7051	95.5743	76.4175
453	383	0.5444	3005.09	0.7782	90.0961	79.1638
468	398	0.6048	2997.89	0.8492	85.4319	81.6036
483	413	0.6653	2991.38	0.9184	81.3148	83.7056
498	428	0.7258	2985.36	0.9858	77.3340	85.1844
513	443	0.7863	2979.83	1.0517	73.4035	86.0067
528	458	0.8468	2974.66	1.1161	69.5444	86.2429
543	473	0.9073	2969.8	1.1791	66.3466	86.7080
558	488	0.9677	2965.31	1.2408	63.4730	87.0949
573	503	1.0282	2961.19	1.3013	60.8420	87.3693
588	518	1.0887	2957.39	1.3607	58.8561	88.1975
603	533	1.1492	2953.69	1.4191	56.8551	88.6831
618	548	1.2097	2950.18	1.4764	55.0244	89.1352
633	563	1.2702	2946.91	1.5327	53.8640	90.4339
648	578	1.3306	2943.63	1.5882	52.6167	91.3900
663	593	1.3911	2940.51	1.6427	51.4835	92.3554
678	608	1.4516	2937.56	1.6965	50.7106	93.8108
693	623	1.5121	2934.61	1.7494	49.7298	94.7392
708	638	1.5726	2931.86	1.8017	49.6557	97.2945
723	653	1.6331	2929.15	1.8531	48.6506	97.9276
738	668	1.6935	2926.45	1.9039	48.4515	100.0812
753	683	1.7540	2923.97	1.9541	47.7553	101.1252
768	698	1.8145	2921.17	2.0036	47.2388	102.4532
783	713	1.8750	2918.83	2.0524	46.8454	103.9698
798	728	1.9355	2916.29	2.1007	46.6843	105.9432
813	743	1.9960	2913.98	2.1484	46.8087	108.5332
828	758	2.0565	2911.65	2.1956	45.8166	108.4638

TABLE 17B.2 Partial Table of $\sqrt{t}$, p_w, $dp/d\sqrt{t}$, and $\sqrt{t}\ dp/d\sqrt{t}$ for Example 17.1

Total Time, t (secs)	Elapsed Time, t (secs)	$\sqrt{t_D}$	p_w (psi)	$dp/d\sqrt{t}$	$\sqrt{t}\ dp/d\sqrt{t}$
318	248	15.7480	3123.09	—	—
333	263	16.2173	3098.72	46.1012	747.6358
348	278	16.6733	3080.28	37.0900	618.4147
363	293	17.1172	3065.26	31.7287	543.1083
378	308	17.5499	3052.42	31.3258	549.7664
393	323	17.9722	3040.86	27.4284	492.9481
408	338	18.3848	3030.73	24.7891	455.7420
423	353	18.7883	3021.44	22.7671	427.7555
438	368	19.1833	3012.94	21.1294	405.3313
453	383	19.5704	3005.09	19.6370	384.3035
468	398	19.9499	2997.89	18.4017	367.1122
483	413	20.3224	2991.38	17.2816	351.2040
498	428	20.6882	2985.36	16.2405	335.9863
513	443	21.0476	2979.83	15.2422	320.8107
528	458	21.4009	2974.66	14.3061	306.1636
543	473	21.7486	2969.8	13.5357	294.3823
558	488	22.0907	2965.31	12.8410	283.6660
573	503	22.4277	2961.19	12.2079	273.7954
588	518	22.7596	2957.39	11.6933	266.1344
603	533	23.0868	2953.69	11.2066	258.7255
618	548	23.4094	2950.18	10.7732	252.1931
633	563	23.7276	2946.91	10.4696	248.4176
648	578	24.0416	2943.63	10.1883	244.9427
663	593	24.3516	2940.51	9.9224	241.6258
678	608	24.6577	2937.56	9.7214	239.7081
693	623	24.9600	2934.61	9.5097	237.3627
708	638	25.2587	2931.86	9.4338	238.2841
723	653	25.5539	2929.15	9.2131	235.4301
738	668	25.8457	2926.45	9.1439	236.3305
753	683	26.1343	2923.97	8.9734	234.5129
768	698	26.4197	2921.17	8.8769	234.5241
783	713	26.7021	2918.83	8.7681	234.1266
798	728	26.9815	2916.29	8.7163	235.1790
813	743	27.2580	2913.98	8.7130	237.4985
828	758	27.5318	2911.65	8.4982	233.9718

APPENDIX

17C BCA AND ACA PLOTS DATA FOR EXAMPLE 17.1

TABLE 17C.1 Gas Properties Including Pseudo-pressures for Example 17.1

Pressure p (psi)	Reduced Pressure p_r	Gas Compressibility Factor, z	Gas Viscosity μ (cp)	$2p/(\mu z)$	Δp	Pseudo-pressure p_p (psi²/cp)	Normalized Pseudo-pressure p_{pn} (psi)*
2100	3.12	0.861	0.0168	290327	25	3.63E+06	11.74
2125	3.16	0.861	0.0169	292449	25	1.09E+07	35.32
2150	3.20	0.861	0.0170	294539	25	1.83E+07	59.06
2175	3.23	0.861	0.0170	296599	25	2.56E+07	82.97
2200	3.27	0.860	0.0171	298833	25	3.31E+07	107.06
2225	3.31	0.860	0.0172	300830	25	4.06E+07	131.31
2250	3.34	0.860	0.0173	302796	25	4.81E+07	155.73
2275	3.38	0.860	0.0174	304731	25	5.57E+07	180.30
2300	3.42	0.860	0.0174	306635	25	6.34E+07	205.03
2325	3.46	0.860	0.0175	308509	25	7.10E+07	229.91
2350	3.49	0.860	0.0176	310352	25	7.88E+07	254.95
2375	3.53	0.860	0.0177	312165	25	8.66E+07	280.13
2400	3.57	0.860	0.0178	313948	25	9.44E+07	305.45
2425	3.60	0.860	0.0179	315700	25	1.02E+08	330.92
2450	3.64	0.860	0.0179	317423	25	1.10E+08	356.53
2475	3.68	0.860	0.0180	319116	25	1.18E+08	382.28
2500	3.72	0.860	0.0181	320779	25	1.26E+08	408.16
2525	3.75	0.860	0.0182	322413	25	1.34E+08	434.18
2550	3.79	0.860	0.0183	324017	25	1.42E+08	460.33
2575	3.83	0.860	0.0184	325592	25	1.50E+08	486.60
2600	3.86	0.860	0.0185	327138	25	1.59E+08	513.01
2625	3.90	0.860	0.0186	328655	25	1.67E+08	539.53
2650	3.94	0.860	0.0187	330143	25	1.75E+08	566.18
2675	3.98	0.860	0.0188	331602	25	1.83E+08	592.95
2700	4.01	0.860	0.0189	333033	25	1.92E+08	619.83
2725	4.05	0.860	0.0189	334435	25	2.00E+08	646.83
2750	4.09	0.860	0.0190	335810	25	2.08E+08	673.94
2775	4.12	0.860	0.0191	337156	25	2.17E+08	701.16

Pressure p (psi)	Reduced Pressure p_r	Gas Compressibility Factor, z	Gas Viscosity μ (cp)	$2p/(\mu z)$	Δp	Pseudo-pressure p_p (psi²/cp)	Normalized Pseudo-pressure p_{pn} (psi)*
2800	4.16	0.860	0.0192	338474	25	2.25E+08	728.49
2825	4.20	0.860	0.0193	339764	25	2.34E+08	755.93
2850	4.24	0.860	0.0194	341027	25	2.42E+08	783.46
2875	4.27	0.860	0.0195	342262	25	2.51E+08	811.10
2900	4.31	0.861	0.0196	343308	25	2.59E+08	838.83
2925	4.35	0.861	0.0197	344491	25	2.68E+08	866.65
2950	4.38	0.861	0.0198	345647	25	2.76E+08	894.57
2975	4.42	0.861	0.0199	346776	25	2.85E+08	922.58
3000	4.46	0.863	0.0200	347575	25	2.94E+08	950.66
3025	4.50	0.863	0.0201	348656	25	3.02E+08	978.83
3050	4.53	0.863	0.0202	349710	25	3.11E+08	1007.07
3075	4.57	0.864	0.0203	350594	25	3.20E+08	1035.40
3100	4.61	0.864	0.0204	351599	25	3.29E+08	1063.80
3125	4.64	0.865	0.0205	352439	25	3.38E+08	1092.28
3150	4.68	0.867	0.0206	353120	25	3.46E+08	1120.82
3175	4.72	0.867	0.0207	354056	25	3.55E+08	1149.46
3200	4.76	0.867	0.0208	354967	25	3.64E+08	1178.11
3225	4.79	0.868	0.0209	355723	25	3.73E+08	1206.85
3250	4.83	0.868	0.0210	356587	25	3.82E+08	1235.67
3275	4.87	0.880	0.0209	355910	25	3.91E+08	1264.49
3300	4.90	0.880	0.0210	356757	25	4.00E+08	1293.31

*Normalized pseudo-pressures based at reservoir pressure of 2332 psia (Craig and Blasingame[7])

TABLE 17C.2 Partial Table of Δt, p_w, p_{pnw}, Δp_{pn}, and $\Delta t\, d\Delta p_{pn}/d\Delta t$ for Example 17.1

Total Time, t (secs)	Elapsed Time, t (secs)	Δt	p_w (psi)	p_{pnw} (psi)	Δp_{pn}	$\Delta t\, d\Delta p_{pn}/d\Delta t$
381	311	0.00	3050.00	1006.29	—	—
393	323	11.86	3040.86	996.49	9.80	9.80
408	338	26.86	3030.73	985.63	20.65	20.65
423	353	41.86	3021.44	975.68	30.61	30.61
438	368	56.86	3012.94	966.57	39.72	34.02
453	383	71.86	3005.09	958.15	48.13	39.17
468	398	86.86	2997.89	950.44	55.85	43.50
483	413	101.86	2991.38	943.46	62.82	46.98
498	428	116.86	2985.36	937.01	69.27	49.71
513	443	131.86	2979.83	931.08	75.20	51.70
528	458	146.86	2974.66	925.54	80.74	53.12
543	473	161.86	2969.80	920.33	85.95	54.48
558	488	176.86	2965.31	915.52	90.76	55.57
573	503	191.86	2961.19	911.11	95.18	56.41
588	518	206.86	2957.39	907.03	99.25	57.33
603	533	221.86	2953.69	903.07	103.22	58.04
618	548	236.86	2950.18	899.31	106.98	58.70
633	563	251.86	2946.91	895.80	110.48	59.79
648	578	266.86	2943.63	892.29	114.00	60.85
663	593	281.86	2940.51	888.94	117.34	61.77
678	608	296.86	2937.56	885.78	120.50	62.91
693	623	311.86	2934.61	882.62	123.66	63.89
708	638	326.86	2931.86	879.67	126.61	65.56
723	653	341.86	2929.15	876.77	129.51	66.19
738	668	356.86	2926.45	873.88	132.41	67.79
753	683	371.86	2923.97	871.22	135.07	68.51
768	698	386.86	2921.17	868.22	138.07	69.83
783	713	401.86	2918.83	865.71	140.57	70.85
798	728	416.86	2916.29	862.99	143.30	72.30
813	743	431.86	2913.98	860.51	145.77	74.09
828	758	446.86	2911.65	858.02	148.27	73.99
843	773	461.86	2909.41	855.61	150.67	75.24
858	788	476.86	2907.03	853.06	153.22	76.10
873	803	491.86	2904.76	850.63	155.65	77.44
888	818	506.86	2902.63	848.35	157.94	78.81
903	833	521.86	2900.59	846.16	160.12	79.56

TABLE 17C.3 Partial Table of p_{pnw}, $p_{pnw} - p_{pni}$, F_L^2, and $F_L^2\, d\Delta p_{pnw}/dF_L^2$ for Example 17.1

Total Time, t (secs)	Elapsed Time, t (secs)	p_w (psi)	p_{pnw} (psi)	$p_{pnw} - p_{pni}$ (psi)	F_L^2	$F_L^2 d\Delta p_{pnw}/dF_L^2$
1803	1733	2805.53	744.29	432.463	1.000	—
1818	1748	2804.31	742.98	431.156	0.885	—
1833	1763	2803.19	741.78	429.955	0.840	—
1848	1778	2801.85	740.34	428.519	0.807	—
1863	1793	2800.62	739.02	427.201	0.779	35.055
1878	1808	2799.50	737.82	426.001	0.756	32.753
1893	1823	2798.24	736.47	424.650	0.735	32.553
1908	1838	2796.94	735.08	423.257	0.716	32.519
1923	1853	2795.83	733.89	422.068	0.699	33.787
1938	1868	2794.63	732.60	420.782	0.683	34.151
1953	1883	2793.45	731.34	419.517	0.669	35.326
1968	1898	2792.27	730.08	418.252	0.655	35.609
1983	1913	2791.13	728.85	417.031	0.642	36.366
1998	1928	2790.01	727.65	415.830	0.630	37.073
2013	1943	2788.82	726.38	414.555	0.619	36.884
2028	1958	2787.74	725.22	413.398	0.608	39.292
2043	1973	2786.63	724.03	412.208	0.598	39.714
2058	1988	2785.55	722.87	411.051	0.588	40.200
2073	2003	2784.49	721.74	409.915	0.579	41.992
2088	2018	2783.26	720.42	408.596	0.570	39.498
2103	2033	2782.21	719.29	407.471	0.561	40.590
2118	2048	2781.18	718.19	406.367	0.553	41.041
2133	2063	2780.04	716.97	405.145	0.545	41.937
2148	2078	2779.14	716.00	404.181	0.537	42.486
2163	2093	2778.07	714.86	403.034	0.530	43.614
2178	2108	2777.08	713.80	401.973	0.522	43.981
2193	2123	2775.93	712.56	400.741	0.516	43.149
2208	2138	2775.07	711.64	399.819	0.509	44.738
2223	2153	2773.97	710.46	398.640	0.502	44.767
2238	2168	2773.02	709.45	397.622	0.496	44.921
2253	2183	2772.01	708.36	396.540	0.490	46.444
2268	2198	2771.08	707.37	395.543	0.484	46.144
2283	2213	2770.04	706.25	394.428	0.478	46.374
2298	2228	2769.14	705.29	393.464	0.473	48.497
2313	2243	2768.07	704.14	392.317	0.467	46.456
2328	2258	2767.22	703.23	391.406	0.462	48.471

TABLE 17C.4 | Partial Table of F_L and p_{pnw} Plotted in Figure 17.28 of Example 17.1

Total Time, t (secs)	Elapsed Time, t (secs)	p_w (psi)	F_L	p_{pnw} (psi)
11838	11768	2556.29	0.2507	477.18
11898	11828	2555.72	0.2501	476.57
11958	11888	2555.18	0.2494	475.99
12018	11948	2554.61	0.2487	475.38
12078	12008	2554.02	0.2481	474.74
12138	12068	2553.56	0.2474	474.25
12198	12128	2553.06	0.2468	473.71
12258	12188	2552.54	0.2461	473.16
12318	12248	2551.97	0.2455	472.55
12378	12308	2551.38	0.2449	471.91
12438	12368	2550.79	0.2443	471.28
12498	12428	2550.27	0.2436	470.72
12558	12488	2549.72	0.2430	470.13
12618	12548	2549.35	0.2424	469.74
12678	12608	2548.77	0.2418	469.12
12738	12668	2548.17	0.2412	468.47
12798	12728	2547.75	0.2406	468.02
12858	12788	2547.24	0.2400	467.48
12918	12848	2546.72	0.2394	466.92
13218	13148	2544.31	0.2365	464.34
13518	13448	2542.02	0.2338	461.88
13818	13748	2539.76	0.2311	459.46
14118	14048	2537.47	0.2285	457.01
14418	14348	2535.33	0.2260	454.71
14718	14648	2533.29	0.2235	452.53
15018	14948	2531.25	0.2212	450.34
15318	15248	2529.39	0.2189	448.35
15618	15548	2527.33	0.2167	446.14
15918	15848	2525.58	0.2146	444.26
16218	16148	2523.82	0.2125	442.38
16518	16448	2522.09	0.2105	440.52
16818	16748	2520.34	0.2085	438.65
17118	17048	2518.61	0.2066	436.79
17418	17348	2516.98	0.2047	435.05
17718	17648	2515.59	0.2029	433.56
18018	17948	2514.07	0.2012	431.93
18318	18248	2512.62	0.1994	430.37

TABLE 17C.5 Partial Table of F_R, Horner time, and p_{pnw} for Example 17.1

Total Time, t (secs)	p_w (psi)	Elapsed Time, t (secs)	F_R	Horner time $(t + t_p)/t_p$	p_{pnw} (psi)
30318	2474.69	30248	0.0235	1.0082	389.7253
30918	2473.33	30848	0.0230	1.0080	388.2678
31518	2471.82	31448	0.0226	1.0079	386.6495
32118	2470.38	32048	0.0222	1.0077	385.1062
32718	2469.03	32648	0.0217	1.0076	383.6595
33318	2468.32	33248	0.0213	1.0075	382.8985
33918	2467.72	33848	0.0210	1.0073	382.2555
34518	2467.03	34448	0.0206	1.0072	381.5161
35118	2466.31	35048	0.0202	1.0071	380.7444
35718	2465.62	35648	0.0199	1.0070	380.0050
36318	2464.97	36248	0.0196	1.0068	379.3083
36918	2464.2	36848	0.0192	1.0067	378.4831
37518	2463.56	37448	0.0189	1.0066	377.7973
38118	2462.82	38048	0.0186	1.0065	377.0042
38718	2462.19	38648	0.0183	1.0064	376.3290
39318	2461.53	39248	0.0181	1.0063	375.6217
39918	2460.81	39848	0.0178	1.0062	374.8501
40518	2460.32	40448	0.0175	1.0061	374.3249
41118	2459.54	41048	0.0173	1.0060	373.4890
41718	2459.05	41648	0.0170	1.0060	372.9639
42318	2458.43	42248	0.0168	1.0059	372.2994
42918	2457.62	42848	0.0165	1.0058	371.4314
43518	2456.92	43448	0.0163	1.0057	370.6812
44118	2456.51	44048	0.0161	1.0056	370.2418
44718	2455.96	44648	0.0159	1.0056	369.6523
45318	2455.31	45248	0.0156	1.0055	368.9557
45918	2454.57	45848	0.0154	1.0054	368.1627
46518	2454.15	46448	0.0152	1.0053	367.7126
47118	2453.5	47048	0.0150	1.0053	367.0160
47718	2452.99	47648	0.0148	1.0052	366.4694
48318	2452.49	48248	0.0147	1.0051	365.9335
48918	2451.84	48848	0.0145	1.0051	365.2369
49518	2451.42	49448	0.0143	1.0050	364.7868
50118	2450.92	50048	0.0141	1.0050	364.2510
50718	2450.41	50648	0.0140	1.0049	363.7044
51318	2449.81	51248	0.0138	1.0048	363.0614

CHAPTER

18 Immiscible Fluid Displacement

18.1 Introduction

Most of the oil and gas recovered from reservoirs is displaced immiscibly by water and/or gas. The displacement could be in the form of solution gas drive, gas cap expansion, water influx from aquifers or injection of water and/or gas. Solution-gas drive, gas cap expansion, and water influx from aquifers are essentially natural processes that supply energy to the reservoir for hydrocarbon recovery. Gas and water injection are designed and installed to artificially supply energy to the reservoir and thereby improve hydrocarbon recovery. It is important to understand the fundamental processes that occur when reservoir fluids are displaced immiscibly by gas or water. The displacement process is affected by the wettability of the rock, and the mobility ratio between the displaced and the displacing fluids. The total efficiency of the displacement process is measured in terms of the effectiveness of water or gas in displacing the reservoir fluids, and the proportion of the reservoir actually contacted by the displacing fluids. In this chapter, basic concepts in immiscible fluid displacement are presented. These are then followed with the presentation of the fractional flow equation, the Buckley-Leverett equation, and the Welge method for estimating average water saturation in a water displacement process. These equations are presented to familiarize the engineer with some of the classical developments in the analysis of immiscible displacement processes before the advent and widespread application of reservoir simulation techniques. This approach is intended to enable the engineer to become conversant with some of the terms generally used in the industry to analyze and discuss the results from reservoir simulation when applied to immiscible displacement processes.

18.2 Basic Concepts in Immiscible Fluid Displacement

In this section, basic concepts in the immiscible fluid displacement process are presented. These concepts are useful in discussion and understanding of immiscible displacement processes. Some of the concepts are also used in discussions of secondary and enhanced hydrocarbon recovery processes in Chapters 19 to 21. The concepts that are presented in this section are: rock wettability, capillary pressure, relative permeability, mobility and mobility ratio, fluid displacement efficiency, volumetric displacement efficiency, and total recovery efficiency.

18.2.1 Rock Wettability

Rock wettability is the tendency of either the water phase or the oil phase to preferentially maintain contact with the rock surface in a multiphase fluid system. Thus, the surface of a water-wet rock, preferentially maintains contact with water, while the surface of an oil-wet rock will preferentially maintain contact with oil in an oil-water system. The most common method of determining rock wettability is by measurement of the contact angle, θ, between the rock surface and the fluid system.[1] For an oil-water system in contact with a solid surface, the contact angle, θ, is the angle between the fluid-solid interface measured through the water phase. The rock surface is considered to be water-wet when $\theta < 90°$, and oil-wet when $\theta > 90°$ (Figure 18.1). When $\theta \approx 90°$, the rock surface is considered to be intermediate- or neutral-wet. There are other methods for determining rock wettability based on displacement studies. These are the Amott[2] wettability test and the United States Bureau of Mines (USBM)[3] wettability test. The Amott and USBM methods are relatively fast, with tests completed in a few days, in comparison to the contact angle method, which may require several months for completion. Additional methods for determining rock wettability are: Imbibition rate test,[4] Hysteresis of relative permeability curve,[5] and Nuclear Magnetic Relaxation (NMR).[6]

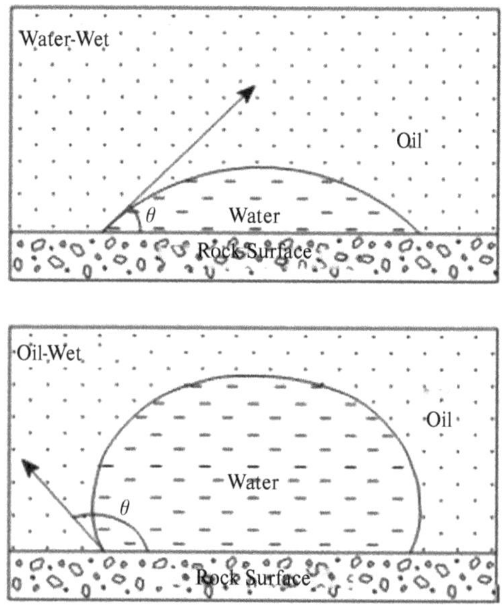

▲ FIGURE 18.1 Water-wet rock and oil-wet rock

Rock wettability is classified into uniform or homogeneous wetting, and non-uniform or heterogeneous wetting. Uniform wetting, in contrast to non-uniform wetting, assumes that the entire reservoir rock has a single form of wettability. Non-uniform wettability is characterized by distinct zones within the reservoir rock system that are preferentially water-wet or oil-wet. The distribution and extent of wetting heterogeneity is determined by the variation of the reservoir rock minerals. Heterogeneous wettability consists of wetting states that have been described as mixed wettability, fractional wettability, "Dalmatian" wetting, and speckled wetting.[7]

The composition of the reservoir oil affects the wettability of the rock. The wetting state of reservoir rock is affected by the presence of polar compounds such as asphaltenes, film forming components, and high molecular weight paraffins. Other factors that may affect rock wettability include the type of minerals present in the rock, the reservoir rock type (quartz, silica, calcite, etc.), and salinity of the connate water.

Rock wettability is usually measured from core plugs or whole length cores. The wettability of the core samples can be altered during preparation of the core samples before testing. Furthermore, in-situ rock wettability may be different from laboratory measured rock wettability due to the following reasons:

1. Flash vaporization of reservoir fluids from cores due to depressurization as the cores are retrieved from the reservoir.
2. Invasion of mud filtrate.
3. Asphaltenes deposition or wax precipitation in the cores due to pressure and temperature changes.
4. Drying of cores during storage.

It is important to ensure that the wettability of the core samples is not altered during handling. In cases where the wettability of the core may have been altered, procedures should be undertaken to restore it as close as possible to its native state before initiating the wettability tests.

18.2.2 Capillary Pressure

Capillary pressure, p_c, is commonly defined as the difference in the pressure of the non-wetting phase and the pressure of the wetting phase. This is represented as:

$$p_c = p_{nw} - p_w \tag{18.1}$$

In Eq. (18.1), p_{nw} = pressure in the non-wetting phase; and p_w = pressure in the wetting phase. As defined in Eq. (18.1), it is evident that capillary pressure is related to the wettability of the rock as discussed in Section 18.2.1. For a water-wet rock in an oil/water system, the capillary pressure derived from Eq. (18.1) is:

$$p_c = p_o - p_w \tag{18.2}$$

Similarly, for an oil-wet rock in an oil/water system, the capillary pressure is:

$$p_c = p_w - p_o \tag{18.3}$$

By convention, the capillary pressure in a water-wet rock as represented in Eq. (18.2) is designated as positive. Hence, capillary pressure in an oil-wet rock is designated as negative.

The phenomenon of capillarity in reservoirs can be discussed in terms of capillary pressure as measured in capillary tubes. For a capillary tube, capillary pressure is determined as:

$$p_c = \frac{2\sigma \cos\theta}{r} \tag{18.4}$$

In Eq. (18.4), p_c = capillary pressure, dynes/cm^2; σ = the interfacial tension between the two immiscible phases, dynes/cm; θ = contact angle, degrees; and r = radius of the capillary tube, cm. As shown by Eq. (18.4), capillary pressure is inversely proportional to the radius of the capillary tube. This means that capillary tubes with smaller radii will result in higher capillary pressures if other terms in Eq. (18.4) remain the same. A reservoir can be considered as a porous medium consisting of pores of different sizes. Since the sizes of these pores are at dimensions of capillary tubes, it is easily deduced by application of Eq. (18.4) that capillary pressure will be higher in pores with smaller radii. Capillary pressure can also be related to height above the free water level (FWL) in a reservoir. The FWL is defined as the depth at which oil-water capillary pressure is zero. Applying these definitions, capillary pressure in an oil/water system can be represented also as:

$$p_c = \Delta\rho g H \tag{18.5}$$

In Eq. (18.5), $\Delta\rho = \rho_w - \rho_o$, where ρ_w is the density of water, and ρ_o is the density of oil; g = acceleration due to gravity; and H = height above the FWL. By applying Eqs. (18.4) and (18.5), it can be deduced that reservoirs with relatively high capillary pressure also have large transition zone. Transition zone is defined as height between the FWL and irreducible connate water saturation. The effects of capillary pressure on the interpretation of FWL and transition zone were presented in Chapter 6 (Section 6.5.3).

The capillary pressure curves for a strongly water-wet rock are shown in Figure 18.2. Curve 1 is the imbibition curve and Curve 2 is the drainage curve. Imbibition is described as the process that results in the increase of the saturation of the wetting phase, while drainage is the process that leads to decrease in the saturation of the wetting phase. Figure 18.2 illustrates the path-dependent hysteresis that exists between the imbibition and drainage capillary pressure curves. The capillary pressure curves of an oil-wet rock are shown in Figure 18.3. Note that the capillary pressures are negative by convention. Also, note the hysteresis between the imbibition and drainage capillary pressure curves. The capillary pressure curves for a neutral-wet rock are shown in Figure 18.4.

One of the important applications of capillary pressure curves is the distribution of initial fluid saturations in the reservoir. This application utilizes the dependence of capillary pressure on fluid saturations, the wettability of the reservoir rock, pore distribution, and the sizes of the pores and the pore throat. It is common to use the drainage capillary pressure curve in this application since the reservoir is assumed to be initially filled with water before the migration of oil into it. This essentially assumes a water-wet rock system. This assumption could be in error since it is possible for the wettability of the rock to change with time if the oil contains polar compounds, such as asphaltenes. Capillary pressure data from various core samples can be correlated with the J-function. The J-function is represented as:

$$J(S_w) = \frac{0.217 p_c}{\sigma \cos\theta} \sqrt{\frac{k}{\phi}} \tag{18.6}$$

In Eq. (18.6), $J(S_w)$ = dimensionless function dependent on saturation of the wetting phase; p_c = capillary pressure, psi; σ = interfacial tension, dynes/cm; θ = contact angle, degrees; k = rock permeability, md; and ϕ = porosity, fraction. The J-function can be used to distribute initial fluid saturations in the reservoir as part of the process of reservoir characterization and building a reservoir model for simulation.

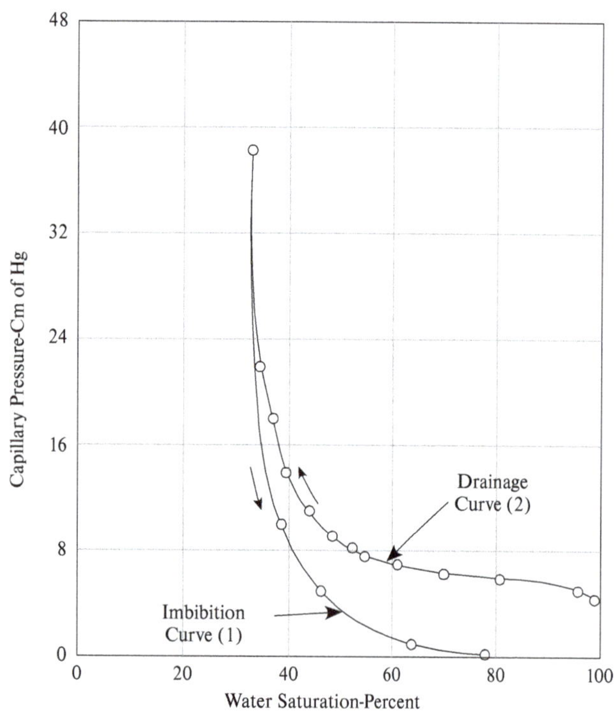

▲ **FIGURE 18.2** Capillary pressure curve for strongly water-wet rock (from Killins et al.[8])

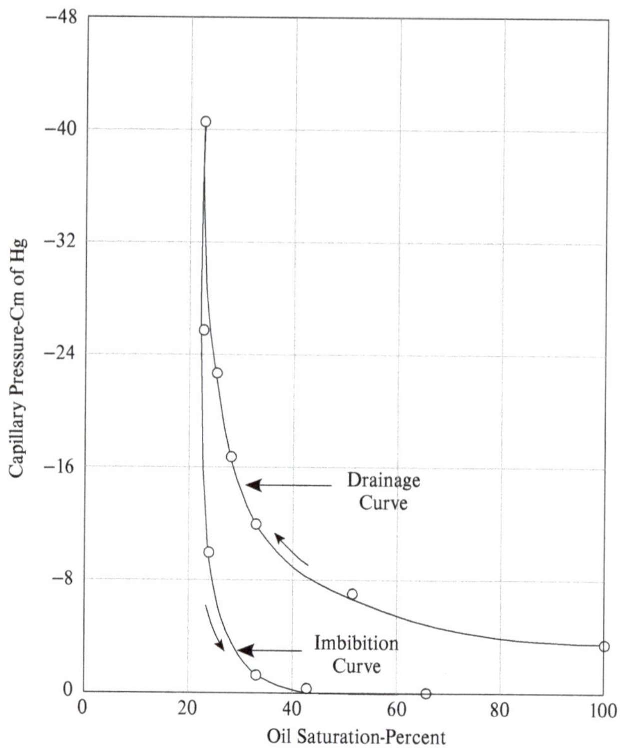

▲ **FIGURE 18.3** Capillary pressure curve for oil-wet rock (from Killins et al.[8])

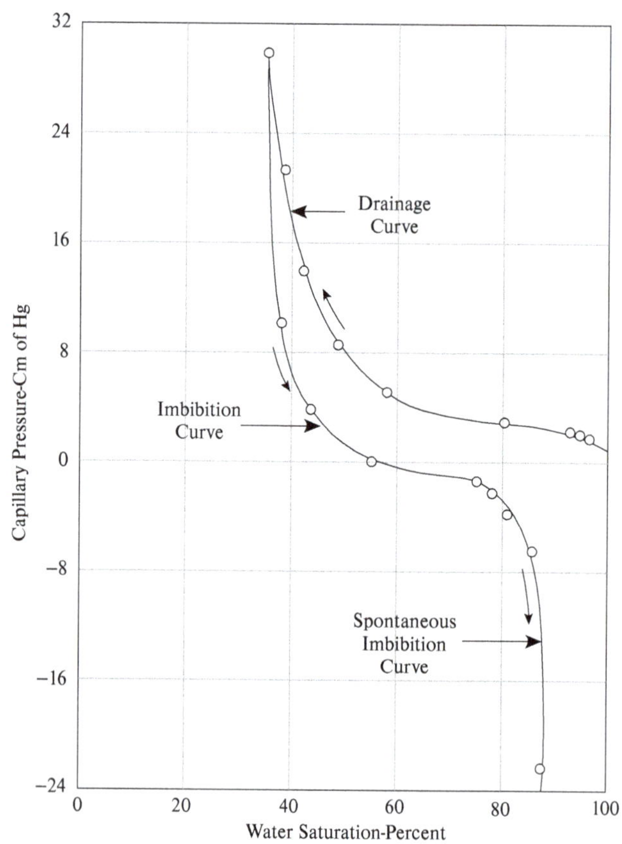

▲ **FIGURE 18.4** Capillary pressure curve for neutral-wet rock (from Killins et al.[8])

18.2.3 Relative Permeability

Relative permeability of a fluid is the ratio of its effective permeability to the absolute permeability of the porous medium. Basic concepts in the application of relative permeability curves were presented in Chapter 3, Section 3.3. Figures 3.4 and 3.5 show the relative permeability curves for an oil-water system, and gas-oil system, respectively. Note that relative permeability curves are affected by the wettability of the rock. One of the most important factors that affect oil recovery in an immiscible displacement process is the magnitude of the relative permeability of the displacing fluid with respect to the displaced fluid. This is shown by the fractional flow equation developed in this chapter in Section 18.3.

18.2.4 Mobility and Mobility Ratio

The mobility of a fluid, λ, is defined as the ratio of its effective permeability to its viscosity. For a fluid, i, mobility, λ_i, is given by:

$$\lambda_i = \frac{k_i}{\mu_i} \tag{18.7}$$

In Eq. (18.7), k_i = effective permeability of the fluid; and μ_i = viscosity of the fluid. For instance, the mobility of water is $\lambda_w = k_w / \mu_w$. Similarly, the mobility of oil is $\lambda_o = k_o / \mu_o$.

Mobility ratio is the ratio of the mobility of the displacing phase to the mobility of the displaced phase. Mobility ratio, M, is represented as:

$$M = \frac{\lambda_D}{\lambda_d} \tag{18.8}$$

In Eq. (18.8), λ_D − mobility of the displacing phase; and λ_d = mobility of the displaced phase. Mobility ratios are defined with reference to saturations at specific locations of the flood front. In this book, mobility ratio is defined with reference to endpoint saturations assuming piston-like displacement. Using this convention, the mobility ratio for water displacing oil in a waterflood is defined as:

$$M = \frac{\left(k_w / \mu_w\right)_{S_{or}}}{\left(k_o / \mu_o\right)_{S_{wi}}} \tag{18.9}$$

In terms of relative permeability, Eq. (18.9) can be written as:

$$M = \frac{\left(k_{rw} / \mu_w\right)_{S_{or}}}{\left(k_{ro} / \mu_o\right)_{S_{wi}}} \tag{18.10}$$

In Eq. (18.10), $\left(k_{rw}\right)_{S_{or}}$ = water relative permeability at residual oil saturation, S_{or}; and $\left(k_{ro}\right)_{S_{wi}}$ = oil relative permeability at initial water saturation, S_{wi}.

From Eq. (18.8), it is evident that when $M \leq 1$, the displacement process is considered favorable because the displaced fluid is more mobile than the displacing fluid. When M is much greater than one, the displacement process is considered to be unfavorable. High mobility ratios could lead to viscous fingering, resulting in significant amounts of bypassed hydrocarbon reserves.

18.2.5 Fluid Displacement Efficiency

Fluid displacement efficiency, E_D, is sometimes called microscopic displacement efficiency. It is defined as the volume of oil displaced from the invaded region divided by the volume of oil initially-in-place in the invaded region. Fluid displacement efficiency is affected by rock wettability, capillary pressure, relative permeability, and mobility ratios of the fluids. If reservoir pressure is maintained at initial conditions and saturation within the invaded region is reduced to residual oil saturation, the displacement efficiency can be estimated with the expression:

$$E_D = 1 - \frac{S_{or}}{S_{oi}} \tag{18.11}$$

In Eq. (18.11), S_{or} = residual oil saturation in the swept (invaded) region; and S_{oi} = initial oil saturation in the swept region.

18.2.6 Volumetric Displacement Efficiency

Volumetric displacement efficiency, E_V, is sometimes termed macroscopic displacement efficiency. It is the fraction of the reservoir volume swept by the displacing fluid. It is composed of two parts, namely, areal sweep efficiency, E_A, and vertical sweep efficiency, E_I. Areal sweep efficiency is the fraction of the reservoir area contacted by the displacing fluid. Vertical sweep efficiency is the fraction of thc vcrtical reservoir cross-section contacted by the displacing fluid. Volumetric displacement efficiency is the product of areal and vertical sweep efficiencies:

$$E_V = E_A \times E_I \tag{18.12}$$

18.2.7 Total Recovery Efficiency

Total recovery efficiency, E_R, is the volume of oil displaced divided by the initial volume of oil in place in the swept portion of the reservoir. Thus, total recovery efficiency is given by:

$$E_R = E_D \times E_V \tag{18.13}$$

18.3 Fractional Flow Equations

The fractional flow equation is used to calculate the flow rate of a fluid as a fraction of the total fluid flow rate when only two fluids are flowing in the reservoir. The flow rate of the fluid at any point in the reservoir depends on its saturation at that point. Since relative permeability of the fluid is dependent on saturation, it follows then that the flow rate of the fluid is dependent on its relative permeability at that point in the reservoir. Fractional flow of a fluid in a reservoir is primarily dependent on its relative permeability but can be affected by capillary and gravity forces. The fractional flow equations developed in this book are based on the assumption of linear flow.

18.3.1 Fractional Flow Equation for Oil Displaced by Water

The fractional flow equation developed in this section is for water displacing oil in an oil-water reservoir.

Applying Eq. (12.16), the flow rate of water in reservoir barrels is represented as:

$$q_w = -1.127 \times 10^{-3} \frac{k_w A}{\mu_w} \left[\frac{\partial p_w}{\partial l} + 0.00694 \rho_w \sin\alpha \right] \tag{18.14}$$

The oil flow rate is represented as:

$$q_o = -1.127 \times 10^{-3} \frac{k_o A}{\mu_o} \left[\frac{\partial p_o}{\partial l} + 0.00694 \rho_o \sin\alpha \right] \tag{18.15}$$

In Eqs. (18.14) and (18.15), q_w, q_o = flow rates of water and oil, respectively, RB/D; k_w, k_o = effective permeabilities of water and oil, respectively, md; μ_w, μ_o = viscosities of water and oil, respectively, cp; p_w, p_o = phase pressures in water and oil, respectively, psia; A = cross-sectional area for fluid flow, ft^2; l = flow path length, ft; ρ_w, ρ_o = densities of water and oil, respectively, at reservoir conditions, lbm/ft^3; and α = dip angle of the reservoir measured counter clockwise from the horizontal to the positive flow path (Figure 18.5).

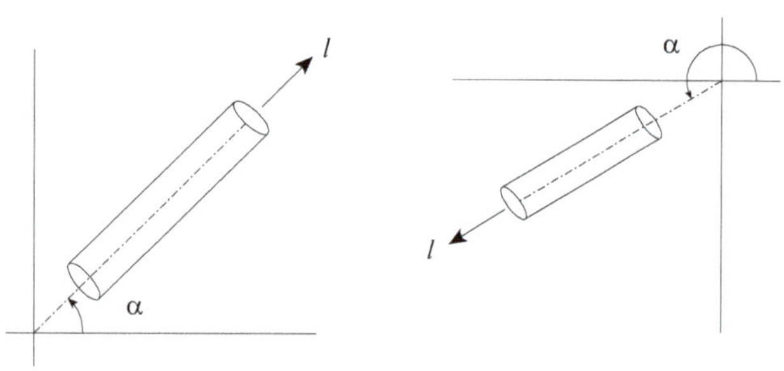

▲ **FIGURE 18.5** Flow path from the horizontal to the positive path

The total reservoir liquid flow rate, q_t, is the sum of the water flow rate, q_w, and the oil flow rate, q_o, expressed as:

$$q_t = q_w + q_o \tag{18.16}$$

The fractional flow of water, f_w, as a fraction of the total flow rate is:

$$f_w = \frac{q_w}{q_t} \tag{18.17}$$

The fractional flow of oil, f_o, as fraction of the total flow rate is:

$$f_o = \frac{q_o}{q_t} \tag{18.18}$$

Rearranging Eqs. (18.17) and (18.18) yields:

$$f_w q_t = q_w \tag{18.19}$$

and

$$(1 - f_w) q_t = q_o \tag{18.20}$$

Substituting Eq. (18.14) into Eq. (18.19) and rearranging gives the following equations:

$$f_w q_t = -1.127 \times 10^{-3} \frac{k_w A}{\mu_w} \left[\frac{\partial p_w}{\partial l} + 0.00694 \rho_w \sin\alpha \right] \tag{18.21}$$

$$-\frac{f_w q_t \mu_w}{k_w A} = 1.127 \times 10^{-3} \left[\frac{\partial p_w}{\partial l} + 0.00694 \rho_w \sin\alpha \right] \tag{18.22}$$

Substituting Eq. (18.15) into Eq. (18.20) and rearranging gives the following equations:

$$(1 - f_w) q_t = -1.127 \times 10^{-3} \frac{k_o A}{\mu_o} \left[\frac{\partial p_o}{\partial l} + 0.00694 \rho_o \sin\alpha \right] \tag{18.23}$$

$$-\frac{(1 - f_w) q_t \mu_o}{k_o A} = 1.127 \times 10^{-3} \left[\frac{\partial p_o}{\partial l} + 0.00694 \rho_o \sin\alpha \right] \tag{18.24}$$

Subtracting Eq. (18.22) from Eq. (18.24) yields:

$$-\frac{(1 - f_w) q_t \mu_o}{k_o A} + \frac{f_w q_t \mu_w}{k_w A} = 1.127 \times 10^{-3} \left[\frac{\partial p_o}{\partial l} - \frac{\partial p_w}{\partial l} + 0.00694 (\rho_o - \rho_w) \sin\alpha \right] \tag{18.25}$$

Applying Eq. (18.2), $p_c = p_o - p_w$, and taking derivatives gives:

$$\frac{\partial p_c}{\partial l} = \frac{\partial p_o}{\partial l} - \frac{\partial p_w}{\partial l} \tag{18.26}$$

Substituting Eq. (18.26) into Eq. (18.25) and simplifying yields:

$$-\frac{q_t\mu_o}{k_oA}+\frac{f_wq_t}{A}\left(\frac{\mu_o}{k_o}+\frac{\mu_w}{k_w}\right) = 1.127\times10^{-3}\left[\frac{\partial p_c}{\partial l}+0.00694(\rho_o-\rho_w)\sin\alpha\right] \tag{18.27}$$

Rearranging Eq. (18.27) gives:

$$\frac{f_wq_t}{A}\left(\frac{\mu_o}{k_o}+\frac{\mu_w}{k_w}\right)=\frac{q_t\mu_o}{k_oA}+1.127\times10^{-3}\left[\frac{\partial p_c}{\partial l}+0.00694(\rho_o-\rho_w)\sin\alpha\right] \tag{18.28}$$

Solving for f_w from Eq. (18.28) gives:

$$f_w=\frac{\frac{\mu_o}{k_o}}{\left(\frac{\mu_o}{k_o}+\frac{\mu_w}{k_w}\right)}+\frac{1.127\times10^{-3}\frac{A}{q_t}\left[\frac{\partial p_c}{\partial l}+0.00694(\rho_o-\rho_w)\sin\alpha\right]}{\left(\frac{\mu_o}{k_o}+\frac{\mu_w}{k_w}\right)} \tag{18.29}$$

Multiplying the right side of Eq. (18.29) by k_o/μ_o yields:

$$f_w=\frac{1}{\left(1+\frac{\mu_w}{k_w}\times\frac{k_o}{\mu_o}\right)}+\frac{1.127\times10^{-3}\frac{k_oA}{\mu_oq_t}\left[\frac{\partial p_c}{\partial l}+0.00694(\rho_o-\rho_w)\sin\alpha\right]}{\left(1+\frac{\mu_w}{k_w}\times\frac{k_o}{\mu_o}\right)} \tag{18.30}$$

Equation (18.30) is the complete form of the fractional flow equation for water flowing linearly in an oil-water reservoir. A quick examination of Eq. (18.30) shows that it has all the factors that affect the flow of water in an oil-water reservoir. These factors are fluid properties (viscosities, densities), rock properties (effective permeabilities, saturations, capillary pressure), total flow rate, and structural inclination of the reservoir (dip angle, α). If it is assumed that fluid properties and total flow rate are constant, then fractional flow of water is only a function of water saturation in the reservoir. A plot of f_w vs. S_w shown in Figure 18.6 is commonly called the fractional flow curve. The fractional flow curve can be used to calculate the flow rate of water in the reservoir at any water saturation.

The effect of capillary pressure can be assumed to be negligible. In such cases, Eq. (18.30) reduces to the following form:

$$f_w=\frac{1}{\left(1+\frac{\mu_w}{k_w}\times\frac{k_o}{\mu_o}\right)}+\frac{7.821\times10^{-6}\frac{k_oA}{\mu_oq_t}(\rho_o-\rho_w)\sin\alpha}{\left(1+\frac{\mu_w}{k_w}\times\frac{k_o}{\mu_o}\right)} \tag{18.31}$$

Equation (18.30) can be simplified further by assuming that both capillary and gravity effects are negligible to give:

$$f_w=\frac{1}{\left(1+\frac{\mu_w}{k_w}\times\frac{k_o}{\mu_o}\right)} \tag{18.32}$$

In terms of relative permeabilities, Eq. (18.32) can be written as:

$$f_w=\frac{1}{\left(1+\frac{\mu_w}{k_{rw}}\times\frac{k_{ro}}{\mu_o}\right)} \tag{18.33}$$

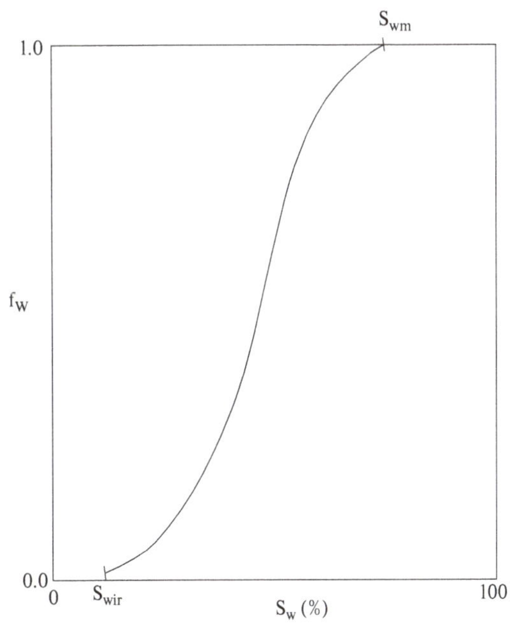

▲ **FIGURE 18.6** Fractional flow curve for water in an oil-water system

18.3.2 Fractional Flow Equation for Oil Displaced by Gas

For reservoirs in which oil is displaced by gas, a comparable set of equations can be written for fractional flow of gas. The fractional flow equation for gas flow is:

$$f_g = \frac{1}{\left(1+\frac{\mu_g}{k_g}\times\frac{k_o}{\mu_o}\right)} + \frac{1.127\times10^{-3}\frac{k_o A}{\mu_o q_t}\left[\frac{\partial p_c}{\partial l}+0.00694\left(\rho_o-\rho_g\right)\sin\alpha\right]}{\left(1+\frac{\mu_g}{k_g}\times\frac{k_o}{\mu_o}\right)} \quad (18.34)$$

Equation (18.34) shows that the flow of a gas in a reservoir is affected by fluid properties, rock properties, total flow rate, and structural inclination of the reservoir. If it is assumed that fluid properties and total flow rate are constant, then fractional flow of gas is only a function of gas saturation in the reservoir. A fractional flow curve of, f_g versus S_g, similar to Figure 18.6 can be generated for gas flow in a reservoir.

If the effects of capillary pressure are assumed to be negligible, Eq. (18.34) reduces to:

$$f_g = \frac{1}{\left(1+\frac{\mu_g}{k_g}\times\frac{k_o}{\mu_o}\right)} + \frac{7.821\times10^{-6}\frac{k_o A}{\mu_o q_t}\left(\rho_o-\rho_g\right)\sin\alpha}{\left(1+\frac{\mu_g}{k_g}\times\frac{k_o}{\mu_o}\right)} \quad (18.35)$$

If the effects of both capillary pressure and gravity are assumed to be negligible, Eq. (18.34) simplifies to the form:

$$f_g = \frac{1}{\left(1+\frac{\mu_g}{k_g}\times\frac{k_o}{\mu_o}\right)} \quad (18.36)$$

In terms of relative permeability, Eq. (18.36) is written as:

$$f_g = \frac{1}{\left(1+\frac{\mu_g}{k_{rg}}\times\frac{k_{ro}}{\mu_o}\right)} \quad (18.37)$$

18.4 The Buckley-Leverett Equation

The Buckley-Leverett[9] equation is based on the principle of conservation of mass for linear flow of a fluid (water or gas) through a reservoir at constant total flow rate. To illustrate the derivation of the Buckley-Leverett equation, the case of water displacing oil is used. Note that the same equation can be developed representing the case for gas displacing oil.

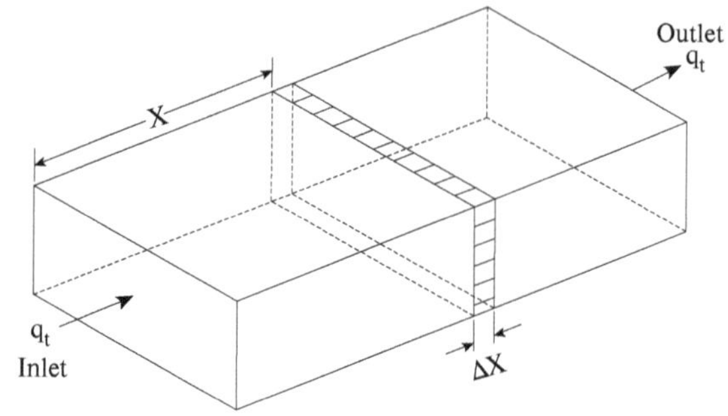

▲ **FIGURE 18.7 Linear reservoir model**

Consider a volume element of a linear reservoir model shown in Figure 18.7. Let the thickness of the element be represented as Δx and located at a distance x from the inlet face of the linear model. A volumetric balance in terms of the water phase (assuming density of water is constant) for the element of the reservoir model can be written as:

$$\begin{bmatrix}\text{Vol. of water flowing into}\\ \text{element in time, }\Delta t\end{bmatrix} - \begin{bmatrix}\text{Vol. of water flowing out of}\\ \text{element in time, }\Delta t\end{bmatrix} = \begin{bmatrix}\text{Accumulation of water}\\ \text{in element, }\Delta x\end{bmatrix} \quad (18.38)$$

Equation (18.38) can be expressed algebraically as:

$$\left[f_w q_t \Delta t\right]_x - \left[f_w q_t \Delta t\right]_{x+\Delta x} = \left[\frac{\phi A \Delta x}{5.615}\Delta S_w\right] \quad (18.39)$$

In Eq. (18.39), f_w = fraction of water in flow stream; q_t = constant total flow rate, RB/D; Δt = time interval, days; ϕ = porosity, fraction; A = cross-sectional area of flow, ft^2; and S_w = water saturation in the element. Rearranging Eq. (18.39) gives:

$$\left[\frac{\phi A}{5.615 q_t}\frac{\Delta S_w}{\Delta t}\right] = -\frac{\left[f_w\right]_{x+\Delta x} - \left[f_w\right]_x}{\Delta x} \quad (18.40)$$

Taking limits as $\Delta t \to 0$ and $\Delta x \to 0$ yields the continuity equation:

$$\frac{\phi A}{5.615 q_t}\left(\frac{\partial S_w}{\partial t}\right)_x = -\left(\frac{\partial f_w}{\partial x}\right)_t \quad (18.41)$$

As stated in Section 18.3.1, the fractional flow of water is a function of water saturation only if fluid properties and total flow rate are constant. By application of chain rule, $f_w = f_w(S_w)$, can be expressed as:

$$\left(\frac{\partial f_w}{\partial x}\right)_t = \left(\frac{\partial f_w}{\partial S_w}\right)_t\left(\frac{\partial S_w}{\partial x}\right)_t \quad (18.42)$$

Substituting Eq. (18.42) into Eq. (18.41) and rearranging gives:

$$\left[\frac{\partial S_w}{\partial t}\right]_x = -\frac{5.615q_t}{\phi A}\left[\left(\frac{\partial f_w}{\partial S_w}\right)_t\left(\frac{\partial S_w}{\partial x}\right)_t\right] \quad (18.43)$$

Equation (18.43) gives water saturation as a function of time at a given location. A more useful equation expressing water saturation as a function of location at a given time can be developed from Eq. (18.43). For any displacement, the distribution of water saturation is a function of both location and time. This is represented as:

$$S_w = S_w(x,t) \quad (18.44)$$

The total derivative of S_w is then:

$$dS_w = \left(\frac{\partial S_w}{\partial x}\right)_t dx + \left(\frac{\partial S_w}{\partial t}\right)_x dt \quad (18.45)$$

Since the focus is on a fixed water saturation, then $dS_w = 0$. And Eq. (18.45) becomes:

$$0 = \left(\frac{\partial S_w}{\partial x}\right)_t dx + \left(\frac{\partial S_w}{\partial t}\right)_x dt \quad (18.46)$$

By rearrangement, Eq. (18.46) becomes:

$$\left(\frac{dx}{dt}\right)_{S_w} = -\frac{\left(\frac{\partial S_w}{\partial t}\right)_x}{\left(\frac{\partial S_w}{\partial x}\right)_t} \quad (18.47)$$

Substituting Eq. (18.47) into Eq. (18.43) gives:

$$\left(\frac{dx}{dt}\right)_{S_w} = \frac{5.615q_t}{\phi A}\left(\frac{\partial f_w}{\partial S_w}\right)_t \quad (18.48)$$

Since the total flow rate is assumed to be constant, then fractional flow of water is independent of time. Hence,

$$\left(\frac{\partial f_w}{\partial S_w}\right)_t = \frac{df_w}{dS_w} \quad (18.49)$$

Equation (18.48) then becomes:

$$\left(\frac{dx}{dt}\right)_{S_w} = \frac{5.615q_t}{\phi A}\frac{df_w}{dS_w} \quad (18.50)$$

Equation (18.50) is the Buckley-Leverett equation. It is also called the frontal advance equation. Integration of Eq. (18.50) yields a useful form of the Buckley-Leverett equation:

$$x = \frac{5.615q_t t}{\phi A}\left(\frac{df_w}{dS_w}\right)_{S_w} \quad (18.51)$$

In Eq. (18.51), x = distance travelled by a fixed saturation in time t, feet; q_t = total flow rate, RB/D; t = time interval, days; ϕ = porosity, fraction; A = cross-sectional area of flow, ft^2; and $(df_w/dS_w)_{S_w}$ = slope of the fractional flow curve at S_w.

A similar equation can be written for gas displacement of oil as follows:

$$x = \frac{5.615q_t t}{\phi A}\left(\frac{df_g}{dS_g}\right)_{S_g} \quad (18.52)$$

In Eq. (18.52), f_g = fractional flow of gas; and S_g = gas saturation. All previous definitions of terms in Eq. (18.51) apply.

Equation (18.51) can be used to calculate the distribution of water saturation as a function of time in a linear reservoir under water injection or aquifer influx. The distance travelled by a given saturation in a specified time interval is proportional to the slope of the fractional flow curve at that saturation assuming the total flow rate and reservoir properties are constant. Using this approach, the distribution of water saturation in the reservoir as a function of time can be calculated by determining the slope of the fractional flow curve at that saturation. However, because of the shape of the fractional flow curve (Figure 18.6), it is possible that two slopes of equal value can exist for two different water saturations. Applying Eq. (18.51), this is interpreted to indicate that two different water saturations can exist at the same location in the reservoir at the same time. The appearance of this contradiction in the application of the frontal advance equation is illustrated in Figure 18.8. Buckley and Leverett[9] recognized that a portion of the saturation distribution curve is imaginary and that the real curve is discontinuous at the flood front. The location of the flood front as determined by material balance is represented in Figure 18.9 by a solid line such that areas A and B are equal. Note the sharp discontinuity of the saturation curve at the flood front as represented in Figure 18.9. This is because capillary and gravity effects were assumed to be negligible. If capillary and gravity effects are considered, the distribution of water saturation at the flood front is more gradual as represented in Figure 18.10.

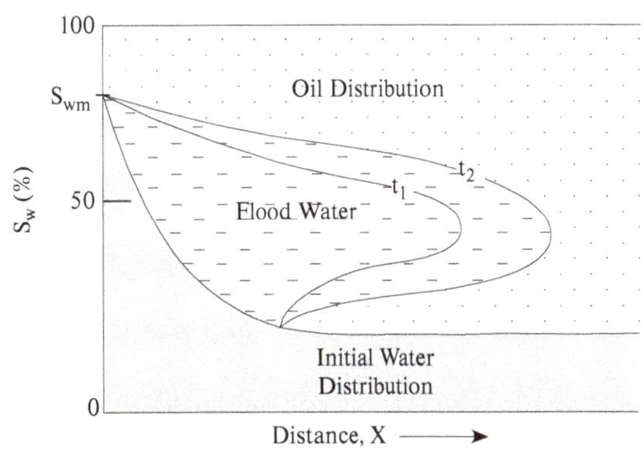

FIGURE 18.8 Saturation distribution based on frontal advance equation

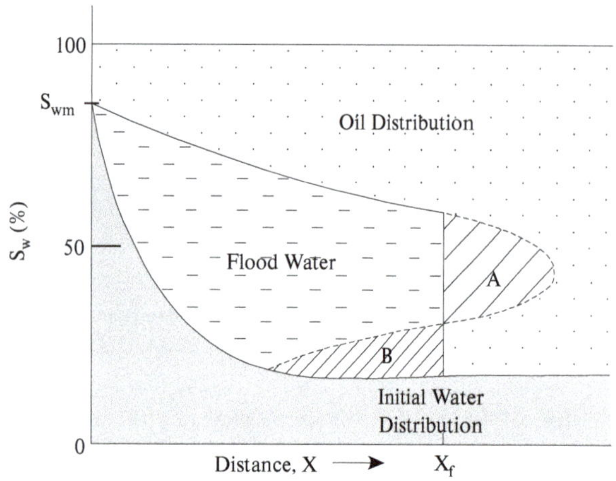

▲ **FIGURE 18.9** **Location of the flood front as determined by material balance**

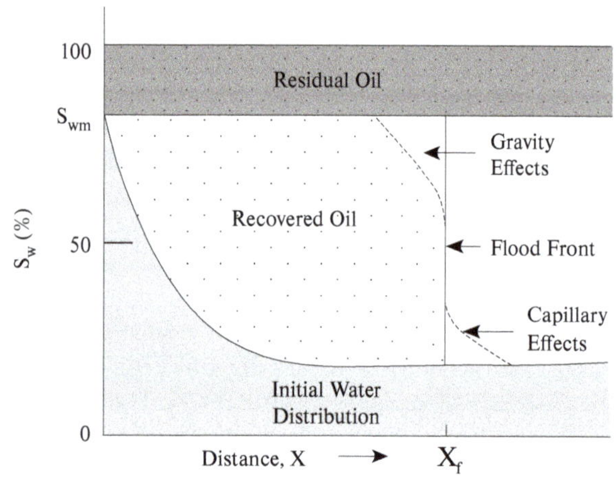

▲ **FIGURE 18.10** **Location of the flood front with capillary and gravity effects**

18.5 The Welge Method

Welge[10] proposed a method for computing oil recovery from gas or water drive that simplified the application of the Buckley-Leverett method. The Welge method is presented with graphical illustrations for the case of water drive in a linear reservoir. The graphical illustrations can be replicated for gas drive by simply replacing f_w and S_w with f_g and S_g, respectively.

18.5.1 Water Saturation at the Flood Front

Water saturation at the flood front, S_{wf}, can be determined graphically using the Welge method by drawing a straight line from initial water saturation tangent to the fractional flow curve as shown in Figure 18.11. If the initial water saturation is greater than the irreducible water saturation (i.e., $S_{wi} > S_{wir}$), the tangent line is drawn from the initial water saturation as shown in Figure 18.12.

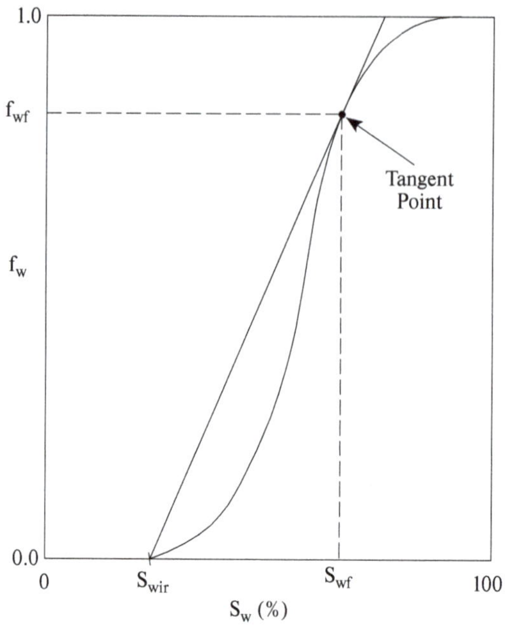

▲ **FIGURE 18.11** **Fractional flow curve with application of the Welge method**

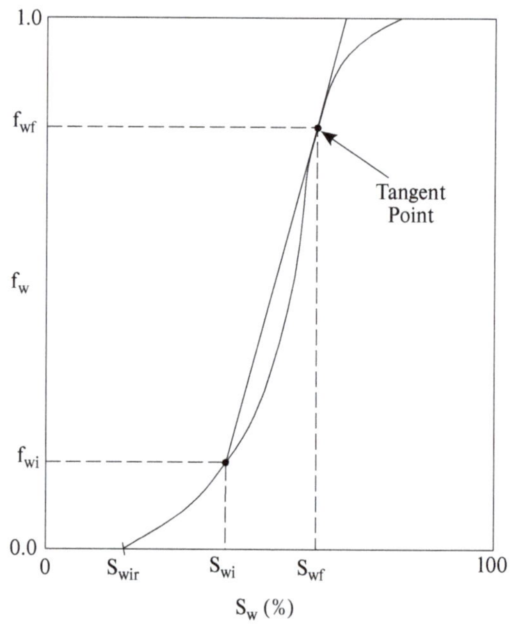

▲ **FIGURE 18.12** **Fractional flow curve with application of Welge method for $S_{wi} > S_{wir}$**

18.5.2 Average Water Saturation Behind the Flood Front

By extending the tangent line drawn to the fractional flow curve as shown in Figures 18.11 or 18.12 to the point where $f_w = 1.0$, the average water saturation, $\bar{S}_w$, behind the flood front can be determined as shown in Figure 18.13. At water breakthrough, $\bar{S}_w = \bar{S}_{wbt}$, where $\bar{S}_{wbt}$ is the average water saturation in the reservoir at water breakthrough.

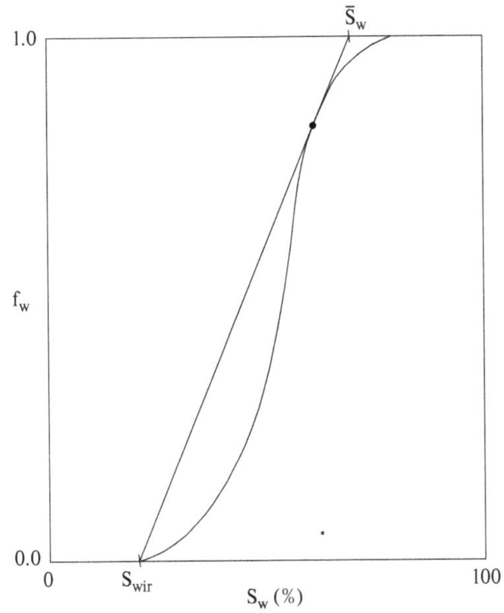

▲ **FIGURE 18.13** **Fractional flow curve with application of Welge method for average water saturation**

18.5.3 Average Water Saturation after Water Breakthrough

The average water saturation after water breakthrough is determined as shown in Figure 18.14 by drawing a tangent line to the fractional flow curve at water saturation, S_{w2}, greater than S_{wf} but less than the maximum water saturation S_{wm} (that is, $S_{wf} < S_{w2} < S_{wm}$). The water saturation, S_{w2}, is the saturation at the outlet end of the linear system after water breakthrough with the corresponding fractional flow of water denoted as f_{w2}. By extending the tangent line to the point where $f_w = 1.0$, the average water saturation, $\overline{S}_{wabt}$, in the system after water breakthrough is determined.

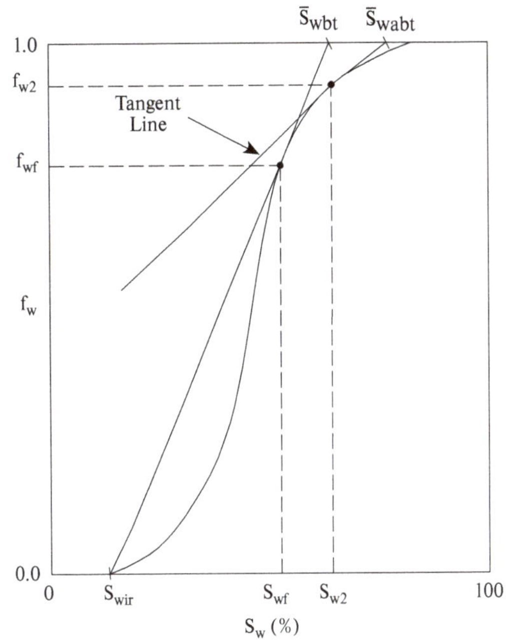

▲ **FIGURE 18.14** **Application of Welge method for average water saturation after water breakthrough**

EXAMPLE 18.1 Calculation of fractional flow curve and water saturations based on the fractional flow curve

Problem

The relative permeability ratio of a linear reservoir as a function of water saturation is given in Table 18.1. Plot the fractional flow curve. Using the fractional flow curve, calculate the following:

1. Average water saturation and fractional flow of water at water breakthrough.
2. Displacement efficiency at water breakthrough.
3. Average water saturation and fractional flow of water after water breakthrough.

Assume capillary pressure and gravity effects are negligible. Other reservoir and fluid properties are as follows:

Porosity, ϕ	0.25
Initial water saturation, S_w	0.20
Viscosity of water, μ_w	0.95 cp
Viscosity of oil, μ_o	1.60 cp

TABLE 18.1 | Data for Example 18.1

S_w	$\frac{k_{ro}}{k_{rw}}$
0.20	∞
0.25	110.20
0.30	34.40
0.35	12.20
0.40	5.60
0.45	2.80
0.50	1.40
0.55	0.75
0.60	0.35
0.65	0.12
0.70	0.02
0.75	0.00

Solution

Step 1: Calculate the fractional flow of water, f_w.

Since capillary and gravity effects are negligible, Eq. (18.33) is used to calculate the fractional flow of water as a function of water saturation. The calculated data are shown in Table 18.2.

Step 2: Plot the fraction flow curve.

The data in Table 18.2 are used to plot the fractional flow curve as shown in Figure 18.15.

Step 3: Determine S_{wf}, f_{wf}, and $\bar{S}_{wbt}$.

Draw a straight line from the initial water saturation tangent to the fractional flow curve as shown in Figure 18.15. Extend this line to the point where, $f_w = 1.0$. Read values for, S_{wf}, f_{wf}, and $\bar{S}_{wbt}$, as shown in Figure 18.15. From Figure 18.15, $S_{wf} = 0.63$, $f_{wf} = 0.88$, and $\bar{S}_{wbt} = 0.68$.

Step 4. Calculate displacement efficiency, E_D, at water breakthrough.

By definition, $E_D = \dfrac{S_{wbt} - S_{wi}}{1 - S_{wi}}$.

$$E_D = \frac{0.68 - 0.20}{1 - 0.20} = 0.60$$

Step 5. Determine fractional water flow and water saturations after water breakthrough.

Expand the fractional flow curve as shown in Figure 18.16. Draw tangents at saturations greater than S_{wf}. Read the fractional flow of water and average saturations as shown in Figure 18.16. For instance, at $S_{w2} = 0.68$, $f_{w2} = 0.97$, and $\bar{S}_{wabt} = 0.71$. Note that several average water saturations after water breakthrough can be determined by drawing additional tangents at saturations greater than S_{wf}.

TABLE 18.2 Plot Data for Example 18.1

S_w	f_w
0.20	0.00
0.25	0.02
0.30	0.05
0.35	0.12
0.40	0.23
0.45	0.38
0.50	0.55
0.55	0.69
0.60	0.83
0.65	0.93
0.70	0.99
0.75	1.00

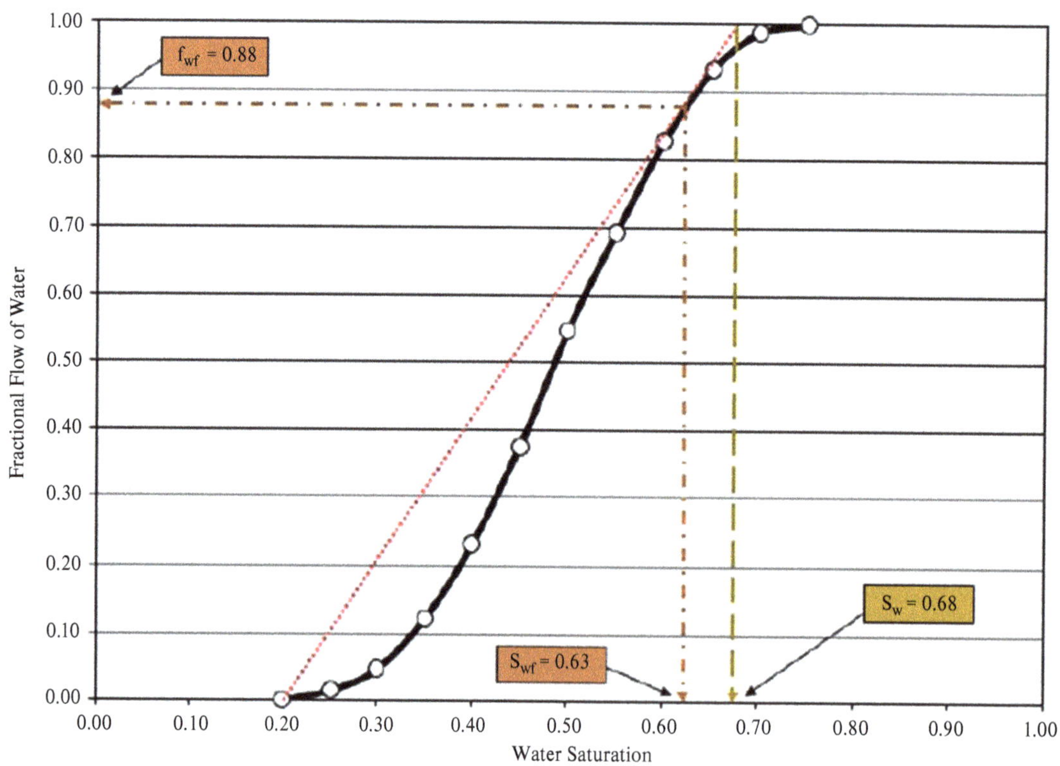

▲ **FIGURE 18.15** Fractional flow curve for Example 18.1

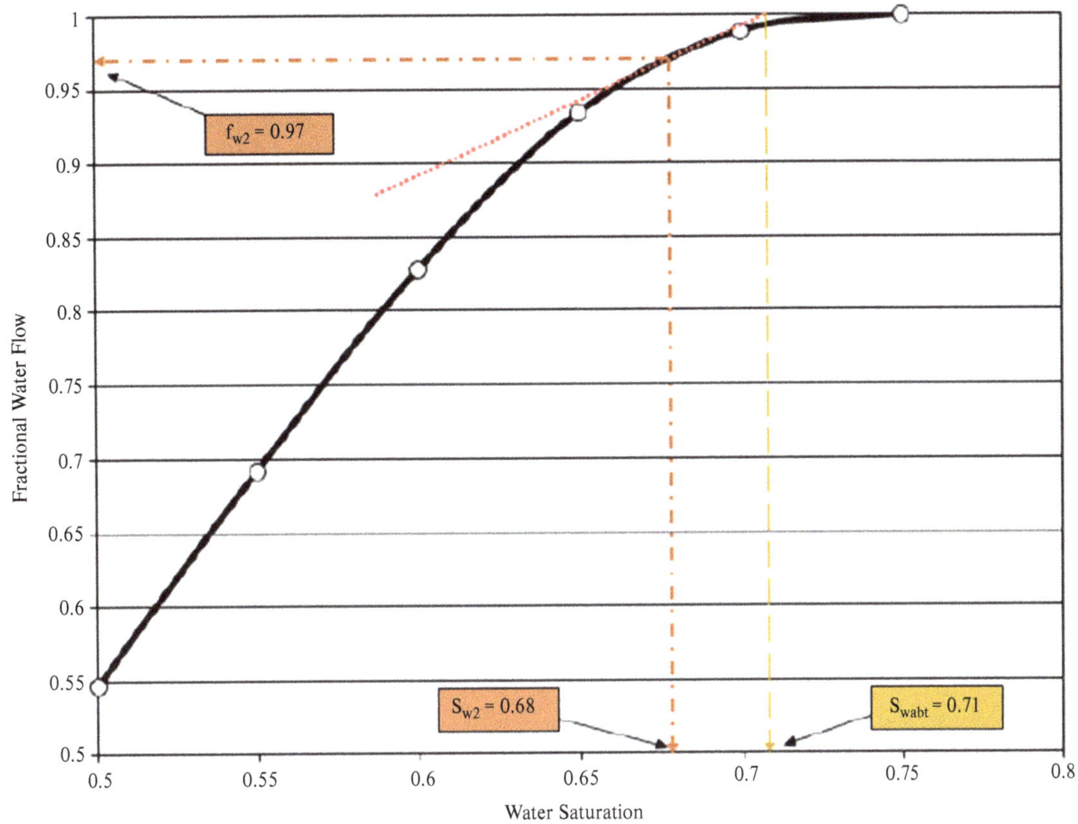

▲ **FIGURE 18.16** **Expanded fractional flow curve for Example 18.1**

End of Example 18.1

18.6 Summary

The treatment of immiscible displacement in this chapter is by no means exhaustive. The main objective was to introduce the engineer to basic concepts and some of the historical theoretical developments in immiscible displacements. In the process, several important terms were introduced in discussing the importance of rock wettability, capillary pressure, relative permeability, mobility ratio, and displacement efficiency in immiscible displacements. The fractional flow equation was developed to convey the impact of fluid and rock properties and reservoir geometry on immiscible displacement. In fact, this relatively simple equation can be used to discuss the fractional flow of water or gas in many waterflood or gasflood projects. Finally, the Welge method was introduced to illustrate determination of several terms, such as saturations at the flood front, average saturations behind the flood front, saturations at breakthrough, and saturations after breakthrough. These terms are useful in discussing performance of waterflooding or gasflooding on a wider scale that may involve application of reservoir simulation. For this reason, there was no attempt made in this chapter to present immiscible displacements in more rigorous and quantitative terms.

Nomenclature

A	cross-sectional area to flow, ft^2
E_A	areal sweep efficiency
E_D	fluid displacement efficiency
E_I	vertical sweep efficiency
E_R	total recovery efficiency
E_v	volumetric sweep efficiency
f_g	fractional flow of gas
f_o	fractional flow of oil
f_w	fractional flow of water
f_{wf}	fractional flow of water at the flood front
f_{wf2}	fractional flow of water at the outlet after water breakthrough
g	acceleration due to gravity
H	height above the free water level
k	absolute permeability, md
k_g	effective gas permeability, md
k_o	effective oil permeability, md
k_w	effective water permeability, md
k_{rg}	relative gas permeability
k_{ro}	relative oil permeability
k_{rw}	relative water permeability
l	flow path length, ft.
M	mobility ratio
p_c	capillary pressure, psia
p_{nw}	pressure in the non-wetting phase, psia
p_w	pressure in the wetting phase or water, psia
p_o	pressure in the oil phase, psia
q_o	oil flow rate, RB/D
q_t	total flow rate, RB/D
q_w	water flow rate, RB/D
r	radius of the capillary tube, cm
S_{oi}	initial oil saturation, fraction
S_{or}	residual oil saturation, fraction
S_w	water saturation, fraction
S_{w2}	water saturation at the outlet after water breakthrough, fraction
S_{wi}	initial water saturation, fraction
S_{wm}	maximum water saturation, $S_{wm} = 1 - S_{or}$, fraction
S_{wir}	irreducible water saturation, fraction
$\bar{S}_w$	average water saturation, fraction
$\bar{S}_{wabt}$	average water saturation after water breakthrough, fraction
$\bar{S}_{wbt}$	average water saturation at water breakthrough, fraction
S_{wf}	water saturation at the flood front, fraction
t	time, days
α	dip angle of reservoir or formation, degrees
ϕ	porosity, fraction
μ	fluid viscosity, cp
μ_g	gas viscosity, cp
μ_o	oil viscosity, cp
μ_w	water viscosity, cp
ρ	fluid density, lbm/ft^3
ρ_g	gas density, lbm/ft^3
ρ_o	oil density, lbm/ft^3
ρ_w	water density, lbm/ft^3
θ	contact angle, degrees
λ	fluid mobility, md/cp
σ	interfacial tension, dynes/cm

REFERENCES

1. Treiber, L.E., Archer, D.L., and Owens, W.W.: "A Laboratory Evaluation of the Wettability of Fifty Oil-Producing Reservoirs," *SPEJ* (Dec. 1972) 531-540.
2. Amott, E.: "Observations Relating to the Wettability of Porous Rock," *Trans.* AIME, (1959) 216, 156-192.
3. Donaldson, E.C., Thomas, R.D., and Lorenz, P.B.: "Wettability Determination and Its Effect on Recovery Efficiency," *SPEJ* (March 1969) 13-20.
4. Morrow, N.R.: "Wettability and Its Effect on Oil Recovery," *JPT* (Dec. 1990) 1476-1484.
5. Owens, W.W., and Archer, D.L.: "The Effect of Rock Wettability on Oil-Water Relative Permeability Relationships," *JPT* (July 1971) 873-878.
6. Brown, R.J., and Fatt, I.: "Measurements of Fractional Wettability of Oilfield Rocks by Nuclear Magnetic Relaxation Method," *Trans.* AIME (1956) 207, 262-264.

7. Agbalaka, C., Dandekar, A.Y., Patil, S.L., Khataniar, S., and Hemsath, J.R.: "The Effect of Wettability on Oil Recovery: A Review," paper SPE 114496 presented at the 2008 SPE Asia Pacific Oil & Gas Conference and Exhibition, Perth, Australia, (October 20-22, 2008).
8. Killins, C.R., Nielson, R.F., and Calhoun, J.C., Jr.: "Capillary Desaturation and Imbibition in Rocks," Prod. Monthly (Feb. 1953) 18, 2, 30-39.
9. Buckley, S.E., and Leverett, M.C.: "Mechanism of Fluid Displacements in Sands," *Trans.* AIME (1942) 107, 107-116.
10. Welge, H.J.: "A Simplified Method for Computing Oil Recovery by Gas or Water Drive," *Trans.* AIME (1952) 195, 91-98.

ADDITIONAL REFERENCE

- Willhite, G.P.: *Waterflooding,* SPE Textbook Series, Vol. 3, Richardson, Texas (1986).

PROBLEMS

18-1. The interfacial tension between oil and water in a core plug was measured to be 2.65 dynes/cm. If the contact angle between the fluids and the rock surface was measured to be 65 degrees, and the average radius of the pores was measured to be 0.254 cm, calculate the capillary pressure in the core plug.

18-2. In a waterflood project, the rock and fluid properties of reservoir oil and injection water were measured to be:

Oil viscosity, μ_o	1.75 cp
Water viscosity, μ_w	1.15 cp
Oil relative permeability at initial water saturation, $(k_{ro})_{S_{wi}}$	0.70
Water relative permeability at residual oil saturation, $(k_{rw})_{S_{or}}$	0.65

Calculate:

a. The mobility ratio of the displacement process

b. Is the mobility ratio favorable or unfavorable?

18-3. Consider an experiment where capillary tubes of various diameters were placed in a large beaker of water and the other ends of the capillary tubes were open to air. The surface tension between air and water was measured to be 72 dynes/cm and the contact angle between air and water was measured to be zero degree. Assume water density was 1 gm/cm^3. The radii of the capillary tubes were designated as shown below:

Tube	Radius of Capillary Tube (cm)
A	1.0000
B	0.0500
C	0.0010
D	0.0005
E	0.0001

Calculate:

a. The height water will rise in each capillary tube in feet

b. The capillary pressure in each tube in psi

c. If the contact angle between air and water was 100 degrees, what is the height of water in Tube C in inches?

18-4. Five core samples from a sandstone reservoir were used in a laboratory capillary pressure measurement. The capillary pressures measured on the five core samples versus water saturation are given in Table 18.3. In the laboratory experiments, the interfacial tension of the air-brine system was measured to be 73 dynes/cm and the interfacial tension of the reservoir oil-water system was measured to be 32.5 dynes/cm. The contact angle for the air-brine system was measured to be zero degrees and for the reservoir oil-water system was measured to be 30 degrees. Using the data in Table 18P.1, prepare the following plots:

a. Semi-log plot of permeability versus water saturation at each capillary pressure

b. Capillary pressure curve versus water saturation at reservoir conditions at geometric mean permeability of 200 md.

TABLE 18P.1 Data for Problem 18-4

Core Sample	Permeability (md)	Capillary Pressure				
		75 psi	50 psi	25 psi	10 psi	5 psi
		Water Saturation (%)	Water Saturation (%)	Water Saturation (%)	Water Saturation (%)	Water Saturation (%)
A	470	18.5	22.0	29.0	39.0	49.5
B	300	22.5	25.5	34.0	45.5	56.0
C	115	30.0	34.0	41.0	53.5	65.0
D	50	36.0	40.5	51.0	64.0	77.0
E	27	41.0	44.0	55.0	69.0	81.5

18-5. Suppose a sandstone reservoir was waterflooded for many years such that the producing water-cut is 100%. Calculate the following efficiencies for the waterflood process:

a. Fluid displacement efficiency, if the initial oil saturation at the start of the waterflood is 75% and the residual oil saturation at the end of the waterflood is 20%.

b. Volumetric displacement efficiency, if the vertical and areal sweep efficiencies of the waterflooding process are 70% and 78%, respectively.

c. Total recovery efficiency.

18-6. A horizontal reservoir is being waterflooded. The relative permeability data for the oil-water system are given in Table 18P.2. If oil viscosity is 5 cp and water viscosity is 0.5 cp, calculate and plot the fractional flow curve of water for the displacement process. Assume capillary pressure and gravity effects are negligible.

TABLE 18P.2 Relative Permeability Data for Problem 18-6

Water Saturation, S_w(%)	Water Relative Permeability, k_{rw}	Oil Relative Permeability, k_{ro}
20	0.000	0.800
25	0.002	0.610
30	0.009	0.470
35	0.020	0.370
40	0.033	0.285
45	0.051	0.220
50	0.075	0.163
55	0.100	0.120
60	0.132	0.081
65	0.170	0.050
70	0.208	0.027
75	0.251	0.100
80	0.300	0.000

18-7. The relative permeability data for an oil/water system in a reservoir are given in Table 18P.3. Other rock and fluid data for the reservoir are given as follows:

Absolute permeability, k_a	45 md
Formation porosity, ϕ	0.18
Initial water saturation, s_{wi}	0.30
Water viscosity, μ_w	0.62 cp
Oil viscosity, μ_o	2.48 cp
Oil FVF, B_o	1.37 RB/STB
Water FVF, B_w	1.04 RB/STB
Oil gravity, γ_o	0.80
Water gravity, γ_w	1.03
Density of pure water, ρ_w	62.4 lbm/ft^3
Total flow rate, q_t	10,000 RBPD
Length of the reservoir, L	1625 feet
Area of the reservoir, A	500,000 ft^2
Reservoir dip angle, α	30 degrees

Calculate and plot the water fractional flow curve for the reservoir assuming capillary pressure gradients are negligible.

TABLE 18P.3 Relative Permeability Data for Problem 18-7

Water Saturation, S_w(%)	Oil Relative Permeability, k_{ro}	Water Relative Permeability, k_{rw}
30	0.940	0.000
40	0.800	0.040
50	0.440	0.110
60	0.160	0.200
70	0.045	0.300
80	0.000	0.440

18-8. Using the water fractional flow curve plotted in Problem 18-7, calculate:

a. Fractional water flow at breakthrough

b. Average water saturation at breakthrough

c. Displacement efficiency at breakthrough

d. Cumulative volume of water injected at breakthrough

e. Fractional water flow and average water saturation after water breakthrough at waterfront saturation of 70%.

18-9. Applying the reservoir data given in Problem 18-7 and the fractional flow curve plotted from the same problem, calculate:

a. Time required for water breakthrough assuming injection only at the inlet end of the reservoir.

b. Time required for water saturation front at the outlet face of the reservoir (end of the reservoir) to reach 70%.

18-10. A waterflood experiment was conducted on a core sample from a sandstone reservoir. The water fractional flow data calculated from the flood test are given in Table 18P.4. The flood test was conducted with the core sample saturated with refined oil whose viscosity was measured to be 10.4 cp. The viscosity of injected water was 1.04 cp. Later, the viscosity of the actual reservoir oil was measured to be 4.5 cp. Assuming negligible capillary pressure and gravity effects, calculate:

a. The modified water fractional flow data for the waterflood test based on the viscosity of the actual reservoir oil.

b. What is the error in calculated oil recovery (as a fraction of reservoir pore volume) at water breakthrough if the modified fractional flow curve was not used?

TABLE 18P.4 Water fractional flow data for Problem 18-10

Water Saturation, (%)	Fractional Flow Curve, f_w
10	0.000
20	0.067
30	0.187
40	0.408
50	0.752
60	0.905
70	0.978
80	1.000

CHAPTER

19 Secondary Recovery Methods

19.1 Introduction

Secondary recovery methods are defined as processes that can be used to increase hydrocarbon recovery from the reservoir beyond primary recovery. Typical secondary recovery methods are considered to be intervention methods implemented during the primary recovery period to improve projected low hydrocarbon recovery from the primary process. The presentation of secondary recovery methods in this chapter is focused on waterflooding and gasflooding. Waterflooding of hydrocarbon reservoirs is generally an immiscible displacement process since water is virtually immiscible with hydrocarbons even at high pressures. Gasflooding can be an immiscible or miscible displacement process, depending on pressure and temperature of the reservoir, the type of injection gas, the composition of the injection gas, the composition of the reservoir fluid, the nature of reservoir heterogeneities, and other factors. Waterflooding and gasflooding are presented in this chapter as immiscible displacement processes. Gasflooding, as a miscible displacement process, is presented in Chapter 21 as an enhanced oil recovery process.

In this chapter, secondary recovery methods are presented in descriptive terms under the categories of well patterns, well surveillance, reservoir performance monitoring, and project management. The application of prediction methods such as the Stiles[1] method, Dykstra-Parsons[2] method, Craig-Geffen-Morse[3] method, and other methods for the calculation of recoveries from waterflooding are not presented in this book. Similar methods, such as the Welge[4] method for gasflooding are also not presented. For flood design and calculation of flood performance, these analytical methods have been rendered virtually obsolete by faster and more robust methods based on reservoir simulation. It may be necessary to use these analytical methods to gain a fundamental understanding of immiscible displacement processes, but the same knowledge can be gained by using very simple reservoir models to explore key factors (such as rock and fluid properties, reservoir heterogeneities, well pattern, injection rates, production rates, etc.) that influence hydrocarbon recovery under secondary recovery processes. The application of reservoir simulation for predicting reservoir performance under secondary recovery processes is presented in Chapter 23.

19.2 Waterflooding

Waterflooding is the most common secondary recovery method used worldwide to improve oil recovery from reservoirs. The performance of a waterflood project in terms of oil recovery depends on many factors, such as rock wettability, rock and fluid properties, formation heterogeneities, flood patterns, composition of injection water, water injection rates, fluid production rates, etc. These factors should be considered in the design and installation of a waterflood project.

19.2.1 Waterflood Patterns

Waterflood patterns are based on the arrangement of injectors and producers in a waterflooded reservoir. The patterns are generally designed to improve areal sweep efficiency. It is important to note that waterflood patterns evolved from conversion of onshore oil fields developed with regular well spacing from primary production to secondary production by waterflooding. The criteria for converting these oil fields to waterflooding were to minimize drilling of new wells, maximize water injection, and improve oil recovery. Consequently, the waterflood pattern installed in these oil fields depended on the density of existing wells, reservoir properties, and projected rate of oil recovery. Generally, pattern flooding is not efficient for development of deepwater reservoirs or reservoirs found in similar environments with high development costs. In these high cost environments, the ratio of water injectors to producers, and the location of these wells should be optimized through the use of reservoir simulation models.

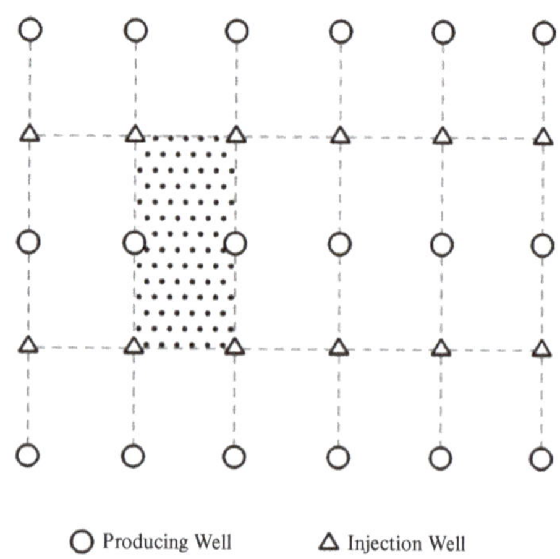

▲ FIGURE 19.1 Direct line drive pattern

Some of the common waterflood patterns are the direct line drive, the staggered line drive, the five-spot, and the nine-spot patterns. The direct line drive pattern is shown in Figure 19.1. For this pattern, the injectors and producers offset each other directly. This could lead to early water breakthrough if the distances between the injectors and producers are small. The impact of early water breakthrough on sweep efficiency for the direct line drive pattern can be reduced somewhat by using the staggered line drive pattern, as shown in Figure 19.2. As seen in Figure 19.2, the alternate rows of injectors and producers are displaced so that the wells do not offset each other directly. Note that the ratio of injectors to producers in both patterns is one. The five-spot pattern is shown in Figure 19.3. The five-spot pattern has been used widely in many field applications and model studies. This pattern generally results in good areal sweep efficiency. The ratio of injectors to producers in a five-spot pattern is also one. The nine-spot pattern is shown in Figure 19.4. This pattern is used in low permeability reservoirs where high water injection volumes are desired. The ratio of injectors to producers in a nine-spot pattern is three. There are numerous variations of these basic flood patterns, such as inverted five-spot and inverted nine-spot patterns. It is not necessary to replicate these other patterns in this book since the engineer can devise flood patterns that are best suited to the development of a particular reservoir with modern reservoir simulation tools.

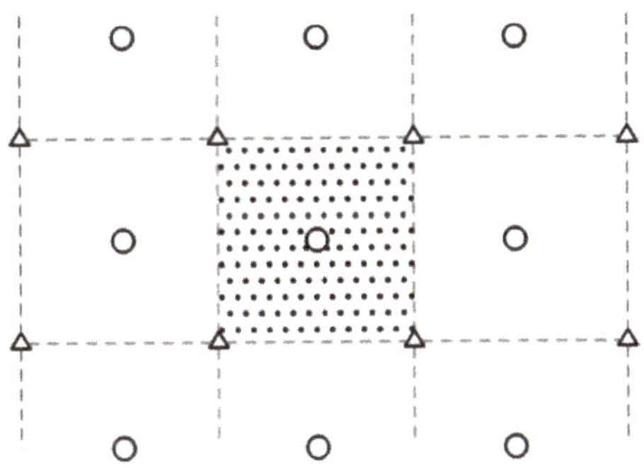

▲ **FIGURE 19.2 Staggered line drive pattern**

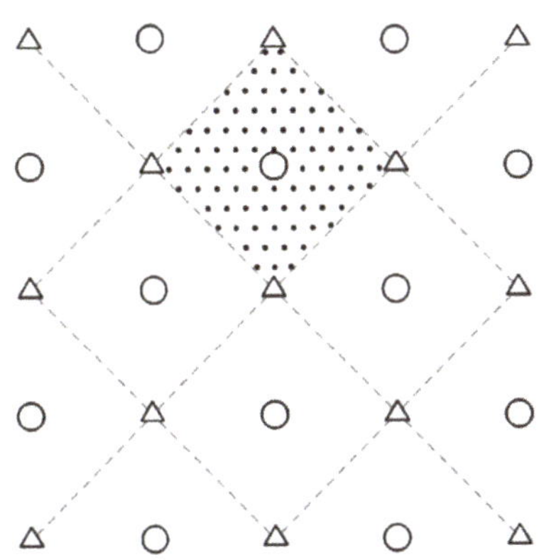

▲ **FIGURE 19.3 Normal five-spot pattern**

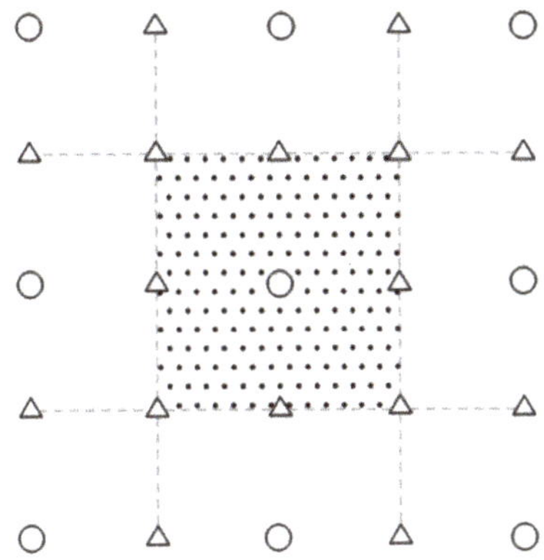

▲ **FIGURE 19.4 Normal nine-spot pattern**

Peripheral flooding is developed when injectors are located around part or the entire boundary of the reservoir, as shown in Figure 19.5. If the injectors are located along one side of the reservoir or the middle of the reservoir, it is called a line flood. Peripheral flooding is generally used in reservoirs where the objectives are to maintain reservoir pressure, minimize water production, and improve oil recovery.

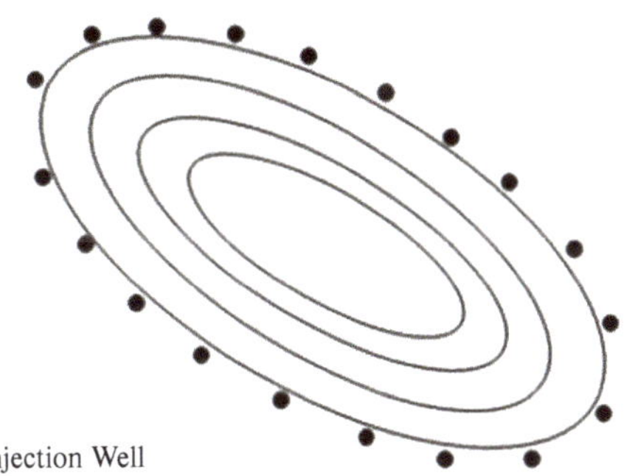

▲ **FIGURE 19.5 Peripheral flood pattern**

19.2.2 Waterflood Design

The steps recommended in this book for the design of a waterflood are based entirely on the application of numerical reservoir simulation models. This approach reflects the experience of the author that numerical reservoir modeling of a waterflood is superior to all the earlier prediction methods (Stiles method, Dykstra-Parsons method, Craig-Geffen-Morse method, etc.) for waterflood design. Each of these earlier prediction methods made key assumptions (such as linear flow, layered system, no crossflow, piston-like displacement, constant sweep efficiency, etc.) which limit their applicability in the design of a waterflood. Furthermore, none of these methods can incorporate the impact of reservoir heterogeneities, such as fracturing, faults, permeability barriers, spatial variations of fluid and rock properties, etc. in a waterflood design. Numerical reservoir simulation of the waterflooding process can be applied without any of these assumptions, while incorporating any geologic features that may impact the performance of the waterflood. Numerical reservoir modeling also allows the waterflood design to be subjected to more rigorous tests on uncertainty and sensitivity of key variables that could impact the performance of the waterflood.

19.2.3 Recommended Steps in Waterflood Design

1. Construct a geologic model of the reservoir or project area.
   a. Identify and include all faults and other structural features that may affect fluid flow in a geologic framework model of the reservoir.
   b. Identify and include all reservoir heterogeneities, such as permeability barriers, reservoir unconformities, etc. in the geologic model.
   c. Perform characterization of the geologic model to include areal and vertical variation of reservoir properties such as facies, net pay, porosity, permeability, and saturations. Additional details on geologic modeling and reservoir characterization are presented in Chapter 22.
2. Analyze rock/fluid properties data.
   a. Determine mineralogy of reservoir rocks.
   b. Conduct studies on compatibility of injection water with reservoir rocks.
   c. Determine Pressure-Volume-Temperature (PVT) properties of reservoir fluids, including saturation pressures and oil viscosity.
3. Construct reservoir flow model with data from Steps 1 and 2. (Note: Reservoir simulation concepts are presented in Chapter 23.)
   a. If the reservoir had prior production history, history-match reservoir model to obtain the current depleted state of the reservoir before the start of waterflooding.
   b. At the completion of Step 3a, determine gas cap size if a gas cap is present and extent of aquifer influx if reservoir has an active aquifer.
   c. Compare pressure distribution in model after Step 3a to actual pressure data. Identify state of reservoir depletion.
   d. Explore distribution of fluid saturation in the model after Step 3a to identify potential undepleted areas of the reservoir that would be targeted for waterflooding.
4. Run prediction cases.
   a. Run a base prediction case assuming continuation of current depletion strategy.
   b. Run several predictive cases assuming the reservoir is waterflooded with different numbers and locations of water injectors and producers.
   c. Compare results from Steps 4a and 4b. If the waterflood cases indicate substantial improvement in total oil recovery, then proceed to Step 5.
5. Optimize waterflood design.
   a. Choose the cases from Step 4b with the "best" reservoir performance and optimize the numbers and locations of injectors and producers.
   b. Optimize injection and production rates for each case.
   c. Rank the cases by incorporating project economics. Economic evaluation methods are presented in Chapter 25.
6. Perform sensitivity analyses.
   a. Select two or three cases from Step 5 and perform sensitivity analyses on key reservoir and operational variables of the waterflood design.
   b. Repeat economic analyses of the entire project based on results from Step 6a.
7. Conduct a pilot waterflood project.
   a. For large waterflood projects, it is recommended that a pilot project should be installed. The pilot waterflood project would provide data, such as water injection rates, that could be used to adjust the waterflood design and improve predictions before implementation of the project over the entire reservoir.

19.2.4 Waterflood Management

In many cases, proper management of a waterflood project is as important as designing the waterflood. A well-designed waterflood can become an engineering and economic failure if several important surveillance practices are not implemented in the management of the project. Monitoring and surveillance practices are presented for injection wells, injection water, production wells, production water, facilities, and the reservoir.

A. **Injection Wells.** One of the key determinants of a successful waterflood project is achieving and maintaining the desired levels of water injection rates. Water injection wells may exhibit reduced injectivity over time due to scale buildup and plugging of the formation from solid deposits. It is important to test the injection wells periodically to ensure that injectivity had not declined substantially. The most common test that can be used to gauge well injectivity is the fall-off test. Pressure fall-off tests on injectors are similar and are conducted in the same manner as buildup tests on producers. Data from pressure fall-off tests are analyzed with the same techniques demonstrated in Chapter 13 for buildup tests. Some caution should be exercised in analyzing pressure fall-off test data since multiple fluid boundaries may exist around the injector and create complications in the analysis.[5-7]

   The condition of an injection well can be monitored periodically by preparing a Hall[8] plot on the performance of the injector. The Hall plot can be prepared from data routinely collected as part of monitoring the performance of the injector. These data include cumulative water injection and injection pressure versus time. The Hall method is based on the radial form of the Darcy flow equation expressed as:

$$q_w = \frac{0.00708 k_w h \left(p_{inj} - \overline{p}\right)}{\mu_w \ln\left(\frac{r_e}{r_w}\right)} \tag{19.1}$$

In Eq. (19.1), q_w = water injection rate, BWPD; k_w = effective permeability to water, md; h = formation thickness, ft; p_{inj} = bottomhole injection pressure, psia; $\overline{p}$ = average reservoir pressure, psia; μ_w = water viscosity, cp; r_e = external radius, ft; and r_w = wellbore radius, ft. The cumulative water injection, W_i, can be expressed as:

$$W_i = \int_0^t q_w dt \tag{19.2}$$

Substituting Eq. (19.1) into Eq. (19.2) and rearranging yields:

$$\int_0^t \left(p_{inj} - \overline{p}\right) dt = \left(\frac{141.2 \mu_w \ln\left(\frac{r_e}{r_w}\right)}{k_w h}\right) W_i \tag{19.3}$$

In integrated form, Eq. (19.3) can be expressed as:

$$\sum \Delta\left(p_{inj} - \overline{p}\right) \times \Delta t = \left(\frac{141.2 \mu_w \ln\left(\frac{r_e}{r_w}\right)}{k_w h}\right) W_i \tag{19.4}$$

If it is assumed that all terms on the right hand side of Eq. (19.4) are constant except W_i, then Eq. (19.4) can be written in an equivalent form:

$$\sum \Delta\left(p_{inj} - \overline{p}\right) \times \Delta t = m_H W_i,$$

$$\text{where } m_H = \left(\frac{141.2 \mu_w \ln\left(r_e / r_w\right)}{k_w h}\right) \tag{19.5}$$

Equation (19.5) implies that a plot of $\sum \Delta\left(p_{inj} - \overline{p}\right) \times \Delta t$ versus W_i should yield a straight line if m_H is constant. The parameter m_H would not be constant if permeability is reduced due to plugging or increased due to stimulation or fracturing. A characteristic Hall plot is shown in Figure 19.6. Note the changes in the shape of the curve due to effects of various injection conditions as indicated in Figure 19.6.

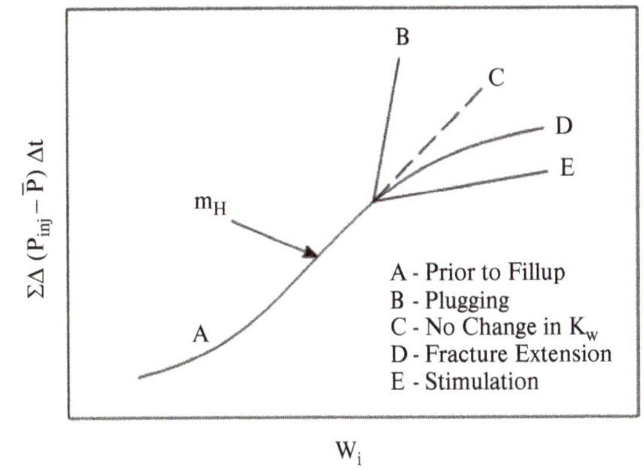

▲ **FIGURE 19.6** **Characteristic Hall plot**

To maximize water injection, it is sometimes necessary to determine the maximum injection pressure at which the formation would be fractured. This could be done using the step rate test.[9] A step rate test is a sequential measurement of stabilized water injection rates at injection pressures recorded at specified time intervals. A plot of injection pressures versus injection rates would indicate a discontinuity if the fracture pressure was reached, as shown in Figure 19.7. The point of intersection of straight lines drawn on data points on either side of the discontinuity represents the fracture pressure. After establishing the fracture pressure, water injection is maximized by maintaining injection pressure as high as possible without exceeding the indicated fracture pressure.

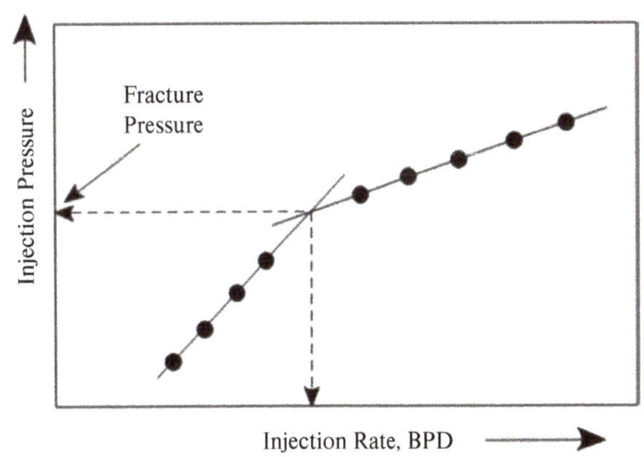

▲ **FIGURE 19.7** **Step rate test plot**

It is good waterflood management practice to run injection profile logs on water injection wells to determine the intervals of the formation that are actually taking injected water. It is recommended that injection profile logs should be run at least once a year on water injectors. Injection profile logs are useful in determining the formation intervals that are being waterflooded and could be used to plan future remedial work to improve distribution of water injection into the formation.

EXAMPLE 19.1 Preparation of the Hall Plot from Well Injection Data

Problem

Prepare a Hall plot from the data for an injection well shown in Table 19.1. Determine the condition of the well from the Hall plot. Estimate permeability of the well. Other well and reservoir data are as follows:

Formation thickness, h	50 ft
Reservoir radius, r_e	1000 ft
Well radius, r_w	0.5 ft
Water viscosity, μ_w	1.15 cp

TABLE 19.1 Injection Well Data for Example 19.1

N	Δt (days)	Injection Pressure, p_{inj} (psia)	Average Pressure, $\bar{p}$ (psia)	Injection Volume, (bbls)
1	25	4260	3205	25,000
2	20	4295	3215	19,800
3	30	4285	3210	29,250
4	22	4279	3211	22,220
5	28	4295	3225	27,500
6	30	4305	3230	25,500
7	31	4322	3232	24,025
8	29	4328	3233	22,765
9	24	4343	3227	19,200
10	27	4338	3229	21,465
11	30	4353	3234	23,400
12	21	4360	3236	16,695

Solution

Step 1: Prepare data for Hall plot.

Applying Eq. (19.5), use the data from Table 19.1 to prepare the data for Hall plot as shown in Table 19.2.

Step 2: Make Hall plot.

Use the data in Table 19.2 to make a Hall plot as shown in Figure 19.8. Note that the y-axis is represented as Y-Function in Figure 19.8 where the Y-Function = $\sum(\Delta p \times \Delta t)$.

Step 3: Evaluate the Hall plot.

Review the Hall plot diagnostically using the characteristic shapes shown in Figure 19.6. For this example, there is no change in the slope of the plot after fill-up of the area around the injector. For this reason, the injector does not appear to be damaged.

Step 4: Estimate the permeability of the formation around the water injector.

From Figure 19.8, the slope of the line, $m_H = 1.4152$

From Eq. (19.5),

$$m_H = \frac{141.2\mu_w \ln \frac{r_e}{r_w}}{k_w h}$$

Therefore,

$$k_w = \frac{141.2\mu_w \ln \frac{r_e}{r_w}}{m_H h}$$

$$= \frac{141.2 \times 1.15 \times \ln \frac{1000}{0.5}}{1.4152 \times 50} = 17.44 \text{ md}$$

This is the average effective permeability to water within the area taking water from the injector.

TABLE 19.2 Plot Data for Example 19.1

N	Injection Pressure, p_{inj} (psia)	Average Pressure, $\bar{p}$ (psia)	Δp (psia)	Δt (days)	$\Sigma(\Delta p \times \Delta t)$	Injection Volume, (bbls)	Cumulative Injection, W_i (bbls)
1	4260	3205	1055	25	26,375	25,000	25,000
2	4295	3215	1080	20	47,975	19,800	44,800
3	4285	3210	1075	30	80,225	29,250	74,050
4	4279	3211	1068	22	103,721	22,220	96,270
5	4295	3225	1070	28	133,681	27,500	123,770
6	4305	3230	1075	30	165,931	25,500	149,270
7	4322	3232	1090	31	199,721	24,025	173,295
8	4328	3233	1095	29	231,476	22,765	196,060
9	4343	3227	1116	24	258,260	19,200	215,260
10	4338	3229	1109	27	288,203	21,465	236,725
11	4353	3234	1119	30	321,773	23,400	260,125
12	4360	3236	1124	21	345,377	16,695	276,820

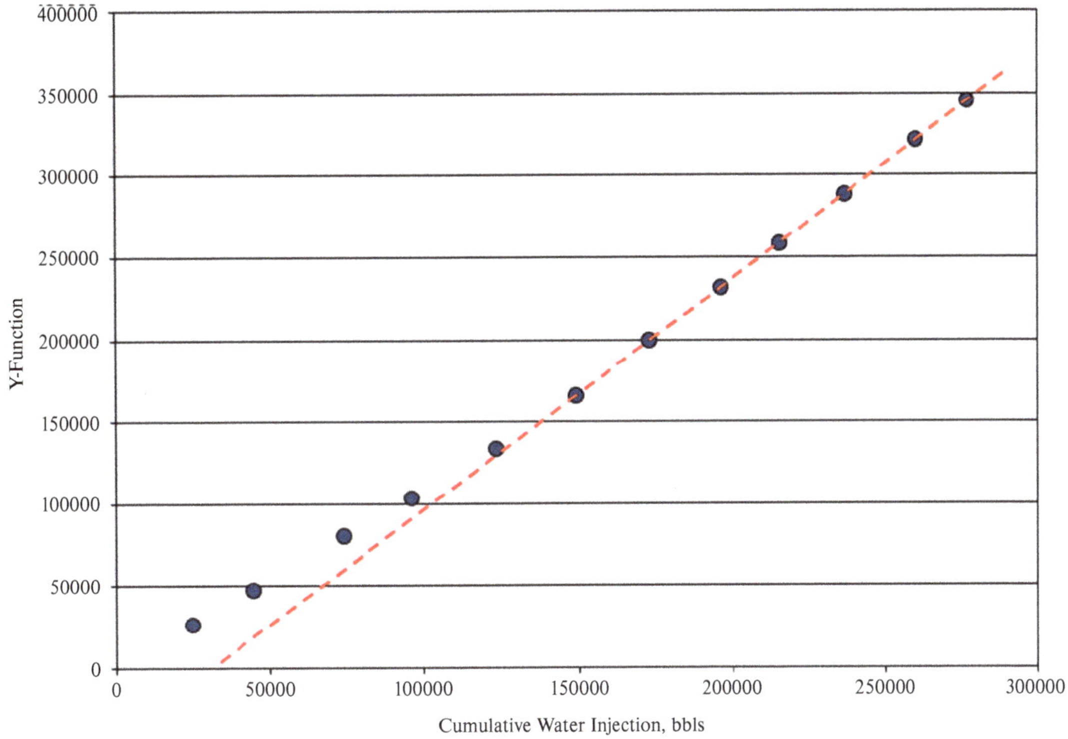

▲ FIGURE 19.8 Hall plot for Example 19.1

End of Example 19.1

B. Injection Water. The source and quality of injection water is another key component in the management of a waterflood project. The source of injection water should be sufficient to supply as much water as needed during the early stages of the waterflood project. Later on, as the project matures, produced water could be recycled and used to make up the required volumes of injection water. The quality of injection water should be determined by performing complete water analysis on all sources of water. The water analysis process should evaluate the compatibility of source water with reservoir formation water, swelling clays or other minerals in the reservoir rock, the type of injection facility (open or closed), and the extent of water treatment required before the water can be used safely for injection. Extensive water treatment should be avoided because it could be expensive, and adversely impact the economics of the project. Water analysis should be performed regularly throughout the life of the waterflood project to monitor the quality of injection water.

Injection water, if not treated, could cause corrosion, solid precipitation, and promote growth of bacteria. The major cause of corrosion in injection water is dissolved gases in the form of oxygen, carbon dioxide, and hydrogen sulfide. Oxygen in injection water is very corrosive, and may cause precipitation of ferrous hydroxide, if the injection water contains iron. If oxygen is present in injection water, it should be removed by stripping or other processes. Solid precipitation can be avoided by minimizing the presence of iron, manganese, and barium in the injection water. Water containing barium is likely to precipitate barium sulfate, if injected into a reservoir containing sulfate. Barium sulfate can cause severe plugging of production wells and production facilities.[10,11] Bacteria growth should be controlled by treating injection water with chlorine or other bactericides. The most important bacteria that should be controlled are the iron bacteria and sulfate reducing bacteria. The iron bacteria cause precipitation of iron and manganese hydroxides which can plug the formation. The sulfate reducing bacteria react with sulfate ion to create iron sulfide which can also plug the formation. Hydrogen sulfide is also generated as a by-product of the reaction. Hydrogen sulfide is a very corrosive and toxic substance that can pose serious problems on the operation of the waterflood project. The generation of hydrogen sulfide by sulfate reducing bacteria should be controlled and eliminated.

C. Production Wells. Production wells must be monitored regularly in a waterflood project just as injection wells. They are equally important for achieving engineering and economic success on the waterflood project. Each production well should be tested frequently (depending on the availability of test equipment) to measure fluid production rates. The measured production rates should be plotted on various graphs to monitor the response of the well to water injection. Typical graphs commonly used are plots of fluid production rates versus time, percent oil cut versus cumulative oil produced, or water-oil ratio versus cumulative oil produced. There are numerous other types of plots that can be used to monitor well performance. These plot types can be tailored to suit the particular needs of the project.

If the production well appears to be damaged, a pressure buildup test should be conducted on the well to determine the extent of damage. Methods for analysis of pressure buildup tests were presented in Chapters 13 to 15. A common cause of reduction in well productivity is deposition of scale on the sandface and in the production string. Some types of scale can be removed by simulation of the wells with acid.

Profile (fluid entry) logs should be run periodically on production wells. These profile logs are extremely important in monitoring the intervals that are contributing to production. Historical reviews of profile production logs can be used to determine intervals that have been flooded, and intervals that were not affected by water injection. These reviews are useful in reallocation of water injection and targeting of specific formation intervals for water injection.

D. **Produced Water.** Water analysis should be performed regularly on produced water as a means of monitoring the progress of the waterflood and tracking potential problems with solids deposition caused by mixing of source water and formation water. At early stages of the waterflood project, produced water consists mostly of formation water. Later, as production wells are affected by water breakthrough, produced water will contain increasing amounts of injected water. The proportion of injected water in produced water is sometimes used as an indicator in monitoring the location of the floodfront.

In many waterflood projects, some of the produced water is recycled and used as injection water. This reduces the amount of produced water to be disposed, thereby reducing the environmental impact of the project. Produced water recycled for re-injection should be treated to remove harmful substances, and improve its quality, as was the case for the source injection water. In some operations, produced water is also treated to remove harmful substances before disposal.

E. **Facilities.** The facilities used in a waterflood project should be able to withstand potential threat from massive corrosion caused by the water. Protection of injection and production facilities from corrosion begins with careful water analysis, and selection of equipment with proper metallurgy. It is recommended that major waterflood projects should retain the services of a corrosion prevention department. The corrosion engineers should routinely monitor all facilities used in the project for corrosion and implement preventative measures to forestall onset of corrosion in these facilities.

19.2.5 Management of Waterflooded Reservoirs

Several elements of proper management of waterflooded reservoirs were presented previously under management of injection wells, injection water, production wells, produced water, and facilities. This approach was taken to emphasize the importance of these key sectors in the successful operation of a waterflood project. In addition to these key sectors, the waterflooded reservoir should be managed as a whole. This means that the reservoir should be monitored regularly to determine the location of the floodfront, evaluate the distributions of reservoir pressure and fluid saturations, and exploit opportunities for re-alignment of injectors and producers to improve sweep efficiency and hydrocarbon recovery.

Successful management of waterflooded reservoirs must include sustained programs for gathering data on the reservoir.[12-16] Pressure data should be obtained regularly from producers, injectors, and observation wells, if available. The pressure data should be analyzed and used to track the movement of injected water in the reservoir. The pressure data are also used to monitor reservoir voidage balance and ensure that reservoir pressure is maintained for waterflood projects designed primarily for pressure maintenance. In addition to pressure data, fluid saturations can be monitored with carbon/oxygen logs.[17] Injection and production profile logs should be obtained regularly and assembled with the fluid saturations data to develop a comprehensive view of the status of the project.

4D seismic surveys are very effective tools for management of waterflooded reservoirs. A baseline survey should be taken before start of the waterflood project. This should be followed subsequently with periodic 4D seismic surveys. Data from 4D seismic surveys can be analyzed and used to locate floodfronts, bypassed reserves, water breakthroughs, etc. It is highly recommended to incorporate 4D seismic surveys as essential tools for management of waterflood projects if acquisition costs of the seismic data are not prohibitive.

A reservoir model is typically used to design the waterflood project as discussed in Section 19.2.2. This model should be used to manage the waterflood project. The model should be updated with pressure data, injection and production data, and other data acquired as part of monitoring program. In some cases, it might be necessary to reconstruct the reservoir model entirely, if new data indicate substantial new knowledge on reservoir geology and architecture. The functional reservoir model should be used to locate infill wells, allocate water injection, re-align wells, and predict future performance of the waterflood project. In general, waterflood projects should adopt the principles stated in Chapter 24 for sound management of petroleum reservoirs.

19.3 Gasflooding

Gasflooding is presented in this chapter as an immiscible displacement process. Immiscible displacement occurs in a displacement process where a distinct interface (or boundary) exists between the displaced fluid and the displacing fluid. This includes displacement processes that are described as near-miscible. The key factors that determine whether a gasflooding process is immiscible or miscible are reservoir pressure, reservoir temperature, composition of injection gas, composition of reservoir fluid, and reservoir heterogeneities such as faults and permeability barriers. The impact of each factor can be determined with laboratory measurements and modeling of the displacement process. Gas miscible displacement processes are presented under enhanced oil recovery methods in Chapter 21.

The injection gas in immiscible gasflooding could be nitrogen, hydrocarbon gas, flue gas, carbon dioxide, or any other gas mixtures. In a typical displacement process, sufficient quantities of the gas are injected into the reservoir to displace the reservoir fluid to producing wells. Gasflooding is sometimes used instead of waterflooding because the fluid displacement (microscopic) efficiency of gas is superior in some applications. In addition, some of the injection gas (such as hydrocarbon gas and carbon dioxide) can dissolve in the reservoir fluid to reduce its viscosity and density, thereby improving its mobility in the reservoir. On the other hand, the volumetric sweep efficiency (composed of areal and vertical sweep efficiencies) of gas displacement processes is generally low due to gravity over-ride, and viscous fingering caused by higher mobility ratios of gas in comparison to liquids. This effect can be compounded in low permeability reservoirs, or reservoirs with severe heterogeneities, such as faults and permeability barriers. Mobility ratios of gas displacement processes can be reduced by alternately injecting water and gas. This is described as a WAG (Water-Alternating-Gas) displacement process. Volumetric sweep efficiency can be improved by application of WAG displacement process. This has been shown to improve oil recovery based on field results.[18]

19.3.1 Applications of Gasflooding

The primary applications of gas injection into a hydrocarbon reservoir are pressure maintenance, gas cycling, and fluid displacement. In many gas injection projects, these three primary functions are achieved at the same time to some degree.[19] Pressure maintenance by gas injection is routinely implemented in many reservoirs. Reservoirs that are good candidates for pressure maintenance may have experienced steep, rapid pressure decline during primary depletion accompanied with the development of a secondary gas cap, or enlargement of a primary gas cap. For such reservoirs, gas is injected into the gas cap at sufficient volumes to reduce or stop the decline of reservoir pressure. In some cases, gas injection for pressure maintenance is initiated early close to the start of production to prevent any decline in reservoir pressure. This practice is used, in some cases, to keep reservoir pressure from falling below bubble point for oil reservoirs, or dew point for gas reservoirs. Gas cycling is used to improve recovery from condensate reservoirs by vaporizing fluids that may have condensed in the reservoir with injected gas. Gas cycling can be used also to improve hydrocarbon recovery from secondary gas caps in oil reservoirs. In many steeply dipping reservoirs, gas is injected at the crest of the structure to assist in displacing oil toward downdip wells.[19] The volumetric sweep efficiency can be high for such gravity stable processes.[20] Gasflooding in horizontal reservoirs can lead to gravity over-ride, where the injected gas rises to the top of the reservoir because it is less dense than oil. Severe cases of gravity over-ride can lead to low volumetric sweep efficiencies leaving large portions of the reservoir unswept by injected gas. The effects of gravity over-ride and viscous fingering (early gas breakthrough) can be mitigated by using mobility control measures such as WAG as part of the injection strategy.

19.3.2 Gasflood Design

This section on gasflood design begins with an overview of two main factors that determine the performance of a gas flood. The first key factor in gasflood design is obtaining a fluid property model that represents the interactions that occur between the reservoir fluids and the injected gas. These interactions can be both physical and chemical in nature. For instance, lean hydrocarbon gas is capable of physically vaporizing light hydrocarbons from reservoir oil.[21] Carbon dioxide can dissolve chemically into reservoir oil to reduce its density and viscosity and thereby improve its ability to flow. These changes can be measured in the laboratory by conducting PVT studies, slim-tube tests, and core floods with the injection gas and the reservoir fluids.[22] The data from these studies can be used to modify the parameters of an Equation of State (EOS) which can then be used to simulate and reproduce the phase behavior resulting from the interaction between the injected gas and reservoir fluids. A procedure for regression of the parameters of an EOS was presented in Chapter 7.

The second key factor in gasflood design is proper geologic modeling and characterization of the reservoir that is the candidate for gasflooding.[23] Geologic modeling and characterization of a reservoir for gasflooding should begin with the construction of a geologic model that incorporates all available geologic data. The geologic model is the framework that includes all available structural, stratigraphic, and lithological data. It is important that the geologic model should include all known faults, fractures, and reservoir heterogeneities, including variation of rock properties such as permeability, porosity and saturations.[24,25] The geologic model should represent all reservoir conditions and properties that could impact the gasflooding process. A general procedure for geologic modeling and reservoir characterization is presented in Chapter 22.

The geologic model is then coupled with the EOS fluid model to create a compositional simulation model for the gasflooding process. Note that the compositional simulation model as recommended here can be used to investigate the performance of the gasflooding process under immiscible as well as miscible conditions.

19.3.3 Recommended Steps in Gasflood Design

1. Perform comprehensive PVT studies.
   a. Obtain PVT data on the reservoir fluids.
   b. Measure PVT data on mixtures of reservoir fluids with the injection gas at different compositions.
   c. Conduct multicontact miscibility tests to establish the conditions at which immiscible displacement can occur.
2. Conduct slim tube and core flood tests.
   a. Establish displacement efficiencies with slim tube tests.
   b. Verify results of slim tube tests with core flood tests if whole core samples of the reservoir are available.
   c. Obtain relative permeability data from the slim tube or the core flood tests.
3. Build a fluid properties predictor from an EOS.
   a. Use the data from Step 1 to build a fluid properties predictor with an EOS. The procedures for regression of the parameters of an EOS from PVT data are provided in Chapter 7.
   b. Enhance the predictive capabilities of the EOS with the data from slim tube tests and the core flood tests conducted in Step 2.
4. Perform geologic modeling and characterization of the reservoir.
   a. Construct a geologic model of the reservoir using all available data. Include all faults and other geologic structures that can affect fluid flow.
   b. Define reservoir stratigraphic framework. Incorporate lithological data from petrophysical analysis. Detailed analysis should include definition of facies and assignment of rock properties based on facies.
   c. Apply geostatistics to populate reservoir properties.
   d. Generate several realization of the geologic model using uncertainty analysis. (Note: Detailed geologic modeling and reservoir characterization procedures are presented in Chapter 22.)
5. Construct compositional reservoir flow models with fluid properties model from Step 3 and geologic models from Step 4. (Note: Reservoir simulation is presented in Chapter 23.)
   a. If the reservoir had prior production history, history-match flow models to obtain current state of the reservoir.
   b. Review pressure and saturation distributions in history-matched models. Compare results to actual data.
   c. Select two or more flow models for prediction of gasflood performance.
6. Run prediction cases.
   a. Make prediction runs assuming several gas injection rates.
   b. Make prediction runs by changing location and number of gas injection wells.
   c. Make prediction runs by applying mobility control measures such as the WAG injection strategy.
   d. Compare results and select the cases with the optimal hydrocarbon recoveries.
7. Rank gasflood design.
   a. Rank cases from Step 6d after performing economic analysis. (Economic evaluation of petroleum projects is presented in Chapter 25).
8. Perform sensitivity analyses.
   a. Select two or more cases from Step 7 and conduct sensitivity analyses on key reservoir and operational variables of the gasflood project.
   b. Incorporate economic analyses in the sensitivity runs.
   c. Select a prediction case for project implementation.
9. Conduct a pilot gasflood project.
   a. A pilot gasflood project is recommended, especially for large projects. The pilot project is essential for testing the data and the results generated with the reservoir models. It also provides the opportunity to check the hydrocarbon recoveries predicted for the project before implementing the project on the entire reservoir. A pilot project is necessary in major projects as a field test that can be used to avoid disastrous outcomes in terms of technical and economic performance of gasflood projects. More details on applications of pilot floods are presented in Chapter 21.

19.3.4 Gasflood Management

The main requirement for sound management of gasflood projects is extensive surveillance and monitoring of gas injectors, producers, gas processing facilities, and the reservoir.[26] The gas injectors, producers, and processing facilities should be maintained in good condition for the project to be successful. The reservoir should be monitored closely to ensure that the gasflood is achieving recoveries as projected in designing the process. The surveillance and monitoring practices for gas injectors, producers, processing facilities, and the reservoir are presented.

A. **Gas Injectors.** It is essential to maintain the gas injectors in good conditions so that the required gas volumes are injected. This is particularly important for pressure maintenance projects. Gas injectivity may decline with time due to plugging of perforations from deposition of wax, asphaltene, and hydrates. Loss of gas injectivity can be remediated by stimulating the well with acids, solvents, or other agents for removal of well damage. It is also important to run injection profile logs periodically on gas injectors to evaluate the distribution of the injected gas over the perforated interval. If the profile injection logs indicate unfavorable distribution of injection gas across the perforated interval, then the gas injector should undergo remedial work to direct injected gas to the intended formation intervals. If the injected gas is carbon dioxide or flue gas, the injector should be completed with equipment that can resist corrosion, and the well should be inspected regularly for corrosion. Inspection of well equipment for corrosion also applies to wells used in WAG operations.

B. **Producers.** Producers in gasflood projects should be monitored regularly to prevent excessive production of injected gas which could result from early gas breakthrough due to channeling, or gravity over-ride.[27] If injected gas is allowed to flow to producers without contacting unswept portions of the reservoir, the volumetric sweep efficiency will be reduced resulting in poor hydrocarbon recovery. Gas over-rides or channeling is particularly detrimental to gas cycling or gas displacement projects. This problem can be remedied at the producers by isolating intervals showing high gas production with packers, or by squeezing them with cement. As recommended for injectors, producers should be under surveillance regularly for corrosion, especially if the injection gas is carbon dioxide or flue gas. In most gasfloods containing carbon dioxide or other corrosive injectants, producers are outfitted with subsurface and surface equipment that are resistant to corrosion.

C. **Processing Facilities.** All facilities used in the injection of gas and processing of produced fluids are grouped under the umbrella of processing facilities. These are gas injection facilities, production separators and stock tanks, flow lines, gas plants (for processing, separation, and recycling of injection gas), and storage tanks. These facilities must be fitted to resist corrosion and inspected regularly to avoid the initiation of corrosion in the equipment, especially if the injection gas contains carbon dioxide or other corrosive agents.

19.3.5 Management of Gasflood Reservoirs

Reservoirs under gasflood require extensive program of monitoring and surveillance. The program depends on the type of the gasflood project. For instance, if the gas injection project is for pressure maintenance, extensive program for gathering pressure data across the reservoir should be implemented. In addition, gas production should be monitored and controlled to limit excessive production of injected gas. If gas cycling is the objective of the project, then the surveillance program should ensure that the injected gas is controlled to sweep a large portion of the reservoir. This will require monitoring of the intervals taking gas at the injectors and measuring the liquid content of produced gas at the producers. The surveillance and monitoring program should be adapted to meet the stated objectives of the gasflood project.

It is also highly recommended to incorporate 4D seismic surveys as part of the management program for gasflooded reservoirs as discussed earlier for waterflooded reservoirs.

The progression of a gasflood project is rapid because of the high mobility of gas. Consequently, it is highly recommended that the gasflood project should be managed with a reservoir model.[28] The reservoir model should be the same model used in designing the project or an updated version of it. The reservoir model should be updated and refined with data gathered from the monitoring and surveillance programs. If actual performance of the gasflood is substantially different from model predictions, the entire project should be re-evaluated. The reservoir model should be used to adjust injection and production volumes and predict future performance of the gasflood project. In general, gasflood projects should adopt the principles stated in Chapter 24 for sound management of petroleum reservoirs.

Nomenclature

h	formation thickness, ft	r_e	reservoir radius, ft
k_w	water effective permeability, md	r_w	wellbore radius, ft
$\bar{p}$	average reservoir pressure, psia	t	time, days
p_{inj}	injection pressure, psia	W_i	cumulative water injection, bbls
q_w	water injection rate, STB/D	μ_w	viscosity of water, cp

Abbreviations

EOS	Equation of State	WAG	Water-Alternating-Gas
PVT	Pressure-Volume-Temperature		

REFERENCES

1. Stiles, W.E.: "Use of Permeability Distribution in Waterflood Calculations," *Trans.* AIME (1949) 186, 9-13.
2. Dykstra, H., and Parsons, R.L.: "The Prediction of Oil Recovery by Waterflood," Secondary Recovery of Oil in the United States, 2nd Ed., API (1950) 160-174.
3. Craig, F.F., Jr., Geffen, T.M., and Morse, R.A.: "Oil Recovery Performance of Pattern Gas or Water Injection Operations from Model Tests," *Trans.* AIME (1955) 204, 7-15.
4. Welge, H.J.: "A Simplified Method for Computing Oil Recovery by Gas or Water Drive," *Trans.* AIME (1952) 195, 91-98.
5. Earlougher, R.C., Jr.: *Advances in Well Test Analysis,* Monograph Series, SPE, Richardson, Texas, (1977).
6. Hazebroek, P., Rainbow, H., and Mathews, C.S.: "Pressure Falloff in Water Injection Wells," *Trans.* AIME (1958) 213, 250-260.
7. Merrill, L.S., Kazemi, H., and Gogarty, W.B.: "Pressure Falloff Analysis in Reservoirs with Fluid Banks," *JPT* (July 1974) 809-818.
8. Hall, H.N.: "How to Analyze Injection Well Performance," *World Oil* (October 1963) 128-130.
9. Felsenthal. M.: "Step Rate Tests Determine Safe Injection Pressures in Floods," *The Oil and Gas Journal* (October 28, 1974) 49-54.
10. Vetter, O.J.: "Oilfield Scale–Can We Handle It?" *JPT* (December 1976) 1402-1408.
11. Vetter, O.J., Kandarpa, V., and Harauka, A.: "Prediction of Scale Problems Due to Injection of Incompatible Waters," *JPT* (February 1982) 273-284.
12. Stiles, L.H.: "Optimizing Waterflood Recovery in a Mature Waterflood, The Fullerton Clearfork Unit," paper SPE 6198 presented at the 51st SPE Annual Technical Conference and Exhibition, New Orleans, (October 3-6, 1976).
13. Ghauri, W.K.: "Production Technology Experience in a Large Carbonate Waterflood, Denver Unit, Wasson San Andres Field," *JPT* (September 1980) 1493-1502.
14. Langston, E.P., Shirer, J.A., and Nelson, D.E.: "Innovative Reservoir Management–Key to Highly Successful Jay/LEC Waterflood," *JPT* (May 1981) 783-791.
15. Ezekwe, N., Smith, S., Wilson, M., Humphrey, M., and Murphy, J.: "Performance of a Peripheral Waterflood Project in the Main Body 'B' Reservoirs (Stevens) at the Elk Hills Oil Field, California," paper SPE 21759 presented at the SPE Western Regional Meeting, Long Beach, California, (March 20-21, 1991).
16. Kenawy, F.A., Bibars, O.A., Hanafy, H.H., and Rezk, A.S.: "Waterflood Strategy, Challenges and Innovations," paper SPE 53136 presented at the 1999 Middle East Oil Show, Bahrain, (February 20-23, 1999).
17. Weiland, J., Mikulencak, D., Fox, P., Frisch, G., and Azari, M.: "Waterflood Surveillance in the Mars Field Deepwater GOM: Mississippi Canyon Block 807," paper SPE 115365 presented at the 2008 Annual Technical Conference and Exhibition, Denver, Colorado, (September 21-24, 2008).
18. Christensen, J.R., Stenby, E.H., and Skauge, A.: "Review of WAG Field Experience," *SPEREE* (April 2001) 97-106.
19. Wei, M.H., Yu, J.P., Moore, D.M., Ezekwe, N., Querin, M.E., and Williams, L.L.: "Case History of Pressure Maintenance by Crestal Gas Injection in the 26R Gravity Drainage Reservoir," paper SPE 24035 presented at the SPE Western Regional Meeting, Bakersfield, California, (March 30-April 1, 1992).
20. Naylor, P., and Frorup, M.: "Gravity-Stable Nitrogen Displacement of Oil," paper SPE 19641 presented at the 64th Annual Technical Conference and Exhibition of SPE, San Antonio, Texas, (October 8-11, 1989).

21. Shyeh, J.J., and Stadler, M.P.: "Effect of Injectant Composition and Pressure on Displacement of Oil by Enriched Hydrocarbon Gases," *SPERE* (May 1995) 109-115.

22. Egemann, P., Lombard, J.-M., Modavi, A., and Kalam, M.Z.: "Gas Process Displacement Efficiency Comparisons on a Carbonate Reservoir," *SPEREE* (December 2006) 621-629.

23. Thawer, R., Alhendi, A., Al Mazroui, Y., Boyd, D., Masuzawa, T., Sugawara, Y., Hollis, C., and Lowden, B.: "Controls on Vertical and Horizontal Flow in a Carbonate Reservoir that Impact Gasflooding and Waterflooding," paper SPE 87237 presented at the 9th Abu Dhabi International Petroleum Exhibition and Conference, (October 15-18, 2000).

24. Wylie, P.L., and Mohanty, K.K.: "Effect of Wettability on Oil Recovery by Near-Miscible Gas Injection," *SPEREE* (December 1999) 558-563.

25. Wylie, P., and Mohanty, K.K.: "Effect of Water Saturation on Oil Recovery by Near-Miscible Gas Injection," *SPERE* (November 1997) 264-268.

26. Young, R.E., Fairfield, W.H., and Dykstra, H.: "Performance of a High-Pressure Gas Injection Project, Swanson River Field, Alaska," *JPT* (February 1977) 99-104.

27. Ruhovets, N., and Wyatt, D.F.: "Quantitative Monitoring of Gasflooding in Oil-Bearing Reservoirs by Use of a Pulsed Neutron Tool," *SPEFE* (December 1995) 255-258.

28. Lawrence, J.J., Teletzke, G.F., Hutfliz, J.M., and Wilkinson, J.R.: "Reservoir Simulation of Gas Injection Processes," paper SPE 81459 presented at the SPE 13th Middle East Oil Show & Conference, Bahrain, (April 5-8, 2003).

ADDITIONAL REFERENCES

- Brice, B.W., and Renouf, G.: "Increasing Oil Recovery from Heavy Oil Waterfloods," paper SPE 117327 presented at the 2008 SPE International Thermal Operations and Heavy Oil Symposium, Calgary, Alberta, Canada, (October 20-23, 2008).
- Burger, J.E., and Mohanty, K.K.: "Mass Transfer from Bypassed Zones during Gas Injection," *SPERE* (May 1997) 124-130.
- Jackson, R.: "Why Waterfloods Fail," *World Oil* (March 1968) 65.
- Kumar, M., Hoang, V., Satik, C., and Rojas, D.: "High-Mobility-Ratio-Waterflood Performance Prediction: Challenges and New Insights," *SPEREE* (February 2008) 186-196.
- Rossen, W.R., and Shen, C.: "Gravity Segregation in Gas-Injection IOR," paper SPE 107262 presented at the SPE Europe/EAGE Annual Conference and Exhibition, London, United Kingdom, (June 11-14, 2007).
- Shirer, J.A., Ainsworth, W.J., and White, R.W.: "Selection of a Waterflood Pattern for the Jay-Little Escambia Creek Fields," paper SPE 4978 presented at the 49th Annual Fall Meeting of SPE, Houston, Texas, (October 6-9, 1974).
- Willhite, G.P.: *Waterflooding,* SPE Textbook Series, Vol. 3, Richardson, Texas (1986).

PROBLEMS

19-1. Discuss the differences and similarities between pattern waterflooding and peripheral waterflooding as reservoir management strategies.

19-2. Injection water (sourced from water wells, streams, rivers, seawater, etc.) can be incompatible when mixed with formation water in the reservoir. List chemical compounds that can cause formation damage in the reservoir and injection facilities or pose harmful health threats to personnel from mixing of incompatible injection and formation waters. What are the procedural steps that should be undertaken to prevent or mitigate these threats to successful operation of a waterflood project?

19-3. Is a miscible gasflood project always preferable to an immiscible gasflood project? Discuss reservoir and economic conditions that could justify application of immiscible gasflooding instead of miscible gasflooding.

19-4. A step rate test was conducted on an ultra-deepwater well at a total depth of 24,000 feet. The step rate data are provided in Table 19P.1. Ignore frictional losses from the injection and assume the specific gravity of injection water is 1.01. Using the step rate data, determine:

a. The fracture pressure for the well

b. The fracture gradient for the well

c. Recommended water injection rate for the well

TABLE 19P.1 Step rate data for Problem 19-4

Date	Water Injection Rate (BWPD)	Wellhead Pressure (psia)
3-Mar-19	10,700	8650
4-Mar-19	10,150	8383
5-Mar-19	10,222	8425
6-Mar-19	10,295	8460
7-Mar-19	10,370	8512
8-Mar-19	10,450	8534
9-Mar-19	10,517	8567
10-Mar-19	10,595	8615
11-Mar-19	10,668	8640
12-Mar-19	10,750	8656
13-Mar-19	10,810	8665

19-5. The injection data of a well over a period of 10 months are provided in Table 19P.2. Other well and reservoir data are given as follows:

Formation thickness, h	43 ft
Reservoir radius, r_e	744.73 ft
Wellbore radius, r_w	0.454 ft
Water viscosity, μ_w	1.06 cp

Applying the data provided, determine:

a. Permeability of the formation around the water injector

b. The condition of the formation (damaged or stimulated) around the water injection

TABLE 19P.2 Data for Problem 19-5

N	Δt (Days)	Injection Pressure (psia)	Average Reservoir Pressure (psia)	Water Injection Volumes (bbls)
1	14	3235	2290	14,000
2	29	3262	2295	29,000
3	31	3280	2292	30,000
4	30	3310	2294	28,000
5	22	3302	2296	23,000
6	31	3312	2360	32,000
7	27	3309	2357	29,000
8	25	3314	2362	24,000
9	30	3310	2361	29,000
10	31	3314	2365	31,500

CHAPTER

20 Low Salinity Waterflooding

20.1 Introduction

Low salinity waterflooding (LSWF) is the injection of low salinity water (LSW) containing less than 5000 Total Dissolved Solids (TDS) into a reservoir to achieve incremental oil recovery ranging from 6% to 80% of original oil-in-place (OOIP) (based on corefloods[1]) in comparison to injection of high salinity water (HSW) regarded as containing TDS greater than 5000 ppm. The classification of LSW and HSW in terms of TDS boundaries is arbitrary and only serves as rule-of-thumb for general purposes. Actual salinity of water considered to be low salinity (LS) or high salinity (HS) depends on the rock formation, source of the water, and application. As will be discussed later, determination of the range of LSW suitable for injection is a very rigorous process which is necessary to avoid damaging the reservoir due to swelling and deflocculation of salinity-sensitive clays present in some clastic rocks. Generally, there are two types of low salinity effect (LSE) defined as incremental oil recovery due to injection of LSW into sandstone or carbonate rocks. The two types of LSE depend on the state of water saturation in the cores or reservoirs at the start of LSWF. When the cores or reservoirs are at initial connate water saturation, (S_{wi}), the LSE is categorized as secondary mode LSWF. The LSE at S_{wi} is shown in Figure 20.1. In some applications of LSWF, the cores or reservoir formations were initially flooded with HSW to reach a residual oil saturation, (S_{or}). This is then followed by flooding with LSW. The LSE in this case is categorized as tertiary mode LSWF. The LSE at S_{or} is shown in Figure 20.2.

The primary objective of LSWF is to increase oil recovery beyond the level of oil recovery that could be achieved by high salinity waterflooding (HSWF). However, installation of LSWF requires facilities to generate LSW that provide other secondary benefits that enhance the performance of the waterflood. These ancillary benefits associated with LSWF are as follows:

1. Extended production of relatively dry oil (lower water cut) due to delayed water breakthrough.
2. Removal of sulfate from seawater or produced water that reduces formation damage and/or plugging of downhole equipment due to formation of sulfate scales such as Barium Sulfate and Strontium Sulfate.
3. Reservoir souring is also mitigated due to removal of sulfate.
4. Improved control of sulfate-reducing bacteria in the injection facilities thereby reducing the risk of Hydrogen Sulfide (H_2S) induced corrosion.
5. Better filtration generates cleaner water that reduces formation plugging and improves water injectivity.

This chapter on LSWF has been placed between the previous chapter on waterflooding (Chapter 19) and the next chapter on enhanced oil recovery (Chapter 21) because it can be regarded as an improved waterflooding method or an enhanced oil recovery method, depending on definitions of secondary and enhanced oil recovery processes. However, irrespective of any particular viewpoint, it is generally accepted that LSWF is an improved oil recovery method.

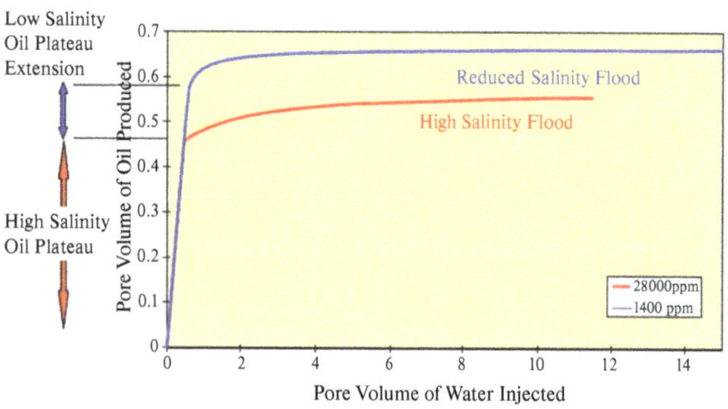

▲ **FIGURE 20.1 Secondary coreflood response (LSE at S_{wi})**

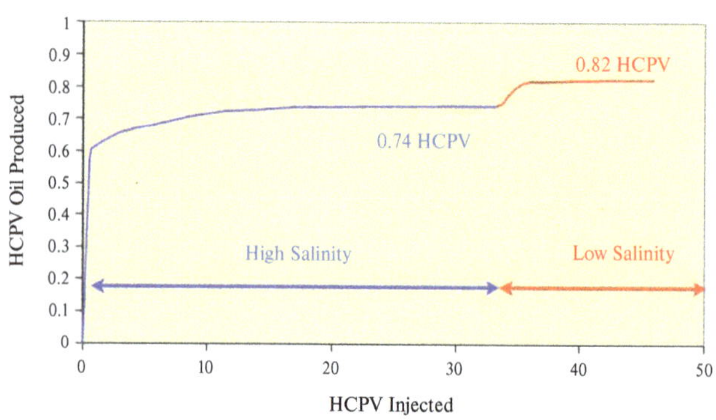

▲ **FIGURE 20.2 Tertiary coreflood response (LSE at S_{or})**

20.2 Laboratory Displacement Experiments Using Sandstone Cores

Increased oil recovery by waterflooding sandstone cores with fresh water or LSW was reported by Bernard[2] in 1967. However, seminal works reported in the literature that showed increased oil recovery by injection of LSW into clayey clastic cores originated from Jadhunandan,[3] Jadhunandan and Morrow,[4,5] Yildiz and Morrow,[6] and Tang and Morrow.[7-9] These reports demonstrated that the LSE occurred in both corefloods and spontaneous imbibition tests with sandstone cores. Other researchers that reported on LSE include Webb et al.,[10,11] Zang and Morrow,[12] Zang et al.,[13,14] Lager et al.,[15-17] and Ligthelm et al.[18] Since the reports of these investigators, there has been a remarkable increase in the number of reported LSE in the literature covering corefloods on clastic and carbonate cores under laboratory and reservoir conditions.[1] Most of these corefloods reported increased oil recovery due to injection of various concentrations of LSW.

20.3 Field Tests in Sandstone Reservoirs

The first field trial of LSE was reported by Webb et al.[19] as a Log-Inject-Log test in 2003. Results from these tests showed that residual oil saturation was reduced by 25% to 50% within the near-wellbore region of the well. This was followed by Single Well Chemical Tracer Tests (SWCTTs) reported by McGuire et al.[20] on four sandstone wells in Alaska's North Slope. In these SWCTTs, remaining oil saturations were reduced in the range of 6% to 12% of OOIP. These translated to incremental waterflood recovery of 8% to 19%. Further SWCTTs in the Endicott Field, Alaska, were reported by Seccombe et al.[21] in 2008. Incremental oil recoveries from these tests that averaged 26% were in good agreement with coreflood results. The first field interwell test was conducted in the same Endicott Field and reported in 2010 by Seccombe et al.[22] The test was conducted on a producer-injector pair about 1040 feet apart. Results from the interwell test indicated a drop in watercut from 95% to 92% and an incremental oil recovery equal to 10% of the swept total pore volume. Furthermore, the results from the interwell tests were consistent with the results from Endicott coreflood tests and SWCTTs. Many other field tests have been conducted by other investigators but were not reported in the literature. These field tests reported by British Petroleum (BP) were the trailblazers in demonstrating that LSWF also worked at field scale.

20.4 Laboratory Displacement Experiments Using Carbonate Cores

Early evidence that indicated that injecting sea water (with low salinity) into carbonate reservoirs could lead to high oil recovery was observed in fractured chalk reservoirs in the North Sea (Sylte et al.,[23] Hallenbeck et al.[24]). About 15 years after LSE was reported in the literature for clastic reservoirs, seminal laboratory works were reported in the literature documenting evidence of increased oil recovery from injection of seawater of different salinity and ionic concentrations into Stevns Klint chalk (Austad et al.,[25] Zhang and Austad,[26,27] Zhang et al.,[28] Strand et al.,[29] Zhang et al.,[30] Strand and Austad[31]). The increased oil recovery reported in these studies were attributed to the presence of specific ions and their relative concentrations in the seawater. These ions identified as Ca^{2+}, Mg^{2+}, and SO_4^{2-} were called Potential Determining Ions (PDIs). The part played by these ions in promoting increased recovery will be discussed in Section 20.6.2, "Mechanisms for LSE in Carbonate Cores/Reservoirs." Other researchers have reported results that demonstrated that increased oil recoveries were observed by flooding carbonate cores from other geographic regions with water of varying salinities and ionic strength (Webb et al.,[32] Yousef et al.,[33] Al-Harrasi et al.,[34] Sepehnoori et al.[35]). On the basis of these laboratory corefloods reported in the literature, it is generally accepted that modified water/seawater with varying salinities and ionic strengths can achieve increased oil recoveries from carbonate cores as were observed with sandstone cores even though the underlying mechanisms could be considered somewhat different.

20.5 Field Tests in Carbonate Reservoirs

A field test on a Saudi Arabian carbonate reservoir was reported by Yousef et al.[36] in 2012. These tests were SWCTTs conducted on two wells, Well A and Well B. In Well A, seawater was injected to achieve a residual oil saturation. This was then followed with the injection of "SmartWater" which was 10 times diluted seawater. The test in Well A resulted in additional seven (7) saturation units reduction in residual oil saturation over seawater injection. The test in Well B achieved reductions in residual oil saturations from injection of field seawater followed by injection of two (2) times diluted seawater, and then ten (10) times diluted seawater. The second stage (2 times diluted seawater) showed three (3) saturation units reduction in residual oil saturation, and the third stage also showed three (3) saturation units reduction in residual oil saturation. The total reduction in residual oil saturation from the two stages was six (6) saturation units over seawater injection. The results from the field tests were in general agreement with reported research results (Yousef et al.[37]) and pre-test simulation studies. The authors stated that field trials involving multi-well demonstration pilot were planned to evaluate impact of "SmartWater" on ultimate recovery and reserves. The results from these field trials have not yet been published till date.

20.6 Mechanisms for LSE

Incremental oil recovery due to LSE has been observed in clastic and carbonate rocks. Although the mechanisms assumed to cause LSE in clastic and carbonate rocks are considered to be somewhat different, studies reported in the literature suggest that LSE is attributable to changes in wettability from mostly oil-wet to mostly water-wet in both types of rocks. The changes in rock wettability have been traced to physical and chemical changes occurring at the rock surfaces due to alterations in the salinity and ionic strength of injected water. Several mechanisms have been proposed to explain the LSE on clastic and carbonate rocks. Many of these mechanisms seem plausible, but it is important to note that no single mechanism have been accepted in the literature to explain the phenomenon of LSE in either clastic or carbonate rocks. For this reason, this book will present only the mechanisms widely used to explain LSE in clastic and carbonate rocks. There is little doubt that some of the mechanisms presented in this book will be discarded in the future as more research are completed and additional knowledge gained on LSE in both clastic and carbonate rocks.

20.6.1 Mechanisms for LSE in Sandstone Cores and Reservoirs

As stated earlier, LSE in sandstone cores and reservoirs is usually accompanied with changes in wettability of the rocks from mostly oil-wet to mixed-wet or water-wet. Generally, changes in wettability of rock surfaces from oil-wet to water-wet are accompanied with lowering of interfacial tensions. Wettability changes in clastic rocks are dependent on following factors as reported by Berg et al.,[38] Doust et al.,[39] Vledder et al.,[40] Austad et al.,[41] Rivet et al.,[42] Jackson et al.[43]:

1. pH and composition of the water
2. Amount of clay in the rock
3. Characteristics of the rock surface
4. Reservoir temperature
5. Composition of the in-situ crude oil

Several mechanisms proposed in the literature that can lead to LSE due to changes in wettability are discussed separately in the following sections.

20.6.1.1 Migration of Fines

The migration of fines as a mechanism responsible for increased oil recovery due to injection of LSW was first proposed by Tang and Morrow[8] in 1999. They inferred from the results of floods on Berea cores that mobilization of the detached clay particles exposed the rock surfaces such that their wettability was changed from oil-wet to water-wet. Furthermore, migration of the fine particles plugged-off small pore throats thereby causing formation damage and indirectly promoting microscopic displacement efficiencies. Al-Shalabi and Sepehrnoori[44] reported that fine migration occurs when the LSW contains insufficient concentrations of cations especially Ca^{2+} and Mg^{2+}. It is important to note that LSE has been observed in corefloods without any production of clay fines (Jerauld et al.[45]) or no LSE with production of fines (Jackson et al.[43]). Consequently, migration of fines is considered most likely to be an effect but not the cause of LSE.

20.6.1.2 Increase in pH/Lower Interfacial Tension

McGuire et al.[20] reported that LSE mechanism could be caused by lower interfacial tension due to pH increase as observed in many corefloods that they performed over several years. They claimed that the saponification effects of LSW lead to increasingly water-wet rock surfaces similar to the effects of alkaline/surfactant flooding. Increase in pH as the mechanism responsible for LSE has been disputed in the literature (Lager et al.[46]) based on these observations:

1. No direct correlation has been established between LSE and the acid number of the crude oil as seen with alkaline flooding.
2. LSE has been observed in corefloods with relatively no change in pH.
3. None of the experiments showing changes in pH were performed at reservoir conditions with live reservoir fluids.
4. Proton buffering due to desorption of proton from oxides and organic matter would prevent pH from increasing in an actual reservoir.
5. Presence of carbon dioxide in the majority of petroleum reservoirs would act also as buffers and prevent pH from increasing in most reservoirs.

20.6.1.3 Multicomponent Ionic Exchange

The multicomponent ionic exchange (MIE) mechanisms for LSE was proposed by Lager et al.[16] The MIE mechanism is based on the theory that certain mechanisms are responsible for the adsorption of organic matter on clay surfaces. In the presence of LSW, MIE will occur by displacing organic polar compounds and organo-metallic complexes from clay surfaces and substituting them with simpler cations. The MIE process was described by Austad et al.[41] as illustrated in Figure 20.3. As shown in Figure 20.3, injection of LSW destabilizes the rock-brine equilibrium due to its lower ionic composition. As a result, Ca^{2+} cations are released from the clay surface which is then balanced by adsorption of H^+ ions in the brine. The surplus OH^- ions created by the process increases the pH of the solution, resulting in reduction of interfacial tension with attendant changes in wettability of the clay surface. To reestablish balance, the OH^- ion will take a H^+ ion from the oil component thereby releasing the oil complex from the clay surface. Similar mechanism is depicted for acidic oil components as shown in Figure 20.3.

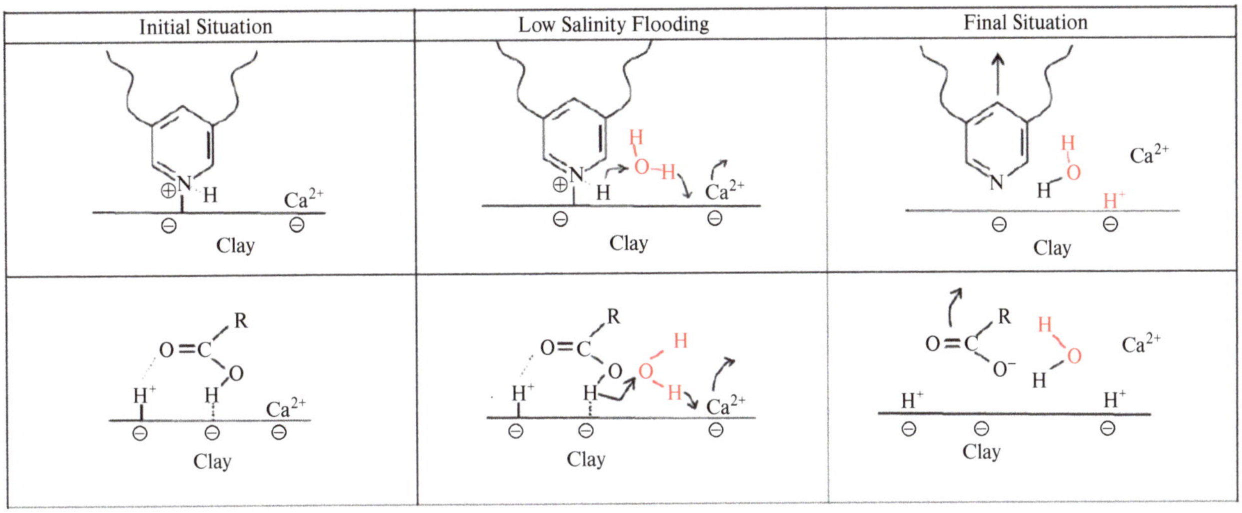

▲ **FIGURE 20.3 Proposed MIE mechanism for LSE (from Austad[41])**

20.6.1.4 Electrical Double Layer Expansion

Electrical double layer (EDL) expansion is the mechanism proposed for explanation of LSE because the wettability of the rock surface is altered in the process (Ligthelm et al.[18]). The EDL is electrically neutral with a particle and a double layer as shown in Figure 20.4. The EDL consists of a layer of ions that are strongly bound (the Stern layer) and a second layer with ions weakly bound to the clay surface. It has been reported by Lee et al.[47] and Jackson et al.[43] that the thickness of the EDL is increased by LSW. The increased EDL thickness causes an increase in the zeta-potential of the EDL (Zeta potential is the potential at the shear plane of the EDL (Figure 20.4)). The increased zeta-potential is exhibited by increased electrostatic repulsive forces on the clay surface. If the EDL-induced electrostatic forces become stronger than the forces binding the oil to the clay surface, the oil is released and then recovered. The EDL expansion as a dominant mechanism that represents increased oil recovery by LSW has been demonstrated in a study by Nasralla and Nasr-Ef-Din.[48]

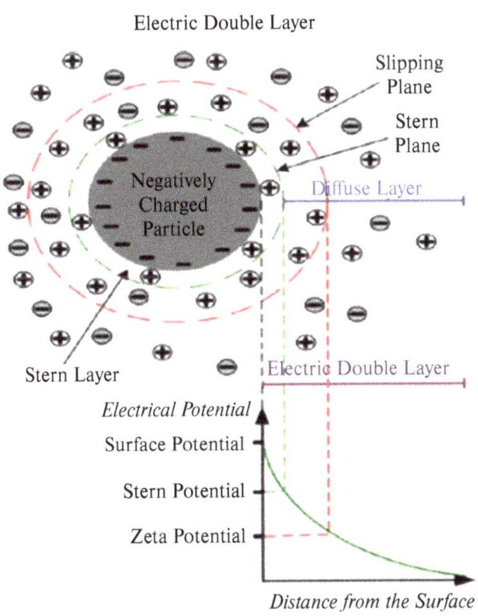

▲ **FIGURE 20.4 Schematic of the EDL (from Nasralla and Nasr-El-Din[48])**

20.6.2 Mechanisms for LSE in Carbonate Cores/Reservoirs

In carbonate reservoirs, increased oil recovery has been linked to mechanisms that change the wettability of carbonate rocks from oil-wet to water-wet. Laboratory studies show that the wettability of carbonate rocks can be modified by two major means:

1. Increasing the concentrations of surface acting anions which are SO_4^{2-}, BO_3^{3-}, and PO_4^{3-}.
2. Reducing the ionic strength of the injected water.

As discussed in Section 20.4, Austad and his co-workers demonstrated that increased oil recovery from carbonate cores was promoted by the presence of PDIs (Ca^{2+}, Mg^{2+}, and SO_4^{2-}) in the injected seawater. The suggested mechanism was that the sulfate ion (SO_4^{2-}) adsorbed to the positive charged sites on the carbonate surface and thereby lower its positive surface charge. As a result, the repulsive electrostatic positive surface charge was reduced which then allowed excess Ca^{2+} ions in solution to get closer to the carbonate surface and react with the adsorbed oil polar compound. The newly formed calcium carboxylate compound (oil) was then released from the carbonate surface. Zhang et al.[30] reported that at high temperatures, the Ca^{2+} ion would be replaced by the Mg^2 ion in the oil complex. Gupta et al.[49] reported that borate (BO_3^{3-}) and phosphate (PO_4^{3-}) ions could be used in place of sulfate ions to increase oil recovery by the same mechanism. These surface acting ions increased oil recovery by altering the wettability of the carbonate rock surface from oil-wet to water-wet. Note that addition of sulfate ions to injection water can cause formation of insoluble scale (Barium Sulfate and Strontium Sulfate) that can damage the reservoir and/or plug-off wells if the formation water contains Barium and Strontium ions. It can also cause produced fluids to become sour. It is advisable to test the composition of formation water for presence of barium and strontium ions or implement mitigating remedies before adding sulfate ions to injection water as a way to increase oil recovery.

Experimental work reported by various researchers showed also that the wettability of carbonate rocks can be altered by lowering the ionic strength of the injected water (Almehaideb et al.,[50] Yousef et al.,[33] Alotaibi and Yousef[51]). Lowering the ionic strength of injected water can be used as an alternative means of increasing oil recovery from carbonate reservoirs. The mechanism associated with lowering the ionic strength of injected water can be explained with the EDL expansion mechanism (Alotaibi and Yousef[51]).

20.7 Guidelines for Planning, Designing, and Installing LSWF Projects

There are several sequential activities that should be carefully scheduled and executed in planning, designing, and finally installing a LSWF project in an oil field. These activities typically could take several years and involve different functional groups in any organization. **It is important to note that injection of LSW in any reservoir (clastic or carbonate) could have very adverse consequences if tedious and exhaustive laboratory tests were not conducted with actual core samples from the reservoir to demonstrate that the salinity of the water will achieve higher oil recovery and not cause irreparable damage to the reservoir.** LSWF is an emerging technology and has not been proven like common HSWF. Note that it is not advisable to adopt any shortcuts in conducting tests with actual reservoir cores at reservoir conditions and with live reservoir crude samples before proceeding to field implementation. If economically feasible, field tests with pilot floods should be considered.

To plan, design, and install an LSWF project in a reservoir, specific steps are proposed here as guides. These steps serve as guides and can be modified as necessary to adapt to a particular reservoir or project.

Step 1: Screening of Potential Candidate Reservoirs

This is the first step to determine whether the reservoir is a good candidate for LSWF. Some of the recommended screening criteria are:

1. Salinity of the connate water. Connate water salinity higher than 10,000 ppm is preferred.
2. Wettability of the reservoir rocks, if known. Oil-wet rock is required.
3. Available source of large quantities of water for injection. If reservoir is offshore, the source will be seawater. The composition of the seawater should be obtained and analyzed for potential use as source for LSW. If reservoir is onshore, the water source could be water wells. As with seawater, the composition of the water should be analyzed for use as the source for LSW. In either case, analysis of the source water should focus on whether it contains elements or chemical compounds that would be difficult or expensive to remove by filtration before the required composition of LSW is achieved.
4. The reservoir should have an existing HSWF or be a good candidate for HSWF. Remember that LSWF is an improvement over HSWF due to its ability to increase oil recovery higher than HSWF. Hence, a reservoir with an existing HSWF or approved for HSWF is automatically a good candidate for evaluation for LSWF if fluid and rock properties meet certain criteria as discussed earlier.
5. The potential size of incremental oil recovery should be large enough to justify additional capital expenditure required for installation of LSWF.
6. Generating LSW requires additional equipment for power generation, filtration, and processing of HSW. This additional equipment has space and weight requirements. For offshore installations, space and weight requirements are critical determinants on whether LS projects are feasible especially for existing platforms. Even for new offshore projects, the cost of additional space and weight of LS equipment becomes a very important decisive factor. Many potential offshore LS projects have not advanced beyond the screening stage because the costs of retrofitting existing platforms for LS equipment were high. These costs are very high due to lost production especially if the platform will need to be shut down for several weeks or months for retrofitting.
7. Conduct x-ray diffraction (XRD) analyses of core samples from the reservoir. If sandstone reservoir, determine the clay content of the samples. Sandstone reservoirs with high clay content are considered good candidates for LSWF. If carbonate reservoir, determine the type of carbonate especially if anhydrites are present.

Carbonate rocks containing anhydrites have shown higher oil recovery under LSWF (Austad et al.[52]).

Step 2: Conduct Injectivity and Special Core Analyses (SCAL) Tests

This is probably the most important step in the entire process of screening potential candidates for LSWF. It begins with selecting many core samples representing different rock types, zones, and areas of the reservoir. It is important to select at least a core sample from each section of the reservoir that would be subjected to LSWF. Some of the cores will be used for injectivity tests, and others for SCAL tests. It is also important to select reputable laboratories that will perform the tests and agree on protocols that will be used to conduct the tests. Major oil companies have research facilities that have considerable experience in conducting these tests with established protocols. For small independent oil companies that use commercial laboratories, it is important that protocols that would be used to conduct these tests are reviewed thoroughly and documented.

The injectivity tests should be conducted with brines covering a wide range of salinities representing brines expected to be processed at the injection facilities. Typical salinities to be tested could range from source water salinity to 100 ppm water salinity. It is important to make sure that injectivity tests are conducted at the specific range of salinities expected to be used in the project (e.g., 1000–3000 ppm). The injectivity tests should be designed to determine the potential reduction in permeability that would result from injection of brines with progressively lower salinities. If severe reductions in permeability (more than 50% reduction from original permeability) are observed, this should raise serious concerns that injection of LSW could impair the reservoir due to swelling of clays and drastically reduce its productivity. Such reservoirs may not be good candidates for LSWF.

Special core analyses (SCAL) involving corefloods are other laboratory experiments designed to evaluate incremental oil recovery and reductions in permeability due to injection of low salinity brines. These tests should be conducted with representative core samples at reservoir conditions with live reservoir oils. For secondary LSE, the cores should be flooded with HSW, restored to initial state, and then flooded with LSW. The oil recoveries from both floods are then compared. For tertiary LSE, the cores should be flooded to residual oil saturation with HSW, and then flooded with LSW. The incremental oil recovery from flooding with LSW after HSW is then measured. SCAL corefloods to evaluate LSE is a very specialized operation that requires experienced laboratory personnel. Careful attention should be devoted to the protocols used to conduct the tests. Very specialized equipment is required to conduct a successful test. As was the case with injectivity tests, major oil companies have in-house capabilities to conduct these corefloods with well-established protocols. Independent oil companies should seek reputable commercial laboratories with experience on the subject to perform these tests as long as well documented protocols are adopted. The results from SCAL should include oil recoveries, permeability data, relative permeability data, and capillary data recorded from the tests.

SCAL corefloods are very tedious, laborious, and time-consuming tests that may take 6 months or more to reach completion. It is important to plan and schedule them ahead at the start of the screening process. Since the most important decision on the screening process rely on them, it is necessary to recognize that SCAL tests are on the critical path on the timely progression of the LSWF project.

Step 3: Modeling and Uncertainty Analysis

The key to modeling LSE is matching the results of corefloods with a reservoir model and scaling up the results from cores to reservoir. By scaling up the coreflood results, relative permeability curves representing the LSE will be generated for use in the fullfield models or sector models. Reservoir simulators that have incorporated algorithms that account for grid block compositional changes due to flooding with LSW are available either in-house in major oil companies or commercially with simulator vendors. It is very important to understand how the LSE has been implemented in these simulators before using them to generate production forecasts for LSWF projects.

Typically, production forecasts are generated for both the HSWF cases and the LSWF cases. The differences between the forecasts are the incremental oil recovery due to LSE. The incremental oil production profile becomes the input for economic evaluation of the LSWF project.

Evaluation of an LSWF project should include extensive uncertainty analysis. In this exercise, uncertainty parameters to be investigated should include variations in the salinity of injection brine, relative permeability data, porosity and permeability distributions, connate fluid properties, brine injection rates, etc. Incorporating variations in these parameters in an uncertainty analyses would generate a range of outcomes that would present a realistic picture of potential outcomes from the LSWF project.

Step 4: Selection of Water Injection Facilities

The source of water used in waterflooding is usually determined by the location of the reservoir. This is also dictated by convenience and economics. It is usually cheaper to use water from nearby sources for waterflooding than to transport or pipe water from far away locations. Most giant reservoirs discovered around the world are located in remote arid regions or offshore in deep water. In either case, waterflooding these giant reservoirs require sourcing and processing large quantities of water to achieve water quality suitable for injection into the reservoir. The reservoir can be irreversibly damaged if water that does not meet quality specifications is injected over a sustained period of time. An example of injection water quality specification is shown in Table 20.1. It is important that water injection facilities are designed to meet these water quality specifications most of the time to avoid impairment of water injectivity and potential reservoir damage from loss of permeability.

The facilities required for water injection will be presented based on design for offshore installation. This is justified due to the fact that most giant reservoirs under development are offshore and the challenges of designing water injection facilities are more acute for offshore installations. Note that the designs for offshore installation can be readily modified and adapted for onshore installations. A block diagram for an offshore conventional water injection facility is shown as Figure 20.5. Note that Figure 20.5 includes a seawater lift system, coarse filtration units, and micro/ultra-filtration units. The micro/ultra-filtration units are not included in most installations, but it is advisable to use them depending on the permeability of the reservoir and the specified water quality required to prevent reservoir damage. The block diagram for LSW is shown in Figure 20.6. Note that Figure 20.6 is very similar to Figure 20.5 except for the inclusion of reverse osmosis (RO) units. RO was selected as the preferred desalination technology in comparison to other desalination methods (e.g., mechanical vapor compression, multi-stage flash distillation, multi-effect distillation, electrodialysis, forward osmosis, and freeze distillation) because of the following reasons:

1. It is a matured technology; has been used in many onshore and offshore freshwater plants.
2. Less footprint on the platform.
3. Less weight on the platform.
4. Less power requirements.
5. Very flexible skid configuration.

It is important to note that RO was selected as the technology for generation of LSW based on current knowledge and state of water processing technology. It is likely that in the future superior means of generating LSW could be developed based on advancements of membrane filtration technology. Another superior technology if developed, would have to overcome the advantages listed above for RO.

In some waterflood, installation of a sulfate removal unit (SRU) is compared to installation of an RO unit. This can be achieved readily in design by replacing the RO unit with an SRU as shown in Figure 20.6 (red dashed box). The RO unit provides all the benefits of the SRU plus the incremental oil recovery due to LSE. However, the RO unit involves additional equipment and operation costs that should be considered in an incremental economic analysis. For instance, RO has 50% rejection rate of feed water compared to 25% for SRU. Table 20.2 compares the similarities and differences between RO units and SRUs.

During this step, the reservoir engineer should pay particular attention to the water quality specifications and ensure that those specifications would be met by the equipment selected to deliver water for injection. The reservoir engineer should lead the team that have been tasked to develop the water quality specifications. The main contribution of the reservoir engineer is to ensure that the water quality specified for the injection system will not damage the reservoir and should meet the salinity range designated for the project. This responsibility continues during construction, installation, and commissioning of the water injection facility.

TABLE 20.1 Quality Specifications for Injection Water

Parameter	Seawater	Sulfate Removal Water	Low Salinity Water
Total dissolved solids	Ambient	No specification	Can achieve < 500 ppm
pH	Ambient	No specification	5.5-6.5
Sulfate reduction	Ambient	<20 mg/L	<5 mg/L
Total suspended solids	<5 mg/L	<5 mg/L	<2 mg/L
Absolute particle size	<1 μm	<1 μm	<1 μm
Dissolved oxygen	<10 ppb	<10 ppb	<10 ppb
Sulfate reducing bacteria	<1 CFU/L	<1 CFU/L	<1 CFU/L

TABLE 20.2 SRU vs. RO: Similarities and Differences

Parameter	Sulfate Removal Unit (SRU)	Reverse Osmosis (RO) Unit
Filtration	Needs coarse and ultra-filtration	Same
Operations and maintenance	Needs continuous chemical injection and periodic chemical cleaning	Same
Salinity reduction	Minimal	Can achieve < 500 ppm
Sulfate reduction	<20 mg/L	<5 mg/L
Feed water reject rate	25% (Feed = 1.33 × Inj. required)	50% (Feed = 2 × Inj. required)
Operating pressure	350-450 psi	1000-1200 psi
Number of stages	2	1

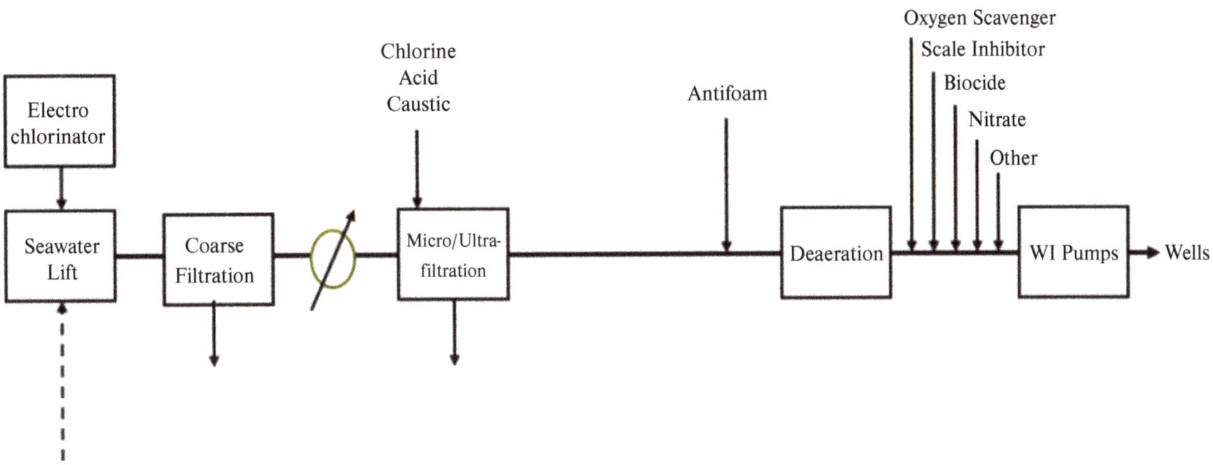

▲ **FIGURE 20.5** **Block flow diagram for offshore seawater injection facility**

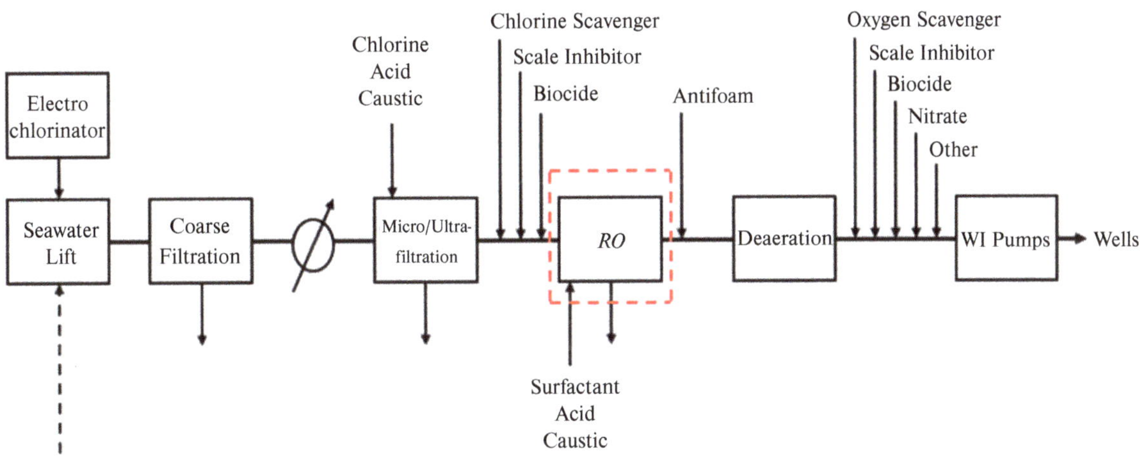

▲ **FIGURE 20.6** **Block flow diagram for offshore low salinity water injection facility**

Step 5: Optimize Injection Water Salinity/Composition by Blending

The four types of most common swelling clays in the reservoir are Smectite (includes Montmorillonite), Illite, Smectite/Illite mixed layer, and Kaolinite. In this group, smectite clays are the most sensitive to swelling from freshwater. In 1990, Scheuerman and Bergersen[53] published flocculation salinity diagrams for the four clay types. These diagrams are reproduced as Figure 20.7. Scheuerman and Bergersen identified four zones in these diagrams to indicate potential effect of water composition on each clay type. These four zones are demarcated in Figure 20.7 as:

- Zone A - Compatible water (no impairment)
- Zone B - Impairing water
- Zone C - Some impairment
- Zone D - Unknown- need more data

To use these diagrams, injection water composition is plotted on the flocculation salinity diagram of the most water-sensitive clay in the reservoir. The compatibility of the water with the particular clay is determined based on the zone that the composition of the injection water falls. The flocculation salinity diagram for mixed clay (Smectite/Illite) is expanded in Figure 20.8 for clarity. It is important to note that it is possible to devise water compositions that can be verified to be compatible in Zone D through experiments with cores from the candidate reservoir.

Permeate water from RO is low in salinity (in some cases less than 500 ppm) and divalent ions. This can cause clays to swell and deflocculate leading to impairment of the reservoir and loss of permeability. There are opportunities to mitigate the risks from clay swelling/deflocculation by blending the RO permeate with intermediate streams from the water processing facilities. A partial flow schematic diagram for processing of

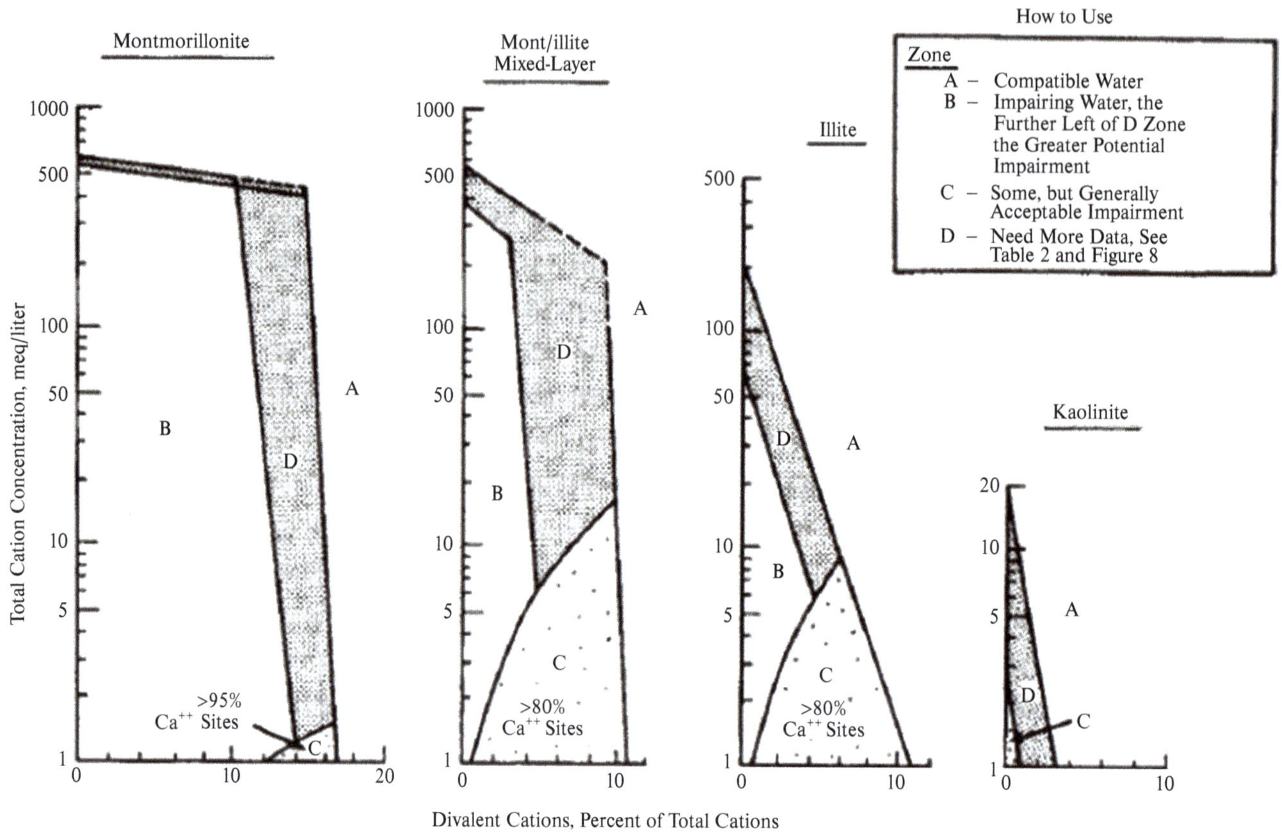

▲ **FIGURE 20.7** **Salinity compatibility criteria for various types of clay (from Scheuerman and Bergersen[53])**

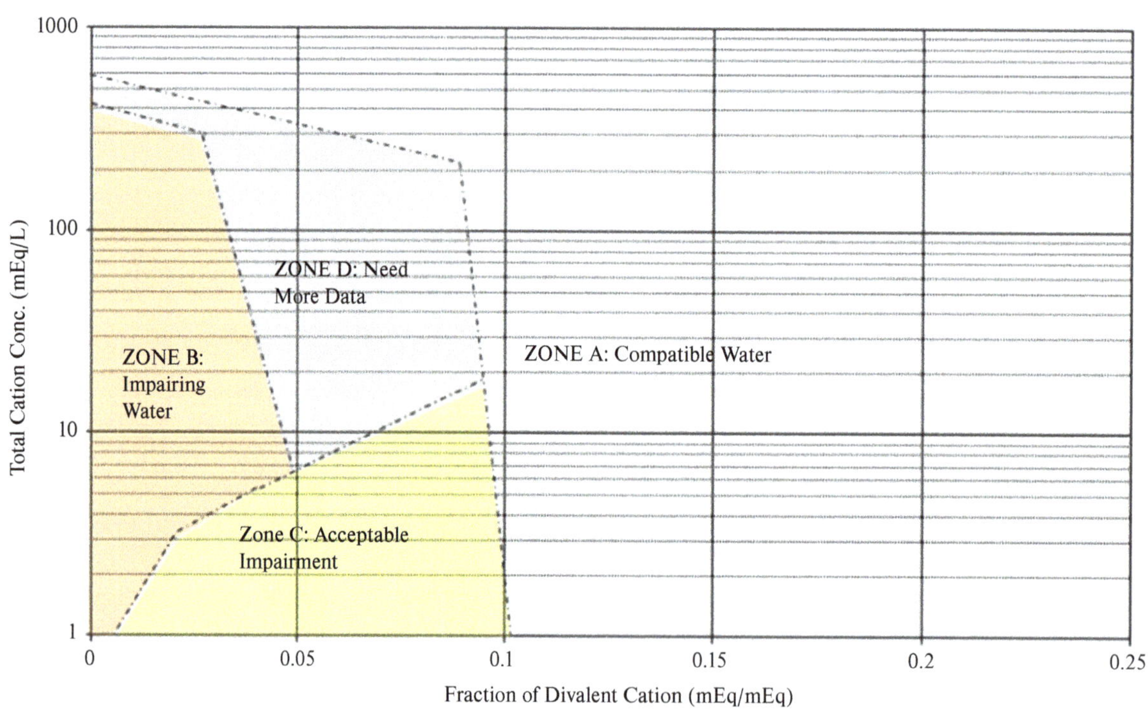

▲ **FIGURE 20.8** **Salinity compatibility criteria for mixed clays (Smectite and Illite)**

seawater for blending is shown in Figure 20.9. Note that other configurations of the water processing facility are possible. Figure 20.9 is used only to illustrate potential blending paths for the injection water. Figure 20.9 shows five sources of water that can be used to blend RO permeate as follows:

1. Filtered seawater
2. Nano-filtration (NF) Array 1 reject
3. NF Array 2 reject
4. NF-combined permeate (or RO feed)
5. RO reject

The compositional pathways for blending these five sources of intermediate streams with RO permeate are shown in Figure 20.10 for Smectite clays (Ayirala et al.[54]). Note that Figure 20.10 shows the four zones (Zones A, B, C, and D) of water effect on clays as illustrated in Figures 20.7 and 20.8. Also, note that Figure 20.10 shows the water salinity lines of 1000 ppm and 4000 ppm required for the injection water produced from the blending process. The required injection water salinity range could vary depending on requirements for specific reservoirs. Figure 20.10 shows dilution paths for the blending of filtered seawater, NF array 1 reject, NF array 2 reject, NF-combined permeate, and RO reject with RO permeate water. Note that all compositions of NF-combined permeate and RO reject water blended with RO permeate water will be in Zone B (reservoir impairment zone) even though some compositions fall within the desired salinity range. On the other hand, blending the RO permeate water with filtered seawater (3% vol) or NF array 1 reject (2% vol), or NF array 2 reject (1% vol) would generate blended water that falls in Zone A (no reservoir impairment zone) and also satisfy

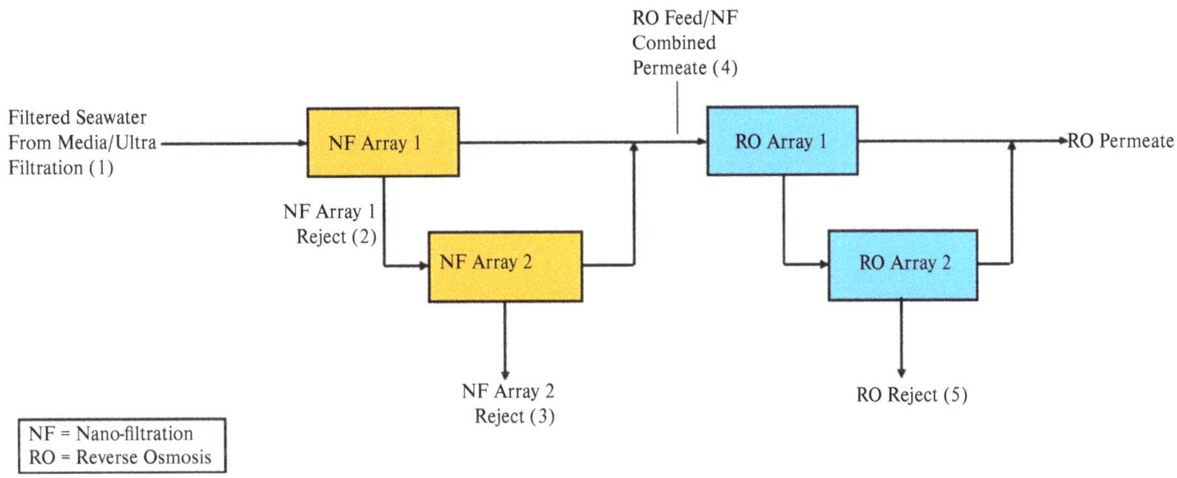

▲ **FIGURE 20.9** **Flow schematic showing sources of intermediate streams for blending**

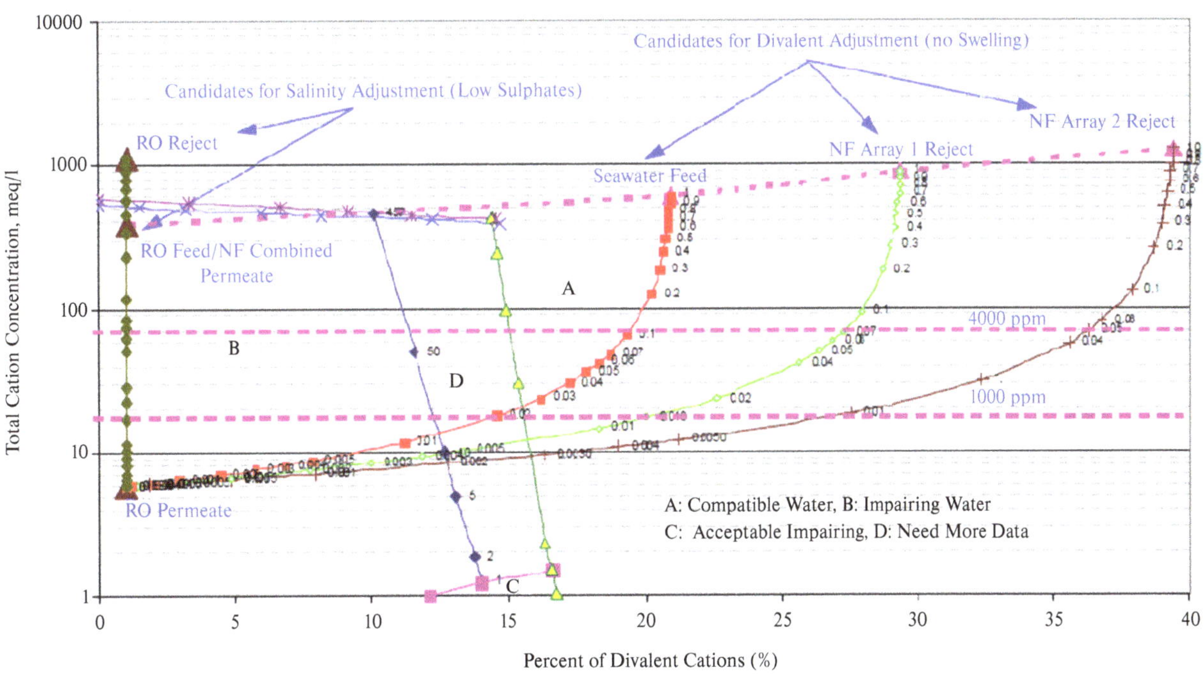

▲ **FIGURE 20.10** **Effect of blended water on flocculation salinity of injection water for Smectite clays (from Ayirala et al.[54])**

the salinity range requirement. Note that blending with filtered seawater could introduce risk of scale formation and reservoir souring if the sulfate concentration is not reduced below certain levels (e.g., <40 ppm). Similar compositional pathways for water blending are shown in Figure 20.11 for Illite clay (Ayirala et al.[54]). Note that this clay is less sensitive to freshwater swelling than Smectite so the impairment zone (Zone B) is much smaller in comparison. Blending with RO permeate water to achieve the stated salinity range would require the same volumes of seawater (3% vol), or NF array 1 reject (2% vol), or NF array 2 reject (1% vol) as was the case for Smectite clay but the composition of the blended water is located further away in the no impairment areas of Zone A (Figure 20.11).

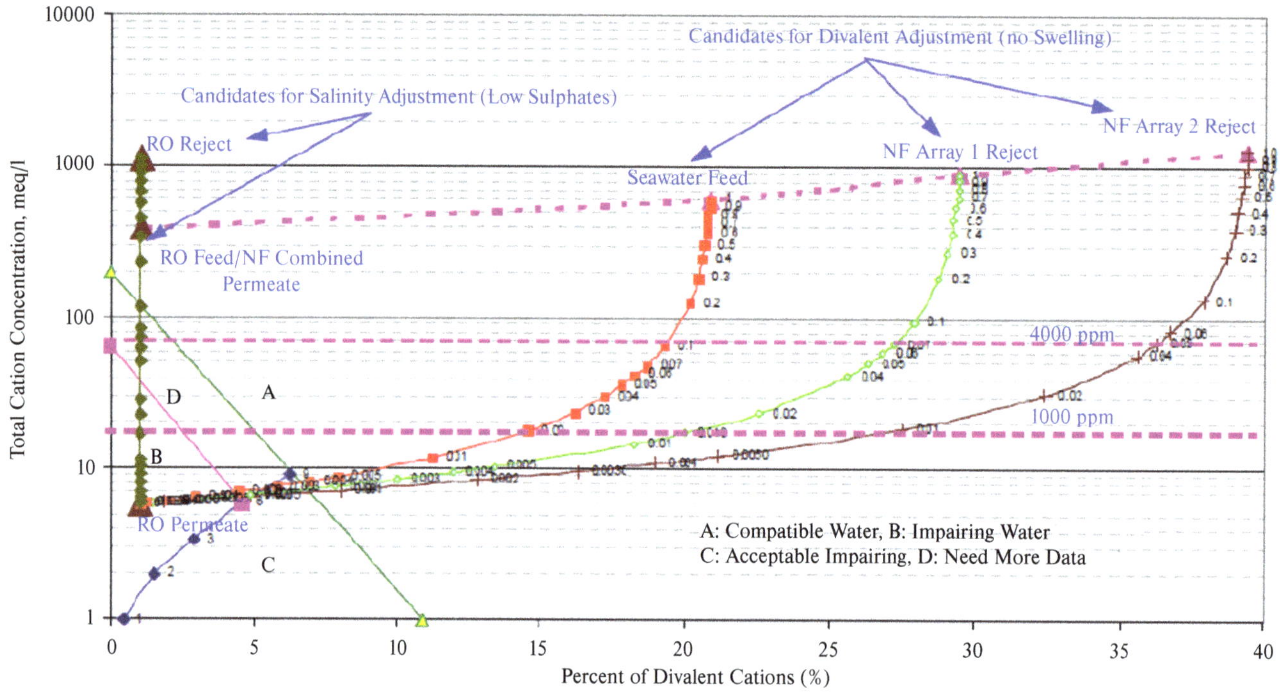

▲ **FIGURE 20.11** **Effect of blended water on flocculation salinity of injection water for Illite clays (from Ayirala et al.[54])**

Step 6: Project Economics

Economic evaluation of LSWF project is conducted as incremental with the HSWF as the base case. Consequently, all the costs including capital expenditure (CAPEX) and operational expenditures (OPEX) are treated as incremental. Extreme care should be exercised in the determination and assignment of these costs so that the LSWF project is not burdened with costs that normally should be carried by the HSWF as the base case. Some costs that should be assigned to the LSWF project include the cost of desalination units, nano-filtration units, extra pumping units, and additional power-generation units needed to operate these units. The incremental operating expenses required to run all the equipment associated with LSW generation should be identified and assigned to the incremental LSWF case.

The production profiles (oil, water, and gas) are also incremental over the HSWF base case. It is important that incremental hydrocarbon recovery associated with LSWF is carefully determined and fully accounted for in the production profile. The incremental recoveries are sometimes marginal and imprecise accounting of the production profile could render the LSWF case uneconomical.

The criteria for determining the economic feasibility of incremental projects varies across many companies. Many companies may use net present value (NPV), profitability index (PI), pay-out (PO), etc. Note that economic evaluation of LSWF project should be determined most effectively as an incremental case. Economic evaluations of petroleum projects using these criteria are presented in Chapter 25.

Step 7: Risk Assessments and Mitigations

The final step in planning, designing, and installing an LSWF project is to conduct a preliminary exercise of risk assessments and mitigations. This step is important because it raises the awareness of all participants that expectations from LSF may not turn out as planned or designed. And that it is necessary to consider remedies if adverse consequences are to be avoided. To initiate this step, some examples of risk assessments and mitigation are given in Table 20.3, but it should be noted that these examples are not exhaustive nor complete for any LSWF project. The project team should explore, discuss, and develop other potential risks specific to the project and devise ways to mitigate them.

TABLE 20.3 | LSWF Risks Assessments and Mitigations

RISKS	MITIGATIONS
LSW outside specified operating water quality range	• Install automated monitoring systems on water generating plant. Plans to monitor permeate salinity, water hardness, ionic concentration with probes. Validate with periodic lab tests. • Dump water not meeting specifications overboard if offshore • Reduce/shut-in water injection • Reduce/shut-in production • Convert to SRU as last option if problem(s) cannot be resolved
Formation swelling causing injectivity loss due to permeability reductions	• Isolate and monitor injectivity of some injection wells at start-up • Install automated monitoring systems on water generating plant. Monitor permeate salinity, water hardness, ionic concentration with probes. Validate with periodic lab tests. • Install equipment for continuous monitoring of injection well performance. Monitor salinity at wellhead, measure injection rates, and injection pressures with wellhead and downhole gauges. • Conduct additional injectivity tests with new cores
Incremental oil recovery due to LSW injection below forecast	• Review results from laboratory corefloods. Ensure results were interpreted accurately • Review simulation of LSE in reservoir models. Modify as necessary with field performance data • Perform injection and production logs to check vertical and areal sweep efficiencies between injectors and producers. Re-distribute injection coverage if necessary. • Conduct additional SCAL tests with new cores
Water treatment equipment and/or membrane failure (downtime)	• Over-inject during uptime to build voidage bank • Have standby units as insurance • Reduce/shut-in production

20.8 Brief Case History of Claire Ridge LSWF Project

British Petroleum (BP) announced first oil production from the LSWF project in parts of the Claire Ridge reservoir on November 23, 2018. A brief review of this first offshore implementation of an LSWF project is considered necessary to illustrate that the technology has moved from laboratory and field tests to actual field installation. Even though the full results and experiences from this field implementation have not been published in the petroleum literature to date (November 2018), it is still a notable development and readers are well-advised to monitor the industry literature for such publications which are expected not to be available for some time.

20.8.1 An Overview of Clair Ridge Field

The Clair is the largest oil reservoir in the United Kingdom continental shelf with over 6 billion barrels of OOIP. The reservoir consists of Devonian sandstones with low formation permeability (averaging 50 md) and contains oil with relatively high viscosity of 3.2 cP as average. This combination of low formation permeability and high viscous oil resulted in estimates of potentially non-commercial oil production rates (less than 1000 bopd) in the harsh North Atlantic offshore environment. It was later discovered in 1991 that the formations contain interconnected open fractures which could be exploited with horizontal wells to achieve oil production at commercial rates. This led to the development of Clair Phase 1 project that achieved first oil in February 2005. Five years of production history from Clair Phase 1 project indicated that oil is being produced from the matrix by several drive mechanisms that included viscous displacement, gravity drainage, and water imbibition from fracture into the matrix rock. The success with Clair Phase 1 led to development plans for Clair Phase 2 which will focus on an area of the field called Clair Ridge. After considering gas flooding and carbon dioxide flooding, LSWF was selected for the Clair Ridge project (Robbana et al.[55]). The Clair Ridge project is projected to recover 640 million barrels of oil through 2050. First oil was announced on November 23, 2018, with peak production projected to reach 120,000 bopd later. As of November 2018, the partners in the Clair Field are BP (28.6%), Shell (28%), Chevron (19.4%), and Conoco-Phillips (24%). BP is the operator of the field.

20.8.2 Low Salinity Coreflood Tests

As recommended in Step 2 of Section 20.7, corefloods were conducted at reservoir temperature and pressure with live reservoir fluids on core samples obtained from the Main Field area (Well A02) and the Ridge area (Wells 13Z and 14/14Z). The incremental oil recovery over high salinity corefloods range from 9% to 13.1% (Robbana et al.[55]).

20.8.3 Reservoir Modeling and Analysis

Sector models with 48 different "reservoir description types" based on combinations of "possible joint and sub-seismic fracture descriptions" were used in the simulation studies (Robbana et al.[55]). For the sector model of each reservoir description type, a high salinity waterflood and a low salinity waterflood were conducted. Incremental oil recovery due to low salinity effect was calculated as the difference in total oil recovery between the two types of waterflooding. The average incremental oil recovery due to low salinity flooding for the 48 sector models was 14.5%.

20.8.4 Selection of Water Injection Facilities

The desalination method selected for generation of low salinity water was reverse osmosis. The facility design shown as Figure 20.12 is essentially similar to Figure 20.6 as discussed in Step 4 of Section 20.7. It is worthy to note that reverse osmosis was selected for the Clair Ridge project because it is a proven technology for generation of low salinity water, has less footprint, and weighs less in comparison to other methods for generating low salinity water as noted in Step 4 of Section 20.7.

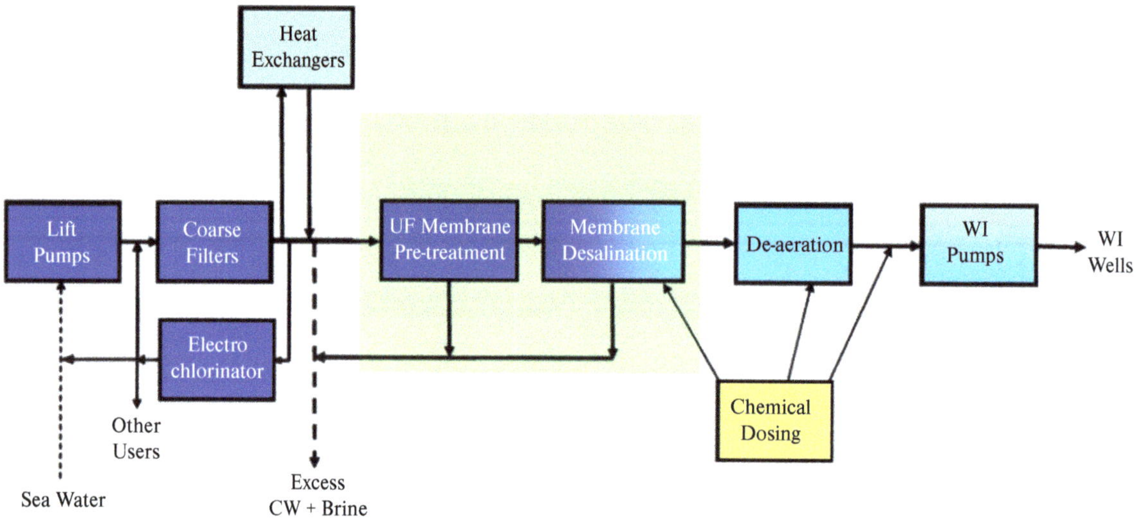

▲ FIGURE 20.12 **Schematic diagram of Clair Ridge water injection facility**

Nomenclature

S_{or}	Residual oil saturation
S_{wi}	Initial water saturation

Abbreviations

BP	British Petroleum
CAPEX	Capital Expenditure
EDL	Electric Double Layer
HS	High Salinity
HSW	High Salinity Water
HSWF	High Salinity Waterflooding
LS	Low Salinity
LSE	Low Salinity Effect
LSW	Low Salinity Water
LSWF	Low Salinity Waterflooding
MIE	Multicomponent Ionic Exchange
NF	Nano-Filtration
OOIP	Original Oil-In-Place
OPEX	Operational Expenditure
PDI	Potential Determining Ions
RO	Reverse Osmosis
SCAL	Special Core Analysis
SRU	Sulfate Removal Unit
SWCTT	Single Well Chemical Tracer Test
TDS	Total Dissolved Solids

REFERENCES

1. Morrow, N., and Buckley, J.: "Improved Oil Recovery by Low-Salinity Waterflooding," paper SPE 129421, Distinguished Author Series, *JPT* (May 2011) 106-112.
2. Bernard, G.G.: "Effect of Floodwater Salinity on Recovery of Oil from Cores Containing Clays," paper SPE 1725 presented at the 38th Annual California Regional Conference, Los Angeles, California, (October 26-27, 1967).
3. Jadhunandan, P.: *Effects of Brine Composition, Crude Oil and Aging Conditions on wettability and Oil Recovery*, PhD Dissertation, PRRC, New Mexico Tech, 1990.
4. Jadhunandan, P., and Morrow, N.R.: "Spontaneous Imbibition of Water by Crude Oil/Brine/Rock Systems," *In Situ*, 15 (4), (1991), 319-345.
5. Jadhunandan, P., and Morrow, N.R.: "Effect of Wettability on Waterflood Recovery for Crude Oil/Brine/Rock Systems," *SPERE*, 10(1), (1995), 40-46.
6. Yildiz, H.O., and Morrow, N.R.: "Effect of Brine Composition Recovery by Waterflooding Moutray Crude Oil," *J. Petrol. Sci. and Eng.*, 14, (1996), 159-168.
7. Tang, G.Q., and Morrow, N.R.: "Salinity, Temperature, Oil Composition and Oil Recovery by Waterflooding," SPE 36680-PA, *SPE Journal*, (November 1, 1997).
8. Tang, G.-Q., and Morrow, N.R.: "Influence of Brine Composition and Fines Migration on Crude Oil/Brine/Rock Interactions and Oil Recovery," *J. Petrol. Sci. and Eng.* (1999) 24, 99-111.
9. Tang, G., and Morrow, N.R.: "Oil Recovery By Waterflooding and Imbibition: Invading Brine Cation Valency and Salinity," paper SCA9911 presented at the International Symposium of the Society of Core Analysts held in Golden, Colorado, (August 1-4, 1999).
10. Webb, K.J., Black, C.J.J., and Edmonds, I.J.: "Low Salinity Oil Recovery: The Role of Reservoir Condition Corefloods," paper C18 presented at the 13th European Symposium on Improved Oil Recovery, Budapest, Hungary, (2005).
11. Webb, K.J., Black, C.J.J., and Tjetland, G.: "A Laboratory Study Investigating Methods for Improving Oil Recovery in Carbonates," International Petroleum Technology Conference (IPTC), Doha, Qatar, (2005).
12. Zhang, Y., and Morrow, N.R.: "Comparison of Secondary and Tertiary Recovery with Change in Injection Brine Composition for Crude Oil/Sandstone Combinations," paper SPE 99757 presented at the 2006 SPE/DOE Symposium on Improved Oil Recovery held in Tulsa, Oklahoma, (22-26 April 2006).
13. Zhang, P., Tweheyo, M.T., and Austad, T.: "Wettability Alteration and Improved Oil Recovery by Spontaneous Imbibition of Seawater Into Chalk: Impact Of The Potential Determining Ions: Ca^{2+}, Mg^{2+} and SO_4^{2-}. Colloids and Surfaces A: *Physicochem. Eng. Aspects*, (2007) 301, 199-208.
14. Zhang, Y., Xie, X., and Morrow, N.R.: "Waterflood Performance by Injection of Brine with Different Salinity for Reservoir Cores," paper SPE 109849 presented at the 2007 SPE Annual Technical Conference and Exhibition held in Anaheim, California, (November 11-14, 2007).
15. Lager, A., Webb, K.J., and Black, C.J.J.: "Impact of Brine Chemistry on Oil Recovery," paper A24 presented at the 14th European Symposium on Improved Oil Recovery, Cairo, Egypt, (2007).
16. Lager, A., Webb, K.J., Black, C.J.J., Singleton, M., and Sorbie, K.S.: "Low Salinity Oil Recovery: an Experimental Investigation," *Petrophysics* (2008) 49(1), 28-35.
17. Lager, A., Webb, K.J., Collins, I.R., and Richmond, D.M.: "LoSal™ Enhanced Oil Recovery: Evidence of Enhanced Oil Recovery at The Reservoir Scale," paper SPE 113976 presented at the 2008 SPE/DOE Improved Oil Recovery Symposium, Tulsa, Oklahoma,
18. Ligthelm, D.J., Gronsveld, J., Hofman, J., Brusee, N., Marcelis, F., and van der Linde, H.: "Novel Waterflooding Strategy by Manipulation of Injection Brine Composition," paper SPE 119835 presented at the 2009 SPE EUROPEC/EAGE Annual Conference and Exhibition held in Amsterdam, The Netherlands, (June 8-11, 2009).
19. Webb, K.J., Black, C.J.J., and Al-Jeel, H.: "Low Salinity Oil Recovery - Log-Inject-Log." paper SPE 81460 presented at the SPE 13th Middle East Oil Show and Conference, Bahrain, (April 5-8, 2003).
20. McGuire, P.L., Chatham, J.R., Paskvan, F.K., Sommer, D.M., and Carini, F.H.: "Low Salinity Oil Recovery: An Exciting New EOR Opportunity for Alaska's North Slope," paper SPE 93093 presented at the 2005 SPE Western Regional Meeting, Irvine, California, (March 30-April 1, 2005).
21. Seccombe, J.C., Lager, A., Webb, K.J., Jerauld, G., and Fueg, E.: "Improving Waterflood Recovery: LowSal™ EOR Field Evaluation," paper SPE 113480 presented at 2008 SPE/DOE Improved Oil Recovery Symposium, Tulsa, Oklahoma, (April 19-23, 2008).
22. Seccombe, J., Lager, A., Jerauld, G., et al.: "Demonstration of Low-Salinity EOR at Interwell Scale, Endicott Field, Alaska," paper SPE 129692 presented at the 2010 SPE Improved Oil Recovery Symposium, Tulsa, Oklahoma, (April 24-28, 2010).
23. Sylte, J.E., Hallenbeck, L.D., and Thamas, L.K.: "Ekofisk Formation Pilot Waterflood," paper SPE 18276 presented at SPE Annual Technical Conference and Exhibition, Houston, Texas, (October 2-5, 1988).
24. Hallenbeck, L.D., Sylte, J.E., Ebbs, D.J., Thamas, L.K.: "Implementation of the Ekofisk Field Waterflood," *SPE Form Eval.* (1991) 6, 284-290.
25. Austad, T., Strand, S., Hognesen, E. J., and Zhang, P.: "Seawater as IOR Fluid in Fractured Chalk," paper SPE 93000 presented at the International Symposium on Oilfield Chemistry, The Woodlands, Texas, (February 2-4, 2005).
26. Zhang, P., and Austad, T: "Waterflooding in Chalk: Relationship between Oil Recovery, New Wettability Index, Brine Composition, and Catonic Wettability Modifier," paper SPE 94209 presented at the SPE Europec/EAGE Annual Conference, Madrid, Spain, (June 13-16, 2005).
27. Zhang, P., and Austad, T.: "Wettability and Oil Recovery from Carbonates: Effects of Temperature and Potential Determining Ions," *Colloid Surface A* (2006) 279, 1-3, 179-187.
28. Zhang, P., Tweheyo, M.T., and Austad, T.: "Wettability Alteration and Improved Oil Recovery in Chalk: The Effect of Calcium in the Presence of Sulfate," *Energ. Fuel.* (2006), 20(5), 2056-2062.

29. Strand, S., Hognesen, E.J., and Austad, T.: "Wettability Alteration of Carbonates: Effects of Potential Determining Ions (Ca^{2+} and SO_4^{2-}) and Temperature," *Colloid Surface A* (2006), 275(1-3), 1-10.
30. Zhang. P., Tweheyo, M.T., and Austad, T.: "Wettability Alteration and Improved Oil Recovery by Spontaneous Imbibition of Seawater into Chalk: Impact of the Potential Determining Ions Ca^{2+}, Mg^{2+}, and SO_4^{2-}," *Colloid Surface A* (2007), 301(1), 199-208.
31. Strand, S., and Austad, T.: "Effect of Temperature on Enhanced Oil Recovery from Mixed-wet Chalk Cores by Spontaneous Imbibition and Forced Displacement using Seawater," *Energ. Fuel.* (2008), 22(5), 3222-3225.
32. Webb, K.J., Black, C.J.J., and Tjetland, G.: "A Laboratory Study Investigating Methods for Improving Oil Recovery in Carbonates," paper IPTC 10506 presented at the International Petroleum Technology Conference in Doha, Qatar, (November 21-23, 2005).
33. Yousef, A.A., Al-Saleh, S., A-Kaabu, A., and Al-Jawfi, M.: "Laboratory Investigation of the Impact of Injection-Water Salinity and Ionic Content on Oil Recovery from Carbonate Reservoirs," *SPE Form Eval.* (2011), 14(15), 578-593.
34. Al Harrasi, A., Al-maamari, R.S., and Masalmeh, S.K.: "Laboratory Investigation of Low Salinity Waterflooding for Carbonate Reservoirs," paper SPE 161468 presented at the Abu Dhabi International Petroleum Conference and Exhibition, Abu Dhabi, (November 11-14, 2012).
35. Sepehrnoori, K., Pope, G., and Al-shalabi, E.W.: "Geochemical Interpretation of Low Salinity Water Injection in Carbonate Oil Reservoirs," paper SPE 169101 presented at the SPE Improved Oil Recovery Symposium, Tulsa, (April 12-16, 2014).
36. Yousef, A. A., Liu, J., Blanchard, G., et al.: "SmartWater Flooding: Industry's First Field Test in Carbonate Reservoirs," paper SPE 159526 presented at the SPE Annual Technical Conference and Exhibition, San Antonio, Texas, (October 8-10, 2012).
37. Yousef, A.A., Al-Saleh, S.H., and Al-Jawfi, M.S.: "SmartWater Flooding for Carbonate Reservoirs: Salinity and Role of Ions," paper SPE 141082 presented at the SPE Middle East Oil and Gas Show and Conference, Manama, Bahrain, (September 25-28, 2011).
38. Berg, S., Cense, A.W., Jansen, E., and Bakker, K.: "Direct Experimental Evidence of Wettability Modification by Low Salinity," paper SCA2009-12 presented at the International Symposium of the Society of Core Analysts, Noordwijk, The Netherlands, (September 27-30, 2009).
39. Doust, A.R., Puntervold, T., Strand, S., and Austad, T.: "Smart Water as Wettability Modifier in Carbonate and Sandstone: A Discussion of Similarities/Differences in the Chemical Mechanisms," *Energy Fuels* (2009), 23, 4479-4485.
40. Vledder, P., Gonzalez, I.E., Fonseca, C., Cesar, J., Terrence, W., and Ligthelm, D.J.: "Low Salinity Water Flooding: Proof of Wettability Alteration on A Field Wide Scale," paper SPE 129564 presented at the SPE Improved Oil Recovery Symposium, Tulsa, Oklahoma, (April 24-28, 2010).
41. Austad, T., Doust, A.R., and Puntervold, T.: "Chemical Mechanisms of Low Salinity Water Flooding in Sandstone Reservoirs," paper SPE 129767 presented at the SPE Improved Oil Recovery Symposium, Tulsa, Oklahoma, (April 24-28, 2010).
42. Rivet, S.M., Lake, W., and Pope, G.A.: "A Coreflood Investigation of Low-Salinity Enhanced Oil Recovery," paper SPE 134297 presented at the SPE Annual Technical Conference and Exhibition, Florence, Italy, (September 19-22, 2010).
43. Jackson, M.D., Vinogradov, J., Hamon, G., and Chamerois, M.: "Evidence, Mechanisms And Improved Understanding of Controlled Salinity Waterflooding Part 1: Sandstones," *Fuel* (2016), 185, 772-793.
44. Al-Shalabi, E.W., and Sepehrnoori, K.: "A Comprehensive Review of Low Salinity/Engineered Water Injections and Their Application in Sandstone and Carbonate Rocks," *J. Petrol. Sci. Eng.* (2016), 139, 137-16.
45. Jerauld, G.R., Lin, C.Y., Webb, K.J., and Seccombe, J.C.: "Modeling Low-Salinity Waterflooding," *SPEREE*, December 2008; SPE102239-PA.
46. Lager, A., Webb, K.J., Black, C.J.J., Singleton, M., and Sorbie, K.S.: "Low-Salinity Oil Recovery-An Experimental Investigation," paper SCA2006-36 presented at the International Symposium of the Society of Core Analysts, Trondheim, Norway, (September 12-16, 2006).
47. Lee, S.Y., Webb, K.J., Collins, I.R., et al.: "Low Salinity Oil Recovery: Increasing Understanding of the Underlying Mechanisms," paper SPE 129722 presented at the SPE improved Oil Recovery Symposium, Tulsa, Oklahoma, (April 24-28, 2010).
48. Nasralla, R.A., and Nasr-El-Din, H.A.: "Double-Layer Expansion: Is it a Primary Mechanism of Improved Oil Recovery by Low-Salinity Waterflooding?" *SPEREE*, February 2014; SPE 154334-PA.
49. Gupta, R., Smith, G.G., Willingham, T., Lo Cascio, M., Shyeh, J.J., and Harris, C.: "Enhanced Waterflood for Middle East Carbonate Cores: Impact of Injection Water Composition," paper SPE 142668 presented at SPE Middle East Oil and Gas Show and Conference, Manama, Bahrain, (September 25-28, 2011).
50. Almehaideb, R.A., Ghannam, M.T., and Zekri, A.Y.: "Experimental Investigation of Contact Angles of Crude Oil-Microbial Solution on Carbonate Rocks," *Pet. Sci. & Technol.* (2004) 22(3-4), 423-438.
51. Alotaibi, M.B., and Yousef, A.A.: "The Role of Individual and Combined Ions in Waterflooding Carbonate Reservoirs: Electrokinetic Study," SPEREE, February 2017; SPE 177983-PA.
52. Austad, T., Shariatpanahi, S.F., Strand, S., Black, C.J.J., and Webb, K.J.: "Conditions for a Low-Salinity Enhanced Oil Recovery (EOR) Effect in Carbonate Oil Reservoirs," *Energy Fuels* (2011), 26, 1, 569-575.
53. Scheuerman, R.F., and Bergersen, B.M.: "Injection-Water Salinity, Formation Pretreatment, and Well-Operations Fluid-Selection Guidelines," *JPT* (July 1990) 836-845; SPE 18461-PA.
54. Ayirala, S., Uehara-Nagamine, E., Matzakos, A., Chin, R., Doe, P., and Van den Hoek, P. J.: "A Designer Water Process for Offshore Low Salinity and Polymer Flooding Applications," paper SPE 129926 presented at the SPE Improved Oil Recovery symposium, Tulsa, Oklahoma, (April 24-28, 2010).
55. Robbana, E., Buikema, T., Mair, C., et al.: "Low Salinity Enhanced Oil Recovery - Laboratory to Day One Field Implementation - *Losal* EOR into the Clair Ridge Project," paper SPE 161750 presented at the Abu Dhabi International Petroleum Exhibition and Conference, Abu Dhabi, UAE, (November 11-14, 2012).

ADDITIONAL REFERENCES

- Cissokho, M., Bertin, H., Boussour, S., Cordier, P., and Hamon, G.: "Low Salinity Oil Recovery on Clayey Sandstone: Experimental Study," *Petrophysics* (2010), 51.
- Kumar, M., Fogden, A., Morrow, N.R., and Buckley, J.S.: "Mechanisms of Improved Oil Recovery from Sandstone by Low Salinity Flooding," *Petrophysics-SPWLA-Journal of Formation Evaluation and Reservoir Description* 52, no. 6 (2011); 428.
- Law, S., and Sutcliffe, P.: "Secondary Application of Low Salinity Waterflooding to Forties Sandstone Reservoirs," paper SPE 170725-MS presented at the SPE Annual Technical Conference and Exhibition, Amsterdam, The Netherlands, (October 27-29, 2014).
- Myint, P.C., and Firoozabadi, A.: "Thin liquid films in improved oil recovery from low-salinity brine," *Curr. Opin. Colloid. In.* 20.2 (2015); 105-114.
- Romanuka, J., Hofman, J., Ligthelm, D.J., et al.: "Low Salinity EOR in Carbonates," paper SPE 153869-MS presented at the Eighteenth SPE Improved Oil Recovery Symposium, Tulsa, Oklahoma, (April 14-18, 2012).
- Rotondi, M., Callegaro, C., Masserano, F., and Bartosek, M.: "Low Salinity Water Injection: eni's Experience," paper SPE 171794-MS presented at the Abu Dhabi International Petroleum Exhibition and Conference, Abu Dhabi, UAE, (November 11-13, 2014).
- Shehata, A.M., and Nasr-El-Din, H.A.: "Laboratory Investigations to Determine the Effect of Connate-Water Composition on Low-Salinity Waterflooding in Sandstone Reservoirs," *SPE RE&E* (February 2017); 171690-PA.
- Shehata, A.M., and Nasr-El-Din, H.A.: "The Role of Sandstone Mineralogy and Rock Quality in the Performance of Low-Salinity Waterflooding," *SPE RE&E* (February 2017); 181754-PA.
- Sorbie, K.S., and Collins, I.R.: "A Proposed Pore-Scale Mechanism for How Low Salinity Waterflooding Works," paper SPE 129833 presented at the 2010 SPE Improved Oil Recovery Symposium, Tulsa, Oklahoma, (April 24-28, 2010).
- Webb, K., Lager, A., and Black, C.: "Comparison of High/Low Salinity Water/Oil Relative Permeability," International Symposium of the Society of Core Analysts, Abu Dhabi, UAE. Vol. 29, 2008.

PROBLEMS

20-1. The low salinity effect (LSE) that increases oil recovery from sandstone and carbonate rocks has been described by different mechanisms. Discuss the mechanism that appears most likely responsible for LSE. Could LSE result from a combination of mechanisms?

20-2. An oil reservoir operated by your company has been identified as a potential candidate for low salinity waterflooding (LSWF). Write a detailed proposal on the activities/work that should be completed in screening the reservoir for application of LSWF.

20-3. Suppose the reservoir in Problem 20-2 meets all the screening criteria as an LSWF candidate, develop a scope of work for conducting injectivity and special core analysis on core samples from the reservoir. What measures (permeability reduction, oil recovery, etc.) should be established to determine success or failure of the core tests?

20-4. An offshore deepwater reservoir has been selected as a candidate for installation of low salinity water injection. Discuss potential barriers that could be encountered in the design and installation of low salinity water injection equipment on:

a. An existing platform

b. New platform

c. What are the potential impact of economics on the installation of the injection equipment on either an existing platform or a new platform?

20-5. Low salinity waterflood projects are very sensitive to oil prices. List factors that could improve the economic feasibility of low salinity waterflood projects in a low oil price environment. (Low oil price assumed as less than \$40 per bbl.)

CHAPTER

21 Enhanced Oil Recovery

21.1 Introduction

Enhanced oil recovery (EOR) is sometimes called tertiary oil recovery or improved oil recovery. In this book, EOR processes are defined to include **all** processes that increase oil recovery **beyond** primary or secondary recovery processes. This definition was broadened to include EOR processes that have been used to improve oil recovery beyond the primary stage of reservoir depletion, such as steam flooding, or at the end of secondary recovery processes that involve waterflooding. This definition of EOR processes might appear to be in conflict with the inclusion of immiscible gasflooding as a secondary recovery process in Chapter 19. Immiscible gasflooding was presented in Chapter 19 as a matter of convenience since the immiscible process of waterflooding was also presented in that chapter. A case can be made that immiscible gasflooding is an EOR process. Such classification is also acceptable. Also, note that low salinity waterflooding presented in Chapter 20 may be considered an EOR process.

A typical EOR process involves injecting a fluid other than water into the reservoir. These EOR injectants usually have economic values either as marketable fluids or the cost incurred to procure and process them for injection into the reservoir. Consequently, EOR processes are very sensitive to oil prices. The price of oil on a sustainable basis must exceed the cost of the injectant plus operating costs by a sizeable margin for an EOR process to be considered economical. For this reason, an EOR process must be **efficient** in terms of cost per barrel of oil recovered and also **effective** in substantially increasing the volume of oil recovered beyond the current recovery process. Economic evaluation is the key important step in the selection of an EOR process and is emphasized throughout the selection process.

EOR processes are important as technologies that could help meet the growing demand for oil in the world. It is estimated that roughly 65% of the original oil-in-place (OOIP) remains in the reservoir after primary and secondary recoveries for conventional reservoirs. The remaining oil in unconventional reservoirs can be as high as 90% of OOIP. This remaining oil can be recovered by applying suitable EOR processes. The potential for EOR processes is clearly substantial and is responsible for the growth of EOR projects in all oil producing regions of the world.[1] In this chapter, the three main classes (gas injection, chemical, and thermal) of EOR processes are presented with sub-processes that are grouped under each class. This is followed with a summary of criteria for selection of EOR processes based on reservoir properties. Each EOR process is presented later with more details in terms of the mechanism that facilitates oil recovery from the reservoir, and the reservoir conditions that are suitable for its application. Finally, the steps to be followed in the evaluation, implementation, and management of an EOR project are presented.

21.2 EOR Processes

EOR processes can be classified under three main groups. These are Miscible Gas Injection Processes, Chemical Processes, and Thermal Processes. Under these main groups are specific types of applications of the EOR processes which are defined and named as follows:

A. Miscible Gas Injection Processes
   1. Nitrogen injection
   2. Hydrocarbon (HC) gas injection
   3. Carbon dioxide (CO_2) injection
   4. Sour gas, flue gas, etc., injection

B. Chemical Processes
   1. Polymer flooding
   2. Polymer/surfactant flooding
   3. Alkali-Surfactant-Polymer (ASP) flooding
   4. Low Salinity Waterflooding (Chapter 20)
   5. Microbial

C. Thermal Processes
   1. In-Situ Combustion (ISC) or High Pressure Air Injection (HPAI)
   2. Steam/Hot water injection
   3. Steam-Assisted Gravity Drainage (SAGD)

The total number of EOR projects in operation around the world is summarized in Figure 21.1 from 1971 to 2017. In 2017, there were 375 EOR projects in operation in the world. These EOR projects produced around 2 MMBOPD which is about 2% of global oil production. Most EOR projects have been concentrated in North America. In 2013, 75% of all EOR projects were in operation in North America. In 2017, this number had declined to 40% with EOR projects located in other countries such as United Arab Emirates, Saudi

Arabia, China, India, Colombia, Kuwait, and Venezuela. Production from EOR projects is expected to grow in future years as shown in Figure 21.2. Oil production from EOR projects is projected to grow from 2.7 MMBOPD in 2025 to 4.5 MMBOPD in 2040, accounting for 4% of global oil production in 2040.[1]

It can be deduced from Figure 21.2 that the global growth in EOR oil production is mainly from CO_2-EOR and other gas-EOR projects through 2040. CO_2-EOR is growing rapidly globally because it serves the dual purpose of improving oil recovery and providing means for carbon dioxide sequestration or storage.[2] Carbon capture, utilization, and storage (CCUS) is the driver as many governmental agencies around the world promote policies that encourage the application of CO_2-EOR. In the United States of America (USA), CO_2-EOR is projected to contribute 740,000 BOPD and account for 12% of total oil production by 2040.

An expanding frontier for the application of EOR processes is unconventional reservoirs. It was stated in Chapter 1 (Section 1.2.3) that tight oil production will

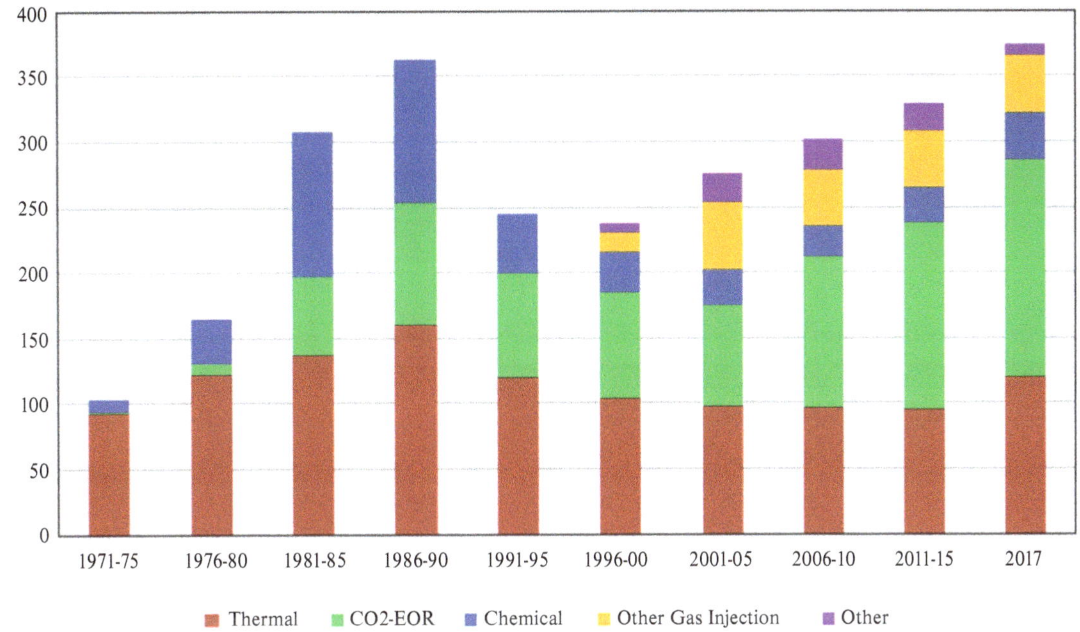

▲ **FIGURE 21.1** **Number of EOR projects in operation worldwide[1]**

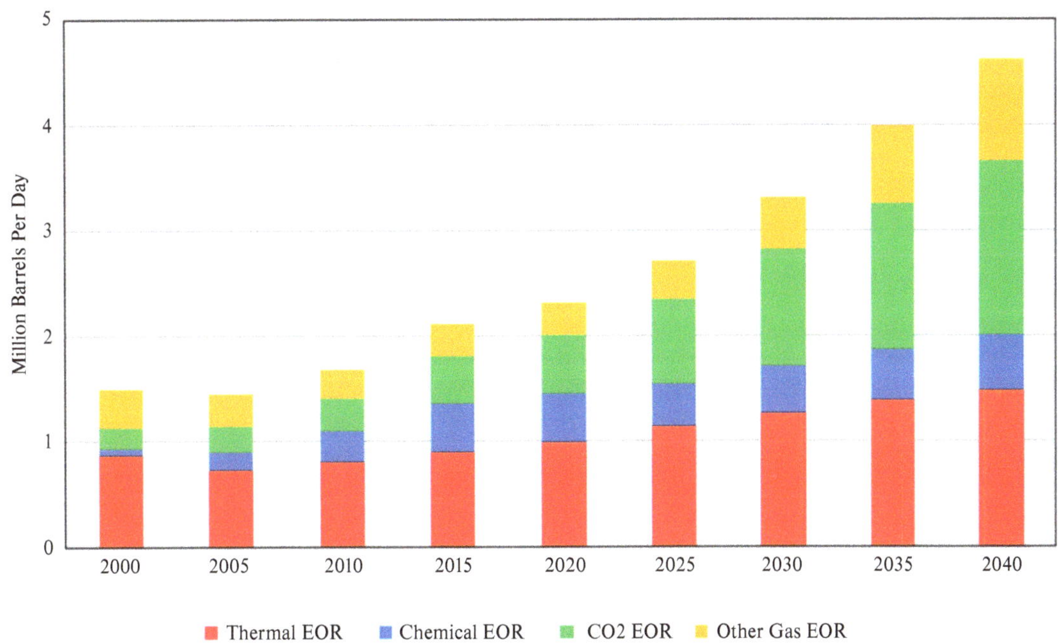

▲ **FIGURE 21.2** **Oil production from EOR projects worldwide[1]**

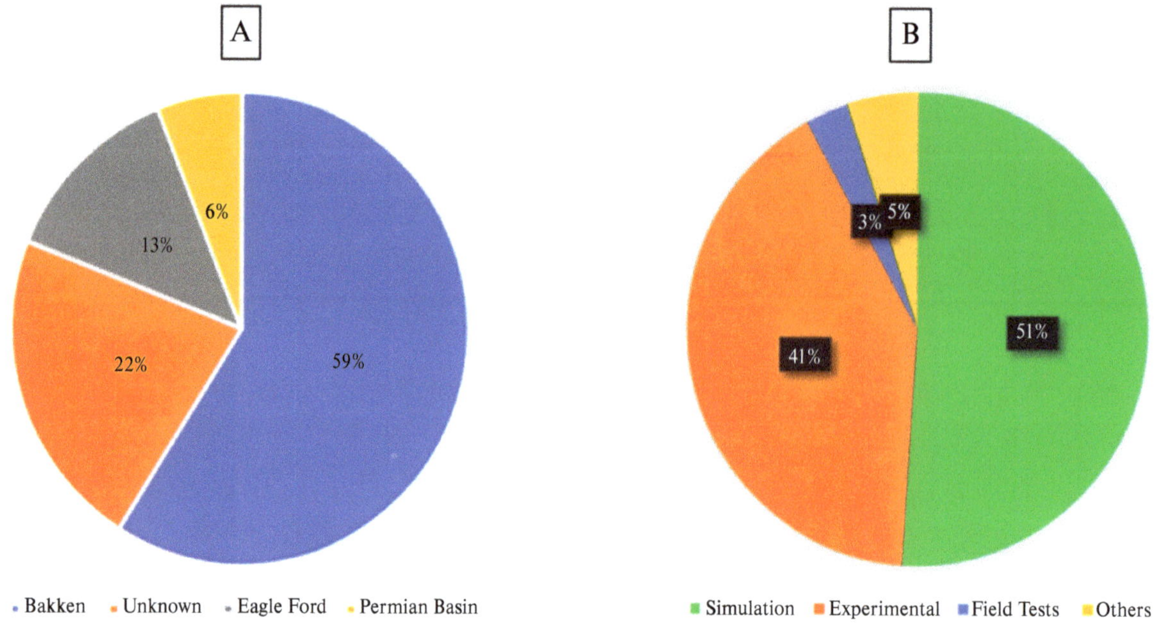

▲ **FIGURE 21.3 A and B** **A. Formations studied for EOR in USA[4]; B. Techniques used in studies[4]**

account for 65% of the cumulative oil production in the USA from 2017 to 2050. Most of this production is from primary depletion. The primary oil recovery factor for unconventional reservoirs ranges from 7% to 10%.[3] Consequently, a large portion of the oil-in-place in unconventional reservoirs are unrecovered at the end of primary depletion and would be a huge resource base to be targeted with EOR processes. In the USA, many unconventional reservoirs have been investigated for application of EOR processes. The proportions of formations that have been studied for EOR processes are shown in Figure 21.3A while the proportions of the different techniques that have been used in these studies are shown in Figure 21.3B. As Figure 21.3B shows, 92% of these studies were either laboratory experiments or simulation studies. Only 3% of these studies were field pilot tests.[4] The EOR processes that have been tried on unconventional reservoirs are gas injection, water injection, and surfactant injection. The gases used in the gas injection projects are carbon dioxide, nitrogen, and natural gas. Most of the gas injection projects used carbon dioxide as shown in Figure 21.4.

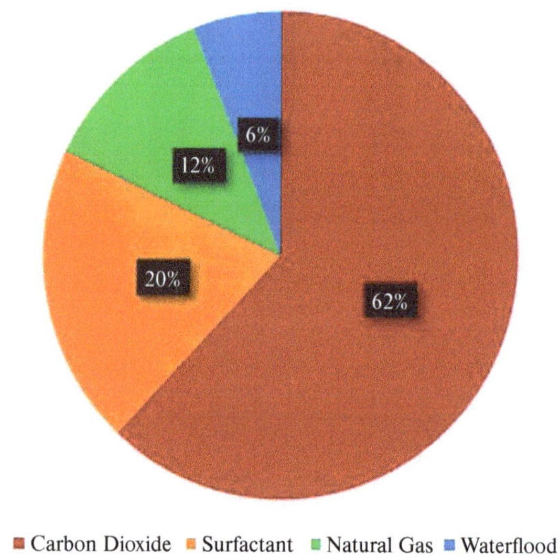

▲ **FIGURE 21.4** **Most applied EOR process in unconventional reservoirs in USA[4]**

Consequently, EOR-gas injection especially CO_2-EOR injection, has substantial potential for both conventional and unconventional reservoirs as the vehicle to supply more oil to the world and at the same time assist in curbing climate change caused by carbon dioxide emissions.

21.3 EOR Screening Criteria

The application of EOR processes are both reservoir-specific and reservoir fluid-specific. This literally means that each EOR process must be specifically evaluated before it can be applied to a reservoir. The evaluation process is typically extensive and may include laboratory work, geologic and reservoir modeling, economic analyses, and in many cases field trial in the form of a pilot test. The selection criteria presented here are meant to serve as the first-pass screening procedures that compare the candidate reservoir with other reservoirs that have been produced with an EOR process. They should not replace the rigorous evaluation procedure that each EOR process must undergo before it is actually implemented in the field. Note that the criteria presented in the following section apply mainly to conventional reservoirs. These criteria are not applicable and should not be used for unconventional reservoirs.

21.3.1 EOR Screening Criteria for Miscible Gas Injection Processes

The EOR screening criteria for miscible gas injection processes are presented in Table 21.1. The injection gases are nitrogen/flue gas, hydrocarbon gas, and carbon dioxide. The screening criteria for nitrogen/flue gas and carbon dioxide are based on all the successful field projects reported in the 2008 Worldwide EOR Survey.[5] The criteria for hydrocarbon gas injection are based on the 2008 Worldwide EOR Survey[5] and a survey of North Sea EOR projects reported by Awan et al.[6] The screening criteria for miscible gas injection shown in Table 21.1 indicate that most light oil reservoirs meet the criteria for any of these injection gases. The most important factors that determine the choice of injection gas appear to be availability (gas supply) and cost.

TABLE 21.1 | EOR Screening Criteria for Miscible Gas Injection Processes

Gas Injection Process (miscible)	Fluid Properties			Reservoir Properties				
	Gravity (°API)	Viscosity (cp)	Temp. (°F)	Porosity (%)	Perm. (md)	Oil Sat. at start (% PV)	Lithology	Depth (ft)
Nitrogen/flue gas	>30	<0.5	>250	>10	>30	>50	Carbonate or Sandstone	>7000
Hydrocarbon (HC) gas	21–57	0.1–1.3	136–290	4–26	10–5000	30–98	Carbonate or Sandstone	4000–14500
Carbon dioxide (CO_2)	28–44	0.4–3.0	100–250	4–26	2–500	25–90	Carbonate or Sandstone	2000–12000

21.3.2 EOR Screening Criteria for Chemical Flooding Processes

The EOR screening criteria for chemical flooding processes are shown in Table 21.2. Note that the same screening criteria are recommended for polymer/surfactant flooding, alkali/surfactant/polymer flooding, and polymer flooding. This is as a result of the screening criteria being based on successful field projects reported in the 2008 Worldwide EOR Survey.[5] Few projects were reported for polymer/surfactant flooding and alkali/surfactant/polymer flooding in the survey. In any case, chemical flooding using polymers, surfactants, and alkalis require extensive laboratory testing and design of the chemicals to suit the properties of the reservoir rock. Hence, the screening criteria for these chemicals serve basically as guides. The criteria for microbial EOR are based on the criteria reported by Lazar,[7] and Bryant and Lindsey.[8] Again, as with all chemical EOR processes, microbial EOR (MEOR) requires extensive laboratory and field testing.

TABLE 21.2 | EOR Screening Criteria for Chemical Flooding Processes

Chemical Flooding Process	Fluid Properties			Reservoir Properties				
	Gravity (°API)	Viscosity (cp)	Temp. (°F)	Porosity (%)	Perm. (md)	Oil Sat. at start (% PV)	Lithology	Depth (ft)
Polymer/surfactant flooding	14–34	5–80	80–160	20–30	170–900	60–75	Sandstone	1300–4600
Alkali/surfactant/polymer (ASP) flooding	14–34	5–80	80–160	20–30	170–900	60–75	Sandstone	1300–4600
Polymer flooding	14–34	5–80	80–160	20–30	170–900	60–75	Sandstone	1300–4600
Microbial (MEOR)*	—	5–50	<176	≥20	>50	—	Sandstone or Carbonate	<7700

*Includes water salinity <150,000 ppm.

21.3.3 EOR Screening Criteria for Thermal Processes

The EOR screening criteria for thermal processes are shown in Table 21.3. The criteria for thermal processes are based on successful field projects reported in the 2008 Worldwide EOR Survey.[5] The criteria for steam flooding also apply to hot water injection and the application of steam flooding described as steam-assisted gravity drainage (SAGD). Note that steam flooding has been applied successfully to a wide range of heavy oil sandstone reservoirs.

TABLE 21.3 | EOR Screening Criteria for Thermal Flooding Processes

Thermal Process	Fluid Properties			Reservoir Properties				
	Gravity (°API)	Viscosity (cp)	Temp. (°F)	Porosity (%)	Perm. (md)	Oil Sat. at start (% PV)	Lithology	Depth (ft)
ISC or HPAI	19–33	2–660	110–230	17–32	10–1265	50–94	Sandstone or Carbonate	400–8300
Steam	8–30	50–500,000	45–290	15–65	100–10000	44–90	Sandstone* or Tripolite	200–3600

*Includes unconsolidated sandstone.

21.4 Miscible Gas Injection Processes

A miscible displacement process is defined as a fluid displacement process where there is **no** interface (or boundary) between the displacing fluid and the displaced fluid. In this chapter, all EOR gas injection processes that are presented assume that reservoir conditions exist that miscibility is achieved between the injected gas and the reservoir fluid. This is in contrast to the immiscible gasflooding processes presented in Chapter 19. However, since the main distinction between miscible and immiscible gasflooding depends largely on the amount of hydrocarbon fluids recovered, most of the material (screening criteria, project design and implementation, etc.) presented for miscible gasflooding are also applicable to immiscible gasflooding processes.

21.4.1 Basic Concepts on Miscibility for Gas Displacement Processes

The concept of miscibility in a gas displacement process can be demonstrated with a ternary phase diagram. A ternary phase diagram for the methane/n-butane/decane[9] system at 2500 psia and 160 °F is shown in Figure 21.5. As shown in Figure 21.5, this system has a two-phase region which is enclosed by a bubble point line and a dew point line that meet at the plait (or critical) point. At 2500 psia and 160°F, mixtures of this system with compositions within the two-phase envelope are immiscible, while mixtures of the system with composition outside the two-phase envelope are miscible. The relative size of the two-phase envelope in Figure 21.5 depends on temperature and pressure of the system. For instance, at constant temperature of 160°F, increasing the pressure of the system from 2500 to 3250 psia will shrink the two-phase envelope, as shown in Figure 21.6. This also means that at higher pressures more compositions of the methane/n-butane/decane system are miscible at 160°F. For simplicity, complex hydrocarbon systems can be represented with a pseudoternary diagram, as shown in Figure 21.7. Note that in a pseudoternary diagram, pseudocomponents are used to represent the multicomponent hydrocarbon system, as shown in Figure 21.7. As was the case for simple ternary diagrams, a two-phase region can exist in a multicomponent system formed by a bubble point line and a dew point line that meet at the plait (or critical) point. Consequently, compositions of the system that fall within the two-phase envelope are considered to be immiscible while compositions outside the two-phase envelope can be miscible. These concepts are used in the descriptions of miscible conditions in gas displacement processes.

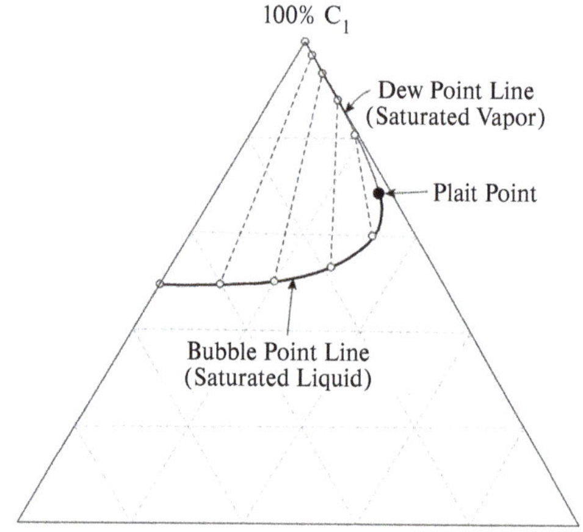

▲ **FIGURE 21.5 Ternary phase diagram of methane/n-butane/decane system at 2500 psia and 160°F (from Hutchinson and Braun[9])**

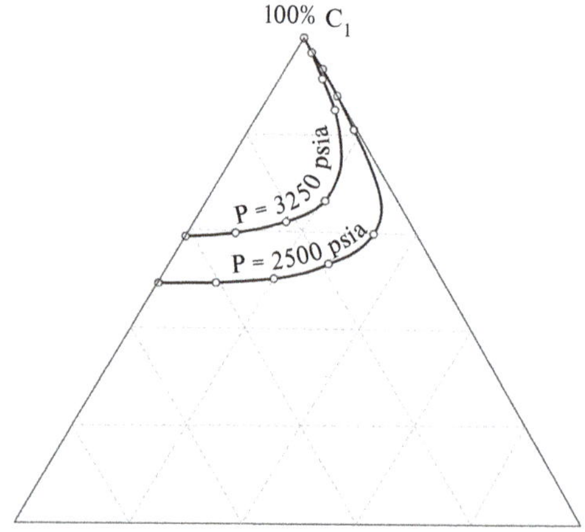

▲ **FIGURE 21.6 Effect of pressure on the two-phase envelope of methane/n-butane/decane system at 160°F (from Hutchinson and Braun[9])**

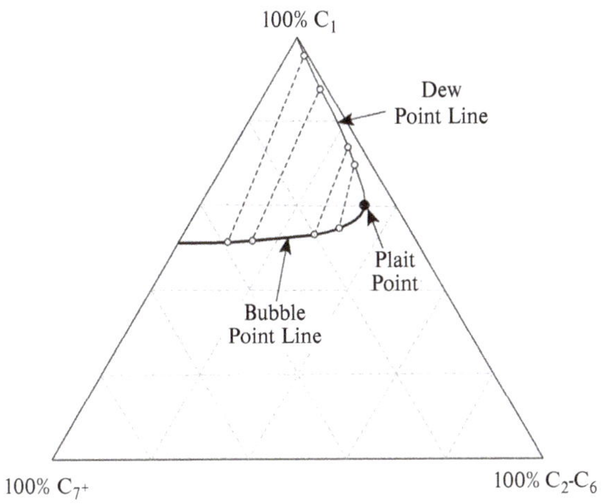

▲ **FIGURE 21.7 Pseudoternary phase diagram for a complex hydrocarbon system (from Hutchinson and Braun[9])**

21.4.2 First-Contact Miscibility (FCM)

The phenomenon of first-contact miscibility (FCM) occurs when the injected fluid is completely miscible with the reservoir fluid under the existing reservoir pressure and temperature. In FCM, the injected fluid completely mixes with the reservoir fluid to create a transition fluid that miscibly displaces the reservoir fluid. FCM can be considered an instantaneous process of complete mixing of the injected fluid with the reservoir fluid. The factors that determine FCM are reservoir temperature and pressure, composition of the injected fluid, and composition of the reservoir fluid. Suppose Figure 21.8 is the pseudoternary diagram for a CO_2-hydrocarbon mixture. Line AB is the critical tie line drawn through the critical point of the phase envelope. As a simple illustration of FCM, pure CO_2 will achieve FCM with reservoir fluid in the shaded region of Figure 21.8.

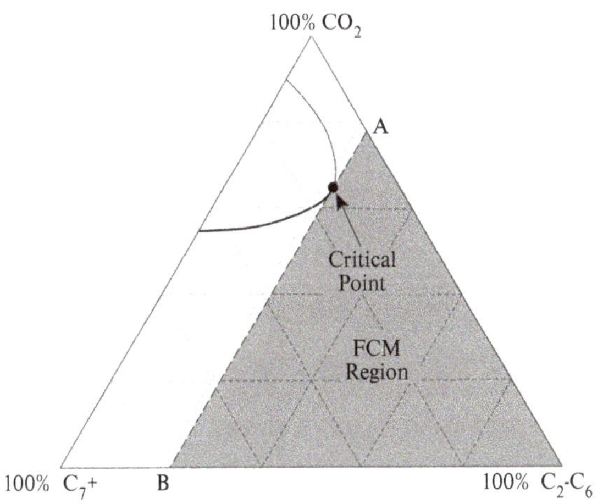

▲ **FIGURE 21.8 Pseudoternary diagram for CO_2-hydrocarbon system**

21.4.3 Multiple-Contact Miscibility (MCM)

As the term suggests, multiple-contact miscibility (MCM) is achieved as the result of repeated contacts in the reservoir between the reservoir oil and the injected fluid, and also fluids generated in-situ by interactions between the injected fluid and the reservoir oil.[10,11] There are two main processes by which MCM is achieved in the reservoir. These are the vaporizing gas drive and the condensing gas drive. In the vaporizing gas drive, the composition of the injected fluid is progressively modified by vaporization of the intermediate hydrocarbon components (C_2 to C_6) in the reservoir oil until the modified fluid becomes ultimately miscible with the original reservoir oil. In the condensing gas drive, intermediate hydrocarbon components in the injection fluid condense in the reservoir oil to generate a modified fluid that becomes miscible with the injection fluid. Zick[12] and Stalkup[13] reported that some MCM processes involve a combination of the vaporizing gas drive and the condensing gas drive. This third MCM process is termed the combined condensing and vaporizing (CV) gas drives. The type of MCM process that is dominant in a displacement drive depends on the composition of the reservoir oil, the composition of the injection fluid, and the temperature and pressure of the reservoir. These processes are illustrated with pseudoternary phase diagrams in the following sections.

21.4.4 Vaporizing Gas Drive MCM Process

The mechanism for achieving MCM by vaporizing gas drive is best illustrated with a pseudoternary diagram, as shown in Figure 21.9. The pseudoternary diagram is simplified to represent a system composed of methane (C_1)/ethane-hexane (C_2-C_6)/Heptane-plus (C_{7+}). Suppose a lean gas composed entirely of methane is injected into a reservoir containing oil with composition represented at point O in Figure 21.9. The methane gas will vaporize the intermediate components of the oil to form a fluid with composition represented at point 1 in the two-phase region. This fluid consists of a vapor phase shown as V_1, and a liquid phase shown as L_1, on the phase envelope. As the displacement drive progresses, the vapor phase, V_1, advances to contact additional original reservoir oil, and again vaporize more intermediate components to form another fluid with composition represented at point 2 within the two-phase envelope. This fluid separates into a vapor phase, V_2, and a liquid phase, L_2. The vapor phase, V_2, advances to contact more original oil to form a third fluid with composition represented at point 3 within the two-phase envelope. The vapor phase, V_3, from the fluid at point 3 contacts more original oil to form a fluid represented at point 4, whose vapor phase, V_4, achieves miscibility with original oil as shown in Figure 21.9. As described, the vaporizing gas drive MCM process is a mechanism whereby the vapor phase of the displacing fluid is progressively enriched through vaporization of the intermediate components in the original reservoir oil by repeated contacts until a fluid miscible with the original reservoir oil is achieved. Note that reservoir

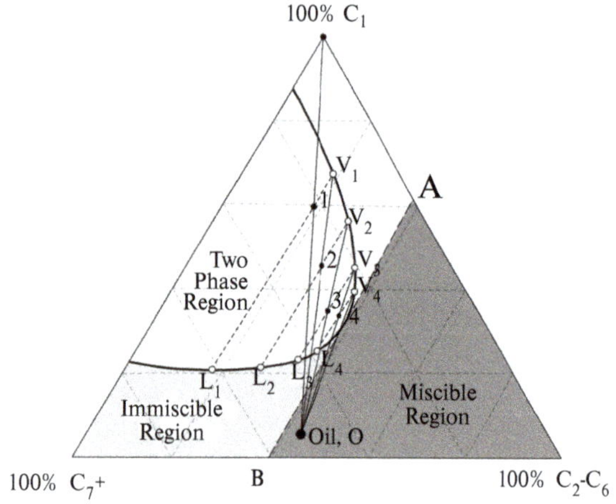

▲ **FIGURE 21.9** **MCM by vaporizing gas mechanism**

temperature and pressure are assumed to be constant, and local thermodynamic equilibrium is assumed to have been achieved between the intermediate phases during this process.

In Figure 21.9, line AB drawn through the critical point on the two-phase envelope is called the critical tie-line. For the fluid system at fixed conditions represented in Figure 21.9, reservoir oil compositions on the critical tie-line or to the right of the critical tie-line are miscible with injection gas composition to the left of point A on the critical tie-line. Furthermore, Figure 21.9 shows that reservoir oil composition to the left of the critical tie-line will be immiscible with injection gas composition to the left of point A on the critical tie-line. Note that these descriptive representations on the pseudoternary diagram are applicable only to vaporizing gas drive processes. The critical tie-line can be used also to define the minimum miscibility pressure (MMP) for a fluid system at fixed conditions. The MMP is defined as the minimum pressure at which the reservoir oil composition lies on a critical tie-line extension for a specified fluid system. The MMP represents the minimum pressure at which MCM can be achieved in the reservoir for the fluid system. At pressures less than MMP, the displacement process will be immiscible.

21.4.5 Condensing Gas Drive MCM Process

The mechanism of attaining MCM by condensing gas drive is illustrated with a pseudoternary diagram shown as Figure 21.10. In Figure 21.10, the injected fluid is represented by point G, and the reservoir oil is shown as point O. To initiate the multiple contact process, the intermediate components in the enriched injection gas will condense in the reservoir oil to form a fluid within the two-phase envelope with composition represented at point 1. This fluid is composed of a vapor phase, V_1, and a liquid phase, L_1. As the displacement advances, the liquid phase, L_1, is contacted by more injection gas to form a fluid with composition shown at point 2. This fluid is composed of a vapor phase, V_2, and a liquid phase, L_2. The liquid phase, L_2, is contacted by more injection gas to form a fluid shown at point 3 composed of a vapor phase, V_3, and liquid phase, L_3. Further contact between the liquid, L_3, and the injection gas produces a fluid which is ultimately miscible with the injection gas. The condensing gas drive MCM process can be described as the mechanism of progressive enrichment of the condensed liquid phase with intermediate hydrocarbons in the injection gas resulting from multiple contacts that ultimately produces a fluid that is miscible with the injection gas. Note that this description of the condensing gas drive MCM process assumes that reservoir conditions are fixed, and local thermodynamic equilibrium is attained between the intermediate liquid and vapor phases.

In Figure 21.10, the critical tie-line is shown as line CD. Any injection fluid composition to the left of point D will not achieve miscibility with the oil composition to the left of point C. The oil composition to the left of point C can achieve MCM with injection fluid composition at point D or to the right of point D. For the condensing gas drive, the MMP is defined as the minimum pressure at which the injection fluid composition lies on the extension of a critical tie-line. For the condensing gas drive, the injection fluid composition must lie on the critical tie-line or to the right of the critical tie-line and the oil composition must lie on the opposite (left) side of the critical tie-line as shown in Figure 21.10 for MCM to be achieved. Another important parameter that can be used to define the condensing gas MCM process is the minimum miscibility enrichment (MME). This describes the process of enriching the composition of the injection gas with intermediate hydrocarbons (C_2 to C_6) to achieve miscibility with the reservoir oil. The MME at a given pressure is defined as the minimum enrichment of the injection gas composition, such that the gas composition lies on an extension of a critical tie-line. Thus, for a condensing-gas drive process, miscibility can be achieved either by increasing pressure or by enriching the composition of the injection fluid with intermediate hydrocarbons.

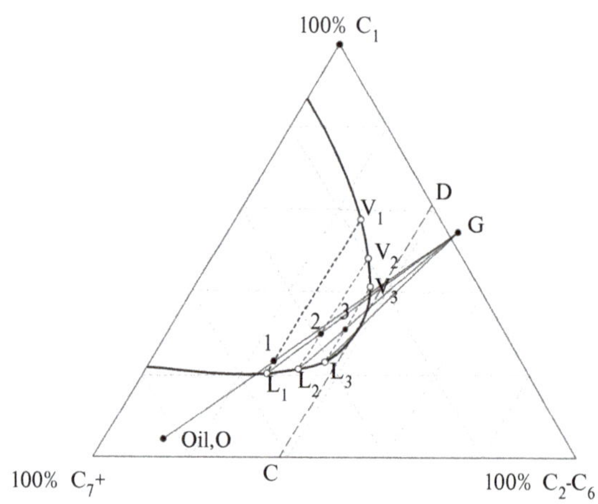

▲ **FIGURE 21.10** **MCM by condensing gas mechanism**

21.4.6 Combined Condensing/Vaporizing (CV) Gas Drive MCM Process

Zick[12] reported that a combined condensing/vaporizing gas drive mechanism rather than the traditional condensing gas drive alone is the key mechanism in the displacement of real reservoir oil by enriched fluids. This observation was supported by the work done by Stalkup.[13] The importance of the observation is that MMPs and MMEs estimated by pseudoternary methods are different from MMPs and MMEs observed for combined CV drives. This development should be expected since pseudoternary diagrams are inadequate for representation of a complex fluid system, such as reservoir hydrocarbon systems.

21.5 Methods for Determination of MMP or MME for Gasfloods

The two main techniques used in the determination of MMP or MME can be broadly classified as analytical techniques and experimental methods. Analytical techniques, which are based on numerical computations, can be subdivided further into empirical correlations and compositional simulations. Empirical correlations are based on fitting equations to MMP data obtained from experimental methods, and/or computed from compositional simulations. Compositional simulations methods generally use equations of state to predict MMP or MME from the compositions of the injection gas and the reservoir oil. The three common analytical techniques that use compositional simulations are the tie-line method, the 1-D slimtube simulation, and the mixing cell method. The two most common experimental methods are the slimtube method and the rising bubble apparatus method.

21.5.1 Analytical Techniques for Estimation of MMP or MME

Analytical techniques for the estimation of MMPs or MMEs include all computational methods that do not involve laboratory experiments. Analytical techniques can be subdivided into empirical correlations and compositional simulations. Compositional simulations usually require the use of an equation of state for prediction of phase equilibria, and are comprised of the tie-line method, the 1-D slimtube simulation method, and the multiple mixing-cell method.

21.5.1.1 Empirical Correlations for MMP

Empirical correlations are sets of equations generated from MMP data measured by experimental methods and/or MMP data calculated from compositional simulations. Empirical correlations are used to predict the MMPs for reservoir oils with various types of gas injection. These correlations provide quick estimates of MMPs which can be useful during screening of various gas injection processes for the reservoir. The MMPs calculated from empirical correlations can have large errors and should not replace MMPs obtained from experimental or compositional simulation methods. Empirical correlations should be used to predict MMPs during the early stages of screening the reservoir for various types of miscible gas injection processes. Empirical correlations for prediction of MMPs have been developed on the basis of type and composition of the injection gas. This classification is used in their presentation.

MMP Correlations for Hydrocarbon Injection Gases. A set of equations for prediction of MMPs for hydrocarbon injection gases was presented by Glasø.[14] These equations were derived from the graphical correlations developed by Benham et al.[10] The Glasø correlations state MMP as a function of reservoir temperature, molecular weight of the C_{7+} in the oil, the mole percent of methane in the injection gas, and the molecular weight of the intermediates (C_2 through C_6) in the injection gas. The Glasø correlations for predicting MMP for hydrocarbon gas/oil systems are as follows:

$$MMP_{x=34} = 6329.0 - 25.410y - z(46.745 - 0.185y) + T\left(1.127\times10^{-12}y^{5.258}e^{319.8zy^{-1.703}}\right) \tag{21.1}$$

$$MMP_{x=44} = 5503.0 - 19.238y - z(80.913 - 0.273y) + T\left(1.700\times10^{-9}y^{3.730}e^{13.567zy^{-1.058}}\right) \tag{21.2}$$

$$MMP_{x=54} = 7437.0 - 25.703y - z(73.515 - 0.214y) + T\left(4.920\times10^{-14}y^{5.520}e^{21.706zy^{-1.109}}\right) \tag{21.3}$$

In Eqs. (21.1) to (21.3), MMP = minimum miscibility pressure, psig; x = molecular weight of C_2 through C_6 in the injection gas, lbm/lb-mole; y = corrected molecular weight of C_{7+} in the stock-tank oil where $y = \left(2.622/\gamma_{o,C_{7+}}^{-0.846}\right)^{6.588}$ and $\gamma_{o,c_{7+}}$ is the specific gravity of stock-tank oil; z = methane in injection gas, mole percent; and T = reservoir temperature, °F. Glasø[14] reported a deviation of 2 to 8 percent between the MMP calculated from Eqs. (21.1) to (21.3) and experimentally determined MMP data.

EXAMPLE 21.1 Calculation of MMP from the Glasø Correlations for Hydrocarbon Gas Injection

Problem

Calculate the MMP for a reservoir oil and hydrocarbon gas injection mixture with the following properties:

Molecular weight of C_2 through C_6 in injection gas	32 lbm/lb-mole
Mole percent of methane in injection gas	65.2
Corrected molecular weight of C_{7+} in stock-tank oil	224 lbm/lb-mole
Reservoir temperature	160°F

Solution

Using Eq. (21.1),

$$MMP_{x=34} = 6329.0 - 25.410y - z(46.745 - 0.185y) + T\left(1.127\times10^{-12} y^{5.258} e^{319.8zy^{-1.703}}\right)$$

$$= 6329.0 - 25.410\times224 - 65.2(46.745 - 0.185\times224) + 160\left(1.127\times10^{-12}\times224^{5.258} e^{319.8\times65.2\times224^{-1.703}}\right)$$

$$= 3557.55 \text{ psig}$$

Using Eq. (21.2),

$$MMP_{x=44} = 5503.0 - 19.238y - z(80.913 - 0.273y) + T\left(1.700\times10^{-9} y^{3.730} e^{13.567zy^{-1.058}}\right)$$

$$= 5503.0 - 19.238\times224 - 65.2(80.913 - 0.273\times224) + 160\left(1.70\times10^{-9}\times224^{3.730} e^{13.567\times65.2\times224^{-1.058}}\right)$$

$$= 2749.74 \text{ psig}$$

By extrapolation, $MMP_{x=32}$ = 3719.26 psig.

The experimental MMP reported in the literature for this system by Glasø[14] is 3700 psig. This represents an error of +0.52%.

End of Example 21.1

MMP Correlations for Nitrogen Injection Gases. Two correlations are presented for prediction of MMP when the injection gas is nitrogen. These correlations were developed by Glasø[14] and Firoozabadi and Aziz.[15] The Glasø correlation for nitrogen/oil systems are based on limited experimental data and should be used with caution. The MMP correlations for nitrogen gas injection by Glasø are expressed as functions of molecular weight of the C_{7+} in the stock-tank oil, mole percent of the intermediates (C_2 through C_6) in the reservoir oil, and reservoir temperature. The Glasø MMP correlations for nitrogen gas injection are as follows:

For molecular weight of $C_{7+} > 160$ and mole percent of intermediates >28, the correlation to use is:

$$MMP = 6364.0 - 12.090M_{C_{7+}} + T\left(1.127\times10^{-12} M_{C_{7+}}^{5.258} e^{23,025.0M_{C_{7+}}^{-1.703}} - 20.80\right) \tag{21.4}$$

For molecular weight of $C_{7+} < 160$ and mole percent of intermediates >28, the correlation to use is:

$$MMP = 7695.1 - 12.090M_{C_{7+}} + T\left(1.127\times10^{-12} M_{C_{7+}}^{5.258} e^{23,025.0M_{C_{7+}}^{-1.703}} - 39.77\right) \tag{21.5}$$

If the mole percent of intermediates is <28, the correlation to use is:

$$MMP = 9364.0 - 12.090M_{C_{7+}} + T\left(1.127\times10^{-12} M_{C_{7+}}^{5.258} e^{23,025.0M_{C_{7+}}^{-1.703}} - 20.80\right) - 99.3C_{2-6} \tag{21.6}$$

In Eqs. (21.4) to (21.6), MMP = minimum miscibility pressure, psig; $M_{C_{7+}}$ = molecular weight of the C_{7+} in the stock-tank oil, lbm/lb-mole; T = reservoir temperature, °F; and C_{2-6} = mole percent of the intermediates (C_2 through C_6) in the reservoir oil.

The Firoozabadi and Aziz[15] correlation for prediction of MMP for nitrogen/oil systems is based on experimental data and compositional simulation using the Peng-Robinson equation of state. This correlation can also be used to predict MMP for lean-gas injection. (A lean gas is composed mostly of methane gas.) There was no accuracy reported for this correlation in comparison with experimental data. This correlation should be used with caution. The Firoozabadi and Aziz[15] correlation for prediction of MMP for nitrogen or lean gas injection is:

$$MMP = 9433 - 188\times10^3\left(\frac{C_{2-5}}{M_{C_{7+}}T^{0.25}}\right) + 1430\times10^3\left(\frac{C_{2-5}}{M_{C_{7+}}T^{0.25}}\right)^2 \tag{21.7}$$

In Eq. (21.7), MMP = minimum miscibility pressure, psia; $M_{C_{7+}}$= molecular weight of C_{7+}, lbm/lb-mole; C_{2-5} = mole percent of C_2 through C_5 including CO_2 and H_2S in the reservoir fluid; and T = reservoir temperature, °F.

EXAMPLE 21.2 Calculation of MMP from the Glasø, and Firoozabadi and Aziz Correlations for Nitrogen Gas Injection

Problem

Calculate the MMP for a reservoir oil and nitrogen gas injection mixture with the following properties:

Mole percent of C_2 through C_5 in reservoir oil	25.17
Molecular weight of C_{7+} in reservoir oil	193.3 lbm/lb-mole
Reservoir temperature	164°F
Bubble point pressure of reservoir oil	4000 psi

Solution

To apply the Glasø correlations to this nitrogen/oil system, Eq. (21.4) was selected because the molecular weight of the C_{7+} fraction was >160 lbm/lb-mole and the mole percent of the intermediates was assumed >25.17. Note that the Glasø correlations are based on intermediate composition of C_2 through C_6, whereas the Firoozabadi and Aziz correlation is based on intermediates composition of C_2 through C_5.

Using Eq. (21.4) of the Glasø correlation:

$$MMP = 6364.0 - 12.090 M_{C_{7+}} + T\left(1.127\times10^{-12} M_{C_{7+}}^{5.258} e^{23{,}025.0 M_{C_{7+}}^{-1.703}} - 20.80\right)$$

$$= 6364.0 - 12.090\times193.3 + 164\left(1.127\times10^{-12}\times193.3^{5.258}\times e^{23{,}025\times193.3^{-1.703}} - 20.8\right)$$

$$= 4295.19 \text{ psig}$$

Applying the Firoozabadi and Aziz correlation, Eq. (21.7) gives:

$$MMP = 9433 - 188\times10^{3}\left(\frac{C_{2-5}}{M_{C_{7+}} T^{0.25}}\right) + 1430\times10^{3}\left(\frac{C_{2-5}}{M_{C_{7+}} T^{0.25}}\right)^{2}$$

$$= 9433 - 188\times10^{3}\left(\frac{25.17}{193.3\times164^{0.25}}\right) + 1430\times10^{3}\left(\frac{25.17}{193.3\times164^{0.25}}\right)^{2}$$

$$= 4485.62 \text{ psia}$$

The reported experimental MMP for this system is 4280 psi. For screening purposes, both correlations gave very good MMP estimates for this nitrogen/reservoir oil system, although the Glasø correlation appears to be superior for this system.

End of Example 21.2

MMP Correlations for Pure Carbon Dioxide Injection Gases. Many correlations have been published in the literature for the prediction of MMP for pure carbon dioxide (CO_2)-oil systems.[16] Three of these correlations were selected on the basis of simplicity and accuracy as reported by their authors for presentation in this book. The three correlations that are recommended for prediction of MMP for pure CO_2-Oil systems are the Glasø[14] correlation, the Emera and Sarma[16] correlation, and the Yuan et al.[17] correlation. Glasø[14] correlated MMP for pure CO_2-Oil systems as a function of molecular weight of the C_{7+} fraction in the stock-tank oil, the mole percent of the intermediates (C_2 through C_6) in the oil, and reservoir temperature. For mole composition of C_2 through C_6 greater than 18%, the Glasø correlation for pure CO_2-Oil systems is:

$$MMP_{pure} = 810.0 - 3.404 M_{C_{7+}} + T\left(1.700\times10^{-9} M_{C_{7+}}^{3.730} e^{786.8 M_{C_{7+}}^{-1.058}}\right) \quad (21.8)$$

For mole composition of C_2 through C_6 less than 18%, the Glasø correlation for pure CO_2-Oil systems is:

$$MMP_{pure} = 2947.9 - 3.404 M_{C_{7+}} + T\left(1.700\times10^{-9} M_{C_{7+}}^{3.730} e^{786.8 M_{C_{7+}}^{-1.058}}\right) - 121.2 C_{2-6} \quad (21.9)$$

In Eqs. (21.8) and (21.9), MMP_{pure} = minimum miscibility pressure for pure CO_2 gas injection, psig; $M_{C_{7+}}$ = molecular weight of the C_{7+} in the stock-tank oil, lbm/lb-mole; T = reservoir temperature, °F; and C_{2-6} = mole percent of the intermediates (C_2 through C_6) in the reservoir oil. For a pressure range of 900 to 4400 psig, Glasø[14] reported an average deviation of −4.41% and standard deviation of 11.68% for MMP predicted with these correlations and experimental MMP data reported in the literature.

The Emera[16] correlation for pure CO_2 injection is represented as:

$$MMP_{pure} = 5.0093\times10^{-5}\times(1.8T+32)^{1.164} \times M_{C_{5+}}^{1.2785}\times\left(\frac{C_{C_1+N_2}}{C_{C_{2-4}+H_2S+CO_2}}\right)^{0.1073} \quad (21.10)$$

If the bubble point pressure (p_b) of the oil is less than 50 psi, the Emera[16] correlation for pure CO_2 injection becomes:

$$MMP_{pure} = 5.0093\times10^{-5}\times(1.8T+32)^{1.164}\times M_{C_{5+}}^{1.2785} \quad (21.11)$$

Note that for the Emera correlations, if the MMP is less than p_b, then $MMP_{pure} = p_b$. In Eqs. (21.10) and (21.11),

MMP_{pure} = minimum miscibility pressure for pure CO_2, MPa; $M_{C_{5+}}$ = molecular weight of the C_{5+} in the stock-tank oil, lbm/lb-mole; T = reservoir temperature, °C; $C_{C_1+N_2}$ = mole fraction of the volatiles (C_1 and N_2) in the reservoir oil; and $C_{C_{2-4}+H_2S+CO_2}$ = mole fraction of the intermediates (C_2, C_3, C_4, H_2S, and CO_2) in the reservoir oil. Emera and Sarma[16] reported an average deviation of 5.86% and standard deviation of 7.96% for MMP predicted with their correlations when compared with experimental MMP data available in the literature.

The Yuan et al.[17] correlation is based on analytical MMPs calculated using the tie-line method.[18] The expression of the Yuan correlation for pure CO_2 gas injection is given by:

$$MMP_{pure} = a_1 + a_2 M_{C_{7+}} + a_3 P_{C_{2-6}} + \left(a_4 + a_5 M_{C_{7+}} + a_6 \frac{P_{C_{2-6}}}{M_{C_{7+}}^2} \right) T + \left(a_7 + a_8 M_{C_{7+}} + a_9 M_{C_{7+}}^2 + a_{10} P_{C_{2-6}} \right) T^2 \quad (21.12)$$

The coefficients in Eq. (21.12) are:

$a_1 = -1.4364\text{E}+03$; $a_2 = 0.6612\text{ E}+01$;

$a_3 = -4.4979\text{ E}+01$; $a_4 = 0.2139\text{ E}+01$;

$a_5 = 1.1667\text{ E}-01$; $a_6 = 8.1661\text{ E}+03$;

$a_7 = -1.2258\text{ E}-01$; $a_8 = 1.2883\text{ E}-03$;

$a_9 = -4.0152\text{ E}-06$; $a_{10} = -9.2577\text{ E}-04$.

In Eq. (21.12), MMP_{pure} = minimum miscibility pressure for pure CO_2 gas injection, psia; $M_{C_{7+}}$ = molecular weight of C_{7+} fraction in reservoir oil, lbm/lb-mole; $P_{C_{2-6}}$ = mole percent of C_2 to C_6 in reservoir oil; and T = reservoir temperature, °F. The range of the molecular weight of the C_{7+} of all the reservoir oils used in the regression of the correlations is 139 to 319 lbm/lb-mole. The mole composition of C_2 to C_6 in the oil has a range of 2.0 to 40.3%, and the reservoir temperature ranged from 71 to 300°F. Yuan et al.[17] reported an average absolute error of 11.9%, when MMPs computed with the correlation are compared with 41 slimtube MMPs. Also, the Yuan correlation is applicable over a wider temperature range than other existing correlations.[17]

EXAMPLE 21.3 Calculation of MMP for Pure CO_2 Gas Injection Using the Glasø, Emera, and Yuan Correlations

Problem

Calculate the MMP for pure CO_2 gas injection into a reservoir containing oil with the following properties:

Mole percent of C_1 in reservoir oil	4.4
Mole percent of C_2 through C_6 in reservoir oil	24.0
Molecular weight of C_{5+} in reservoir oil	185 lbm/lb-mole
Molecular weight of C_{7+} in reservoir oil	234 lbm/lb-mole
Reservoir temperature	180°F

Solution

For the Glasø correlation, since mole percent of C_2 to C_6 is >18, using Eq. (21.8) gives:

$$MMP_{pure} = 810.0 - 3.404 M_{C_{7+}} + T\left(1.700\times10^{-9} M_{C_{7+}}^{3.730} e^{786.8 M_{C_{7+}}^{-1.058}}\right)$$

$$= 810.0 - 3.404\times234 + 180\left(1.700\times10^{-9}\times234^{3.730} e^{786.8\times234^{-1.058}}\right)$$

$$= 2451.75 \text{ psig}$$

For the Emera correlation, substituting the given properties in Eq. (21.10) gives:

$$MMP_{pure} = 5.0093\times10^{-5}\times(1.8T+32)^{1.164}\times M_{C_{5+}}^{1.2785}\times\left(\frac{C_{C_1+N_2}}{C_{C_{2-4}+H_2S+CO_2}}\right)^{0.1073}$$

$$= 5.0093\times10^{-5}\times(180)^{1.164}\times(185)^{1.2785}\times\left(\frac{4.4}{24}\right)^{0.1073}$$

$$= 13.9457 \text{ MPa}$$

$$= 2022.63 \text{ psia}$$

For the Yuan correlation, substituting the given properties in Eq. (21.12) gives:

$$MMP_{pure} = a_1 + a_2 M_{C_{7+}} + a_3 P_{C_{2-6}} + \left(a_4 + a_5 M_{C_{7+}} + a_6 \frac{P_{C_{2-6}}}{M_{C_{7+}}^2} \right) T + \left(a_7 + a_8 M_{C_{7+}} + a_9 M_{C_{7+}}^2 + a_{10} P_{C_{2-6}} \right) T^2$$

$$= -1436.4 + 6.612\times234 - 44.979\times24 + \left(2.139 + 0.11667\times234 + 8166.1\times\frac{24}{(234)^2}\right)\times180 + [-0.12258 + 0.0012883\times234 - 4.0152\text{E}-06(234)^2 - 9.257\text{E}-04\times24](180)^2$$

$$= 2927.30 \text{ psia}$$

The experimental MMP from slimtube measurements for this system was reported by Yuan et al.[17] as 3250 psia. The Yuan correlation prediction of 2927.30 psia is the closest to the experimental result of 3250 psia.

End of Example 21.3

MMP Correlations for Impure Carbon Dioxide Injection Gases. In actual field applications, CO_2 injection gas is not available at 100% purity. CO_2 injection gas contains impurities such as H_2S, SO_X, N_2, O_2, and hydrocarbons (C_1-C_4). Some of the impurities such as H_2S and SO_X could lower the MMP for the CO_2-Oil system, while impurities such as N_2 and CH_4 could increase the MMP for the system. For these reasons, it is important to account for the effect of impurities present in the CO_2 injection gas on achieving miscibility for the CO_2-Oil system. This impact is reflected in the calculation of MMPs for impure CO_2 injection gas. Numerous correlations have been published for the calculation of MMP for impure CO_2-Oil systems. These include correlations by Yuan et al.,[17] Johnson and Pollin,[18] Alston et al.,[19] Sebastian et al.,[20] Kovarik,[21] Orr and Silva,[22] Enick et al.,[23] Eakin and Mitch,[24] Dong,[25] and Emera and Sarma.[26] Only three of these correlations were selected on the basis of simplicity, and accuracy as reported by their authors for presentation in this book. The recommended correlations for calculation of MMP for impure CO_2 gas injection mixtures are the Sebastian et al.[20] correlation, Yuan et al.[17] correlation, and the Emera and Sarma[26] correlation. The Sebastian correlation was selected because of its simplicity and ability to handle CO_2-rich injection gas containing up to 55 mol% of impurities such as H_2S, N_2, and hydrocarbons (C_1-C_4). The Yuan correlation was selected because it covers a wider temperature range, especially when applied with the Yuan correlation for pure CO_2. Note however that the Yuan correlation was derived for impure CO_2 containing only methane as the impurity. The Emera and Sarma correlation was selected because it can be used for gas mixture containing up to 78 mol% of impurities such as H_2S, SO_X, N_2, O_2, and hydrocarbons (C_1-C_4), and was claimed to be more accurate than all the other correlations by its authors.

The Sebastian correlation is expressed as a function of pseudocritical temperature, $\bar{T}_{cM}$, of the drive gas mixture in the following form:

$$\left(\frac{MMP_{impure}}{MMP_{pure}}\right) = 1.0 - 2.13\times10^{-2}\left(\bar{T}_{cM} - 304.2\right) + 2.51\times10^{-4}\left(\bar{T}_{cM} - 304.2\right)^2 - 2.35\times10^{-7}\left(\bar{T}_{cM} - 304.2\right)^3 \quad (21.13)$$

The pseudocritical temperature, $\bar{T}_{cM}$, is expressed as:

$$\bar{T}_{cM} = \sum_i x_i T_{ci} \quad (21.14)$$

In Eq. (21.13) and (21.14), MMP_{impure} = minimum miscibility pressure for impure CO_2, psi; MMP_{pure} = minimum miscibility pressure for pure CO_2, psi; $\bar{T}_{cM}$ = pseudocritical temperature for impure CO_2 injection gas mixture, °K; x_i = mole fraction of component, i, in the injection gas mixture; and T_{ci} = critical temperature of component, i, in the injection gas mixture, °K. Values of critical temperatures for various gases are given in Table 21.4. Note that the critical temperature for H_2S was reduced from the actual value by Sebastian et al.[20] to improve the accuracy of the correlation. The temperature range of the experimental data used to develop the Sebastian correlation is 100 to 160°F, and the pressure range is 1000 to 3000 psi. The accuracy of the correlation is an average absolute deviation of 12% and a standard deviation of 17%, when the MMPs predicted with the correlation are compared with experimentally measured MMPs as reported by Sebastian et al.[20]

TABLE 21.4 Critical Temperature of Gases for the Sebastian et al.[20] Correlation

Component	Critical Temp., T_c (°K)
Hydrogen	33.2
Nitrogen	126.2
Oxygen	154.6
Carbon monoxide	132.9
Methane	190.6
Carbon dioxide	304.2
Ethane	305.2
Hydrogen sulfide	325.0*
Propane	369.8
i-Butane	408.1
n-Butane	425.2
i-Pentane	433.8
n-Pentane	469.6
n-Heptane	507.4

*Reduced from actual to improve correlation fit to data.

The Yuan correlation, which is based on 119 MMPs calculated by the tie-line method, is given as:

$$\left(\frac{MMP_{impure}}{MMP_{pure}}\right) = 1 + m\left(P_{CO_2} - 100\right) \quad (21.15)$$

The term, m, is represented by:

$$m = a_1 + a_2 M_{C_{7+}} + a_3 P_{C_{2\text{-}6}} + \left(a_4 + a_5 M_{C_{7+}} + a_6 \frac{P_{C_{2\text{-}6}}}{M^2_{C_{7+}}}\right)T + \left(a_7 + a_8 M_{C_{7+}} + a_9 M^2_{C_{7+}} + a_{10} P_{C_{2\text{-}6}}\right)T^2 \quad (21.16)$$

The coefficients in Eq. (21.16) are:

$a_1 = -6.5996\text{E}-02$; $a_2 = -1.5246\ \text{E}-04$;

$a_3 = 1.3807\ \text{E}-03$; $a_4 = 6.2384\ \text{E}-04$;

$a_5 = -6.7725\ \text{E}-07$; $a_6 = -2.7344\ \text{E}-02$;

$a_7 = -2.6953\ \text{E}-06$; $a_8 = 1.7279\ \text{E}-08$;

$a_9 = -3.1436\ \text{E}-11$; $a_{10} = -1.9566\ \text{E}-08$.

Note that Eq. (21.16) is strictly applicable for CO_2-rich gas injection mixtures containing up to 40 mol% of methane only. In Eqs. (21.15) and (21.16), P_{co_2} = mole percent of CO_2 in the injection gas. All other terms are as previously defined in Eq. (21.12). The maximum average error reported for this correlation is 9% for the MMP ratio for an injection gas containing 10 to 20 mol% methane.

The Emera correlation, which can be used for CO_2-rich gas injection mixtures containing H_2S, SO_X, N_2, O_2, and hydrocarbons (C_1-C_4), is expressed as:

$$\left(\frac{p_{r,impure}}{p_{r,pure}}\right) = 3.406 + 5.786\times\left(\frac{1.8T_{cw}+32}{1.8T_{c,CO_2}+32}\right) - 23.0\times\left(\frac{1.8T_{cw}+32}{1.8T_{c,CO_2}+32}\right)^2 + 20.48\times\left(\frac{1.8T_{cw}+32}{1.8T_{c,CO_2}+32}\right)^3 - 5.7\times\left(\frac{1.8T_{cw}+32}{1.8T_{c,CO_2}+32}\right)^4 \quad (21.17)$$

In Eq. (21.17),

$$p_{r,impure} = \frac{MMP_{impure}}{p_{cw}} \quad (21.18)$$

$$p_{cw} = \sum_{i=}^{n} w_i p_{ci} \quad (21.19)$$

$$p_{r,pure} = \frac{MMP_{pure}}{p_{c,CO_2}} \quad (21.20)$$

$$T_{cw} = \sum_{i=1}^{n} w_i T_{ci} MF_i \quad (21.21)$$

Values for the multiplying factors, MF_i, are given in Table 21.5. In Eqs. (21.17) to (21.21), T_{cw} = weight averaged pseudocritical temperature adjusted with, MF_i, °C; T_{c,CO_2} = pure CO_2 critical temperature, °C; MMP_{impure} = MMP for impure CO_2-Oil, MPa; MMP_{pure} = MMP for pure CO_2-Oil, MPa; w_i = weight fraction of component i, fraction; p_{ci} = critical pressure of component i, MPa; T_{ci} = critical temperature of component i, °C; and MF_i = multiplying factors. The Emera correlation is valid for impure CO_2 containing up to 78 mol% of impurities and up to 20 mol% of N_2. The temperature range of the data used to develop the correlation is 105.4 to 234.0°F. The range of molecular weight of C_{5+} fraction in the oils used to develop and test the correlation is 166.2 to 267.5. The average error reported for the correlation is 4.7% with a standard deviation of 6.3%, when compared with experimental data.

TABLE 21.5 Multiplication Factors for the Emera and Sarma[26] Correlation

Component	Multiplication Factor, *MF*
Sulfur dioxide	0.30
Hydrogen sulfide	0.59
Carbon dioxide	1.00
Ethane	1.10
Methane	1.60
Nitrogen	1.90
All other gases	1.00

EXAMPLE 21.4 Calculation of MMP for Impure CO_2 Gas Injection Using the Sebastian, Yuan, and Emera Correlations

Problem

Calculate the MMP for impure CO_2 gas injection into reservoir oil with the following properties:

Mole percent of CO_2 in injection gas mixture	90
Mole percent of CH_4 in injection gas mixture	10
Mole percent of C_2 to C_6 in the oil	23.62
Molecular weight of C_{7+} fraction in the oil	240 lbm/lb-mole
Reservoir temperature	106°F

Solution

Step 1: Calculate MMP for pure CO_2 gas injection.

The Yuan correlation is used to calculate the MMP for pure CO_2 gas injection because it appears to be superior to other correlations for this purpose as shown in Example 21.3.

Substituting the properties given for the oil into Eq. (21.12) for the Yuan correlation, MMP_{pure} = 1890.32 psia or 13.03 MPa.

Step 2: Calculate MMP for impure CO_2 gas injection using the Sebastian correlation.

Using Eq. (21.14) to calculate the mole average critical temperature for the injection gas mixture:

$$\bar{T}_{cM} = \sum_i x_i T_{ci}$$

$$= 0.9 \times 304.2 + 0.1 \times 190.6 = 292.84\ ^\circ\text{K}$$

Values for T_{ci} were taken from Table 21.4.

Substituting the value for $\bar{T}_{cM}$ in Eq. (21.13) gives:

$$\left(\frac{MMP_{impure}}{MMP_{pure}}\right) = 1.0 - 2.13\times10^{-2}\left(\bar{T}_{cM} - 304.2\right) + 2.51\times10^{-4}\left(\bar{T}_{cM} - 304.2\right)^2 - 2.35\times10^{-7}\left(\bar{T}_{cM} - 304.2\right)^3$$

$$= 1.0 - 2.13\times10^{-2}(292.84 - 304.2) + 2.51\times10^{-4}(292.84 - 304.2)^2 - 2.35\times10^{-7}(292.84 - 304.2)^3$$

$$= 1.2747$$

Using the MMP for pure CO_2 gas injection calculated in Step 1, MMP_{impure} = 2409.59 psia.

Step 3: Calculate MMP for impure CO_2 gas injection using the Yuan correlation.

Substituting the given properties of the oil into Eq. (21.16) gives:

$$m = a_1 + a_2 M_{C_{7+}} + a_3 P_{C_{2-6}} + \left(a_4 + a_5 M_{C_{7+}} + a_6 \frac{P_{C_{2-6}}}{M^2_{C_{7+}}}\right)T + \left(a_7 + a_8 M_{C_{7+}} + a_9 M^2_{C_{7+}} + a_{10} P_{C_{2-6}}\right)T^2$$

$$= -0.065996 - 0.00015246\times240 + 0.0013807\times23.62 + \left(\begin{array}{l}0.00062384 - 6.7725E-07\\ \times240 - 0.027344\times\dfrac{23.62}{(240)^2}\end{array}\right)\times106$$

$$+\left[-2.6953E-06 + 1.7279E-08\times240 - 3.1436E-11\times(240)^2 - 1.9566E-08\times23.62\right](106)^2$$

$$= -0.0315$$

For the Yuan correlation, Eq. (21.15) gives:

$$\left(\frac{MMP_{impure}}{MMP_{pure}}\right) = 1 + m\left(P_{CO_2} - 100\right)$$

$$= 1 - 0.0315(90 - 100)$$

$$= 1.315$$

MMP_{impure} = 2485.77 psia.

Step 4: Calculate MMP for impure CO_2 gas injection using the Emera correlation.

The critical properties for carbon dioxide and methane were obtained from Table 5.1. Thus, p_{c,co_2} = 1070 psia; T_{c,co_2} = 87.76 °F; p_{c,c_1} = 667 psia; and T_{c,c_1} = −116.66 °F.

Using Eq. (21.19) to calculate weight average pseudocritical pressure for the impure CO_2 injection gas gives:

$$p_{cw} = \sum_{i=}^{n} w_i p_{ci}$$

$$= 0.9611\times7.3774 + 0.03892\times4.5988$$

$$= 7.2694 \text{ MPa}$$

Also, using Eq. (21.21) and Table 21.5 to calculate the weight average pseudocritical temperature for the impure CO_2 injection gas yields:

$$T_{cw} = \sum_{i=1}^{n} w_i T_{ci} MF_i$$

$$= 0.9611\times30.98\times1.0 + 0.0389\times-82.59\times1.6$$

$$= 24.6345\ ^\circ\text{C}$$

From Eq. (21.20) and MMP_{pure} = 13.03 MPa from Step 1;

$$p_{r,pure} = MMP_{pure} / p_{c,CO_2}$$

$$= 13.03/7.38 = 1.77$$

Substituting Eq. (21.17) with T_{cw} = 24.6284°C and T_{c,co_2} = 30.98°C yields:

$$\left(\frac{p_{r,impure}}{p_{r,pure}}\right) = 3.406 + 5.786\times\left(\frac{1.8T_{cw}+32}{1.8T_{c,CO_2}+32}\right) - 23.0\times\left(\frac{1.8T_{cw}+32}{1.8T_{c,CO_2}+32}\right)^2 + 20.48\times\left(\frac{1.8T_{cw}+32}{1.8T_{c,CO_2}+32}\right)^3 - 5.7\times\left(\frac{1.8T_{cw}+32}{1.8T_{c,CO_2}+32}\right)^4$$

$$= 3.406 + 5.786\times\left(\frac{1.8\times24.63+32}{1.8\times30.98+32}\right) - 23.0\times\left(\frac{1.8\times24.63+32}{1.8\times30.98+32}\right)^2 + 20.48\times\left(\frac{1.8\times24.63+32}{1.8\times30.98+32}\right)^3 - 5.7\times\left(\frac{1.8\times24.63+32}{1.8\times30.98+32}\right)^4$$

$$= 1.2524$$

$$\left(\frac{p_{r,impure}}{p_{r,pure}}\right) = 1.2524;\ p_{r,impure} = 2.2167 \text{ and from}$$

Eq. (21.18),

$$MMP_{impure} = p_{r,impure} \times p_{cw} = 2.2167 \times 7.2692$$
$$= 16.1139 \text{ MPa or } 2337.11 \text{ psia.}$$

$$MMP_{impure} / MMP_{pure} = 1.2342.$$

The results from the three correlations are compared in Table 21.6 to actual experimental data reported for the CO_2 gas injection mixture by Sebastian et al.[20] Note that the Yuan correlation predicted the impure MMP that was closest to the experimental value. However, the Yuan correlation can be applied to impure CO_2 gas injection mixtures containing only methane. The Sebastian and Emera correlations can be applied to CO_2 mixtures containing other gases, such as H_2S, SO_x, N_2, O_2, and hydrocarbons (C_1-C_4). For this reason, either the Sebastian correlation or the Emera correlation is recommended for the calculation of MMP for impure CO_2 mixtures, especially if the mixture contains impurities other than methane.

TABLE 21.6 Comparison of Impure CO_2 MMPs Calculated in Example 21.4

Correlation Type	MMP_{impure}/ MMP_{pure}	MMP_{impure} (psia)	Error (%)
Sebastain, et al.[20]	1.2747	2409.6	–4.2
Yuan, et al.[17]	1.3149	2485.6	–1.1
Emera and Sarma[26]	1.2342	2333.0	–7.2
Experimental	1.3300	2514.1	+0.0

End of Example 21.4

21.5.1.2 Compositional Simulations

Compositional simulations are used to predict MMP or MME from the compositions of the reservoir oil and the injection gas. The three main methods for prediction of MMP or MME by compositional simulations are the tie-line method, the 1-D slimtube simulation, and the multiple mixing-cell method. Compositional simulations require the use of an equation of state (EOS) to represent the phase behavior of the reservoir oil and the injection gas. The ability of an EOS to adequately represent the phase behavior of a complex mixture of reservoir oil and injection gas depends on the characterization of the oil and "tuning" of the EOS to experimental data. Fluid characterization of the reservoir oil and tuning of the EOS to experimental data are crucial in the development of an EOS that can be used reliably in compositional simulations to predict MMPs or MMEs.[27,28] In Chapter 7, Section 11, a procedure for characterization of fluid systems and tuning of an EOS was presented. This procedure is recommended for characterizations of oil and gas systems and tuning of the EOS used in compositional simulations to predict MMPs or MMEs.

The Tie-Line Method. The tie-line method, sometimes called the analytical method, is used to calculate MMP by compositional simulation along the displacement path of the reservoir oil by the injected fluid. The tie-line method for calculating MMP is fast in comparison with other methods, and the computed MMP is not affected by either physical or numerical dispersion. The tie-line method is based on the model for one-dimensional flow of a multicomponent mixture without the influence of dispersion.[29,30] The analytical theory, as developed by Johns and Orr[31] and Wang and Orr[32] showed that the displacement path is controlled by $n_c - 1$ tie lines. These include the tie line for the oil, the tie line for the injection fluid, and $n_c - 3$ crossover tie lines. In the tie-line method, MMP is calculated as the minimum pressure at which one of the tie lines becomes a critical tie line. A critical tie line is defined as the tie line that has zero length and is tangent to the critical locus. If the oil tie line becomes the critical tie line, the displacement type is vaporizing gas drive. If the gas (for the injection fluid) tie line becomes the critical tie line, the displacement type is condensing gas drive. If the development of MCM is controlled by one of the crossover tie lines, then the displacement type is combined condensing-vaporizing gas drive. Methods for calculation of MMPs using the tie-line method have been developed by Johns and Orr,[31] Wang and Orr,[32] Jessen et al.,[33] and Yuan and Johns.[34]

1-D Slimtube Simulation. MMP can be determined by conducting compositional simulation of the process in a 1-D reservoir model. In 1-D slimtube simulation, oil recovery at 1.2 HCPV (hydrocarbon pore volume) of gas injection is plotted against displacement pressure. MMP is determined at the bend or break-over (also called the "knee") of the plot as shown in Figure 21.11. The MMP determined by 1-D compositional simulation is affected by numerical dispersion from truncation error. The magnitude of numerical dispersion is directly proportional to the number of grid blocks in the model. Thus, the effect of numerical dispersion on MMP can be eliminated by increasing the number of grid blocks in the 1-D simulation model. The dispersion-free MMPs calculated by the tie-line method are close to MMPs determined by the 1-D slimtube simulation for most oil-gas injection systems. MMPs based on the tie-line method are about 100 to 200 psi lower than MMPs from slimtube simulations.[28]

Multiple Mixing-Cell Method. The multiple mixing-cell method can be used to estimate MMP. It is similar to other computational methods because it relies on an EOS to predict the equilibrated compositions of the mixtures from flash calculations. There have been many variations of the multiple mixing-cell method published in the literature.[35] However, the basic model of the method involves analytically mixing injection gas with reservoir oil, separating the equilibrated mixture into vapor and liquid phases, and further analytically mixing fresh oils with either the vapor phase or the liquid phase. To simulate vaporizing gas drive, the equilibrated vapor phase from the previous cell is contacted with fresh oil in the next cell. This process of multiple contacts is repeated for several steps until the resultant vapor phase is miscible with fresh oil. Similarly, for condensing gas drive, the equilibrated liquid phase from the previous cell is contacted with fresh injection gas. Miscibility is achieved when the condensed liquid phase is miscible with injection gas after several steps in the multi-cell mixing process. This procedure can be used to predict reliable MMPs for either purely vaporizing gas drive or condensing gas drive. However, Zick[12] and Stalkup[13] have demonstrated that most real oil displacements are combined condensing-vaporizing gas drive processes. Ahmadi and Johns[35] have published a multiple mixing-cell method that calculates MMPs for condensing-vaporizing gas drives using a modification of the tie-line method. They claim that the accuracy of MMP predicted with their method is comparable to MMP predicted by the tie-line method.

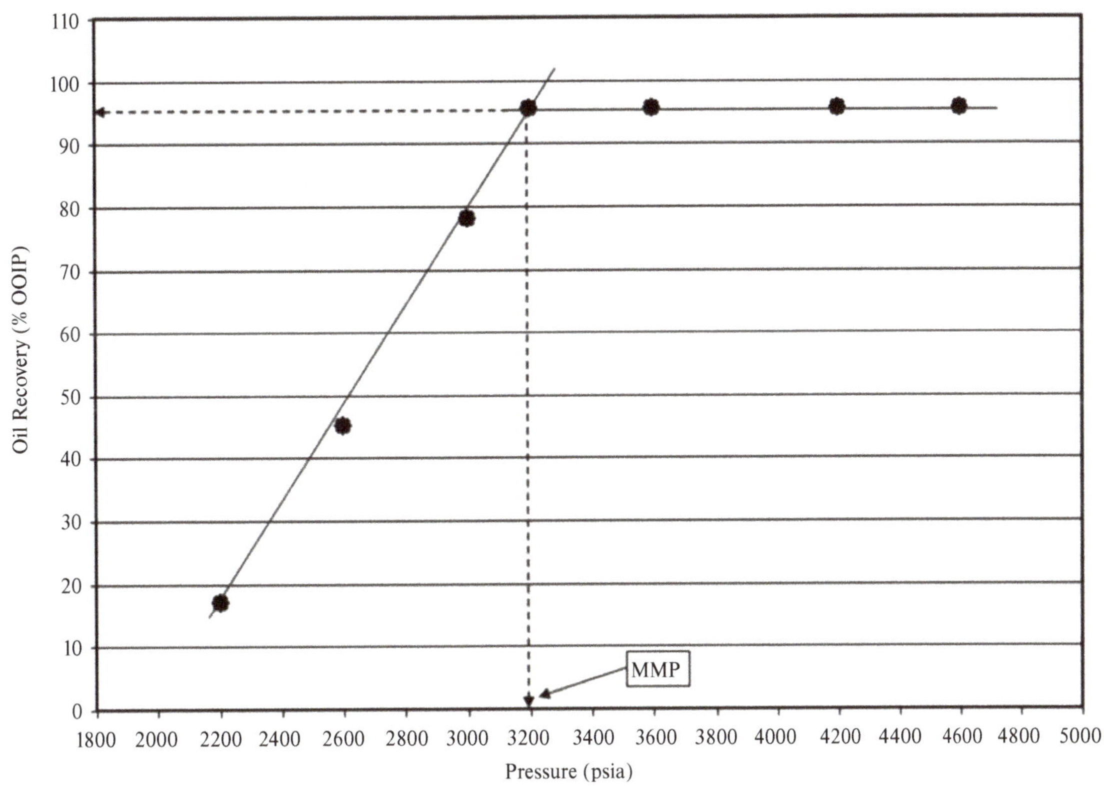

▲ **FIGURE 21.11** Determination of MMP from 1-D slimtube simulation or experiments

21.5.2 Experimental Methods

Two experimental methods are used in the measurement of MMP or MME. These are slimtube and the rising-bubble apparatus (RBA) methods. The slimtube method is by far more widely used than the RBA method. However, the RBA method can be used to obtain comparable MMP data in much less time than the slim tube method.

21.5.2.1 Slimtube Method

The slimtube equipment consists of a 40 to 60 feet long, 0.25 inch internal diameter, stainless steel tubing, packed with either fine sands or glass beads of a size in the range of 100 to 200 mesh. This tubing is spirally coiled to render gravity effects insignificant in the displacement process. Also, viscous fingering is considered negligible because of the small diameter of the tubing. Fluids stored in stainless cylinders are injected into the coiled tubing at a constant rate by an injection pump. Pressure across the coiled tubing is controlled by a backpressure regulator. A sight glass is used to monitor fluids flowing out of the coiled tube. The effluent fluids are measured with simple equipment that range from a wet-test meter to more elaborate equipment that may include a gas chromatograph. Most of the slimtube apparatus is enclosed in an air-bath maintained at any desired constant temperature. A schematic diagram of a typical slimtube apparatus is shown in Figure 21.12. Slimtube measurements are affected by the length and diameter of the tube, the type of packing material, particle size of the packing material, and the rate of displacement.[36-38]

In a typical slimtube experiment, the packed medium in the coiled tube is saturated with the oil that is the subject of the experiment. The apparatus is maintained at the desired temperature and pressure set for the experiment. The displacement fluid is then injected at constant rate until a predetermined pore volume of the fluid has been injected. This is typically set at injection of 1.2 pore volumes of displacement fluid. The oil recovery at this pore volume of injection and pressure is recorded. The experiment is repeated at different pressures with all the other variables (such as injection rate, temperature, etc.) kept constant. A plot of oil recovery versus displacement pressure similar to Figure 21.11 will yield the MMP at the "bend" or break-over of the plot. Slimtube experiments can be used to screen reservoirs for miscible processes and should be included in the design and evaluation of reservoirs for gasfloods.

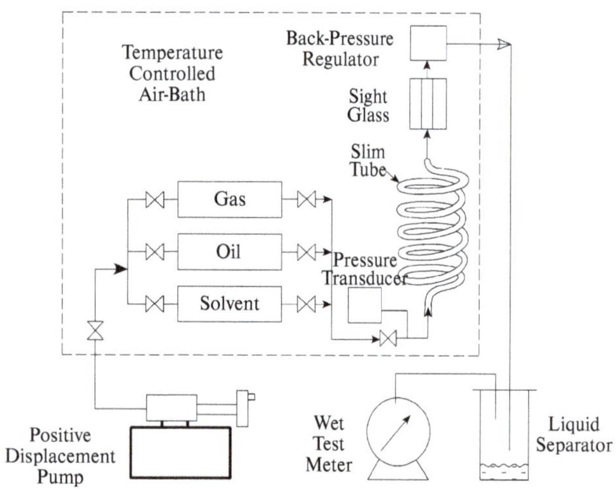

▲ FIGURE 21.12 Schematic diagram of slimtube apparatus

21.5.2.2 Rising-Bubble Apparatus (RBA) method

Since it was introduced in the early 1980s, the rising-bubble apparatus (RBA) has gained acceptance in the petroleum industry as an alternative method for measuring MMP because it is quicker than the slimtube method. The essential parts of the RBA consist of a high pressure visual cell equipped with a camera to record observations during the experiment. Inside the visual cell, there is a flat glass tube into which the reservoir fluid is charged before the experiment is initiated. The glass tube is made flat so that bubbles rising in the oil can be seen more clearly. The internal dimensions of the glass tube are typically 0.04 by 0.2 inches and about 8 inches long. A schematic diagram of the RBA is shown in Figure 21.13. A detailed description and operational procedure for the RBA is provided by Novosad et al.,[39] Christiansen and Haines,[40] Elsharkawy et al.,[41] and Thomas et al.[42] Measurements of MMP with RBA are very dependent on monitoring and interpretation of the shapes of bubbles formed as the gas bubbles travel through the oil column in the glass tube. Data reported in the literature by Novosad et al.,[39] Elsharkawy et al.,[41] Thomas et al.,[42] Zhou and Orr,[43] and Bon et al.[44] show that MMPs measured with RBA are comparable in magnitude with those measured with slimtubes. The key advantage of the RBA is that it can be used to determine MMP at approximately one-tenth of the time required for a slimtube experiment.

21.6 Types of Miscible Gas Flooding

Miscible displacement can be achieved by injecting different types of gases or gas mixtures, depending on the composition of fluids in the reservoir, reservoir conditions, and injection pressure. The types of gases used for miscible flooding include nitrogen, flue gas, light hydrocarbons (C_1-C_6), and carbon dioxide. Flue gas is generated from power plants, internal combustion engines, etc. and contains a mixture of gases including a high percentage of nitrogen (>66%), carbon dioxide (9%-12%), and other gases, such as sulfur dioxide and oxygen in smaller amounts. The MMP required for flue gas injection is close to the requirements for nitrogen injection. For this reason, flue gas is grouped with nitrogen in the discussion of miscible gas flooding. Note however that flue gas is very corrosive because it contains carbon dioxide and sulfur dioxide. The manner by which miscibility is achieved depends on the type of injection gas. Generally, nitrogen and flue-gas achieve MCM by the vaporizing gas drive mechanism. Hydrocarbon gas and carbon dioxide can achieve MCM by vaporizing gas drive, condensing gas drive or combined condensing-vaporizing gas drive mechanisms, depending on the compositions of the reservoir fluid and injection gas, and injection pressure. In this section, the nature of miscible gas flooding with the different gases is presented. This is followed by a presentation on different types of gas injection strategies.

21.6.1 Nitrogen/Flue-gas Miscible Gas Flooding

Nitrogen/flue-gas is sometimes selected as the injection gas for miscible flooding because it is relatively cheap and abundantly available in comparison with other potential injection gases such as hydrocarbon (HC) gas or carbon

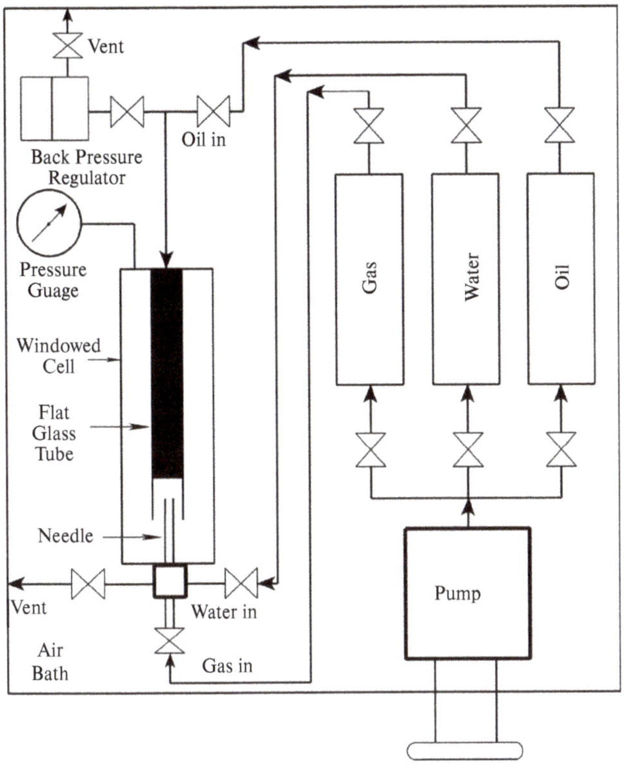

▲ **FIGURE 21.13** **Schematic diagram of rising-bubble apparatus**

dioxide. Nitrogen/flue-gas can achieve miscibility at higher MMP than either HC gas or CO_2 at the same reservoir conditions. This condition limits the application of nitrogen/flue gas to reservoirs with relatively high pressures. The mechanism by which nitrogen/flue-gas achieves miscibility with oil is through vaporization of the light intermediate components present in the oil. This is essentially the mechanism of vaporizing gas drive. For this reason, nitrogen/flue-gas is suitable for light oil reservoirs as candidates for miscible flooding. The criteria for screening reservoirs for miscible flooding with nitrogen were given in Table 21.1. The correlations for predicting MMP with nitrogen as the injection gas were provided as Eqs. (21.4) to (21.7). Evaluation of nitrogen for use as an injection gas should include the cost of removing nitrogen from the produced gas before it can be sold. Note that flue-gas is corrosive and may not be suitable for application in many reservoirs without preventative procedures implemented to combat corrosion in processing equipment.

21.6.2 Hydrocarbon (HC) Miscible Gas Flooding

In hydrocarbon (HC) miscible gas flooding, light hydrocarbons (C_1-C_6) are injected into the reservoir at pressures sufficiently high to achieve miscibility with the reservoir oil. Many pressure maintenance projects with HC gas injection involve some aspects of miscible gas flooding, if reservoir pressures are maintained above MMP. HC gas injection can be classified broadly into three categories. The first category involves the injection of liquefied petroleum gas (LPG), such as propane at pressures sufficiently high to achieve FCM. The second category involves the injection of enriched gas (containing C_2 to C_6 at enriched levels) to achieve MCM by predominantly condensing gas drive mechanism. The third category entails the injection of lean gas (containing mostly methane) at high pressures to achieve MCM by vaporizing the light components (C_2-C_6) in the oil using predominantly the vaporizing drive mechanism. In most real applications, the drive mechanisms for the second and third categories involve some degree of combination of condensing-vaporizing gas drives. The criteria for screening reservoirs for miscible flooding with HC gas were given in Table 21.1. The correlations for predicting MMP with light hydrocarbons as the injection gas were provided as Eqs. (21.1) to (21.3). In addition to enhanced oil recovery due to miscible conditions, HC gas flood improves oil recovery by swelling the oil, reducing oil viscosity, and physical displacement of the oil from the pore spaces. The application of HC gas flooding is limited by poor sweep efficiencies due to potential for viscous fingering and gravity over-ride, and the high economic value of hydrocarbon gases.

21.6.3 Carbon Dioxide Gas Flooding

Carbon dioxide (CO_2) gas is a very effective agent for miscible displacement of oil. In most cases, CO_2 achieves miscibility with reservoir oil by extracting light to intermediate (C_1 to C_6) components of the oil in an MCM, vaporizing gas drive mechanism. Also, CO_2 improves oil recovery by swelling the oil, reducing its viscosity, and lowering interfacial tension between the oil and the CO_2-enriched fluid. The criteria for screening reservoirs for miscible flooding with CO_2 gas were given in Table 21.1. The correlations for predicting MMP with CO_2 as the injection gas were provided as Eqs. (21.8) to (21.12) for pure CO_2, and Eqs. (21.13) to (21.21) for CO_2 containing impurities. CO_2 can cause extensive corrosion and should be used in a field project after extensive corrosion preventative measures have been implemented. Carbon dioxide can be obtained from naturally occurring CO_2 reservoirs or separated from flue gases for injection. There is a persistent clamor around the world to reduce the amount of carbon dioxide in the atmosphere which is considered as a major contributor to climate change. The amount of carbon dioxide emitted into the atmosphere can be reduced by removing carbon dioxide from flue gas and injecting it into reservoirs to improve oil recovery. This is a novel proposal which can contribute to solving the dangers posed to the entire world by climate change, and at the same time contribute to meeting the increasing need around the world for additional oil supply. Oil and gas reservoirs are good candidates for storage or sequestration of carbon dioxide because these reservoirs had geologic seals that originally trapped the hydrocarbons provided the seals had not been breached by field operations.[45,46] It is anticipated that large scale operations in the future will combine the use of carbon dioxide for oil recovery and sequestration of large volumes of carbon dioxide in the same reservoir and its associated aquifers. A model that estimates the cumulative oil recovery from CO_2 flooding and CO_2 storage potential for a given reservoir has been provided by Wood et al.[47] This model can be used to screen many reservoirs to determine the best candidates for CO_2 flooding and storage.

21.6.4 Types of Miscible Gas Injection Strategies

Mobility control is a major concern in miscible gas injection processes because of the relatively low viscosity of the injected gas. The mobility ratio between injected gas and displaced reservoir oil is unfavorable in most applications. This causes viscous fingering and gravity over-ride which creates low volumetric sweep efficiency. To remedy this condition, most miscible gas injection processes, inject gas and water to improve volumetric sweep of the reservoir. If water and gas are injected in alternating cycles, the process is called water-alternating gas (WAG) injection strategy. In some application, water and gas are injected simultaneously. This injection strategy is called simultaneous-water-and-gas (SWAG). SWAG injection strategy is considered to be a variant of WAG. WAG injection strategy is used in most miscible flood projects around the world. A WAG ratio of 1:1 (pore volume of water injected versus pore volume of gas injected) is used in most field applications. Note that application of WAG results in loss of injectivity of water and gas over time. Some miscible gas floods are initiated with the strategy of continuous gas injection (CGI) which can be switched later to WAG or SWAG to achieve better mobility control. The volume of gas injected can be reduced as the flood progresses to reduce cost and improve mobility control. This injection strategy is described as tapering. Tapering is used in most carbon dioxide and hydrocarbon injection projects to reduce the cost of injecting these gases into reservoirs for hydrocarbon recovery.

21.7 Chemical Flooding Processes

Chemical flooding processes are displacement processes that improve oil recovery from reservoirs with the aid of chemicals that are generally called surfactants, polymers, and alkalis. These chemicals are combined in various proportions to create injection systems which are termed polymer/surfactant (micellar) process, and alkali/surfactant/polymer (ASP) process. Only polymers are injected for mobility control and improved oil recovery in polymer flooding. Chemical flooding processes improve oil recovery by lowering the interfacial tension (IFT) between oil and water, solubilization of the oil in the micelles, emulsification of the oil and water, alteration of the wettability of the rock, and enhancement of the mobility of the displacing fluids. For each application of chemical flooding, the chemicals must be formulated and tailored to the properties of the reservoir rock and its fluid system. This usually requires extensive testing of various combinations of the chemicals in laboratory coreflood tests followed with a pilot test in the field.[48] There are few field-wide chemical flooding projects in the world in comparison with either CO_2 or HC gas injections projects. The decline of chemical flooding can be attributed to the high cost of the chemicals and higher risk associated with achieving predicted oil recoveries with field-wide projects. The chemical flooding processes that have been applied in the field are surfactant/polymer floods, ASP floods, and polymer floods. In some fields, microbial flooding has been tested. These chemical flooding processes are discussed in more detail in the following sections.

21.7.1 Polymer/Surfactant Flooding

In a typical polymer/surfactant flood (also called micellar/polymer flood), a slug of micellar solution (consisting of water, surfactant, electrolytes, etc.) is injected into the reservoir. The size of the micellar slug can vary from 3% to 30% of the flood pattern pore volume (PV) depending on the flood design. The micellar slug is followed with a mobility buffer slug made up of polymer and water for mobility control. The entire micellar/polymer slug is then chased with drive water. The polymer solution sometimes contains a biocide to control microorganisms that can degrade the polymer and reduce the viscosity of the solution. The micellar slug is formulated such that a favorable mobility ratio exists between the slug and the oil bank that is formed by the flooding process. This sometimes requires the addition of small amounts of polymers in the composition of the micellar solution. The micellar slug is usually the most expensive solution used in micellar/polymer flooding. The surfactants in the micellar slug are degraded with the advancement of the flood due to adsorption on rock surfaces and trapping of some of the micellar solution as a residual phase. Adsorption of surfactants on rock surfaces can be reduced by preflushing the flood zone with fresh water. Preflushing with fresh water also reduces incompatibilities between the surfactant and formation brine. The application of micellar/polymer floods have declined because of the high cost of micellar solution and limited economic success of the process in field tests.[49]

21.7.2 Alkali/Surfactant/Polymer (ASP) Flooding

Alkali/surfactant/polymer (ASP) flooding is more widely used than micellar/polymer flooding in field applications. ASP formulation utilizes the favorable mechanisms of its key components (alkali, surfactant, and polymer) to improve oil recovery.[50] The main functions of the alkaline component are to promote emulsification of the crude oil, reduce IFT, reduce adsorption of the surfactants, and regulate phase behavior of the mixtures. The main function of the surfactants is to reduce IFT between the oil and the injected slugs. The polymers are used for mobility control and improvement of sweep efficiencies. The favorable attributes of ASP formulations have been tested and proven in many laboratory and field tests.[51-53] ASP flooding has been applied in most cases to sandstone reservoirs, but laboratory and pilot tests suggest that it can also be applied to carbonate reservoirs.[54,55]

21.7.3 Polymer Flooding

Polymer flooding is the most widely used chemical flooding process for both sandstone and carbonate reservoirs. Polymer flooding can be considered as a waterflooding process in which polymers have been added to the solution to improve mobility control. Thus, polymer flooding is able to achieve higher oil recovery than ordinary waterflooding due to improved volumetric sweep efficiency resulting from better mobility control. Factors affecting the application of polymer flooding are the cost of the polymers, the permeability of the reservoir, and the temperature of the reservoir. The cost of polymers limits the concentration and slug size of the polymers that can be used to achieve higher oil recovery. Low rock permeability can limit the use of more viscous polymers to improve mobility control and can cause low injectivity. Polymers are not stable at reservoir temperatures higher than 175 °F. Many field applications of polymer flooding have been reported as technical and economic success.[56-58]

21.7.4 Microbial Enhanced Oil Recovery (MEOR)

Microbial activities in the reservoir are believed to increase oil recovery by altering the wettability of the rock, reducing interfacial tension, reducing oil viscosity, generation of gases such as carbon dioxide, production of surfactants, etc. These mechanisms, which combine to improve oil recovery by the MEOR process, are complex and not yet fully understood. Consequently, after many years of laboratory and pilot tests, MEOR processes have not been applied on a large scale in the field in comparison with other EOR processes. There are no active MEOR projects in the United States in 2008,[5] and small-scale MEOR projects are in operation in China,[5] and the North Sea.[6] The screening criteria for microbial processes are presented in Table 21.2. Generally, MEOR processes are applicable to reservoirs with injected and connate water salinities <150,000 ppm; oil viscosity in the 5 to 50 cp range; reservoir temperature <176°F; rock porosity ≤20%; rock permeability ≥50 md.; and reservoir depth <7700 ft. Most MEOR tests have been on sandstone reservoirs, although there have been tests in carbonate reservoirs.[6] The bacteria source in MEOR processes can be indigenous or exogenous. Indigenous bacteria are already present in the reservoir while exogenous bacteria are cultivated and injected into the reservoir. Microbial processes in the reservoir can be aerobic or anaerobic. For aerobic processes, oxygen must be injected to maintain the required concentration of bacteria. Injection of oxygen into the wellbore may cause severe corrosion problems. Anaerobic processes require injection of large amounts of nutrients in the form of sugars to sustain the growth of bacteria. This will increase the cost of the MEOR process due to the additional costs of purchasing, processing, and injecting the nutrients. Large-scale field utilization of MEOR processes have not occurred in the industry because of limited knowledge of the process, and lack of models to predict potential oil recovery from application of the technology.

21.8 Thermal Processes

Thermal EOR processes are defined to include all processes that supply heat energy to the rocks and fluids contained in a reservoir thereby enhancing the ability of oil (including other fluids) to flow by primarily reducing its viscosity. The oil caused to flow by the supply of thermal energy is produced through nearby wells. The two main thermal processes that are presented in this section are steamflooding and in-situ combustion. Steamflooding has achieved substantial commercial successes around the world and is ranked as one of the top EOR processes. In-situ combustion process, by comparison, has attained limited commercial success, and is actively applied in few reservoirs around the world. The main obstacle to the application of in-situ combustion technology is ability to manage and control the advancement and progression of the combustion process. Steamflooding has been used on highly viscous, shallow reservoirs at depths of 200 to 3600 feet while in-situ combustion can be applied to less viscous, deeper reservoirs at depths reaching 8300 feet (Table 21.3). In principle, in-situ combustion is the only viable thermal process for deeper reservoirs where steam cannot be used due to thermodynamic limitations of steam (critical pressure of steam = 3200.1 psia, critical temperature of steam = 705.1°F from Table 5.1).

The number of thermal projects has declined in the United States from 1990 to 2004 but appears to be increasing from 2006 to 2008 (Figure 21.1). In the United States, oil production from thermal processes has declined from a peak production of 480,000 BOPD in 1986 to 293,000 BOPD in 2008 (Figure 21.2).

Thermal EOR processes have significant impact on oil recovery in other regions of the world. Canada has substantial oil reserves deposited as heavy oil or in tar sands which are good candidates for thermal oil recovery processes. China has heavy oil deposits of more than 1.9 billion tons of oil reserves.[59] First commercial development in China started in 1982. Between 1993 and 2004, oil production from thermal processes in China reached the level of 10 to 13×10^6 tons per year.[59] Venezuela has many thermal projects in the Tia Juana field.[5] The Duri steamflood project in Indonesia is the largest thermal project in the world averaging about 190,000 BOPD in 2008.[5]

In this section, more emphasis is placed on steamflooding, since it is by a wide margin the most commercially applied thermal recovery process. The presentation includes various methods of steamflooding, description of steamflood models, and management of steamflood projects. This is followed with a descriptive presentation of in-situ combustion.

21.8.1 Steamflooding Methods

Steamflooding is an established EOR technique that has been applied successfully on many heavy oil reservoirs around the world. The process started in early 1960 with cyclic steam injection in the Tia Juana Field in Venezuela, the Schoonebeek Field in the Netherlands, and the Yorba Linda Field in California.[60,61] The three main methods of steamflooding are cyclic steam stimulation (CSS), steam drive (SD), and steam-assisted gravity drainage (SAGD). SAGD can be considered as a modification of SD for heavy oil reservoirs including tar sands. This section concludes with a discussion of water-alternating-steam process (WASP) as a steam injection strategy for mobility control.

21.8.1.1 Cyclic Steam Stimulation (CSS)

Cyclic steam stimulation (CSS) is a simple and cheap method of applying thermal recovery process on a reservoir. It involves injecting steam into a well for several weeks, shutting the well in as long as necessary to allow the steam to heat the oil in the areas around the well, and putting the well back on production to recover the heated oil (Figure 21.14). This process is repeated when the production from the well declines to a low level. The cycle is repeated many times until the ratio of oil produced to steam injected termed oil-steam ratio (OSR) drops to a level that is considered uneconomic. CSS was the first steamflooding technique used in heavy oil reservoirs. It can still be used in a limited pilot test on a well to investigate whether steam injection can enhance oil recovery in a particular field before undertaking more expensive pattern pilot tests that will require more wells and more equipment.

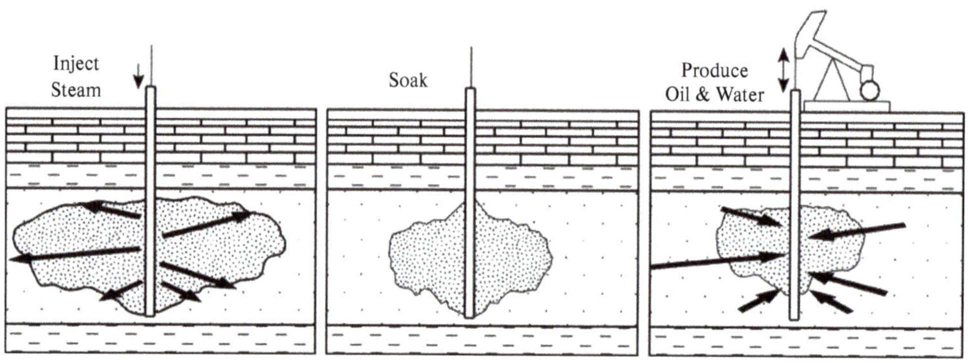

▲ **FIGURE 21.14** **Cyclic steam stimulation (CSS) process**

21.8.1.2 Steam Drive (SD)

In steam drives, steam is injected continuously at injectors with the aim of driving oil towards producers. Typically, in most projects steam injection is organized in patterns. For instance, in a normal five-spot pattern (Figure 19.3), the producer is located in the middle of the pattern and surrounded by four steam injectors. Various pattern configurations of injectors and producers have been used in many fields. The selection of the pattern that is best suited for a particular reservoir is an essential part of the design and pilot testing process, before field implementation of the project. Steam drive was initially described as essentially a frontal displacement process similar to waterflooding, whereby steam displaces oil towards a producer. Later, experimental and field evidence showed that the mechanism of oil recovery by steam is dominated by gravity drainage caused by steam rising toward the top of the formation due to gravity over-ride (Figure 21.15). The accumulation of steam at the top of the formation creates a steam zone which

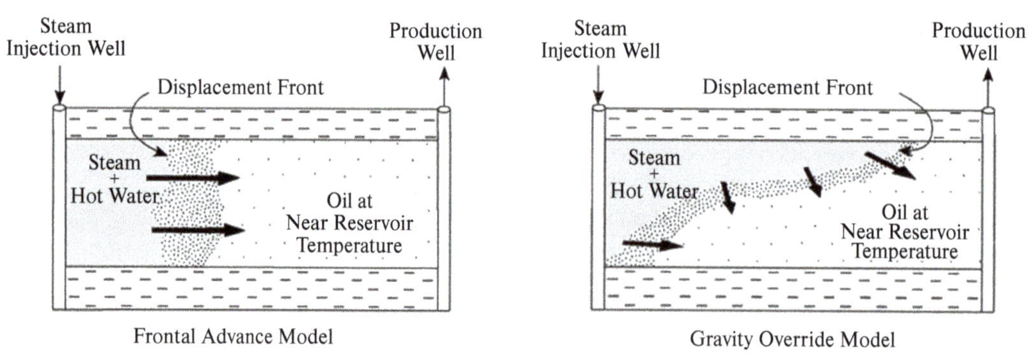

▲ **FIGURE 21.15** **Steam drive models**

continues to expand with more steam injection. This steam zone can propagate and breakthrough at the producers to cause inefficient utilization of the energy from the steam due to cycling. The main challenges of managing a steam drive project are monitoring the size and growth of the steam zone, monitoring producers for early steam breakthrough, adjusting steam inject rates, and measuring the trend of oil-steam ratio for each pattern.

21.8.1.3 Steam-Assisted Gravity Drainage (SAGD)

The concept of steam-assisted gravity drainage (SAGD) was tested and shown to be successful by the Alberta Oil Sands Technology and Research Authority (AOSTRA) at the Underground Test Facility in Fort McMurray, Alberta, Canada, between 1987 and 2004.[62,63] In 2008, more than 10 commercial SAGD projects were in operation in the Athabasca area of Canada, and SAGD has quickly become the preferred thermal recovery process for oil sands in the area.[63] The SAGD process is applicable to other heavy oil and oil sands deposits in other countries such as China, Venezuela, and Russia. The SAGD process consists of two horizontal wells about 15 feet apart located close to the bottom of the formation (Figure 21.16). Steam is injected into the top horizontal well, while the horizontal well below it functions as the producer. The steam creates an expanding steam chamber around the injector as more steam is injected. Within the steam chamber and at its boundaries, as the viscosity of the oil is reduced, its mobility increases causing it to drain under gravity toward the production well. The SAGD process should be applied to reservoirs with formation thickness greater than 50 feet, good vertical permeability, and absence of thief zones.[63] In some SAGD projects, a hydrocarbon solvent (such as methane) is added to the steam to improve oil recovery. This variation of the SAGD process has been described as the expanding solvent-SAGD (ES-SAGD).[64] In another variation, vertical wells are used for steam injection, while the producer remains a horizontal well as shown in Figure 21.17. This configuration of injectors and producer is termed Cross SAGD (XSAGD).[65] Oil recoveries with SAGD in pilot tests can reach 65% of the oil-in-place in the project area in comparison with CSS that achieves oil recovery in the range of 25% to 30%.[66] The heavy oil/bitumen in-place in Canada is estimated to be 1.6 trillion barrels.[66] With this huge resource base, SAGD is positioned to be a very important technology that is bound to be used to meet future energy needs in the world.

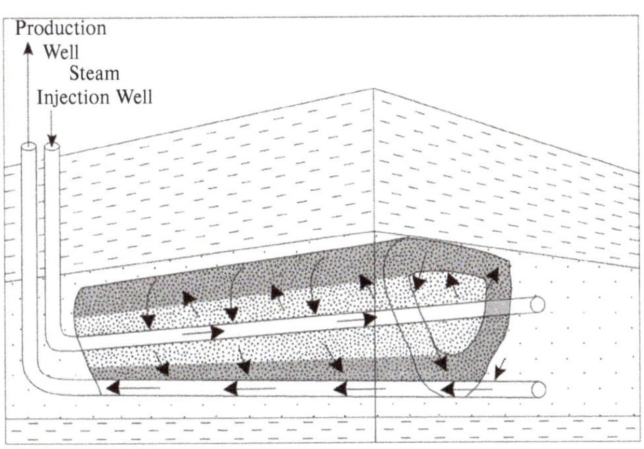

▲ FIGURE 21.16 SAGD process diagram

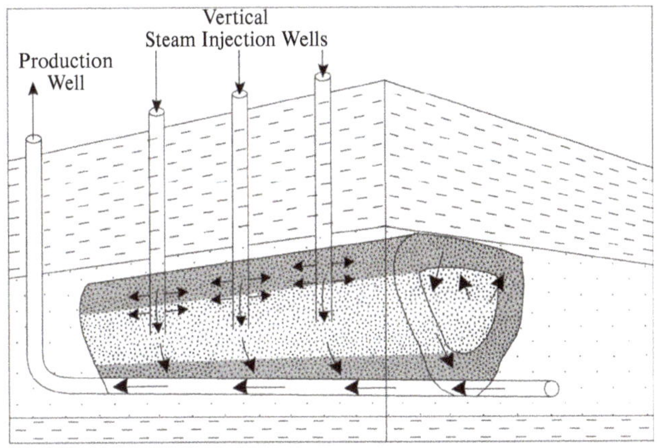

▲ FIGURE 21.17 XSAGD process diagram

21.8.1.4 Water-Alternating Steam Process (WASP)

Water-alternating gas (WAG) was described in Section 21.6.4 as an injection strategy used in gasfloods to reduce viscous fingering, improve vertical sweep, and thereby increase oil recovery. This concept is also used in steamflooding by injecting water alternating with steam to reduce fingering of steam and improve its vertical sweep efficiency. This strategy is called water-alternating-steam process or WASP. The main benefit of WASP over continuous steam injection is reduction or elimination of premature steam breakthrough at producing wells. Early steam breakthrough at producers wastes heat energy and reduces productivity of the wells.

21.8.2 Steamflood Models

Steamflood models are used to predict the performance of the process in terms of oil production rates, oil-steam ratios, cumulative oil recovery, steam zone size, etc. Consequently, steamflood models are essential tools used in the screening, pilot testing, field implementation, and management of steamflood projects. The two main mathematical steamflood models widely used in the industry are analytical models and numerical simulation models. Analytical steamflood models are based on heat flow and energy balance equations, assuming uniform reservoir properties and shape of the steam zone. Numerical steamflood simulation models are based on simultaneous solutions of material balance and energy balance equations, including possibly an equation of state for component phase equilibrium calculations.

Numerical simulation models can incorporate most of the known geologic features (such as structures, faults, permeability barriers, etc.) and spatial variation of rock and fluid properties for the reservoir. Numerical models are more rigorous than analytical models in representing the physical processes of steamflooding and achieving results that include the impact of geological heterogeneities on the performance of steamfloods. Consequently, numerical simulation models are recommended as the preferred method for prediction of steamflood performance.

21.8.2.1 Analytical Steamflood Models

With the introduction of steamflooding technology in 1960, analytical models were developed to predict the performance of steamfloods. The complexity of published analytical models has grown over many years with the spread of steamflooding. The first steamflood analytical model was published by Marx and Langenheim in 1959.[67] Analytical or semi-analytical models published by other authors in chronological order are as follows: Boberg and Lantz, 1966;[68] Mandl and Volek, 1967;[69] Farouq Ali, 1970;[70] Myhill and Stegemeir, 1978;[71] Gomaa, 1980;[72] Rhee and Doscher, 1980;[73] Jones, 1981;[74] Aydelotte and Pope, 1983;[75] van Lookeren, 1983;[76] Vogel, 1984;[77] Neuman, 1985;[78] Miller and Leung, 1985;[79] and Chandra and Mamora, 2007.[80] This list is not meant to include all analytical steamflood models ever published in the petroleum literature, but serve to illustrate the historical development work performed on this application. The theoretical development and applications of any of these analytical models are not presented in this book. This is because the utility of these models has been surpassed by numerical steamflood simulation models as pointed out in the previous section.

21.8.2.2 Numerical Steamflood Simulation Models

Early works on the development of numerical simulators for steamfloods were published by Shutler,[81,82] Abdalla and Coats,[83] Weinstein et al.,[84] Coats et al.,[85] Coats,[86,87] and Grabowski et al.[88] The application of numerical simulators on steamfloods was boosted with the publication of the Fourth SPE Comparative Solution Project on steamflood simulators in 1987.[89] Because of the complexities of the physical processes involved in steamflooding, numerical simulation of steamfloods require smaller grid sizes and large computation times, in comparison with black oil simulators. However, recent advances in computer hardware and improvements in formulation of solution algorithms have made it possible to construct very heterogeneous reservoir models for simulation of steam floods and achieve considerable reduction in computation times. For instance, parallel computing and dynamic gridding are some of the advances in application of numerical steamflood simulators.[90,91] The development of material and energy balance equations representing numerical simulators for steamfloods are not presented in this book. However, the interested reader is encouraged to consult published literature that is referenced in this book and other books that have been published on numerical simulations. In addition, there are several competent thermal numerical simulators available commercially in the petroleum industry which can be used to model any of the described steamflood processes.

21.8.3 Management of Steamflood Projects

Management of a steamflood project offers unique challenges because the heat energy supplied to the reservoir must be utilized efficiently and effectively to maximize oil recovery. The management of steamflood projects can be subdivided into the management of three interrelated areas: surface equipment management, well management, and reservoir management. These areas are presented briefly in the next sections.

21.8.3.1 Surface Equipment Management

The key objective of surface equipment management in steamfloods is to minimize heat losses and optimize the process of generating steam. To minimize heat losses, most projects use insulated tubing and coupling joints coated with Teflon. Many projects have also incorporated cogeneration of steam and power using cogeneration plants as a cost effective means of improving the economics of steamflooding.

21.8.3.2 Well Management

The well system in a typical steamflood project consists of steam injectors, oil producers, and observation wells. Management of steam injectors usually requires profile control so that the steam can reach the targeted intervals. This is achieved by using surfactants or high temperature polymers to reduce mobility of steam and re-direct steam to unswept intervals, thereby improving sweep efficiency. Another practice is the use of multizone completion techniques to isolate and inject steam into separate target zones. Steam injectors can be remediated by cementing existing perforated intervals and recompleting the well in other intervals that has not been flooded with steam. Similar remedial practice is conducted at the producers by either shutting-in updip producers as steam breaks through their perforated intervals or recompleting them to produce from lower intervals. Producers should be monitored regularly to measure fluid production rates and detect steam breakthrough. One of the important activities in managing steamfloods is the use of observation wells to monitor the temperature and growth of the steam zone. This can be accomplished by running frequent temperature surveys or equipping the observation wells with fiber-optic distributed temperature sensing (DTS) technology.[92,93]

21.8.3.3 Steamflood Reservoir Management Strategy

Steamflooded reservoirs require higher levels of monitoring and surveillance because of the close spacing of injectors and producers, the potential for early steam breakthrough from gravity over-ride, and the need to ensure that the energy supplied by the steam is effectively used in heating the reservoir to produce oil. Heat management in steamfloods have evolved over several years.[94,95] The recommended practice is to manage the steamflooding process with a numerical thermal simulator with the goal of optimal utilization of the heat energy from steam. The simulator should be used to adjust steam injection rates, monitor the steam zone, adjust or modify the flood patterns, and predict future flood performance.

21.8.4 In-Situ Combustion (ISC)/High Pressure Air Injection (HPAI)

In-situ combustion (ISC)/High pressure air injection (HPAI) are EOR processes in which compressed air is injected into heavy and light oil reservoirs under conditions, such that the oxygen in the injected air reacts with a small fraction of the crude oil at high temperatures to create a combustion front. The distinction between ISC and HPAI processes is blurred, although both processes are definitely related. Turta et al.[96] state that an ISC process is associated with high peak temperatures (350°C to 600°C), and may require ignition to initiate it, whereas an HPAI process may not have high peak temperature and does not require ignition. The ISC process appears to be more applicable to heavy oil reservoirs, whereas HPAI is better suited for light oil reservoirs. In this book, both processes are considered to be air injection processes (AIP). A schematic diagram of an air injection displacement process is shown in Figure 21.18. The displacement path consists of an air zone behind the combustion/reaction zone. The combustion/reaction zone is preceded by a coke zone, a condensation zone, and mobilized oil bank. High oil recoveries have been reported from laboratory and field tests of AIP.[97-99] However, there are few air injection projects around the world. In the United States, the remaining few air injection projects are based on relatively small reservoirs located in North and South Dakota.[5] Steamflooding have surpassed air injection as the preferred EOR process even though both processes can satisfy the screening criteria for the same light or heavy oil reservoirs. This is the result of air injection being considered as riskier than steamflooding because the propagation of its combustion/reaction front is more difficult to predict and control.

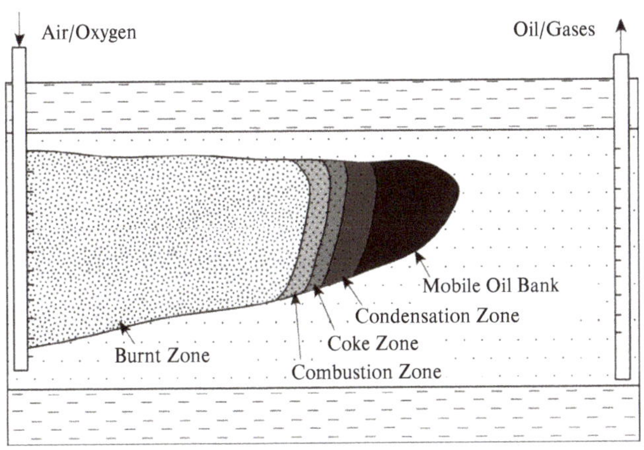

▲ **FIGURE 21.18** **Air injection process diagram**

21.9 Implementation of EOR Projects

Taking an EOR project from initial screening to field implementation is a long, labor-intensive process.[100] It is one of the most elaborate processes in project management and execution that requires the involvement of all the major areas of petroleum engineering. EOR projects require input from reservoir engineers, geologists, petrophysicists, geophysicists, drilling engineers, production engineers, facilities engineers, field personnel, economists, etc., for proper evaluation and implementation of the process. Assembling these professionals in a functional team is more efficient and effective for execution of the project.

The procedure for screening, testing, and implementing an EOR process is divided into stages. The stages are organized sequentially as follows:

1. Process screening and selection.
2. Quick economic evaluation of selected processes.
3. Geologic and reservoir modeling of selected processes.
4. Expanded economic evaluation of processes.
5. Pilot testing, if necessary.
6. Update/upgrade simulation models with pilot test data/results.
7. Detailed economic evaluation of selected process.
8. Field implementation of process.
9. Project management.

These stages should not be followed rigidly as enumerated. The exchange of data and knowledge among the stages as the evaluation progresses should be incorporated as an integral part of the process. The stages for evaluation and implementation of an EOR process have been organized into a flowchart shown as Figure 21.19. Additional suggestions and guidance on work required at each stage are provided in the following sections. These suggestions and guidance are not exhaustive by any measure. Potential users are advised to use them as starting points for more detailed investigation at each stage of the process.

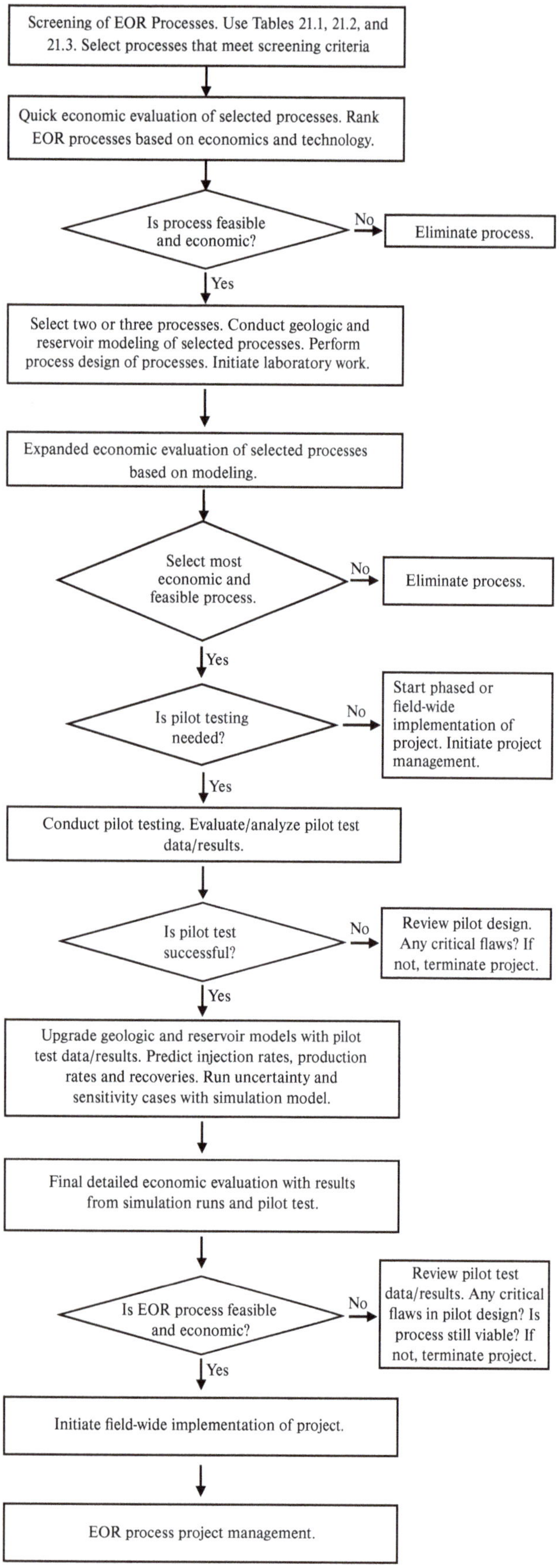

▲ **FIGURE 21.19** **Flow chart for EOR process implementation**

21.9.1 Process Screening and Selection

EOR processes tend to be specific to geologic characteristics of the reservoir and the type of fluid present in the reservoir. These observations are evident by reviewing the screening criteria presented in Tables 21.1 to 21.3 for various EOR processes. For instance, generally thermal processes apply to heavy oils in shallow reservoirs, while miscible gas injection processes apply to lighter oils in deeper reservoirs. To use the screening criteria presented in Tables 21.1 to 21.3, key fluid and reservoir properties for the candidate reservoir should be averaged and compared with the ranges listed in the tables. If the average rock and fluid properties of the candidate reservoir meet the screening criteria, the process is selected for further evaluation.

As part of the screening process, it is important to assemble all data (including geologic, rock properties, fluid properties, pressure transients, production, injection, static and flowing pressure, production logs, etc.) on the candidate reservoir. These data should be reviewed and evaluated to determine the volume of oil to be targeted with the EOR process, and the approximate distribution of the targeted oil in the reservoir. Other issues to address at this stage are whether they are known geologic features existing in the reservoir, such as high permeability layers or permeability barriers that may cause the potential EOR process to fail. Some questions that can help with the selection process are as follows:

1. Do the rock and fluid properties of the reservoir match the screening criteria for the selected process as listed in Tables 21.1 to 21.3?
2. Are there known geologic features such as faults, high permeability layers, permeability barriers, thief zones, etc. in the reservoir that might be detrimental to the selected process?
3. Is the target oil volume in the reservoir sufficient for this process?
4. Can the selected process recover the targeted oil?
5. What additional data are needed to improve the evaluation of the process?

Addressing these and many other questions will lay the groundwork necessary for evaluation of the EOR processes before proceeding to the next stage of the evaluation process.

21.9.2 Quick Economic Evaluation of Selected Processes

In this stage, the selected EOR processes are subjected to quick economic evaluation. The objective of this exercise is to check which processes are economically viable. Some of the data needed for the economic evaluation are:

1. Sources and costs of the gas, chemicals, and steam to be used for injection.

2. Estimates of expected hydrocarbon recoveries.
3. Market prices for the hydrocarbons recovered from the process.
4. Cost of processing produced fluids.
5. Cost of facilities for the EOR process.

Most of the required data can be estimated from analogue fields, if available. Alternatively, industry rule-of-thumb estimates can be used. The goal at this stage is to quickly eliminate processes that are clearly uneconomic and do not merit further evaluations.

21.9.3 Geologic and Reservoir Modeling of Selected Processes

Geologic and reservoir modeling of selected processes is a very important stage of the evaluation process. Geologic models of the reservoir are constructed from all available geologic data including structural, stratigraphic, seismic, rock properties, fluid data, etc. The geologic models should be reviewed in recognition of historical production data. The geological models should be characterized and populated with reservoir properties. Representative geologic models are then converted to reservoir models for simulation. The reservoir models should be subjected to a history-matching process, if the reservoir has historical production data. The history-matching process should serve as a test of the geologic models and could be used to rank the geologic models in terms of how closely they match historical reservoir performance.

The most representative geologic model is then used to evaluate the selected EOR processes. This requires incorporating laboratory data that are specific to the EOR process in constructing the simulation model. Using the simulation models, each selected EOR processes is evaluated in terms of injection rates, production rates, displacement efficiency, cumulative recovery, etc., depending on the process design. The output from this stage is then used in an expanded economic evaluation in the next stage.

21.9.4 Expanded Economic Evaluation of Selected Processes

In the expanded economic evaluation, the outputs from the design of the simulated EOR processes are used. These outputs are primarily injection rates and production rates from process design. Development costs, such as the costs to install flood patterns, drill new wells (producers, injectors, and observation wells), install specialized facilities for fluid processing, etc., are included. At this stage, the cost of injectants and selling prices of produced hydrocarbons are upgraded to reflect market rates. All the relevant costs that could significantly impact the economic viability of the processes must be included in the analyses at this stage. At the conclusion of the stage, one process is chosen for pilot testing or field implementation, if pilot testing is not needed. Pilot testing may not be necessary, if the EOR process has been applied successfully in an analogue field or if it was decided that conducting a pilot test will not reduce the risks and uncertainties associated with the project. In that case, a phased field implementation of the project should be used to mitigate risks.

Note that methods for economic evaluation of petroleum projects are presented in Chapter 25.

21.9.5 Pilot Testing

Pilot testing is very important in the evaluation of an EOR process. Pilot tests are typically designed to reduce the risks and uncertainties associated with an EOR process. Some of these risks and uncertainties may have been identified during simulation of the EOR process. Risks and uncertainties associated with EOR processes are typically elevated because of limited data on the impact of well spacing, pattern configuration, injection and production rates, injection strategy, vertical and areal sweep efficiencies, gravity over-ride, viscous fingering, thief zones, mobility control, etc. Pilot testing can be used to assess the impact of these factors on the EOR process before embarking on a field-wide implementation. For a pilot test to be successful, the objectives of the test must be clearly defined.[101] Some key objectives of a pilot test that should be considered are:

1. Acquisition of geologic and performance data to improve the process simulation model developed in the earlier stage.
2. Evaluation of the impact of geology on volumetric sweep efficiency in the target area.
3. Assessment of the efficiency of the selected process in increasing hydrocarbon recovery.
4. Validation of predicted injection and production rates.
5. Testing processing facilities and identifying potential problems with handling injection fluids or processing produced fluids.

The pilot test should be designed with the simulation model in terms of selection of well type, well spacing, pattern configuration, well completion, injection strategy, injection and production rates, injection volumes, etc. The data acquired from the pilot should be compared with predictions from the simulation model. This comparison should serve as guide on the reliability of the simulation model and the level of confidence that could be placed on its predictions.

Pilot tests can be classified according to well configurations.[101] The four common configurations are:

1. Non-producing pilot.
2. Small-scale unconfined pilot.
3. Small-scale confined pilot.
4. Multi-pattern producing pilot.

A non-producing pilot consists of one injector and is used primarily for injectivity tests. If observation wells are added to the configuration, the system can be used for vertical conformance tests. Small-scale unconfined/confined and multi-pattern pilots have producing wells (and in many cases observation wells) in their configurations. These types of pilot tests yield valuable data on process efficiency, reservoir conformance, performance data, and overall direct assessment of the EOR process.[101]

Acquisition of accurate and reliable data during pilot testing is important for assessment of the success of the EOR process and future utilization of the data in field implementation. Frequent and regular monitoring of all wells involved in the pilot testing should include measurements of injection and production rates, bottomhole pressure surveys, running of injection and production profiles, pressure transient tests, temperature surveys, time-lapse logging, coring, use of chemical or radioactive tracers, etc. The nature and frequency of the well tests depend on the type of EOR process. For instance, installation of permanent downhole gauges for continuous monitoring of reservoir pressure may be desirable for miscible injection processes. The surveillance and monitoring program for the pilot test should be as extensive as necessary to acquire the data needed to meet the stated objectives for the test.

21.9.6 Upgrade Geologic and Reservoir Models with Pilot Test Data/Results

The geologic and reservoir models used in designing the pilot test should be upgraded with the data and results from the pilot test. In practice, this activity should be conducted almost simultaneously with the pilot test. As soon as data are collected from the pilot test, comparisons should be conducted between predictions from the simulation model and the results from the test. Severe divergence between model predictions and pilot test results should be investigated thoroughly. This may be an indication that the reservoir model is not a good representation of the actual reservoir and should be reviewed. Actually, there are many other issues that could cause the difference between predictions and actual results other than the reservoir model. These issues must be resolved before using the model for future predictive work. Future predictive work should include extensive uncertainty analysis on critical parameters identified by the pilot test. This should be accompanied with sensitivity analysis on injectant costs and product prices.

21.9.7 Final Detailed Economic Evaluation

The final detailed economic evaluation incorporates the data from the pilot test and predicted rates from the upgraded simulation model. Further detailed costs of equipment and facilities, projected costs of injectants, future product prices, production costs, etc., should be included in the final economic analysis. This is the last economic evaluation of the entire project before proceeding to field implementation.

21.9.8 Field-Wide Project Implementation

Commercial field implementation of an EOR process is capital intensive. It demands careful planning and budgeting so that the project is installed in a timely manner. It is advantageous to use the team that installed and managed the pilot test for the field implementation. Also, field personnel trained in the operation of the pilot should be retained for the field implementation phase.

21.9.9 EOR Process Project Management

Proper management of EOR projects requires an intensive program of monitoring and surveillance of reservoir performance. The monitoring and surveillance practices that were proven during the pilot testing phase should be adopted for the field implementation of the EOR process. Well surveillance and monitoring should be conducted regularly by a team dedicated to this activity. The data gathered by the team should be used in frequent well and project reviews.

The simulation model history-matched with the results from the pilot tests should be used in the management of the project. Data gathered from field surveillance of the project should be used to update the model as necessary. The simulation model should be used in a predictive mode to adjust injection and production rates, pattern configuration, well locations, injection strategy, and other activities which will optimize the management of the EOR process.

Nomenclature

$M_{C_{5+}}$	molecular weight of C_{5+} fraction, lbm/lb-mole	T_{ci}	critical temperature of component *i* in the gas mixture
$M_{C_{7+}}$	molecular weight of C_{7+} fraction, lbm/lb-mole	γ_o	specific gravity of stock tank oil
T	reservoir temperature, °F		

Abbreviations

AIP	air injection process	ISC	in-situ combustion
ASP	alkali-surfactant-polymer	MCM	multiple-contact miscibility
CGI	continuous gas injection	MEOR	microbial enhance oil recovery
CSS	cyclic steam stimulation	MME	minimum miscibility enrichment
CV	combined condensing and vaporizing	MMP	minimum miscibility pressure
DTS	distributed temperature sensing	OOIP	original oil-in-place
EOS	equation of state	OSR	oil-steam ratio
EOR	enhanced oil recovery	RBA	rising-bubble apparatus
ES-SAGD	expanding solvent steam-assisted gravity drainage	SAGD	steam-assisted gravity drainage
FCM	first-contact miscibility	SD	steam drive
HC	hydrocarbon	SWAG	simultaneous water-and-gas
HCPV	hydrocarbon pore volume	WAG	water-alternating gas
HPAI	high pressure air injection	WASP	water-alternating-steam process
IFT	interfacial tension	XSAGD	cross-steam-assisted gravity drainage

REFERENCES

1. McGlade, C., Sondak, G., and Han, Mei: "Commentary: Whatever Happened to Enhanced Oil Recovery?" https://www.iea.org/newsroom/news/2018/november/whatever-happened-to-enhanced-oil-recovery.html. Accessed on July 05, 2019.
2. Melzer, L.S.: "Carbon Dioxide Enhanced Oil Recovery (CO_2 EOR): Factors Involved in Adding Carbon Capture, Utilization and Storage (CCUS) to Enhanced Oil Recovery." *Center for Climate and Energy Solutions* (2012).
3. Sheng, J.J.: "Critical Review of Field EOR Projects in Shale and Tight Reservoirs." *J. Petrol. Sci. Eng.* (2017), 159, 654-665.
4. Alfarge, Dheiaa Khafief Khashan. "Integrated Study on the Applicability of CO_2-EOR in Unconventional Liquids Rich Reservoirs." (2018).
5. Koottungal, L: "2008 Worldwide EOR Survey," *Oil and Gas Journal* (April 21, 2008) 47-59.
6. Awan, A.R., Teigland, R., and Kleppe, J.: "A Survey of North Sea Enhanced-Oil-Recovery Projects Initiated During the Years 1975 to 2005," *SPEREE* (June 2008) 497-512.
7. Lazar, I.: "MEOR Field Trials Carried Out Over the World During the Last 35 Years," International Conference on MEOR, Norman, Oklahoma, May 27-June 1, 1990.
8. Bryant, R.S., and Lindsey, R.P.: "World-Wide Applications of Microbial Technology for Improving Oil Recovery," paper SPE/DOE 35356 presented at the SPE/DOE Improved Oil Recovery Symposium, Tulsa, Oklahoma, (April 21-24, 1998).
9. Hutchinson, C.A., and Braun, P.H.: "Phase Relations of Miscible Displacement in Oil Recovery," *AICHEJ* (March 1961) 7, 64-72.
10. Benham, A.L., Dowden, W.E., and Kunzman, W.J.: "Miscible Fluid Displacement—Prediction of Miscibility," *Trans.,* AIME (1960) 219, 229-237.
11. Stalkup, F.I.: "Status of Miscible Displacement," *JPT* (April 1983) 815-826.
12. Zick, A.A.: "A Combined Condensing/Vaporizing Mechanism in the Displacement of Oil by Enriched Gases," paper SPE 15493 presented at the SPE 61st Annual Technical Conference and Exhibition, New Orleans, Louisiana, (October 5-8, 1986).

13. Stalkup, F.I.: "Displacement Behavior of the Condensing/Vaporizing Gas Drive Process," paper SPE 16715 presented at the SPE 62nd Annual Technical Conference and Exhibition, Dallas, Texas, (September 27-30, 1987).
14. Glasø, Ø.: "Generalized Minimum Miscibility Pressure Correlation," *SPEJ* (December 1985) 927-934.
15. Firoozabadi, A., and Aziz, K.: "Analysis and Correlation of Nitrogen and Lean-Gas Miscibility Pressure," *SPERE* (November 1986) 575-582.
16. Emera, M.K., and Sarma, H.K.: "Use of Genetic Algorithm to Estimate CO_2-Oil Minimum Miscibility Pressure–A Key Parameter in Design of CO_2 Miscible Flood," *Journal of Petroleum Science & Engineering* (2005) 46, 37-52.
17. Yuan, H., Johns, R.T., Egwuenu, A.M., and Dindoruk, B.: "Improved MMP Correlations for CO_2 Floods Using Analytical Gasflooding Theory," *SPEREE* (October 2005) 418-425.
18. Johnson, J.P., and Pollin, J.S.: "Measurement and Correlation of CO_2 Miscibility Pressures," paper SPE 9790 presented at the SPE/DOE Enhanced Oil Recovery Symposium, Tulsa, Oklahoma, (April 5-8, 1981).
19. Alston, R.B., Kokolis, G.P., and James, C.F.: "CO_2 Minimum Miscibility Pressure: A Correlation for Impure CO_2 Streams and Live Oil Systems," *SPEJ* (April 1985) 268-274.
20. Sebastian, H.M., Wenger, R.S., and Renner, T.A.: "Correlation of Minimum Miscibility Pressure for Impure CO_2 Streams," *JPT* (November 1985) 2076-2082.
21. Kovarik, F.S.: "A Minimum Miscibility Pressure Study Using Impure CO_2 and West Texas Oil Systems: Data Base, Correlations, and Compositional Simulation," paper SPE 14689 presented at the SPE Production Technology Symposium, Lubbock, Texas, (November 11-12, 1981).
22. Orr, F.M., Jr., and Silva, M.K.: "Effect of Oil Composition on Minimum Miscibility Pressure–Part 2: Correlation," *SPERE* (November 1987) 479-491.
23. Enick, R.M., Holder, G.D., and Morsi, B.I.: "A Thermodynamic Correlation for the Minimum Miscibility Pressure in CO_2 Flooding of Petroleum Reservoir," *SPERE* (February 1988) 81-92.
24. Eakin, B.E., and Mitch, F.J.: "Measurement and Correlation of Miscibility Pressures of Reservoir Oils," paper SPE 18065 presented at the SPE Annual Technical Conference and Exhibition, Houston, Texas, (October 2-5, 1988).
25. Dong, M.: Potential of Greenhouse Gas Storage and Utilization Through Enhanced Oil Recovery–Task 3: Minimum Miscibility Pressure Studies. Final Report (SRC Publication No. P-110-468-C-99), September 1999.
26. Emera, M.K., and Sarma, H.K.: "A Reliable Correlation to Predict the Change in Minimum Miscibility Pressure When CO_2 Is Diluted with Other Gases," *SPEREE* (August 2006) 366-373.
27. Egwuenu, A.M., and Johns, R.T.: "Improved Fluid Characterization for Miscible Gas Floods," *SPEERE* (August 2008) 655-665.
28. Yuan, H., Chopra, A., Marwah, V., and Stalkup, F.: "Fluid Characterization for Miscible Floods," paper SPE 114913 presented at the SPE Annual Technical Conference and Exhibition, Denver, Colorado, (September 21-24, 2008).
29. Monroe, W.W., Silva, M.K., Larsen, L.L., and Orr, F.M., Jr.: "Composition Paths in Four-Component Systems: Effect of Dissolved Methane on 1D CO_2 Flood Performance," *SPERE* (August 1990) 423-432.
30. Johns, R.T., Dindoruk, B., and Orr, F.M., Jr.: "Analytical Theory of Combined Condensing/Vaporizing Gas Drive," *SPE Advanced Technical Series* (1993) 1, No. 2.
31. Johns, R.T., and Orr, F.M., Jr.: "Miscible Gas Displacement of Multicomponent Oils," *SPEJ* (March 1996) 39-50.
32. Wang, Yun, and Orr, F.M., Jr.: "Calculation of Minimum Miscibility Pressure," paper SPE 39683 presented at the SPE/DOE Improved Oil Recovery Symposium, Tulsa, Oklahoma, (April 19-22, 1998).
33. Jessen, K., Michelsen, M.L., and Stenby, E.H.: "Effective Algorithm for Calculation of Minimum Miscibility Pressure," paper SPE 50632 presented at the SPE European Petroleum Conference, Hague, The Netherlands, (October 20-22, 1998).
34. Yuan, H., and Johns, R.T.: "Simplified Method for Calculation of Minimum Miscibility Pressure or Enrichment," *SPEJ* (December 2005) 416-425.
35. Ahmadi, K., and Johns, R.T.: "Multiple Mixing-Cell Method for MMP Calculations," paper SPE 116823 presented at the SPE Annual Technical Conference and Exhibition, Denver, Colorado, (September 21-24, 2008).
36. Orr, F.M., Jr., Silva, M.K., Lien, C.L., and Pelletier, M.T.: "Laboratory Experiments to Evaluate Field Prospects for CO_2 Flooding," *JPT* (April 1982) 888-898.
37. Randall, T.E., and Bennion, D.B.: "Laboratory Factors Influencing Slim Tube Results," *The Journal of Canadian Petroleum Technology* (July-August, 1989) 28, 4, 60-70.
38. Flock, D.L., and Nouar, A.: "Parametric Analysis on the Determination of the Minimum Miscibility Pressure in Slim Tube Displacements," *The Journal of Canadian Petroleum Technology* (September-October, 1984) 80-88.
39. Novosad, Z., Sibbald, L.R., and Costain, T.G.: "Design of Miscible Solvents for a Rich Gas Drive–Comparison of Slim Tube and Rising Bubble Tests," *The Journal of Canadian Petroleum Technology* (January-February, 1990) 29, 1, 37-42.
40. Christiansen, R.L., and Haines, H.K.: "Rapid Measurement of Minimum Miscibility Pressure with the Rising-Bubble Apparatus," *SPERE* (November 1987) 523-527.
41. Elsharkawy, A.M., Poettmann, F.H., and Christiansen, R.L.: "Measuring Minimum Miscibility Pressure: Slim-Tube or Rising-Bubble Method?" paper SPE/DOE 24114 presented at the SPE/DOE Eight Symposium on Enhanced Oil Recovery, Tulsa, Oklahoma, (April 22-24, 1992).
42. Thomas, F.B., Zhou, X.L., Bennion, D.B., and Bennion, D.W.: "A Comparative Study of RBA, P-x, Multicontact, and Slim Tube Results," *The Journal of Canadian Petroleum Technology* (February 1994) 33, 2, 17-26.
43. Zhou, D., and Orr, F.M., Jr.: "Analysis of Rising-Bubble Experiments to Determine Minimum Miscibility Pressures," *SPEJ* (March 1998) 19-25.
44. Bon, J., Sarma, H.K., and Theophilos, A.M.: "An Investigation of Minimum Miscibility Pressure for CO_2-Rich Injection Gases with Pentanes-Plus Fraction," paper SPE 97536 presented at the SPE International Improved Oil Recovery Conference in Asia Pacific, Kuala Lumpur, Malaysia, (December 5-6, 2005).

45. Orr, F.M., Jr.: "Storage of Carbon Dioxide in Geologic Formations," SPE 88842, Distinguished Author Series (September 2004) 90-97.

46. Qi, R., Laforce, T.C., and Blunt, M.J.: "Design of Carbon Dioxide Storage in Oilfields," paper SPE 115663 presented at the Annual Technical Conference and Exhibition, Denver, Colorado, (September 21-24, 2008).

47. Wood, D.J., Lake, L.W., Johns, R.T., and Nunez, V.: "A Screening Model for CO_2 Flooding and Storage in Gulf Coast Reservoirs Based on Dimensionless Groups," *SPEERE* (June 2008) 513-520.

48. Flaaten, A.K., Nguyen, Q.P., Pope, G.A., and Zhang, J.: "A Systematic Laboratory Approach to Low-Cost, High Performance Chemical Flooding," paper SPE 113469 presented at the 2008 SPE/DOE Improved Oil Recovery Symposium, Tulsa, Oklahoma, (April 19-23, 2008).

49. Manrique, E.J., Muci, V.E., and Gurfinkel, M.E.: "EOR Field Experiences in Carbonate Reservoirs in the United States," *SPEREE* (December 2007) 667-686.

50. Arihara, N., Yoneyama, T., Akita, Y., and XiangGuo, L.: "Oil Recovery Mechanisms of Alkali-Surfactant-Polymer Flooding," paper SPE 54330 presented at the SPE Asia Pacific Oil and Gas Conference and Exhibition, Jakarta, Indonesia, (April 20-22, 1999).

51. Vargo, J., Turner, J., Vergnani, B., Pitts, M.J., Wyatt, K., Surkalo, H., and Patterson, D.: "Alkaline-Surfactant-Polymer Flooding of the Cambridge Minnelusa Field," *SPEREE* (December 2000) 552-558.

52. Hernández, C., Chacón, L.J., Anselmi, L., Baldonedo, A., Qi, J., Dowling, P.C., and Pitts, M.J.: "ASP System Design for an Offshore Application in La Salina Field, Lake Maracaibo," *SPEREE* (June 2003) 147-156.

53. Jiechang, C., Xu, D., and Bai, W.: "Commercial ASP Flood Test in Daqing Oil Field," paper SPE 117824 presented at the Abu Dhabi International Petroleum Exhibition and Conference, Abu Dhabi, United Arab Emirates, (November 3-6, 2008).

54. Zhang, D.L., Liu, S., Yan, W., Puerto, M., Hirasaki, G.J., and Miller, C.A.: "Favorable Attributes of Alkali-Surfactant-Polymer Flooding," paper SPE 99744 presented at the SPE/DOE Symposium on Improved Oil Recovery, Tulsa, Oklahoma, (April 22-26, 2006).

55. Zubari, H.K., and Babu, V.C.: "Single Well Tests to Determine the Efficiency of Alkaline-Surfactant Injection in a Highly Oil-Wet Limestone Reservoir," paper SPE 81464 presented at the SPE Middle East Oil Show and Exhibition, Bahrain, (April 5-8, 2003).

56. Taber, J.J., Martin, F.D., and Seright, R.S.: "EOR Screening Criteria Revisited—Part 1: Introduction to Screening Criteria and Enhanced Recovery Field Projects," *SPERE* (August 1997) 189-198.

57. Wang, D., Jiechang, C., Junzheng, W., and Gang, W.: "Experiences Learned After Production More Than 300 Million Barrels of Oil by Polymer Flooding in Daqing Oil Field," paper SPE 77693 presented at the SPE Annual Technical Conference and Exhibition, San Antonio, Texas, (September 29-October 2, 2002).

58. Wang, D., Seright, R.S., Shao, Z., and Wang, J.: "Key Aspects of Project Design for Polymer Flooding at the Daqing Oil Field," *SPEREE* (December 2008) 1117-1124.

59. Shouliang, Z., Yitang, Z., Shuhong, W., Shangqi, L., Xiuluan, L., and Songlin, L.: "Status of Heavy-Oil Development in China," Paper SPE 97844 presented at the 2005 SPE International Thermal Operations and Heavy Oil Symposium, Calgary, Alberta, (November 1-3, 2005).

60. Matthews, C.S.: "Steamflooding," *JPT* (March 1983) 465-471.

61. Blevins, T.R.: "Steamflooding in the U.S.: A Status Report," *JPT* (May 1990) 548-554.

62. Nasr, T.N., and Ayodele, O.R.: "Thermal Techniques for the Recovery of Heavy Oil and Bitumen," paper SPE 97488 presented at the SPE International Improved Oil Recovery Conference in Asia Pacific, Kuala Lumpur, Malaysia, (December 5-6, 2005).

63. Jimenez, J.: "The Field Performance of SAGD Projects in Canada," paper SPE 12860 presented at the International Petroleum Technology Conference, Kuala Lumpur, Malaysia, (December 3-5, 2008).

64. Das, S.: "Improving the Performance of SAGD," paper SPE 97921 presented at the SPE International Thermal Operations and Heavy Oil Symposium, Calgary, Alberta, Canada, (November 1-3, 2005).

65. Stalder, J.L.: "Thermal Efficiency and Acceleration Benefits of Cross SAGD (XSAGD)," paper SPE 117244 presented at the SPE International Thermal Operations and Heavy Oil Symposium, Calgary, Alberta, Canada, (October 20-23, 2008).

66. Gupta, S.: "Unlocking the Planet's Heavy Oil and Bitumen Resources–A Look at SAGD," SPE 108799, Distinguished Lecture Series, 2005.

67. Marx, J.W., and Langenheim, R.H.: "Reservoir Heating by Hot Fluid Injection," *Trans.* AIME (1959) 216, 312-315.

68. Boberg, T.C., and Lantz, R.B.: "Calculation of the Production Rate of a Thermally Stimulated Well," *JPT* (December 1966) 1613-1623.

69. Mandl, G., and Volek, C.W.: "Heat and Mass Transport in Steam-Drive Processes," paper SPE 1896 presented at the 42nd SPE Annual Fall Meeting, Houston, Texas, (October 1-4, 1967).

70. Farouq Ali, S.M.: "Graphical Determination of Oil Recovery in a Five-Spot Steamflood," paper SPE 2900 presented at the SPE Rocky Mountain Regional Meeting, Casper, Wyoming, (June 8-9, 1970).

71. Myhill, N.A., and Stegemeier, G.L.: "Steam-Drive Correlation and Prediction," *JPT* (February 1978) 173-182.

72. Gomaa, E.E.: "Correlations for Predicting Oil Recovery by Steamflood," *JPT* (February 1980) 325-332.

73. Rhee, S.W., and Doscher, T.M.: "A Method for Predicting Oil Recovery by Steamflooding Including the Effects of Distillation and Gravity Override," *SPEJ* (August 1980) 249-266.

74. Jones, J.: "Steam Drive Model for Hand-Held Programmable Calculators," *JPT* (September 1981) 1583-1598.

75. Aydelotte, S.R., and Pope, G.A.: "A Simplified Predictive Model for Steamdrive Performance," *JPT* (May 1983) 991-1002.

76. van Lookeren, J.: "Calculation Methods for Linear and Radial Steam Flow in Oil Reservoirs," *SPEJ* (June 1983) 427-439.

77. Vogel, J.V.: "Simplified Heat Calculations for Steamfloods," *JPT* (July 1984) 1127-1136.
78. Neuman, C.H.: "A Gravity Override Model of Steamdrive," *JPT* (January 1985) 163-169.
79. Miller, M.A., and Leung, W.K.: "A Simple Gravity Override Model of Steamdrive," paper SPE 14241 presented at the 60th SPE Annual Technical Conference and Exhibition, Las Vegas, Nevada, (September 22-25, 1985).
80. Chandra, S., and Mamora, D.D.: "Improved Steamflood Analytical Model," *SPEREE* (December 2007) 638-643.
81. Shutler, N.D.: "Numerical Three-Phase Simulation of the Linear Steamflood Process," *SPEJ* (June 1969) 232-246.
82. Shutler, N.D.: "Numerical Three-Phase Model of the Two-Dimensional Steamflood Process," *SPEJ* (December 1970) 405-417.
83. Abdalla, A., and Coats, K.H.: "A Three-Phase, Experimental and Numerical Simulation Study of the Steamflood Process," paper SPE 3600 presented at the SPE-AIME 46th Annual Fall Meeting, New Orleans, Louisiana, (October 3-6, 1971).
84. Weinstein, H.G., Wheeler, J.A., and Woods, E.G.: "Numerical Model for Steam Stimulation," paper SPE 4759 presented at the SPE-AIME Third Improved Oil Recovery Symposium, Tulsa, Oklahoma, (April 22-24, 1974).
85. Coats, K.H., George, W.D., Chu, Chieh, and Marcum, B.E.: "Three-Dimensional Simulation of Steamflooding," *SPEJ* (December 1974) 573-592.
86. Coats, K.H.: "Simulation of Steamflooding with Distillation and Solution Gas," *SPEJ* (October 1976) 235-247.
87. Coats, K.H.: "A Highly Implicit Steamflood Model," *SPEJ* (October 1978) 369-383.
88. Grabowski, J.W., Vinsome, P.K., Lin, R.C., Behie, A., and Rubin, B.: "A Fully Implicit General Purpose Finite-Difference Thermal Model for In Situ Combustion and Steam," paper SPE 8396 presented at the 54th Annual Fall Technical Conference and Exhibition, Las Vegas, Nevada, (September 23-26, 1979).
89. Aziz, K., Ramesh, A.B., and Woo, P.T.: "Fourth SPE Comparative Solution Project: Comparison of Steam Injection Performance," *JPT* (December 1987) 1576-1584.
90. Kumar, M., Satik, C., and Hoang, V.: "New Developments in Steamflood Modeling," paper SPE 97719 presented at the 2005 SPE International Thermal Operations and Heavy Oil Symposium, Calgary, Alberta, Canada, (November 1-3, 2005).
91. Shi, X., Chang, Y., Muller, M., Obi, E., and Lim, K.: "A General Unstructured Grid, Parallel, Thermal Simulator and its Application for Large Scale Thermal Models," paper SPE 119172 presented at the 2009 SPE Reservoir Simulation Symposium, The Woodlands, Texas, (February 2-4, 2009).
92. Johnson, D.O., Sugianto, R., Mock, P.H., and Jones, C.H.: "Identification of Steam-Breakthrough Intervals with DTS Technology," *SPEPF* (February 2004) 41-48.
93. Nath, D.K., Sugianto, R., and Finley, D.: "Fiber-Optic Distributed-Temperature-Sensing Technology Used for Reservoir Monitoring in an Indonesian Steamflood," *SPEDC* (June 2007) 149-156.
94. Hanzlik, E.J., and Mims, D.S.: "Forty Years of Steam Injection in California–The Evolution of Heat Management," paper SPE 84848 presented at the SPE International Improved Oil Recovery Conference in Asia Pacific, Kuala Lumpur, Malaysia, (October 20-21, 2003).
95. Ziegler, V.M., Crookston, R.B., Sanford, S.J., and Merrell, J.M.: "Recommended Practices for Heat Management of Steamflood Projects," paper SPE 25808 presented at the International Thermal Operations Symposium, Bakersfield, California, (February 8-10, 1993).
96. Turta, AT., and Singhai, A.K.: "Reservoir Engineering Aspects of Light-Oil Recovery by Air Injection," *SPEREE* (August 2001) 336-344.
97. Fassihi, M.R., Yannimaras, D.V., and Kumar, V.K.: "Estimation of Recovery Factor in Light-Oil Air-Injection Projects," *SPERE* (August 1997) 173-178.
98. Kumar, V.K., Gutierrez, D., Moore, R.G., and Mehta, S.A.: "Case History and Appraisal of the West Buffalo Red River Unit High-Pressure Air Injection Project," paper SPE 107715 presented at the 2007 SPE Hydrocarbon Economics and Evaluation Symposium, Dallas, Texas, (April 1-3, 2007).
99. Gutierrez, D., Taylor, A.R., Kumar, V.K., Ursenbach, M.G., Moore, R.G., and Mehta, S.A.: "Recovery Factors in High-Pressure Air Injection Projects Revisited," *SPEREE* (December 2008) 1097-1106.
100. Hite, J.R., Avasthi, S.M., and Bondor, P.L.: "Planning EOR Projects," paper SPE 92006 presented at the 2004 SPE International Petroleum Conference, Puebla, Mexico, (November 8-9, 2004).
101. Teletzke, G.F., Wattenbarger, R.C., and Wilkinson, J.R.: "Enhanced Oil Recovery Pilot Testing Best Practices," paper SPE 118055 presented at the 2008 Abu Dhabi International Petroleum Exhibition and Conference, Abu Dhabi, UAE, (November 3-6, 2008).

ADDITIONAL REFERENCES

- Abou-Kassem, J.H., and Aziz, K.: "Sensitivity of Steamflood Model Results to Grid and Timestep Sizes," *SPEJ* (February 1984) 65-74.
- Akin, S.: "Mathematical Modeling of Steam-Assisted Gravity Drainage," *SPEREE* (October 2005) 372-376.
- Chen, Q., Gerritsen, M.G., and Kovscek, A.R.: "Effects of Reservoir Heterogeneities on the Steam-Assisted Gravity-Drainage Process," *SPEREE* (October 2008) 921-932.
- Collins, P.M.: "Geomechanical Effects on the SAGD Process," *SPEREE* (August 2007) 367-375.
- Dindoruk, B., Orr, F.M., Jr., and Johns, R.T.: "Theory of Multicontact Miscible Displacement with Nitrogen," *SPEJ* (September 1997) 268-279.
- Gates, I.D., Kenny, J., Hernandez-Hdez, I.L., and Bunio, G.L.: "Steam-Injection Strategy and Energetics of Steam-Assisted Gravity Drainage," *SPEREE* (February 2007) 19-34.

- Hammershaimb, E.C., and Kuuskraa, V.A.: "Recovery Efficiency of Enhanced Oil Recovery Methods: A Review Of Significant Field Tests," paper SPE 12114 presented at the 58th Annual Technical Conference and Exhibition, San Francisco, California, (October 5-8, 1983).
- Holm, L.W., and Josendal, V.A.: "Effect of Oil Composition on Miscible-Type Displacement by Carbon Dioxide," paper SPE 8814 presented at the SPE-DOE Enhanced Oil Recovery Symposium, Tulsa, Oklahoma, (April 20-23, 1980).
- Holm, L.W., and Josendal, V.A.: "Mechanism of Oil Displacement by Carbon Dioxide," *JPT* (December 1974) 1427-1436.
- Hong, K.C.: "Recent Advances in Steamflood Technology," paper SPE 54078 presented at the 1999 International Thermal Operations and Heavy Oil Symposium, Bakersfield, California, (March 17-19, 1999).
- Jensen, T.B., Harpole, K.J., and Østhus, A.: "EOR Screening for Ekofisk," paper SPE 65124 presented at the SPE European Petroleum Conference, Paris, France, (October 24-25, 2000).
- Johns, R.T., Yuan, H., and Dindoruk, B.: "Quantification of Displacement Mechanisms in Multicomponent Gasfloods," *SPEJ* (September 2004) 314-321.
- Jones, J.: "Why Cyclic Steam Predictive Models Get No Respect," *SPERE* (February 1992) 67-74.
- Neuman, C.H.: "A Mathematical Model of the Steam Dive Process–Derivation," paper SPE 5495.
- Rathmell, J.J., Stalkup, F.I., and Hassinger, R.C.: "A Laboratory Investigation of Multiple Displacement by Carbon Dioxide," paper SPE 3483 presented at the SPE 46th Annual Fall Meeting, New Orleans, Louisiana, (October 3-6, 1971).
- Satik, C., Kumar, M., Defrancisco, S., Hoang, V., and Basham, M.: "Important Modeling Parameters for Predicting Steamflood Performance," *SPEREE* (October 2007) 508-513.
- Stalkup, F.: "Predicting the Effect of Continued Gas Enrichment Above the MME on Oil Recovery in Enriched Hydrocarbon Gas Floods," paper SPE 48949 presented at the SPE Annual Technical Conference and Exhibition, New Orleans, Louisiana, (September 27-30, 1998).
- Stalkup, F.I.: "Carbon Dioxide Miscible Flooding: Past, Present, and Outlook for the Future," *JPT* (August 1978) 1102-1112.
- Stalkup, F.I.: "Status of Miscible Displacement," *JPT* (April 1983) 815-826.
- Taber, J.J., Martin, F.D., and Seright, R.S.: "EOR Screening Criteria Revisited—Part 2: Applications and Impact of Oil Prices," *SPERE* (August 1997) 199-205.
- Wu, R.S, and Batycky, J.P.: "Evaluation of Miscibility from Slim Tube Tests," *The Journal of Canadian Petroleum Technology* (November-December 1990) 29, 6, 63-70.
- Yellig, W.F., and Metcalfe, R.S.: "Determination and Prediction of CO_2 Minimum Miscibility Pressure, *JPT* (January 1980) 160-168.

PROBLEMS

21-1. A ternary diagram is shown in Figure 21P.1. Using the ternary diagram, determine:

a. The composition of the mixture at X_1

b. The composition of the mixture at X_2

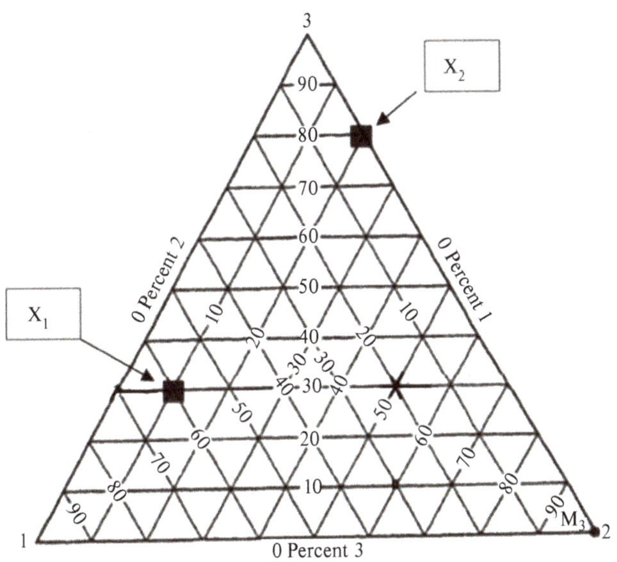

▲ **FIGURE 21P.1** **Ternary diagram for Problem 21-1**

21-2. The pseudoternary diagram of a mixture consisting of methane (C_1), ethane (C_2) to hexane (C_6), and heptanes⁺ (C_{7+}) is represented in Figure 21P.2. Using the ternary diagram, demonstrate the development of multiple-contact miscibility (MCM) between pure methane and crude composition shown in the diagram. Assume vaporizing gas drive MCM is the dominant process.

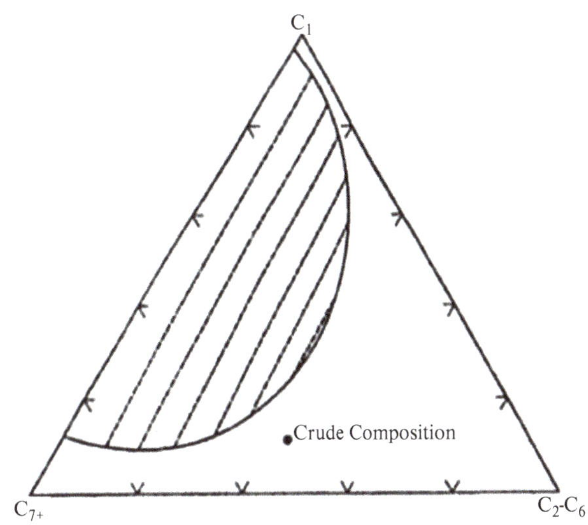

▲ **FIGURE 21P.2** **Ternary diagram for Problem 21-2**

21-3. The pseudoternary diagram of a fluid system consisting of carbon dioxide (CO_2), ethane (C_2) to hexane (C_6), and heptanes plus (C_{7+}) is shown in Figure 21P.3. The composition of the oil in the reservoir is shown in the ternary diagram at point O. If the composition of the injection fluid is 96% CO_2 and 4% C_2 to C_6, can miscible displacement be achieved between the injection fluid and the oil at point O? Explain your answer with the aid of the ternary diagram.

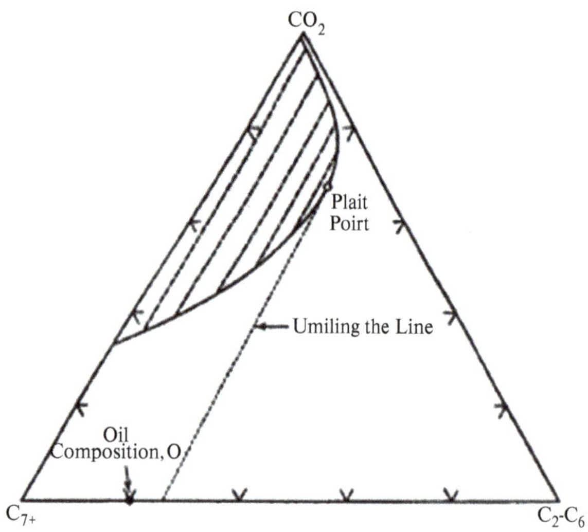

▲ **FIGURE 21P.3 Ternary diagram for Problem 21-3**

21-4. Consider the pseudoternary diagram shown in Figure 21P.4. The composition of the reservoir oil is shown at point C. Suppose an injection fluid with a composition shown as point A on the ternary diagram is to be used. Use the ternary diagram to demonstrate how MCM could be achieved assuming vaporizing gas drive process. If the reservoir oil composition is at point B instead, could MCM be achieved also? Explain your answer with the aid of the ternary diagram.

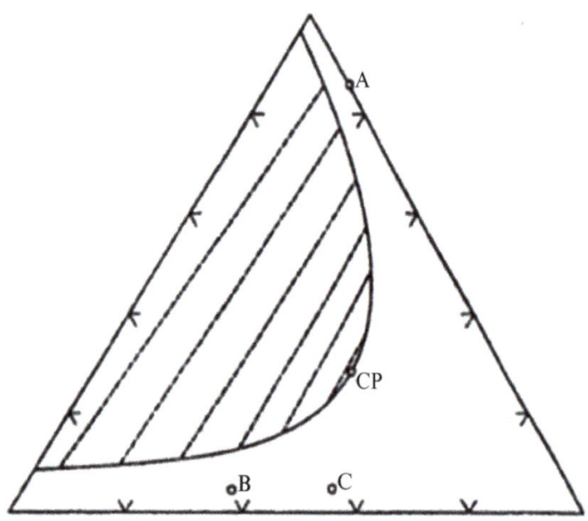

▲ **FIGURE 21P.4 Ternary diagram for Problem 21-4**

21-5. The composition of hydrocarbon injection gas mixture and reservoir oil are given below as:

Molecular weight of C_2 to C_6 components in injection gas	33 lbm/lb-mole
Molecular percent of methane in injection gas	69.2
Corrected molecular weight of C_{7+} in stock tank oil	226 lbm/lb-mole
Reservoir temperature	165°F

Calculate the minimum miscibility pressure (MMP) using the Glasø correlation.

21-6. Calculate the minimum miscibility pressure (MMP) for pure CO_2 gas injection into a reservoir containing oil with the following properties using the Glasø, Emera, and Yuan correlations.

Mole percent of C_1 in reservoir oil	25
Mole percent of C_2 to C_6 in reservoir oil	37.93
Molecular weight of C_{5+} in reservoir oil	94.83 lbm/lb-mole
Molecular weight of C_{7+} in reservoir oil	114.07 lbm/lb-mole
Reservoir temperature	200°F

21-7. Calculate the minimum miscibility pressure (MMP) for impure CO_2 gas injection into a reservoir containing oil with the following properties using the Sebastian, Yuan, and Emera correlations.

Mole percent of CO_2 in injection gas	81
Mole percent of methane in injection gas	19
Mole percent of C_2 to C_6 in reservoir oil	25.6
Molecular weight of C_{7+} in reservoir oil	245 lbm/lb-mole
Reservoir temperature	110°F

CHAPTER

22 Geologic Modeling and Reservoir Characterization

22.1 Introduction

Geologic modeling is the art of constructing a structural and stratigraphic model of a reservoir from analyses and interpretations of seismic data, log data, core data, fluid production data, etc., plus inclusion of knowledge on geologic structures acquired by the geoscientists from basin sedimentology and outcrop studies of similar formations. Reservoir characterization can be described as the process by which various rock definitions and petrophysical properties (such as facies, flow units, porosity, permeability, net sand, saturations, etc.) are distributed in the geologic model in a consistent and logical manner to obtain equally probable representations of the reservoir. Geologic modeling and reservoir characterization are processes that require extensive knowledge and experience before they can be applied properly on a reservoir. The tools and techniques used in geologic modeling and reservoir characterization are numerous, varied, and intricate. Due to these reasons, this chapter serves mainly as an introduction to these subjects.

The basic concepts in geologic modeling and reservoir characterization presented in this chapter include the types and sources of data, scale and integration of data, and data quality and control. This is followed with a general procedure for conducting geologic modeling and characterization of a reservoir. The chapter concludes with examples on the application of geologic modeling and characterization on two reservoirs.

22.2 Sources of Data for Geologic Modeling and Reservoir Characterization

There are many sources of data for geologic modeling and reservoir characterization. These depend on the stages of exploration, appraisal, and development of the reservoir. At the exploration stage, data sources are limited to seismic, analogue reservoirs, and outcrops. In the appraisal stage, data sources expand to include cores, logs, fluids, and other subsurface data acquired from wells drilled into the reservoir. During the development stage, additional data may be available from fluid production tests, and other tests designed to assess the extent and productive potential of the reservoir. These sources of data should be integrated at each stage to construct and characterize reservoir models that are based on **all** available data. A brief description of the various sources of data for geologic modeling and reservoir characterization is presented in this section. This list is not considered to be complete and exhaustive. It is provided as a guide to some of the data that are generally used for geologic modeling and reservoir characterization. Advancements in the technology for data acquisition from reservoirs and transformation of these data into quantities that are useful for geologic modeling and reservoir characterization are making significant progress in the industry. It is expected that the list of data sources presented in this book will continue to grow as more data sources become available in the future.

22.2.1 Seismic Data

One of the important sources of data for geologic modeling and reservoir characterization is seismic data. Seismic data can be used in geologic modeling to map the structure and stratigraphy of the reservoir, as well as designate the locations of faults, if any are present. The applications of seismic data in geologic modeling are especially important in deepwater reservoirs, where due to few well penetrations, seismic data are used almost exclusively in the construction of the geologic models.[1] Another important use of seismic data is in reservoir characterization. Seismic impedance data obtained from seismic inversion processes are widely used in reservoir characterization.[2,3] The seismic inversion data are typically integrated with well log and other petrophysical data to improve reservoir characterization.

22.2.2 Outcrop and Basin Studies

Knowledge and experience gained from outcrop and basin studies by the geoscientist generally aid in the construction of the geologic model and characterization of the reservoir. During field trips, geoscientists gain valuable visual description of the structure and stratigraphy of outcrops which could be analogous to subsurface structures that are the candidates for geologic modeling. Another practice that aids geologic modeling is basin studies. Generally, the depositional environment described in basin studies can aid in defining the structure and stratigraphy of geologic structures located in similar basins.[4,5]

22.2.3 Well Log Data

Well log data are obtained from running tools in wells drilled in the reservoir or analogue reservoirs. These include well log data obtained from logging-while-drilling (LWD), wireline logs, and any other means of running tools in wells. These log data may include gamma ray, resistivity, density, neutron, sonic, nuclear magnetic resonance (NMR), image logs, vertical seismic profile, etc. The log data are correlated, processed, and interpreted. They are then used in the determination of the structure and stratigraphy of the reservoir in concert with other data, such as seismic data. Further petrophysical analyses of the log data provide data for facies description, and other rock properties data (such as porosity, permeability, water saturation, formation thickness, etc.) for reservoir characterization. Well log data are very important in the process of geologic modeling and reservoir characterization.

22.2.4 Core Data

Samples of the reservoir rock are obtained from whole and sidewall cores. From these core samples, direct data on many rock properties can be measured in the laboratory. Core plugs taken from whole core samples can be used in routine core analysis (RCA) programs to measure basic rock properties such as porosity, permeability, and water saturation. Some of the core plugs are used in special core analysis (SCAL) programs to measure capillary pressures and relative permeability data. In addition, high-density profile permeability data can be obtained from whole core samples at close intervals using the mini-permeameter tool. Other data obtained from core samples include detailed core descriptions, geomechanical data, x-ray diffraction data, sand particle size distribution, geochemical data, etc. All these data are integrated and used in geologic modeling and characterization of the reservoir.

22.2.5 Formation Pressures and Fluid Properties Data

Formation pressures and fluid samples are obtained from the reservoir using wireline formation tester (WFT) tools. More details on the use of WFT to measure formation pressures and obtain fluid samples are presented in Chapter 6, Section 6.5. Formation pressures can be used in pressure-depth plots to determine fluid contacts, identify existence of reservoir heterogeneities or permeability barriers, and evaluate the presence of reservoir compartments and/or multiple reservoirs. These data can supplement other data in geologic modeling of the reservoir. Fluid properties, such as bubble point pressure, formation volume factor, solution gas-oil ratio, oil viscosity, oil gravity, etc., can be measured from fluid samples. Vertical and spatial variation of fluid properties can be incorporated in the characterization of the reservoir.

22.2.6 Pressure Transient Test Data

Pressure transient test data include data obtained from drill stem tests (DST), and conventional/unconventional drawdown and buildup tests. Procedures for the analysis of pressure transient test data are presented in Chapters 13 to 17. Analyses of pressure transient tests generally provide data on formation permeability, location of faults, and presence of reservoir boundaries. The permeability data obtained from pressure transient tests are at different scale from permeability data obtained from well logs and core measurements. These permeability data should be reconciled before they can be used in the characterization of the reservoir. Further discussions on scales of different data sources are presented in Section 22.4.

22.2.7 Reservoir Performance Data

Reservoir performance data include all fluid production and pressure data measured after the reservoir has been placed on production. These include data from well production rates, flowing and static bottomhole pressures, production logs, injection logs, etc. These data are compiled and compared with assumptions made during construction of the geologic model and characterization of the reservoir. For instance, performance data can be used to verify whether a fault assumed to be a sealing fault in the geologic model is acting as a sealing fault based on actual performance data. If it appears that the fault is not sealed on the basis of performance data, the geologic model should be modified.

22.3 Data Quality Control and Quality Assurance

Data quality control and quality assurance (QC/QA) are essential for execution of successful geologic modeling and reservoir characterization programs. The data acquired or collected at all stages of exploration, appraisal, development, and fluid production should be subjected to rigorous QC/QA procedures. These data should be organized and stored in a data management system that is accessible to a multidisciplinary team. The QA/QC program applied to each data set varies with the type of data. It is important to emphasize that a comprehensive QA/QC program should be devised for each data type discussed earlier, and the program should be under the management of a professional with experience in the acquisition, collection, and processing of that particular type of data.

Techniques in data analytics should be used in descriptive, diagnostic, predictive, and prescriptive processing of all datasets. Application of data analytics on datasets will facilitate efficient utilization of all data in geologic modeling and reservoir characterization.

22.4 Scale and Integration of Data

Petrophysical data used in reservoir characterization are measured at different scales. For instance, petrophysical data measured on core plugs are at a different scale from similar data obtained from well logs. Typically, the dimensions of a core plug are 2 inch diameter by 2 inch length to give a core volume of 0.0036 ft.[3] Most modern well logs measure petrophysical properties as an average over an interval in the range from 2 to 10 feet along the wellbore. It is rather obvious then that the volume of rock sampled by well logs is several orders of magnitude larger than the volume of rock sampled by core plugs. The difference in scale is even higher when data from well tests and seismic interpretations are included. The dependence of petrophysical data on scale is described as the support effect. The support for core, well log, well test, and seismic data ranges from a few inches to thousands of feet. These differences in scale should be considered before petrophysical data from these sources are integrated and used for reservoir characterization.[6–8]

Many techniques have been used to reconcile petrophysical data with different scales or supports. Frykman and Deutsch[7] used volume-variance scaling techniques to integrate data of different scales. Al-Ali and Worthington[8] used averaging techniques to reconcile core plug and well log data. Shafer and Ezekwe[9] applied regression analysis in the integration of core and well log permeability data. The methods that have been reported to be successful appear to apply to specific data types and data sets. Consequently, no particular method for reconciling petrophysical data with different supports is recommended. Different methods should be tried and reviewed to find a method that is applicable to the data set. It is important to recognize that petrophysical data with different scales or supports should be integrated before using them in characterizing the reservoir.

22.5 General Procedure for Geologic Modeling and Reservoir Characterization

A general procedure for geologic modeling and reservoir characterization is presented in this section. This general procedure is designed to function as a guide in devising workflows for geologic modeling and reservoir characterization. The approach recognizes that the quantity, quality, and type of data available for the process vary from one reservoir to the other. For this reason, all the steps in the general procedure are described in generic terms and may not apply in all cases. The workflow captures all the essential steps that are applicable in a geologic modeling and reservoir characterization process. The workflow could be modified and changed as necessary to meet specific requirements of a project. The general procedure recommended for geologic modeling and reservoir characterization is presented under the following processes:

1. Generation of geologic surfaces or horizons
2. Structural modeling
3. Stratigraphic modeling
4. Correlation and assignment of well log data
5. Property modeling
6. Uncertainty analysis
7. Upscaling of geologic model to reservoir flow model

The application of the general procedure for geologic modeling and reservoir characterization is illustrated with two examples at the end of the chapter.

22.5.1 Generation of Geologic Surfaces or Horizons

Geologic surfaces or horizons can be mapped from seismic data or correlation of formation tops from well logs. In some cases, seismic and well log data are used together to map geologic surfaces that define the geologic structure. The number of geologic surfaces generated depends on the complexity of the structure. In simple structures, two geologic surfaces representing the top and base of the structure may be sufficient. In complex geologic structures, additional surfaces may be added to capture geologic features (such as faults, permeability barriers, unconformities, truncations, etc.) that are necessary in the construction of the geologic model. Additional surfaces may also be added to serve as guides in defining the stratigraphy of the geologic model. The number of geologic surfaces used in developing the geologic model is typically determined by the geoscientists responsible for the construction of the model.

22.5.2 Structural Modeling

Structural modeling involves the design and definition of geologic features within the geologic model. Structural modeling generally includes modeling of faults and other geologic structures that were observed or mapped from seismic data, well log data, and performance data, if available. Fault modeling is a very important process under structural modeling. The faults represented in the geologic model may include vertical, inclined, curved, or reverse faults. The fault system may consist of a single fault, crossing faults, branching faults, truncated faults, eroded faults, etc. The complexity of the fault system is a major consideration in the determination of the appropriate fault models to use in the representation of the faults. The potential impact of the faults on reservoir performance is also a major consideration in the selection of the fault models. The capability of the software available for the fault modeling exercise is part of the consideration. Selection of fault models and modeling of faults are processes that are highly dependent on the particular geologic model, features available in the fault modeling software, and experience of the modeler. For these reasons, no specific fault modeling process is recommended. However, the geoscientists and engineers should work closely together to devise fault models that represent the fault system structurally as envisioned by the geoscientists, while including the impact of the faults on reservoir performance as recognized by the engineers.

22.5.3 Stratigraphic Modeling

The stratigraphy of the geologic model is determined from seismic and well log data. Seismic data can be used to define the stratigraphy of the geologic structure, depending on the quality of the data. The most reliable means of defining the stratigraphy of the structure is correlation of formation tops determined from well logs. By correlating formation tops, the stratigraphy of the structure can be defined and categorized into zones. The geologic zones should represent geologic units separated by recognizable geologic features, such as correlated shale intervals. These geologic zones can be subdivided further into geologic layers either to represent flow units or facies. The use of stratigraphic modeling to create geologic zones which can be subdivided further into geologic layers should precede the definition of facies or flow (hydraulic) units during the reservoir characterization step.

22.5.4 Correlation and Assignment of Well Log Data

Well log data are processed with petrophysical models to generate rock properties, such as porosity, permeability, water saturation, formation thickness, etc., before they are loaded into the geologic model. Also, the well logs may have been used to define and assign facies or flow units to specific formation intervals. Classification of the formation into facies or flow units are based on well log data, studies of mud log data, and analogous studies of outcrops and depositional basins.

The processed well log data, comprising rock property data and facies designations, are loaded into the geologic model. The processed well log data are tied to well tops previously used in defining the stratigraphy of the geologic model. Consequently, the processed well log data are correlated in depth to the geologic zones created in the geologic model. The processed well log data can then be used in characterizing the geologic model as discussed in property data modeling.

22.5.5 Property Data Modeling

In property data modeling, some of the methods that can be used to populate the geologic model with rock, fluid, or any property data are presented. There are many different property data modeling techniques. These include techniques as simple as direct assignment of properties to the gridblocks in the geologic model to sophisticated methods based on geostatistics. Geostatistics can be described as the application of mathematical methods based on statistics in analyzing, integrating, interpreting, and distributing geologic, petrophysical, or any other property-based data in a geologic model. Geostatistics is presented in this book as a methodology widely used in the petroleum industry for reservoir characterization. An introduction to basic concepts in geostatistics, such as variogram analysis, stochastic, and object modeling are presented. The reader is advised to consult text books[10-12] on geostatistics for more detailed and comprehensive treatments of these topics. Property modeling is generally based on the grid system applied on the geologic model. In this section, model gridding is presented followed by the application of geostatistical methods.

22.5.5.1 Model gridding

One of the critical processes in property modeling is the design of the grid system for the geologic model. The design of the grid system for the model should be the joint effort of the geoscientists and reservoir engineers involved in the project. The key factors that should be considered in the design of the grid system are:

1. Resolution of the geologic and petrophysical data
2. Size of the static geologic model
3. Limitations on the size of the reservoir flow model
4. Potential uses of the reservoir model
5. Upscaling (also termed upgridding or scaleup) techniques to convert the static geologic model into a dynamic reservoir model

1. **Resolution of the geologic and petrophysical data.** One of the key factors in the design of the model grid system is the resolution or scale of the available geologic and petrophysical data. The grid system should be designed at a scale that reflects the scale of the data to be used in the construction of the geologic model. This approach enables data used in the modeling process to be captured at the scale of the original data, while avoiding potential "smoothing" effects caused by applying averaging techniques to the original data in order to fit the grid system. For instance, in designing the areal grid system, the grid dimensions should be selected to ensure that important geologic features, such as faults, permeability barriers, unconformities, etc., are represented adequately in the geologic model. If the areal grid dimensions are too large, these geologic features could be poorly represented or actually absent in the geologic model. Similarly, the vertical grid dimensions should be as close as possible to the scale of the available vertical data. For instance, if log data are available at half-foot intervals, the vertical grid dimensions should be at this resolution or as close as possible to this scale to avoid excessive smoothing of the vertical data caused by averaging to fit a much larger vertical grid scale.

2. **Size of the static geologic model.** The selection of the grid system is sometimes determined by the desired size of the static geologic model. Some conditions that may cause the size of the geologic model to be limited include potential applications of the model, quality, and quantity of available data, resources assigned to the project and time constraints. For instance, if the geologic model is a conceptual model and very little data are available, there is little or no need to construct a geologic model with millions of gridblocks in this case, since a geologic model with few thousand gridblocks is sufficient for the application. Similarly, if limited human resources and time are available for work on the project, a smaller geologic model with few thousand gridblocks could be adequate due to limited resources. The key point is that the grid system should be designed to fit the intended applications of the geologic model and at the same time accommodate utilization of available resources.
3. **Limitations on the size of the reservoir flow model.** In most cases, geologic models are converted into flow models for simulation of reservoir processes. This should be taken into consideration in designing a grid system for the geologic model. A large geologic model will convert into a large reservoir flow model, if excessive upscaling is not applied. Excessive upscaling will lead to loss of the detailed geologic features present in the geologic model. The key message is that if the objective is to obtain a reservoir flow model with a few thousand gridblocks, it is adequate to construct a geologic model that is at most five times the expected size of the reservoir flow model.
4. **Potential uses of the reservoir model.** Reservoir simulation models are typically designed to achieve specific objectives. These objectives generally define the potential uses of the reservoir model which may include investigations of reservoir drive mechanisms, effect of rock and fluid properties on hydrocarbon recovery, impact of faults on reservoir performance, incremental recoveries from secondary and/or enhanced oil recovery processes, etc. These potential uses of the reservoir model can directly influence the design of its grid system. For instance, a reservoir model designed to evaluate the impact of fluid properties on hydrocarbon recovery may require a finer grid system than a reservoir model designed to investigate the effects of faults on reservoir performance. Since reservoir models are built from geologic models, the potential uses of the reservoir model should be considered in the design of the grid system for the geologic models.
5. **Upscaling (also termed upgridding or scaleup) techniques to convert the static geologic model into a reservoir flow model.** A general rule of thumb is to minimize the extent of upscaling required to convert the static geologic model to a reservoir flow model. Upscaling generally results in some loss of heterogeneities built into the larger geologic model. This should be borne in mind in designing a grid system for the geologic model. Irrespective of the upscaling techniques used to convert the geologic model into a reservoir model, the objective should be to select a grid system for the geologic model that should be retained as much as feasible in the reservoir model. Additional details on upscaling are presented in this chapter as part of the workflow in geologic modeling and reservoir characterization.

22.5.5.2 Geostatistical methods

Geostatistical methods are presented in this book as part of the workflow in geologic modeling and reservoir characterization. Some prior knowledge and exposure to geostatistical methods is assumed. It is recommended that the reader consult textbooks[10-12] for detailed presentations of geostatistical methods. In this section, an overview on spatial analysis using semivariograms (called variograms hereafter) is presented, followed by a brief presentation of geostatistical estimation and conditional simulation methods.

1. **Spatial analysis with variograms.** Many geostatistical methods are based on variograms. The variogram is a geostatistical tool widely used for spatial analysis of the variability of facies, porosity, permeability, and any other petrophysical and geologic data. The semivariogram function can be defined as one-half of the squared difference between random variables separated by a distance. It is represented mathematically as:

$$\gamma(\mathrm{h}) = \frac{1}{2N(\mathrm{h})}\sum_{i=1}^{N(\mathrm{h})}\left[z(x_i) - z(x_i + \mathrm{h})\right]^2 \qquad (22.1)$$

In Eq. (22.1), $\gamma(\mathrm{h})=$ the semivariogram at the lag distance, h; $N(\mathrm{h})$ is the number of sample pairs at lag distance, h; $z(x_i)$ and $z(x_i + \mathrm{h})$ are the sample values at locations x_i, and $(x_i + \mathrm{h})$, respectively. Note that by definition, the semivariogram, $\gamma(\mathrm{h})$, is one-half of the variogram, $2\gamma(\mathrm{h})$. But the term variogram is used in spatial analysis generally instead of the more correct term semivariogram. In many practical applications of Eq. (22.1), the number of pairs at a given lag distance may not be statistically significant. To increase the number of sample pairs at any lag distance, a tolerance, $(\pm\Delta\mathrm{h})$, is often added to the lag distance. The modified semivariogram with the added tolerance is represented as:

$$\gamma(\mathrm{h}\pm\Delta\mathrm{h}) = \frac{1}{2N(\mathrm{h})}\sum_{i=1}^{N(\mathrm{h})}\left[z(x_i) - z(x_i + \mathrm{h})\right]^2 \qquad (22.2)$$

As Eq. (22.2) indicates, sample pairs are collected within a band represented by the lag distance, h ±Δh. This is illustrated in Figure 22.1 in which four sample pairs were formed by using the tolerance, Δh.

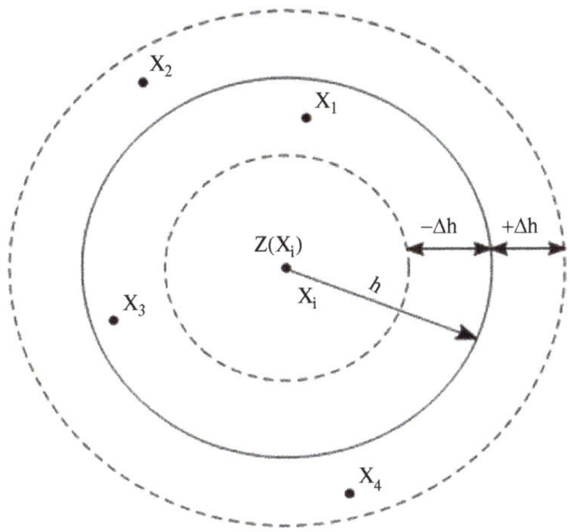

▲ **FIGURE 22.1 Sample pairs formed by lag distance h and lag tolerance Δh**

A diagram showing all the features of a simple variogram is shown in Figure 22.2. The features shown in Figure 22.2 are the sill, the range, and the nugget. The point at which the variability between the sample pairs becomes flat is called the sill. The sill is equal to the variance of the sample data. The distance at which the sill is reached is called the range. Beyond the range, the sample data are no longer correlated. The nugget is the point at which the variogram intersects the y-axis above the origin. The nugget effect is caused by the combination of geologic microstructures and measurement errors. Sparse data may also cause nugget effects. The presence of nuggets in the variogram has significant effect on the simulation of distributed properties.[13]

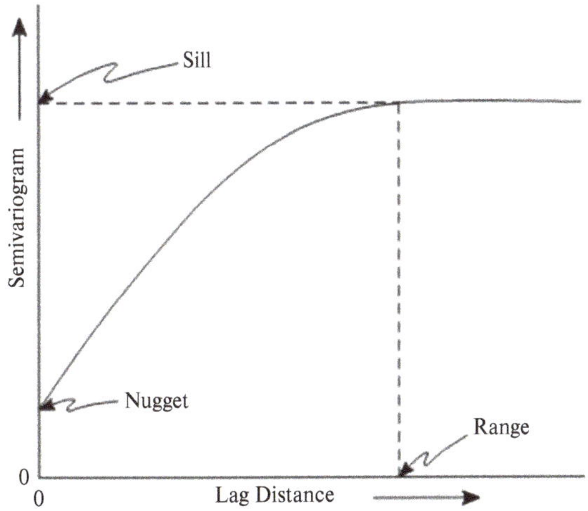

▲ **FIGURE 22.2 A variogram showing sill, range, and nugget**

By plotting the variance for each sample pair against the lag distance as calculated from Eq. (22.1) or Eq. (22.2), an experimental variogram is generated. If the experimental variogram represents spatial relationship in all directions, it is called an omnidirectional or isotropic variogram. In some cases, different experimental variograms define spatial relationships in different directions. These experimental variograms are called directional or anisotropic variograms. Anisotropic variograms are functions of lag distances as well as directions. There are two types of anisotropic variograms. These are geometric anisotropy and zonal anisotropy. Geometric anisotropy is observed where the variograms in different directions have different ranges, but the same sill as shown in Figure 22.3. This type of anisotropy is generally observed when analyzing areal data in different directions. Zonal anisotropy is observed where the variograms in different directions have different ranges and sills as shown in Figure 22.4. This type of anisotropy is present where variograms in vertical direction are compared to horizontal variograms in the areal direction. Calculation of anisotropic variograms can be a very tedious process. However, most geostatistical software packages offer visual graphical support for the calculation of anisotropic variograms.

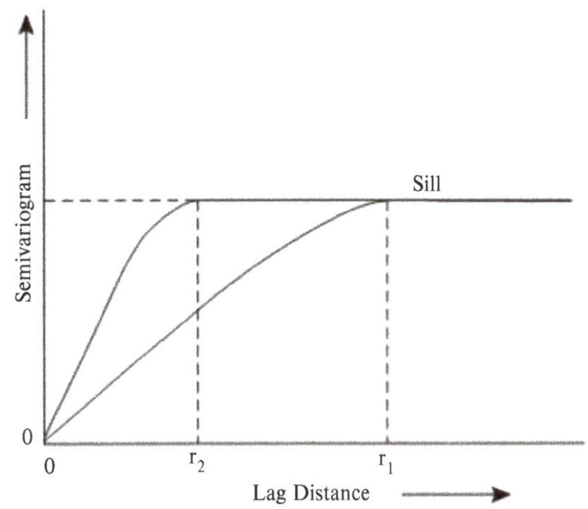

▲ **FIGURE 22.3 A variogram showing geometric anisotropy**

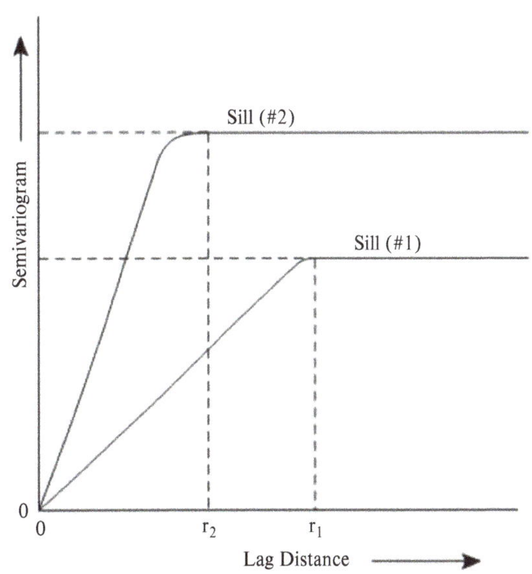

▲ **FIGURE 22.4 A variogram showing zonal anisotropy**

2. **Variogram models.** Variogram models, expressed as mathematical functions, are divided into two groups. The first group is variogram models with sills, also called transition models. The second group is variogram models without sills, also called non-transition models.

   A. **Transition models.** The spherical model is an example of a transition model. If the range is represented as, r, and the sill is normalized to 1, the spherical model can be expressed as:

$$\gamma(\mathrm{h}) = \begin{cases} \frac{3}{2}\frac{\mathrm{h}}{r} - \frac{1}{2}\left(\frac{\mathrm{h}}{r}\right)^3 & \text{if } \mathrm{h} \le r \\ 1 = \text{sill} & \mathrm{h} \ge r \end{cases} \tag{22.3}$$

   Another transitional model is the exponential model. The exponential model is represented as:

$$\gamma(\mathrm{h}) = 1 - \exp\left(-\frac{3\mathrm{h}}{r}\right) \tag{22.4}$$

   The Gaussian model is a transitional model represented as:

$$\gamma(\mathrm{h}) = 1 - \exp\left(-\frac{3\mathrm{h}^2}{r^2}\right) \tag{22.5}$$

   In Eqs. (22.3) to (22.5), h = lag distance; and the sill of the sample data has been normalized to 1. The shapes of the spherical, exponential, and Gaussian models are shown in Figure 22.5. Note that the spherical model is linear near the origin, while the exponential and Gaussian models are parabolic near the origin. For practical purposes, the ranges of these models are defined as the lag distances at which the variograms reach 95 percent of their respective sill values.

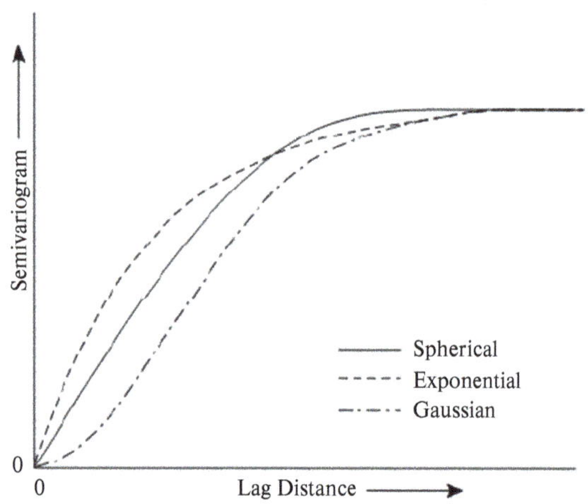

▲ **FIGURE 22.5** **Shapes of spherical, exponential, and Gaussian variogram models**

   B. **Non-transition models.** The most common example of a non-transition model is the Power Law model. The Power Law model is expressed as:

$$\gamma(\mathrm{h}) = \mathrm{h}^{\theta}, \text{ where, } \theta, \text{ is between 0 and 2 excluding these limits} \tag{22.6}$$

   The shapes of the Power Law model are shown in Figure 22.6. Other non-transition models include the cosine model and the log normal model. These models are rarely used in modeling experimental variograms.

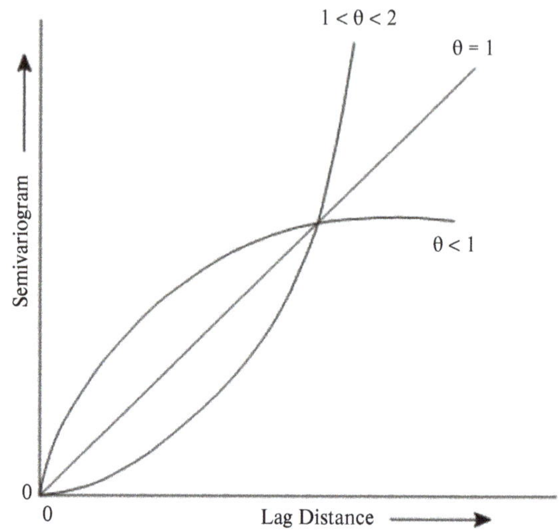

▲ **FIGURE 22.6** **Shape of Power Law variogram model**

   C. **Modeling of experimental variograms.** After the experimental variogram has been calculated with either Eq. (22.1) or (22.2), the next step in spatial analysis is modeling of the experimental variograms with one or combinations of the variogram models. The requirement for the use of any function as a variogram model is that the function must be conditionally positive definite. This condition ensures the existence of a unique solution for the estimation of a variable at an unsampled location. The transition and non-transition models described earlier satisfy this condition. In modeling experimental variograms, it is important to capture major features such as the sill, range, and nugget.[14] In some cases, any linear combinations of variogram models to create a nested model are permissible. However, it is better to use simple models and avoid nested structures as much as possible. In a sense, modeling of experimental variograms is a curve-fitting exercise involving systematic fitting of several parameters, namely the sill, the range, and the nugget. Most geostatistical software packages provide visual and interactive graphical applications to assist in the modeling of experimental variograms.

   The choice of variogram models used in modeling (or fitting) the experimental variogram have considerable impact on the resulting simulated realizations using the same sample data as input.[7] Figures 22.7a to 22.7c show the 2D-simulated realizations using exponential, spherical, and Gaussian variogram models with the same sill and range values. The exponential variogram model gives a realization that appears "noisy," while the Gaussian model produces a smoother realization.[7] The realization with the spherical variogram model lies in-between these two realizations. These figures are used to illustrate the impact of variogram models on simulated realizations and emphasize that the choice of variogram models during modeling of experimental variograms should be guided by knowledge of spatial variability of the sample data.

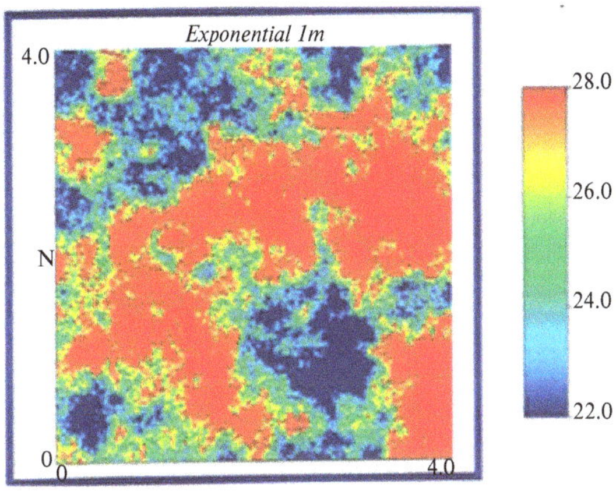

▲ FIGURE 22.7a 2D-simulated property using exponential variogram model (from Frykman and Deutsch[7])

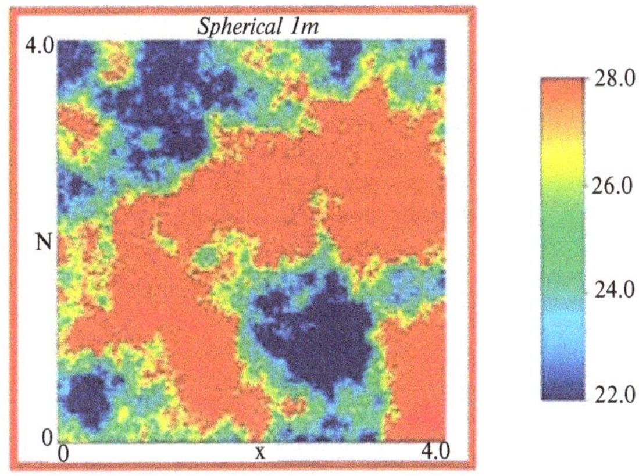

▲ FIGURE 22.7b 2D-simulated property using spherical variogram model (from Frykman and Deutsch[7])

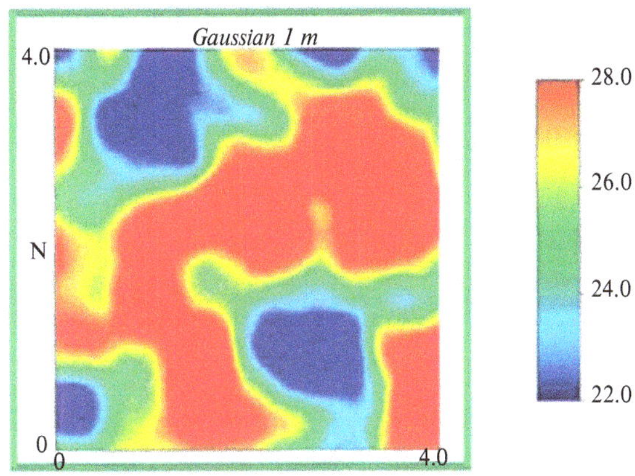

▲ FIGURE 22.7c 2D-simulated property using Gaussian variogram model (from Frykman and Deutsch[7])

EXAMPLE 22.1 Calculation of an Experimental Variogram as a Function of Lag Distance

Problem

Calculate an experimental variogram at lag distances of 1 to 12 feet for the data shown in Table 22.1.

Solution

The experimental variogram at all lag distances were calculated using Eq. (22.1).

At lag distance, h = 1 ft.:

Substituting Eq. (22.1) gives:

$$\gamma(1) = \frac{\begin{array}{c}(19.0-20.2)^2+(20.2-21.4)^2 \\ +(21.4-19.7)^2+\ldots+(21.2-20.9)^2 \\ +(20.9-21.5)^2+(21.5-21.8)^2\end{array}}{2\times 33}$$
$$= 0.5073$$

At lag distance, h = 2 ft.:

TABLE 22.1 Data for Example 22.1

Depth (feet)	Porosity (%)	Depth (feet)	Porosity (%)
6050	19.00	6067	20.90
6051	20.20	6068	21.40
6052	21.40	6069	18.20
6053	19.70	6070	18.10
6054	22.00	6071	18.40
6055	21.00	6072	18.70
6056	21.70	6073	19.10
6057	21.40	6074	19.60
6058	19.80	6075	20.40
6059	19.30	6076	20.80
6060	18.90	6077	21.40
6061	19.70	6078	21.00
6062	18.30	6079	21.50
6063	19.50	6080	21.20
6064	19.90	6081	20.90
6065	20.20	6082	21.50
6066	20.70	6083	21.80

Substituting Eq. (22.1) gives:

$$\gamma(2)=\frac{\begin{array}{l}(19.0-21.4)^2+(21.4-22.0)^2\\+(22.0-21.7)^2+\ldots+(20.8-21.0)^2\\+(21.0-21.2)^2+(21.2-21.5)^2\end{array}}{2\times 32}$$

$$=0.7458$$

The variograms calculated at higher lag distances are shown in Table 22.2. The results in Table 22.2 were plotted as shown in Figure 22.8. A simple computer program can be developed for the calculation of experimental variograms.

TABLE 22.2 Variogram Data Calculated from Data in Table 22.1

Lag Distance, h (ft)	Variogram, γ(h)
1	0.5073
2	0.7458
3	1.0752
4	1.4928
5	1.6659
6	1.9164
7	1.7024
8	1.7160
9	1.3986
10	1.3554
11	1.1913
12	1.2559

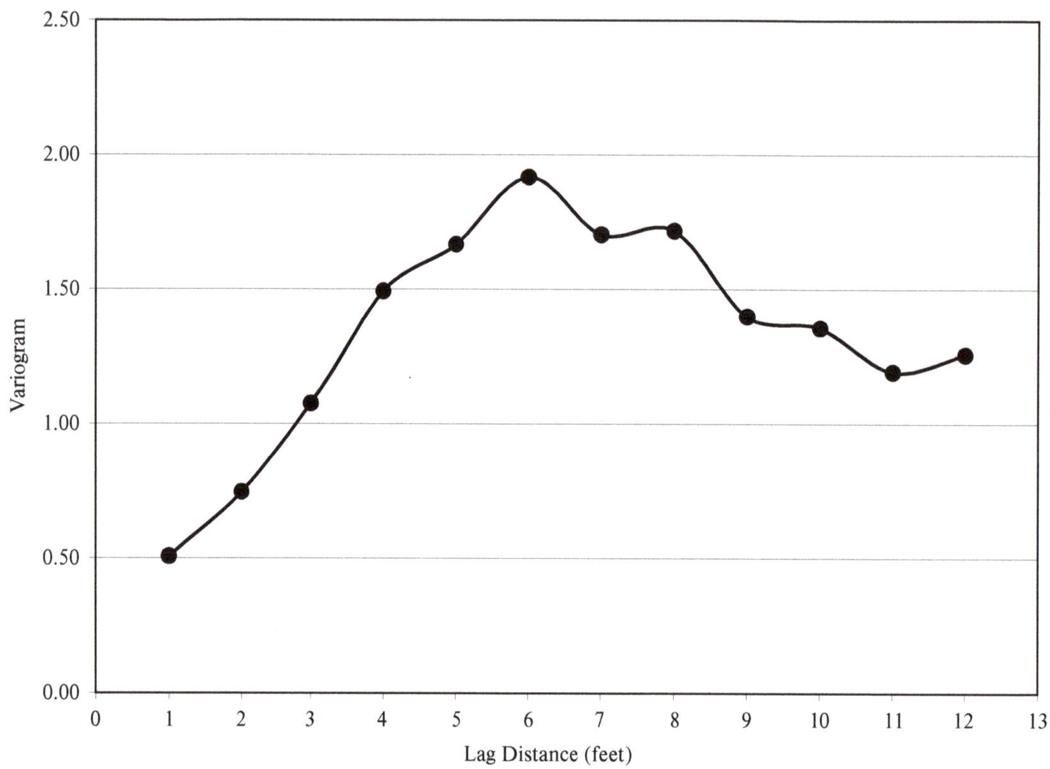

▲ **FIGURE 22.8** **Plot of the experimental variogram calculated in Example 22.1**

End of Example 22.1

3. **Geostatistical estimation and conditional simulation methods.** There are many estimation and conditional simulation methods that can be used to assign values at unsampled locations. For reservoir characterization, kriging is the most applied estimation method. Kriging is different from conditional simulation in three key areas:

   a. Kriging yields the best local estimate of the unsampled value without considering the resulting spatial variability. In conditional simulation, global features and spatial statistics take precedence over local accuracy.

   b. Kriging tends to produce a "smoothing" effect unlike conditional simulation.

   c. Kriging produces a single model (realization) of the attribute. Conditional simulation produces many alternate models (probable realizations) of the attribute.

The development of estimation and conditional simulation methods involve complex statistical and mathematical equations. Furthermore, implementations of these methods for analysis of real data require in-depth knowledge of geostatistics. Due to these reasons, the statistical theories and mathematics that are the bases of estimation and conditional simulation methods are not presented. These topics are available in several geostatistical books.[10-12] For the purposes of this book, estimation methods are limited to a listing of commonly used kriging methods. Similarly, conditional simulation methods are limited to a listing of methods commonly used in reservoir characterization.

A. **Kriging methods.** Kriging is a minimum variance estimator based on variograms. It is an unbiased estimator that accounts for the configuration of the sampled data. There are many variation of the kriging method. The kriging methods most used in reservoir characterization are:
   - Ordinary kriging
   - Kriging with external drift
   - Cokriging
   - Collocated cokriging
   - Indicator kriging

B. **Conditional simulation methods.** Conditional simulation methods are stochastic processes of building alternative, equiprobable models (realizations) using the spatial distribution of the attribute. The simulation process is described as conditional if it produces realizations that honor sample data at their locations. The attributes simulated can be either continuous or categorical. Conditional simulation methods commonly used in reservoir characterization are:
   - Sequential Gaussian simulation (SGS)
   - Sequential Indicator simulation (SIS)
   - Object-based modeling

22.5.6 Uncertainty Analysis

There is some degree of uncertainty inherent in any geologic model because of the differences (or errors) between the true and sampled values of data used in building the model. If the amount of sample data available for constructing the model is sparse, the level of uncertainty is expected to be high. Conversely, if the amount of sample data available is large and well distributed across the volume of interest, the level of uncertainty is comparatively less. In geologic model building, the two main sources of uncertainties are structural data and petrophysical data. Structural uncertainties may result from processing and interpretation of seismic data, mapping of faults, structural and stratigraphic correlation of log data, "picking" of structure tops from seismic and/or log data, description of the depositional environment, etc. Petrophysical uncertainties may derive from measurements of porosity, permeability, and water saturation data, determination of gross and net thickness data, definition and assignment of facies, location of fluid contacts, etc. Practically most of the data used in the construction of geologic models have some degree of uncertainty associated with them. Consequently, it is important to develop methodologies for assessing and quantifying the uncertainty associated with geologic models built from these data.

One of the methods that can be used to assess quantitatively the degree of uncertainty associated with a geologic model is by generation of multiple, equiprobable realizations of the model. Even though the realizations are equally probable (equiprobable), there are differences from one realization to the next realization. These differences are measures of the uncertainties present in the models. As a result, the realizations can serve as the input for uncertainty analysis. By application of statistical analysis, several measures of uncertainty can be generated from the set of multiple realizations. These statistical metrics may include the mean, standard deviation, and the cumulative probability distribution of variables such as total pore volumes and oil-in-place volumes which can be calculated from multiple realizations. A typical cumulative probability curve representing the distribution of oil-in-place volumes based on multiple realizations is shown in Figure 22.9. The nominal classification based on P10, P50, and P90 quantiles are shown for illustration purposes in Figure 22.9. For instance, the P10 quantile represents realizations with volumes of oil-in-place equal to or larger than the designated volume. Note that in some organizations, the designations of magnitudes of P10, P50, and P90 quantiles are reversed from the designations used in this book. From the cumulative probability curve calculated for a variable, the probability of achieving any level of outcome for that variable can be estimated. The cumulative probability curve can also be used to select specific realizations (geologic models) at the P10, P50, and P90 quantiles for further analyses in reservoir flow simulations.

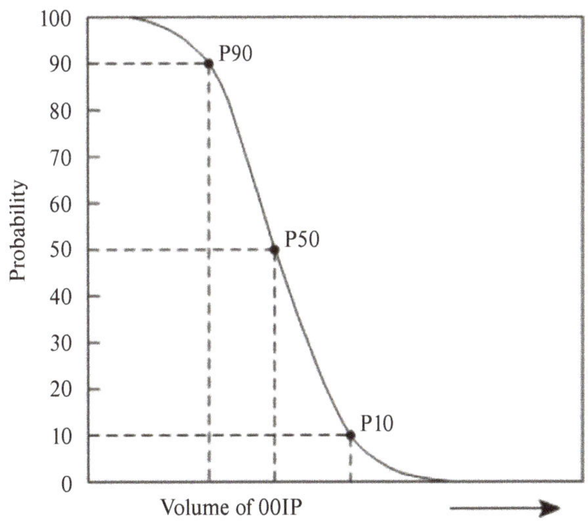

▲ **FIGURE 22.9 Cumulative probability curve for OOIP based on multiple realizations showing P10, P50, and P90 quantiles**

22.5.7 Upscaling of Geologic Model to Reservoir Flow Model

The term Upscaling (also called Upgridding or Scaleup) is used to describe the systematic process of building a coarse-gridded reservoir flow (simulation) model from a fine-gridded geologic model.[15] The objective in the upscaling process is to reduce the number of gridblocks in the geologic model by a factor of two or more in converting it into a reservoir flow model, and still retain important characteristics of the geologic model. Upscaling is sometimes necessary because geologic model with multimillion gridblocks are routinely constructed in many projects with the advent of advanced techniques for geologic modeling and reservoir characterization. The sizes of these geologic models are in some cases too large to simulate as reservoir flow models because of the need to reduce model run times, evaluate alternative reservoir management strategies, or conduct iterative simulation runs in a history matching process. Other reasons for scaling up geologic models include limitations on computer memory storage required for large simulation models, and the cost of processing the output from large models. However, it is relevant to observe that recent advances in computer hardware and simulation software technologies indicate that the practice of upscaling geologic models to simulation models is expected to decline with time. In other words, it is expected that geologic models will be converted into reservoir simulation models without the need for upscaling as computer hardware and simulation software technologies advance over time. This view is supported by the construction and simulation of billion-cell reservoir models as reported in the literature.[16-18]

22.5.7.1 Recommended steps for upscaling a geologic model to a reservoir model

The following steps are recommended for upscaling a geologic model to a reservoir model:

1. Grid construction for the reservoir model
2. Generation of properties for reservoir model gridblocks
3. Verification of the upscaling process

Step 1: Grid construction for the reservoir model

The goal in designing a grid system for the reservoir simulation model is to retain the major geologic features and heterogeneities built into the geologic model. It is also important to consider the potential uses of the reservoir model in designing the grid system. Major geologic features that should be preserved in the reservoir model include faults, fault geometry, permeability barriers, location of fluid contacts, stratigraphy of the geologic model, and variability of rock and fluid properties. Potential applications of the reservoir model that can depend on the grid system include monitoring rapid changes in fluid saturations and pressures, fluid flow around wells, and microscopic displacement processes as observed in enhanced oil recovery processes.

The areal grid system determines the location of geologic features in the X, Y coordinates. Consequently, it is important to select an areal grid system for the reservoir model that closely mirrors the areal grid system of the geologic model. This approach ensures that the location of major geologic features in the geologic model is maintained in the simulation model. A common practice that is recommended is to use the same areal grid system of the geologic model for the reservoir model.[19] If it is necessary to use a different areal grid system for the reservoir model, it is recommended that the new areal grid system should be devised to encompass consistent multiples of the geologic model gridblocks, and have the same orientation as the areal grid of the geologic model.

By using the same or similar areal grid system for the reservoir model as present in the geologic model, most of the upscaling exercise is focused on selecting the vertical grid system or layering for the reservoir model. In upscaling the vertical grids from the geologic model to the reservoir model, it is important to maintain the stratigraphy developed for the geologic model. The recommended practice is to reduce the number of layers in the reservoir model by combining similar layers in the geologic model **within** the same zone, while retaining geologic layers that mark the location of permeability barriers such as identifiable shale layers. Proper upscaling of the vertical grid system can be achieved by close inspection of the geologic model coupled with a thorough understanding of the stratigraphy used in its construction.

The reservoir engineers and the geoscientists who constructed the geologic model should work closely together to obtain a representative reservoir model from the geologic model. This cooperative effort should begin at the initial construction of the grid system for the geologic model. With proper input from the reservoir engineer, the areal grid system for the geologic model should be devised with the intention that the same grid system is retained for the reservoir model. Similarly, the vertical grid system for the geologic model should be devised to capture the high resolution of data at the geologic level, while retaining the ease of combining vertical layers that exhibit similar characteristics during the upscaling process.

Step 2: Generation of properties for reservoir model gridblocks

The reservoir model gridblocks are populated with rock properties, such as porosity, net sand thickness, water saturation, permeability, rock types or facies based on the properties of the gridblocks in the geologic model. The method of upscaling rock properties from the geologic model to the reservoir model depends on the type of rock property to be upscaled. Static rock properties such as porosity, net sand thickness, and water saturation are scaled-up using averaging techniques for the corresponding properties in the geologic model. For instance, gridblock porosities in the reservoir model can be upscaled as the bulk volume weighted averages of the porosities of corresponding geologic model gridblocks. Net sand thickness of gridblocks in the reservoir model can be upscaled as the sum of the net sand thickness of the corresponding gridblocks in the geologic model. Water saturations can be upscaled as pore volume weighted averages of the corresponding gridblock water saturations in the geologic model.

There are no simple techniques for upscaling rock types or facies, since a gridblock in the geologic model may represent more than one rock type or facies. An approach that is recommended is to assign rock type or facies based on the proportion of non-shale rock type or facies in the corresponding geologic model gridblocks.[15] The rock type or facies with the highest proportion is assigned to the corresponding reservoir model gridblock.

The best practice for upscaling permeability is by applying flow-based scale averaging (FBSA).[15] This method uses the Darcy flow equation to calculate the effective permeability due to the flow of a single phase fluid in the fine grid geologic model under set pressures, flow rates, and boundary conditions. The effective permeabilities assigned to the reservoir model gridblocks have been shown to achieve the same flow characteristic as the corresponding geologic model gridblocks.[20] Many geologic modeling software provide permeability upscaling techniques based on FBSA. It is strongly recommended that this technique should be used in upscaling permeabilities. Geometric and harmonic averaging techniques should not be used since these averaging techniques generally yield lower gridblock permeabilities in the reservoir model.

Step 3: Verification of the upscaling process

Verification of the upscaling process is a quality control step to ensure that the reservoir model upscaled from the geologic model is a **reasonable** representation of the geologic model. This process involves an intensive interrogation of the upscaled reservoir model. Every aspect of the reservoir model should be investigated and compared with the geologic model. Visual and graphic displays should be used as much as possible in comparing the two models.

The first activity in the verification process is to compare the two models and ascertain that major geologic features such as faults, permeability barriers, etc. present in the geologic model are correctly reproduced in the reservoir model. The location and geometry of all faults in the two models should be compared and shown to be correctly represented in the reservoir model. Other geologic heterogeneities and features also should be compared in the same manner.

This should then be followed by comparisons of the gridblock properties, such as porosities, net sand thickness, saturations, permeabilities, etc. Cross-sections of the geologic model displaying these properties should be compared with the same cross-section of the reservoir model. The comparisons should be conducted in a systematic and thorough manner. Another way of comparing the petrophysical properties of the two models is by plotting the histograms of each property in the two models together. The histograms should show similar trends. Most geologic modeling software has applications that facilitate these comparisons. These applications can be used to calculate and compare statistical metrics such as mean, standard deviation, maximum and minimum values of any petrophysical parameter in the geologic and reservoir models.

It is also useful to calculate and compare gross rock volumes, pore volumes, and hydrocarbon-in-place volumes, of the geologic and reservoir models. These volumetric comparisons should also be conducted for major geologic zones as defined in the stratigraphy. The gross rock volume of the geologic and reservoir models should agree within a margin of 1 to 2 percent.[15] The hydrocarbon-in-place volumes between the two models should agree within a margin of 1 to 5 percent.

At the end of the verification process, if no major deficiencies or errors are found in the upscaled reservoir model, then the upscaling effort can be considered to be successful. The upscaled reservoir model can then be used for flow simulation work.

EXAMPLE 22.2 Geologic Modeling and Characterization of Reservoir "A"

Problem

Reservoir "A" consists of two main sand formations, designated here as 1A and 1B sands. The 1A sands are separated from the 1B sand by a shale interval as shown in Figure 22.10. The deepest sand formation designated as the A2 sand is not included in the model. The data available for building and characterizing the geologic model are:

1. Top and base structural contour maps for 1A and 1B sands
2. Net sand maps for 1A and 1B sands
3. Fault maps for the major faults
4. Processed log data from five wells
5. Core data from one well

Using the above data, build and characterize a geologic model for the reservoir.

Solution

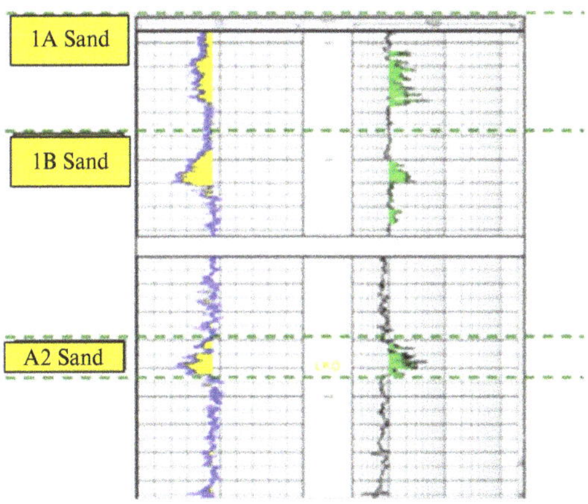

▲ FIGURE 22.10 Type log of Reservoir "A" in Example 22.2

Step 1: Generate geologic surfaces or horizons

From the top and base structural contour maps, surface maps were generated for the top of 1A sand and base of 1B sand using facilities available in a commercial geologic modeling software. The surfaces generated by this process are shown in Figure 22.11. Also, net sand map surfaces were generated from net sand contour maps for 1A and 1B sands. These are shown as Figures 22.12 and 22.13, respectively.

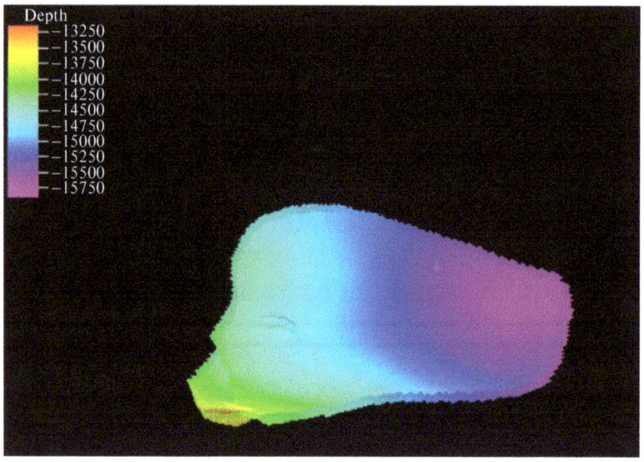

▲ **FIGURE 22.11** **Top and base surface maps for Reservoir "A" in Example 22.2**

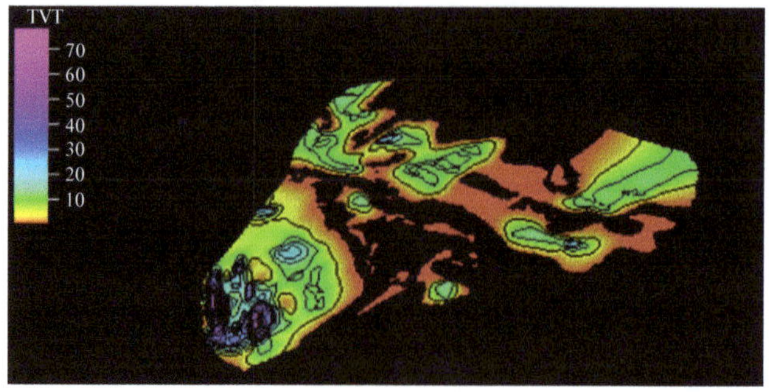

▲ **FIGURE 22.12** **Net sand surface map of 1A sand in Example 22.2**

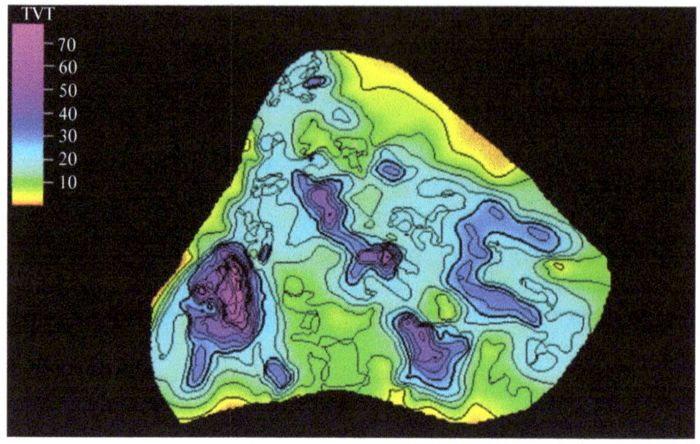

▲ **FIGURE 22.13** **Net sand surface map of 1B sand in Example 22.2**

Step 2: Build the structural model

For this reservoir, the main geologic feature that must be represented in the geologic model is the main fault that divides the structure into a southwestern section and a northeastern section. The southwestern section has gas in the 1A sand and the 1B sand is wet. The northeastern section has oil in both the 1A and 1B sands. The juxtapositions of the sands across the main fault are shown in Figure 22.14. The main fault was considered to be a sealing fault at discovery.

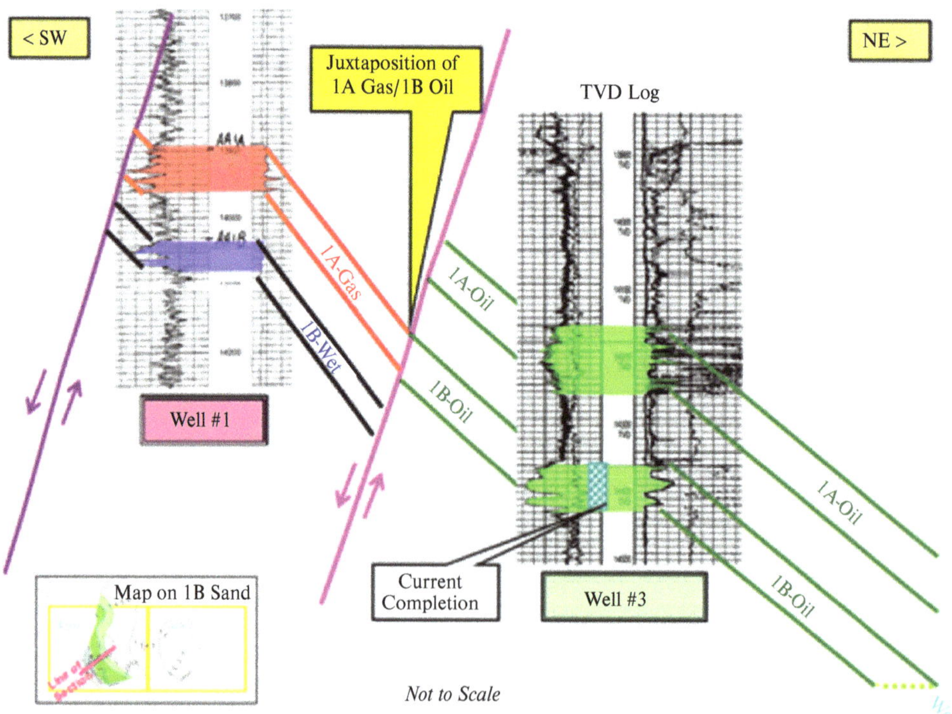

▶ **FIGURE 22.14** **Cross-section through the main fault in Reservoir "A" of Example 22.2**

The faults were digitized from structure maps into polygons. From the polygons, fault surfaces were generated and are shown in Figure 22.15. The fault surfaces were treated as vertical faults and used in the construction of the geologic model.

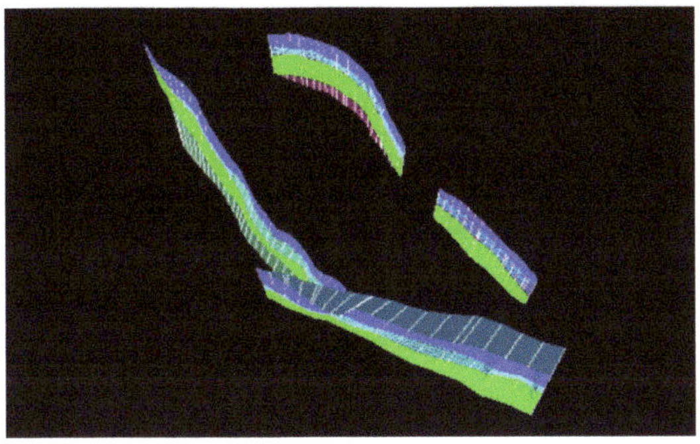

▲ **FIGURE 22.15 Fault surfaces in Reservoir "A" of Example 22.2**

Step 3: Develop the model stratigraphy

The stratigraphy of the geologic model is relatively simple. It consists of three zones. Zone 1 is assigned to the 1A sand. Zone 2 designated the shale interval between 1A and 1B sands. And Zone 3 is assigned to the 1B sand.

Step 4: Load and assign well log data

Processed log data were available from five wells. The log data include porosity, water saturation, and permeability. The log data were calibrated with core data from one of the wells. The log data were uploaded and assigned to corresponding gridblocks in the geologic model.

Step 5: Conduct property modeling

To start property modeling, a grid system has to be created for the geologic model. The grid system designed for the model is 91 × 104 × 34 for a total of 321,776 gridblocks. The dimensions of the gridblocks on the average are 197.6 ft × 200.5 ft × 5.8 ft. The areal grid is shown as Figure 22.16. For the vertical layering, 10 layers were assigned to 1A sand and 20 layers were assigned to the 1B sand. The shale interval between the two sands was assigned 4 layers. This grid system was considered to be sufficiently fine to capture the heterogeneity of the rock properties and could still be used for flow simulation without the need to upscale the geologic model.

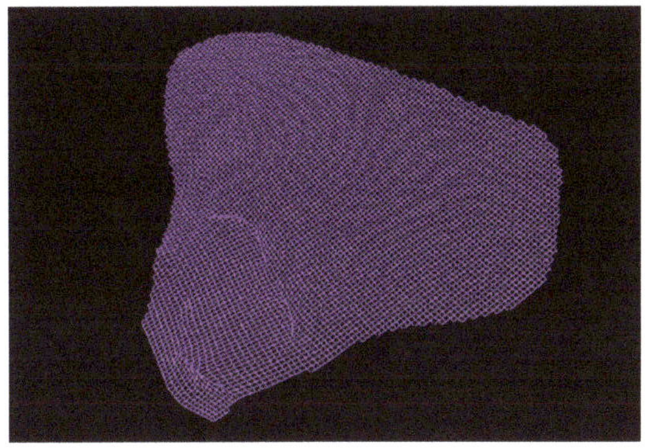

▲ **FIGURE 22.16 Areal grid system for the geologic model in Example 22.2**

The next phase in property modeling is data analysis using variogram modeling. Data analysis was conducted on porosity, permeability, and water saturation data upscaled from well logs. A variogram model was fitted to the particular data for the corresponding sand. For example, a spherical variogram model was fitted to the porosity data for 1A sand as shown in Figure 22.17.

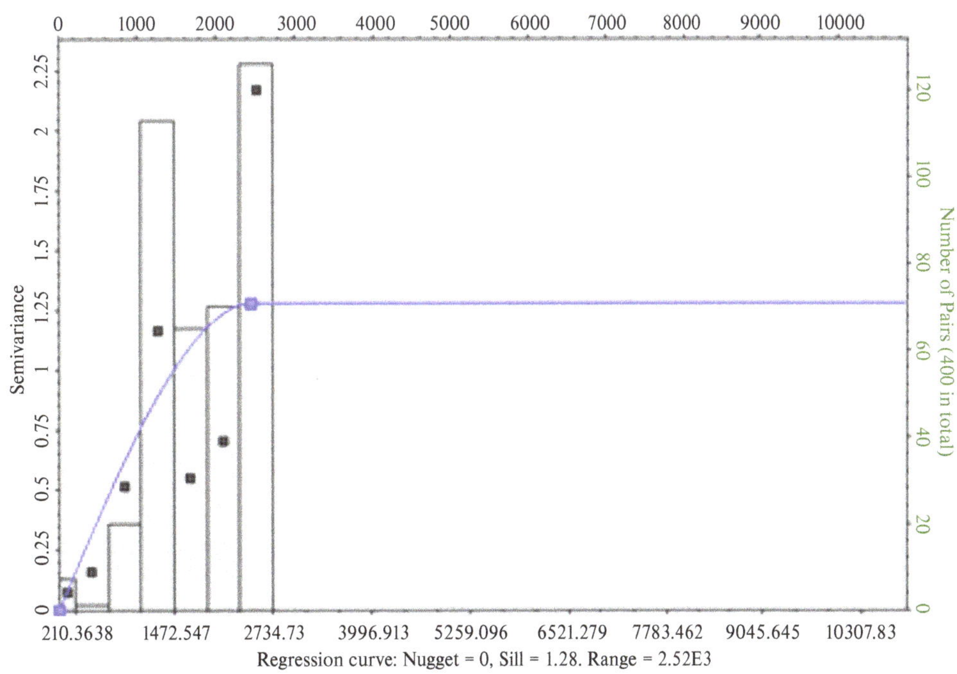

▶ **FIGURE 22.17 Spatial variogram (spherical model) of porosity in the 1A sand (nugget = 0, sill = 1.28, range = 2520 ft) of Example 22.2**

Similar variogram models were obtained for other petrophysical data. These variogram models were then used in conditional simulation of the petrophysical data.

In conditional simulation of petrophysical data for the geologic model, porosity was distributed in the model using the method of Sequential Gaussian Simulation (SGS) with collocated cokriging on net sand data. Figure 22.18 shows the geologic model with the distribution of porosity generated by this process. Similar distributions of permeability and water saturation were generated using SGS with collocated cokriging on distributed porosity data.

▲ FIGURE 22.18 Porosity distribution for Reservoir "A" of Example 22.2

End of Example 22.2

EXAMPLE 22.3 Geologic Modeling and Characterization of Reservoir "B"

Problem

The reservoir in this example is much larger and has more complex lithology than the reservoir in Example 22.2. The data available for building and characterizing the geologic model for Reservoir "B" are:

1. Seismic data
2. Horizons from seismic data
3. Fault maps from seismic data
4. Processed log data from three wells
5. Core data from two wells
6. Analogs from outcrop studies

Using the above data, build and characterize a geologic model for Reservoir "B."

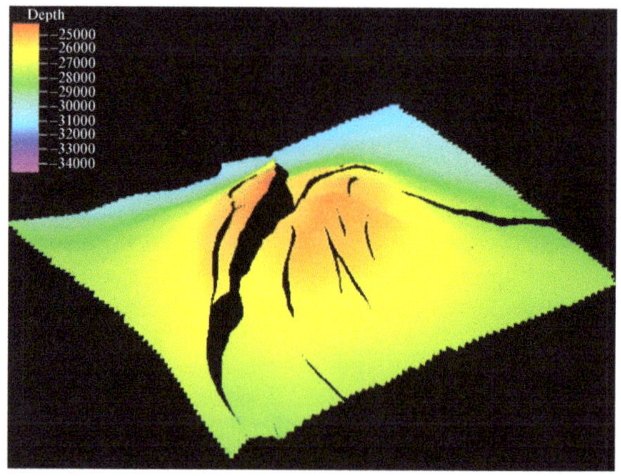

▲ FIGURE 22.19 Horizon map on top of structure of Reservoir "B" in Example 22.3

Solution

Step 1: Generate geologic surfaces or horizons

Since this is a very large reservoir with over 2500 feet of gross thickness, several horizons were "picked" from seismic data. Only the surface maps generated for the top and base of the structure are shown in Figures 22.19 and 22.20.

Step 2: Build the structural model

It is evident in the surface maps shown in Figures 22.19 and 22.20 that Reservoir "B" has a complex fault system. The challenge in constructing a structural model for this reservoir is proper representation of the faults in the geologic model. The faults were mapped from seismic data. The fault model used in the

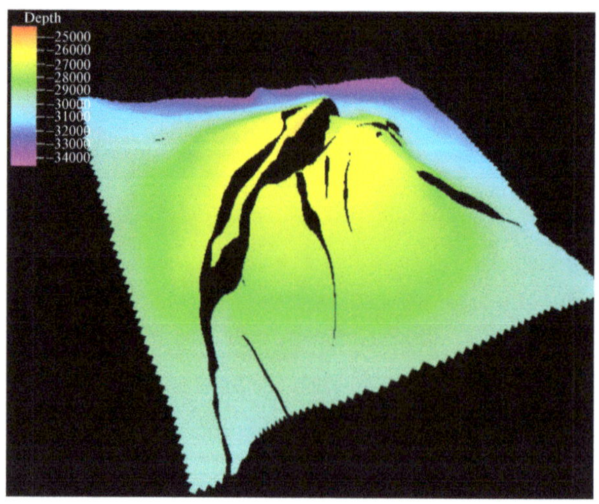

▲ FIGURE 22.20 Horizon map on base of structure of Reservoir "B" in Example 22.3

structural model is shown in Figure 22.21. All faults were treated as vertical faults in the fault model.

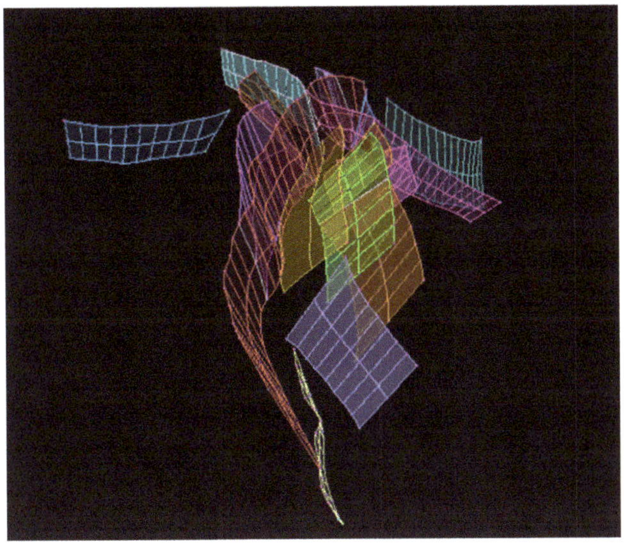

▲ **FIGURE 22.21** **Fault surfaces of Reservoir "B" in Example 22.3**

Step 3: Develop the model stratigraphy

The stratigraphy of the reservoir was defined in the geologic model with 45 zones. This required the generation of 91 conformable horizons. It was necessary to refine the stratigraphy as much as possible so that vertical variations of lithology are captured in the geologic model.

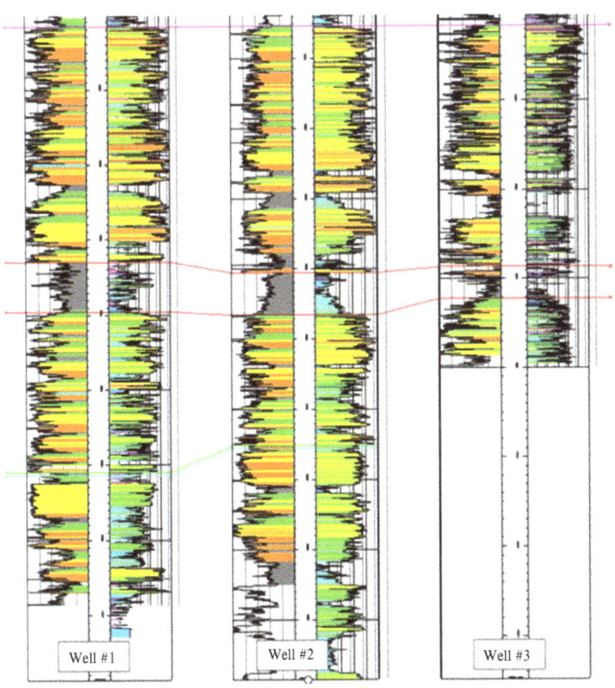

▲ **FIGURE 22.22** **Correlation of well log data in Example 22.3**

Step 4: Load and assign well log data

Processed log data were available from three wells. An example of correlation of well log data is shown in Figure 22.22. The well log data were also used to define facies in the model. Five facies were designated for the various flow units present in the model. Figure 22.23 shows the classification of the facies in the model.

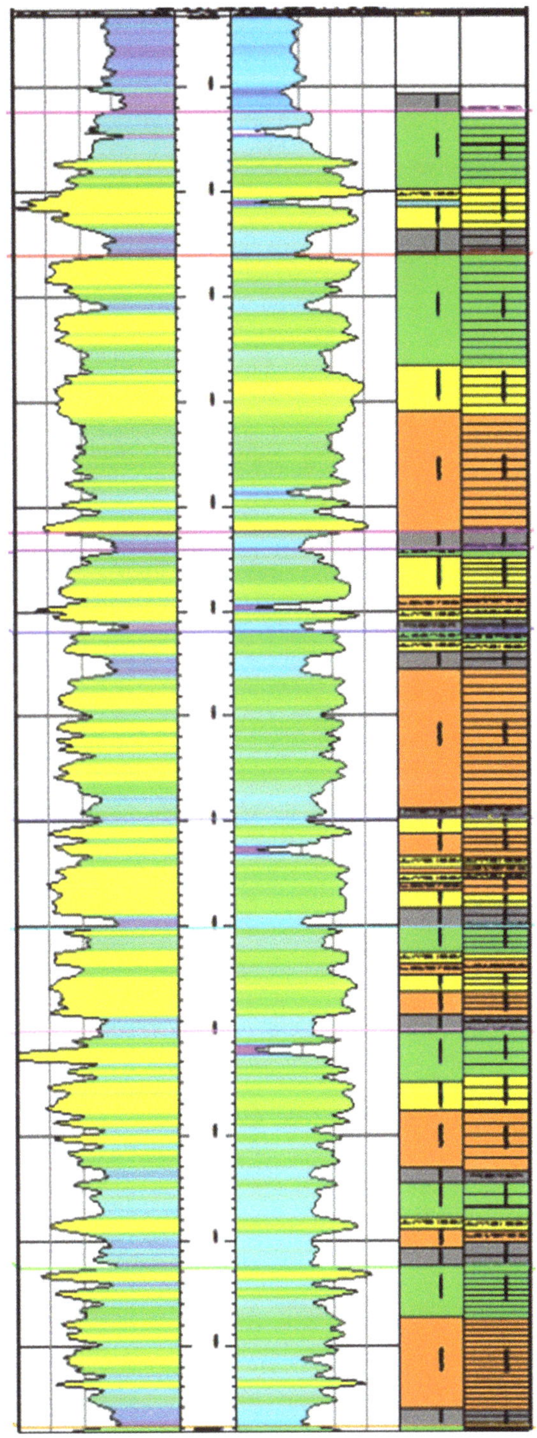

▲ **FIGURE 22.23** **Facies definition and upscaling of a well in Example 22.3**

Step 5: Conduct property modeling

The grid system designed for the model is 113 × 112 × 557 for a total of 7.0 million gridblocks. The dimensions of the gridblocks on the average are 400 ft × 404 ft × 3.6 ft. The areal grid is shown as Figure 22.24. The areal grid as designed is expected to be retained in the flow model without further need for upscaling. The vertical resolution of the grid system was chosen so that well data are captured at the scale of the log data.

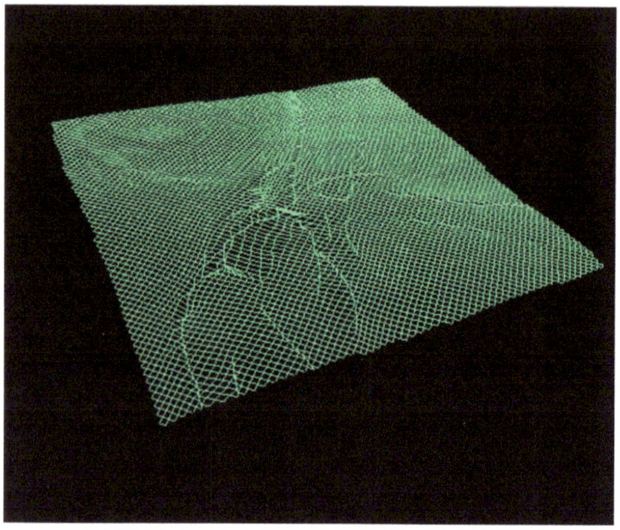

▲ **FIGURE 22.24 Areal grid system of the geologic model in Example 22.3**

The variogram models were based on various data sources as follows:

- Directional variograms were based on seismic data. An example of variogram map based on seismic amplitude is shown as Figure 22.25.
- Vertical variogram models were based on well log data. Vertical trends and relative proportions of facies in the geologic model were derived from logs.
- Input variograms were based on published outcrop studies.

The geologic model was populated stochastically using the variogram models derived as described earlier. Figure 22.26 shows a realization of facies distribution in one layer of the geologic model. Figure 22.27 is a cross-section showing vertical distribution of facies in the geologic model. Porosity distribution in the geologic model was by facies type. Figure 22.28 shows porosity distribution for a realization of the geologic model. Net-to-gross ratio was also distributed by facies. Permeability distribution by facies was based on cloud transform from porosity/permeability crossplots. The porosity/permeability crossplot for all the wells is shown in Figure 22.29. The permeability distribution for a realization of the geologic model is shown in Figure 22.30.

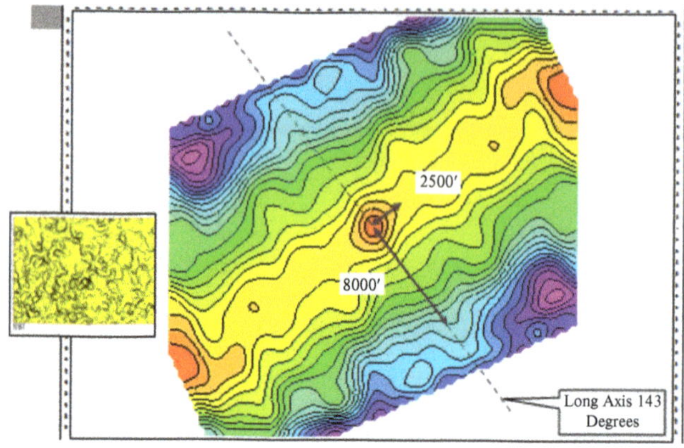

▲ **FIGURE 22.25 Variogram map based on seismic amplitude for Example 22.3**

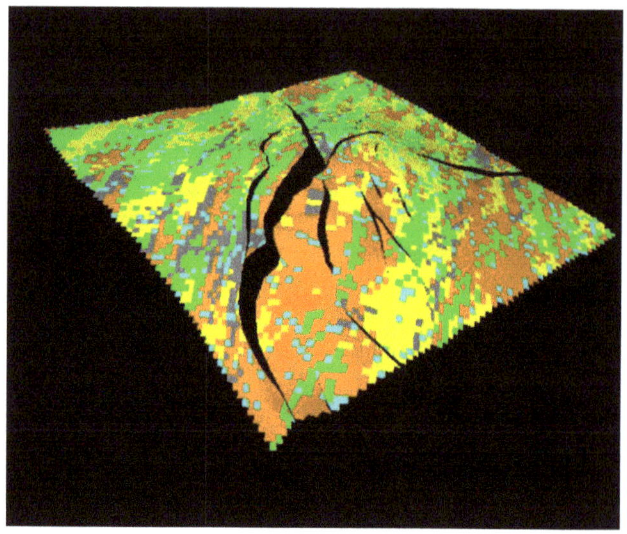

▲ **FIGURE 22.26 A realization of facies distribution in one layer of Reservoir "B" in Example 22.3**

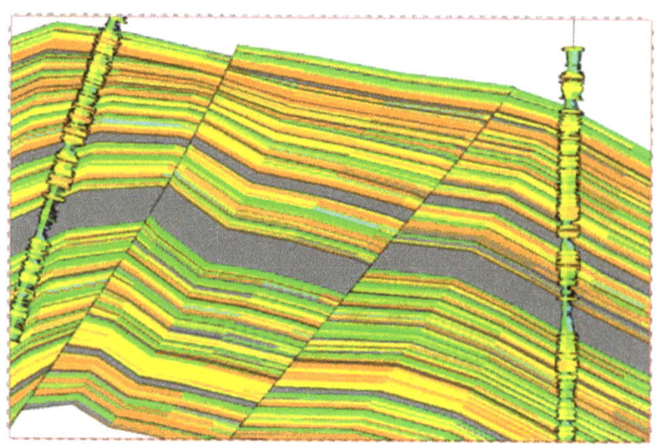

▲ **FIGURE 22.27 Cross-section showing a vertical realization of facies in Example 22.3**

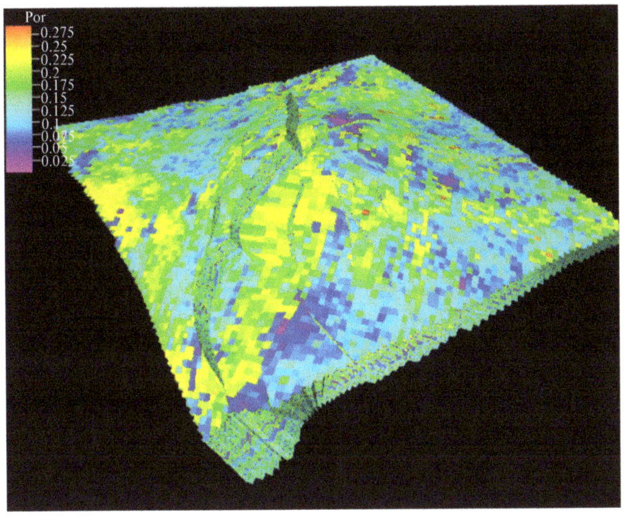

▲ **FIGURE 22.28** **A porosity realization of Reservoir "B" in Example 22.3**

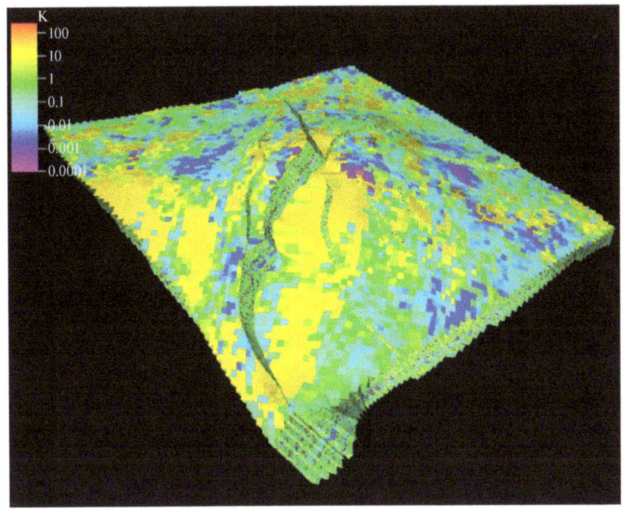

▲ **FIGURE 22.30** **A permeability realization of Reservoir "B" in Example 22.3**

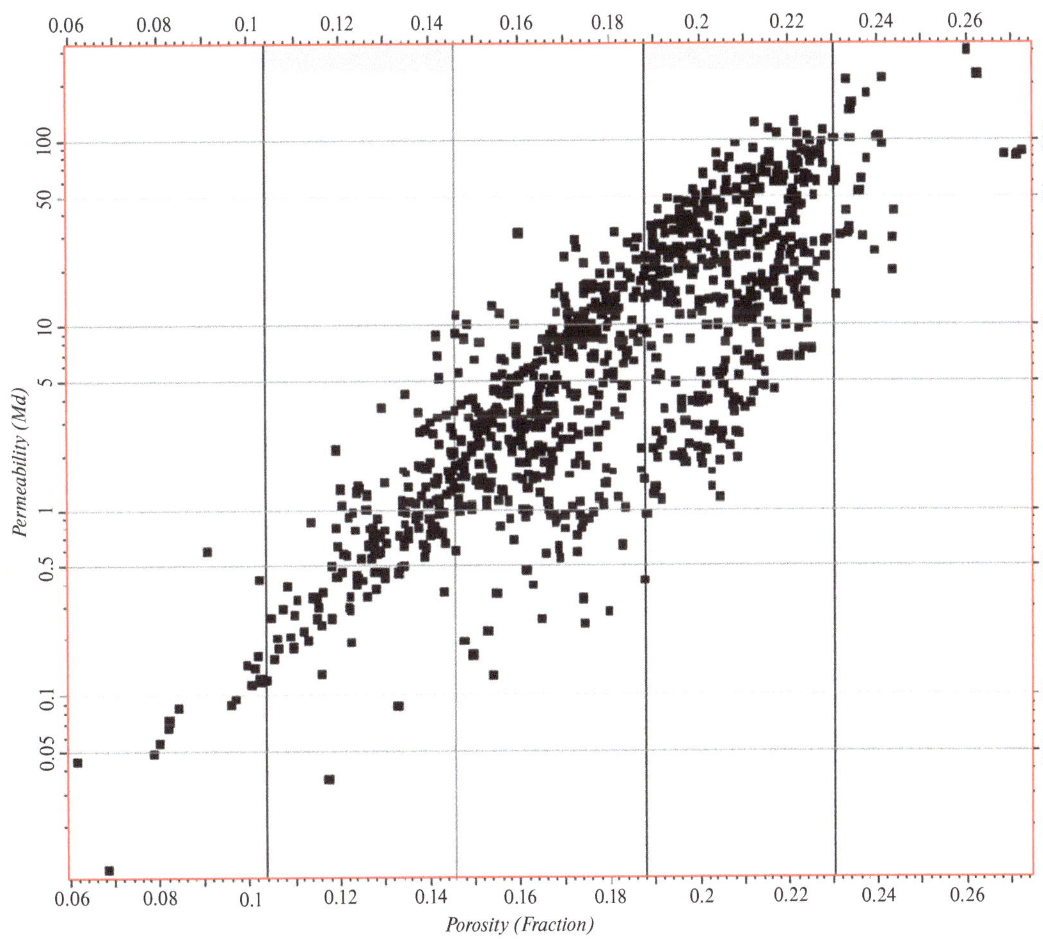

▲ **FIGURE 22.29** **Porosity/permeability cross-plot for all wells in Reservoir "B" of Example 22.3**

End of Example 22.3

Nomenclature

h	lag distance
x_i	location of sample points
$z(x_i)$	sample values at location x_i
N(h)	number of sample pairs at lag distance, h
r	range of a semivariogram
γ(h)	semivariograms at lag distance, h
Δh	lag tolerance

Abbreviations

DST	drill stem test
FBSA	flow-based scale averaging
LWD	logging-while-drilling
NMR	nuclear magnetic resonance
QA	quality assurance
QC	quality control
RCA	routine core analysis
SCAL	special core analysis
SGS	sequential Gaussian simulation
SIS	sequential indicator simulation
WFT	wireline formation tester

REFERENCES

1. Ezekwe, J.N., and Filler, S.L.: "Modeling Deepwater Reservoirs," paper SPE 95066 presented at the Annual Technical Conference and Exhibition, Dallas, Texas, (October 9-12, 2005).
2. McCarthy, P., Brand, J., Paradiso, B., Ezekwe, J., Wiltgen, N., Bridge, A., Willingham, R., and Bogaards, M.: "Using Geostatistical Inversion of Seismic and Borehole Data to Generate Reservoir Models for Flow Simulations of Magnolia Field, Deepwater Gulf of Mexico," *SEG Technical Program Abstracts,* (2005) 1351-1354.
3. Wu, Heyong, Wu, Haibo, Wang, D., Nian, J., Sun, Z., and Dusseault, M.: "Combined Seismic Inversion Methods in Reservoir Model Development," paper SPE 109626 presented at the 2007 Annual Technical Conference and Exhibition, Anaheim, California, (November 11-14, 2007).
4. Wickens, H.D., and Bouma, A.H.: "The Tanqua Fan Complex, Karoo Basin, South Africa–Outcrop Analog for Fine-Grained, Deepwater Deposits," *in* A.H. Bouma and C.G. Stone, eds., *Fine-grained turbidite systems*, AAPG Memoir 72/SEPM Special Publication 68, Chapter 14, (2000).
5. Carr, M., and Gardner, M.H.: "Portrait of a Basin-Floor Fan for Sandy Deepwater Systems, Permian Lower Brushy Canyon Formation, West Texas," *in* A.H. Bouma, C.E. Stelting, and C.G. Stone, eds., *Fine-grained turbidite systems*, AAPG Memoir 72/SEPM Special Publication 68, Chapter 20, (2000).
6. Panda, M.N., Mosher, C., and Chopra, A.K.: "Reservoir Modeling Using Scale-Dependent Data," *SPEJ* (June 2001) 157-170.
7. Frykman, P., and Deutsch, C.V.: "Practical Application of Geostatistical Scaling Laws for Data Integration," *Petrophysics* (May-June 2002) 153-171.
8. Al-Ali, H.A., and Worthington, P.F.: "Applications of Petrophysical Scale Reconciliation to Saudi Arabian Reservoirs," paper IPTC 10726 presented at the International Petroleum Technology Conference, Doha, Qatar, (November 21-23, 2005).
9. Ezekwe, N.J., and Shafer, J.L.: "Methods for Upscaling Diverse Rock Permeability Data for Reservoir Characterization and Modeling," paper SPE 109128 presented at the Annual Technical Conference and Exhibition, Anaheim, CA, (November 11-14, 2007).
10. Isaaks, E.H., and Srivastava, R.M.: *An Introduction to Geostatistics,* Oxford University Press, Oxford, United Kingdom, (1989).
11. Deutsch, C.: *Geostatistical Reservoir Modeling*, 1st Ed., Oxford University Press, New York, (2002).
12. Kelkar, M., and Perez, G.: *Applied Geostatistics for Reservoir Characterization*, SPE, Richardson, Texas, (2002).
13. Yarus, J.M., and Chambers, R.L.: "Practical Geostatistics–An Armchair Overview for Petroleum Reservoir Engineers," paper SPE 103357, Distinguished Author Series, *JPT* (November 2006) 78-87.
14. Gringarten, E., and Deutsch, C.V.: "Methodology for Variogram Interpretation and Modeling for Improved Reservoir Characterization," paper SPE 56654 presented at the 1999 SPE Annual Technical Conference and Exhibition, Houston, Texas, (October 3-6, 1999).
15. Stern, D.: "Practical Aspects of Scaleup of Simulation Models," paper SPE 89032, Distinguished Author Series (September 2005) 74-81.

16. Dogru, A.H., Fung, L.S.K., Middya, U., Al-Shaalan, T.M., Pita, J.A., Kumar, K.H., Su, H.J., Tan, J.C.T., Hoy, H., Dreiman, W.T., Hahn, W.A., Al-Harbi, R., Al-Youbi, A., Al-Zamel, N.M., Mezghani, M., and Al-Mani, T: "A Next-Generation Parallel Reservoir Simulator for Giant Reservoirs," paper SPE 119272 presented at the SPE Reservoir Simulation Symposium, The Woodlands, Texas, (February 2-4, 2009).

17. Dogru, A.H, Fung, L.S.K., Al-Shaalan, T.M., Middya, U., and Pita, J.A.: "From Mega-Cell to Giga-Cell Reservoir Simulation," paper SPE 116675 presented at the SPE Annual Technical Conference and Exhibition, Denver, Colorado, (September 21-24, 2008).

18. Saggaf, M.M.: "A Vision for Future Upstream Technologies," paper SPE 109323, Distinguished Author Series (March 2008) 94-98.

19. Brown, J.M.G., and Ezekwe, N.: "Improved Reservoir Management Using Geostatistical Reservoir Description, 24Z Stevens Reservoir, Elk Hills Oilfield, California," SPE 46232 presented at the Western Regional Meeting in Bakersfield, California, (May 10-13, 1998).

20. Peaceman, D.W.: "Effective Transmissibilities of a Gridblock by Upscaling–Comparison of Direct Methods with Renormalization," paper SPE 36722, *SPEJ* (September 1997) 338-349.

ADDITIONAL REFERENCES

- Al Maskeen, A.M.A.: "Advanced Geological Modeling and Uncertainty Analysis in a Complex Clastic Gas Reservoir from Saudi Arabia," paper SPE 109275 presented at the 2007 Asia Pacific Oil & Gas Conference and Exhibition, Jakarta, Indonesia, (October 30-November 1, 2007).
- Al-Aziz, S.A., Al-Rowyeh, A., AL-Enezi, T., Chetri, H.B., and AL-Anzi, E.: "Sabiriyah Upper Burgan Geological Model and Reservoir Characterization," paper SPE 120380 presented at the 2009 SPE Middle East Oil & Gas Show and Conference, Bahrain, (March 15-18, 2009).
- Aniekwena, A.U., McVay, D.A., Ahr, W.M., and Watkins, J.S.: "Integrated Characterization of the Thin-Bedded 8 Reservoir, Green Canyon 18, Gulf of Mexico," paper SPE 84051 presented at the 2003 SPE Annual Technical Conference and Exhibition, Denver, Colorado, (October 5-8, 2003).
- Bogan, C., Johnson, D., Litvak, M., and Stauber, D.: "Building Reservoir Models Based on 4D Seismic & Well Data in Gulf of Mexico Fields," paper SPE 84370 presented at the 2003 SPE Annual Technical Conference and Exhibition, Denver, Colorado, (October 5-8, 2003).
- Caers, J., Asveth, P., and Mukerji, T.: "Geostatistical integration of rock physics, seismic amplitudes and geological models in North-Sea turbidite systems," paper SPE 71321 presented at the 2001 SPE Annual Technical Conference and Exhibition, New Orleans, Louisiana, (September 30-October 3, 2001).
- Camey, S., Chai, S.N., Abu Bakar, M.F., Barr, D., and Leal, L.: "Reservoir Characterization of the Miocene Reservoirs of the North Malay Basin, Malaysia-Thailand Joint Development Area (MTJDA)," paper SPE 116311 presented at the 2008 SPE Asia Pacific Oil & Gas Conference and Exhibition, Perth, Australia, (October 20-22, 2008).
- Christie, M.A.: "Tenth SPE Comparative Solution Project: A Comparison of Upscaling Techniques," SPEREE (August 2001) 308-317.
- Christie, M.A.: "Upscaling for Reservoir Simulation," *JPT* (November 1996) 1004-1010.
- Delfiner, P.: "Three Statistical Pitfalls of Phi-K Transforms," paper SPE 102093 presented at the 2006 SPE Annual Technical Conference and Exhibition, San Antonio, Texas, (September 24-27, 2006).
- Lerat, O., Nivlet, P., Doligez, B., Lucet, N., Roggero, F., Berthet, P., Lefeuvre, F., and Vittori, J.: "Construction of a Stochastic Geological Model Constrained by High-Resolution 3D Seismic Data–Application to the Girassol Field, Offshore Angola," paper SPE 110422 presented at the 2007 SPE Annual Technical Conference and Exhibition, Anaheim, California, (November 11-14, 2007).
- Salazar, M.O., and Villa, J.R.: "Permeability Upscaling Techniques for Reservoir Simulation," paper SPE 106679 presented at the 2007 SPE Latin America and Carribean Petroleum Engineering Conference, Buenos Aires, Argentina, (April 15-18, 2007).
- Shew, R.D., Tiller, G.M., Hackbarth, C.J., Rollins, D.R., and White, C.D.: "Characterization and Modeling of Channel and Thin-Bedded Turbidite Prospects in the Gulf of Mexico: Integration of Outcrops, Modern Analogs, and Subsurface Data," paper SPE 30535 presented at the 1995 SPE Annual Technical Conference and Exhibition, Dallas, Texas, (October 22-25, 1995).
- Shmaryan, L.E., and Deutsch, C.V.: "Object-Based Modeling of Fluvial/Deepwater Reservoirs with Fast Data Conditioning: Methodology and Case Studies," paper SPE 56821 presented at the 1999 SPE Annual Technical Conference and Exhibition, Houston, Texas, (October 3-6, 1999).
- Slatt, R.M., Browne, G.H., Davis, R.J., Clemenceau, G.R., Colbert, J.R., Young, R.A., Anxionnaz, H., and Spang, R.J.: "Outcrop-Behind Outcrop Characterization of Thin-bedded Turbidites for Improved Understanding of Analog Reservoirs: New Zealand and Gulf of Mexico," paper SPE 49563 prepared for presentation at the 1998 SPE Annual Technical Conference and Exhibition, New Orleans, Louisiana, (September 27-30, 1998).
- Stewart, J., Dunn, P., Lyttle, C., Campion, K., Oyerinde, A., and Fischer, B.: "Improving Performance Prediction in Deep-Water Reservoirs: Learning From Outcrop Analogues, Conceptual Models and Flow Simulation," paper IPTC 12892 presented at the International Petroleum Technology Conference, Kuala Lumpur, Malaysia, (December 3-5, 2009).
- Strebelle, S., Payrazyan, K., and Caers, J.: "Modeling of a Deepwater Turbidite Reservoir Conditional to Seismic Data Using Multiple-Point Geostatistics," paper SPE 77425 presented at the 2002 SPE Annual Technical Conference and Exhibition, San Antonio, Texas, (September 29-October 2, 2002).
- Worthington, P.F.: "The Effect of Scale on the Petrophysical Estimation of Intergranular Permeability," SPWLA 44th Annual Logging Symposium, (June 22-25, 2003).

PROBLEMS

22-1. Calculate an experimental vertical variogram for the water saturation data given in Table 22P.1.

TABLE 22P.1 Data for Problem 22-1

Depth (feet)	Water Saturation (%)	Depth (feet)	Water Saturation (%)
6050	25.2	6067	27.5
6051	24.3	6068	27.8
6052	22.1	6069	25.4
6053	25.8	6070	24.3
6054	21.5	6071	23.9
6055	20.9	6072	23.5
6056	26.0	6073	22.4
6057	23.3	6074	21.0
6058	19.8	6075	19.6
6059	22.9	6076	19.8
6060	25.4	6077	20.4
6061	23.4	6078	22.8
6062	22.8	6079	21.9
6063	24.7	6080	21.5
6064	20.4	6081	20.5
6065	19.8	6082	19.5
6066	26.8	6083	19.8

22-2. Calculate an experimental vertical variogram for the porosity data given in Table 22P.2.

TABLE 22P.2 Data for Problem 22-2

Depth (feet)	Porosity (%)	Depth (feet)	Porosity (%)
5832	20.7	5846	18.4
5833	21.4	5847	17.4
5834	18.7	5848	18.2
5835	17.4	5849	18.9
5836	17.4	5850	18.9
5837	17.2	5851	17.4
5838	18.4	5852	18.2
5839	18.4	5853	18.4
5840	18.4	5854	19.4
5841	18.9	5855	20.4
5842	18.7	5856	20.7
5843	17.9	5857	21.4
5844	18.2	5858	21.7
5845	18.4	—	—

22-3. The case history of Red Hawk reservoirs was presented in Chapter 9, Section 9.6.1. The structure maps for the Surculus interval and Tamalis interval were provided as Figures 9.16 and 9.17, respectively. Average rock properties for the intervals were given in Table 9.9. Applying all the data given in the case history, construct:

a. A geologic model of the Surculus and Tamalis intervals. Assume constant porosity and permeability, water saturation, and net pay for the grid cells in the model

b. Adjust geologic model if necessary, to match given data.

(Hint: Use a geologic modeling software to solve this problem.)

22-4. The case history of the West Cameron reservoirs was presented in Chapter 9, Section 9.6.2. The structure maps for the M-4/M-4A and M-6 reservoirs were given as Figures 9.24 and 9.25, respectively. Average rock properties of the reservoirs were in Table 9.13. The average net-to-gross ratio for both reservoirs is 0.7834. Using the data given in the case history, construct:

a. A geologic model of the M-4/M-4A and M-6 reservoirs. Assume constant porosity, permeability, net-to-gross, and water saturation values for the grid cells.

b. Adjust the geologic model to match the total productive volumes reported in Table 9.13 of the case history as necessary.

(Hint: Use a geologic modeling software to solve this problem.)

22-5. The case history of the Manatee reservoirs was presented in Chapter 10, Section 10.8. The structure maps and the net sand maps were also provided with the case history in Figures 10.8 to 10.11. Other average reservoir petrophysical data were given in Table 10.6, Applying all the data given in the case history, construct:

a. A geologic model of the S10A2 and S10B reservoirs. Assume constant porosity, permeability, and water saturation values for the grid cells.

b. Adjust the geologic model to match the planimetered reservoir volumes reported in the case history as necessary.

(Hint: Use a geologic modeling software to solve this problem.)

CHAPTER

23 Reservoir Simulation

23.1 Introduction

The term "reservoir simulation" is generally used to describe the activities involved in the building and execution of a model that represents the reservoir, such that the behavior of the model mirrors or "simulates" as much as possible the observed behavior of the actual reservoir. The model built in this case consists of sets of mathematical equations that represent material balance, fluid flow, and other physical processes occurring in the reservoir, subject to some defined constraints and conditions. Many forms of the mathematical model that represent the reservoir can be developed, depending on the nature of the reservoir, physical data available to describe the reservoir, processes occurring within the reservoir, and the objectives of simulating the behavior of the reservoir. In this book, three-dimensional, three-phase mathematical reservoir models (called black oil models) are used almost exclusively to describe the general principles and practice of reservoir simulation. This approach was taken to limit the scope of the material covered on reservoir simulation in this book, and also in recognition of the extensive body of work and books already published on reservoir simulation. These books[1-3] and literature[4-6] are recommended as additional sources of material for acquiring more knowledge on the principles and applications of petroleum reservoir simulation.

Why is it necessary to conduct numerical simulations of petroleum reservoirs? The answers to this cogent question can be approached in a variety of ways. But the simplest, most straightforward answer to the question was provided by Coats[7] with this statement: "Whereas the field can be produced only once–and at considerable expense–a model can be produced or 'run' many times at low expense over a short period of time." In other words, reservoir simulation provides a powerful tool for evaluating alternative reservoir management strategies, thereby making it possible to select an optimum management strategy, based on current/future reservoir and operating conditions. Since the above statement was made by Coats[7] in 1969, the application of reservoir simulation as a tool for reservoir management, and for assessments of uncertainty and risks associated with reservoir development, has become widespread in the petroleum industry. There is hardly any recent major field development or project that is not evaluated or assessed with some form of reservoir simulation. The reach of project evaluation with reservoir simulation is quite extensive from the development of deepwater or unconventional reservoirs to the improvement of hydrocarbon recovery with secondary or enhanced oil recovery processes. The relevant question in the industry is no longer why simulate, but rather how to effectively and efficiently apply the technology of reservoir simulation to develop reservoir management strategies, optimize reservoir development, and improve hydrocarbon recovery.

Reservoir simulation (or modeling) is applied on the evaluation or assessment of key reservoir management issues, problems or operating questions such as:

1. What are the expected hydrocarbon recoveries under current and alternative reservoir management strategies? Can the expected hydrocarbon recovery be improved?
2. Is the reservoir a good candidate for pressure maintenance? Should gas or water injection be used for pressure maintenance?
3. For gas condensate reservoirs, will gas cycling improve total hydrocarbon recovery from the reservoir?
4. What is the optimum well count and well locations for development of new conventional or unconventional reservoirs?
5. Is the development of a new discovery economic under prevailing drilling and completion costs, operating costs, and product prices for conventional/unconventional reservoirs? What management strategies can be used to change or improve the economic outcomes of new projects?
6. What is the optimum producing rate for each well in the reservoir? What is the effect of producing rate on hydrocarbon recovery? Can the reservoir sustain production at levels necessary to maintain profitability?
7. Is the reservoir a good candidate for secondary recovery processes? If so, when is the proper time to initiate the secondary recovery process? What type of secondary recovery process to implement; for instance, pattern waterflooding versus peripheral (flank) waterflooding? What is the expected incremental hydrocarbon recovery from secondary recovery processes?
8. What types of enhanced oil recovery processes will be suitable for the reservoir? What is the ranking of various enhanced oil recovery processes when applied to the reservoir? What is the expected incremental recovery and profitability of the selected enhanced oil recovery process?

9. How is the reservoir performing under current operating strategy? Are they indications from current performance data of future problems? How can those problems be avoided or mitigated by modifying current practices and/or management strategies?
10. What are the uncertainties and risks associated with current or planned reservoir management strategies? Are those uncertainties and risks acceptable? How can those uncertainties and risks be mitigated by adopting alternative reservoir management strategies?

These are some of the problems of reservoir management that can be evaluated with reservoir simulation for conventional or unconventional reservoirs. Clearly, there are many more problems of reservoir management that can be addressed with reservoir simulation. Consequently, reservoir simulation has developed over many years to finally become an indispensable tool in modern reservoir engineering practice.

The material balance equations for three-dimensional, three-phase immiscible flow of oil, water, and gas were derived by Muskat.[8] These material balance equations are the fundamental equations used in black oil simulators. There are no analytical solutions of these non-linear, partial differential equations. Solutions of these equations are generally based on numerical methods. Advancement of numerical reservoir simulation has depended on the development of numerical methods to solve these partial differential equations, and availability of high-speed digital computers for implementation of the solution techniques. Douglas et al.[9] proposed an Implicit (simultaneous) method for solving two-dimensional, two-phase flow problems. This method was used later by Coats et al.[10] on three-dimensional, two-phase flow problems, and by Coats[11,12] on three-dimensional, three-phase flow of incompressible and compressible fluids, respectively. An alternative solution technique described as Implicit-Pressure, Explicit Saturation (IMPES) method was proposed by Sheldon et al.[13] and Stone and Garder.[14] The method was later used on a two-dimensional, multiphase flow simulator by Fagin and Stewart[15] in 1966. The differences between these two solution methods (Implicit and IMPES) are discussed later. The above account is meant to provide a historical perspective and is not considered an accurate historical account of the development of numerical reservoir simulators. Many other pioneers contributed in various ways to the development of this technology. It is not feasible to provide a complete list of these pioneers, but their contributions to the development of reservoir simulation is hereby acknowledged.

The use of numerical methods to solve the equations formulated to represent material balance and fluid flow in reservoirs leads to non-exact solutions. Moreover, application of different numerical solution techniques on the same set of equations will generally yield solutions that are slightly different, including associated errors. In addition, solutions can be achieved by making different sets of assumptions in the process of formulating the equations for solution. The point is that numerical simulators generally yield results that may vary when applied to the same reservoir problem, depending on the formulation of the equations and the numerical technique used in solving these equations. For many years, the non-uniqueness of solutions provided by numerical simulators undermined its acceptability as a valuable tool for reservoir analyses. These skepticisms on the applicability of numerical reservoir simulators have been erased gradually over the years in the petroleum industry through publication of sponsored comparative studies using reservoir simulators built by various groups,[16-25] and availability of a large body of work on reservoir simulation in the literature.[1-7, 26-29]

Numerical reservoir simulators are classified on the basis of reservoir type, reservoir process, and formulation of the simulator. The following types of reservoir simulators are commercially available in the petroleum industry:

1. Black oil simulators: oil, water, and gas are treated as separate immiscible phases. This simulator can be used on dry gas, and black oil reservoirs (as defined in Chapter 5), and immiscible recovery processes (Chapters 18 to 20). The black oil simulator is the "workhorse" of the petroleum industry with respect to reservoir simulation.
2. Compositional simulators: oil and gas are represented as hydrocarbon components. Water is present as a phase. This type of simulator is used on volatile oil, and gas condensate reservoirs (as defined in Chapter 5), and miscible gas enhanced oil recovery processes (Chapter 21).
3. Dual-porosity simulators: used on naturally fractured reservoirs that exhibit dual porosity behavior and unconventional reservoirs (Chapters 11, 15, and 17)
4. Thermal simulators: used for simulating thermal processes such as cyclic steam injection, steam flooding, and in-situ combustion (Chapter 21).
5. Chemical flood simulators: applicable to enhanced oil recovery processes using alkalis, surfactants, and polymers (Chapter 21).
6. Streamline simulators: This class of simulators is widely used in the industry for upscaling of large geologic models, modeling of waterflooding, and other uncomplicated reservoir processes. Streamline simulators can be used in conjunction with the other grid-based simulators and are not considered a substitute for them.[30] Streamline simulators are not covered in this book.

Reservoir simulation is presented in this book with more emphasis on its use as a tool for reservoir analysis and evaluation. Theoretical development and formulation of equations are used where necessary to illustrate basic concepts such as solution techniques, stability issues, material balance errors, etc. Detailed presentations of various formulations of the basic material balance equations and solution techniques are not covered in this book. These are available in other books written exclusively on reservoir simulation.[1-3] In this chapter, the basic material balance and flow equations are presented for a black oil simulator. This is then followed with presentation of basic terms, concepts, and solution methods encountered in the practice of reservoir simulation. The chapter concludes with the presentation of the basic structure of data input generally encountered in most commercial simulators.

23.2 Derivation of the Continuity Equation in Rectangular Form

Consider a rectangular block with fluid flow in three directions, x, y, and z as shown in Figure 23.1. During time interval, Δt, the principle of mass conservation can be applied on the block to obtain the following mass balance equation:

$$\begin{aligned}&\text{Mass flowing into the block in time, } \Delta t\\&-\text{Mass flowing out of the block in time, } \Delta t \quad (23.1)\\&=\text{Mass accumulation in the block in time, } \Delta t\end{aligned}$$

Let mass flow rate per unit cross-sectional area normal to the direction of flow be denoted as F. Using this designation, fluid flow into the block in the x-direction at x is $(F_x)_x$ and flow out of the block at $x+\Delta x$ is $(F_x)_{x+\Delta x}$. Similarly, flow into the block in the y and z directions are designated as $(F_y)_y$ and $(F_z)_z$, respectively. And flow out of the block in the y and z directions are designated as $(F_y)_{y+\Delta y}$ and $(F_z)_{z+\Delta z}$, respectively. The length, width, and height of the block are designated as Δx, Δy, and Δz, respectively, as shown in Figure 23.1. Using these definitions, it can be written from Eq. (23.1) that:

Mass flowing into block in time, Δt

$$=\left[\left(F_x\right)_x \Delta y\Delta z+\left(F_y\right)_y \Delta x\Delta z+\left(F_z\right)_z \Delta x\Delta y\right]\Delta t \tag{23.2}$$

Mass flowing out of block in time, Δt

$$\begin{aligned}&=\left[\left(F_x\right)_{x+\Delta x} \Delta y\Delta z+\left(F_y\right)_{y+\Delta y} \Delta x\Delta z+\left(F_z\right)_{z+\Delta z} \Delta x\Delta y\right]\Delta t\\&+q\Delta x\Delta y\Delta z\Delta t\end{aligned} \tag{23.3}$$

Mass accumulation in block in time, Δt

$$=\left[\left(C_p\right)_{t+\Delta t}-\left(C_p\right)_t\right]\Delta x\Delta y\Delta z \tag{23.4}$$

In Eq. (23.3), q is a source/sink term that represents mass flow into or out of the block primarily through a well. For a well that is a producer, $q>0$ (or positive), and for a well that is an injector, $q<0$ (or negative) by convention. In Eq. (23.4), C_p is the concentration of any phase (oil, water, and gas) in the block defined in mass per unit volume.

Substituting Eqs. (23.2) to (23.4) into Eq. (23.1) gives:

$$\begin{aligned}&\left[\left(F_x\right)_x \Delta y\Delta z+\left(F_y\right)_y \Delta x\Delta z+\left(F_z\right)_z \Delta x\Delta y\right]\Delta t\\&-\left[\left(F_x\right)_{x+\Delta x} \Delta y\Delta z+\left(F_y\right)_{y+\Delta y} \Delta x\Delta z+\left(F_z\right)_{z+\Delta z} \Delta x\Delta y\right]\Delta t\\&-q\Delta x\Delta y\Delta z\Delta t=\left[\left(C_p\right)_{t+\Delta t}-\left(C_p\right)_t\right]\Delta x\Delta y\Delta z\end{aligned} \tag{23.5}$$

Dividing Eq. (23.5) with $\Delta x\Delta y\Delta z\ \Delta t$ and collecting like terms yields:

$$\begin{aligned}&-\frac{\left(F_x\right)_{x+\Delta x}-\left(F_x\right)_x}{\Delta x}-\frac{\left(F_y\right)_{y+\Delta y}-\left(F_y\right)_y}{\Delta y}\\&-\frac{\left(F_z\right)_{z+\Delta z}-\left(F_z\right)_z}{\Delta z}-q=\frac{\left(C_p\right)_{t+\Delta t}-\left(C_p\right)_t}{\Delta t}\end{aligned} \tag{23.6}$$

Taking limits as Δx, Δy, Δz, and Δt approach zero gives:

$$-\frac{\partial F_x}{\partial x}-\frac{\partial F_y}{\partial y}-\frac{\partial F_z}{\partial z}-q=\frac{\partial C_p}{\partial t} \tag{23.7}$$

Equation (23.7) is the continuity equation for fluid flow in rectangular coordinates (three directions). Equation (23.7) is similar in form to Eq. (12.52) developed in Chapter 12 as the continuity equation in radial form.

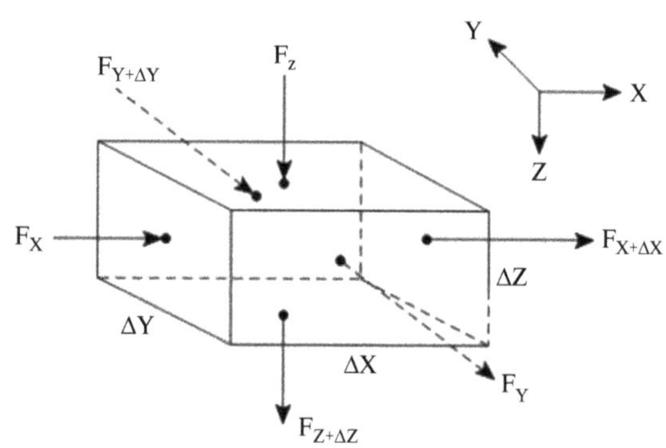

▲ **FIGURE 23.1 A gridblock showing fluid flow in three directions**

23.3 Flow Equations for Three-Phase Flow of Oil, Water, and Gas

The mass flow rate per unit area in a given direction for any fluid is the product of the density of the fluid and its velocity in that direction. If oil, water, and gas are denoted with the subscripts o, w, and g, respectively, then $(\vec{F})_o$, $(\vec{F})_w$, and $(\vec{F})_g$, respectively, can be represented as:

$$\left(\vec{F}\right)_o=\frac{\rho_{osc}}{B_o}\vec{v}_o \tag{23.8}$$

$$\left(\vec{F}\right)_w = \frac{\rho_{wsc}}{B_w}\vec{v}_w \tag{23.9}$$

$$\left(\vec{F}\right)_g = \frac{\rho_{gsc}}{B_g}\vec{v}_g + \frac{R_{so}\rho_{gsc}}{B_o}\vec{v}_o + \frac{R_{sw}\rho_{gsc}}{B_w}\vec{v}_w \tag{23.10}$$

In Eqs. (23.8) to (23.10), $\vec{F}$ is the mass flow rate in x, y, and z directions; $\vec{v}$ is the fluid velocity in x, y, and z directions; ρ_{osc}, ρ_{wsc}, and ρ_{gsc} are oil, water, and gas densities at standard conditions, respectively; R_{so} and R_{sw} are gas solubilities in oil and water, respectively, scf/STB; and B_o, B_w, and B_g are formation volume factors of oil, water, and gas, respectively, RB/STB. For oil, water, and gas, the concentration of each phase, C_p, is defined respectively as:

$$C_o = \frac{\phi\rho_{osc}S_o}{B_o} \tag{23.11}$$

$$C_w = \frac{\phi\rho_{wsc}S_w}{B_w} \tag{23.12}$$

$$C_g = \phi\rho_{gsc}\left[\frac{S_g}{B_g} + R_{so}\frac{S_o}{B_o} + R_{sw}\frac{S_w}{B_w}\right] \tag{23.13}$$

In Eqs. (23.11) to (23.13), ϕ is the porosity of the block; and S_o, S_w, and S_g are oil, water, and gas saturations, respectively, in the block. A constraint applied to fluid saturations in the block is the requirement that the sum of the fluid saturations is equal to unity:

$$S_o + S_w + S_g = 1 \tag{23.14}$$

Substituting Eqs. (23.8) to (23.13) into Eq. (23.7) with vector expansions gives a mass balance equation for each phase as:

For oil:

$$-\left[\frac{\partial}{\partial x}\left(\frac{\rho_{osc}}{B_o}v_{xo}\right) + \frac{\partial}{\partial y}\left(\frac{\rho_{osc}}{B_o}v_{yo}\right) + \frac{\partial}{\partial z}\left(\frac{\rho_{osc}}{B_o}v_{zo}\right)\right] - q_o = \frac{\partial}{\partial t}\left(\frac{\phi\rho_{osc}S_o}{B_o}\right) \tag{23.15}$$

For water:

$$-\left[\frac{\partial}{\partial x}\left(\frac{\rho_{wsc}}{B_w}v_{xw}\right) + \frac{\partial}{\partial y}\left(\frac{\rho_{wsc}}{B_w}v_{yw}\right) + \frac{\partial}{\partial z}\left(\frac{\rho_{wsc}}{B_w}v_{zw}\right)\right] - q_w = \frac{\partial}{\partial t}\left(\frac{\phi\rho_{wsc}S_w}{B_w}\right) \tag{23.16}$$

For gas:

$$\begin{aligned} &-\frac{\partial}{\partial x}\left(\frac{\rho_{gsc}}{B_g}v_{xg} + \frac{R_{so}\rho_{gsc}}{B_o}v_{xo} + \frac{R_{sw}\rho_{gsc}}{B_w}v_{xw}\right) \\ &-\frac{\partial}{\partial y}\left(\frac{\rho_{gsc}}{B_g}v_{yg} + \frac{R_{so}\rho_{gsc}}{B_o}v_{yo} + \frac{R_{sw}\rho_{gsc}}{B_w}v_{yw}\right) \\ &-\frac{\partial}{\partial z}\left(\frac{\rho_{gsc}}{B_g}v_{zg} + \frac{R_{so}\rho_{gsc}}{B_o}v_{zo} + \frac{R_{sw}\rho_{gsc}}{B_w}v_{zw}\right) \\ &-q_g = \frac{\partial}{\partial t}\left[\phi\rho_{gsc}\left(\frac{S_g}{B_g} + \frac{R_{so}S_o}{B_o} + \frac{R_{sw}S_w}{B_w}\right)\right] \end{aligned} \tag{23.17}$$

Eqs. (23.15), (23.16), and (23.17) can be simplified by canceling out ρ_{osc}, ρ_{wsc}, and ρ_{gsc}, respectively to get:

For oil:

$$-\left[\frac{\partial}{\partial x}\left(\frac{v_{xo}}{B_o}\right) + \frac{\partial}{\partial y}\left(\frac{v_{yo}}{B_o}\right) + \frac{\partial}{\partial z}\left(\frac{v_{zo}}{B_o}\right)\right] - \frac{q_o}{\rho_{osc}} = \frac{\partial}{\partial t}\left(\frac{\phi S_o}{B_o}\right) \tag{23.18}$$

For water:

$$-\left[\frac{\partial}{\partial x}\left(\frac{v_{xw}}{B_w}\right) + \frac{\partial}{\partial y}\left(\frac{v_{yw}}{B_w}\right) + \frac{\partial}{\partial z}\left(\frac{v_{zw}}{B_w}\right)\right] - \frac{q_w}{\rho_{wsc}} = \frac{\partial}{\partial t}\left(\frac{\phi S_w}{B_w}\right) \tag{23.19}$$

For gas:

$$\begin{aligned} &-\frac{\partial}{\partial x}\left(\frac{v_{xg}}{B_g} + \frac{R_{so}v_{xo}}{B_o} + \frac{R_{sw}v_{xw}}{B_w}\right) \\ &-\frac{\partial}{\partial y}\left(\frac{v_{yg}}{B_g} + \frac{R_{so}v_{yo}}{B_o} + \frac{R_{sw}v_{yw}}{B_w}\right) \\ &-\frac{\partial}{\partial z}\left(\frac{v_{zg}}{B_g} + \frac{R_{so}v_{zo}}{B_o} + \frac{R_{sw}v_{zw}}{B_w}\right) \\ &-\frac{q_g}{\rho_{gsc}} = \frac{\partial}{\partial t}\left[\phi\left(\frac{S_g}{B_g} + \frac{R_{so}S_o}{B_o} + \frac{R_{sw}S_w}{B_w}\right)\right] \end{aligned} \tag{23.20}$$

In vector notation, Eqs. (23.18) to (23.20) can be written as:

For oil:

$$-\nabla\cdot\frac{\vec{v}_o}{B_o} - \frac{q_o}{\rho_{osc}} = \frac{\partial}{\partial t}\left(\frac{\phi S_o}{B_o}\right) \tag{23.21}$$

For water:

$$-\nabla\cdot\frac{\vec{v}_w}{B_w} - \frac{q_w}{\rho_{wsc}} = \frac{\partial}{\partial t}\left(\frac{\phi S_w}{B_w}\right) \tag{23.22}$$

For gas:

$$-\nabla\cdot\left(\frac{\vec{v}_g}{B_g}+\frac{R_{so}\vec{v}_o}{B_o}+\frac{R_{sw}\vec{v}_w}{B_w}\right)-\frac{q_g}{\rho_{gsc}} = \frac{\partial}{\partial t}\left[\phi\left(\frac{S_g}{B_g}+\frac{R_{so}S_o}{B_o}+\frac{R_{sw}S_w}{B_w}\right)\right] \quad (23.23)$$

In vector notation,

$$\nabla\cdot\vec{v}=\frac{\partial}{\partial x}v_x+\frac{\partial}{\partial y}v_y+\frac{\partial}{\partial z}v_z \quad (23.24)$$

If the fluid velocities in Eqs. (23.21) to (23.23) are based on Darcy's law, then the fluid velocities for each phase can be represented in terms of potential as:

$$\vec{v}_p=-\vec{\vec{K}}\lambda_p\nabla\Phi_p \quad (23.25)$$

In Eq. (23.25), $\vec{v}$ is the fluid velocity of phase, p, in vector form; $\vec{\vec{K}}$ represents permeability as a tensor; λ_p is the mobility of the phase, where $\lambda_p = k_{rp}/\mu_p$; and Φ_p is the potential of the phase, where $\Phi_p = p_p - (\rho_p z)/144$, assuming $g = g_c$.

Substituting Eq. (23.25) into Eqs. (23.21) to (23.23) gives:

For oil:

$$\nabla\cdot\frac{\vec{\vec{K}}\lambda_o}{B_o}\cdot\nabla\Phi_o-\frac{q_o}{\rho_{osc}}=\frac{\partial}{\partial t}\left(\frac{\phi S_o}{B_o}\right) \quad (23.26)$$

For water:

$$\nabla\cdot\frac{\vec{\vec{K}}\lambda_w}{B_w}\cdot\nabla\Phi_w-\frac{q_w}{\rho_{wsc}}=\frac{\partial}{\partial t}\left(\frac{\phi S_w}{B_w}\right) \quad (23.27)$$

For gas:

$$\nabla\cdot\vec{\vec{K}}\left(\frac{\lambda_g}{B_g}\nabla\Phi_g+\frac{R_{so}\lambda_o}{B_o}\nabla\Phi_o+\frac{R_{sw}\lambda_w}{B_w}\nabla\Phi_w\right)-\frac{q_g}{\rho_{gsc}}=\frac{\partial}{\partial t}\left[\phi\left(\frac{S_g}{B_g}+\frac{R_{so}S_o}{B_o}+\frac{R_{sw}S_w}{B_w}\right)\right] \quad (23.28)$$

Equations (23.26) to (23.28) are the basic equations for immiscible flow of oil, water, and gas, respectively. When written in difference form, these are the basic equations solved numerically in black oil simulators. The equations are presented in this book essentially as derived by Fanchi.[31]

The derivations of the continuity equation [Eq. (23.7)] and the flow equations [Eqs. (23.26) to (23.28)] for a black oil simulator were presented to emphasize the point that the development of reservoir simulators are based on the fundamental principles of mass balances and the Darcy flow equation, which are the bedrock of most calculations in petroleum reservoir engineering. In other types of simulators, mass balances are applied on the components within the phases, and other fundamental flow equations may be included in the derivations of the relevant equations. For instance in a compositional simulator, mass balances are applied to the defined components of the hydrocarbon phases (Chapter 7), and the equilibrium between components in the vapor and liquid phases are determined with an equation of state.[27,32] For thermal simulators, an energy balance on the system in addition to mass balances on water, steam, and oil are included in the relevant equations.[33,34] The point here is to reassure potential users of reservoir simulation that reservoir simulators are based on fundamental principles of mass balance, energy balance, Darcy flow equations, etc. that form the basis of most analytical reservoir engineering calculations. With proper geologic description and reservoir characterization, application of reservoir simulation should yield results that are comparable, if not superior, to analytical methods.

23.4 Basic Concepts, Terms, and Methods in Reservoir Simulation

In this section, basic concepts, terms, and methods that are commonly used in the practice of reservoir simulation are presented. This section is intended for readers who have little or no prior knowledge of reservoir simulation. The subjects covered in this section are relatively basic, and readers with prior knowledge of reservoir simulation may elect to skip this section entirely.

23.4.1 Grid Systems

The discretization in space of the difference forms of the flow equations [such as Eqs. (23.26) to (23.28)] in reservoir simulators requires the use of grids. The key factors that influence the design of grid systems for reservoir simulation were discussed extensively in Chapter 22, Section 22.5.5.1 under Model Gridding. These factors are summarized here to re-emphasize their importance in the design of grid systems for reservoir simulation as follows:

1. Resolution of the geologic and petrophysical data
2. Size of the static geologic model
3. Limitations on the size of the flow reservoir model
4. Potential uses of the reservoir model
5. Upscaling (also termed upgridding or scaleup) techniques to convert the static geologic model into a reservoir flow model

In this section, the different types of grid systems and their potential effects on reservoir simulation results are presented.

23.4.1.1 Cartesian Grids

The most common type of grid systems used in reservoir simulation are based on Cartesian coordinates of x, y, and z. Cartesian grids can be based on a block-centered grid or a point-distributed grid systems (Figure 23.2). The block-centered grid is better for calculating mass accumulation terms in the flow equations, while the point-distributed grid is more accurate for calculating flows between the gridblocks.[35] The differences between the two grid systems (block-centered vs point-distributed) are insignificant, if the Cartesian grid systems are uniform. Most commercial simulators use block-centered grids. Cartesian grid systems can be classified as one-dimensional (1D), two-dimensional (2D), and three-dimensional (3D). Figure 23.3 shows a 1D grid system in the x-direction. Figure 23.4 shows a 2D grid system for an areal model (x, y), and a cross-sectional model (x, y). Figure 23.5 shows a 3D grid system for a reservoir model (x, y, z). The discretization in space of the simulator flow equations assumes that the gridblocks are orthogonal. This means that each column of gridblocks is at right angles to each row of gridblocks. In designing a Cartesian grid system for reservoir simulation, it is important to maintain the gridblocks orthogonal to each other. Otherwise, the results from the simulation may not be accurate, if the gridblocks are severely non-orthogonal. Most commercial simulators incorporate features to ensure orthogonality of the gridblocks for irregular grid systems, especially if the static geologic model is intended to be used in reservoir simulation.

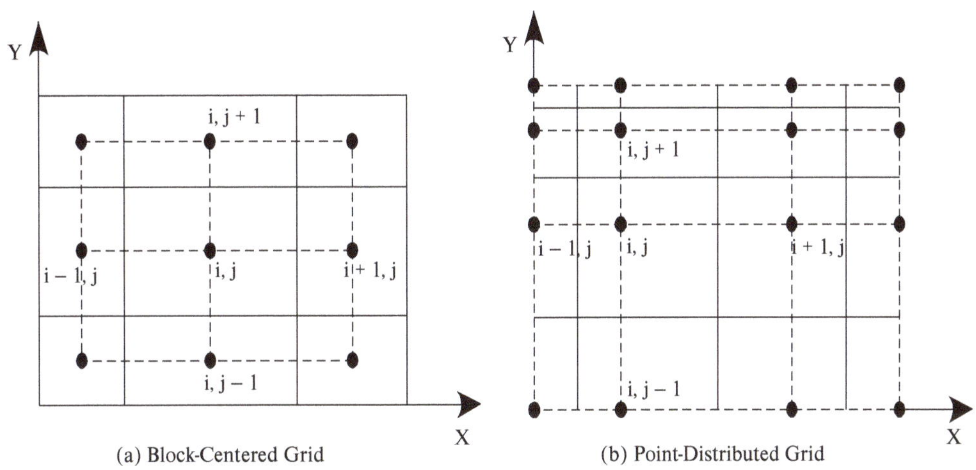

▲ **FIGURE 23.2** **Block-centered and point-distributed grids (from Aziz[35] © 1993 SPE, Reproduced with permission)**

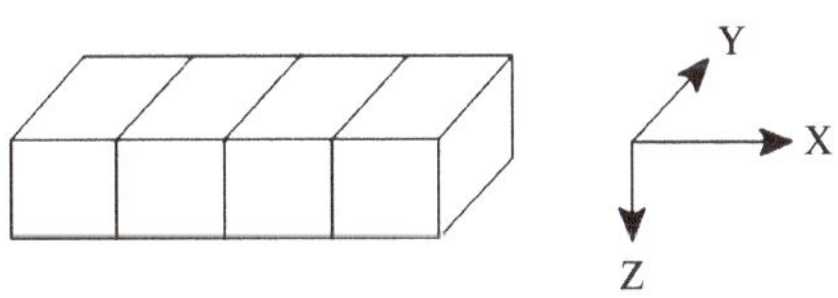

▲ **FIGURE 23.3** **1D grid model**

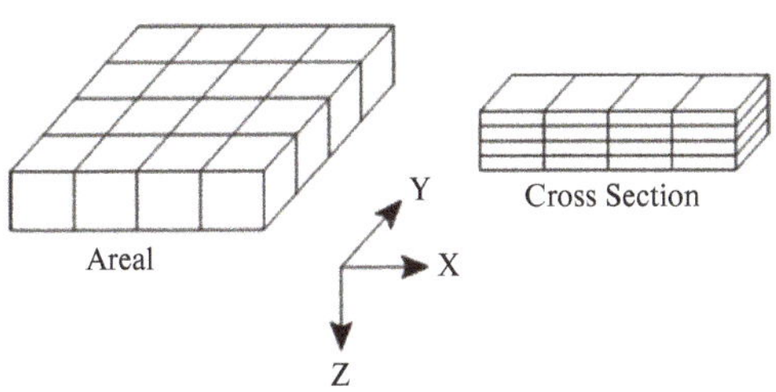

▲ **FIGURE 23.4** **2D grid model**

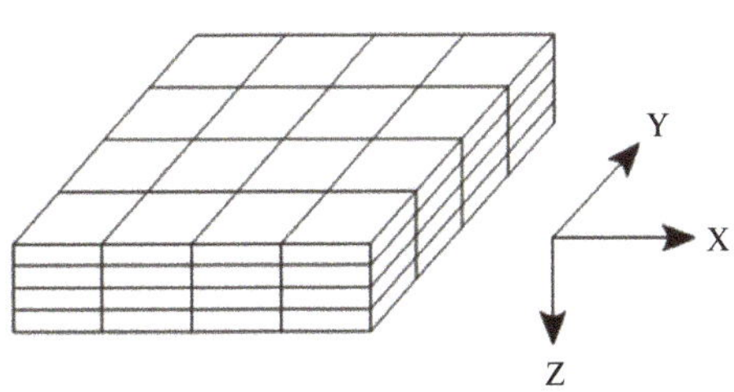

▲ **FIGURE 23.5** **3D grid model**

23.4.1.2 Radial Grids

Grid systems are also based on radial coordinates of r, θ, and z. An example of a grid system based on radial coordinates is shown in Figure 23.6. Radial grid systems are used mainly in the representation of wells in models that show rapid changes in fluid saturations near the wellbore as in gas cusping or water coning studies. In some applications, the wells are represented with radial grids while the remaining parts of the reservoir have Cartesian grids, thereby creating a **hybrid grid**[36] system as shown in Figure 23.7.

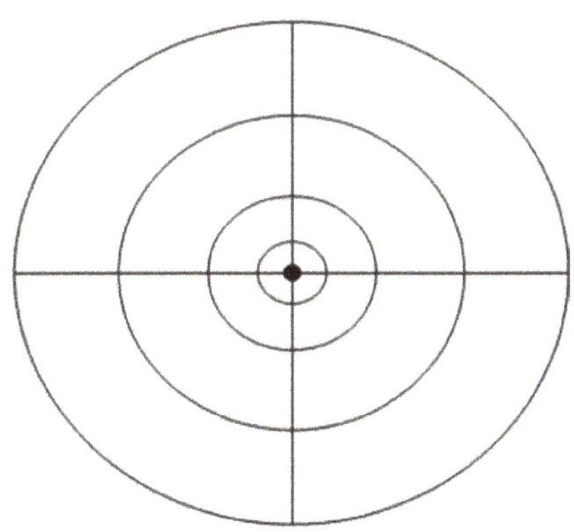

▲ **FIGURE 23.6** **Radial grid model**

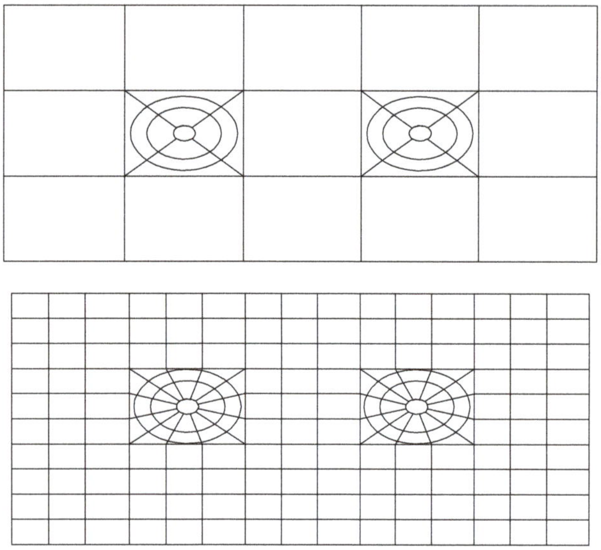

▲ FIGURE 23.7 **Hybrid grid model (from Aziz[35] © 1993 SPE, Reproduced with permission)**

23.4.1.3 Corner-Point Geometry

Corner-point geometry is based on specifying the locations of all the eight corners of each gridblock as shown in Figure 23.8. This gridding system is useful for representing complex geologic geometries such as faults, permeability barriers, irregular boundaries, sand channels, etc. However, they could generate highly irregular grids that may not be orthogonal. As pointed out earlier, non-orthogonal grids, especially from corner-point geometry, could yield erroneous flow simulation results.[35]

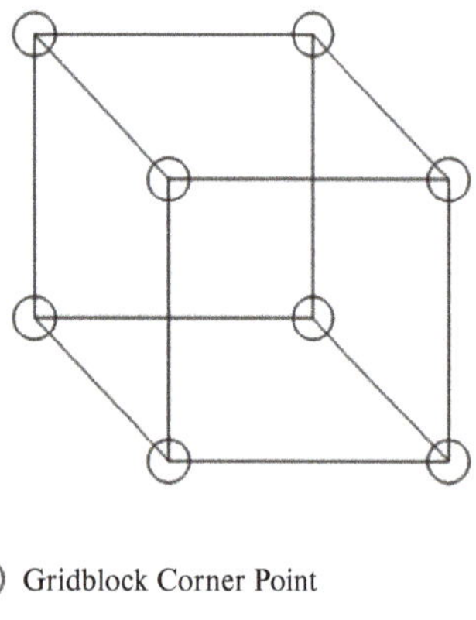

▲ FIGURE 23.8 **Gridblock showing corner-point geometry**

23.4.1.4 Voronoi or Perpendicular Bisection (PEBI) Grids

Voronoi grid, named after Voronoi,[37] is very flexible and locally orthogonal. It is also called perpendicular bisection (PEBI) grid in the petroleum industry. A Voronoi gridblock is defined as the region of space that is closer to its gridpoint than to any other gridpoint.[38] Voronoi grids are based on a generalization of the point-distributed grid system (Figure 23.2). Voronoi gridblocks are locally orthogonal because a line that joins gridpoints of any two connected gridblocks is perpendicular to the gridblock boundary between these two gridpoints and is divided into two equal parts by the boundary.[35] Voronoi grids are very flexible for representing complex well geometries, such as horizontal wells, and major geological features, such as faults, permeability barriers, etc. Examples of Voronoi grids are shown in Figure 23.9. The flexibility of Voronoi grid is illustrated in Figure 23.10. Wide applications of Voronoi grid in the petroleum industry are still limited primarily because many engineers are more familiar with using Cartesian-based grid systems, especially for analyses, visualization, and reporting of simulation results.

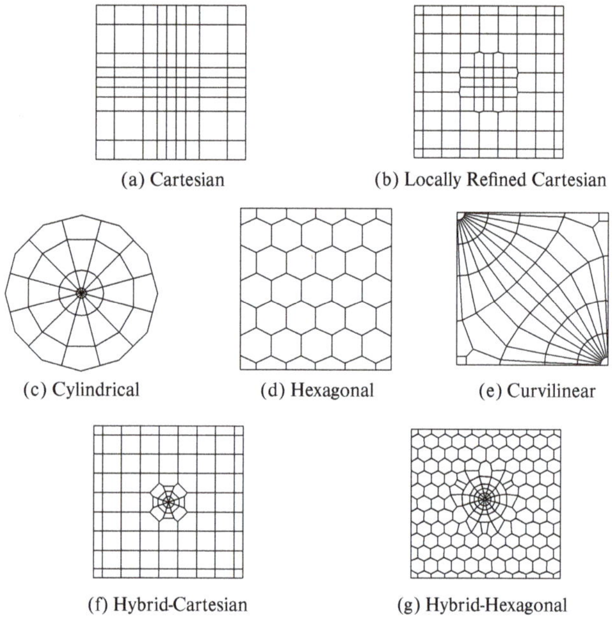

▲ FIGURE 23.9 **Examples of Voronoi grids (from Aziz[35] © 1993 SPE, Reproduced with permission)**

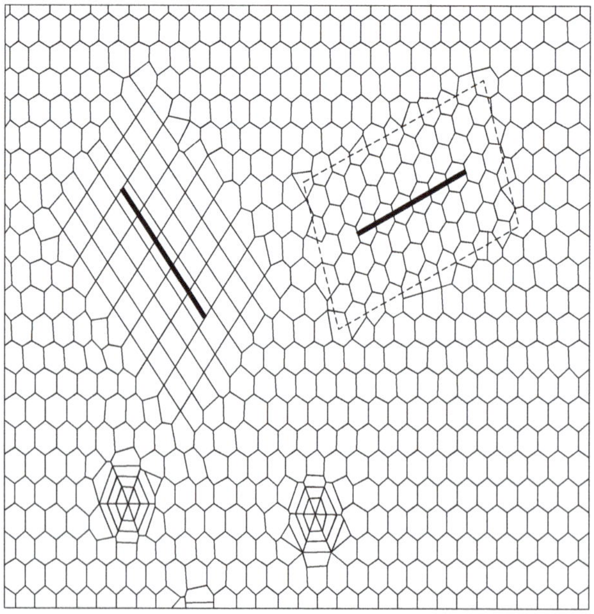

▲ FIGURE 23.10 **Flexibility of Voronoi grids (from Aziz[35] © 1993 SPE, Reproduced with permission)**

23.4.1.5 Local Grid Refinement (LGR)

Local grid refinement (LGR) is a technique used to increase the grid density in regions of the reservoir where rapid changes in saturations and pressures are occurring so as to represent those changes more accurately. LGRs are used mainly around well locations or to monitor saturation changes at fluid boundaries. An example of LGR around well locations is shown in Figure 23.11 for Cartesian grids. LGR should be used with caution due to two main potentially adverse consequences from indiscriminate application of the technique. The first adverse result of LGR is fluid flow between a large gridblock and a small grid block in an LGR area, which could cause large oscillations in computed results. The second adverse effect is a significant increase in model run times, if many LGRs are present in a flow simulation model. A reverse form of LGR is **Grid Coarsening**. Grid coarsening involves reduction of the number of grid blocks in portions of the reservoir that do not have rapid changes in saturation or pressures, such as in aquifers surrounding the hydrocarbon portions of the reservoir (Figure 23.12). Grid coarsening is beneficial in terms of reducing computation and storage memory required for a given flow simulation model.

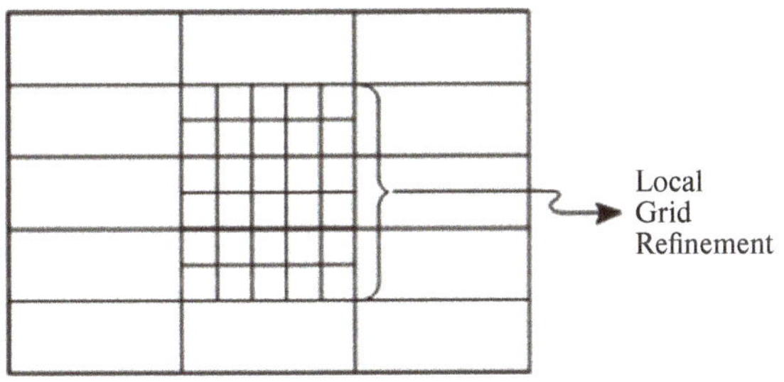

▲ FIGURE 23.11 Local grid refinement

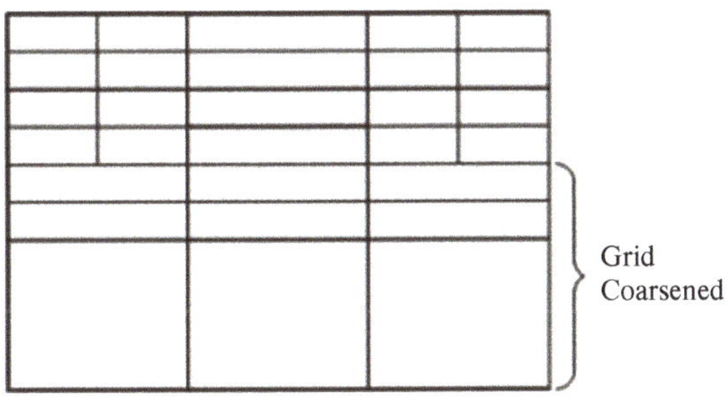

▲ FIGURE 23.12 Grid coarsening

23.4.1.6 Grid Orientation Effects

The orientation of Cartesian grids can have strong effects on the shapes of the displacement front and fluid breakthrough times. These effects are known as grid orientation effects. Strong grid orientation effects were reported by Todd et al.[39] for highly adverse mobility waterfloods and by Coats et al.[40] for pattern steamfloods. The impact of grid orientation effects can be reduced by using nine-point finite differencing[41] as shown in Figure 23.13 instead of five-point finite differencing shown in Figure 23.14. In nine-point finite difference formulation, flow is allowed between a gridblock and all its eight neighbor gridblocks including gridblocks diagonally adjacent to it (Figure 23.13). Five-point finite difference formulation allows flow between the gridblock and its four adjacent gridblocks (Figure 23.14). Note that nine-point formulation requires more computation time than five-point formulation. Most commercial simulators provide the option of selecting either five-point or nine-point finite differencing schemes. The recommendation is to use nine-point finite difference formulation, if the process being simulated involves adverse mobility ratios, such as in waterfloods, gasfloods, or steamfloods.

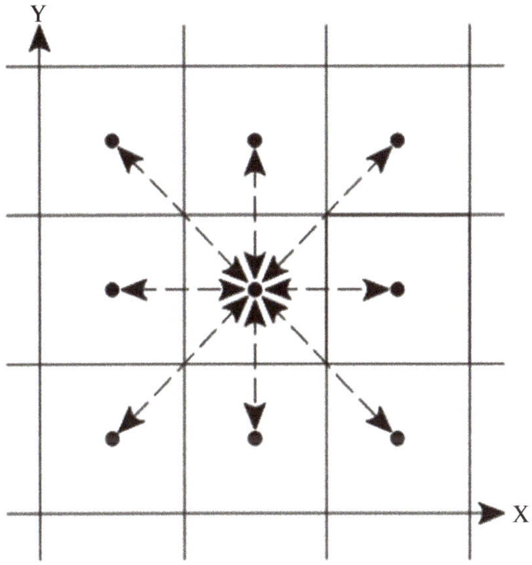

▲ FIGURE 23.13 Nine-point differencing

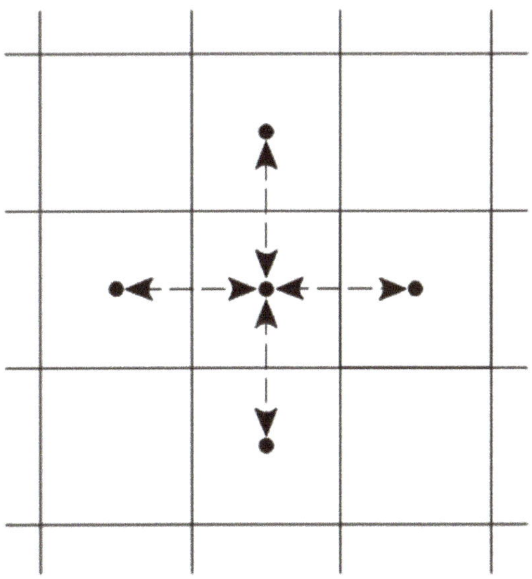

▲ FIGURE 23.14 Five-point differencing

23.4.2 Timesteps

The difference equations resulting from the discretization in time of the flow equations in reservoir simulators are solved in a number of timesteps. Timesteps are progressive increments of time used to advance simulation of the process from initial time to the stated end time. As the solution progresses in time, the selection of timestep sizes plays a very important role in ensuring that solutions of the difference equations are accurate, efficient, and stable. Most commercial simulators feature sophisticated automatic time step control that is based on pre-determined set of default parameters and tolerances. For many simulator applications, these automatic time step selection criteria are adequate, and no further adjustments are needed. However, for specialized simulator applications, the user may prefer to use manual timestep control. The procedure for implementing manual timestep control is typically provided in the technical documentation of the commercial simulator.

Simulator timestep sizes are influenced by simulator startup, timing of input and output of data, and large changes in reservoir pressures, production rates, and injection rates. At startup, the timestep size for most commercial simulators is automatically set at about 1 day or less depending on the application. As the simulation progresses and the solution stabilizes, large time steps are taken, depending on other controls applied on the simulator. One of such controls applied on the simulator is data input and output. Most simulators adjust their timestep sizes as necessary to precisely comply with data input and/or output specified by time (or date). Frequent data input or output could slow down the simulator and increase run times, thereby increasing computation costs. Large changes in pressures, production rates, and injection rates may also cause reduction in timestep sizes. One technique that could be used to avoid sudden large changes in reservoir conditions (pressures, saturations, etc.) is to incorporate additional times (or dates) for data input so as to "smooth-out" or reduce the impact of the large changes (such as production rates, injection rates, etc.) on the progress of the simulation.

Except for experienced users of simulators, it is recommended that automatic time step selection should be used as provided in commercial simulators. In general, automatic time step control in commercial simulator will yield reduced run times and computation costs.

23.4.3 Formulations of Simulator Equations

Three main methods are used in the formulation of simulator equations for solution. These are the **IMPES** (Implicit Pressure, Explicit Saturation) method, the **Sequential** method, and the **Implicit** method. The IMPES method was first applied to black oil systems by Sheldon et al.,[13] Stone and Garder,[14] and Fagin and Stewart.[15] For the IMPES formulation, the flow equations are combined to eliminate unknown saturations, thereby generating a single pressure equation. The pressure equation is solved implicitly for pressures at each gridblock at the current timestep designated as $n+1$, using parameters at the previous (or old) timestep designated as n. This is then followed by the explicit substitution of the pressures into the corresponding flow equations to calculate saturations at the current timestep, $(n+1)$, for each gridblock.[42] The advantages of the IMPES method are that it requires less computing time per timestep and less computer memory for storage. The main disadvantage of the IMPES method is that it is relatively unstable at large timesteps. There are techniques for improving the stability of the IMPES method as reported by Coats.[43,44] To improve the stability of IMPES formulations, MacDonald and Coats[45] introduced a modification of the IMPES formulation which was later named the **Sequential** method by Spillette et al.[46] The Sequential method is a two-step solution process. In the first step, the pressure equation is solved implicitly for pressures at each gridblock as was done in the IMPES method. This is then followed by a second step in which the saturation equations derived with the aid of fractional flow equations are also solved implicitly. However, the Sequential method may have material balance problems in areas of the gridblocks with large saturation (or composition) changes. The formulation method that is widely used in many simulators (black oil, compositional, thermal, chemical, etc.) is the **Implicit** formulation. In the Implicit method, the flow equations are formulated to be solved simultaneously at the current timestep, $n+1$. A fully implicit formulation was first published by Blair and Weinaug[47] in 1969. This was followed by authors that used Implicit formulations in black oil,[48] compositional,[32] and thermal[34,49] simulators. The Implicit formulation requires more computer memory for storage and higher computation costs than either the IMPES or the Sequential methods. However, the Implicit formulation is considerably more stable and can take larger timestep than either IMPES or Sequential methods, especially for flow problems with large saturation (or composition) changes in the gridblocks. It is noteworthy to discuss a hybrid solution method that cleverly combines the strengths of the IMPES and Implicit methods in one solution method. This hybrid method is called the **Adaptive Implicit Method** (AIM). The AIM approach uses Implicit method in gridblocks or groups of gridblocks with rapid pressure and saturation changes, while the IMPES method is used in the rest of the gridblocks with slower (less rapid) changes in pressures and saturations.[50,51] The switch between IMPES and Implicit methods under AIM is automatic, and is applied on gridblocks or groups of gridblocks when needed and as long as required to achieve optimum balance on stability and cost of the computation process. However, with the advancement of computer hardware especially in terms of computation speed and cheap memory for storage, most commercial simulators use Implicit formulation as the preferred formulation for solution of most flow problems. Consequently, experienced users are required by most commercial simulators to specifically select the other solution methods (IMPES, SEQUENTIAL, AIM), if these methods are better suited for the solution of the flow problems.

23.4.4 Material Balance Errors and Other Convergence Criteria

One criterion used in all simulators to check the accuracy and stability of the solution at each timestep is the material balance error (MBE). A low value for material balance error by itself is not sufficient measure of the accuracy of the solution. Material balance error is used, in conjunction with other criteria, to determine convergence of the solution at each timestep. Commercial simulators use different methods to calculate and normalize material balance errors. Irrespective of the method used, the criterion applied in most simulators is that the material balance error for each gridblock at each timestep should be close to zero. One method for calculating material balance error is the summation of mass accumulation for each phase at the beginning and at the end of the timestep for all gridblocks. The net mass accumulation for each phase should be equal to net influx through the wells during the timestep. For a black oil simulator, the material balance error is calculated as follows:

For the oil phase:

$$\text{MBE}_o = \frac{\sum_{j=1}^{m}\left[PV\left(\frac{S_o}{B_o}\right)\right]_j^n - \sum_{j=1}^{m}\left[PV\left(\frac{S_o}{B_o}\right)\right]_j^{n+1}}{q_o \Delta t} - 1 \tag{23.29}$$

For the water phase:

$$\text{MBE}_w = \frac{\sum_{j=1}^{m}\left[PV\left(\frac{S_w}{B_w}\right)\right]_j^n - \sum_{j=1}^{m}\left[PV\left(\frac{S_w}{B_w}\right)\right]_j^{n+1}}{q_w \Delta t} - 1 \tag{23.30}$$

For the gas phase:

$$\text{MBE}_g = \frac{\sum_{j=1}^{m}\left[PV\left(\frac{S_g}{B_g}\right)\right]_j^n - \sum_{j=1}^{m}\left[PV\left(\frac{S_g}{B_g}\right)\right]_j^{n+1}}{q_g \Delta t} - 1 \tag{23.31}$$

In Eqs. (23.29) to (23.31), PV = pore volume of the gridblock, j; S_o, S_w, and S_g, are saturations of oil, water, and gas, respectively, in gridblock, j; B_o, B_w, and B_g are formation volume factors for oil, water, and gas, respectively, in gridblock, j; q_o, q_w, and q_g are net influx through wells for oil, water, and gas, respectively, in gridblock, j; Δt = length of timestep; m = total number of gridblocks in reservoir model; n = previous timestep; and $n+1$ = current timestep. Many commercial simulators consider material balance errors to be sufficiently small for all the phases, if less than 1×10^{-7}.

Other measures that are used to check stability and accuracy of the solutions are changes in saturations of the phases in iteration at each timestep. The iteration is considered to have converged, if maximum changes in the saturation of each phase for each gridblock is less than a preset maximum saturation change for the timestep. Many commercial simulators set the largest convergence error for any phase in any grid block at less than 0.001. If the conditions for the material balance error for each phase and the convergence error for phase saturations are not met for the timestep size, the calculations and iterations are repeated with a reduction in the size of the timestep until convergence is achieved. The amount of reduction of timestep size applied automatically due to convergence failure varies among commercial simulators. Some commercial simulators automatically reduce the timestep size by half after a convergence failure. This can drastically slow down the speed of the simulation run. Note that the amount of cut applied to timestep size due to convergence failure can be specified by the user in most commercial simulators, in addition to changes on the convergence criteria. However, these options should be used as necessary with considerable caution and only by very experienced users of simulators because they can lead to simulation results that are unstable and erroneous.

23.4.5 Numerical Dispersion

Numerical dispersion is a term generally used to describe the false appearance of smeared spatial gradients of saturation or concentration arising from time and space discretization of the flow equations in simulators based on finite differences.[52] In simpler terms, numerical dispersion describes the smearing of saturation or concentration of the displacing phase ahead of the actual front. Numerical dispersion generally increases with increases in the areal sizes of the model gridlocks, (Δx, Δy), and increases in timestep sizes, Δt. Consequently, for the same areal grid sizes, numerical dispersion is likely to be larger with Implicit solution methods that use larger timestep sizes than IMPES solution methods that use smaller timestep sizes. For both IMPES and Implicit solution techniques, numerical dispersion can be reduced by increasing the number of areal gridblocks. This approach is expected to create larger simulation models. However, if accurate prediction of water, steam, or solvent breakthrough is an objective for simulation of the displacement process, then it is necessary to use as many areal gridblocks as necessary in order to minimize the effects of numerical dispersion. Other methods for reducing numerical dispersion include use of pseudofunctions as demonstrated by Kyte and Berry,[53] and mobility weighting techniques such as the two-point, upstream mobility weighting scheme proposed by Todd et al.[39]

23.4.6 Well Model

A well model is used in simulation to represent the flow between the wellbore and the reservoir gridblock in which the well is located. Assuming the well is producing from a single gridblock, the flow rate of a phase, p(oil, water, or gas), into the well is given by:

$$q_{p,j} = \frac{WI_j}{\mu_p}\left(p_j - p_{wf}\right) \tag{23.32}$$

In Eq. (23.32), $q_{p,j}$ = the flow rate of phase, p, in gridblock, j; WI_j = well index; μ_p = viscosity of phase, p; p_j = gridblock pressure; and p_{wf} = flowing bottomhole pressure of the well. For the sign convention, the flow is considered positive from the gridblock into the well, and negative from the well into the gridblock. For Cartesian grids, the well index, WI_j (also called connection transmissibility factor), was derived by Peaceman[54] as:

$$WI_j = \frac{2\pi\sqrt{k_x k_y}\,h}{\ln\dfrac{r_o}{r_w} + s} \tag{23.33}$$

In Eq. (23.33), WI_j = well index for gridblock, j; k_x, k_y = permeabilities in the x and y directions, respectively; h = net thickness of the gridblock; r_o = equivalent radius of the gridblock; r_w = wellbore radius; and s = skin factor. The equivalent radius of the gridblock, r_o, is defined as the distance from the well at which the flowing wellbore pressure is equal the average pressure of the gridblock.[55,56] According to Peaceman,[56] equivalent wellbore radius, r_o, for an anisotropic reservoir in Cartesian grids is given by:

$$r_o = 0.28\frac{\sqrt{\left(k_y/k_x\right)^{1/2}\Delta x^2 + \left(k_x/k_y\right)^{1/2}\Delta y^2}}{\left(k_y/k_x\right)^{1/4} + \left(k_x/k_y\right)^{1/4}} \tag{23.34}$$

In Eq. (23.34), r_o = equivalent wellbore radius; k_x, k_y = permeabilities in the x and y directions, respectively; Δx, Δy = x and y dimensions of the gridblock, respectively. For an isotropic reservoir, the equivalent wellbore radius, r_o, in Cartesian grids is given by:

$$r_o = 0.14\sqrt{\left(\Delta x^2 + \Delta y^2\right)} \tag{23.35}$$

It is important to note that the well index, WI_j, for a gridblock is not the same as the productivity (or injectivity) index of the well. The productivity index, J, of a well is defined as:

$$J = \frac{q_p}{\bar{p} - p_{wf}} \tag{23.36}$$

In Eq. (23.36), q_p = production rate of phase, p(oil, water, or gas); $\bar{p}$ = average reservoir pressure or pressure at the drainage radius of the well; and p_{wf} = flowing bottomhole pressure of the well. Assuming steady-state, radial Darcy flow within the drainage radius of the well, a relationship between the well index and the productivity index of the well is expressed as:

$$J = \sum_j\left[\frac{WI_j}{\mu_p}\left\{\frac{\ln\left(r_o/r_w\right) + s}{\ln\left(r_d/r_w\right) + s}\right\}\right] \tag{23.37}$$

In Eq. (23.37), WI_j = well index for gridblock, j; μ_p = viscosity of phase, p; r_o = equivalent wellbore radius; r_w = wellbore radius; s = skin factor; and r_d = drainage radius of the well. The summation is over all the gridblocks connected to the well. If the drainage radius of the well is assumed to be equal to equivalent wellbore radius, Eq. (23.37) reduces to:

$$J = \sum_j\left[\frac{WI_j}{\mu_p}\right] \tag{23.38}$$

Note that all the equations developed in this section for well models apply only to vertical wells located in rectangular gridblocks in Cartesian grids. They are not applicable to wells located in radial gridblocks or advanced wells. Advanced wells are described as wells with complex trajectories such as horizontal wells and multilateral wells, and wells with downhole control devices (sensors, flow control valves, etc.) called "smart" wells.[57] Peaceman[58] proposed a method for representing a horizontal well in reservoir models with Cartesian grids. The modeling of vertical and horizontal wells in unstructured grids was presented by Palagi and Aziz.[59] A technique for modeling an advanced well that is available in most commercial simulators consists of building a well model by subdividing the wellbore into segments as described by Stone et al.[60] and Holmes et al.[61] Detailed presentations of these techniques for modeling advanced wells in reservoir simulators are not covered in this book. However, since many reservoirs are produced with horizontal wells, multilateral wells, or "smart" wells, it is important to note that these wells require advanced well models for accurate representation of their productivity.

EXAMPLE 23.1 Calculation of Well Index

Problem

Calculate the well index for a vertical well located completely in a single rectangular gridblock in an areal Cartesian grid system. The dimensions and properties of the gridblock are given as follows:

Gridblock dimension, Δx	= 150 ft
Gridblock dimension, Δy	= 100 ft
Gridblock dimension, Δz	= 5 ft
Gridblock permeability, k_x	= 350 md
Gridblock permeability, k_y	= 225 md
Wellbore radius, r_w	= 0.5 ft
Skin factor, s	= 2.3

Assume the net thickness of gridblock, h, is equal to gridblock dimension, Δz.

Solution

Step 1. Calculate equivalent radius of the gridblock, r_o.

Using Eq. (23.34),

$$r_o = 0.28\frac{\sqrt{\left(k_y/k_x\right)^{1/2}\Delta x^2 + \left(k_x/k_y\right)^{1/2}\Delta y^2}}{\left(k_y/k_x\right)^{1/4} + \left(k_x/k_y\right)^{1/4}}$$

$$= 0.28\frac{\sqrt{(225/350)^{1/2}(150)^2 + (350/225)^{1/2}(100)^2}}{(225/350)^{1/4} + (350/225)^{1/4}}$$

$$= 24.3065 \text{ ft}$$

Step 2. Calculate well index for gridblock.

Using Eq. (23.33),

$$WI_j = \frac{2\pi\sqrt{k_x k_y}\,h}{\ln\frac{r_o}{r_w} + s}$$

$$= \frac{2\pi\sqrt{350\times 225}\times 5}{\ln\frac{24.3065}{0.5} + 2.3}$$

$$= 1425.6513 \text{ md-ft}$$

End of Example 23.1

23.4.7 Model Initialization

The initial conditions in the reservoir based on the input data are established during model initialization. These initial conditions are distributions of phase pressures and saturations for each gridblock in the reservoir model. Generally, two methods are used for model initialization. These are simulator generated initial conditions or user specified initial conditions. Under simulator generated initial conditions, the simulator uses the input data to generate equilibrated distributions of phase pressures and saturations at initial conditions. The input data consist of specification of pressure at a given datum depth, depth of fluid contacts (water-oil, gas-oil, or gas-water contacts), capillary pressure at the contacts, fluid densities at datum, and saturation endpoints and capillary pressure data in the relative permeability tables. For the user specified initial conditions, the distributions of pressure and saturations at initial conditions for each gridblock are provided and set by the user. Note that user specified initial conditions may cause a non-equilibrated model at initial conditions. The user should verify that these conditions actually exist in the reservoir at initial conditions as a non-equilibrated model may lead to erroneous results.

After initializing the model, one of the key activities at this stage of model building is to check the quality of the model. A very important measure of model quality is to compare the in-place fluid volumes calculated by the simulator against volumes calculated volumetrically from the geologic models. The fluid-in-place volumes from the two methods should be in close agreement. A good rule-of-thumb is that the deviation between the two methods should be less than 2 percent.

The initialized model is typically the model which is used for simulation of the reservoir process. At this stage it should be reviewed completely before proceeding with the simulation. All the input data should be compared with model data for consistency and reliability. These include checking the model pressure and saturation data, gridblock properties (porosity, permeability, net thickness, etc.), well completion data, production, and injection data, if applicable. The objective of thoroughly checking the model at this stage is to ensure that the initialized model is an accurate representation of the reservoir model before proceeding with the time consuming process of history matching and predictions.

23.4.8 History Matching

History matching is the process of adjusting the properties and parameters of the reservoir model to match past actual performance data measured on the reservoir. The performance data may include pressure data (such as static bottomhole pressures, flowing bottomhole pressures, tubing head pressures, average reservoir pressures, etc.), production and injection data (such as oil, water, and gas production rates and water and gas injection rates), well completion data (such as perforated and isolated intervals), production and injection profile data (from production logs and injection logs), etc. A large variety of performance data can be used in the history matching process. It depends on reservoir type, and methods being used to deplete or improve production from the reservoir. The history matching process is a classic example of an inverse problem which leads to a non-unique solution or response. This means that different history matched models of a reservoir can be achieved by applying the same input data. The history matched models are non-unique solutions or responses to the applied performance data and may yield different predictions of the future performance of the reservoir.

The process of history matching a reservoir model varies from one practitioner to the other. It depends largely on experience and knowledge of reservoir engineering principles. The history matching process also depends considerably on the reservoir model type, the quality and quantity of available performance data, and the objectives of the study. For instance, if the reservoir model was constructed from sparse data, and limited performance data are available, it is reasonable to expect that the history matched model is likely to be limited in its application on prediction of the future performance of the reservoir.

Generally, the history matching process can be divided into two phases, namely the pressure match phase and the saturation match phase. The pressure match phase normally precedes the saturation match phase. During the pressure match phase, the history matching process is focused on matching the net voidage (in reservoir barrels) from the reservoir with the aim of matching pressure distribution in the reservoir over the specified time period. In some cases, only a match of the trend in pressure distribution can be achieved during this phase. The goal during the pressure match phase is to obtain a reservoir model that has approximately the appropriate net volume of fluids in place over the duration of its known history. During the pressure match phase, minimal effort is applied in the validation of the distribution of phases or saturations in the model. At the end of the pressure match phase, the history matching process proceeds to the saturation match phase. During the saturation match phase, attention is then directed toward specifying the actual phases of fluids produced and/or injected into the reservoir. For instance, in history matching black oil reservoir models, oil production rates are typically specified, and the model yields the corresponding water and gas production rates. The water and gas production rates generated from the model are then compared with the actual water and gas production rates. Also, pressures and other performance data generated from the model are compared with actual data. Different forms of the actual data may be used for this comparison during the history matching process. Some practitioners use production rates, cumulative production volumes, producing gas-oil ratios and water-oil ratios, flowing bottomhole pressures, tubing head pressures, average reservoir pressures, etc. The matching of data predicted by the model to actual data can be performed at the level of the field, well groups, or individual well. Again, the choice of actual data to match depends on the quality and quantity of the data, and the objectives of the simulation study. At the end, the saturation match phase generally produces a history matched model with appropriate distribution of pressures and saturations in the reservoir. Generally, the history matching process is terminated at the point where the team engaged in the effort decides that an acceptable history matched model has been obtained. History matched models are ruled to be acceptable, if they are adequate to meet the objectives of the simulation study, and further adjustments to the models do not result in significant changes on the history matched variables.

The following best practices are recommended to enhance the efficiency and effectiveness of the history matching process:

1. Make a log of all activities and changes made on the model during the history matching process. This log should be maintained preferably within the model. If the history match activities are conducted by a team, all members of the team should adhere to these guidelines to ensure that all changes made on the model are documented in this historical record. No changes made on the model should be left out of this record.
2. The changes made on the model during the history match process should be graduated and systematic. Avoid making changes in a haphazard manner. Clear and defined results expected from the changes must be understood before the adjustments to the model are made at each stage.
3. The results from previous changes or adjustments to the model should be analyzed and understood before proceeding with the next set of changes or adjustments. If a team is working cooperatively on the project, it is advantageous for members of the team to reach consensus on the next set of changes or adjustments to be made.
4. The least reliable input data should be modified first during the history match process. For instance, if permeability data are the least reliable data, then permeability data should be changed systematically at the inception of the history match process. If further changes to permeability data lead to unreasonable results, then the next set of input data considered to be unreliable should be changed in a systematic manner.
5. Recognition of the point in the history matching process when further changes on the input data cause insignificant improvement on the state of the history matched model is important. This is the point of diminishing return and should be recognized as the end of the history matching process to save money and resources.

23.4.9 Predictions

The future performance of the reservoir is predicted with the history matched models or with the initialized model in the case that the reservoir has no historical production. For a history matched model, it is usually necessary to adjust individual well productivity index so that a smooth transition is achieved from the history matching phase to the prediction phase. The process of tuning the well productivity indices to achieve the desired smooth transition may require an iterative process. For new reservoirs with no production history, initial well production rates may be adjusted with productivity index from drill stem tests, extended well tests, or production data from analogue reservoirs. The predicted performance of new reservoirs is subject to a much higher level of uncertainty due to lack of historical performance data from the reservoir to guide or condition predicted future performance as is the case with history matched models.[62]

The most effective application of reservoir models is to compare alternative management strategies or to rank competing management strategies. This is actually the understated power of reservoir simulation. The actual reservoir can be produced one time at considerable cost over an extended period of time in many cases, whereas a reservoir model can be "produced" (or run) as many times as needed at a tiny fraction of the cost.[7] With the aid of reservoir models, it is possible to investigate the potential outcome of many reservoir management strategies without conducting actual field or pilot tests. It is reasonable to assert that learned application of reservoir simulation technology should lead to better reservoir management practices when compared with cases where the technology was not applied.

For a typical comparison of alternative reservoir management strategies, it is useful to create a base case that is focused on current operating strategy. For instance, the base case may consist of current operations based on depletion drive mechanism. This can then be compared with other alternative strategies based on fluid injection. The injected fluid could be gas, and/or water. The important point is that predicted performances of the model under various alternative management strategies can be compared with the base case leading to the selection of an optimum management strategy.[63] Note that the use of reservoir models to rank competing strategies were presented in Chapter 19 for secondary recovery methods, and Chapter 21 for enhanced oil recovery methods.

23.4.10 Uncertainty Analysis

The presence of uncertainties in the geologic and petrophysical data used in constructing the geologic model was discussed in Chapter 22, Section 22.5.6. Uncertainty analysis of geologic models was performed by generation of multiple realizations and application of statistical analysis on the realizations to create a cumulative distribution curve. It was suggested then that the cumulative probability curve can be used to select models at the P10, P50, and P90 quantiles for use in reservoir flow simulations. The decision to select geologic models for flow simulation only at the P10, P50, and P90 quantiles is based largely on limitation of resources which can be devoted to the project. In other words, the team may choose to use additional geologic models including the P10, P50, and P90 models, if all the models can be processed at the same time with available resources.

The level of uncertainties present in the geologic models is increased by introducing additional variable parameters during the process of transforming the static geologic model into a reservoir simulation model. These additional variable parameters include fluid properties (oil viscosity, oil gravity, saturation pressures, solution gas-oil ratios, etc.), rock/fluid properties (relative permeability data, rock compressibility data, etc.), fluid contacts (oil-water contact, gas-oil contact, gas-water contact, etc.), well locations, and performance data (production rates, pressures, completion data, etc.). These uncertainties present in the reservoir model from geological data, petrophysical data, fluid properties data, rock/fluid properties data, etc., compound the problem of uncertainty analysis in reservoir simulation models.

One of the processes widely used in the petroleum industry to systematically evaluate the impact of uncertainty on flow simulation outcomes is called **Experimental Design** (ED). Experimental Design is also called **Design of Experiments** (DOE). ED is a statistically based procedure that systematically evaluates the effects of these uncertain variables on dependent outcomes such as hydrocarbon recovery. Commercial software based on Plackett-Burman or Latin Hypercube methods for designing the experiments are readily available in the industry. They are copious documentations of these methods available in the technical manual provided by the vendors of the commercial software. Interested readers are encouraged to consult these technical documents for further details on these methods. It is important to note that the commercial ED software have been configured to facilitate input for most commercial flow simulators.

23.5 General Structure of Reservoir Flow Models

The structures of input data in most reservoir flow simulators are remarkably similar. The general structure of reservoir simulation models is described here to acquaint the reader with sequence of data input in most simulators. Obviously, the actual structure of input data will vary slightly between different simulators. But in many cases, the differences are minor and can be quickly reconciled between simulators from different sources. The purpose of this section is to familiarize the reader with a readily available data structure that can be used to transfer data from one simulator to the other.

23.5.1 Definition of Model and Simulator

The input of data into every simulator begins with definition of the size of the reservoir model and the type of simulator to be used for modeling the reservoir. The key input that defines the size of the reservoir model is the number of gridblocks in the model. For instance, in the Cartesian system, the number of gridblocks in the x, y, and z directions for a 3D model is specified. Also defined in this data input section is the number of wells in the model, the number of tabular data (PVT, relative permeability, equilibration regions, etc.), and the number of initialization regions, etc. The type of simulator to be used including the formulation (solution) type is specified in this section. For instance, a black oil model based on the Implicit formulation may be selected. Most important, the date for start of simulation is specified. This section may be considered as the section in which the scope of the simulation problem is specified for the simulator. In many ways, this section defines the amount of computer memory that will be required to run the reservoir model.

23.5.2 Geologic Model Data

All the structural and petrophysical data in the geological model are typically assembled as data input for the gridblocks in a section of the simulator. The structure of the geologic model is represented by geometrical data on the gridblocks in terms of location and dimensions. This is usually accompanied with separate specifications of the petrophysical data for each gridblock. The petrophysical data usually specified for each gridblock include porosity, permeability, and net sand thickness or net-to-gross ratio data. Initial fluid saturations for each gridblock may also be specified in some models. These data are then followed with modifications to the grid system (such as local grid refinement), and modifications to the petrophysical data specified for the gridblocks.

23.5.3 Fluid Properties Data

This section of the model data set contains data that represent the PVT properties of the fluids present in the reservoir. The PVT data are usually presented in a tabular form for black oil models. For compositional simulators, the PVT data are represented in a compatible form as output generated with an equation of state.

23.5.4 Rock/Fluid Properties Data

Rock/fluid properties data in the form of relative permeability data and capillary pressure data are represented in the model as functions of fluid saturations. These data are usually presented in the simulator in a tabular form. Note that the data in these tables are sometimes used by the simulator to establish initial conditions in the reservoir model, if the option for simulator generated initial conditions is selected.

23.5.5 Model Equilibration Data

Model equilibration data include fluid contact depths (oil-water contact, gas-oil contact, or gas-water contact), capillary pressures at the fluid contacts, and reservoir pressure at a selected datum depth. The model equilibration data are in some cases used by the simulator to establish initial reservoir conditions.

23.5.6 Well Data

In the section of model well data, the locations of the wells in the grid system are specified. Also, the gridblocks in which the wells are completed are specified. The production or injection rates of the wells including the type of fluid produced or injected are specified. The progression of the simulation in terms of time is defined in this section in the form of time steps, cumulative time, or dates. These time-based data are very important because the speed and duration of the simulation are controlled by these data. For reservoirs with production history, the production data are provided at specific time intervals which could be daily, monthly, quarterly, semi-annually, or annually. The frequency of production data entry is totally at the discretion of the user. However, note that higher frequency of production data specifications reduces the speed of the simulator runs during history match. Additional data that may be specified at progressive time periods include introduction of new wells, recompletion of existing wells, and changes to well fluid production or injection rates.

23.5.7 Simulator Data Output

All simulators are capable of generating large amounts of output data, especially for large models. This should be controlled by the user to avoid being overwhelmed by the simulator generated data input and output.[64] In most cases, the amount of simulator data output can be controlled by the user. The minimum output that can be specified by the user may include well performance data and distributions of pressures and saturations over time. Most simulators are equipped with powerful graphical software that can assist in the evaluation and presentation of the simulator output from the model.

Nomenclature

B_g	gas formation volume factor
B_o	oil formation volume factor
B_w	water formation volume factor
C_g	concentration of gas, mass per unit volume
C_o	concentration of oil, mass per unit volume
C_w	concentration of water, mass per unit volume
F	mass flow rate per unit cross-sectional area normal to the direction of flow
$\vec{F}$	mass flow rate per unit cross-sectional area in the x, y, and z directions
h	net thickness of gridblock, j
J	productivity index
k_{rp}	relative permeability of phase, p
k_x	permeability in x-direction
k_y	permeability in y-direction
k_z	permeability in z-direction
$\vec{K}$	permeability tensor
n	timestep
p_j	pressure in gridblock, j
p_{wf}	flowing bottomhole pressure
$\bar{p}$	average reservoir pressure
q	mass flow rate per unit volume
r_d	well drainage radius
r_o	equivalent radius of gridblock
r_w	wellbore radius
R_{so}	gas solubility in oil
R_{sw}	gas solubility in water
s	skin factor
S_g	gas saturation
S_o	oil saturation
S_w	water saturation
$\vec{v}_g$	gas velocity in x, y, and z directions
$\vec{v}_o$	oil velocity in x, y, and z directions
$\vec{v}_w$	water velocity in x, y, and z directions
Δx	grid length in the x-direction
Δy	grid width in the y-direction
Δz	grid height in the z-direction
Δt	time interval
λ_p	mobility of phase, p
μ_p	fluid viscosity of phase, p
ρ_{gsc}	gas density at standard conditions
ρ_{osc}	oil density at standard conditions
ρ_{wsc}	water density at standard conditions
ϕ	porosity
Φ_p	potential of phase, p

Subscripts

g	gas
o	oil
sc	standard conditions
t	time
w	water
x	x-direction
y	y-direction
z	z-direction

Abbreviations

AIM	Adaptive Implicit Method
DOE	Design of Experiments
ED	Experimental Design
IMPES	Implicit Pressure Explicit Saturation
MBE	Material Balance Error
PV	Pore Volume
PVT	Pressure Volume Temperature
WI	Well Index

REFERENCES

1. Aziz, K., and Settari, A.: *Petroleum Reservoir Simulation*, Applied Science Publishers Ltd., London (1979).
2. Mattax, C.C., and Dalton, R.L.: Reservoir Simulation, SPE Monograph Volume 13, Society of Petroleum Engineers, Richardson, Texas (1990).
3. Erketkin, T., Abou-Kassem, J.H., and King, G.R.: "Basic Applied Reservoir Simulation," SPE Textbook Series, Vol. 7, Society of Petroleum Engineers, Richardson, Texas (2001).
4. Coats, K.H.: "Reservoir Simulation: State of the Art," paper SPE 10020, Distinguished Author Series (August 1982) 1633-1642.
5. *Numerical Simulation II,* SPE Reprint Series No. 20, Society of Petroleum Engineers, Richardson, Texas (1986).
6. Dogru, A.H., Fung, L.S.K., Middya, U., Al-Shaalan, T.M., Pita, J.A., Kumar, K.H., Su, H.J., Tan, J.C.T., Hoy, H., Dreiman, W.T., Hahn, W.A., Al--Harbi, R., Al--Youbi, A., Al--Zamel, N.M., Mezghani, M., and Al--Mani, T.: "A Next-Generation Parallel Reservoir Simulator for Giant Reservoirs," paper SPE 119272 presented at the SPE Reservoir Simulation Symposium, The Woodlands, Texas, (February 2-4, 2009).
7. Coats, K.H.: "Use and Misuse of Reservoir Simulation Models," *JPT* (November 1969) 1391-1398.
8. Muskat, M.: *Physical Principles of Oil Production,* McGraw-Hill, New York (1949).
9. Douglas, J., Jr., Peaceman, D.W., and Rachford, H.H., Jr.: "A Method for Calculating Multi-Dimensional Immiscible Displacement," *Trans.* AIME (1959) 216, 297-308.
10. Coats, K.H., Nielsen, R.L., Terhune, M.H., and Weber, A.G.: "Simulation of Three-Dimensional, Two-Phase Flow in Oil and Gas Reservoirs," *SPEJ* (Dec. 1967) 377-388.
11. Coats, K.H.: "An Analysis for Simulating Reservoir Performance Under Pressure Maintenance by Gas and/or Water Injection," *SPEJ* (Dec. 1968) 331-340.
12. Coats, K.H.: "A Mathematical Model for Simulating Three-Dimensional Three-Phase Fluid Flow in Reservoirs," The University of Texas, Austin, Texas, (November 1968).
13. Sheldon, J.W., Harris, C.D., and Bavly, D.: "A Method for General Reservoir Behavior Simulation on Digital Computers," paper 1521-G presented at SPE 35th Annual Fall Meeting, Denver, Colorado, (October 2-5, 1960).
14. Stone, H.L., and Garder, A.O., Jr.: "Analysis of Gas Cap or Dissolved-Gas Drive Reservoirs," *SPEJ* (June 1961) 92-104.
15. Fagin, R.G., and Stewart, C.H., Jr.: A New Approach to the Two-Dimensional Multiphase Reservoir Simulator," *SPEJ* (June 1966) 175-182.
16. Odeh, A.S.: "Comparison of Solutions to a Three-Dimensional Black Oil Reservoir Simulation Problem," *JPT* (Jan. 1981) 13-25.
17. Chappelear, J.E., and Nolen, J.S.: "Second Comparative Solution Project: A Three-Phase Coning Study," *JPT* (March 1986) 345-353.
18. Kenyon, D.E., and Behie, A.: "Third SPE Comparative Solution Project: Gas Cycling of Retrograde Condensate Reservoirs," *JPT* (August 1987) 981-997.
19. Aziz, K., Ramesh, A.B., and Woo, P.T.: "Fourth SPE Comparative Solution Project: Comparison of Steam Injection Simulators," *JPT* (December 1987) 1576-1584.
20. Killough, J., and Kossack, C.: "Fifth SPE Comparative Solution Project: Evaluation of Miscible Flood Simulators," paper SPE 16000 presented at the Ninth SPE Symposium on Reservoir Simulation, San Antonio, Texas, (February 1-4, 1987).
21. Firoozabadi, A., and Thomas, L.K.: "Sixth SPE Comparative Solution Project: Dual-Porosity Simulators," *JPT* (June 1990) 710-715, 762-763.
22. Nghiem, L., Collins, D.A., and Sharma, R.: "Seventh SPE Comparative Solution Project: Modeling of Horizontal Wells in Reservoir Simulation," paper SPE 21221 presented at the 11th SPE Symposium on Reservoir Simulation, Anaheim, California, (February 17-20, 1991).
23. Quandalle, P.: "Eighth SPE Comparative Solution Project: Gridding Techniques in Reservoir Simulation," paper SPE 25263 presented at the 12th SPE Symposium on Reservoir Simulation, New Orleans, Louisiana, (February 28-March 3, 1993).
24. Killough, J.E.: "Ninth SPE Comparative Solution Project: A Reexamination of Black Oil Simulation," paper SPE 29110 presented at the 13th SPE Symposium on Reservoir Simulation, San Antonio, Texas, (February 12-15, 1995).
25. Christie, M.A., and Blunt, M.J.: "Tenth SPE Comparative Solution Project: A Comparison of Upscaling Techniques," *SPEREE* (August 2001) 308-317.
26. Odeh, A.S.: "Reservoir Simulation...What is it?" *JPT* (November 1969) 1383-1388.
27. Coats, K.H., Thomas, L.K., and Pierson, R.G.: "Compositional and Black Oil Reservoir Simulation," *SPEREE* (August 1998) 372-379.
28. Varavei, A., and Sepehrnoori, K.: "An EOS-Based Compositional Thermal Reservoir Simulator," paper SPE 119154 presented at the 2009 SPE Reservoir Simulation Symposium, The Woodlands, Texas, (February 2-4, 2009).
29. Mishev, I., Beckner, B., Fedorova, N., and Terekhov, S.: "Linear Solver Performance Optimization in Reservoir Simulation Studies, paper SPE 119266 presented at the 2009 SPE Reservoir Simulation Symposium, The Woodlands, Texas, (February 2-4, 2009).
30. Datta-Gupta, A.: "Streamline Simulation: A Technology Update," paper SPE 65604, Distinguished Author Series (December 2000) 68-74.
31. Fanchi, J.R.: *Principles of Applied Reservoir Simulation,* Gulf Publishing Co., Houston, Texas (1997).
32. Coats, K.H.: "An Equation of State Compositional Model," *SPEJ* (October 1980) 363-376.
33. Coats, K.H., George, W.D., Chu, Chieh, and Marcum, B.E.: "Three-Dimensional Simulation of Steamflooding," *SPEJ* (December 1974) 573-592.
34. Coats, K.H.: "A Highly Implicit Steamflood Model," *SPEJ* (October 1978) 369-383.

35. Aziz, K.: "Reservoir Simulation Grids: Opportunities and Problems," *JPT* (July 1993) 658-663.

36. Pedrosa, O.A., Jr., and Aziz, K.: "Use of Hybrid Grid in Reservoir Simulation," *SPERE* (November 1986) 611-621.

37. Voronoi, G.: "Nouvelles applications des parameters continues a la teorie des formes quadratiques," *J. Reine Angew. Math.* (1908) 134, 198-287.

38. Palagi, C.L., and Aziz, K.: "Use of Voronoi Grid in Reservoir Simulation," paper SPE 22889 presented at the 1991 SPE Annual Technical Conference and Exhibition, Dallas, Texas, (October 6-9, 1991).

39. Todd, M.R., O'Dell, P.M., and Hirasaki, G.J.: "Methods for Increased Accuracy in Numerical Reservoir Simulators," *SPEJ* (December 1972) 515-530.

40. Coats, K.H., George, W.D., Chu, C., and Marcum, B.E.: "Three-Dimensional Simulation of Steamflooding," *SPEJ* (December 1974) 573-592.

41. Yanosik, J.L., and McCracken, T.A.: "A Nine-Point, Finite-Difference Reservoir Simulator for Realistic Prediction of Adverse Mobility Ratio Displacements," *SPEJ* (August 1979) 253-262.

42. Coats, K.H.: "A Note on IMPES and Some IMPES-Based Simulation Models," *SPEJ* (September 2000), 245-251.

43. Coats, K.H.: "IMPES Stability: Selection of Stable Timesteps," *SPEJ* (June 2003) 181-187.

44. Coats, K.H.: "IMPES Stability: The CFL Limit," *SPEJ* (September 2003) 291-297.

45. MacDonald, R.C., and Coats, K.H.: "Methods for Numerical Simulation of Water and Gas Coning," *SPEJ* (December 1970) 425-436.

46. Spillette, A.G., Hillestad, J.G., and Stone, H.L.: "A High-Stability Sequential Solution Approach to Reservoir Simulation," paper SPE 4542 presented at the SPE 48th Annual Fall Meeting, Las Vegas, Nevada, (September 30-October 3, 1973).

47. Blair, P.M., and Weinaug, C.F.: "Solution of Two-Phase Flow Problems Using Implicit Difference Equations," *SPEJ* (December 1969) 417-424.

48. Bansal, P.P., Harper, J.L., McDonald, A.E., Moreland, E.E., Odeh, A.S., and Trimble, R.H.: "A Strongly Coupled, Fully Implicit, Three Dimensional, Three Phase Reservoir Simulator," paper SPE 8329 presented at the 54th Annual Fall Technical Conference and Exhibition, Las Vegas, Nevada, September 23-26, 1979.

49. Coats, K.H.: "In-Situ Combustion Model," *SPEJ* (December 1980) 533-553.

50. Thomas, G.W., and Thurnau, D.H.: "Reservoir Simulation Using an Adaptive Implicit Method," *SPEJ* (October 1983) 759-768.

51. Thomas, G.W., and Thurnau, D.H.: "The Mathematical Basis of the Adaptive Implicit Method," paper SPE 10495 presented at the Sixth SPE Symposium on Reservoir Simulation, New Orleans, Louisiana, January 31-February 3, 1982.

52. Lantz, R.B.: "Quantitative Evaluation of Numerical Diffusion (Truncation Error)," *SPEJ* (September 1971) 351-320.

53. Kyte, J.R., and Berry, D.W.: "New Pseudo Functions to Control Numerical Dispersion," *SPEJ* (August 1975) 269-276.

54. Peaceman, D. W.: "A New Method for Calculating Well Indexes for Multiple Wellblocks with Arbitrary Rates in Numerical Reservoir Simulation," paper SPE 79687 presented at the SPE Reservoir Simulation Symposium, Houston, Texas, (February 3-5, 2003).

55. Peaceman, D.W.: "Interpretation of Well-Block Pressures in Numerical Reservoir Simulation," *SPEJ* (June 1978) 183-194.

56. Peaceman, D.W.: "Interpretation of Well-Block Pressures in Numerical Reservoir Simulation with Nonsquare Grid Blocks and Anisotropic Permeability," *SPEJ* (June 1983) 531-543.

57. Holmes, J.A.: "Modeling Advanced Wells in Reservoir Simulation," paper SPE 72493, Distinguished Author Series, (November 2001).

58. Peaceman, D.W.: "Representation of a Horizontal Well in Numerical Reservoir Simulation," paper SPE 21217, SPE *Advanced Technology Series*, Vol. 1, No. 1 (1993) 7-16.

59. Palagi, C.L., and Aziz, K.: "Modeling Vertical and Horizontal Wells with Voronoi Grid," *SPERE* (February 1994) 15-21.

60. Stone, T.W., Edmunds, N.R., and Kristoff, B.J.: "A Comprehensive Wellbore/Reservoir Simulator," paper SPE 18419 presented at the 1989 SPE Reservoir Simulation Symposium, Houston, Texas, (February 6-8, 1989).

61. Holmes, J.A., Barkve, T., and Lund, O.: "Application of a Multisegment Well Model to Simulate Flow in Advanced Wells," paper SPE 50646 presented at the 1998 European Petroleum Conference, The Hague, The Netherlands, (October 20-22, 1998).

62. Ezekwe, J.N.: "Comparison of Predictions from Reservoir Models to Actual Production Data: Field Examples," SPE 77972 presented at the Annual Technical Conference and Exhibition in San Antonio, Texas, (September 29-October 2, 2002).

63. Echols, D.P., and Ezekwe, N.: "A Reservoir Management Model for 26R Reservoir, Elk Hills Oil Field, California," SPE 46231 presented at the Western Regional Meeting in Bakersfield, California, (May 10-13, 1998).

64. Gencer, S.C., Ketcherside, B.P., Morrell, G.O., Mulkay, E.L., and Wiegand, K.D.: "Data Management in Reservoir Simulation," paper SPE 106075 presented at the 2007 SPE Reservoir Simulation Symposium, Houston, Texas, (February 25-28, 2007).

ADDITIONAL REFERENCES

- Abou-Kassem, J.H., and Aziz, K: "Analytical Well Models for Reservoir Simulation," *SPEJ* (August 1985) 573-580.
- Chappelear, J.E., and Williamson, A.S.: "Representing Wells in Numerical Reservoir Simulation: Part 2—Implementation," *SPEJ* (June 1981) 339-344.
- Coats, K.H.: "Simulation of Steamflooding with Distillation and Solution Gas," *SPEJ* (October 1976) 235-247.
- Coats, K.H.: "A Highly Implicit Steamflood Model," *SPEJ* (October 1978) 369-383.

- Dogru, A.H.: "Megacell Reservoir Simulation," paper SPE 57907, Distinguished Author Series, (May 2000).
- Ezekwe, N., Echols, D., and Hudgens, P.: "Diagnostic Application of Reservoir Simulation to 31S Structure Stevens Reservoirs–Elk Hills Oil Field, California," SPE 46230 presented at the Western Regional Meeting in Bakersfield, California, (May 10-13, 1998).
- Frost, R.J., Parker, M.J., and Ponting, D.K.: "Scalability and Usability Aspects of Reservoir Simulation," paper SPE 115015 presented at the 2008 SPE Russian Oil & Gas Technical Conference and Exhibition, Moscow, Russia, (October 28-30, 2008).
- Fung, L.S.K., and Dogru, A.H.: "Parallel Unstructured-Solver Methods for Simulation of Complex Giant Reservoirs," *SPEJ* (December 2008) 440-446.
- Hammersley, R.P., and Ponting, D.K.: "Solving Linear Equations in Reservoir Simulation Using Multigrid Methods," paper SPE 115017 presented at the 2008 SPE Russian Oil & Gas Technical Conference and Exhibition, Moscow, Russia, (October 28-30, 2008).
- Iranshahr, D.V., Voskov, D.V., and Tchelepi, H.A.: "Phase Equilibrium Computations Are No Longer the Bottleneck in Thermal Compositional EoS Based Simulation," paper SPE 119166 presented at the 2009 SPE Reservoir Simulation Symposium, The Woodlands, Texas, (February 2-4, 2009).
- Mattax, C.C., and Dalton, R.L.: "Reservoir Simulation," *JPT* (June 1990) 692-695.
- Mehra, R., Hadjitofi, M., and Donnelly, J.K.: "An Automatic Time-Step Selector for Reservoir Models," paper SPE 10496 presented at the Sixth SPE Symposium on Reservoir Simulation, New Orleans, Louisiana, (January 31-February 3, 1982).
- Mishev, I., Beckner, B., Fedorova, N., and Terekhov, S.: "Linear Solver Performance Optimization in Reservoir Simulation Studies," paper SPE 119266 presented at the 2009 SPE Reservoir Simulation Symposium, The Woodlands, Texas, (February 2-4, 2009).
- Najafabadi, N.F., Han, C., Delshad, M., and Sepehrnoori, K.: "Development of a Three Phase, Fully Implicit, Parallel Chemical Flooding Simulator," paper SPE 119002 presented at the 2009 SPE Reservoir Simulation Symposium, The Woodlands, Texas, (February 2-4, 2009).
- Nolen, J.S., and Berry, D.W.: "Tests of the Stability and Time-Step Sensitivity of Semi-Implicit Reservoir Simulation Techniques," *SPEJ* (June 1972) 253-266.
- Peaceman, D.W.: "A New Method for Calculating Well Indexes for Multiple Wellblocks with Arbitrary Rates in Reservoir Simulation," paper SPE 79687 presented at the SPE Reservoir Simulation Symposium, Houston, Texas, (February 3-5, 2003).
- Rietz, D., and Usmani, A.: "Case Studies Illustrating the Use of Reservoir Simulation Results in the Reserves Estimation Process," *SPEREE* (February 2009) 149-158.
- Varavei, A., and Sepehrnoori, K.: "An EOS-Based Compositional Thermal Reservoir Simulator," paper SPE 119154 presented at the 2009 SPE Reservoir Simulation Symposium, The Woodlands, Texas, (February 2-4, 2009).
- Vestergaard, H., Olsen, H., Sikandar, A.S., and Noman, R.: "The Application of Unstructured-Gridding Techniques for Full-Field Simulation of a Giant Carbonate Reservoir Developed with Long Horizontal Wells," *SPEREE* (December 2008) 958-967.
- Williamson, A.S., and Chappelear, J.E.: "Representing Wells in Numerical Reservoir Simulation: Part 1–Theory," *SPEJ* (June 1981) 323-338.
- Wong, T.W., Firoozabadi, A., Nutakki, R., and Aziz, K.: "A Comparison of Two Approaches to Compositional and Black Oil Simulation," paper SPE 15999 presented at the Ninth SPE Symposium on Reservoir Simulation, San Antonio, Texas, (February 1-4, 1997).

PROBLEMS

23-1. A vertical well is located in a gridblock of a reservoir model based on the Cartesian grid system. The dimensions and properties of the gridblock including the well are given as:

Gridblock dimension, Δx	= 250 ft
Gridblock dimension, Δy	= 250 ft
Gridblock dimension, Δz	= 25 ft
Gridblock permeability, k	= 100 md
Wellbore radius, r_w	= 0.454 ft
Skin factor, s	= 0

Assume the net thickness of the gridblock is equal to the gridblock dimension, Δz, and that the gridblock is isotropic. Calculate the well index for the well.

23-2. A vertical well is located in a single gridblock within a reservoir model gridded on the Cartesian system. The dimensions and properties of the gridblock are given as:

Gridblock dimension, Δx	= 175 ft
Gridblock dimension, Δy	= 125 ft
Gridblock dimension, Δz	= 15 ft
Gridblock permeability, k_x	= 150 md
Gridblock permeability, k_y	= 130 md
Wellbore radius, r_w	= 0.25 ft

Assume the gridblock dimension, Δz, is equal to the net thickness of the gridblock.

Calculate the well index of the well:

a. With a skin factor, $s = +2$

b. With a skin factor, $s = -2$

c. Compare and discuss the well indices from (a) and (b).

23-3. A geologic model was constructed for the Red Hawk reservoirs in Chapter 22, Problem 22-3. The relative permeability data for the Surculus and Tamalis intervals can be generated from Corey equations assuming a Corey exponent, N = 3.0. Convert the geologic model into a reservoir model using additional input data given in the case history in Chapter 9, Section 9.6.1. Use the reservoir model to perform the following analyses:

a. Calculate the original gas-in-place (OGIP) for the reservoir models.

b. Compare the OGIP to earlier OGIP calculated in Chapter 9, Problem 9.5a.

c. History-match the reservoir model using the production and pressure data given in Chapter 9, Tables 9.11 and 9.12.

d. What modifications to the reservoir model were made in order to history-match the performance history given in Tables 9.11 and 9.12?

23-4. A geologic model was built for the M-4/M-4A and M-6 reservoirs in Chapter 22, Problem 22-4. The relative permeability data for the M-4/M-4A and M-6 reservoirs are given in Tables 23P.1 to 23P.4, respectively. Build a reservoir model from the geologic model. Then use the reservoir model to conduct the following analyses:

a. Determine the original gas-in-place (OGIP) for the reservoir model.

b. Compare the OGIP for each reservoir to OGIPs calculated earlier for the reservoirs in Chapter 9, Problem 9-6

c. History-match the reservoirs using the production and pressure data given in the case history in Chapter 9, Section 9.6.2.3.

d. What are the modifications that were made to the reservoir model to achieve history match of the performance data?

TABLE 23P.1 M-4/M-4A Oil-Water Relative Permeability Data for Problem 23-4

Sw	Krw	Krow	Pcwo
0.2860	0.0000	0.8000	0
0.3217	0.0043	0.6393	0
0.3574	0.0100	0.5066	0
0.3931	0.0193	0.3969	0
0.4288	0.0324	0.3073	0
0.4645	0.0498	0.2349	0
0.5002	0.0716	0.1771	0
0.5359	0.0981	0.1316	0
0.5716	0.1297	0.0960	0
0.6073	0.1666	0.0687	0
0.6430	0.2092	0.0480	0
0.6787	0.2576	0.0323	0
0.7144	0.3123	0.0206	0
0.7501	0.3735	0.0118	0
0.7858	0.4415	0.0053	0
0.8215	0.5165	0.0006	0
0.8572	0.5990	0.0000	0
0.8929	0.6891	0.0000	0
0.9286	0.7871	0.0000	0
0.9643	0.8935	0.0000	0
1.0000	1.0000	0.0000	0

TABLE 23P.2 M-4/M-4A Gas-Oil Relative Permeability Data for Problem 23-4

Sg	Krg	Krog	Pcgo
0.2100	0.0000	0.8000	0
0.2352	0.0125	0.5488	0
0.2604	0.0240	0.4212	0
0.2856	0.0385	0.3205	0
0.3108	0.0560	0.2421	0
0.3360	0.0765	0.1820	0
0.3612	0.1001	0.1367	0
0.3864	0.1267	0.1028	0
0.4116	0.1562	0.0776	0
0.4368	0.1888	0.0588	0
0.4620	0.2244	0.0444	0
0.4872	0.2630	0.0328	0
0.5124	0.3046	0.0229	0
0.5376	0.3493	0.0141	0
0.5628	0.3969	0.0061	0
0.5880	0.4476	0.0000	0
0.6132	0.5012	0.0000	0
0.6384	0.5579	0.0000	0
0.6636	0.6176	0.0000	0
0.6888	0.6803	0.0000	0
0.7140	0.7400	0.0000	0

TABLE 23P.3 M-6 Oil-Water Relative Permeability Data for Problem 23-4

Sw	Krw	Krow	Pcwo
0.2000	0.0000	0.8000	0
0.2400	0.0043	0.6393	0
0.2800	0.0100	0.5066	0
0.3200	0.0193	0.3969	0
0.3600	0.0324	0.3073	0
0.4000	0.0498	0.2349	0
0.4400	0.0716	0.1771	0
0.4800	0.0981	0.1316	0
0.5200	0.1297	0.0960	0
0.5600	0.1666	0.0687	0
0.6000	0.2092	0.0480	0
0.6400	0.2576	0.0323	0
0.6800	0.3123	0.0206	0
0.7200	0.3735	0.0118	0
0.7600	0.4415	0.0053	0
0.8000	0.5165	0.0006	0
0.8400	0.5990	0.0000	0
0.8800	0.6891	0.0000	0
0.9200	0.7871	0.0000	0
0.9600	0.8935	0.0000	0
1.0000	1.0000	0.0000	0

TABLE 23P.4 M-6 Gas-Oil Relative Permeability Data for Problem 23-4

Sg	Krg	Krog	Pcgo
0.2100	0.0000	0.8000	0
0.2395	0.0125	0.5488	0
0.2690	0.0240	0.4212	0
0.2985	0.0385	0.3205	0
0.3280	0.0560	0.2421	0
0.3575	0.0765	0.1820	0
0.3870	0.1001	0.1367	0
0.4165	0.1267	0.1028	0
0.4460	0.1562	0.0776	0
0.4755	0.1888	0.0588	0
0.5050	0.2244	0.0444	0
0.5345	0.2630	0.0328	0
0.5640	0.3046	0.0229	0
0.5935	0.3493	0.0141	0
0.6230	0.3969	0.0061	0
0.6525	0.4476	0.0000	0
0.6820	0.5012	0.0000	0
0.7115	0.5579	0.0000	0
0.7410	0.6176	0.0000	0
0.7705	0.6803	0.0000	0
0.8000	0.7400	0.0000	0

23-5. A geologic model was constructed for the Manatee reservoirs in Chapter 22, Problem 22-5. The relative permeability data for the S10A2 and S10B reservoirs can be derived from Corey parameters specified in Table 23P.5. Utilize additional data in the case history in Chapter 10, Section 10.8 to transform the geologic model into a reservoir model. Use the reservoir model to perform the following analyses:

a. Determine the original oil-in-place (OOIP) for the S10A2 and S10B reservoirs.

b. Compare the OOIP from the S10A2 reservoir model to the OOIPs calculated in Chapter 10, Problem 10-5 by various other methods.

c. History-match the reservoir model using the performance data given in the case history in Chapter 10, Section 10.8.3.

d. Evaluate the S10A2 reservoir as candidate for further development with another well.

TABLE 23P.5 Corey Parameters for Problem 23-5

Corey Parameter	Value
Oil curve endpoint at initial water saturation, $k_{ro@S_{wi}}$	0.9
Oil curve endpoint at initial gas saturation, $k_{ro@S_{gi}}$	0.8
Water curve endpoint at residual oil saturation, $k_{rw@S_{orw}}$	0.1
Gas curve endpoint at residual oil saturation, $k_{rg@S_{org}}$	1.0
Residual oil saturation to gas, S_{org}	0.3
Residual oil saturation to water, S_{orw}	0.214
Critical gas saturation, S_{gc}	0.035
Corey's exponent, N	3.0

CHAPTER

24 Reservoir Management

24.1 Introduction

Reservoir management is a term which can be described as consisting of operational plans or strategies, based on analyses of current geologic, reservoir, and production data, designed to optimize the development and exploitation of a reservoir with the goal of achieving the maximum efficient **and** economic recovery of its hydrocarbons. The most important ingredient necessary for the creation of sound reservoir management strategies is the collection and analyses of key geologic, reservoir, and performance data.[1] For this reason, this book has emphasized in previous chapters, the collection and analyses of basic geologic, rock, fluid properties and reservoir production data, the construction and characterization of geologic models with these data, and use of reservoir flow models in simulation to devise reservoir management strategies. But depending on the quantity and quality of available data, it is **not** always necessary to use reservoir models in devising effective reservoir management strategies. Other methods such as the general material balance equation (Chapter 8), and other analytical methods (Chapters 9 and 10) can be used in gaining insights into the prevailing mechanisms occurring in the reservoir. These methods can also be used to develop short-term reservoir management strategies, such as curtailing production from high gas-oil ratio wells or high water-cut wells. As additional data are collected from the reservoir, models can then be constructed which could be used to generate long-term management strategies for the reservoir, such as initiation of pressure maintenance by gas and/or water injection.

The collection, analyses, and assimilation of data should be at the center of any sound reservoir management process. The traditional process of creating reservoir management strategies had relied on somewhat regular periods of collecting reservoir and well performance data. These data are then subjected to analyses, and eventually used to generate or modify strategies for managing the reservoir. The petroleum field of current era is rapidly embracing the digital age. This has led to the installation of permanent downhole gauges and control devices in wells, thereby creating the so-called "smart" or "intelligent" wells. Also, petroleum fields are equipped with sensors, real-time data communication systems, massive data storage capacity, flow simulators, and data visualization monitors to create the "intelligent" field.[2] The confluence of "smart" wells and "intelligent" fields is leading the development of a new concept of reservoir management which has been termed "closed-loop reservoir management."[3,4] Closed-loop reservoir management can be described as a process that seeks to maximize reservoir performance in terms of hydrocarbon recovery or net present value for the remaining projected life of the reservoir by changing reservoir management activities based on near real-time processes. In applying closed-loop reservoir management, a combination of reservoir model-based optimization and data assimilation are used. As the use of flow control, flow monitoring, and flow optimization devices continue to spread throughout the petroleum industry, more reservoirs will become candidates for application of closed-loop reservoir management.

To assist reservoir management teams in crafting and implementing sound reservoir management strategies, five reservoir management principles are enunciated. The five reservoir management principles are:[1]

1. Conservation of reservoir energy.
2. Early implementation of simple, proven strategies.
3. Systematic and sustained practice of data collection.
4. Application of emerging technologies for improved hydrocarbon recovery.
5. Long-term retention of staff in multi-disciplinary teams.

In this chapter, the concepts behind the five reservoir management principles are explored further. The chapter is concluded by presenting case histories of three reservoirs where these principles were applied to illustrate implementations of sound and successful reservoir management practices.

24.2 Reservoir Management Principles

The reservoir management principles proposed in this chapter are designed to guide reservoir management teams in the development, implementation, and monitoring of sound reservoir management strategies. The principles are simple, easy to understand, and can be applied to practically all reservoirs. The five principles can be used as a checklist by reservoir management teams to ensure that they are doing all the "right things" in managing their reservoirs. The five reservoir management principles are:

1. Conservation of reservoir energy.
2. Early implementation of simple, proven strategies.

3. Systematic and sustained practice of data collection.
4. Application of emerging technologies for improved hydrocarbon recovery.
5. Long-term retention of staff in multi-disciplinary teams.

24.2.1 Conservation of Reservoir Energy

The central principle of reservoir management is conservation of reservoir energy. This principle applies to controlled and optimized use of the energy stored in the reservoir at discovery or at any stage of depletion to maximize economic and efficient recovery of its reserves. **The principle should not be interpreted to mean that the reservoir should not be produced until the "correct" strategies for managing it have been devised.** It does require however that management strategies applied on the reservoir should avoid depleting reservoir energy inefficiently, especially during the early stages of reservoir development and production. Common production practices that should be avoided because they could deplete reservoir energy include excessive production of gas from the gas cap of a saturated reservoir, high production rates due to excessive pressure drawdown, comingling of production from separate reservoirs, etc. In applying this principle, the reservoir management teams should strive to achieve a balance between conserving reservoir energy and maximizing economic recovery of hydrocarbons from the reservoir.

24.2.2 Early Implementation of Simple, Proven Strategies

The principle of early implementation of simple, proven strategies is directed at supporting the concept of conserving reservoir energy. Simple, proven strategies are reservoir management practices that are known from industry experience to conserve reservoir energy. Some of these simple, proven strategies include some form of pressure maintenance by fluid injection, limited pressure drawdown at production wells, isolation of separate reservoirs at producers, optimal well spacing, selective perforation of productive zones, etc. **The costs of implementing these strategies should be weighed against their expected benefits by the reservoir management teams, such that economic recovery of hydrocarbons is always maintained.**

24.2.3 Systematic and Sustained Practice of Data Collection

The collection, analysis, and assimilation of data can be considered as the foundation on which sound reservoir management strategies can be devised and implemented. The culture of data collection should start at discovery of the reservoir and maintained throughout its life, as long as it is economic. **That is, the value of information gained from the data should outweigh the cost of collecting and analyzing the data.** The data collection process should include geologic, geophysical, petrophysical, pressure, production/injection data, etc. The objectives of data collection, analyses, and assimilation should be to continuously increase knowledge on reservoir hydrocarbon recovery mechanisms and to apply that knowledge on devising improved reservoir management strategies. The petroleum industry is currently adopting the practice of sustained data collection by installing downhole gauges and flow control devices on many key wells. This has led to substantial increase in the amount of data available on many reservoirs. It is expected that this practice will continue to expand in the industry with the ultimate result of key reservoir data available in real time for reservoir management decisions.

24.2.4 Application of Emerging Technologies for Improved Hydrocarbon Recovery

Reservoirs that have been properly managed on the basis of the first three principles of reservoir management discussed earlier are good candidates for application of emerging technologies for improved hydrocarbon recovery. In general, reservoirs whose energies have been conserved with simple, proven strategies and have sufficient data for analyses and assimilation are excellent candidates for application of emerging technologies for improved hydrocarbon recovery. These new technologies for improved hydrocarbon recovery include advances in well architecture, drilling and completion of wells, applications of new chemicals for enhanced recovery, equipment for production operations, etc. For instance, an emerging technology in well architecture that is expected to substantially improve productivity is application of multilateral wells.[5-7] Multilateral wells have been used in many well-managed reservoirs with long production history to improve hydrocarbon recovery. **The reservoir management team should monitor the industry regularly for new emerging technologies and assess the potential of these technologies for improving hydrocarbon recovery from the reservoir.** The realizable benefits of sound reservoir management practices are derived by positioning the reservoir ready to take advantage of new technologies as they emerge in the industry.

24.2.5 Long-Term Retention of Staff in Multi-Disciplinary Teams

In many organizations, reservoir management teams are composed of reservoir engineers, geologists, geophysicists, petrophysicists, production engineers, drilling engineers, facilities engineers, etc. These multi-disciplined teams have been recognized in the industry as being very effective in fashioning sound reservoir management strategies. **This last principle of reservoir management is necessary because it puts emphasis on the importance of keeping members of the team together long enough to gain critical understanding of the geology of the reservoir, and the recovery mechanisms occurring in the reservoir.** After a critical level of knowledge has been achieved by working together on a reservoir, management teams collaborate more efficiently in crafting new strategies or altering existing strategies for improved reservoir management. When staff changes are necessary or mandated, it is recommended that at least one or two key members of the reservoir management team should be retained to train and transfer their knowledge and experience to new team members.

24.3 Case Histories Demonstrating Applications of Reservoir Management Principles

The applications of the reservoir management principles can be demonstrated by reviewing the case histories of three reservoirs, and discussing the strategies used to manage the reservoirs on the basis of these principles. The case histories were selected as examples of reservoirs which have long histories of sound reservoir management practices. Many other case histories reported in the literature or constructed from the histories of reservoirs in many organizations can be used as examples by applying these reservoir management principles.

24.3.1 The Case History of 26R Reservoir (1976–1996)

The reservoir management strategy for the 26R reservoir (1976–1996) can be summarized as follows:

1. Control excessive gas production.
2. Maintain reservoir pressure by gas injection.
3. Assist gravity drainage.
4. Collect data.
5. Improve oil recovery with horizontal wells.

Geology. The 26R reservoir is located within the Elk Hills Field in the southern San Joaquin valley of central California.[8] The reservoir is contained within the steeply, dipping southwestern limb of the 31S Anticline (Figure 24.1). The structure map of the 26R reservoir is shown in Figure 24.2. The 26R structure is roughly three miles long and one mile wide. As shown in Figure 24.2, a non-sealing fault with several hundred feet of throw intersects the structure. The reservoir had an initial oil column of 1800 feet, and a net productive sand thickness of approximately 1150 ft. The 26R reservoir is an Upper Miocene, Stevens turbidite sand channel classified under the Monterey Formation. These sand channels consist of many thin turbidite beds ranging in thickness from a few inches to several feet. Within the beds, sand grain sizes range from very fine to very coarse with infrequent conglomerates. In clean portions of the reservoir, net-to-gross ratios could be as high as 90%. Significant correlatable shale layers exist between turbidite sequences. These continuous shale layers were used to classify five large intervals as mega-units, namely: A-C, C-F, F-K, K-N, and N-P, as shown in Figure 24.3. These mega-units are not considered to be separate reservoirs. Formation beds within the 26R reservoir generally dip from 23 degrees near the top of the structure to 70 degrees at the OWC. The OOIP was estimated at 424 million barrels with an OWC at 6010 feet subsea. The 26R reservoir rock and fluid properties are summarized in Table 24.1.

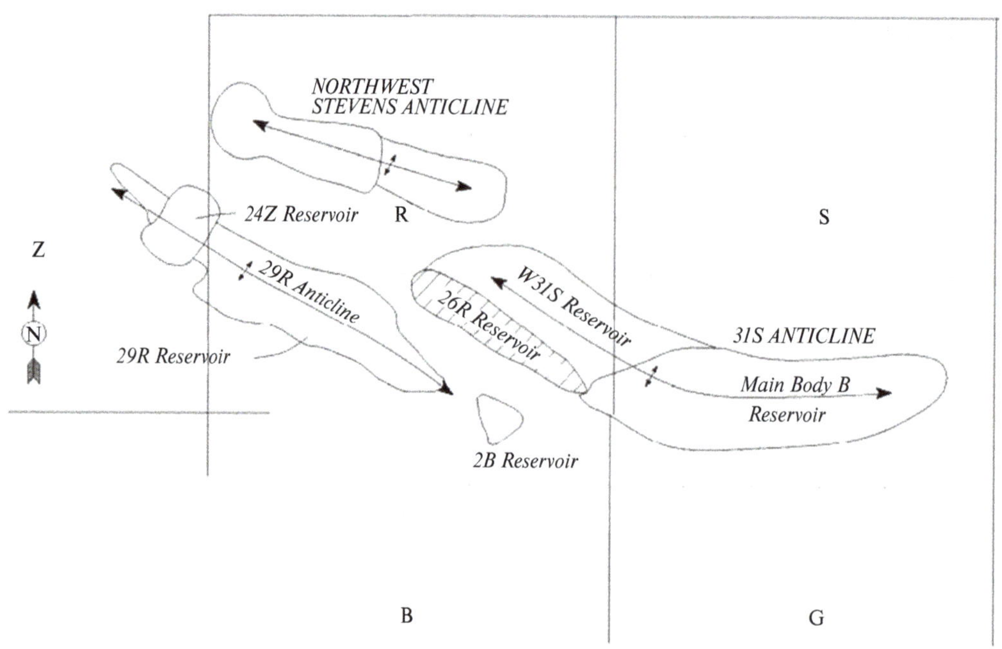

▲ FIGURE 24.1 Map showing anticlines in Elk Hills Field

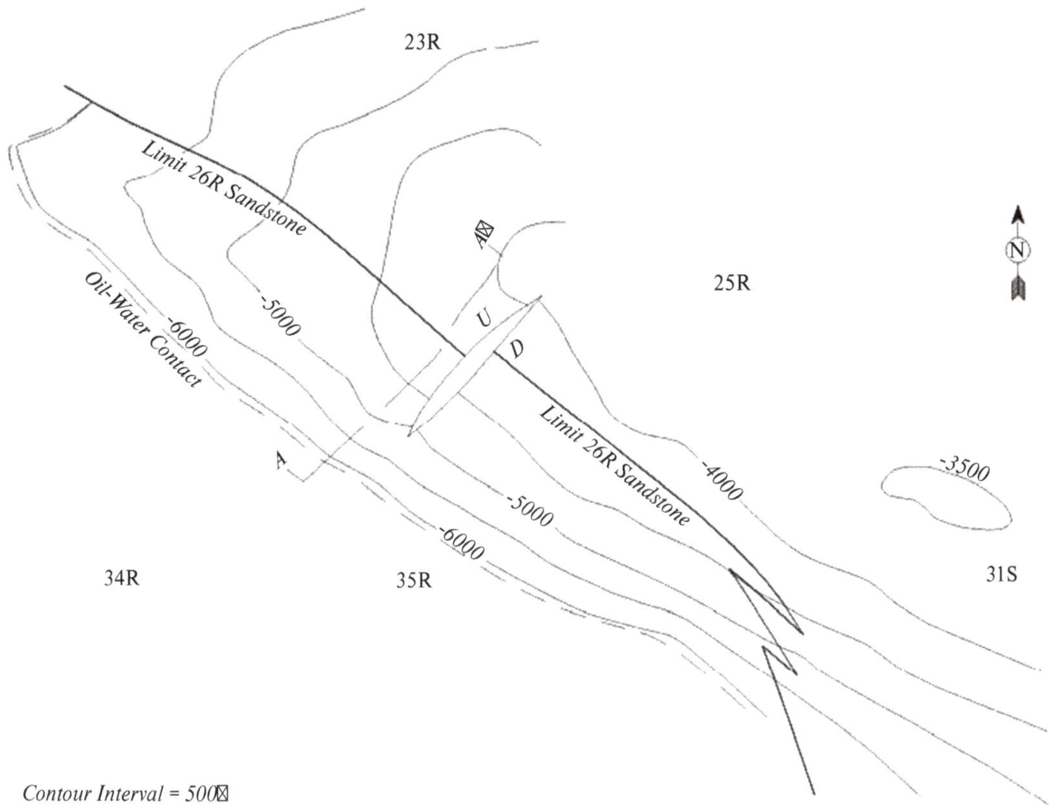

▲ **FIGURE 24.2** **Structure map of 26R reservoir**

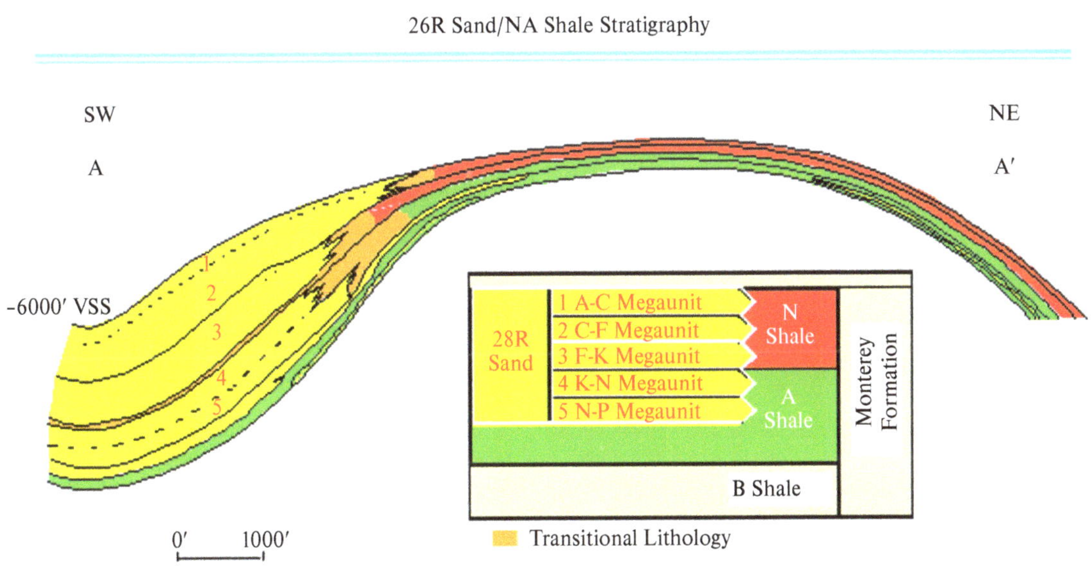

▲ **FIGURE 24.3** **Cross-section of 26R reservoir (along A-A′ in Figure 24.2) showing mega-units**

TABLE 24.1 Summary of 26R Reservoir Rock and Fluid Properties

Property	Parameter
Average porosity	23.6%
Average permeability	88 md
Average water saturation	16%
Initial saturation pressure	3155 psia
Reservoir temperature	210 °F
Reservoir oil viscosity	0.42 cp
Oil gravity	36 °API
Water-oil contact	6010 ft subsea
Residual oil saturation to gas	49%
Residual oil saturation to water	26%
Estimated original oil in place	424 MMBO

Production and Injection History. The 26R reservoir was discovered in 1950. The reservoir was barely produced from 1950 to 1976 because it was mothballed as part of the United States of America Naval Petroleum Reserve to supply oil for naval operations during national emergency. The production history of the 26R reservoir is shown in Figure 24.4. Oil production reached its peak between 1980 and 1983 at average rates of 50,000 BOPD. Gas and water production during this period averaged 72,809 Mcf/D and 1,833 BWPD, respectively. Excessive gas production from the reservoir was controlled by performing gas isolation remedial work with multiple packers and/or shutting-in wells producing at high gas-oil ratios.

Gas injection at the crest of the structure began in October 1976 barely three months after initial production started in July 1976. The management strategy of injecting gas into this reservoir was to balance reservoir voidage, thereby maintaining reservoir pressure. The average pressure history of the 26R reservoir is shown in Figure 24.5. As shown in Figure 24.5, reservoir pressure declined gradually from 1976, in spite of the pressure maintenance program. There is evidence that the inability to maintain pressure in the 26R reservoir was caused by the migration of some of the injected gas into the contiguous and overlaying shale reservoirs (Figure 24.3). However, the gas injection strategy minimized pressure decline and helped to stabilize reservoir pressure.

Displacement Mechanisms. The major fluid displacement mechanisms occurring within the 26R reservoir in order of prominence are gravity drainage, liquids vaporization, oil displacement by gas, and solution gas drive. Gravity drainage in the 26R reservoir is promoted by high dip angles, moderate oil relative permeabilities, and low oil viscosity.[9,10] Liquids vaporization in the secondary gas cap are due to injection and cycling of dry gas as part of the gas injection program.[11] Oil recovery is further assisted by the downward displacement of oil by injected gas. And additional oil recovery is due to expansion of solution gas as reservoir pressure declined gradually. These four mechanisms have worked together to achieve a very efficient system of oil displacement in the 26R reservoir.

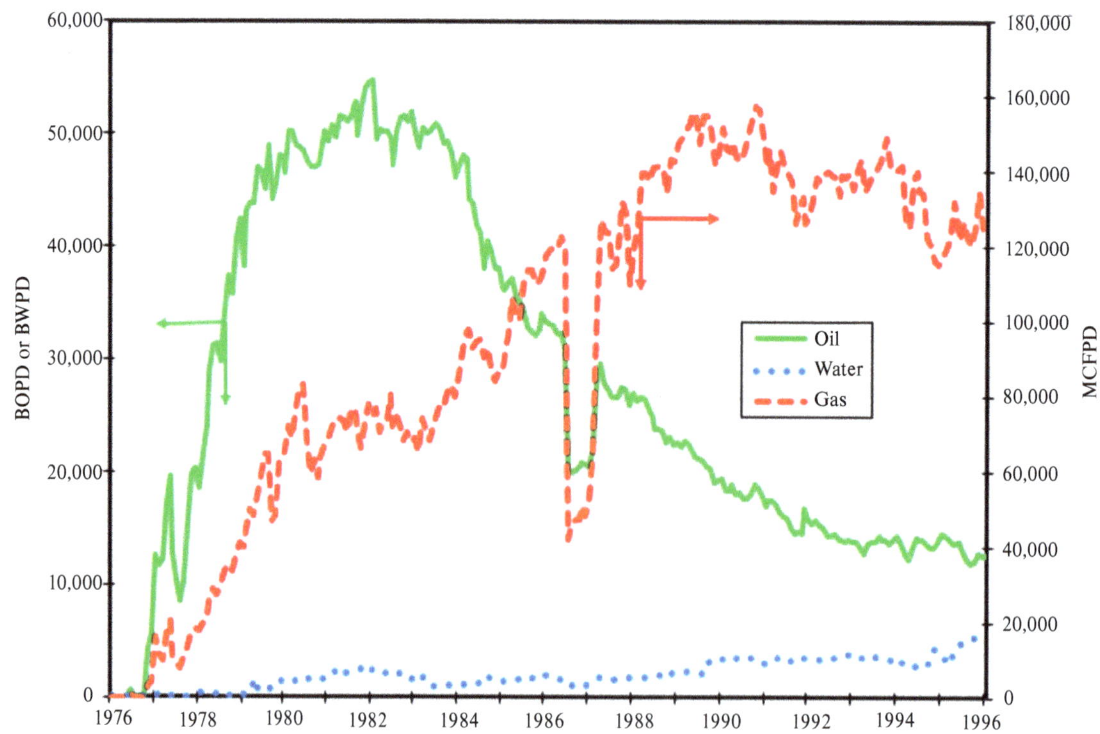

▲ **FIGURE 24.4** Historical production of 26R reservoir

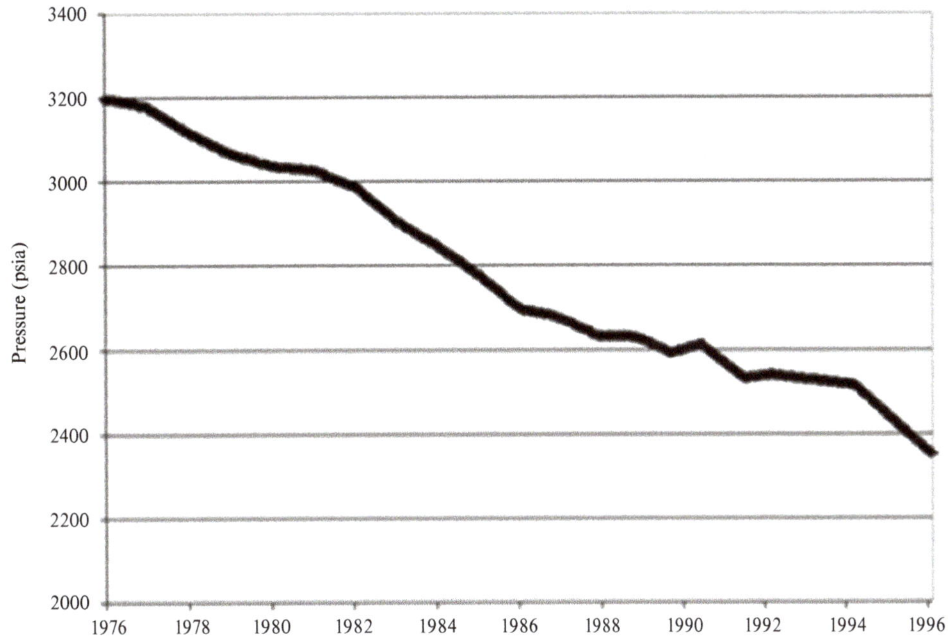

▶ **FIGURE 24.5** **Historical average reservoir pressure of 26R reservoir**

24.3.2 Application of Reservoir Management Principles to 26R Reservoir

Conservation of Reservoir Energy. The reservoir management principle of conservation of reservoir energy was practiced in the 26R reservoir by controlling excessive gas production. This was achieved by installing multiple packers in wells located high on the structure. As the gas-oil contact moved downdip, high gas producing intervals straddled with packers were shut-off. Eventually, these upstructure wells were shut-in as the gas-oil contact reached the lowest packers in the well. The historical gas-oil ratio plot of the 26R reservoir is shown in Figure 24.6. A review of Figure 24.6 indicates that field GOR was controlled in this reservoir for over a 20-year period.

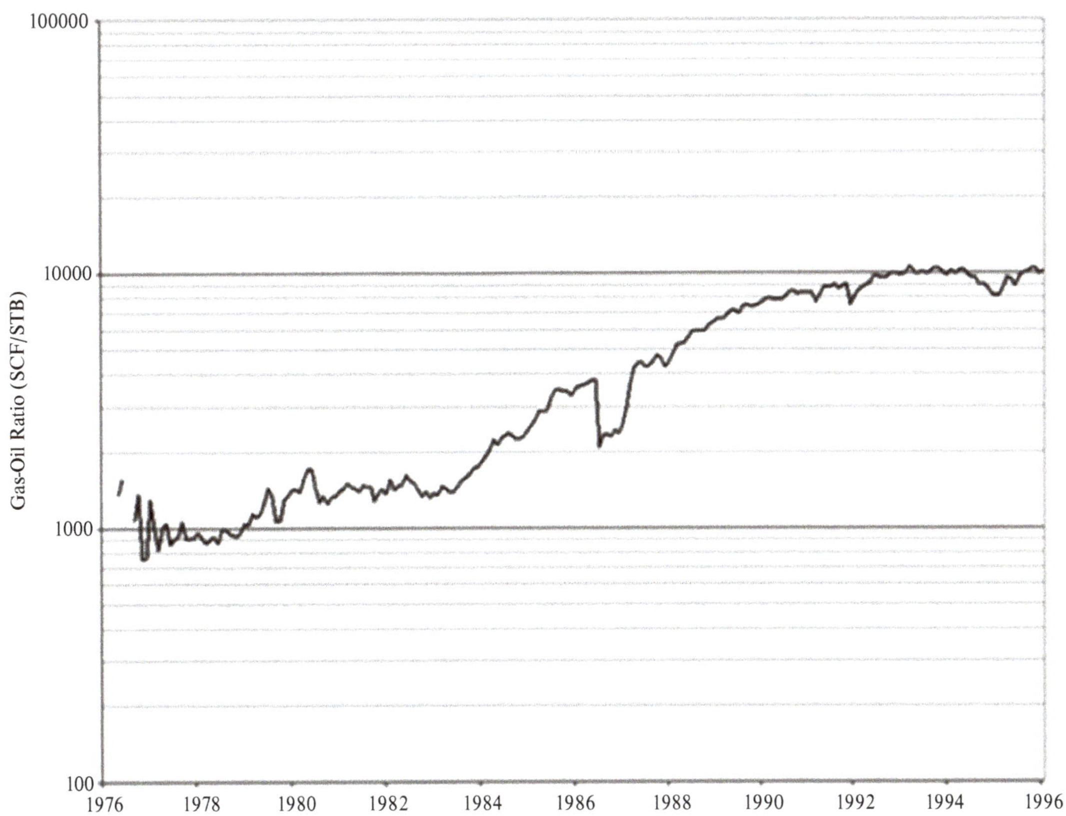

▶ **FIGURE 24.6** **Historical GOR of the 26R reservoir**

Early Implementation of Simple, Proven Strategies. Gas injection into the 26R reservoir for pressure maintenance was started just three months from initial production. Gas was injected into the reservoir at wells located along the crest of the structure. The volume of gas injected was calculated to balance reservoir voidage. Pressure maintenance by gas injection is a simple, proven strategy which can be implemented early in the life of a reservoir to reduce precipitous decline in reservoir pressure. In many reservoirs similar to 26R reservoir, reinjection of produced gas is efficient in terms of pressure maintenance and utilization of excess gas produced from the reservoir. In the case of 26R reservoir, pressure maintenance by gas injection was not entirely successful because reservoir pressure continued to decline gradually over the entire period of gas injection (Figure 24.5). This was due primarily to gas migration out of the 26R reservoir into the overlying shale reservoirs (Figure 24.3). However, it is evident that the pressure decline in this reservoir would have been more severe, if the pressure maintenance program was not installed.

Systematic and Sustained Practice of Data Collection. The collection of reservoir pressure data was very important in the management of the 26R reservoir. Several key wells were designated in the gas cap and the oil leg as pressure monitoring wells. These wells in the gas cap were equipped with permanently installed capillary tubing for regular pressure monitoring. Buildup surveys were conducted on the oil band wells as part of a field wide pressure monitoring program. The field wide pressure monitoring program was conducted every six months. Consequently, substantial pressure data were available on the 26R reservoir. These pressure data were used to adjust gas injection volumes as part of the management strategy of maintaining reservoir pressure. In addition to the pressure data, log data, wireline formation tool pressure data, fluid samples, and whole core samples were collected from all new wells drilled in the reservoir. These data were used to monitor the progression of the gas injection strategy, measure residual oil saturation in the gas cap, and update existing geological and reservoir models. These models[12,13] were used to develop new reservoir management strategies or adjust current strategies.

Application of Emerging Technologies for Improved Hydrocarbon Recovery. The successful implementation of the first three principles of reservoir management positioned the 26R reservoir as a good candidate for application of horizontal well technology. Horizontal well technology was virtually non-existent in the petroleum industry in the 1950s when the 26R reservoir was developed with vertical wells. In the 1980s, as horizontal well technology was emerging in the industry, it was recognized that the 26R reservoir was ideally suited for this technology because of its steeply dipping formation that promoted gravity drainage. It was recognized that horizontal wells could be located within the oil column close to the oil-water contact to produce oil draining to the base of the structure by gravity.[14,15] The first horizontal well was drilled in the 26R reservoir in 1988. By 1996, twenty-two horizontal wells were drilled in the 26R reservoir (Figure 24.7). The horizontal wells were more efficient

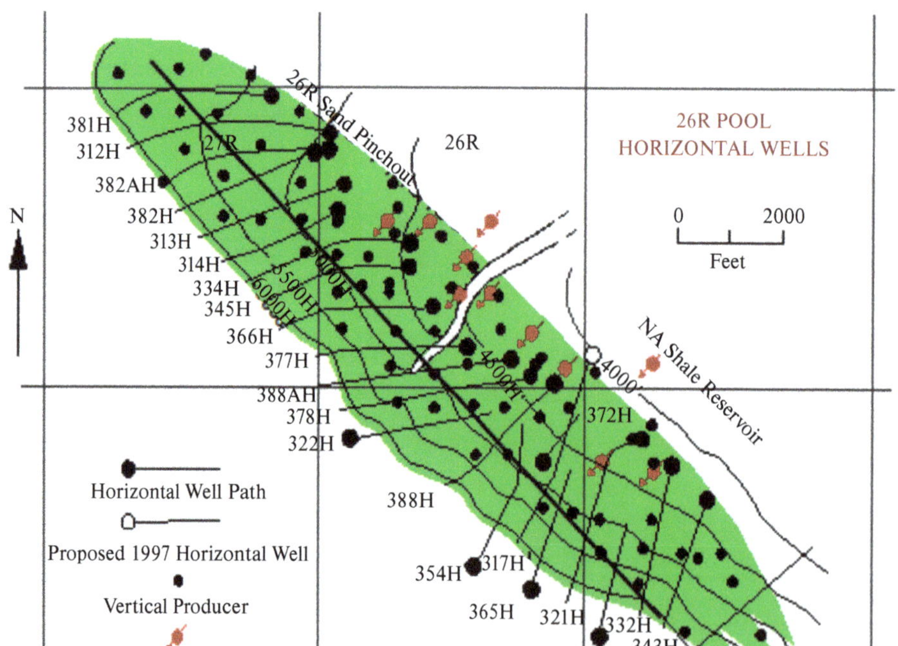

▶ **FIGURE 24.7** **Structure map of 26R reservoir showing horizontal well locations**

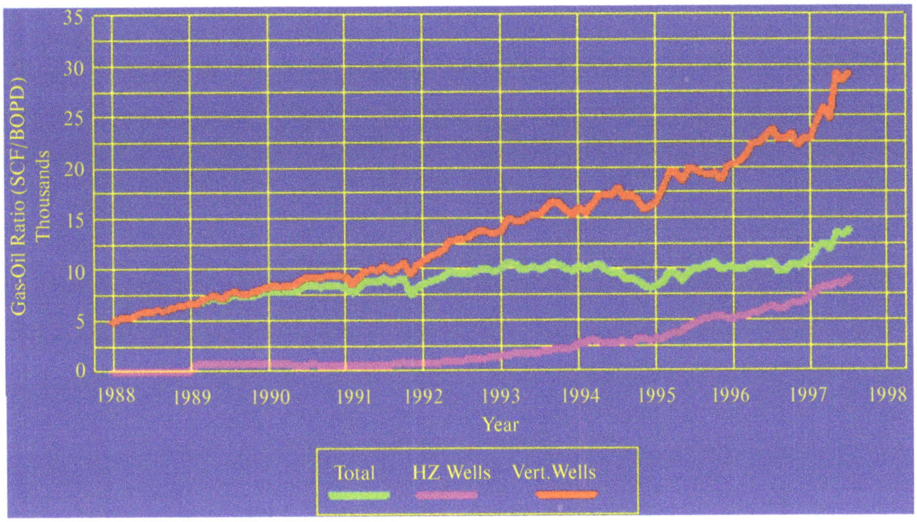

▲ **FIGURE 24.8** **GOR comparison—horizontal versus vertical wells in the 26R reservoir**

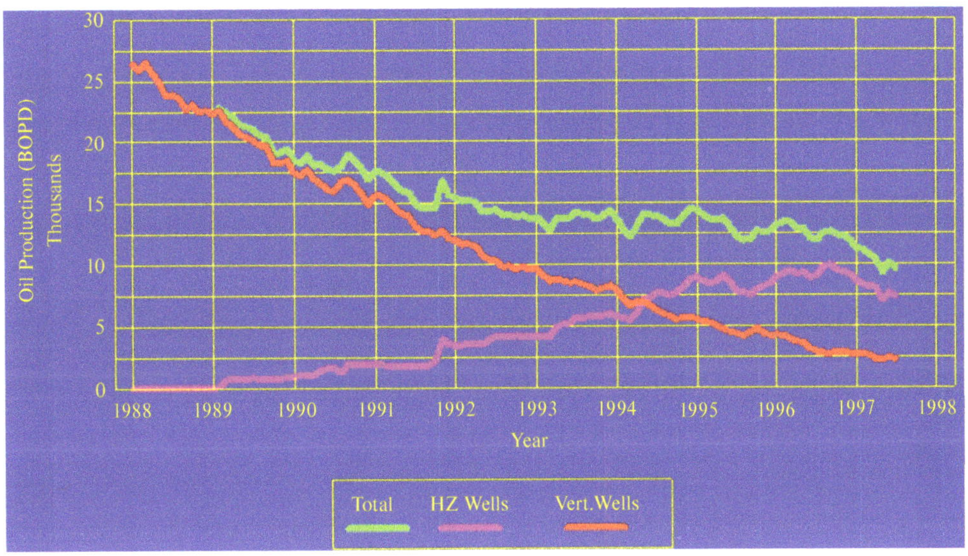

▲ **FIGURE 24.9** **Oil production comparison—horizontal versus vertical wells in the 26R reservoir**

than the vertical wells because they produce at lower GOR, thereby requiring less gas injection for pressure maintenance. In 1998, the horizontal wells produced 70% of the total oil production with one-third of the GOR of the remaining vertical wells (Figures 24.8 and 24.9). In the 1980s, the application of horizontal well technology in the 26R reservoir was the most successful program of its type in the United States of America.

Long-Term Retention of Staff in Multi-Disciplinary Teams. The 26R reservoir management team was composed of staff from geology, petrophysics, reservoir, production, facilities, and other disciplines. Over several years, the staff coalesced into a highly efficient team focused on optimizing hydrocarbon recovery from the reservoir. New reservoir management strategies were devised, or existing strategies were revised as more data were collected, and increased knowledge of reservoir behavior was achieved by the team. For instance, this team was able to adjust gas injection volumes weekly on the basis of pressure data from pressure monitoring wells located in the gas cap and reported total field production volumes. Furthermore, the team was kept generally together for longer than 5 years. When new members were brought in, there were always legacy members who could convey their experience and knowledge of the reservoir to the new members. Keeping the management team of the 26R reservoir stable over a long period contributed to the success achieved in the management of the reservoir.

24.3.3 The Case History of MBB/W31S Reservoirs (1976-1999)

The reservoir management strategy that proved successful in the Main Body "B" (MBB) and Western 31S (W31S) reservoirs can be enumerated as follows:

1. Reduce production of free gas from the gas cap.
2. Maintain reservoir pressure by gas and peripheral water injection.
3. Install extensive program of data collection and monitoring of waterflood performance.
4. Exploit the oil bank created by the waterflood by drilling infill wells.

Geology. The largest of the three anticlines in the Elk Hills Oil Field is the 31S structure (Figure 24.1). The MBB and W31S reservoirs occupy the entire 31S structure which is about 9 miles long and 1.5 miles wide. The 31S structure is an elongated, doubly plunging anticline, which was slightly deflected at its mid region to form two distinct halves, called the MBB and the W31S reservoirs (Figure 24.10). The eastern half of the 31S anticline with formations dipping up to 30 degrees is the MBB reservoir, and the western half of the 31S anticline with formations dipping up to 60 degrees is the W31S reservoir. The two reservoirs were combined for management purposes and called the MBB/W31S reservoirs. The MBB/W31S reservoirs are turbidite, sandstone reservoirs consisting of feldspathic, clay-rich deposits. Core studies show that the turbidite beds are composed of medium to coarse grained dirty sands that grade upward first into fine grained sands, followed by silts and mudstones.[16,17] The complete "Bouma Sequence,"[18] consisting of fining upward cycle of coarse sand, fine sand, silt, and mud were created in a low-density sandy turbidity currents. The Bouma sequences are not complete in all depositional cycles of the MBB/W31S formations. However, where these complete and incomplete Bouma sequences are stacked vertically on top of each other, the resulting formations are very stratified and highly heterogeneous. The Dykstra-Parsons permeability variability coefficients of the MBB/W31S formations range from 0.7 to 0.85.

Several relatively thick shale layers can be correlated over long distances in the MBB and W31S formations. These shale layers have been used to subdivide the MBB and W31S formations into flow units for waterflood management. The flow units in the MBB reservoir are sheet-like sands, whereas the flow units in the W31S reservoir are channelized sands. Some of the flow units in the MBB reservoir can be correlated over distances as long as five miles, whereas the flow units in the W31S reservoir are more sporadic and extend about one-half mile. The MBB and W31S reservoirs overlap in the central portion of the 31S anticline (Figure 24.10).

The OOIP was calculated at 530 million STBO for the MBB reservoir, and 80 million STBO for the W31S reservoir for a total of 610 million STBO for the combined MBB/W31S reservoirs. The range of porosity is 11%–26%. The geometric mean air permeability from core samples is 32.3 md, with a range of 10 to 250 md. The range of initial water saturation is 30%–45%. The mobility ratio for water displacement is favorable at 0.6, with residual oil saturation to water of 25%. The MBB/W31S reservoir rock and fluid properties are summarized in Table 24.2.

TABLE 24.2 Summary of MBB/W31S Reservoir Rock and Fluid Properties

Property	Parameter
Porosity range	11%-26%
Air permeability range	10-250 md
Initial water saturation range	30%-45%
Initial average reservoir pressure	3150 psia
Initial bubble point pressure	2965 psia
Reservoir temperature	210°F
Reservoir oil viscosity	0.40 cp
Oil gravity	36°API
Mobility ratio	0.6
Residual oil saturation to water	25%
Estimated original oil-in-place	610 MMBO

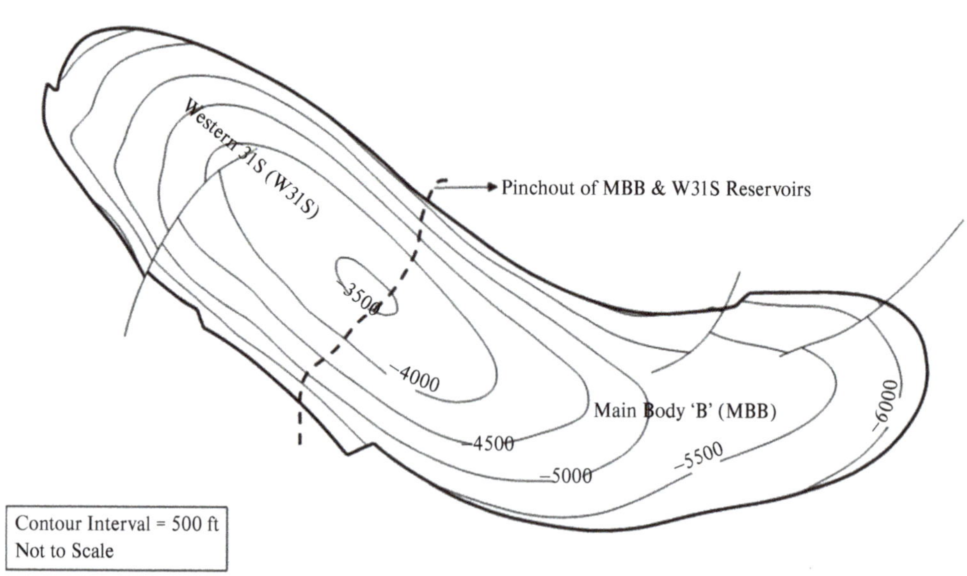

FIGURE 24.10
Structure map of MBB/W31S reservoirs

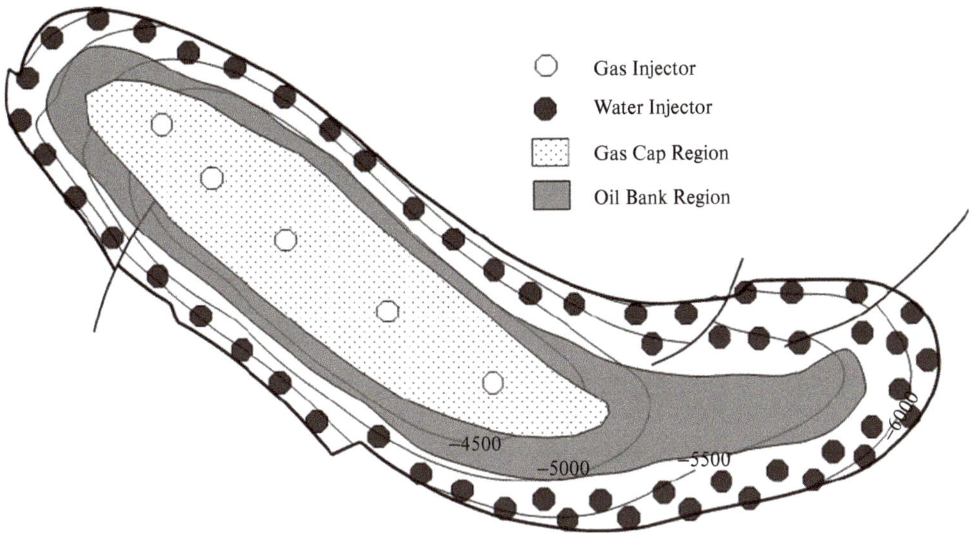

▲ **FIGURE 24.11** **Map of MBB/W31S reservoirs showing the peripheral waterflood project**

Production and Injection History. Before the energy crisis in mid-1970s, the MBB/W31S reservoirs were partially developed and preserved as part of the United States of America Naval Petroleum Reserve to supply oil for naval operations in case of national emergency. Cumulative oil production from initial production in January 1942 through December 1975 was just 1.15 million barrels of oil. Consequently, oil production could be considered to have started in the MBB/W31S reservoirs in June 1976 under primary depletion. Total oil production rate rose rapidly to 25,900 BOPD in May 1977 but was later reduced to 10,765 BOPD toward the end of that year due to rapid decline of reservoir pressure. A pressure maintenance program by gas injection at the crest of the structure was initiated in October 1976. In June 1978, a pilot peripheral waterflood project was started on the southeastern nose of the MBB reservoir. The pilot waterflood project was designed to evaluate the effectiveness of peripheral water injection in displacing oil toward the first line of in-board producers. The pilot waterflood project was very successful because the in-board producers responded with higher oil production rates. This led to the expansion of the peripheral water injection around the entire MBB/W31S reservoirs from 1982 to 1985 (Figure 24.11). The gas injection program was ended in 1988 with cumulative gas injection volume of 117 BCF. Gas injection was terminated because increased water injection was considered sufficient for pressure maintenance.

Peripheral water injection resulted in the development of a significant oil bank within the MBB/W31S reservoirs (Figure 24.11). The oil bank was created updip in areas of the structure where a large secondary gas cap had developed because of the decline in reservoir pressure and the gas injection program. As the oil bank developed, infill wells were drilled on the leading edge of the oil bank. Most wells drilled in the oil bank produced at GORs that were close to solution GOR. Development drilling of infill wells reduced well spacing in the MBB/W31S reservoirs from 20 acres to 10 acres. The production history of MBB/W31S reservoirs is shown in Figure 24.12. The historical gas and water injection rates are shown in Figure 24.13. The historical gas-oil ratio and water-oil ratio for the reservoirs are shown in Figure 24.14. The historical average reservoir pressure of MBB/W31S reservoirs is shown in Figure 24.15.

Displacement Mechanisms. At the start of production, the dominant drive mechanisms in the MBB/W31S reservoirs were by fluid expansion and solution gas drive. These were assisted by gas displacement from expansion of the secondary gas cap. Later, the main mechanism that assisted oil recovery in the MBB/W31S reservoirs is oil displacement by water injected at the periphery of the reservoirs. Oil displacement by waterflooding in the MBB/W31S reservoirs had a favorable mobility ratio of 0.6. Peripheral waterflooding in the MBB/W31S reservoirs over time created an oil bank and shrunk its secondary gas cap as shown in Figure 24.11.

24.3.4 Application of Reservoir Management Principles to MBB/W31S Reservoirs

Conservation of Reservoir Energy. After the MBB/W31S reservoirs were opened up for production in June 1976, there was a rapid decline in reservoir pressure. To conserve reservoir energy, oil production was curtailed from 25,900 BOPD to 10,765 BOPD in 1977. This early intervention measure reduced the growth of the secondary gas cap that formed after average reservoir pressure initially at 3150 psia

▲ **FIGURE 24.12** **Historical production of MBB/W31S reservoirs**

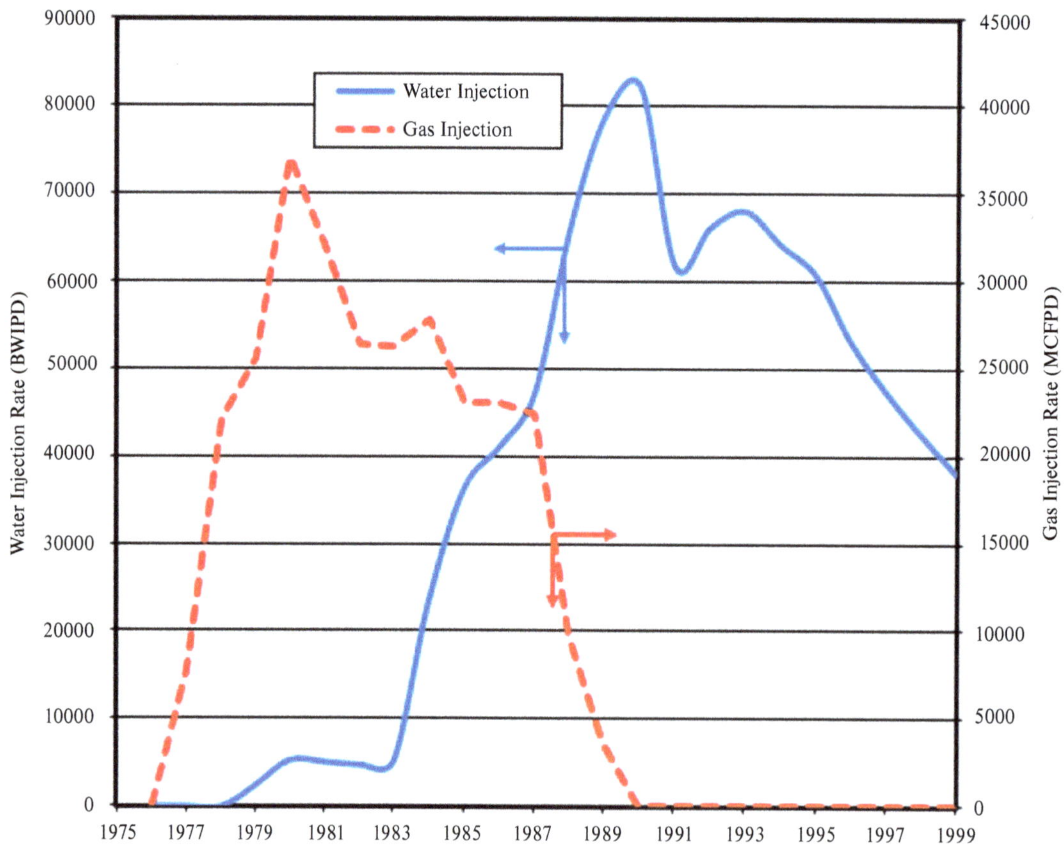

▲ **FIGURE 24.13** **Historical gas and water injection in the MBB/W31S reservoirs**

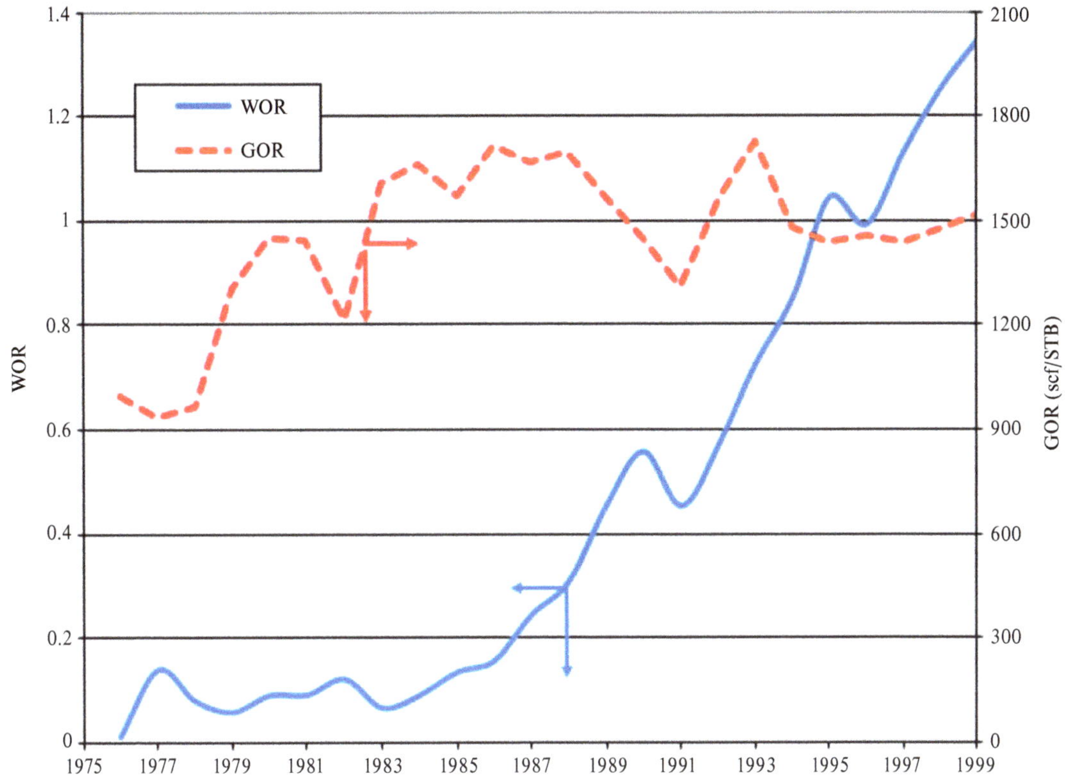

▲ **FIGURE 24.14** Historical GOR and WOR for MBB/W31S reservoirs

▲ **FIGURE 24.15** Historical average reservoir pressure of MBB/W31S reservoirs

declined about 450 psi below the bubble point pressure at 2950 psia. By shutting in high gas-oil ratio wells and increasing water injection, the producing gas-oil ratio of the reservoirs was maintained relatively stable as shown in Figure 24.14.

Early Implementation of Simple, Proven Strategies. Gas injection at the crest of the structure was started in October 1976. This was followed with the start of a pilot water injection project in June 1978. The gas injection project was primarily to maintain reservoir pressure. The pilot peripheral water injection was designed to evaluate oil displacement with water, and also to maintain reservoir pressure. Gas and water injection are simple, proven strategies that can be used to maintain reservoir pressure and improve oil recovery. In the case of MBB/W31S reservoirs, peripheral water injection strategy was selected because there were concerns about premature water breakthrough at the producers for the highly heterogeneous MBB/W31S sands. Also, peripheral water injection was cheaper to install compared with pattern water flood and could be discontinued quickly, if adverse results were observed. Early implementation of these two simple strategies improved oil recovery in the MBB/W31S reservoirs substantially, when compared with the performance of almost identical reservoirs in the area.[19,20]

Systematic and Sustained Practice of Data Collection. An extensive data collection program was undertaken as an integral part of the reservoir management strategy for the peripheral waterflood project in the MBB/W31S reservoirs. This data collection system included pressure monitoring in wells located in key areas of the reservoir, wireline formation test pressures, core samples, and complete log suites from new infill wells, production logs from producing wells, injection profile logs from water injectors, fluid entry surveys with downhole videos, and gamma ray surveys to identify layers showing significant water production. For instance as of 1996, the MBB/W31S reservoirs had a database consisting of conventional core analysis on 9000 samples, and special core analysis on 37 samples.[21] Included in this database are logs for 725 wells, including 413 wells with modern log suites.[22] In the area of pressure data, the MBB reservoir had about 7000 static pressure data, 80 buildups/falloffs, and 43 wells with wireline formation test points.[22] In addition, a well was drilled behind the floodfront for the sole purpose of evaluating the potential for bypassed reserves in the less permeable layers of the MBB reservoir in 1988. As a result of this massive database, the management of the MBB/W31S waterflood project was constantly updated and improved to take advantage of the advancement of the floodfront into unflooded areas of the reservoir.

Application of Emerging Technologies for Improved Hydrocarbon Recovery. The emerging technologies that were used to improve hydrocarbon recovery in the MBB/W31S reservoirs were mainly in the area of monitoring and surveillance of the waterflood project. New tools for fluid entry surveys were routinely used to determine the layers that were being flooded. This assisted the reservoir management team to adjust the water injection profile by installation of multiple packers in water injection wells. For instance, downhole video cameras were especially useful in identifying water producing layers for remedial work to isolate or shut-off water production.

Long-Term Retention of Staff in Multi-Disciplinary Teams. The successful peripheral waterflooding of the MBB/W31S reservoirs can be linked to the formation of a management team composed of staff from diverse disciplines. This team worked together on the management of this project for many years. The team was able to devise new management strategies or modify existing strategies based on analysis of extensive data collected on the waterflood project.

24.3.5 The Case History of the Shaybah Field

The Shaybah field in Saudi Arabia is a low permeability carbonate reservoir overlain by a large gas cap and underlain by a relatively weak aquifer. The Shaybah field was developed initially with massive application of single lateral horizontal wells. Later to improve well productivity, multilateral wells with maximized reservoir contact (MRC) were used to develop the reservoir. The use of horizontal wells and later multilateral wells with MRC as reservoir management strategies were successful in this reservoir due to the following reasons:

1. The formations in the reservoir have low permeability and porosity.
2. Ability to control and reduce producing gas-oil ratio.
3. Ability to control and reduce producing water cut.
4. The need to improve well productivity index.
5. The requirement of achieving production targets.
6. Reduction of unit development cost.
7. Use of smart completion technology.

Geology. The Shaybah field in the Rub' al-Khali desert of Saudi Arabia is about 40 miles long and 8 miles wide.[23] The structure of the field is described as a gently folded northeast/southwest-trending anticline. The formation consists primarily of Cretaceous age sandstones, shales, and carbonates with rudist buildups that change laterally into barrier and shelf slope facies.[23] The average matrix porosity is 25% with no lateral variations. The matrix permeability is facies-dependent and has spatial variations. In south Shaybah, matrix permeability has a range of 5 to 10 md, whereas in the north Shaybah the range of matrix permeability is from 50 to 200 md. Three-dimensional seismic data show that the Shu'aiba carbonate reservoir in the Shaybah field contains a number of faults. These faults have been identified from openhole logs and are more prevalent in the northern part of the reservoir. The Shu'aiba formation contains Arabian Extra Light oil with average API of 42°, and solution gas-oil ratio of 750 scf/STB. The Shaybah field was discovered in 1968 and placed on production in July 1998.

Development History. The surface terrain of the Rub' al-Khali desert, where the Shaybah field is located, comprises of salt flats area called *sabkhahs* and mountainous sand dunes which are as high as 650 feet. Due to the rugged character of the terrain, the Shaybah field was developed from the flat sabkhahs. This required the use of highly directional wells to reach formation targets. The Shaybah field was developed in 1996 with 1-km long, single-lateral horizontal wells designed to drain the oil effectively, while reducing cusping of gas from the gas cap. From 1998 to 2001, reservoir contact was increased by drilling single-lateral wells with lengths of 2 to 3 km based on well and field performance of the 1-km horizontal wells. The longer, single-lateral horizontal wells resulted in significant improvements in well performance, in terms of higher productivity indices, lower drawdowns, and further delays of gas cusping into wells.[23-25] To capitalize on the success of drilling these extra long, single-lateral horizontal wells, the concept of drilling multilateral (ML) and maximum reservoir contact (MRC) wells was developed in Saudi Aramco.[23,26,27]

An MRC well is defined as a well with a minimum aggregate reservoir contact of 5 km either as a single-lateral or a multilateral configuration.[25] The first MRC well was drilled in the Shaybah field in 2002. This was a trilateral well with total reservoir contact of 8.5 km. This was followed by two other MRC wells with reservoir contacts of 5.8 km and 12.3 km, respectively. These three MRC wells had openhole completions. By the end of 2007, a total of 50 ML/MRC wells have been completed in the Shaybah field.[25] MRC wells in the Shaybah field showed reductions in water cut and gas-oil ratio, increases in oil production and well productivity index, and reduction in unit development cost.[23,25]

Displacement Mechanisms. The Shaybah field was described as a low permeability carbonate reservoir overlain by a large gas cap and underlain by a relatively weak aquifer. From this description, it is evident that the primary drive mechanisms for hydrocarbon recovery are gas cap expansion, solution gas drive, and weak water drive from the aquifer. When these drive mechanisms are coupled with the low permeability conditions present in the reservoir, it becomes quite comprehensible that reservoir development with wells that can reduce drawdown, minimize free gas and water production, and achieve high target production rates are desirable. These observations undoubtedly led to the decision to use long single-lateral horizontal wells for the initial development of the reservoir in 1996.

24.3.6 Application of Reservoir Management Principles to the Shaybah Field

It is instructive to apply the five principles of reservoir management to the Shaybah Field and illustrate how the reservoir management strategies devised for this field may have contributed to the success reported for the field.

Conservation of Reservoir Energy. The reservoir energy in the Shaybah field was conserved by the management strategy of using horizontal wells instead of vertical wells for the initial development of the reservoir. The use of horizontal wells minimized production of free gas from the overlying primary gas cap, thereby conserving reservoir energy. It also reduced coning of water from the aquifer which could have reduced hydrocarbon recovery potentials of the wells. The use of horizontal wells minimized the potential for early gas breakthrough, while maintaining production at economic rates.[23]

Early Implementation of Simple, Proven Strategies. Horizontal well technology emerged in the petroleum industry in the early 1980s. In 1996 alone, about 3700 horizontal wells were drilled around the world and the total number of horizontal wells in the world was about 38,000.[28] The same year 1996, the reservoir management team chose to develop the Shaybah field with horizontal wells. This is a clear example of early implementation of simple, proven strategy to maximize hydrocarbon recovery. By 1996 when horizontal well technology was implemented for the development of the Shaybah field, the technology has been proven around the world to be efficient, effective, and economic for developing low permeability reservoirs with large gas caps.

Systematic and Sustained Practice of Data Collection. The database of the Shaybah field was initially created from core and log data of 29 delineation wells drilled before 1976. The database was used to build a facies-based geological model in 1995.[29] As horizontal wells were drilled during the development phase from 1996 to 2002, additional data were obtained from cores and logs leading to the update of the geological model in 1998, and in 2002. Consequently, it appears that a culture of systematic and sustained practice of data collection was a part of the management strategy of the Shaybah field. During drilling of MRC wells, the motherbore was logged with a logging-while-drilling (LWD) tool, and a measurement-while-drilling (MWD) log was run on the laterals. These tools enabled the use of real-time geosteering to improve the efficiency of drilling the ML/MRC wells.[26] Furthermore, the data collected from the closely spaced laterals of the ML/MRC wells improved the distribution of facies in the Shaybah geologic model.[29]

Application of Emerging Technologies for Improved Hydrocarbon Recovery. The Shaybah field is a very good example of the application of emerging technologies for improvement of hydrocarbon recovery. In this case, the management team of the Shaybah field developed the technology of ML/MRC wells to improve hydrocarbon recovery from the field.[23-28] ML/MRC wells were designed for the Shaybah field to take advantage of its low permeability and relatively low porosity, minimize cusping gas from the large overlying primary gas cap, and avoid coning water from the underlying weak aquifer, while maintaining desired target oil production rates. In comparison to single-lateral horizontal wells, ML/MRC wells produced at lower GORs (Figure 24.16), and higher oil rates resulting in much higher cumulative oil productions (Figures 24.17 and 24.18). The productive indices (PIs) of ML/MRC wells were higher than those of single-lateral horizontal wells as shown in Figure 24.19. The costs of ML/MRC wells defined in terms of unit-development costs (drilling + completion + hookup costs per B/D of initial production) are lower when compared with single-lateral horizontal wells. Figure 24.20 shows normalized (relative to 1 km horizontal well) unit-development costs in terms of reservoir contact for single-lateral horizontal wells and ML/MRC wells in South Shaybah field. Consequently, in terms of performance and costs, ML/MRC wells are considerably superior to single-lateral wells justifying the application of this advanced well technology for the development of the Shaybah field.

The management team of the Shaybah field has continued to apply emerging technologies to improved hydrocarbon recovery.[30] This is reflected in the use of smart technologies to improve control of new ML/MRC wells and conversion of existing single-lateral horizontal wells into ML/MRC wells. The smart technologies deployed in the Shaybah field include downhole flow control systems (DHFCS) and permanent downhole monitoring systems (PDHMS). The DHFCS are surface-controlled hydraulic inflow control valves (ICV) and the PDHMS consists of surface-based pressure and temperature monitoring systems. Smart completions have been applied successfully with expandable liners as shown in Figure 24.21 and in open-holes with swellable packers as shown in Figure 24.22. In addition, production equalizer is being used in the Shaybah field as an advanced completion technology to control and reduce GORs of wells.[30] The application of these new technologies clearly demonstrate that the management team of the Shaybah field are proactive in using emerging new technologies to improve hydrocarbon recovery from the field.

Long-Term Retention of Staff in Multi-Disciplinary Teams. Evidence of the application of this principle in the management of the Shaybah field is revealed by reviewing the technical literature published on the Shaybah field.[23-31] It is worthy to note that the names of the same authors have reoccurred in most of these publications. It can be readily deduced from these publications that these authors have worked consistently on the Shaybah field from the drilling of the single-lateral horizontal wells to drilling of ML/MRC wells and application of smart well technologies. There is little doubt that this long-term retention of staff on the Shaybah field has contributed to its resounding success.

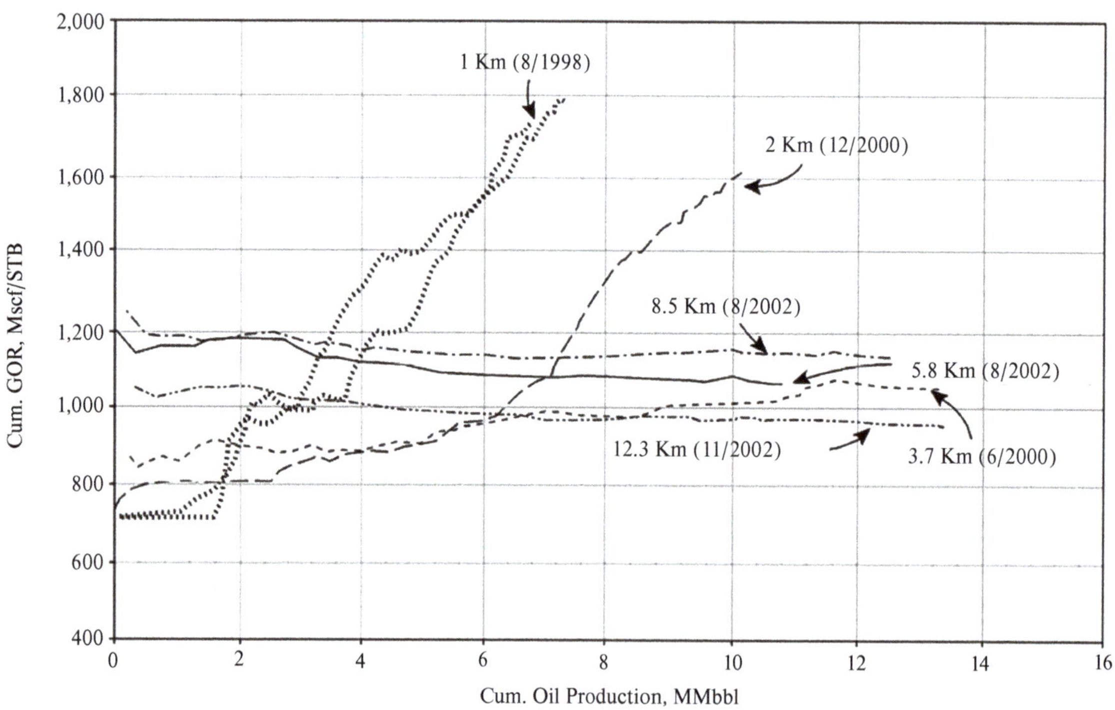

▲ **FIGURE 24.16** Shaybah Field GOR plot based on reservoir contact-South area wells (from Salamy et al.[25] © 2008 SPE, Reproduced with permission)

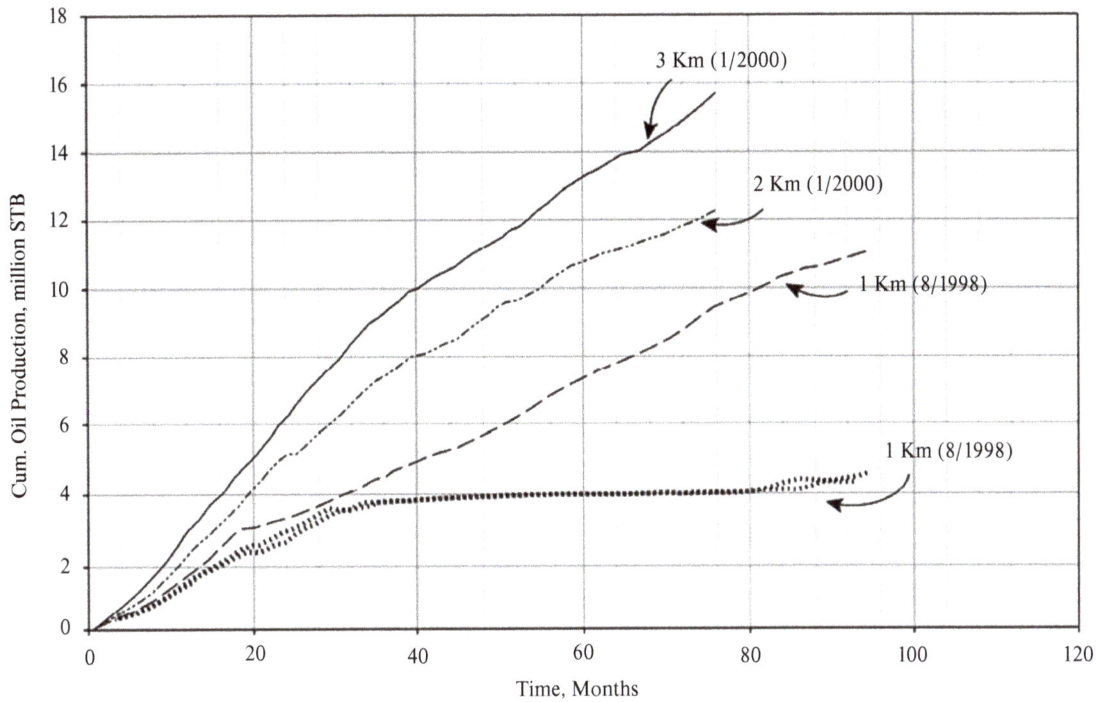

▲ **FIGURE 24.17** **Shaybah Field cumulative oil production vs. time based on reservoir contact-North Area wells (from Salamy et al.[25] © 2008 SPE, Reproduced with permission)**

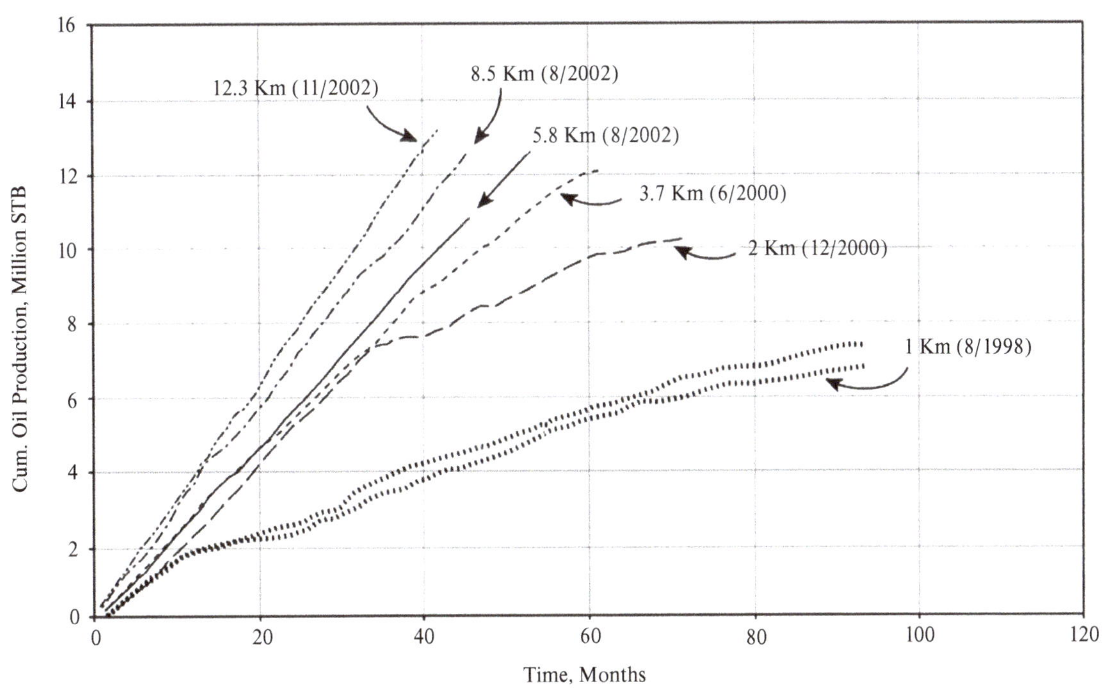

▲ **FIGURE 24.18** **Shaybah Field cumulative oil production vs. time based on reservoir contact-South Area wells (from Salamy et al.[25] © 2008 SPE, Reproduced with permission)**

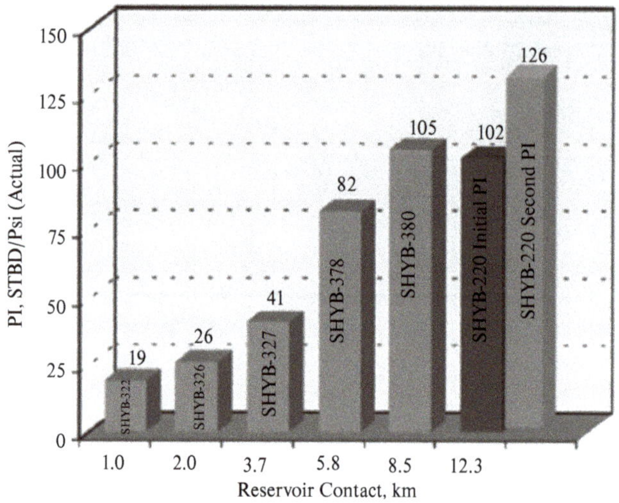

▲ **FIGURE 24.19** **Productivity index versus reservoir contact in South Shaybah (from Saleri et al.[23] © 2008 SPE, Reproduced with permission)**

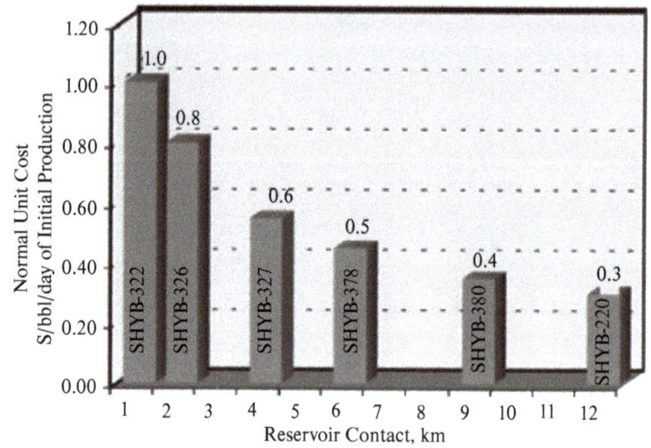

▲ **FIGURE 24.20** **Normalized unit cost versus reservoir contact in South Shaybah (from Saleri et al.[23] © 2008 SPE, Reproduced with permission)**

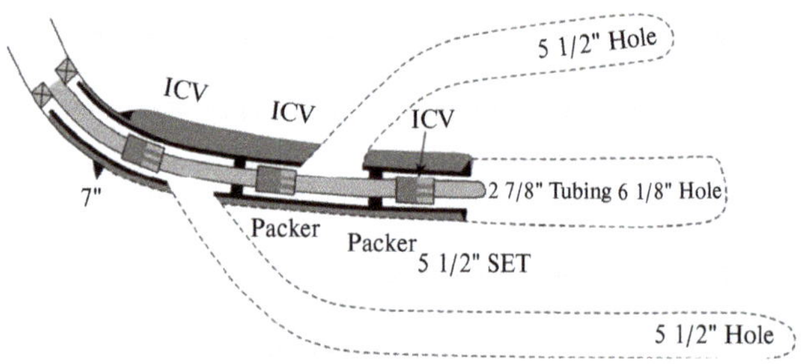

▲ **FIGURE 24.21** **Smart completion inside 5 1/2" expandable liner (from Salamy et al.[25] © 2008 SPE, Reproduced with permission)**

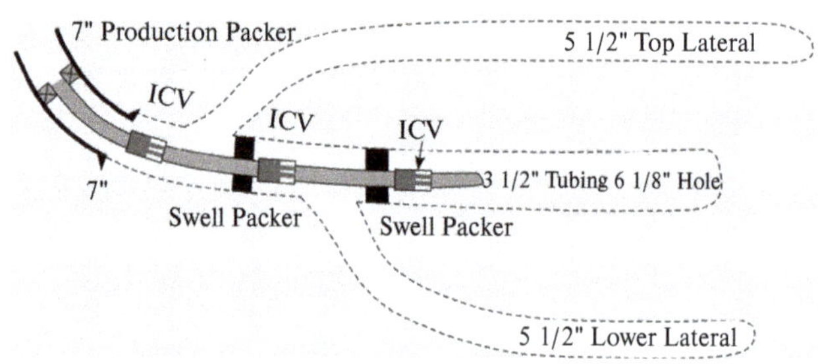

▲ **FIGURE 24.22** **Swell-packer in openhole smart completion (from Salamy et al.[25] © 2008 SPE, Reproduced with permission)**

REFERENCES

1. Ezekwe, J.N.: "Applied Reservoir Management Principles with Case Histories," paper SPE 84148 presented at the SPE Annual Technical Conference and Exhibition, Denver, Colorado, (October 5-8, 2003).
2. Konopczynski, M.R., and Ajayi, A.: "Reservoir Surveillance, Production Optimisation and Smart Workflows for Smart Fields–A Guide for Developing and Implementing Reservoir Management Philosophies and Operating Guidelines in Next Generation Fields," paper SPE 112244 presented at the 2008 SPE Intelligent Energy Conference and Exhibition, Amsterdam, The Netherlands, (February 25-27, 2008).
3. Mubarak, S.M.: "Real-Time Reservoir Management from Data Acquisition Through Implementation: Closed-Loop Approach," paper SPE 111717 presented at the 2008 SPE Intelligent Energy Conference and Exhibition, Amsterdam, The Netherlands, (February 25-27, 2008).
4. Jansen, J.D., Douma, S.D., Brouwer, D.R., Van den Hof, P.M.J., Bosgra, O.H., and Heemink, A.W.: "Closed-Loop Reservoir Management," paper SPE 119098 presented at the 2009 SPE Reservoir Simulation Symposium, The Woodlands, Texas, (February 2-4, 2009).
5. Meader, T., Allen, F., and Riley, G.: "To the Limit and Beyond-The Secret of World-Class Extended-Reach Drilling Performance at Wytch Farm," paper IADC/SPE 59204 presented at the 2000 IADC/SPE Drilling Conference, New Orleans, Louisiana, (February 23-25, 2000).
6. Yaliz, A., Chapman, T., and Downie, J.: "Case Study of a Quad-Lateral Horizontal Well in the Lennox Field: A Triassic Oil-Rim Reservoir," paper SPE 75249 presented at the SPE/DOE Improved Oil Recovery Symposium, Tulsa, Oklahoma, (April 13-17, 2002).

7. Mercado, O., Velez, J., and Fipke, S.: "Multilateral Wells in the Castilla Field of Eastern Colombia: A Case History of the Guadalupe Reservoir," paper SPE/IADC 119458 presented at the SPE/IADC Drilling Conference and Exhibition, Amsterdam, The Netherlands, (March 17-19, 2009).
8. Wei, M.H., Yu, J.P., Moore, D.M., Ezekwe, N., Querin, M., and Williams, L.L.: "Case History of Pressure Maintenance by Crestal Gas Injection in the 26R Gravity Drainage Reservoir," paper SPE 24035 presented at the Western Regional Meeting, Bakersfield, California, (March 20-April 1, 1992).
9. Essley, P.L., Jr., Hancock, G.L., Jr., and Jones, K.E.: "Gravity Drainage Concepts in a Steeply Dipping Reservoir," paper SPE 1029-G, SPE of AIME, Petroleum Conference on Production and Reservoir Engineering, Tulsa, Oklahoma, (March 20-21, 1958).
10. Lewis, J.O.: "Gravity Drainage in Oil Fields," *Trans. AIME* (1944) 155, 133-154.
11. Fleckenstein, W.W., Bouck, L.S., Hudgens, P.D., Querin, M.E., and Williams, L.L.: "An Application of Oil Vaporization Evaluation Methods," paper SPE 24041 presented at the 1992 Western Regional Meeting, Bakersfield, California, (March 30-April 1, 1992).
12. Echols, D.P., and Ezekwe, N.: "A Reservoir Management Model for 26R Reservoir, Elk Hills Oil Field, California," paper SPE 46231 presented at the 1998 SPE Western Regional Meeting, Bakersfield, California, (May 10-13, 1998).
13. Ezekwe, N., Echols, D., and Hudgens, P.: "Diagnostic Application of Reservoir Simulation to 31S Structure Stevens Reservoirs–Elk Hills Oil Field, California paper SPE 46230 presented at the 1998 SPE Western Regional Meeting, Bakersfield, California, (May 10-13, 1998).
14. Gangle, F.J., Schultz, K.L., McJannet, G.S., and Ezekwe, N.: "Improved Oil Recovery Using Horizontal Wells at Elk Hills, California," *SPE Drilling and Completion* (March 1995) 27-33.
15. Mut, D.L., Moore, D.M., Thompson, T.W., and Querin, E.M.: "Horizontal Well Development of a Gravity Drainage Reservoir, 26R Pool, Elk Hills Field, California," paper SPE 35710 presented at the 1996 SPE Western Regional Meeting, Anchorage, Alaska, (May 22-24, 1996).
16. Ezekwe, N., Smith, S., Wilson, M., Humphrey, M., and Murphy, J.: "Performance of a Peripheral Waterflood Project in the Main Body 'B' Reservoirs (Stevens) at the Elk Hills Oil Field, California" paper SPE 21759 presented at the Western Regional Meeting, Long Beach, California, (March 20-22, 1991).
17. Walker, T., Kerns, S., Scott, D., White, P., Harkrider, J., Miller, C., and Singh, T.: "Fracture Stimulation Optimization in the Redevelopment of a Mature Waterflood, Elk Hills Field, California," paper SPE 76723 presented at the SPE Western Regional/AAPG Pacific Section Joint Meeting, Anchorage, Alaska, (May 20-22, 2002).
18. Bouma, A.H.: *Sedimentology of Some Flysch Deposits: A Graphical Approach to Facies Interpretation,* Elsevier, Amsterdam (1962).
19. Underdown, D.R., Schultz, D.J., Marino, A., and Miranda, R.: "Optimizing Production by Fluid/Rock Studies for the Stevens Sands, North Coles Levee Field, Kern County, California," paper SPE 21785 presented at the Western Regional Meeting, Long Beach, California, (March 20-22, 1991).
20. MacAllister, D.J.: "Evaluation of CO_2 Flood Performance: North Coles Levee CO_2 Pilot, Kern County, California," paper SPE 15499 presented at the 61st Annual Technical Conference and Exhibition, New Orleans, Louisiana, (October 5-8, 1986).
21. Starcher, M.G., Engineer, R., and Wilson, M.: "Case History of the 31S Peripheral Waterflood Project, Stevens Zone, Elk Hills Field, California," paper SPE 35673 presented at the 1996 SPE Western Regional Meeting, Anchorage, Alaska, (May 22-24, 1996).
22. Hampton, T.J., Danielli, C., McChesney, D., and Singh, S.P.: "Fullfield Numerical Simulation Using Geostatistical Reservoir Model of Upper Main Body "B" Reservoir, Monterey Formation, Elk Hills, California," paper SPE 68879 presented at the SPE Western Regional Meeting, Bakersfield, California, (March 26-30, 2001).
23. Saleri, N.G., Salamy, S.P., Mubarak, H.K., Sadler, R.K., Dossary, A.S., and Muraikhi, A.J.: "Shaybah-220: A Maximum-Reservoir-Contact (MRC) Well and its Implications for Developing Tight-Facies Reservoirs," *SPEREE* (August 2004) 316-321.
24. Saleri, N.G.: " 'Learning Reservoirs': Adapting to Disruptive Technologies," *JPT* (March 2002) 57-60.
25. Salamy, S.P., Al-Mubarak, H.K., Ghamdi, M.S., and Hembling, D.: "Maximum-Reservoir-Contact-Wells Performance Update: Shaybah Field, Saudi Arabia," *SPE Production & Operation* (November 2008) 439-443.
26. Dossary, A.S., and Mahgoub, A.A.: "Challenges and Achievements of Drilling Maximum Reservoir Contact (MRC) Wells in Shaybah Field," paper SPE 85307 presented at the SPE/IADC Middle East Drilling Technology Conference and Exhibition, Abu Dhabi, UAE, (October 20-22, 2003).
27. Saleri, N.G., Salamy, S.P., and Al-Otaibi, S.S.: "The Expanding Role of the Drill Bit in Shaping the Subsurface," *JPT* (December 2003) 53-58.
28. Salamy, S.P.: "A New Generation of Wells for Developing Tight Reservoir Facies," Paper SPE 108806, *SPE Distinguished Lecture Series Program,* 2004-2005.
29. Al-Awami, F., Hemanthkumar, K., Salamy, S., and Sadler, R.: "A Mega-Cell Simulation Model of the Shaybah Field," paper IPTC 10240 presented at the International Petroleum Technology Conference, Doha, Qatar, (November 21-23, 2005).
30. Salamy, S.P., Al-Mubarak, H.K., Hembling, D.E., and Al-Ghamdi, M.S.: "Deployed Smart Technologies Enablers for Improving Well Performance in Tight Reservoirs–Case: Shaybah Field, Saudi Arabia," paper SPE 99281 presented at the SPE Intelligent Energy Conference and Exhibition, Amsterdam, The Netherlands, (April 11-13, 2006).
31. Salamy, S.P., Al-Mubarak, S., Al-Mubarak, H., Al-Dawood, N., and Al-Alawi, A.: "Maximum Reservoir Contact Wells: Six Years of Performance-Lessons Learned and Best Practices," paper SPE 118030 presented at the 2008 Abu Dhabi International Petroleum Exhibition and Conference, Abu Dhabi, UAE, (November 3-6, 2008).

ADDITIONAL REFERENCES

- Al-Hussainy, R., and Humphreys, N.: "Reservoir Management: Principles and Practices," *JPT* (December 1996) 1129-1135.
- Hembling, D., Salamy, S., Qatani, A., Carter, N., and Jacob, S.: "Swell Packers: Enabling Openhole Intelligent and Multilateral Well Completions for Enhanced Oil Recovery," paper IADC/SPE 100824 presented at the IADC/SPE Asia Pacific Drilling Technology Conference and Exhibition, Bangkok, Thailand, (November 13-15, 2006).
- Hill, A.D., Zhu, D., and Economides, M.J.: *Multilateral Wells,* SPE, Richardson, Texas (2008).
- Mubarak, S., Dawood, N., and Salamy, S.: "Lessons Learned from 100 Intelligent Wells Equipped with Multiple Downhole Valves," paper SPE 126089 presented at the 2009 SPE Saudi Arabia Section Technical Symposium, Al Khobar, Saudi Arabia, (May 9-11, 2009).
- Saputelli, L., Nikolaou, M., and Economides, M.J.: "Self-Learning Reservoir Management," *SPEREE,* (December 2005) 534-547.
- Wang, C., Li, G., and Reynolds, A.C.: "Production Optimization in Closed-Loop Reservoir Management," paper SPE 109805 presented at the 2007 SPE Annual Technical Conference and Exhibition, Anaheim, California, (November 11-14, 2007).

PROBLEMS

24-1. A concise case history of the Agbami Field is presented to assist in the discussion of implemented reservoir management strategies. Review the case history and discuss the questions following the case history.

Location. The Agbami Field straddles Blocks OML (Oil Mining Lease) 127 and OML 128 located approximately 70 miles offshore Nigeria in the Gulf of Guinea in about 4800 feet of water (Figure 24P.1). After unitization of OML 127 and 128, the proportional unit ownerships as of June 2016 are: FAMFA (Chevron) 49.6%, Petrobras 12.4%, Statoil 17.54%, and Chevron 20.46%. Chevron is the operator of Agbami Field.

Geology. The Agbami geologic structure consists of a doubly plunging anticlinal structure cut by a major thrust fault along its crestal axis with several other minor faults that further partitioned the structure into smaller fault blocks (Figure 24P.2). The reservoirs are deepwater turbidite sand deposits (Figure 24P.3), and the structure trends in the northwest-southeast direction. The Agbami structure is dominated by four major stacked reservoir units identified sequentially as 13MY, 14MY, 16MY, and 17MY. OOIP is estimated at 2.5 billion barrels with about 80% contained in the 17MY reservoir. A stratigraphic cross-section of the 17MY reservoir is shown in Figure 24P.4.

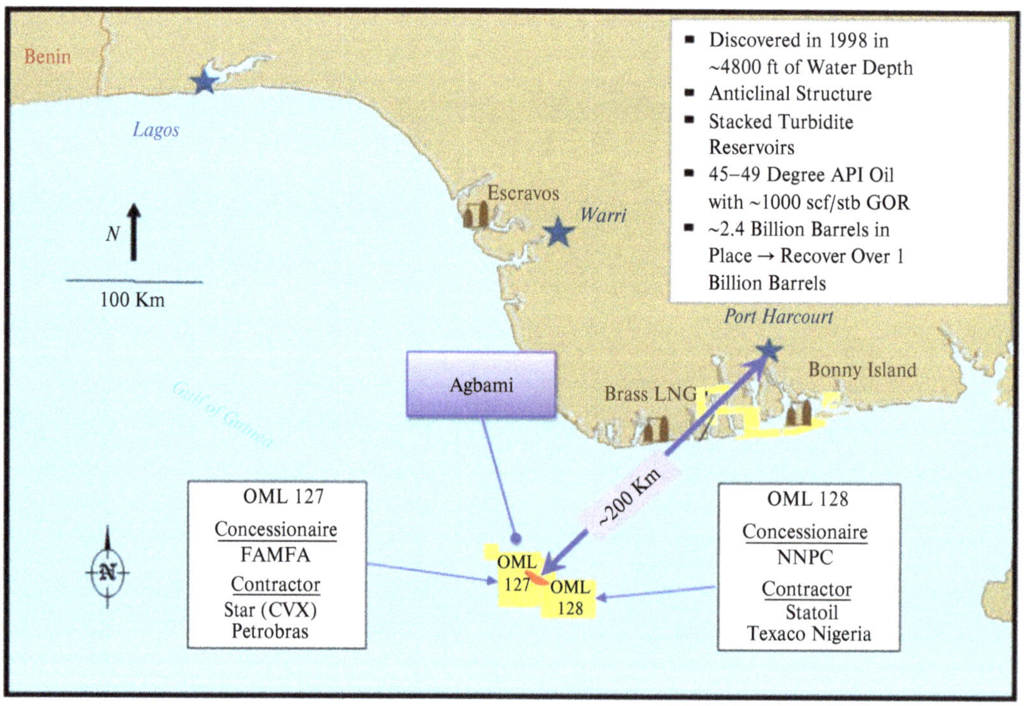

▲ **FIGURE 24P.1** **Location map for Agbami Reservoir, Offshore Nigeria**

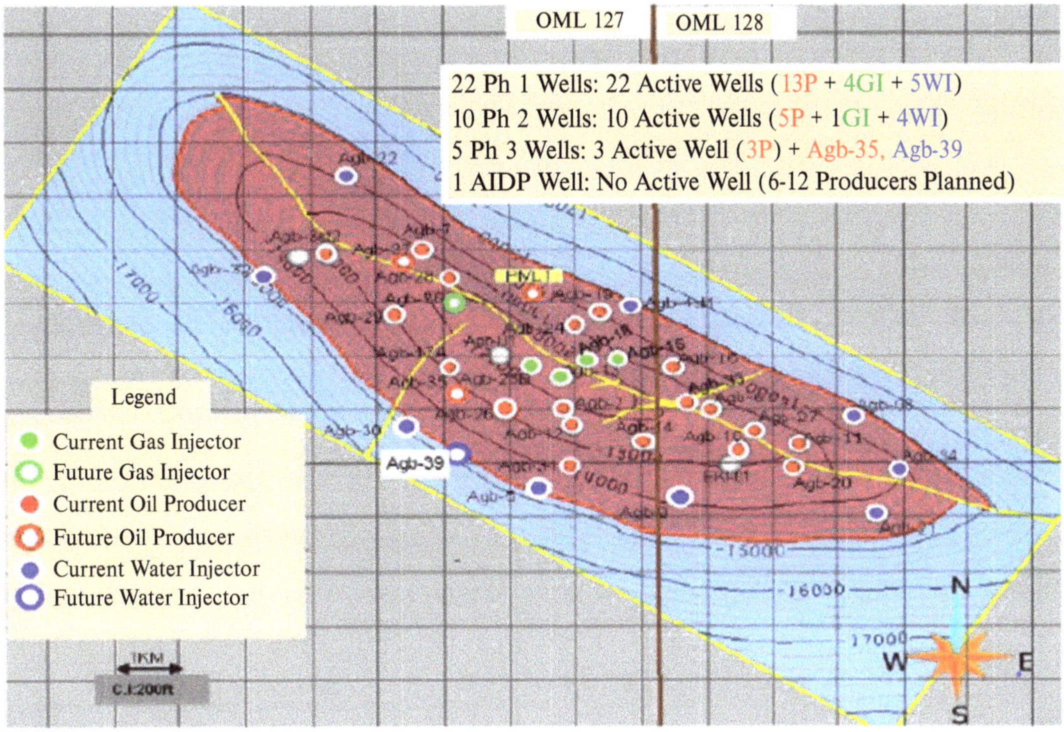

▲ **FIGURE 24P.2** **Agbami 17MY top structure map showing well locations (June 2016)**

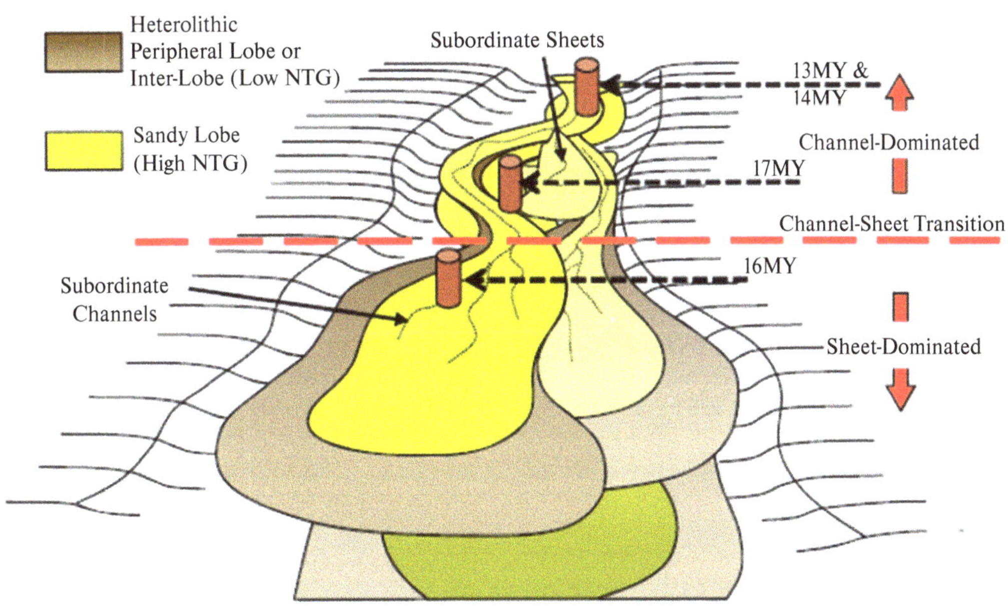

▲ **FIGURE 24P.3** **Agbami conceptual depositional model**

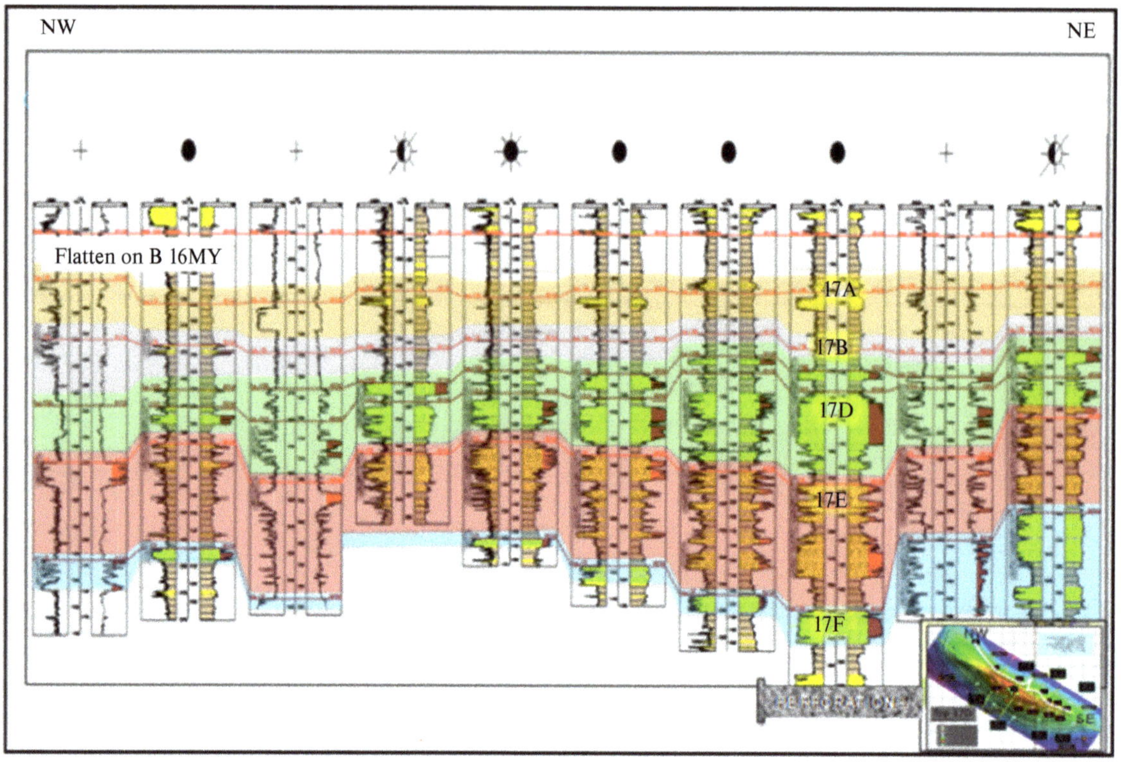

▲ FIGURE 24P.4 **Stratigraphic cross-section through 17MY reservoir**

Rock and Fluid Properties. The Agbami reservoirs have excellent rock and fluid properties that justify ranking them as one of the best and most prolific reservoirs in the world. A summary of the rock and fluid properties is provided in Table 24P.1. Note that the hydrocarbon fluids in the reservoirs are highly undersaturated. This thermodynamic condition is highly conducive to the reservoir management strategy implemented since initial production in 2008. This reservoir management strategy is primarily full pressure maintenance by crestal gas re-injection and peripheral water injection to retain most of the reservoir in undersaturated state and limit the growth of a secondary gas cap.

Development and Production History. Agbami reservoirs were discovered with Agbami No. 1 well drilled in late 1998. An appraisal program consisting of five wells and one sidetrack well was conducted from late 1999 to middle of 2001. After the appraisal program, a Field Development Plan (FDP) was approved. The FDP consisted of a totally subsea infrastructure tied to a Floating Production Storage Offloading (FPSO) vessel through manifolds, flowlines, and umbilicals with a Taught Anchor Leg Mooring (TALM) buoy offloading system. The original FDP proposed a total of 38 subsea wells that consisted of 22 oil producers, 11 water injectors, and 5 gas injectors. The FDP was a three-phased development plan. Phase 1 included the FPSO, subsea and topsides infrastructure, and 22 wells. Phases 2 and 3 planned 10 and 5 development wells, respectively. Since the original FDP, a Phase 4 Stage 1 (called AIDP- Agbami Infill Well Development Program) consisting of 6 infill wells is currently in progress (as of June 2016). This is expected to be followed by a Phase 4 Stage 2 with additional 6 wells made up of 4 producers, 1 water injector and 1 gas injector. A summary of the wells in the FDP is shown in Table 24P.2.

TABLE 24P.1 Average Rock/Fluid Properties for Agbami Reservoirs

Rock/Fluid Property	Average Property Value
Permeability	1045 mD
Initial water saturation	24%
Porosity	21%
Net-to-gross	83%
Net Pay	95.1 ft
Oil API gravity	45 degree
Oil viscosity	0.40 cp
Initial formation volume factor	1.62 rb/STB
Initial solution gas-oil ratio	1224 scf/STB
Bubble point pressure	3130 psi
Initial reservoir pressure	5687 psi

TABLE 24P.2 Summary of wells for Agbami Field FDP (June 2016)

	Phase 1	Phase 2	Phase 3	Phase 4 (Stages 1 & 2)	Total
Producer	13	5	4	10	32
Water Injector	5	4	1	1	11
Gas Injector	4	1	—	1	6
Total	**22**	**10**	**5**	**12**	**49**

Most of the producers and injectors in Agbami reservoirs are equipped with intelligent well appliances that enabled systematic and sustained data collection. An intelligent well completion (IWC) of a producer is shown in Figure 24P.5. IWCs allowed for focused monitoring of wells and rapid adjustments of well performance.

4D seismic technology is currently used in Agbami Field to monitor spatial movement of reservoir and injected fluids, and to optimize placement of infill wells. The data collected from IWCs and 4D seismic have been used to progressively update various generations of Agbami Field reservoir models. Since first oil in 2008, about six generations of Agbami Field reservoir models have been built or updated as more knowledge is gained on the reservoirs.

The Agbami Field posted first oil at initial total production rate of about 95,000 bopd on July 28, 2008. Peak total production rate of 250,000 bopd was reached on August 13, 2009. The field has been maintained at close to peak production since this date. This is a remarkable feat for any reservoir anywhere in the world. Cumulative oil production from Agbami field through January 2018 is approximately 776 million barrels of oil. The historical average monthly oil production rate through January 2018 is shown in Figure 24P.6.

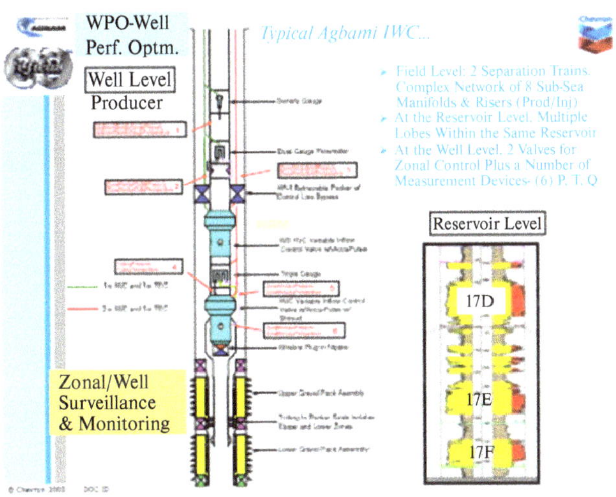

▲ **FIGURE 24P.5** **Example of Intelligent well completion for an Agbami well**

Based on the case history of Agbami reservoirs presented above and additional information in the references cited below:

a. Identify the main oil displacement mechanisms occurring in the Agbami reservoirs.

b. Categorize the reservoir management strategies implemented in the Agbami Field since initial production under the five main principles of reservoir management presented in this book.

c. Discuss potential reservoir management strategies that could be implemented in future to maintain or improve reservoir performance in the Agbami Field.

Note: Additional references on the Agbami Field:

- Adeyemo, K.M., Aigbe, C., Chukwumaeze, I.C., Meinert, D., and Shyrock, S.: "Intelligent Well Completions in Agbami: A Review of the Value Added and Execution Performance," paper OTC 20191 (2009).
- Jeboda, O., Bodunrin, A., Atuanya, C., and Ogidi, E.: "Reservoir Management Best Practices and Lessons Learned from the Agbami Field," paper SPE 193471 (2018).
- Narahara, G.M., Spokes, J.J., Brennan, D.D., Maxwell, G., and Base, M.: "Incorporating uncertainties in Well-Count Optimization with Experimental Design for the Deepwater Agbami Field," SPE RE&E December 2005; SPE 91012-PA.
- Odusote, F.: "Deepwater Nigeria Field Development: Challenges, Best Practices and Lessons Learned from the Agbami Field," paper SPE 167534 presented at the SPE Nigeria Annual International Conference and Exhibition, Lagos, Nigeria, (August 5-7, 2013).
- Olatunde, F., Adeyinka, A., Lawal, O., Iyiola, S., Faparusi, D., Bodunrin, A., Sustakoski, R., Ebo, H.: "Time Lapse Seismic Survey: Adding Value in the Agbami Field," paper SPE 189104 (2017).
- Sunmonu, O., Ogbuagu, F., Ebo, H., and Odusote, O.: "Material Balance Validation of Subsurface Uncertainties in a Deepwater Turbidite System: A Case Study of the Agbami Field," paper SPE 172839-MS presented at SPE Nigeria Annual International Conference and Exhibition, Lagos, (August 05-07, 2014).

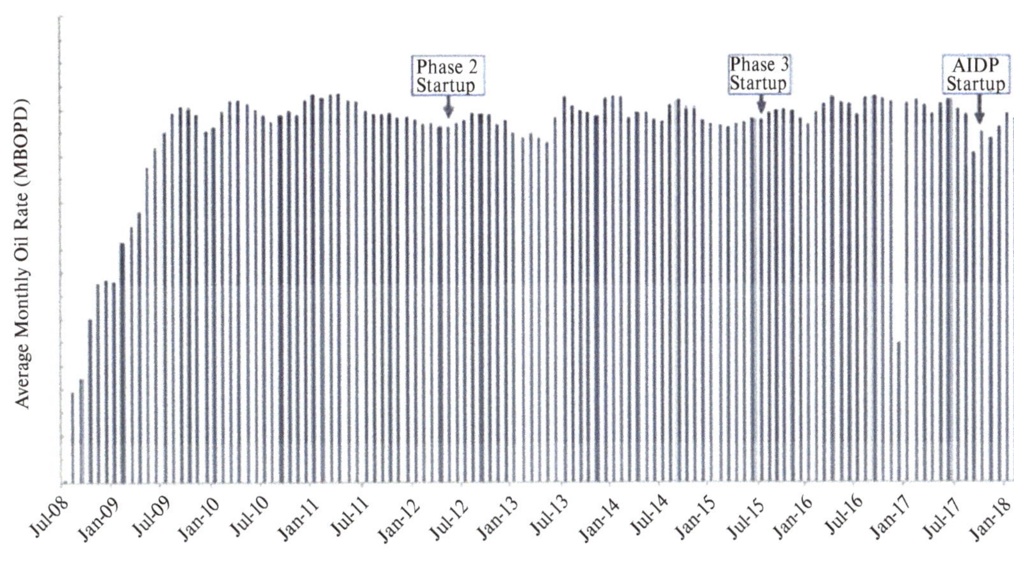

▶ **FIGURE 24P.6** **Agbami Field monthly oil production history (through January 2018)**

24-2. An abbreviated case history of the Akpo Field is presented to assist in the discussion of applied reservoir management strategies. Review the case history and discuss the questions following the case history.

Location. The Akpo field is located in OML 130 about 125 miles offshore south of Port Harcourt in water depth ranging from 3640 to 5576 feet (Figure 24P.7). The proportional unit ownerships of OML 130 are CNOOC (45%), Total (24%), Petrobras (16%), and SAPETRO (15%) as of June 2016. The operator of Akpo Field is Total (TUPNI).

Geology. The Akpo Field is a large anticlinal structure consisting of turbidite channels and lobes deposited in three distinct accumulations (eastern, central, and western) as shown in Figure 24P.8. The lithology is sandstones of Upper Miocene and Middle Miocene age. Five main reservoir series have been identified from seismic interpretations and well correlations. These main reservoir series are: "A" reservoir consisting of "A" Upper and "A" Lower, "BC" reservoir, "D" reservoir, "EF" reservoir, and "G" reservoir. The distribution of these reservoir series across the structure is shown in a cross-section in Figure 24P.9. A structure map of the Akpo field with well locations is shown in Figure 24P.10.

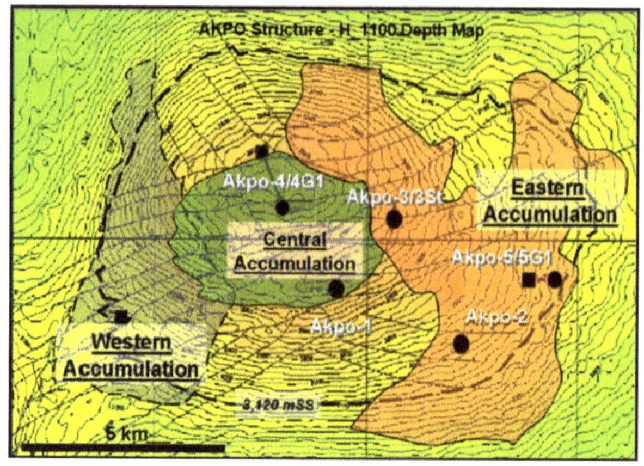

▲ FIGURE 24P.8 Overview of Accumulations in Akpo Structure

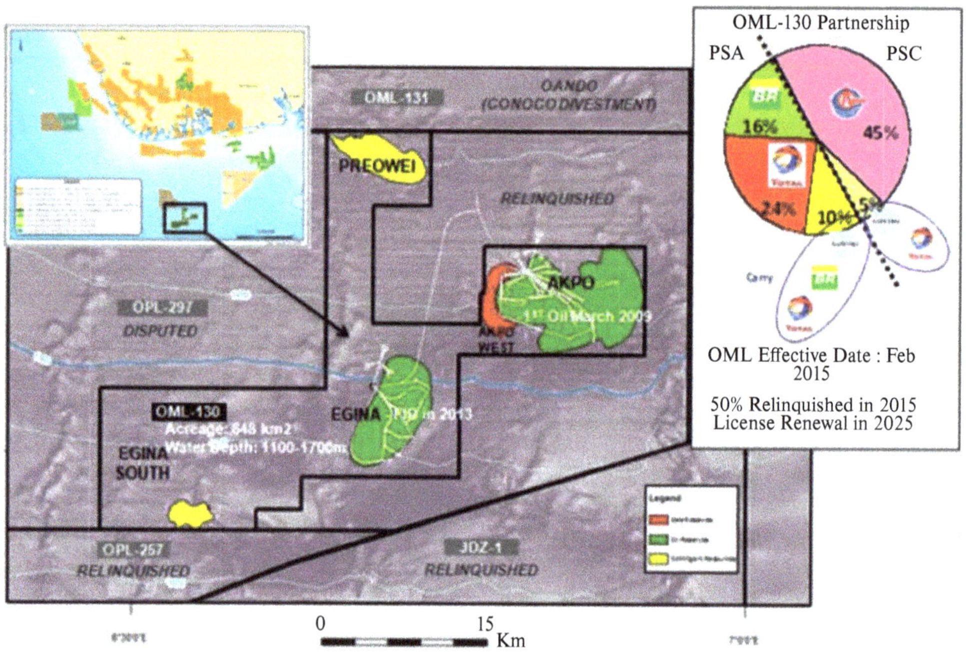

▶ FIGURE 24P.7 Map showing Akpo location (June 2016)

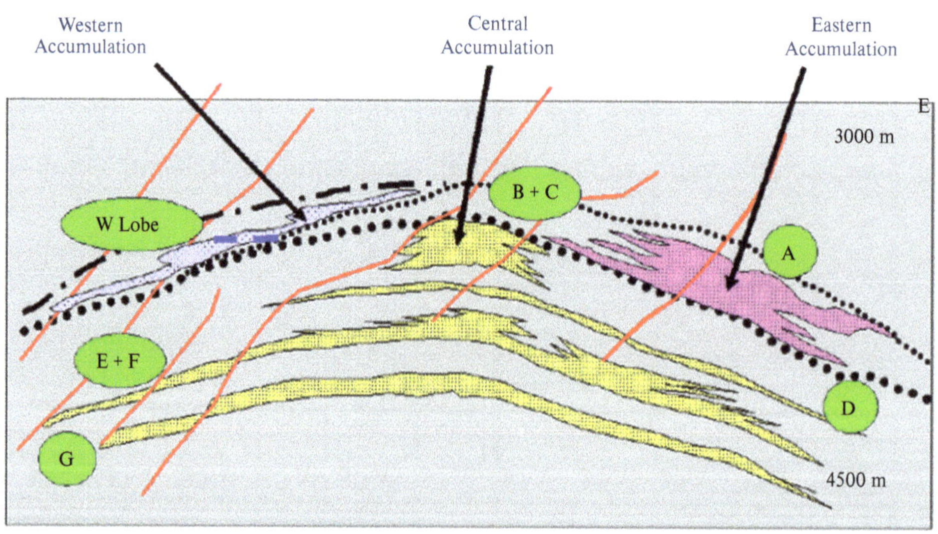

▶ FIGURE 24P.9 An East-West Cross-section through Akpo Structure

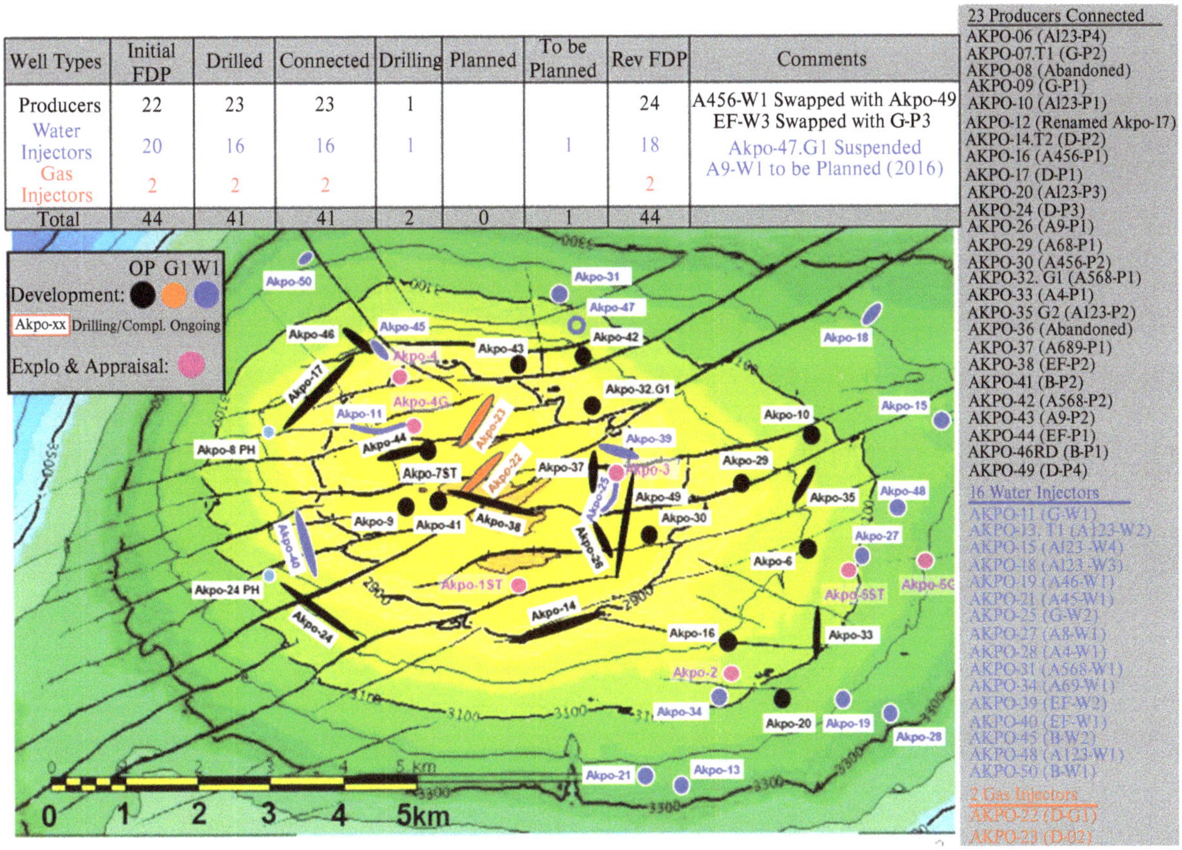

▲ **FIGURE 24P.10** **Akpo Structure Map showing Well Locations (June 2016)**

Rock and Fluid Properties. The fluids in Akpo reservoirs exist in a critical thermodynamic state. In common terms, this means that the fluids exist neither in a liquid nor gaseous state in the reservoirs. This has led to the misidentification of Akpo reservoirs as condensate reservoirs. Akpo reservoir fluids produce condensate at surface or separator conditions. But the fluids existed in critical states under initial reservoir conditions. Due to the nature of fluids in Akpo reservoirs, the fluid properties data vary significantly with depth, temperature, and pressure. This has been reflected by providing these data in the form of a range as shown in Table 24P.3. Total OOIP in all reservoirs is estimated at 1.293 billion barrels based on separator conditions.

Development and Production History. Akpo Field was discovered in April 2000 by Akpo-1 exploration well. An appraisal program was conducted by drilling four additional wells between June 2000 and July 2002. A Field Development Plan (FDP) was approved and started in 2003. This FDP (rev 0) was based on 44 subsea wells with water injection in reservoirs A, BC, EF, and G, gas re-injection in Reservoir D, and export of remaining gas to Bonny NLNG through Amenam. The distribution of wells in the FDP (rev 0) and subsequent revision in FDP (rev 1) are shown in Table 24P.4 and Figure 24P.10. The FDP was completed in February 2016. Note that in FDP (rev 1), two wells designated as water injectors in FDP (rev 0) were re-assigned as oil producers (Figure 24P.10). An infill well program targeting five potential candidates was initiated in 2016. The first well (E-P3) of the program was drilled in 2016. The remaining four wells have not yet been approved. The FDP (rev 0) is a fully subsea infrastructure tied to a spread-moored Floating Production Storage Offloading (FPSO) vessel through manifolds, flowlines, and umbilicals. The FPSO has the

TABLE 24P.3 Rock/Fluid Properties Data for Akpo Reservoirs

Akpo Rock/Fluid Property	Property Value Range
Permeability	400-1200 mD
Initial water saturation	10%-25%
Porosity	17%-25%
Net-to-gross	40%-80%
Net Pay	56–216 ft
Oil API gravity	44–50 degree
Oil viscosity	0.06–0.15 cp
Initial formation volume factor	2.0–4.20 rb/STB
Initial gas-liquid ratio	1965–5615 scf/STB
Saturation pressure	4438–5642 psi
Initial reservoir pressure	4743–6048 psi

following capacities: 2 million barrels of oil storage, 175,000 bopd average oil production, 235,000 bopd total liquids processing, 530 MMcfpd produced fluids, 215 MMcfpd gas injection, 320 MMcfpd export gas, 150,000 bwpd produced water, and 420,000 bwpd water injection. The capacities of the FPSO are summarized in Figure 24P.11.

The critical state of fluids in the Akpo reservoirs has led to the implementation of full pressure maintenance to ensure that the fluids remain at this critical state to avoid condensation in the reservoirs. Water is being injected into the "A," "BC," "EF," and "G" reservoirs, and gas is re-injected into the "D" reservoir, primarily for pressure maintenance with some benefit of secondary recovery. Field production started in March 2009. The field is producing with increasing water-cut especially from the "A" reservoirs.

4D seismic technology is currently applied in Akpo reservoirs to monitor spatial movement of reservoir fluids and injected water, and to optimize placement of infill wells. 4D seismic data have been very beneficial in locating floodfronts and bypassed reserves particularly in the "A" reservoirs. The data collected from 4D seismic have been used to progressively update various generations of Akpo reservoir models. Total cumulative production through August 2016 is 406 million barrels of oil. The production/injection history through March 2016 is shown in Figure 24P.12.

TABLE 24P.4 | Well distributions in Akpo Field (FDP rev 1, June 2016)

	Reservoir Unit	Phase 1	Phase 2	Total
Producers	A upper	3	2	5
	A Lower	2	6	8
	BC	–	2	2
	D	3	1	4
	EF	–	2	2
	G	2	1	3
	Sub Total	10	14	24
Water Injectors (2 WI waiting for results of the M2)	A upper	4	1	5
	A Lower	3	2	5
	BC	–	2	2
	D	–	–	0
	EF	–	2	2
	G	2	–	2
	Unassigned		2	2
	Sub Total	9	9	18
Gas Injectors	A upper	–	–	0
	A Lower	–	–	0
	BC	–	–	0
	D	2	–	2
	EF	–	–	0
	G	–	–	0
	Sub Total	2	0	2
Total		**21**	**23**	**44**

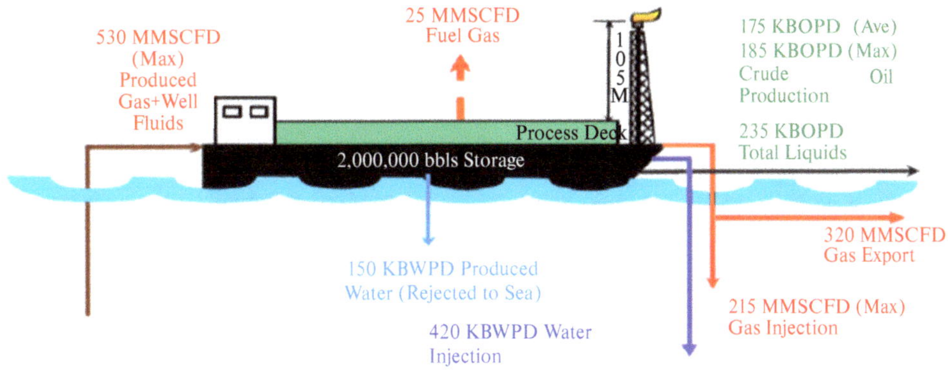

▲ **FIGURE 24P.11** **Akpo FPSO capacities**

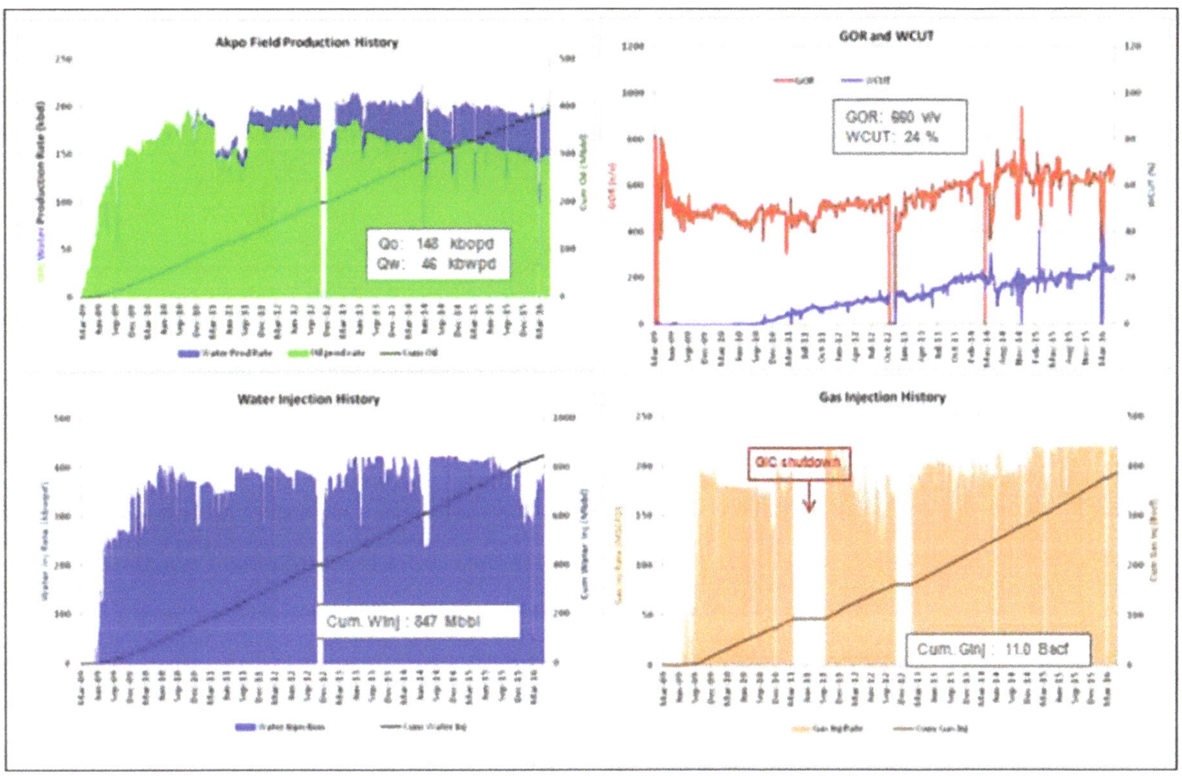

▲ **FIGURE 24P.12** **Akpo Production/Injection History (Through March 2016)**

Based on the case history of Akpo reservoirs presented above and additional information in the references cited below:

a. Discuss the significance of the thermodynamic state of fluids in the Akpo reservoirs on the selection of reservoir management strategies for the reservoirs.

b. Identify the main fluid displacement mechanisms occurring in the Akpo reservoirs.

c. Categorize the reservoir management strategies implemented since initial production under the five main principles of reservoir management presented in this book.

d. Discuss potential reservoir management strategies that could be implemented in future to maintain or improve reservoir performance in the Akpo reservoirs.

Note: References on the Akpo reservoirs:

- Itiola, T.O., Michel, B., and Bez, M.: "Akpo, Nigeria: From Seismic Interpretation to Geomodel," paper SPE 111912-MS (January 1, 2007).
- Piovesanel, E., Claudel, G., Atewologun, A., Adeyemi, A., and Yuh, S.: "AKPO Condensate Field Case Study, Nigeria: 4D Seismic Monitoring of a Complex Deepwater Turbidite Reservoir," paper IPTC 17035 presented at the International Petroleum Technology Conference, Beijing, China, (March 26-28, 2013).
- Rafin, F., Allain, L., and Ludot, B.: "AKPO: A Giant Deep Offshore Development," *SPE Projects, Facilities & Construction*, (June 2008); SPE 118892-PA.
- Rafin, F., and Allain, L.: "AKPO: Early Completion of a Giant Nigerian Deep Offshore Development," paper OTC 20989 presented at the 2010 Offshore Technology Conference, Houston, Texas, (May 3-6, 2010).

24-3. A short case history of the Greater Plutonio Field is presented to assist in the discussion of applied reservoir management strategies. Review the case history and discuss the questions following the case history.

Location. The Greater Plutonio Field, consisting of five smaller fields (Cobalto, Cromio, Galio, Paladio, and Plutonio), is located in Block 18 about 99 miles offshore Angola, in water depths of between 3960 to 4620 feet (Figure 24P.13). The unit ownership of Greater Plutonio Field is 50% BP and 50% Sonangol Sinopec International. Sonangol Sinopec International is a joint venture owned by China Petroleum & Chemical Corporation and Angola's national oil company, Sonangol. In 2018, BP and Sonangol agreed to extend the production license for Block 18 to 2032 with Sonangol taking an 8 percent equity interest in the block, pending government approval. BP is the operator of Greater Plutonio Field.

Geology: The Greater Plutonio fields are Upper Oligocene turbidite reservoirs. The seismic data obtained in the area were of very high quality and enabled imaging of both subsurface geologic structures and reservoir fluid distributions (Figure 24P.14). These high-quality seismic images coupled with high cost of the deepwater environment led to the decision to proceed to field development phase with very little appraisal drilling phase in the Greater Plutonio Field.

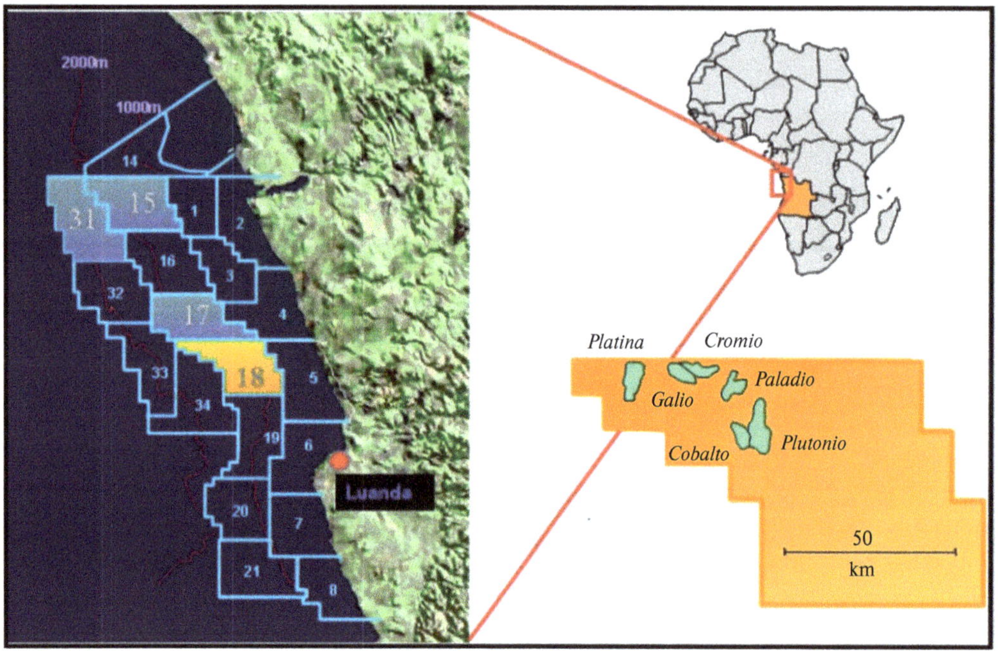

▲ FIGURE 24P.13 Location map showing Greater Plutonio fields, offshore Angola

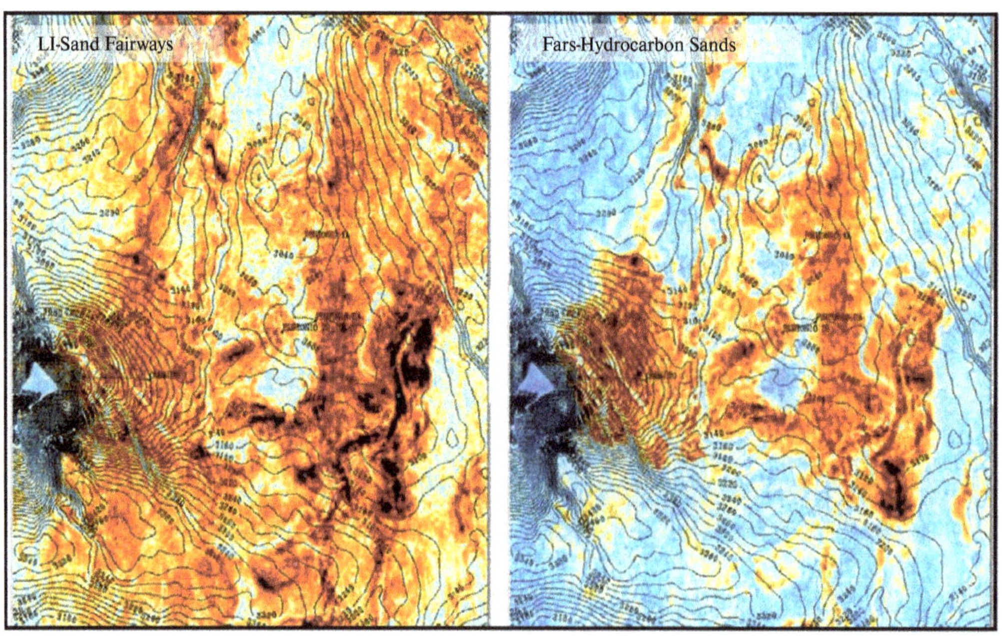

▲ FIGURE 24P.14 Seismic volumes over Plutonio and Cobalto fields

Rock and Fluid Properties. The formations that make-up the Greater Plutonio reservoirs are semi-consolidated rocks. The reservoir rocks have high porosities with high permeabilities that exceed 1 darcy. The Greater Plutonio oil has a low sulfur content and has API gravity of about 33.2.

Development History. The Greater Plutonio fields were discovered between 1999 and 2000. The development of Greater Plutonio Field was approved in early 2004. The entire field development plan (FDP) had a total of 43 wells consisting of 20 oil producers, 20 water injectors, and 3 gas injectors. Field development is 100% subsea tied back to an FPSO. The FPSO has storage capacity of 1.74 million barrels of oil, processing capacity of 240,000 bopd, gas handling capacity of 400 MMscfpd, and water injection rate of 450,000 bwpd. The subsea layout comprised of 94 miles of flowline, 9 manifolds, and 69 miles of instrument and control umbilicals. A single riser tower of about 4151 feet connects the FPSO to the subsea system. Oil production started on October 1, 2007 after about 3.5 years of development. Oil production quickly reached 200,000 bopd after just 3 months of start-up. Water and gas injection reached 300,000 bwpd and 300 MMcfpd respectively, within 12 months of start-up. The Greater Plutonio Field was operated as one of BP's Field of the Future flagship using advanced reservoir management technologies.

The application of advanced reservoir management technologies on the Greater Plutonio Field required that:

1. Permanent downhole pressure/temperature gauges were installed in all wells for continuous measurement of pressure and temperature data.
2. Multi-phase Flow Meters (MPFM) were used for well testing and metering thereby eliminating the need for surface separators.
3. Acoustic sand detectors were installed in all open-hole gravel pack completions for early detection of sand production.
4. Interval control valves (ICVs) were installed in water injectors to direct water injection into specific zones (Figure 24P.15). This reduced the number of water injectors required for the project.
5. Field-wide monitoring with 4D seismic data acquisition program was implemented.
    - Base line survey was conducted in 2000
    - Monitor 1 survey was completed in 2009
    - Monitor 2 survey was completed in 2011
    - Monitor 3 survey was completed in 2013
        - ▹ Difference maps and volumes from the 4D seismic data were used to locate infill well, simulate reservoir performance, and modify reservoir management strategies
        - ▹ 4D seismic data were used to identify advancing waterfront in the Galio reservoir (Figure 24P.16) and shrinking gas cap in the Plutonio reservoir (Figure 24P.17)

At the conclusion of the initial field development plan in two phases, a third-development phase has added four oil producers and two water injectors in 2015. The Greater Plutonio Field overall reservoir management strategy is full pressure maintenance by water injection.

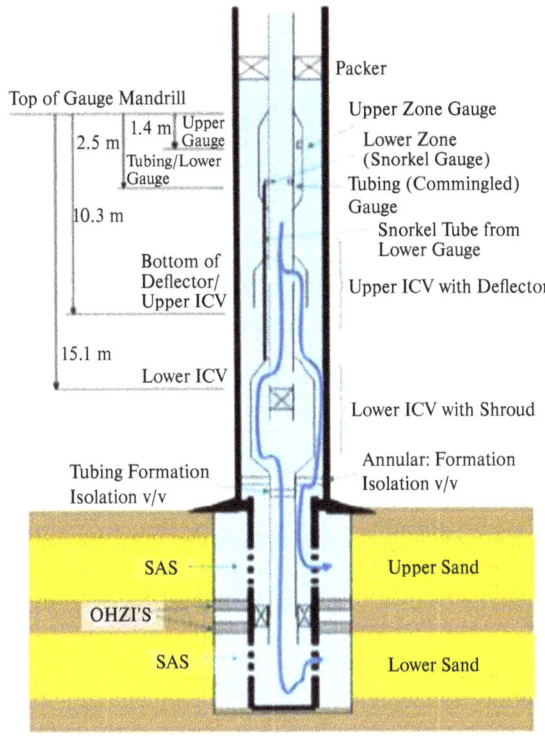

▲ **FIGURE 24P.15** **ICVs installed on a typical water injector**

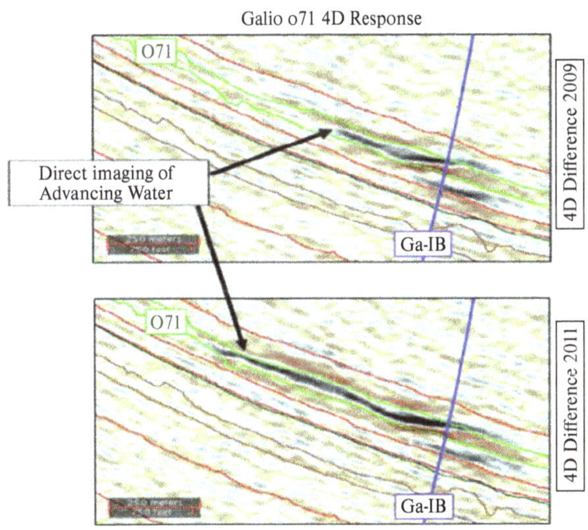

▲ **FIGURE 24P.16** **4D response showing advancing water floodfront in the Galio reservoir**

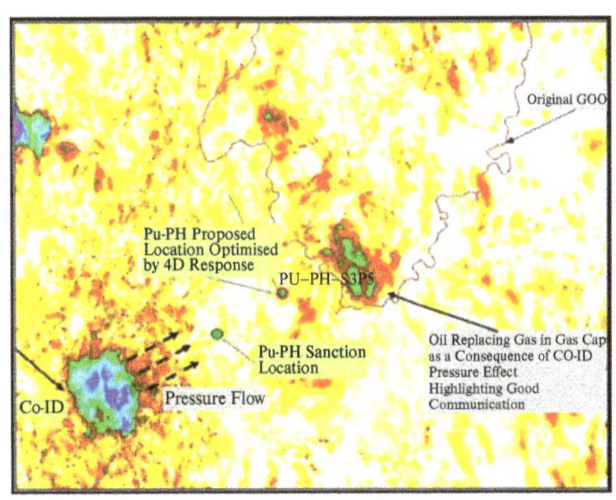

▲ **FIGURE 24P.17** **4D response showing shrinking gas cap in the Plutonio reservoir**

Based on the case history of the Greater Plutonio Field presented above and additional information in the references cited below:

a. Identify the main oil displacement mechanisms occurring in the Greater Plutonio reservoirs.

b. Categorize the reservoir management strategies implemented since initial production under the five main principles of reservoir management presented in this book.

c. Discuss potential reservoir management strategies that could be implemented in future to maintain or improve reservoir performance in the Greater Plutonio fields.

Note: References on the Greater Plutonio Field:

- Booth, D., and Sebastian, P.M.: "Greater Plutonio - Real Time Reservoir Management in a High Cost, Deepwater Environment," paper SPE 128542-MS (January 1, 2010).
- Dickens, J., Latin, D. M., Verra, G., Blosser, W. R., Edmonds, G. A., and Grimshaw, G.: "The BP Field of the Future Programme: The Continuing Mission To Deliver Value," paper SPE 128672-MS (January 1, 2010).
- Riviere, M. "A Review of 4D Seismic at Greater Plutonio and PSVM," *First EAGE/ASGA Petroleum Exploration Workshop*, Luanda, Angola, (October 2-4, 2017).

24-4. Construct the case history of a field (or reservoirs) from published literature, internal company documents or company vault files, etc. Use case histories presented in this chapter as examples for the types of data to search for about the field or reservoir(s) in creating the case history. Using the case history as source:

a. Identify the main oil displacement mechanisms occurring in the field or reservoirs.

b. Categorize the reservoir management strategies implemented since initial production under the five main principles of reservoir management presented in this book.

c. Discuss potential reservoir management strategies that could be implemented in future to maintain or improve reservoir performance in your candidate field or reservoirs.

CHAPTER

25 Economic Evaluation of Petroleum Projects and Property

25.1 Introduction

Economic evaluation of petroleum projects and property is one of the fundamental tasks required of all engineers working in the petroleum industry. Engineers should be skilled in various methods used to conduct economic evaluation. They should lead the economic evaluation process or at least be a key member of the evaluation team. Ultimately, a project is executed on the basis of potential economic rewards that it could generate for the organization independent of its technical justification. Petroleum companies or organizations succeed and thrive on the basis of the cumulative economic rewards generated from various projects executed within the companies or organizations. These projects originate from the desks of engineers and must be economically justified. The role of the engineer in proposing projects that must be justified economically for execution has a very wide range. Some of these projects are listed as:

1. Well remedials
2. Drilling of new or infill wells as producers and injectors
3. Well/Field abandonments
4. Field/Lease developments
5. Installation of improved/enhanced petroleum recovery projects
6. Sale and acquisition of leases/property, etc.
7. Mergers with or acquisitions of competitors
8. Entry into new geographical regions/basins/countries
9. Exit from geographical region/basins/countries
10. Bidding for onshore/offshore leases
11. "Farming" into new/existing property
12. Evaluation of new prospects

This list is not comprehensive but serves to show some of the activities an engineer could be expected to support with some measure of economic evaluation before a decision can be made by managers in the company or organization. A successful engineer must be very resourceful in assembling required input data (reserves, production rates, capital expenditures, operating costs, revenues, etc.) and very skilled in performing the required economic evaluations. Finally, it is equally important that the engineer should know how to present economic yardsticks that justify viability of the project in a concise and readily comprehensible format to the management of the company or organization for approval.

25.2 Classification of Oil and Gas Resources

Economic evaluation of petroleum projects and property should begin with a clear understanding of the classification of petroleum resources on the basis of in-place volumes, production, and commerciality. One of the important functions of a reservoir engineer is the classification of oil and gas resources. In many companies and organizations, the practice of classifying oil and gas resources is guided by rules and procedures proposed by the Society of Petroleum Engineers (SPE)[1] in conjunction with these other organizations:

1. World Petroleum Council (WPC)
2. American Association of Petroleum Geologists (AAPG)
3. Society of Petroleum Evaluation Engineers (SPEE)
4. Society of Exploration Geologists (SEG)
5. Society of Petrophysicists and Well Log Analysts (SPWLA)
6. European Association of Geoscientists and Engineers (EAGE)

This classification known as Petroleum Resources Management System (PRMS)[1] as proposed by SPE is shown in Figure 25.1. The definition of terms used in the PRMS will be discussed in following sections. For financial reports, the United States Securities and Exchange Commission (SEC)[2] has instituted procedures and regulations that are required in the reporting and classification of oil and gas resources. These rules and regulations are mandatory and carry substantial penalties if violated. To avoid any subjective interpretation of these rules and regulations, readers are directed to the SEC website[2] for direct comprehension of these SEC directives.

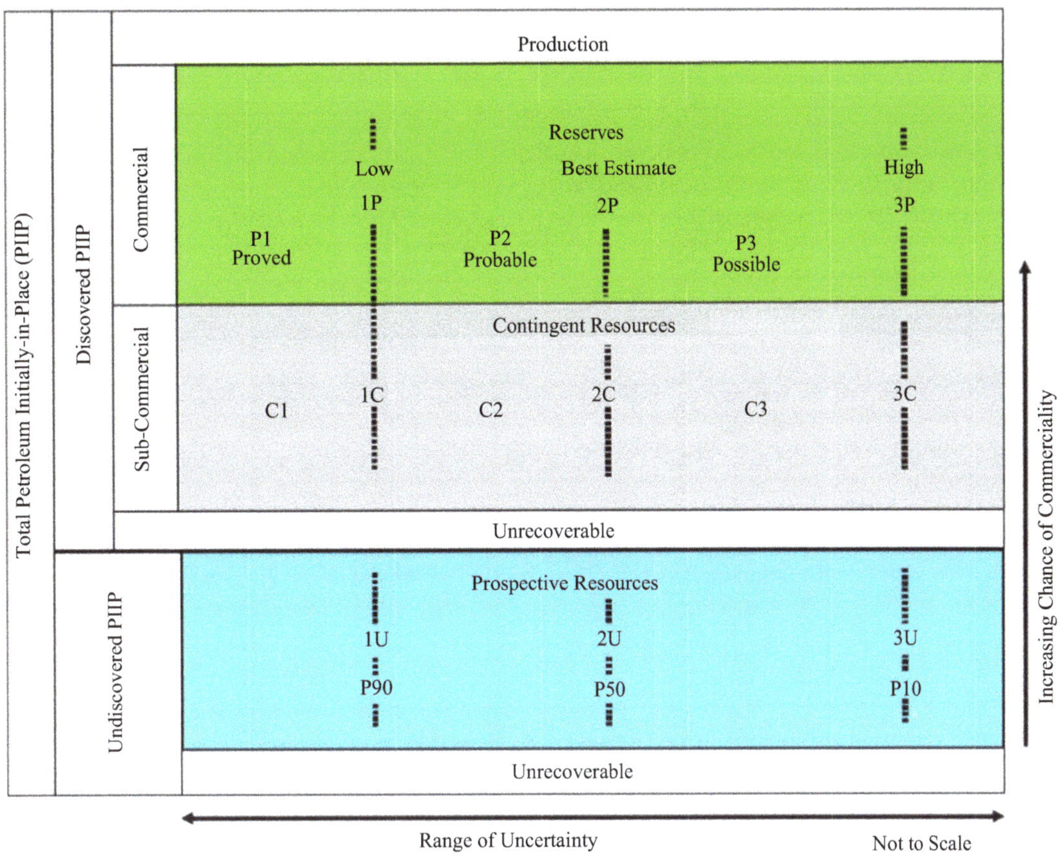

▲ **FIGURE 25.1** Resources classification framework (from SPE[1])

25.2.1 Definition of Terms in the SPE PRMS

- **Total Petroleum-Initially-in-Place (PIIP):** This is the total quantity of petroleum resources estimated to exist originally in natural accumulations before any production. It includes discovered and undiscovered accumulations.
- **Discovered PIIP:** The quantity of petroleum estimated to be contained within known accumulations before production as of a given date.
- **Undiscovered PIIP:** The quantity of petroleum estimated to be undiscovered within known accumulations as of a given date.
- **Commercial:** A project is considered as commercial if there is evidence of a firm commitment to progress the project to development and production within a reasonable period of time.
- **Sub-Commercial:** A project is considered to be uncommercial if there is no evidence of firm commitment to proceed with its development and production within a reasonable period of time. Sub-commercial projects are categorized as contingent resources.
- **Production:** Production is the cumulative quantities of hydrocarbons that have been recovered over a given time period. Production can be reported based on product sales quantities. However, engineering analyses and project economic evaluations require total production quantities (sales and non-sales) to be used.
- **Reserves:** Reserves are quantities of petroleum expected to be commercially recoverable through application of development projects to known accumulation from a stated date going forward under stated conditions. Four criteria must be met in the definition of reserves: (1) Discovered (2) Recoverable (3) Commercial and (4) Remaining, all on the basis of development projects employed. Reserves are further classified based on range of uncertainty as Proved, Probable, and Possible.
  - **Proved reserves** are "quantities of Petroleum that, by analysis of geoscience and engineering data, can be estimated with reasonable certainty to be commercially recoverable from known reservoirs and under defined technical and commercial conditions."[1] Proved reserves are denoted as 1P reserves. This is equal to P1 reserves that denote a low estimate of reserves.
  - **Probable reserves** are "those additional Reserves which analysis of geoscience and engineering data indicate are less likely to be recovered than Proved Reserves but more certain to be recovered than

Possible Reserves."[1] Probable reserves are denoted as P2. Note that 2P reserves is the sum of Proved plus Probable reserves. It represents the best estimate of reserves.

    - **Possible reserves** are "those additional reserves that analysis of geoscience and engineering data suggest are less likely to be recoverable than Probable Reserves."[1] Possible reserves are denoted as P3 reserves. Note that 3P reserves is the sum of Proved plus Probable plus Possible reserves. It denotes the high estimate of reserves.

- **Contingent Resources:** Contingent resources are quantities of petroleum estimated, as of a stated date, to be potentially recoverable from known accumulations, using development projects that are currently not considered to be commercial due to certain contingencies. Contingent resources are further classified based on a range of uncertainty as C1, C2, and C3 resources.
    - **C1 Resources** are low estimates of contingent resources. C1 is equal to 1C resources.
    - **C2 Resources** represents contingent resources at a technical level of probable but not commercially matured to be considered reserves. 2C resources are the sum of C1 plus C2 resources. 2C resources are the best estimates of contingent resources.
    - **C3 Resources** represent contingent resources at a technical level of possible but not commercially matured to be considered reserves. 3C resources are the sum of C1 plus C2 plus C3 resources. 3C resources are the high estimates of contingent resources.
- **Prospective Resources:** Prospective resources are quantities of petroleum estimated, as of a stated date, to be potentially recoverable from undiscovered accumulations employing future development projects. Prospective resources are further classified on the basis of uncertainty as unrisked low estimate (1U), unrisked best estimate (2U), and unrisked high estimate (3U).
- **Unrecoverable Resources:** Unrecoverable resources are quantities of either discovered or undiscovered PIIP estimated, as of a stated date, to be unrecoverable by current known projects. A portion of unrecoverable resources may be recoverable in future as commercial environment change, new technologies emerge, or more data are obtained. Some portion may never be recoverable due to limitations imposed by subsurface interactions of rocks and fluids.

25.2.2 Estimation of Recoverable Resources

The estimation of recoverable resources is a major process that is routinely undertaken in many companies and organizations annually or in some cases continuously throughout the year to meet budgetary, reporting, and regulatory requirements. Some organizations, especially small or independent oil companies, retain outside consultant groups to perform resource determination functions for them. Major oil companies typically have departments that co-ordinate and administer the determination process of resources to ensure consistent guidelines and methods are applied across various classes of assets managed by different asset teams. Whether the estimation of recoverable resources was conducted by an external party or internally, the desired goal for all these organizations is to achieve estimates of recoverable resources that are supported by documented data-based evidence that are defensible before financial and regulatory agencies. In the United States of America (USA) for instance, filing false or misleading reserves with the SEC is a prosecutable violation of the law that carries serious consequences including incarceration if convicted.[2]

The reservoir engineer plays a central role in the process of estimating recoverable resources for the assets under his/her assignment. For large assets, this responsibility may fall on a team of reservoir engineers supported by geoscientists. If the company uses outside consultants to conduct determination of resources, then the reservoir engineer would be responsible for assembling subsurface and performance data usually requested by these firms. Even under these circumstances, it is advisable for the engineer to perform his/her own analyses so that he/she would be able to critically review the results submitted by the outside consultants. For organizations that conduct internal resources determinations, the reservoir engineer would collect and analyze all subsurface and performance data needed for resource determination usually with the support and guidance of the internal resource estimation expert. The resource estimation expert of the organization is the final arbiter on the determination and classification of resources. It is the primary responsibility of the resource estimation expert to ensure that the entire processes meet the guidelines established by the organization and also are in conformity with the guidelines and rules established by regulatory agencies such as the SEC.

25.2.3 Analytic Methods for Resource Determination

The estimation of resources using analytic methods is typically performed using the following techniques:

1. Analogy
2. Volumetric methods
3. Material balance
4. Decline curve analyses
5. Rate transient analyses
6. Reservoir simulation

25.2.3.1 Analogy

Analogy is frequently used in resource determination by assuming that the performance of the subject reservoir would be similar to the performance of another reservoir in similar geologic formation or basin. Analogy is typically used during exploration, discovery, or early development stages of a reservoir because little or no direct data are available on the subject reservoir or prospect. The analogous reservoirs are considered to possess properties similar to the subject reservoir in the following areas:

1. **Geology:** Similar depositional environment and structure. Located in the same basin or possesses similar geologic features.
2. **Rock properties:** Comparable ranges of porosity, permeability, saturation, net pay, and net-gross ratios.
3. **Fluid properties**: Comparable fluid properties such as viscosity, density, and fluid composition
4. **Thermodynamic properties**: Comparable subsurface temperatures and pressures
5. **Drive mechanisms**: Similar drive mechanisms such as solution-gas drive, aquifer drive, gravity drainage, fracture-dominated production, etc.
6. **Development strategy**: Comparable well type (vertical vs. horizontal), well spacing, well counts, well completion methods, artificial lift systems, fluid processing, sales products, etc.
7. **Recovery methods**: Similar recovery methods such as waterflooding, gasflooding, EOR methods, etc.

It is not necessary that the analog reservoir should meet all the criteria suggested to qualify as an analog. However, it is important that specific areas of analogy should be identified and documented before such application can be justified.

25.2.3.2 Volumetric Methods

The volume of PIIP can be calculated using volumetric methods. Volumetric methods require input of average values of porosity, saturation, net pay, area of the reservoir, and formation volume factors in consistent units. Volumetric methods for calculations of PIIP are presented in Chapter 9 for various types of gas reservoirs and in Chapter 10 for various types of oil reservoirs. The key parameters that affect calculation of PIIP using volumetric methods are:

1. **Reservoir area:** The area of the reservoir is affected by heterogeneities, unconformities, and other geologic features. Consequently, the reservoir area to be used to calculate PIIP should be obtained from competent geologic studies.
2. **Petrophysical properties**: Petrophysical properties such as porosity, net pay, and fluid saturation should be obtained from competent petrophysical studies.
3. **Fluid contacts:** The location of fluid contacts (oil-water contact, gas-water contact, gas-oil contact, lowest known hydrocarbon, etc.) has significant impact on volumetric calculations. These contacts must be determined as accurately as possible from available data.
4. **Fluid distribution:** Spatial distribution of fluid saturation is important factor that could influence PIIP calculations. This should be recognized and incorporated in PIIP estimations.

The calculation of PIIP using volumetric methods can be conducted using either deterministic or probabilistic methods. The methods discussed in Chapters 9 and 10 are deterministic. In probabilistic method, all or any of the input parameters of area, net pay, porosity, and saturations are represented by a probability distribution. The probability distributions reflect the uncertainties associated with the input parameters.

After calculating the PIIP, the resource is calculated by applying a recovery factor which can be based on analogy or industry experience depending on the development strategy planned for the reservoir. Note that the recovery factor applied can be arbitrary and can lead to overestimation or underestimation of resources.

25.2.3.3 Material Balance Methods

The General Material Balance Equation (GMBE) was derived in Chapter 8. The GMBE is the basic equation for application of material balance methods. Estimations of recoverable resources using material balance methods were shown in Chapter 9 for gas reservoirs and Chapter 10 for oil reservoirs. It is important to note that successful application of material balance methods depends on availability of reliable production and pressure data over a reasonable length of time such that reservoir drive mechanisms could be determined with some degree of confidence.

25.2.3.4 Decline Curve Analyses

Decline curve analyses (DCA) based on Arps' decline equations (exponential, harmonic, and hyperbolic equations) are the workhorse of the petroleum industry for production forecasting and estimation of resources. The theory and applications of DCA were presented in Chapter 11, "Production Forecasting for Conventional and Unconventional Reservoirs." The use of type curves (such as Fetkovich type curves) for production forecasting was discussed in Chapter 11.

25.2.3.5 Rate Transient Analyses

Rate transient analyses (RTA) are typically used to perform production forecasting for unconventional reservoirs. These were presented in Chapter 11 for modified Arp's model, Duong model, and Stretched Exponential Decline (SEDM) model. Examples are also presented in Chapter 11 on application of these models.

25.2.3.6 Reservoir Simulation

Reservoir simulation has gained acceptance by regulatory agencies (such as SEC) as a tool for determination of resources.[3] Reservoir models are effective tools for forecasting reservoir performance because they provide flexibility in the investigation of many development plans and varied depletion strategies. The techniques of model construction and reservoir simulation were presented in Chapters 22 and 23, respectively.

25.2.4 Resource Assessment Techniques

Resource assessment techniques can be classified generally on the basis of techniques used in the process as:

1. Deterministic
2. Geostatistical
3. Probabilistic
4. Integrated
5. Aggregation

25.2.4.1 Deterministic Technique

In deterministic technique for resource assessment, discrete values are assigned to input parameters such as porosity, fluid saturation, net pay thickness, area, fluid contacts, recovery factors, etc. This will result in the calculation of a discrete result for the PIIP or recoverable resources. Generally, application of deterministic techniques leads to estimation of a low case, a best case, and a high case. These cases reflect the confidence assigned by the evaluator to the input parameters used in the calculation. For instance, the best case implies that the evaluator assigned most likely input parameters in the determination of the result for the case. The deterministic technique assigns no chances of occurrence (or probabilities) to any of the cases.

25.2.4.2 Geostatistical Technique

The application of geostatistical techniques for resource assessment was presented in Chapter 22, "Geologic Modeling and Reservoir Characterization." In Section 22.5.6, "Uncertainty Analysis," the generation of equiprobable models (realizations) using geostatistical methods was discussed. The equiprobable models reflect uncertainties which can be used as input in a statistical analysis to generate a cumulative distribution function. The cumulative distribution function can then be used to assess the probability of a particular outcome such as PIIP or recoverable resource.

25.2.4.3 Probabilistic Technique

Probabilistic technique for resource assessment is applied by assuming that most input parameters can be represented by a probability distribution function within an assigned range. The probability distribution function for any input parameter is sampled randomly to generate a complete distribution of potential PIIP or recoverable resource. This technique is represented by the application of Monte Carlo Simulation or similar methods in generating the distributions of PIIP or recoverable resources.

25.2.4.4 Integrated Technique

Integrated technique involves the combination of deterministic, geostatistical, and probabilistic techniques in the evaluation of PIIP or recoverable resources at different levels of the evaluation process to improve the reasonableness of the estimates. For instance, the area of interest can be established using deterministic techniques while all other input parameters are based on probabilistic techniques.

25.2.4.5 Aggregation Technique

Aggregation technique is a method of reporting resources at different levels in an organization. The first level of categorizing resources is the reservoir level. Resources at reservoir level are summed to report resources at field, project, property, or organizational level. Further summations may result in reports at geographic, country, subsidiary, or company level, depending on the structure of the organization.

25.3 Time Value of Money

The worth of money changes over time. For instance, the value of $1000 today is not the same as 1 year from today. This can be demonstrated easily by offering to give someone $1000 today compared with $1000 a year from today. Applying objective reasoning, one would prefer to receive $1000 today than wait to receive the same amount 1 year later. Clearly by making this choice, the value of money has been intuitively associated with time. In other words, the value of $1000 today is considered to be worth more than $1000 a year later. However, if the offer is $1000 today compared with $1100 a year later, the choice could be different depending on the value of the additional $100 to the person receiving the offer. Is the additional $100 sufficient reward for waiting 1 year to receive the $1000? The additional $100 is the "interest" or inducement for waiting 1 year before receiving the $1000 offered initially. This offer could be rejected depending on the person's willingness to "risk" receiving $1000 today compared with $1100 a year later. In other words, interest can be considered to be compensation for the risk that was assumed in postponing payment today for some other time in the future. By linking interest to risk in this way, it is evident that higher risk would be associated with higher interest. If the offer of $1000 today is changed to $1500 a year later, most rational investors would accept the offer to receive $1500 a year later. Clearly, the additional $500 is sufficient inducement or interest for the risk of waiting 1 year before receiving the initial \$1000. These simple propositions illustrate that money has value associated with time. And this time value of money is reflected by the interest we are willing to accept as reward for postponing receipt today compared with some other time in the future. For the earlier simple propositions, \$1000 today compared to \$1000 a year later has 0% interest; \$1000 today compared with \$1100 a year later has 10% interest; \$1000 today compared with \$1500 a year later has 50% interest. Interest is sometimes expressed as a rate. For instance, the interest for the propositions discussed here can be expressed as 0% per year, 10% per year, and 50% per year, respectively. Interest can be expressed per period as interest per day, interest per month, interest per quarter, interest per 6 months, etc. Simply stated, interest rate is calculated by dividing the amount of interest paid per period by the principal amount owed at the beginning of the period. For companies, interest rate may be called discount rate, "hurdle" rate, etc. Interest rates or discount rates are applied in the same manner in economic evaluations. Note that compound interest describes the concept where the interest earned in the previous period is added to the principal at the beginning of the period to form the principal for the current period. In most economic evaluations that exceed 1 year, the interest is compounded.

In this section, several formulas for evaluating the time value of money will be presented. All the formulas are based on the concept of equivalence. The concept of equivalence implies that cashflows (representing revenues and costs) received over a specified time frame are equivalent if converted with the same rules of interest rate and compounding periods. For instance, an offer of \$1000 today compared with \$1100 a year later is considered equivalent if the receiver is indifferent to the choice. In this example, the interest is 10% and the period is 1 year. The formulas presented in this section are useful for calculating equivalence between various forms of cashflows as will be illustrated later.

To facilitate derivations and uses of these formulas, certain symbols common to all the formulas are defined as follows:

- $P \equiv$ Present lump-sum of money. Normally refers to sum of money at time zero but may refer to any time-based reference point.
- $F \equiv$ Future lump-sum of money received at a future designated date (or n periods from start date).
- $A \equiv$ Amount of equal payments (or investments) in a uniform series of payments (or investments) made at each period. Assumes payments are made at the end of the period.
- $B \equiv$ Amount of equal payments (or investments) in a uniform series of payments (or investments) made at each period. Assumes payments are made at the beginning of the period.
- $C \equiv$ Amount of equal payments (or investments) in a uniform series of payments (or investments) made at each period. Assumes payments are made at middle of period (or mid-period).
- $i \equiv$ Interest rate per period.
- $i_n \equiv$ Nominal interest rate per year.
- $i_e \equiv$ Effective interest rate per year.
- $i_{mon} \equiv$ Effective interest rate per month.
- $t \equiv$ Number of years.
- $m \equiv$ Number of compounding periods per year.
- $n \equiv$ Total number of compound interest periods, $n = m \times t$.

A time diagram is useful in displaying the symbols as designated earlier and facilitating solution of the problems. An example of a time diagram is shown in Figure 25.2. The time diagram is similar to a cash flow diagram which will be presented later. In Figure 25.2, P represents the present sum of money at time zero; F represents the future lump-sum of money at the end of period, n; A designates uniform series of equal payments at the end of each compounding period; and the compounding interest periods are designated as 1, 2, 3,......, $n-1$, n.

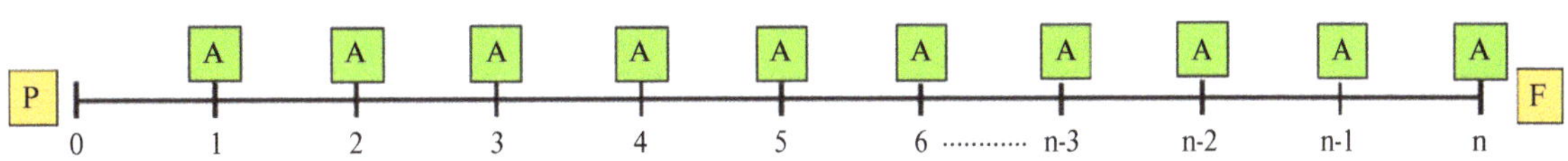

▲ FIGURE 25.2 Time diagram

25.3.1 Future Worth (Value) of a Single Payment

Suppose a lump-sum of money, P, is invested at an interest rate, i, for one period. At the end of the period, the amount of money, F, generated by the investment is given by:

$$F = P(1+i) \tag{25.1}$$

Since the investment is for first period, Eq. (25.1) can be represented as:

$$F_1 = P(1+i) \tag{25.2}$$

At the end of the second period, interest is paid on the amount of money at the beginning of the period which is, F_1. So the amount of money at the end of the second period is:

$$\begin{aligned} F_2 &= F_1(1+i) \\ &= P(1+i)(1+i) = P(1+i)^2 \end{aligned} \tag{25.3}$$

Similarly, at the end of the third period the amount of money generated is given by:

$$
\begin{aligned}
F_3 &= F_2(1+i) \\
&= F_1(1+i)(1+i) \\
&= P(1+i)(1+i)(1+i) = P(1+i)^3
\end{aligned}
\tag{25.4}
$$

From these equations, it can be deduced that for, n, compounding periods, the general form of the equation for calculating future worth, F, of a lump-sum of money, P, invested at the beginning of the period is:

$$F = P(1+i)^n \tag{25.5}$$

The expression, $(1+i)^n$, in Eq. (25.5) is called the single payment compound amount factor.

EXAMPLE 25.1 Calculation of the Future Value of a Single Lump-Sum Investment or Payment

Problem

Calculate the future value of $10,000 invested at interest rate of 12% compounded annually at the end of 7 years.

Solution

$P = \$10{,}000$

$i = 12\%$

$n = 7$

$F = ?$

Substituting these values into Eq. (25.5) gives:

$$
\begin{aligned}
F &= \$10{,}000 \times (1+0.12)^7 \\
&= \$22{,}106.81
\end{aligned}
$$

End of Example 25.1

25.3.2 Present Worth (Value) of a Single Payment

Suppose it is desired to calculate the present worth, P, of a future lump-sum of money, F, at interest rate, i, over n compounding periods, Eq. (25.5) can be rearranged to give:

$$P = \frac{F}{(1+i)^n} \tag{25.6}$$

The expression, $1/(1+i)^n$, in Eq. (25.6) is called the single payment present worth factor.

EXAMPLE 25.2 Calculation of the Present Value of a Future Single Lump-Sum Payment

Problem

Calculate the present value of $22,106.81 to be received at the end of 7 years at interest rate of 12% compounded annually.

Solution

$F = \$22{,}106.81$

$i = 12\%$

$n = 7$

$P = ?$

Substituting these values into Eq. (25.6) gives:

$$
\begin{aligned}
P &= \frac{\$22{,}106.81}{(1+0.12)^7} \\
&= \$10{,}000.00
\end{aligned}
$$

End of Example 25.2

25.3.3 Uniform Series Compound Amount

As defined previously, suppose the symbol A represents the amount of equal payments (or investments) in a uniform series over n periods at a compound interest i per period, the future value, F, of the payments (or investments) can be derived with the aid of the time diagram shown as Figure 25.3. Note that each payment (or investment) has the same compound interest for different number of periods (Figure 25.3). Equation (25.5) can be used to represent the future values of these series of payments (or investments) as follows:

$$
\begin{aligned}
F = {} & A(1+i)^0 + A(1+i)^1 + A(1+i)^2 + \ldots. \\
& + A(1+i)^{n-3} + A(1+i)^{n-2} + A(1+i)^{n-1}
\end{aligned}
\tag{25.7}
$$

Multiplying both sides of Eq. (25.7) by $(1+i)$ gives:

$$
\begin{aligned}
F(1+i) = {} & A(1+i)^1 + A(1+i)^2 + A(1+i)^3 + \ldots. \\
& + A(1+i)^{n-2} + A(1+i)^{n-1} + A(1+i)^n
\end{aligned}
\tag{25.8}
$$

Subtracting Eq. (25.7) from Eq. (25.8) gives:

$$F(1+i) - F = A(1+i)^n - A \tag{25.9}$$

By rearrangement, Eq. (25.9) gives:

$$F = \frac{A\left[(1+i)^n - 1\right]}{i} \quad (25.10)$$

The factor, $\left[(1+i)^n - 1\right]/i$, is called the uniform series compound amount factor. The factor is used to calculate a future single amount, F, that is equivalent to a uniform series of payments (or investments), A. Note that the series of payments (or investments) are made at the end of the periods.

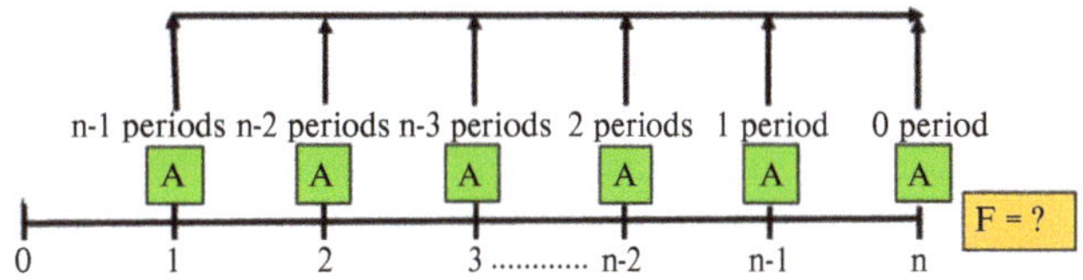

▲ **FIGURE 25.3** **Time diagram for uniform series compound amount derivation**

EXAMPLE 25.3 Calculation of the Future Value of a Series of Uniform Payments (or Investments) Made at the End of Each Period

Problem

Calculate the future value of a uniform series of \$5000 invested at the end of each year for 20 years assuming an interest rate of 10% per year compounded annually.

Solution

$A = \$5000$

$i = 10\%$

$n = 20$

$F = ?$

Substituting these values into Eq. (25.10) gives:

$$F = \frac{\$5000\left[(1+0.10)^{20} - 1\right]}{0.10}$$

$$= \$286,375.00$$

End of Example 25.3

25.3.4 Sinking Fund Deposit Amount

A sinking fund deposit amount is a series of equal payments (or investments), A, made at the end of each period that is expected to yield a future amount, F, at the end of n periods of such payments (or investments) at a given compounded interest rate, i, per period. Rearranging Eq. (25.10) gives:

$$A = F\left[\frac{i}{(1+i)^n - 1}\right] \quad (25.11)$$

The factor, $i/\left[(1+i)^n - 1\right]$, is called the sinking fund deposit factor. It is used to calculate equal end of period payments (or investments), A, that are equivalent to a future lump-sum payment, F, at a compounded interest rate, i, per period over n periods.

EXAMPLE 25.4 Calculation of the Amount of a Series of Uniform Payments (or Investments) to be Made at the End of Each Period to Accumulate a Future Lump-Sum Amount

Problem

Calculate the amount of a uniform series invested at the end of each year for 30 years assuming an interest rate of 8% per year compounded annually to accumulate a lump-sum of \$3,000,000.

Solution

$F = \$3,000,000$

$i = 8\%$

$n = 30$

$A = ?$

Substituting these values into Eq. (25.11) gives:

$$A = \$3,000,000\left[\frac{0.08}{(1+0.08)^{30} - 1}\right]$$

$$= \$26,482.30$$

End of Example 25.4

25.3.5 Capital Recovery Amount

The amount of a uniform series of equal end of period payments (or investments), A, over n periods at a compounded interest rate per period, i, that is equivalent to a present value single payment, P, can be derived by combining Eq. (25.5) and Eq. (25.11) as follows:

Starting with Eq. (25.11):

$$A = F\left[\frac{i}{(1+i)^n - 1}\right]$$

Substituting Eq. (25.5) gives:

$$A = P(1+i)^n\left[\frac{i}{(1+i)^n - 1}\right] \tag{25.12}$$

Equation (25.12) can be rearranged to give:

$$A = P\left[\frac{i(1+i)^n}{(1+i)^n - 1}\right] \tag{25.13}$$

The factor, $\left[\frac{i(1+i)^n}{(1+i)^n - 1}\right]$, is called the capital recovery factor. It can be used to calculate equal end of period payments (or investments), A, that are equivalent to a present value single payment, P, at a compound interest rate per period, i, over n periods. The capital recovery factor is sometimes expressed in an equivalent form as: $\left[\frac{i}{(1+i)^n - 1} + i\right]$.

EXAMPLE 25.5 Calculation of the Amount of a Series of Uniform Payments (or Investments) to Be Made at the End of Each Period to be Equivalent to a Present Lump-Sum Amount.

Problem

Suppose that a retiree has \$3,000,000 to invest in an annuity account. If the annuity offers 8.0% annual interest rate over a 20-year period, what is the annuity payment that the retiree will receive at the end of each year over the 20-year period.

Solution

$P = \$3{,}000{,}000$

$i = 8\%$

$n = 20$

$A = ?$

Substituting these values into Eq. (25.13) gives:

$$A = \$3{,}000{,}000\left[\frac{0.08(1+0.08)^{20}}{(1+0.08)^{20} - 1}\right]$$

$$= \$305{,}556.63$$

End of Example 25.5

25.3.6 Uniform Series Present Worth Amount

The present worth, P, of a series of uniform equal payments (or investments), A, made at the end of the period for n periods at compound interest rate per period, i, can be determined by rearranging Eq. (25.13) to give:

$$P = A\left[\frac{(1+i)^n - 1}{i(1+i)^n}\right] \tag{25.14}$$

EXAMPLE 25.6 Calculation of the Present Lump-Sum Amount That Is Equivalent to a Series of Uniform Payments (or Investments) Made at the End of Each Period.

Problem

Suppose that a retiree was offered equal uniform series of annual payments of \$3000 over a 20-year period at a compound interest rate of 7% per year. What is the present value of the annuity?

Solution

$A = \$3000$

$i = 7\%$

$n = 20$

$P = ?$

Substituting these values into Eq. (25.14) gives:

$$P = \$3000\left[\frac{(1+0.07)^{20} - 1}{0.07(1+0.07)^{20}}\right]$$

$$= \$31{,}782.04$$

End of Example 25.6

25.3.7 Timing of Payments (or Investments) Considerations

It has been assumed in all the compound interest equations presented earlier that payments (or investments) occurred at the end of the period. This is not always the case. In some financial transactions, payments (or investments) are assumed to occur at the beginning of the period or at mid-point of the period. In such cases, Eqs. (25.5) to (25.14) can still be used if particular care is exercised in determining the correct number of periods to be used in the calculations using time diagrams. This will be illustrated later in Example 25.11.

25.3.7.1 Compound Interest Equations for Beginning-of-Period

For ease of applications, some forms of the compound interest equations presented earlier have been derived for beginning-of-period calculations. Payments (or investments) due at the beginning of the period are referred to as lease or annuity-due payments (or investments), B. An equal uniform series of lease payments, B, is related to present worth, P, at compound interest, i, for n periods by the equation:

$$B = P\left[\frac{i(1+i)^{n-1}}{(1+i)^{n}-1}\right] \tag{25.15}$$

The future worth, F, of payments made at the beginning of the period is represented by the equation:

$$B = F\left[\frac{i(1+i)^{-1}}{(1+i)^{n}-1}\right] \tag{25.16}$$

Equation (25.16) is called modified sinking fund.

EXAMPLE 25.7 Calculation of the Amount of a Series of Uniform Payments (or Investments) to Be Made at the Beginning of Each Period to Be Equivalent to a Present Lump-Sum Amount

Problem

Suppose that a retiree has \$3,000,000 to invest in an annuity account. If the annuity offers 8.0% annual interest rate over a 20-year period, what is the annuity payment that the retiree will get at the beginning of each year over the 20-year period.

Solution

P = \$3,000,000

i = 8%

n = 20

B = ?

Substituting these values into Eq. (25.15) gives:

$$B = \$3,000,000\left[\frac{0.08(1+0.08)^{20-1}}{(1+0.08)^{20}-1}\right]$$

$$= \$282,922.80$$

End of Example 25.7

EXAMPLE 25.8 Calculation of the Amount of a Series of Uniform Payments (or Investments) To Be Made at the Beginning of Each Period to Accumulate a Future Lump-Sum Amount

Problem

Calculate the amount of a uniform series invested at the beginning of each year for 30 years assuming an interest rate of 8% per year compounded annually to accumulate a lump-sum of \$3,000,000.

Solution

F = \$3,000,000

i = 8%

n = 30

B = ?

Substituting these values into Eq. (25.16) gives:

$$B = \$3,000,000\left[\frac{0.08(1+0.08)^{-1}}{(1+0.08)^{30}-1}\right]$$

$$= \$24,520.65$$

End of Example 25.8

25.3.7.2 Compound Interest Equations for Mid-Period

Also, some forms of the compound interest equations have been derived for mid-period calculations. An equal uniform series of lease payments, *C*, made at mid-period is related to present worth, *P*, at compound interest, *i*, for *n* periods by the equation:

$$C = P\left[\frac{i(1+i)^{n-0.5}}{(1+i)^n - 1}\right] \tag{25.17}$$

The future worth, *F*, of payments (or investments) made at the mid-period is represented by the equation:

$$C = F\left[\frac{i(1+i)^{-0.5}}{(1+i)^n - 1}\right] \tag{25.18}$$

EXAMPLE 25.9 Calculation of the Amount of a Series of Uniform Payments (or Investments) to Be Made at the Mid-Period to Be Equivalent to a Present Lump-Sum Amount

Problem

Suppose that a retiree has $3,000,000 to invest in an annuity account. If the annuity offers 8.0% annual interest rate over a 20-year period, what is the annuity payment that the retiree will receive at the mid-year over the 20-year period.

Solution

P = $3,000,000

i = 8%

n = 20

C = ?

Substituting these values into Eq. (25.17) gives:

$$C = \$3{,}000{,}000\left[\frac{0.08(1+0.08)^{20-0.5}}{(1+0.08)^{20} - 1}\right]$$

$$= \$294{,}022.00$$

End of Example 25.9

EXAMPLE 25.10 Calculation of the Amount of a Series of Uniform Payments (or Investments) to Be Made at Mid-Period to Accumulate a Future Lump-Sum Amount

Problem

Calculate the amount of a uniform series invested at each mid-year for 30 years assuming an interest rate of 8% per year compounded annually to accumulate a lump-sum of $3,000,000.

Solution

F = $3,000,000

i = 8%

n = 30

C = ?

Substituting these values into Eq. (25.18) gives:

$$C = \$3{,}000{,}000\left[\frac{0.08(1+0.08)^{-0.5}}{(1+0.08)^{30} - 1}\right]$$

$$= \$25{,}482.61$$

The impact of timing of receipts or investment are summarized in Table 25.1. Note that Table 25.1 shows the annuities to be received for the same present value or future value, interest rate, and periods. The only difference between the calculations is whether the payments were received at the beginning of the year, middle of the year, or end of the year. As should be expected, middle of the year payments is valued between beginning of the year payments and end of the year payments (Table 25.1).

TABLE 25.1 Summary Showing Impact of Timing of Periodic Payments on Annuities

Periodic Timing of Payments	Annuities based on $P = \$3 \times 10^6$, $i = 8\%$ per yr., $n = 20$	Annuities based on $F = \$3 \times 10^6$, $i = 8\%$ per yr., $n = 20$
Beginning-of-year (or period)	$282,922.80 (Example 25.7)	$24,520.65 (Example 25.8)
Middle-of-year (or period)	$294,022.00 (Example 25.9)	$25,482.61 (Example 25.10)
End-of-year (or period)	$305,556.63 (Example 25.5)	$26,482.30 (Example 25.4)

End of Example 25.10

25.3.8 Non-Uniform Series of Payments (or Investments)

The equations that have been presented earlier assumed a series of uniform payments (or investments). The present or future values of these payments (or investments) were calculated with equations derived to facilitate those computations. However, in many situations the payments (or investments) are not uniform over the years (or periods). They are variable. In such cases, calculation of present value, future value, or any other equivalent values can be accomplished more easily with the aid of time diagrams. The use of time diagrams to solve these problems is illustrated in Example 25.11.

EXAMPLE 25.11 Calculation of the Present Values, Future Values, and Annuities with the Aid of Time Diagrams of Non-Uniform Payments (or Investments) at End of the Period, Beginning of the Period, and Middle of the Period

Problem

Suppose you have received payments of \$400 at the end of years 1, 2, and 3; \$600 at end of years 4, 5, 6, and 7; \$1000 at the end of years 8, 9, and 10. For these payments, calculate the following equivalent values assuming 9% interest rate compounded annually:

a. The present value of the payments at year zero

b. The future value of the payments at end of year 10

c. Equivalent equal uniform payments assuming payments are received at the end of the year

d. Equivalent equal uniform payments assuming payments are received at the beginning of the year

e. Equivalent equal uniform payments assuming payments are received at the middle of the year

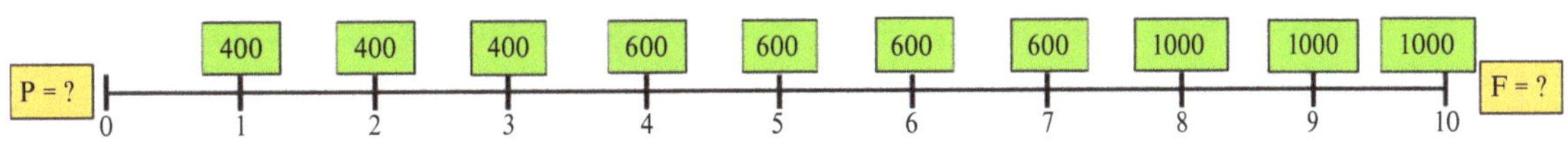

▲ FIGURE 25.4 Time diagram: end of year payments for Example 25.11

Solution

$i = 0.09$

$(1+i) = 1.09$

The time diagram for the payments is shown as Figure 25.4 as end-of-year payments.

a. Using Figure 25.4 and Eq. (25.6), the present value of the payments at year zero is:

$$P = \frac{400}{(1.09)^1} + \frac{400}{(1.09)^2} + \frac{400}{(1.09)^3} + \frac{600}{(1.09)^4} + \frac{600}{(1.09)^5} + \frac{600}{(1.09)^6} + \frac{600}{(1.09)^7} + \frac{1000}{(1.09)^8} + \frac{1000}{(1.09)^9} + \frac{1000}{(1.09)^{10}}$$

$$= \$3898.22$$

Alternate solution: Using Figure 25.4, Eqs. (25.6) and (25.14), the present value of the payments at year zero is:

$$P = 400\left[\frac{(1.09)^3 - 1}{0.09(1.09)^3}\right] + 600\left[\frac{(1.09)^4 - 1}{0.09(1.09)^4}\right]\left[\frac{1}{(1.09)^3}\right] + 1000\left[\frac{(1.09)^3 - 1}{0.09(1.09)^3}\right]\left[\frac{1}{(1.09)^7}\right] = \$3898.22$$

b. Using Figure 25.4 and Eq. (25.5), the future value of the payments at year 10 is:

$$F = 400(1.09)^9 + 400(1.09)^8 + 400(1.09)^7 + 600(1.09)^6 + 600(1.09)^5 + 600(1.09)^4 + 600(1.09)^3 + 1000(1.09)^2 + 1000(1.09)^1 + 1000$$

$$= \$9228.50$$

Alternate solution: Using Figure 25.4 and Eqs. (25.5) and (25.10), the future value of the payments at year 10 is:

$$F = 400\frac{\left[(1.09)^3 - 1\right]\times(1.09)^7}{0.09} + 600\frac{\left[(1.09)^4 - 1\right]\times(1.09)^3}{0.09} + 1000\frac{\left[(1.09)^3 - 1\right]}{0.09}$$

$$= \$9228.50$$

c. Equivalent equal uniform payments assuming payments are received at the end of the year.

$P = \$3898.22$

$i = 0.09$

$n = 10$

Substituting these values into Eq. (25.13) gives:

$$A = 3898.22\left[\frac{0.09(1.09)^{10}}{(1.09)^{10}-1}\right] = \$607.42$$

d. Equivalent equal uniform payments assuming payments are received at the beginning of the year

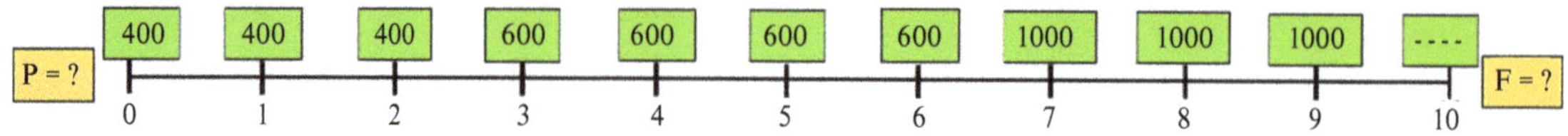

▲ **FIGURE 25.5** **Time diagram: beginning of year payments for Example 25-11**

The time diagram for the payments is shown in Figure 25.5 as beginning-of-year payments.

Using Figure 25.4 and Eq. (25.6), the present value of the payments at year zero is:

$$P = 400 + \frac{400}{(1.09)^1} + \frac{400}{(1.09)^2} + \frac{400}{(1.09)^3} + \frac{400}{(1.09)^4}$$

$$+ \frac{400}{(1.09)^5} + \frac{400}{(1.09)^6} + \frac{400}{(1.09)^7} + \frac{400}{(1.09)^8} + \frac{400}{(1.09)^9}$$

$$= \$4249.06$$

Alternate solution: Using Figure 25.5 and Eqs. (25.6) and (25.14), the present value at year zero is:

$$P = 400 + 400\left[\frac{(1.09)^2-1}{0.09(1.09)^2}\right] + 600\left[\frac{(1.09)^4-1}{0.09(1.09)^4}\right]$$

$$\times\left[\frac{1}{(1.09)^2}\right] + 1000\left[\frac{(1.09)^3-1}{0.09(1.09)^3}\right]\left[\frac{1}{(1.09)^6}\right]$$

$$= \$4249.06$$

Note the future value calculations are not shown here but it is also an alternate solution.

Equivalent equal uniform payments assuming payments are received at the beginning of the year.

$P = \$4249.06$

$i = 0.09$

$n = 10$

Substituting these values into Eq. (25.15) gives:

$$B = 4249.06\left[\frac{0.09(1.09)^9}{(1.09)^{10}-1}\right]$$

$$= \$607.42$$

e. Equivalent equal uniform payments assuming payments are received at the middle of the year.

The time diagram for the payments is shown in Figure 25.6 as middle-of-year payments.

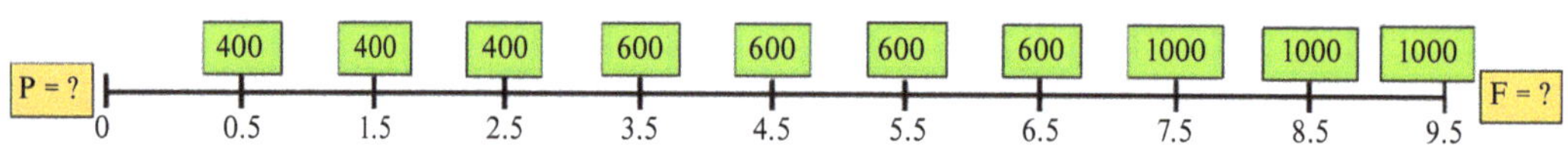

▲ **FIGURE 25.6** **Time diagram: middle of year payments for Example 25.11**

Using Figure 25.6 and Eq. (25.6), the present value of the payments at year zero is:

$$P = \frac{400}{(1.09)^{0.5}} + \frac{400}{(1.09)^{1.5}} + \frac{400}{(1.09)^{2.5}} + \frac{400}{(1.09)^{3.5}}$$

$$+ \frac{400}{(1.09)^{4.5}} + \frac{400}{(1.09)^{5.5}} + \frac{400}{(1.09)^{6.5}}$$

$$+ \frac{400}{(1.09)^{7.5}} + \frac{400}{(1.09)^{8.5}} + \frac{400}{(1.09)^{9.5}}$$

$$= \$4069.86$$

Alternate solution: Using Figure 25.6 and Eqs. (25.6) and (25.14), the present value at year zero is:

$$P = \frac{400}{(1.09)^{0.5}} + 400\left[\frac{(1.09)^2-1}{0.09(1.09)^2}\right]\left[\frac{1}{(1.09)^{0.5}}\right]$$

$$+ 600\left[\frac{(1.09)^4-1}{0.09(1.09)^4}\right]\left[\frac{1}{(1.09)^{2.5}}\right]$$

$$+ 1000\left[\frac{(1.09)^3-1}{0.09(1.09)^3}\right]\left[\frac{1}{(1.09)^{6.5}}\right]$$

$$= \$4069.86$$

Note the future value calculations are not shown here but it is also an alternate solution.

Equivalent equal uniform payments assuming payments are received at the beginning of the year.

$P = \$4069.86$

$i = 0.09$

$n = 10$

Substituting these values into Eq. (25.17) gives:

$$C = 4069.86\left[\frac{0.09(1.09)^{9.5}}{(1.09)^{10}-1}\right] = \$607.42$$

Note the equivalency of the uniform payments received irrespective of the timing of the payments.

End of Example 25.11

25.3.9 Period, Nominal, and Effective Interest Rates

In previous sections and example problems, interest rates have been presented on an annual basis. These are generally referred to as nominal annual interest rates which are denoted with the symbol, i_n. The interest rate per period is denoted with the symbol, i. Suppose the number of compounding periods per year is denoted by the symbol, m, then the relationship between the nominal annual interest rate and the period interest rate is given by the expression:

$$i_n = i \times m \tag{25.19}$$

If the number of years is represented as t, the total number of periods is given by the expression:

$$n = m \times t \tag{25.20}$$

The annual effective interest rate, i_e, is the annual interest rate which is equivalent to the period interest rate, i, compounded m times per year. The relationship between annual effective interest and period interest rates is expressed as:

$$i_e = (1+i)^m - 1 \tag{25.21}$$

The American mortgage lending industry uses Annual Percentage Rate (APR) to represent effective interest rate for loans that include deductions based on points and other fees. This is illustrated in Example 25.12.

Interest rates can be compounded on a monthly, quarterly, or semi-annual basis. If interest rate is compounded monthly, the effective monthly interest rate per month, i_{mon}, can be derived from Eq. (25.21) as:

$$i_{mon} = (1+i_n)^{1/12} - 1 \tag{25.22}$$

Sometimes, interest rates are compounded on a continuous basis. On continuous compounding, the period interest rates become increasingly smaller as the number of periods approach infinity. For continuous compounding, this is expressed as:

$$\lim_{m\to\infty}\left(1+\frac{i_n}{m}\right) = e^{i_n} \tag{25.23}$$

For continuous compounding, substituting Eq. (25.23) into Eq. (25.21) gives:

$$i_e = e^{i_n} - 1 \tag{25.24}$$

Note that period, nominal, and effective interest rates can be used consistently as long as the number of compounding periods per year is applied correctly. Also for continuous compounding, use Eq. (25.24) to calculate the applicable effective interest rate per year. The applications of period, nominal, effective interest rates with discrete and continuous compounding of interest will be illustrated in Example 25.13.

EXAMPLE 25.12 Calculation of Annual Percentage Rate (APR)

Problem

Suppose you are buying a home for an agreed price of \$250,000 with a loan offered at 4% interest rate per year. The mortgage payment at the end of each month is scheduled for 30 years and the lender is asking for 2 points (2% fee of loan value) in addition for the loan. What is the APR for the loan?

Solution

Using Eq. (25.19), period interest rate, $i = 4\%/12 = 0.3333\%$ or 0.003333

From Eq. (25.20), number of periods, $n = 12 \times 30 = 360$

$P = \$250{,}000$

$i = 0.003333$

$n = 360$

Substituting these values into Eq. (25.13),

$$\text{Monthly payments} = 250000\left[\frac{0.003333(1.003333)^{360}}{(1.003333)^{360}-1}\right] = \$1193.48$$

$$\text{Actual loan amount received by borrower} = \$250{,}000 - 0.02(\$250000) = \$245{,}000$$

Using Eq. (25.13),

$$\left[\frac{i(1+i)^{360}}{(1+i)^n - 1}\right] = \frac{\$1193.48}{\$245000} = 0.004871 \equiv \text{"Term"}$$

At 4% nominal interest rate, $i = 0.003333$,

"Term" $= 0.004774$

At i% nominal interest rate, $i = ?$,

"Term" $= 0.004871$

At 5% nominal interest rate, $i = 0.004167$,

"Term" $= 0.005368$

By interpolation,

$$\frac{i - 0.003333}{0.004167 - 0.003333} = \frac{0.004871 - 0.004774}{0.005368 - 0.004774}$$

$i = 0.003469$ or 0.3469% per month

Using Eq. (25.21),

$$i_e = (1 + 0.003469)^{12} - 1 = 0.0424 \text{ or } 4.24\% = \text{APR}$$

End of Example 25.12

EXAMPLE 25.13 Calculation of Present Values with Period and Effective Interest Rates Using Discrete and Continuous Compounding of Interest Rates

Problem

Suppose a retiree will receive payments of \$5000 each year at the end of years 1, 2, 3, 4, and 5. Assuming nominal rate of 12% per year, calculate:

a. Present value at year 0 if nominal annual interest rate is compounded quarterly
b. Present value at year 0 if nominal annual interest rate is compounded continuously

Solution

a. Present value at quarterly compounding

Total number of periods, $n = m \times t = 4 \times 5 = 20$

From Eq. (25.19), period interest rate, $i = 12\%/4 = 3\%$

Substituting period interest rate and number of periods into Eq. (25.6) gives:

$$P = \frac{5000}{(1.03)^4} + \frac{5000}{(1.03)^8} + \frac{5000}{(1.03)^{12}} + \frac{5000}{(1.03)^{16}} + \frac{5000}{(1.03)^{20}}$$

$= \$17{,}780.59$

Alternate solution: Using effective interest rate per year.

From Eq. (25.21), effective interest rate is given by:

$$i_e = (1 + 0.03)^4 - 1 = 0.1255 \text{ or } 12.55\%$$

Substituting effective interest rate per year and number of years into Eq. (25.6) gives:

$$P = \frac{5000}{(1.1255)^1} + \frac{5000}{(1.1255)^2} + \frac{5000}{(1.1255)^3} + \frac{5000}{(1.1255)^4} + \frac{5000}{(1.1255)^5}$$

$= \$17{,}780.98$

Alternate solution: Using Eq. (25.14) gives:

$$P = 5000\left[\frac{(1.1255)^5 - 1}{0.1255(1.1255)^5}\right]$$

$= \$17{,}780.98$

b. Present value at continuous compounding

From Eq. (25.24), effective interest rate at continuous compounding is given by:

$$i_e = e^{i_n} - 1 = e^{0.12} - 1 = 0.1275 \text{ or } 12.75\%$$

Substituting effective interest rate per year and number of years into Eq. (25.6) gives:

$$P = \frac{5000}{(1.1275)^1} + \frac{5000}{(1.1275)^2} + \frac{5000}{(1.1275)^3} + \frac{5000}{(1.1275)^4} + \frac{5000}{(1.1275)^5}$$

$= \$17{,}693.96$

Alternate solution: Substituting into Eq. (25.14) gives:

$$P = 5000\left[\frac{(1.1275)^5 - 1}{0.1275(1.1275)^5}\right]$$

$= \$17{,}693.96$

End of Example 25.13

25.3.10 The Rule of 72

The Rule of 72 is an approximation that can be used to quickly calculate how long it will take for an investment of money to double at a given interest rate. The Rule of 72 is represented as:

$$i \times n = 72 \qquad (25.25)$$

In Eq. (25.25), i = interest rate per period in percent, and n = number of periods. Application of the Rule of 72 is illustrated with Example 25.14.

EXAMPLE 25.14 Application of the Rule of 72

Problem

If you invested \$50,000 at 8% interest rate per year, how many years will it take for your investment to double?

Solution

$P = \$50{,}000$

$F = \$100{,}000$

$i = 8\%$

$n = ?$

Substituting these values into Eq. (25.5) gives:

$$\$100000 = \$50000(1.08)^n$$

$$2 = (1.08)^n$$

$$n = \frac{\ln 2}{\ln 1.08} = 9.0065 \text{ years}$$

Using Eq. (25.25),

$$n = \frac{72}{8}$$

$$= 9.0000 \text{ years}$$

In Example 25.14, the error in using the Rule of 72 to calculate the number of years it will take for the investment to double is −0.072%. This is sufficiently close for most applications. However, the error associated with using the Rule of 72 becomes significant as the period interest rate becomes large (greater than 15%) or the number of compounding periods becomes small (less than 3).

End of Example 25.14

25.3.11 The Rule of 78

The Rule of 78 is a method that is sometimes used to calculate the amount of accrued interest if a loan is paid off before the due date. It usually results in a higher effective interest rate than the loan interest late. Lenders benefit from applying the Rule of 78 but borrowers should avoid it in negotiating the terms of the loan. The Rule of 78 is also used in calculating refund for prepaid insurance premium that is cancelled before the end of the insurance policy.

If the number of periods is equal to, n, then the sum of compounding periods is represented as:

$$\text{Sum of Compounding Periods (SCP)} = \frac{n \times (n+1)}{2} \quad (25.26)$$

The SCP in a year if compounded every month is therefore:

$$\text{SCP} = \frac{12 \times 13}{2} = 78$$

This is the source of the term "Rule of 78." The cumulative fraction interest earned through any period, j, is calculated with the generalized formula:

Cumulative Fraction Interest Earned

$$= \frac{j \times n - \text{SCP}(j-1)}{\text{SCP}(n)} \quad (25.27)$$

Table 25.2 provides factors that can be applied to a 1-year loan due at the end of year if the loan is paid-off at the end of any month. Application of Table 25.2 is illustrated with Example 25.15.

The Rule of 78 is sometimes used to compute the amount of refund due on prepaid insurance if the policy is terminated before the end of its term. If the termination is effective at the end of period, j, and the total period of the insurance is n, then the fraction to be refunded is given by:

$$\text{Fraction to be refunded} = \frac{\text{SCP}(n-j)}{\text{SCP}(n)} \quad (25.28)$$

In some transactions, "flat" interest is applied to the initial principal at each compounding period. This is different from compound interest which is applied to unpaid principal and accrued interest at each compounding period. The cumulative flat interest is given by:

$$\text{Cumulative Flat Interest} = P \times i \times n \quad (25.29)$$

Application of flat interest and Rule of 78 is illustrated in Example 25.17.

TABLE 25.2 Prepayment Factors for 1-Year Loans Using the Rule of 78

Month	Cum. Frac. Int. Earned	Monthly Interest	Prepayment Factor	Multiplier Factor
1	0.1538	0.1538	1 + 0.1538 × i	1.8462
2	0.2949	0.1410	1 + 0.2949 × i	1.7692
3	0.4231	0.1282	1 + 0.4231 × i	1.6923
4	0.5385	0.1154	1 + 0.5385 × i	1.6154
5	0.6410	0.1026	1 + 0.6410 × i	1.5385
6	0.7308	0.0897	1 + 0.7308 × i	1.4615
7	0.8077	0.0769	1 + 0.8077 × i	1.3846
8	0.8718	0.0641	1 + 0.8718 × i	1.3077
9	0.9231	0.0513	1 + 0.9231 × i	1.2308
10	0.9615	0.0385	1 + 0.9615 × i	1.1538
11	0.9872	0.0256	1 + 0.9872 × i	1.0769
12	1.0000	0.0128	1 + 1.0000 × i	1.0000

EXAMPLE 25.15 Application of the Rule of 78 on a Bank Loan

Problem

Suppose you borrowed \$20,000 from a bank at interest rate of 12% for 1 year to be paid back at the end of the year under the terms of Rule of 78. If you decide to pay off the loan at the end of 7 months, what is your payoff amount? What is the effective interest rate of the loan if paid off at end of 7 months?

Solution

$P = \$20,000$

$i = 12\%$

$F = ?$

Using Table 25.2, pay off amount at 7 months is:

$$F = P(1+0.8077\times i)$$
$$= \$20000(1+0.8077\times 0.12)$$
$$= \$21,938.48$$

Using Table 25.2, the effective interest rate of the loan if paid off at end of 7 months is:

$$\text{Effective interest rate at end of 7 months} = \text{Multiplier factor}\times 12\%$$
$$= 1.3846\times 12\% = 16.6152\%$$

End of Example 25.15

EXAMPLE 25.16 Application of the Rule of 78 on Prepaid Insurance Premium

Problem

If you paid \$1200 insurance premium on your car for 1 year and decided to cancel the policy after 5 months, how much is your refund from the insurance company?

Solution

$j = 5$

$n = 12$

Using Eq. (25.27),

$$\text{Cumulative fraction of premium earned} = \frac{5\times 12-\text{SCP}(5-1)}{\text{SCP}(12)}$$
$$= \frac{60-10}{78} = 0.6410$$

$$\text{Amount of the refund} = \$1200(1-0.6410)$$
$$= \$430.80$$

Alternate solution:

Using Eq. (25.28),

$$\text{Fraction to be refunded} = \frac{\text{SCP}(12-5)}{\text{SCP}(12)}$$
$$= 0.3590$$

$$\text{Amount to be refunded} = \$1200\times 0.3590$$
$$= \$430.80$$

End of Example 25.16

EXAMPLE 25.17 Application of the Rule of 78 on a Bank Loan Lasting More Than 1 Year

Problem

Suppose you borrowed \$15,000 from a bank at 10% interest rate per year to be paid back in 5 years at equal payments of \$4500 each year. What is your repayment schedule if the loan was based on the Rule of 78?

Solution

$P = \$15,000$

$i = 10\%$

$n = 5$

Using Eq. (25.29):

$$\text{Cumulative flat interest} = \$15000\times 0.1\times 5$$
$$= \$7500$$

Applying Eq. (25.27) on the basis of Rule of 78, the repayment schedule is shown in Table 25.3.

TABLE 25.3 | Schedule of Repayment for Example 25.17

Year	Cum. Fraction Interest Earned	Cum. Interest ($)	Interest Paid Per Year ($)	Principal Paid Per Year ($)	Total Paid Per Year ($)
1	0.3333	2500.00	2500.00	2000.00	4500.00
2	0.6000	4500.00	2000.00	2500.00	4500.00
3	0.8000	6000.00	1500.00	3000.00	4500.00
4	0.9333	7000.00	1000.00	3500.00	4500.00
5	1.0000	7500.00	500.00	4000.00	4500.00
Total			**7500.00**	**15000.00**	**22500.00**

End of Example 25.17

25.3.12 Cash Flow Diagram

The time diagram was shown in Figure 25.2 as a way of graphically illustrating the flow of investments, payments, annuities, etc., with the time periods of the events. Since then, time diagrams have been used to illustrate solutions to several example problems. However, there is an alternative time diagram that is called cash flow diagram which can be used as well to illustrate the time flow of money. The cash flow diagram is equivalent to the time diagram in all respects. The choice between use of time diagram or cash flow diagram is entirely dependent on individual preferences. In the petroleum industry, the cash flow diagram is preferred in most cases. A sample of the cash flow diagram is shown as Figure 25.7. In cash flow diagrams, investments or any cash outflow is represented as negative, and net revenues or any cash inflows are shown as positive. For instance, in Figure 25.7 investment at time zero (P) is shown as negative. Net cash flows (NCF) (i.e., revenues-expenses) within the span of the project are assumed to be positive, and the net cash outflow at the end of the project is assumed to be negative. Note that Figure 25.7 is a simplified version of a cash flow diagram and it is used here to illustrate the essential concepts of a cash flow diagram. Only cash flow diagrams will be used to illustrate cash flows in petroleum projects for the remainder of this chapter.

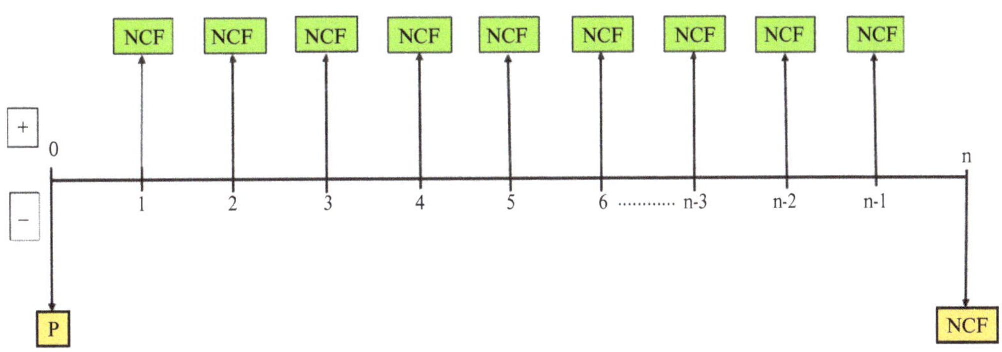

▲ FIGURE 25.7 Cash flow diagram

EXAMPLE 25.18 Application of the Cash Flow Diagram on a Project

Problem

An independent oil company invested $8,500,000 in drilling, completing, and placing an oil well on production. The expected annual net cash flows based on production from the well over its projected life of 7 years are shown in Figure 25.8. If the discount rate for the company is 12% per year, what is the net present value (NPV) of the investment. Assume the net cash flows shown in Table 25.4 occur at the end of the year. The cash flow diagram representing the investment is shown as Figure 25.8.

TABLE 25.4 | Net Cash Flow for Example 25.18

Year	Net cash flow ($)
0	-8,500,000
1	4,000,000
2	3,500,000
3	3,000,000
4	2,500,000
5	2,000,000
6	1,500,000
7	-2,000,000

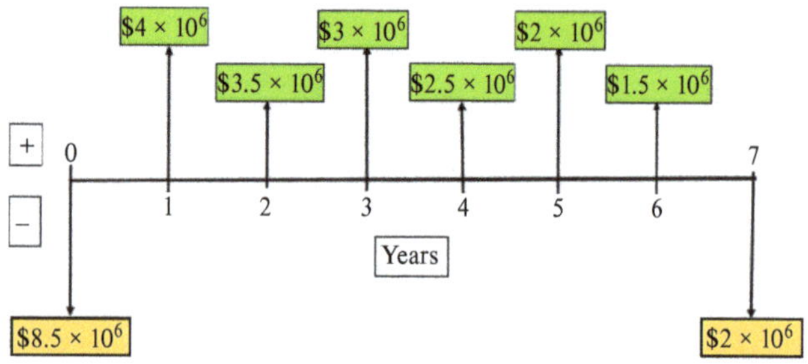

▲ **FIGURE 25.8** **Net cash flow diagram for Example 25.18**

Solution

$i = 12\%$

Using Eq. (25.6),

$$\begin{aligned} NPV &= -\$8.5\times10^6 + \frac{\$4\times10^6}{(1.12)} + \frac{\$3.5\times10^6}{(1.12)^2} \\ &\quad + \frac{\$3.0\times10^6}{(1.12)^3} + \frac{\$2.5\times10^6}{(1.12)^4} + \frac{\$2.0\times10^6}{(1.12)^5} \\ &\quad + \frac{\$1.5\times10^6}{(1.12)^6} - \frac{\$2.0\times10^6}{(1.12)^7} \\ &= \$2,575,845.04 \end{aligned}$$

End of Example 25.18

25.4 Economic Valuation Parameters

Various methods of accounting for the time value of money have been presented in Section 25.2. These methods dealt with calculating the present value (worth) or future value (worth) of a lump-sum of money or a series of payments (or incomes) at any specified time period. In many cases, the time period was chosen to be time zero for convenience. In all the examples shown in Section 25.2, straight or compound interest rates were either specified or calculated without any discussions on the relevance or sources of these discount rates. However, in many companies or organizations, projects are evaluated economically based on discount rates specified by the company or organization. These discount rates are computed by the financial department of the company or organization to meet certain expected growth in the wealth of the company or organization over ten or more years in the future. The two types of rates generally used in most companies or organizations to evaluate projects are:

1. Average cost of capital
2. Average investment opportunity rate

The average cost of capital is typically calculated as the average cost of debt (bonds, etc.) and the cost of equity (stocks, etc.) weighted on the basis of the debt/equity ratio of the company. The average investment opportunity rate is the rate of return that the company would expect if available funds were invested in other opportunities available to the company. Generally, companies would prefer to invest funds at the investment opportunity rate rather than at cost of capital since the former rate would lead to faster growth of the wealth of the company. Average cost of capital could be looked at as just investing funds at the cost of borrowing the funds from debtors and shareholders. Since the primary goal of the company is to maximize shareholder wealth, investing at higher rates offered by other opportunities would generate faster growth of the shareholder wealth. It should be evident that either the average cost of capital or average investment opportunity rate is the minimum rate of return acceptable to the company for funding of projects using any time value of money investment criteria. Average costs of capital and average investment opportunity rates are internally generated and specific to the growth needs of each company. Some companies use a "hurdle rate" which is based on some internal objectives established by the company. Note that "hurdle rates" are subjective and specific to the needs of each company. Some economic evaluators suspect that risks are included in hurdle rates used by many companies. However, it is generally accepted by industry practitioners that risk should not be included in rates used in economic evaluations. These rates should reflect either the average cost of capital or the average investment opportunity rate of the company. The risks associated with any project should be handled in a separate risk analysis exercise. When the risks associated with a project are embedded in the discount rates, the potential dangers posed by the project to the company are hidden and may not be fully addressed. A project that passed all the economic evaluation criteria may still not be pursued by the company if it is judged to be too risky by management.

Generally, there are two main types of economic valuations that are conducted on projects. These are:

1. Valuation parameters not based on time value of money
2. Valuation parameters based on time value of money

Economic valuation parameters **not** based on time value of money presented in this book are:

a. Undiscounted pay-back (pay-out)
b. Undiscounted profitability index

Economic valuation parameters based on the time value presented in the book are:

a. Discounted pay-back (pay-out)
b. Discounted profitability index
c. Net present value (NPV)
d. Net present value profile
e. Internal rate of return (IRR)

25.4.1 Valuation Parameters *Not* Based on the Time Value of Money

Some economic valuation parameters are not based on the time value of money. These parameters are easy to calculate and offer quick measure on the economic profitability of the project. They provide a quick yardstick which can be used to decide whether to proceed with more robust economic evaluation of the project.

25.4.1.1 Undiscounted Pay-Back (Pay-Out)

Undiscounted pay-back (also termed pay-out) is a yardstick based on time required to recover the initial investment. It is a quick parameter that can be calculated readily from initial investments and projected net cash flows. Undiscounted pay-back should not be used as a parameter for decision on the economic profitability of a project. Some of the main limitations of undiscounted pay-back as a valuation parameter are:

1. It is not based on time value of money.
2. It ignores cash flows beyond the pay-back period.
3. It does not measure the profitability of the project.
4. It cannot be used for ranking of projects.

EXAMPLE 25.19 Calculation of Pay-Back

Problem

Calculate the pay-back period for a project with the net cash flow shown in Table 25.5.

Solution

Using the cumulative cash flow column of Table 25.5,

$$\text{Pay-back} = 4 + \frac{30000}{60000} = 4.5 \text{ years}$$

TABLE 25.5 Net Cash Flow for Example 25.19

Year	Net Cash Flow ($)	Cumulative Net Cash Flow ($)
0	−480,000	−480,000
1	140,000	−340,000
2	125,000	−215,000
3	105,000	−110,000
4	80,000	−30,000
5	60,000	+30,000
6	50,000	+80,000
7	40,000	+120,000

End of Example 25.19

25.4.1.2 Undiscounted Profitability Index

Undiscounted profitability index is also called undiscounted profit-to-investment ratio. This is calculated by dividing the undiscounted sum of net operating income (NOI) or net cash flow (NCF) by the undiscounted sum of investments (INV). If undiscounted sum of net operating income is used in the calculation, the break-even value is equal to one which indicates that all the investments are recovered. For this case, PI values less than one are unprofitable while values higher than one are considered profitable. If the undiscounted sum of net cash flow is used in the calculation, the break-even value is zero indicating that all the investments are recovered. For this case, PI values less than zero are unprofitable and values higher than zero are considered profitable. Note that the profitability index for undiscounted net operating income can be calculated from the profitability index of net cash flow by adding a value of 1.0 to the result. Undiscounted profitability index for NOI can be calculated with the expression:

$$\text{Undiscounted PI} = \frac{\text{Sum of undiscounted NOI}}{\text{Sum of undiscounted INV}} \tag{25.30}$$

Alternatively, the undiscounted profitability index for net cash flow can be calculated with the expression:

$$\text{Undiscounted PI} = \frac{\text{Sum of undiscounted NCF}}{\text{Sum of undiscounted INV}} \tag{25.31}$$

EXAMPLE 25.20 Calculation of Undiscounted Profitability Index (PI)

Problem

Calculate the profitability indices for the undiscounted net operating income and undiscounted net cash flow from the cash flows shown in Table 25.6.

Solution

Using Eq. (25.30) and Table 25.6, undiscounted PI for net operating income is:

$$\text{Undiscounted PI} = \frac{\$600,000}{\$480,000} = 1.25$$

Alternatively, using Eq. (25.31) and Table 25.6, undiscounted PI for net cash flow is:

$$\text{Undiscounted PI} = \frac{\$120,000}{\$480,000} = 0.25$$

Note that the PI for net operating income could be calculated from the PI for net cash flow by adding one to the PI of net cash flow as discussed earlier.

TABLE 25.6 Cash flow and calculations for Example 25.20

Year	Investment ($)	Net Operating Income (NOI) ($)	Net Cash Flow (NCF) ($)
0	480,000	0	−480,000
1		140,000	140,000
2		125,000	125,000
3		105,000	105,000
4		80,000	80,000
5		60,000	60,000
6		50,000	50,000
7		40,000	40,000
Total	**480,000**	**600,000**	**120,000**
	PI	**1.25**	**0.25**

End of Example 25.20

25.4.2 Valuation Parameters Based on the Time Value of Money

Economic valuation parameters based on the time value of money are the best methods to measure the profitability of a project. They can be used effectively to rank and compare projects based on common criteria. Earlier in this chapter, several measures of time value of money such as present values and future values were presented and illustrated with examples. These measures based on discounting money to a specific time period such as zero time will be used in the calculation of these economic valuation parameters.

25.4.2.1 Discounted Pay-Back (Pay-Out)

Discounted pay-back or pay-out is similar in all respects to the undiscounted version except that the net cash flow is discounted at the average investment opportunity rate or average cost of capital. As an economic valuation parameter, it is still limited in its applicability by all the factors listed earlier for the undiscounted pay-back parameter.

EXAMPLE 25.21 Calculation of Pay-back with Discounted Cash Flow

Problem

Calculate the discounted pay-back period for a project with the net cash flow at end of the year shown in Table 25.7. Assume average investment opportunity rate of 5% per year.

Solution

Using the cumulative cash flow column of Table 25.7:

$$\text{Discounted Pay-back} = 5 + \frac{29757.27}{37310.77}$$
$$= 5.80 \text{ years}$$

By comparing Examples 25.19 and 25.21, note that pay-back increased by 1.3 years by discounting the same net cash flow at 5% annual investment opportunity rate.

TABLE 25.7 Net Cash Flow at End of the Year for Example 25.21

Year	Cash Flow (NCF) ($)	Present Value NCF Discounted at 5% ($)	Cumulative PV of NCF ($)
0	−480,000	−480,000.00	−480,000.00
1	140,000	133,333.33	−346,666.67
2	125,000	113,378.68	−233,287.98
3	105,000	90,702.95	−142,585.03
4	80,000	65,816.20	−76,768.84
5	60,000	47,011.57	−29,757.27
6	50,000	37,310.77	7553.50
7	40,000	28,427.25	35,980.76

End of Example 25.21

25.3.2.2 Discounted Profitability Index

Discounted profitability index is also called discounted profit-to-investment ratio. It is calculated by dividing the discounted sum of net operating income (NOI) or net cash flow (NCF) by the discounted sum of investments (INV). If the discounted sum of net operating income is used in the calculation, the break-even value is equal to one which indicates that all the investments are recovered. For this case, PI values less than one are economically unprofitable while PI values greater than one are economically profitable. If the discounted sum of net cash flow is used in the calculation, the break-even value is zero indicating that all the investments are recovered. For this case, PI values less than zero are classified as uneconomic and PI values greater than one are classified as economically profitable. Discounted profitability index for net operating income (NOI) can be calculated with the expression:

$$\text{Discounted PI} = \frac{\text{Sum of discounted NOI}}{\text{Sum of discounted INV}} \tag{25.32}$$

Alternatively, the discounted profitability index for net cash flow can be calculated with the expression:

$$\text{Discounted PI} = \frac{\text{Sum of discounted NCF}}{\text{Sum of discounted INV}} \tag{25.33}$$

EXAMPLE 25.22 Calculation of Discounted Profitability Index (PI)

Problem

Calculate the profitability indices for the discounted net operating income and discounted net cash flow shown in Table 25.8. Assume average investment opportunity rate of 5% per year.

Solution

Using Eq. (25.32) and Table 25.8, the discounted PI for NOI is:

$$\text{Discounted PI} = \frac{\$515{,}980.76}{\$480{,}000} = 1.07$$

Alternatively, using Eq. (25.33) and Table 25.8, the discounted PI for NCF is:

$$\text{Discounted PI} = \frac{\$35{,}980.76}{\$480{,}000} = 0.07$$

On the basis of the PIs calculated for the example project, it is obvious that this project is barely profitable or economic. Note the effect of compounding by comparing the undiscounted PIs to discounted PIs calculated in Examples 25.20 and 25.22, respectively.

TABLE 25.8 Cash Flow for Example 25.22

Year	Investment ($)	Net Operating Income (NOI) ($)	Net Cash Flow (NCF) ($)	PV of NOI Discounted at 5% ($)	PV of NCF Discounted at 5% ($)
0	480,000	0	-480,000	0	-480,000.00
1		140,000	140,000	133,333.33	133,333.33
2		125,000	125,000	113,378.68	113,378.68
3		105,000	105,000	90,702.95	90,702.95
4		80,000	80,000	65,816.20	65,816.20
5		60,000	60,000	47,011.57	47,011.57
6		50,000	50,000	37,310.77	37,310.77
7		40,000	40,000	28,427.25	28,427.25
Total	**480,000**	**600,000**	**120,000**	**515,980.76**	**35,980.76**
PI	—	—	—	**1.07**	**0.07**

End of Example 25.22

25.4.2.3 Net Present Value (NPV)

Net present value (NPV) is the most common and widely accepted economic valuation parameter used in the petroleum industry. NPV is the sum of the discounted future net cash flows of a project to a specified time period typically chosen to be time zero. Note that the future net cash flows can be discounted to any other chosen time period other than time zero. NPV calculation is represented by the expression:

$$NPV = \sum_{j=0}^{n} \frac{(NCF)_j}{(1+i)^j} \tag{25.34}$$

In Eq. (25.34), NCF_j, is the net cash flow for period, j; i, is the compounding discount rate per period; and n, is the number of periods equivalent to expected life or duration of the project. The economic valuation criteria using NPV as the parameter are as follows:

- $NPV < 0$, project is uneconomic and unacceptable at the applied discount rate.
- $NPV = 0$, project is at break-even at the applied discount rate.
- $NPV > 0$, project is economic and acceptable at the applied discount rate.

Note that the applied discount rate used in the calculation could be the average investment opportunity rate, the average cost of capital or the "hurdle" rate generally applied to similar projects in the company or organization.

EXAMPLE 25.23 Calculation of Net Present Value (NPV)

Problem

Calculate the NPVs of a project with the annual net cash flow shown in Table 25.9 using the average investment opportunity rates of 5% and 8% per year.

Solution

Using Eq. (25.34) and Table 25.9, the NPV at 5% rate of return is +$35,980.76. At 8% rate of return, the NPV is –$5365.16. Note that for this example, the project is economic and acceptable when evaluated at 5% rate of return but turns uneconomic and unacceptable when evaluated at 8% rate of return. The results of the calculations are also shown in Table 25.9.

TABLE 25.9 Net Cash Flow for Example 25.23

Year	Cash Flow (NCF) ($)	Present Value NCF Discounted at 5% ($)	Present Value NCF Discounted at 8% ($)
0	-480,000	-480,000.00	-480,000.00
1	140,000	133,333.33	129629.63
2	125,000	113,378.68	107167.35
3	105,000	90,702.95	83352.39
4	80,000	65,816.20	58802.39
5	60,000	47,011.57	40834.99
6	50,000	37,310.77	31508.48
7	40,000	28,427.25	23339.62
	NPV	**35,980.76**	**-5365.16**

End of Example 25.23

25.4.2.4 Net Present Value Profile

In some cases, it is necessary to compare the net present value profiles of two or more mutually exclusive projects. The process involves calculating the net present values of the projects at different discount rates. By making a plot of the NPVs at different discount rates, it is instructive to observe the impact of net cash flows over time on selection of the most economically profitable project. This observation is best illustrated in Example 25.24.

EXAMPLE 25.24 Comparison of Mutually Exclusive Projects with Net Present Value Profile

Problem

Projects A and B are two mutually exclusive projects with net cash flows shown in Table 25.10. Prepare a plot of net present values at discount rates of 0%, 2%, 5%, 8%, and 12% for the Projects A & B. Compare the results.

Solution

Calculation of NPVs for Projects A & B at the stated discount rates are shown in Table 25.10. A plot of NPVs versus discount rates for Project A & B is shown as Figure 25.9.

Projects A & B have the same initial investment of $480,000. By examining Table 25.10, it is observed that the net cash flows of Project A are larger in the early years and taper off at the later years while the cash flows for Project B are smaller in the early years and larger in the later years. Comparing the NPVs in Figure 25.9, it is observed that at discount rates less than 4.7%, the NPVs of Project B are higher than the NPVs of Project A. This means that Project B is more profitable than Project A and should be selected at discount rates less than 4.7%. However, at discount rates greater than 4.7%, Project A has higher NPVs and should be selected. This illustrates the importance of using present value profiles to compare mutually exclusive projects instead of calculating their NPVs at a single discount rate.

TABLE 25.10 NPV Profiles for Projects A & B in Example 25.24

Project A						
Year	Net Cash Flow (NCF) ($)	Present Value NCF Discounted at 0% ($)	Present Value NCF Discounted at 2% ($)	Present Value NCF Discounted at 5% ($)	Present Value NCF Discounted at 8% ($)	Present Value NCF Discounted at 12% ($)
0	-480,000	-480,000	-480,000.00	-480,000.00	-480,000.00	-480,000.00
1	140,000	140,000	137,254.90	133,333.33	129629.63	125000.00
2	125,000	125,000	120,146.10	113,378.68	107167.35	99649.23
3	105,000	105,000	98,943.85	90,702.95	83352.39	74736.93
4	80,000	80,000	73,907.63	65,816.20	58802.39	50841.45
5	60,000	60,000	54,343.85	47,011.57	40834.99	34045.61
6	50,000	50,000	44,398.57	37,310.77	31508.48	25331.56
7	40,000	40,000	34,822.41	28,427.25	23339.62	18093.97
Total	**120,000**	**120,000**	**83,817.30**	**35,980.76**	**-5365.16**	**-52,301.26**

Project B						
Year	Net Cash Flow (NCF) ($)	Present Value NCF Discounted at 0% ($)	Present Value NCF Discounted at 2% ($)	Present Value NCF Discounted at 5% ($)	Present Value NCF Discounted at 8% ($)	Present Value NCF Discounted at 12% ($)
0	-480,000	-480,000	-480,000.00	-480,000.00	-480,000.00	-480,000.00
1	40,000	40,000	39,215.69	38,095.24	37,037.04	35,714.29
2	50,000	50,000	48,058.44	45,351.47	42,866.94	39,859.69
3	60,000	60,000	56,539.34	51,830.26	47,629.93	42,706.81
4	80,000	80,000	73,907.63	65,816.20	58,802.39	50,841.45
5	115,000	115,000	104,159.04	90,105.51	78,267.07	65,254.09
6	145,000	145,000	128,755.85	108,201.23	91,374.60	73,461.51
7	160,000	160,000	139,289.63	113,709.01	93,358.46	72,375.87
Total	**170,000**	**170,000**	**109,925.62**	**33,108.92**	**-30,663.57**	**-99,786.28**

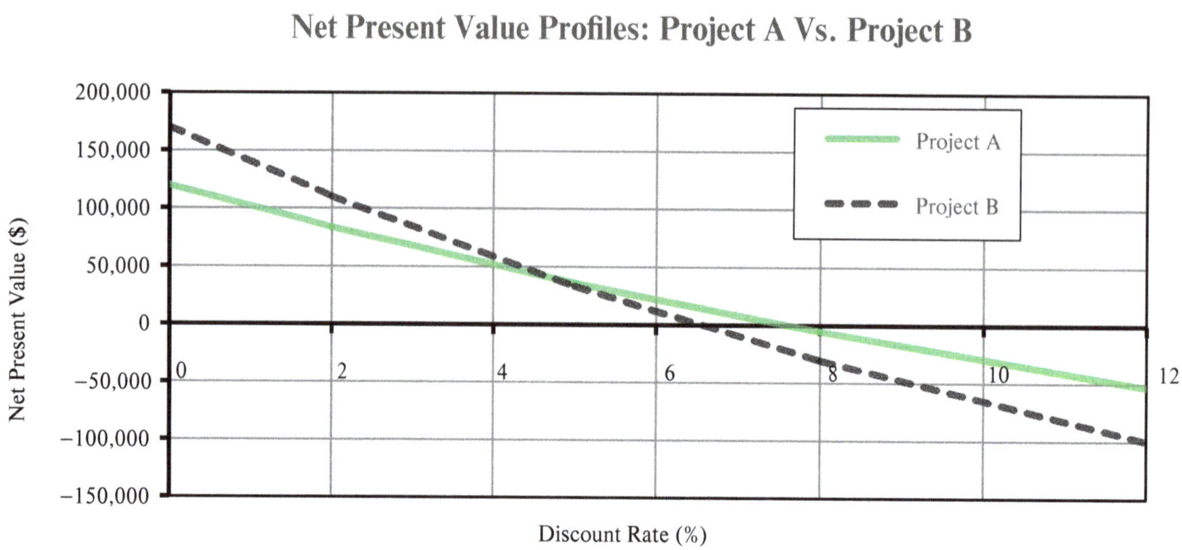

▲ **FIGURE 25.9** **Comparison plot of NPVs versus discount rates for Projects A & B**

End of Example 25.24

25.4.2.5 Internal Rate of Return (IRR)

Internal rate of return (IRR) is another measure of economic profitability. It is calculated as the discount rate that equates the sum of all net cash flows to zero. IRR is called by other names such as discounted cash flow rate of return, discounted rate of return, rate of return, internal yield, effective yield to maturity, return on investment, etc. The equation for calculation of IRR is represented as:

$$\sum_{j=0}^{n} \frac{(\mathrm{NCF})_j}{(1+\mathrm{IRR})^j} = 0 \tag{25.35}$$

In Eq. (25.35), $(\mathrm{NCF})_j$ is the net cash flow for period j; IRR is the internal rate of return (or discount rate) per period; and n is the number of periods equivalent to the expected life or duration of the project. As Eq. (25.35) indicates, IRR can be calculated by trial-and-error technique or by using root-finding methods such as the Newton-Raphson method. Another form of Eq. (25.35) can be derived by noting that:

$$(\mathrm{NCF})_j = (\mathrm{NOI})_j - (\mathrm{INV})_j \tag{25.36}$$

In Eq. (25.36), $(\mathrm{NOI})_j$ is the net operating income for period j, and $(\mathrm{INV})_j$ is the investment for period j. Substituting Eq. (25.36) into Eq. (25.35) gives:

$$\sum_{j=1}^{n} \frac{(\mathrm{NOI})_j}{(1+\mathrm{IRR})^j} - \sum_{j=0}^{n} \frac{(\mathrm{INV})_j}{(1+\mathrm{IRR})^j} = 0 \tag{25.37}$$

EXAMPLE 25.25 Calculation of Internal Rate of Return (IRR)

Problem

Calculate the internal rate of return for Project A net cash flow in Table 25.11.

Solution

For the simple net cash flows shown in Table 25.11, the straightforward way to calculate IRR for Project A is to calculate the NPV at different discount rates as shown in the same table. Then make a plot of NPV versus discount rate as shown in Figure 25.10. The IRR is at the point the curve crosses the x-axis. By trial-and-error, calculate the discount rate that will give NPV close to zero as shown in Table 25.11. This discount rate which is the IRR is 7.586%.

TABLE 25.11 Calculation of IRR for Example 25.25

Project A								
Year	Net Cash Flow (NCF) ($)	PV NCF Discounted at 0% ($)	PV NCF Discounted at 2% ($)	PV NCF Discounted at 5% ($)	PV NCF Discounted at 7% ($)	PV NCF Discounted at 7.586% ($)	PV NCF Discounted at 8% ($)	PV NCF Discounted at 12% ($)
0	-480,000	-480,000	-480,000.00	-480,000.00	-480,000.00	-480,000.00	-480,000.00	-480,000.00
1	140,000	140,000	137,254.90	133,333.33	130,841.12	130,128.46	129629.63	125000.00
2	125,000	125,000	120,146.10	113,378.68	109,179.84	107,993.72	107167.35	99649.23
3	105,000	105,000	98,943.85	90,702.95	85,711.28	84,318.33	83352.39	74736.93
4	80,000	80,000	73,907.63	65,816.20	61,031.62	59,712.73	58802.39	50841.45
5	60,000	60,000	54,343.85	47,011.57	42,779.17	41,626.74	40834.99	34045.61
6	50,000	50,000	44,398.57	37,310.77	33,317.11	32,243.00	31508.48	25331.56
7	40,000	40,000	34,822.41	28,427.25	24,909.99	23,975.61	23339.62	18093.97
Total	**120,000**	**120,000**	**83,817.30**	**35,980.76**	**7770.13**	**-1.41**	**-5365.16**	**-52,301.26**

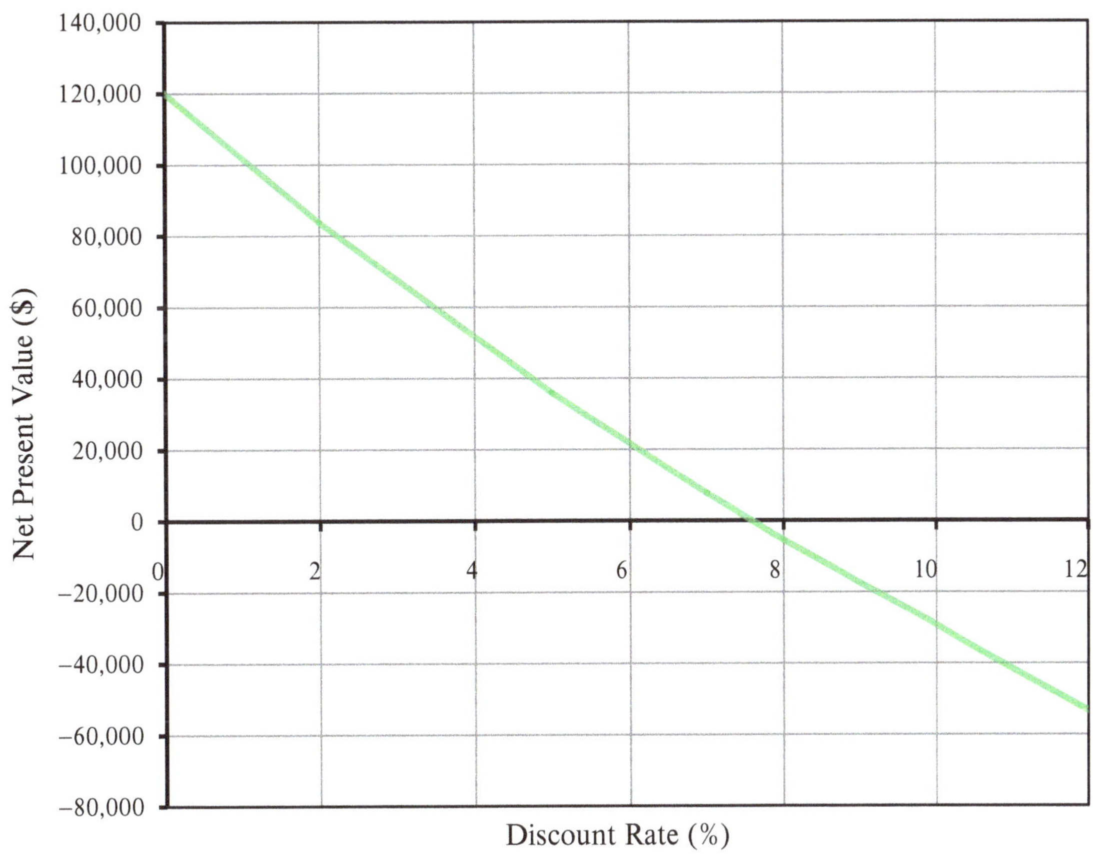

▲ FIGURE 25.10 **Calculation of IRR for Example 25.25**

25.4.2.6 Multiple Internal Rates of Return (IRR)

The equations for the calculation of IRR shown as Eqs. (25.35) or (25.37) implies that there is only one IRR that satisfies the requirement that the sum of discounted cash flows is equal to zero. This is not always the case. Multiple rates of return can occur when there are more than one IRR that can result from the sum of discounted cash flow being equal to zero. This means essentially that solutions to Eq. (25.35) or (25.37) results in more than one root. This can occur when there is additional significant investment in a project sometime later in its life after the initial investment. This can cause the sign of net cash flow to change more than once on a cash flow diagram. Consider the cash flow diagrams shown in Figure 25.11 (a and b). Figure 25.11a is a typical cash flow diagram for a project with an initial investment at time zero followed by positive net cash flows for all future time periods. In this case, the net cash flow changes sign once from negative at time zero to positive at time period 1 and subsequent time periods. For this case shown in Figure 25.11a and similar cases where the cash flow changes sign once, only one root representing the IRR will be calculated. Figure 25.11b is an example of project with significant investment in time period 5. This investment is shown to cause the net cash flow to turn negative in time period 5 from positive in time period 4. Consequently, Figure 25.11b has two sign changes. The first sign change is from time zero to time period 1 representing the initial investment, and the second sign change is from time period 4 to time period 5 representing the second investment at time period 5. For the case shown in Figure 25.11b and similar cases where there is more than one sign change, it is likely that more than one root representing the IRR will be calculated. The calculation of more than one IRR presents a special problem regarding the "correct" IRR to use in making financial decisions on the project. The answer is that neither of the two or more IRRs is the "correct" IRR to use for economic decision. In fact, it can be argued that the results do not represent true IRR as represented in Eq. (25.35) or (25.37). When two or more IRRs are calculated for a project, the recommended approach is to use NPV method for evaluating the project. An alternative approach is to use the Modified IRR method that will be discussed in Section 25.4.2.7.

In the petroleum industry, projects that may require substantial investments later after initial development investments include infill well drilling, secondary recovery projects such as waterflooding and gasflooding, pressure maintenance projects, installation of artificial lift systems, improved/enhanced oil recovery projects, etc. Some of these projects could be classified as acceleration projects. Acceleration projects are regarded as projects that result in production of accelerated reserves **not** incremental reserves. Acceleration of reserves means that total recoverable reserves remains the same, but the reserves are produced faster and earlier than originally planned. Incremental reserves mean that total recoverable reserves are increased by the amount of the incremental reserves. Example 25.26 illustrates the application of NPV and IRR methods on an acceleration project.

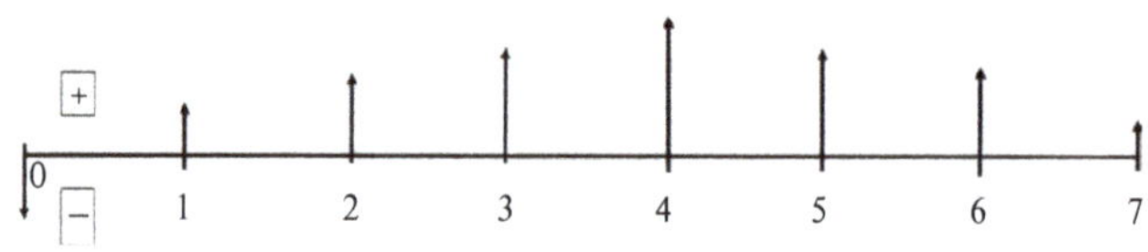

▲ FIGURE 25.11a Cash flow diagram for single IRR

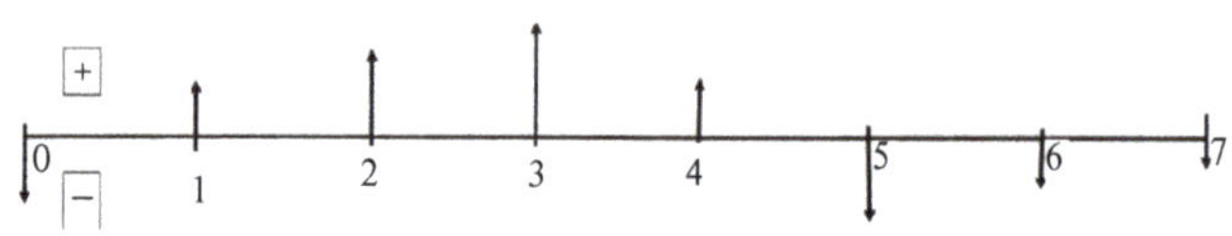

▲ FIGURE 25.11b Cash flow diagram for multiple IRR

EXAMPLE 25.26 Application of NPV and IRR Methods on an Acceleration Project

Problem

Suppose Project A is a field that has been developed and is producing with the net cash flow shown in Table 25.12. The operator of the field is proposing to drill two infill wells at the cost of $20,000 each to accelerate recovery of reserves. This proposal is termed Project B. The accelerated net cash flow from Project B is shown in Table 25.13. Should the operator get the approval to perform Project B?

Solution

NPVs for Project A and Project B at different discount rates are calculated as shown in Tables 25.12 and 25.13, respectively. A plot of NPVs versus discount rates for the two projects is shown in Figure 25.12. Examining Figure 25.12 indicates that Project A will be preferred if the average investment opportunity rate of the field is less than 4.77%. Project B should be approved if the average investment opportunity rate is more than 4.77% but less than 313.3%. Project A should be retained if the average investment opportunity rate of the field is greater than 313.3%. It is reasonable to expect that the average investment opportunity rate for the field should be higher than 4.77% but less than 313.3%. Hence, the operator should get the approval to proceed with the project.

An alternative solution to the problem is to take the difference between the net cash flows of Project B and Project A. Then calculate the NPVs of the net cash flow difference as shown in Table 25.14. A plot of these NPVs versus discount rates is also shown in Figure 25.12. Table 25.15 shows a summary of the calculations for the example. The alternative solution gives the same results and could be used to reach the same decisions as explained earlier.

TABLE 25.12 NCF and NPV Calculations for Project A in Example 25.26

Project A													
Year	INV ($)	NOI ($)	NCF ($)	PV NCF Disc. at 0% ($)	PV NCF Disc. at 10% ($)	PV NCF Disc. at 20% ($)	PV NCF Disc. at 40% ($)	PV NCF Disc. at 60% ($)	PV NCF Disc. at 80% ($)	PV NCF Disc. at 100% ($)	PV NCF Disc. at 200% ($)	PV NCF Disc. at 400% ($)	PV NCF Disc. at 600% ($)
0	0	0	0	0	0	0	0	0	0	0	0	0	0
1	0	115,000	115,000	115,000	104,545	95,833	82,143	71,875	63,889	57,500	38,333	23,000	16,429
2	0	110,000	110,000	110,000	90,909	76,389	56,122	42,969	33,951	27,500	12,222	4400	2245
3	0	90,000	90,000	90,000	67,618	52,083	32,799	21,973	15,432	11,250	3,333	720	262
4	0	80,000	80,000	80,000	54,641	38,580	20,825	12,207	7621	5000	988	128	33
5	0	75,000	75,000	75,000	46,569	30,141	13,945	7153	3969	2344	309	24	4
6	0	65,000	65,000	65,000	36,691	21,768	8633	3874	1911	1016	89	4	1
7	0	65,000	65,000	65,000	33,355	18,140	6166	2421	1062	508	30	1	0
Total	**0**	**600,000**	**600,000**	**600,000**	**434,329**	**332,935**	**220,633**	**162,472**	**127,834**	**105,117**	**55,304**	**28,277**	**18,974**

TABLE 25.13 NCF and NPV Calculations for Project B in Example 25.26

Project B													
Year	INV ($)	NOI ($)	NCF ($)	PV NCF Disc. at 0% ($)	PV NCF Disc. at 10% ($)	PV NCF Disc. at 20% ($)	PV NCF Disc. at 40% ($)	PV NCF Disc. at 60% ($)	PV NCF Disc. at 80% ($)	PV NCF Disc. at 100% ($)	PV NCF Disc. at 200% ($)	PV NCF Disc. at 400% ($)	PV NCF Disc. at 600% ($)
0	-40,000	0	-40,000	-40,000	-40,000	-40,000	-40,000	-40,000	-40,000	-40,000	-40,000	-40,000	-40,000
1	0	260,000	260,000	260,000	236,364	216,667	185,714	162,500	144,444	130,000	86,667	52,000	37,143
2	0	195,000	195,000	195,000	161,157	135,417	99,490	76,172	60,185	48,750	21,667	7800	3980
3	0	100,000	100,000	100,000	75,131	57,870	36,443	24,414	17,147	12,500	3704	800	292
4	0	45,000	45,000	45,000	30,736	21,701	11,714	6866	4287	2813	556	72	19
5	0	0	0	0	0	0	0	0	0	0	0	0	0
6	0	0	0	0	0	0	0	0	0	0	0	0	0
7	0	0	0	0	0	0	0	0	0	0	0	0	0
Total	**-40,000**	**600,000**	**560,000**	**560,000**	**463.388**	**391,655**	**293,361**	**229,952**	**186,063**	**154,063**	**72,593**	**20,672**	**1433**

TABLE 25.14 NCF and NPV for Project B Minus Project A in Example 25.26 – Alternative Solution

Project B minus Project A													
Year	INV ($)	NOI ($)	NCF ($)	PV NCF Disc. at 0% ($)	PV NCF Disc. at 10% ($)	PV NCF Disc. at 20% ($)	PV NCF Disc. at 40% ($)	PV NCF Disc. at 60 %($)	PV NCF Disc. at 80% ($)	PV NCF Disc. at 100% ($)	PV NCF Disc. at 200% ($)	PV NCF Disc. at 400% ($)	PV NCF Disc. at 600% ($)
0	-40,000	0	-40,000	-40,000	-40,000	-40,000	-40,000	-40,000	-40,000	-40,000	-40,000	-40,000	-40,000
1	0	0	145,000	145,000	131,818	120,833	103,571	90,625	80,556	72,500	48,333	29,000	20,714
2	0	0	85,000	85,000	70,248	59,028	43,367	33,203	26,235	21,250	9444	3400	1735
3	0	0	10,000	10,000	7513	5787	3644	2441	1715	1250	370	80	29
4	0	0	-35,000	-35,000	-23,905	-16,879	-9111	-5341	-3334	-2188	-432	-56	-15
5	0	0	-75,000	-75,000	-46,569	-30,141	-13,945	-7153	-3969	-2344	-309	-24	-4
6	0	0	-65,000	-65,000	-36,691	-21,768	-8633	-3874	-1911	-1016	-89	-4	-1
7	0	0	-65,000	-65,000	-33,355	-18,140	-6166	-2421	-1062	-508	0	-1	0
Total	**-40,000**	**0**	**-40,000**	**-40,000**	**29,059**	**58,720**	**72,728**	**67,481**	**58,229**	**48,945**	**17,318**	**-7605**	**-17,542**

TABLE 25.15 Summary of NPVs for Example 25.26

Rate (%)	NPV Project A ($)	NPV Project B ($)	NPV Project B minus Project A ($)
0	600,000	560,000	-40,000
10	434,329	463,388	29,059
20	332,935	391,655	58,720
40	220,633	293,361	72,728
60	162,472	229,952	67,481
80	127,834	186,063	58,229
100	105,117	154,063	48,945
200	55,304	72,593	17,318
400	28,277	20,672	-7605
600	18,974	1433	-17,542

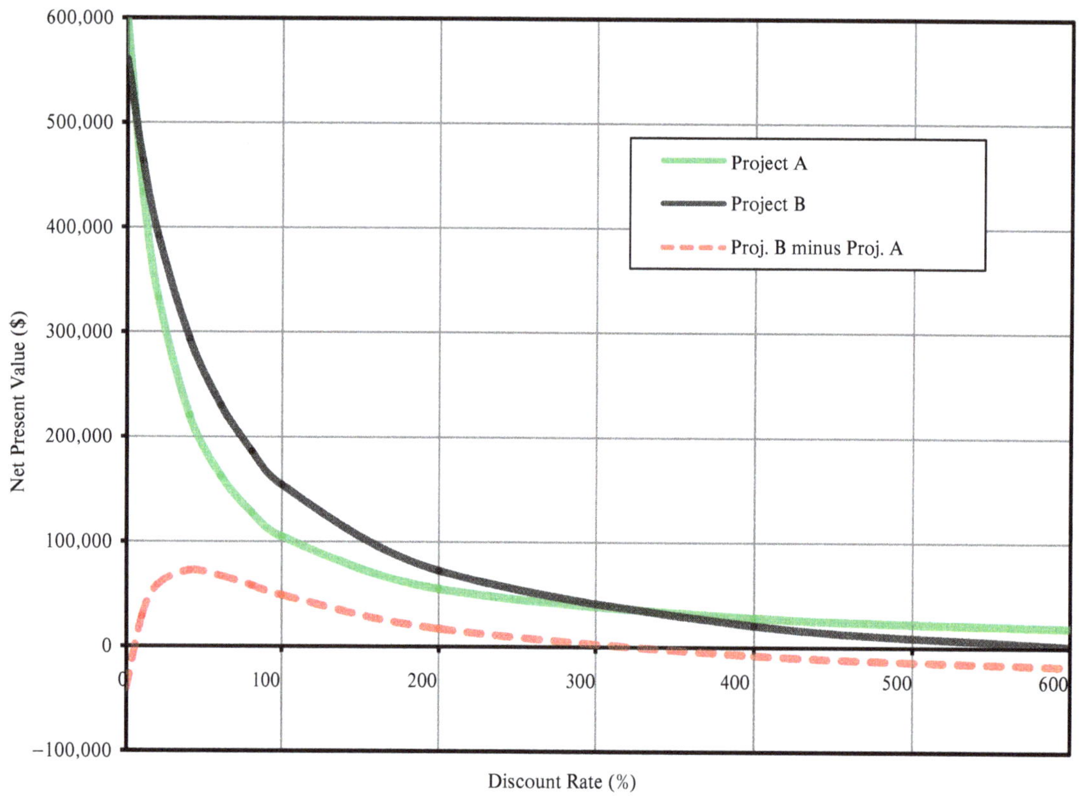

▲ FIGURE 25.12 **Plot of NPVs versus discount rates for Example 25.26**

End of Example 25.26

25.4.2.7 Modified Internal Rate of Return (IRR)

The modified internal rate of return method is used to calculate IRR for projects with additional investments later in the life of the project. The modified method ensures that a single IRR is calculated. It is also described as IRR using the escrow approach. For the modified IRR, future investments are discounted to time zero using the average investment opportunity rate or the average cost of capital. The expression for the modified IRR method is represented by the equation:

$$\sum_{j=1}^{n} \frac{(\mathrm{NOI})_j}{(1+\mathrm{IRR})^j} - \sum_{j=0}^{n} \frac{(\mathrm{INV})_j}{\left(1+i^*\right)^j} = 0 \qquad (25.38)$$

In Eq. (25.38), i^* is the average investment opportunity rate or the average cost of capital.

EXAMPLE 25.27 Calculation of Modified Internal Rate of Return (IRR)

Problem

Calculate the modified internal rate of return for the cash flow shown in Table 25.16 with initial investment of $480 in year zero and additional investment of $45,000 in year 5. The operating net incomes per year are also shown in the table. Assume the average cost of capital is 4% per year. Repeat the calculation using conventional IRR method using the same cash flow in Table 25.17.

Solution

Using the cash flows and NPV method as shown in Table 25.16, the modified IRR was calculated to be 4.932%. In Table 25.17, the conventional IRR was calculated to be 5.055%. The results are very close even though the modified IRR was slightly lower than the conventional IRR.

Note that the cash flow for this example is the same as for Example 25.25 except additional investment of $45,000 in year 5. This additional investment was made simply to maintain the cashflow forecasted for the original investment in Example 25.25. The consequence of the additional investment is that the return on the project is reduced from the original IRR of 7.586% in Example 25.25 to 4.932% as calculated in this example.

TABLE 25.16 Cash Flow for Modified IRR Calculation in Example 25.27

Project A										
Year	INV ($)	PV INV Disc. at 4% ($)	NOI ($)	PV NOI Disc. at 0% ($)	PV NOI Disc. at 2% ($)	PV NOI Disc. at 4.93% ($)	PV NOI Disc. at 5% ($)	PV NOI Disc. at 6% ($)	PV NOI Disc. at 7% ($)	PV NOI Disc. at 8% ($)
0	-480,000	-480,000	0	0	0	0	0	0	0	0
1	0	0	140,000	140,000	137,255	133,420	133,333	132,075	130,841	129,630
2	0	0	125,000	125,000	120,146	113,526	113,379	111,250	109,180	107,167
3	0	0	105,000	105,000	98,944	90,879	90,703	88,160	85,711	83,352
4	0	0	80,000	80,000	73,908	65,987	65,816	63,367	61,032	58,802
5	-45,000	-36,987	60,000	60,000	54,344	47,164	47,012	44,835	42,779	40,835
6	0	0	50,000	50,000	44,399	37,456	37,311	35,248	33,317	31,508
7	0	0	40,000	40,000	34,822	28,556	28,427	26,602	24,910	23,340
Total	**-525,000**	**-516,987**	**600,000**	**600,000**	**563,817**	**516,988**	**515,981**	**501,538**	**487,770**	**474,635**
ΣNOI − ΣINV				83,013	46,831	2	-1006	-15,448	-29,217	-42,352

TABLE 25.17 Cash Flow for Conventional IRR Calculation in Example 25.27

Project A										
Year	INV ($)	NOI ($)	NCF ($)	PV NCF Disc. at 0% ($)	PV NCF Dis. at 2% ($)	PV NCF Disc. at 5% ($)	PV NCF Disc. at 5.05% ($)	PV NCF Disc. at 6% ($)	PV NCF Disc. at 7% ($)	PV NCF Disc. at 8% ($)
0	-480000	0	-480,000	-480,000	-480,000	-480,000	-480,000	-480,000	-480,000	-480,000
1		140,000	140,000	140,000	137,255	133,333	133,264	132,075	130,841	129,630
2		125,000	125,000	125,000	120,146	113,379	113,260	111,250	109,180	107,167
3		105,000	105,000	105,000	98,944	90,703	90,561	88,160	85,711	83,352
4		80,000	80,000	80,000	73,908	65,816	65,678	63,367	61,032	58,802
5	-45000	60,000	15,000	15,000	13,586	11,753	11,722	11,209	10,695	10,209
6		50,000	50,000	50,000	44,399	37,311	37,194	35,248	33,317	31,508
7		40,000	40,000	40,000	34,822	28,427	28,323	26,602	24,910	23,340
Total	**-525,000**	**600,000**	**75,000**	**75,000**	**43,059**	**722**	**2**	**-12,088**	**-24,314**	**-35,991**

End of Example 25.27

25.4.3 Recommended Practice for Economic Valuations

The Society of Petroleum Evaluation Engineers (SPEE) has recommended practices for economic evaluation of petroleum projects. These recommended evaluation practices (REPs) cover elements that should be included in economic evaluation reports, and methods for discounting cash flows. The REPs can be accessed at the SPEE website.[4] SPEE recommended that month-end discounting should be applied for calculations based on monthly cash flows and that mid-year discounting should be applied for calculations based on annual cash flows. SPEE also recommended that monthly discounting should use effective monthly interest rates as defined in Eq. (25.22). Application of effective monthly interest rates in discounting monthly cash flows yield results that are very close to results calculated by discounting annual cash flows with mid-year discounting. SPEE further recommended that cash flows based on monthly calculations should be discounted no earlier than month-end and cash flows based on annual calculations should be discounted at mid-year.

These REPs were used in the evaluation of the drilling prospect problem presented by Wright and Thompson.[5] The input data and results of the calculations are provided in Appendix 25A. The methods used in the economic calculations shown in Appendix 25A should be noted as it applied most of the techniques covered in this chapter.

25.5 Depreciation, Depletion, and Amortization

Investments by individuals, businesses, companies, and organizations in certain types of property used for the production of income can be recovered in before-tax dollars by application of depreciation, depletion, and amortization methods. The cost basis by which depreciation, depletion, and amortization are computed is usually the acquisition cost of the property. Other cost basis could be established depending on the applicable tax laws of the country in which the property is located. Examples of costs that may be recovered through depreciation over the useful life of the assets include buildings, equipment, machinery, etc. These types of physical property are considered to be tangible assets. The costs to acquire oil and gas leases, mineral rights, and other natural rights are examples of investment costs that can be recovered by depletion. Certain business acquisition costs such as costs of acquiring franchises, research and development costs, registration cost for trademarks, patents, copyrights, etc., are examples of investment costs that may be recovered by amortization. These types of property are grouped as intangible assets.

25.5.1 Depreciation Methods

The four basic depreciation methods to be discussed in this section are:

1. Straight Line method
2. Declining Balance method
3. Declining Balance switching to Straight Line method
4. Units of Production method

25.5.1.1 Straight Line Method

This is the simplest method of calculating the annual depreciable amount of a property. It is also the slowest method of recovering the cost of a property through its useful life. The annual depreciation amount of a property is calculated by multiplying the cost basis of the property by the straight-line depreciation rate. The straight-line depreciation rate is given by the expression:

$$\text{Straight-line depreciation rate} = \frac{1}{n} \tag{25.39}$$

In Eq. (25.39), n = allowable depreciable life in years. The annual depreciation amount is given by:

$$\text{Annual straight-line depreciation} = \frac{\text{Cost Basis}}{n} \tag{25.40}$$

EXAMPLE 25.28 Calculation of Straight-Line Depreciation

Problem

Assume that you purchased an equipment for \$500,000. The equipment has a useful life of 10 years. What is the annual depreciation deduction for the equipment using straight line depreciation method assuming that the equipment has no salvage value at the end of 10 years?

Solution

Using Eq. (25.40),

Annual straight-line depreciation amount

$$= \frac{\$500{,}000}{10} = \$50{,}000$$

Note that in Example 25.28, it was assumed that the equipment has no salvage value at the end of 10 years. If the equipment has a salvage value, this is deducted from the cost of the equipment to calculate the cost basis in some countries. For instance, if the salvage value of the equipment is \$100,000 at the end of 10 years, then the cost basis is \$400,000 and the annual straight-line depreciation amount is \$40,000. In the USA, asset useful life and salvage value are not considered in the calculation of annual depreciation amount under the MACRS (Modified Accelerated Cost Recovery System) tax law.[6] If the straight-line depreciation method is used to calculate the annual depreciation amount under MACRS, the result will be based on the depreciable life of the equipment as assigned under its category and the timing convention established by the system. It is important that applicable laws and regulations of any particular country should be observed before applying any property depreciation methods. Consequently, only the basic forms of the depreciation methods are presented in this book for general guidance.

End of Example 25.28

25.5.1.2 Declining Balance Method

The declining balance method is considered an accelerated method of calculating annual depreciation amounts. This means that the depreciation amount is largest in the first year and then declines in subsequent years. The declining balance depreciation concept is applicable as various rates of declining depreciation. Some of the most common rates of declining balance depreciation are:

1. 200% declining balance referred to as double declining balance because the rate is two times the straight-line rate or $2/n$, where n is the depreciable life of the property in years.
2. 150% declining balance or $1.5/n$.
3. 125% declining balance or $1.25/n$.

The annual depreciation amount is calculated by multiplying the adjusted basis with the declining balance rate as represented in the expression:

$$\begin{aligned}&\text{Annual declining balance depreciation}\\&= \text{Declining balance rate} \times \text{Adjusted Basis}\end{aligned} \tag{25.41}$$

and the Adjusted Basis is calculated as:

$$\begin{aligned}&\text{Adjusted Basis} = \text{Cost Basis} - \text{Cumulative}\\&\text{Depreciation Already Taken}\end{aligned} \tag{25.42}$$

Adjusted basis is sometimes called book value or tax book value.

EXAMPLE 25.29 Calculation of Declining Balance Depreciation

Problem

Assume that you purchased an equipment for \$500,000. The equipment has a useful life of 10 years. What are the annual depreciation deductions for the equipment using double declining balance depreciation method assuming that the equipment has no salvage value at the end of 10 years?

Solution

For double declining balance, the annual decline rate is $2/n$. Substituting with $n = 10$ gives annual decline rate as 20%. The annual depreciation amounts are calculated as shown in Table 25.18. Note that the cumulative depreciation amount taken at the end of 10 years is \$446,313. This is less than the \$500,000 initial cost of the equipment. This means that the equipment is not fully depreciated at the end of 10 years. This the main drawback of the declining balance depreciation method. It will take infinite years to fully depreciate a property using the declining balance depreciation method. This is the main reason that it is allowed in some cases to switch from declining balance to straight-line depreciation to fully recover the investment in an asset.

TABLE 25.18 | Annual Declining Balance Amounts for Example 25.29

Year	Double Declining Balance Rate	Adjusted Basis (\$)	Depreciation Amount (\$)
1	0.2	500,000	100,000
2	0.2	400,000	80,000
3	0.2	320,000	64,000
4	0.2	256,000	51,200
5	0.2	204,800	40,960
6	0.2	163,840	32,768
7	0.2	131,072	26,214
8	0.2	104,858	20,972
9	0.2	83,886	16,777
10	0.2	67,109	13,422
Total	–	–	**446,313**

End of Example 25.29

25.5.1.3 Declining Balance Switching to Straight Line Method

As noted in Example 25.29, the cost of a property may not be fully recovered over its depreciable life using the declining balance method. For this reason, investors are allowed in some cases to switch from declining balance to straight-line depreciation to fully depreciate an asset within its depreciable life. This is essentially a combination of the declining balance and straight-line methods. It is desirable to switch when the straight-line rate is equal or higher than the declining balance rate. Note that the remaining adjusted basis is kept constant and depreciated with straight-line method for the remaining depreciable life of the asset.

EXAMPLE 25.30 Calculation of Declining Balance Switching to Straight-Line Depreciation

Problem

Assume that you purchased an equipment for $500,000. The equipment has a useful life of 10 years. What are the annual depreciation deductions for the equipment using double declining balance depreciation switching to straight line depreciation assuming that the equipment has no salvage value at the end of 10 years?

Solution

Switch to occur in year 6 when straight-line rate is equal to double declining balance rate. Remaining depreciable life at year 6 is 5 years. Straight-line rate is 1/5 or 20%. The annual depreciation amounts are calculated as shown in Table 25.19. Note that the cost of the equipment is fully depreciated by the end of year 10 as shown in Table 25.19.

TABLE 25.19 Annual Depreciation Amounts for Example 25.30

Year	Method	Double Declining Balance Rate	Adjusted Basis ($)	Depreciation Amount ($)
1	DDB	0.2	500,000	100,000
2	DDB	0.2	400,000	80,000
3	DDB	0.2	320,000	64,000
4	DDB	0.2	256,000	51,200
5	DDB	0.2	204,800	40,960
6	SL	0.2	163,840	32,768
7	SL	0.2	163,840	32,768
8	SL	0.2	163,840	32,768
9	SL	0.2	163,840	32,768
10	SL	0.2	163,840	32,768
Total	—	—	—	**500,000**

DDB, double declining balance; SL, straight line.

End of Example 25.30

25.5.1.4 Units of Production Method

In Units-of-Production method, annual depreciation amount is calculated by multiplying the cost basis of an asset by the ratio of the units produced in a depreciable year and the total units expected to be produced by the asset over its useful life. Expressed in the form of equation, this is represented as:

$$\text{Annual depreciation} = \frac{\text{Units produced in year}}{\text{Total units to be produced over useful life}} \times \text{Cost Basis} \qquad (25.43)$$

The concept of units-of-production method is based on the total number of units the asset is estimated to produce over its depreciable life rather than on a specified period of useful life. It is based on $ per unit rather than $ per year.

EXAMPLE 25.31 Calculation of Depreciation Using Units of Production Method

Problem

Assume that you purchased an equipment for \$500,000. The equipment has a useful life of 10 years. What are the annual depreciation deductions for the equipment using annual production units shown in Table 25.20 if the equipment is expected to produce 250,000 units over its productive life of 10 years with no salvage value?

Solution

For Year 1, using Eq. (25.43) depreciation amount is:

$$\text{Depreciation amount, Year 1} = \frac{52,000}{250,000} \times \$500,000$$
$$= \$104,000$$

The depreciation expense for the remaining years are shown in Table 25.20.

TABLE 25.20 Annual Depreciation Amounts for Example 25.31

Year	Expected Total Production Units	Production Units	Depreciation Rate	Adjusted Basis (\$)	Depreciation Amount (\$)
1	250,000	52,000	0.2080	500,000	104,000
2	250,000	42,000	0.1680	500,000	84,000
3	250,000	34,000	0.1360	500,000	68,000
4	250,000	27,600	0.1104	500,000	55,200
5	250,000	22,480	0.0899	500,000	44,960
6	250,000	22,480	0.0899	500,000	44,960
7	250,000	17,360	0.0694	500,000	34,720
8	250,000	13,520	0.0541	500,000	27,040
9	250,000	10,640	0.0426	500,000	21,280
10	250,000	7920	0.0317	500,000	15,840
Total	—	**250,000**	—	—	**500,000**

End of Example 25.31

25.5.2 Depletion Methods

Depletion methods are applied to mineral deposits, oil and gas wells, and standing timber. Oil, gas, and mineral depletion can be calculated by two methods, namely:

1. Cost depletion
2. Percentage depletion

Cost depletion is represented in equation form as:

$$\text{Cost depletion/year} = \frac{\text{Units produced and sold during the year}}{\text{Units recoverable at the begining of the year}} \times \text{Adjusted Basis} \tag{25.44}$$

In Eq. (25.44),

$$\text{Adjusted Basis} = \text{Cost Basis} \pm \text{Adjustments} - \text{Cumulative Depletion} \tag{25.45}$$

Percentage depletion is determined as a specified percentage of gross income minus royalties from the sale of minerals recovered from the mineral asset within the tax year. Cost and percentage depletion methods are governed by complex tax rules and laws in the USA. These rules and laws are modified regularly. Current rules and laws are available at the United States Internal Revenue Service (IRS) website.[6]

25.5.3 Amortization Methods

Amortization methods apply to intangible asset costs while depreciation methods apply to tangible asset costs. Intangible asset costs include franchise costs, research and development costs, registration costs for trademarks, patents, copyrights, business start-up costs, etc. Amortization allows recovery of such costs with calculations similar to straight-line depreciation over 5 years or other specified life depending on the asset. Certain rules and laws applicable to amortization of intangible assets are available at the IRS website.[6]

25.6 Economic Evaluation of a Petroleum Property

The economic evaluation of a petroleum property depends on the fiscal regime in effect in the country at the time the agreement or contract between the parties was consummated. Petroleum fiscal regime of a country is a compendium of regulations, agreements, and laws that were devised for the administration of economic benefits obtained from the exploration, development, and production of petroleum in that country. The fiscal regimes are used to regulate financial transactions between the participating or contracting entities. The structures of fiscal regimes vary from one country to the other. In the USA, onshore mineral rights (including petroleum rights) are owned generally by individuals, families, corporations, organizations, associations, governments (local, state, and federal), etc., while offshore mineral rights are owned by the federal government. In other parts of the world (African countries, some European countries, China, India, Middle East, Far East, Russia, etc.), all mineral rights are owned exclusively by the state. Consequently, the fiscal regime that governs petroleum property depends on the country in which the asset is located. In general, the fiscal regime applicable to a petroleum property determines the structure of its economic evaluation procedures. For this reason, two economic evaluation procedures will be presented in this book. The first economic evaluation model will apply in general to onshore petroleum property in the USA. The second economic evaluation model will be based on fiscal regimes generally used in other parts of the world. The two economic evaluation models are similar in structure and can be modified and adapted to other fiscal regimes as necessary.

25.6.1 Economic Evaluation of an Onshore Petroleum Property in the USA

It is necessary that this topic should be presented by using terms that are familiar to petroleum operators in the USA. Some of these terms are in use in other parts of the world but their origins can be traced to petroleum transactions in the USA.

In the USA, the holder of mineral interest to a property owns the rights to minerals (including petroleum) underneath the property. The ground surface above the minerals may be owned by another entity. Although the mineral interest owner has legal access to the minerals, it is expected that the ground surface owner would be compensated for any damages caused during operations to recover the minerals. The mineral interest owner can grant a leasehold interest in the form of a working interest to a second party to explore, develop and produce the minerals in return for royalty payments. For instance, a mineral interest owner may grant 100% working interest to the leaseholder in return for 12.5% royalty interest. In this case, the leaseholder (lessee) will pay 100% of the cost to explore, develop and produce the minerals, and takes 87.5% of the revenues. The mineral interest owner (lessor) gets 12.5% of the revenues. This is translated as the lessor having 12.5% net revenue interest (NRI) and the lessee having an 87.5% NRI. In due course, the leaseholder may choose to "farmout" a portion of the lease to a third party. A farmout agreement is an agreement between the leaseholder (termed the Farmor) and another party (termed the Farmee) whereby the farmor agrees to assign all or part of the working interest to the farmee in exchange for certain contractual obligations that usually includes drilling and testing of wells.[7] After the contractual obligations have been fulfilled, the farmee is said to have "earned" an assignment on the lease. The farmor typically reserves an overriding royalty interest (ORRI) with the option to elect to convert the ORRI to a portion of the working interest on the lease upon payout of the drilling, completion, and production costs. This option is termed back-in after payout. A back-in option is the same as a reversionary interest depending on the exact terms of the agreement. The decision to convert an ORRI depends whether the farmor is willing to assume additional risks and costs in exchange for potentially higher interest in the net revenues. If the farmor decides not to back-in, then the ORRI is retained as is. If the farmor decides to revert, then the ORRI is converted to a pre-agreed working interest which will be proportionately reduced.

There are several incentives that motivate farmors and farmees to join in farmout agreements. Some of these incentives that may interest farmors are:

1. To preserve the lease.
2. To salvage the lease if the lease has been idled or abandoned.
3. Share risk of exploration and drilling costs.
4. Obtain geologic information that could be valuable for other nearby prospects.

Some of the incentives that may motivate the farmees are:

1. Acquire acreage quickly.
2. Obtain acreage without lease capital expenditures.
3. Gain access to a prospective area.
4. Opportunity to operate in a new area.

A simplified cash flow model for a petroleum property in the USA is presented to illustrate some of transaction terms discussed earlier in this section as follows:

Total Gross Production

 multiplied by Net Revenue Interest (NRI or Royalty)

= Net Production

 multiplied by Price ($/unit)

= Gross Revenues ($)

 minus operating costs

 minus Severance and Ad Valorem taxes

 minus other costs

= Net Operating Income Before Income Tax (NOI BFIT) ($)

minus Investments ($)

= Net Cash Flow Before Income Tax (NCF BFIT) ($)

minus Federal Income Tax (FIT) ($)

= Net Cash Flow After Income Tax (NCF AFIT) ($)

This cash flow model is illustrated with a multi-well development project in Appendix 25B.

25.6.2 Economic Evaluation of a Petroleum Property in Other Parts of the World

The permission to explore, develop, and produce oil and gas in other parts of the world is granted through agreements and contracts. Typically, the agreement/contract is negotiated between a foreign oil company, designated here as an International Oil Company (IOC) and the state (host country) represented by the government or one of its agencies such as the National Oil Company (NOC). For the rest of this section, IOC will be used to represent foreign oil companies, and NOC will be used to represent the country, state, government or any of its agencies. The four main types of contracts negotiated or bid between the IOCs and NOCs are as follows:

1. Concession
2. Joint Venture (JV)
3. Production Sharing Contract or Agreement (PSC or PSA)
4. Service Contract (SC)

The differences between these types of contracts are conceptual in nature and depend largely on the degree of control granted to the IOCs, the sharing of risks and rewards, and the level of involvement by the NOCs in the exploration, development, and operation of the petroleum property. A summary of the risks and rewards for the four main types of contracts are shown in Table 25.21.

TABLE 25.21 | Risk and Reward for the Four Main Types of Contracts

Contract Type	IOC	NOC
Concession	Takes all risks and all rewards	Limited reward depends on production and product price
Joint Venture	Proportional share in risk and reward	Proportional share in risk and reward
PSC	Assumes all exploration risks. Shares in reward based on terms.	No exploration risks. Shares in reward based on terms.
Service Contract	Limited risk. Limited reward.	Most of the risk and reward

25.6.2.1 Concession

The earliest type of contract between IOCs and NOCs is the concession contract.[8] The early concession contracts have four main characteristics, namely:

1. Covered vast areas of the country
2. Long duration sometimes as long as 75 years
3. Contractor own all the resources produced from the concession area
4. Contractor was granted complete control over development and production of resources

In these early concession contracts, the NOC received royalties, one-time bonuses, and other types of compensation. However, it is necessary to note that these early concession contracts have been renegotiated with terms more favorable to the host country. Modern concession contracts incorporate more favorable terms for the host country. These favorable terms include shorter contract duration, work obligations (means specific work that the IOC must accomplish within a specified period of time such as acquisition and analysis of seismic data, drilling of exploratory wells, etc.), relinquishment clause (release all or parts of the concession area if certain conditions are not met), higher royalties, and bonus payments, involvement of NOC in certain aspects of the program, etc. Concession contracts are mostly phased out and no longer used in most parts of the world.

25.6.2.2 Joint Venture (JV)

A Joint Venture (JV) is a contract between the IOC and the State (represented by the NOC) to jointly engage in the exploration, development, and operation of a prospect assuming commercial quantities of petroleum are discovered. The JV partners agree on the proportional ownership of the production depending on the contracted working interest. Note that in this form of JV, the NOC shares some of the exploration risks. Although the JV is sometimes structured such that the NOC is "carried" (IOC pays all the costs) through the exploration phase until commercial discovery is made. JVs can be established in the from of equity or contract. As an equity company, the JV partners own a specified percentage of the company. For the contract form of JVs, a Joint Operating Agreement (JOA) is signed between the IOC and NOC specifying the percentage of production owned by each partner. It should be noted that JVs are used by many host countries as vehicles for technology transfer from the IOCs to the NOCs to enable the NOCs to eventually achieve capacities to independently conduct exploration, development, and operational activities.

25.6.2.3 Production Sharing Contract or Agreement (PSC or PSA)

Production Sharing Contract (PSC) or Agreement (PSA), collectively referred to as PSC hereafter, is an agreement between the IOCs and a host country (typically represented by the NOC) whereby each party receives a certain percentage of the production after the IOC have recovered a specified amount of costs and expenses. PSCs were introduced in the petroleum industry in Indonesia in 1966.[8,9] They have become the most common types of contractual agreement between IOCs and NOCs for the exploration, development, and operation of petroleum property in the world. The terms of PSCs are established between IOCs and NOCs either through negotiation or bidding. PSCs are distinguished from other types of contracts by certain basic conditions which are:

1. The IOC bears all the exploration risk, i.e., if no oil is found in commercial quantities, the IOC receives no compensation.
2. If the discovery is commercial, the state owns the resources and the installations after development of the property.

The basic form of a PSC has the following four features:

1. **Royalty:** The IOC pays a specified royalty as a percentage of gross production to the NOC. Sometimes, royalty is not fixed at a certain percentage but is made variable based on average oil production per day. The general range of royalty is 6% to 15% of gross production.
2. **Cost oil:** The IOC receives a pre-specified share of the production to recover exploration and development costs and operational expenses after deduction of royalty payments. Most PSCs have a cost-oil recovery limit which is set at a certain percentage of production (e.g., 60%) per period while allowing unrecovered costs to be carried forward and recovered in the next period if the limit is not exceeded. Note that some PSCs may not have any limit on the amount of cost oil that could be deducted per period.
3. **Profit oil:** The remaining oil production (after deducting cost oil) is then shared between the NOC and the IOC at stipulated proportion such as 60/40 (meaning NOC receives 60% of the profit oil and the IOC receives 40% of the profit oil).
4. **Income tax:** IOC then pays income tax to the NOC on its profit oil share at a rate stipulated in the contract or based on local tax laws.

The basic form of a PSC is shown in Figure 25.13. Note that PSCs have changed substantially over the years from this basic form. Some of the new features introduced in more recent PSCs are summarized as follows:[9]

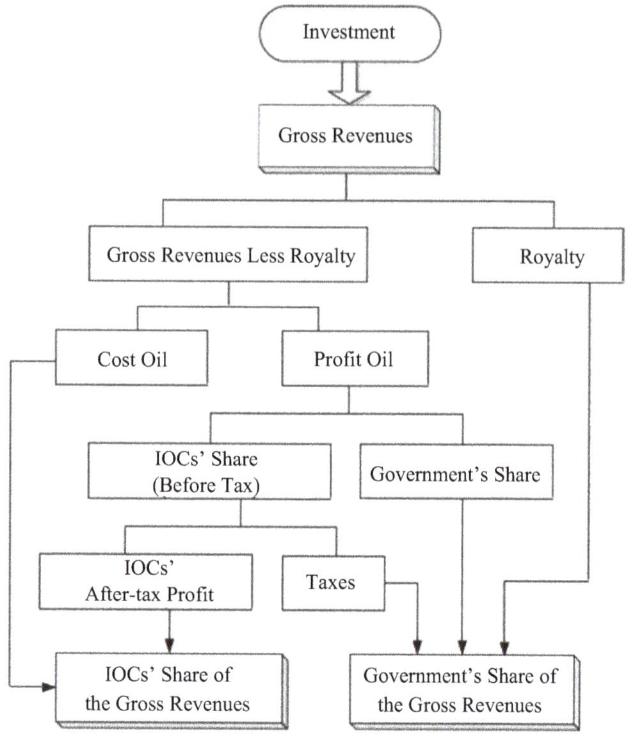

▲ **FIGURE 25.13** Allocation chart of revenues in a basic PSC (from Feng et al.[10])

1. **Bonuses:** Sometimes, PSCs include bonuses to be paid to the State (NOC) once certain milestones have been reached. Types of bonuses include signature bonus, discovery bonus, production bonus, etc. Production bonuses are usually structured to be due after certain milestones have been reached in terms of production rate per day or cumulative production volume.
2. **Tax holidays:** Some PSCs provide tax holidays as further incentive to the IOCs. These tax holidays could be 5 years of no taxes from start of the contract or from start of production if commercial discovery is made.
3. **Work program:** The work program is an important part of the PSC. It stipulates commitment of the IOC to complete certain activities (such as acquisition and processing of seismic data, drilling of exploratory wells, employment and/or training of local workforce, etc.) within specified time periods.
4. **Fixed and sliding scales:** Royalties, cost oil, profit oil, and production bonuses can be based as a fixed percentage of production or can be based on a sliding scale tied to production milestones. The production milestones could be attaining certain level of daily production rate or cumulative production volume. The use of sliding scales to determine payments is becoming very common and can be complex depending on the PSC. However, application of sliding scales adds a degree of flexibility to the PSC by providing the opportunity for the State (NOC) to increase its share of profits if the project becomes more profitable.

5. **Participation:** Most PSCs have terms that grant NOCs the option (at their discretion) to participate in the venture if the discovery is commercial. Under these terms, the IOCs still bear the costs and risks during the exploration phase. In other words, the NOCs are "carried" through the exploration phase. The amount of working interest that the NOC has the option to take if the discovery is commercial depends on the PSC and varies from 5% to 50%.
6. **Contract duration and commerciality:** Typically, minimum exploration period with extensions are stipulated in the PSC. If the discovery is commercial, the production period starts after a work program has been approved by the NOC. For example, the PSC may stipulate an exploration period of 5 years with two extensions of 2 years each followed by a production period of 25 years with a possible 5-year extension. Note that in some PSCs commerciality of a discovery is declared by the IOC with approval by the NOC.

A worksheet illustrating an economic model for a PSC with a fixed scale is shown in Appendix 25C.

25.6.2.4 Service Contracts

Service contracts are forms of petroleum fiscal regimes whereby IOCs provide services and know-how (expertise) to NOCs in return for fees or a share of the production. Service contracts can be divided into pure service contracts and risk service contracts on the basis of risk allocation between the host government (NOCs) and IOCs. In pure service contracts, IOCs are compensated with a flat fee for their services and expertise. Under risk service contracts, IOCs are compensated on the basis of oil revenues generated in the project. In either case, IOCs have no equity in the project and are obligated to transfer operation of the oilfield including surface infrastructure to the host country (NOCs) after start of production. Buyback is variant of risk service contracts whereby the IOCs are paid back for services rendered by allocation of a proportion of oil produced in the project.[10] An allocation chart of revenues for a basic buyback contract is shown in Figure 25.14. Various types of buyback contracts are widely used in Iran and Iraq.

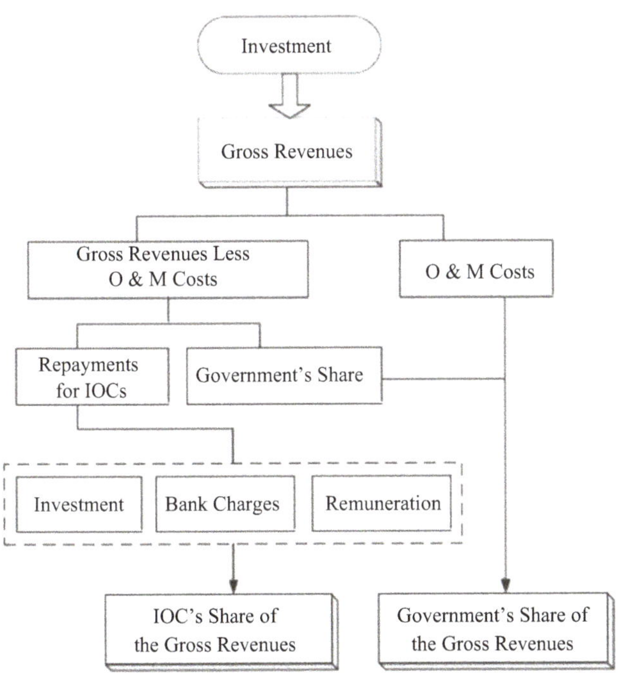

▲ **FIGURE 25.14 Allocation chart of revenues in a basic buyback contract (from Feng et al.[10])**

25.7 Decision and Risk Analysis Applied to Petroleum Evaluations

This chapter on economic evaluation of petroleum projects and property has been presented thus far assuming the absence of uncertainty and risk in the business of exploring, developing, and producing petroleum resources. In reality, the petroleum business is full of uncertainties and risks in every stage starting from exploration to production at the wellhead. It is the responsibility of petroleum professionals to manage these uncertainties and risks to achieve and maintain a viable and profitable business. **Uncertainty** can be defined as the outcome of an event which cannot be reliably predicted nor controlled. And **risk** can be defined as an event that could adversely impact the outcome if it were to occur. Uncertainty analysis is recommended in the petroleum economic evaluations instead of using single-valued outcomes.[11] It is an integral part of decision and risk analyses. Uncertainty analysis can be conducted with the aid of tools such as Monte Carlo simulation.[12] Monte Carlo simulation techniques have been discussed extensively in other books.[13] A more robust method for incorporating uncertainty analysis in reservoir modeling was described in Chapter 22, "Geologic Modeling and Reservoir Characterization," Section 22.5.6, "Uncertainty Analysis." As described in the section, geostatistical modeling can be used to incorporate uncertainty in model building leading to generation of multiple realizations of potential outcomes.

According to Newendorp and Schuyler,[13] decision and risk analysis can be divided into five steps as follows:

1. Identify decision alternatives.
2. Identify possible outcomes. Calculate NPV for each outcome.
3. Assess the probability of each outcome. This is the risk analysis step.
4. Calculate the expected monetary value (EMV) for each alternative.
5. Select the alternative with the highest EMV.

Several methods are available that can be used to conduct decision and risk analyses. One of the common methods is application of Decision Trees.[13] A decision tree can be described as a graph that uses a tree-like model to illustrate all possible outcomes, including probabilities pertaining to a decision. Monte Carlo simulation can also be applied in decision and risk analysis. Several companies and organizations have incorporated more advanced methods such as Portfolio Analysis[14] and Real Option[15] in their decision and risk analysis process. Clearly, presentation of these methods is outside the scope of this book.

Decision and risk analyses as applied to petroleum projects have been presented briefly in this book because the subjects are very complex and could not be adequately addressed here due to the limited scope of the book. However, decision and risk analyses should be included as important and integral parts of any significant economic analyses. Readers are advised to advance their knowledge in this field by accessing other publications on this subject.[13,16] Decision and risk analyses are recommended as standard practice for evaluation of major petroleum projects.

Nomenclature

A	amount of equal payments in a uniform series of payments made at end of period
B	amount of equal payments in a uniform series of payments made at beginning of period
C	amount of equal payments in a uniform series of payments made at mid-period
F	future lump-sum of money or future value
i	interest rate per period
i_e	effective interest rate per year
i_{mon}	effective interest rate per month
i_n	nominal interest rate per year
m	number of compounding periods per year
n	total number of compound interest periods
P	present lump-sum of money or present value
t	time in years

Abbreviations

AFIT	After Federal Income Tax
BFIT	Before Federal Income Tax
DCA	Decline Curve Analysis
DDB	Double Declining Balance
EMV	Expected Monetary Value
GMBE	General Material Balance Method
INV	Investment
IOC	International Oil Company
IRR	Internal Rate of Return
JV	Joint Venture
MACRS	Modified Accelerated Cost Recovery System
NCF	Net Cash Flow
NOC	National Oil Company
NOI	Net Operating Income
NPV	Net Present Value
NRI	Net Revenue Interest
ORRI	Over-Ride Revenue Interest
PI	Profitability Index
PIIP	Petroleum Initially In-Place
PRMS	Petroleum Resources Management System
PSA	Production Sharing Agreement
PSC	Production Sharing Contract
RTA	Rate Time (Transient) Analysis
SC	Service Contract
SEC	Securities Exchange Commission
SL	Straight Line

REFERENCES

1. Society of Petroleum Engineers (SPE)-Petroleum Resources Management Systems (PRMS)- 2018 Update. https://www.spe.org/en/industry/petroleum-resources-management-system-2018/?aliId=368702674. Accessed October 4, 2018.
2. Securities and Exchange Commission (SEC): Reserves Definitions and Interpretations. https://www.sec.gov/divisions/corpfin/guidance/cfactfaq.htm. Accessed October 4, 2018.
3. Rietz, D., and Usmani, A.: "Case Studies Illustrating the Use of Reservoir Simulation Results in the Reserves Estimation Process, *SPE RE&E*, February 2009; SPE 110066-PA.
4. Society of Petroleum Evaluation Engineers: Recommended Engineering Practices (REPs). https://secure.spee.org/resources/recommended-evaluation-practices-reps. Accessed November 29, 2018.
5. Wright, J.D., and Thompson, R.S.: "A Comparative Analysis of 12 Economic Software Programs," paper SPE 68588 presented at the SPE Hydrocarbon Economics and Evaluation Symposium, Dallas, Texas, (April 2-3, 2001).
6. United States of America: Internal Revenue Service. https://www.irs.gov/. Accessed December 1, 2018.
7. Lowe, J.S.: "Analyzing Oil and Gas Farmout Agreements." *SW LJ* 41 (1987), 759.
8. Radon, J.: "The ABCs of Petroleum Contracts: License-Concession Agreements, Joint Ventures, and Production-sharing Agreements." *A Reporter' s Guide to Energy and Development of Open Society Institute* (2005).
9. Bindemann, K.: *Production-Sharing Agreements: An Economic Analysis*. Oxford Institute for Energy Studies, October 1999.
10. Feng, Z., Shui-Bo Z., and Ying G.: "On Oil Investment and Production: A Comparison of Production Sharing Contracts and Buyback Contracts." *Energy Economics* (2014) 42, 395-402.
11. Hultzsch, P., Lake, L.W., and Gilbert, R.B.: "Uncertainty Estimation through the Life of the Field." *Energy Exploration & Exploitation* (2008) 26, 3, 175-195.
12. Murtha, J.A.: "Monte Carlo Simulation: Its Status and Future." *JPT* (1997), 361-373: SPE 37932.
13. Newendorp, P., and Schuyler, J.: *Decision Analysis for Petroleum Exploration Planning Press.* Colorado, (2000).
14. Edwards, R.A., and Hewett, T.A.: "Applying Financial Portfolio Theory to the Analysis of Producing Properties," paper SPE 26392 presented at the SPE Annual Technical Conference and Exhibition, Houston, Texas, (October 3-6, 1993).
15. Lohrenz, J., and Dickens, R.N.: "Option Theory for Evaluation of Oil and Gas Assets: The Upsides and Downsides," SPE 25837 presented at the SPE Hydrocarbon Economics and Evaluation Symposium, Dallas, (March 29-30, 1993).
16. Suslick, S.B., Schiozer, D., and Rodriguez, M.R.: "Uncertainty and Risk Analysis in Petroleum Exploration and Production." *Terrae* (2009), 6, 1.

PROBLEMS

25-1. If you deposited $20,000 in a bank that promised to pay you 5% interest rate per year compounded annually, what is the value of your deposit at the end of 10 years?

25-2. A bank in your town is offering 7% interest rate per year compounded annually. How much money would you deposit in your bank account if you expect to have $40,000 at the end of 12 years?

25-3. A bank customer deposited $11,000 at the end of each year for 7 years into a bank account. How much money is in the bank account as a result of these deposits only if the bank paid 6% interest per year compounded annually?

25-4. Company XYZ offers a retirement investment plan for employees. The retirement plan guarantees a 9% rate of return per year compounded annually. If an employee plans to retire with $2,000,000 in the retirement plan at the end of 25 years, how much money should the employee invest in the retirement account at the end of each year?

25-5. If the employee in Problem 25-4 plans to retire with $2,000,000 in the retirement plan at the beginning of the 25th year, how much should the employee invest in the retirement account at the beginning of each year?

25-6. If the employee in Problem 25-4 plans to retire with $2,000,000 in the retirement account in the middle of the 25th year, how much should the employee deposit into the retirement account in the middle of each year over the 25-year period?

25-7. As a recent retiree, you have $2,500,000 that you will like to invest in an annuity for a period of 17 years. A company sells an annuity to you that guarantees 6.5% annual interest rate over the 17-year period. What is the annuity payment that you are guaranteed to receive at the end of each year over the duration of the annuity investment?

25-8. The employee in Problem 25-7 elects to receive the annuity payments at the beginning of each year. What is the annuity payment at the beginning of each year over the 17-year life of the annuity?

25-9. The retiree in Problem 25-7 elects to receive the annuity payments at the middle of the year. What is the annuity payments at the middle of the year over the 17-year life of the annuity?

25-10. An employee of a company had a 12-year investment plan. This plan was to invest $2,000 at the end of years 1, 2, 3, and 4; $4,000 at the end of years 5, 6, and 7; and $6,000 at the end of years 8, 9, 10, 11, and 12. Calculate the following equivalent values of these investments assuming 7.5% interest rate compounded annually:

a. The present value of the investments at year zero.
b. The future value of the investments at the end of year 12.
c. Equivalent equal uniform investments assuming the investments were made at the end of the year.

d. Equivalent equal uniform investments assuming the investments were made at the middle of the year.

e. Equivalent equal uniform investments assuming the investments were made at the beginning of the year.

25-11. A homeowner agreed to buy a new home at the price of $750,000, A mortgage company offered to lend the money for the purchase of the home at 5% annual interest rate with 1% origination fees plus 2% loan discount fees (2% fee of loan value). The loan repayments are scheduled at the end of each month for 30 years. Calculate the annual percentage rate (APR) of the loan.

25-12. An investor receives annuity payments of $9300 at the end of each year for 10 years. If the nominal annual interest rate is 8%, calculate:

a. The present value of the payments at year zero if the nominal annual interest rate is compounded semi-annually.

b. The present value of the payments at year zero if the nominal interest rate is compounded continuously.

25-13. The net cash flows of a project at the end of the year are given in Table 25P.1 over a period of 10 years. Calculate:

a. The undiscounted pay-back for the project.

b. The discounted pay-back for the project assuming the average cost of capital for the company is 5.5% per year.

TABLE 25P.1 Net Cash Flow for Problem 25-13

Year	Net Cash Flow ($)
0	−1,500,000
1	450,000
2	350,000
3	300,000
4	250,000
5	225,000
6	200,000
7	175,000
8	150,000
9	100,000
10	50,000

25-14. The investment for the project in Table 25P.1 is $1,500,000. Calculate:

a. Undiscounted profitability index for net operating income

b. Undiscounted profitability index for net cash flow

c. Discounted profitability index for net operating income assuming cost of capital is 5.5%

d. Discounted profitability index for net cash flow assuming cost of capital is 5.5%

25-15. Calculate the net present value of the project with net cash flows given in Table 25P.1 at average cost of capital of 5.5% per year and 12% per year. Is the project profitable at both costs of capital?

25-16. The net cash flows of Projects A and B at the end of the year are given in Table 25P.2. Projects A and B are mutually exclusive. Using the net cash flows in Table 25P.2, calculate:

a. The net present values of Projects A and B at annual discount rates of 0%, 2%, 5%, 8%, 12%, and 14%.

b. Make a plot of the net present values for Projects A and B at the discount rates.

c. Determine the discount rate at which the net present values of Projects A and B are equivalent.

d. Discuss the range of discount rates at which either project could be preferable to the other based on net present value criteria.

TABLE 25P.2 Net Cash Flow for Problem 25-16

Year	Net Cash Flow ($)	
	Project A	Project B
0	−1,500,000	−1,500,000
1	450,000	75,000
2	350,000	125,000
3	300,000	155,000
4	250,000	205,000
5	225,000	235,000
6	200,000	280,000
7	175,000	315,000
8	150,000	340,000
9	100,000	365,000
10	50,000	405,000

25-17. The net cash flows for Project B are given in Table 25P.2. Calculate the internal rate of return for Project B.

25-18. The primary drive mechanism of an oil reservoir has been classified as solution-gas drive based on production data. The net cash flows from producing the reservoir as a solution-gas drive reservoir are shown under Project X in Table 25P.3 over a period of 10 years. The engineer assigned to the reservoir proposes to drill four wells at the total cost of $550,000 for pressure maintenance. The pressure maintenance project will only accelerate recovery of reserves. The net cash flow for the proposed pressure maintenance project are given under Project Y in Table 25P.3 Evaluate the proposal and determine whether Project Y should be approved on the basis of the average investment opportunity rate criteria of the company.

TABLE 25P.3 Net Cash Flow for Problem 25-18

Year	Net Cash Flow ($)	
	Project X	Project Y
0	0	-550,000
1	1,049,975	2,800,050
2	999,995	2,150,250
3	799,965	947,401
4	699,975	397,416
5	649,945	197,401
6	549,925	97,176
7	549,915	0
8	440,000	0
9	435,000	0
10	415,000	0

25-19. Calculate the modified internal rate of return for the cash flows given in Table 25P.4. Assume the average cost of capital for the company is 4.5%.

TABLE 25P.4 Cash Flow for Problem 25-19

Year	Investment ($)	Net Operating Income ($)
0	-1,250,000	0
1	0	73,500
2	0	126,000
3	0	156,000
4	0	204,000
5	0	236,000
6	-430,000	281,000
7	0	317,000
8	0	342,000
9	0	367,000
10	0	407,000

25-20. A company purchased an equipment at a cost of $1,700,000. The equipment is expected to be used for 15 years with zero salvage value at the end of 15 years.

a. Calculate the annual depreciable amounts using double declining balance method switching over to straight-line method to fully recover the investment

b. What is the straight-line depreciation rate and the year to switch to straight-line method?

25-21. A manufacturing company purchased an equipment for $1,250,000. The equipment was expected to last 12 years with no salvage value. The annual production units estimated for the equipment are given in Table 25P.5. Calculate the annual depreciable deductions for the equipment.

TABLE 25P.5 Production Units for Problem 25-21

Year	Production Units
1	142,500
2	120,000
3	102,000
4	82,800
5	75,000
6	67,500
7	45,000
8	37,500
9	31,920
10	23,760
11	15,000
12	7020

25-22. A small independent oil company plans to drill a development oil well. The well will be drilled and completed in 1 month at total cost of $1,800,000. Production from the development well will start on the first day of the following month after completion of the well. Initial oil production rate from the well is estimated to be 190 bopd declining at effective rate of 48% per year. The total operating cost for the well is $10,000 per month escalated at a rate of 2% per year. The company has 100% working interest and 80% net revenue interest on the well. Oil price is estimated to start at $60 per barrel escalated at the rate of 2% per year. Severance and Ad Valorem tax rate is 12%. The expected economic life of the well is 6 years. Assume end of month compounding. Calculate:

a. Gross reserves in barrels for the well.

b. Pay-out for the project.

c. Net present value at 0% and 10% rates of return.

d. Profitability index at 10% rate of return.

e. Internal rate of return for the project.

f. Prepare an annual summary of monthly net cash flow for the project.

25-23. An oil company is planning a five well development project. The parameters of the project are summarized as follows:

a. Number of wells	= 5
b. Initial oil production rate per well	= 1850 bopd
c. Effective production decline rate for each well per year	= 50%
d. Total operating costs per month escalated at 3.5% per year	= $200,000
e. Oil price per barrel escalated at 3.5% per year	= $50
f. Working interest	= 100%
g. Net revenue interest	= 85%
h. Severance and Ad Valorem tax rate	= 15%
i. Federal income tax rate	= 20%
j. Days in 1 year	= 365.25

Assumptions: One month to drill and complete each well. Wells to be placed on production at 6-month intervals. Each well produces at initial rate. Use exponential decline for the wells (see Chapter 11). Production starts at month 1. Project life is 7 years.

Calculate:

a. Gross reserves in barrels for project.

b. Production rate for each month.

c. Pay-out for the project.

d. Net present values at 10% and 20% rates of return.

e. Profitability indices at 10% and 20% rates of return.

f. Internal rate of return for project.

g. Prepare an annual summary of monthly net cash flow for project.

25-24. The production data, operating and overhead costs for an oil field developed under a profit sharing contract (PSC) are provided in Table 25P.6. Other terms of the PSC are given as follows:

a. Royalty payments are fixed at 20% of gross oil revenues per year

b. Cost oil is limited to 45% maximum of net oil revenues per year

c. Profit oil is shared at fixed scale of 60% for the country and 40% for the International Oil Company (IOC)

d. Discovery bonus of $5,000,000

e. First oil bonus of $15,000,000

f. Country tax rate of 30%

g. Annual discount rate of 15%

h. Oil price of $59 per bbl escalated at rate of 2.5% per year

i. Total development cost of field is $40,200,000

j. PSC expires at the end of 15 years

TABLE 25P.6 | Input Data for Problem 25-24

Year	Production (Mbo)	Operating Costs ($M)	Overhead ($M)
1	1480.52	606.91	400.00
2	2330.06	561.32	450.00
3	1651.85	635.28	475.00
4	1171.04	932.27	550.00
5	830.19	1368.09	530.00
6	588.54	2007.65	520.00
7	417.24	2946.19	510.00
8	295.79	3549.22	480.00
9	209.69	3051.17	395.00
10	148.66	2607.98	285.00
11	105.39	1848.87	190.00
12	74.71	1310.72	150.00
13	52.97	929.21	115.00
14	37.55	658.74	60.00
15	26.62	467.00	35.00

Using these PSC terms:

a. Develop a worksheet as an economic model for the field.

b. Calculate the net present value of the field for the IOC at the end of the PSC.

c. Calculate the net present value of the field for the country at the end of the PSC.

APPENDIX

25A ECONOMIC EVALUATION OF A DRILLING PROSPECT FROM PAPER SPE 68588*

Appendix 25A: Summary

Input Parameters		Economic Input Parameters:			
Effective decline rate per year =	0.5	Working Interest =	100%		
Arps decline rate per year =	0.6931	Net Revenue Interest =	85%		
Arps decline rate per month =	0.0578	Oil Price per bbl ($) =	17.50		
Days in one year =	365.25	Well cost per month ($) =	2000		
Days in one month =	30.438	Escalation factor per year =	3%		
Project Life, years =	5	Effective Escalation factor/mo.	0.002466		
Nominal interest/yr =	10%	Severance & Ad Valorem taxes	10%		
Effective interest rate per month =	0.7974%	Drilling Cost ($) =	425,000		
Nominal interest/yr =	20%				
Effective interest rate per month =	1.5309%	**Summary of Results**			
Nominal interest/yr =	34.30%	Gross Reserves, bbls =	50,080	PI Undiscounted (NOI) =	1.38
Effective interest rate per month =	2.4880%	Pay-out, years =	1.50	PI Undiscounted (NCF) =	0.38
		Internal rate of return (%) =	34.30	PI Discounted @10% (NOI) =	1.23
Assumptions:		Net Present Value ($) @ 0% =	160,369	PI Discounted @10% (NCF) =	0.23
1. Revenues from oil sale received at end of next month		Net Present Value ($) @ 10% =	99,366	PI Discounted @20% (NOI) =	1.12
2. End of month compounding		Net Present Value ($) @ 20% =	51,948	PI Discounted @20% (NCF) =	0.12

Appendix 25A: Annual Summary of Monthly Calculations

Year	Average BOPD	Gross Prod bbl	Net Prod bbl	Avg. Oil Price $/bbl	Gross Revenues ($)	Sev. + Ad Val. Taxes ($)	Operating Costs ($)	NOI BFIT ($)	NCF ($)
2001	76	24,781	21,063	17.76	373,581	37,358	22,273	313,950	-111,050
2002	39	13,957	11,863	18.27	216,398	21,640	24,996	169,762	169,762
2003	20	6,978	5,932	18.82	111,445	11,145	25,746	74,554	74,554
2004	10	3,489	2,966	19.38	57,394	5,739	26,519	25,136	25,136
2005	6	875	744	19.79	14,722	1,472	11,283	1,967	1,967
Total		**50,080**	**42,568**		**773,541**	**77,354**	**110,818**	**585,369**	**160,369**

*Wright, J.D., and Thompson, R.S.: "A Comparative Analysis of 12 Economic Software Programs," paper SPE 68588 presented at the SPE Hydrocarbon Economics and Evaluation Symposium, Dallas, Texas, April 2–3, 2001.

Appendix 25A-1: Monthly Calculations

Date	Months	BOPD	Gross BOPM	Net BOPM	Oil Price $/bbl	Gross Revenues ($)	Sev. + Ad Val. Taxes ($)	Operating Costs ($)	NOI BFIT ($)	Cum. Net Cash Flow (S)	NCF ($)	NPV at 10% ($)	NPV at 20% ($)	NPV at 34.30% ($)
1/1/2001	1	0	0	0	0	0	0	0	0	0	-425,000	-425,000	-425,000	-425,000
2/1/2001	2	100	2,958	2,514	17.54	44,101	4,410	2,000	37,691	387,309	37,691	37,097	36,563	35,884
3/1/2001	3	94	2,792	2,373	17.59	41,729	4,173	2,005	35,551	351,758	35,551	34,714	33,967	33,024
4/1/2001	4	89	2,635	2,240	17.63	39,484	3,948	2,010	33,526	318,232	33,526	32,477	31,549	30,387
5/1/2001	5	84	2,487	2,114	17.67	37,360	3,736	2,015	31,609	286,623	31,609	30,378	29,297	27,954
6/1/2001	6	79	2,347	1,995	17.72	35,350	3,535	2,020	29,795	256,828	29,795	28,409	27,199	25,710
7/1/2001	7	75	2,216	1,883	17.76	33,448	3,345	2,025	28,079	228,749	28,079	26,560	25,246	23,641
8/1/2001	8	71	2,091	1,778	17.80	31,649	3,165	2,030	26,454	202,295	26,454	24,826	23,426	21,732
9/1/2001	9	67	1,974	1,678	17.85	29,946	2,995	2,035	24,917	177,378	24,917	23,198	21,732	19,973
10/1/2001	10	63	1,863	1,584	17.89	28,335	2,834	2,040	23,462	153,916	23,462	21,670	20,155	18,350
11/1/2001	11	59	1,759	1,495	17.94	26,811	2,681	2,045	22,085	131,832	22,085	20,237	18,686	16,854
12/1/2001	12	56	1,660	1,411	17.98	25,368	2,537	2,050	20,782	111,050	20,782	18,892	17,318	15,474
1/1/2002	13	53	1,567	1,332	18.03	24,004	2,400	2,055	19,548	91,502	19,548	17,631	16,045	14,202
2/1/2002	14	50	1,479	1,257	18.07	22,712	2,271	2,060	18,381	73,121	18,381	16,447	14,859	13,030
3/1/2002	15	47	1,396	1,186	18.11	21,490	2,149	2,065	17,276	55,844	17,276	15,336	13,755	11,950
4/1/2002	16	45	1,317	1,120	18.16	20,334	2,033	2,070	16,231	39,614	16,231	14,294	12,728	10,954
5/1/2002	17	42	1,243	1,057	18.20	19,240	1,924	2,075	15,241	24,373	15,241	13,316	11,772	10,036
6/1/2002	18	40	1,174	998	18.25	18,205	1,821	2,080	14,304	10,068	14,304	12,399	10,882	9,191
7/1/2002	19	37	1,108	942	18.29	17,226	1,723	2,086	13,418	-3,349	13,418	11,538	10,053	8,412
8/1/2002	20	35	1,046	889	18.34	16,299	1,630	2,091	12,579	-15,928	12,579	10,731	9,282	7,694
9/1/2002	21	33	987	839	18.38	15,422	1,542	2,096	11,784	-27,712	11,784	9,974	8,565	7,033
10/1/2002	22	31	932	792	18.43	14,593	1,459	2,101	11,032	-38,744	11,032	9,264	7,898	6,425
11/1/2002	23	30	879	747	18.47	13,808	1,381	2,106	10,321	-49,065	10,321	8,597	7,277	5,864
12/1/2002	24	28	830	705	18.52	13,065	1,306	2,111	9,647	-58,712	9,647	7,973	6,699	5,349
1/1/2003	25	26	783	666	18.57	12,362	1,236	2,117	9,009	-67,721	9,009	7,387	6,162	4,874
2/1/2003	26	25	739	628	18.61	11,697	1,170	2,122	8,405	-76,126	8,405	6,837	5,662	4,437
3/1/2003	27	24	698	593	18.66	11,068	1,107	2,127	7,834	-83,960	7,834	6,322	5,198	4,035
4/1/2003	28	22	659	560	18.70	10,472	1,047	2,132	7,293	-91,253	7,293	5,839	4,766	3,665
5/1/2003	29	21	622	528	18.75	9,909	991	2,138	6,780	-98,033	6,780	5,385	4,364	3,325
6/1/2003	30	20	587	499	18.80	9,376	938	2,143	6,295	-104,328	6,295	4,961	3,991	3,012
7/1/2003	31	19	554	471	18.84	8,871	887	2,148	5,836	-110,164	5,836	4,562	3,644	2,724
8/1/2003	32	18	523	444	18.89	8,394	839	2,153	5,401	-115,566	5,401	4,189	3,322	2,460
9/1/2003	33	17	493	419	18.94	7,942	794	2,159	4,990	-120,555	4,990	3,839	3,022	2,217
10/1/2003	34	16	466	396	18.98	7,515	752	2,164	4,600	-125,155	4,600	3,511	2,744	1,995

Date	Months	BOPD	Gross BOPM	Net BOPM	Oil Price $/bbl	Gross Revenues ($)	Sev. + Ad Val. Taxes ($)	Operating Costs ($)	NOI BFIT ($)	Cum. Net Cash Flow (S)	NCF ($)	NPV at 10% ($)	NPV at 20% ($)	NPV at 34.30% ($)
11/1/2003	35	15	440	374	19.03	7,111	711	2,169	4,230	-129,385	4,230	3,204	2,486	1,790
12/1/2003	36	14	415	353	19.08	6,728	673	2,175	3,881	-133,266	3,881	2,916	2,246	1,602
1/1/2004	37	13	392	333	19.12	6,366	637	2,180	3,550	-136,816	3,550	2,646	2,023	1,430
2/1/2004	38	13	370	314	19.17	6,024	602	2,185	3,236	-140,052	3,236	2,393	1,817	1,272
3/1/2004	39	12	349	297	19.22	5,700	570	2,191	2,939	-142,991	2,939	2,156	1,625	1,127
4/1/2004	40	11	329	280	19.26	5,393	539	2,196	2,658	-145,648	2,658	1,934	1,447	994
5/1/2004	41	11	311	264	19.31	5,103	510	2,202	2,391	-148,039	2,391	1,726	1,282	873
6/1/2004	42	10	293	249	19.36	4,828	483	2,207	2,139	-150,178	2,139	1,532	1,130	762
7/1/2004	43	9	277	235	19.41	4,569	457	2,213	1,899	-152,077	1,899	1,350	988	660
8/1/2004	44	9	261	222	19.46	4,323	432	2,218	1,673	-153,750	1,673	1,179	857	567
9/1/2004	45	8	247	210	19.50	4,090	409	2,223	1,458	-155,208	1,458	1,020	736	482
10/1/2004	46	8	233	198	19.55	3,870	387	2,229	1,254	-156,462	1,254	870	624	405
11/1/2004	47	7	220	187	19.60	3,662	366	2,234	1,061	-157,523	1,061	731	520	334
12/1/2004	48	7	207	176	19.65	3,465	347	2,240	879	-158,402	879	600	424	270
1/1/2005	49	7	196	166	19.70	3,279	328	2,245	705	-159,107	705	478	335	212
2/1/2005	50	6	185	157	19.74	3,102	310	2,251	541	-159,648	541	364	253	158
3/1/2005	51	6	174	148	19.79	2,935	294	2,257	385	-160,034	385	257	178	110
4/1/2005	52	6	165	140	19.84	2,777	278	2,262	238	-160,271	238	157	108	66
5/1/2005	53	5	155	132	19.89	2,628	263	2,268	98	-160,369	98	64	44	27
6/1/2005	54	5	147	125	19.94	2,487	249	2,273	-35	-160,334				
7/1/2005	55	5	138	118	19.99	2,353	235	2,279	-161	-160,172				
8/1/2005	56	4	131	111	20.04	2,226	223	2,285	-281	-159,891				
9/1/2005	57	4	123	105	20.09	2,107	211	2,290	-394	-159,497				
10/1/2005	58	4	116	99	20.14	1,993	199	2,296	-502	-158,995				
11/1/2005	59	4	110	93	20.19	1,886	189	2,301	-604	-158,391				
12/1/2005	60	4	104	88	20.24	1,785	178	2,307	-701	-157,690				
1/1/2006	61	3												
TOTAL			**50,080**	**42,568**					**585,369**		**160,369**	**99,366**	**51,948**	**7**

APPENDIX

25B CASH FLOW MODEL FOR A MULTI-WELL DEVELOPMENT PROJECT

Appendix 25B-1: Summary

Input Parameters		**Economic Input Parameters:**		**Note: For modified IRR, investments discounted at 10%**	
Effective decline rate per year =	0.5	Working Interest =	100%		
Arps decline rate per year =	0.6931	Net Revenue Interest =	85%		
Arps decline rate per month =	0.0578	Oil Price $ per bbl =	55.00		
Days in one year =	365.25	Well cost per month ($M) =	24		
Days in one month =	30.438	Escalation factor per year =	3.5%		
Project Life, years =	10	Effective Escalation factor/mo. =	0.002871		
Total no. of wells at initial rate of 2500 bopd/well =	4	Severance & Ad Valorem taxes =	15%		
Nominal interest/yr =	10%	Federal Income Tax Rate =	20%		
Effective interest rate per month =	0.7974%	Drilling & Completion Cost/Well ($M) =	4,500		
Nominal interest/yr =	20%	**Summary of Results**			
Effective interest rate per month =	1.5309%	Gross Reserves, bbls =	5,258,985	PI Undiscounted (NOI BFIT) =	12.36
Nominal interest/yr =	624.50%	Pay-out, months =	6.5	PI Undiscounted (NCF BFIT) =	11.36
Effective interest rate per month =	17.9424%	Modified Internal Rate of Return (%) =	624.5	PI Undiscounted (NCF AFIT) =	9.04
Assumptions:		NPV NCF AFIT @ 0% (M$) =	162,682.62	PI NCF(AFIT) & INV Discounted @10% =	7.72
1. Revenues from oil sale received at end of next month		NPV NCF AFIT @ 10% (M$) =	129,559.60	PI NCF (AFIT) & INV discounted @20% =	6.76
2. End of month compounding		NPV NCF AFIT @ 20% (M$) =	106,735.86		

Appendix 25B-2: Annual Summary of Monthly Calculations

Year	Average (BOPD)	Gross Prod (bbl)	Net Prod (bbl)	Avg. Oil Price ($/bbl)	Gross Revenues ($M)	Sev. + Ad Val. Taxes ($M)	Operating Costs ($M)	Investments ($M)	NOI BFIT ($M)	NCF BFIT ($M)	Federal Income Tax ($M)	NCF AFIT ($M)	NPV NCF AFIT at 10% ($M)	NPV NCF AFIT at 20% ($M)
2017								4,500.00		-4,500.00		-4,500.00	-4,500.00	-4,500.00
2018	2,943	1,044,527	887,848	56.04	49,840.25	7,476.04	292.59	9,000.00	42,071.62	33,071.62	6,614.32	26,457.30	24,906.19	23,583.68
2019	5,183	1,839,624	1,563,681	58.00	90,705.28	13,605.79	302.83	4,500.00	76,796.65	72,296.65	14,459.33	57,837.32	49,509.16	42,987.62
2020	3,361	1,192,646	1,013,749	60.03	60,734.76	9,110.21	313.43	0.00	51,311.12	51,311.12	10,262.22	41,048.89	32,141.02	25,724.41
2021	1,680	596,323	506,875	62.13	31,430.24	4,714.54	324.40	0.00	26,391.30	26,391.30	5,278.26	21,113.04	15,029.03	11,026.59
2022	840	298,162	253,437	64.30	16,265.15	2,439.77	335.76	0.00	13,489.62	13,489.62	2,697.92	10,791.70	6,984.04	4,697.37
2023	420	149,081	126,719	66.55	8,417.21	1,262.58	347.51	0.00	6,807.13	6,807.13	1,361.43	5,445.70	3,204.35	1,975.85
2024	210	74,540	63,359	68.88	4,355.91	653.39	359.67	0.00	3,342.85	3,342.85	668.57	2,674.28	1,430.98	809.06
2025	105	37,270	31,680	71.30	2,254.18	338.13	372.26	0.00	1,543.80	1,543.80	308.76	1,235.04	601.21	311.79
2026	53	18,635	15,840	73.79	1,166.54	174.98	385.29	0.00	606.27	606.27	121.25	485.02	215.10	102.45
2027	28	8,177	6,950	76.15	528.55	79.28	331.35	0.00	117.91	117.91	23.58	94.33	38.54	17.03
TOTAL		5,258,985	4,470,137		265,698.07	39,854.71	3,365.09	18,000.00	222,478.28	204,478.28	41,795.66	162,682.62	129,559.60	106,735.86

Appendix 25B-3: Monthly Calculations of Cash Flow Model for a Multi-Well Development Project

Date	Months	BOPD	Gross BOPM	Net BOPM	Oil Price $/bbl	Gross Revenues ($M)	Sev. + Ad Val. Taxes ($M)	Operating Costs ($M)	Investments ($M)	NOI BFIT ($M)	NCF BFIT ($M)	Federal Income Tax ($M)	NCF AFIT ($M)	NPV NCF AFIT at 10% ($M)	NPV NCF AFIT at 20% ($M)
12/01/17	0	0	0	0	0	0.00	0.00	0.00	4,500.00	0.00	–4,500.00		–4,500.00	–4,500.00	–4,500.00
01/01/18	1	2500	73938	62,847	55.16	3,466.52	519.98	24.00	0.00	2,922.54	2,922.54	584.51	2,338.03	2,301.18	2,268.05
02/01/18	2	2360	69,788	59,320	55.32	3,281.35	492.20	24.07	0.00	2,765.08	2,765.08	553.02	2,212.06	2,159.98	2,113.50
03/01/18	3	2227	65,871	55,990	55.48	3,106.07	465.91	24.14	0.00	2,616.02	2,616.02	523.20	2,092.82	2,027.37	1,969.42
04/01/18	4	2102	62,174	52,848	55.63	2,940.16	441.02	24.21	0.00	2,474.93	2,474.93	494.99	1,979.94	1,902.85	1,835.10
05/01/18	5	1984	58,684	49,882	55.79	2,783.11	417.47	24.28	0.00	2,341.36	2,341.36	468.27	1,873.09	1,785.92	1,709.89
06/01/18	6	1873	55,391	47,082	55.95	2,634.44	395.17	24.35	4,500.00	2,214.93	–2,285.07	-457.01	-1,828.05	-1,729.19	–1,643.62
07/01/18	7	4268	126,220	107,287	56.11	6,020.38	903.06	24.42	0.00	5,092.91	5,092.91	1,018.58	4,074.33	3,823.50	3,608.01
08/01/18	8	4028	119,136	101,265	56.28	5,698.80	854.82	24.49	0.00	4,819.49	4,819.49	963.90	3,855.59	3,589.60	3,362.83
09/01/18	9	3802	112,449	95,582	56.44	5,394.39	809.16	24.56	0.00	4,560.67	4,560.67	912.13	3,648.54	3,369.96	3,134.26
10/01/18	10	3589	106,138	90,217	56.60	5,106.24	765.94	24.63	0.00	4,315.68	4,315.68	863.14	3,452.54	3,163.70	2,921.17
11/01/18	11	3387	100,181	85,154	56.76	4,833.49	725.02	24.70	0.00	4,083.77	4,083.77	816.75	3,267.01	2,970.01	2,722.51
12/01/18	12	3197	94,558	80,374	56.93	4,575.30	686.30	24.77	4,500.00	3,864.24	–635.76	–127.15	–508.61	-458.71	-417.45
01/01/19	13	5518	163,189	138,710	57.09	7,918.75	1,187.81	24.84	0.00	6,706.10	6,706.10	1,341.22	5,364.88	4,800.30	4,336.92
02/01/19	14	5208	154,030	130,925	57.25	7,495.77	1,124.36	24.91	0.00	6,346.49	6,346.49	1,269.30	5,077.19	4,506.95	4,042.47
03/01/19	15	4916	145,385	123,577	57.42	7,095.37	1,064.31	24.98	0.00	6,006.08	6,006.08	1,201.22	4,804.87	4,231.47	3,767.96
04/01/19	16	4640	137,225	116,641	57.58	6,716.37	1,007.45	25.05	0.00	5,683.86	5,683.86	1,136.77	4,547.09	3,972.77	3,512.04
05/01/19	17	4379	129,523	110,094	57.75	6,357.61	953.64	25.13	0.00	5,378.84	5,378.84	1,075.77	4,303.07	3,729.83	3,273.46
06/01/19	18	4134	122,253	103,915	57.91	6,018.01	902.70	25.20	4,500.00	5,090.11	590.11	118.02	472.09	405.96	353.71
07/01/19	19	6402	189,330	160,930	58.08	9,346.64	1,402.00	25.27	0.00	7,919.37	7,919.37	1,583.87	6,335.50	5,404.97	4,675.33
08/01/19	20	6042	178,703	151,898	58.25	8,847.38	1,327.11	25.34	0.00	7,494.93	7,494.93	1,498.99	5,995.94	5,074.82	4,358.03
09/01/19	21	5703	168,673	143,372	58.41	8,374.79	1,256.22	25.42	0.00	7,093.16	7,093.16	1,418.63	5,674.52	4,764.78	4,062.22
10/01/19	22	5383	159,207	135,326	58.58	7,927.44	1,189.12	25.49	0.00	6,712.84	6,712.84	1,342.57	5,370.27	4,473.63	3,786.45
11/01/19	23	5081	150,271	127,730	58.75	7,503.99	1,125.60	25.56	0.00	6,352.83	6,352.83	1,270.57	5,082.26	4,200.22	3,529.35
12/01/19	24	4796	141,837	120,561	58.92	7,103.16	1,065.47	25.64	0.00	6,012.05	6,012.05	1,202.41	4,809.64	3,943.46	3,289.66
01/01/20	25	4527	133,876	113,795	59.09	6,723.74	1,008.56	25.71	0.00	5,689.47	5,689.47	1,137.89	4,551.57	3,702.35	3,066.21

Date	Months	BOPD	Gross BOPM	Net BOPM	Oil Price $/bbl	Gross Revenues ($M)	Sev. + Ad Val. Taxes ($M)	Operating Costs ($M)	Investments ($M)	NOI BFIT ($M)	NCF BFIT ($M)	Federal Income Tax ($M)	NCF AFIT ($M)	NPV NCF AFIT at 10% ($M)	NPV NCF AFIT at 20% ($M)
02/01/20	26	4273	126,362	107,408	59.26	6,364.58	954.69	25.78	0.00	5,384.11	5,384.11	1,076.82	4,307.29	3,475.93	2,857.90
03/01/20	27	4033	119,270	101,380	59.43	6,024.61	903.69	25.86	0.00	5,095.06	5,095.06	1,019.01	4,076.05	3,263.30	2,663.69
04/01/20	28	3806	112,576	95,690	59.60	5,702.80	855.42	25.93	0.00	4,821.45	4,821.45	964.29	3,857.16	3,063.62	2,482.64
05/01/20	29	3593	106,258	90,319	59.77	5,398.18	809.73	26.01	0.00	4,562.45	4,562.45	912.49	3,649.96	2,876.12	2,313.85
06/01/20	30	3391	100,294	85,250	59.94	5,109.83	766.47	26.08	0.00	4,317.28	4,317.28	863.46	3,453.82	2,700.03	2,156.49
07/01/20	31	3201	94,665	80,465	60.11	4,836.89	725.53	26.16	0.00	4,085.20	4,085.20	817.04	3,268.16	2,534.68	2,009.80
08/01/20	32	3021	89,352	75,949	60.28	4,578.52	686.78	26.23	0.00	3,865.51	3,865.51	773.10	3,092.41	2,379.40	1,873.05
09/01/20	33	2852	84,337	71,686	60.46	4,333.95	650.09	26.31	0.00	3,657.56	3,657.56	731.51	2,926.04	2,233.58	1,745.56
10/01/20	34	2692	79,603	67,663	60.63	4,102.45	615.37	26.38	0.00	3,460.70	3,460.70	692.14	2,768.56	2,096.65	1,626.71
11/01/20	35	2540	75,135	63,865	60.80	3,883.32	582.50	26.46	0.00	3,274.36	3,274.36	654.87	2,619.49	1,968.06	1,515.91
12/01/20	36	2398	70,918	60,281	60.98	3,675.88	551.38	26.53	0.00	3,097.97	3,097.97	619.59	2,478.38	1,847.31	1,412.62
01/01/21	37	2263	66,938	56,897	61.15	3,479.53	521.93	26.61	0.00	2,930.99	2,930.99	586.20	2,344.80	1,733.92	1,316.33
02/01/21	38	2136	63,181	53,704	61.33	3,293.67	494.05	26.69	0.00	2,772.94	2,772.94	554.59	2,218.35	1,627.43	1,226.57
03/01/21	39	2016	59,635	50,690	61.51	3,117.74	467.66	26.76	0.00	2,623.31	2,623.31	524.66	2,098.65	1,527.44	1,142.89
04/01/21	40	1903	56,288	47,845	61.68	2,951.20	442.68	26.84	0.00	2,481.68	2,481.68	496.34	1,985.34	1,433.54	1,064.88
05/01/21	41	1796	53,129	45,159	61.86	2,793.56	419.03	26.92	0.00	2,347.61	2,347.61	469.52	1,878.09	1,345.37	992.16
06/01/21	42	1696	50,147	42,625	62.04	2,644.34	396.65	26.99	0.00	2,220.69	2,220.69	444.14	1,776.56	1,262.57	924.37
07/01/21	43	1600	47,332	40,233	62.22	2,503.09	375.46	27.07	0.00	2,100.55	2,100.55	420.11	1,680.44	1,184.82	861.18
08/01/21	44	1511	44,676	37,974	62.39	2,369.38	355.41	27.15	0.00	1,986.83	1,986.83	397.37	1,589.46	1,111.80	802.27
09/01/21	45	1426	42,168	35,843	62.57	2,242.82	336.42	27.23	0.00	1,879.17	1,879.17	375.83	1,503.34	1,043.24	747.36
10/01/21	46	1346	39,802	33,831	62.75	2,123.02	318.45	27.30	0.00	1,777.26	1,777.26	355.45	1,421.81	978.86	696.17
11/01/21	47	1270	37,568	31,933	62.93	2,009.62	301.44	27.38	0.00	1,680.79	1,680.79	336.16	1,344.63	918.40	648.45
12/01/21	48	1199	35,459	30,140	63.11	1,902.27	285.34	27.46	0.00	1,589.47	1,589.47	317.89	1,271.57	861.63	603.97
01/01/22	49	1132	33,469	28,449	63.29	1,800.66	270.10	27.54	0.00	1,503.02	1,503.02	300.60	1,202.42	808.32	562.51
02/01/22	50	1068	31,591	26,852	63.48	1,704.48	255.67	27.62	0.00	1,421.18	1,421.18	284.24	1,136.95	758.27	523.87
03/01/22	51	1008	29,818	25,345	63.66	1,613.43	242.01	27.70	0.00	1,343.72	1,343.72	268.74	1,074.97	711.26	487.84
04/01/22	52	952	28,144	23,922	63.84	1,527.25	229.09	27.78	0.00	1,270.38	1,270.38	254.08	1,016.30	667.12	454.26

Date	Months	BOPD	Gross BOPM	Net BOPM	Oil Price $/bbl	Gross Revenues ($M)	Sev. + Ad Val. Taxes ($M)	Operating Costs ($M)	Investments ($M)	NOI BFIT ($M)	NCF BFIT ($M)	Federal Income Tax ($M)	NCF AFIT ($M)	NPV NCF AFIT at 10% ($M)	NPV NCF AFIT at 20% ($M)
05/01/22	53	898	26,564	22,580	64.02	1,445.67	216.85	27.86	0.00	1,200.96	1,200.96	240.19	960.77	625.68	422.96
06/01/22	54	848	25,073	21,312	64.21	1,368.45	205.27	27.94	0.00	1,135.24	1,135.24	227.05	908.19	586.76	393.79
07/01/22	55	800	23,666	20,116	64.39	1,295.35	194.30	28.02	0.00	1,073.03	1,073.03	214.61	858.42	550.22	366.60
08/01/22	56	755	22,338	18,987	64.58	1,226.16	183.92	28.10	0.00	1,014.13	1,014.13	202.83	811.31	515.91	341.25
09/01/22	57	713	21,084	17,922	64.76	1,160.66	174.10	28.18	0.00	958.38	958.38	191.68	766.71	483.69	317.63
10/01/22	58	673	19,901	16,916	64.95	1,098.66	164.80	28.26	0.00	905.60	905.60	181.12	724.48	453.43	295.61
11/01/22	59	635	18,784	15,966	65.14	1,039.98	156.00	28.34	0.00	855.64	855.64	171.13	684.51	425.03	275.09
12/01/22	60	599	17,730	15,070	65.32	984.42	147.66	28.42	0.00	808.34	808.34	161.67	646.67	398.36	255.96
01/01/23	61	566	16,735	14,224	65.51	931.84	139.78	28.50	0.00	763.56	763.56	152.71	610.85	373.31	238.14
02/01/23	62	534	15,795	13,426	65.70	882.07	132.31	28.59	0.00	721.17	721.17	144.23	576.94	349.80	221.53
03/01/23	63	504	14,909	12,672	65.89	834.95	125.24	28.67	0.00	681.04	681.04	136.21	544.83	327.72	206.04
04/01/23	64	476	14,072	11,961	66.08	790.35	118.55	28.75	0.00	643.05	643.05	128.61	514.44	306.99	191.62
05/01/23	65	449	13,282	11,290	66.27	748.13	112.22	28.83	0.00	607.08	607.08	121.42	485.66	287.53	178.17
06/01/23	66	424	12,537	10,656	66.46	708.17	106.23	28.92	0.00	573.03	573.03	114.61	458.42	269.25	165.64
07/01/23	67	400	11,833	10,058	66.65	670.34	100.55	29.00	0.00	540.79	540.79	108.16	432.63	252.09	153.97
08/01/23	68	378	11,169	9,494	66.84	634.54	95.18	29.08	0.00	510.27	510.27	102.05	408.22	235.99	143.09
09/01/23	69	356	10,542	8,961	67.03	600.64	90.10	29.17	0.00	481.38	481.38	96.28	385.10	220.86	132.95
10/01/23	70	336	9,950	8,458	67.22	568.56	85.28	29.25	0.00	454.02	454.02	90.80	363.22	206.66	123.50
11/01/23	71	318	9,392	7,983	67.42	538.19	80.73	29.33	0.00	428.13	428.13	85.63	342.50	193.33	114.70
12/01/23	72	300	8,865	7,535	67.61	509.44	76.42	29.42	0.00	403.61	403.61	80.72	322.88	180.82	106.50
01/01/24	73	283	8,367	7,112	67.80	482.23	72.33	29.50	0.00	380.39	380.39	76.08	304.31	169.07	98.86
02/01/24	74	267	7,898	6,713	68.00	456.47	68.47	29.59	0.00	358.41	358.41	71.68	286.73	158.04	91.75
03/01/24	75	252	7,454	6,336	68.19	432.09	64.81	29.67	0.00	337.60	337.60	67.52	270.08	147.69	85.12
04/01/24	76	238	7,036	5,981	68.39	409.01	61.35	29.76	0.00	317.90	317.90	63.58	254.32	137.97	78.94
05/01/24	77	225	6,641	5,645	68.59	387.16	58.07	29.84	0.00	299.24	299.24	59.85	239.39	128.84	73.19
06/01/24	78	212	6,268	5,328	68.78	366.48	54.97	29.93	0.00	281.58	281.58	56.32	225.26	120.28	67.83
07/01/24	79	200	5,917	5,029	68.98	346.90	52.04	30.01	0.00	264.85	264.85	52.97	211.88	112.24	62.84
08/01/24	80	189	5,584	4,747	69.18	328.37	49.26	30.10	0.00	249.02	249.02	49.80	199.21	104.69	58.19

Date	Months	BOPD	Gross BOPM	Net BOPM	Oil Price $/bbl	Gross Revenues ($M)	Sev. + Ad Val. Taxes ($M)	Operating Costs ($M)	Investments ($M)	NOI BFIT ($M)	NCF BFIT ($M)	Federal Income Tax ($M)	NCF AFIT ($M)	NPV NCF AFIT at 10% ($M)	NPV NCF AFIT at 20% ($M)
09/01/24	81	178	5,271	4,480	69.38	310.83	46.62	30.19	0.00	234.02	234.02	46.80	187.22	97.61	53.86
10/01/24	82	168	4,975	4,229	69.58	294.23	44.13	30.27	0.00	219.82	219.82	43.96	175.86	90.96	49.83
11/01/24	83	159	4,696	3,992	69.78	278.51	41.78	30.36	0.00	206.38	206.38	41.28	165.10	84.72	46.08
12/01/24	84	150	4,432	3,768	69.98	263.64	39.55	30.45	0.00	193.64	193.64	38.73	154.91	78.87	42.58
01/01/25	85	141	4,184	3,556	70.18	249.55	37.43	30.53	0.00	181.59	181.59	36.32	145.27	73.37	39.33
02/01/25	86	134	3,949	3,356	70.38	236.22	35.43	30.62	0.00	170.17	170.17	34.03	136.13	68.21	36.30
03/01/25	87	126	3,727	3,168	70.58	223.60	33.54	30.71	0.00	159.35	159.35	31.87	127.48	63.37	33.48
04/01/25	88	119	3,518	2,990	70.78	211.66	31.75	30.80	0.00	149.11	149.11	29.82	119.29	58.83	30.86
05/01/25	89	112	3,321	2,822	70.99	200.35	30.05	30.89	0.00	139.41	139.41	27.88	111.53	54.57	28.41
06/01/25	90	106	3,134	2,664	71.19	189.65	28.45	30.98	0.00	130.23	130.23	26.05	104.18	50.57	26.14
07/01/25	91	100	2,958	2,515	71.39	179.52	26.93	31.06	0.00	121.53	121.53	24.31	97.22	46.82	24.03
08/01/25	92	94	2,792	2,373	71.60	169.93	25.49	31.15	0.00	113.29	113.29	22.66	90.63	43.30	22.06
09/01/25	93	89	2,636	2,240	71.80	160.86	24.13	31.24	0.00	105.48	105.48	21.10	84.39	40.00	20.23
10/01/25	94	84	2,488	2,114	72.01	152.26	22.84	31.33	0.00	98.09	98.09	19.62	78.47	36.90	18.53
11/01/25	95	79	2,348	1,996	72.22	144.13	21.62	31.42	0.00	91.09	91.09	18.22	72.87	33.99	16.95
12/01/25	96	75	2,216	1,884	72.42	136.43	20.46	31.51	0.00	84.45	84.45	16.89	67.56	31.27	15.48
01/01/26	97	71	2,092	1,778	72.63	129.14	19.37	31.60	0.00	78.17	78.17	15.63	62.53	28.71	14.11
02/01/26	98	67	1,974	1,678	72.84	122.25	18.34	31.69	0.00	72.21	72.21	14.44	57.77	26.32	12.84
03/01/26	99	63	1,864	1,584	73.05	115.72	17.36	31.79	0.00	66.57	66.57	13.31	53.26	24.07	11.66
04/01/26	100	59	1,759	1,495	73.26	109.53	16.43	31.88	0.00	61.23	61.23	12.25	48.98	21.96	10.56
05/01/26	101	56	1,660	1,411	73.47	103.68	15.55	31.97	0.00	56.16	56.16	11.23	44.93	19.98	9.54
06/01/26	102	53	1,567	1,332	73.68	98.15	14.72	32.06	0.00	51.36	51.36	10.27	41.09	18.13	8.59
07/01/26	103	50	1,479	1,257	73.89	92.90	13.94	32.15	0.00	46.82	46.82	9.36	37.45	16.40	7.71
08/01/26	104	47	1,396	1,187	74.10	87.94	13.19	32.24	0.00	42.51	42.51	8.50	34.00	14.77	6.90
09/01/26	105	45	1,318	1,120	74.32	83.24	12.49	32.34	0.00	38.42	38.42	7.68	30.74	13.24	6.14
10/01/26	106	42	1,244	1,057	74.53	78.80	11.82	32.43	0.00	34.55	34.55	6.91	27.64	11.81	5.44
11/01/26	107	40	1,174	998	74.74	74.59	11.19	32.52	0.00	30.88	30.88	6.18	24.70	10.48	4.79
12/01/26	108	37	1,108	942	74.96	70.60	10.59	32.62	0.00	27.40	27.40	5.48	21.92	9.22	4.18

Date	Months	BOPD	Gross BOPM	Net BOPM	Oil Price $/bbl	Gross Revenues ($M)	Sev. + Ad Val. Taxes ($M)	Operating Costs ($M)	Investments ($M)	NOI BFIT ($M)	NCF BFIT ($M)	Federal Income Tax ($M)	NCF AFIT ($M)	NPV NCF AFIT at 10% ($M)	NPV NCF AFIT at 20% ($M)
01/01/27	109	35	1,046	889	75.17	66.83	10.02	32.71	0.00	24.10	24.10	4.82	19.28	8.05	3.62
02/01/27	110	33	987	839	75.39	63.26	9.49	32.80	0.00	20.97	20.97	4.19	16.78	6.95	3.11
03/01/27	111	32	932	792	75.61	59.88	8.98	32.90	0.00	18.00	18.00	3.60	14.40	5.92	2.63
04/01/27	112	30	880	748	75.82	56.68	8.50	32.99	0.00	15.19	15.19	3.04	12.15	4.95	2.18
05/01/27	113	28	830	706	76.04	53.66	8.05	33.09	0.00	12.52	12.52	2.50	10.02	4.05	1.77
06/01/27	114	26	784	666	76.26	50.79	7.62	33.18	0.00	9.99	9.99	2.00	7.99	3.21	1.39
07/01/27	115	25	740	629	76.48	48.08	7.21	33.28	0.00	7.59	7.59	1.52	6.07	2.42	1.04
08/01/27	116	24	698	593	76.70	45.51	6.83	33.37	0.00	5.31	5.31	1.06	4.25	1.68	0.72
09/01/27	117	22	659	560	76.92	43.08	6.46	33.47	0.00	3.15	3.15	0.63	2.52	0.99	0.42
10/01/27	118	21	622	529	77.14	40.78	6.12	33.56	0.00	1.10	1.10	0.22	0.88	0.34	0.14
11/01/27	119	20	587	499	77.36	38.60	5.79	33.66	0.00	–0.85	-0.85				
12/01/27	120	19	554	471	77.58	36.54	5.48	33.76	0.00	–2.70	-2.70				
01/01/28	121	18	523	445	77.81	34.59	5.19	33.85	0.00	–4.46	-4.46				
02/01/28	122	17	494	420	78.03	32.74	4.91	33.95	0.00	–6.12	-6.12				
03/01/28	123	16	466	396	78.25	30.99	4.65	34.05	0.00	–7.71	-7.71				
04/01/28	124	15	440	374	78.48	29.33	4.40	34.15	0.00	–9.21	-9.21				
05/01/28	125	14	415	353	78.70	27.77	4.17	34.24	0.00	–10.64	-10.64				
06/01/28	126	13	392	333	78.93	26.28	3.94	34.34	0.00	–12.00	-12.00				
07/01/28	127	13	370	314	79.16	24.88	3.73	34.44	0.00	–13.29	-13.29				
08/01/28	128	12	349	297	79.38	23.55	3.53	34.54	0.00	–14.52	-14.52				
09/01/28	129	11	329	280	79.61	22.29	3.34	34.64	0.00	–15.69	-15.69				
10/01/28	130	11	311	264	79.84	21.10	3.17	34.74	0.00	–16.80	-16.80				
11/01/28	131	10	293	249	80.07	19.97	3.00	34.84	0.00	–17.86	-17.86				
12/01/28	132	9	277	235	80.30	18.91	2.84	34.94	0.00	–18.87	-18.87				
01/01/29	133	9	261	222	80.53	17.90	2.68	35.04	0.00	–19.83	-19.83				
02/01/29	134	8	247	210	80.76	16.94	2.54	35.14	0.00	–20.74	-20.74				
03/01/29	135														
TOTAL			**5,258,985**	**4,470,137**		**265,698**	**39,855**	**3,365**	**18,000**	**222,478.28**	**204,478.28**	**41,795.66**	**162,682.62**	**129,559.60**	**106,735.86**

APPENDIX

25C WORKSHEET FOR A PSC WITH FIXED SCALE

Appendix 25C

Worksheet for a Production Sharing Contract with Fixed Scale		
Fiscal Regime		**Notes:**
Royalty =	15%	Royalty is fixed at 15% of gross oil revenues per year
Cost oil limit per year =	40%	Cost oil is limited at maximum of 40% of net oil revenues per year
Profit Oil share for NOC =	55%	Profit oil is shared at fixed scale of 55% for NOC and 45% for IOC
Profit Oil share for IOC =	45%	Discovery bonus of $2,500,000
Country Income Tax rate =	25%	Start of production bonus of $5,000,000
Annual Discount Rate =	15%	

Year	Development Costs ($M)	Bonuses ($M)	Production (mbo)	Oil Price ($/bbl)	Gross Revenues ($M)	Royalty ($M)	Net Revenue ($M)	Operating Costs ($M)	Overhead ($M)	Total Recoverable Costs ($M)	Annual Cost Limit ($M)	Annual Costs Recovery ($M)	Remaining Unrecovered Costs ($M)
1	23,000	2,500							200	23,200		0	23,200
2		5,000	1,071	55	58,796	8,819	49,976	439	200	639	19,991	19,991	3,848
3			1,886	57	107,176	16,076	91,099	454	250	704	36,440	4,553	0
4			1,222	59	71,915	10,787	61,128	470	250	720	24,451	720	0
5			611	61	37,216	5,582	31,634	487	200	687	12,653	687	0
6			306	63	19,259	2,889	16,370	504	175	679	6,548	679	0
7			153	65	9,967	1,495	8,472	521	150	671	3,389	671	0
8			76	68	5,158	774	4,384	540	125	665	1,754	665	0
9			38	70	2,669	400	2,269	458	100	558	908	558	0
10			19	72	1,381	207	1,174	278	55	333	470	333	0
11			8	75	625	94	532	147	30	177	213	177	0
Total	**23,000**	**7,500**	**5,390**		**314,162**	**47,124**	**267,038**	**4,298**	**1,735**	**29,033**		**29,033**	

Year	Total Profit Oil ($M)	NOC Share @55% ($M)	IOC Share @45% ($M)	IOC NCF BFIT ($M)	INCOME TAX @25% ($M)	IOC NCF AFIT ($M)	NPV IOC NCF @ 15% ($M)	NOC NCF ($M)	NPV NOC NCF @ 15% ($M)
1				–25,700	0	–25,700	–22,348	2,500	2,174
2	29,986	16,492	13,494	27,845	3,373	24,472	18,504	24,866	18,802
3	86,547	47,601	38,946	42,794	9,737	33,058	21,736	57,337	37,700
4	60,408	33,224	27,183	27,183	6,796	20,388	11,657	40,020	22,882
5	30,947	17,021	13,926	13,926	3,482	10,445	5,193	20,502	10,193
6	15,692	8,630	7,061	7,061	1,765	5,296	2,290	10,396	4,494
7	7,800	4,290	3,510	3,510	878	2,633	990	5,168	1,943
8	3,720	2,046	1,674	1,674	418	1,255	410	2,464	806
9	1,710	941	770	770	192	577	164	1,133	322
10	841	463	379	379	95	284	70	557	138
11	355	195	160	160	40	120	26	235	50
Total	**238,005**	**130,903**	**107,102**	**99,602**	**26,776**	**72,827**	**38,692**	**165,178**	**99,504**

INDEX

abnormally pressured gas reservoirs. *See* geopressured gas reservoirs
Abou-Kassem. *See* DAK correlation
absolute permeability, 33, 37, 42, 337, 438
ACA (After Closure Analysis), 408, 420
ACA Horner Plot, 419f
acentric factors, 8, 128, 501
acoustic logs. *See* logs
Adaptive Implicit Method (AIM), 545, 552
AFIT (After Federal Income Tax), 627, 635
afterflow, 301, 310, 338, 400
Agarwal equivalent time, 349, 355, 386
AIM. *See* Adaptive Implicit Method
alkali/surfactant/polymer (ASP) flooding, 484, 499
Al-Marhoun correlation, 79, 82
altered zone, permeability of, 298, 340
Amott wettability test, 435
Amortization Methods, 622
Analyses of Diagnostic Fracture Injection Tests (DFITs), 401-411
analytical steamflood models, 502, 503
analytical techniques (for estimation of MME/MMP), 488
angle of contact, 96, 105, 273
Angus and Troika reservoir,194, 205
anisotropic reservoir model, 374,376
Annual Percentage Rate (APR), 602
API Recommended Practice 44: Sampling Petroleum Reservoir Fluids, 93
API (American Petroleum Institute), 229
application of emerging technologies for improved hydrocarbon recovery, 559, 560, 565, 571-573. *See also* reservoir management principles
approximation models (water influx models), 143
aquifer models. *See also* water influx
  Carter-Tracy aquifer model, 143, 146-148, 175, 190t,
  Fetkovich aquifer model, 143-145, 146t, 175, 211, 212f,
  Schilithuis aquifer model, 143
  van Everdingen-Hurst aquifer model, 143, 264, 265, 274
aquifers
  finite-acting, 205, 264
  infinite-acting, 313, 318, 320, 327, 338
  linear, 143, 144f, 187, 194
  productivity indices, 145, 145t
  radial, 143f, 145, 273, 273f
Archie equations, 48
Archie parameters, 48-50
Arp's Decline Equations, 209-210
ARI (Advanced Resources International), 20
ASP flooding. *See* alkali/surfactant/polymer flooding
asphaltenes, 94, 436
average gas gravity, 78
average net pay thickness
  Surculus interval, 158
  Tamalis interval, 158
average net sand thickness, 203
average permeability
  of parallel beds, 257
  of serial concentric segments, 258
average reservoir pressures
  M-4/M-4A reservoirs, 168t,
  M-6 reservoirs, 170t
average rock properties
  M-4/M-4A reservoirs, 158, 166t
  M-6 reservoirs, 166t
  Surculus interval, 158t
  Tamalis interval, 158t
average water saturation (Welge method)
  after water breakthrough, 443
  behind flood front, 443
averaging techniques, 444
Aziz and Firoozabad correlation, 489, 490. *See also* MMP
Aziz and Wichert correlations, 63, 73

Barnett Shale, 222
basin studies, 515
BCA (Before Closure Analysis), 407, 420
Bcf/D (Billion cubic feet of gas per day), 20
BDF (Boundary Dominated Flow), 229
Beggs-Robinson correlation, 80-82
Bergman-Sutton correlation, 80, 82
BFIT (Before Federal Income Tax), 627
BHFP (Bottom Hole Flowing Pressure), 229
BHP (Bottomhole Pressure), 420

bilinear flow
  defined, 364
  fractured well test analysis (straightline methods) and, 364
binary interaction parameter (BIP), 119, 126
binary systems, 56-57
  defined, 56
  P-T (pressure-temperature) phase diagram of, 57f
BIP (binary interaction parameter), 119
black oil reservoirs, 58, 89, 92
  saturated, 90
  undersaturated, 90
black oil sample (reservoir fluid study), 100-103, 109
  calculated analysis, 107
  differential liberation, 103, 110
  gas differently liberated, 111
  pressure-volume properties (CCE), 110
  reservoir oil flash liberation tests (comparison), 111
  summary, 112
  viscosity data, 112
black oil simulators, 503, 538, 541
bottomhole sampling. *See* subsurface
  sampling bottom-water drive reservoirs, 144f,
BOPD (barrels of oil per day), 20
BP (British Petroleum), 20, 477
boundary effects, 338
bounded cylindrical reservoir
  constant pressure (case 3), 265
  no flow (case 2), 264
Bourdet
  calculation of pressure derivatives, 362
  derivative type curves, 345, 345f, 347, 349
  Gringarten type curves (example), 352
Boyle's law, 59, 117
Brouweri sands, 156
bubble point
  calculations (example), 81-83
  pressure, 77
bubble point pressure calculation, 117
Buckley-Leverett equation, 441- 443
buildup well tests
  with constant production rate before shut-in, 301-303
  with constant production rate before shut-in (gas wells), 331
  data (Gringarten-Bourdet type curves example), 347
  defined, 297, 301,
  Horner's time approximation and, 316, 317, 333
  MDH plot and, 315, 316
  permeability/reservoir pressure/skin factor (example) and, 317, 318
  time regions, 300, 301f
bulk density, 25, 31
bulk volume, 23, 24
bulk volume irreducible (BVI), 42
bulk volume irreducible water, 35
BVI. *See* bulk volume irreducible

calculation from interpretation model parameters (type-curve matching stage), 351, 352-368
capillary displacement pressure, 97, 105
capillary end effects, 38
capillary pressure, 41, 96-100
  defined, 96
  immiscible fluid displacement and, 435-447
capillary tubes, 436, 448
CAPEX (Capital Expenditure), 477
Capital Recovery Amount, 596
capillary viscometer, 103
carbonate rocks, 465, 467
carbon dioxide (physical constants), 61t
carbon dioxide injection gases, correlations for, 490-492,
carbon dioxide miscible gas flooding, 512
carbon monoxide (physical constants), 61t
Carter-Tracy aquifer model, 143, 146-148, 175, 190t,
Cartesian grids, 542-544
case histories, 154, 559
  Manatee reservoirs, 194-202
  MBB/W31S reservoirs, 566-571
  Red Hawk gas reservoir, 154-163
  reservoir management, 504, 537, 559-561
  Shaybah field, 571-575
  26R reservoir, 561-566
  West Cameron 580 reservoir, 164-172
cash flow diagram, 606, 607
Cation-Exchange-Capacity (CEC), 50, 53
CBM (coalbed methane), 8, 9f, 10f, 20
CCE. *See* Constant Composition Expansion
CEC (Cation-Exchange-Capacity), 50, 53
cementation factor, 29, 30, 53
cementation factor for shaly sands, 53
centrifugal method, 35, 50
channeling, 38, 374, 399

characteristic time constant, NMR, 35, 42, 50, 53
Charle's law, 59, 60
chemical flooding processes, 484, 499, 500
  ASP flooding, 500
  EOR screening criteria for, 483, 484t
  MEOR and, 484, 500
  polymer flooding, 500
  polymer/surfactant flooding, 481, 499
chemical flood simulators, 538
chemical potential, 55, 84, 119, 128, 129. *See also* Gibbs energy
chromatograph analysis, 159
  of flash gas, 159
  of separator gas, 112
  of separator liquid, 113
  slimtube method and, 496
Claire Ridge LSWF Project (case history), 476-477
classification of oil and gas resources, 589
clay, volume of, 23, 30
clay-bound water, volume of, 23, 30
clay cation-exchange-capacity, 50, 53
clay water associated with clay counterions, 53
clean sands, 23, 48, 49
closed loop reservoir management, 559
cloud transforms, 37
Coates-Denoo equation, 35
Coates equation, 34, 35, 36f
coefficient of isothermal compressibility
  of formation water, 71-72
  of gas, 72
  of oil, 79, 81
cokriging, 524, 529
Cole Plot, 142, 142f
collocated cokriging, 524, 529
complex hydrocarbon system (pseudoternary phase diagram), 485, 486f
composite reservoir or multiple region reservoir model, 374
compositional simulations (MMP/MME), 488
  multiple mixing-cell method, 488, 495
  1-D slimtube simulation, 488, 495, 496, 496f
  tie-line method, 488, 491, 492, 495, 496
compositional simulators, 538, 551
compounds, physical constants of, 61t
compressibility of wellbore fluid, 340, 348
compressible fluids, 254
  radial diffusivity equation and, 261-262
  straightline methods and, 297-339
Concession contract, 624
Conventional Reservoirs Versus Unconventional Reservoirs, 1
condensate gases. *See also* gas-condensate reservoirs
  correlation method, 65-67
  properties, 59-61
  recombination method and, 64-65, 65t, 66t
condensing gas drive MCM process, 487
condensing/vaporizing (CV) gas drive MCM process, 488
conditional simulation methods
  object-based modeling, 524
  SGS, 524, 529, 533
  SIS, 524, 533
conditional stimulation methods, 524
conductivity
  of clay-bound water, 50, 53
  of formation or rock, 29, 30, 51, 53
  of formation water, 29, 30, 48, 49, 51, 53
  of fully water-saturated formation or rock, 53
  of shale, 51, 53
connate water volume, 132, 134
conservation of reservoir energy, 559, 560, 564, 568, 572. *See also* reservoir management principles
Constant Composition Expansion (CCE), 100
  black oil sample, 109
  experimental diagram, 100f
  gas condensate sample, 113
constant pressure lines, 56
constant production rate before shut-in (buildup tests), 312-314, 331-335
constant single rate drawdown well tests
  defined, 300, 300f
  for gas wells, 320-326
  slightly compressible fluids and, 207, 253-254
constant temperature lines, 56
constant terminal pressure solution, 265-266
  cumulative oil production for well (example) and, 266
  superposition principle and, 270-276
constant terminal rate solution
  bounded cylindrical reservoir, constant pressure (case 3), 265
  bounded cylindrical reservoir, no flow (case 2), 264
  line source well, infinite reservoir (case 1), 262-263
  superposition principle and, 270-276

constant volume depletion (CVD), 100, 101
continuity equation, radial, 261
continuity equation in rectangular form, 539
conventional bottomhole samplers, 93–94
core data, 516
core plugs, 24
  length of, 30
  NMR logs *v.*, 34
  permeability data and, 34
  radius of, 24, 30
  rock wettability and, 436
core porosity, NMR porosity *v.*, 28, 28f
core samples, permeability from, 34
Corey equations, 38–40
Corey's exponent, 39, 40, 42, 557
corner-point geometry, 543
correlation method, 65–68
correlations
  Al-Marhoun, 79, 82
  Beggs-Robinson, 82
  Bergman-Sutton, 82
  for calculation of oil PVT properties, 77–82
  for calculation of water PVT properties, 83
  DAK, 68–69
  Emera, 490, 492, 493, 494, 495
  Firoozabadi and Aziz, 328, 489
  for gas compressibility factor, 68–69
  Glasø, 488, 489, 490,
  for hydrocarbon injection gases (MMP), 488
  for impure carbon dioxide injection gases (MMP), 492–495
  Lucas, 70
  Ng-Egbogah, 80, 82
  for nitrogen injection gases (MMP), 489–490
  Petrosky and Farshad, 78, 81
  for pure carbon dioxide injection gases (MMP), 490–491
  for residual gas saturations in gas reservoirs under water influx, 177
  Sarma, 490, 491, 492, 493
  Sebastian, 492, 495
  Sutton, 62, 63, 64, 68
  Valko and McCain, 77, 78, 81, 214
  Wichert and Aziz, 63, 73
  Yuan, 490, 491, 492, 495
corresponding states, principle of, 61, 68, 71
Craig-Geffen-Morse method, 451, 452
cricondenbar, 56, 57
cricondentherm, 57, 58, 59
critical compressibility factor, 128
critical gas saturation, 37, 42, 185, 557
critical point, 55
critical pressures (for compounds), 61t
critical temperatures (for compounds), 61t
critical volumes (for compounds), 61t
critical water saturation, fraction, 37, 42
Cross SAGD (XSAGD), 502, 508
cross-sectional area, cm^2, 33
CSG (coal seam gas), 20
CSS (cyclic steam simulation), 501, 508
cubic EOS, roots from, 120
cumulative gas injection, 132, 134, 568
cumulative gas produced, 134, 142, 173
cumulative oil production, 131, 183, 185, 203, 208, 340, 580
cumulative oil production for well (example), 11, 134, 580
cumulative production gas-oil ratio, 134, 186, 203,
cumulative reservoir voidage, 186, 203
cumulative water influx, 134, 144, 146, 266, 273, 274, 276
cumulative water influx (van Everdingen-Hurst method), 143–148
cumulative water injection, 134, 203, 454, 461
cumulative water production, 131, 134, 203
cutoff criteria, 47, 51, 52, 52f
CVD. *See* constant volume depletion
CV (condensing/vaporizing) gas drive MCM process, 486
cyclic steam simulation (CSS), 501, 502, 508

DAK (Dranchuk and Abou-Kassem) correlation, 68–69
Dalmation wetting, 436
damage ratio (DR), 299, 341
darcy, 2, 33, 34
Darcy fluid flow equation, 255
  average permeability of parallel beds, 257, 257f
  average permeability of serial concentric segments, 258–259
  PSS state/incompressible fluids, 259–260
  radial forms of, 256–260
  SS flow, incompressible fluids, 259–260
  SS flow/compressible fluids, 256–257
  well production rate in SS flow, 516, 250, 257

data (for geologic modeling and reservoir characterization). *See also specific types of data*
  core data, 516
  fluid properties data, 516, 551
  quality control/quality assurance, 516
  scale/integration of, 517
  sources of, 515-516
  well log data, 516
DCA (Decline Curve Analysis), 229, 627
DDB (Double Declining Balance), 627
Decision and Risk Analysis: Applied to Petroleum Evaluations, 626
dead-oil viscosity, 80, 82, 84
Dean-Starks method, 35, 50
deconvolution concepts (well test analysis), 389-395. *See also* well test analysis
  defined, 389
  pressure-rate deconvolution model and, 390
  von Schroeter deconvolution algorithm and, 390-391
density, 84
  of formation water, 83
  of gas, 70
  reduced, 450, 458
density-derived porosity, 25, 30
density-derived shale porosity, 30
density logs, 25
  crossover in gas interval, 26, 27f
depletion study (CCD), 101
depreciation, depletion, and amortization, 619-622
Depletion Methods, 622
depth below OWC, 105
derivative type curves. *See* type curves
Design, Planning, and Execution of DFIT, 399-401
Design of Experiments (DOE), 550
de-watering practices, 142
dew point, 56, 57
dew point pressure calculation (EOS), 123
DFIT (Diagnostic Fracture Injection Test), 420
Diagnostic Plots, 213
differential liberation (DL)
  black oil sample, 100, 103
  defined, 101
  experimental diagram, 102f
dimensionless pressure (for finite/infinite aquifers), 178-179
dimensionless variables, for hydraulically fractured wells, 363
dimensionless wellbore storage coefficient, 302, 348, 350, 351, 386
directional fractures, NFR with, 376f
direct line drive pattern, 452, 451f
Discounted Pay-Back (Pay-Out), 609
Discounted Profitability Index, 610
displacement mechanisms
  MBB/W31S reservoirs, 567
  Shaybah field, 571
  26R reservoir, 561-566
distributed temperature sensing (DTS), 503, 508
DL. *See* differential liberation
DOE. *See* Design of Experiments
double porosity behavior, NFRs and, 381-382
double porosity model, 377
DPR method, 68
DR (damage ratio), 299, 341
drainage area, pressure function of, 313, 340
drainage displacements, 37, 40
Dranchuk and Abou-Kassem (DAK) correlation, 68
drawdown, 92, 253
drawdown well tests, 300-301
  constant single rate, 300, 300f, 327
  multi-rate, 306, 306f, 307, 327
  time regions, 301, 301f
  two-rate, 310
drillstem tests (DST), 34, 297
drive index, 182
  formation and connate water compressibility drive index, 182
  formation and water compressibility drive index, 182
  gas cap drive index, 182
  oil drive index, 182
  water drive index, 182
dry gas, 58, 59, 89, 133, 137, 154
dry gas reservoirs, 137, 158
  characteristics, 137
  dry gas behaviors (phase diagram), 485, 485f
  volumetric, material balance for, 592
  volumetric calculations for, 29
Dry Gas Well, 216
DST. *See* drillstem tests
DTS (distributed temperature sensing), 503, 508
dual-porosity simulators, 538
Dual-Water model, 47, 50

Dykstra-Parsons method, 452, 567
Dykstra-Parsons permeability variability coefficient, 567

early implementation of simple, proven strategies, 559, 560, 565, 571, 572. *See also*
reservoir management principles
early time region. *See* ETR
economic evaluation (EOR projects), 481, 482, 504
economic evaluation of petroleum projects and property, 589
economic evaluation of petroleum property, 623, 624
ECS well logs. *See* elemental capture spectroscopy well logs
ED. *See* Experimental Design
EDL (Electric Double Layer), 477
edge-water drive reservoirs, 143, 144f
effective compressibility, reservoir, 184, 203
effective conductivity (of shaly sands), 49, 53
effective permeability
  for fluid, 33, 42, 340
  for gas, 33, 42, 33
  for oil, 33, 42
  for water, 33, 42, 461
effective porosity, 23, 25, 28, 42
effective wellbore radius, 229, 297, 300
EIA (Energy Information Administration), 20
electromagnetic (EM) viscometer, 103
elemental capture spectroscopy (ECS) well logs, 23, 30
Elk Hills Oil Field, 567, 555. *See also* MBB/W31S reservoirs; 26R reservoir
Emera correlation, 490, 493 . *See also* MMP emerging technologies for improved hydrocarbon recovery, application of, 560, 565, 571, 572
empirical correlations. *See also* correlations for
  MMP, 38-40,
  relative permeability data and, 37-40
empirical models, Stone's, 41
EM (electromagnetic) viscometer, 103
EMV (Expected Monetary Value), 627
Energy trade for USA, 15f
energy conservation (reservoirs), 559, 560, 564, 568, 572. *See also* reservoir management principles
enhanced oil recovery. *See* EOR
EOR (enhanced oil recovery), 41, 481-508
  chemical flooding processes, 484, 499-500
  defined, 481
  improved oil recovery and, 499
  introduction to, 481, 465
  microbial, 481, 484, 500
  miscible gas injection processes, 485-488, 505
  processes, 481-483
  tertiary oil recovery and, 481
  thermal processes, 481, 500-504
EOR projects
  economic evaluation and, 504, 505
  geologic/reservoir modeling and, 506
  implementation of, 504-508
  management, 507
  oil prices *v.*, 481
  oil production from, 482, 482f, 566f
  pilot testing and, 506-507
  process screening/selection, 505
  total number of, 481
EOR screening criteria
  for chemical flooding processes, 484, 484t
  for miscible gas injection processes, 484, 484t
  for thermal EOR processes, 484, 485t
EOS (equations of state), 117-128
  bubble point pressure calculation and, 123, 183
  cubic, roots from, 117-118
  dew point pressure calculation and, 123
  historical introduction to, 117
  hydrocarbon plus fractions and, 123-125
  ideal gas, 117
  phase equilibrium of mixtures and, 119-120
  phase equilibrium predictions with, 125-127 PR, 117, 119, 120, 122
  preparing for use (in analyses of PVT data), 125-127
  real gas, 117
  RK, 118-119
  SRK, 118-119
  tuning of, 118, 126, 495
  two-phase flash calculation and, 122-123
  vdW, 117-118, 120
  volume translation and, 121-122
equations of state. *See* EOS
equilibrium constant, 122, 128
ES (expanding solvent)-SAGD, 481, 484, 502
estimation methods, 523
ethane (physical constants), 121t
ethane/n-heptane mixtures (P-T phase diagram), 57, 57f
ETR (early time region), 301, 341, 349
examples

average permeability of parallel beds, 257
average permeability of serial concentric segments, 258
bubble point calculations, 123, 183
constant rate drawdown test, 304-306
cumulative oil production for well, 183, 568, 573,
cumulative water influx (van Everdingen-Hurst method), 274
experimental variogram, 520-523
flowing bottomhole pressure of gas well, 547, 552
fractional flow curve calculation, 444-446
fracture and reservoir parameters from type curves in NFR, 381
gas gravity/total reservoir fluid production rate (calculation), 60-61, 67
gas in place, 19, 131, 137, 138
gas mixture pseudo-critical properties (SSBV mixing rules), 62-64
geologic modeling, 515-532
Gringarten/Bourdet type curves, 347-348
Hall plot, 454
Horner's time approximation, 313, 409
hydraulically fractured well (type curves), 368-374
multi-rate drawdown test, 306-309
non-Darcy flow coefficient, 327-331
OGIP (for volumetric dry gas reservoir), 137
OOIP (undersaturated oil reservoir with water influx), 186
radius of investigation, 297
recombination method (wet gas gravity), 64, 66t
reservoir "A," 526-529
reservoir "B," 529-532
reservoir pressure calculation, 263
skin factor/effective wellbore radius, 300
total pressure drop of well in multi-well reservoir, 303-304
two-rate drawdown test, 310-312
von Schroeter deconvolution algorithm, 390, 391, 395
water influx calculation (Carter-Tracy method), 147, 148t
water influx calculation (Fetkovich method), 144-145, 146t
water saturation calculation (Archie equation), 48
wellbore pressure calculation, 265
well index calculation, 548
well production rate in PSS flow, 260
well production rate in SS flow, 259

EUR (estimated ultimate recovery), 20
exothermic samplers, 94
expanding solvent (ES)-SAGD, 502, 508
Experimental Design (ED), 550
experimental method, 496
experimental methods (for MME/MMP measurement), 488
RBA method, 497
    slimtube method, 496-497
experimental variogram models, 521-522
exponential integral, 262, 270, 281
    chart for, 281f
exponential variogram model, 522, 522f
extensive properties, 55

facies (rock types), 29
facilities
    gasflood management, 459
    waterflood projects, 453
Farshad and Petrosky correlation, 78, 81
FBSA (flow-based scale averaging), 526, 533
FCP (Fracture Closure Pressure), 420
FCM (first-contact miscibility), 486, 508
Fetkovich aquifer model, 143, 144-146, 146t
FFI. *See* free-fluid index
field data, relative permeability data and, 38
Field Tests:
    Carbonate reservoirs, 466
    Sandstone reservoirs, 466
finite-acting aquifers, 143, 148
finite conductivity vertical fracture
    defined, 364
    type curve for, 369f
finite reservoirs. *See* van Everdingen-Hurst entries
Firoozabadi and Aziz correlation, 489, 490. *See also* MMP
first-contact miscibility. *See* FCM
five-point differencing, 544, 544f
five-spot pattern,452, 452f, 501
flash gas (chromatograph analysis), 159
flow-based scale averaging (FBSA), 526, 533
flow efficiency, 299, 340
flow equations for three-phase flow of oil/water/gas, 539-541
flowing bottomhole pressure of gas well (example), 210, 213, 547, 552
flow models, of hydraulically fractured wells, 364-365

flow periods (NFRs), 380, 381, 391
flow rate, cc/sec, 33, 42
Flow Regimes, 207
flow regimes (in hydraulically fractured wells), 363
    bilinear flow, 364, 365
    formation linear flow, 364, 366-367
    fracture linear flow, 364
    pseudo-radial flow, 364, 368
flow simulators (general structure), 550-551
flow zone indicator (FZI), 36
fluid composition (reservoir fluid study), 100
fluid density, 25, 30, 96
fluid displacement efficiency, 438
fluid flow (in petroleum reservoirs), 253-277
    Darcy fluid flow equation, 255
    PSS flow, 259-260
    radial continuity equation and, 261
    radial diffusivity equation and, 261
    real gas pseudo-pressure concept and, 267-269
    SS flow, 260
    superposition principle and, 270-276
    transient flow, 254
    well injectivity index and, 277
    well productivity index and, 276
fluid properties data, 453, 516. *See also* formation pressures; reservoir rock/fluid properties
fluids
    compressible, 254
    incompressible, 253
    slightly compressible, 253-254, 261, 262, 304
fluid transit time, 25, 26, 30
FMB Flowing Material Balance, 212, 229
formation and connate water compressibility drive index, 182
formation and water compressibility drive (FWCD) index, 182, 203
formation compressibility factor, 153, 182
formation conductivity, 29, 30, 51
formation factor, 48, 53
formation interval transit time, 26, 30
formation linear flow
    defined, 364
    fractured well test analysis (straightline methods) and, 365-368
formation pressures (from WFTs), 96
    capillary pressure and, 96
    fluid properties data and, 516
    supercharging and, 99
formation resistivity factor, 48
formation water
    coefficient of isothermal compressibility of, 83
    conductivity, 29, 30, 51
    density, 83
    viscosity of, 83
formulas (for compounds), 61t
formulations, of simulator equations, 545-546
FPP (Fracture Propagation Pressure), 420
fractional flow curve, 440, 442, 443,444
    calculation (example), 444-446
    defined, 440
fractional flow equations, 438-441
    for oil displaced by gas, 440-441
    for oil displaced by water, 439-440
fractional wettability, 436
Fracture Height Recession/Fracture Model, 401
Fracture Tip Extension, 404
fracture and reservoir parameters from type curves in NFR (example), 381-385
fractured well flow models, 364-365. *See also* hydraulically fractured wells
    finite conductivity vertical fracture, 364
    infinite conductivity vertical fracture, 364, 365f
    uniform flux vertical fracture, 365
fracture linear flow, 364
fracture permeability, 363, 377, 379, 383, 386
free-fluid index (FFI), 28, 35
free-fluid porosity, 28
Free Water Level (FWL), 97, 105, 436
frontal advance equation, 442. *See also* Buckley-Leverett equation
fugacity, 119, 120, 121, 128
fugacity coefficient, 119, 120, 128
function to account for extra conductivity of shale, 49, 53
future worth, 594
FWCD (formation and water compressibility drive) index, 182, 203
FWL. *See* Free Water Level
FZI (flow zone indicator), 36

gamma distribution model, 124
gamma function, 124, 214
gas cap drive index, 182
gas compressibility, 71-77

gas compressibility factor, 60, 62, 70
calculation of, 103-104
charts, 69f
correlations for, 68-70
defined, 60
gas condensate. *See* condensate gases
gas-condensate reservoirs,58, 59, 92, 133. *See also* retrograde gas condensate reservoirs
gas cycling (designing) for, 59
management of, 59
retrograde behavior of, 58-59
saturated, 93
undersaturated, 92
gas condensate sample (reservoir fluid study), 112-115
chromatograph analysis of separator gas, 112
hydrocarbon analyses of produced wellstream, 115
pressure-volume properties (CCE), 114
reservoir fluid composition, 113
retrograde condensation during gas depletion, 115
saturation pressures from stepwise recombinations, 113
summary, 112
gas constant, 60, 84, 117, 128
gas density, 70
gas deviation factor, 60
gas diffusivity equation
real gas pseudo-pressure concept and, 267-270
real gas pseudo-pressure/pseudo-time and, 319
gases
properties, 62-64
real gas equation, 60
gas factor, 60
gasflooding, 451, 458-460. *See also* immiscible fluid displacement
applications of, 459
design steps for, 458-459
immiscible, 481
management of, 459-460
miscible, types of, 485
MMP/MME and, 488-495
gasflood reservoirs, 457
gas flow equations
normalized pseudo-pressure/pseudo-time, 319
pressure/pressure-squared as variables in, 320
real gas pseudo-pressure/pseudo-time, 319
gas formation volume factor, 70, 74, 84, 103, 131, 137, 260
gas gravity
defined, 60-61
total reservoir fluid production rate (calculation), 67-68
gas injectors (gasflood management), 459
gas in place (example), 192
gas mixtures (pseudo-critical properties), 61-64
known composition, 62-63
non-hydrocarbon gas impurities and, 63
SSBV mixing rules (example), 72-74
unknown composition, 63-64
gas-oil ratio (GOR), 78. *See also* solution gas-oil ratios
gas-oil relative permeability curves, 37f, 40, 41f
gas pseudo-pressure concept (real gas), 267-270
gas pseudo-pressure/pseudo-time
normalized, 319
real, 319
gas PVT data (S10A2/S10B reservoirs), 200t
gas reservoirs, 137-173
case histories of, 154-172
classification of, 137
Cole plot for, 142, 142f
geopressured, 149-154
GMBE for, 133-134
Havlena-Odeh straight line method for, 143-144, 143f
overpressured, 137
Red Hawk gas reservoir, 154-164
volumetric, 137-141
with water influx, 141-143
water influx models and, 143-148
West Cameron 580 reservoir, 164-172
gas saturation, fraction, 37, 42, 47, 137, 142, 173, 177, 185, 441
gas specific gravity (of compounds), 61t
gas viscosity, 70-71
gas wells
constant rate drawdown tests for, 320-327
de-watering practices and, 142
flowing bottomhole pressure of, 213, 266, 268-269
multi-rate drawdown tests for, 306-309
Gaussian quadrature, 124, 125t, 126
Gaussian simulation, sequential, 524, 529, 533
Gaussian variogram model, 521, 521f
General Material Balance Equation. *See* GMBE
generation of geologic surfaces/horizons, 517

geologic model data, 551
geologic modeling (and reservoir characterization), 515
- examples, 526–533
- gasflood design and, 458
- general procedure for, 517–533
- object-based, 524
- property data modeling, 518–523
- QC/QA programs, 516
- reservoir "A" (example), 526–529
- reservoir "B" (example), 529–533
- scale/integration of data, 516
- sources of data for, 515–516
- stratigraphic modeling, 518
- structural modeling, 517
- uncertainty analysis and, 524

geologic/reservoir modeling (EOR projects), 506
geologic surfaces/horizons, generation of, 517
geology. *See* reservoir geology
geometric anisotropy, variogram and, 520f
geopressured gas reservoirs, 149
- Ramagost and Farshad method and, 150–152
- Roach method and, 152–154

geopressured zones (along Gulf Coast), 149, 149f
geostatistical methods, 519–524
- conditional stimulation methods, 523, 524
- estimation methods, 523
- spatial analysis with variograms, 519–521
- variogram models, 521–523

Gibbs energy, 119, 120, 128. *See also* chemical potential
Gibbs' phase rule, 56
Glasø correlations, 488, 489. *See also* MMP
GMBE (General Material Balance Equation), 131–134
- application of, 133–134
- derivation of, 131–133
- for gas reservoirs, 133
- Ramagost and Farshad method and, 150–152
- saturated oil reservoirs with water influx and, 192
- undersaturated oil reservoirs (with water influx) and, 186–192
- volumetric gas reservoirs and, 137–141
- volumetric saturated oil reservoirs and, 192–193

GMBE (General Material Balance Method), 627
GOR. *See* gas-oil ratio
gravity drainage mechanism (oil reservoirs), 182
grid coarsening, 544, 544f
grid orientation effects, 544
grid systems, 541
- Cartesian grids, 542
- corner-point geometry, 543
- LGR technique, 544
- PEBI grids, 543
- radial grids, 542–543
- Voronoi, 543

Gringarten/Bourdet type curves (example), 352–354
Gringarten type curves, 347–348
gross rock thickness, 52, 52f
Gulf Coast, geopressured zones along, 149, 149f
Gulf of Mexico. *See also* Red Hawk gas reservoir; retrograde gas condensate reservoirs; WC 580 reservoir
- condensate gas reservoir in, 59, 137
- deepwater reservoirs in, 80, 99, 253, 363, 451
- Manatee reservoirs in, 194–203
- oil PVT properties calculations and, 77
- Red Hawk reservoir in, 155f
- WC 580 reservoir in, 164

Hall plot, 454, 454f, 455
Havlena-Odeh straight line method, 142–143, 143f
HCPV. *See* hydrocarbon pore volume
heterogeneous (non-uniform) wetting, 436
high hydrocarbon saturations, 47
highly conductive fault, NFR and, 376f, 376f
high water injection volumes, 452
high water saturations, 47
historical liquids and water production (Well 1ST2, Well 2ST2), 163f
historical performance data
- for Well 1ST2, 160t
- for Well 2ST2, 161t

historical pressure and gas production
- for Well 1ST2, 162f
- for Well 2ST2, 162f

historical production data
- M-4/M-4A reservoirs, 166t–168t, 171f
- M-6 reservoirs, 164, 165f, 166t, 169t, 170t, 172f
- S10A2 reservoir, 196f, 197f, 200, 201t, 202f
- S10B reservoir, 194, 196, 198f, 199f, 200, 200t

history matching, 548–549
homogeneous reservoir model, 374–375
homogeneous (uniform) wetting, 436
horizontal permeability data, 34
Horner's pseudo-producing time approximation, 317, 333

Horner's time approximation, 333-336
Hurst aquifer model. *See* van Everdingen-Hurst aquifer model
HS (High Salinity), 477
HSW (High Salinity Water), 477
HSWF (High Salinity Waterflooding), 477
hydraulically fractured well flow models, 364-365
    finite conductivity vertical fracture, 364
    infinite conductivity vertical fracture, 365, 365f
    uniform flux vertical fracture, 365
hydraulically fractured wells (well test analysis), 365-368. *See also* well test analysis
    dimensionless variables for, 363
    flow regimes in, 363-364
    procedure steps for, 366
    straightline methods and, 365-368
    type-curve matching and, 368-374
    type curves (example) and, 347-355
hydraulic units, 36
hydrocarbon analyses of produced wellstream (gas condensate sample), 115
hydrocarbon injection gases, correlations for, 488-489
hydrocarbon miscible gas flooding, 498
hydrocarbon plus fractions, 123-125
hydrocarbon pore volume (HCPV), 29, 30, 47, 132, 137
hydrocarbon recovery, emerging technologies for, 560, 565, 571, 572
hydrocarbons
    volume shift parameter correlation coefficients for, 122t
    volume shift parameter for, 121t
hydrocarbon saturation, 47
hydrocarbon saturation, fraction, 47, 53
hydrogen (physical constants), 61t
hydrogen sulfide (physical constants), 61t
hysteresis, 37, 435

IARF. *See* infinite-acting radial flow
ideal gas EOS, 117
ideal gas equation, 59
identification of flow periods, 381
identification of interpretation model (type-curve matching stage), 351, 368
IFT. *See* interfacial tension
imbibition curve, 436, 437f
imbibition displacement, 37
imbibition process, 39 imbibition rate test, 435
immiscible fluid displacement, 435-447. *See also* capillary pressure; relative permeability; rock wettability
    basic concepts in, 435-438
    Buckley-Leverett equation and, 441-443
    fluid displacement efficiency and, 438
    fractional flow equations and, 440-441
    gasflooding and, 458-461
    introduction to, 435
    mobility/mobility ratio and, 438
    relative permeability and, 438
    total recovery efficiency and, 438
    volumetric displacement efficiency and, 438
    Welge method and, 443-446
IMPES (Implicit Pressure, Explicit Saturation) method, 545
implementation of simple, proven strategies, 560, 565, 571, 572. *See also*
    reservoir management principles
implicit constraint encoding, 390
Implicit method, 545
Implicit Pressure, Explicit Saturation method. *See* IMPES method
improved hydrocarbon recovery, emerging technologies for, 565, 571, 572
improved oil recovery, 481. *See also* EOR
impure carbon dioxide injection gases, correlations for, 492-495
incompressible fluids, 253. *See also* radial forms of Darcy fluid flow equation
indicator kriging, 524
infinite-acting aquifers, 143, 148
infinite-acting radial flow (IARF), 313, 347, 349, 355
infinite conductivity vertical fracture
    defined, 364
    type curve for, 369f
infinite reservoirs. *See* line source well, infinite reservoir; van Everdingen-Hurst entries
initial reservoir pressure, 58, 144
initial reservoir temperature, 59, 70, 77, 84
initial water saturation, fraction, 39, 137, 438, 443
injection history. *See* production/injection history
injection water (waterflood projects), 456
injection wells
    tests for, 297
    waterflood management and, 457
    well injectivity index and, 277
in-situ combustion (ISC)/high pressure air injection (HPAI), 504

intelligent fields, 559
intensive properties, 55
interest rates
  annual percentage rate (APR), 602
  period, nominal, and effective, 602
  The Rule of 72, 602, 603
  The Rule of 78, 604, 605
interfacial tension (IFT), 96, 105, 437, 467, 499
interporosity flow, 377
interporosity flow coefficient, 377, 379, 380
Internal Rate of Return (IRR), 613, 627
  Modified Internal Rate of Return (IRR), 617
  Multiple Internal Rates of Return (IRR), 614
INV (Investment), 627
IOC (International Oil Company), 627
irreducible water saturation, fraction, 35, 42, 50, 443
ISC (in-situ combustion)/HPAI (high pressure air injection), 504
ISIP (Instantaneous Shut-In Pressure), 420
isobars, 56
isobutane (physical constants), 61t
isochronal flow tests, 328–331
isopentane (physical constants), 61t
isothermal gas compressibility, 60, 62, 68, 71, 84, 320
isothermal oil compressibility, 79–80
isothermal oil compressibility at bubble point, 79, 81
isothermal water compressibility, 83, 134, 182
isotherms, 56

JV (Joint Venture), 624, 627

Katz chart. *See* Standing and Katz chart Kay's mixing rule, 62, 75
KB (Kelly Bushing), 229
Kessler-Lee equations, 62, 125
Klinkenberg constant, 33, 42
Klinkenberg effect, 33
Kozeny-Carman equation, 35
Kozeny constant, 35
kriging methods, 524
  collocated cokriging, 524, 529

lag distance, 519, 520, 521
lag tolerance, 520, 533
largest pore throat radius, 97, 105
late time region. *See* LTR
Late Transient Flow (LTF) regime, 255
length, cm, 24, 30, 33
LGR (local grid refinement) technique, 544
linear aquifers, 143, 144f, 187, 194
line source well, infinite reservoir
  (case 1), 262–263
  with skin factor, 298–299
liquid phase mole fraction, 62, 128, 491, 492
liquid specific gravity (for compounds), 61t
LNG (liquefied natural gas), 20
local grid refinement (LGR) technique, 544
logging-while-drilling (LWD) tools, 30, 516, 572
long term retention of staff in multi-disciplinary teams, 559, 560–561, 566, 571, 573. *See also*
  reservoir management principles
low hydrocarbon saturations, 47
low water saturations, 47
LS (Low Salinity), 477
LSE (Low Salinity Effect), 477
LSE, mechanisms, 467
LSW (Low Salinity Water), 477
LSWF (Low Salinity Waterflooding), 477
LSWF Projects: Planning, Designing and Installing, 469-476
LTF (Late Transient Flow) regime, 255
LTR (late time region), 301, 302, 309
Lucas method, 70
LWD tools. *See* logging-while-drilling tools

MACRS (Modified Accelerated Cost Recovery System), 627
M-4/M-4A reservoirs
  average reservoir pressures for, 168t, 171f
  average rock properties, 166, 166t
  fluid properties for, 166t
  historical production data for, 167t, 171f
  structure data for, 164t
  structure map of, 165f
M-6 reservoirs
  average reservoir pressures for, 170t, 172f
  average rock properties, 166, 166t
  fluid properties for, 166t
  historical production data for, 169t, 172f
  structure data for, 164t
  structure map of, 165f
management principles. *See* reservoir management principles

Manatee reservoirs, 194-202
geology, 195-199
map of, 196f
reservoir pressure/production data, 200-202
review questions, 204
type log for, 195f
material balance errors (MBEs), 546
material balance methods, 592. *See also* GMBE for gas reservoir with water influx, 133
Roach method and, 152-154
for saturated oil reservoirs with water influx, 192
for undersaturated oil reservoirs (with water influx), 186-192
for volumetric gas reservoirs, 131-141
for volumetric saturated oil reservoirs, 193
material balance model (for oil reservoir), 131f
matrix block model, 377, 377f
matrix permeability, 377, 379, 386
NFR and, 377f
Shaybah field and, 571
MBB/W31S reservoirs
displacement mechanisms, 571, 572
historical gas/water injection in, 569f
historical production of, 569f
management principles applied to, 568-571
map of, 568f
MBB/W31S reservoirs (case history), 566-568
geology, 567
production/injection history, 568
reservoir rock and fluid properties, 579t
structure map of, 562f
MBEs. *See* material balance errors
McCain correlations. *See* Valko and McCain correlations
Mcf (Thousand cubic feet of gas), 20
MCM (multiple contact miscibility)
condensing gas drive MCM process, 487
CV (condensing/vaporizing) gas drive MCM process, 487
defined, 487
vaporizing gas drive MCM process, 488
MCN (Multiple-Carbon-Number) groups, 123, 128
MDH (Miller, Dyes, Hutchinson) plot, 315-316
mean flowing pressure, 33, 42
mean-square in water saturation regression, 49, 53
melting curve, 55
MEOR. *See* microbial EOR methane (physical constants), 61t
methane/n-butane/decane system
effect of pressure on two-phase envelope of, 485, 485f
ternary phase diagram for, 485-486, 485f
microbial EOR (MEOR), 484, 500
middle time region. *See* MTR
MIE (Multicomponent Ionic Exchange), 477
Miller, Dyes, Hutchinson (MDH) plot, 315-316
millidarcies, 36
minimum miscibility enrichment. *See* MME
minimum miscibility pressure. *See* MMP
miscible gas flooding
carbon dioxide, 498-499
hydrocarbon, 498
nitrogen/flue-gas, 497-498
types, 497-499
miscible gas injection processes, 484, 499, 485-488
basic concepts, 485
condensing gas drive MCM process, 487
CV gas drive MCM process, 488
EOR screening criteria for, 484, 485t
FCM and, 486
MCM and, 487
types of, 485
vaporizing gas drive MCM process, 486-487
mixed wettability, 436
mixing rules
Kay's, 62, 75
SBV, 62,
SSBV, 62-64
MME (minimum miscibility enrichment)
analytical techniques for estimation of, 488-496
defined, 488
experimental methods in measurement of, 496-497
methods for determination of, 488-497
MMP (minimum miscibility pressure)
analytical techniques for estimation of, 488-496
correlations for hydrocarbon injection gases, 488
correlations for impure carbon dioxide injection gases, 492-495
correlations for nitrogen injection gases, 489-490
correlations for pure carbon dioxide injection gases, 490-491
defined, 488
empirical correlations for, 488-495

experimental methods in measurement of, 496-497
methods for determination of, 488-497
mobility, 438
mobility ratio, 438
Modeling DFIT with a Hydraulic Fracture Model, 401
model equilibration data, 551
model gridding, 518-519
modeling. *See* geologic modeling; reservoir simulation
model initialization (reservoir simulation), 548
models. *See also* aquifer models
anisotropic reservoir model, 376
approximation models (water influx models), 143
double porosity model, 377
Dual-Water model, 50
gamma distribution model, 124
Gaussian variogram model, 521, 522
homogeneous reservoir model, 374
hydraulically fractured well flow models, 364-365
hydraulically fractured wells, 368-385
material balance model (for oil reservoir), 131f
matrix block model, 377, 377f
multiple region or composite reservoir model, 375
naturally fractured reservoir models, 374-378
porosity model, for shaly sand reservoir, 24f
pressure-rate deconvolution model, 390
pseudosteady-state models (water influx models), 143
single fracture model, 376
slab model, 377, 377f
steamflooding, 500
Stone's empirical models, 41
total least squares error model, 390
variogram, 519-532
water influx models, 143-147
Waxman-Smits model, 47, 49-50
well, 547
modified fugacity, 119, 128
molar entropy, 119, 128
molar volume, 119, 121, 128
molecular weight, lbm/lb mole, 62, 64, 65
molecular weights (for compounds), 61t
mole fraction of component *i*
in liquid phase, 84
in vapor phase, 84
mole fraction of vapor in mixture, 62, 122, 128, 492,
MTR (middle time region), 297, 301, 304, 305, 313, 317, 337, 338, 341, 347, 350
MTR line, slope of, 305, 313, 316, 334
multicomponent systems, 57-58
multi-disciplinary teams, 559, 560-561, 566, 571, 573. *See also* reservoir management principles
multilayered reservoirs, 339, 339f
multiphase flow, 336-338
Multiple-Carbon-Number (MCN) groups, 123, 128
multiple contact miscibility. *See* MCM
multiple mixing-cell method, 488, 495
multiple region or composite reservoir model, 375
multi-rate drawdown well tests, 306-309
for gas wells, 320

naturally fractured reservoir models, 374-378
anisotropic reservoir model, 374, 376
double porosity model, 377
homogeneous reservoir model, 374-375
multiple region or composite reservoir model, 375
single fracture model, 376
naturally fractured reservoirs (NFRs), 374-378. *See also* well test analysis with directional fractures, 376f
double porosity behavior and, 381-382
flow periods and, 380, 381, 391
fracture and reservoir parameters from type curves (example) in, 381-385
highly conductive fault and, 376f
matrix block model for, 377, 377f
matrix permeability and, 377, 379, 386
semilog pressure response of, 369, 369f
slab model for, 377, 377f
straightline methods and, 365-368
type curves and, 380-381
n-butane (physical constants), 61t
n-decane (physical constants), 61t
near wellbore zone. *See* wellbore
NCF (Net Cash Flow), 627
net confining stress, 34
net pay isopach maps
S10A2 reservoir, 196f, 197f, 200, 201t, 202f
S10B reservoir, 194, 196, 198f, 199f, 200, 200t
net pay thickness, 47, 52, 52f
Net Present Value Profile, 611
net present value (NPV), 611-613
net reservoir, 52, 52f

net sand, 52, 52f
neutron-derived porosity, 26, 30
  in clay, 26, 30
  in shale, 26, 30
neutron porosity logs, 26–28
  crossover in gas interval, 26, 27f
Newton-Raphson method, 68, 122, 613
NF (Nano-Filtration), 477
NFP (Net Fracture Pressure), 420
NFRs. *See* naturally fractured reservoirs
Ng-Egbogah correlation, 80, 82
n-heptane (physical constants), 61t
n-hexane (physical constants), 61t
Nigerian saturation equation, 50
nine-point differencing, 544f
nine-spot pattern, 452, 452f
nitrogen (physical constants), 61t
nitrogen/flue-gas miscible gas flooding, 497
nitrogen injection gases, correlations for, 489
NMR characteristic time constant, 35, 42, 50, 53
NMR (Nuclear Magnetic Relaxation) Method, 435
NMR (nuclear magnetic resonance) porosity logs, 50
  core plugs *v.*, 34
  core porosity *v.*, 28, 28f
  water saturations from, 50–51
NOC (National Oil Company), 627
NOI (Net Operating Income), 627
Normal Leak-off, 402, 403f
Normalized Rates and Pressures, 212
n-nonane (physical constants), 61t
n-octane (physical constants), 61t
non-Archie rocks, 49, 51. *See also* shaly sands
non-Darcy flow coefficient, 319, 327-331
non-hydrocarbon gas impurities, 63
non-transition variogram models, 521
non-uniform (heterogeneous) wetting, 436
normal five-spot pattern, 452, 452f, 501
normalization parameter in gamma function, 124, 128
normalized gas pseudo-pressure/pseudo-time, 319
normalized Gibbs energy, 120
normalized saturation of the wetting phase, 39, 42
normalized water saturation, 39, 42
normal nine-spot pattern, 452, 452f
n-pentane (physical constants), 61t
NPV (Net Present Value), 611, 627
NRI (Net Revenue Interest), 627
Nuclear Magnetic Relaxation (NMR) Method, 435
nuclear magnetic resonance porosity logs. *See* NMR porosity logs
numerical dispersion, 546-547
numerical reservoir simulators, 538. *See also* reservoir simulation
numerical steamflood simulation models, 503

object-based modeling, 524
OBM (oil-based mud) systems, 94, 126
Odeh method. *See* Havlena-Odeh straight line method
OFA (Optical Fluid Analyzer), 95, 105
OGIP. *See* original gas in place
oil
  effective permeability, 33, 37, 38, 42, 337
  oil compressibility, 79–80
  oil compressibility factor (calculation), 104
  oil displaced by gas, fractional flow equation for, 440–441
  oil displaced by water, fractional flow equation for, 439–440
  oil drive index, 182
  oil prices, EOR projects *v.*, 481
  oil production (from EOR projects), 482, 482f, 566f
  oil PVT data (S10A2/S10B reservoirs), 200t
  oil reservoir drive mechanisms, 181-182
  oil saturation, fraction, 39, 47, 203
  oil viscosity, 80–82
  relative permeability, 33, 37, 438
oil-based mud systems. *See* OBM systems
oil formation volume factor, 78–79, 103, 105
  at bubble point pressure, 77–78
  solution GOR, calculation of, 78
oil reservoirs, 92, 181–202. *See also* black oil reservoirs; volatile oil reservoirs, 58, 89
  gravity drainage mechanism, 182
  material balance model for, 131f
  saturated, 192
  undersaturated, 183
Oil-Water Contact (OWC), 97, 143
oil-water relative permeability curves, 41, 37, 37f,
1-D slimtube simulation, 488, 495, 496, 496f
OOIP (original oil in place)
  GMBE and, 131
  in undersaturated oil reservoir, 186–192
  in undersaturated reservoirs with water influx, 186

OOIP (Original Oil-In-Place), 477
OPEX (Operational Expenditure), 477
Optical Fluid Analyzer (OFA), 95, 105
original gas in place (OGIP), 131, 137, 138, 150, 152
original oil in place. *See* OOIP
ORRI (Over-Ride Revenue Interest), 627
outcrop studies, 515
overpressured gas reservoirs, 137. *See also* geopressured gas reservoirs
OWC. *See* Oil-Water Contact
oxygen (physical constants), 61t

parallel beds, average permeability of, 257, 257f
PDI (Potential Determining Ions), 477
p_D and t_D for radial flow (van Everdingen and Hurst)
  finite reservoirs with closed outer boundary, 283, 293t
  finite reservoirs with constant pressure outer boundary, 285t
  infinite reservoirs with constant terminal rate at inner boundary, 291t
PEBI (perpendicular bisection) grids, 543
Peng-Robinson (PR) EOS. *See* PR EOS
peripheral flood pattern, 452, 452f
permeability, 33-42. *See also* effective permeability; relative permeability
  absolute, 33
  applications of, 41-42
  from core samples, 34
  fracture permeability, 229, 363, 376f, 377, 379
  gas-oil relative permeability curves, 37, 40, 41f
  horizontal, 34
  matrix permeability, 377, 379, 386
  measured with air, 33, 42
  from pressure transient tests, 34, 363, 507
  sources of, 38-41
  vertical, 34, 502
  vertical permeability data, 34
  from well logs, 34-37
permeability of the altered zone, 298, 340
Permian Basin, 216
perpendicular bisection (PEBI) grids, 543
petroleum reservoir fluid flow regimes. *See* fluid flow
Petrosky and Farshad correlation, 78, 81
phase, 55
phase diagrams, 55-59
  binary systems and, 56-57
  multicomponent systems and, 57-58
  pseudoternary phase diagram for complex hydrocarbon system, 485, 486f
  reservoir fluid types and, 57, 58f
  retrograde behavior and, 58-59
  single component systems and, 55-56
  ternary phase diagram for methane/n-butane/decane system, 485-486, 485f
  wet gas/dry gas behaviors, 485, 485f
phase envelopes, 56, 57, 58, 59, 127f, 137, 485
phase equilibrium of mixtures, 119-120
phase equilibrium predictions (with EOS), 125-127
physical constants of compounds, 61t
PI (Profitability Index), 627
PIIP (Petroleum Initially In-Place), 627
pilot testing (EOR projects), 506-507
pistonned bottomhole samplers, 94
Pitzer acentric factor, 118
Pliocene
  lower Pliocene, 196
  upper Pliocene sands, 154
PKN (Perkins-Kern-Nordgren), 420
PMP (Pressure Match Point), 350, 369, 380
polymer flooding, 500
polymer/surfactant flooding, 499
pore throat radius, 97, 105
pore volume, 29, 30, 47, 132, 137
pore volume-to-grain volume ratio, 36, 42
porosity, 23-14
  effective, 23-24, 28, 30
  free-fluid, 28
  primary, 23
  secondary, 23
  total, 23-24
porosity data
  applications of, 28-29
  direct measurements of, 24-25
  indirect measurements of, 25-28
  sources of, 24-28
porosity model, for shaly sand reservoir, 24f
porosity-permeability crossplot, 37, 37f
porosity-permeability transforms (in reservoir characterization), 29, 29f
Potential Resources in Unconventional Reservoirs in the World, 19, 20
Poupon-Leveaux equation, 47, 50

Power Law variogram model, 521, 521f
predictions, reservoir simulation and, 537
PR (Peng-Robinson) EOS, 117, 119
cubic form, roots of, 119
present worth, 595
pressure data applications (WFT), 516
pressure derivative calculations (Bourdet), 362, 347
pressure difference, atm, 33, 34, 401, 407
Pressure-Dependent Leakoff, 402
pressure function of drainage area, 257, 313
Pressure Match Point. *See* PMP
pressure/pressure-squared as variables in gas flow equations, 320
pressure/production data. *See* reservoir pressure/production data pressure-rate deconvolution model, 390
pressure response, von Schroeter deconvolution algorithm and, 390
pressure-temperature diagrams. *See* P-T phase diagrams
pressure transient test data, 34, 516
pressure transient tests,34, 363, 507. *See also* well test analysis
pressure-volume phase diagrams. *See* P-V phase diagrams
pressure-volume-temperature properties. *See* PVT properties
primary porosity, 23
principle of corresponding states, 61, 68, 71
PRMS (Petroleum Resources Management System), 627
processing facilities (gasflood management), 459
process screening/selection (EOR projects), 505
Procedure for DFIT Analyses, 411-419
produced water (waterflood projects), 457
producers (gasflood management), 459
production/injection history
MBB/W31S reservoirs, 566-571
26R reservoir, 561-566
production/pressure data. *See* reservoir pressure/production data
production wells (waterflood projects), 456
propane (physical constants), 61t
property data modeling, 518-524
geostatistical methods and, 519-524
model gridding in, 518-519
PSA (Production Sharing Agreement), 625, 627
PSC (Production Sharing Contract), 625, 627
PSS (Pseudosteady-state), 229
pseudo-critical pressure, 61, 71, 84
pseudo-critical properties, of gas mixtures, 62-64
pseudo-critical temperature, 61, 71, 84
pseudo-pressure/pseudo-time
normalized gas, 319
real gas, 267-269
pseudo-radial flow
defined, 364, 368
fractured well test analysis (straightline methods) and, 365-368
radial flow *v.*, 209, 364
pseudo-reduced compressibility, 71, 72f
pseudo-reduced isothermal compressibility, 83, 84
pseudo-reduced temperature/pressure, 61, 71
pseudosteady-state (PSS) flow, 207, 255
incompressible fluids (Darcy flow equation), 256, 259-260
well production rate in, 260
pseudosteady-state models (water influx models), 143
pseudoternary phase diagram for complex hydrocarbon system, 485, 486f
pseudo time. *See* real gas pseudo-pressure/pseudo-time
PSS flow. *See* pseudosteady-state flow
P-T (pressure-temperature) phase diagrams, 55, 55t, 56, 57
of binary system, 56, 57f
for ethane/n-heptane mixtures, 58-59, 58f
pure carbon dioxide injection gases, correlations for, 490-491
P-V (pressure-volume) phase diagrams, 56, 56f
PVT laboratory measurements (reservoir fluid studies), 89-104
applications of, 99-100
black oil sample, 109-111
gas condensate sample, 112-116
guidelines, 90t
PVT (pressure-volume-temperature) properties (reservoir fluids), 55-83
from EOS, 117-128
water, 200
QC (quality control)/QA (quality assurance) programs (geologic modeling), 516
Q_D and t_D for radial flow (van Everdingen-Hurst)
finite reservoirs with closed outer boundaries, 283, 293t
infinite reservoirs with constant terminal pressure, 291t
quality control/quality assurance for data, 516

radial aquifers, 143f, 273
radial continuity equation, 261

radial diffusivity equation, 261–262. *See also* gas diffusivity equation
compressible fluids and, 304
slightly compressible fluids and, 207, 253–254
radial flow. *See also* radial forms of Darcy fluid flow equation; van Everdingen-Hurst aquifer model
IARF, 313, 347, 349, 355
incompressible fluids and, 256, 259–260
parallel beds and, 257, 257f
pseudo-radial flow *v.*, 364, 368
serial concentric segments and, 258
radial forms of Darcy fluid flow equation, 256–261
average permeability of parallel beds, 257, 257f
average permeability of serial concentric segments, 258–259
PSS state/incompressible fluids, 259–260
SS flow, incompressible fields, 259–260
SS flow/compressible fluids, 256–257
well production rate in SS flow, 516, 250, 257
radial grids, 542–543
radius of capillary tube, 97, 436, 447
radius of investigation, 297
Ramagost and Farshad method, 150–152
Raphson method. *See* Newton-Raphson method
R/P (reserves-to-production ratio), 20
RBP (Retrievable Bridge Plug), 420
RBA (rising-bubble apparatus) method, 497
RCAL (routine core analysis), 34
real gas EOS, 117
real gas equation, 60
real gas pseudo-pressure concept, 267–269
real gas pseudo-pressure/pseudo-time, 319
recombination method, 64–65, 65t, 66t
Red Hawk gas reservoir, 154-156
drilling history (summary) for, 155t
location of, 155f
reservoir pressure/production data, 158
reservoir rock/fluid properties, 158, 158t
review questions, 175
rock properties of intervals in, 158t
Surculus interval, 157f, 159
Tamalis interval, 156, 157f, 158, 159
type log for, 156f
wellbores in, 158f
Redlich-Kwong EOS. *See* RK EOS
reduced density, 84
reduced pressure, 61, 68, 84
reduced pseudo-pressure, 84, 61
reduced temperature, 61, 68,
regression analysis, 350
regularization based on curvature, 390
relative permeability, 33, 37, 438
applications of, 41–42
defined, 33, 438
for fluid, 33, 37
for gas, 33, 37
immiscible fluid displacement and, 438
of non-wetting phase, 33, 39, 42
for oil, 33, 37
sources of, 33–37
three-phase, 41
two-phase, 41
for water, 33, 37
of wetting phase, 33, 39, 42
relative permeability curves, 38, 41, 438, 470
gas-oil system, 37, 37f, 438
oil-water system, 37, 37f, 438
reservoir "A" (example), 526–529
reservoir area, acres, 137, 173
reservoir "B" (example), 529–533
reservoir characterization, 515. *See also* geologic modeling
defined, 515
geologic modeling and, 515–532
porosity-permeability transforms in, 29, 29f
reservoir effective compressibility, 184, 203
reservoir energy conservation,560, 559. *See also* reservoir management principles
reservoir fluids. *See also* PVT properties
PVT laboratory measurements, 89–104
types (phase diagram), 55, 55f
reservoir fluid sampling, 89–105
errors associated with, 90, 91, 113
guidelines, 90t
subsurface sampling, 93–94
surface sampling, 89, 91
uses of, 99, 103
reservoir fluid saturations. *See* water saturations
reservoir fluid studies. *See* PVT laboratory measurements
reservoir geology
Manatee reservoirs, 194-200
MBB/W31S reservoirs, 566–571

Red Hawk gas reservoir, 154-156
Shaybah field, 571
26R reservoir, 561-566
West Cameron 580 reservoir, 164-172
reservoir management, 504, 537, 559-561
case histories, 154, 559
closed loop, 559,
defined, 154-158
introduction to, 154, 196
MBB/W31S reservoirs, 566-571
problems, reservoir simulation and, 555-557
Shaybah field, 571
26R reservoir, 561-566
reservoir management principles, 559-561
application of emerging technologies for improved hydrocarbon recovery, 559, 560, 565, 571-573
applied to 26R reservoir, 561-566
applied to MBB/W31S reservoirs, 551-561
applied to Shaybah field, 571
conservation of reservoir energy, 559, 560, 564, 568, 572
early implementation of simple, proven strategies, 559, 560, 565, 571, 572
long term retention of staff in multi-disciplinary teams, 559, 560-561, 566, 571, 573
systematic and sustained practice of data collection, 560, 565, 571, 572
reservoir oil flash liberation tests (comparison), 111
reservoir or formation thickness, feet, 257, 262, 266, 268, 273, 302, 328, 360
reservoir performance data, 516
reservoir pressure calculation (example), 263
reservoir pressure/production data
Manatee reservoirs, 194-203
Red Hawk gas reservoir, 154-164
West Cameron 580 reservoir, 164-172
reservoir productive intervals, 51-52
reservoir quality index (RQI), 36
reservoir rock/fluid properties
MBB/W31S reservoirs, 567tt
Red Hawk gas reservoir, 155t
S10A1/S10B reservoirs, 200t
26R reservoir, 561-566
West Cameron 580 reservoir, 164-172
reservoirs. *See also* black oil reservoirs; dry gas reservoirs; gas-condensate reservoirs; gas reservoirs; oil reservoirs; volatile oil reservoirs; *specific reservoirs*
multilayered, 339, 339f
Saudi Arabian, 34
single layer, 339
types, 6-11
reservoir simulation, 537-552
concepts/terms/methods in, 541-550
continuity equation in rectangular form, 539
flow equations for three-phase flow of oil/water/gas, 539-541
general structure of flow simulators, 550-551
GMBE *v.*, 131
grid systems and, 541-544
history matching and, 548-549
introduction to, 537
MBEs and, 546
model initialization and, 548
numerical dispersion and, 546-547
predictions and, 550
reservoir management problems and, 555-558
reservoir performance and, 559
timesteps and, 545
types of, 538-539
uncertainty analysis and, 550
well model and, 547
reservoir simulation models (data structure), 550-551
reservoir temperature, 59, 70, 77, 84
residual gas saturations in gas reservoirs under water influx (correlations), 177
residual oil saturation to gas, 37, 40, 565
residual oil saturation to water, 39, 438, 567
residual saturation of the non-wetting phase, 39, 42
residual saturation of the wetting phase, 38, 39, 42
resistivity index, ratio, 48, 53
resistivity of formation or rock, 48, 53
resistivity of formation water, 48, 53,
resistivity of fully water-saturated formation or rock, 48, 53
Resource Assessment Techniques, 593
Resources Triangle, 2f
retention of staff in multi-disciplinary teams, 559, 560-561, 566, 571, 573. *See also* reservoir management principles
retrograde condensation, 58, 59, 336
retrograde condensation during gas depletion (gas condensate sample), 115
retrograde gas condensate reservoirs, 58, 59, 59f, 118
characterization of, 138
gas gravity (example), 64-65

gas in place (example), 138
volumetric, material balance for, 139-140
volumetric calculations for, 138-139
rising-bubble apparatus (RBA) method, 497
RK (Redlich-Kwong) EOS, 117, 118
cubic form, roots of, 119 Roach method, 152-154
Robinson correlation. *See* Beggs-Robinson correlation
rock/fluid properties data, 516, 551. *See also* reservoir rock/fluid properties
rock matrix, 26, 28, 374
density, 28, 30
transit time, 26, 30
rock shale volume, fraction, 23, 53, 50
rock types. *See* facies
rock wettability, 97, 435
classification of, 436
core plugs and, 436
defined, 435
immiscible fluid displacement and, 435-436
roots from cubic EOS, 120
routine core analysis (RCAL), 34
RO (Reverse Osmosis), 477
RTA (Rate Time (Transient) Analysis), 627
RQI (reservoir quality index), 36

S10A1 reservoir, 194
gas PVT data for, 200t
oil PVT data for, 200t
reservoir rock/fluid properties for, 200, 200t
S10A2 reservoir, 196f, 197f, 200, 201t, 202f
historical production data for, 201t, 202f
net pay isopach map for, 196f, 197f, 200, 201t, 202f
structure map for, 196f
S10B reservoir, 194, 196, 200, 200t
gas PVT data for, 200t
historical production data for, 201t, 202f
net pay isopach map for, 199f
oil PVT data for, 200t
reservoir rock/fluid properties for, 200, 200t
structure map for, 198f
SAGD (steam-assisted gravity drainage), 502
salinity, mg/L, 83, 84
Sarma correlation, 492. *See also* MMP saturated gas-condensate reservoirs, 58
saturated oil reservoirs, 192
volumetric, 192
with water influx (material balance approach), 193
saturation exponent, 29, 48, 50, 51
saturation exponent for shaly sands, 53
saturation of the wetting phase, 39, 42
Saudi Arabian reservoirs, 34
SBV mixing rules, 62-64
SC (Service Contract), 626, 627
SCAL (Special Core Analysis), 477
SCAL (special core analysis), 470, 516
scaleup. *See* upscaling
scaling method, 470, 517, 519, 525
Schilithuis aquifer model, 143
SCN (single carbon number), 123, 124
screening criteria. *See* EOR screening criteria
SDs (steam drives), 501-502
Sebastian correlation,492, 495. *See also* MMP
secondary porosity, 23
secondary recovery methods, 451-460
defined, 451
gasflooding, 458
introduction to, 451
waterflooding, 451
SEC (Securities Exchange Commission), 627
SEDM (Stretched Exponential Decline Model), 216, 229
seismic data, 515
semi-continuous thermodynamics, 124, 126
semilog pressure response, of NFR, 369, 369f
semivariograms, 519, 533. *See also* variograms
separator/stock-tank fluids (compositions), 65t
separator tests, 102, 102f
sequential Gaussian simulation (SGS), 524, 529
sequential indicator simulation (SIS), 524, 529
Sequential method, 545
serial concentric segments, average permeability of, 258-259
SGS (sequential Gaussian simulation), 524, 529
shale (function to account for extra conductivity of shale), 51, 53
shale effect, 481
Shale Gas/Oil Reservoirs, 8, 9f
shale porosity, 30
shale volume, fraction, 23, 53, 50
shaly sand reservoir, porosity model for, 24f
shaly sands, 53

cementation factor for, 53
effective conductivity of, 49, 53
saturation exponent for, 53
shape factor, 35, 42
shape parameter in gamma function, 128, 214
Shaybah field (case history), 571
development history, 572
displacement mechanisms, 572
geology, 571
management principles applied to, 572-573
Simandoux equation, 47, 50
Simple Fracture Orientation Models, 398
simple, proven strategies, 560. *See also* reservoir management principles
simulations. *See also* reservoir simulation
compositional (MMP/MME), 495-496
SGS, 524, 529
SIS, 524
simulator data output, 551
simulator equations, 545-546
simultaneous-water-and-gas (SWAG), 499, 508
single carbon number (SCN), 123, 124
single component systems, 55-56
single fracture model, 376
single layer reservoir, 339
single-phase samplers, 94
Sinking Fund Deposit Amount, 596
SIS (sequential indicator simulation), 524
SK chart. *See* Standing and Katz chart
skimming method, 95, 95f
skin, 298
skin factor
buildup well test (example) and, 313-315
constant rate drawdown test (example) and, 304-306
defined, 298
effective wellbore radius (example), 300
Horner's time approximation (examples) and, 316-318
line source well, infinite reservoir solution and, 262
MDH plot and, 315-316
multi-rate drawdown test (example) and, 306-309
two-rate drawdown test (example) and, 310-312
SL (Straight Line), 627
slab model (for NFRs), 377, 377f
slightly compressible fluids, 207, 253-254
radial diffusivity equation and, 261-262
straightline methods and, 365-368
slimtube method, 496-497
slippage, 33
slope of MTR line, 305, 313, 316, 334
smart wells, 547
Soave-Redlich-Kwong (SRK) EOS, 118-119
cubic form, roots of, 119
solution gas-oil ratios, 78
at bubble point, 77, 78, 79, 84, 101
oil FVF, calculation of, 78
scf/STB, 78, 80, 104, 132
sonic-derived porosity, 26, 30
in clay, 26
in shale, 26
sonic (acoustic) logs, 26
spatial analysis with variograms, 519-521
special core analysis (SCAL), 470, 516
specific conductivity of exchangeable cations, 49, 53
speckled wetting, 436
spherical variogram model, 522, 522f
spin-echo signal decay, 35, 50
SRK EOS. *See* Soave-Redlich-Kwong EOS
SSBV mixing rules, 62-64
defined, 62
gas mixture (pseudo-critical properties) example, 62-64
SS flow. *See* steady-state flow
staggered line drive pattern, 452, 452f
Standing and Katz (SK) chart, 68
Standing correlation, 78
Standing equation, 63
Starling equation, 68
steady-state (SS) flow
compressible fluids (Darcy equation), 260, 262
defined, 255
incompressible fluids (Darcy equation), 256, 259-260
well production rate in, 257, 259
steady-state methods, relative permeability data and, 38
steady-state models (water influx model), 143-148
steam-assisted gravity drainage (SAGD), 502
steam drives (SDs), 501-502
steamflooding methods, 501-502
CSS, 501
SAGD, 502
SD, 501
WASP, 502

steamflooding models, 502–503
steamflood projects management, 503–504
steamflood reservoir management strategy, 504
step rate test plot, 454, 454f
Stiles method, 452
stock-tank oil gravity, 64, 67, 77, 79, 80
stock-tank oil specific gravity, 64, 78, 81
stock-tank/separator fluids (compositions), 65t
Stone's empirical models, 41
straightline methods, 365–368. *See also* well test analysis
  compressible fluids, 319–327
  historical perspective, 304
  hydraulically fractured well test analysis, 363–386
  NFRs and, 374–378
  slightly compressible fluids, 207, 253–254
stratigraphic modeling, 518
streamline simulators, 538
structural modeling, 517
structure data (M-6, M-4/M-4A reservoirs), 164t
structure maps
  M-4/M-4A reservoirs, 165f
  M-6 reservoirs, 165f
  of MBB/W31S reservoirs, 568f
  S10A2 reservoir, 196f
  S10B reservoir, 196f
  Surculus interval, 157f, 159
  Tamalis interval, 156, 157f, 158, 159
  26R reservoir, 561-566
SRU (Sulfate Removal Unit), 477
sublimation curve, 55
subsurface sampling (bottomhole sampling), 93–94
  errors associated with, 91
  methods/tools, 93–94
subtraction method, 95
superposition principle, 270–276
  constant terminal pressure solution and, 265–266
  constant terminal rate solution and, 270
  total pressure drop of well in multi-well reservoir (example), 270–271
Surculus interval, 157f, 158, 158t, 159. *See also* Well 1ST2
  average net pay thickness of, 158
  average rock properties of, 158t
  reservoir pressure and production data for, 158
  structure map, 157f
surface area, 35, 42
surface equipment management (steamflood projects), 503
surface sampling, 89, 91
sustained and systematic practice of data collection, 560, 565, 571, 572. *See also* reservoir management principles
Sutton-Bergman correlation, 80, 82
Sutton correlation, 62, 63, 68, 70, 71, 74, 80
SWCTT (Single Well Chemical Tracer Test), 477
SWAG (simultaneous-water-and-gas), 499, 508
systematic and sustained practice of data collection, 560, 565, 571, 572. *See also* reservoir management principles

Tamalis interval, 156, 157f, 158, 159. *See also* Well 1ST2; Well 2ST2
  average net pay thickness of, 158
  average rock properties of, 158t
  reservoir pressure and production data for, 158
  structure map, 157f
TBP (True-Boiling-Point) analyses, 123, 128
temperature-volume phase diagrams. *See* T-V phase diagrams
ternary phase diagram for methane/n-butane/decane system, 485–486, 485f
Tcf (Trillion cubic feet of gas), 20
TDS (Total Dissolved Salts), 477
Terms in the SPE PRMS, Defined, 590
tertiary oil recovery, 481. *See also* EOR
thermal EOR processes, 484, 485t
  EOR screening criteria for, 484, 484t
  steamflooding methods, 501–502
  steamflooding models, 502–503
  steamflood projects management, 503–504
thermal simulators, 538
thermodynamic equilibrium, 55
thermodynamics, semi-continuous, 124, 126
thickness-averaged formation permeability, 339, 340
three-parameter gamma function, 124
three-phase relative permeability, 41
three-stage separator test experimental diagram, 102f
tie-line method, 488, 491, 492, 495, 496
Time Match Point (TMP), 350, 351, 355, 370, 372, 381
time regions
  buildup well tests, 301–302
  drawdown tests, 300, 300f
  ETR, 301, 341, 349
  LTR, 301, 302, 309
  MTR, 297, 301, 304, 305, 313, 317, 337, 338, 341, 347, 350

time value of money, 593
timesteps, 545
Timur equation, 35
Tight Gas/Oil Reservoirs, 6, 7f
TMP. *See* Time Match Point
tortuosity, 35, 42
TOC (total organic carbon), 20
Transient Decline Models, 214
Transverse Storage, 406
tortuosity factor, 48, 51, 58
total least squares error model, 390
total liberated gas in DL, 104, 105
total mole fraction of component in mixture, 128
total pore volume, 132, 134, 466
total porosity, 23, 24, 26, 29, 35, 49
total pressure drop of well in multi-well reservoir (example), 303-304
total recovery efficiency, 438
transient flow, 254
transition variogram models, 521
triple point, 55
Troika reservoir, 194, 194f
TRR (technically recoverable resources), 20
True-Boiling-Point (TBP) analyses, 123, 128
T-V (temperature-volume) phase diagrams, 55, 55t, 56, 57
26R reservoir (case history), 561-566
    cross-section of, 562f
    displacement mechanisms, 563
    geology, 561
    historical average reservoir pressure of, 564f
    historical production of, 563f
    management principles applied to, 564-566
    production/injection history, 563
    rock and fluid properties, 563t
    structure map, 562f
two-phase envelope, 485, 485f
two-phase flash calculation (EOS), 122-123
two-phase formation volume factor, 134
two-phase relative permeability, 37
two-rate drawdown tests, 310-312
type-curve matching
    calculation from interpretation model parameters (stage), 352
    defined, 350,
    hydraulically fractured well test analysis and, 363-386
    identification of interpretation model (stage), 351
    procedures for manual application of, 350-351
    stages of procedures, 351-352
    validation of interpretation model results (stage), 352
type curves (well test analysis), 367, 445-474. *See also* well test analysis
    Agarwal equivalent time and, 349, 355, 386
    boundary effects and, 338
    Bourdet derivative, 347, 349
    characteristic shapes of, 351, 357-358
    definition of, 347
    for finite conductivity vertical fracture, 364
    Gringarten, 347-348
    hydraulically fractured well (example) and, 363
    for infinite conductivity vertical fracture, 364, 365f
    NFRs and, 374-378
type log
    Manatee reservoirs, 194-203
    Red Hawk gas reservoir, 154-156
Type Curves for Analysis of Production Data, 211
Typical DFIT Profile, 397

uncertainty analysis
    geologic modeling and, 524
    reservoir simulation and, 550
Unconventional Reservoirs, 1, 5
undersaturated gas condensate reservoirs, 92
undersaturated oil reservoirs, 92
    volumetric, 137-141
undersaturated oil reservoirs with water influx, 186-192
    example,187
    GMBE and, 131-134
    material balance method and, 186-187
    OOIP and, 131,183, 187, 465
    volume method and, 186
Undiscounted Pay-Back (Pay-Out), 608
Undiscounted Profitability Index, 608
uniform flux vertical fracture, 365
uniform (homogeneous) wetting, 436
Uniform Series Compound Amount, 595
United States Bureau of Mines wettability test, 435
Unit Slope Line (USL), 190, 348, 350
universal gas constant, 117, 128
unsteady-state methods, relative permeability data and, 38
unsteady-state models (water influx models), 143

upgridding. *See* upscaling
upper Pliocene sands, 154
upscaling (upgridding, scaleup), 519, 525
USL (Unit Slope Line), 190, 348, 350

validation of interpretation model results (type-curve matching stage), 352
Valko and McCain correlations, 77, 78, 81, 214
van der Waals (vdW) EOS, 117-118
  cubic form, roots of, 117
van Everdingen-Hurst aquifer model, 143, 264, 265, 274
van Everdingen-Hurst–p_D and t_D for radial flow
  finite reservoirs with closed outer boundary, 283, 293t
  finite reservoirs with constant pressure outer boundary, 285t
  infinite reservoirs with constant terminal rate at inner boundary, 291t
van Everdingen-Hurst–Q_D and t_D for radial flow
  finite reservoirs with closed outer boundaries, 283, 293t
  infinite reservoirs with constant terminal pressure, 291t
vapor phase mole fraction, 121, 128
vapor pressure curve, 55
variogram models, 521-523
  experimental, 521
  exponential, 522, 522f
  non-transition models, 521
  Power Law, 214, 521, 521f
  spherical, 521, 521f
  transition models, 521
variograms
  Gaussian variogram model, 521, 521f
  geometric anisotropy and, 520f
  semivariograms and, 519, 533
  spatial analysis with, 519-521
  zonal anisotropy and, 520, 520f
vdW EOS. *See* van der Waals EOS
vertical permeability data, 34, 182
viscometers, 103
viscosities
  data, black oil sample, 103
  dead-oil, 80, 82, 84
  of formation water, 83
  gas, 70-71
  measurements, 103
  oil, 80-82
  viscosity, cp, 84
  water, 83, 84
viscous fingering, 38, 458, 499, 502
volatile oil reservoirs, 58, 89
  saturated, 192
  undersaturated, 183
volume method
  undersaturated oil reservoirs with water influx, 186-192
  volumetric saturated oil reservoirs, 137-141
volume shift parameter, 121t
  correlation coefficients for hydrocarbons, 122t
  for hydrocarbons, 121t, 122t
volume translation (EOS), 121-122
volumetric approach (gas reservoirs with water influx), 141-143
volumetric calculation, 28
volumetric displacement efficiency, 438
volumetric gas reservoirs, 137-141
  dry, OGIP for, 137
  material balance equations for, 139-140
volumetric saturated oil reservoirs, 192-193
volumetric sweep efficiency, 458, 460, 499, 506
volumetric undersaturated oil reservoirs, 137-141
  volume calculations (above bubble point pressure), 183-185
  volume calculations (below bubble point pressure), 185-186
volume-variance scaling techniques, 517
von Schroeter deconvolution algorithm, 390-391
  application examples for, 391-395
  guidelines for application of, 395
  implicit constraint encoding, 390
  pressure response and, 390
  regularization based on curvature, 390
  total least squares error model, 390
Voronoi grid system, 543

W31S reservoir. *See* MBB/W31S reservoirs
WAG (water-alternating-gas), 458, 461, 459, 499
WASP (water-alternating-steam process), 502
water (physical constants), 61t
water-alternating-gas. *See* WAG
water-alternating-steam process (WASP), 502
water-based mud (WBM), 97, 98, 105
water drive index, 182

waterflooded reservoirs, 457
waterflooding, 451
waterflood patterns, 451-452
direct line drive pattern, 451, 451f
five-spot pattern, 452, 452f
nine-spot pattern, 452, 452f
peripheral flood pattern, 452, 452f
staggered line drive pattern, 452, 452f
waterflood projects
design steps for, 453
facilities and, 453
injection water and, 456
management of, 453
produced water and, 457
production wells and, 456
water formation volume factor, 83
water influx. *See also* aquifer models
calculation (Carter-Tracy method), 147, 148t
calculation (Fetkovich method), 144-145, 146t
gas reservoirs with, 177
models, 143-148
residual gas saturations in gas reservoirs (correlations), 177
saturated oil reservoirs with (material balance approach), 192
undersaturated oil reservoirs with, 92
water injection rates, 453, 454, 461, 568
water PVT properties, 83
water saturation, fraction, 37, 42
water saturation at the flood front (Welge method), 443-446
water saturations, 47-52
carbonate rocks and, 50
clean sands and, 48
defined, 47
estimation of, 47
from NMR logs, 50
shaly sands and, 29, 49
uncertainties in estimation of, 51
water viscosity, 83, 84
Watson characterization factor, 125, 128
Waxman-Smits model, 47, 49-50
WBM. *See* water-based mud
WC (West Cameron) 580 reservoir, 164-172. *See also* M-4/M-4A reservoirs; M-6 reservoirs
cross section of, 164f
reservoir geology, 164
reservoir pressure/production data, 168, 168t, 172f
reservoir rock/fluid properties, 166, 166t
Welge method, 443-446
average water saturation after water breakthrough, 444
average water saturation behind flood front, 443-444
water saturation at the flood front, 443
Well 1ST2, 158. *See also* Surculus interval; Tamalis interval
historical liquids and water production, 163f
historical performance data, 160t
historical pressure and gas production, 162f
Well 2ST2, 158
historical liquids and water production, 163f
historical performance data, 161t
historical pressure and gas production, 162f
wellbore (near wellbore zone), 299
formation damage near wellbore, 298, 299f
zone with altered permeability, 298, 298f
wellbore fluid, compressibility of, 340, 348
wellbore phase redistribution effects, 338, 338f
wellbore pressure calculation (example), 321
wellbores, in Red Hawk reservoir, 155f
wellbore storage, 302, 348, 350, 351, 386
effects, 338
wellbore storage coefficient, 302, 340, 348, 350, 351, 380, 386
wellbore storage coefficient, dimensionless, 350, 351, 355
well conditioning, 92
well data, 551
well index (WI), 547
well index calculation (example), 548
well injectivity index, 277
well log data, 516
correlation/assignment of, 518
well logs
ECS, 23, 30
permeability from, 34-37
well management (steamflood projects), 503-504
well model, 547
well production rate
in PSS flow, 260
in SS flow, 259
well productivity index, 276
wellstream gas gravity, 65, 84
well test analysis. *See also* deconvolution concepts; hydraulically fractured wells; naturally fractured reservoirs; straightline methods; type curves

advancements in, 367–368
basic concepts in, 297–302
boundary effects, 338
deconvolution concepts, 389–395
historical account, 297
hydraulically fractured wells, 363
multilayered reservoirs, 339, 339f
multiphase flow, 336–338
naturally fractured reservoirs, 374–378
pressure transient tests, 34, 363, 507
radius of investigation, 297
special topics in, 336–339
straightline methods, 365–368
type curves, 368–374
wellbore phase redistribution effects, 338, 338f
wellbore storage effects, 338
West Cameron 580 reservoir. *See* WC 580 reservoir
wet gas, 58, 64, 66t
correlation method, 65–68
defined, 58
recombination method and, 64–65, 65t
wet gas reservoirs, 89, 137
volumetric, material balance for, 592
volumetric calculations for, 29
wet gas behaviors (phase diagram), 55, 58f
Well Test Analysis, 397
wettability. *See* rock wettability
WFTs (wireline formation testers), 94–100
formation pressures from, 96
OBM systems, 94–95
pressure data applications, 516
WI (well index), 547, 548
Wichert and Aziz correlations, 63, 73
wireline formation testers. *See* WFTs
Wyllie-Rose correlation, 35

XSAGD (Cross SAGD), 502, 508

Yuan correlation, 491, 492. *See also* MMP

zonal anisotropy, variogram and, 520f

www.ingramcontent.com/pod-product-compliance
Lightning Source LLC
LaVergne TN
LVHW071344210926
848486LV00003B/225
9781733389006